Membrane Handbook

Membrane Handbook

Edited by

W. S. Winston Ho, Ph.D.
and
Kamalesh K. Sirkar, Ph.D.

VAN NOSTRAND REINHOLD
New York

Disclaimer

To the best of the authors' and publisher's knowledge, the information contained in this handbook is accurate. However, the publisher and authors assume no responsibility nor liability for any consequences arising from the use of the information contained herein. Final determination of the suitability of any information for use contemplated by any user is the sole responsibility of the user. In addition, the citation of commercial products, tradenames, or names of manufacturers is not to be construed as an endorsement or recommendation for use.

Copyright © 1992 by Van Nostrand Reinhold

Library of Congress Catalog Card Number 91-43661
ISBN 0-442-23747-2

All rights reserved. No part of this work covered by the copyright hereon may be reproduced or used in any form or by any means—graphic, electronic, or mechanical, including photocopying, recording, taping, or information storage and retrieval systems—without written permission of the publisher.

Manufactured in the United States of America

Published by Van Nostrand Reinhold
115 Fifth Avenue
New York, NY 10003

Chapman and Hall
2-6 Boundary Row
London, SE1 8HN, England

Thomas Nelson Australia
102 Dodds Street
South Melbourne 3205
Victoria, Australia

Nelson Canada
1120 Birchmount Road
Scarborough, Ontario M1K 5G4, Canada

16 15 14 13 12 11 10 9 8 7 6 5 4 3 2 1

Library of Congress Cataloging-in-Publication Data

Membrane handbook/editors, W. S. Winston Ho and Kamalesh K. Sirkar.
    p. cm.
  Includes index.
  ISBN 0-442-23747-2
  1. Membranes (Technology)—Handbooks, manuals, etc. I. Ho, W. S.
Winston, 1943-   , II. Sirkar, Kamalesh K., 1942-
TP159.M4M4444 1992
660'.2842—dc20   91-43661
                                                                                              CIP

To Annie Ho and Keka Sirkar
 for their support
 and
To the scientists and engineers
 for advancing membrane science and technology

Applications of membrane processes are rapidly expanding in fields ranging from chemical engineering, environmental science, and water treatment to food science and electronic engineering. As a result, there is a critical need for a central resource on these processes that bridges the gap between fundamentals and everyday industrial uses. This new work answers that need by offering a comprehensive description of principles, membranes, membrane modules, process design, applications, cost estimates, and data for membrane processes.

The Membrane Handbook thoroughly examines today's commercialized membrane processes. It also details new membrane processes under development. Each process is described in significant detail, with coverage of theory, design considerations, and cost-effective operation.

The handbook first deals with commercialized but not always well-understood membrane processes:
• gas permeation • pervaporation • dialysis • electrodialysis • reverse osmosis • ultrafiltration • microfiltration • emulsion liquid membranes.

It then examines new membrane processes under development, as well as membrane-based controlled release. You'll find information on: • membrane-based solvent extraction • hollow-fiber contained liquid membrane • membrane reactors • facilitated transport • electrostatic pseudo-liquid-membrane • membrane-based gas absorption and stripping • membrane distillation • perstraction • controlled release.

Enhanced by more than 400 illustrations, the handbook serves as an invaluable tool for scientists and engineers in the chemical, petroleum, petrochemical, paper, textile, pharmaceutical, and electronic industries. It offers sound practical guidance for food and medical technologists, as well as for engineers involved in pollution control, water treatment, biotechnology, and chemical processing. Moreover, this handbook provides a solid foundation for specialized graduate courses in separations, membrane processes, membrane reactors, controlled release using synthetic membranes, and diffusion in thin films.

About the Editors

W.S. Winston Ho, Ph.D., is a Senior Engineering Associate in Corporate Research of Exxon Research and Engineering Company in Annandale, New Jersey. He has written numerous papers and several book chapters on membranes and separation processes, including a chapter on membrane processes in Perry's Chemical Engineers' Handbook. A co-inventor of membrane solvent extraction and Exxon's FLEXSORB® gas treating technology and New Jersey Inventor of the Year of 1991, Dr. Ho holds about 30 U.S. patents in membranes and separation processes.

Kamalesh K. Sirkar, Ph.D., holds a Sponsored Chair in membrane separations and biotechnology and is a Professor of Chemical Engineering at New Jersey Institute of Technology in Newark, New Jersey. He was with Stevens Institute of Technology. He has authored many papers and several book chapters on membrane processes and jointly edited New Membrane Materials and Processes for Separation. He is on the editorial board of Journal of Membrane Science. Holder of several U.S. patents and a consultant to industry, Dr. Sirkar is the inventor of contained liquid membrane and solvent extraction technologies based on microporous membranes.

List of Contributors

Mauro A. Accomazzo, Ph.D., Vice President, Research & Development, IonPure Technologies Corporation, 10 Technology Drive, Lowell, MA 01851 (Part VIII, Microfiltration: Chapter 34, Deadend Microfiltration: Applications, Design, and Cost). Previous affiliation: Millipore Corporation.

Dibakar Bhattacharyya, Ph.D., Professor, Department of Chemical Engineering and Center of Membrane Sciences, University of Kentucky, Lexington, KY 40506 (Part VI, Reverse Osmosis: Chapters 21–24, Introduction and Definitions, Theory, Design, and Selected Applications)

Robert H. Davis, Ph.D., Associate Professor, Department of Chemical Engineering, University of Colorado, Boulder, CO 80309-0424 (Part VIII, Microfiltration: Chapters 31–33, Definitions, Theory for Deadend Microfiltration, and Theory for Crossflow Microfiltration)

Anthony J. DiLeo, Ph.D., Manager, Advanced Separations Group, Millipore Corporation, 80 Ashby Road, Bedford, MA 01730 (Part VIII, Microfiltration: Chapter 34, Deadend Microfiltration: Applications, Design, and Cost)

Josef Draxler, Dr.-Ing., Assistant Professor, Institute for Chemical Engineering and Environmental Technologies, Technical University of Graz, Inffeldgasse 25, A-8010 Graz, Austria (Part IX, Emulsion Liquid Membranes: Chapters 39 and 40, Applications and Capital and Operating Costs)

Gregory K. Fleming, Ph.D., Research Engineer, E. I. Du Pont de Nemours & Company, Inc., Willow Bank Plant, 305 Water Street, Newport, DE 19804 (Part II, Gas Permeation: Chapters 2–6, Definitions, Theory, Design of Gas Permeation Systems, Applications, and Economics)

Hubert L. Fleming, Ph.D., Vice President, Marketing and Commercial Development, Zenon Environmental, Inc., 13 Estates Drive, Sussex, NJ 07461 (Part III, Pervaporation: Chapters 7–10, Definitions and Background, Theory, Design, and Applications and Economics)

Edward W. Funk, Ph.D., Manager of Process Technology, Engineered Products and Process Technology, Allied Signal, Inc., 50 East Algonquin Road, Des Plaines, IL 60017 (Part VII, Ultrafiltration: Chapters 26–30, Introduction and Definitions, Theory and Mechanistic Concepts, Membranes, Module and Process Configuration, and Applications and Economics)

Vinay Goel, Ph.D., Vice President, Membrane Research, Millipore Corporation, 80 Ashby Road, Bedford, MA 01730 (Part VIII, Microfiltration: Chapters 34 and 35, Deadend Microfiltration: Applications, Design, and Cost and Crossflow Microfiltration: Applications, Design, and Cost)

Donald C. Grant, M.S., Research Manager, FSI International, 322 Lake Hazeltine Drive, Chaska, MN 55318 (Part VIII, Microfiltration: Chapter 32, Theory for Deadend Microfiltration)

Zhongmao Gu, B.S., Separations Group Leader, China Institute of Atomic Energy, P.O. Box 275-93, Beijing 102413, China (Part IX, Emulsion Liquid Membranes: Chapter 38, Design Considerations; Part X, New Membrane Processes under Development: Chapter 45, Electrostatic Pseudo-Liquid-Membrane)

Scott M. Herbig, M.S., Director, Controlled Release Pharmaceuticals, Bend Research, Inc., 64550 Research Road, Bend, OR 97701 (Part XI, Chapter 47, Controlled Release)

W. S. Winston Ho, Ph.D., Senior Engineering Associate, Corporate Research, Exxon Research and Engineering Company, Route 22 East, Annandale, NJ 08801 (Part I, Chapter 1, Overview; Part IX, Emulsion Liquid Membranes: Chapters 36–38, Definitions, Theory, and Design Considerations)

Robert Kaiser, Sc.D., President, ARGOS Associates, Inc., 12 Glengarry Road, Winchester, MA 01890 (Part VIII, Microfiltration: Chapters 34 and 35, Deadend Microfiltration: Applications, Design, and Cost and Crossflow Microfiltration: Applications, Design, and Cost)

Stephen B. Kessler, M.Eng., Manager, Product Development, Sepracor Inc., 33 Locke Drive, Marlborough, MA 01752 (Part IV, Dialysis: Chapters 11–15, Definitions, Theory, Design, Applications, and Cost Estimates)

Elias Klein, Ph.D., Professor, Kidney Disease Program, University of Louisville, Louisville, KY 40292 (Part IV, Dialysis: Chapters 11–15, Definitions, Theory, Design, Applications, and Cost Estimates)

Sudhir S. Kulkarni, Ph.D., Scientist, Chemical Engineering Division, National Chemical Laboratory, Pune 411008, India (Part VII, Ultrafiltration: Chapters 26–30, Introduction and Definitions, Theory and Mechanistic Concepts, Membranes, Module and Process Configuration, and Applications and Economics)

Norman N. Li, Sc.D., Director, Engineered Products and Process Technology, Allied Signal, Inc., 50 East Algonquin Road, Des Plaines, IL 60017 (Part VII, Ultrafiltration: Chapters 26–30, Introduction and Definitions, Theory and Mechanistic Concepts, Membranes, Module and Process Configuration, and Applications and Economics; Part IX, Emulsion Liquid Membranes: Chapters 36–38, Definitions, Theory, and Design Considerations)

Sudipto Majumdar, Ph.D., Research Associate, Center for Membranes and Separation Technologies, Department of Chemistry and Chemical Engineering, Stevens Institute of Technology, Hoboken, NJ 07030 (Part X, New Membrane Processes under Development: Chapter 42, Hollow-Fiber Contained Liquid Membrane). Present address: Department of Chemical Engineering, Chemistry and Environmental Science, New Jersey Institute of Technology, University Heights, Newark, NJ 07102

Rolf J. Marr, Dr.-Ing., Professor, Institute for Chemical Engineering and Environmental Technologies, Technical University of Graz, Inffeldgasse 25, A-8010 Graz, Austria (Part IX, Emulsion Liquid Membranes: Chapters 39 and 40, Applications and Capital and Operating Costs)

Stephen L. Matson, Ph.D., President, Arete Technologies, Inc., 15 Withington Lane, Harvard, MA 01451 (Part X, New Membrane Processes under Development: Chapter 43, Membrane Reactors)

Scott B. McCray, Ph.D., Director of Membrane Development, Bend Research, Inc., 64550 Research Road, Bend, OR 97701 (Part VI, Reverse Osmosis: Chapters 23 and 24, Design and Selected Applications)

Peter M. Meier, Ph.D., U.S. Marketing Manager, Food and Beverage, Millipore Corporation, 80 Ashby Road, Bedford, MA 01730 (Part VIII, Microfiltration: Chapter 34, Deadend Microfiltration: Applications, Design, and Cost)

Stephen L. Michaels, M.S., Marketing Manager, Process Systems Division, Millipore Corporation, 80 Ashby Road, Bedford, MA 01730 (Part VIII, Microfiltration: Chapter 35, Crossflow Microfiltration: Applications, Design, and Cost)

Leon Mir, Ph.D., Vice President & General Manager, Process Systems Division, Millipore Corporation, 80 Ashby Road, Bedford, MA 01730 (Part VIII, Microfiltration: Chapter 35, Crossflow Microfiltration: Applications, Design, and Cost)

Richard D. Noble, Ph.D., Professor, Department of Chemical Engineering, Center for Separations Using Thin Films, University of Colorado, Boulder, CO 80309-0424 (Part X, New Membrane Processes under Development: Chapter 44, Facilitated Transport)

Aldo Pitt, M.S., Consulting Scientist, Millipore Corporation, 80 Ashby Road, Bedford, MA 01730 (Part VIII, Microfiltration: Chapter 34, Deadend Microfiltration: Applications, Design, and Cost)

Malcolm Pluskal, Ph.D., Consulting Scientist, Millipore Corporation, 80 Ashby Road, Bedford, MA 01730 (Part VIII, Microfiltration: Chapter 34, Deadend Microfiltration: Applications, Design, and Cost)

Ravi Prasad, Ph.D., Senior Development Engineer, Separations Products Division, Hoechst Celanese Corporation, Charlotte, NC 28373 (Part X, New Membrane Processes under Development: Chapter 41, Membrane-Based Solvent Extraction)

John A. Quinn, Ph.D., Robert D. Bent Professor, Department of Chemical Engineering, University of Pennsylvania, Philadelphia, PA 19104-6393 (Part X, New Membrane Processes under Development: Chapter 43, Membrane Reactors)

Roderick J. Ray, Ph.D., Vice President and Chief Operating Officer, Bend Research, Inc., 64550 Research Road, Bend, OR 97701 (Part VI, Reverse Osmosis: Chapters 23–25, Design, Selected Applications, and Cost Estimates)

Amitava Sengupta, Ph.D., Chemical Engineer, Chemical Engineering Laboratory, SRI International, 333 Ravenswood Avenue, Menlo Park, CA 94025 (Part X, New Membrane Processes under Development: Chapter 42, Hollow-Fiber Contained Liquid Membrane)

Kamalesh K. Sirkar, Ph.D., Professor, Center for Membranes and Separation Technologies, Department of Chemistry and Chemical Engineering, Stevens Institute of Technology, Hoboken, NJ 07030 (Part I, Chapter 1, Overview; Part X, New Membrane Processes under Development: Chapters 41, 42, and 46, Membrane-Based Solvent Extraction, Hollow-Fiber Contained Liquid Membrane, and Other New Membrane Processes). Present address: Department of Chemical Engineering, Chemistry and Environmental Science, New Jersey Institute of Technology, University Heights, Newark, NJ 07102

C. Stewart Slater, Ph.D., Associate Professor, Department of Chemical Engineering, Manhattan College, Manhattan College Parkway, Riverdale, NY 10471 (Part III, Pervaporation: Chapters 7–10, Definitions and Background, Theory, Design, and Applications and Economics)

Kelly L. Smith, M.S., Director of Research, Controlled Release, Bend Research, Inc., 64550 Research Road, Bend, OR 97701 (Part XI, Chapter 47, Controlled Release)

Heinrich Strathmann, Dr.-Ing., Professor, Institut für Chemische Verfahrenstechnik, Universität Stuttgart, Böblinger Strasse 72, 7000 Stuttgart 1, Germany (Part V, Electrodialysis: Chapters 16–20, Introduction and Definitions, Theory, Ion-Exchange Membranes, Design and Cost Estimates, and Applications)

J. Douglas Way, Ph.D., Assistant Professor, Department of Chemical Engineering, Oregon State University, Corvallis, OR 97331-2702 (Part X, New Membrane Processes under Development: Chapter 44, Facilitated Transport)

Michael E. Williams, M.S., Ph.D. Student, Department of Chemical Engineering and Center of Membrane Sciences, University of Kentucky, Lexington, KY 40506 (Part VI, Reverse Osmosis: Chapters 21–24, Introduction and Definitions, Theory, Design, and Selected Applications)

Raymond R. Zolandz, Ph.D., Senior Research Engineer, Polymer Products Department, E. I. Du Pont de Nemours & Company, Inc., Experimental Station, P.O. Box 80323, Wilmington, DE 19880-0323 (Part II, Gas Permeation: Chapters 2–6, Definitions, Theory, Design of Gas Permeation Systems, Applications, and Economics)

Contents

Foreword		xi
Preface		xiii
Part I.	Overview	1
Chapter 1.	Overview	3
Part II.	Gas Permeation	17
Chapter 2.	Definitions	19
Chapter 3.	Theory	25
Chapter 4.	Design of Gas Permeation Systems	54
Chapter 5.	Applications	78
Chapter 6.	Economics	95
Part III.	Pervaporation	103
Chapter 7.	Definitions and Background	105
Chapter 8.	Theory	117
Chapter 9.	Design	123
Chapter 10.	Applications and Economics	132
Part IV.	Dialysis	161
Chapter 11.	Definitions	163
Chapter 12.	Theory	167
Chapter 13.	Design	186
Chapter 14.	Applications	206
Chapter 15.	Cost Estimates	212
Part V.	Electrodialysis	217
Chapter 16.	Introduction and Definitions	219
Chapter 17.	Theory	223
Chapter 18.	Ion-Exchange Membranes	230
Chapter 19.	Design and Cost Estimates	246
Chapter 20.	Applications	255
Part VI.	Reverse Osmosis	263
Chapter 21.	Introduction and Definitions	265
Chapter 22.	Theory	269
Chapter 23.	Design	281
Chapter 24.	Selected Applications	312
Chapter 25.	Cost Estimates	355
Part VII.	Ultrafiltration	391
Chapter 26.	Introduction and Definitions	393
Chapter 27.	Theory and Mechanistic Concepts	398
Chapter 28.	Membranes	408
Chapter 29.	Module and Process Configuration	432
Chapter 30.	Applications and Economics	446
Part VIII.	Microfiltration	455
Chapter 31.	Definitions	457
Chapter 32.	Theory for Deadend Microfiltration	461
Chapter 33.	Theory for Crossflow Microfiltration	480
Chapter 34.	Deadend Microfiltration: Applications, Design, and Cost	506
Chapter 35.	Crossflow Microfiltration: Applications, Design, and Cost	571
Part IX.	Emulsion Liquid Membranes	595
Chapter 36.	Definitions	597
Chapter 37.	Theory	611
Chapter 38.	Design Considerations	656
Chapter 39.	Applications	701
Chapter 40.	Capital and Operating Costs	718
Part X.	New Membrane Processes under Development	725
Chapter 41.	Membrane-Based Solvent Extraction	727
Chapter 42.	Hollow-Fiber Contained Liquid Membrane	764
Chapter 43.	Membrane Reactors	809
Chapter 44.	Facilitated Transport	833
Chapter 45.	Electrostatic Pseudo-Liquid-Membrane	867
Chapter 46.	Other New Membrane Processes	885
Part XI.	Controlled Release	913
Chapter 47.	Controlled Release	915
Index		936

Foreword

Membrane processes have wide industrial applications covering many existing and emerging uses in the chemical, petrochemical, petroleum, environmental, water treatment, pharmaceutical, medical, food, dairy, beverage, paper, textile, and electronic industries. The existing applications include: (1) dialysis for the purification of human blood (the artificial kidney), (2) electrodialysis for the desalination of brackish water to produce potable water, (3) reverse osmosis for the desalination of seawater, (4) ultrafiltration for the concentration of large protein molecules from cheese, casein whey, and milk, and (5) microfiltration for the sterilization of pharmaceutical and medical products, beer, wine, and soft drinks. Since membrane processes generally have low capital investment, as well as low energy consumption and operating cost, there are a number of emerging applications, which include: (1) gas permeation for the removal of acid gases from natural gas, (2) pervaporation for the dehydration of alcohols and organics and the separation of organics, (3) emulsion liquid membranes for wastewater treatment, (4) novel contactors for gas absorption/stripping, and (5) membrane reactors which combine chemical reaction and separation.

Although there is a great body of literature covering various aspects of membrane science and technology, to date there has been no single handbook covering the entire field and the full range of applications. Such a source is welcome.

This handbook reviews the published literature, presents an in-depth description of commercialized membrane processes, and gives a state-of-the-art review of new membrane process concepts under development. It is intended to be a single source of underlying principles, membranes, membrane modules, process design, applications, and cost estimates. It is also a first attempt to bridge the gap between the theory and practice.

There are several groups which may benefit from this handbook. It can be used as educational material for industrial personnel engaged in membrane separations. For scientists and engineers active in research and development in synthetic membranes, it will serve as a single source of reference for the entire field. Engineers evaluating separation processes will find this handbook to be a guide to allow membrane approaches to be compared with other separation processes. To students examining separation processes, membrane separations, membrane reactors, membrane-based controlled release, and diffusion in thin films, it should be a valuable sourcebook.

Frank B. Sprow, Ph.D.

Vice President
Corporate Research
Exxon Research and
Engineering Company
Annandale, New Jersey

Preface

This handbook deals with processes using synthetic membranes. It consists of eleven parts: overview, gas permeation, pervaporation, dialysis, electrodialysis, reverse osmosis, ultrafiltration, microfiltration, emulsion liquid membranes, new membrane processes under development, and controlled release. The membrane processes have been arranged in the following order: (1) commercialized membrane separation processes (Parts II–IX), (2) new membrane processes under development (Part X), and (3) controlled release (Part XI). Processes using biological membranes have not been included.

Part I, consisting of a chapter, provides an overview of this handbook. This overview gives a definition of a membrane and a membrane process, and it serves as an introduction to the remaining parts of this handbook. It discusses the characteristics of eight commercialized membrane separation processes with respect to the following seven aspects: (1) separation goal, (2) nature of species retained (size of the species), (3) nature of species transported through membrane, electrolytic or volatile, (4) minor or major species of feed solution transported through membrane, (5) driving force, (6) mechanism for transport/selectivity, and (7) phase of feed and permeate streams. It includes additional considerations required for the selection of a membrane process or a hybrid process. This chapter also discusses the characteristics of new membrane processes for separation under development with respect to these seven aspects.

Parts II–IX treat the eight commercialized membrane separation processes in significant detail with an individual part devoted to each process. Each part has a detailed description of the following five aspects of each process: (1) definitions, (2) theory, (3) design, (4) applications, and (5) cost estimates. In each part, an individual chapter is devoted to each of these five aspects, with a few exceptions. The goal is to have a comprehensive coverage with respect to practical application and fundamental understanding.

For the new membrane processes under development covered in Part X, an individual chapter is devoted to each of the following processes: (1) Membrane-Based Solvent Extraction (Chapter 41), (2) Hollow-Fiber Contained Liquid Membrane (Chapter 42), (3) Membrane Reactors (Chapter 43), (4) Facilitated Transport (Chapter 44), and (5) Electrostatic Pseudo-Liquid-Membrane (Chapter 45). These have been followed by Chapter 46, Other New Membrane Processes, which includes: (1) membrane-based gas absorption and stripping, (2) membrane distillation, and (3) perstraction. The five aspects mentioned above for the commercialized processes have also been used to describe each of these new processes except for their cost information which, however, is rarely available.

The last part of this handbook consisting of a chapter describes controlled release processes using membranes. This chapter is structured after the pattern used for the other membrane processes. However, attention is given to the description of typical controlled release technologies since these are commercialized processes.

There is a wide variation in the nomenclature used in the literature of membrane science and technology. To avoid such a problem, a list of general notation used throughout this handbook has been provided in the very beginning. Specialized notation, if needed, has been added to the end of particular chapters.

W. S. Winston Ho
Annandale, New Jersey

Kamalesh K. Sirkar
Newark, New Jersey

Acknowledgments

We wish to thank the contributors for their efforts in the preparation of their chapters in this handbook and their willingness to share their expertise and insights with those involved with membranes through these chapters. We also wish to thank Ms. Marianne Kane for many hours of typing throughout the preparation of this handbook. Many thanks are also due to the reviewers, to whom the publisher sent the manuscripts of all chapters, for their invaluable comments and suggestions.

The following copyright owners are acknowledged for granting permission for the use of their tables and figures: Academic Press (Figure 47-3), American Chemical Society (Figures 42-5, 42-18, and 44-7), American Institute of Chemical Engineers (Figures 4-14, 29-6, 29-8, 30-3, 37-6, 37-7, 41-4, 41-6, 41-8, 42-3, 42-4, 42-6, 42-7, 42-11, 42-13, 42-14, 42-16, and 42-17), American Journal of Medicine and ALZA Corporation (Figures 47-9 and 47-16), Bakish Materials Corporation (Figures 10-14, 10-15, 10-16, 10-17, and 10-23), Baxter Healthcare Corporation (Figures 13-3 and 13-4), Bend Research, Inc. and R. J. Ray (Tables 25-1 through 25-15, 25-17, and 25-18, Figures 25-1 through 25-14, and Appendix Worksheets in Chapter 25), Business Communications Co., Inc. (Table 29-3), Canon Communications, Inc. (Figure 34-17), Chemical Industry Press (Table 45-3 and Figures 45-1, 45-3, 45-4, 45-5, 45-6, and 45-7), Chemical Processing and Coors Brewing Company (Figure 30-2), Childwall University Press Ltd. (Figure 38-2), China Institute of Atomic Energy (Table 45-1), China Ocean Press and Water Treatment (Table 45-4 and Figure 38-3), CRC Press, Inc. (Table 30-3), Marcel Dekker, Inc. (Table 30-2 and Figures 38-7, 38-8, 38-9, 38-10, 38-11, and 39-12), E. I. Du Pont de Nemours & Co., Inc., and G. L. Poffenbarger (Tables 5-2, 5-4, and 5-5 and Figures 5-1, 6-1, and 6-2), Elsevier Science Publishers (Tables 3-3, 3-4, 38-1, and 45-2 and Figures 3-7, 3-8, 3-13, 3-14, 4-3, 38-1, 41-12, 41-14, 42-9, 42-10, 42-12, 44-3, 45-2, 45-8, 46-6, 47-11, and 47-15), GFT and H. L. Fleming (Table 10-3 and Figures 7-7, 9-11, 9-13, 10-6, 10-8, 10-10, 10-11, 10-18, 10-19, and 10-20), Hoechst Celanese Corporation (Table 5-3 and Figures 10-32, 41-7, 41-10, and 42-15), International Scientific Communications, Inc. (Figure 34-23), Journal of Environmental Science (Figures 32-4, 32-7, 32-8, 32-9, and 32-10), Koch Membrane Systems (Figures 29-1 and 29-2), McGraw-Hill, Inc. (Tables 35-1, 35-3, and 35-4 and Figures 34-2, 35-1, 35-2, 35-6, and 44-2), Membrane Technology and Research, Inc. (Figures 7-8, 10-25, 10-27, 10-28, 10-29, 10-30, and 10-33), Millipore Corporation (Table 29-2 and Figures 29-3, 30-1, 34-1, 34-3, 34-4, 34-7 through 34-10, 34-12 through 34-14, 34-16, 34-18 through 34-22, 35-3, 35-4, 35-7, and 35-9 through 35-14), Millipore Corporation and G. Larrabee (Table 34-16), Millipore Corporation and Texas Instruments Corporation (Figure 34-11), Nadir Separations (Hoechst) (Tables 28-4 and 28-8), The New York Academy of Sciences (Figures 43-1, 43-2, 43-3, 43-11, 43-12, 43-13, 43-14, and 43-23), New York State Energy Research and Development Authority (Figure 42-19), Noyes Publications (Figure 34-15), Parenteral Drug Associates (Figures 34-5 and 34-6), PCI Membrane Systems (Table 28-5), Plenum Publishing Corporation (Figures 37-1, 37-2, 37-3, 37-4, 37-5, 38-5, 38-6, 44-6, and 47-5), Royal Society of Chemistry (Figures 5-6 and 5-7), W. B. Saunders Company (Figures 12-11 and 13-8), Sepracor Inc. (Figure 41-9), Solid State Technology (Figure 32-6), Swiss Contamination Control (Figures 32-15, 32-16, 32-17, and 32-18), Ube Industries (Table 4-2), U.S. Department of Energy (Figures 5-4 and 5-5), and John Wiley & Sons (Tables 3-2, 3-3, and 3-4 and Figures 3-1, 3-3, 3-4, 3-5, 3-6, 3-7, 3-8, 3-9, 3-10, 3-11, 3-12, 12-2, 12-8, 12-12, and 13-9).

Finally, we wish to acknowledge support from the Management of Exxon Research and Engineering Company and the Administration of Stevens Institute of Technology and those of the contributors. The contributors and editors are solely responsible for the content of this handbook.

W. S. Winston Ho
Kamalesh K. Sirkar

General Notation

Units are given in terms of physical quantities, length (L), mass (M), time (t), temperature (T), amount of substance (mol, mole), electric current (A, ampere), electric potential (V, volt), energy ($E = ML^2/t^2$), and pressure ($p = M/Lt^2$). Special notation is also present in some chapters.

a	activity, various units or dimensionless, or particle radius, L
A	area, L^2
A	species A
A_i	plasticization coefficient for species i, L^3/mol or L^3/M
A_{ij}	interaction coefficient between species i and j, L^3/mol or L^3/M
b	channel height, L, or Langmuir affinity constant, Lt^2/M or p^{-1}, or the ratio of frictional force of solute to the bulk solution, dimensionless
B	species B
B_{ij}	interaction coefficient between species i and j, dimensionless
c	concentration, mol/L^3 [or cm^3 (STP)/cm^3 polymer] or M/L^3
c_{Ai}	concentration of species A at the inside radius of membrane, mol/L^3 [or cm^3 (STP)/cm^3 polymer] or M/L^3
c_{Ao}	concentration of species A at the outside radius of membrane, mol/L^3 [or cm^3 (STP)/cm^3 polymer] or M/L^3
c_H'	Langmuir sorption capacity, L^3 (STP)/L^3 polymer [cm^3 (STP)/cm^3 polymer]
C	heat capacity, L^2/t^2T or E/MT
C_s	reflection constant, 1.14 for air, dimensionless
d	diameter, L
d_f	diameter of filter fibers, L
d_h	hydraulic diameter, L
d_i	inside diameter, or molecular diameter of species i, L
d_o	outside diameter, L
d_p	pore diameter, or particle diameter, L
d_s	stirrer or impeller diameter, L
d_t	tube diameter, L
D	diffusion coefficient, L^2/t
$\hat{D}$	dimensionless diffusivity, $\eta D/r^2 \tau_{wo}$
D_0	infinite dilution diffusivity, or characteristic diffusivity, L^2/t
Da	Damköhler number, $k_f c^{n-1} l^2/D$ (n = order of reaction), dimensionless
D_{AB}	diffusion coefficient for solute-carrier complex, L^2/t
D_b	Brownian diffusion coefficient, L^2/t
D_{eff}	effective diffusion coefficient, L^2/t
D_i	diffusivity in the internal phase, L^2/t
D_m	diffusivity in the membrane phase, L^2/t
D_s	shear-induced diffusion coefficient, L^2/t
D_T	thermodynamic diffusion coefficient, L^2/t
E	energy, ML^2/t^2 or E, or activation energy, ML^2/t^2 mol or E/mol, or enhancement factor ($F - 1$), dimensionless
$\Delta \bar{E}_a$	activation energy, ML^2/t^2 mol or E/mol
f	Fanning friction factor, or inertia lift function, dimensionless
f_i	fugacity of component i, M/Lt^2 or p
F	facilitation factor (ratio of flux with carrier present to flux without carrier), dimensionless, or $F = D_H/D_D$, convenient dimensionless group in the dual-mode sorption model
F	Faraday's constant (9.652 × 10^4 amp · s/g-equivalent), At/g-equivalent
g	gravitational acceleration, L/t^2

xvii

g_c	gravitational conversion factor, dimensionless [or ft · lb/(lb force · s^2) in the fps system]	K_{di}	diffusion constant relating diffusivity to concentration for species i, L^5/mol t or L^5/Mt
G	Gibbs free energy, ML^2/t^2 or E	K_{eq}	equilibrium constant (for extraction reaction), dimensionless or various units
GPU	gas permeation unit, 10^{-6} cm^3 (STP)/(cm^2 · s · cm Hg)	K_i	distribution coefficient for species i, a_i/a_i^1, dimensionless
H	enthalpy, ML^2/t^2 or E	K_{req}	equilibrium constant for stripping or re-extraction reaction, dimensionless or various units
H	Henry's law constant ($c = Hp$), t^2 mol/L^2M [or cm^3 (STP)/(cm^3 polymer · atm)], or t^2/L^2	K_s	solubility product, various units, or sorption coefficient, mol/L^3 or M/L^3
H_o	channel half height, L	K_t	tortuosity correction factor for bubble point, dimensionless
$\Delta \hat{H}_v$	latent heat of vaporization (per unit mass), L^2/t^2 or E/M	l	membrane thickness, L
i	electric current density, A/L^2	L	membrane length, or height of a column extractor, L
$i_{\lim}$	limiting current density, A/L^2	LRV	filter log reduction value, dimensionless
I	electric current, A	m	molal concentration, mol/M
j	mass transfer or permeation rate, mol/t (or kg-equivalent/s)	m_i	distribution coefficient for species i, a_i/a_i^1, dimensionless
j_k	mass transfer rate due to diffusional transport, mol/t (or kg-equivalent/s) or M/t	M	mass, M
j_Φ	mass transfer rate due to breakage, mol/t (or kg-equivalent/s) or M/t	M	molarity, mol/L^3 (g-mole/liter or kg-mole/m^3)
J	flux, mol/$L^2 t$ or $M/L^2 t$, or volumetric flux, L/t	$\tilde{M}$	molecular weight, M/mol
$\langle J \rangle$	area-averaged volumetric flux, L/t	n	number of moles
J_i	flux of species i, mol/$L^2 t$ or $M/L^2 t$	n_p	number of pores per unit membrane area, L^{-2}
$\langle J_L \rangle$	area-averaged volumetric flux from inertial lift model, L/t	N	number of hollow fibers, or number of emulsion globules
$\langle J_s \rangle$	area-averaged volumetric flux from shear-induced diffusion model, L/t	N	normal (g-equivalent/liter or kg-equivalent/m^3)
J_v	volumetric flux, L/t	$\tilde{N}$	Avogadro's number, 6.023×10^{23} molecules/mol
k	thermal conductivity, $ML/t^3 T$ or E/LtT, or Boltzmann's constant [1.38×10^{-16} g · cm^2/(s^2 · °K)], $ML^2/t^2 T$ or E/T, or mass transfer coefficient, L/t	Δp	pressure drop, M/Lt^2 or p
		p	pressure or partial pressure, M/Lt^2 or p
		P	permeability coefficient, various units, or fractional penetration, dimensionless
k_f	forward reaction rate constant, (mol/L^3)$^{1-n}/t$, n = order of reaction, various units	Pe	Peclet number, vd/D, or $v_e L/D_{Le}$, dimensionless
k_i	mass transfer coefficient for species i, L/t	q	heat flux, M/t^3 or $E/L^2 t$
k_r	reverse reaction rate constant, (mol/L^3)$^{1-n}/t$, n = order of reaction, various units	Q	flow rate or permeation rate, mol/t or M/t, or volumetric flow rate, L^3/t
K	distribution coefficient, or hydrodynamic constant, dimensionless, or $K=C_H^1 b/H$, convenient dimensionless group in the dual-mode sorption model	Q_e	volumetric flow rate of the external phase, L^3/t
		Q_i	volumetric flow rate of the internal phase, L^3/t
K_c	cake permeability, L^2	Q_m	volumetric flow rate of the membrane phase, L^3/t

Q'	excess particle flux in boundary layer, L^2/t	t_i	transport number of ionic species i, dimensionless
$\hat{Q}$	dimensionless excess particle flux	T	temperature, T
$\hat{Q}_{cr}$	critical excess particle flux for cake formation, dimensionless	T_b	normal boiling point, T
		T_c	critical temperature, T
Q_v	volumetric flow rate, L^3/t	T_g	glass transition temperature, T
r	radius, or radial coordinate, L	u_i	absolute mobility of species i, t mol/M or L^2 mol/Et
r_d	drum radius, L		
r_i	inside radius, L	v	velocity, L/t
r_o	outside radius, or emulsion globule radius, L	v_e	velocity of the external phase, L/t
		v_L	inertial lift velocity for constricted tube or channel, L/t
r_p	pore radius, L		
r_t	tube radius, L	$v_{L,0}$	inertial lift velocity for unconstricted tube or channel, L/t
R	gas constant, ML^2/t^2T mol, or rejection of solute, or recycle ratio, dimensionless		
		v_s	superficial face velocity, L/t
		V	volume, L^3
R_c	cake resistance, M/L^2t (or pt/L), or L^{-1}	V_e	volume of the external phase, L^3
		V_f	fractional free volume, dimensionless
$\hat{R}_c$	specific cake resistance per unit thickness, L^{-2}	V_i	total volume of the internal phase, L^3
		V_m	volume of the membrane phase, L^3
R_c'	specific cake resistance (mass basis), L/M	V_s	suspension volume, L^3
		$\bar{V}_i$	partial molar volume of species i, L^3/mol
R_e	electrical resistance, V/A		
Re	Reynolds number, $dv\rho/\eta$, dimensionless	w	cake mass per unit membrane area, M/L^2
R_f	reaction rate for extraction per interfacial area, mol/L^2t (or kg-equivalent/m^{2}s)		
		w_i	mass fraction of species i, dimensionless
R_i	rejection of solute species i, or interception parameter, dimensionless	w_1	penetrant mass fraction, dimensionless
		W	channel width, or drum width, or width of membrane sheet, L
R_m	membrane resistance, M/L^2t (or pt/L), or L^{-1}		
		W_e	total consumption amount of the reagent in the external phase, kg-equivalent
R_{mf}	final membrane resistance, L^{-1}		
R_{mi}	initial membrane resistance, L^{-1}	We	Weber number, $\omega^2 d_s^3 \rho_e/\gamma$, dimensionless
R_r	reaction rate for stripping per interfacial area, mol/L^2t (or kg-equivalent/m^{2}s)		
		W_i	total consumption amount of the reagent in the internal phase, kg-equivalent
s	cake compressibility factor, dimensionless		
		x_{cr}	critical distance for cake formation, L
S	entropy, ML^2/t^2T or E/T, or shape factor for flux correction in a cylindrical geometry, dimensionless	x_i	mole fraction of species i, dimensionless
		y_i	mole fraction of species i in permeate, dimensionless
S_c	specific surface area (pore surface area/solids volume) in cake, L^{-1}		
		z	valence, dimensionless, or z coordinate, L
Sc	Schmidt number, $\eta/\rho D$, dimensionless		
Sh	Sherwood number, $k_i L/D$, dimensionless	z_i	valence of species i, dimensionless

Greek Letters

S_m	specific surface area (pore surface area/solids volume) in membrane, L^{-1}
t	time, t
$t_{1/2}$	half-life, t

α	selectivity, or mobility ratio ($D_{AB}c_T/D_A c_{A0}$) for facilitated transport, or

xx General Notation

	flow distribution parameter, dimensionless	θ_f	angle subtended by submerged drum, dimensionless
α_{ij}	separation factor, $(y_i''/y_j'')/(y_i'/y_j')$, $y_i = c_i$, p_i, w_i, x_i, etc., dimensionless	λ	ratio of species diameter to pore diameter, or parameter of the advancing front model, dimensionless
$\Delta\alpha$	difference in polymer thermal expansion coefficients above and below glass transition, T^{-1} (or $°K^{-1}$)	Λ	reduced filter coefficient, dimensionless
		μ	chemical potential, ML^2/t^2 mol or E/mol
β	matrix model coefficient in diffusion expression, L^3 polymer/L^3 (STP) [or cm^3 polymer/cm^3 (STP)]	v	kinematic viscosity, L^2/t
		v_i	number of ions per molecule of electrolyte i
$\Delta\beta$	difference in polymer compressibility above and below glass transition, Lt^2/M or p^{-1} (or atm^{-1})	ξ	current utilization, or flow redistribution decrement parameter, dimensionless
γ	surface or interfacial tension, M/t^2 or E/L^2, or plasticization parameter in the free-volume model, dimensionless	π	osmotic pressure, M/Lt^2 or p
		ρ	density or mass concentration, M/L^3
γ_i	activity coefficient of species i, dimensionless	ρ_0	pure fluid density, M/L^3
		ρ_e	density of the external phase, M/L^3
$\dot\gamma$	shear rate, t^{-1}	ρ_m	density of the membrane phase, M/L^3
$\dot\gamma_0$	shear rate at edge of boundary layer, t^{-1}	ρ_s	cake solids density, M/L^3
Γ	tubesheet length, L	σ	reflection coefficient, or inhomogeneity factor, dimensionless
$\Gamma°$	inlet ratio of the internal reagent equivalents to the feed solute equivalents, dimensionless	σ_c	matrix model coefficient in solubility expression, L^3 polymer/L^3 (STP) [cm^3 polymer/cm^3 (STP)]
Γ_i	ratio of the internal reagent equivalents to the external phase solute equivalents for stage i, dimensionless	σ_0	infinite dilution solubility coefficient in matrix model, L^3 (STP)/(L^3 polymer · M/Lt^2) (or cm^3 (STP)/(cm^3 polymer · atm)]
δ	polarization layer thickness, or stagnant film thickness, L, or solubility parameter (with subscript), $(E/L^3)^{1/2}$	τ	tortuosity factor, or number of pore volumes of fluid passed through filter, dimensionless
δ_c	cake thickness, L	τ_c	cake fouling time constant, t
ϵ	porosity, or void fraction, or inhomogeneity factor, dimensionless	τ_m	membrane fouling time constant, t
ϵ_s	suspension porosity, dimensionless	τ_w	wall shear stress for constricted tube or channel, M/Lt^2
ε	inverse Damköhler number, Da^{-1}, dimensionless	τ_{w0}	wall shear stress for unconstricted tube or channel, M/Lt^2
η	viscosity, M/Lt, or efficiency, dimensionless	ϕ	volume fraction, or solids volume fraction, dimensionless
η_0	pure fluid viscosity, M/Lt		
η_d	single fiber efficiency due to diffusion, dimensionless	ϕ_s	suspension solids volume fraction, dimensionless
η_e	viscosity of the external phase, M/Lt		
η_i	single fiber efficiency due to interception, dimensionless	ϕ_w	solids volume fraction at cake surface, dimensionless
η_m	viscosity of the membrane phase, M/Lt	ψ	electric potential, V
η_s	solvent viscosity, M/Lt, or single collector efficiency, dimensionless	ψ_s	solvent association factor, dimensionless
θ	diffusion time lag, or time constant, t, or stage cut, or angle, dimensionless	ω	stirring rate or impeller speed, rpm
		Ω	angular velocity, t^{-1}

Diacritical Marks

~	per mole
^	per unit mass
‾	average value
·	time rate of change

Superscripts

′	value in the feed stream, on the upstream side or on the high-pressure side of the membrane, or value in the phase external to the membrane
″	value in permeate, product or extract, or value on the downstream side or on the low-pressure side of the membrane value in the membrane indicated by absence of superscript
°	standard reference state
*	ideal case

Subscripts

0	initial value
A, B	particular species
b	bulk
c	cake
f	feed
i	general species index or solute species i
in	inlet
int	interface
j	species j
l	liquid
lm	logarithmic mean
m	membrane
out	outlet
p	product, permeate, permeant, or polymer
r	retentate or reject
s	solution or shell side
t	tube or tube side
T	total
v	vapor
w	water or solvent
x,y,z	three coordinate directions

I
Overview

1
Overview

W. S. Winston Ho
Exxon Research and Engineering Company

Kamalesh K. Sirkar*
Stevens Institute of Technology

INTRODUCTION
DEFINITION OF A MEMBRANE AND A MEMBRANE PROCESS
MEMBRANE PROCESSES
  Commercialized Membrane Separation Processes
New Membrane Processes under Development
Controlled Release
REFERENCES

INTRODUCTION

Membranes are primarily used for separation, and membrane processes are generally separation processes. Over the last 30 years, such processes have been widely adopted by different industries. Large-scale commercial uses of membrane separations have displaced conventional separation processes. More are expected in the future. Membrane separation processes are often more capital and energy efficient when compared with conventional separation processes. Membrane devices and systems are almost always compact and modular. In addition, membrane processes can sometimes achieve totally novel results. Membrane processes treated in this handbook are, therefore, important from current use and future development points of view.

This chapter provides an overview of this membrane handbook. The succeeding chapters provide detailed descriptions of individual membrane processes including those already commercialized and new ones under development. This overview will facilitate the development of a perspective on membrane processes treated in this handbook. To that end, this overview covers (1) the definition of a membrane and a membrane process and (2) membrane processes. A notation section for the whole handbook has been provided before this overview. *Specialized notation, if needed, has been provided at the end of particular chapters.*

DEFINITION OF A MEMBRANE AND A MEMBRANE PROCESS

A membrane process requires two bulk phases physically separated by a third phase, the membrane. The *membrane* is an interphase between the two bulk phases. It is either a homogeneous phase or a heterogeneous collection of phases. The membrane phase may be any one or a combination of the following: nonporous solid, microporous or macroporous solid with a fluid

---
*Present affiliation: New Jersey Institute of Technology

(liquid or gas) in the pores, a liquid phase with or without a second phase, or a gel. The membrane phase is almost always thin when compared with the dimensions of the bulk phases in at least two other directions. Hollow fibers and emulsion liquid membranes provide exceptions where the membrane thickness is of the order of the dimensions of one of the bulk phases. (For a variety of definitions of membrane, see Lonsdale 1989.)

The membrane phase interposed between two bulk phases controls the exchange of mass between the two bulk phases in a membrane process. In membrane separation processes, the bulk phases are mixtures. One of the species in the mixture is allowed to be exchanged in preference to others. The membrane is selective to one of the species. One bulk phase is enriched in one of the species while the other is depleted of it. A *membrane process* then allows selective and controlled transfer of one species from one bulk phase to another bulk phase separated by the membrane. The definition of every membrane process treated in this handbook is provided at the beginning of every process description.

The movement of any species across the membrane is caused by one or more driving forces. These driving forces arise from a gradient of chemical potential or electrical potential. A gradient in chemical potential may be due to a concentration gradient or pressure gradient or both. The transmembrane flux of any species per unit driving force is proportional to the permeability of the species. If the driving force is described by the use of a partial pressure difference (Δp_i) or a concentration difference (Δc_i) across the membrane for species i, then

Transmembrane flux of species i

$$= \left(\frac{\text{permeability of species } i}{\text{effective membrane thickness}} \right) (\Delta p_i \text{ or } \Delta c_i). \quad (1\text{--}1)$$

The ratio, permeability of species i/effective membrane thickness, is sometimes called the permeance (which is also called permeability in some special processes, e.g., dialysis and microfiltration). The membrane selectivity between any two species can be defined in a number of ways. A common definition called separation factor α_{ij} for two species i and j is

$$\alpha_{ij} = \frac{c_i''/c_j''}{c_i'/c_j'}, \quad (1\text{--}2)$$

where the prime and double prime superscripts denote the upstream bulk phase (feed or retentate) and the downstream bulk phase (permeate), respectively. The separation factor is equal to the ratio of the two species' permeabilities under conditions in which the downstream pressure or concentration is negligible in comparison to the upstream pressure or concentration. Such selectivities are reduced or destroyed if the membrane has gross defects through which nonselective hydrodynamic flow of one bulk phase into the other bulk phase occurs.

The membrane control of the transmembrane transport rates of species is established by two effects: (1) different transport rates due to different membrane/solute/solvent interactions and (2) the partitioning or exchange of the species at the two interfaces on two sides of the membrane, bulk phase 1/membrane interface and membrane/bulk phase 2 interface. The second effect is, in general, more important than the first. The existence of two interfaces is in contrast to conventional equilibrium-based separation processes in which there is only one interface, the bulk phase 1/bulk phase 2 interface, where the partitioning or exchange of the species occurs. Further, bulk phases 1 and 2 must be immiscible in equilibrium-based processes. In a membrane process with the membrane interposed between the two bulk phases, bulk phase 1 may be miscible or immiscible with bulk phase 2.

Thus, for a *solid membrane,* bulk phase 1 and bulk phase 2 can be any combination of miscible or immiscible liquid phases and gaseous phases. The following combinations are commercial processes: gas 1/solid membrane/gas 2 (gas permeation), liquid 1/solid membrane/gas (vapor) 2 (pervaporation), and liquid 1/solid membrane/miscible liquid 2 (dialysis, electrodialysis, reverse osmosis, ultrafiltration, and microfiltration). The other combinations, gas 1/solid membrane/liquid 2 (e.g., CO_2 removed from a breathing mixture by a solid membrane swept by seawater on the other side, Lokhandwala and Stern 1990) and liquid 1/solid membrane/immiscible liquid 2 [membrane-based solvent extraction (Ho, Lee, and Liu 1976; Lee, Ho, and Liu 1976a, 1976b)], are also feasible. Since diffusional rates through a solid are very low, at most one of the bulk phases can be a solid (e.g., controlled release systems) unless extremely slow transport rates are acceptable for the process.

For a *liquid membrane*, bulk phase 1 and bulk phase 2 can be any combination of gaseous phases, immiscible liquid phases, and a solid phase. Here bulk liquid phases must be immiscible with the liquid membrane phase. The following combinations are feasible: gas 1/liquid membrane/gas 2, gas 1/liquid membrane/liquid 2, liquid 1/liquid membrane/gas 2 (Chapters 42 and 44), and liquid 1/liquid membrane/liquid 2 (Chapters 36 through 40, 42, 44, and 45). Some of them have been either commercialized (e.g., liquid 1/liquid membrane/liquid 2, Chapters 36 through 40) or demonstrated in the laboratory. Processes with one bulk solid phase and a liquid membrane are also realizable.

For a *gas membrane*, the two bulk phases could be any combination of liquid and solid phases. The liquid phases may or may not be miscible. However, as mentioned before, the use of bulk solid phases has not been reported yet.

In all such membrane processes, the membrane phase may be thin but it does not allow direct contact between the two bulk phases. A class of newly emerging processes involves a microporous or macroporous membrane that functions primarily to immobilize the interface of two immiscible bulk fluid phases that contact each other directly inside the membrane pores or at a membrane/bulk phase boundary (Kiani, Bhave, and Sirkar 1984; Sirkar 1988). When such a process is used for separation, there are no functional and conceptual differences between it and a classical equilibrium-based separation process. To distinguish them from rate-governed membrane separation processes, they are identified as membrane-based equilibrium or contacting processes for separation. The use of microporous or macroporous membranes, however, confers other advantages to membrane-based contacting processes over their conventional counterparts.

MEMBRANE PROCESSES

A large number of processes use synthetic membranes (Hwang and Kammermeyer 1975; Meares 1976; Lacey and Loeb 1979; Belfort 1984; Ho and Li 1984; Kesting 1985; Sourirajan and Matsuura 1985; Bungay, Lonsdale, and dePinho 1986; Cheryan 1986; Cabasso 1987; Gutman 1987; Rautenbach and Albrecht 1989; Porter 1990). Processes using biological membranes have not been included in this handbook. Is the principal objective of the membrane process the separation of a mixture? For the industrial or commercial user, an additional aspect of great utility is whether the process has been commercialized or not. A combination of these two concerns has led to the following order of presentation of the membrane processes in this handbook: (1) commercialized membrane separation processes, (2) new membrane processes under development, and (3) controlled release.

Commercialized Membrane Separation Processes

In all membrane separation processes, the feed is separated into a stream that goes through the membrane, the permeate, and a fraction of feed that does not go through the membrane, the retentate, the reject, or the concentrate. In a few processes, e.g., dialysis, electrodialysis, and emulsion liquid membrane, a wash or strip stream is needed on the permeate side.

Eight commercialized membrane separation processes have been treated in detail in this handbook: gas permeation, pervaporation, dialysis, electrodialysis, reverse osmosis, ultrafiltration, microfiltration, and emulsion liquid membrane (the nanofiltration process is covered under reverse osmosis). Large-scale or widespread commercial uses were developed earliest for microfiltration, dialysis (hemodialysis), electrodialysis, and reverse osmosis. Ultrafiltration was adopted for commercial uses immediately afterward. The first large-scale gas separation modules were developed by Du Pont as early as 1970; but the first successful commercial membrane-gas-separation processes were announced by Monsanto in the late 1970s. Many more have appeared since then. Significant commercialization of pervaporation separation began during the period 1985–1990. Commercialization of emulsion liquid membranes for the removal of zinc and phenol from wastewaters was also achieved in Austria and China, respectively, around 1986.

Although all commercialized membrane separation processes, except gas permeation and microfiltration of gaseous feeds, are capable of separating liquid solutions or liquid phase feeds, virtually all of them are principally used for separating aqueous solutions. The solutes may be microsolutes (inorganic or organic), macrosolutes, proteins, polymers, colloids, whole cells, precipitates, dirt particles, etc. An exception is pervaporation whose principal commercial use at this time is for removing small amounts of water from organic liquids such as alcohols. Ultrafiltration has also been used for treating waste aqueous streams with a very high volume fraction of organics. Microfiltration separation of organic liquids is also practiced.

Characteristics and Selection of Commercialized Membrane Separation Processes

The principal characteristics of the eight commercialized membrane separation processes can be specified based on the following seven aspects (Kesting 1985):

1. Separation goal
2. Nature of species retained (size of the species)
3. Nature of species transported through membrane, electrolytic or volatile
4. Minor or major species of feed solution transported through membrane
5. Driving force
6. Mechanism for transport/selectivity
7. Phase of feed and permeate streams.

Table 1–1 identifies such characteristics for each commercialized membrane separation process and thereby allows a preliminary selection of the membrane process relevant to the separation goal. This table is self-explanatory. However, a brief discussion of a few separation problems capable of being solved by different membrane processes is useful.

Consider a dilute aqueous solution of an electrolytic microsolute. Recovering the solvent or concentrating the electrolyte may be achieved by either reverse osmosis or electrodialysis. Selection of a process will require consideration of several factors. For example, reverse osmosis is more economical than electrodialysis at higher electrolyte concentrations. However, at lower electrolyte concentrations, electrodialysis is more economical (Chapters 19 and 20).

On the other hand, if the dilute aqueous solution also contains proteins as a key product, electrodialysis may be a convenient way of eliminating the microsolute, the electrolyte. Ultrafiltration as well as dialysis can eliminate the microsolute in such a problem. The nature of the proteins, the fouling tendency of membranes, the colloidal characteristics of the solution, and ultimately the cost will guide the process selection among the three competing processes, electrodialysis, ultrafiltration, and dialysis. An additional consideration of relevance now is the nature of the electrolyte-containing waste stream produced. Electrodialysis will produce a waste stream concentrated in the electrolyte whereas dialysis will yield a stream more dilute than the feed stream. The waste stream in ultrafiltration can at the most be as concentrated as the feed stream.

If a minor amount of a volatile microsolute is present in a particular aqueous stream, competing membrane processes for obtaining purer water are reverse osmosis and pervaporation. If the microsolute has to be recovered simultaneously, pervaporation can produce a stream of almost pure microsolute from the permeate, whereas reverse osmosis can produce a concentrated aqueous solution of the microsolute. On the other hand, if the membrane has a very high selectivity, the reverse osmosis permeate, water, is likely to be sufficiently pure. A high membrane area is required in pervaporation to produce an equivalent quality of water since the driving force for the small amount of the volatile solute to be removed is low.

Ionizable solutes, which may be acids, bases, cations, or anions of metallic salts, or electrolytes, may be removed from an aqueous solution and concentrated in the receiving phase by emulsion liquid membranes. Often reverse osmosis membranes are able to reject such solutes very efficiently, especially if the solution pH is in the required range. The characteristics of the membrane processes identified in Table 1–1 are insufficient for the selection of the required process for such solutes, e.g., phenol, organic acids, etc. In the absence of a heuristical guide, detailed analysis of the competing processes would be needed. Nonmembrane separation processes (e.g., carbon adsorption for organic solutes) have to be analyzed often along with the competing membrane processes.

A single membrane process is sometimes inefficient in solving a given separation problem. Hybrid processes have been developed for such cases. A hybrid process may combine a membrane process with a nonmembrane process or combine a membrane process with another membrane process; the hybrid can achieve an efficiency level not achievable by either process used alone. No focused treatment of membrane-based hybrid processes has been provided in this handbook. However, examples of hybrid processes have been included in individual parts of the handbook [e.g., Chapter 5 for air separation via membrane permeation with pressure swing adsorption and cryogenics and CO_2 separation via membrane permeation with amine treating and cryogenics, Chapter 10 for pervaporation with distillation and pervaporation with condensation and phase separation, Chapter 14 for hemodialysis with hemofiltration, Chapter 24 for a hybrid of reverse osmosis (RO) with multistage flash distillation, etc.]. Other hybrid processes, for example, nanofiltration-RO (Rautenbach and Gröschl 1990), RO with electrodialysis, ultrafiltration with RO, and RO with liquid membranes, are of increasing interest.

In the examples cited above, the two processes operated separately in two separate units so that the product from one process was the feed to the next process. The two processes may be further connected via a recycle stream. Processes exist, however, in which two different separation mechanisms are operating simultaneously and locally in a single device. An example is hemodiafiltration (Chapter 14) where diffusion (the basis for dialysis) and convection (the basis for ultrafiltration) are simultaneously present. Other examples of some interest include combining an electric field with ultrafiltration.

Not included in Table 1–1 is Donnan dialysis (treated in Chapters 14 and 20 according to current practice); it is a membrane-based equilibrium process that is totally different from those included in this table. Although solute partitioning occurs in such a process at two aqueous liquid/membrane interfaces on two sides of an ion-exchange membrane, it is not a rate-governed membrane separation process. The separation achieved between feed and strip solutions is not lost even if the assembly of feed solution/ion-exchange membrane/strip solution is closed to the surrounding, a feature characteristic of equilibrium separation processes such as absorption, solvent extraction, distillation, etc. On the other hand, in simple rate-governed membrane separation processes such as gas permeation, dialysis, etc., all separation achieved is lost after some time if the permeation cell is closed to the surrounding.

Table 1–1 does not identify the structural characteristics of the membranes used in particular membrane separation processes. The relevant details are provided in each of the suc-

TABLE 1-1. Principal Characteristics of Commercialized Membrane Separation Processes.

Separation Process	Separation Goal	Nature of Species Retained (Size)	Nature of Species Transported through Membrane	Minor/Major Species Transported	Driving Force	Mechanism for Transport/Selectivity	Phase of Feed and Permeate Streams
Gas permeation	Stream/streams enriched or depleted in a particular species	Larger species retained unless highly soluble	Gaseous. Smaller species/more soluble species	Either	Concentration gradient (partial pressure difference)	Solution-diffusion	Gaseous
Pervaporation	Same as above	Same as above	More soluble/smaller/more volatile nonelectrolytes	Preferably minor species	Concentration gradient, temperature gradient	Solution-diffusion	Liquid feed, gaseous permeate
Dialysis	Macrosolute solution free of microsolute, microsolute solution free of macrosolute	>0.02 μm retained, >0.005 μm retained in hemodialysis	Microsolute, smaller solute	Minor species. Solvent transported under osmotic unbalance	Concentration gradient	Sieving, hindered diffusion in microporous membranes	Liquid
Electrodialysis	Solution free of microions, concentrated solution of microions, fractionation of microions	Co-ions, macroions[a] and water retained	Microionic species	Minor ionic species, small amounts of water by electroosmosis	Electrical potential gradient, electro-osmosis (minor amount)	Counter-ion transport via ion exchange membranes	Liquid

Process	Stream/streams enriched or depleted in a particular species	Size	Species retained/permeated	Major species	Driving force	Mechanism	Phase
Reverse osmosis	Solvent free of all solutes, concentrated solution of microsolutes	1- to 10-Å microsolute species	Solvent. Species retained may be electrolytic or volatile	Major species solvent	Hydrostatic pressure gradient vs. osmotic pressure gradient	Preferential sorption/capillary flow (solution-diffusion-imperfection)	Liquid
Ultrafiltration	Solution free of macrosolute, macrosolute solution free of microsolute, macrosolute fractionation	10- to 200-Å macrosolute species	Solution of microsolutes	Major solvent, minor microsolutes	Hydrostatic pressure gradient vs. small osmotic pressure gradient	Sieving	Liquid
Microfiltration	Solution free of particles, gas free of particles	0.02- to 10- μm particles	Solution/gas free of particles	Major solvent, minor microsolutes/macrosolutes	Hydrostatic pressure gradient	Sieving	Liquid or gas
Emulsion liquid membrane	Stream/streams enriched or depleted in a particular species	Generally not size-selective except in host-guest chemistry	Species with high solubility in liquid membranes	Minor species. Can be major species in organic mixture separation	Concentration gradient, pH gradient	Solution-diffusion, facilitated transport	Generally liquid feed, emulsion containing permeate

[a]Macromolecules with charge.

ceeding parts of this handbook. Although most of the membranes referred to in this handbook are polymeric in nature, other membranes, e.g., ceramic, metal, etc., are also included. However, a few common features of membranes are dictated by the mechanism for transport/selectivity. For example, only nonporous or solvent membranes can allow the solution-diffusion mechanism. Therefore, the species-selective layer in gas permeation and pervaporation membranes is solvent-like, i.e., nonporous. Sieving by a membrane presupposes the existence of pores. Thus, membranes in dialysis, ultrafiltration, and microfiltration have pores; depending on solute size, they are either microporous or macroporous. By IUPAC classification, pores of less than 20 Å in diameter are called *micropores*, pores between 20 and 500 Å are called *mesopores*, and pores larger than 500 Å are called *macropores*. There is no such agreed practice *vis-à-vis* membranes. Membranes for ultrafiltration and microfiltration are generally identified in industry as microporous.

Membranes containing negatively charged groups fixed to the membrane matrix allow the passage of cations but reject anions; they are called cation-exchange membranes (Chapter 18). On the other hand, anion-exchange membranes with positively charged groups fixed to the membrane matrix allow the passage of anions but not cations. Combining a cation-exchange membrane with an anion-exchange membrane in a manner such that one is adjacent to the other results in a bipolar membrane. Bipolar membranes can be used for splitting water into hydrogen and hydroxyl ions and then converting salts into acids and bases (Chapters 18 and 20).

The following aspects of each of the eight commercialized membrane separation processes are covered in significant detail in this handbook with an individual part devoted to each process (Parts II–IX for these eight processes): (1) definitions, (2) theory, (3) design, (4) applications, and (5) cost estimates. An individual chapter is devoted to each of these five aspects of each process with a few exceptions. The goal is to provide comprehensive coverage from both practical application and fundamental understanding points of view.

New Membrane Processes under Development

As indicated above, after describing the commercialized membrane separation processes, the remaining parts of the handbook cover new membrane processes under development (Chapters 41 through 46) and controlled release processes using membranes (Chapter 47). The new membrane processes under development can be divided conveniently into three categories: (1) new membrane-based equilibrium separation processes, (2) new membrane separation processes, and (3) membrane reactors.

New Membrane-Based Equilibrium Separation Processes

New membrane-based equilibrium separation processes for solvent extraction, gas absorption, and stripping are covered in significant detail in this handbook. Although vacuum membrane distillation is a membrane-based equilibrium separation process, it is discussed under membrane distillation. Of these, membrane gas absorption and stripping have been commercialized for some time for blood oxygenation using microporous membranes. This commercial use should have catapulted the membrane gas absorption and stripping process to the category of commercialized processes; however, there is no such recognition in the literature (Kesting 1985). Literature investigations have also been quite limited for potentially important areas such as membrane gas scrubbing or stripping. Substantial efforts, however, are currently being devoted to commercializing membrane stripping, membrane solvent extraction, and membrane gas absorption using microporous membranes. Hence they have been included here.

These processes open an additional window for the use of membranes. A true membrane separation process has, in general, practical limitations to achieving very high levels of purification of fluid streams being processed. Conventional equilibrium-based separation processes, such as solvent extraction, absorption, stripping, distillation, etc., suffer from no such limitations. Membrane-based equilibrium or contacting processes are as capable of purification as their nonmembrane counterparts. Membrane-based equilibrium processes have many additional benefits including the fact that they are nondispersive.

New Membrane Separation Processes

The new membrane separation processes under development and included in this handbook are hollow-fiber contained liquid membrane, facilitated transport, electrostatic pseudo-liquid-membrane, membrane distillation, and perstraction. The first three processes provide alternative methods of using liquid membranes. Facilitated transport also covers separation through polymeric membranes with fixed carriers. Membrane distillation is similar to osmotic distillation, and both use gas membranes (located in the pores of a microporous membrane). Studied first in the 1960s, perstraction may have potential for separating organic liquid mixtures. Piezodialysis has not been included due to lack of any developmental effort and very poor quality membranes.

Additional membrane processes exist that could be termed gas membrane processes. Here a volatile solute (e.g., a vapor or a gas) is transferred from one liquid on one side of a microporous or macroporous membrane to another liquid on the other side of the membrane through gas-filled or vapor-filled pores of the membrane. The gas gap or vapor gap is assumed to act as a membrane. Due to the limited literature on such processes and their close relation to membrane gas absorption and stripping and membrane distillation, they are covered briefly in the chapters on the above-mentioned processes.

Membrane Reactors

Chemical reaction processes generally require the separation of products. Of considerable use also is the separation of a mixture to provide purer reactants to the reactor. Incorporation of a membrane or two in the reactor can provide considerable benefits of synergy via reaction-separation and/or separation-reaction processes. These benefits may be due to equilibrium shift, the reduction of product inhibition, etc. New processes being developed using membrane reactors are included in Chapter 43. The principal characteristics of the separation required for membrane reactors can be the same as those of the other membrane processes for separation (see Tables 1–1, 1–2, and 1–3).

Characteristics of New Membrane Processes for Separation under Development

As mentioned above, the two classes of new membrane processes for separation under development are membrane-based equilibrium separation processes and membrane separation processes. The principal characteristics of these two classes of processes can be specified in Tables 1–2 and 1–3, respectively.

Table 1–2 on membrane-based equilibrium separation processes identifies the membrane only as an interface stabilizer or a phase barrier. The driving forces, transport mechanisms, and basis for selectivity are to be sought in the two contacting phases, liquid/liquid and gas/liquid (King 1980). The membrane has little to do with the process selectivity, which is governed by the partitioning characteristics between the two phases. The achievement of nondispersive contacting of the two phases via the membrane for satisfactory operation, however, depends significantly on the membrane pore size and wetting characteristics. Under certain conditions, the membrane pore size can also influence solute selectivity (Chapter 41).

In all processes of Table 1–2, there is only one interface, either a liquid/liquid or a gas/liquid interface, where solute partitioning occurs when microporous membranes are used. Thus, these are not true membrane separation processes where solute partitioning occurs at two interfaces on the two sides of the membrane. Although vacuum membrane distillation is considered under membrane distillation (Chapter 46), which is nominally a membrane separation process, it is included in Table 1–2 to emphasize that there is only one interface in vacuum membrane distillation.

As described in Table 1–3, three new mem-

TABLE 1-2. Principal Characteristics of New Membrane-Based Equilibrium Separation Processes under Development.

Separation Process	Separation Goal	Nature of Species Retained (Size)	Nature of Species Transported through Membrane	Minor/Major Species Transported	Driving Force	Mechanism for Transport/ Selectivity	Phase of Feed and Permeate Streams
Membrane solvent extraction	Solute-free feed solution, solute-enriched extractant solution	Microsolute and macrosolute not extracted	Species preferentially extracted, nonionic or ionic	Minor. Membrane as phase interface stabilizer	Concentration gradient, different solubilities in extractant	Selective partitioning between extract and feed phases	Liquid feed and extract, both immiscible
Membrane gas absorption	Solute-free feed gas, solute-enriched absorbent	Gaseous species not absorbed	Gaseous species preferentially absorbed	Usually minor. Membrane as phase barrier for absorbent	Concentration gradient, different solubilities in absorbent	Selective absorption between absorbent/feed	Gaseous feed, liquid absorbent, both immiscible
Membrane stripping	Solute-free feed liquid, solute enriched in stripping gas	Liquid species not stripped	Volatile species stripped from feed	Usually minor. Membrane as phase barrier for feed	Concentration gradient, different volatilities	Higher volatility between feed/ stripping gas	Liquid feed, stripping gas, both immiscible
Vacuum membrane distillation	Solute-free feed liquid, solute-free feed liquid evaporated/ condensed	Microsolute/ solvent not evaporated	Volatile species with higher vapor pressure	Either. Membrane as phase barrier for feed	Equilibrium vapor pressure minus permeate side pressure	Higher volatility	Liquid feed, permeate vapor at low pressure

TABLE 1-3. Principal Characteristics of New Membrane Separation Processes under Development.

Separation Process	Separation Goal	Nature of Species Retained (Size)	Nature of Species Transported through Membrane	Minor/Major Species Transported	Driving Force	Mechanism for Transport/ Selectivity	Phase of Feed and Permeate Streams
Hollow-fiber contained liquid membrane	Stream/streams enriched or depleted in a particular species	Generally not size-selective except in host-guest chemistry	Species with high solubility in liquid membrane	Minor species. Can be major species in organic mixture or gas separation	Concentration gradient, pH gradient	Solution-diffusion, facilitated transport	Liquid/liquid, gas/gas, gas/liquid, liquid/gas
Facilitated transport	Same as above	Same as above	Species with high complexing ability in membrane	Minor species. Can be major species	Concentration gradient, pH gradient	Facilitated transport	Gas/gas, gas/liquid, liquid/liquid, liquid/gas
Electrostatic pseudo-liquid-membrane	Same as above	Same as above	Species with high solubility in liquid membrane	Minor species (generally)	Concentration gradient, pH gradient	Solution-diffusion, facilitated transport	Liquid/liquid (miscible)
Membrane distillation	Removal of volatile liquid species from feed	Nonvolatile microsolute and solvent	Volatile species with higher vapor pressure	Either	Partial/vapor pressure difference due to temperature gradient	Higher volatility, diffusion through gas membrane	Liquid/liquid (miscible)
Osmotic distillation	Removal of species (H_2O) from solution to concentrate microsolutes	Nonvolatile microsolute and solvent	Volatile species (usually H_2O vapor)	Major	Partial pressure difference of H_2O in opposite direction to osmotic pressure gradient	Higher volatility, diffusion through gas membrane	Liquid feed, liquid product
Gas membrane	Species removed from a solution to another solution	Same as above	Volatile species	Minor	Partial pressure difference	Solubility in product liquid, strippability from feed	Liquid feed, liquid product
Perstraction	Feed enriched in one species, sweep enriched in other species	Generally not size-selective	More soluble species	Either	Concentration gradient	Solution-diffusion	Usually miscible feed/sweep liquids

brane separation processes under development, hollow-fiber contained liquid membrane, facilitated transport, and electrostatic pseudo-liquid-membrane, use liquid membranes. The other three new processes described in this table, membrane distillation, osmotic distillation, and gas membrane, use gas membranes. None of these three gas membrane processes can process conventional nonpolar organic liquid mixtures and are primarily limited to aqueous solutions or polar organics, which do not wet the hydrophobic microporous membranes used. Among the processes identified in this table, only the perstraction process uses a solid membrane. In addition, perstraction is one of the few membrane processes (including those via pervaporation and liquid membranes) capable of separating organic liquid mixtures.

For the new membrane processes under development, an individual chapter is devoted to each of the following processes:

1. Membrane-Based Solvent Extraction (Chapter 41)
2. Hollow-Fiber Contained Liquid Membrane (Chapter 42)
3. Membrane Reactors (Chapter 43)
4. Facilitated Transport (Chapter 44)
5. Electrostatic Pseudo-Liquid-Membrane (Chapter 45).

These have been followed by Chapter 46, "Other New Membrane Processes," which covers in order the following processes: (1) membrane-based gas absorption and stripping, (2) membrane distillation, and (3) perstraction.

The basic structure followed earlier in the treatment of commercialized membrane separation processes, namely, definitions, theory, design, applications, and cost, has also been followed here with the following differences. Only one chapter instead of five is devoted to a particular process with general description, theory, design, applications, etc., each spanning a section of a chapter. Cost information is rarely available for new processes under development. Potential applications have been identified for each process. In Chapter 46 on other new membrane processes, each section covering a particular process or processes has separate subsections dealing with the general description, theory, design, applications, etc.

Controlled Release

The last chapter of the handbook describes controlled release processes using membranes. Unlike almost all processes described in Chapters 41 through 46, membrane-based controlled release processes have been commercialized for more than a decade. In comparison with the membrane processes covered in the rest of the handbook, the size of commercialized devices for controlled release is smaller by orders of magnitude. In addition, there is no specific separation goal in controlled release.

Whenever a chemical, biochemical, or polymeric species needs to be administered or supplied at low levels continuously or intermittently for an extended period in any application, controlled release technologies are used. In a variety of controlled release products, the controlled release of the species is often mediated by a membrane. The membrane controls the rate at which the species is supplied from a reservoir to the environment where it is to be delivered. The species may be present in pure form or in a mixture, which may be a liquid solution, liquid dispersion, solid dispersion, etc. The species transported may be a microsolute or a macrosolute, nonvolatile or volatile, electrolytic or nonelectrolytic. The membrane may be nonporous or microporous. Usually, the driving force is a concentration difference and the mechanism is solution-diffusion; however, osmotic pressure difference has been used to create a hydrostatic pressure difference to effect drug release in osmotic pump devices.

Other control release technologies are available that do not use membranes for the rate control of active species delivery. These include the erosion of a polymer-drug matrix by chemical or biological means, magnetic systems, liposomes, etc. Only membrane-based controlled release processes will be covered in Chapter 47. Such processes are increasingly used in human health care, animal husbandry, agricultural activities, and consumer products.

This chapter is structured after the pattern followed for the other membrane processes. However, attention is paid to the description of typical controlled release technologies because these are commercialized processes.

There are processes used in industry that employ membranes but are neither separation nor controlled release processes. Typical examples are sensors using membranes as one of a number of important components. Such processes have not been covered in this handbook.

REFERENCES

Belfort, G. 1984. *Synthetic Membrane Processes*. New York: Academic Press.

Bungay, P. M., H. K. Lonsdale, and M. N. dePinho. 1986. *Synthetic Membranes: Science, Engineering and Applications*. Dordrecht, Holland: D. Reidel Publishing Co.

Cabasso, I. 1987. Membranes. *Encycl. Polym. Sci. Eng.* 9:509–579.

Cheryan, M. 1986. *Ultrafiltration Handbook*. Lancaster, PA: Technomic Publishing Co.

Gutman, R. G. 1987. *Membrane Filtration: The Technology of Pressure-Driven Crossflow Processes*. Bristol, U.K.: Adam Hilger.

Ho, W. S., L. T. C. Lee, and K. J. Liu. 1976. Membrane hydrometallurgical extraction process. U.S. Patent 3,957,504.

Ho, W. S., and N. N. Li. 1984. Membrane processes. In *Perry's Chemical Engineers' Handbook*, ed. R. H. Perry and D. W. Green, 6th ed., pp. 17-14–17-35. New York: McGraw-Hill Book Co.

Hwang, S. T., and K. Kammermeyer. 1975. *Membranes in Separations*, Vol. VII in *Techniques of Chemistry*, ed. A. Weissberger. New York: Wiley Interscience.

Kesting, R. E. 1985. *Synthetic Polymeric Membranes: A Structural Perspective*, 2nd ed. New York: John Wiley & Sons.

Kiani, A., R. R. Bhave, and K. K. Sirkar. 1984. Solvent extraction with immobilized interfaces in a microporous hydrophobic membrane. *J. Membr. Sci.* 20:125–145.

King, C. J. 1980. *Separation Processes*, 2nd ed. New York: McGraw Hill Book Co.

Lacey, R. E., and S. Loeb. 1979. *Industrial Processing with Membranes*, reprinted ed. Huntington, New York: Robert E. Krieger Publishing Co.

Lee, L. T. C., W. S. Ho, and K. J. Liu. 1976a. Membrane solvent extraction. U.S. Patent 3,956,112.

Lee, L. T. C., W. S. Ho, and K. J. Liu. 1976b. Novel high diffusivity membranes. U.S. Patent 3,951,789.

Lokhandwala, K. A., and S. A. Stern. 1990. A membrane process for removal of CO_2 from diving atmosphere. In *Proc. 1990 Intl. Congress on Membranes and Membrane Processes*, 20–24 August 1990, Chicago, IL, Vol. II, pp. 845–847.

Lonsdale, H. K. 1989. What is a membrane? Part II. *J. Membr. Sci.* 43:1–3.

Meares, P. 1976. *Membrane Separation Processes*. Amsterdam: Elsevier Scientific Publishing Co.

Porter, M. C. 1990. *Handbook of Industrial Membrane Technology*. Park Ridge, NJ: Noyes Publications.

Rautenbach, R., and R. Albrecht. 1989. *Membrane Processes*. New York: John Wiley & Sons.

Rautenbach, R., and A. Gröschl. 1990. Separation potential of nanofiltration membranes. *Desalination* 77:73–84.

Sirkar, K. K. 1988. Immobilized interface solute transfer apparatus. U.S. Patent 4,789,468.

Sourirajan, S., and T. Matsuura. 1985. *Reverse Osmosis/Ultrafiltration Process Principles*. Ottawa, Canada: National Research Council Canada Publications.

II
Gas Permeation

2.	Definitions	19
3.	Theory	25
4.	Design of Gas Permeation Systems	54
5.	Applications	78
6.	Economics	95

2

Definitions

Raymond R. Zolandz and Gregory K. Fleming
E. I. Du Pont de Nemours & Company, Inc.

THE BASIC MEMBRANE PROCESS
  FOR GAS SEPARATION
TERMINOLOGY

DRIVING FORCE FOR GAS
  SEPARATION
NOTATION
REFERENCES

THE BASIC MEMBRANE PROCESS FOR GAS SEPARATION

The concept of separating gases with polymeric membranes is more than 100 years old, but the widespread use of gas separation membranes has occurred only within the last 10 to 15 years. Commercialization depended on the development of membranes with sufficient productivity and separating ability to make them economically attractive in industrial applications.

Keller (1987) estimated that membranes are about one-third as developed as traditional unit operations such as distillation. Therefore, it is not surprising that the number of available membranes and the number of economically feasible separations have both increased in recent years, and should continue to do so.

A typical membrane process for gas separation is shown in Figure 2–1. The stream to be separated is fed to the membrane device at an elevated pressure, where it passes across one side of the membrane. The opposite side of the membrane is held at a lower pressure. The pressure difference across the membrane provides the driving force for the diffusion of gas across the membrane.

Separation is achieved because of differences in the relative transport rates of the feed components. Components that diffuse more rapidly become enriched in the low-pressure *permeate* stream, while the slower components are concentrated in the *retentate* or *residue* stream. The degree to which components are separated is governed by the ability of the membrane to discriminate between those components, as well as by the relative driving force of each component. The fact that the membrane process separates components based on their relative rates of permeation distinguishes it from equilibrium processes such as distillation or extraction.

In equilibrium processes, multicomponent thermodynamics determines the extent of separation achieved between two distinct phases. In the case of gas separation with membranes, the

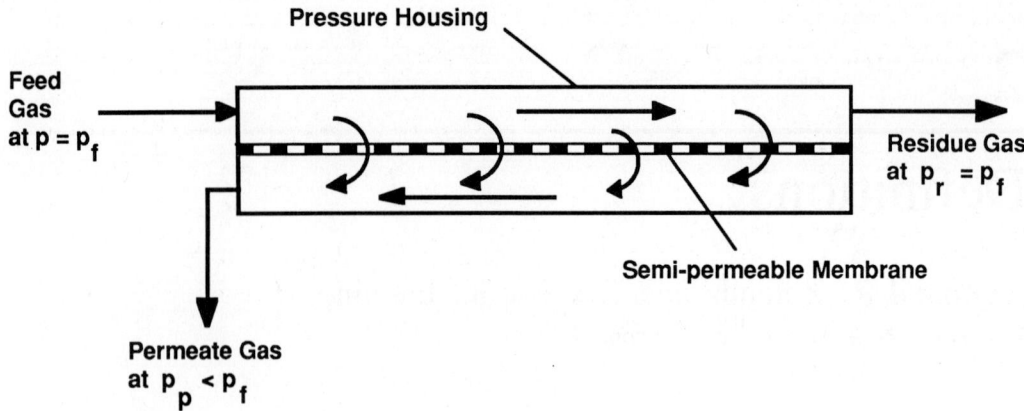

FIGURE 2–1. Schematic diagram for the membrane separation of gases.

feed and product streams are all gases, and no phase change occurs. If a membrane system were permitted to go to equilibrium, permeation would continue to occur until the pressure and composition of gas on both sides of the membrane were equal, with no separation being achieved. Separation is achieved only if the system is maintained away from the equilibrium state. An analogy, however, can be drawn between gas separation by membranes and batch distillation.

In a batch distillation process, the first vapor that emerges is the most highly enriched in the low-boiling component: a very small stage cut yields the purest possible overhead product. As distillation continues, the liquid becomes increasingly enriched in the high boiler, as does the vapor in equilibrium with it. The overhead product becomes less pure as a larger fraction of the initial charge is boiled over.

Similarly, in a membrane unit, the first gas that permeates is the most highly enriched in the rapidly diffusing component(s). A very small stage cut across the membrane yields the purest permeate product. As larger stage cuts are taken, the feed gas becomes enriched in the gases which permeate more slowly, and both their driving force and their concentration in the permeate are increased. The permeate becomes less pure as a larger fraction of the feed gas is permeated. The trade-off between product purity and product recovery is inherent in the membrane process.

TERMINOLOGY

A membrane will separate gases only if some components pass through the membrane more rapidly than others. This requirement places constraints on the structure of the membrane's separating layer, as depicted in Figure 2–2.

If the membrane contains pores large enough to allow convective flow, separation will not occur. If the size of the pores is smaller than the mean free path of the gas molecules, then convective flow is replaced by *Knudsen diffusion*. In this case gas molecules interact with the pore walls much more frequently than with one another. Low molecular weight gases are able to diffuse more rapidly than heavier ones, and separation occurs. In the limit of zero permeate pressure, the difference in transport rates of two components is inversely proportional to the square root of the ratio of their molecular weights. This type of diffusion process does not provide sufficient separation in most instances, although the technique has been used in the large-scale separation of uranium isotopes.

In some cases transport rates through Knudsen diffusion membranes can be enhanced by other mechanisms, such as surface diffusion, in which molecules adsorb to the surface of the pores and then diffuse along that surface. High selectivity can be achieved in cases for which preferential adsorption of one component occurs.

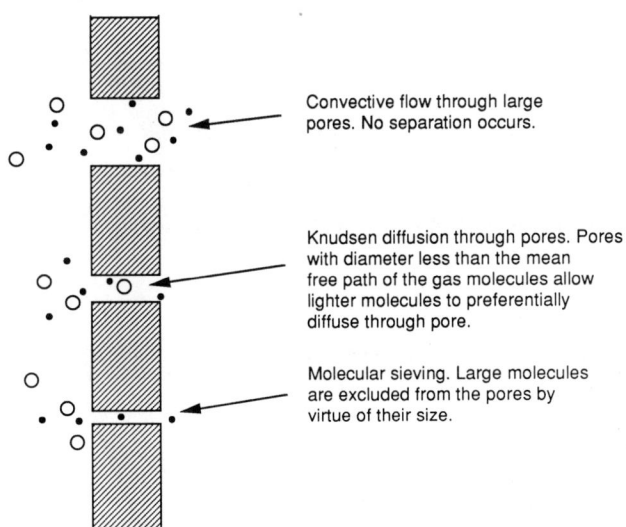

FIGURE 2–2. Mechanisms for gas separation using membranes.

If the pores are small enough, large molecules are unable to pass through them and are excluded by the membrane. This *molecular sieving* is potentially useful in separating molecules of different sizes.

Commercial application of porous membranes for gas separations has not yet been realized. It is difficult to fabricate membranes with the narrow pore-size distribution that is required for molecular sieving, and current materials are fragile. In addition, water and other vapors can condense within the fine pores and block them, greatly affecting transport rates across the membranes.

Research on microporous membranes is, however, continuing. Ceramic membranes (Hsieh, Bhave, and Fleming 1988; Niwa et al. 1988) and the carbon molecular sieve membranes (Koresh and Soffer 1987; Soffer and Koresh 1987) are receiving the most attention, but materials such as polyphosphazenes and porous metallic membranes are also being studied (McCaffrey et al. 1987).

The membranes currently used in most commercial applications are *solution-diffusion membranes*. These membranes are so named because transport occurs when gas molecules dissolve into the membrane and then diffuse across it. While solution-diffusion membranes can be made of a liquid layer supported on a porous support, such membranes have drawbacks that limit their application. Virtually all of today's commercial solution-diffusion membranes are made of polymeric materials, and hence most of our attention will be given to membranes of this type.

Polymeric solution-diffusion membranes for gas separation have been described as having four *structural levels,* each of which influences the ultimate performance of the membrane (Hoehn 1985). These structural levels are as follows:

Level I   Chemical composition of the polymer that forms the selective membrane layer
Level II Steric relationships in repeat units of the selective polymer

Level III   Morphology of the membrane's separating layer

Level IV   The overall membrane structure, including structural relationships between the separating layer and the rest of the membrane.

The first two structural levels involve the chemistry of the polymer and its influence on the rate at which gas molecules diffuse through the membrane. These are discussed further in Chapter 3.

At the third structural level, membranes can be classified as being *symmetric* or *asymmetric*, depending on whether or not the morphology of the membrane is the same across the entire thickness of the membrane. Symmetric (homogeneous) membranes have a uniform density across their thickness, while asymmetric membranes do not.

Commercial solution-diffusion membranes are of the asymmetric type depicted in Figure 2–3. They possess a porous *support layer*, which provides mechanical strength and allows the membrane to tolerate both the manufacturing process and the pressure differentials imposed during operation of the membrane. On one side of this support layer is a thin layer of material, which governs the molecular transport rate and achieves the separation. This thin region is often referred to as the membrane's *active layer* or *skin*. The rate of gas transport across the membrane is inversely proportional to the thickness of the active layer.

The active layer has a defined level IV structure. If the active and support layers are formed in a single operation from a single material the membrane is said to have an *integrally skinned* structure. On the other hand, if the active layer is a coating on the support layer, or if a coating is applied over the active layer, one has what is referred to as a *composite membrane*.

In addition to the four structural levels of membranes, three higher *levels of organization* are required in commercially viable membrane separations. The first of these is the *geometry* of the membrane. Two geometries are commonly used, namely, *flat films* and *hollow fibers*.

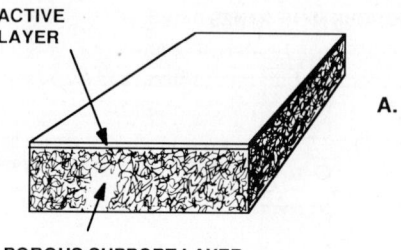

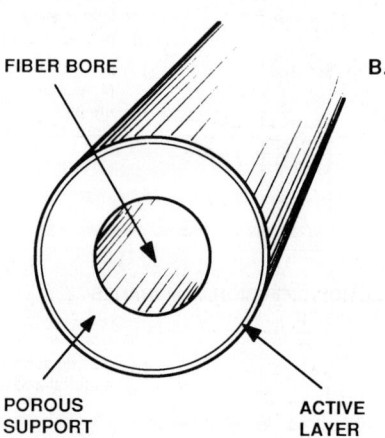

FIGURE 2–3.   Schematic of flat-film (A) and hollow-fiber (B) membranes.

The membrane geometry influences the next higher organizational level, which is the manner in which the membrane is *packaged*. The final package, consisting of the membranes assembled into their pressure containment, is referred to as a *membrane module*, *permeator*, or *membrane separator*.

The highest level of organizational structure is the membrane *system*, which is all of the hardware required along with the permeators in order to achieve a given separation.

The levels of organization defined above are discussed in more detail in Chapter 4.

DRIVING FORCE FOR GAS SEPARATION

Gas separation with membranes occurs via pressure-driven diffusion. This differentiates it from membrane processes such as ultrafiltration in which convective flow plays a role.

From thermodynamics we know that a diffusion process is spontaneous only if the change in chemical potential of the diffusing species is negative. For a component i in an ideal gas mixture, the chemical potential is given by the expression

$$\mu_i = \mu_i^\circ(T) + RT \cdot \ln(p_i), \quad (2\text{–}1)$$

where $\mu_i^\circ(T)$ is the chemical potential of pure component i at the system temperature T and the reference pressure of 1 atm, and p_i is the partial pressure of component i. The change in chemical potential across the membrane from the feed f to the permeate p is thus given as

$$\mu_{ip} - \mu_{if} = [\mu_i^\circ(T) + RT \cdot \ln(p_{ip})]$$

$$- [\mu_i^\circ(T) + RT \cdot \ln(p_{if})]$$

$$= RT \cdot \ln\left(\frac{p_{ip}}{p_{if}}\right). \quad (2\text{–}2)$$

Since the difference in chemical potential must be negative for spontaneous diffusion to occur, $\ln(p_{ip}/p_{if})$ must also be negative, which requires that $p_{ip} < p_{if}$. Therefore, the difference in partial pressure (or, in the case of a real gas, the difference in fugacity) provides the driving force for transport across the membrane.

Classical thermodynamics, however, does not address the rate at which the diffusion process occurs, since diffusion is a kinetic process. Fick's law of diffusion provides a description of the kinetics. This is discussed further in Chapter 3.

From Fick's law the equation for the flux of a given component i across the membrane is derived:

$$J_{vi} = (P_i/l)(p_{if} - p_{ip}), \quad (2\text{–}3)$$

where J_{vi} is the volumetric flux of i across the membrane, P_i is the permeability of i in the membrane polymer, l is the thickness of the active layer, and p_{if} and p_{ip} are, respectively, the partial pressures of i on the feed f and permeate p sides of the membrane. Note that the flux of any component is proportional to the difference in its partial pressures on opposite sides of the membrane, and it can be increased by an increase in the pressure differential across the membrane.

The *relative* transport rate of two components i and j is obtained directly from the above equation as

$$\frac{J_{vi}}{J_{vj}} = \frac{(P_i/l)(p_{if} - p_{ip})}{(P_j/l)(p_{jf} - p_{jp})}. \quad (2\text{–}4)$$

This expression can be rewritten as

$$\frac{J_{vi}}{J_{vj}} = K_1 p_f \left(\frac{x_{if} - (1/K_2)x_{ip}}{x_{jf} - (1/K_2)x_{jp}}\right), \quad (2\text{–}5)$$

where x indicates the mole fraction of the subscripted component in the indicated stream, K_1 is the ratio P_i/P_j, and K_2 is the ratio of the feed pressure to the permeate pressure ($K_2 = p_f/p_p$). From Eq. (2–5) we see that the relative flux of the more rapidly permeating component (assumed to be component i) compared to the other component will increase as the term in brackets increases. Since, as a result of the separation process, both $x_{ip} > x_{if}$ and $x_{jp} < x_{jf}$, increasing K_2 gives the desired increase in the relative flux of component i. Thus, the *pressure ratio* between the feed and permeate streams influences the relative permeation rates and is therefore important in determining the extent of separation that can be achieved.

While partial pressure difference provides the driving force for most gas separations with membranes, externally applied fields can provide an additional driving force for separation. In cases where an electrochemical reaction is used to drive the separation, the additional driving force is $z_i F \Delta\varphi$, where z_i is the charge on the specie, F is Faraday's constant, and $\Delta\varphi$ is the electrical potential across the separation cell. As noted by Winnick (1990), a potential difference of only 60 mV acting on a species with a charge of ± 2 electrons provides the equivalent of 100 atm of pressure driving force. This makes it feasible to separate electrochemically gases that have a low feed concentration. Carbon dioxide, oxygen, and sulfur oxides have been separated in the laboratory by this technique.

A magnetic field can also be used to separate a paramagnetic gas such as oxygen from nonparamagnetic gases such as nitrogen (Yamamoto, Ishizu, and Ichihara 1987), although this technology has yet to be commercialized.

NOTATION

General Notation

See the General Notation section at the beginning of this handbook.

Special Notation

K_1 defined locally as the ratio of component permeabilities, P_i/P_j

K_2 defined locally as the ratio of feed and permeate pressures, p_f/p_p

REFERENCES

Hoehn, H. H. 1985. Aromatic polyamide membranes. *ACS Symp. Ser.* 269:81–98.

Hsieh, H. P., R. R. Bhave, and H. L. Fleming. 1988. Microporous alumina membranes. *J. Membr. Sci.* 39(3):221–242.

Keller, G. E. 1987. *Separations, New Directions for an Old Field*. AIChE Monograph Series No. 17. New York: American Institute of Chemical Engineers.

Koresh, J. E. and A. Soffer. 1987. The carbon molecular sieve membranes: general properties and the permeability of CH_4/H_2 mixture. *Sep. Sci. Technol.* 22(2&3):973–982.

McCaffrey, R. R., R. E. McAtee, A. E. Grey, C. A. Allen, D. G. Cummings, A. D. Appelhans, R. B. Wright, and J. G. Jolley. 1987. Inorganic membrane technology. *Sep. Sci. Technol.* 22(2&3):873–887.

Niwa, M., H. Ohya, Y. Tanaka, N. Yoshikawa, K. Matsumoto, and N. Negishi. 1988. Separation of gaseous mixtures of CO_2 and CH_4 using a composite microporous glass membrane on ceramic tubing. *J. Membr. Sci.* 39(3):301–314.

Soffer, A. and J. E. Koresh. 1987. Separating device. U.S. Patent 4,685,940.

Winnick, J. 1990. Electrochemical membrane gas separation. *Chem. Eng. Prog.* 86(1):41–46.

Yamamoto, A., T. Ishizu, and K. Ichihara. 1987. Method and apparatus for separating gases. U.S. Patent 4,704,139.

3

Theory

Raymond R. Zolandz and Gregory K. Fleming
E. I. Du Pont de Nemours & Company, Inc.

INTRODUCTION
TRANSPORT IN RUBBERY
  POLYMERS
TRANSPORT IN GLASSY
  POLYMERS
  Dual-Mode Model
  Matrix Model
MIXED GAS TRANSPORT

EFFECT OF CONCENTRATION AND
  PRESSURE ON DIFFUSIVITY
FREE-VOLUME DIFFUSION MODEL
EFFECT OF TEMPERATURE
MEMBRANE MATERIALS AND
  PREDICTION OF PERMEABILITY
NOTATION
REFERENCES

INTRODUCTION

Gas diffusion through nonporous membranes is a concentration gradient driven process, which is generally well described by Fick's first law (Crank 1975),

$$J = -D\nabla c, \tag{3-1}$$

where D is the diffusion coefficient and c refers to the local gas or penetrant concentration. For unidirectional diffusion through a flat membrane, Eq. (3–1) can be written for species i as (Crank 1975)

$$J_i = -D_i(c_i)\frac{dc_i}{dx}, \tag{3-2}$$

where $D_i(c_i)$ indicates that the diffusion coefficient can be dependent on the local composition of penetrant i.

The permeability coefficient is defined in terms of the steady-state flux J_i and the pressure or fugacity driving force, Δp_i or Δf_i, normalized by the membrane thickness l,

$$P_i = \frac{J_i}{\Delta f_i/l}. \tag{3-3}$$

The common unit of P_i is Barrer, 10^{-10} cm^3 (STP) · cm/(cm^2 · s · cm Hg).

Substitution of the expression for steady-state flux, Eq. (3–2), into Eq. (3–3) yields

$$P_i = -D_i(c_i)\frac{dc_i}{dx}\frac{l}{\Delta f_i}, \tag{3-4}$$

where Δf_i refers to the difference in fugacity for a given species at the upstream conditions, $x = 0$, and at the downstream conditions, $x = l$, respectively, and dc_i/dx is the local concentration gradient at a given position in the membrane. The product of $D_i(c_i)$ and dc_i/dx must be a constant at each point in a flat membrane, since the steady-state permeability is constant for fixed upstream and downstream conditions (Chern et al. 1985). For the special case in which the concentration of penetrant is zero at the downstream face of the membrane, Eq. (3–4) can be rearranged and integrated for the appropriate boundary conditions (Koros and Chern 1987), namely,

$$\int_0^l \frac{P_i}{l} dx = \int_{c_{il}}^{c_{io}} \frac{D_i(c_i)\, dc_i}{(f_i' - f_i'')}. \qquad (3\text{--}5)$$

This equation can be rearranged to the form

$$P_i = \underbrace{\frac{1}{(f_i' - f_i'')} \int_{f_i''}^{f_i'} [D_i(c_i)\, dc_i]}_{\text{kinetic factor}} \underbrace{\left(\frac{df_i}{dc_i}\right)}_{\text{thermodynamic factor}}.$$

$$(3\text{--}6)$$

Thus, Eq. (3–6) indicates that permeability is a function of the product of kinetic and thermodynamic factors. While the fugacity driving force used in these equations is more thermodynamically correct, pressure is used in many of the following equations. For many permanent gases at moderate pressures (≤ 50 atm), the difference between pressure and fugacity is small.

A more common way of expressing Eq. (3–6) is by the relation

$$P_i = D_i S_i, \qquad (3\text{--}7)$$

where D_i and S_i are the diffusivity and solubility coefficients for component i, respectively. The quantity S_i is thermodynamic in nature and is affected by polymer-penetrant interactions as well as excess interchain gaps in glassy polymers. The average diffusion coefficient D_i is kinetic in nature and largely determined by polymer-penetrant dynamics. Typically, the permeability and solubility are measured independently, and the diffusion coefficient is calculated by Eq. (3–7). In this case, the diffusion coefficient is an average measure of the penetrant diffusivity in the membrane at the upstream and downstream concentrations.

The ability of the membrane to separate gases is characterized by the separation factor

$$\alpha_{ij} = \frac{y_i/y_j}{x_i/x_j}, \qquad (3\text{--}8)$$

where y_i, x_i and y_j, x_j refer to the mole fraction of components i and j in the product and feed streams, respectively.

When the downstream pressure is negligible compared to the upstream pressure, the separation factor is determined by

$$\alpha_{ij}^* = \frac{P_i}{P_j}. \qquad (3\text{--}9)$$

For the special case in which the absolute downstream pressure is close to zero, the above ratio of permeabilities is referred to as the "ideal" separation factor and is so denoted by an asterisk (*). In the absence of plasticizing effects due to high levels of penetrant sorption, the separation factor for a mixed gas case, α_{ij}, can be closely approximated by a ratio of pure gas permeabilities for the individual components i and j (Kim 1988).

For cases in which the downstream pressure cannot be ignored, the true separation factor defined by Eq. (3–8) is related to the ratio of the permeability coefficients by (Barbari, Koros, and Paul 1989; King 1980):

$$\alpha_{ij} = \alpha_{ij,\text{inh}} \left(\frac{x_i(\alpha_{ij} - 1) + 1 - r\,\alpha_{ij}}{x_i(\alpha_{ij} - 1) + 1 - r} \right), \qquad (3\text{--}10)$$

where

$$\alpha_{ij,\text{inh}} = \frac{P_i}{P_j} \qquad (3\text{--}11)$$

and r is the ratio of the downstream and upstream pressures (p_i''/p_i'). For permeation experiments for which the downstream pressure is essentially zero, r is zero and Eq. (3–10) is equivalent to Eq. (3–9).

Also, a useful way to gain a better understanding of how the membrane material performs the separation is by dividing the separation factor into its diffusivity and solubility components. If Eq. (3–7) is substituted into Eq. (3–9), the ideal separation factor can be expressed as (Koros and Chern 1987)

$$\alpha^*_{ij} = \left(\frac{D_i}{D_j}\right)\left(\frac{S_i}{S_j}\right), \quad (3-12)$$

$$\alpha^*_{ij} = \begin{pmatrix}\text{mobility}\\ \text{selectivity}\end{pmatrix}\begin{pmatrix}\text{solubility}\\ \text{selectivity}\end{pmatrix}. \quad (3-13)$$

By understanding how molecular structure within a polymer family affects the two components of the separation factor, we can conceivably optimize the polymer structure for a desired gas separation. Recently, significant advances have been made in the understanding of how polymer structure, especially for glassy polymers, affects the separation ability of materials. This subject is covered later in this chapter in the section "Membrane Materials and Prediction of Permeability."

TRANSPORT IN RUBBERY POLYMERS

A rubbery polymer is an amorphous polymeric material that is above its softening or glass transition temperature under the conditions of use (Billmeyer 1971). The sorption of low molecular weight penetrants in rubbery materials is typically described by Henry's law for cases in which the sorbed concentrations are low (Stannett et al. 1979):

$$c = k_D p, \quad (3-14)$$

where c is the gas concentration in the polymer, cm^3 (STP)/(cm^3 polymer), k_D is the Henry's law coefficient, cm^3 (STP)/(cm^3 polymer atm), and p is the penetrant pressure. For rubbery polymers and with low concentrations of penetrant, the diffusion coefficient in Eq. (3–2) is typically constant and the permeability is independent of the feed pressure as indicated by Eq. (3–15) (Stannett et al. 1979):

$$P = k_D D. \quad (3-15)$$

For rubbery polymers in the presence of high-activity gases or vapors, deviations from simple Henry's law sorption are observed. Sorption and permeation isotherms in such cases are typically strongly convex to the pressure axis. In the absence of strong polymer-penetrant interactions, the Flory-Huggins expression represented by Eq. (3–16) provides a satisfactory description of penetrant solubility for such behavior (Flory 1969; Fleming and Koros 1986):

$$\ln(p/p^o) = \ln(1 - \phi_p) + \phi_p + \chi\phi_p^2, \quad (3-16)$$

where p is the penetrant pressure, p^o is the vapor pressure of the penetrant, ϕ_p is the volume fraction of the polymer, and χ is the so-called Flory-Huggins parameter. For such cases the polymer is likely to be highly swollen, which may result in a concentration-dependent diffusion coefficient. Concentration-dependent diffusion is covered later in this chapter in the section "Mixed Gas Transport."

TRANSPORT IN GLASSY POLYMERS

A glassy polymer is an amorphous polymeric material that is below its softening or glass transition temperature under the conditions of use (Billmeyer 1971). Due to the more restricted segmental motions in glassy polymers, these materials offer enhanced "mobility selectivity" as compared to rubbery polymers (Stannett et al. 1979; Koros and Chern 1987). Because they are inherently more size and shape selective than rubbery materials, glassy polymers are more commonly used as the selective layer in gas separation membranes. Table 3–1 shows the diameters of several common gases.

Glassy polymers are able to discriminate effectively between extremely small differences in molecular dimensions of common gases (e.g., 0.2 to 0.5 Å). For example, Table 3–2 shows a comparison of permeabilities and selectivities for both a rubbery and a glassy polymer. Solubility often dominates diffusional characteristics for transport in rubbery polymers (e.g., $P_{He} < P_{C_2H_4}$). On the other hand, transport in glassy polymers is most often governed

TABLE 3–1. Kinetic Sieving Dimensions of Penetrants Based on Zeolite Sorption Cutoffs (Breck 1974).

Molecule	He	H_2	NO	CO_2	Ar	O_2	N_2	CO	CH_4
Kinetic diameter (Å)	2.6	2.89	3.17	3.3	3.4	3.46	3.64	3.76	3.8

Molecule	C_2H_4	C_3H_8	$n\text{-}C_4$	CF_2Cl_2	C_3H_6	CF_4	$i\text{-}C_4$
Kinetic diameter (Å)	3.9	3.96	4.3	4.3	4.4	4.5	4.7

by penetrant size, although some exceptions have been observed with unique glassy materials that contain large amounts of free volume, such as poly(1-trimethylsilyl-1-propyne) (Ichiraku, Stern, and Nakagawa 1987).

As noted above, transport through rubbery polymers is typically described by a solution-diffusion process, whereby solution of low molecular weight penetrants into rubbery polymers is similar to penetrant sorption into low molecular weight liquids. Consequently, gaseous sorption at low concentrations in rubbery polymers can be described by a linear Henry's law relation as expressed by Eq. (3–14) and illustrated in Figure 3–1(a) (Koros and Chern 1987). For high penetrant uptake in rubbery polymers, sorption isotherms are described in Eq. (3–16) and illustrated in Figure 3–1(b) (Koros and Chern 1987). Gas sorption in glassy polymers is more complex, and a typical sorption isotherm is presented in Figure 3–1(c) for gases up to moderate pressures (20 to 30 atm) (Koros and Chern 1987).

The physical characteristic of glassy polymers that is commonly linked to the complex sorption isotherm is the "unrelaxed" volume locked into these materials when they are quenched below the glass transition. As illus-

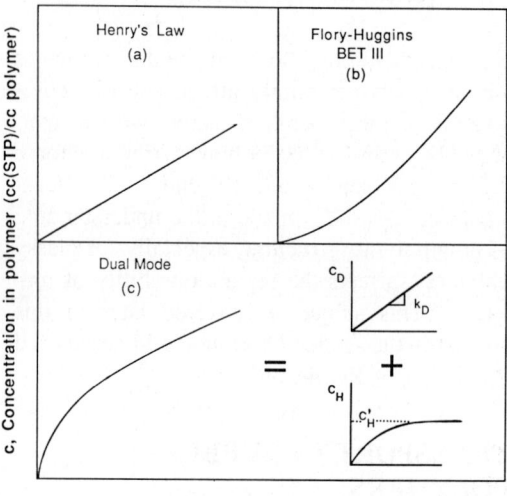

FIGURE 3–1. Typical gas sorption isotherm forms for polymeric media (reprinted from Koros and Chern 1987 with permission).

trated in Figure 3–2, there is a break in the volume versus temperature plot for a polymer as the temperature is lowered below the glass transition T_g. Below the glass transition, the excess or "unrelaxed volume" is $V_g - V_l$. This excess volume is thought to be the result of trapped nonequilibrium chain conformations in

TABLE 3–2. Permeabilities and Selectivities of Various Gas Pairs in Silicone Rubber (Stern, Shah, and Hardy 1987) and Polycarbonate (Koros, Chan, and Paul 1977; Jordan 1988).

Polymer	T (°C)	P_{He} (Barrer)	P_{He}/P_{CH_4}	$P_{He}/P_{C_2H_4}$	P_{CO_2} (Barrer)	P_{CO_2}/P_{CH_4}	$P_{CO_2}/P_{C_2H_4}$	P_{O_2} (Barrer)	P_{O_2}/P_{N_2}
PDMS	35	561	0.41	0.15	4550	3.37	1.19	933	2.12
PC	35	14	50	33.7[a]	6.5	23.2	14.6[a]	1.48	5.12

PDMS is poly(dimethyl siloxane). Values for PDMS at 100 psia.
PC is bisphenol-A-polycarbonate. Values for PC at 10 atm (147 psia).
[a]Reported by Jordan (1988).

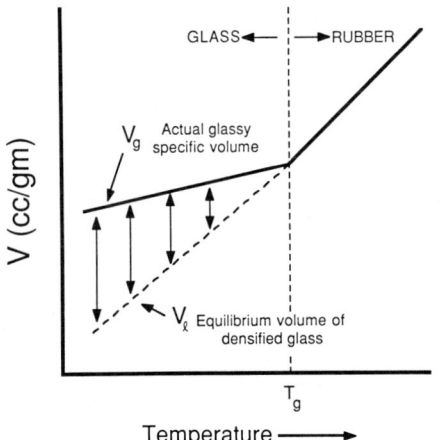

FIGURE 3–2. Polymer specific volume as a function of temperature.

quenched glasses, which result from the extraordinarily long relaxation times for segmental motions in the glassy state. The excess volume present in glassy polymers allows for the accommodation of additional penetrant above that observed in low molecular weight liquids and rubbers (Chern et al. 1983b).

Dual-Mode Model

Gas solubility in glassy polymers can be accurately described by the so-called dual-mode sorption model. Inherent in the dual-mode sorption model is the concept of sorption into two idealized environments (Barrer, Barrie, and Slater 1958; Michaels, Vieth, and Barrie 1963). One population of sorption is viewed as arising from uptake into a dissolved environment similar to sorption in low molecular weight liquids and rubbery polymers, and is described by a Henry's law relation. The second population of sorption is viewed as being due to uptake into the unrelaxed volume or "microvoids" present in glassy polymers. This population of sorption is described as a Langmuir "hole-filling" process. The total sorption is the sum of these two populations as follows:

$$c = c_D + c_H, \quad (3-17)$$

$$c = k_D p + \frac{c'_H b p}{1 + bp}, \quad (3-18)$$

where k_D is the Henry's law constant, p is the penetrant pressure, and c'_H and b are the Langmuir capacity constant and affinity constant, respectively. The component parts that make up the dual-mode sorption model are illustrated in Figure 3–1(c). The success of the dual-mode model in describing penetrant sorption in glassy polymers is due to the physical significance that can be related to model parameters. This subject has been covered in detail in previous studies (Barrer, Barrie, and Slater 1958; Michaels, Vieth, and Barrie 1963; Paul and Koros 1976; Chan, Koros, and Paul 1978; Koros and Paul 1978a, 1978b; Stannett et al. 1979; Chern et al. 1983b; Koros and Chern 1987).

The companion transport model to the dual-mode sorption model expresses the local flux J in terms of a two-part contribution (Paul and Koros 1976; Koros, Chan, and Paul 1977; Chan, Koros, and Paul 1978):

$$J = -D_D \frac{dc_D}{dx} - D_H \frac{dc_H}{dx}, \quad (3-19)$$

where D_D and D_H refer to the mobility of the dissolved and Langmuir sorbed components, respectively. Due to differences in the energetics of diffusional jumps in these two environments, D_D is typically much larger than D_H except for noncondensable gases such as helium or hydrogen. The transport coefficients obey Arrhenius expressions, and the activation energy tends to be larger for D_H than for D_D (Paul and Koros 1976; Koros, Chan, and Paul 1977; Chan, Koros, and Paul 1978).

For the special case in which the downstream pressure is effectively zero, the appropriate expression derived from the dual-mode sorption and transport models for steady-state permeability of a pure component in a glassy polymer is given by (Koros and Paul 1978b):

$$P = k_D D_D \left(1 + \frac{FK}{1 + bp}\right), \quad (3-20)$$

where $F = D_H/D_D$ and $K = c'_H b/k_D$ are convenient dimensionless groups and p is the upstream driving pressure. The first term in Eq. (3–20) describes transport in the Henry's law

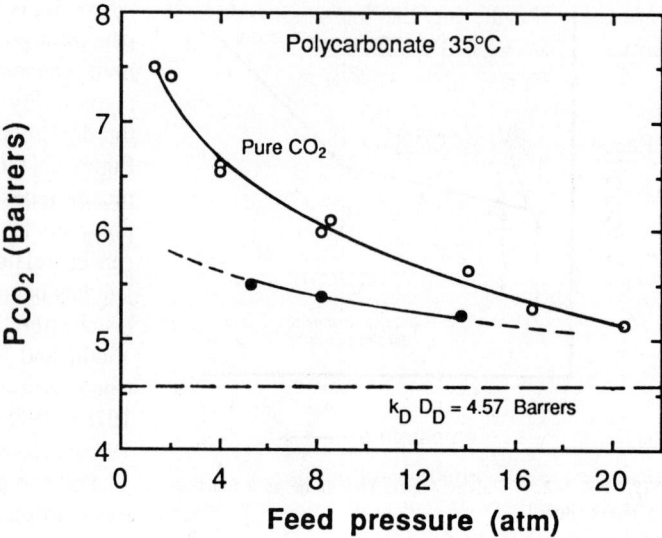

FIGURE 3-3. Pressure dependence of CO_2 transport in glassy polycarbonate. Also depicted is the effect of a low partial pressure of isopentane (117 mm Hg) on CO_2 flux (reprinted from Chern et al. 1983a with permission).

environment, while the second term is related to the Langmuir environment. Equation (3–20) approaches a limiting value equal to $k_D D_D$ at high pressures. In the dual-mobility formalism this asymptotic approach is related to saturation of the Langmuir capacity at high pressures, and as the pressure increases additional flux contributions are primarily due to the Henry's law term, $k_D D_D$. This concept is illustrated in Figure 3–3, which shows the CO_2 permeability in glassy polycarbonate. As the CO_2 pressure is increased, the permeability asymptotically approaches the $k_D D_D$ limit.

The above expression for dual-mode transport accounts for two types of diffusion jumps. The terms D_D and D_H represent diffusional jumps in the Henry's law and Langmuir environment, respectively. Barrer (1984) extended the dual-mode transport model to account for the effect of neighboring gas molecules on transport, and it is assumed in this treatment that a gas molecule can execute four distinct diffusional jumps. If the Henry's law and Langmuir environments are denoted as D and H, respectively, the four possible jumps are depicted by

$D \rightarrow D$ (dissolved to dissolved)
$H \rightarrow H$ (hole to hole)
$D \rightarrow H$ (dissolved to hole)
$H \rightarrow D$ (hole to dissolved)

where the four diffusion coefficients are represented by D_{DD}, D_{HH}, D_{DH}, and D_{HD}. The resulting expression for dual-mode transport developed by Barrer is

$$P = k_D D_{DD} + \frac{c'_H b(D_{HH} + D_{HD}) - k_D D_{DH}}{1 + bp}$$

$$+ 2k_D D_{DH} \frac{\ln(1 + bp)}{bp}. \qquad (3\text{–}21)$$

While Eq. (3–20) allows for determination of D_D and D_H from a linear regression of P plotted versus $1/(1 + bp)$, determination of the diffusion coefficients in the more complex expression above [Eq. (3–21)] requires nonlinear regression techniques.

Matrix Model

Another model describing gas sorption and transport in glassy polymers has been proposed (Raucher and Sefcik 1983a, 1983b). The so-called matrix model was developed based on the assertion that gas molecules exist in a glassy material as a single population. The peculiar form of the sorption and permeability concentration dependence is said to be due to gas/polymer interactions. These gas/polymer interactions were proposed to be strong enough to change the polymer's structural and dynamic properties by altering the interchain potential energy (Raucher and Sefcik 1983a).

In the matrix model, the concentration of sorbed penetrant is related to gas pressure by (Raucher and Sefcik 1983b):

$$c = \sigma_0 p \, \exp(-\sigma_c c), \quad (3\text{--}22)$$

which is most often simplified to the form

$$c = \frac{\sigma_0 p}{1 + \sigma_c c}, \quad (3\text{--}23)$$

where σ_c is said to indicate the ability of the penetrant to affect main chain polymer motions. The diffusion coefficient is written as (Raucher and Sefcik 1983b):

$$\bar{D} = D_0 \, \exp(\beta c), \quad (3\text{--}24)$$

which for small values of βc is closely approximated by

$$\bar{D} = D_0 \, (1 + \beta c), \quad (3\text{--}25)$$

where D_0 is the infinite dilution diffusion coefficient and β is said to be a measure of penetrant-induced depression of the polymer glass transition.

The resulting expression for the permeation coefficient is obtained by substituting Eqs. (3–23) and (3–25) into Eq. (3–7) (Raucher and Sefcik 1983b):

$$P = D_0 \sigma_0 \frac{(1 + \beta c)}{(1 + \sigma_c c)}. \quad (3\text{--}26)$$

Little work has been done to relate the parameters of the matrix model to physically meaningful measures of polymer and penetrant properties. A study (Barbari, Koros, and Paul 1988) of gas sorption in various polymers indicated that σ_c in Eq. (3–23) was larger for lower sorbing gases such as N_2 than for more highly sorbing gases such as CO_2. Because N_2 is not expected to affect the main chain motions of a glassy polymer more than CO_2, Barbari, Koros, and Paul (1988) concluded that the parameters of the model did not possess easily definable physical significance.

MIXED GAS TRANSPORT

Transport of gas mixtures in rubbery polymers can ideally be represented by the product of the Henry's law constant and diffusion coefficient as illustrated by Eq. (3–15). In the absence of plasticization or hydrostatic compression effects, gas transport in rubbery polymers is unaffected by the presence of other penetrants (Shakespear 1918; Yi-Yan, Felder, and Koros 1980).

On the other hand, the presence of secondary components can have profound effects on the transport rate of penetrants in glassy polymers (Robeson 1969; McCandless 1972; Pye, Hoehn, and Panar 1976a, 1976b; Antonson et al. 1977; Chern et al. 1983a, 1983c, 1984). In general, the transport rate of a penetrant A in a glassy polymer is depressed by the presence of an additional penetrant B. This flux depression for A is thought to be due to a reduction in the penetrant concentration or solubility, namely, Eq. (3–7). The total sorption level for A is depressed by the second component B due to exclusion of A from Langmuir sorption sites that would be available in the pure gas case.

Studies by Sanders (Sanders et al. 1983; Sanders and Koros 1986) have provided excellent support for the above rationalization for "competitive" sorption. Figure 3–4 clearly illustrates the progressive reduction in CO_2 sorption level as the ethylene pressure is increased in the presence of a constant CO_2 partial pressure for poly(methyl methacrylate) (PMMA) at 35°C.

Likewise, Figure 3–3 shows how the transport rate of CO_2 in polycarbonate is depressed by the presence of even a low level of a heavy hydrocarbon such as isopentane (Chern et al. 1983a). Increases in the isopentane level could eventually eliminate the Langmuir component of transport, reducing the CO_2 flux to the $k_D D_D$ limit.

Koros and coworkers (Koros 1980; Koros et al. 1981) have used the physical rationale inherent in the dual-mode formalism to extend the

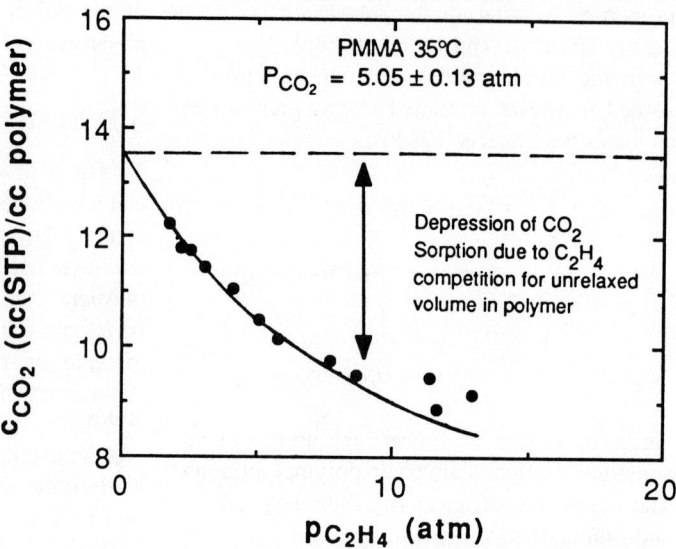

FIGURE 3–4. Depression of CO_2 sorption in PMMA below the pure component level by the presence of C_2H_4. The CO_2 sorption depression is a result of exclusion from unrelaxed volume sorption sites with increasing C_2H_4 partial pressure (reprinted from Sanders and Koros 1986 with permission).

pure component expressions given in Eqs. (3–18) and (3–20) to account for mixed penetrants:

$$c_A = k_{DA}p_A + \frac{c'_{HA}b_A p_A}{1 + b_A p_A + b_B p_B}, \quad (3\text{–}27)$$

$$P_A = k_{DA}D_{DA}\left(1 + \frac{F_A K_A}{1 + b_A p_A + b_B p_B}\right). \quad (3\text{–}28)$$

In these expressions, A refers to the primary component while B refers to the secondary component. Note that the above expressions can be extended to account for additional penetrants by adding the appropriate $b_i p_i$ terms to the Langmuir portion of these equations. Both the permeability depression in Figure 3–3 and the sorption depression in Figure 3–4 are predicted very well by Eqs. (3–27) and (3–28).

Studies have also extended the dual-mode expression for mixed gas permeability [Eq. (3–28)] to account for interdiffusion between the Henry's law D and Langmuir sites H (Sada et al. 1988; Story and Koros 1989). These treatments are based on the work of Barrer (1984) for the pure gas case. The interested reader is referred to the references for further details.

EFFECT OF CONCENTRATION AND PRESSURE ON DIFFUSIVITY

For transport in both rubbery and glassy polymers, the flux can be written in terms of Fick's law with an effective diffusion coefficient $D_{\text{eff}}(c)$ that is dependent on local concentration:

$$J = -D_{\text{eff}}(c)\frac{dc}{dx}. \quad (3\text{–}29)$$

The concentration dependence of the diffusion coefficient for penetrant transport in polymers can be evaluated from steady-state permeability and equilibrium solubility data without any reference to a particular model. Rearrangement of Eq. (3–6) results in the expression (Koros, Chan, and Paul 1977; Koros and Chern 1987):

$$P p_0 = -\int_{c_0}^{c_l} D_{\text{eff}}(c)\, dc, \quad (3\text{–}30)$$

where $c_l = 0$ is assumed for the case where the penetrant pressure is zero at the downstream face. Then, by applying the Liebnitz rule,

$$D_{\text{eff}}(c_0) = p_0\left.\frac{dP}{dc}\right|_{c_0} + P\left.\frac{dp}{dc}\right|_{c_0}, \quad (3\text{–}31)$$

which can be simplifed to the form (Koros, Chan, and Paul 1977; Koros and Chern 1987):

$$D_{\text{eff}}(c_0) = \left.\frac{dp}{dc}\right|_{p_0}\left(p_0\left.\frac{dP}{dp}\right|_{p_0} + \left.P\right|_{p_0}\right). \quad (3\text{–}32)$$

Hence, the concentration dependence of the local diffusion coefficient $D_{\text{eff}}(c_0)$ can be de-

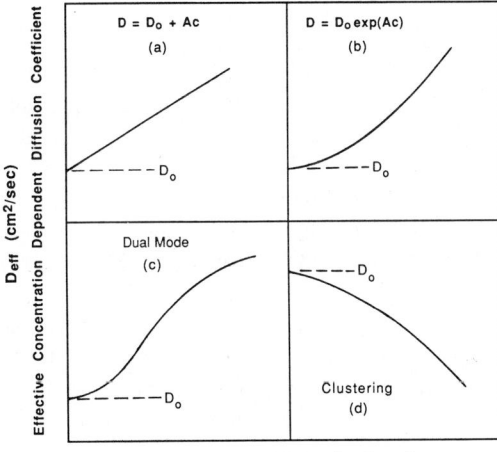

FIGURE 3–5. Typical forms for concentration-dependent diffusion coefficients in polymeric media (reprinted from Koros and Chern 1987 with permission).

termined by the pressure dependence of both the permeability and solubility isotherms.

For low sorbing gases such as N_2 and even CH_4 in rubbery polymers, both P and dc/dp are essentially constant, leading to a constant diffusion coefficient. On the other hand, when concentration dependence is observed, the form will generally take the shape of one of the curves in Figure 3–5, as illustrated by Koros and Chern (1987). Transport of penetrants that have a strong affinity for rubbery polymers can be characterized by a linearly increasing diffusion coefficient as shown in Figure 3–5(a). In cases for which the penetrant has a strong plasticizing effect on the rubbery polymer, an exponentially increasing diffusion coefficient is found, as illustrated in Figure 3–5(b) (Rogers 1965; Fujita 1968; Vrentas and Duda 1977a; Stern and Frisch 1981).

Clustering of penetrant molecules is often observed for diffusion of hydrogen bonding species in relatively nonpolar materials (Barrie 1968). When such a clustering effect occurs, groups of penetrant molecules transport together, the effective diameter for the penetrant is increased, and the effective diffusion coefficient is reduced. As shown in Figure 3–5(d), the clustering effect usually leads to a decreasing diffusion coefficient because clus-

tering becomes more severe at higher concentrations.

For gas transport in glassy polymers, the diffusion coefficient is usually found to be concentration dependent even in the absence of a strong affinity of the penetrant for the material. As shown in Figure 3–5(c), the diffusion coefficient for a gas in a glassy polymer is usually convex to concentration at low sorption levels, and is concave to concentration at higher sorption levels (Koros, Chan, and Paul 1977).

This seemingly complicated dependence of the diffusion coefficient can be independently estimated by substitution and differentiation of the expressions for dual-mode sorption and transport using Eq. (3–32). The resulting expression for $D_{\text{eff}}(c)$ is shown in Eq. (3–33) in terms of the local concentration of dissolved penetrant c_D as well as other dual-mode terms defined above (Koros, Chan, and Paul 1977, Chern et al. 1983b):

$$D_{\text{eff}}(c) = D_D \left\{ \frac{1 + [FK/(1 + \alpha c_D)^2]}{1 + [K/(1 + \alpha c_D)^2]} \right\}, \quad (3\text{–}33)$$

where $\alpha = b/k_D$.

As indicated previously, the concentration dependence of the diffusion coefficient can also be evaluated independently of a transport model by examination of the basic sorption and permeation data via Eq. (3–32).

Figure 3–6 shows a plot of the concentration dependence of D_{eff} as determined by Eq. (3–32)

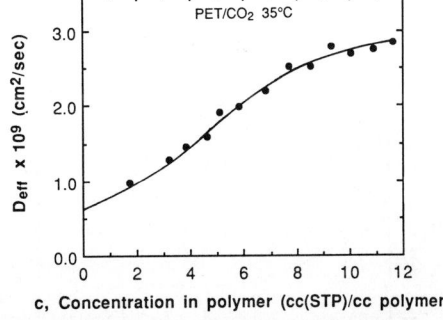

FIGURE 3–6. Concentration-dependent diffusion coefficient for CO_2 in poly(ethylene terephthalate). The data points are determined by Eq. (3–32), and the solid line is calculated using Eq. (3–33), which is based on dual-mode assumptions (reprinted from Koros and Paul 1978b with permission).

versus the predicted dependence calculated by Eq. (3–33) for CO_2 transport in poly(ethylene terephthalate) (PET) (Koros, Chan, and Paul 1977). Here the concentration dependence of the diffusion coefficient is shown to be predicted very well by the dual-mode approximation and agrees with the expected dependence shown in Figure 3–5(c).

In some membrane applications, feed streams contain very high levels of strongly interacting gases such as CO_2 and H_2S. Exposure to high levels of these gases results in a large depression of the effective glass transition temperature T_g for some polymeric materials (Chiou, Barlow, and Paul 1985; Sanders 1988). The glass transition depression is thought to increase the size of segmental motions that participate in the diffusion process, and the increased mobility of the polymeric chains usually results in a drop in the selectivity.

The effect of CO_2 on glassy polymer plasticization has received much attention (Chiou, Barlow, and Paul 1985; Jordan, Koros, and Fleming 1987; Sanders 1988; Jordan, Fleming, and Koros 1990). The CO_2 pressure necessary to induce plasticization is widely different and dependent on material properties. Koros and Hellums (1989a, 1989b) recently presented results for a series of well-characterized polycarbonates. Figure 3–7 shows CO_2 permeability results for silicone rubber and a series of polycarbonates. While the permeability for the rubbery material steadily increases with CO_2 pressure, permeability in the polycarbonates is found to reach a minimum followed by an upswing. The minima shown here in the permeability isotherms closely coincide with a decline in selectivity for the material.

The local diffusion coefficient $D_{\text{eff}}(c)$ is a useful measure of transport plasticization. The local diffusion coefficient was determined from

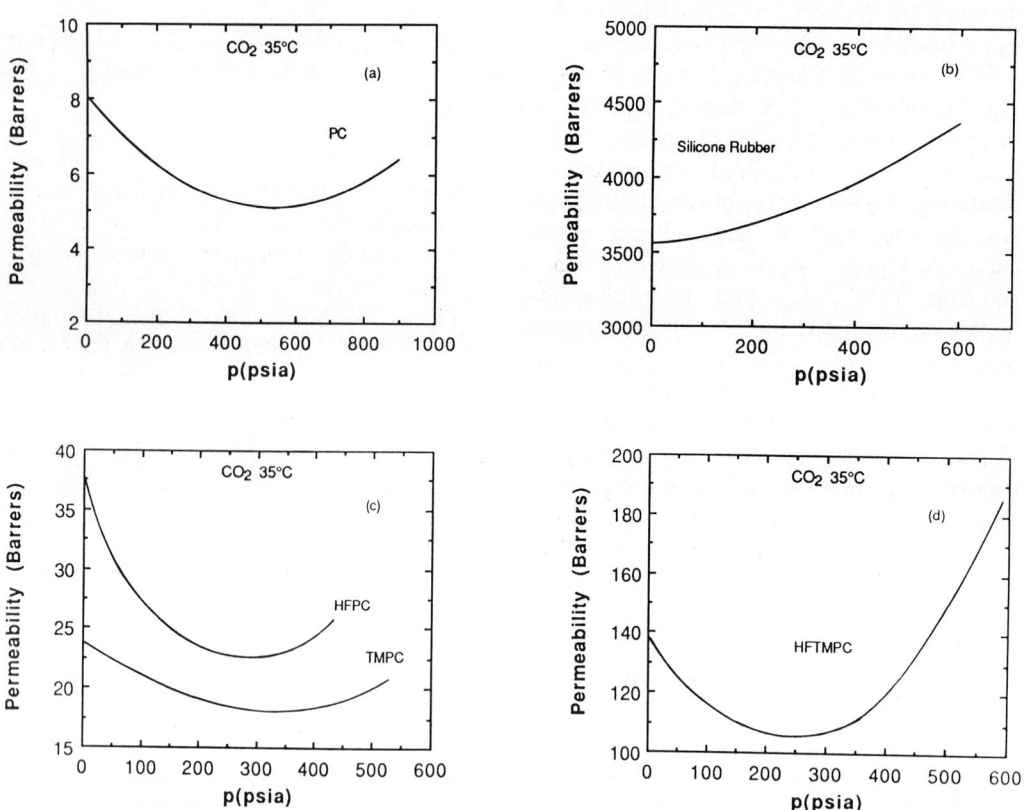

FIGURE 3–7. Carbon dioxide permeability in silicone rubber and polycarbonates at 35°C. Structures for the polycarbonates are illustrated in Table 3–3 (reprinted from Koros and Hellums 1989a, 1989b with permission).

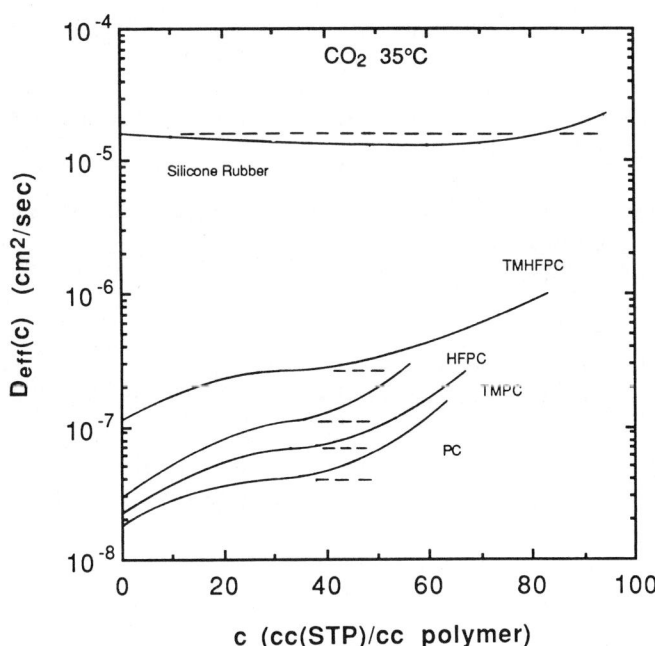

FIGURE 3–8. Local concentration-dependent diffusion coefficients for CO_2 in silicone rubber and polycarbonate materials in Table 3–3(a–d) (reprinted from Koros and Hellums 1989a, 1989b with permission).

the CO_2 permeability and sorption isotherms for these materials using Eq. (3–32), and the resulting relations are shown in Figure 3–8 (Koros and Hellums 1989a, 1989b). At moderate concentrations, $D_{eff}(c)$ rises to a plateau level. Such behavior is similar to that presented earlier in Figure 3–6 for PET/CO_2 and is not indicative of plasticization. However, the upward deviation from the plateau level (e.g., dashed lines) at higher concentrations does indicate the onset of transport plasticization. The pressure at which this plasticization condition is reached roughly corresponds to the minima in the permeability isotherms in Figure 3–7. This shows that the useful operating range for these materials is widely variable and is dependent on a complex interrelation between the penetrant solubility and polymer chain rigidity. Currently, suitable process conditions for a material cannot be predicted *a priori*, so tests must be performed to determine these limits for new materials.

FREE-VOLUME DIFFUSION MODEL

The free-volume theory of diffusion in polymers was first developed mainly by Fujita (Fujita, Kishimoto, and Matsumoto 1960; Fujita 1961, 1968) and since its origination several versions have evolved (Frisch, Klempner, and Kwei 1971; Vrentas and Duda 1976, 1977b; Stern, Kulkarni, and Frisch 1983). The premise for free-volume diffusion in polymers is free-volume transport in liquids. Useful forms of the free-volume theory of diffusion have been developed primarily through the works of Duda and Vrentas (Vrentas and Duda 1976, 1977b) and Stern and Frisch (Frisch, Klempner, and Kwei 1971; Stern, Kulkarni, and Frisch 1983). Although simple and useful versions of the theory have been developed, the primary challenge in applying the theory is providing precise physical definitions for free-volume parameters.

The free-volume theory of transport postulates that movement of molecules depends on the free volume available as well as the availability of energy sufficient to overcome polymer-polymer attractive forces (Vrentas and Duda 1986). Vrentas and Duda (1986) propose that the specific volume of the polymer and polymer-penetrant mixture is comprised of three components: (1) the occupied volume, which is the volume of the equilibrium liquid at $0°K$, (2) the interstitial free volume, which is small and distributed uniformly throughout the material, and (3) the hole free volume, which is large enough to facilitate molecular transport. Redistribution of the interstitial free volume requires a large energy input. On the other hand, redistribution of the hole free volume requires no additional energy, so this volume randomly migrates throughout the polymer matrix. Random mobility of the hole free volume may facilitate both a slow interdiffusion of polymer chains as well as penetrant transport.

Stern has proposed that the flux of penetrant can be written in terms of the local volume fraction of penetrant in the polymer, ϕ, and the mutual diffusion coefficient, D (Stern, Kulkarni, and Frisch 1983):

$$J = -\frac{D}{1-\phi}\frac{d\phi}{dx}, \qquad (3\text{-}34)$$

where the diffusion coefficient D is assumed to be a function of the local concentration as discussed above in the section "Effect of Concentration and Pressure on Diffusivity."

If the partial molar volume of penetrant can be approximated by the gaseous molar volume in organic liquids, the volume fraction of the penetrant can be estimated from the penetrant sorption level. Fleming and Koros (1986) have discussed the similarity of partial molar volumes of gases in low molecular weight liquids and rubbery polymers. Because rubbery polymers are essentially high molecular weight liquids, a reasonable estimate of the partial molar volume of a penetrant in a rubbery polymer is the infinite dilution partial molar volume of the penetrant in organic liquids (Horiuti 1931; Chueh and Prausnitz 1967).

The relationship between the thermodynamic diffusion coefficient D_T and the mutual diffusion coefficient D is given by

$$D_T = \left(\frac{D}{1-\phi}\right)\left[\frac{d\ln a}{d\ln \phi}\right]_T^{-1} \qquad (3\text{-}35)$$

Because gases exhibit a relatively low solubility in polymers, the term $(d\ln a/d\ln \phi)_T$ is ~ 1.0 (Stern, Kulkarni, and Frisch 1983); hence, the flux can be expressed by

$$J = -D_T \frac{d\phi}{dx}. \qquad (3\text{-}36)$$

The fractional free volume V_f in a polymer has been expressed as a linear function of temperature, pressure, and concentration (Stern, Fang, and Frisch 1972; Stern, Kulkarni, and Frisch 1983):

$$V_f = \phi_{fs}^\circ + \Delta\alpha(T - T_s) - \Delta\beta(p - p_s) + \gamma\phi.$$

$$(3\text{-}37)$$

Here ϕ_{fs}° is the fractional free volume of the pure polymer at the reference temperature and pressure, T_s (or T_g) and p_s. Usually the reference temperature T_s is chosen as the glass transition temperature T_g, and the reference pressure p_s is 1 atm. The terms $\Delta\alpha$ and $\Delta\beta$ are typically taken as the differences in the thermal expansion coefficients and compressibilities above and below T_g, respectively. The parameter γ is used to characterize the ability of the penetrant to plasticize the polymer.

According to Fujita, Kishimoto, and Matsumoto (1960) and Fujita (1961), the self-diffusion coefficient is dependent on the fractional free volume according to the relationship

$$D_i = RTA_d \exp(-B_d/V_f) = u_iRT, \qquad (3\text{-}38)$$

where A_d and B_d are related to the size and shape of the penetrant molecule and must be determined empirically, and V_f is given by Eq. (3-37). The term u_i is the so-called "mobility" of the penetrant and is the inverse of the "resistance coefficient" of the medium. The resistance coefficient is equal to the product of the effective viscosity of the medium and the effective diameter of the penetrant (Bearman 1961).

The mutual diffusion coefficient is related to the mobility and mass fraction of the penetrant (w_i) by the relation (Koros and Hellums 1989b)

$$D_{ij} = RTu_i(1 - w_i)\frac{d\ln a_i}{d\ln w_i}. \qquad (3\text{-}39)$$

For rubbery polymers at low pressures (<300 psia), it has been shown that the solubility of gases can be represented by Henry's law (Fleming 1987). In these cases, the term $d\ln a_i/d\ln w_i$ is equal to unity. Combination of Eqs. (3-37) through (3–39) provides an expression that relates the mobility of the penetrant to the free volume of the polymer:

$$u_i = A_d \exp\left(\frac{-B_d}{\phi_{fs}^{\circ} + \Delta\alpha(T - T_s) - \Delta\beta(p - p_s) + \gamma\phi_i}\right)$$

$$\cong \frac{D_{ij}}{RT(1 - w_i)}. \quad (3\text{–}40)$$

A recent study (Jordan and Koros 1990) used a multivariable nonlinear least-squares fit of Eq. (3–40) to determine A_d, B_d, and γ for transport of various gases in silicone rubber. Parameters determined in their analysis supported the physically observed effects of hydrostatic compression and plasticization as illustrated in sorption, transport, and volume dilation measurements (Fleming 1987; Jordan and Koros 1990). For low sorbing gases such as He and N_2, mobilities were found to decrease with increasing pressure because hydrostatic effects dominate plasticization. On the other hand, highly condensable penetrants such as CO_2 and C_2H_4 were found to act as plasticizers as was evidenced by increases in mobility within the pressure range studied. A complex interplay between plasticization and compression was found for gas mixtures of low sorbing (e.g., N_2) and highly condensable gases (e.g., CO_2). Figure 3–9 illustrates the behavior of CO_2 mobility in silicone rubber for different gas compositions as calculated by Jordan and Koros (1990) using Eq. (3–40). As the N_2 content increases, the CO_2 mobility is depressed due to the compressive nature of N_2 on silicone rubber.

EFFECT OF TEMPERATURE

Temperature can have a large effect on the transport rate of small penetrants in polymeric media. The transport of small molecules in rubbery polymers is viewed as an activated process and obeys an Arrhenius relationship (Barrer 1937; Van Amerongen 1964):

$$D = D^{\circ} \exp(-E_d/RT), \quad (3\text{–}41)$$

where E_d is the activation energy, or the energy required to facilitate a diffusional jump for a penetrant molecule. The preexponential factor, D°, is often related to the entropy of activation and the diffusional jump length.

The solubility of gases in rubbery polymers usually exhibits a similar behavior (Van Amerongen 1964):

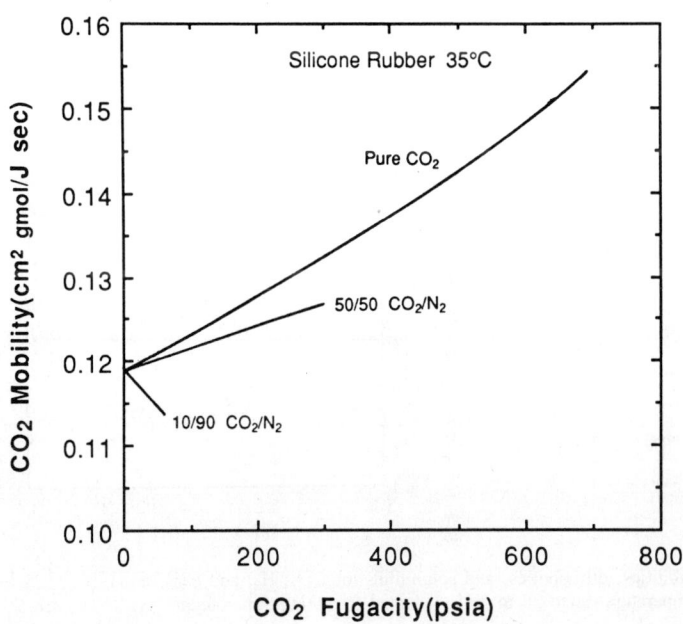

FIGURE 3–9. Carbon dioxide mobility calculated using Eq. (3–40) for pure CO_2 and mixtures with N_2 (reprinted from Jordan and Koros 1990 with permission).

$$S = S° \exp(-E_s/RT), \quad (3\text{-}42)$$

where E_s is the enthalpy of sorption.

In accordance with Eqs. (3–41) and (3–42), the temperature dependence of transport and solubility parameters for gases in rubbery polymers is usually expressed in semilogarithmic form as a function of $1/T$. Shown in Figure 3–10 are the permeability, diffusivity, and solubility coefficients for gases in natural rubber as a function of inverse temperature (Van Amerongen 1964; Koros and Chern 1987). Figure 3–10 shows that the diffusion coefficient is most strongly dependent on temperature and is shown to change by 200 to 300% over the temperature range of 25 to 50°C. Solubility coefficients are found to vary less than 30% within this same range. The permeability, which is made up of the diffusivity and solubility coefficients, changes within the observed ranges for the individual factors.

The temperature dependence of gas transport in glassy polymers is an activated process similar to that observed for rubbers. Shown in Figure 3–11 are solubility, diffusivity, and permeation coefficients for poly(ethylene terephthalate) (PET)/CO_2 as measured by Koros (Koros and Paul 1978a, 1978b). By means of Eq. (3–7), the effective diffusion coefficient was calculated from the sorption and permeation measurements of Koros at 20 atm.

Inflections near the polymer glass transition in Arrhenius plots of D were observed in early studies by Meares (1954) and led to the first quantified studies of dual-mode sorption (Barrer, Barrie, and Slater 1958; Michaels, Vieth, and Barrie 1963). Although the transport process for gases in glassy and rubbery polymers

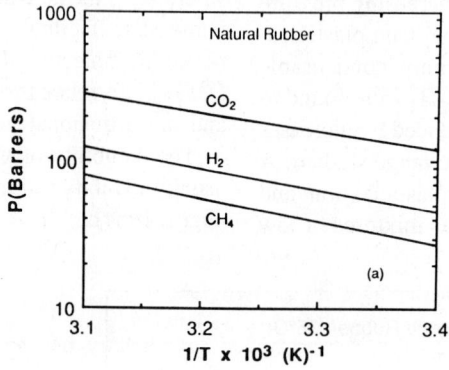

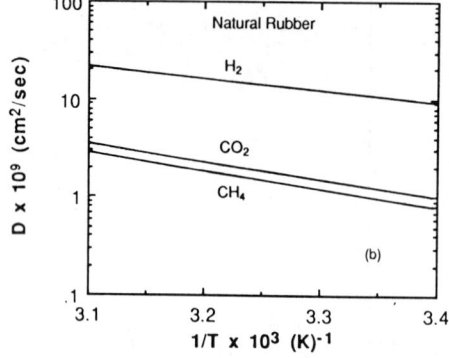

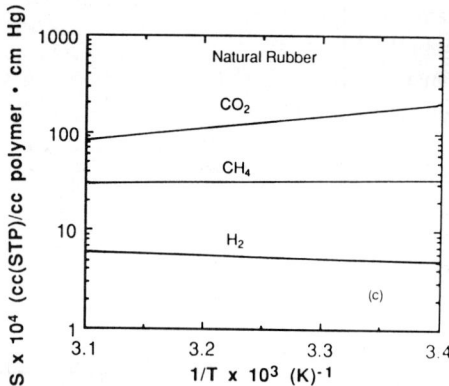

FIGURE 3–10. Permeabilities, diffusivities, and solubilities for CO_2, H_2, and CH_4 in natural rubber over the temperature range 25 to 50°C (adapted from Van Amerongen 1964; Koros and Chern 1987).

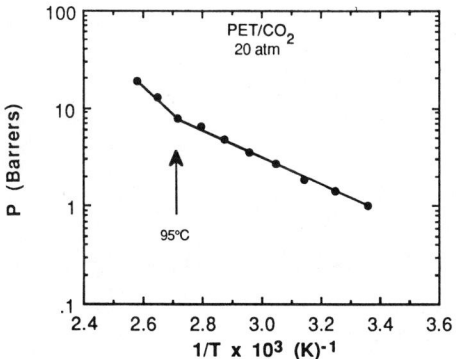

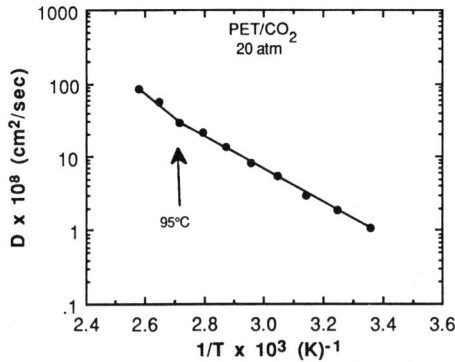 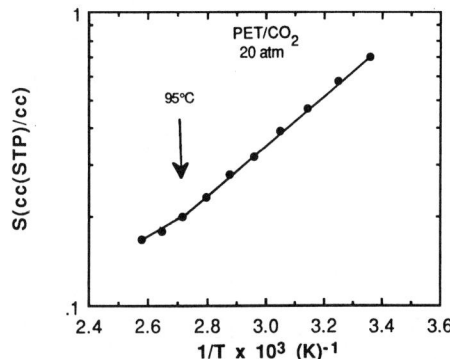

FIGURE 3–11. Permeability, diffusivity, and solubility for CO_2 in PET over the temperature range 25 to 115°C (adapted from Koros and Paul 1978a, 1978b). The glass transition for PET is 95°C (Koros and Paul 1978b).

can be similarly treated as an activated process, Figure 3–11 shows that the activation energy (namely, the slope) is smaller in the glassy state. The lower activation energy for the glassy state is thought to be due to the abrupt change in mobility for the polymer matrix as the T_g is transversed. Presumably, in the much less mobile environment of the glassy polymer, smaller segments are involved in the diffusion process. Also, because these smaller segments govern penetrant diffusion jumps, more subtle discrimination in penetrant sizes is obtained, which leads to higher selectivities.

Converse to the behavior observed for the diffusion and permeation coefficients, the solubility coefficient is much more strongly dependent on temperature below T_g. Below T_g, the sorption level is thought to be strongly affected by sorption into the Langmuir-like environment. As shown in Figure 3–12, the Langmuir capacity has been found to be strongly dependent on temperature and is generally predicted to disappear at the glass transition (Koros and Paul 1981; Chern et al. 1983b).

MEMBRANE MATERIALS AND PREDICTION OF PERMEABILITY

An important result of the recent and rapidly growing interest in gas separation membranes has been the development of a host of new materials specifically designed to enhance gas transmission and permselectivity. While early gas membranes used polymers readily available or borrowed from liquid separation membranes, the current trend is to employ materials with more advanced molecular structures constructed specifically for gas separation.

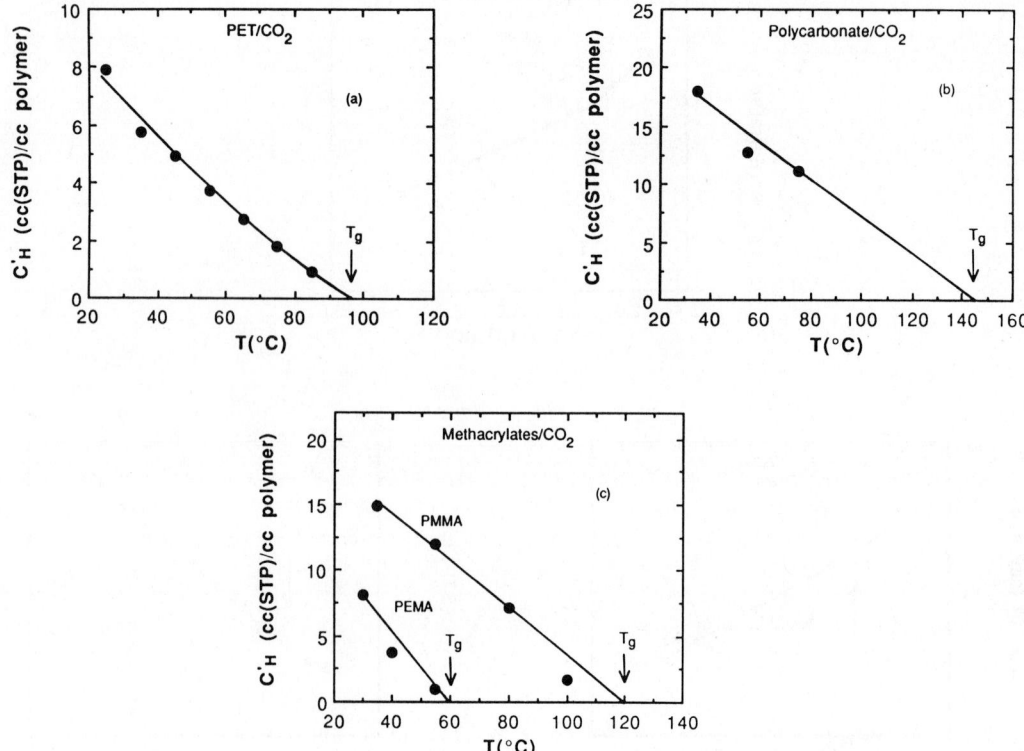

FIGURE 3-12. Temperature dependency of Langmuir sorption capacity c'_H for CO_2 sorption in glass polymers. Langmuir sorption capacity is found to approach zero at the polymer glass transition T_g (reprinted from Koros and Paul 1981 with permission).

Recently, several studies (Kim et al. 1988; Koros and Hellums 1989; Stern, Mi, and Yamamoto 1989) have reported results for families of polymers that show favorable deviations from standard permeability versus selectivity correlations found for commercially available materials. Shown in Table 3-3 are structures for two classes of materials, polyimides and polycarbonates, where molecular design concepts have been applied. Table 3-4 shows the permeability, solubility, and selectivities for these materials, and Figure 3-13 illustrates the permselectivity of a few of these materials versus "standard" trade-off curves.

Gas membrane manufacturers have also vigorously pursued new materials that possess superior properties. For instance, shown in Figure 3-14 is the H_2/CH_4 permeability and selectivity trade off for a family of polyaramides developed by Du Pont (Ekiner and Vassilatos 1990). Du Pont has also developed other materials that possess unique gas separation characteristics (Hayes 1987, 1988a, 1988b, 1989).

Dow has recently patented a series of halogenated polycarbonates and polyestercarbonates that have high O_2/N_2 selectivities (Anand et al. 1989a, 1989b; Jeanes 1989). Dow has also patented a process to form asymmetric hollow fibers from tetrabrominated polycarbonate, which is thought to be the basis for the recently introduced Generon II® membranes (Sanders et al. 1988; Sanders, Clark, and Jensvold 1989). This fiber-spinning patent is unique in that it claims to describe a method whereby the separating layer is between porous inner and outer surfaces. The intrinsic properties of tetrabrominated polycarbonate have been reported (Muruganandam, Koros, and Paul

TABLE 3–3. Well-characterized Families of Polycarbonates(a–f) (Koros and Hellums 1989a, 1989b) and Polyimides (1–8) (Kim et al. 1988).

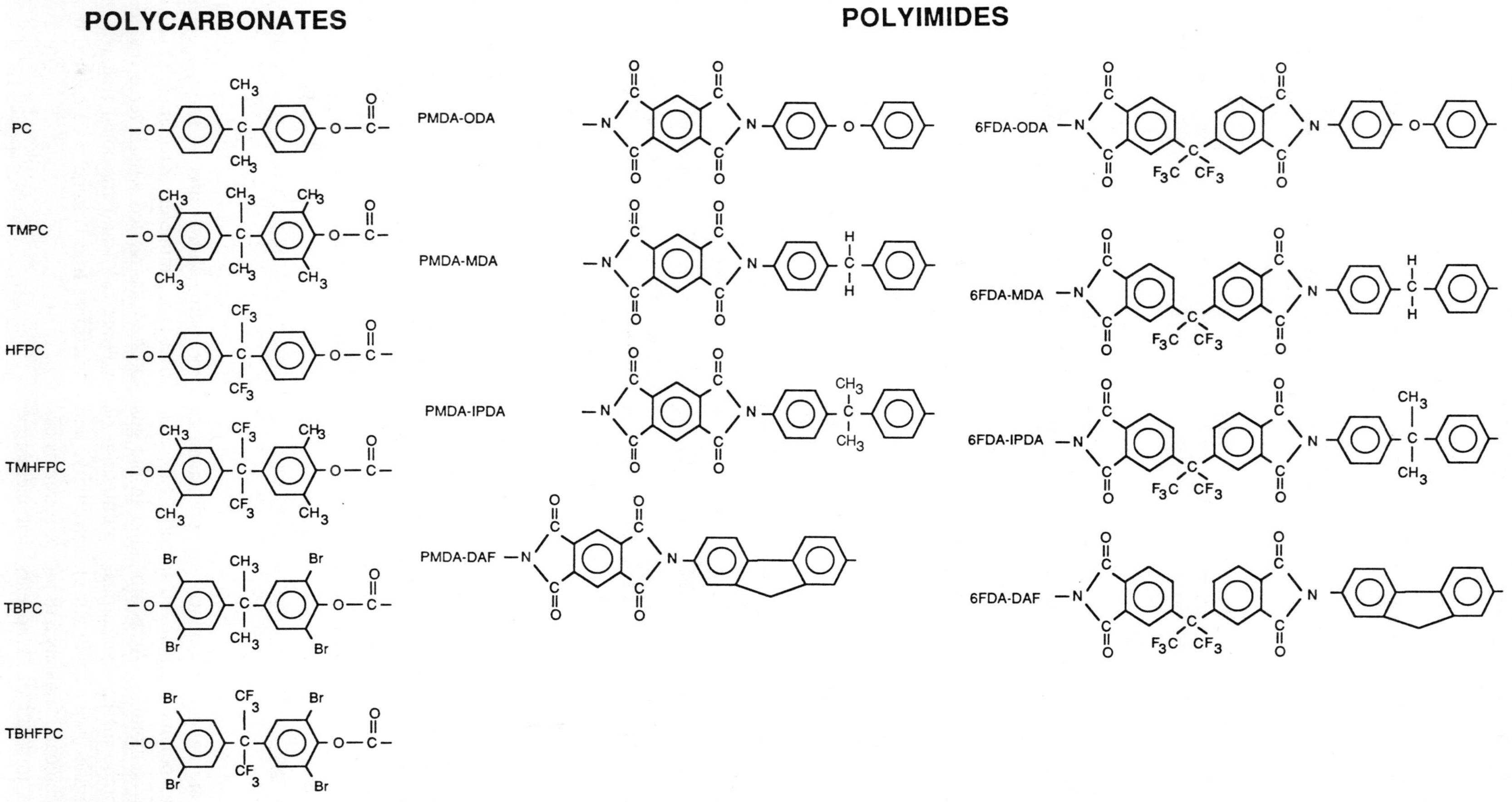

POLYCARBONATES

PC

TMPC

HFPC

TMHFPC

TBPC

TBHFPC

POLYIMIDES

PMDA-ODA

PMDA-MDA

PMDA-IPDA

PMDA-DAF

6FDA-ODA

6FDA-MDA

6FDA-IPDA

6FDA-DAF

TABLE 3–4. Permeabilities and Selectivities of Polycarbonates (Koros and Hellums 1989a, 1989b) and Polyimides (Kim et al. 1988).

Polymer	Permeabilities in Barrer at 35°C			Ideal Selectivities at 35°C		
	He 10 atm	O_2 2 atm	CO_2 10 atm	He/CH_4 10 atm	O_2/N_2 2 atm	CO_2/CH_4 10 atm
Polycarbonates						
(a) PC	13	1.6	6.8	35	4.8	19
(b) TMPC	46	5.6	18.6	50	5.1	21
(c) HFPC	60	6.9	24	57	4.1	23
(d) TMHFPC	206	32	111	44	4.1	24
(e) TBPC	18	1.4	4.2	140	7.5	34
(f) TBHFPC	100	9.7	32	112	5.4	36
(g) TB/TBHF-co-PC	49	4.9	16	110	6.2	34
Polyimides						
(1) PMDA-ODA	8.0	0.61	2.71	134.9	6.1	45.9
(2) PMDA-MDA	9.4	0.98	4.03	94	4.9	42.9
(3) PMDA-IPDA	37.1	7.1	26.8	41.1	4.7	29.7
(4) PMDA-DAF	1.9	—	0.15	921	—	71.6
(5) 6FDA-ODA	51.5	4.34	23	135.4	5.2	60.5
(6) 6FDA-MDA	50	4.6	19.3	117.1	5.7	44.9
(7) 6FDA-IPDA	71.2	7.53	30	102.1	5.6	42.9
(8) 6FDA-DAF	98.5	7.85	32.2	156.3	6.2	51.1

1987) and are shown in Table 3–4 and Figure 3–13(c).

Like Du Pont, Ube also has developed new materials specifically for H_2/CH_4 separations. Their product bulletin lists impressive asymmetric properties for two different membranes as shown in Table 3–5 (Ube Industries 1989). The polymers used by Ube are claimed to be aromatic polyimides based on biphenyltetracarboxylic dianhydride (BPDA) and aromatic diamines. Two studies (Hakuta et al. 1986; Tanaka et al. 1989) provide at least a partial list of aromatic diamines available to Ube.

Permeabilities and permselectivities of many of the gas membrane materials developed in recent academic and industrial efforts have been summarized in Table 3–6. Also included in this table are common "engineering" polymers, as well as polymers employed in early gas separation membranes. As competition in the gas membrane market intensifies, optimization of membrane productivity (flux), selectivity, and environmental resistance to process streams becomes imperative. It appears that membrane suppliers are currently targeting the nitrogen-enriched air (NEA) market. The size and importance of this market are represented by the abundance of O_2/N_2 measurements in the patent literature for new materials.

Our current understanding of the effects of polymer structure does not allow for the design of a good gas separation membrane polymer without initial screening tests. Generalized correlations cutting across different polymer families are not available, but trends have been identified within these families. With the methods discussed below, testing of some candidate materials may be eliminated by initially screening two or three polymers within a family to develop a rough correlation.

Figures 3–15(a) and 3–15(b) illustrate a useful correlation tool reported by Hellums (1990) for permeability and diffusivity. Here, gas transport parameters for a family of polycarbonates are correlated with inverse fractional free volume (l/V_f) and fractional free volume. The fractional free volume is calculated by:

$$V_f = \frac{V - V_0}{V}, \qquad (3\text{–}43)$$

Theory 43

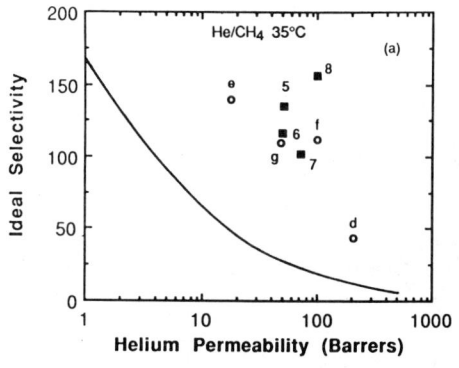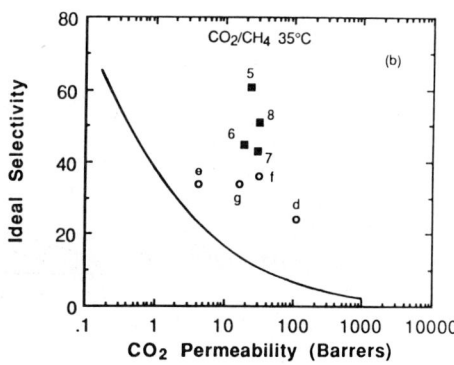

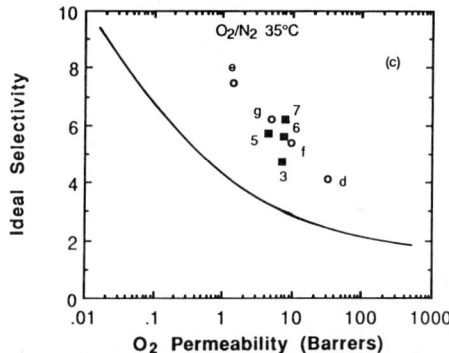

FIGURE 3–13. Permeability–selectivity relationships for novel polycarbonates (adapted from Koros and Hellums 1989a) and polyimides (adapted from Kim et al. 1988) as compared to the trade-off for standard commercial polymers. Polycarbonate (a–g) and polyimide (1–8) structures are shown in Table 3–3 and permeability data are listed in Table 3–4.

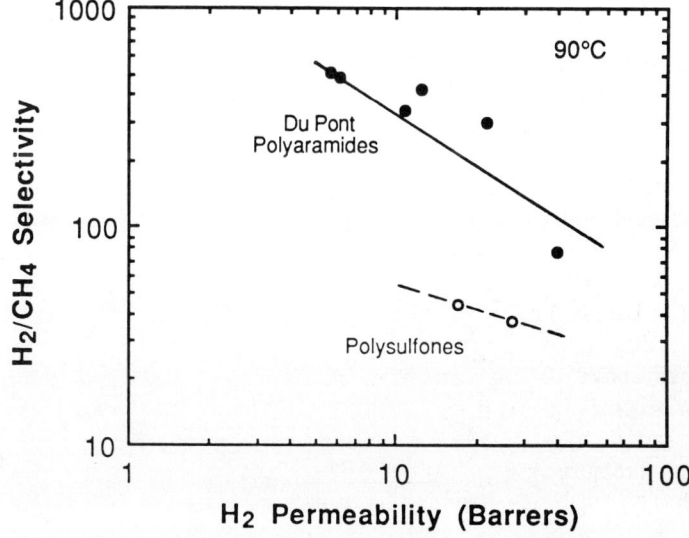

FIGURE 3–14. Permeability–selectivity relationship for H_2/CH_4 separation with novel polyaramides developed by Du Pont (adapted from Ekiner and Vassilatos 1990).

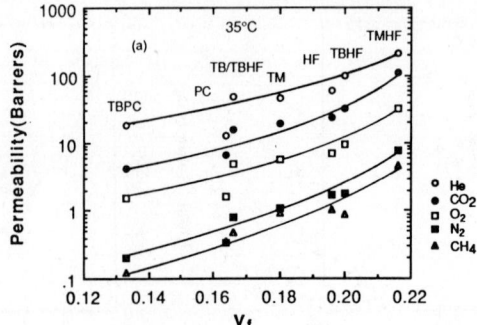
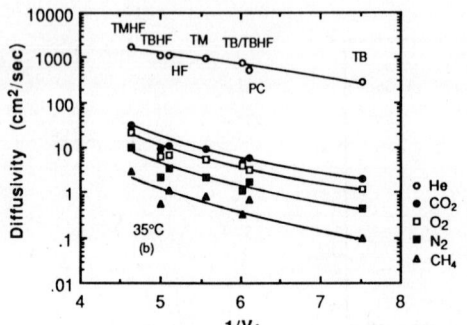
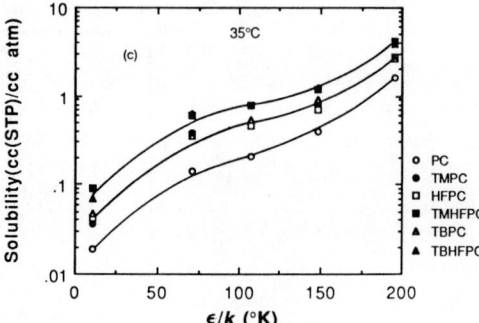

FIGURE 3–15. Permeability correlation with fractional free volume (V_f), diffusivity correlation with inverse fractional free volume ($1/V_f$), and solubility correlation with Lennard-Jones well potential (ϵ/k) for a series of polycarbonates at 35°C (Hellums 1990).

where V is the inverse of the measured polymer density, and V_0 is the polymer occupied specific volume. The occupied specific volumes were estimated by the method of Lee (1980) using the group contribution correlation of Van Krevelen (Van Krevelen and Hoftyzer 1976) for calculation of the van der Waals volumes.

Generalized correlations of gas solubility in different families of glassy polymers are also not available. Several studies have shown solubility for gases in rubbery polymers correlates well with ϵ/k or other measures of condensability (e.g., T_c, T_b) (Michaels and Bixler 1961; Van Amerongen 1964). As discussed above, gas solubility in glassy polymers is thought to be governed by both a Langmuir component as well as a Henry's law dissolution component. At penetrant pressures above 10 atm, the total sorption population becomes increasingly dominated by Henry's law dissolution. Hence, at these penetrant pressures,

TABLE 3–5. Ube Polyimide H_2 Membrane Performance (Ube Industries 1989).

	Membrane Type	
Gas Component	A	B-H
H_2	100 GPU[a]	500 GPU[a]
	H_2/Gas Component Selectivity	
H_2O	0.10	0.25
He	1.11	1.25
CO_2	10	3.3
H_2S	25	10
O_2	33	17
Ar	67	50
CO	100	56
N_2	170	83
C_2H_4	200	250
CH_4	250	125
C_2H_6	1000	590

[a]$GPU = \dfrac{cm^3 \ (STP)}{cm^2 \cdot s \cdot cm \ Hg} \times 10^{-6}$.

TABLE 3-6. Permeabilities (Barrer) and Selectivities of Various Gases in Dense Polymer Films.

Polymer	T (°C)	He	H_2	CO_2	O_2	He/CH_4	H_2/CH_4	CO_2/CH_4	O_2/N_2	N_2/CH_4	Ref.
Rubbery Polymers											
Natural rubber	25	30.3	49	134	24	1.05	1.63	4.7	2.76	0.30	1
Poly(4-methyl-1-pentene)	35	114	97.8	83	27.3	7.5	8.7	6.3	4.2	0.50	2
Silicone rubber (PDMS)	35	561	—	4553	933	0.41	—	3.37	2.12	0.33	3
Glassy Polymers											
Miscellaneous											
TMSP	25	6510	16200	33100	10000	0.41	1.01	2.07	1.48	0.42	4
TMSP	35	—	—	28000	7730	—	—	2.15	1.56	0.38	5
Polysulfone	35	13.0	14.0	5.6	1.4	49	53	22	5.6	1.0	6
Polycarbonate	35	14.0	—	6.5	1.48	50	—	23.2	5.12	0.93	7
PPO	35	105	—	61	16.8	24.4	—	14.2	4.41	0.95	8,9
PEI (Ultem®)	35	9.4	—	1.33	0.41	261	—	36.9	8.04	1.42	10
Cellulose Acetate[a] (~2.5 DS)	35	13.6	5.0	5.5	0.68	68	48	27.5	3.4	0.73	11
Cellulose Acetate[a] (2.45 DS)	35	16.0	12.0	4.75	0.82	107	80	32	5.5	1.0	12
Ethyl Cellulose	30	35.6	4.9	47.5	11.2	4.75	0.66	6.34	3.4	0.44	13
Polycarbonates											
TBBA-PC	30	—	—	3.60	0.85	—	—	35	7.4	1.1	14
TBBA:BA(70:30)-PC	30	—	—	—	0.93	—	—	—	6.7	1.1	14
TBBA:BA(50:50)-PC	30	—	—	6.4	0.98	—	—	29	6.4	—	14
TBBA:BA(30:70)-PC	30	—	—	3.6	0.80	—	—	32	6.4	1.1	14
TBBA:TMBA(1:1)-PC	30	—	—	7.0	1.87	—	—	30	6.9	1.17	14
TCBA-PC	30	—	—	2.6	1.45	—	—	25	6.3	2.2	14
TCBA:BA(70:30)-PC	30	—	—	—	1.34	—	—	—	6.1	—	14
TCBA:BA(50:50)-PC	30	—	—	5.5	1.24	—	—	25	5.4	1.05	14
TCBA:BA(30:70)-PC	30	—	—	3.6	0.80	—	—	32	6.1	1.15	14
TMBA-PC	30	—	—	16.3	3.9	—	—	27	5.0	1.3	14
TMBA:BA(50:50)-PC	30	—	—	5.7	1.40	—	—	24	6.0	—	14
TMBA:BA(30:70)-PC	30	—	—	5.7	1.30	—	—	29	5.8	1.08	14

TABLE 3-6. (continued)

Polymer	T (°C)	He	H_2	CO_2	O_2	He/CH_4	H_2/CH_4	CO_2/CH_4	O_2/N_2	N_2/CH_4	Ref.
Polyestercarbonates											
TBPEC(20% ester) (T:I = 1:1)	30	10.8	—	3.55	0.97	121	—	40	7.2	1.6	15
TBPEC(50% ester) (T:I = 1:9)	30	11.8	—	3.73	0.96	114	—	39	7.2	1.4	15
TBPEC(67% ester) (T:I = 8:2)	30	10.3	—	4.75	1.08	108	—	33	7.7	1.14	15
TBPEC(67% ester) (T:I = 0:1)	30	—	—	—	1.1	—	—	—	7.7	—	15
TBPEC(80% ester) (T:I = 8:2)	30	11.4	—	5.26	1.23	81	—	39	7.2	1.2	15
TBPEC(95% ester) (T:I = 1:1)	30	—	—	—	1.4	—	—	—	6.7	—	15
TBPEC(95% ester) (T:I = 2:8)	30	—	—	—	1.4	—	—	—	8.0	—	15
Poly(pyrrolone)											
6FDA-TADPO(PYRR)	35	89	—	27.6	7.9	165	—	51.1	6.5	2.4	16
Poly(imides)											
6FDA-6FpDA	35	—	—	63.9	16.3	—	—	39.9	4.7	2.17	17
6FDA-6FmDA	35	—	—	5.1	1.8	—	—	63.8	6.9	3.26	17
3,3' ODA-PMDA	35	—	3.6	0.5	0.13	—	450	62	7.2	2.5	18
4,4' ODA-PMDA	35	—	3.0	1.14	0.22	—	115	43	4.5	1.8	18
3,3' ODA-6FDA	35	—	14.0	2.10	0.68	—	437	64	6.8	3.0	18
4,4' ODA-6FDA	35	—	52.5	22.0	5.05	—	97	41	5.4	1.7	18
MPD-6FDA	35	—	20.3	8.23	2.61	—	145	58	7.2	2.6	19
PPD-6FDA	35	—	23	11.8	2.10	—	128	65	5.5	1.9	19
2,4-DAT-6FDA	35	—	87.2	28.6	7.44	—	124	41	5.7	1.9	19
2,6-DAT-6FDA	35	—	107	42.5	11.0	—	115	46	5.2	2.3	19
4,4'-ODA-BPDA	50	—	5.2	0.87	—	—	173	29	—	—	20
DDS-BPDA	50	—	11.3	2.57	—	—	133	30.2	—	—	20
DDBT-BPDA	50	—	31.2	8.20	—	—	130	34.2	—	—	20
DAD-6FDA	25	530	—	381	73	23	—	25	3.2	—	21

DAD:DAM(1:1)-6FDA	25	—	320	54	—	—	24	3.6	—	21
DAM-6FDA	35	—	691	—	—	—	14.2	—	—	21
DAM-6FDA	25	—	—	—	—	—	—	3.6	—	21
DAD-BTDA:6FDA(1:19)	25	396	—	120	11	12	—	3.4	—	22
DAM-BTDA:6FDA(1:19)	25	293	340	141	11	6.8	—	3.5	—	22
DAM-BTDA:6FDA(1:3)	25	265	—	120	13	—	—	4.0	1.8	22
DAM-BTDA:6FDA(1:1)	25	179	221	59	10	15	—	4.2	1.34	22
DAM-BTDA:6FDA(1:1)	25	223	106	38	25	20	—	4.1	1.7	22
DAM:MPD(4:1)-6FDA	25	83	146	43	21	23	—	3.9	—	22
DAM:PPD(1:1)-6FDA	25	—	178	42	—	—	—	3.6	—	22
DAM-1,5 ND(9:1)-6FDA	25	340	—	100	14	—	—	4.5	—	23
MEMA-6FDA	25	—	—	17	—	—	—	4.6	—	23
MDIPA-BTDA	25	—	—	15	—	—	—	5.1	—	23
MMIPA-BTDA	25	—	—	10	—	—	—	4.7	—	23
MMIPA-BPDA	25	—	—	9.0	—	—	—	5.9	—	24
MIA-6FDA	25	—	—	8.21	—	—	—	4.7	—	24
MBA-6FDA	25	—	—	19.0	—	—	—	5.27	—	24
FFA-6FDA	25	—	—	12.2	—	—	—	5.10	—	24
FIA-6FDA	25	—	—	16.8	—	—	—	4.18	—	24
FMIA-6FDA	25	—	—	85.9	—	—	—			24

[a] Permeability measurements for cellulose acetate polymers are sensitive to film preparation method as well as degree of acetyl substitution (DS).

Nomenclature

Polymers

PDMS	Polydimethylsiloxane
PEI	Polyetherimide
PEC	Polyestercarbonate
PC	Polycarbonate
PPO	Poly(2,6-dimethylphenyleneoxide)
PYRR	Poly(pyrrolone)
TMSP	Poly[1-trimethylsilyl-1-propyne]

Diamines

DAD	2,3,5,6-tetramethyl-1,3-phenylenediamine; diaminodurene
DAM	2,4,6-trimethyl-1,3-phenylenediamine; diaminomesitylene
2,4-DAT	2,4-diaminotoluene
2,6-DAT	2,6-diaminotoluene
DDBT	dimethyl-3,7-diaminodibenzothiophene-5,5'dioxide (mixture of methyl isomers)
DDS	4,4'-diaminodiphenylsulfone
FFA	4,4'-(9-fluorenylidene)bis(2-fluoroaniline)
6FmDA	4,4'-(9-fluorenylidene)bis(2-isopropylaniline)
6FpDA	4,4'-(9-fluorenylidene)bis(2-methyl-6-isopropylaniline)
MCDEA	4,4'-methylenebis(3-chloro-2,6-diethylaniline)
MDIPA	4,4'-methylenebis(2,6-diisopropylaniline)
MEMA	4,4'-methylenebis(2-ethyl-6-methylaniline)

TABLE 3-6. (*continued*)

Nomenclature (*continued*)

MMIPA   4,4'-methylenebis(2-methyl-6-isopropylaniline)
MIA   4,4'-methylenebis(2-isopropylaniline)
MBA   4,4'-methylenebis(2-t-butylaniline)
MDA   4,4'-methylenedianiline
MPD   1,3-phenylenediamine; metaphenylenediamine
PPD   1,4-phenylenediamine; paraphenylenediamine
1,5-ND   1,5-naphthalenediamine
3,3'-ODA   3,3'-oxydianiline
4,4'-ODA   4,4'-oxydianiline
TADPO   2,2',3,3'-tetraaminodiphenylether

Dianhydrides

BPDA   3,3'4,4' biphenyl tetracarboxylic dianhydride
BTDA   3,3'4,4' benzophenone tetracarboxylic dianhydride
6FDA   4,4' (hexafluoro isopropylidene)bis(phthalic anhydride), hexafluorodianhydride
PMDA   1,2,4,5 benzene tetracarboxylic dianhydride, pyromellitic dianhydride

Bisphenol A

TBBA   3,3'5,5' tetrabromobisphenol A
TCBA   3,3'5,5' tetrachlorobisphenol A
TMBA   3,3'5,5' tetramethylbisphenol A

1. Bixler and Sweeting (1971).
2. Mohr, Paul, Misna and Lagow (1991).
3. Stern, Shah, and Hardy (1987).
4. Tien et al. (1989).
5. Ichiraku, Stern and Nakagawa (1987).
6. McHattie, Koros, and Paul (1991).
7. Koros, Chan, and Paul (1977).
8. Toi, Morrel, and Paul (1982).
9. Hwang, Choi, and Kammermeyer (1974).
10. Barbari, Koros, and Paul (1989).
11. Schell (1975).
12. Puleo, Paul, and Kelly (1989).
13. Hwang, Choi, and Kammermeyer (1974).
14. Anand et al. (1989a).
15. Anand et al. (1989b).
16. Walker and Koros (1990).
17. Coleman and Koros (1990).
18. Stern, Mi, and Yamamoto (1989).
19. Yamamoto, Mi, and Stern (1990).
20. Tanaka et al. (1989).
21. Hayes (1987).
22. Hayes (1988b).
23. Hayes (1989).
24. Langsome and Burgoyne (1990).

solubility correlation methods used for rubbery polymers are expected to work well for glassy polymers.

Figure 3–15(c) shows the correlation of solubility coefficients with Lennard-Jones potential well depth ϵ/k for the polycarbonates. The nonlinearity of this correlation suggests that additional factors are at play. Also, it is interesting that the behavior is independent of structural variations within the polycarbonate family. Similar solubility correlations have also been reported (Kim et al. 1988) for the series of polyimides listed in Tables 3–3 and 3–4.

NOTATION

General Notation

See the General Notation section at the beginning of this handbook.

Special Notation

a	activity, dimensionless
A_d	free-volume diffusivity parameter, L^2 mol/Et or cm^2 · mol/(J · s)
B_d	free-volume diffusivity parameter, dimensionless
b	Langmuir affinity constant, p^{-1} or atm^{-1}
c	sorbed penetrant concentration in polymer, mol/L^3 or cm^3 (STP)/(cm^3 polymer)
D	average diffusion coefficient based on upstream and downstream concentrations, L^2/t or cm^2/s
D_D, D_H	mobility of dissolved and Langmuir sorbed components, L^2/t or cm^2/s
D_0	infinite dilution diffusion coefficient in matrix model, L^2/t or cm^2/s
D_{ij}	mutual diffusion coefficient, L^2/t or cm^2/s
E_d	activation energy for penetrant diffusion, E/mol or kcal/mol
E_s	activation energy for penetrant sorption, E/mol or kcal/mol
F	D_H/D_D, convenient dimensionless group in dual mobility transport model
J_i	diffusive flux of component i, mol/$L^2 t$ or cm^3 (STP)/(cm^2 · s)
k_D	Henry's law coefficient for gas sorption in polymers, mol/$L^3 p$ or cm^3 (STP)/(cm^3 polymer · atm)
K	$c_H' b/k_D$, convenient dimensionless group in dual mobility transport model
$p°$	vapor pressure, p or atm or psia
p_s	reference pressure, 1 atm
P	permeability = (diffusive flux)(membrane thickness)/(partial pressure difference), Barrer = 10^{-10} cm^3 (STP) · cm/(cm^2 · s · cm Hg)
S	equilibrium solubility coefficient, mol/$L^3 p$ or cm^3 (STP)/(cm^3 polymer · atm)
T_s	reference temperature, T_g, T or °C
V_0	polymer occupied specific volume, L^3/M or cm^3/g

Greek Letters

α_{ij}^*	ideal separation factor, applies to the case in which the downstream pressure is effectively zero, dimensionless
$\alpha_{ij,\text{inh}}$	inherent separation factor, the permeability ratio of component i relative to component j for the case in which the permeate pressure is greater than zero, dimensionless
γ	plasticization parameter in free-volume model, dimensionless
ϕ	volume fraction of sorbed penetrant, dimensionless
ϕ_p	volume fraction of polymer, dimensionless
$\phi_{fs}°$	fractional free volume of pure polymer at the reference temperature and pressure, dimensionless

REFERENCES

Anand, J. N., S. E. Bales, D. C. Feay, and T. O. Jeanes. 1989a. Tetrabromo bisphenol based polycarbonate membranes and method of using. U.S. Patent 4,840,646.

Anand, J. N., D. C. Feay, S. E. Bales, and T. O. Jeanes. 1989b. Semi-permeable membranes consisting predominantly of polycarbonates derived from tetrahalobisphenols. U.S. Patent 4,818,254.

Antonson, G. R., R. J. Gardner, C. F. King, and D. Y. Ko. 1977. Analysis of gas separation by permeation of hollow fibers. *Ind. Chem. Process Des. Dev.* 16(4):463.

Barbari, T. A., W. J. Koros, and D. R. Paul. 1988.

Gas sorption in polymers based on bisphenol-A. *J. Polym. Sci., Polym. Phys. Ed.* 26:729.

Barbari, T. A., W. J. Koros, and D. R. Paul. 1989. Polymeric membranes based on bisphenol-A for gas separations. *J. Membr. Sci.* 42:69–86.

Barrer, R. M. 1937. Nature of diffusion in polymers. *Nature* 140:106.

Barrer, R. M., J. A. Barrie, and J. Slater. 1958. Sorption and diffusion in ethyl cellulose. Part III. Comparison between ethyl cellulose and rubber. *J. Polym. Sci.* 27:177.

Barrer, R. M. 1984. Diffusivities in glassy polymers for the dual mode sorption model. *J. Membr. Sci.* 18:25.

Barrie, J. A. 1968. Water in polymers. Chap. 8 in *Diffusion in Polymers*, ed. J. Crank and G. S. Park. New York: Academic Press.

Bearman, R. J. 1961. On the molecular basis of some current theories of diffusion. *J. Phys. Chem.* 65:1961.

Billmeyer, F. W. 1971. *Textbook of Polymer Science*, 2nd ed. New York: Wiley-Interscience.

Bixler, H. J., and O. J. Sweeting. 1971. *The Science and Technology of Polymer Films*, Vol. II, p. 85, ed. O. J. Sweeting. New York: John Wiley and Sons.

Breck, D. W. 1974. *Zeolite Molecular Sieves*. New York: John Wiley and Sons.

Chan, A. H., W. J. Koros, and D. R. Paul. 1978. Analysis of hydrocarbon gas sorption and transport in ethyl cellulose using the dual mode sorption/partial immobilization models. *J. Membr. Sci.* 3:117.

Chern, R. T., W. J. Koros, H. B. Hopfenberg, and V. T. Stannett. 1983a. Reversible isopentane-induced depression of carbon dioxide permeation through polycarbonate. *J. Polym. Sci., Polym. Phys. Ed.* 21:753.

Chern, R. T., W. J. Koros, E. S. Sanders, S. H. Chen, and H. B. Hopfenberg. 1983b. Implications of the dual mode sorption and transport models for mixed gas permeation. In ACS Symp. Ser. No. 233, *Industrial Gas Separations*, ed. T. E. Whyte, C. M. Yon, and E. H. Wagener, pp. 47–73. Washington, DC: American Chemical Society.

Chern, R. T., W. J. Koros, E. S. Sanders, and R. E. Yui. 1983c. Second component effects on sorption and permeation of gases in glassy polymers. *J. Membr. Sci.* 15:157.

Chern, R. T., W. J. Koros, H. B. Hopfenberg, and V. T. Stannett. 1984. Selective permeation of CO_2 and CH_4 through Kapton® polyimide: Effects of penetrant competition and gas phase nonidealities. *J. Polym. Sci., Polym. Phys. Ed.* 22(6):1061.

Chern, R. T., W. J. Koros, H. B. Hopfenberg, and V. T. Stannett. 1985. Material selection for membrane-based gas separations. In ACS Symp. Ser. No. 269, *Materials Science of Synthetic Membranes,* ed. D. R. Lloyd, pp. 25–45. Washington, DC: American Chemical Society.

Chiou, J. S., J. W. Barlow, and D. R. Paul. 1985. Plasticization of glassy polymers by CO_2. *J. Appl. Polym. Sci.* 30:2633.

Chueh, P. L., and J. M. Prausnitz. 1967. Vapor-liquid equilibria at high pressures: Calculation of partial molar volumes in nonpolar liquid mixtures. *AIChE J.* 13(6):1099.

Coleman, M. R., and W. J. Koros. 1990. Isomeric polyimides based on fluorinated dianhydrides and diamines for gas separation applications. *J. Membr. Sci.* 50:285.

Crank, J. 1975. *The Mathematics of Diffusion*. 2nd ed. Clarendon: Oxford.

Ekiner, O. M., and G. Vassilatos. 1990. Polyaramide hollow fibers for hydrogen/methane separation: spinning and properties. *J. Membr. Sci.* 53:259–273.

Erb, A. J., and D. R. Paul. 1981. Gas sorption and transport in polysulfone. *J. Membr. Sci.* 8:11.

Fleming, G. K. 1987. Dilation of silicone rubber and glassy polycarbonates due to high pressure gas sorption. Ph.D. diss., University of Texas at Austin.

Fleming, G. K., and W. J. Koros. 1986. Dilation of polymers by sorption of carbon dioxide at elevated pressures: 1. Silicone rubber and unconditioned polycarbonate. *Macromol.* 19:2285–2291.

Flory, P. J. 1969. *Principles of Polymer Chemistry*. Ithaca, NY: Cornell University.

Frisch, H. L., D. Klempner, and T. Kwei. 1971. Modified free volume theory of penetrant diffusion in polymers. *Macromol.* 4:237.

Fujita, H. 1961. Diffusion in polymer diluent systems. *Fortschr. Hochpolym. Forsch.* 3:1.

Fujita, H. 1968. Organic vapors above the glass transition temperature. In *Diffusion in Polymers*, ed. J. Crank and G. S. Park. New York: Academic Press.

Fujita, H., A. Kishimoto, and K. Matsumoto. 1960. Concentration and temperature dependence of diffusion coefficients for systems polymethyl acrylate and n-alkyl acetates. *Trans. Faraday Soc.* 56:424.

Hakuta, T., K. Haray, K. Obata, Y. Shindo, N. Ito, and H. Yoshitome. 1986. The use of membranes in the Japanese "C_1" chemistry programme. In *Proc. 4th BOC Priestley Conf.*, p. 281.

Hayes, R. A. 1987. Polyimide gas separation membranes. U.S. Patent 4,705,540.

Hayes, R. A. 1988a. Polyimide gas separation membranes. U.S. Patent 4,717,393.

Hayes, R. A. 1988b. Polyimide gas separation membranes. U.S Patent 4,717,394.

Hayes, R. A. 1989. Polyimide gas separation membranes. U.S. Patent 4,838,900.

Hellums, M. W. 1990. Gas sorption and permeation in a series of polycarbonates. Ph.D. diss., University of Texas at Austin.

Hoshay, A., and L. M. Robeson. 1976. Sulfonated polysulfone. *J. Appl. Polym. Sci.* 20:1885.

Horiuti, J. 1931. On the solubility of gas and coefficient of dilation by absorption. *Sci. Papers Inst. of Phy. and Chem. Res. (Tokyo)* 19:1655.

Hwang, S. T., C. K. Choi, and K. Kammermeyer. 1974. Gaseous transfer coefficients in membranes. *Sep. Sci.* 9:461.

Ichiraku, Y., S. A. Stern, and T. Nakagawa. 1987. An investigation of the high gas permeability of poly(1-trimethylsilyl-1-propyne). *J. Membr. Sci.* 34:5.

Jeanes, T. O. 1989. Gas separation membranes derived from polycarbonates, polyesters, polyestercarbonates containing tetrafluoro bisphenol-F. U.S. Patent 4,851,014.

Jordan, S. M. 1988. The effects of carbon dioxide exposure on the permeability behavior of silicone rubber and glassy polycarbonates. Ph.D. diss., University of Texas at Austin.

Jordan, S. M., G. K. Fleming, and W. J. Koros. 1990. Permeability of carbon dioxide at elevated pressures in substituted polycarbonates. *J. Polym. Sci., Polym. Phys. Ed.* 28:2305.

Jordan, S. M., W. J. Koros, and G. K. Fleming. 1987. The effects of CO_2 exposure on pure and mixed gas permeation behavior: comparison of glassy polycarbonate and silicone rubber. *J. Membr. Sci.* 30:191.

Jordan, S. M., and W. J. Koros. 1990. Permeability of pure and mixed gases in silicone rubber at elevated pressures. *J. Polym. Sci., Polym. Phys. Ed.* 28:795.

Kim, T. H. 1988. Gas sorption and permeation in a series of aromatic polyimides, Ph.D. diss., University of Texas at Austin.

Kim, T. H., W. J. Koros, G. R. Husk, and K. C. O'Brien. 1988. Relationship between gas separation properties of aromatic polyimides. *J. Membr. Sci.* 37:45.

King, C. J. 1980. *Separation Processes*, 2nd ed. New York: McGraw-Hill Book Co.

Koros, W. J. 1980. Model for sorption of mixed gases in glassy polymers. *J. Polym. Sci., Polym. Phys. Ed.* 18:981.

Koros, W. J., A. H. Chan, and D. R. Paul. 1977. Sorption and transport of various gases in polycarbonates. *J. Membr. Sci.* 2:165.

Koros, W. J., and R. T. Chern. 1987. Separation of gaseous mixtures using polymer membranes. In *Handbook of Separation Process Technology*, ed. R. W. Rousseau, pp. 862–953. New York: John Wiley & Sons.

Koros, W. J., R. T. Chern, H. B. Hopfenberg, and V. T. Stannett. 1981. A model for permeation of mixed gases and vapors in glassy polymers. *J. Polym. Sci., Polym. Phys. Ed.* 19:1513.

Koros, W. J., and M. W. Hellums. 1989a. Gas separation membrane material selection criteria: differences for weakly and strongly interacting feed components. *Fluid Phase Equilibria* 53:339–354.

Koros, W. J., and M. W. Hellums. 1989b. Transport properties. *Encyclopedia of Polymer Science and Engineering*. 2nd. ed. (supplement volume), pp. 724–802. New York: John Wiley & Sons.

Koros, W. J., and D. R. Paul. 1978a. CO_2 sorption in poly(ethylene terephthalate) above and below the glass transition. *J. Polym. Sci., Polym. Phys. Ed.* 16:1947.

Koros, W. J., and D. R. Paul. 1978b. Transient and steady state permeation in poly (ethylene terephthalate) above and below the glass transition. *J. Polym. Sci., Polym. Phys. Ed.* 16:2171.

Koros, W. J., and D. R. Paul. 1981. Observation concerning the temperature dependence of the Langmuir sorption capacity of glassy polymers. *J. Polym. Sci., Polym. Phys. Ed.* 19:1655.

Lansome, M., and W. F. Burgoyne. 1990. Effects of ortho substituent volume on gas permeability of polyimides. In *Proc. 1990 Intl. Conf. on Membranes and Membrane Processes*, 20–24 August 1990, Chicago, IL, Vol. II, p. 809.

Lee, W. M. 1980. Selection of barrier materials from molecular structure. *Polym. Eng. Sci.* 20:65.

McCandless, F. P. 1972. Separation of binary mixtures of CO and H_2 by permeation through polymeric films. *Ind. Chem. Process Des. Dev.* 11:470.

McHattie, J. S., W. J. Koros, and D. R. Paul. 1991. Gas transport properties of polysulfones, part 1: role of symmetry of methyl group placement on biphenol rings. *Polymer* 32(5):840.

Meares, P. 1954. Diffusion of gases through polyvinyl acetate. *J. Am. Chem. Soc.* 76:3415.

Michaels, A. S., and H. J. Bixler. 1961. Solubility of gases in polyethylene. *J. Polym. Sci.* 50:393.

Michaels, A. S., W. R. Vieth, and J. A. Barrie. 1963. Diffusion and solution of gases in poly(ethylene terephthalate). *J. Appl. Phys.* 34(1):13.

Mohr, J. M., D. R. Paul, T. E. Mlsna, and R. J. Lagow. 1991. Surface fluorination of composite membranes. part I. transport properties. *J. Membr. Sci.* 55:131.

Muruganandam, N., W. J. Koros, and D. R. Paul. 1987. Gas sorption and transport in substituted polycarbonates. *J. Polym. Sci., Polym. Phys. Ed.* 25:1999.

Paul, D. R., and W. J. Koros. 1976. Effect of partially immobilizing sorption on permeability and diffusion time lag. *J. Polym. Sci., Polym. Phys. Ed.* 14:675.

Puleo, A. C., D. R. Paul, and S. S. Kelly. 1989. The effect of degree of acetylation on gas sorption and transport behavior in cellulose acetate. *J. Membr. Sci.* 47:301.

Pye, D. G., H. H. Hoehn, and M. Panar. 1976a. Measurement of gas permeability of polymers. I. Permeabilities in constant volume/variable pressure apparatus. *J. Appl. Polym. Sci.* 20:1921.

Pye, D. G., H. H. Hoehn, and M. Panar. 1976b. Measurement of gas permeability of polymers. II. Apparatus of determination of mixed gases and vapors. *J. Appl. Polym. Sci.* 20:287.

Raucher, D., and M. D. Sefcik. 1983a. Gas transport and cooperative main chain motions in glassy polymers. In ACS Symp. Ser. No. 233, *Industrial Gas Separations,* ed. T. E. Whyte, C. M. Yon, and E. H. Wagener, pp. 89–110. Washington, DC: American Chemical Society.

Raucher, D., and M. D. Sefcik. 1983b. Sorption and transport in glassy polymers. In ACS Symp. Ser. No. 233, *Industrial Gas Separations,* ed. T. E. Whyte, C. M. Yon, and E. H. Wagener, pp. 111–124. Washington, DC: American Chemical Society.

Robeson, L. M. 1969. The effect of antiplasticization on secondary loss transitions and permeability of polymers. *Polym. Eng. Sci.* 9:277.

Rogers, C. E. 1965. Solubility and diffusivity. In *Physics and Chemistry of the Organic State,* ed. D. Fox, M. M. Labes, and A. Weissberger, Vol. II, pp. 510–635. New York: Wiley-Interscience.

Sada, E., H. Kumazawa, P. Xu, and M. Nishigaki. 1988. Mechanism of gas permeation through glassy polymer films. *J. Membr. Sci.* 37:165.

Sanders, E. S. 1988. Penetrant-induced plasticization and gas permeation in glassy polymers. *J. Membr. Sci.* 37:63.

Sanders, E. S., D. O. Clark, and J. A. Jensvold. 1989. Semi-permeable membranes with an internal discriminating region. U.S. Patent 4,772,392.

Sanders, E. S., D. O. Clark, J. A. Jensvold, H. N. Breck, G. G. Libscomb, and F. L. Coan. 1988. Process for preparing POWADIR membranes from tetrahalo bisphenol-A polycarbonates. U.S. Patent 4,772,392.

Sanders, E. S., and W. J. Koros. 1986. Sorption of CO_2, C_2H_4, N_2O and their binary mixtures in poly(methyl methacrylate). *J. Polym. Sci., Polym. Phys. Ed.* 24:175.

Sanders, E. S., W. J. Koros, H. B. Hopfenberg, and V. T. Stannett. 1983. Pure and mixed gas sorption of carbon dioxide and ethylene in poly(methyl methacrylate). *J. Membr. Sci.* 13:161.

Schell, W. J. 1975. Separation of coal hydrogasification gases by permselective membranes. *ACS Div. Fuel Chemical Preprints.* 20:253.

Shakespear, G. A. 1918. Reports of the advisory committee on aeronautics. T. 1164.

Stannett, V. T., W. J. Koros, D. R. Paul, H. K. Lonsdale, and R. W. Baker. 1979. Recent advances in membrane science and technology. *Adv. Polym. Sci.* 32:69–121.

Stern, S. A., S. M. Fang, and H. L. Frisch. 1972. Effect of pressure on gas permeability coefficients. A new application of free volume theory. *J. Polym. Sci.;* A-2. 10:201.

Stern, S. A., and H. L. Frisch. 1981. The selective permeation of gases through polymers. *Ann. Rev. Mater. Sci.* 11:523.

Stern, S. A., S. S. Kulkarni, and H. L. Frisch. 1983. Test of a free volume model of gas permeation through polymer membranes. I. Pure CO_2, CH_4, C_2H_4 and C_3H_8 in polyethylene. *J. Polym. Sci., Polym. Phys. Ed.* 21:467.

Stern, S. A., Y. Mi, and H. Yamamoto. 1989. Structure/permeability relationships of polyimide membranes. Applications to the separation of gas mixtures. *J. Polym. Sci., Polym. Phys. Ed.* 27:887.

Stern, S. A., V. M. Shah, and B. J. Hardy. 1987. Structure-permeability relationships in silicone polymers. *J. Polym. Sci., Polym. Phys. Ed.* 25:1263.

Story, B. J., and W. J. Koros. 1989. Comparison of three models for permeation of CO_2/CH_4 mixtures in poly(phenylene oxide). *J. Polym. Sci., Polym. Phys. Ed.* 27:1927.

Tanaka, K., H. Kita, K. Okamato, A. Nakamura,

and Y. Kusuki. 1989. Gas permeability and permselectivity in polyimides based on 3,3', 4,4' biphenyltetracarboxylic dianhydride. *J. Membr. Sci.* 47:203.

Tien, C. F., A. C. Savoca, A. D. Surnamer, and M. Langsam. 1989. Chemical structure/permeation relationship for polysilylpropynes. *Proc. ACS Div. Polym. Mater. Sci. Eng.* 61:507.

Toi, K., G. Morel, and D. R. Paul. 1982. Gas sorption and transport in poly(phenylene oxide) and comparisons with other glassy polymers. *J. Appl. Polym. Sci.* 27:2997.

Ube Industries. 1989. Ube gas separation system by polyimide membranes. Product brochure.

Van Amerongen, G. J. 1964. Diffusion in elastomers. *Rubber Chem. Technol.* 37(5):1065.

Van Krevelen, D. W., and P. J. Hoftyzer. 1976. *Properties of Polymers.* Chap. 4, p. 129. Amsterdam: Elsevier.

Vrentas, J. S., and J. L. Duda. 1976. Diffusion of small molecules in amorphous Polymers. *Macromol.* 9:785.

Vrentas, J. S., and J. L. Duda. 1977a. Diffusion in polymer-solvent systems. I. reexamination of the free volume theory. *J. Polym. Sci., Polym. Phys. Ed.* 15:403; Diffusion in polymer-solvent systems. II. A predictive theory for the dependence of diffusion coefficients on temperature, concentration and molecular weight. *J. Polym. Sci., Polym. Phys. Ed.* 15:417.

Vrentas, J. S., and J. L. Duda. 1977b. Solvent and temperature effects on diffusion in polymer-solvent systems. *J. Appl. Polym. Sci.* 21:1715.

Vrentas, J. S., and J. L. Duda. 1986. Diffusion. In *Encyclopedia of Polymer Science,* ed. J. I. Kroschwitz, 2nd ed., Vol. 5, pp. 36–68. New York: John Wiley & Sons.

Walker, D. R. B., and W. J. Koros. 1991. Transport characterization of a polypyrrolone for gas separations. *J. Membr. Sci.* 55:99.

Yamamoto, H., Y. Mi, and S. A. Stern. 1990. Structure/permeability relationships of polyimide membranes. II. *J. Polym. Sci., Polym. Phys. Ed.* 28:2291.

Yi-Yan, N., R. M. Felder, and W. J. Koros. 1980. Selective permeation of hydrocarbon gases in poly(tetrafluoroethylene) and poly(fluoroethylene/propylene copolymer). *J. Appl. Polym. Sci.* 25:1755.

4

Design of Gas Permeation Systems

Raymond R. Zolandz and Gregory K. Fleming
E. I. Du Pont de Nemours & Company, Inc.

GENERAL DESIGN
  CONSIDERATIONS
MEMBRANE DESIGN
  Membrane Structure
  Chemical Compatibility
  Pressure and Temperature Effects
  Liquid Membranes
MEMBRANE MODULES
  Flow Patterns
  Other Design Considerations
  Spiral-Wound Permeators

Hollow-Fiber Modules
Choosing Flow Patterns in Hollow-
  Fiber Modules
Modeling the Performance of Gas
  Permeation Modules
CONFIGURATIONS USED IN GAS
  SEPARATIONS
  Arrangement of Gas Separators
  Novel Device Configurations
NOTATION
REFERENCES

GENERAL DESIGN CONSIDERATIONS

In Chapter 2 we stated that membranes can be considered to have four structural levels and that three additional organizational levels exist between the membrane itself and the final membrane process. The design at all of these levels must be aimed at the ultimate goal of providing a cost-effective, robust separation technique.

To this end, the membrane polymers must have good permeabilities and permselectivities, be compatible with the process environment in which they will be used, and be amenable to fabrication into useful membranes.

The membranes must be mechanically strong so that they can be incorporated into permeators without being damaged. They must also be able to withstand the pressure differentials imposed on them during operation.

Membrane separators (permeators) are designed to incorporate large amounts of membrane area per unit volume of pressure housing and to provide for effective contact of gas with the membrane surface. Provisions must be taken to minimize deleterious pressure drops within the permeator.

Like the membranes, the permeator components need to be robust so that they can tolerate process pressures and temperatures, and the materials of construction must be chemically compatible with the process application.

The systems into which the permeators are placed are generally composed of standard chemical process industry components. Necessary piping, valves, heat exchangers, liquid knockouts, controls, and instrumentation are generally industry standard and "off the shelf." For this portion of the membrane-related equipment, good engineering practice will lead to the "minimum essential" system design.

Membrane-based processes are most suitable (Schell 1983) in applications for which the following are true:

1. The stream to be processed is of low to moderate volume (150 KSCFD to 200 MSCFD, where KSCFD = 10^3 standard cubic feet per day and MSCFD = 10^6 SCFD), since membrane costs scale approximately linearly with throughput.
2. The feed contains moderate concentrations (10 to 85 mol%) of the more permeable gas, so that a reasonable feedside partial pressure of the more permeable gas is readily achieved.
3. The feed is already at moderate to high pressure (250 to 2000 psig) and moderate temperature (30 to 150°F), in order to reduce or avoid compression and heating or cooling costs.
4. The product gas(es) are not required at absolute purity or 100% recovery, since these are not readily achieved with membrane technology.

Membranes are also favored in cases for which the nonpermeate (retentate) is used as a product at elevated pressure, as in natural gas sweetening and production of nitrogen-enriched air, since the nonpermeate is delivered at elevated pressure. Modularity, simplicity, and light weight will also favor membranes in particular applications.

Membranes alone are generally not suitable when very high product purity and recovery are required. However, membranes may be coupled with other unit operations to achieve an effective solution to such separation problems.

MEMBRANE DESIGN

Membrane Structure

The commercial use of membranes for gas separations did not begin until after other membrane applications such as reverse osmosis and ultrafiltration were well established. This was due in large part to the fact that membranes suitable for gas separations are more difficult to fabricate. To be commercially viable, the membranes must have a very thin separating layer that is also free from defects. Such defects can allow significant bypass of feed gas to the permeate stream, leading to poor separation.

Membranes are generally fabricated by casting films or spinning fibers from polymer solutions. Polymer concentrations of 20 to 40% by weight are commonly employed. The polymer content is generally such that the solutions are close to the point of phase separation. In most cases the solution is cast or extruded into a quench medium, which consists of nonsolvent for the polymer. Nonsolvent diffuses into the nascent membrane, causing spinodal decomposition of the polymer solution into polymer-rich and polymer-poor regions, which ultimately become the walls and pores of the membrane. Temperature may also be used to induce this decomposition or to influence its rate. An evaporation step is usually provided prior to the quench, during which solvent evaporates from what ultimately will be the active layer of the membrane. Details of these and other membrane formation processes can be found in the excellent text by Kesting (1985) as well as in the monograph edited by Lloyd (1985).

The quench liquid and any solvent remaining from the membrane formation process must be removed from the membrane before it can be used for gas processing. For some membranes this step is accomplished by water washing and a simple evaporative drying step. For other membranes the quench liquid must be displaced with solvent(s), which are then evaporated. Such solvent displacement is needed in these cases in order to preserve the morphology of the membrane. Drying without the use of the intermediate solvent would result in collapse or densification of the asymmetric structure, which can greatly reduce membrane productivity.

Additional thermal or chemical treatments can be applied to the membranes prior to their use in gas separations. In some membranes these treatments are required in order to "heal" defects in the membrane surface.

To minimize the surface area required, the throughput of the membrane must be maximized. From Eqs. (3–3) and (3–15) of Chapter 3, and using partial pressure instead of fugacity, the following relationship is obtained for transport of a pure component across the active layer of the membrane:

$$J_v = Q_v/A = P(p_f - p_p)/l. \qquad (4-1)$$

From this expression, we see that for a given driving force large values of J_v are obtained when the polymer permeability P is large and the thickness of the membrane l is small. (Permeability values for a number of common polymers are given in Table 3–6 of Chapter 3). For commercial membranes, l is typically of the order of 1000 to 2000 Å, with 400 to 1000 Å attainable in more finely tuned membranes. Figure 4–1 shows a cross section of a typical hollow-fiber membrane. The fiber structure compares well with the idealized version depicted in Figure 2–3 of Chapter 2, in which the active layer is backed by a porous layer that provides mechanical support.

The membrane skin offers a resistance R_1 to the flow of gas across the membrane, which is in series with the resistance R_2 of the porous support. By analogy to electrical circuits, two transport resistances R_1 and R_2 in series have an effective resistance R_{eff} given as

$$R_{\text{eff}} = R_1 + R_2. \qquad (4-2)$$

In most cases the resistance of the porous layer is very much lower than that of the dense layer. Thus, R_{eff} is essentially equal to R_1, and the support layer does little to impede transport through the membrane. In the case of a highly permeable active layer and a relatively nonporous support layer, the overall transport rate has been observed to be significantly affected by the support layer (Pinnau et al. 1988).

The active separating region is often more than a simple layer of a single polymer. For example, the "resistance model" composite membranes developed by Henis and Tripodi (1981) consist of a base polymer formed into a membrane whose active layer has some defects. An occluding layer of a highly permeable but low-selectivity rubbery polymer is used to caulk these defects. Even though the caulking material is highly permeable, the area of the pores is typically small compared to the area of the intact membrane (less than 0.01%). The filled pores and the intact skin can be considered as resistances in parallel, so that the skin resistance R_1 is calculated from the resistance of the

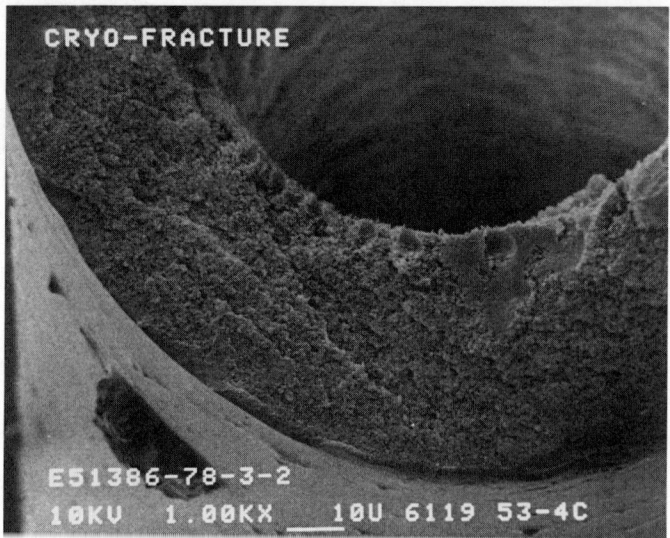

FIGURE 4–1. Photomicrograph of an experimental asymmetric hollow-fiber membrane. (Courtesy of E. I. Du Pont de Nemours & Company, Inc.)

intact skin (A) and of the filled pores (B) by the relationship

$$1/R_1 = 1/R_A + 1/R_B \quad (4-3)$$

so that

$$(P_{\text{eff}} A_{\text{tot}})/l = (P_A A_A)/l + (P_B A_B)/l, \quad (4-4)$$

where A_{tot} is the total membrane area, equal to the sum of A_A and A_B, and the thicknesses for both the intact skin and the filled pores are assumed to be the same and equal to l. The flux through the active layer is then given as

$$\begin{aligned} J_v &= (p_F - p_P)(P_{\text{eff}} A_{\text{tot}}/l) \\ &= (p_F - p_P)(P_A A_A/l + P_B A_B/l). \end{aligned} \quad (4-5)$$

Even if the permeability P_B is high, when the area of plugged pores (A_B) is small, the contribution of the term $P_B A_B/l$ to the effective permeability parameter $P_{\text{eff}} A_{\text{tot}}/l$ will be small. The bulk of the permeating gas will therefore go through the selective membrane polymer rather than through the less selective plugging material.

In practice, the material used to plug the defects also coats the entire surface of the membrane. The resistance offered by this coating, like the resistance of the underlying porous support, acts in series with the resistance of the separating layer. However, referring to Eq. (4-2), if a coating material with a very high permeability is used, the overall transport through the membrane is hardly affected.

Other techniques have been developed to overcome defects in the active layer of membranes. One such method is that of Cabasso and Lundy (1986) in which a permeable, nonselective "gutter layer" is applied to a highly porous substrate. A selective coating is then applied over the gutter layer. While defects are normally present in the selective coating, the gutter layer prevents convective flow through the membrane.

Commercially available membranes show a wide variation in properties, based on both the range of polymers used and the membrane structures achieved. An example showing this variability was presented in Table 3–5 of Chapter 3, which gave permeation properties for two very different hydrogen separation membranes from the same manufacturer. Most manufacturers do not disclose their performance properties, so the specific permeation properties of most membranes on the market cannot be readily determined. Table 4–1, however, gives reasonable ranges of properties that might be assumed for current (1990) polymeric membranes. In general, membranes with higher permeability coefficients will possess lower selectivities.

TABLE 4–1. Estimated Ranges of Membrane Properties in Commercially Available Gas Separation Membranes.

Hydrogen Separation Membranes	
Hydrogen permeability coefficient	30 to 500 GPU
Hydrogen/methane selectivity	30 to 250
Carbon Dioxide Separation Membranes	
Carbon dioxide permeability coefficient	10 to 200 GPU
Carbon dioxide/methane selectivity	5 to 30
Air Separation Membranes	
Oxygen permeability coefficient	5 to 250 GPU
Oxygen/nitrogen selectivity	3 to 7

Note: 1 GPU = 10^{-6} cm^3 (STP)/cm$^2 \cdot$ s $\cdot$ cm Hg.

Chemical Compatibility

For membranes to operate properly in a commercial application, they must not come in contact with either bulk liquids or with chemical species that are detrimental to the membrane polymer. Each of these situations is considered in turn below.

Commercial polymeric membranes are tolerant of liquid contaminants to varying degrees. Liquids can affect the functioning of the membrane in several ways. First, any liquid film that coats the membrane acts as an additional resistance to the diffusion of gases. This increased resistance results in lower transport rates and generally lower selectivities, because "fast" gases are usually slowed to a greater degree than "slow" gases. Irreversible fouling usually occurs with nonvolatile liquids such as oil, since the temperatures required to evaporate the liquid are normally deleterious to the membrane. Liquid fouling can be reversible for volatile liquids, although membrane performance will sometimes be permanently affected.

Removal of liquids from the microporous region of the membrane by simple evaporation can be detrimental to the structure of the membrane. In the evaporation process, surface tension forces can be generated that lead to *membrane compaction,* which is the collapse of part of the porous support layer adjacent to the active layer. This increases both the effective thickness of the active layer and the membrane's resistance to gas transport. Compaction is usually irreversible. Even if compaction does not occur, the forces generated during solvent evaporation can lead to cracking of the active layer, considerably lowering the effective selectivity of the membrane.

The compaction and cracking problems tend to become greater as the pore size decreases or the polymer wettability increases. Thus, hydrophobic polymers are more affected by nonpolar liquids (e.g., hydrocarbons), while hydrophilic polymers are more affected by polar compounds (e.g., water).

Specific chemical interactions between condensed liquids or gaseous feed components and the membrane polymer will also affect membrane performance. Solvents for the membrane polymer can plasticize it, causing it to soften. Once the polymer is softened it is easier to distort the "metastable" membrane morphology. This will tend to accelerate the rate of membrane compaction and shorten membrane life.

As noted in Chapter 3, gases that are highly soluble in the membrane polymer will also affect permeation. Carbon dioxide in particular is very soluble in many polymers. Increasing the CO_2 partial pressure of the feed gas will often increase the CO_2 permeability of the membrane, with a concomitant decrease in the selectivity for CO_2 over other gases (Sanders 1988). Such effects, if known from laboratory or field measurements, can be accounted for when designing a membrane system. In extreme cases, however, CO_2 can highly plasticize the membrane, leading to the compaction problem noted above for liquid solvents.

Finally, the performance of the membrane will be affected if chemical reactions of components in the feed alter the chemistry of the polymer. As an example, hydrochloric acid was found to be a problem with cellulose acetate membranes in an application where hydrogen was to be recovered from a butamer unit purge (Anon. 1985). It was necessary to scrub the HCl upstream of the permeators to ensure good membrane life.

Each membrane polymer has its unique chemical resistance properties. Table 4–2 gives chemical compatibility information for a particular commercial aromatic polyimide membrane. Membrane manufacturers can be expected to have chemical resistance data for their membrane polymers, especially for commonly encountered species.

Pressure and Temperature Effects

A key operational challenge for membranes is to withstand the large pressure differentials imposed on them. As might be expected, polymers with a higher modulus produce membranes having a greater capability to withstand large transmembrane pressure differences (Ekiner and Vassilatos 1990).

The productivity of a permeation system will normally decrease with time as the membrane densifies (compacts) under the applied pressure differential. As noted above, compaction results in reduced membrane productivity, since the membrane resistance increases as previously porous regions of the membrane structure are compressed and densified, increasing the effective skin thickness. A number of factors will influence the rate at which densification occurs, including the tendency of the polymer to undergo creep and the presence of condensable agents in the feed, which can plasticize the membrane material. The decrease in membrane productivity is often linear when

TABLE 4–2. Chemical Compatibility of a Commercial Ube Polyimide Membrane (reprinted from Ube Industries 1989 with permission).

Component	Volume % in Feed Gas					
	0.01	0.1	1	3	5	10
H_2O	A	A	A	A	A	A
H_2S	A	A	A	A	B	B
HCl	A	A	A	A	B	B
NH_3	A	B	C	C	C	C
CH_3OH	A	A	A	A	A	B
CH_3OCH_3	A	A	A	A	A	B
CH_3CHO	A	A	A	A	B	B
BTX	A	A	A	B	B	B
Gasoline	A	A	A	A	A	B

Note: BTX = benzene/toluene/xylenes.

A: The membrane can be used satisfactorily for an extended period of time.

B: The membrane can withstand the condition in this range for a short period of time. However, long-term operation in this range is not recommended, otherwise some performance decline will result.

C: Operation in this range is to be avoided. The membrane will be irreversibly damaged.

plotted against the logarithm of time. Decline rates for commercial membranes are generally between 1 and 10% per year in typical applications. Membrane selectivities often improve slightly when compaction occurs.

As demonstrated in Figure 4–2, certain morphologies can provide more stable performance than others. In this case the membranes were hollow fibers which, though made from very similar polymers, had different morphological (levels III and IV) structures. Because of its superior structure, the membrane formed by the wet-spinning process showed higher initial hydrogen permeability and better permeability retention than the structure produced by the dry-jet wet process (Coats, Wilkens, and

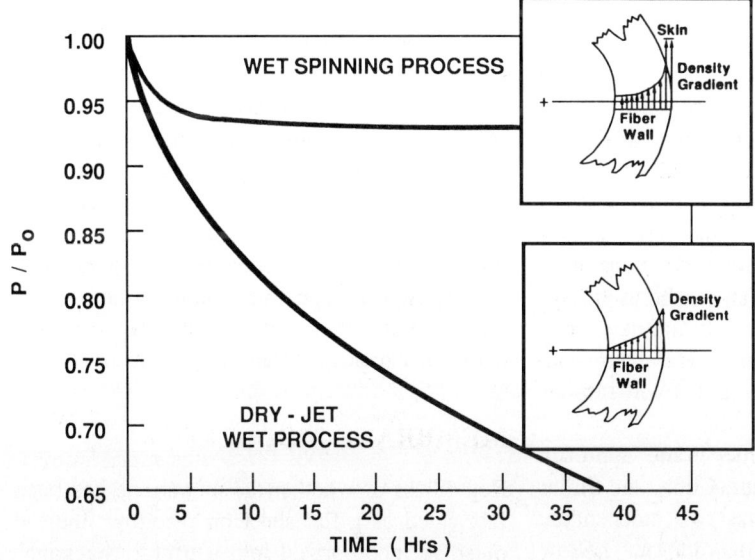

FIGURE 4–2. Influence of membrane structure on performance stability (Coats, Wilkens, and Zolandz 1987).

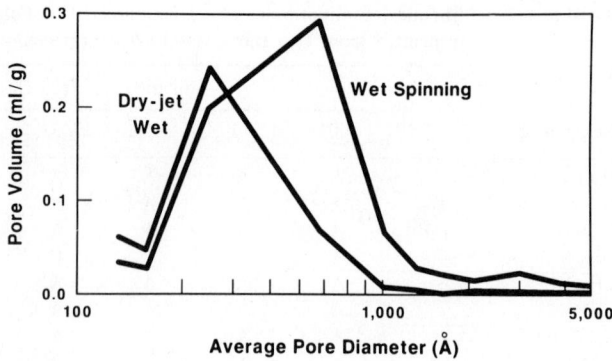

FIGURE 4–3. The effects of spinning process on membrane pore volume (reprinted from Weinberg 1988 with permission).

Zolandz 1987). As seen in Figure 4–3, the wet-spun membrane has both a greater porosity and a larger mean pore diameter than the dry-jet membrane (Weinberg 1988).

The temperature limits for membranes are related to the glass transition temperature of the polymer. As the glass transition temperature is approached, the polymer chains become more mobile and the membrane is increasingly susceptible to creep and compaction. Rigid polymers with high glass transition temperatures (such as the aromatic polyimides and polyamides) are preferred materials over first-generation polymers such as cellulose acetate and polysulfone in applications for which elevated operating temperatures are required.

Liquid Membranes

Liquid films on porous supports also form solution-diffusion membranes that can be used for gas separations. The liquids, which are held in place by capillary forces, naturally form defect-free layers. (See Chapter 42 for a review of related techniques.) This reliance on capillary forces severely limits the allowable transmembrane pressure difference that can be used. Applications for liquid membranes that have been tested in the laboratory include natural gas separation using supported liquid hydrocarbons (Brennan, Fane, and Fell 1986).

An alternative liquid membrane approach has been taken by Majumdar, Guha, and Sirkar (1988). Their permeator has two independent but intermingled sets of microporous hollow fibers. The liquid membrane resides in the shell of the permeator rather than in the walls of the hollow fibers. The effective thickness of the liquid membrane is made small by packing the fibers close together. Feed is introduced via the tubeside of one set of hollow fibers, transport occurs across the liquid in the shell, and the permeate is removed via the tubeside of the other set of fibers.

The above device is somewhat limited by the fact that current microporous membranes cannot support large pressure differentials without disruption of the liquid membrane layer. If the separation of interest cannot be carried out using the available pressure differential, then a "sweep" fluid must be utilized on the permeate side of the membrane. The sweep fluid, maintained at or near the pressure of the feed, is made to flow past the membrane in order to convey permeating molecules away from the membrane.

Inclusion of a "carrier" molecule in a liquid membrane can result in so-called "facilitated transport." The "carrier" molecule selectively and reversibly binds to specific component(s) of the feed mixture, then diffuses with it across the membrane, leading to enhanced transport rates.

Hollow-fiber-contained liquid membranes and facilitated transport are discussed in further detail in Chapters 42 and 44, respectively.

MEMBRANE MODULES

Regardless of whether the membrane has been fabricated as a flat sheet or a hollow fiber, it must be incorporated into a useful "package" that is readily usable in commercial processes.

Design of Gas Permeation Systems   61

The two membrane geometries lead to different permeator designs, although the same design criteria apply in both cases. The permeators need to contain as much surface area per unit volume as possible, but still allow good flow distribution and efficient contact of the feed gas with the membrane.

Spiral-wound modules will typically contain approximately 1000 ft^2 of membrane area per cubic foot of pressure vessel volume. Hollow-fiber permeators have approximately three times as much area per unit volume as the spiral-wound units. In early hollow-fiber modules the area advantage was offset by the fact that the thickness of the active layer was two to three times greater for hollow fibers than for flat films. Improvements by hollow-fiber manufacturers have resulted in membrane thicknesses comparable to those found in flat-film membranes.

Permeator housings for both spiral-wound and hollow-fiber membranes are normally manufactured using standard sized pipe. Housing sizes range from approximately 4 to 12 in. in diameter and from 4 to 20 ft in length. Materials of construction for the housing depend on the applications, with carbon steel and aluminum being suitable in many cases.

Flow Patterns

The performance of a membrane separator will be affected by the relative directions of feed and permeate flow in the vicinity of the active layer of the membrane and/or the relative flow directions of the bulk feed and permeate streams.

As shown in Figure 4–4, the feed and permeate streams may be directed cocurrent or countercurrent to one another. Crossflow permeation, with the permeate stream per-

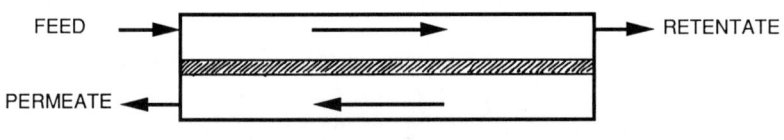

a. COUNTERCURRENT FLOW

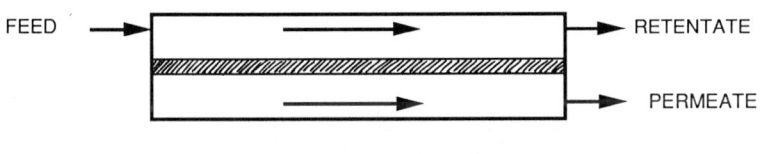

b. COCURRENT FLOW

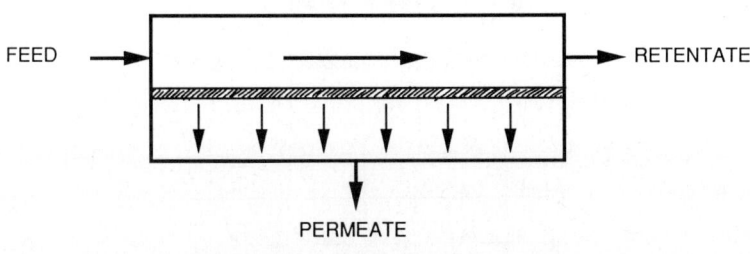

c. CROSSFLOW PERMEATION

FIGURE 4–4.
Flow patterns for gas separation membranes.

pendicular to the membrane, may also be practiced.

For co- and countercurrent permeation, the gas in contact with the downstream side of the membrane consists of gas that has just permeated through the membrane plus the bulk permeate that is flowing past it. As discussed later in this chapter, a set of differential equations relates the concentration profiles on the feed and permeates side of the membrane to each other.

For crossflow permeation, on the other hand, the permeate gas that is in contact with the active layer consists entirely of gas that has just passed through the membrane. There is no flow of permeate gas past the membrane from adjoining regions on the permeate side of the membrane, so the local permeate composition is not "coupled" to the permeate composition elsewhere along the membrane. In the case of two-component crossflow permeation, the following relationship holds at all points on the permeate side of the membrane:

$$\frac{x_{1p}}{x_{2p}} = \frac{P_1(x_{1f}p_f - x_{1p}p_p)}{P_2(x_{2f}p_f - x_{2p}p_p)}. \quad (4\text{-}6)$$

The local permeate composition is thus determined by the selectivity of the membrane, the feed composition and pressure, and the permeate pressure.

Note that the relative directions of the *bulk* feed and permeate stream are *not always related* to the *local* flow directions in the vicinity of the active layer. For symmetric (homogeneous) membranes, which possess no porous support region, the directions of the bulk and local flows will always be the same. This is not necessarily the case with asymmetric membranes, due to the presence of the porous support layer.

Consider, for example, an asymmetric membrane operated in the countercurrent mode with the feed gas flowing adjacent to the active layer of the membrane. Even though the bulk permeate stream flows countercurrent to the feed, the gas that permeates through the active layer can continue to flow perpendicular to the surface through the porous layer until it enters the bulk permeate stream, as shown in Figure 4–5. When this occurs *local* crossflow permeation results, and the permeator behaves as a crossflow device. This situation is most likely to occur in cases for which the permeate flow rate is high due to high membrane permeabilities (Pan 1986). Some early flat-film membranes had high enough permeabilities for this crossflow phenomenon to occur, while early hollow-fiber membranes did not.

For asymmetric hollow-fiber membranes with moderate permeabilities, such crossflow permeation is not expected to occur, and the local and bulk flow directions can be taken as identical. Sidhoum, Sengupta, and Sirkar (1988), for example, showed that the performance of their asymmetric hollow fibers operated with shellside feed correlated well with that expected from cocurrent or countercurrent flow. No indications of local crossflow were found in this study.

While delivery of the feed gas to the side of the membrane with the active layer is common, this is not always necessary. Instead it can be contacted with the porous support side. For

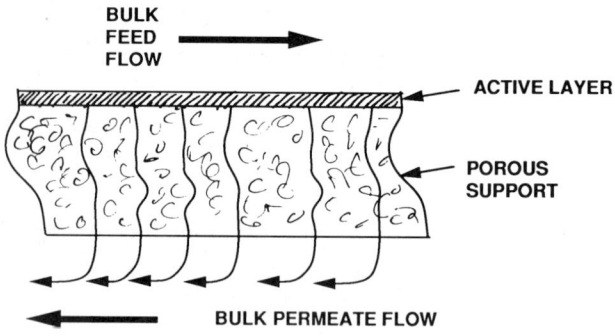

FIGURE 4–5. Crossflow permeation with asymmetric membranes.

typical hollow fibers, the back-diffusion of nonpermeating gases is rapid enough so that concentration polarization (the buildup of the concentration of "slow" gas at the feed/membrane interface) does not occur (Gollan 1988). Comparison of modeling calculations with experimental data for cellulose acetate fibers confirmed that standard co- and countercurrent models are adequate to predict device performance in such cases (Sidhoum, Sengupta, and Sirkar 1988).

A potential problem with having the pressurized feed in contact with the support side of the membrane is that in this configuration there is no support layer acting as a backing on the low-pressure side. The active layer is thus under tension rather than compression, making it more susceptible to damage when a pressure differential is applied. In practice, this is usually not a serious problem because the applications where such operation is useful (as in tube-side feed for air separations) involve operating pressures of only several hundred pounds per square inch.

Other Design Considerations

In an effective permeator design, pressure drop on the feed side should be minimized in order to avoid significant loss of driving force. Minimizing the pressure drop on the feed side of the membrane also reduces compression costs when the retentate is a product that will be used at higher pressures. Flow restrictions leading to undue pressure buildup on the permeate side need to be avoided for similar reasons.

Stagnant zones on the feed side of the membrane must be avoided. Slowly permeating gases will become concentrated in these stagnant regions, resulting in a localized increase in their partial pressure. The increased partial pressures result in increased transport of these components across the membrane, which lowers the efficiency of the separation.

Other factors influencing permeator design include ease of fabrication, manufacturing costs, and permeator robustness during operation. The designs arrived at by taking the above constraints into consideration are discussed in the following sections.

Spiral-Wound Permeators

A schematic of a spiral-wound permeator made from flat membrane films is given in Figure 4–6. The module is constructed by taking two membrane sheets and sealing them together at three edges to form a membrane *leaf,* which is somewhat like a large, open envelope. A spacer is provided inside of the leaf in order to maintain a gap between the two membrane sheets. The open end of the leaf is then sealed to a perforated tube. When the permeator is operated, the pressurized feed gas is brought into contact with the outer surface of the leaf. Gas diffuses through the walls of the leaf and then flows along the inside of the envelope to the collection tube, which is held at a pressure lower than that of the feed.

The maximum length of the leaf is determined by the allowable pressure drop for the permeate gas as it flows to the collection tube, and by manufacturing considerations. Multiple leaves can be attached to a single collection tube. Because a cylindrical pressure vessel is the simplest and cheapest design, the leaf or leaves are rolled up into a spiral configuration, forming a cylindrical cartridge that can then be inserted into a tubular pressure vessel. A common practice is to install more than one spiral-wound cartridge in a single pressure vessel, which reduces the number of piping connections needed for a given membrane area.

Note that in spiral-wound modules the bulk feed and permeate flows are neither cocurrent nor countercurrent to each other. Instead the flows at any point in the spiral are in approximately parallel planes but at right angles to one another. However, since the permeability of film-type membranes is generally high enough so that local crossflow permeation occurs, the relative directions of the feed and permeate matter little unless there is severe pressure buildup on the permeate side of the membrane.

Spiral-wound permeators are operated as depicted in Figure 4–6, with the feed gas contact-

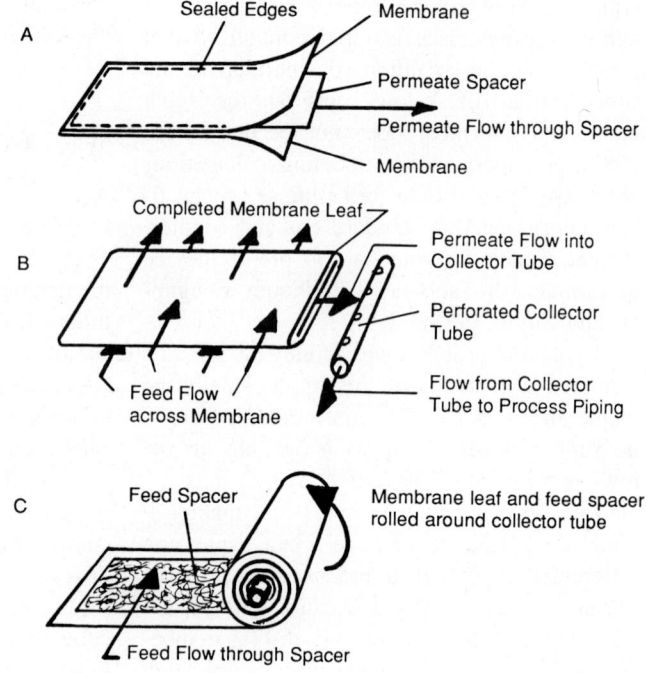

FIGURE 4-6. Schematic of a spiral-wound permeator assembly.

ing the outer surface of the membrane leaf. This arrangement yields the simplest module construction because otherwise an additional distributor would be required to deliver feed gas into the leaf. Also, it is easier to provide adequate mechanical support for the membrane and to maintain the integrity of the seals at the leaf edges when the high pressure is applied to the outer surface.

The detailed design of the elements associated with a spiral-wound permeator, such as spacers and collection tubes, will influence permeator performance. Thus, Cooley and Dethloff (1985) noted how improvements in permeator design led to better module performance at low feed rates and better removal of trace components from the feed gas. It was also reported that the performance of the improved modules more closely followed model calculations of device performance.

Hollow-Fiber Modules

Commercial hollow-fiber modules are operated in the three different configurations depicted in Figure 4-7. These are:

1. Shellside feed, with countercurrent flow of feed and permeate
2. Shellside feed, with crossflow of the feed gas
3. Tubeside feed, with countercurrent flow of feed and permeate.

For all of these flow patterns the permeators are constructed somewhat like shell and tube heat exchangers. The fibers, which are generally between 100 and 500 μm in diameter, are arranged parallel to one another and pass through tubesheets at either one or both ends of the device. Seals are provided between the tubesheet exterior and the pressure vessel in order to isolate the high-pressure feed from the low-pressure permeate.

Shellside Feed Permeators

In shellside feed permeators, the feed is brought in contact with the outer surface of the membrane fibers. Gas permeates into the fiber and then flows down the fiber bore, from which it passes through the tubesheet and out of the pressure vessel. Two different flow con-

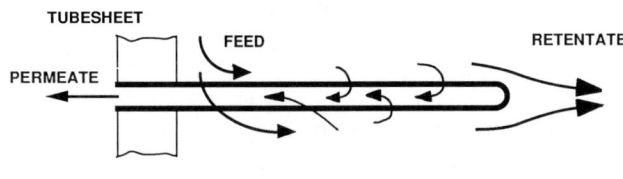

A. SHELLSIDE FEED, COUNTERCURRENT FLOW

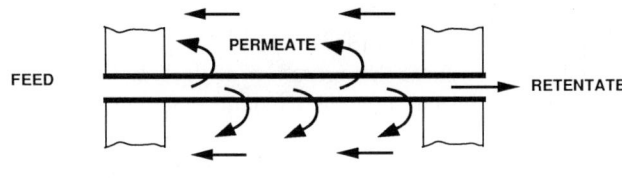

B. TUBESIDE FEED, COUNTERCURRENT FLOW

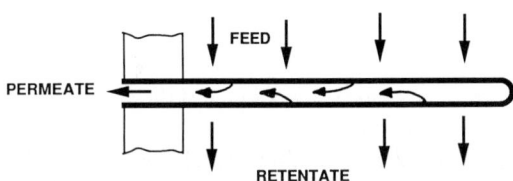

C. CROSSFLOW SHELLSIDE FEED

FIGURE 4-7. Flow configurations for hollow-fiber permeators.

figurations, countercurrent and radial crossflow, are used in commercial devices with shellside feed.

In countercurrent permeators the feed and permeate streams flow parallel to the fibers but in opposite directions. In a radial crossflow permeator, on the other hand, the feed gas flows in a radial direction, perpendicular to the length of the fibers (Figure 4-8).

As discussed further below, an ideal countercurrent permeator will give better performance than a crossflow feed device having membranes with the same permeation properties. However, crossflow feed devices with advanced membranes are competitive with countercurrent devices. Crossflow feed devices are, in addition, readily engineered to achieve uniform feed gas distribution and are more readily scaled up to larger diameter devices.

For all permeators with shellside feed, provisions must be made to distribute the feed gas evenly and achieve uniform flow through the bundle. In shellside feed countercurrent permeators, gas is generally introduced through a distributor at one end of the device. The feed gas first flows radially into the bundle, then turns and flows parallel to the fibers. The distributor is designed to deliver gas as uniformly as possible across the diameter of the permeator. The fibers must be uniformly packed throughout the length of the permeator in order to maintain proper flow distribution.

In crossflow feed permeators, the feed can flow either radially inward or outward. Inward crossflow has an advantage over outward flow in terms of maintaining good flow distribution because the molar and volumetric flows of the feed decrease as it flows through the permeator, due to permeation. For permeators with inward flow, the cross-sectional area through which flow occurs decreases along with the decreasing flow rate. Flow velocities are maintained, and the tendency toward stagnant zones is reduced. Stagnant zones are also minimized in crossflow

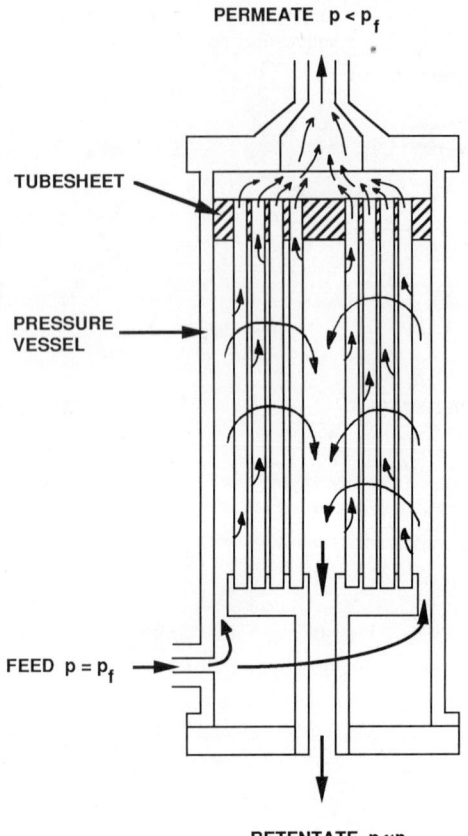

FIGURE 4-8. Schematic of a crossflow feed permeator.

permeators because the fibers themselves serve as flow distributors.

Radial crossflow feed must be distinguished from the definition for crossflow permeation introduced previously. The former term refers to the bulk flow pattern within a permeator, whereas the latter term refers to flow of permeate gas perpendicular to the membrane's active layer. Note that a radial crossflow permeator can experience crossflow permeation if it contains fibers with very high permeability coefficients (P/l).

Modules with crossflow feed can have tubesheets at either one or both ends. For a module with two tubesheets, an axial symmetry exists about the center of the permeator, and the device behaves exactly like two permeators with half the overall length. This reduces the pressure drop in the fiber bores, which leads to improved performance at the expense of requiring a second tubesheet. Permeators with two tubesheets have been used in CO_2 separations (Parro 1985).

Pressure drop can be minimized in single tubesheet crossflow devices by limiting the length of the membrane fibers. Shorter fiber lengths, however, will result in less surface area being packaged in a single pressure vessel unless multiple permeator bundles are placed into a single pressure vessel, as in the design of Edwards (1987).

Note that crossflow devices, unlike countercurrent modules, require a central perforated tube and an annular space between the bundle of fibers and the inner wall of the pressure vessel. One acts as a gas distributor, the other as a collector. Both these spaces represent pressure vessel volume that is not filled with membrane area, and so are made as small as practical.

An interesting radial crossflow design has been described by Coplan (1987) in which several permeate cuts are taken from separate sets of fibers housed in a single module. Such a permeator could be used in the two-stage recovery of hydrogen from ammonia purge gas, which is described in Chapter 5.

Tubeside Feed Permeators

Permeators in which feed gas is delivered to the bore of the hollow fibers are also used commercially. Two tubesheets are required, with feed gas being introduced through one tubesheet while nonpermeating gas exits from the other. The permeate stream is removed countercurrent to the feed because this is the most efficient method of operation. The first commercial-scale hollow-fiber gas permeators described in the literature were in fact tubeside feed devices that were tested in various applications, including hydrogen recovery in the ammonia process and H_2/CO adjustment (Gardner, Crane, and Hannan 1977).

Choosing Flow Patterns in Hollow-Fiber Modules

The choice between shellside and tubeside feed in hollow-fiber modules is generally made based on the separation to be performed. Hollow-fiber permeators for hydrogen separations, for example, will normally utilize shell-side feed. Such a design minimizes the pressure drop occurring between the feed and the retentate streams, an important consideration in applications such as H₂/CO adjustment for synthesis gas (syngas) plants. Also, most H₂ recovery applications involve large transmembrane pressure drops, and the hollow fibers have better resistance to pressure when they are under compression, which occurs with shellside feed. Finally, since permeate pressures in hydrogen applications are usually well above atmospheric pressure, the actual volumetric flow of permeate gas is low and so boreside permeate pressure drops will be minimal.

In contrast, a number of factors favor tubeside feed when air is separated to produce nitrogen-enriched air (NEA). First, transmembrane pressure differentials are relatively small, so that undue mechanical stresses are not generated. Also, about half of the molar flow fed to the permeator exits the device as permeate at atmospheric pressure. Less pressure drop will occur if this relatively large volume of permeate is removed on the shellside, and so better device performance is realized when tubeside feed is used.

The difference between shellside and tubeside feed for NEA production is demonstrated in Figure 4–9. The curves show predicted ideal performance for permeators with identical surface areas and tubesheet lengths operated under the same conditions in the two feed modes. Conditions used in the calculations are as follows: 100 psig feed pressure, 0 psig permeate pressure, 40°C, P_{O_2} = 30 GPU, P_{N_2} = 6 GPU, active fiber length = 4 ft, tubesheet length = 4 in., fiber outer diameter = 300 μm, and fiber inner diameter = 150 μm. Note that 1 GPU (gas permeation unit) = 10^{-6} cm³ (STP)/cm² · s · cm Hg.

At any nitrogen purity, operation with tubeside feed recovers a larger fraction of the feed gas. Likewise, at a fixed recovery, a higher purity product is delivered by the tubeside feed device.

In applications where shellside feed per-

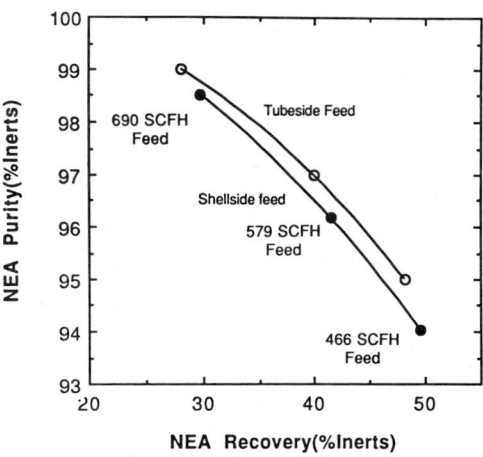

FIGURE 4–9. Comparison of tubeside and shellside feed hollow-fiber devices. Parameters are cited in the text.

meators are used, such as in hydrogen recovery, the choice between countercurrent and crossflow devices [see Figures 4–7(a) and 4–7(c)] is not obvious. If the countercurrent and crossflow devices contain the same fiber geometry and the same number of fibers, in the ideal case a countercurrent device will always outperform a crossflow device with the same fiber properties. However, the membrane properties of products from different vendors are not the same. Crossflow devices with superior membranes can outperform countercurrent devices in many applications. This is demonstrated for an idealized binary separation in Figure 4–10.

The two lower curves of Figure 4–10 show hydrogen product purity as a function of hydrogen recovery under a fixed set of operating conditions for countercurrent and crossflow devices with identical membrane properties (a hydrogen permeability of 50 GPU and a H₂/CH₄ selectivity of 50). At any specified hydrogen recovery, the countercurrent device yields a purer hydrogen product.

If the crossflow permeator, however, has membranes with twice the H₂ permeability and three times the H₂/CH₄ selectivity as the other permeators, it would, at any given hydrogen recovery, deliver a higher purity product, as shown by the upper dashed curve in Figure 4–10.

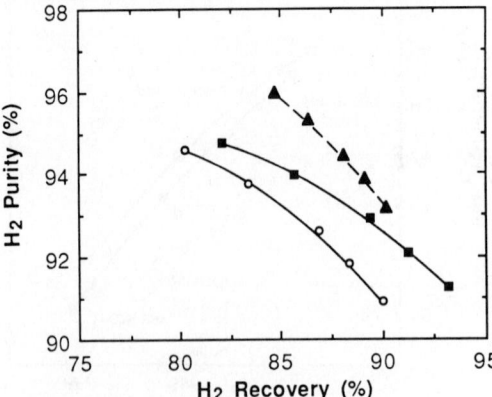

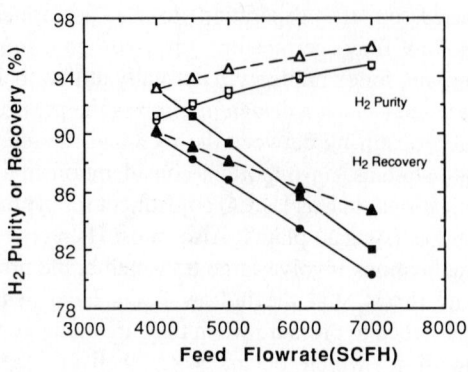

FIGURE 4-10. Comparison of shellside feed permeators with countercurrent and crossflow feed. Parameters used in the calculations: feed pressure = 1100 psig, permeate pressure = 400 psig, active fiber length = 4 ft, tubesheet length = 4 in., fiber outer diameter = 300 μm, fiber inner diameter = 150 μm. The curves are for the following cases: squares: Countercurrent flow, P_{H_2} = 50 GPU, P_{H_2}/P_{CH_4} = 50; circles: Crossflow feed, P_{H_2} = 50 GPU, P_{H_2}/P_{CH_4} = 50; and triangles: Crossflow feed, P_{H_2} = 100 GPU, P_{H_2}/P_{CH_4} = 150.

FIGURE 4-11. Comparison of shellside feed permeators with countercurrent and crossflow feed as a function of feed flow rate. Data and symbols are the same as Figure 4-10.

The same purity and recovery data can be plotted against the amount of feed gas processed by the permeator, as is done in Figure 4-11. Here it is seen that, for a given feed flow rate, the crossflow and countercurrent permeators with the same fiber properties produce approximately the same product purity, but the crossflow device operates at a lower recovery. For the superior membrane operated in crossflow mode, the purity of the permeate recovered at any given feed flow is always higher than that for the other membrane. The hydrogen recovery, however, will be greater or less than that for the countercurrent module depending on the feed flow rate.

Modeling the Performance of Gas Permeation Modules

Simulation models for gas separation membranes have been developed by a number of investigators. Membrane suppliers have proprietary programs that model the performances of their particular products. Any such permeator model must include

1. A relationship describing transport across the membrane
2. Mass balance equations
3. Relationships or assumptions for the pressure drops that occur on both sides of the membrane due to convective flow of gas
4. Boundary conditions that reflect the configuration and operation of the permeator being modeled.

Transport across the membrane is generally described by Eq. (4-1). This relationship can be used in developing the mass balance equation for the feed side of the membrane. The result is given in Eq. (4-7), which expresses the change in molar flow of a component i in a differential element (see Figure 4-12):

$$-\frac{d(x_{if}Q)}{dA} = \frac{P_i p_0 (p_f x_{if} - p_p x_{ip})}{RT_0 l}. \quad (4-7)$$

Note that dA is an incremental element of area related to membrane geometry. For a single

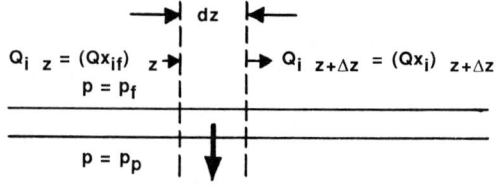

FIGURE 4-12. Differential mass balance for membrane permeation.

rectangular membrane $dA = W\,dz$, where W is the width of the sheet. For a hollow asymmetric fiber, $dA = \pi d_o\,dz$, while for a symmetric fiber $dA = \pi d_{lm}\,dz$ with $d_{lm} = (d_o - d_i)/\ln(d_o/d_i)$, where d_o and d_i are, respectively, the outer and inner fiber diameters.

Rigorous models will account for pressure drops within the modules, although these are often ignored. For flat-film modules, the pressure drop relationships will depend on the type of spacers used within and between the membrane leafs.

For hollow-fiber permeators, the Hagen-Poiseuille equation for flow through tubes can be used to describe pressure drop on the tube side of the fiber:

$$\frac{dp_t}{dz} = \frac{-8\eta v_b}{g_c r_i^2}. \quad (4-8)$$

Pressure drop on the shell side of the fibers is often neglected, although empirical relationships or correlations for flow through porous media or banks of tubes can be used.

The boundary conditions and methods of solution for the permeator model will vary depending on the relative flow directions of the feed and permeate streams and the assumptions used in a given formulation. Simplified models such as that of Rautenbach and Dahm (1986) may be used during initial design stages, but fairly rigorous models are required to obtain good performance estimates.

Antonson et al. (1977) presented an early set of equations developed specifically for commercial-scale hollow-fiber permeators and examined a number of hollow-fiber configurations for binary separations. A description of their model (presented in dimensional form) is informative because it shares many elements with other formulations. Their model assumes that

1. The ideal gas law applies.
2. The operation is isothermal.
3. Viscosity of the mixed gas is the sum over all components: $\eta_{\text{mixture}} = \Sigma_i x_i \eta_i$.
4. The integrated Hagen-Poiseuille relationship describes the tubeside pressure drop.
5. Pressure on the shell side is constant.
6. Axial diffusion can be neglected.
7. Permeability coefficients are independent of pressure and composition.

For *tubeside feed* (Figure 4–13), the component balances are

$$\frac{dQ_{1t}}{dz} = \pi d_{lm} N \left(\frac{P_1 P_0}{lRT_0}\right)\left(\frac{Q_{1t} P_t}{Q_{1t} + Q_{2t}} - x_{1s} P_s\right), \quad (4-9)$$

$$\frac{dQ_{2t}}{dz} = \pi d_{lm} N \left(\frac{P_2 P_0}{lRT_0}\right)\left(\frac{Q_{2t} P_t}{Q_{1t} + Q_{2t}} - x_{2s} P_s\right), \quad (4-10)$$

and the pressure drop relationship on the tube side is

$$\frac{dp_t}{dz} = \frac{128 RT(\eta_1 Q_{1t} + \eta_2 Q_{2t})}{\pi g_c d_i^4 N p_t}. \quad (4-11)$$

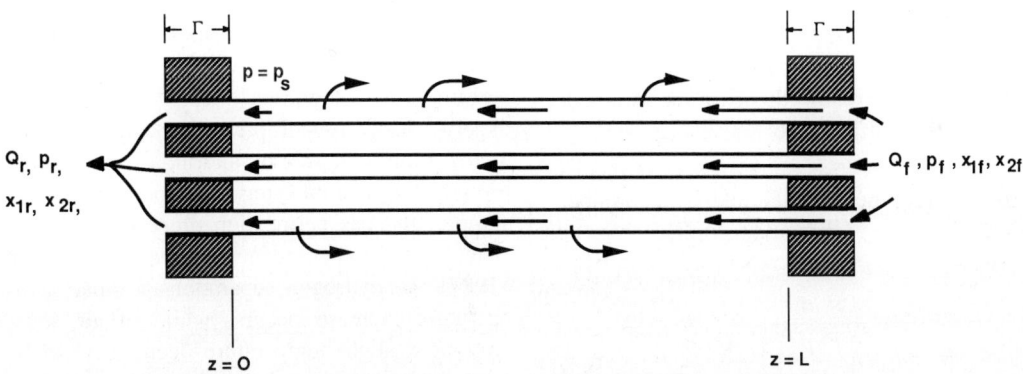

FIGURE 4–13. Idealized tubeside feed permeator for mathematical modeling.

The latter relationship derives from Eq. (4–8), using the ideal gas law to relate the volumetric flow to the molar flow.

Boundary conditions at the feed end of the permeator ($z = L$) are:

$$Q_1 = x_{1f}Q_f, \qquad (4-12)$$

$$Q_2 = x_{2f}Q_f, \qquad (4-13)$$

$$p_t = \left[p_f^2 - 2\left(\frac{128\eta_f Q_f RT}{\pi g_c d_i^4 N}\right)\right]^{1/2}. \qquad (4-14)$$

The latter condition derives from the fact that the pressure of the feed gas drops as the gas flows from the face of the tubesheet to the beginning of the region where permeation occurs.

On the shell side, in the case of countercurrent flow, the composition at the residue end ($z = 0$) is determined strictly by the ratio of the permeabilities of the two components, since at $z = 0$ crossflow permeation occurs:

$$\frac{x_{1s}}{x_{2s}} = \frac{P_1(x_{1t}p_t - x_{1s}p_s)}{P_2(x_{2t}p_t - x_{2s}p_s)} \qquad z = 0. \qquad (4-15)$$

Note that this last shellside condition requires knowledge of the tubeside pressure p_t, making it necessary to solve the problem iteratively. Residue gas conditions are assumed, and then the feed gas conditions are calculated by integration of the equations. The calculated feed values are compared to actual values, the assumed residue conditions are adjusted, and the iteration process continued until one converges on a solution.

For *shellside feed* entering at $z = L$, the component balances resemble Eqs. (4–9) and (4–10), except that in this case the shell side is at the elevated pressure:

$$\frac{dQ_{1s}}{dz} = \pi d_{lm} N \left(\frac{P_1 p_0}{lRT_0}\right)\left(\frac{Q_{1s}p_s}{Q_{1s} + Q_{2s}} - x_{1t}p_t\right),$$
$$(4-16)$$

$$\frac{dQ_{2s}}{dz} = \pi d_{lm} N \left(\frac{P_2 p_0}{lRT_0}\right)\left(\frac{Q_{2s}p_s}{Q_{1s} + Q_{2s}} - x_{2t}p_t\right),$$
$$(4-17)$$

Since at any point in the countercurrent flow permeator $Q_{it} = Q_{is} - Q_{ir}$, the tubeside pressure drop is given by

$$\frac{dp_t}{dz} = -\left(\frac{128RT}{\pi g_c d_i^4 N p_t}\right)[\eta_1(Q_{1s} - Q_{1r})$$
$$+ \eta_2(Q_{2s} - Q_{2r})], \qquad (4-18)$$

where Q_{1r} and Q_{2r} are the molar flows of the respective components in the shellside residue stream exiting the permeator. The boundary conditions with shellside feed at $z = L$ are

$$Q_{1s} = Q_{1f}x_{1f}, \qquad (4-19)$$

$$Q_{2s} = Q_{2f}x_{2f}. \qquad (4-20)$$

The boundary condition for the tubeside pressure is calculated from p_p, the pressure at which the permeate is collected, by accounting for the pressure drop experienced by the permeate gas as it flows through the tubesheet:

$$p_t = \left[p_p^2 + 2\left(\frac{128\eta_p Q_p RT}{\pi g_c d_i^4 N}\right)\right]^{1/2}, \quad z = L. \quad (4-21)$$

The solution to the shellside feed problem is obtained by making initial guesses for the residue composition and the tubeside pressure at the residue end of the permeator, performing the integration, then adjusting the guesses as necessary.

This model and its extensions provide a useful approach to modeling actual hollow-fiber permeators because, unlike some other models, all of the permeator dimensions have fixed values, rather than being variables that are determined by integration.

The basic model presented above can be made more complex by altering or eliminating some of the simplifying assumptions. For example, the gas permeation process is actually isenthalpic rather than isothermal (Gorissen 1987). In hydrogen separations a small temperature increase occurs, while in air separations a slight temperature decrease is to be expected. Only for very nonideal gases such as carbon dioxide is there a significant temperature effect.

As mentioned above, Rautenbach and Dahm (1986) proposed a simplified bicomponent model for a countercurrent hollow-fiber permeator that can be used for initial performance estimates. To arrive at an analytical solution, it is assumed that the partial pressure of the fast gas in the permeate is a constant.

Having made this assumption, equations describing the flow within the permeator and the concentration can be derived as follows:

$$\frac{Q_f}{Q_{in}} = \left(\frac{x_f - X_1}{x - X_1}\right)^{\mathcal{H}_1} \left(\frac{x - X_2}{x_f - X_2}\right)^{\mathcal{H}_2}, \quad (4-22)$$

$$\frac{\Phi z}{L} = \frac{(X_1 - x_f)(X_1 - X_2)}{(1 - \alpha)(X_2 + a_1)} \left\{ \left(1 + \frac{(X_2 + X_1)(x_f - X_2)}{(X_1 + a_1)(X_1 + x_f)}\right) - \left(\frac{(x - X_2)(x_f - X_1)}{(x_f - X_2)(x - X_1)}\right)^{\mathcal{H}_2} \right.$$

$$\left. \times \left[1 + \frac{(X_2 + a_1)(x_f - X_2)}{(X_1 + a_1)(X_1 - x_f)} \left(\frac{(x - X_2)(x_f - X_1)}{(x_f - X_2)(x - X_1)}\right)^{\mathcal{H}_3}\right]\right\}, \quad (4-23)$$

where

$\Phi = A \cdot p_{1f} \cdot (P_1/l)/Q_f$
$\mathcal{H}_1 = (X_1 + a_1)/(X_1 - X_2)$
$\mathcal{H}_2 = (X_2 + a_1)/(X_1 - X_2)$
$\mathcal{H}_3 = (X_1 + a_1)/(X_2 + a_1)$
$X_1 = a_2/2 + [(a_2/2)^2 - a_3]^{1/2}$
$X_2 = a_2/2 - [(a_2/2)^2 - a_3]^{1/2}$
$a_1 = (1 - p_p/p_f)\alpha/(1 - \alpha) - p_p x_p/p_f$
$a_2 = 1 + p_p x_p/p_f + (p_p/p_f)\alpha/(1 - \alpha)$
$a_3 = (p_p x_p/p_f)/(1 - \alpha)$
$\alpha = P_2/P_1$

and where x is the local mole fraction of fast gas on the feedside of the membrane, x_f and x_p are the mole fractions of fast gas in the feed and bulk permeate streams, respectively, Q_f is the feed flow rate at position z, and Q_{in} is the initial feed flow rate. The position coordinate z is defined to be zero at the end of the fiber from which permeate is withdrawn, with the tubesheet understood to have zero length. Other variables are as previously defined.

This simplified model compared favorably to the numerical solution obtained when variation in permeate partial pressure is considered, and with the series solution of Boucif, Majumdar, and Sirkar (1984). Deviations from the numerical solution were found to be greatest at high stage cuts and when the permeability factor Φ was large.

Models of spiral-wound permeators are similar to those for hollow-fiber permeators. Equation (4-7) still defines the differential mass balance for the permeation process, and ap-

propriate pressure drop relations, though sometimes ignored, can be incorporated as desired.

As mentioned above, local crossflow permeation (Figure 4-5) can occur with highly permeable asymmetric membranes. In the presence of such crossflow permeation, Eq. (4-6) applies at all points on the membrane surface. As a result, if the permeate pressure is constant (as often occurs to a good approximation in flat-film permeators), the relative flow directions of the bulk feed and permeate streams do not affect permeator performance (Pan 1983). In cases for which the permeate pressure does vary, performance differences due to flow directions are less pronounced when local crossflow permeation does occur compared to when it does not occur.

A simplified model that can be used to estimate the required suface area for spiral-wound permeators has been presented by Hogsett and Mazur (1983). The area is calculated from the relationship

$$A = \frac{x_{1p} Q_p}{(P_1 p_0 / lRT_0)(p_f x_{1lm} - p_p x_{1p})}, \quad (4-24)$$

where $x_{1lm} = (x_{1f} - x_{1r})/\ln(x_{1f}/x_{1r})$. It was recommended that when the concentration change between the feed and residue is large then the calculation should be done for several different composition increments. As an example, the suggestion was made that when the composition from feed to residue changes from 50 to 6% CO_2, the calculation should be performed for the increments from 50 to 25%, 25 to 13%, and 13 to 6%, and the results summed.

Koros and Chern (1987) discuss a number of other published models, which are not discussed here. As noted by them, analytical solutions are only possible in cases where significant simplifying assumptions have been made. Many of these models consider only two-component systems. Models of multicomponent permeation for various flow patterns have been developed (Li, Acharya, and Hughes 1990; Shindo et al. 1985).

CONFIGURATIONS USED IN GAS SEPARATIONS

Arrangement of Gas Separators

The manner in which gas permeators are arranged with respect to one another is determined by both the module design and the process application, which we will consider in turn. Configurations of membranes with other unit operations are discussed in Chapter 5.

Permeators are designed to operate within a range of gas flow rates, and it is good practice to stay within this range. Operating a permeator below the minimum design flow rate can result in poor distribution of the feed gas, while operating above the recommended maximum flow can cause excessive pressure drop between the feed and retentate ports. Operation below the recommended feed range is least desirable because feed maldistribution can lead to stagnant zones, which result in both diminished permeate purity and lower permeator productivity.

In a given membrane stage the proper flow rates are achieved by piping permeators in parallel and splitting the feed stream equally among them. If the feed flow rate will vary with time, a stage can consist of parallel banks of permeation modules. The required number of banks, or of individual modules within banks, is then brought on-line as needed.

A single membrane stage operating at fixed feed and permeate pressures is limited as to the extent of separation that it can achieve. For membranes with fixed selectivities, the maximum achievable purity for a single membrane stage, obtainable in the limit of zero stage cut (where stage cut is the ratio of the molar permeate flow to the molar feed flow), depends on the feed concentration. The purity of the permeate stream decreases from this maximum as the percent recovery of the more permeable component is increased, as demonstrated, for example, in Figure 4–10. This trade-off between purity and recovery is intrinsic to the membrane process.

This same limitation is experienced by the simplest multiple-stage membrane system in which two stages are piped in series, with the retentate from the first stage being fed directly to the second stage, with no recycle of permeate gas. This type of staging is useful in cases for which a large fraction of the feed gas is to be withdrawn as the permeate: The first stage contains a larger number of permeators to handle the initial feed volume, whereas the second stage, with fewer permeators, processes the reduced volume of the first-stage retentate. This arrangement allows maintenance of the proper gas velocities in each individual permeator.

The banks of modules in the series configuration discussed above need not be operated at the same permeate pressure. Monsanto's process for hydrogen recovery from ammonia purge gas, for example, runs the second stage at a lower permeate pressure than the first. The lower permeate pressure in the second stage increases both the available driving force and the pressure ratio between the feed and permeate streams. This serves to improve both the purity and yield of recovered hydrogen, at the expense of some additional recompression costs.

While the process mentioned above operates at two different permeate pressures, stages can be operated in series with interstage compression. This serves to increase the partial pressure of the more permeable components prior to the downstream stage.

To increase product recovery beyond what can be provided by the once-through systems discussed above, the membranes must be operated with recycle. Recycle of permeate causes the feed composition to the membrane to increase. This results in the production of a higher product purity at any given recovery. Recompression of the recycled permeate adds both investment and operating cost to the membrane system, although membranes can still be competitive in these cases. Note that recompressed permeate can be recycled to the same membrane stage or to a previous stage in a multistage process.

Analyses of permeation systems with recycle have been performed by a number of investigators, including Weller and Steiner (1950), Stern, Perrin, and Naimon (1984), and Majumdar et al. (1987). Among the descriptions of multistage permeation processes are those of Stookey, Graham, and Pope (1984) and of Mazur and Chan (1982) for natural gas processing and of Schendel (1982) for CO_2 recovery in enhanced oil recovery projects.

Various other permeator configurations, though not generally practiced commercially, are of some interest. For example, recycle strippers and enrichers have both been studied (Kao 1988). In the recycle stripper the feed is brought into the module in the usual manner and a portion of the permeate stream is recycled back to the feed stream. In the recycle enricher the feed is brought in as a sweep gas on the permeate side of the membrane and a portion of the permeate is compressed and fed to the high-pressure side of the membrane. The recycle stripper was found to outperform the enricher in terms of both separating effectiveness and compressor requirements.

In a general analysis, Pan and Habgood (1978a, 1978b) examined a membrane cascade in which permeate is recompressed and passed on to the next membrane stage while the nonpermeate is recycled back to the previous stage. With the feed introduced at an intermediate stage, both enricher and stripper sections exist. Their results show that, for a given separation at constant membrane permeability, an optimum selectivity exists that leads to a minimum membrane area.

Another multimodule configuration that has received considerable attention is the so-called "continuous membrane column" proposed by Hwang and Thorman (1980). The column, shown in Figure 4–14, consists of an enriching and a stripping section, with feed being introduced between them. At the "top" of the enriching section, a portion of the permeate

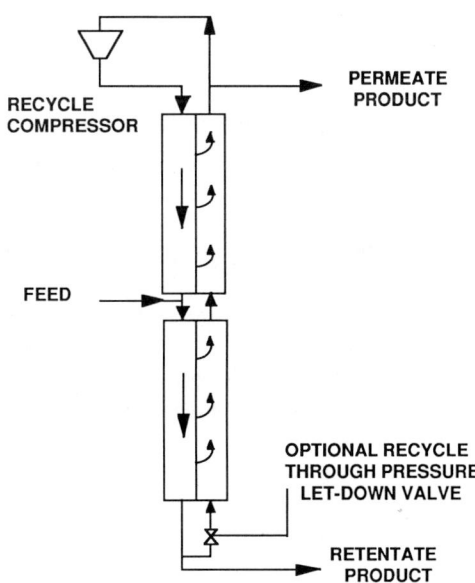

FIGURE 4–14. Schematic of a continuous membrane or membrane rectification column (reprinted from Hwang and Thorman 1980 with permission).

stream is compressed and returned to the high-pressure side of the membrane. Even with a low-selectivity membrane, high degrees of separation can be achieved with sufficient recycle: This is analogous to operating a distillation column at high reflux ratios in cases where the component volatilities are similar.

The performance of a continuous membrane column has been compared to that of two membrane stages in series (Kao, Qiu, and Hwang 1989). In this study, the permeate from the second stage was recycled to the first stage. The series arrangement was preferable for cases of lower stage cut and higher permeate enrichment, particularly when using membranes with high selectivity. The continuous-membrane column, on the other hand, could produce a more highly enriched product than the series configuration, even with low-selectivity membranes.

Another study of these columns (Schulz and Werner 1984) showed that if backmixing occurred, then the efficiency of the separation would be seriously decreased.

Laboratory-scale separations examined on continuous membrane columns include methane recovery from mixtures with CO_2 and/or nitrogen (Hwang and Ghalchi 1982).

The performance of a continuous membrane column can be analyzed using techniques similar to those used for distillation columns. For example, McCabe-Thiele diagrams have been developed for continuous membrane columns (Schulz, Michele, and Werner 1982a, 1982b). The method uses a "quasi-equilibrium" curve that represents the maximum purity achievable in the permeate at a given feed composition. This curve varies depending on membrane selectivity and operating pressure ratio. Mass balance equations provide operating lines, and then the required number of "stages" of separation can be stepped off.

While it is possible to operate membrane rectification schemes using standard hollow-fiber countercurrent permeators, the need for recompression of the top reflux stream adds cost to the overall membrane system and makes such an approach unattractive in many applications.

Novel Device Configurations

Membrane modules incorporating two different membrane materials have been demonstrated in the laboratory. Perrin and Stern (1986), for example, reported enhanced separation of helium-methane mixtures using a combination of cellulose triacetate and silicone rubber membranes. Helium permeates more rapidly through the former and methane more rapidly through the latter.

In a "normal" single-membrane permeator, the fast gas (helium) is stripped across the membrane and becomes depleted in the retentate stream. As the fast gas becomes depleted, its concentration in the gas permeating the membrane is diminished.

By adding the second membrane, the slow component is also stripped from the retentate. Because slow gas is removed across the second membrane, the fast gas concentration in the retentate does not become as depleted as it does in the single-membrane permeator, and purer fast-gas product can be obtained from the first membrane.

Two-membrane permeators have also been used for the separation of ternary mixtures (Sengupta and Sirkar 1987, 1988). In this case the feed contains two "fast" gases, with each type of membrane being more permeable to a different fast gas. Three streams, each enriched in a different component, are produced. The two-membrane permeator was compared to two separate permeators in series, each containing one of the membrane types used in the two-membrane permeator (Sengupta and Sirkar 1988). Performance for the series permeators varies depending on which type of permeator is placed first in the series. For recovery of either fast component individually, the performance of the two-membrane permeator was intermediate to that of the two series configurations. However, in terms of overall recovery of the fast components, the two-membrane device generally showed the better performance.

Although adding a second set of fibers to a permeator increases the complexity of its design, such a device is not impractical. In fact, a patent for a commercial-scale permeator incorporating two types of membranes has been issued (Perrin 1988).

A different module concept that employs two sets of the same fiber type to produce improved separations is the "internally staged" permeator (Sidhoum et al. 1988; Sidhoum, Majumdar, and Sirkar 1989). In this device gas is fed into the tubeside of one set of hollow fibers. Gas permeates out of these fibers into the shell of the device, which is held at an intermediate pressure. Permeation then occurs from the shell side into the second set of fibers, which is held at a lower pressure. Two retentate streams and a single low-pressure permeate stream are produced. Experiments and modeling studies both showed that this configuration provided a better separation than a conventional cascade. The same complexities involved in the two-membrane permeators apply to the internally staged permeator.

NOTATION

General Notation

See the General Notation section at the beginning of this handbook.

Special Notation

A	membrane area, L^2 or cm^2
A_A	surface area of intact skin of porous "resistance model" membrane, L^2 or cm^2
A_B	surface area of pores on "resistance model" membrane skin, L^2 or cm^2
A_{tot}	total unit surface area of membrane, L^2 or cm^2
d_i	fiber inner diameter, L or μm
d_{lm}	log mean diameter of fiber, $(d_o - d_i)/\ln(d_o/d_i)$, L or μm
d_o	fiber outer diameter, L or μm
g_c	gravitational conversion factor, ft · lb/(lb force · s^2)
J_v	volumetric flux, L/t or cm/s
N	total number of hollow fibers in the hollow fiber module
P	permeability = (diffusive flux)(membrane thickness)/(partial pressure difference), Barrer = 10^{-10} cm^3 (STP) · cm/(cm^2 · s · cm Hg)
P_{eff}	effective permeability of "resistance model" membrane, cm^3 (STP) · cm/(cm^2 · s · cm Hg)
P_A	permeability of intact skin of porous membrane, cm^3 (STP) · cm/(cm^2 · s · cm Hg)
P_B	permeability of caulking agent used in "resistance model" membrane, cm^3 (STP) · cm/(cm^2 · s · cm Hg)
p_0	standard pressure, 1 atm
p_s	shellside gas pressure, p or atm or psia
p_t	tubeside gas pressure, p or atm or psia
Q	molar rate, mol/t or mol/s
Q_v	volumetric flow rate, L^3/t or cm^3/s
R	universal gas constant, ML^2/t^2T mol or cal/(mol · K)
R_{eff}	effective transport resistance, l/P, cm^2 · s · cm Hg/cm^3 (STP)
R_A	transport resistance of intact skin of "resistance model" membrane, cm^2 · s · cm Hg/cm^3 (STP)
R_B	transport resistance of filled pores in "resistance model" membrane, cm^2 · s · cm Hg/cm^3 (STP)
R_1	transport resistance of membrane skin, cm^2 · s · cm Hg/cm^3 (STP)
R_2	transport resistance of membrane support, cm^2 · s · cm Hg/cm^3 (STP)
T_0	standard temperature, 273.15°K
z	axial dimension, L or cm

Greek Letters

η	viscosity, M/Lt or cp
Γ	tubesheet length, L or cm

Subscripts

p	permeate side of hollow fiber or flat membrane
s	shell side of hollow fiber

REFERENCES

Anon. 1985. Results confirm membrane's H_2 recovery role. *Oil Gas J.* 83(43):121.

Antonson, C. R., R. J. Gardner, C. F. King, and D. Y. Ko. 1977. Analysis of gas separation by permeation in hollow fibers. *Ind. Eng. Chem. Process Des. Dev.* 16(4):463–469.

Boucif, N., S. Majumdar, and K. K. Sirkar. 1984. Series solution for a gas permeator with countercurrent and cocurrent flow. *Ind. Eng. Chem. Fundam.* 23(4):470–480.

Brennan, M. S., A. G. Fane, and C. J. D. Fell. 1986. Natural gas separation using supported liquid membrane. *AIChE J.* 32(9):1558–1560.

Cabasso, I., and K. A. Lundy. 1986. Method of making membranes for gas separation and the composite membranes. U.S. Patent 4,602,922.

Coats, S. G., J. A. Wilkens, and R. R. Zolandz. 1987. Improved polyaramide membranes for hydrogen separations. Poster presented at Conference on Emerging Technologies in Materials, 18–20 August 1987, Minneapolis, MN.

Cooley, T. E., and W. L. Dethloff. 1985. Field tests show membrane processing attractive. *Chem. Eng. Prog.* 81(10):45–50.

Coplan, M. J. 1987. Module for multistage gas separation. U.S. Patent 4,676,808.

Edwards, D. W. 1987. Multiple bundle fluid separation apparatus. U.S. Patent 4,670,145.

Ekiner, O. M., and G. Vassilatos. 1990. Polyaramide hollow fibers for hydrogen/methane separation: spinning and properties. *J. Membr. Sci.* 53:259–273.

Gardner, R. J., R. A. Crane, and J. F. Hannan. 1977. Hollow fiber permeator for separating gases. *Chem. Eng. Prog.* 73(10):76–78.

Gollan, A. 1988. Gas separating. U.S. Patent 4,734,106.

Gorissen, H. 1987. Temperature changes involved in gas separation membranes. *Chem. Eng. Process.* 22(2):63–67.

Henis, J. M. S., and M. K. Tripodi. 1981. Composite hollow fiber membranes for gas separation: the resistance model approach. *J. Membr. Sci.* 8(3):233–246.

Hogsett, J. E., and W. H. Mazur. 1983. Estimate membrane system area. *Hydrocarbon Process.* 62(8):52–54.

Hwang, S. T., and S. Ghalchi. 1982. Methane separation by a continuous membrane column. *J. Membr. Sci.* 11(2):187–198.

Hwang, S. T., and J. M. Thorman. 1980. The continuous membrane column. *AIChE J.* 26(4):558–566.

Kao, Y.-K. 1988. A parametric study of recycle membrane separators. *J. Membr. Sci.* 39(2):143–156.

Kao, Y. K., M.-M. Qiu, and S. T. Hwang. 1989. Critical evaluations of two membrane gas permeator designs: continuous membrane column and two strippers in series. *Ind. Eng. Chem. Res.* 28(10):1514–1520.

Kesting, Robert E. 1985. *Synthetic Polymeric Membranes*. New York: John Wiley and Sons.

Koros, W. J., and R. T. Chern. 1987. Separation of gaseous mixtures using polymer membranes. In *Handbook of Separation Process Technology*, ed. R. W. Rousseau, pp. 862–953. New York: John Wiley and Sons.

Li, K., D. R. Acharya, and R. Hughes. 1990. Mathematical modelling of multicomponent membrane permeators. *J. Membr. Sci.* 52(2):205–219.

Lloyd, Douglas R., ed. 1985. *Materials Science of Synthetic Membranes. ACS Symp. Ser. No. 269*. Washington, DC: American Chemical Society.

Majumdar, S., A. K. Guha, and K. K. Sirkar. 1988. A new liquid membrane technique for gas separation. *AIChE J.* 34(7):1135–1145.

Majumdar, S., L. B. Heit, A. Sengupta, and K. K. Sirkar. 1987. An experimental investigation of oxygen enrichment in a silicone capillary permeator with permeate recycle. *Ind. Eng. Chem. Res.* 26:1434–1440.

Mazur, W. H., and M. C. Chan. 1982. Membranes for natural gas sweetening and CO_2 enrichment. *Chem. Eng. Prog.* 78(10):38–43.

Pan, C. Y. 1983. Gas separation by permeators with high-flux asymmetric membranes. *AIChE J.* 29(4):545–552.

Pan, C. Y. 1986. Gas separation by high-flux asymmetric hollow-fiber membranes. *AIChE J.* 32(12):2020–2027.

Pan, C. Y., and H. W. Habgood. 1978a. Gas separation by permeation: part I. calculation methods and parametric analysis. *Can. J. Chem. Eng.* 56(4):197–209.

Pan, C. Y., and H. W. Habgood. 1978b. Gas separation by permeation: part II. effect of permeate pressure drop and choice of permeate pressure. *Can. J. Chem. Eng.* 56(4):210–217.

Parro, D. 1985. Membrane CO_2 separation. *Energy Prog.* 5(1):51–54.

Perrin, J. 1988. Hollow fiber multimembrane cells and permeator. U.S. Patent 4,880,440.

Perrin, J. E., and S. A. Stern. 1986. Separation of a helium-methane mixture in permeators with two types of polymer membranes. *AIChE J.* 32(11):1889–1901.

Pinnau, I., J. G. Wijmans, I. Blume, T. Kuroda, and K.-V. Peinemann. 1988. Gas permeation through composite membranes. *J. Membr. Sci.* 37(1):81–88.

Rautenbach, R., and W. Dahm. 1986. Simplified calculation of gas-permeation hollow-fiber modules for the separation of binary mixtures. *J. Membr. Sci.* 28(3):319–327.

Sanders, E. S. 1988. Penetrant induced plasticization and gas permeation in glassy polymers. *J. Membr. Sci.* 37(1):63–80.

Schell, W. J. 1983. Membrane use/technology growing. *Hydrocarbon Process.* 62(8):43–46.

Schendel, R. L. 1982. EOR + CO_2 = a gas processing challenge. *Oil Gas J.* 80(43):158, 163–166.

Schulz, G., H. Michele, and U. Werner. 1982a. New process developments in gas separation with membranes. *J. Membr. Sci.* 11(2):311–319.

Schulz, G., H. Michele, and U. Werner. 1982b. Membrane rectification columns for gas separation and determination of the operating lines using the McCabe-Thiele diagram. *J. Membr. Sci.* 12(2):183–194.

Schulz, G., and U. Werner. 1984. Gas separation using the membrane rectification technique. *Desalination* 51(1):123–133.

Sengupta, A., and K. K. Sirkar. 1987. Ternary gas mixture separation in two-membrane permeators. *AIChE J.* 33(4):529–539.

Sengupta, A., and K. K. Sirkar. 1988. Ternary gas separation using two different membranes. *J. Membr. Sci.* 39(1):61–77.

Shindo, Y., T. Hakuta, H. Yoshitome, and H. Inoue. 1985. Calculation method for multicomponent gas separation by permeation. *Sep. Sci. Technol.* 20(5&6):445–459.

Sidhoum, M., S. Majumdar, R. R. Bhave, A. Sengupta, and K. K. Sirkar. 1988. Experimental behaviour of asymmetric CA membranes and its use in novel separation schemes. *AIChE Symp. Ser.* 84(261):102–112.

Sidhoum, M., S. Majumdar, and K. K. Sirkar. 1989. An internally staged hollow-fiber permeator for gas separation. *AIChE J.* 35(5):764–774.

Sidhoum, M., A. Sengupta, and K. K. Sirkar. 1988. Asymmetric cellulose acetate hollow fibers: studies in gas permeation. *AIChE J.* 34(3):417–425.

Stern, S. A., J. E. Perrin, and E. J. Naimon. 1984. Recycle and multimembrane permeators for gas separations. *J. Membr. Sci.* 20(1):25–43.

Stookey, D. J., T. E. Graham, and W. M. Pope. 1984. Natural gas processing with PRISM separators. *Environmental Prog.* 3(3):212–214.

Ube Industries. 1989. Preliminary information on Ube gas separations system. Product brochure.

Weinberg, M. 1988. The use of pore parameters for monitoring process and property variations in gas separation membranes. *J. Membr. Sci.* 38 (3):237–246.

Weller, S., and W. A. Steiner. 1950. Engineering aspects of separation of gases. Fractional permeation through membranes. *Chem. Eng. Prog.* 46(11):585–590.

5

Applications

Raymond R. Zolandz and Gregory K. Fleming
E. I. Du Pont de Nemours & Company, Inc.

INTRODUCTION
HYDROGEN RECOVERY
   H_2/CO Adjustment
   Refinery Applications
   Petrochemical Applications
OXYGEN/NITROGEN SEPARATION
   Commercial Membranes for Air Separation
   Hybrid Systems for Air Separation
HELIUM RECOVERY
REMOVAL OF ACID GASES FROM LIGHT HYDROCARBONS
   Carbon Dioxide Recovery in Petroleum Applications
   Hybrid Systems for Carbon Dioxide Separations
   Biogas Processing
OTHER APPLICATIONS
REFERENCES

INTRODUCTION

About forty years ago Weller and Steiner (1950) considered membrane processes for the separation of hydrogen from hydrogenation tail gas, enrichment of refinery gas, air separation, and helium recovery from natural gas. However, commercial membranes capable of performing these separations economically have only become available within the last 15 years.

A list of commercial membrane suppliers is given in Table 5–1, along with the applications for which their membranes are used. Each of these applications is considered in this chapter. Comparisons of the membrane processes to alternative technologies appear in both this chapter and the following chapter on economics.

HYDROGEN RECOVERY

Gas separation membranes saw their first widespread commercial use in hydrogen recovery. Hydrogen recovery is applicable to a number of processes, with most applications falling into one of three broad categories:

1. Synthesis gas ratio adjustment (H_2/CO adjustment)
2. H_2 recovery from hydroprocessing purge streams
3. Hydrogen recovery from ammonia plant purge streams and other petrochemical plant streams.

Membranes compete with pressure swing adsorption and cryogenic systems in these hy-

TABLE 5–1. Commercial-Scale Membrane Suppliers (adapted and updated from Spillman 1989).

Company	CO_2	H_2	O_2	N_2	Other[a]
A/G Technology (AVIR)	X		X	X	
Asahi Glass (HISEP)			X	X	
Cynara (Dow)	X				
Dow (Generon)			X	X	
Du Pont/L'Air Liquide (MEDAL)		X		X	
Grace Membrane Systems	X	X			X
Hoechst-Celanese (Separex)	X	X			X
International Permeation	X				X
Membrane Technology and Research					X
Nippon Kokan K.K.					X
Osaka Gas			X		
Oxygen Enrichment Co.			X		
Perma Pure					X
Permea (Air Products)	X	X	X	X	X
Techmashexport (USSR)			X		
Teijin Ltd.			X		
Toyobo			X		
Ube Industries		X	X	X	X
Union Carbide (Linde)	X	X	X		
UOP/Union Carbide	X				

[a]Includes solvent recovery, dehumidification, and/or helium recovery membranes.

drogen recovery applications and, as seen in Table 5–2, are competitive with these technologies over a wide range of operating conditions. Membrane systems have the advantage of low capital cost and ease of operation. Competing systems, however, often deliver the purified hydrogen at or near the same pressure as the feed gas, which results in lower recompression costs than those for membrane systems, where the hydrogen product is always at a pressure lower than that of the original feed.

H_2/CO Adjustment

The use of membranes for H_2/CO adjustment in synthesis gas (syngas) was given early consideration. For example Gardner, Crane, and Hannan (1977) discussed the potential for membrane use in an oxo-alcohol plant.

Syngas contains hydrogen and carbon monoxide along with impurities such as CO_2, CH_4, N_2, and water. The H_2/CO ratio of the feed will vary depending on its source, whereas the desired H_2/CO ratio for the product (which is used as reactor feedstock) depends on the synthesis in which the gas will be used. Syngas made via steam-methane reforming has a 3:1 H_2/CO ratio, whereas typical applications require a ratio of between 0 and 2.1. Membranes are ideally suited for stripping hydrogen out of the syngas in order to reduce the H_2/CO ratio. The syngas is generally already under pressure, so feed compression is not required. A typical mass balance around a single-stage Separex unit (Schott et al. 1987) producing a 2:1 ratio gas is given in Table 5–3.

The product stream from H_2/CO adjustment is usually the high-pressure nonpermeate stream, which, except for the pressure drop through the permeator, is at the feed pressure. Even this small pressure drop, however, can be an important consideration, particularly in plants where membranes are to be added to an existing process. In such plants available pressure differentials for fluid flow can be limited. While tubeside feed hollow-fiber devices can and have been used in this application, shellside hollow-fiber or spiral-wound modules are generally preferable, since they have a lower pressure drop through the permeator from feed

TABLE 5–2. Comparison of Hydrogen Recovery Systems (reprinted from Poffenbarger and Gastinne 1989 with permission).

	Membrane	PSA	Cryogenics
Relative investment	1	1 to 3	2 to 3
Maximum operating pressure (psia)	2000	600	1000
Minimum hydrogen content in feed (%)	15 to 20	50	20
Maximum hydrogen purity (%)	99	99.999	98.5
Maximum hydrogen recovery (%)	95	85	95
Product pressure/feed pressure	Lower	Same	Same
Retentate pressure/feed pressure	Same	Lower	Lower
Fractionation of heavies	No	No	Yes
Modularity	Yes	No	No
Ease of operation	Very easy	Average	Average

to retentate. Because the product is delivered at high pressure, expensive recompression is not required.

Production of high-purity CO from a H_2/CO mixture via membranes is also possible, although multiple membrane stages are generally required.

TABLE 5–3. Mass Balance Around a Separex Membrane in a Syngas Ratio Adjustment (reprinted from Schott et al. 1987 with permission).

	Mole % in Stream		
Component	Feed	Syngas	H_2 Fuel
CO	24.18	32.51	1.57
H_2	73.32	64.32	97.76
N_2	0.10	0.14	0.004
CH_4	2.14	2.89	0.11
H_2O	0.26	0.14	0.55
CO_2 (ppm)	100	115	70
Total feed flow, MSCFD (10^6 SCFD)		18.5	
Pressure, psig		325	
Temperature, °F		110	

Refinery Applications

In petroleum refining, hydrogen is increasingly needed to process heavy and sour crudes. Various purge and offgases are produced, as listed in Table 5–4. All are potential feeds for hydrogen recovery units. The hydrogen content of these streams ranges from 15 to 80%, whereas 90 to 95% hydrogen purity is required to recycle the hydrogen to a processing unit.

The exact configuration of the overall hydrogen recovery system will vary from refinery to refinery, but the membrane process is basically the same in all cases. As an example, consider the case of hydrogen recovery from a hydrocracker unit.

In a typical hydrocracker, hydrogen and oil are introduced to the reactor, where some of the hydrogen reacts with the oil to produce light hydrocarbons. Liquids and gas exit the reactor and are disengaged in a high-pressure separator. The gas phase contains hydrogen, which can be compressed and returned to the reactor.

A purge must be taken on the recycle stream to prevent the buildup of light hydrocarbons in the reactor loop. (These light hydrocarbons act as inerts, which reduce the efficiency of the reactor by lowering the hydrogen partial pressure in the gas phase.)

TABLE 5–4. Typical Hydrogen Membrane Performance of Polyamide Membranes in Refining Applications (reprinted from Poffenbarger and Gastinne 1989 with permission).

Process Stream	Primary Separation	Typical Values for Hydrogen Membrane Recovery System		
		Feed Purity (%)	Permeate Purity (%)	Recovery (%)
Catalytic reformer offgas	H_2/CH_4	70 to 80	90 to 97	75 to 95+
Fluid catalytic cracker offgas	H_2/CH_4	15 to 20	80 to 90	70 to 80
Hydroprocessing unit purge gas	H_2/CH_4	60 to 80	85 to 95	80 to 95
Pressure swing adsorption offgas	H_2/CH_4	50 to 60	80 to 90	65 to 85

Traditionally these inerts are removed either by means of a direct high-pressure purge or by using an oil absorber to concentrate the inerts prior to purging them. The former process has the disadvantage of removing 4 mol of hydrogen from the process for every mole of hydrocarbon removed. The latter has a high capital cost and still loses about 1 mol of hydrogen per mole of hydrocarbon. Membranes can be used in lieu of the absorber system, at a reduced capital cost and better process efficiency.

In a process described by Bollinger, Long, and Metzger (1984) a portion of such reactor offgases is treated in a membrane system (in this case a Prism™ system from Permea) to produce a hydrogen-rich permeate stream at low pressure. This hydrogen-rich stream is then repressurized so that it can be introduced into the reactor. The retentate stream, which is enriched in hydrocarbons, is purged from the reactor loop and used as fuel gas. This purge contains only about 20% hydrogen, versus 50% from an oil absorber and 80% in a high-pressure purge.

Once the membrane unit is in place, several strategies can be pursued for operation of the overall system. One strategy is to operate the process at the same reactor H_2 partial pressure as before, thereby minimizing the need for makeup hydrogen. Alternatively, operation can take place at the same makeup hydrogen rate, which, because of lower H_2 losses, increases the H_2 partial pressure in the loop. Operating in the latter mode, heavier crudes can be processed at the same operating conditions or the same crudes can be processed with greater throughput or at milder operating conditions.

Recovery of hydrogen from other gases associated with hydrotreating units is similar to that described above for a hydrocracker. Both the concentration of hydrogen in the feed and the feed pressure will vary depending on the unit. The hydrogen recovered by the membrane system is either recycled to the original reactor or sent to another one.

An example in which gas from one reactor is processed by a membrane to produce feed gas for a second reactor is shown in Figure 5–1 (Grotewold 1990). In the original flowsheet, gas containing 71% hydrogen exiting from a gas oil hydrotreater (GOHDT) at 1050 psig was processed by splitting the stream between a cryogenic unit and a light cycle oil hydrodesulfurizer (LCOHDS) operating at 430 psig, as shown in Figure 5–1(a). The hydrogen sent to the latter unit was originally used on a once-through basis.

A membrane unit was installed to take advantage of the existing pressure difference between the two reactors. The membrane produces a hydrogen-enriched gas that can be sent to the LCOHDS without compression. Because the feed gas to the LCOHDS has been enriched in hydrogen, the offgas is now pure enough so that it can be recycled to the gas oil hydrotreater via the makeup compressor. The efficiency of the associated cryogenic unit is also improved, since it now processes a stream that has been enriched in the heavier components. The Du Pont permeation system used in this application

FIGURE 5–1. Membrane recovery of refinery hydrogen (reprinted from Grotewold 1990 with permission).

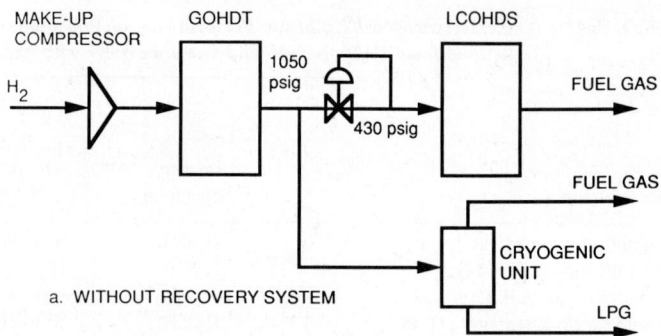

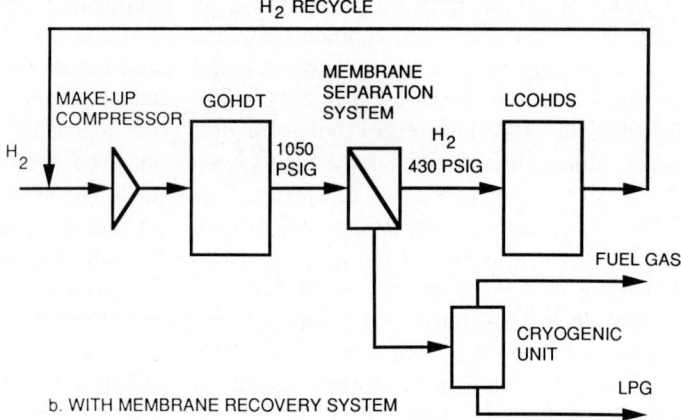

is shown in Figure 5–2. The design specifications called for processing 12 million SCFD of feed, with 75% hydrogen recovery at a minimum 95% hydrogen purity. The aromatic polyamide membranes met or exceeded all process specifications. The calculated payback period of 1.7 yr did not reflect several intangible or difficult to estimate benefits such as longer catalyst life.

Recovery of hydrogen from a catalytic reformer at Seibu Oil's Onoba City refinery was the first announced commercial unit using Ube technology (Anon. 1986). The polyimide membranes used in this application have also found use in other refinery applications.

The polyimides mentioned above and other high-selectivity membranes such as the aromatic polyamides can be used to produce high-purity hydrogen from moderately pure streams.

For example, Ube membranes have been used to upgrade a stream at 360 psia containing 98.8% H_2 to one at 165 psia with 99.99% hydrogen purity (Ube Industries 1989). A 70% recovery of hydrogen was cited, at a total purification cost of $1.5 per MSCF of hydrogen.

Cellulose acetate membranes have been tested in recovering hydrogen from a reformer unit (Yamashiro et al. 1985) and a butamer unit (Cooley and Dethloff 1985). In the latter application the hydrogen purge stream from the butamer unit was passed through a caustic wash to remove the HCl that is formed in the reactor. Unwashed gas, with 2000 to 4000 ppm HCl, was found to be detrimental to membrane performance. The purge stream, at 295 psig and containing 61.1% hydrogen, was processed to recover 53% of the hydrogen at a purity of

FIGURE 5–2. A commercial hydrogen separation system. (Courtesy of E. I. du Pont de Nemours & Company, Inc.)

95.4%. This gas was recompressed and returned to the butamer unit. A commercial unit to recover 97% of the hydrogen at greater than 90% purity was proposed, which presumably would involve staging or recycle. Other cellulose acetate membranes have also been tested for this application (Crull 1983).

In another refinery application for membranes, CO_2 is removed from reformer offgas. This application (Schell, Houston, and Hopper 1983) requires a membrane that permeates CO_2 more rapidly than hydrogen. An example is given for a two-stage membrane system that utilizes membranes with a CO_2/H_2 selectivity of 5 and a CO_2/CO selectivity of 10. The CO_2 content of a 250-psig feed stream is lowered from 6 to 0.2% CO_2, with 50% retention of H_2 in the nonpermeate stream.

Petrochemical Applications

Several nonrefinery processes also have streams containing recoverable hydrogen. Table 5–5 lists a number of such applications.

In the petrochemical industry, membranes have found extensive use in the recovery of hydrogen from ammonia purge gas. This was the first widespread use of gas separation membranes, using Permea's Prism™ system.

In the ammonia process, hydrogen and nitrogen are reacted directly to form ammonia. The hydrogen is generally produced via a steam-reforming process, which uses water, air, and a light hydrocarbon such as methane. Reagent nitrogen enters with the air, as do inerts such as argon.

Ammonia is produced along with a gas that

TABLE 5–5. Typical Hydrogen Membrane Performance of Polyamide Membranes in Petrochemical Applications (reprinted from Poffenbarger and Gastinne 1989 with permission).

Process Stream	Primary Separation	Typical Values for Hydrogen Membrane Recovery System		
		Feed Purity (%)	Permeate Purity (%)	Recovery (%)
Ammonia purge gas	H_2/CH_4	60 to 70	80 to 95	85 to 95
Methanol purge gas	H_2/CH_4	50 to 85	90 to 95	80 to 95
Benzene recycle gas	H_2/CH_4	50 to 60	90 to 95	85 to 95
Cyclohexane feed gas	H_2/CH_4	60 to 70	90 to 95	90 to 95
Synthesis gas	H_2/CO	60 to 80	90 to 95	80 to 95

contains approximately 60% H_2, 5% Ar, 15% CH_4, and 20% N_2. This gas can be recycled to the reactor, but a purge is required to prevent the buildup of inerts. Without some separation process, all of the hydrogen in this purge stream is lost from the process and must be replaced by hydrogen generated from methane or obtained from other sources.

The purge gas is available at about 2000 psig (136 bars) and can be processed conveniently in a two-stage membrane system. Both stages are operated at the 2000-psig feed pressure, but with different permeate pressures (Graham and MacLean 1979; MacLean, Prince, and Chae 1980). The first stage operates with a permeate pressure of approximately 1000 psi (68 bars), producing a high-purity hydrogen product that can be sent back to an intermediate stage of the synthesis gas compressor.

Since the nonpermeate stream from the first membrane stage has been partially depleted in hydrogen, it has a lower partial pressure of hydrogen than the original feed. This gas is fed to the second membrane stage, where a lower permeate pressure (350 psi, 24 bars) is used to provide additional driving force for hydrogen transport. The lower permeate pressure in the second stage also provides a larger pressure ratio between the feed and permeate streams, which helps to increase product recovery. The low-pressure permeate is sent back to the front of the compression train for recycle to the ammonia reactor.

The two-stage process can achieve 85 to 90% or better hydrogen recovery at hydrogen purities of 90 to 95 mol %, depending on the selectivity of the membranes used.

If the membranes are sensitive to ammonia, a water wash must be provided for ammonia removal upstream of the membrane unit.

The membrane process described above is often more economical than alternative technologies (Baldus and Tillman 1986). Compared to the best alternative process, cryogenic removal of inerts, the membrane process recovers more of the hydrogen at a lower investment cost. The cryogenic process, however, does consume less energy since all of the hydrogen product is produced at approximately 1000 psig (70 to 80 bars); the smaller recompression cost offsets the cost of the required refrigeration. A pure, pressurized nitrogen stream can also be recovered in the cryogenic process. This will affect process economics in cases for which the plant's air compression capacity is limited.

Note that a one-stage membrane process operating at the lower of the two permeate pressures could achieve equivalent or improved separation with less membrane area. However, since the hydrogen product would all be at a lower pressure in the one-stage system, economics favor the two-stage system due to the savings in recompression costs.

Membranes can also be used to recover hydrogen from methanol plant purge gas, the recovered hydrogen being recycled to the methanol reactor. Recovery of hydrogen in this application can simultaneously yield higher methanol production and lower operating costs (Burmaster and Carter 1983).

Metallic membranes made with palladium can be used for hydrogen recovery in applications where a high-temperature feed gas is available. One commercial application (Anon. 1985) of such metallic membranes is in a process for morpholine, in which hydrogen serves as both a carrier gas and as a catalyst regenerant. The hydrogen is recovered at high purity and recycled to the reactor feed stream. Because only hydrogen permeates the membrane to any degree, extremely high purity hydrogen (99.9995+%) can be achieved.

In cases where a high-purity hydrogen product must be produced at high recoveries from a low-purity feed, polymeric membranes can be used in conjunction with other technologies. One such process (Doshi 1987) combines membranes and pressure-swing adsorption (PSA). The membranes provide an initial bulk separation, producing a moderately pure permeate stream that is further processed in the PSA unit. The PSA waste gas is compressed and recycled back to the membrane separator. Additional membrane stages can be introduced in the process to enhance product recovery further.

There are occasional applications where the hydrocarbon-enriched nonpermeate is the desired product and the hydrogen is the by-product stream. In these cases the hydrogen can be burned as fuel, with the hydrocarbon-rich product collected at elevated pressure.

OXYGEN/NITROGEN SEPARATION

One membrane process that is becoming increasingly important is the separation of air into nitrogen- and oxygen-enriched streams. Either or both of these streams can be the desired product. Since current membranes are all more permeable to oxygen, the nitrogen-rich stream is recovered at essentially the feed pressure, while the oxygen-rich stream is produced at reduced pressure.

In comparison to hydrogen recovery processes, air separations are relatively straightforward. Hydrogen separations involve feed gases from a number of sources at different feed compositions and pressures and with different impurities. For air separations there is a single source of feed at fixed composition with essentially no impurities. Furthermore, standard air compression systems operate in the range of approximately 100 to 190 psig (8 to 14 bars absolute), making this a convenient and economical operating range.

Pretreatment of the feed gas is minimal in air separations because aggressive components are normally absent. The major pretreatment requirement is to avoid introduction of liquid water or compressor oil into the permeator. Fouling with water can slow transport either reversibly or irreversibly, depending on the membrane. Oil can coat the membrane and foul it irreversibly, increasing the resistance to gas transport. Standard technology can be applied to solve both of these potential problems.

Oxygen-enriched air (OEA) from membranes has, to date, generally been limited to medical uses. Membrane systems for this application can be made lighter and more inexpensively than alternatives such as small-scale PSA units or bottled oxygen (Golan and Kleper 1987). Commercial-scale use of oxygen-enriched gas from membranes (such as enhanced combustion applications) has yet to reach its full potential.

Nitrogen-enriched air (NEA) has been the more successful product to date for a number of reasons. Membrane selectivities do not need to be high in order to produce a relatively pure nitrogen stream. For instance, under typical operating conditions a membrane with an O_2/N_2 selectivity of 3 can produce 95% pure NEA at 30% recovery. If the selectivity is too low, however, nitrogen losses into the oxygen-rich permeate become excessive and compression costs escalate. Since the nitrogen product is the nonpermeate, it remains at essentially the original feed pressure and can, therefore, be delivered at typical inert gas pressures of 70 to 90 psig without additional compression.

Membranes compete with nitrogen produced either cryogenically or via pressure-swing adsorption, as well as with oxygen-free gas produced in inert gas generators. In many instances membranes enjoy a cost advantage over the other technologies for inert gas production (see Chapter 6). In addition, they provide the following advantages over other nitrogen sources (Generon Systems 1989; Bhat and Beaver 1988):

1. A 99.99% pure nitrogen is not always required; therefore, one need not pay for excess purity.
2. Tank rental and delivery expenses are reduced or eliminated.
3. Long-term cost stability is ensured.
4. The nitrogen supply is secure, with little or no need to store reserves.
5. Membrane units are small and portable, so they can be used in areas with limited space and relocated as necessary.
6. Membrane operation is flexible and allows easy adjustment of product flow or purity.
7. The modular nature of the membranes simplifies future expansion or contraction of capacity.

Typical NEA purities range from 95 to 99% nitrogen, with the delivered purity being governed by the requirements of the application. As in other gas separations, a single-stage membrane system will produce higher purity gas only at the expense of product recovery. Figure 5–3 shows such purity/recovery curves for

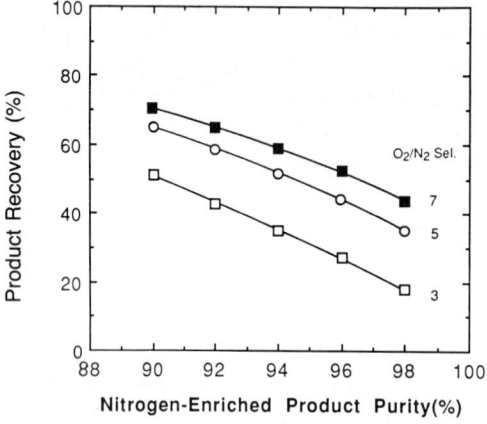

FIGURE 5–3. Purity versus recovery curves for production of nitrogen-enriched air.

membranes with different O_2/N_2 selectivities under typical operating conditions. Note that improved O_2/N_2 selectivity results in increased NEA recovery and hence lower compression costs. (Calculations were performed for tube-side feed, with the feed pressure = 100 psig, permeate pressure = 0 psig, fiber active length = 36 in., tubesheet length = 3 in., fiber outer diameter = 400 μm, fiber inner diameter = 200 μm, and oxygen permeability = 30 GPU, where 1 GPU = 10^{-6} cm^3 (STP)/cm$^2 \cdot$ s $\cdot$ cm Hg).

Membrane selectivity can be important in making the membrane system competitive with respect to alternative technologies. For example, when a nitrogen production system was compared to a PSA system (Baldus and Tillman 1986), investment costs for the two technologies were found to be similar and therefore operating costs were the important factor in distinguishing the two technologies. Thus, at a given purity of nitrogen product, the technology with the highest product recovery would have the advantage. In this study, PSA recoveries were midway between those for membranes with O_2/N_2 selectivities of 4.5 and 9.0, indicating that there is a crossover point at an intermediate selectivity where the two technologies have equal costs.

An interesting membrane process for offshore production of inert gas has been described by Beaver and Paton (1987). The inert gas is injected into geologic formations that are subsiding due to withdrawal of natural gas. The nitrogen is used to repressurize the formations and reduce subsidence. Turbine exhaust rather than air was chosen as the feed in this application, since much of the initial oxygen has been converted into the more permeable CO_2. The exhaust gas is compressed, then processed in a membrane unit to produce a nonpermeate with about 2% oxygen, which is reduced to 5 ppm by a catalytic system.

Because membranes do not need to remain vertical, as do distillation towers, the above membrane process could be installed on a ship instead of on the offshore platforms, where space and weight constraints are important.

Commercial Membranes for Air Separation

Advances in membrane-based air separation have been rapid. Both Permea and Dow unveiled their first membrane systems for air separation in 1985, and both have introduced second-generation products since then.

In the Permea system, the second-generation membrane (Prism Alpha) uses the first-generation polymer, but has a thinner active layer. The Alpha membranes have a twofold to threefold productivity increase over their predecessor as a result of the improved morphology.

The upgraded Dow system (Generon II) employs a new membrane polymer. The new polymer was chosen for its good combination of productivity and selectivity. For 99.5% pure NEA the product flow rates for Generon II are 10 times those of the first-generation membranes. For 95% pure NEA the product recovery is double that of the original membrane system (Heitz and Milne 1990).

Other manufacturers have also entered the NEA production market. Some of these new suppliers have installed large-scale nitrogen production plants.

A Linde NitroGen™ system, for example, produces 17,000 ft^3/h of 97% pure nitrogen for use in a pharmaceutical facility. The unit, operated and maintained by Linde, was claimed to be the most economical supply for the size and purity of delivered nitrogen (Berolatti 1987). An even larger capacity system has been installed by MEDAL (Du Pont/L'Air Liquide) for an undisclosed application (Barry 1989).

For low-volume applications a system such as the Avir™ system from A/G Technology offers a compact, lightweight gas source. Weighing less than 30 lb, the units operate at room temperature and 100-psig feed pressure, and supply up to 300 SCFH of 95% N_2 or 1200 SCFH of 35% oxygen-enriched air (A/G Technology Corporation 1987).

Hybrid Systems for Air Separation

The combination of membrane-based air separation with other technologies in hybrid systems is sometimes advantageous. Membranes, for example, can be used to provide a bulk cut ahead of an adsorption system, reducing the size of the adsorber. Such a combination is useful in gas generation systems for aircraft (Beaver, Bhat, and Sarcia 1988). The membrane performs an initial separation on the feed air, and the nitrogen-rich retentate that is produced is used to blanket the fuel tanks. The oxygen-rich permeate is compressed and sent to a PSA unit, which produces breathing oxygen for the crew.

Membranes can be combined with cryogenics in both large-scale (Agrawal and Auvil 1986) and small-scale systems. One small hybrid system has been reported that produces 5 L per day of liquid nitrogen with a 2-in.-diameter × 3-ft-long hollow-fiber module coupled with a refrigerated cold head (Beaver, Bhat, and Sarcia 1988).

Membranes can also be used in conjunction with catalytic oxygen removal (Beaver, Bhat, and Sarcia 1988). The membrane provides an initial separation ahead of the polishing unit. The membrane reduces the oxygen content of the feed gas to around 0.5% O_2. This gas is introduced into the catalytic unit, in which the oxygen is reacted with hydrogen to produce a final inert gas with less than 5 ppm oxygen.

HELIUM RECOVERY

Helium permeabilities are normally slightly higher than hydrogen permeabilities, making helium separation from natural gas a feasible application for hydrogen-separation membranes. If the feed contains moderate to high helium concentrations, a one-stage system similar to a typical hydrogen recovery unit may be used.

Unfortunately, in many cases, such as in much of the natural gas that contains helium, the helium concentration is low. In such cases either very high He/CH_4 selectivities (500+) are necessary or a staged system with intermediate permeate recompression must be used in order to recover the bulk of the helium while minimizing losses of the natural gas.

Helium is used as a diluent in breathing gas

mixtures used by deep sea divers. Membranes may be used to recover helium from the spent gas. Membranes with He/N_2 selectivities of at least 50 are needed in this application, with selectivities in excess of 100 being desirable (Peinemann, Ohlrogge, and Knauth 1986).

REMOVAL OF ACID GASES FROM LIGHT HYDROCARBONS

Carbon Dioxide Recovery in Petroleum Applications

A potentially large market for carbon dioxide separation systems is the treatment of gas produced in oil fields where enhanced oil recovery (EOR) via CO_2 injection is practiced. In the EOR process, CO_2 is injected into the oil-bearing strata at pressures in excess of 1000 psig (70 bars). The carbon dioxide dissolves the oil and carries it through the underground strata to the producing wells. Once the CO_2/oil phase is brought above ground, the pressure is lowered and most of the hydrocarbons drop out as liquids. The overhead offgas consists of CO_2, CH_4 and other hydrocarbons, nitrogen, and often H_2S.

For economic and environmental reasons, removal of the CO_2 from the offgas and reinjection into the oil field is desirable. The recycled gas generally needs to have a CO_2 purity of at least 95%. This minimum level of purity is necessary to maintain the solvent power of the CO_2. The presence of inerts raises the "miscibility pressure," which is the pressure at which the CO_2 and hydrocarbons form a single fluid phase. High miscibility pressures are undesirable because they lead to higher injection costs and increase the risk of disrupting the underground strata.

Two features make this CO_2 recovery application unique: (1) The volume of produced gas increases greatly over the life of the typical injection project and (2) the CO_2 content of that gas also increases with time. The modularity of membranes allows the addition of processing capacity as needed, a flexibility not available with other technologies.

One of the earliest EOR membrane plants was the SACROC unit, which was started up in 1984 (Parro 1984). The hollow-fiber membrane units are owned and operated by Cynara, a subsidiary of Dow. The gas produced by the wells is processed on a contract basis using a membrane system. The purified CO_2 stream from the membranes is returned to SACROC, which further treats the gas in a previously existing hot potassium carbonate plant prior to reinjecting it into the oil field. The membranes provide a bulk separation that eases the load on the carbonate unit. Before installation of the membranes, the carbonate unit was approaching capacity. The membranes eliminated the need for a second processing plant. The Cynara membrane system processes a total of 70 million SCFD of gas containing 40 to 70% CO_2.

The permeators at SACROC utilize a cross-flow feed design in which permeate is removed at both ends of the fiber bundle. Such operation minimizes the pressure drop in the bore of the hollow fibers. Prior to commissioning the commercial unit, extensive field tests and a four million SCFD pilot facility were run. Two factors found to be important to the operation of the facility are proper gas dehydration upstream of the membrane unit and reduced operating temperatures for the membrane. The reduced operating temperature, achieved by refrigeration, improves both the CO_2 permeation rate and the CO_2/CH_4 selectivity as compared to their values at ambient temperature. The enhanced CO_2 rate is the result of the increased CO_2 solubility in the membrane at the lower temperature. A single membrane stage is used, with multiple banks of permeators in parallel. Plant performance is optimized under varying feed conditions by adjusting the number of permeator banks in operation. Fine-tuning is achieved by valving out permeators within the banks.

Performance of flat-film cellulose acetate (CA) membranes for CO_2 separations has been discussed by Russell and Coady (1982). Their tests used feed streams containing 10 to 80% CO_2. The membranes tested could operate at feed and transmembrane pressures up to 1100 psi, and at temperatures from 45 to 115°F. Feed pretreatment consisted of glycol dehydration and filtration. The feed was limited to no more than 80% CO_2 and less than 7 lb water/MSCFH. It was noted that if water contamination occurred the membranes could be dried in place with only minor loss in the CO_2 permeation coefficient. A programmable controller for process valve control ensured uniform start-up and shutdown sequences, thereby eliminating the risk of damage to the membranes during these operations.

A recent report (U.S. Department of Energy 1989) documents two sets of field tests conducted by UOP in which cellulose acetate membranes were used to treat sour natural gas. Short-term tests at low stage cuts were conducted on two gas streams, one with 4% CO_2 and 0.2% H_2S, the other with 36% CO_2 and 0.4% H_2S. Operating at 900 psi on dehydrated gas, permeate purities of approximately 30 and 90% CO_2, respectively, were obtained. As a result of operating experience, the modules were redesigned to eliminate the high resistance to flow observed on the permeate side of the membrane. Modules containing between 0.5 and 1.5 ft² of surface area produced 12 to 52 CFD of permeate with the low-CO_2 feed and 49 to 164 CFD on the high-CO_2 feed. Over the eight-day test period, membrane flux declined by one-third.

In the second test, CA membranes were used over longer periods, again at low stage cuts. Gas that contained 45% CO_2 and 17% H_2S was fed at 200 psig and 90°F to two membrane units containing a total of 2 ft² of surface area. A schematic of the test unit, typical of small-scale membrane test units, is given in Figure 5–4. Pretreatment of the feed gas consisted of Joule-Thompson cooling followed by liquid knockout and reheating of the gas above its dew point. Despite the high levels of H_2S in the feed, membrane selectivities were maintained (CO_2/CH_4 of 10 to 15, H_2S/CH_4 of 15 to 20).

This same report also mentions tests with facilitated transport membranes. In these tests, a H_2S/CO_2 selectivity of about 3.5 was achieved, which was not deemed high enough for commercial consideration.

While the effect of water vapor on commercial membranes has not been discussed in detail in the literature, water vapor has been

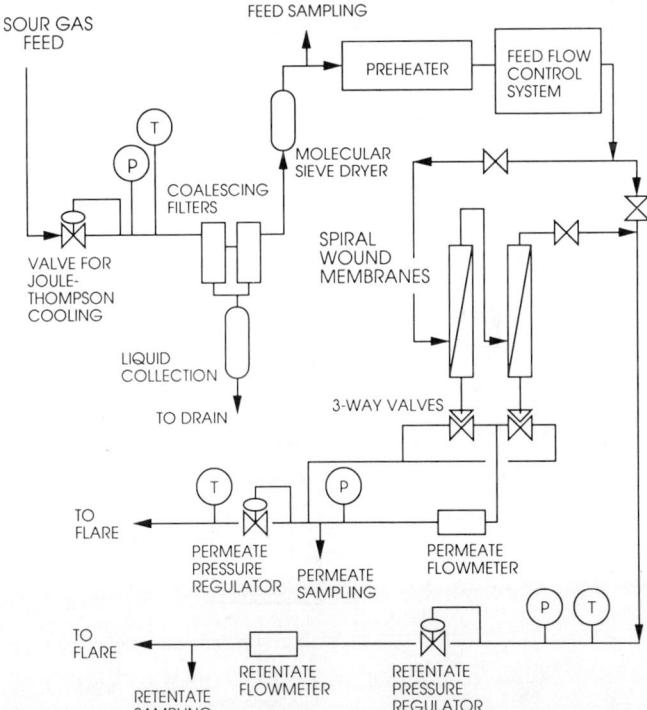

FIGURE 5–4. Schematic of a pilot facility for membrane separation of CO_2 (reprinted from U.S. Department of Energy 1989 with permission).

shown to affect CO_2 and CH_4 permeation through typical membrane polymers (Paulson, Clinch, and McCandless 1983). For the polymers tested, an optimum CO_2 permeation rate was found at a given feed water content, whereas selectivity was found to pass through a minimum. Thus, relative humidity can be used as part of the control strategy for a membrane system, at least in CO_2/CH_4 separations.

Hybrid Systems for Carbon Dioxide Separations

The membrane/potassium carbonate system at SACROC, discussed above, is an example of a hybrid process in which membranes were added to an existing separation train. A number of such hybrids have received attention in the literature. For example, Prism™ permeators have been used ahead of a traditional plant using amine and cryogenic processing units (Backhouse 1986).

Membranes can often be used in combination with multiple unit operations. In the process shown in Figure 5–5, for example, membranes provide an initial separation of acid gases from hydrocarbons (U.S. Department of Energy 1989). The hydrocarbons are sent to an amine unit for final cleanup, and the concentrated acid gases from both the amine and membrane units are fed to a Claus plant for sulfur removal.

A combined membrane/cryogenic process can have advantages over an all-membrane system. In an example discussed by Baldus and Tillman (1986), a feed containing 64% CO_2, 29% hydrocarbons, and 7% nitrogen was to be processed into sales gas with 3% (maximum) CO_2 and another product stream with 97% CO_2. An all-membrane process (Figure 5–6) was quite complex, requiring four membrane stages and three process compressors (one for recycle and two for final compression of the CO_2 product). In the hybrid process (Figure 5–7) the fractionation tower provides a bulk cut between CO_2 and the hydrocarbons, while the membranes strip CO_2 from the hydrocarbon-rich overheads and recycle it back to the cryogenic unit. The hybrid unit was found to offer cost advantages over the membrane unit when the membranes had a CO_2/CH_4 selectivity of 20, which is a typical value for current commercial

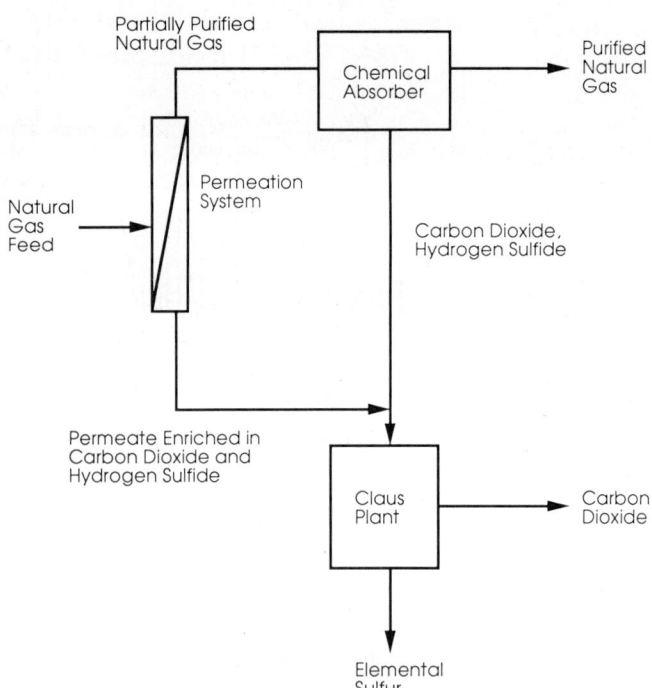

FIGURE 5–5. Process schematic of a membrane system coupled with a Claus reactor and chemical absorber (reprinted from U.S. Department of Energy 1989 with permission).

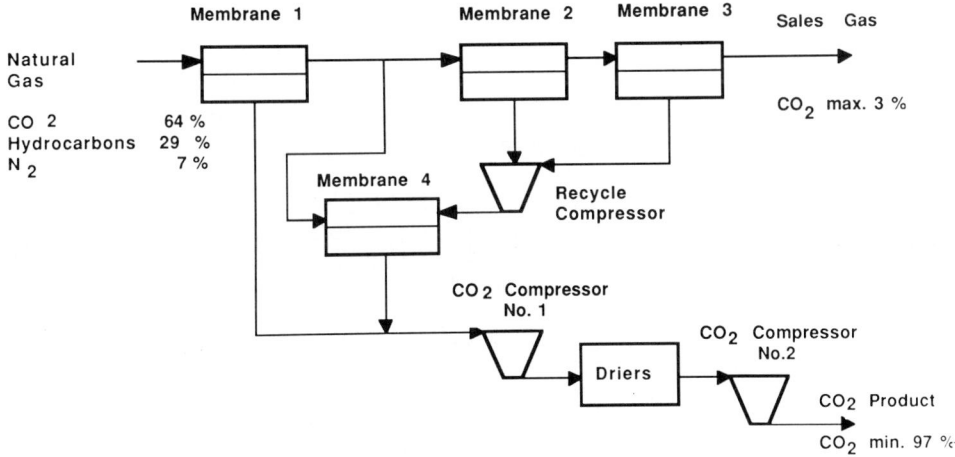

FIGURE 5-6. An all-membrane process for natural gas sweetening (reprinted from Baldus and Tillman 1986 with permission).

membranes. With a membrane selectivity of 50, the all-membrane process was on par with the hybrid unit, although the complexity of the all-membrane process would likely weigh against it.

A membrane-aided distillation process has been discussed by Fluor (Schendel 1984; Schendel and Seymour 1985). A three-column, two-membrane process for treating an 85% CO_2 feed uses an initial permeation system for bulk CO_2 removal followed by a three-column distillation train that separates CO_2 and CH_4 from natural gas liquids (NGLs). Methane is removed in the first column, while the second produces NGLs and a CO_2/C_2 azeotrope. This azeotrope is treated in a second permeation system to remove CO_2 at 95% purity. The nonpermeate is sent on to the third column, which produces NGLs and a CO_2-rich stream, which is recycled back to the second permeation unit.

The separation scheme proposed by Fluor differs from the standard Ryan-Holmes process, which is a multicolumn process that uses a liquid hydrocarbon (typically C_4s) to break the CO_2/C_2 azeotrope. A membrane unit could be used upstream of this process to remove CO_2 from the feed to the Ryan-Holmes unit, thereby reducing the amount of azeotrope that is formed.

Another membrane-aided distillation scheme for CO_2/hydrocarbon separation (Lucadamo 1986) uses the membrane unit after a low-temperature distillation unit rather than before it. The distillation column in this case produces fuel gas, acid gas, and heavy hydrocarbon fractions. The fuel gas is further processed in the membrane separator, where additional CO_2 is removed. This CO_2 is combined with the acid gas from the distillation. A sweeter fuel gas is obtained through use of the membrane, and the recovered CO_2 can be added to the acid gas stream to lower its H_2S content.

Biogas Processing

Membranes can be used to recover methane from biogases such as landfill gas. Such gas is typically 40 to 45% CO_2, 0.5 to 1.0% trace contaminants, and 50 to 56% CH_4 (Schell and Houston 1983). In this straightforward recovery process, the gas is collected, compressed, and then, after any necessary pretreatment, contacted with the membranes. The membrane produces an enriched methane fuel gas at the elevated feed pressure and a low-pressure CO_2 by-product.

The feed pressure required in the membrane process for biogas treatment is governed by the required methane purity. Medium Btu gas (700 to 900 Btu) can be produced at 90 to 96% methane recovery using a feed pressure in the range of 100 to 400 psig, while pressures of 250

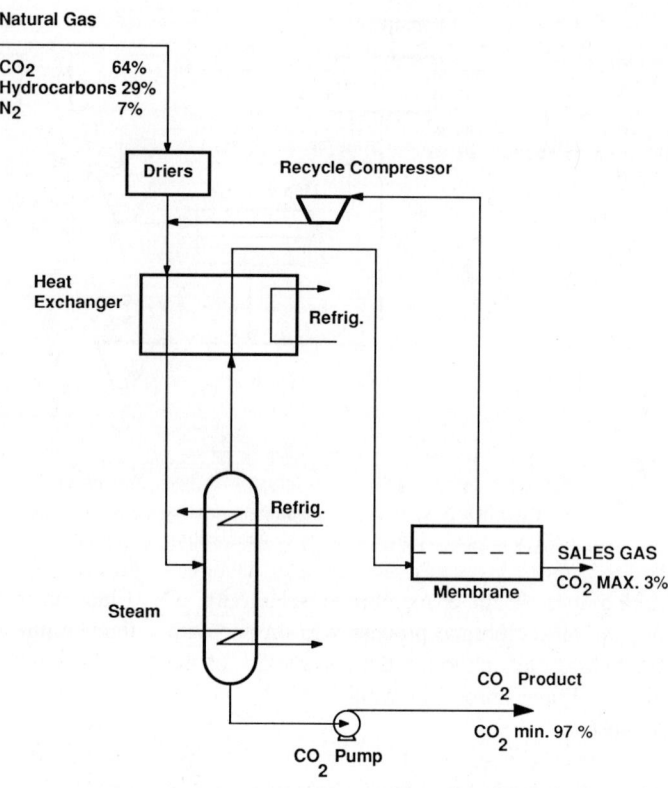

FIGURE 5–7. A hybrid cryogenic/membrane process for natural gas sweetening (reprinted from Baldus and Tillman 1986 with permission).

to 800 psig are needed to produce high-Btu gas at methane recoveries of 82 to 98%. In general, a single-stage system is the most cost-effective recovery system, even though recycle or staging can lead to higher methane recovery.

OTHER APPLICATIONS

Membranes can be used to separate organic vapors from air. In this process (Peinemann, Mohr, and Baker 1986; Baker et al. 1987), solvent-laden air at approximately room temperature is passed across a membrane that is permselective to the solvent. The downstream side of the membrane is held under vacuum in order to provide the driving force for permeation. Solvent is stripped out of the air, and the purified air is vented or reused as appropriate. The effluent stream from the vacuum pump is sent to a condenser, where the solvent is recovered as a liquid.

Rubbery polymers show extremely high permeation rates for many solvents, which results in high solvent-to-air selectivities. They are therefore highly suitable materials for these applications. Butadiene-acrylonitrile rubber, for example, was cited as having a selectivity of about 10^5 for benzene over air. Selectivities greater than 100 to 200 are essential in order to minimize the air in the vacuum and condenser systems. Selectivities greater than 500 to 1000 provide only marginal improvements in solvent recovery as compared to selectivities of 100 to 200. The economics of the process depends largely on the value that can be claimed for the recovered solvent.

Membranes can also be used in place of dessicant dryers for gas dehydration because water is a rapidly permeating gas. Moist pressurized air is fed to one side of the membrane, while dry gas at approximately ambient pressure provides a sweep on the opposite side. A portion of the dry gas produced by the membrane is normally used as the low-pressure sweep. Water diffuses from the feed to this sweep, which is typically discarded.

Dehydration membranes are available from Perma Pure, Permea, and Ube, among others. To produce dry air rather than dry nitrogen-enriched air, membranes with either extremely high water/air selectivities (as in Perma Pure membranes) or low O_2/N_2 selectivities must be used.

There are undoubtedly numerous specialty applications not discussed above in which existing membranes might be used, as well as applications awaiting the development of suitable membranes.

REFERENCES

A/G Technology Corporation. 1987. AVIR™ permeable membrane gas separation technology. Product brochure PB-GS12/87.

Agrawal, R., and S. Auvil. 1986. Process for the generation of gaseous and/or liquid nitrogen. U.S. Patent 4,595,405.

Anon. 1985. Johnson Matthey equipment: récupération d'hydrogène pure dans l'usine Texaco en Grande-Bretagne. *Informations Chemie.* 267(12): 89.

Anon. 1986. A Japanese membrane takes on hydrogen recovery. *Chem. Week* 139(8):73.

Backhouse, I. W. 1986. Recovery and purification of industrial gases using Prism™ separators. In *Membranes in Gas Separation and Enrichment,* Special Publication No. 62, pp. 265–280. London: Royal Society of Chemistry.

Baker, R. W., N. Yoshioka, J. M. Mohr, and A. J. Khan. 1987. Separation of organic vapors from air. *J. Membr. Sci.* 31(2&3):259–272.

Baldus, W., and D. Tillman. 1986. Conditions which need to be fulfilled by membrane systems in order to compete with existing methods in gas separation. In *Membranes in Gas Separation and Enrichment,* Special Publication No. 62, pp. 26–42. London: Royal Society of Chemistry.

Barry, T. R. 1989. Press release. E. I. Du Pont de Nemours & Company, Inc. 11 October 1989.

Beaver, E. P., P. V. Bhat, and D. S. Sarcia. 1988. Integration of membranes with other air separation technologies. *AIChE Symp. Ser.* 84(261): 113–123.

Beaver, E. R., and C. J. Paton. 1987. Membranes for generating nitrogen: gas processing applications. Paper read at AIChE 1987 Spring Meeting, 1 April 1987, at Houston, TX.

Berolatti, L. J. 1987. Press release. Linde/Union Carbide. 23 October 1987.

Bhat, P. V., and E. R. Beaver. 1988. Innovations in nitrogen inerting using membrane systems. *AIChE Symp. Ser.* 84(261):124–129.

Bollinger, W. A., S. P. Long, and T. R. Metzger. 1984. Optimizing hydrocracker hydrogen. *Chem. Eng. Prog.* 80(5):51–56.

Burmaster, B. M., and D. C. Carter. 1983. Increased methanol production using PRISM™ separators. *Energy Prog.* 3(3):158–162.

Cooley, T. E., and W. L. Dethloff. 1985. Field tests show membrane processing attractive. *Chem. Eng. Prog.* 81(10):45–50.

Crull, A. W. 1983. Membranes bring hi-tech to gas processing. *Energy* 8(1):29–30.

Doshi, K. J. 1987. Enhanced gas separation process. U.S. Patent 4,690,695.

Gardner, R. J., R. A. Crane, and J. F. Hannan. 1977. Hollow fiber permeator for separating gases. *Chem. Eng. Prog.* 73(10):76–78.

Generon Systems. 1989. Engineered systems for gaseous nitrogen supply. Product literature.

Golan, A., and M. H. Kleper. 1987. State-of-the-art: permeable membrane gas separation. Paper read at 5th Annual Membrane Technology/Planning Conference, 21 October 1987, at Boston, MA.

Graham, T. E., and D. L. MacLean. 1979. Process for hydrogen recovery from ammonia purge gases. U.S. Patent 4,180,552.

Grotewold, D. 1990. The application of MEDAL's hydrogen membrane technology in petroleum refinery hydroprocessing systems. Paper read at 1990 AIChE Spring National Meeting, 21 March 1990, at Orlando, FL.

Heitz, R., and R. Milne. 1990. Membrane separation research spawning commercial developments. *Chem. Process.* 53(2):62–66.

Lucadamo, G. A. 1986. Membrane-aided distillation for carbon dioxide and hydrocarbon separation. U.S. Patent 4,602,477.

MacLean, D. I., C. E. Prince, and Y. C. Chae. 1980. Energy-saving modifications in ammonia plants. *Chem. Eng. Prog.* 76(3):98–104.

Parro, D. 1984. Membrane CO_2 separation proves out at SACROC tertiary recovery project. *Oil Gas J.* 82(39):85–88.

Paulson, G. T., A. B. Clinch, and F. P. McCandless. 1983. The effects of water vapor on the separation of methane and carbon dioxide by gas permeation through polymeric membranes. *J. Membr. Sci.* 14(2):129–137.

Peinemann, K.-V., J. Mohr, and R. W. Baker. 1986. The separation of organic vapors from air. *AIChE Symp. Ser.* 82(250):19–26.

Peinemann, K.-V., K. Ohlrogge, and H.-D. Knauth. 1986. The recovery of helium from diving gas with membranes. In *Membranes in Gas Separation and Enrichment*, Special Publication No. 62, pp. 329–341. London: Royal Society of Chemistry.

Poffenbarger, G. L., and P. Gastinne. 1989. Hydrogen membrane applications and design considerations. Paper read at AIChE 1989 Spring National Meeting and Petrochemical Refining Exposition, 3 April 1989, at Houston, TX.

Russell, F. G., and A. B. Coady. 1982. Gas permeation process economically recovers CO_2 from heavily concentrated streams. *Oil Gas J.* 80(26):126–134.

Schell, W. J., and C. D. Houston. 1983. Use of membranes for biogas treatment. *Energy Prog.* 3(2):96–100.

Schell, W. J., C. D. Houston, and W. L. Hopper. 1983. Membranes can efficiently separate carbon dioxide from mixtures. *Oil Gas J.* 81(33):52–56.

Schendel, R. L. 1984. Using membranes for the separation of acid gases and hydrocarbons. *Chem. Eng. Prog.* 80(5):39–44.

Schendel, R. L., and J. Seymour. 1985. Take care in picking membranes to combine with other processes for CO_2 removal. *Oil Gas J.* 83(7):84–86.

Schott, M. E., C. D. Houston, J. L. Glazer, and S. P. DiMartino. 1987. Membrane H_2/CO adjustment. Paper read at 1987 AIChE National Meeting, 1 April 1987, at Houston, TX.

Spillman, R. W. 1989. Economics of gas separation membranes. *Chem. Eng. Prog.* 85(1):41–62.

Ube Industries. 1989. Preliminary information on Ube gas separations systems. Product brochure.

United States Department of Energy. 1989. Membrane separation processes for liquid hydrocarbons and gases in the petroleum industry: a technical case study. Publication DE89009464.

Weller, S., and W. A. Steiner. 1950. Engineering aspects of separation of gases: fractional permeation through membranes. *Chem. Eng. Prog.* 46(11):585–590.

Yamashiro, H., M. Hirajo, W. J. Schell, and C. F. Maitland. 1985. Plant uses membrane separation. *Hydrocarbon Process.* 64(2):87–89.

6

Economics

Raymond R. Zolandz and Gregory K. Fleming
E. I. Du Pont de Nemours & Company, Inc.

GENERAL ECONOMIC
 CONSIDERATIONS
ECONOMICS OF GAS SEPARATION
 PROCESSES
 Hydrogen Recovery
 Carbon Dioxide Separation
 Air Separation
REFERENCES

GENERAL ECONOMIC CONSIDERATIONS

The economic analysis of membrane processes for gas separation is intrinsically no different from that for other separations. However, Spillman (1989), in his recent overview of the economics of gas separation with membranes, notes that cost comparisons involving membranes can be particularly difficult for the following reasons:

1. Membranes do not generally perform under the same operating conditions or with the same product split as alternative processes.
2. The costs for membrane-based processes have been changing due to improved membrane performance and increased competition among membrane suppliers.

Evaluating the economics of membrane systems is also difficult because, as noted by Cooley and Dethloff (1985), information on membrane pricing policies is not generally available to potential customers.

Some generalizations regarding the economics of membrane systems can be made, however. Membrane systems often offer lower capital and maintenance costs than competing technologies, but the energy requirements of membrane systems will generally be equivalent to or higher than those for other kinds of separations techniques. As with any technology, costs for a membrane process will increase in cases where compression equipment must be installed.

In designing membrane processes, a number of trade-offs must be considered. In a single-stage process, product recovery can be increased only at the expense of product purity, as discussed in the previous chapter. Operating at increased feed pressure or reduced permeate pressure leads to lower membrane area requirements, but also adds to compression costs when compression is required.

Higher product purities and recoveries can

also be achieved by using multiple membrane stages. This leads to additional capital investment in membranes and/or compressors, which must be balanced against the increased value of the recovered product(s).

For both single- and multiple-stage membrane processes, economic analyses are required in order to arrive at the most cost-effective solution. Membrane suppliers are equipped to perform such analyses for customers and to compare membrane systems to alternative technologies. Some membrane suppliers also market nonmembrane technologies and can therefore recommend *and* provide the customer the lowest cost process to achieve the desired separation, whether or not it involves membranes.

While membrane manufacturers will provide cost estimates based on current membrane performance properties and costs, the potential purchaser can make initial cost estimates. Table 4–1 of Chapter 4 provides ranges of membrane properties, which can be used to estimate the membrane area required for a given separation, using either a simple relationship such as Eq. (4–24) in Chapter 4 or a more complex model. Once the surface area is approximated, an estimate on investment in membranes can be made. A membrane cost of $75/m^2 (Jain 1989) can be used as a reasonable estimate for both flat-film and hollow-fiber membranes. If sensitivity analyses are to be performed, a cost range of $50 to $100 should be considered.

The costs of auxiliary equipment, including feed pretreatment, compression, and instrumentation, can then be estimated as would be done for any other chemical process. Minimal pretreatment would include removal of particulates and mist from the feed stream. In addition, the feed gas must not be saturated in any component in order to avoid condensation on the membrane. Since capillary condensation can occur at temperatures slightly above "normal" dew points, it is standard practice for the feed temperature to be about 20 to 40°C above the feed dew point. This can be accomplished by heating the feed stream, provided the operating temperature of the membrane is not exceeded. Otherwise cooling of the feed gas, condensation and removal of condensables, and reheating of the gas are required.

One must also be sure that the feed temperature is sufficient so that dew points are not exceeded within the permeator. As highly permeable gases such as hydrogen or carbon dioxide are withdrawn across the membrane, the partial pressure of less permeable gases remaining on the retentate side of the membrane will be increased. One must therefore be sure that the residue stream is above saturation conditions. Being 20 to 40°C above the dew point of any condensables in the residue is normally sufficient. Note that water, because it is highly permeable, does not normally concentrate on the retentate side.

ECONOMICS OF GAS SEPARATION PROCESSES

Hydrogen Recovery

The traditional hydrogen recovery processes that membranes compete against include absorption, cryogenic separation, and pressure-swing adsorption.

In comparing membrane and cryogenic processes for hydrogen recovery from ammonia, Tomlinson and Finn (1990) found that both processes can be attractive. Membranes have the advantage of low capital investment and often yield the lowest cost of recovered hydrogen. One estimate (Ube Industries 1989) puts the cost of hydrogen recovered from the ammonia process at 33¢ per thousand standard cubic feet, including all fixed and operating costs.

Membranes, however, do suffer from the fact that the recovered hydrogen is always at a lower pressure than the feed and therefore usually needs to be recompressed prior to use. While cryogenic processes for recovering H_2 in ammonia plants are less energy intensive than membrane processes, they do have associated recompression costs. These costs are incurred because pressure limits in typical heat exchangers require dropping the purge stream from 2000 to 1000 psi prior to chilling the gas.

Estimates of hydrogen recovery costs in other process applications have also been pre-

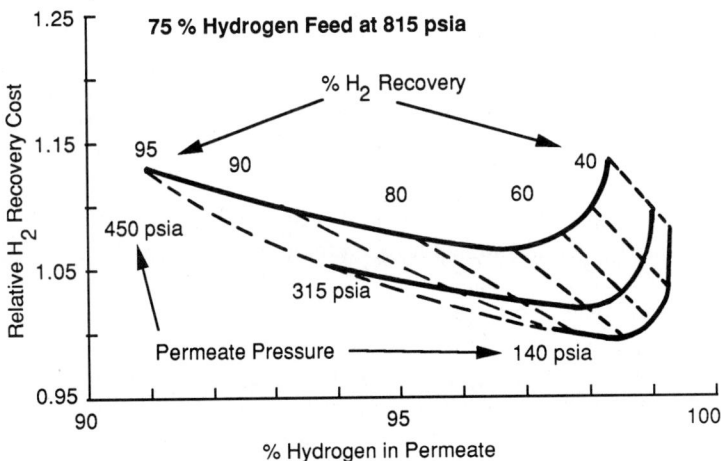

FIGURE 6–1. Hydrogen recovery costs from hydrodesulfurizer purge gas (reprinted from Poffenbarger and Gastinne 1989 with permission).

sented by Ube Industries (1989). The approximate recovery cost in cents per thousand standard cubic feet for those three hydrogen sources follow.

1. Reformer purge, 40¢
2. Hydrocracker purge, 21¢
3. Hydrodesulfurizer offgas, 24¢.

Note that these estimates have been made under a specific set of operating conditions and will vary somewhat in any given application.

Relative hydrogen costs as a function of permeate purity, hydrogen recovery, and permeate pressure have been presented for the recovery of gas from a hydrodesulfurization (HDS) unit (Poffenbarger and Gastinne 1989). The cost of hydrogen, exclusive of recompression, decreases along with the permeate pressure, as seen in Figure 6–1. This occurs because the additional driving force made available by lowering the permeate pressure reduces the required membrane area and membrane investment. For a given permeate pressure an optimum permeate purity and hydrogen recovery exists at which the costs go through a minimum.

Membranes for hydrogen recovery from HDS purge gas are competitive with alternative technologies (Heyd 1986). As seen in Figure 6–2, membranes have a clear advantage in terms of capital costs, with a 30% lower product cost.

The impact of membrane selectivity on the economics of hydrogen recovery has been considered (Weinberg, Wilkens, and Zolandz 1988). Results were presented for the process shown in Figure 6–3, in which feed gas at 800 psig containing 60 mol% hydrogen must be processed to yield a product gas at 900 psig and 95% purity. Since the required product pressure is higher than the available feed pressure, compression of the permeate is required. In this analysis 85% recovery of hydrogen was imposed as a constraint, while the permeate pressure was adjusted to minimize the cost of recovered hydrogen. (As permeate pressure is lowered, the cost of additional recompression

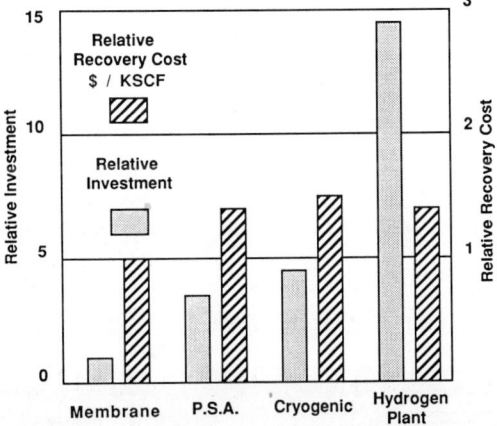

FIGURE 6–2. Comparison of recovery costs for membranes and alternative technologies (reprinted from Heyd 1986 with permission).

FIGURE 6-3. Hydrogen recovery process for purification of a low purity purge gas (Weinberg, Wilkens, and Zolandz 1988).

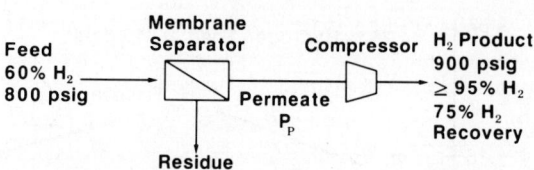

must be balanced against capital investment in permeators.)

The lowest cost membrane system to achieve the desired separation was calculated over a range of membrane selectivities. As shown in Figure 6-4, an optimum hydrogen/methane selectivity of approximately 110 gives the lowest cost hydrogen product. As the selectivity is lowered from the optimum value, the cost of recovered hydrogen increases substantially. Costs are 20% higher when the selectivity drops to 60. Recovery costs increase slightly as the selectivity is increased above the optimal value, but only by 5% as one raises the selectivity from 100 to 300. (Similar results were obtained when the required product purity was 98% rather than 95% hydrogen, except that then the optimum H_2/CH_4 selectivity was found to be approximately 300.)

The cost increase as the selectivity is raised above the optimum is mostly due to added investment in membrane area. This additional area is required because increased selectivity leads to greater hydrogen concentration in the permeate gas and lower hydrogen concentration in the feed gas. The net effect of these concentration changes is a lower partial pressure driving force for hydrogen permeation. Permeator output is reduced because of the decreased driving force; therefore, more membrane area is required.

Remembering that the cost curve in Figure 6-4 is for the lowest cost system operating with membranes of the specified selectivity, it is interesting to note that when the selectivity is at or below the optimum, one just meets the product purity specification. Above the optimum selectivity, the lowest cost system produces hydrogen at a purity in excess of the product specification. Since refineries cannot always quantify the savings accompanying improved hydrogen purity, no economic credit was taken for the additional purity in this analysis. Nonetheless, purer reactor feed gas can be expected to yield cost savings from items such as increased catalyst life.

Carbon Dioxide Separation

Membranes can be cost-effective in separating CO_2 from methane and other hydrocarbons. In an article discussing application of cellulose acetate membranes from Grace it was noted (Cooley and Dethloff 1985) that the manufacturer can "adjust the pricing philosophy to compete across the board with other technologies."

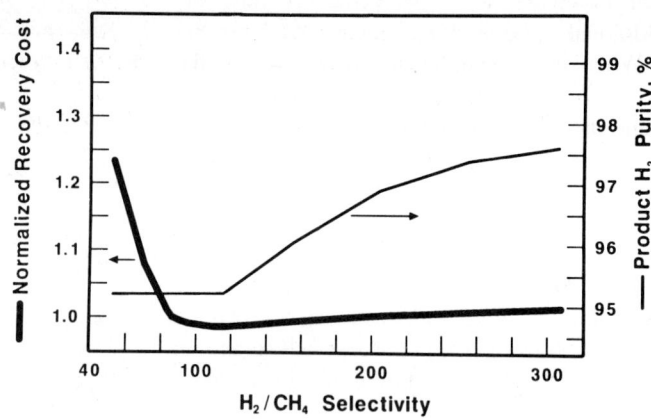

FIGURE 6-4. Effect of membrane selectivity on hydrogen recovery costs (Weinberg, Wilkens, and Zolandz 1988).

Costs for systems to treat three million SCFD of gas containing 8% CO_2 were cited. A "packaged" DEA (diethanolamine) absorption plant for this application cost $460,000, versus $340,000 for a two-stage membrane system. Even though the membrane system required intermediate compression, cited operating costs were 40% lower than those for the amine plant. The capital cost advantage enjoyed by such two-stage systems becomes larger as the CO_2 content of the feed is increased.

Schell (1983) examined the effect of process parameters on gas separation costs. For the two cases of production of 95% CO_2 from a feed gas at 765 psia, lowering the feed composition from 80 to 20% raised costs by a factor of almost 3, since this meant changing from a one-stage to a two-stage system with interstage compression.

In another study (Russell and Coady 1982) cost estimates were presented for processing 20 million SCFD of gas at 800 psig with varying feed CO_2 compositions to produce a hydrocarbon gas with less than 2% CO_2. The application further required that no more than 5% of the hydrocarbons be lost to the CO_2 stream. When the CO_2 content of the feed stream was less than 75%, use of a two-stage cellulose acetate membrane system was necessary in order to simultaneously meet specifications for both product streams. In the staged system the permeate from the first stage was recompressed and processed in a second stage. Capital costs rose from $2.1M (million) for treating a 10% CO_2 feed to $4M to treat a 75% stream. Compression costs varied between one-third and two-thirds of the total cost as the CO_2 concentration went from 10 to 75%.

When the CO_2 content of the feed was above 75% CO_2, the separation could be achieved in a single membrane stage, so the compressor was no longer needed. Capital costs dropped to approximately $1M for the membranes and necessary pretreatment.

Cost comparisons for producing CO_2 for enhanced oil recovery as a function of membrane selectivity highlighted the importance of selectivity on system cost (Schell, Houston, and Hopper 1983). In this example an 80% CO_2 stream at 500 psig was processed to recover 95% pure CO_2 permeate at a fixed CO_2 recovery of 91%. As the selectivity of the membrane was raised, the permeate pressure could be increased while still meeting the objectives of the separation. Higher permeate pressure led to reduced recompression costs and hence lower overall capital cost. Estimated capital costs declined by 31% when the CO_2/CH_4 selectivity was raised from 15 to 45. At present, typical membranes have selectivities of 15 to 25.

The above study also found that capital costs to treat acid gases with 20% CO_2 were 60% lower for membranes as compared to the Selexol absorption process. In the range of 10 to 20% CO_2, membranes were not competitive with DEA treatment, although a combined membrane/DEA system had a 9% cost advantage over DEA alone.

A comparison of CO_2 removal by membranes alone versus removal by either a combined membrane/Benfield potassium carbonate process or a standard TEA (triethanolamine) treatment showed the sensitivity of the economics to the particular details of each separation (Schendel and Seymour 1985). The desired objective was to reduce the CO_2 level of a 44% CO_2 feed to 5% CO_2, which at first glance seems to be an ideal bulk-cut separation for a process involving membranes. TEA, however, was found to be the most economical process, with the all-membrane process being the most expensive.

The costs of the all-membrane system were high in this case because a two-step process with intermediate recompression was required to meet the specified product purity, leading to high capital and operating costs. In the membrane/Benfield scheme the membranes were used to lower the feed gas to 20%, with the Benfield unit providing the final cleanup. Both capital and operating costs were lower than those for the all-membrane system, but evaluated cost was still 10% higher than those for TEA. The TEA system was at about its maximum loading in this case. Had the CO_2 content of the feed been slightly higher, the hybrid system would likely have been the lowest cost process.

Membranes were found to offer a 20 to 60%

lower operating cost than a combined DEA/glycol plant in treating a natural gas stream to reduce CO_2 levels from 6 to 3% for pipeline sales (Markiewicz 1988). The membranes simultaneously removed CO_2 and water from the 18 million SCFD stream, which was at 1180 psig and 95°F. The membranes were arranged in four banks, with a first set of two banks in parallel to each other followed by a similar second set of two banks. Banks of permeators, or individual permeators, could be valved in or out as required.

Air Separation

A graphical technique for estimating the cost of membrane-based air separation has recently been presented (Jain 1989). A simple gas permeator model was used to develop the necessary plots, which depict product recovery and an integral that is proportional to surface area as a function of membrane selectivity and the pressure ratio. With knowledge of the membrane's permeability coefficients, actual surface areas can be calculated. Graphs for both nitrogen-enriched air (NEA, 95 to 99% N_2) and oxygen-enriched air (25 to 35% O_2) are presented.

Golan and Kleper (1986) provided a cost comparison for air separation with several membranes versus pressure-swing adsorption (PSA). For production of 2 to 3 tons of 95% NEA, power requirements for the membrane system were 86 to 118% those of the PSA system. Membrane system costs were 74 to 116% of PSA costs, with the membrane-produced nitrogen costing 1 to 1.6 times the cost of PSA nitrogen. The lower costs in this range of estimates represent costs associated with use of membranes from A/G Technology, whereas the higher costs are an estimate made for first-generation Generon membranes. As noted before, the latter membrane system has been significantly improved from its early version.

In the same study, costs were presented for producing 35% oxygen-enriched air (OEA) by membranes and PSA. The membrane system had 50% lower capital costs and 35% lower power requirements than the PSA system. Delivered oxygen costs from a 10 ton/day plant (based on delivered oxygen) was given as $28/ton via membranes versus $42/ton via PSA.

NEA can be economically produced by membranes for low to moderate volumes (0.1 to 300 SCFH) of NEA at purities up to 99.5% N_2 (Spillman 1989). For higher purities (>99.5% inerts) and larger flow requirements, other technologies offer advantages over current membranes.

Analysis has shown membranes to be competitive with other technologies for producing oxygen-enriched air at purities below 50% and at rates of approximately 10 to 5000 SCFH of delivered oxygen (Beaver, Bhat, and Sarcia 1988). Other technologies are more cost-effective at purities above 50%. Cylinder and liquid oxygen compete at the low and high ends of the flow range, respectively, while PSA competes at all but the lower flow rates.

For oxygen production, as well as the other applications discussed, the region in which membranes are competitive with other technologies will no doubt expand as the capabilities of available membranes are improved.

REFERENCES

Beaver, E. P., P. V. Bhat, and D. S. Sarcia. 1988. Integration of membranes with other air separation technologies. *AIChE Symp. Ser.* 84(261):113–123.

Cooley, T. E., and W. L. Dethloff. 1985. Field tests show membrane processing attractive. *Chem. Eng. Prog.* 81(10):45–50.

Golan, A., and M. H. Kleper. 1986. Membrane-based air separation. *AIChE Symp. Ser.* 82(250):35–47.

Heyd, J. 1986. Hydrogen recovery using membranes in refining applications. Paper read at National Petroleum Refiners Association Meeting, 23–25 March 1986, Los Angeles, CA.

Jain, R. 1989. Method for economic evaluation of membrane-based air separation. *Gas Sep. Purif.* 3(3):123–127.

Markiewicz, G. 1988. Membrane system lowers treating cost at gas plants. *Oil Gas J.* 86(44):71–76.

Poffenbarger, G. L., and P. Gastinne. 1989. Hydrogen membrane applications and design considerations. Paper read at AIChE 1989 Spring

National Meeting and Petrochemical Refining Exposition, 3 April 1989, at Houston, TX.

Russell, F. G., and A. B. Coady. 1982. Gas permeation process economically recovers CO_2 from heavily concentrated streams. *Oil Gas J.* 80(26):126–134.

Schell, W. J. 1983. Membrane use/technology growing. *Hydrocarbon Process.* 62(10):43–46.

Schell, W. J., C. D. Houston, and W. L. Hopper. 1983. Membranes can efficiently separate carbon dioxide from mixture. *Oil Gas J.* 81(33):52–56.

Schendel, R., and J. Seymour. 1985. Take care in picking membranes to combine with other processes for CO_2 removal. *Oil Gas J.* 83(7):84–86.

Spillman, R. W. 1989. Economics of gas separation membranes. *Chem. Eng. Prog.* 85(1):41–62.

Tomlinson, T. R., and A. J. Finn. 1990. H_2 recovery processes compared. *Oil Gas J.* 88(3):35–39.

Ube Industries. 1989. Preliminary information on Ube gas separations systems. Product brochure.

Weinberg, M., J. A. Wilkens, and R. R. Zolandz. 1988. Selectivity and hydrogen separations. Paper read at the Second Annual Meeting of the North American Membrane Society, 1–3 June 1988, Syracuse, NY.

ACKNOWLEDGMENTS

The authors would like to thank their colleagues for their many efforts on our behalf. Special thanks go to Dr. Hugh E. Knipmeyer for his years of support and guidance to the membrane program at Du Pont. The authors are also indebted to Dr. Harvey H. Hoehn and Professor William J. Koros, both of whom have shown deep committment to the advancement of membrane science, and who have willingly shared their insights with us. Dr. Hoehn's recent death is both a personal loss and a loss to the community of membrane scientists.

We would also like to thank our families and friends for their patience and support during the preparation of this work.

Finally, the artistic efforts of L. B. Eichelberger are gratefully acknowledged.

III

Pervaporation

7.	Definitions and Background	105
8.	Theory	117
9.	Design	123
10.	Applications and Economics	132

7

Definitions and Background

Hubert L. Fleming
Zenon Environmental, Inc.

C. Stewart Slater
Manhattan College

INTRODUCTION AND GENERAL
  DESCRIPTION
BACKGROUND

DEFINITIONS AND
  FUNDAMENTALS
NOTATION
REFERENCES

INTRODUCTION AND GENERAL DESCRIPTION

Over the past five years membrane pervaporation has gained acceptance by the chemical industry as an effective process tool for separation and recovery of liquid mixtures. It is currently best identified with dehydration of liquid hydrocarbons to yield high-purity organics, most notably ethanol, isopropyl alcohol, and ethylene glycol. Due to its favorable economics, efficacy, and simplicity, it can be easily integrated into distillation and rectification processes and, depending on the specific process, even replace them. Presently, considerable data are available on industrial-scale processes utilizing pervaporation to evaluate its performance. Chapters 7 through 10 review the historical perspectives of pervaporation, its underlying fundamental principles, design considerations, current commercial applications and processes, and general economics with respect to alternative process technologies.

BACKGROUND

Pervaporation is characterized by the imposition of a barrier (membrane) layer between a liquid and a gaseous phase (Figure 7–1), with mass transfer occurring selectively across the barrier to the gas side. Because of the unique phenomenon of phase change required of the liquid solutes diffusing across the membrane (permselective "evaporation" of the liquid molecules), the process is termed pervaporation. Since different species permeate through the membrane at different rates, a substance at low concentration in the feed stream can be highly enriched in the permeate. Thus, separation occurs, with the efficacy of the separation effect being determined by the physicochemical structure of the membrane.

The early history of development in pervaporation was sporadic with little continuity (Slater and Hickey 1989). Kober (1917) at the New York State Department of Health's research laboratories first mentioned the phenom-

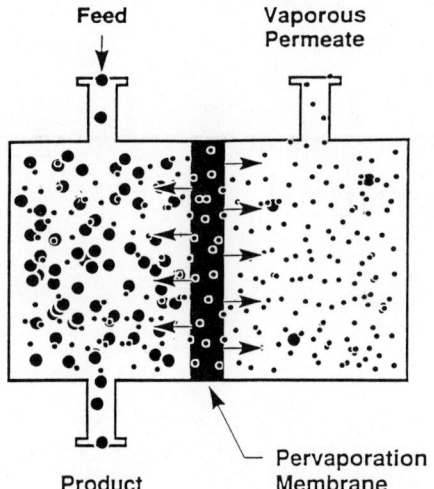

FIGURE 7-1. The pervaporation process.

enon "pervaporation" in a 1917 publication describing experiments with water selective permeation from an albumin/toluene solution through "collodion containers." Farber (1935) at the University of Toronto presented the results of work using pervaporation for concentrating protein solutions. Heisler et al. (1956) published their findings on dewatering ethanol solutions through regenerated cellulose in 1956. Heisler and his associates, working for the U.S. Department of Agriculture's Eastern Regional Laboratory, investigated both basic chemical separations and food applications: ". . . we have used pervaporation to dehydrate mashed potatoes."

The first major research effort in pervaporation was undertaken in the late 1950s by Binning and associates at the American Oil Company (Amoco) in Texas. A paper presented at the 1958 American Chemical Society Meeting, authored by Binning et al., reported the utilization of membrane pervaporation for dehydration of a ternary azeotrope of isopropanol-ethanol-water from the overhead of a distillation column. This paper was followed by several others elucidating their results (Binning and James 1958a, 1958b; Anon. 1958a, 1958b, 1958c). A paper by the group appeared in *Industrial and Engineering Chemistry* describing experiments that separated an equivolume mixture of *n*-heptane and iso-octane through unnamed "thin polymer films" (Binning et al. 1961). Several of the papers discussed azeotropic benzene/methanol separation and reversal of selectivity with different membranes. More than ten patents were issued to Amoco in those early years citing members of the research group as inventors (Binning and Kelly 1959; Binning and Lee 1960a, 1960b, 1960c; Binning 1960; Stuckey 1960; Binning and Stuckey 1960; Binning and Johnson, 1961; Binning 1961; Binning, Jennings, and Martin, 1961, 1962). Most of their studies focused on separation of anhydrous organic mixtures commonly found in petrochemical processing using cellulosic and polyethylene membranes.

Several other early researchers in the field made significant contributions to its development. Sanders and Choo (1960) at Ionics, Inc. (Massachusetts), which became a licensee from Amoco, conducted parallel studies on various types of separations. They published results on the concentration of binary aqueous mixtures of ethanol, isopropanol, methyl ethyl ketone, dioxane, acetonitrile, and formaldehyde. Several hydrocarbon mixtures were also separated, including *n*-heptane/iso-octane and xylene isomers. The types of membranes utilized by Sanders and Choo were not disclosed.

Michaels et al. (1962) at the Massachusetts Institute of Technology investigated polyethylene membranes to separate mixtures of xylene isomers. In a review article, Choo (1962) discussed the phenomenon of membrane pervaporation and its potential for commercialization. Renon and Teyssie (1963) at the French Petroleum Institute investigated organic-organic separations using pervaporation. Carter and Jagannadhaswamy (1964) at the University of Birmingham and the Madras College of Technology, respectively, investigated polyethylene and cellophane membranes for the separation of various alcohols from benzene, carbon tetrachloride, and water mixtures.

Sweeney and Rose (1965) at Applied Sciences Laboratories, Inc., in Pennsylvania used a variety of membranes to examine single solute permeation behavior and binary mixture separation in the pervaporative mode. They documented the separation of binary organic mixtures of ethanol/acetone, acetone/*n*-hexane, and ethanol/*n*-hexane through various membranes (polyethylene, polypropylene, polyvinylchloride, cellulose acetate, cellophane, etc.). Work in the pervaporation field was significant enough to gain mention in a membrane review article by Li, Long, and Henley (1965) and a research paper by Long (1965) of the Esso Research and Engineering Company (now Exxon).

The capacity for separating aqueous-organic and organic-organic mixtures was demonstrated by 1965, but commercial development did not proceed, primarily due to the lack of a market need. Traditional separation technologies including distillation, extraction, and adsorption were deemed sufficient. Further, the membranes then being utilized lacked the high selectivity and permeability necessary to make pervaporation economically attractive. There was much greater interest in the other membrane processes being developed, reverse osmosis and ultrafiltration, in particular.

The energy crisis in the 1970s refocused interest in separation technologies that possessed a high potential for energy savings. Pervaporation was aggressively pursued, primarily in Europe, because of its demonstrated ability to dewater aqueous mixtures for fuel utilization. Further, better membrane materials were being developed in the analogous technologies of reverse osmosis and gas permeation, so that the potential for economically attractive separations by pervaporation was greatly enhanced.

In the mid-1970s, GFT (West Germany) commercialized an economical pervaporation process for dehydrating ethanol and producing high purities that rivaled azeotropic distillation. Following pilot trials in Europe, the first industrial plants were built in Brazil and the Philippines for processes utilizing continuous fermentation of sugar cane, bagasse, and sweet sorghum containing 5 to 7% ethanol, primary distillation to a mash containing 80 to 85% ethanol, with vacuum pervaporation to 96 wt.% (Ballweg et al. 1982). The primary driving forces for the GFT process were:

1. No additives necessary for final separation
2. Reduced energy demand, because only that fraction of the liquid to be evaporated has to be vaporized
3. Only a small vacuum pump is necessary, because condensing permeate continually creates a driving force vacuum
4. Closed-loop operation, with only a small volume of recycled permeate
5. Much lower capital cost
6. Cheap cooling water, with recovery of initial thermal energy supplied as low-pressure steam (<90 psi).

The GFT process is shown schematically in Figure 7–2 for ethanol dehydration. In the late 1970s and continuing into the early 1980s, other integrated distillation/pervaporation plants were built in Europe and Asia. Most of

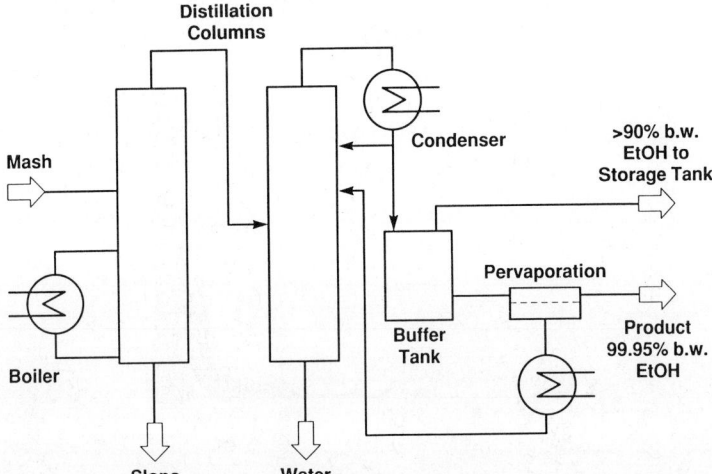

FIGURE 7–2. Integrated distillation/pervaporation plant for recovery from fermenters.

these were of moderate capacity, with typically 1000 to 50,000 L/day of ethanol recovered. As the cost of the permselective membrane module was reduced, and selectivity increased (with the corresponding capital cost decreasing), ethanol purity increased (99.85% was easily attainable) and the integrated process gained industrial acceptance.

In the mid-1980s multiple membrane options were developed, all based on novel proprietary asymmetric composite polymer technology for producing economical, chemically and thermally stable membranes in a flat plate geometry. As shown below, a number of composite membranes were developed, each exhibiting high selectivities (10 to 1000) for specific separations (Brueschke 1988).

1. PVA composites
    water >> MeOH > EtOH >> other organics
2. Silicone composites
    MeOH > EtOH > aldehydes > ketones >> water
    paraffins > olefins
3. Modified cellulose esters
    aromatics > paraffins
    olefins > paraffins
    dienes > olefins
    n-paraffins > branched paraffins
    low MW paraffins > high MW paraffins.

Utilization of GFT's process, which incorporated a vacuum on the permeate, made it industrially feasible to dehydrate and recover solvents such as acetone, ethylene glycol, and tetrahydrofuran (THF), as well as separate mixtures of liquid hydrocarbons. Today, a number of commerical pervaporation plants exist for recovery of solvents, removal of organics from wastewaters, dealcoholization of wines and liquors, as well as many more for ethanol dehydration. In fact, a 150,000 L/day ethanol dehydration plant in Bethenville, France, was started up in early 1988, and became the world's largest pervaporation facility (Rapin 1988).

Both industrial and academic interest in pervaporation has dramatically increased in the past decade. One index of the degree of activity is the number of citations of technical publications and patents in the field. A recent study (Slater and Hickey 1989) identified more than 775 citations spanning the years 1917 to 1989 that referred to pervaporation activities. Of these, 37% were patents, with the remainder composed of technical articles. Approximately 70% of all pervaporation patents and 70% of all pervaporation articles were published from 1984–1989.

Figure 7–3 demonstrates the growth in pervaporation technical activity around the world. Figures 7–4, 7–5, and 7–6 represent the relative breakdown of that growth in the United States, Europe, and Japan, respectively. As shown in these figures, pervaporation research in the 1950s and 1960s was primarily done in the United States. Rapid advancement was made by researchers in Europe in the early 1980s.

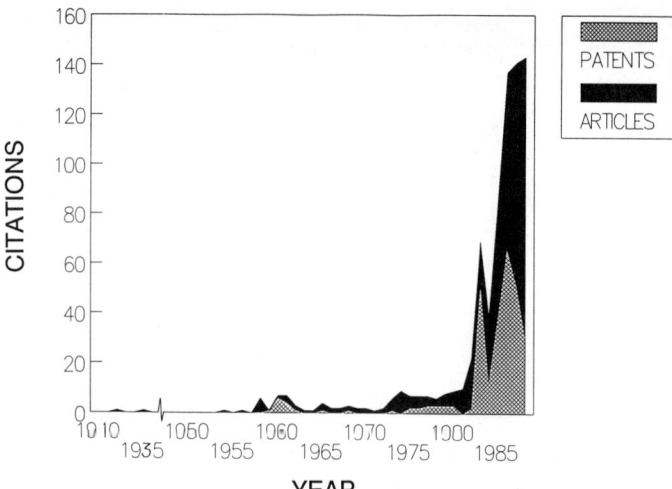

FIGURE 7–3. Chronology of citations worldwide.

Definitions and Background 109

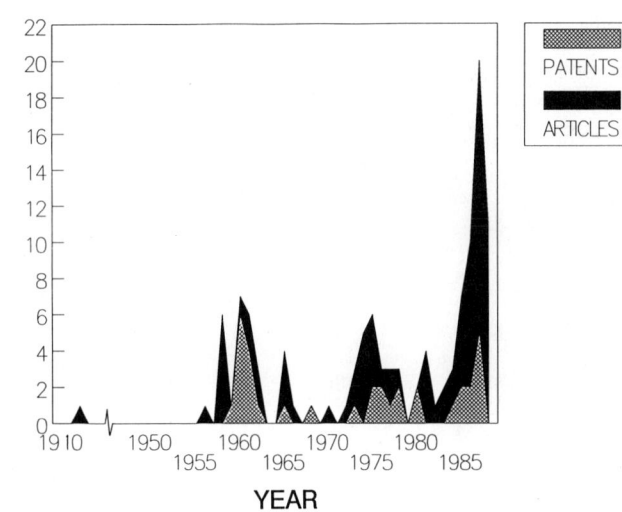

FIGURE 7–4. Chronology of citations in the United States.

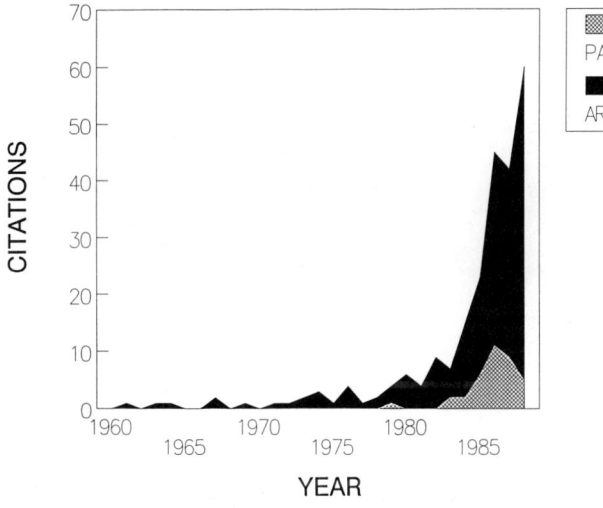

FIGURE 7–5. Chronology of citations in Europe.

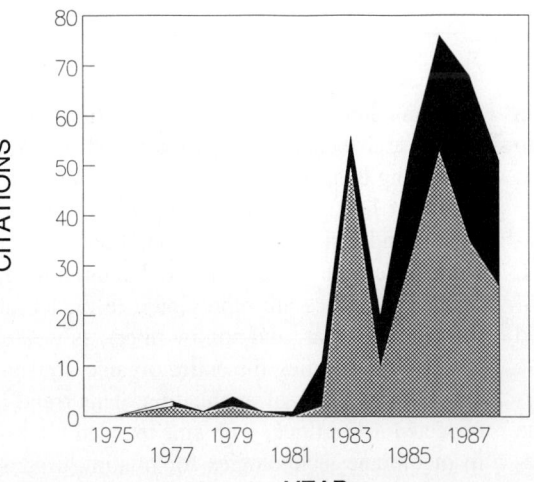
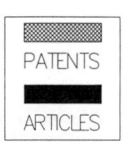

FIGURE 7–6. Chronology of citations in Japan.

TABLE 7-1. Research Groups in Europe.

Research Group	Dates of Recorded Activity	Publications	Patents
Ecole Nat. Super. des Ind. Chim. (ENSIC) (Neel, Aptel, Nguyen et al.)	1969–88	36	1
Reinisch-Westfael Tech. Hochsch. (RWTH) (Aachen Tech. Univ.) (Rautenbach & Albrecht)	1982–88	19	0
GFT (Brueschke, Tusel et al.)	1982–88	13	5
GKSS (Boeddeker, Wenzlaff et al.)	1984–88	10	7
Twente Univ. of Technology (Smolders, Mulder et al.)	1982–88	16	1
Fraunhofer (Strathmann et al.)	1984–88	4	5
Univ. of Paris val de Marne (Brun, Larchet et al.)	1974–86	9	0
Lurgi-Forschung G.m.b.H. (Sander et al.)	1986–88	5	3
Univ. of Heidelberg (Lichtenthaler, Schmittecker et al.)	1986–88	6	0
Delft Univ. of Technology (Groot et al.)	1984–87	4	0
BP Chemicals, Ltd.	1986–88	1	3
Ingenieurhochsch, Koethen (Roedicker et al.)	1976–84	4	0
Lyonnaise des Eaux (Aptel et al.)	1987–89	4	0
Rhone-Poulenc Industries	1979–81	2	1
Hoechst	1976–86	2	1
Hungary Academy of Sciences (Nagy et al.)	1980–86	3	0
University of Aberdeen (Brass and Meares)	1979–82	3	0
Akzo	1985–87	0	3
University of Cologne (Niemoller et al.)	1986–88	3	0

Japanese researchers, with only one citation in 1981, increased that number dramatically to more than 50 in 1983. The output since that time in Japan has continued at a rapid rate.

Currently pervaporation research and development is occurring on a worldwide basis. Japan and Europe account for 46 and 31% of the citations, respectively, with the United States trailing with 15%. The more active research groups are summarized in Tables 7-1 through 7-4 for Europe, Japan, the United States, and the rest of the world, respectively.

Commercialization of pervaporation membranes and technology has expanded as well. Following the lead of GFT, Table 7-5 provides a current listing of seven companies now providing industrial-scale pervaporation systems worldwide. All four of the most common membrane geometries are represented (e.g., spiral, flat plate, tubular, and hollow fiber), as well as companies spanning the entire organic-aqueous spectrum in terms of application. This trend is expected to continue, with an expanded interest in membrane technologies for treating organic streams.

TABLE 7–2. Research Groups in Japan.

Research Group	Dates of Recorded Activity	Publications	Patents
Agency of Industrial Sciences and Technology	1985–89	0	61
Asahi Glass Co.	1983–87	1	35
Nitto Electric Industries	1983–88	0	35
Kuraray Company (Mochizuki et al.)	1983–89	5	12
University of Tokyo (Kimura et al.)	1982–88	15	0
Sophia University (Yoshikawa, Ogata et al.)	1984–88	14	0
Research Institute for Polymers and Textiles (Hirotsu et al.)	1987–89	13	0
Mitsui Engineering & Shipbuilding Co.	1986–88	6	7
Sagami Chem. Research Center	1983–89	8	4
Mitsubishi Heavy Ind., Ltd.	1983–88	0	12
Yamagata University (Suzuki et al.)	1982–88	11	0
Kyoto University (Tanigaaki, Yoshikawa, Eguchi, Masada et al.)	1986–89	11	0
Tokuyama Soda Co.	1984–87	2	9
Showa Denko Co.	1977–86	0	10
Hitachi Co.	1985–88	3	6
Industrial Product Research Institute (Yamada, Hamaya, and Nakagawa)	1977–87	8	0
Ube Industries	1987–88	0	8
Kansai University (Uragami et al.)	1986–89	6	1
Yokohama National University (Matsumoto, Ohya et al.)	1985–88	6	0
Teijin Co.	1983–87	1	5
Sasakura Engineering Co.	1985–89	1	5
Research Assoc. for Basic Polymer Technology	1987–88	6	0
Sanyo Co.	1985	0	4
Toray Industries	1985–89	1	3
Government Industrial Research Institute, Osaka (Murata, Miya et al.)	1985–87	4	0
Waseda University (Itoh, Toya, Terada et al.)	1982–86	4	0

DEFINITIONS AND FUNDAMENTALS

Pervaporation differs from other membrane processes in that the membrane constitutes a barrier between the feed in the liquid phase and the permeate in the gas phase. The driving force that is applied across the membrane creates a chemical potential gradient in the liquid phase, and the selectivity of the membrane is then the determining factor in the relative flow of the different components. In contrast to reverse osmosis, the osmotic pressure is not limiting, because the permeate is kept under low pressure.

Transport across nonporous membranes in pervaporation is generally considered to follow the well-known solution-diffusion model. A

TABLE 7–3.   Research Groups in the United States.

Research Group	Dates of Recorded Activity	Publications	Patents
American Oil Company (Binning et al.)	1958–68	7	14
Monsanto (Lee et al.)	1973–80	2	9
Dow Chemical (Reineke et al.)	1984–88	4	6
University of Maine (Thompson et al.)	1977–88	9	0
State University of New York Polymer Research Institute (Cabasso et al.)	1983–88	7	0
Clemson University (Gooding et al.)	1985–88	7	0
University of Iowa (Hoover, Hwang et al.)	1974–82	4	0
University of Texas at Austin (Paul, Lloyd and Meluch)	1974–85	4	0
Membrane Technology & Research, Inc.	1987–89	3	0
Montana State University (McCandless et al.)	1973–87	3	0
University of California-Los Angeles (Nguyen)	1986–87	3	0
University of Cincinnati (Hwang et al.)	1983–88	3	0
U.S. DOE Solar Energy Research Inst. (Neidlinger, Schissel et al.)	1984–87	1	2
Rensselaer Polytechnic Inst. (Belfort et al.)	1987–89	3	0

TABLE 7–4.   Research Groups in Other Geographic Areas.

Research Group	Dates of Recorded Activity	Publications	Patents
Univ. of Waterloo, Canada (Huang et al.)	1968–89	13	0
Zhejiang Univ., P.R. of China	1983–89	9	5
Univ. Estadual Campinas, Brazil (Schuchardt, Goncalves et al.)	1983–89	3	0
Univ. Federal Rio de Janiero, Brazil	1986–88	3	0
Beijing Inst. Chem., China	1985–88	3	0
Qinghua Univ., China (Xu)	1987–88	3	0
Jadavpur Univ., India (Ghosh, Sanyal, and Mukherjea)	1987–89	3	0
Weizmann Inst. Sci., Israel	1974–89	3	0

Definitions and Background

TABLE 7-5. Current Commercial Producers of Pervaporation Systems.

Organization	Primary Applications	Module Configuration	Membrane Materials
GFT	Dehydration of liquid organics	Plate-and-frame	Polyvinylalcohol composites
MTR	Organic recovery from wastewaters	Spiral wound	Silicones
Lurgi	Dehydration of liquid organics	Plate-and-frame[a]	Polyvinylalcohol composites
	Dehydration of vapors	Plate-and-frame[a]	Polyvinylalcohol composites
Tokuyama Soda	Dehydration of IPA	Hollow-fiber bundles	Chitisan
Kalsep	Dehydration of liquid organics	Tubular bundles	Ion exchange composites
Hoechst Celanese	MeOH/MTBE	Spiral wound	Cellulose acetates
Mitsui	Dehydration of liquid organics	Plate-and-frame[a]	Polyvinylalcohol composites

[a]GFT membranes.

thorough discussion of transport theory and models for pervaporation is given by Aptel and Neel (1986) and others. A summary of the relative points follows.

The driving force for transport is generally recognized as a gradient in chemical potential between the liquid and the vapor. As each component of the feed dissolves in the membrane, then diffuses across the membrane to the permeate side, the flux of each component is given by

$$J_i = -D_i c_i \frac{d(\mu_i/RT)}{dx}. \quad (7\text{-}1)$$

For pervaporation, the activity gradient across the membrane far exceeds the pressure gradient, so that Eq. (7-1) can be rewritten as

$$J_i = -D_i c_i \frac{d(\ln a_i)}{dx}. \quad (7\text{-}2)$$

Equation (7-2) demonstrates that the diffusion coefficient and the activity (i.e., concentration) are the two most important factors in controlling separation.

Selectivity of the pervaporation process α_p is calculated from the liquid compositions of the feed and permeate:

$$\alpha_p = \frac{c_i''/c_j''}{c_i'/c_j'}. \quad (7\text{-}3)$$

This relationship is analogous to that of the selectivity of the sorption process for a swollen membrane:

$$\alpha_s = \frac{c_i/c_j}{c_i'/c_j'}. \quad (7\text{-}4)$$

The corresponding enrichment factor β is defined as

$$\beta = \frac{c_i''}{c_i'}, \quad (7\text{-}5)$$

where c_i is the concentration of each permeable component i and the single prime and double prime indicate feed and permeate, respectively.

Both α and β would be useful if they were constant over most of the useful range of composition. Unfortunately, this is not the case in most systems of industrial interest. Therefore, reporting pervaporation selectivities in the form of a McCabe-Thiele diagram, such as the one shown in Figure 7-7, has become common. Of

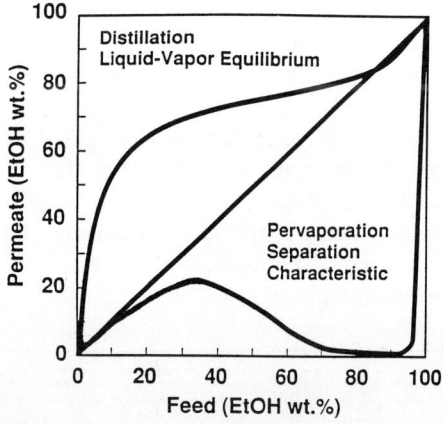

FIGURE 7-7. Separation characteristics of pervaporation: ethanol-water (from Fleming 1989).

FIGURE 7–8. Conventional versus membrane-imposed vapor-liquid equilibria (reprinted from Blume and Baker 1987 with permission).

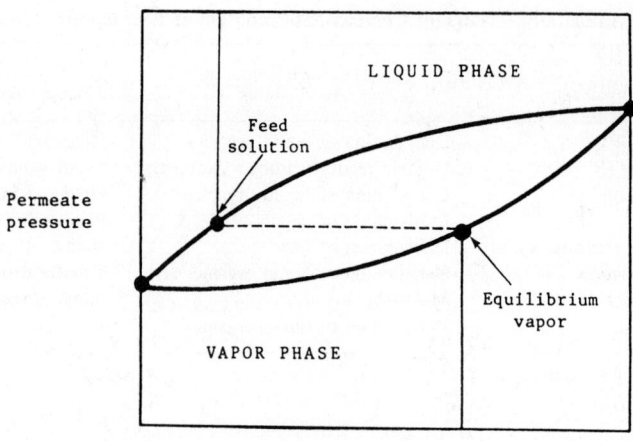

course, such data are then dependent on the process parameters specific to the measurement given.

Separation factors are also often defined in terms of the increase in selectivity due to the membrane, relative to that found simply by thermodynamic evaporation (Blume and Baker 1987; Blume, Wijmans, and Baker 1990). In this reasoning,

$$\alpha_{\text{pervap}} = \frac{p_i''/p_j''}{c_i^l/c_j^l}, \quad (7\text{–}6)$$

$$\alpha_{\text{evap}} = \frac{p_i^l/p_j^l}{c_i^l/c_j^l}, \quad (7\text{–}7)$$

$$\alpha_{\text{mem}} = \frac{p_i''/p_j''}{p_i^l/p_j^l}. \quad (7\text{–}8)$$

Therefore,

$$\alpha_{\text{pervap}} = \alpha_{\text{evap}} \times \alpha_{\text{mem}}. \quad (7\text{–}9)$$

This approach is demonstrated graphically in Figure 7–8 by the well-known Shelden-Thompson plot (Greenlaw et al. 1977; Greenlaw, Shelden, and Thompson 1977). A typical vapor-liquid equilibrium is defined in the upper figure for a two-phase gas-liquid system. As demonstrated in the top figure, separation is determined by volatility differences, with the resulting composition of the two phases defined in the vapor-liquid equilibrium diagram. A feed solution of the represented composition, for example, would exhibit a vapor phase in equilibrium of the composition defined by the diagram. Such a plot would be representative of equilibrium-governed separations such as distillation.

When a pervaporation membrane is imposed, the reasoning is that the new system selectivity is defined as that for evaporation, the conventional equilibrium-based phenomenon defined in the top figure, times that imposed by the intrinsic selectivity of the membrane. This situation is given in the lower plot of Figure 7–8. Intrinsic membrane selectivities greater than 1 enhance the overall selectivity, whereas those <1 decrease separation.

Although physically intuitive, the Shelden-Thompson approach is not particularly helpful in predicting or modeling pervaporation behavior. It also suffers from a lack of thermodynamic consistency.

Pervaporation is a nonequilibrium-based process, for example, being modeled in terms of thermodynamic equilibria.

NOTATION

General Notation

See the General Notation section at the beginning of this handbook.

Special Notation

R gas constant, ML^2/t^2T mol

Greek Letters

α_p pervaporation selectivity (determined from feed and permeate compositions), dimensionless

α_s selectivity of a sorption process, dimensionless

β pervaporation enrichment factor, dimensionless

α_{pervap} pervaporation selectivity determined from α_{evap} and α_{mem}, dimensionless

α_{evap} selectivity of thermodynamic evaporation, dimensionless

α_{mem} selectivity of membrane, dimensionless

REFERENCES

Anon. 18 April 1958a. Plastic film solves tough separations. *Petroleum Week:* 40–42.

Anon. 31 March 1958b. Take a look at permeation. *Chem. Eng. News:* 58–59.

Anon. 5 April 1958c. New way to split a mix. *Chem. Week:* 55–66.

Aptel, P., and J. Neel. 1986. Pervaporation. In *Synthetic Membranes: Science, Engineering, and Application*, ed. P. M. Bungay, H. K. Lonsdale, and M. N. dePinho, pp. 403–436. Dordrecht, Holland: D. Reidel Publishing Company.

Ballweg, A. H., H. E. A. Brueschke, W. H. Schneider, G. F. Tusel, K. W. Boddecker, and A. Wenzlaff. 1982. Pervaporation membranes—an economical method to replace conventional distillation and rectification columns in ethanol distilleries. Paper read at 5th Intl. Symp. on Fuel Alcohol Technology, Auckland, New Zealand.

Binning, R. C. 1960. Organic chemical reactions involving liberation of water. U.S. Patent 2,956,070.

Binning, R. C. 1961. Separation of mixtures. U.S. Patent 2,981,680.

Binning, R. C., and F. E. James. 1958a. Now separate by membrane permeation. *Petrol. Refiner.* 37:214.

Binning, R. C., and F. E. James. 1958b. Permeation: a new way to separate mixtures. *Oil Gas J.* 56(21):104–105.

Binning, R. C., J. F. Jennings, and E. C. Martin. 1961. Separation technique through a permeation membrane. U.S. Patent 2,985,588.

Binning, R. C., R. J. Jennings, and E. C. Martin. 1962. Process for removing water from organic chemicals. U.S. Patent 3,035,060.

Binning, R. C., and W. F. Johnson. 1961. Aromatic separation process. U.S. Patent 2,970,106.

Binning, R. C., and J. T. Kelley. 1959. Hydrocarbon conversion with dialytic separation of the catalyst from the hydrocarbon products. U.S. Patent 2,913,507.

Binning, R. C., and R. J. Lee. 1960a. Prevention of membrane rupture in a separatory process for oil and soluble organic compounds using a nonporous plastic permeation membrane. U.S. Patent 2,923,749.

Binning, R. C., and R. J. Lee. 1960b. Production of high octane alkylate using a permeable membrane separation system. U.S. Patent 2,923,751.

Binning, R. C., and R. J. Lee. 1960c. Separation of azeotropic mixtures. U.S. Patent 2,953,502.

Binning, R. C., R. J. Lee, J. F. Jennings, and E. C. Martin. 1958. Separation of liquid mixtures by permeation. *ACS Div. Petroleum Chem. Preprints* 3(1):131–141.

Binning, R. C., R. J. Lee, J. F. Jennings, and E. C. Martin. 1961. Separation of liquid mixtures by permeation. *Ind. Eng. Chem.* 53:45–50.

Binning, R. C., and J. M. Stuckey. 1960. Method of separating hydrocarbons using ethyl cellulose permselective membrane. U.S. Patent 2,958,657.

Blume, I., and R. W. Baker. 1987. Separation and concentration of organic solvents from water by pervaporation. In *Proceedings of Second International Conference on Pervaporation Processes in the Chemical Industry*, ed. R. A. Bakish, pp. 111–125. Englewood, NJ: Bakish Materials Corporation.

Blume, I., J. G. Wijmans, and R. W. Baker. 1990. The separation of dissolved organics from water by pervaporation. *J. Membr. Sci.* 49(3):253–286.

Brueschke, H. E. A. 1988. State of the art of pervaporation. In *Proceedings of Third International Conference on Pervaporation Processes in the Chemical Industry*, ed. R. A. Bakish, pp. 2–11. Englewood, NJ: Bakish Materials Corporation.

Carter, J. W., and B. Jagannadhaswamy. 1964. Separation of organic liquids by selective permeation through polymeric films. *Brit. Chem. Eng.* 9(8):523–526.

Choo, C. Y. 1962. Membrane permeation. *Adv. Petroleum Chem.* 6(2):73–117.

Farber, L. 1935. Applications of pervaporation. *Science* 82:158.

Fleming, H. L. 1989. Dehydration of organic/aqueous mixtures by membrane pervaporation. In *Proceedings of International Conference on Fuel Alcohols and Chemicals*, ed. W. Kampen. Charlotte, NC: K-Engineering.

Greenlaw, F. W., W. D. Prince, R. A. Shelden, and E. V. Thompson. 1977. Dependence of diffusive permeation rates on upstream and downstream pressures: I. single component permeant. *J. Membr. Sci.* 2:141–151.

Greenlaw, F. W., R. A. Shelden, and E. V. Thompson. 1977. Dependence of diffusive permeation rates on upstream and downstream pressures: II. two component permeant. *J. Membr. Sci.* 2:333–348.

Heissler, E. G., A. S. Hunter, J. Sciliano, and R. M. Treadway. 1956. Solute and temperature effects in the pervaporation of aqueous alcoholic solutions. *Science* 124:77–79.

Kober, P. A. 1917. Pervaporation, perstillation, and percrystallization. *L. Am. Chem. Soc.* 39:944–948.

Li, N. N., R. B. Long, and E. J. Henley. 1965. Membrane separation processes. *Ind. Eng. Chem.* 57(3):18–29.

Long, R. B. 1965. Liquid permeation through plastic films. *Ind. Eng. Chem. Fundam.* 4:445–451.

Michaels, A. S., R. F. Baddour, H. J. Bixler, and C. Y. Choo. 1962. Conditioned polyethylene as a permselective membrane. *Ind. Chem. Proc. Des. Dev.* 1:14–25.

Rapin, J. L. 1988. The Betheniville pervaporation unit—the first large-scale production plant for the dehydration of ethanol. In *Proceedings of Third International Conference on Pervaporation Processes in the Chemical Industry*, ed. R. A. Bakish, pp. 364–378. Englewood, NJ: Bakish Materials Corporation.

Renon, H., and P. Teyssie. 1963. A novel process for separation: permeation. *Rev. Inst. Franc. Petrol.* 18:996–1011.

Sanders, B. H., and C. Y. Choo. June 1960. Latest advances in membrane permeation. *Petrol. Refiner:* 133–138.

Slater, C. S., and P. J. Hickey. 1989. Pervaporation R&D: a chronological and geographic perspective. In *Proceedings of the Fourth International Conference on Pervaporation Processes in the Chemical Industry*, ed. R. A. Bakish, pp. 476–492. Englewood, NJ: Bakish Materials Corporation.

Stuckey, J. M. 1960. Method of separating hydrocarbons using ethyl cellulose permselective membrane. U.S. Patent 2,958,656.

Sweeney, R. F., and A. Rose. 1965. Factors determining rates and separation in barrier membrane permeation. *Ind. Eng. Chem. Proc. Des. Dev.* 4:248–251.

8

Theory

Hubert L. Fleming
Zenon Environmental, Inc.

C. Stewart Slater
Manhattan College

INTRODUCTION
SINGLE-COMPONENT PERMEATION
MULTICOMPONENT PERMEATION
NOTATION
REFERENCES

INTRODUCTION

Pervaporation through a nonporous membrane can be described by the widely accepted solution-diffusion mechanism (Binning and James 1958a, 1958b; Binning et al. 1961). The underlying assumptions of pervaporative transport are:

1. Sorption of the liquid mixture on the feed (upstream) side of the membrane
2. Diffusion through the membrane
3. Desorption on the permeate (downstream) side of the membrane in the vapor phase.

Sorption and diffusion in the polymer represent the important steps in transport of the diffusing component(s). The desorption step is not normally considered the controlling resistance. Although much comparison between the underlying mechanism for pervaporation is possible between gas permeation and reverse osmosis, distinct differences exist. Diffusivity and solubility in pervaporation are highly dependent on composition.

The driving force for transport across the membrane is generally recognized as a chemical potential gradient across the membrane. The chemical potential can be expressed as

$$\mu_i = \mu_i^\circ + RT \ln a_i, \qquad (8\text{-}1)$$

where μ_i° is the standard chemical potential, a_i is the activity of the permeating component, R the universal gas constant, and T the absolute temperature. Component activity is expressed as

$$a_i = \frac{p_i}{p_i^\circ}, \qquad (8\text{-}2)$$

where p_i° is the component saturation pressure of component i and p_i is the vapor pressure. For multicomponent mixtures the component activity is represented by

$$a_i = \gamma_i x_i, \qquad (8\text{-}3)$$

where γ_i is the activity coefficient and x_i is the mole fraction.

SINGLE-COMPONENT PERMEATION

To describe the performance of pervaporation for separating multicomponent mixtures, the transport of pure components must first be understood. Consider the diffusion step in the vacuum permeation of a pure component. The permeation can be described by a Fick's law relationship:

$$J_i = -D_i \frac{dc_i}{dl}, \qquad (8\text{--}4)$$

where J_i is the flux, D_i is the diffusivity, c_i is the concentration in the membrane, and l is the transmembrane distance.

The relationship between diffusion coefficient and concentration has received much attention. Early work by Fujita (1961), which was adapted by Fels and Huang (1971), centered around the free-volume theory. This method is very complex and was found to be difficult to apply to pervaporation. Diffusivity is usually expressed as an exponential function of concentration (Long 1965):

$$D_i = D_{i0} \exp(A_i c_i), \qquad (8\text{--}5)$$

where D_{i0} is the diffusion coefficient at infinite dilution, and A_i is a plasticization coefficient to account for interaction of the particular permeant and polymer. It represents the magnitude of the effect of solvent concentration on solvent mobility in the membrane.

Greenlaw et al. (1977) proposed another relationship between diffusivity and concentration:

$$D_i = D_{i0}(1 + A_i c_i^n) \qquad (8\text{--}6)$$

and found that the simplified expression

$$D_i = kc_i \qquad (8\text{--}7)$$

followed the data from Rogers, Stannett, and Szwarc (1960) for modeling the case of hexane in polyethylene membranes. Rautenbach and Albrecht (1985a, 1985b) also found this simplistic form to be sufficient for basic design calculations. In their analysis of single-component transport, they used a modified form of Greenlaw and coworkers with the exponent n equal to unity.

$$D_i = D_{i0}(1 + A_i c_i). \qquad (8\text{--}8)$$

If the exponential relationship for diffusivity is substituted into Fick's law for diffusion and integrated over the membrane thickness l the permeation equation becomes

$$J_i = \frac{D_{i0}}{A_i l}[\exp(A_i c_{if}) - \exp(A_i c_{ip})], \qquad (8\text{--}9)$$

where c_{if} and c_{ip} represent component concentrations in the membrane at the feed (upstream) and permeate (downstream) sides, respectively. Under vacuum pervaporation or when the permeate side is kept at sufficiently reduced pressure, c_{ip} goes to 0. The equation reduces to

$$J_i = \frac{D_{i0}}{A_i l}[\exp(A_i c_{if}) - 1]. \qquad (8\text{--}10)$$

For a given liquid-polymer system the diffusion coefficient D_{i0}, plasticization coefficient A_i, and membrane thickness l are constant.

The concentration of the permeant in the feed side of the membrane is the variable effecting transport. This interaction between the permeating component and membrane on the feed side produces a swelling phenomenon. The swelling phenomenon increases the membrane thickness. As concentration in this region increases, so does the flux.

The solubility aspects of pervaporative transport are analyzed by examining the basic sorption thermodynamics. If equilibrium conditions are assumed at both the feed and permeate sides of the membrane, then a relationship for equilibrium at the membrane/solution interfaces can be included. A simple expression can be used to relate the concentration to activity using a solubility parameter (Greenlaw et al. 1977; Rogers, Stannett, and Szwarc 1960):

$$c_i = K_{si} a_i. \tag{8-11}$$

In this expression, K_{si} is the sorption coefficient.

An overall permeation equation can be obtained for the pure component case taking into account the feed and permeate streams. The activity of the feed solution is taken as unity and the permeate activity is expressed by the previously presented ratio of downstream pressure to saturation pressure. The relationships for the feed and permeate side conditions can be substituted into Fick's law with the appropriate boundary conditions to obtain an overall permeation equation for the pure component case:

$$J_i = \frac{D_{i0}}{A_i l} \left\{ \exp(A_i K_{si}) - \exp\left[(A_i K_{si}) \left(\frac{p_i''}{p_i^o} \right) \right] \right\}, \tag{8-12}$$

where p_i'' is the downstream pressure and p_i^o is the component saturation pressure. Figure 8-1 represents the concentration and pressure profiles across the membrane.

MULTICOMPONENT PERMEATION

Transport of multicomponent mixtures is difficult to analyze, with most researchers focusing on the binary case. As is the case of pure component permeation, the transport of binary components i and j is dependent on solubility and diffusivity. An interaction occurs between the two components that makes predicting performance based on pure component results difficult. Flux coupling and thermodynamic interaction are two of the recognized phenomena present that affect multicomponent pervaporation separations. Rautenbach and Albrecht (1985a, 1985b, 1989) have reviewed the relevant models on multicomponent transport and have compared them to experimental data for several systems.

The work of Thompson's group has been extensively presented in numerous publications (Greenlaw, Shelden, and Thompson 1977; Greenlaw et al. 1977; Shelden and Thompson 1978, 1984; Knight et al. 1986; Duggal and Thompson 1986). They have endeavored to explain the detailed pervaporative behavior of a variety of mixtures and examined the effects of permeate side and feed side pressure on performance.

Greenlaw, Shelden, and Thompson (1977) and Shelden and Thompson (1978) proposed a simple relationship in which the diffusion coefficients for components i and j are interdependent on both component concentrations:

$$D_i = K_{di}(c_i + B_{ij} c_j), \tag{8-13}$$

$$D_j = K_{dj}(c_j + B_{ji} c_i), \tag{8-14}$$

where K_{di} and K_{dj} are the diffusion constants relating diffusivity to concentration for pure i and j, and the parameters B_{ij} and B_{ji} account for the coupling effects. They proposed a simple relationship between the concentration and activity:

$$c_i = K_{si} a_i, \tag{8-15}$$

$$c_j = K_{sj} a_j. \tag{8-16}$$

The foundation of the model is a diffusive Fickian mechanism for permeation, with constant pressure throughout the membrane equal to the feed side condition, and with sorption equilibrium at both membrane interfaces. Their model contains parameters representing sorption and diffusion coefficients for both components and two flux coupling parameters. They have analyzed the cases for vapor permeate, liquid permeate, and two-phase vapor-liquid permeate (Greenlaw, Shelden, and Thompson 1977; Shelden and Thompson 1978). For the com-

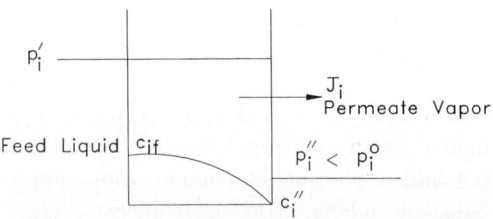

FIGURE 8-1. Pressure and concentration gradients for single-component pervaporation.

monly encountered vapor permeate case, their equation relating the partial pressures of components i and j in the permeate (and hence separation effectiveness) is

$$\frac{p_i''}{p_j''} = \left(\frac{\rho_i \tilde{M}_j}{\rho_j \tilde{M}_i}\right)\left(\frac{K_{di} K_{si}}{K_{dj} K_{sj}}\right)$$

$$\times \frac{\{x_i' - (p_i''/p_i^\circ) \exp[-(\tilde{V}_i/RT)(p' - p_i^\circ)]\}}{\{x_j' - (p_j''/p_j^\circ) \exp[-(\tilde{V}_j/RT)(p' - p_j^\circ)]\}},$$

(8–17)

where the single primes and double primes represent feed side and permeate side conditions, respectively. The concentrations in the feed are represented by the mole fractions x_i', x_j', the saturation pressures are p_i°, p_j°, and the molar volumes are $\tilde{V}_i$, $\tilde{V}_j$. The component properties are density, ρ_i, ρ_j, and molecular weight, $\tilde{M}_i$, $\tilde{M}_j$. The equation can be solved for either p_i'' or p_j'' knowing the total permeate side pressure, from which the permeate composition of both components can be determined.

To develop these expressions the following assumptions were made:

1. The solutions are thermodynamically ideal.
2. The activities of components in the membrane are proportional to their concentration in the membrane.
3. The particular diffusivity-concentration relationship holds.

For the expression presented above the interaction coefficients were set equal to unity.

This relationship holds for some ideal mixtures such as heptane-hexane, but not for nonideal mixtures such as ethanol-water. The group has studied these cases and presented their results (Shelden and Thompson 1984; Knight et al. 1986; Duggal and Thompson 1986). They have developed a numerical method that solves the model equations for thermodynamically nonideal solutions. It also allows the use of a more general relationship for the dependence of component diffusivities and activities on the composition:

$$a_i = f(c_i, c_j), \quad (8\text{--}18)$$

$$a_j = f(c_i, c_j), \quad (8\text{--}19)$$

$$D_i = f(c_i, c_j), \quad (8\text{--}20)$$

$$D_j = f(c_i, c_j). \quad (8\text{--}21)$$

To illustrate their methodology, they use the following relationships (Shelden and Thompson 1984):

$$D_i = D_{i0} + K_{di}(c_i + B_{ij} c_j)^{k_i}, \quad (8\text{--}22)$$

$$D_j = D_{j0} + K_{dj}(c_j + B_{ji} c_i)^{k_j}, \quad (8\text{--}23)$$

where the constants D_{i0}, D_{j0}, K_{di}, K_{dj}, k_i, and k_j depend on individual components and the constants B_{ij}, B_{ji} represent the coupling effects of the interaction of the two components. They have presented the case of the toluene/ethanol (with polyethylene membranes) and water/ethanol with both homogeneous and asymmetric membranes.

Brun et al. (1985) described binary pervaporative transport using an exponential diffusion model that is dependent on individual component concentrations:

$$D_i = D_{i0} \exp(A_{ii} c_i + A_{ij} c_j), \quad (8\text{--}24)$$

$$D_j = D_{j0} \exp(A_{jj} c_j + A_{ji} c_i). \quad (8\text{--}25)$$

This model includes the effects of coupling between permeants. A positive or negative value for the coefficients A_{ij}, A_{ji} means a positive or negative influence of the presence of one component on the rate of transport of the other.

They describe their model as a "six coefficients exponential model" composed of the six D and A coefficients and "well-constrained" parameters, which must be determined empirically. These parameters include the activity coefficients. They have demonstrated their model with benzene/n-heptane and 1,3-butadiene/isobutene mixtures with nitrile-butadiene rubber (NBR) membranes.

Gooding and coworkers (Sferrazza and Gooding 1987; Sferrazza, Escobosa, and Gooding 1988) have analyzed the behavior of binary systems using an approach similar to that of Brun et al. (1985). They use a similar model to describe diffusion:

$$D_i = D_{i0} \exp[A_i(c_i + B_{ij}c_j)], \quad (8\text{--}26)$$

$$D_j = D_{j0} \exp[A_j(c_j + B_{ji}c_i)], \quad (8\text{--}27)$$

where the interaction coefficients B_{ij} and B_{ji} give an indication of the extent of coupling in the membrane. In their model they also use a relationship that is solubility-dependent in the pure component case. They have extensively studied methods for obtaining sorption coefficients needed for the model.

Nguyen (1987) developed a model that describes the permeation behavior of water/tetrahydrofuran/Cuprophan® and water/ethanol/cellulose acetate systems. Similar to the work of Thompson's group, the model describes the effect of downstream pressure on pervaporation behavior. The model includes the evaporation of the permeate from the membrane as another rate-limiting step.

Mulder and Smolders (1984) presented a modified solution-diffusion model that accounts for the coupling of solubility and diffusivity. The model has been applied to the case of ethanol/water/cellulose acetate. Activities of the components in the membrane were determined by polymer/liquid, liquid/liquid interaction parameters, obtained by swelling experiments and from excess free energy of mixing data. Flory-Huggins thermodynamics is used in the calculation of the component activities (Flory 1953).

Mulder, Franken, and Smolders (1985) and Mulder and Smolders (1986) have proposed a preferential sorption versus preferential permeation approach to pervaporative transport. They have found for a wide variety of systems that the component preferentially sorbed by the membrane is also preferentially permeated. Diffusivity is a contributing factor in determining the magnitude of selectivity. They suggest that ideal sorption based on a linear relationship between the concentration of the component in the liquid and that in the membrane is not the actual case. Preferential sorption measurements lead to greater accuracy than the pure component measurements used by others.

NOTATION

General Notation

See the General Notation section at the beginning of this handbook.

Special Notation

a activity, dimensionless
A_{ii} parameter for multicomponent transport ("six coefficients exponential model"), $L^3/$mol or L^3/M
A_{jj} parameter for multicomponent transport ("six coefficients exponential model"), $L^3/$mol or L^3/M
A_{ij} coupling parameter for multicomponent transport ("six coefficients exponential model"), $L^3/$mol or L^3/M
A_{ji} coupling parameter for multicomponent transport ("six coefficients exponential model"), $L^3/$mol or L^3/M
B_{ij} coupling parameter for multicomponent transport, $L^3/$mol or L^3/M
B_{ji} coupling parameter for multicomponent transport, $L^3/$mol or L^3/M
k simplified correlation factor between diffusivity and concentration, $L^5/$mol t or $L^5/M\,t$
k_i exponent in multicomponent transport relationship, dimensionless
k_j exponent in multicomponent transport relationship, dimensionless
K_s sorption coefficient, mol/L^3 or M/L^3
R gas constant, ML^2/t^2T mol

REFERENCES

Binning, R. C., and F. E. James. 1958a. Now separate by membrane permeation. *Petrol. Refiner.* 37:214.

Binning, R. C., and F. E. James. 1958b. Permeation: a new way to separate mixtures. *Oil Gas J.* 56(21):104–105.

Binning, R. C., R. J. Lee, J. F. Jennings, and E. C. Martin. 1961. Separation of liquid mixture by permeation. *Ind. Eng. Chem.* 53:45–50.

Brun, J-P., C. Larchet, R. Melet, and G. Bulvestre.

1985. Modelling of the pervaporation of binary mixtures through moderately swelling, nonreacting membranes. *J. Membr. Sci.* 23:257–283.

Duggal, A., and E. V. Thompson. 1986. Dependence of diffusive permeation rates on upstream and downstream pressures: VI. experimental results for the water/ethanol system. *J. Membr. Sci.* 27:13–30.

Fels, M., and R. Y. M. Huang. 1971. Theoretical interpretation of the effect of mixture composition on separation of liquids in polymers. *J. Macromol. Sci.* 5:89–110.

Flory, P. 1953. *Principles of Polymer Chemistry*. Ithaca, NY: Cornell University Press.

Fujita, H. 1961. Diffusion in polymer diluent systems. *Forschr. Hochpolym. Forsch.* 3:1–47.

Greenlaw, F. W., W. D. Prince, R. A. Shelden, and E. V. Thompson. 1977. Dependence of diffusive permeation rates on upstream and downstream pressures: 1. single component permeant. *J. Membr. Sci.* 2:141–151.

Greenlaw, F. W., R. A. Shelden, and E. V. Thompson. 1977. Dependence of diffusive permeation rates on upstream and downstream pressures: II. two component permeant. *J. Membr. Sci.* 2:333–348.

Knight, K. F., A. Duggal, R. A. Shelden, and E. V. Thompson. 1986. Dependence of diffusive permeation rates on upstream and downstream pressures: V. experimental results for the hexane/heptane (ideal) and toluene/ethanol (non-ideal) systems. *J. Membr. Sci.* 26:31–50.

Long, R. B. 1965. Liquid permeation through plastic films. *Ind. Eng. Chem. Fundam.* 4:445–451.

Mulder, M. H. V., T. Franken, and C. A. Smolders. 1985. Preferential sorption versus preferential permeability in pervaporation. *J. Membr. Sci.* 22:155–173.

Mulder, M. H. V., and C. A. Smolders. 1984. On the mechanism of separation of ethanol/water mixtures by pervaporation: 1. calculation of concentration profiles. *J. Membr. Sci.* 17:289–307.

Mulder, M. H. V., and C. A. Smolders. 1986. Pervaporation solubility aspects of the solution-diffusion model. *Sep. Purif. Methods* 15:1–19.

Nguyen, T. Q. 1987. Modelling of the influence of downstream pressure for highly selective pervaporation. *J. Membr. Sci.* 34:165–183.

Rautenbach, R., and R. Albrecht. 1985a. The separation potential of pervaporation: part 1. discussion of transport equations and comparison with reverse osmosis. *J. Membr. Sci.* 25:1–23.

Rautenbach, R., and R. Albrecht. 1985b. The separation potential of pervaporation: part 2. process design and economics. *J. Membr. Sci.* 25:25–54.

Rautenbach, R., and R. Albrecht. 1989. *Membrane Processes*. New York: John Wiley & Sons.

Rogers, C. E., V. Stannett, and M. Szwarc. 1960. The sorption, diffusion, and permeation of organic vapors in polyethylene. *J. Polymer Sci.* 45:61–82.

Sferrazza, R. A., R. Escobosa, and C. H. Gooding. 1988. Estimation of parameters in a sorption-diffusion model of pervaporation. *J. Membr. Sci.* 35:125–136.

Sferrazza, R. A., and C. H. Gooding. 1987. Determination of sorption selectivity by liquid chromatography. In *Proceedings of the Second International Conference on Pervaporation Processes in the Chemical Industry*, ed. R. A. Bakish, pp. 186–199. Englewood, NJ: Bakish Materials Corporation.

Shelden, R. A., and E. V. Thompson. 1978. Dependence of diffusive permeation rates on upstream and downstream pressures: III. membrane selectivity and implications for separations processes. *J. Membr. Sci.* 4:115–127.

Shelden, R. A., and E. V. Thompson. 1984. Dependence of diffusive permeation rates on upstream and downstream pressures: IV. computer simulation of non-ideal systems. *J. Membr. Sci.* 19:39–49.

9
Design

Hubert L. Fleming
Zenon Environmental, Inc.

C. Stewart Slater
Manhattan College

MODES OF OPERATION
GENERAL DESIGN
  CONSIDERATIONS
MEMBRANES
MEMBRANE MODULES

FEED PRETREATMENT
MEMBRANE CLEANING
NOTATION
REFERENCES

MODES OF OPERATION

In its simplest form, pervaporation is a separation process in which a liquid mixture is contacted with one side of a membrane and one of the components permeates more quickly than the other(s) to the permeate side of the membrane. At the permeate side, the component is evaporated and collected downstream. Generally, a mechanical vacuum is applied to drive this process. This so-called vacuum-driven pervaporation is illustrated in Figure 9–1, with a vacuum pump shown to generate the necessary partial pressure driving force.

A difference in partial vapor pressures (i.e., chemical potentials) may also be realized by a temperature gradient. Sometimes referred to as "thermopervaporation," this fact has been used (Aptel et al. 1976; Aptel and Neel 1986) to design pervaporation systems. As shown in Figure 9–2, temperature gradient-driven pervaporation requires a preheater for increasing the temperature of the feed mixture substantially above that of the permeate. A condenser is also necessary downstream to remove the condensing permeates from the inert sweep gas. Although Figure 9–2 illustrates the use of an outside condenser, several other geometries have been utilized to achieve the necessary condensation. Because of the need to handle large permeate volumes, coupled with low driving forces, this mode of pervaporation has not been

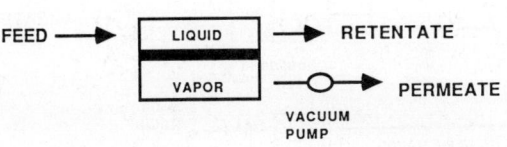

FIGURE 9–1. Vacuum-driven pervaporation.

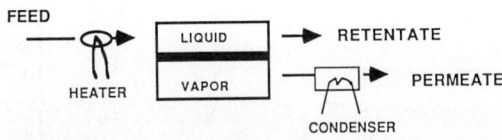

FIGURE 9–2. Temperature gradient-driven pervaporation.

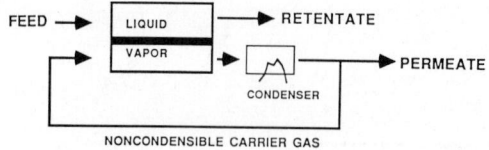

FIGURE 9–3. Carrier gas pervaporation.

demonstrated to be economical in any application.

Rather than impose a vacuum on the permeate side, pervaporation can also be driven simply by the partial vapor pressure generated between a liquid feed and an inert sweep gas on the permeate side. Like temperature gradient-driven pervaporation, this carrier gas mode (Figure 9–3) does not provide sufficient driving force to exhibit fluxes that would be considered economically viable.

In a variation of carrier-moderated pervaporation, an immiscible liquid carrier can be repeatedly vaporized to carry permeate. As shown in Figure 9–4, this process scheme is generally employed when condensable permeates are present. The inert vapor carries the permeating mixture to a condenser. Multiple phases result, with the carrier phase being vaporized and recycled to the pervaporation membrane module. The permeate is collected as a concentrate from the multiphase separator. Again, no commercial processes utilize this scheme. Not only is driving force low, but the entire recovery scheme becomes complex, expensive, and system dependent.

One less complex variation utilizes fractional condensation of the permeate. As shown in Figure 9–5, a permeate-laden vapor is condensed in stages. Here, a vacuum is generated by the condensing permeate. There is no need for a carrier gas or downstream separation. This approach might be considered when multiple products exhibiting different temperatures of condensation are present and can be recovered.

In Figure 9–6 a final mode of operation is considered. The driving force is a vacuum, which is generated by a downstream condenser. The difference is that the condensate is multiphase. Here, a multiphase separator is employed to yield a phase that can be recycled to the feed, and a permeate phase, which is a concentrate. The primary value of this integrated process is that the effect of the multiphase separator dramatically multiplies the overall separation factor for the process.

One final variation worthy of mention is perstraction. As defined in Figure 9–7, perstraction, or permselective extraction, is the membrane analogue of liquid-liquid extraction. Rather than employ an inert gas or vacuum on the permeate side to drive the process, a liquid is imposed (see Chapter 46).

GENERAL DESIGN CONSIDERATIONS

In most commercial processes, a vacuum pump is used to keep the permeate pressure low, supplying a sharp partial pressure driving force.

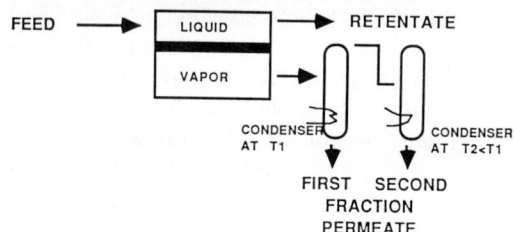

FIGURE 9–5. Pervaporation with fractional permeate condensation.

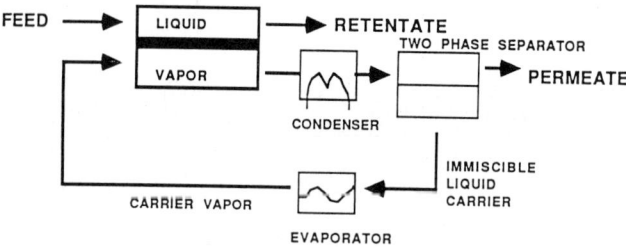

FIGURE 9–4. Pervaporation with condensable, permeate immiscible carrier.

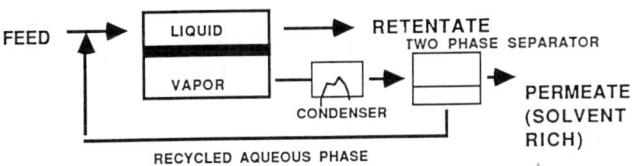

FIGURE 9–6. Pervaporation with two-phase permeate and partial recycle.

The permeate is then condensed, separated, and either recovered or recycled. The process is perpetually driven by condensation of the permeate, creating a significant vacuum and resulting in lower temperatures on the permeate side of the membrane. A typical system is shown in Figure 9–8. The elimination of inert gas sweep, with the associated vacuum capacity and downstream separation inefficiency, has been a major step forward in development of economic pervaporation processes. Although a vacuum pump is still required for startup, it is generally not used in steady-state operation, and it is of very small capacity, low capital cost, and low energy consumption.

Compared with conventional membrane processes such as ultrafiltration or reverse osmosis, fluxes in pervaporation are generally low (<20 kg/m² h). However, selectivities can be extremely high, often exceeding 1000. The result is an unusual membrane process in that concentration polarization is not typical (Figure 9–9). Although reports exist of significant concentration polarization (e.g., Reddy and Reinecke 1988), very few pervaporation processes have demonstrated any effect at all.

Temperature polarization is typical of pervaporation (Figure 9–9). By definition, pervaporation requires the volatilization of a portion of the liquid feed. The enthalpy of vaporization must be consumed by the feed. For a permeate such as water, with $\Delta \hat{H}_v = 1100$ Btu/lb, the resulting temperature drop can be substantial. As shown in Figure 9–10, a large thermal gradient is generally established across the membrane, with continual heat loss to the permeate. As will be shown later, flux is strongly proportional to feed temperature. Typical processes employ interstage heaters to reheat the feed between membrane modules to compensate for this effect.

The temperature drop ΔT in the feed mixture of two components, A and B, between the inlet and the outlet of the pervaporator under adiabatic conditions can be calculated from the following equation (Aptel and Neel 1986):

$$\Delta T = \frac{\overline{J}[\Delta \hat{H}_{v,B} + (\Delta \hat{H}_{v,A} - \Delta \hat{H}_{v,B})\overline{w}_{Ap}]}{Q_f[C_B + (C_A - C_B)w_{Af}]}, \quad (9\text{–}1)$$

where $\overline{J}$ is the average transmembrane flux; $\Delta \hat{H}_{v,A}$ and $\Delta \hat{H}_{v,B}$ are the latent heats of vaporization per unit mass for components A and B, respectively; $\overline{w}_{Ap}$ is the average mass fraction of component A in the permeate; w_{Af} is the inlet mass fraction of component A in the feed; Q_f is the feed flow rate per unit membrane area at the inlet of the pervaporator; and C_A and C_B are the heat capacities for components A and B, respectively. In general, the heats of vaporization are much greater than the heat capacities (about 500 times). The difference between the heats of vaporization is small compared to the individual heats of vaporization, and the difference between the heat capacities is also small compared to the individual heat capacities. Thus, Eq. (9–1) can be approximated as follows (Aptel and Neel 1986):

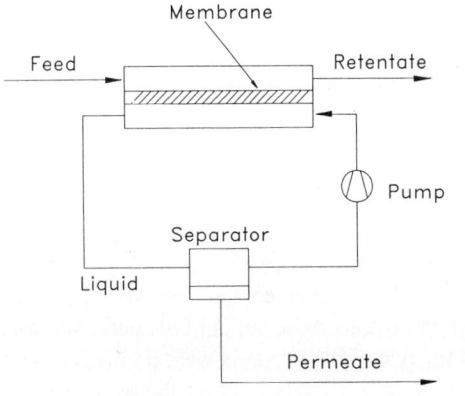

FIGURE 9–7. Perstraction.

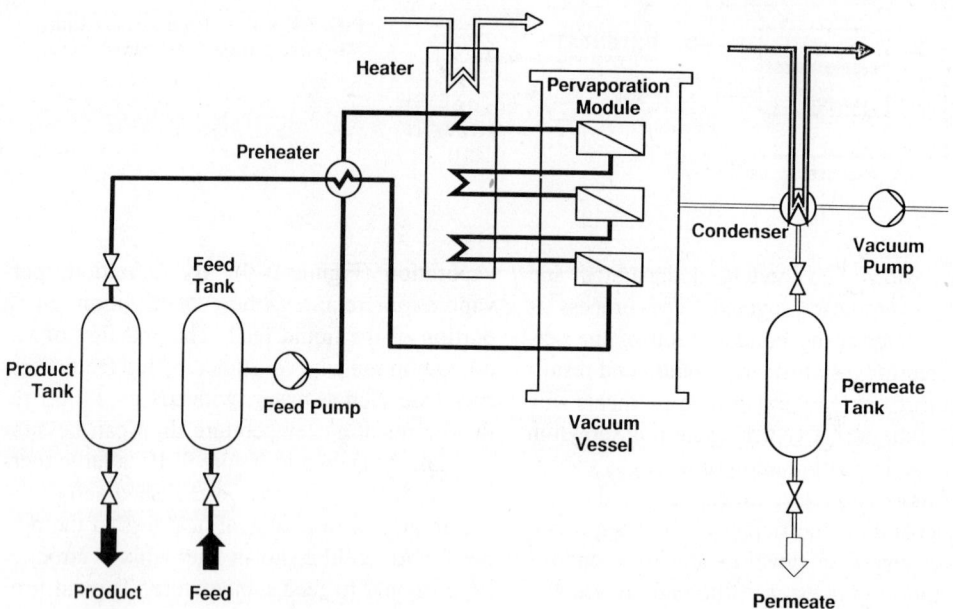

FIGURE 9-8. Pervaporation system schematic.

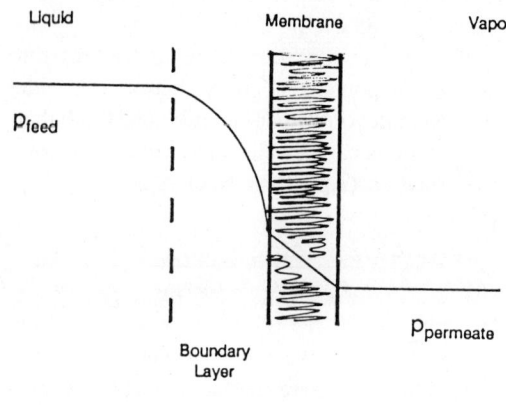

Concentration polarization

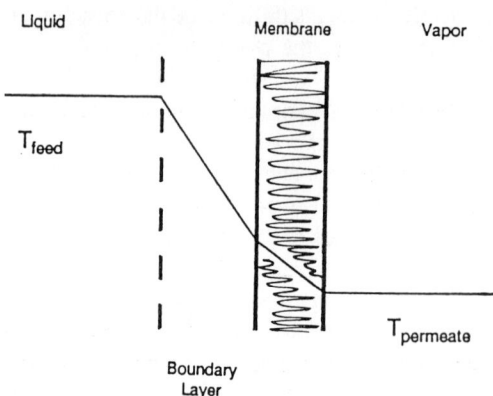

Temperature polarization

FIGURE 9-9. Concentration and temperature polarization.

$$\Delta T \cong \frac{\bar{J}\Delta \hat{H}_{v.B}}{Q_f C_B}. \quad (9-2)$$

Because mass flux, and sometimes selectivity, is enhanced by temperature gradients, the feed temperature is usually elevated as well, with heat recovery from the product liquid. Thus, more than 90% of the energy required for preheating the feed stream is recovered. Further, the necessary energy is only sensible heat, not latent, and is of low quality, typically 1 to 4 bars of available steam.

For a constant transmembrane flux J (also constant enrichment factor, $\beta = w_{Ap}/w_{Af}$) within the range between the inlet concentration of component A in the feed, w_{Af}, and the outlet concentration of component A in the retentate, w_{Ar}, the governing equations for the pervaporation process to separate a two-component feed are (Aptel and Neel 1986)

$$J = Q_f - Q_r, \quad (9-3)$$

$$J\bar{w}_{Ap} = Q_f w_{Af} - Q_r w_{Ar}, \quad (9-4)$$

where Q_r is the retentate flow rate per unit membrane area at the outlet of the pervaporator, and the rest of the symbols were defined earlier. Equations (9-3) and (9-4) are the overall material balances for the total system and component A, respectively.

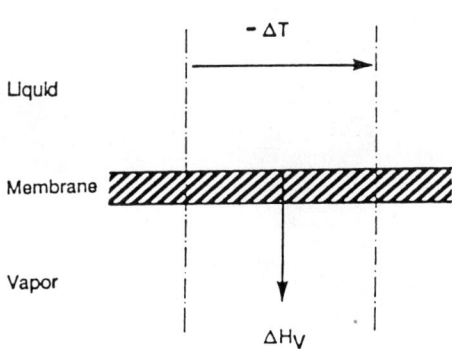

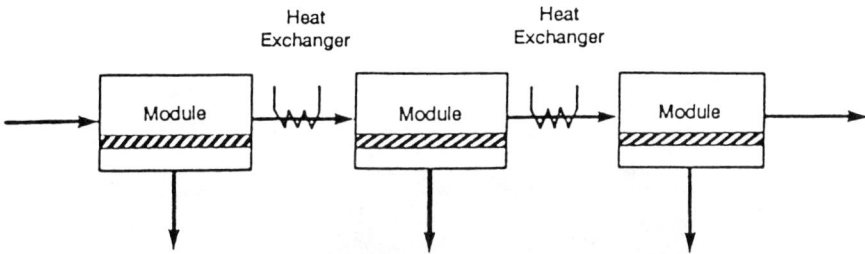

FIGURE 9–10.   Effect of evaporative cooling.

Generally, w_{Ar} is specified as the separation goal for the desired removal of component A from the feed mixture containing an initial, known concentration of w_{Af}. For a given membrane used in the pervaporation process, the transmembrane flux J, corresponding to the permeate composition $\bar{w}_{Ap}$ (or the enhancement factor β), for the concentration range between w_{Af} and w_{Ar}, can be determined experimentally or may be estimated from Eq. (8–12) or Eq. (8–4) with Eqs. (8–13) through (8–17) of Chapter 8. That is, w_{Af}, w_{Ar}, $\bar{w}_{Ap}$, and J are the known variables in Eqs. (9–3) and (9–4). Thus, the two unknowns, Q_f and Q_r, can then be calculated from these two equations. The Q_f term represents the treatment capacity, i.e., the flow rate of feed treated per unit membrane area. The feed flow rate and Q_f give the membrane area required for the process. The Q_r term, which is the flow rate of the retentate produced per unit membrane area, stands for the production capacity of this process.

In cases for which the transmembrane flux varies along the liquid flow path, governing equations for the pervaporation process become complicated and often have to be solved numerically (Rautenbach and Albrecht 1989). The flux variation can be caused by the friction loss of permeate flow (resulting in the increase of permeate pressure).

The key to successful pervaporation lies in the membrane. Selectivity, and flux to a large extent, is essentially controlled by this permselective barrier between the feed liquid and the gaseous permeate. Transport is generally described as a series of three events: (1) preferential sorption of mixture components, (2) diffusion through the membrane, and (3) desorption on the permeate side. Sorption is controlled by specific polymer chemistry and by its interaction with the liquid mixture. Hydrophilic membranes tend to sorb hydrophilic compounds, such as water, for example. Vaporization on the permeate side is generally consid-

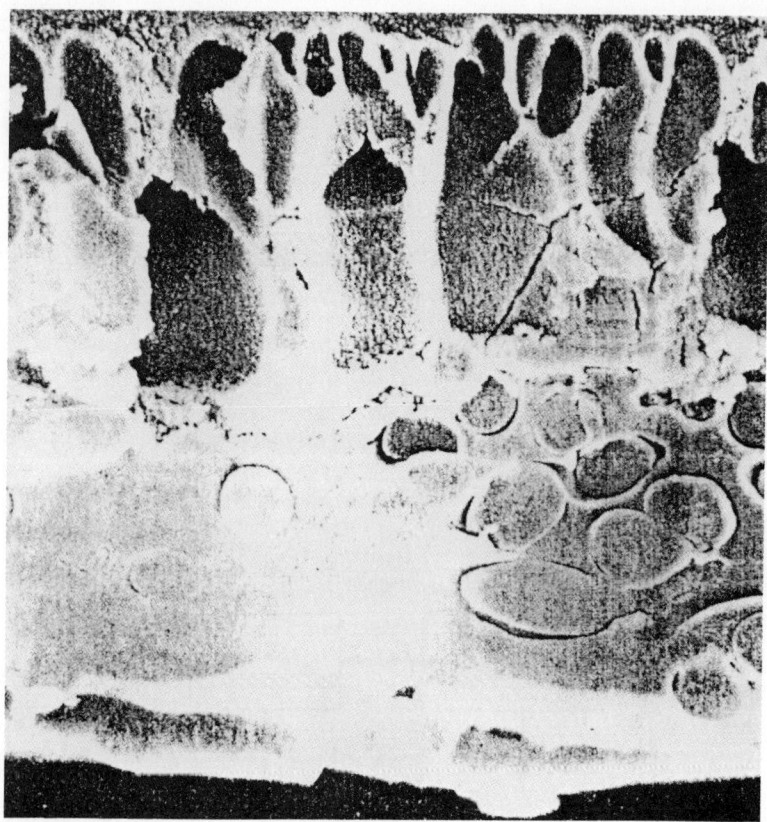

FIGURE 9–11. Photomicrograph of GFT polyvinylalcohol composite membrane (500× magnification). (Courtesy of GFT.)

ered to be a fast, nonselective step. An exception is the isolated instances in which the relative vapor pressures in the membrane, on the permeate side, and in the feed mixture are similar. An example of such a situation is the removal of trace organics from groundwaters (Aptel et al. 1976).

MEMBRANES

The choice of membrane material is critical. The key to commercialization of pervaporation has been the development of asymmetric composite membranes, each layer of which fulfills a specific requirement. The primary family of membranes for water permeation (e.g., ethanol dehydration, solvent recovery) utilizes a supporting layer of nonwoven porous polyester, on which is cast either a polyacrylonitrile (PAN) or polysulfone (PS) ultrafiltration membrane, and finally a 0.1-μm-thick layer of cross-linked polyvinylalcohol (PVA). Figure 9–11 is a photomicrograph of one of the commercial GFT PVA membranes. The PVA provides ultimate separation, with the entire structure being necessary for chemical and thermal stability, and to provide optimum transport properties. Other separations are generally accomplished using the same two sublayers, varying the top layer to modify selectivity. Polydimethylsiloxane (PDMS), for example, cast upon PAN is useful for retarding polar compounds versus nonpolar, such as processes for dealcoholization of liquors. Films of cellulose esters are one example of membrane material that exhibits separation of organics, such as olefin/paraffin and aromatic/paraffin, in petrochemical processing (Aptel et al. 1976).

MEMBRANE MODULES

The cast membranes may be packaged in modules in any of several geometries. For dehydration, GFT utilizes a plate-and-frame module,

dinal temperature drops and an inefficient use of downstream surface area.

FEED PRETREATMENT

As in all membrane processes, the condition of the feed has considerable influence on the performance of the membrane unit. The effect is generally much less for pervaporation relative to the other processes because of the lack of significant concentration polarization. The mass flux through the membrane is small compared to the mass flux parallel to the membrane surface.

In most pervaporation designs the membrane module is operated at linear velocities that result in turbulent flow. The effect of noninteracting solids on the membrane is not severe. Fouling can occur at high solids loadings or in systems where solids are being formed in the operation. An example of the latter case is in dehydration of organic solvents containing salts. As the solvent is dewatered, the solubility for the soluble salts decreases to the point where salts are continuously precipitated on the membrane. Permeate flux is then reduced because of the formation of this resistance layer. A second problem is the formation of solid clogs in the physical constrictions in the module or associated piping. Generally, some form of in-line filtration is employed to minimize solids buildup in the modules.

MEMBRANE CLEANING

Based on previous discussion, the need for membrane cleaning is generally less than for analogous porous membranes. However, membranes do foul and must be treated. Cleaning programs are grouped as (1) clean-in-place and (2) off line. Most pervaporation plants are designed to be cleaned in place. In the case of dehydration membranes, a typical cleaning cycle would utilize one to three bed volumes of clean solvent (isopropyl alcohol or the neat process stream) at an elevated temperature (50 to 80°C) being circulated through the system. This cycle would be implemented once every one to four weeks in most commercial plants.

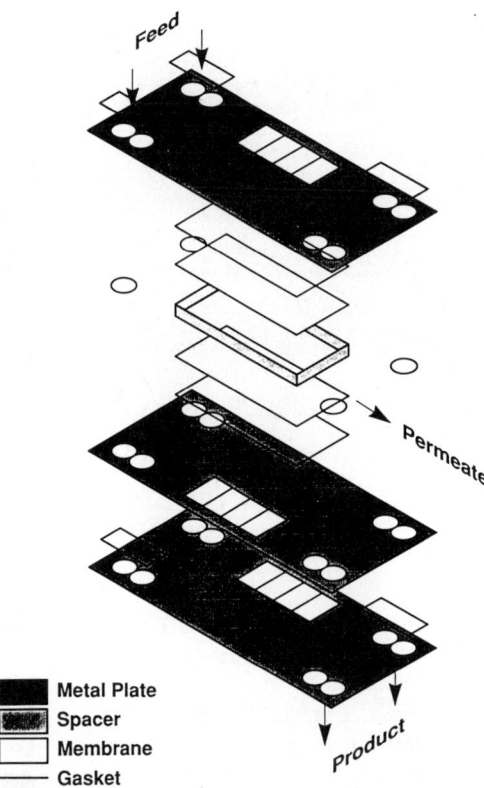

FIGURE 9–12. Pervaporation plate module schematic.

similar to the design used in heat exchangers. A typical module schematic is given in Figure 9–12, and a photograph of a GFT double-stack 50-m^2 module is shown in Figure 9–13. Although other geometries may be employed, the plate-and-frame design has become the module of choice in dehydration processes for two basic reasons. First, this design allows the use of gasketing materials that are resistant to the liquid organic solvents typical in dehydration. Carbon or EPDM (ethylene propylene dimer) compression gaskets are typically required. A spiral-wound unit, for example, is difficult to develop because of the chemical susceptibility of the adhesives required. Second, heat transfer is a prime consideration in optimum system design for dehydration applications. The plate-and-frame module allows high-temperature operation, with efficient interstage heating between stages. Hollow fibers, for example, may have problems with longitu-

FIGURE 9–13. Commercial plate-and-frame pervaporation module. (Courtesy of GFT.)

In the case of organic permeating membranes, the system is generally heat traced and/or contains steam nipples. Steam at a moderate pressure is injected for 2- to 8-h cycles during a cleaning cycle. Deionized water is then circulated to recondition the membrane to its former state.

With few exceptions, the initial performance of commercial pervaporation membranes can be fully recovered. This differs from most other membrane processes where irreversible fouling is significant. Over a period of years pervaporation membranes can degrade, usually by corrosive attack of chemicals in contact with the permselective layer or with the supporting matrix. In general terms, pervaporation membranes exhibit useful lives of 2 to 4 yr, with minimal degradation of flux and selectivity.

NOTATION

General Notation

See the General Notation section at the beginning of this handbook.

Special Notation

C_A heat capacity for component A, L^2/t^2T or E/MT

C_B heat capacity for component B, L^2/t^2T or E/MT

$\Delta\hat{H}_v$ latent heat of vaporization per unit mass, L^2/t^2 or E/M

$\Delta\hat{H}_{v,A}$ latent heat of vaporization per unit mass for component A, L^2/t^2 or E/M

$\Delta\hat{H}_{v,B}$ latent heat of vaporization per unit mass for component B, L^2/t^2 or E/M

J transmembrane flux, M/L^2t

$\bar{J}$ average transmembrane flux, M/L^2t

Q_f feed flow rate per unit membrane area at the inlet of the pervaporator, M/L^2t

Q_r retentate flow rate per unit membrane area at the outlet of the pervaporator, M/L^2t

ΔT temperature drop, T

w_{Af} inlet mass fraction of component A in the feed, dimensionless

w_{Ar} outlet mass fraction of component A in the retentate, dimensionless

w_{Ap} mass fraction of component A in the permeate, dimensionless

$\bar{w}_{Ap}$ average mass fraction of component A in the permeate, dimensionless

Greek Letter

β enrichment factor (w_{Ap}/w_{Af}), dimensionless

REFERENCES

Aptel, P., N. Challard, J. Cuny, and J. Neel. 1976. Application of the pervaporation process to separate azeotropic mixtures. *J. Membr. Sci.* 1:271–287.

Aptel, P., and J. Neel. 1986. Pervaporation. In *Synthetic Membranes: Science, Engineering, and Application,* ed. P. M. Bungay, H. K. Lonsdale, and M. N. dePinho, pp. 403–436. Dordrecht, Holland: D. Reidel Publishing Company.

Rautenbach, R., and R. Albrecht. 1989. *Membrane Processes,* Chap. 12. New York: John Wiley & Sons.

Reddy, D., and C. E. Reineke. 1988. Dehydration with perfluorosulfonic acid ionomer membranes. *AIChE Symp. Ser.* 84(261):84–92.

10

Applications and Economics

Hubert L. Fleming
Zenon Environmental, Inc.

C. Stewart Slater
Manhattan College

INTRODUCTION TO
 PERVAPORATION PROCESSES
WATER REMOVAL FROM LIQUID
 ORGANICS
  Case 1: Ethanol Dehydration
  Case 2: Dehydration of Organics

REMOVAL OF ORGANICS FROM
 WATER
ORGANIC/ORGANIC SEPARATIONS
VAPOR PERMEATION
CONCLUSIONS
NOTATION
REFERENCES

INTRODUCTION TO PERVAPORATION PROCESSES

Over the past few years, the number and variety of industrial pervaporation plants have dramatically increased. At least 20 to 50 plants of a minimum of 5000 L/day product capacity are in operation, with many more in development and pilot phases. In Europe and Asia, the primary driving forces have been (1) reduced energy costs, (2) low overall system capital costs, and (3) superior separations possible, with no limitations imposed by thermodynamic azeotropes relative to azeotropic distillation. In North America, the driving forces have been somewhat different: (1) pollution-free, closed-loop operation, minimum wastewater, and no entrainers, and (2) small, compact units with low capital costs for retrofitting existing plants to increase existing bottlenecked capacity versus distillation and adsorption with molecular sieves. Although the level of energy consumption is considerably less in pervaporation than other competing processes, it is a much less important factor in the United States than is pollution abatement in the selection of pervaporation or integrated pervaporation.

Pervaporation processes can be grouped into the following general categories:

1. Water removal from liquid organics
2. Removal of organics from water
3. Organic/organic separations
4. Vapor permeation.

Although there is considerable activity and interest in all four areas, dehydration processes of the first type are more abundant and account for the majority of plants built to date. To illustrate all categories, two examples of the first will be discussed in depth followed by some discussion of the last three areas.

WATER REMOVAL FROM LIQUID ORGANICS

Case 1: Ethanol Dehydration

Dehydration of fermentation products directly, or following primary distillation, has become the classic example of membrane pervaporation. Because mass transport through the membrane determines the composition of the permeate and, hence, the selectivity, vapor-liquid equilibria with their associated azeotropic effect are irrelevant. Membrane pervaporation is especially effective for separations that are difficult to accomplish via processes governed by thermodynamic equilibrium, such as distillation.

In Table 10–1 many of the more recent references to dehydration of ethanol are given. A number of hydrophilic and ion-exchange polymeric materials have been found that exhibit favorable selectivities and fluxes for water. The predominant commercial membranes are generally those from the polyvinylalcohol (PVA) family in asymmetric composites. Ion-exchange membranes have also been used (Cabasso and Liu 1985).

In dehydration, as in all other pervaporation processes, the driving force for transport is chemical potential. As shown in Figure 10–1 and defined in Eq. (10–1), $\Delta \mu_i$ in the driving force $\Delta \mu_i / l$ can be approximated by Eq. (10–2):

$$\Delta \mu_i = RT \left[\ln(\gamma_i' \, x_i') - \ln\left(\gamma_i'' \frac{p''}{p_i^\circ}\right) \right] \quad (10\text{–}1)$$

$$\cong RT(\ln p_i' - \ln p_i''), \quad (10\text{–}2)$$

as a difference (in terms of natural logarithm) in partial vapor pressures of the permeating component in the liquid feed and the gaseous permeate. Or, expressing the driving force in terms of concentration as in Fick's first law [Eq. (8–4)] and assuming the simple concentration-activity relationship of Eq. (8–11), the driving force can be approximated by Eq. (10–3) as a difference in the partial vapor pressures:

$$\Delta c_i = K_{si}(a_i' - a_i'') = \left(\frac{K_{si}}{p_i^\circ}\right)(p_i' - p_i''). \quad (10\text{–}3)$$

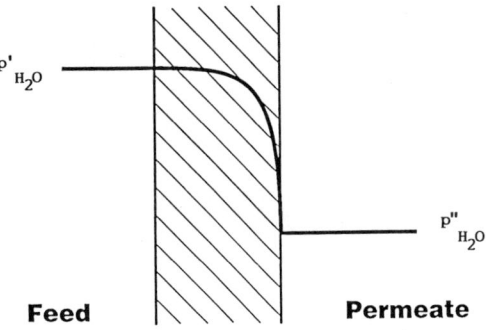

FIGURE 10–1. Driving force in dehydration.

As a practical rule of thumb in dehydration, this difference in partial vapor pressures needs to be at least an order of magnitude in order to exhibit enough flux for an economical process. From the viewpoint of process design, the implication is that this driving force may be realized by (1) increasing feed temperature, (2) decreasing permeate pressure, or (3) decreasing the condenser temperature on the permeate.

Most commercial processes are operated with feed temperatures between 50 and 100°C and permeate pressures of 5 to 20 mbars. The latter is generally achieved with condenser temperatures ranging from –20 up to 30°C. These are easily accessible through employing cooling water or brine chillers. As a rule of thumb, the feed temperature should be as high as possible, up to the normal boiling point. Modified processes are being developed that will run pressurized feeds above the normal boiling point, and run mixed phase feeds (partially vaporized) as well. The condenser temperature will depend on the relative volatility of the permeating mixture. With pure water, the condenser temperature is generally chosen as 3 to 10°C.

A typical ethanol dehydration process is shown in Figure 10–2. The feed is run in a one-pass mode through a preheater, into the membrane modules. The permeate, composed of water with a small amount of ethanol, is condensed and either recycled or disposed. A vacuum pump is situated downstream of the condenser for any noncondensables and for starting the system. The condenser continues to

TABLE 10-1. Selected Water-Permeable Ethanol/Water Pervaporation Membranes.

Membrane Material	Ethanol Feed Concentration (wt.%)	Temperature (°C)	Permeate Pressure (kPa)	Selectivity (α)	Flux (kg/m²-h for H_2O)	Reference
Polyvinylalcohol (GFT)	92 to 100	90 to 100	—	—	0 to 0.9	Sander and Soukup 1988
Polyvinylalcohol (GFT)	0 to 100	60	2.0	High	0 to 4	Wesslein et al. 1988
Polyvinylalcohol (GFT)	60 to 100	75 to 100	0.02–5	50 to 2000	0 to 2	Fleming 1989
Polyvinylalcohol	0 to 100	60	0.01	High	0 to 2.4	Hauser et al. 1987
Polyvinylalcohol	0 to 100	25	0–0.4	5 to 12	0.1 to 0.5	Seok 1987
Cellulose acetate	5 to 95	20	0.01	1 to 3.6	0.3 to 1.2	Changlou 1987
Cellulose triacetate	81 to 95	25	—	2400 to 5900	0.005 to 0.1	Reineke et al. 1987
Carboxymethylcellulose	15 to 95	20 to 50	0.01	3 to 6	—	Changlou 1987
Polysulfone (asymmetric)	0 to 90	40	<0.1	1 to 20	0 to 10	Hirotsu 1988
Acrylic acid-acrylamide GPC	20 to 100	70	—	<1 to 2000	0.5 to 20	Karakane et al. 1988
Polyacrylic acid-polycation	80	70	—	—	1.8	Ellinghorst et al. 1987
Polyvinylfluoride/acrylic acid	0 to 95	70	3.0	—	0 to 6	Niemoller 1988
Polyvinylidenefluoride-N-vinylimidazole						
Nafion™	30 to 98	40	<0.1	Low	<0.5	Kujawski, Nguyen, and Neel 1988

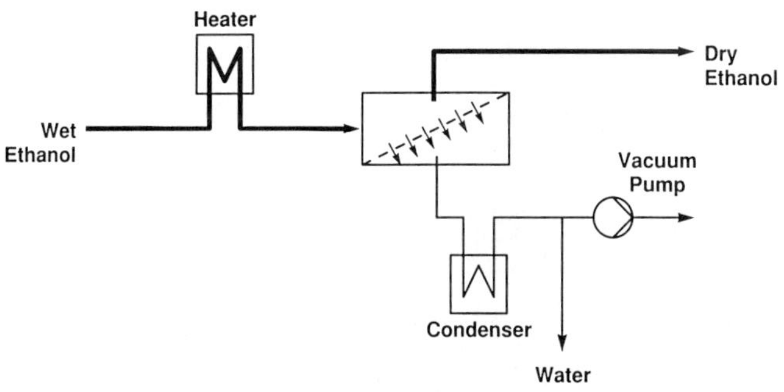

FIGURE 10-2. Ethanol dehydration.

pull vacuum and maintains steady-state operation.

In its applied form (Figure 10–3), multiple-membrane modules are employed in series. One-pass operation is chosen, with the membrane capacity chosen to accomplish the desired degree of dehydration. The feed is reheated between stages, and the membrane modules are generally housed in one or more large vacuum vessels. The permeate is condensed at the bottom of the vacuum vessels and removed. Any degree of dehydration may be chosen in this process, down to <500 ppm water. The feed may contain any degree of water saturation, but generally varies between 5 to 20 wt.% water.

In Figure 7–7 of Chapter 7 permeate compositions for a typical PVA composite pervaporation membrane are plotted over feed composition of ethanol-water mixtures, together with the respective vapor-liquid equilibrium curve at ambient conditions. Note that pervaporation exhibits its highest efficiency in a concentration range of the mixture where distillation is least effective, namely, at high ethanol concentrations. Conversely, at high water levels, distillation is more thermodynamically efficient.

As a result, hybrid processes that integrate pervaporation with distillation have been shown to be the most effective solution in both operating and capital costs. Such a system is shown in Figure 7–2 of Chapter 7. Permeate from the pervaporation membrane unit (containing 5 to 50 wt.% EtOH) is typically recycled back to the distillation columns. The result is an integrated system capable of continuously producing ethanol in multiple purities up to 99.95%, with low energy consumption, almost no wastewater generation, and no chemical additives as entrainers required. No additional heat input is necessary because the still overheads are fed directly to the membranes. The spent mash preheats the recirculating permeate and feed. Because of the continuous recycling, ethanol losses are close to zero (averaging 4% for conventional azeotropic distillation), with virtually no environmental pollution. The cost of the entrainer is also not to be minimized. A 140,000 L/day ethanol plant may use 120,000 to 150,000 L of benzene. At the current price of $0.95/L delivered, a net savings of approximately $120,000 annually is seen on this moderately sized plant (Rapin 1988).

A typical design for an integrated multistage distillation/pervaporation system for ethanol production is shown in Figure 10–4. Rectification and stripping columns follow primary distillation, with permeate recycle fed to the stripping column. Only low-pressure steam (<90 psia, 105 to 110°C) on the order of 0.5 kg per kilogram of ethanol produced is necessary to supply energy to the membrane stack. Not only is this quantity a small fraction of that required for azeotropic distillation or regenerative adsorption (11,200 Btu/gal versus 2000 Btu/gal for membranes), but it is low-quality waste heat, usually available in abundance at little or no charge.

Depending on product requirements and plant size, it is common to optimize the pervaporation system by applying high-flux, lower selectivity membranes early in the train, followed by high-selectivity, lower flux membrane modules at the end. In this manner, total membrane area (e.g., plant capital cost) can be

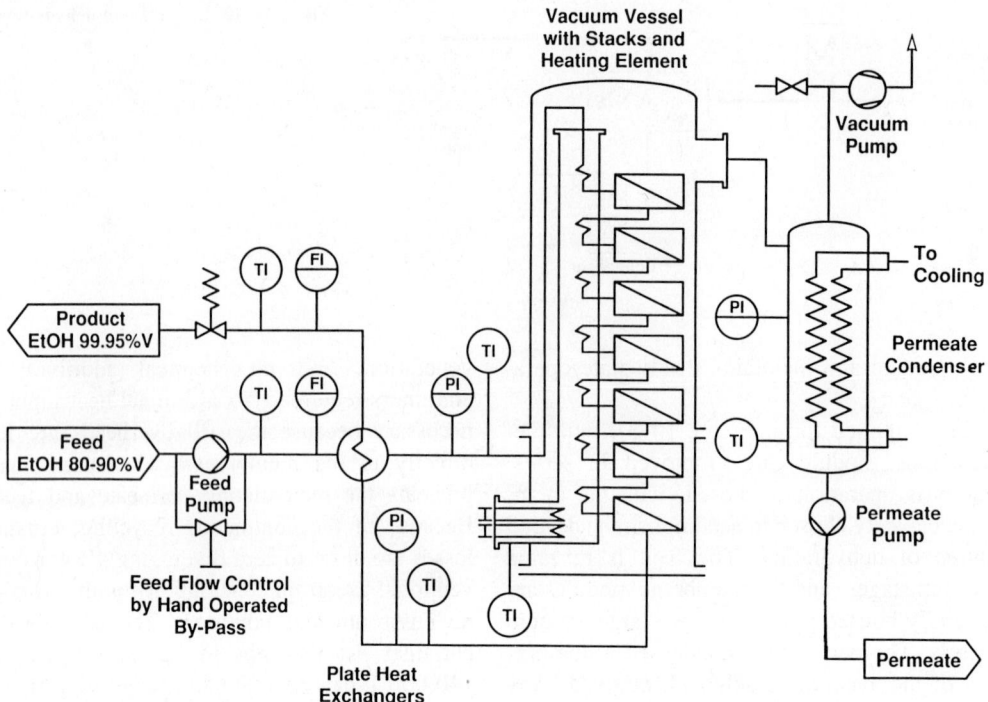

FIGURE 10–3. Pervaporation plant flowsheet for ethanol recovery.

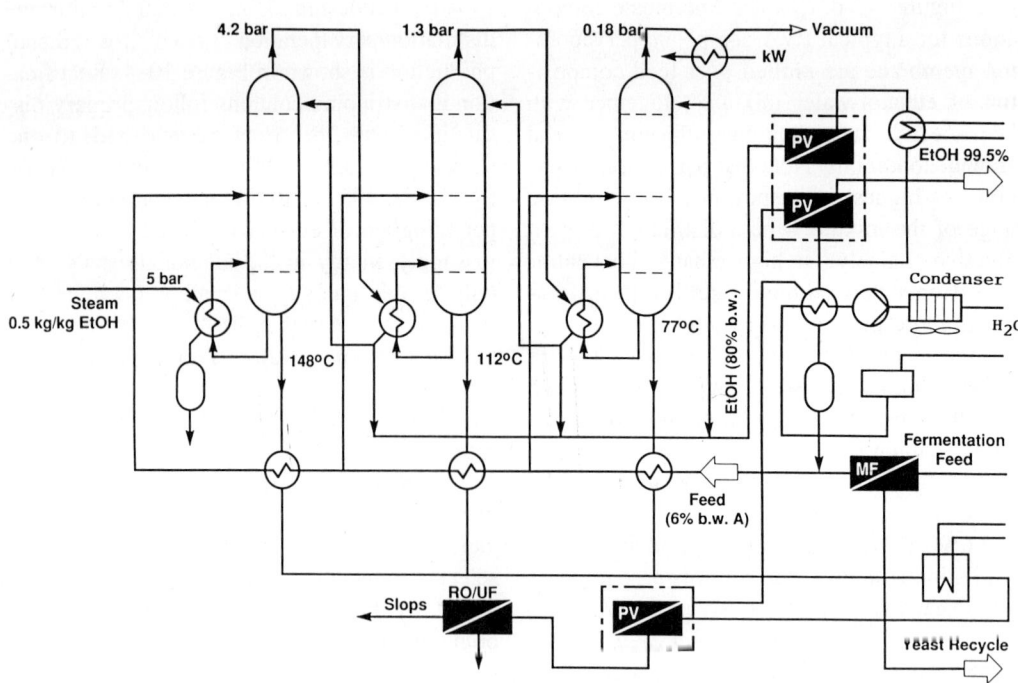

FIGURE 10–4. Integrated distillation/pervaporation plant for ethanol production.

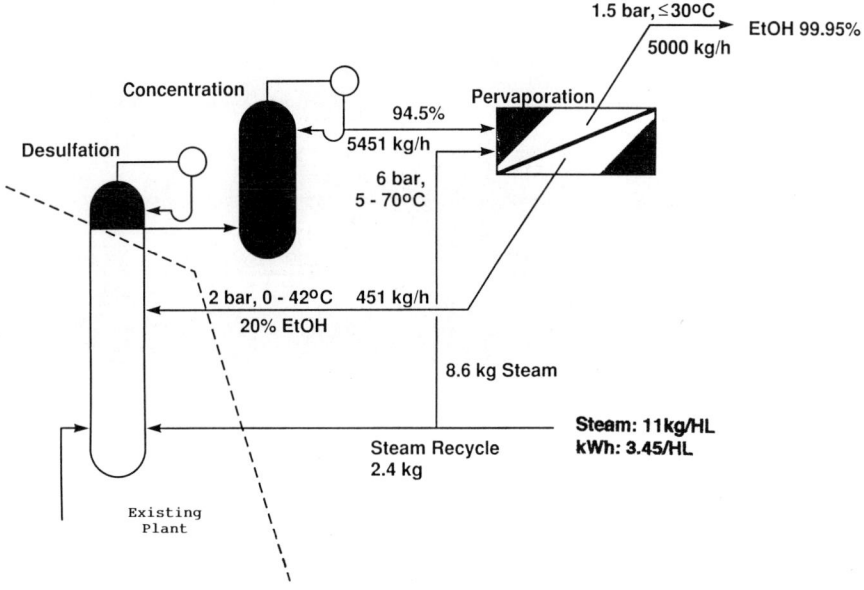

FIGURE 10–5. Betheniville, France, ethanol pervaporation system (from Rapin 1988).

significantly reduced. Because entire families of commercial pervaporation membranes are now available, this optimization process has been a major advancement in the technology.

For example, at the Betheniville, France, ethanol plant built by GFT and shown in Figure 10–5, 2400 m² of pervaporation membranes (to the right) have been retrofitted to existing distillation trains (left side) in order to increase plant capacity significantly, as well as produce higher purity product ethanol from corn, sorghum, and sugar beets (Rapin 1988). Although numerous other options were considered (Table 10–2), integrated distillation/pervaporation coupled with an additional desulphation section resulted in a savings of $0.035/gal over double-effect azeotropic distillation. Energy and labor costs accounted for the bulk of the reduction. As seen in Table 10–2, electricity for the inte-

TABLE 10–2. Comparison of Options for Ethanol Recovery.

Process	Capital Investment Cost (KF)	Wastewater Generated (kL/HL)	Steam Required (kg/HL)	Electricity Required (kWh/HL)	TEP/HL / TEP/AN
1. Actual situation rectification → dehydration purification → liquid-liquid extraction	—	7.04	167	2	0.01135 / 4258
2. Desulfation + 35 stage concentration	2,100	5.71	135	2	0.00927 / 3478
3. Option 2 + double-effect azeotropic distillation	11,100	5.71	110	2	0.00765 / 2869
4. Desulfation + pervaporation	12,650	5.71	11	3.45	0.00157 / 592

Notes: KF: 1000 French francs. HL: hectoliter. TEP: Metric tons equivalent product. AN: annual.

FIGURE 10–6. Betheniville, France, plant for ethanol dehydration. (Courtesy of GFT.)

grated distillation/pervaporation system was 73% greater than the existing system, as well as a projected additional azeotropic distillation unit. However, the huge savings in steam energy (11 versus 167 kg steam/hL of product), and the subsequent 19% reduction in wastewater generated more than compensate for the additional cost of electricity.

An issue not defined in Table 10–2 was replacement cost of membranes. Assuming a life of 1 to 3 yr, with a replacement cost of PVA membrane in existing 304 stainless steel modules at $300/m^2 (Fleming 1989) and a 30% changeout per year, the annual replacement cost would be $189,000. In practice, replacement is far less frequent than this assumption.

Capital costs were 25% lower for the 150,000 L/day ethanol plant, which continuously produces four purities of ethanol. The opened membrane modules of the Betheniville plant are shown in Figure 10–6 with the vacuum chambers sitting to the left, and the plant control package to the right.

Typical capital system costs for skid-mounted, stand-alone pervaporation systems for dehydration of liquid organics, including ethanol, are given in Figure 10–7. Because pervaporation is less mature than many other membrane technologies, very few reports of plant economics have been published. The same situation has been true for published design equations. Although most companies designing pervaporation plants use computer-based algorithms, these tend to be proprietary and system-specific. Relative to other membrane operations, modeling and scale-up of a pervaporation system are much simpler and much more reliable. The algorithms are essentially composed of vapor-liquid equilibria, membrane separation and mass transfer, and conventional capital costing routines.

Figure 10–7 shows that some economy of scale exists for higher volume systems since membranes comprise more of the cost, with ancillary equipment costs decreasing. Capital costs for pervaporation tend to be roughly equal to azeotropic distillation at large-scale green field facilities, and cheaper at reduced scale and retrofits. The same is true of regenerative adsorption, with adsorption capital costs somewhat more expensive at small scale and 10 to 20% cheaper at larger scale than pervaporation. However, these figures for adsorption are misleading because operating costs including energy, adsorbent replacement, and disposal of spent adsorbent are much higher, and are not equivalent for pervaporation.

These economics assume the use of flat-plate, 316 stainless steel shelled membrane modules. This is a fair assumption since most of the current plants utilize such membranes. However, the use of spiral-wound or transverse-flow hollow-fiber modules, such as those from Zenon, Tokuyama Soda, or others, results in module costs of $200 to $500/m^2, a considerable reduction. The net result is large system costs substantially below the cost of using flat-sheet membrane modules. Most of the plants built after 1990 will probably utilize these advanced designs.

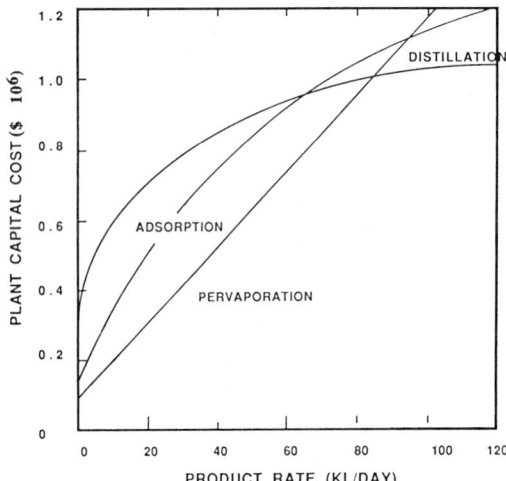

FIGURE 10–7. General economics for solvent dehydration.

TABLE 10–3. Separation Options for Small-Scale Ethanol/Water (Basis: 1000 L/day, 99.5 wt.% ethanol) (from Fleming 1989).

	Pervaporation	Distillation	Adsorption
System cost	$75,000	$140,000	$90,000
Pumps	3 kW	2 kW	2 kW
Steam	45 kg/h at 1.8 bars	70 kg/h at 7.3 bars	90 kg/h at 7.3 bars, 220°C
Entrainer	—	3 L/day	—

An example of key costs for a very small ethanol system is given in Table 10–3. For commercial plants to produce 1000 L/day of 99.5 wt.% ethanol (essentially pilot plants), pervaporation is half the price in capital, with no pollution and low steam requirements. Electricity is greater, due to the need for pump and vacuum energy. An example of a small, 1000 to 6000 L/day pilot pervaporation system is shown in Figure 10–8. The pervaporation pilot plant shown here is an explosion-proof, turnkey, portable unit, is completely automated for unattended operation, and is commercially produced by Zenon (although many others are available) for process evaluation and development.

A typical elevation schematic for a pilot plant for dehydration is shown in Figure 10–9. A specific pilot plant for dehydration developed by GFT is shown for comparison in Figure 10–10, with a photograph of the same unit in Figure 10–11. In this unit with 4 m² of membrane surface area, the vacuum pump is extremely small. Most of the volume is occupied by the vacuum chamber for the membrane modules. As in most membrane processes, the units are very compact relative to conventional equipment to process the same liquid capacity (1000 to 1500 L/day in this case).

FIGURE 10–8. Zenon explosion-proof dehydration pilot plant. (Courtesy of Zenon.)

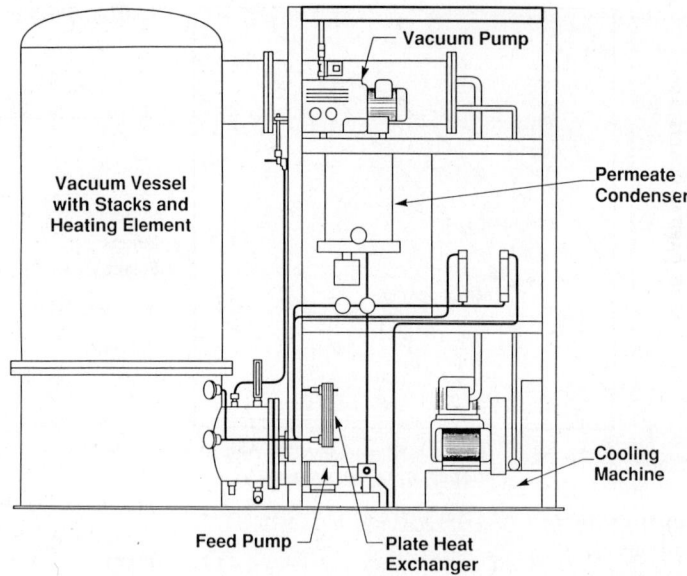

FIGURE 10–9. Elevation drawing for pervaporation pilot plant.

Depending on membrane flux, these figures are roughly accurate for other solvent dehydration applications as well. They may, in fact, cost less because of higher fluxes (less membrane required) than ethanol for the same system size. This is discussed in the next section.

Case 2: Dehydration of Organics

Removal of water from liquid-organic mixtures now accounts for the largest segment of new industrial pervaporation plants. This is even truer in the United States than in Europe or Asia because capital costs and environmental issues are of primary concern, and fewer ethanol plants are being built. In the chemical process industry, there are literally unlimited numbers of organic streams that become contaminated with small (<10%) amounts of water. In most cases, removal of this water to ppm levels prior to the next processing step is economically desirable. Even if distillation is possible (many

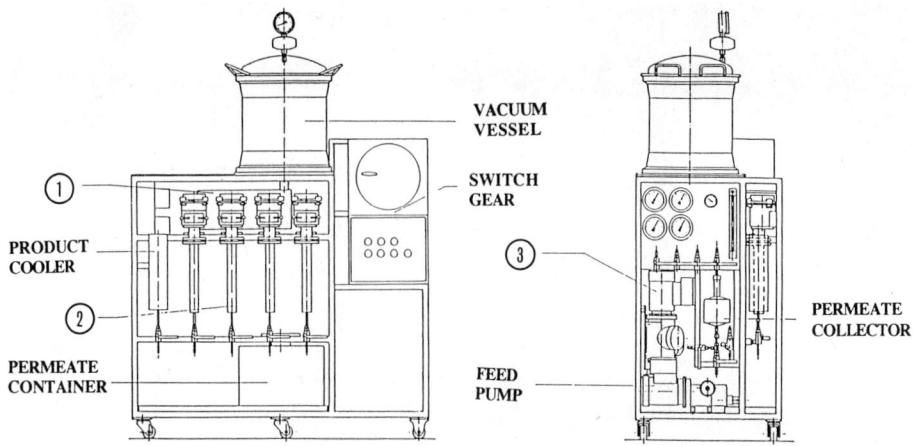

FIGURE 10–10. Schematic of dehydration pilot plant. (Courtesy of GFT.) 1—Condenser; 2—Heat Exchanger; 3—Vacuum Pump.

FIGURE 10–11. GFT dehydration pilot plant. (Courtesy of GFT.)

multicomponent azeotropes formed), it becomes prohibitively expensive to remove small amounts of water. Adsorption with desiccants such as aluminas and zeolites is typically employed. However, regeneration is usually required.

Although the capital costs of adsorption are usually low (assuming existing beds are in place), operating costs including adsorbent replacement, high energy of regeneration, adsorbent disposal, and hazardous gaseous effluents are significant deterrents in many cases. Further, very small volume streams (<10 to 20 gal/min) require small, modular plants because of space requirements, with simple, unattended operation. This feature is considered a strength of pervaporation systems relative to adsorbers. Membrane pervaporation has made major contributions to these applications and is now a preferred process option for dehydration of organics.

A list of organics, which are currently dehydrated and recovered, is given in Table 10–4 to illustrate the strength of pervaporation. The degree of water removal is primarily dependent on the desired economics. Generally, 10 to 20 ppm levels of water in the product are attainable. To approach 1 ppm or less requires much more membrane surface and, in some cases, reduced permeate pressure with a larger vacuum pump. Because of the reduced solubility of water in organics and increased chemical potential driving force, water fluxes tend to be even greater for solvent dehydration than for ethanol. While dependent on feed concentration, 1 to 10 kg/m^2 h is not unusual, resulting in reduced membrane requirements and capital and operating savings. Feed water contents of 0.1 to 10 wt.% are economical with pervaporation. Greater amounts of water are best separated with extraction or other bulk techniques, whereas at lower water content, adsorption may become competitive.

Dehydration of several organics is given in a different form in Table 10–5 (Asada 1987). This table shows that selectivity and water flux may vary widely, even with the same membrane and approximately equivalent process parameters. In general, commercial applications for dehydration can be grouped in two

TABLE 10–4. Industrial Examples of Solvent Dehydration.

Solvent	Water Content	
	Feed (wt.%)	Product (ppm)
1-Butanol	8.4	135
n-Butanol	5.4	800
t-Butanol	10.4	581
THF	0.4	220
Xylene	0.1	140
Methanol	7.1	1650
Methanol/IPA	0.21	300
Caprolactam	10.3	671
Ethanol/IPA	0.6	610
Ethanol/MeOH	2.9	780
Ethanol/benzene	14.1	320
Allylalcohol	4.85	620
Trichlene	0.01	8
MEK	3.8	220
Methylene chloride	0.20	140
Ethylene dichloride	0.22	10
Chlorothene	0.0617	12

TABLE 10–5. Selected Water-Permeable Organic/Water Pervaporation Membranes.

Organic	Water Feed Concentration (wt.%)	Selectivity (α)	Water flux (g/m^2-h)
i-Butanol	8.4	1201	1920
THF/benzene	0.255	805	82
Xylene	0.04	5799	25
Methanol	5.1	58	229
Methanol/BTX[a]	1.1	1823	258
PFP	4.2	22787	1088
Ethanol/benzene	14.1	142	4220
n-Butanol	1.41	929	107
MEK	4.0	3976	907

[a]BTX stands for benzene, toluene, and xylenes.

segments. The first, for dehydration and recycle of solvents in the pharmaceutical and specialty chemicals fields, includes acetone, isopropyl alcohol (IPA), tetrahydrofuran (THF), xylenes, and organic chlorides. Here, the objective is continual removal of a water load resulting from a reaction or other separation step. Volumes may be small and discontinuous, but purity requirements are stringent.

The second group of applications are those generally considered to be large bulk chemical processing. Typical organics include IPA, ethylene glycol, methyl ethyl ketone (MEK), and ethyl acetate. The processes are typically large volume, with fewer requirements on membrane selectivity and more emphasis on flux and versatility of design to accomplish a large turndown ratio (10 to 100%). The objective is generally debottlenecking of existing distillation or adsorption trains.

Chemically, applications tend to be grouped in three segments as follows:

1. Azeotrope forming
2. Nonazeotropic mixtures
3. Azeotropic splitting.

Examples of the first group include most alcohols, esters, and ketones. They were the earliest and most obvious examples or applications of pervaporation because azeotropes do not interfere in this nonequilibrium-based process. The result is a significant advantage over distillation in overcoming the azeotrope. The second group covers more recent applications and includes some alcohols, organic acids, most hydrocarbons, and chlorinated hydrocarbons. The nonazeotropic mixtures also are good candidates for pervaporation, but compete with distillation.

The last group includes amines and some carboxylic acids, among others. Pervaporation is applied here as a retrofit process to overcome an azeotrope, then the products are fed into a distillation or adsorption unit.

A further subgrouping of applications involves organic/aqueous mixtures that are fully miscible versus those with only limited miscibility (miscibility gaps). Although both types are treated, the processes are distinct and different. In the former, the process is much simpler, but the flux tends to be lower due to the higher liquid solubility (i.e., lower chemical activity coefficient) and subsequently lower driving force. In the latter, two phases may be formed on the membrane, and the permeate requires two-phase separation and recycle. Although more complex, the fluxes are generally higher, and the overall separation factor, including two-phase separation, is high for the system. Examples of these two types are MEK and ethyl acetate, respectively.

The versatility of the pervaporation process for solvent dehydration is illustrated in Figure 10–12 for acetone. Vapor-liquid equilibria for this nonazeotropic binary mixtures are given at ambient conditions. The performance of a com-

FIGURE 10–12. Acetone-water vapor-liquid equilibria (from Fleming 1989).

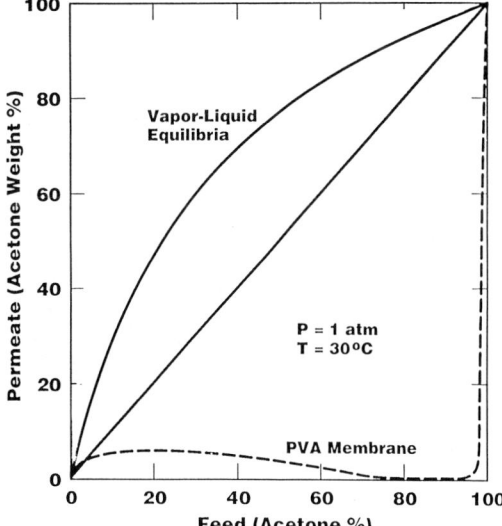

mercial GFT pervaporation membrane is also plotted on the same scale (Fleming 1989). We see that the hydrophilic polyvinylalcohol composite membrane is highly selective for permeating water preferentially over most of the concentration range, with the ability to obtain solvent with extremely low water content. This process is industrially feasible within certain concentration ranges and is more selective than processes such as distillation, which rely on thermodynamic equilibria as their mechanism of separation.

A typical dehydration process employing a two-phase separator with a mixture of limited miscibility is illustrated by purification of dichloroethylene (EDC) in Figure 10–13. In these pilot plant data (Fleming 1989), saturated EDC at 0.2 wt.% from a condenser is preheated and sent to the pervaporation stack containing PVA membranes. Purified EDC containing <10 ppm water is obtained in one pass. The permeate, in this case containing 45 to 50% water, is condensed. Phase separation occurs, with the organic-rich phase recycled to the pervaporation membranes for further purification, and the aqueous phase sent to an existing steam stripper or disposed.

The latter points out another strength of the pervaporation process, namely, the ability to use the limited solubility of water in organics to phase separate the permeate. In many systems, a secondary phase separator in line allows much greater organic recovery, as well as a major reduction in the volume of aqueous water. In

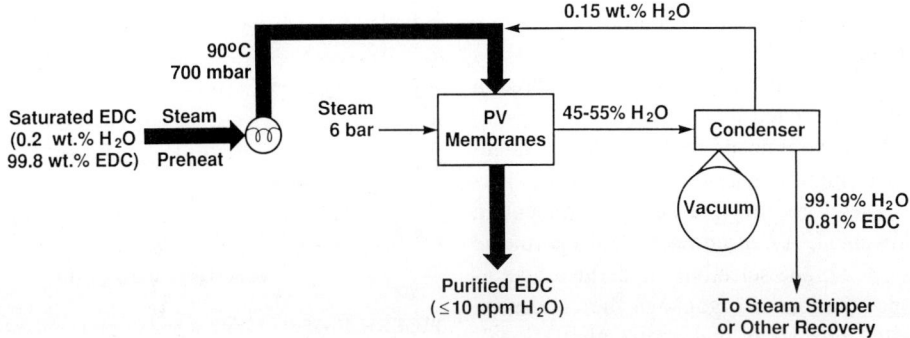

FIGURE 10–13. Purification of dichloroethylene.

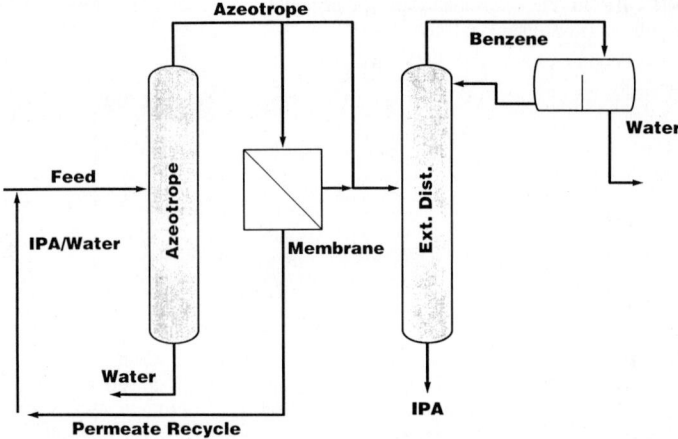

FIGURE 10–14. Isopropyl alcohol separation process (reprinted from Bartels et al. 1988 with permission).

most cases, the aqueous phase is dilute enough to allow disposal in a nonhazardous manner. This example with dichloroethylene is typical for numerous organic systems of industrial interest and is employed extensively. Commercial examples include treatment of halogenated refrigerants and jet engine fuels.

Another example of industrial interest is that reported by Texaco (Bartels et al. 1988) for dewatering isopropyl alcohol. As shown in Figure 10–14, Texaco utilized pervaporation for retrofitting an existing azeotropic/extractive distillation system. In this process pervaporation debottlenecks IPA plant capacity by taking IPA at 85% up to 95% prior to feeding to the extractive distillation column. Only a small pervaporation system is necessary because of the large driving force and low water load. Yet, it eliminates capital requirements for additional extractive distillation capacity. Further, a more desirable entrainer than benzene can be used because of the lower water load on the extraction column.

One further feature of the pervaporation process is illustrated in Figure 10–15. Because of the large volume of feed, Texaco was interested in minimizing solvent loss in the permeate. Therefore, they placed a requirement on the allowable IPA concentration in the permeate at 5%. Membrane selectivity is defined here as the concentration of organic in the condensed permeate. Because of the inverse proportionality of flux and selectivity, this requires a higher selectivity membrane, with reduced flux. Various membrane process configurations are possible, with an economic trade-off in all cases. Optimization is required.

Both the temperature and feed concentration dependence of water flux are illustrated in Figure 10–16. Because both sorption and diffusion are activated processes, there is a strong exponential dependence of flux on feed temperature. In dehydration, fluxes varying by a factor of 2 for every 10°C of temperature are not unusual. The driving force is also strongly dependent on feed concentration. At higher purities, i.e., lower concentrations of the permeating component in the feed, fluxes may drop rapidly.

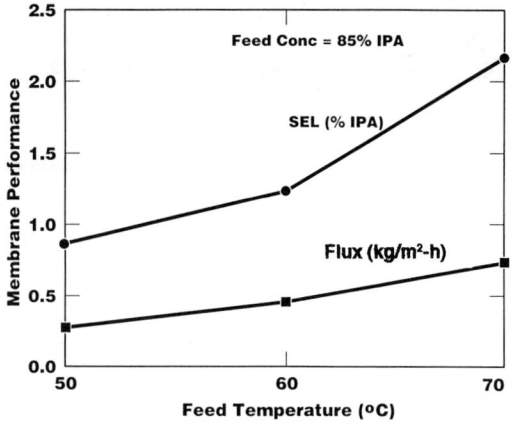

FIGURE 10–15. Effect of temperature on membrane performance (reprinted from Bartels et al. 1988 with permission).

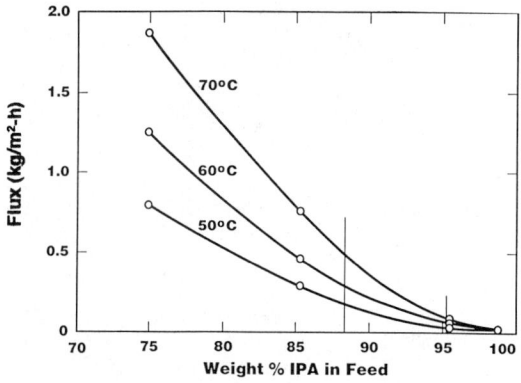

FIGURE 10-16. Dependence of flux on feed concentration (reprinted from Bartels et al. 1988 with permission).

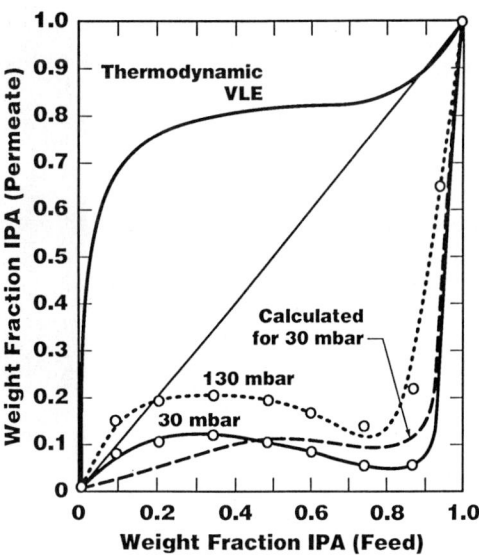

FIGURE 10-17. Separation equilibria for isopropyl alcohol/water at 333 K (reprinted from Rautenbach, Herion, and Franke 1988 with permission).

The effect of permeate pressure on separation is illustrated for IPA dehydration in Figure 10–17. At relatively low permeate pressures (30 mbars and below), water flux (i.e., dynamic selectivity) remains constant (Rautenbach, Herion, and Franke 1988). As permeate pressure is increased, the driving force decreases, eventually to attain the same separation as thermodynamic vapor-liquid equilibria.

A modular pervaporation plant for dehydration is schematically shown in Figure 10–18. This skid-mounted, turnkey plant from GFT contains between 150 and 240 m² of membrane in a number of stages and is typical of the type of unit employing plate-and-frame membrane modules. A variation on this unit is given in Figure 10–19. Here, a two-vacuum chamber system is shown with greater membrane area. The basic unit design remains the same, resulting in scale-up that is essentially linear.

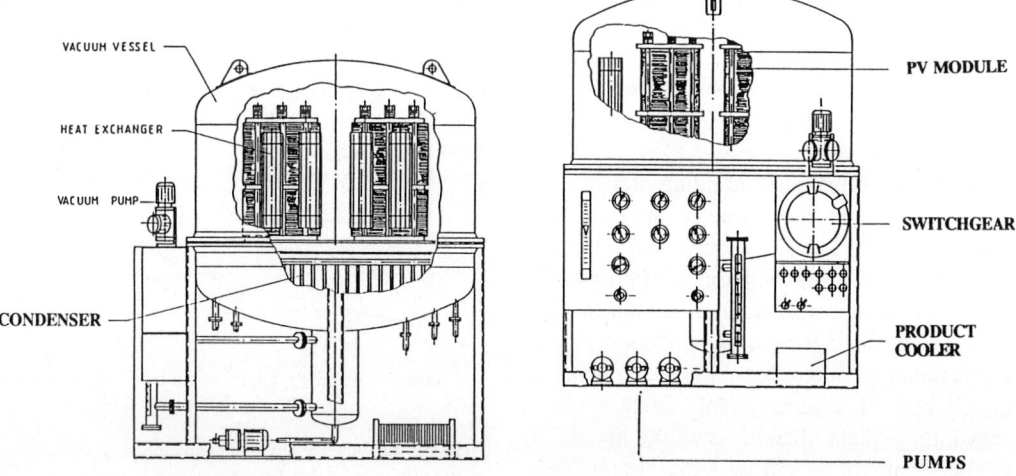

FIGURE 10-18. Schematic of dehydration plant (<240 m²). (Courtesy of GFT.)

FIGURE 10–19. Multistage plants for dehydration/recovery of multiple solvents. (Courtesy of GFT.)

Plant sizing and the economics of dehydration systems largely depend on three parameters: (1) mass to be permeated, (2) driving force (i.e., flux), and (3) the resulting membrane area required. These effects are illustrated in Figure 10–20 for isopropyl alcohol dehydration. The total membrane area required to accomplish a given separation increases relatively linearly with mass of water to be removed. Given the same final concentration, for example, 5 wt.% water requires approximately half the membrane area that 10 wt.% water requires. As the driving force begins to significantly decrease at high organic purity (above 99.8 wt.% in the case of IPA), the amount of membrane area required to accomplish the same separation increases exponentially. This effect is in keeping with the logarithmic decrease in partial vapor pressure driving force. Until this exponential limit is approached, the driving force, i.e., the surface area of membrane required, can be modeled as linear.

As shown in the corresponding plot of plant capital cost with increasing membrane area (Figure 10–21), the cost of dehydration plants above 250 m^2 in area increases in approximately a linear fashion with the total membrane area required. In other words, the membrane modules account for more than 50% of the installed capital cost of a given plant. Scale-up on a dehydration plant then is roughly linear. At smaller plants (0 to 250 m^2 of total membrane area), the cost of ancillary equipment, piping, and instrumentation is more significant, and dominates capital cost.

This effect of module costs controlling plant sizing certainly holds for plate-and-frame modules made of stainless steel. With cheaper materials, or with modules in other geometries, resulting in lower costs per surface area, the economics may be different.

Several other published examples of integrated dehydration by pervaporation with more conventional processes have been considered. Three examples are given to illustrate the capability. Steam stripping of carbon beds results in another toxic gaseous waste, which must be treated. Pervaporation systems may be retrofitted (Figure 10–22) to these beds, the

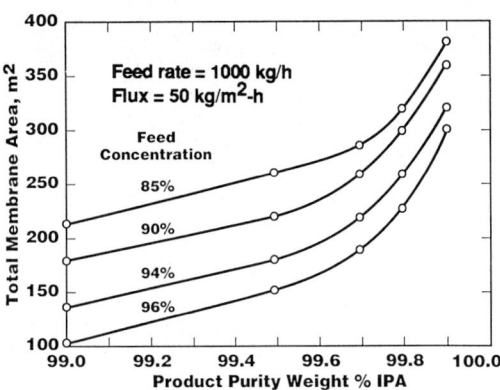

FIGURE 10–20. Required membrane area for isopropyl alcohol dehydration as a function of feed composition (from Fleming 1989).

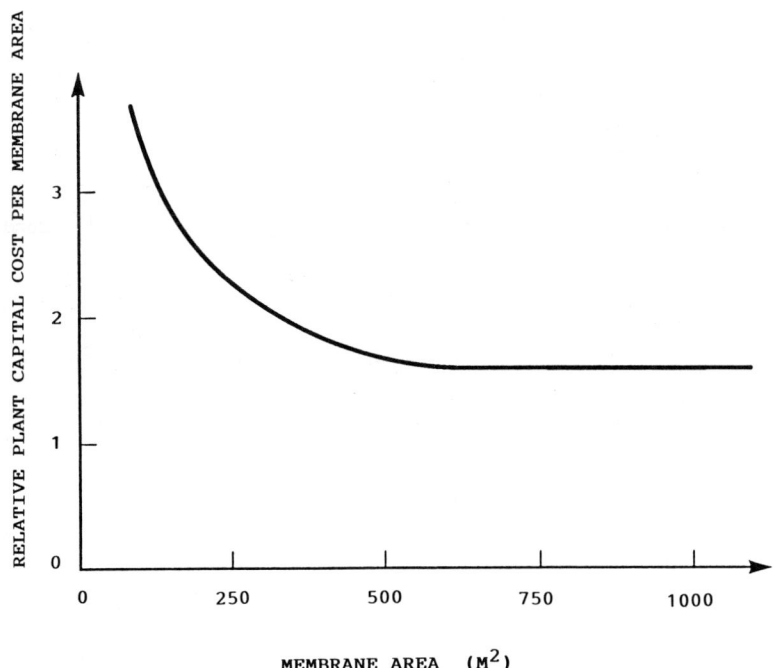

FIGURE 10–21. Capital cost of turnkey pervaporation plants for solvent dehydration.

regenerate condensed, and the waste dehydrated. The solvent is recycled and recovered, as is the condensed steam. In the scheme shown, which is only one of numerous variations, two types of pervaporation modules are utilized—one for dehydration and the other for selective organic permeation.

Rautenbach, Herion, and Franke (1988) have extensively studied a variation on the above scheme. They have considered employing reverse osmosis as a preconcentration step following phase separation. The organic-rich phase is sent to a typical pervaporation dehydration membrane unit for recovery of the solvent fraction for reuse.

Solvent recovery from the vapor phase, such as those processes found in automotive paint lines, may be treated and recovered (Figure

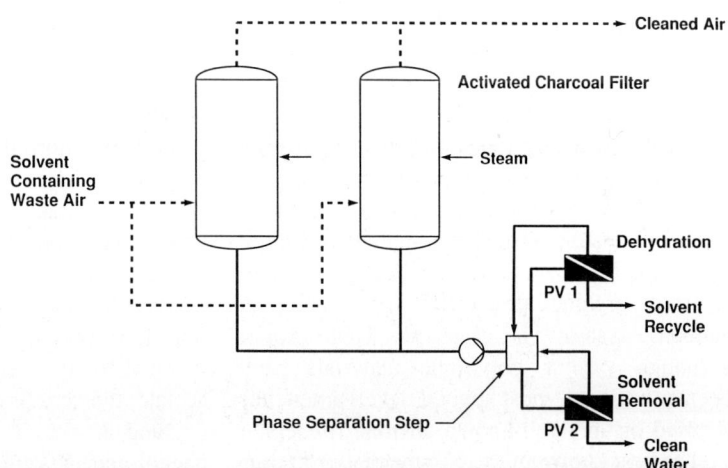

FIGURE 10–22. Integrated process for treatment of adsorbent regeneration stream.

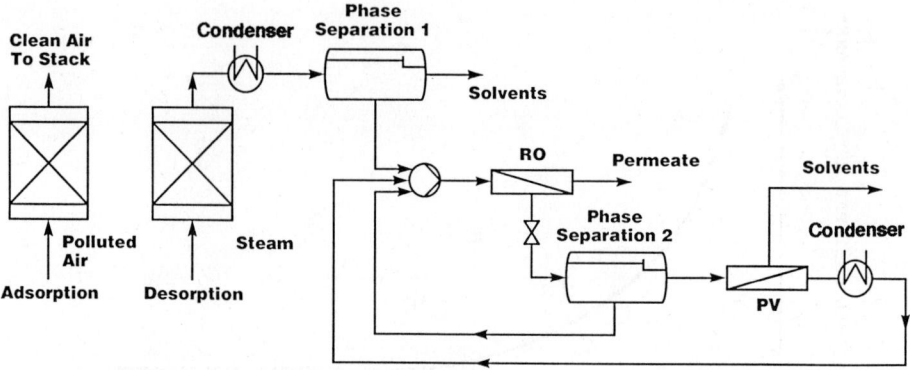

FIGURE 10–23. Integrated reverse osmosis/pervaporation process for processing steam-stripped condensate (reprinted from Rautenbach, Herion, and Franke 1988 with permission).

10–23) by water spray extraction followed by pervaporation. Depending on the type and concentration of organics, either an organic- or water-selective membrane can be employed. One advantage of the water-selective PVA membranes is that they are selective for water over all other species. Multicomponent streams with unknowns or variable compositions can be successfully treated and are not membrane-specific.

REMOVAL OF ORGANICS FROM WATER

Membranes and processes only became commercially available around 1989 for selectively permeating organics from aqueous streams. Because of the associated requirements for good chemical and thermal stability in solvents, most hydrophobic membrane materials, those which are most selective to organics, are not compatible. Currently available composite materials that have been developed for this application are a compromise, with good chemical properties but limited separation abilities. However, some processes are now commercially viable, while others are developing rapidly. A list of selected organic permeable membranes in aqueous systems is given in Table 10–6. Although various polymeric materials have been considered, most applied development has focused on the well-known silicone rubbers.

The best known of the commercial processes is the GFT process for dealcoholization of beers, wines, and liquors. Using polydimethyl silicone-type membranes, reduction or removal of ethanol has been demonstrated in various alcoholic beverages (Escudier et al. 1988). As in the pilot-scale example in Figure 10–24 for beer (Fleming 1989), selective permeation of ethanol is straightforward, with alcohol reduction to 0.7 wt.% for "alcohol-free" beer easily accomplished. Reduction is currently limited to around 0.1 wt.% because membrane selectivity is not as good as with the PVA-based, water-selective materials. Also, numerous contaminants are present. Fusel oils (amyl and propyl alcohol fractions) may also be separated in the process and recovered. Depending on the choice of membrane, permeate quality can be controlled from 15 to 55% ethanol, so that in many cases the permeate is a useful product. In production of low-alcohol wines, for example, the permeate is useful as a salable brandy (Escudier et al. 1988).

Pervaporation of other organics from water is also commercially viable. Removal and recovery of trace organics from groundwaters and industrial wastewaters are under way and are commercially available (Bengston and Boddeker 1988; Kaschemekat et al. 1988). Suggestions have been made that pervaporation is economical for recovery in such streams as ethyl acetate, the various carboxylic acids (citric, lactic, and acetic), as well as aromatics such as phenol and benzene. As seen in Figure 10–25

TABLE 10-6. Selected Organic-Permeable Organic/Water Pervaporation Membranes.

Membrane Material	Organic Feed Concentration (wt.%)	Temperature (°C)	Permeate Pressure (kPa)	Selectivity (α)	Organic Flux (kg/m^2·h)	Reference
Polypropylene	Acetone (45)	30	6.5	3	0.1 to 1.2	Featherstone and Cox 1971
Silicone rubber	Butanol (0 to 8)	30	—	45 to 65	<0.035	Groot et al. 1988
Silicone rubber	IPA (27 to 100)	25	0.33	0.5 to 12	—	Seok 1987
Silicone rubber	IPA (9 to 100)	25	0.67	9 to 22	0.03 to 0.11	Kimura and Nomura 1983
Polyetheramides	Acetic acid (1.5 to 9)	50	<0.2	—	0.18 to 0.28	Bengston and Boddeker 1988
Polyacrylic acid	Acetic acid (48)	15	—	2 to 8	0.4 to 0.55	Huang, Balakrishnan, and Matsuura 1988
Silicone rubber	Ethyl acetate (0.5 to 4)	30	−0.2 to 0.4	High	—	Blume and Baker 1987
GFT ethanol membrane (PDMS)	Ethanol (87 to 100)	60	—	150 to 10,000	0 to 1.6	Kraetz 1988

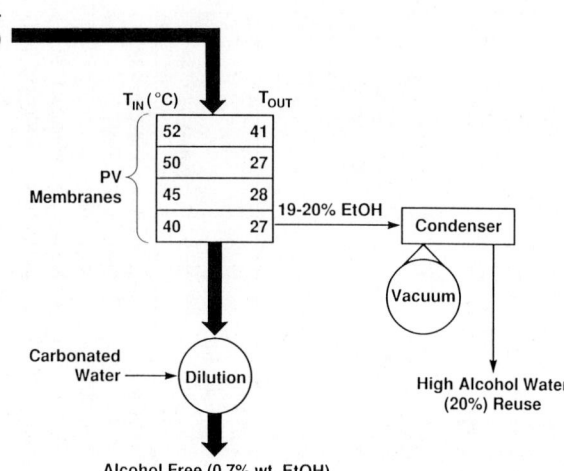

FIGURE 10-24. Pervaporation process for alcohol reduction in beer.

(Blume and Baker 1987; Blume, Wijmans, and Baker 1990), a spiral-wound composite silicone rubber membrane exhibits good selectivity for various organics over water. The flux is seen to be relatively linear with the feed concentration of the organic and selectivity is strongly dependent on the degree of hydrophobicity and water solubility of the organic.

Applications for organic permeation from aqueous streams by pervaporation can be grouped into four categories as follows:

1. Solvent recovery
2. Pollution abatement
3. Concentration of organics (i.e., for disposal)
4. Specialty organic reduction processes.

Examples of the fourth type include that given previously for dealcoholization of beverages, as well as aroma or flavor reduction.

A common example for recovery of solvents by organic permeating pervaporation membranes is illustrated in Figure 10-26. Solvent-laden air streams are treated in many industries with water washers to extract the organics. Conceivably, pervaporation can be utilized to permeate the organics, which can be condensed and recycled to the process stream for reuse. The aqueous retentate, containing minor organics, can be reused in the extraction process. Thus far, no commercial plants of this type have been built. Although detailed economics are not available, it is believed that vapor-phase adsorption of the solvents with activated carbon possesses reasonable economics, so there has been little reason to evaluate alternative technologies.

The best known example of concentration is

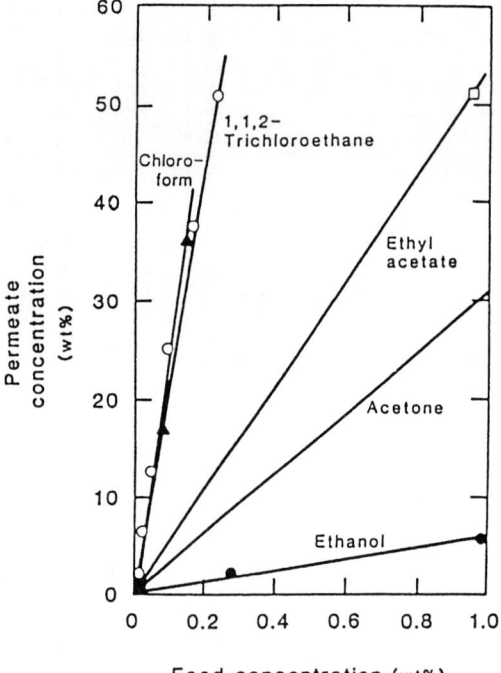

FIGURE 10-25. Permeation of various organics from water (reprinted from Blume and Baker 1987 with permission).

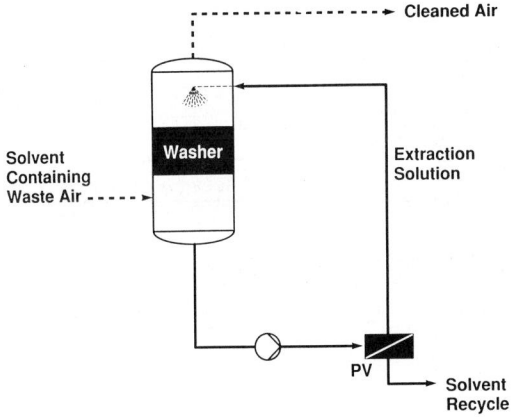

FIGURE 10–26. Removal of solvents from extraction solutions.

that of reducing water volumes in feeds to incinerators. Here, organics are pervaporated (1) to reduce the volume (i.e., the total mass load) to the incinerator and, more importantly, (2) to reduce the energy requirement. A second example in this category that is now being considered is concentration of flavor and aroma compounds in the food and cosmetics industries.

As of 1991, no known commercial plant was utilizing pervaporation of organics from aqueous streams. It is generally agreed that the most likely application of organic permeation to become industrial in the near future is the recovery of valuable organics. An example of such a process is given by Blume and Baker (1987) and Blume, Wijmans, and Baker (1990) for recovery of ethyl acetate. As seen in Figure 10–27, a relatively small membrane system can treat a substantial volume of aqueous feed. Two reasons are given for this: (1) Although the total volume of the stream is large, the total mass flux to be permeated is small, and (2) the membrane modules may be either high surface area/volume spiral wound or hollow fiber. Because the stream is primarily water, with minimal constraints on materials of construction, and because vaporization of organics results in minimal thermal gradients, many more options are available for membrane materials and module configurations than for dehydration.

In the case of ethyl acetate, as with most organics of interest, the partial miscibility of the organic in water is used to advantage. A phase separator is employed on the organic-rich permeate to increase further the overall separation factor in the process. The water phase may be recycled to the feed stream in many cases.

Several requirements must be met if such a process is to be economical. First, the recovery

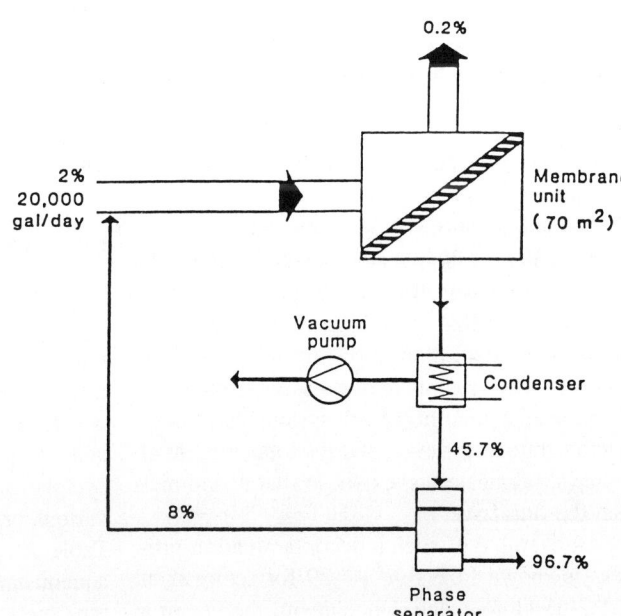

FIGURE 10–27. Ethyl acetate recovery by pervaporation (reprinted from Blume and Baker 1987 with permission).

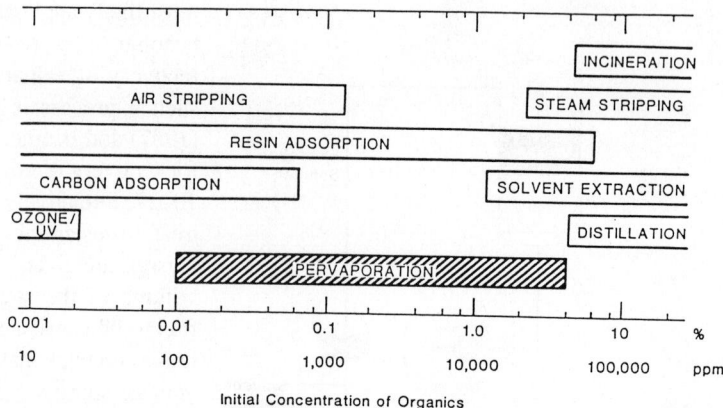

FIGURE 10–28. Commercial methods for removing volatile organic compounds from aqueous streams (reprinted from Blume and Baker 1987 with permission).

value of the solvent must be substantial. The membrane system is not cheap. The payback period must be short, on the order of months. Second, the concentration of the organic in the aqueous feed must be high enough to warrant recovery. Concentrations of 0.1 to 5 wt.% are generally considered good. Higher organic contents usually result in other more conventional treatment technologies (e.g., steam/air stripping, distillation, etc.) being more economically attractive. Streams that are too dilute have low driving force with low flux and high capital cost. It is generally more cost-effective to treat the stream as a waste in this case. The "window of opportunity" for pervaporation is depicted in Figure 10–28.

Finally, the stream must be relatively pure. Because the membranes are generally not greatly selective to specific organics, multiple organic components result in a spectrum of organics in the permeate. The result is a separation process of questionable efficiency and more complicated downstream processing.

Decontamination of groundwaters with dioxins, trihalocarbons, and other pollutants is possible. However, the driving force for separation is low in this case, requiring more membrane area. Also, more sophisticated condenser and recovery systems are necessary because of the permeate volatility. Such systems have been demonstrated, however, and the total amount of material permeated is small, so that economics may be attractive.

An example of such a decontamination process is shown in Figure 10–29 for removal of 1,1,2-trichloroethane. Once again, the use of a phase separator to increase the overall system separation factor proves beneficial. Given the large system costs, the lack of value in recovery of mixed contaminants, and the inability of the membrane unit to achieve very low levels of organic in water, this type of application is likely to be specialized, rather than widespread.

An example of an industrial pilot plant for permeating organics from aqueous streams is shown in Figure 10–30. This unit is manufactured by Membrane Technology and Research (MTR, Menlo Park, CA). The PerVap 100 is one of several pilot units built by MTR for the recovery of volatile organic solvents from aqueous streams. Pilot plants such as these are capable of 99% recovery of solvents, and can be used in batch or continuous mode on streams up to 13,000 gal/day. Several other manufacturers produce such systems, including Zenon Environmental, Inc. (Burlington, Ontario, Canada) and Bend Research (Bend, OR).

ORGANIC/ORGANIC SEPARATIONS

Separations of organic/organic mixtures represent the least developed and largest potential commercial impact for pervaporation. Historically, the earliest interest in pervaporation was focused in this type of application (Binning and James 1958a, 1958b) for benzene/cyclohexane separation. A representative list of various organic mixture separations is given in Table 10–7. Here, unlike the cases of organic/aqueous mixtures, no clear choice of membrane type presents itself. The specific membrane

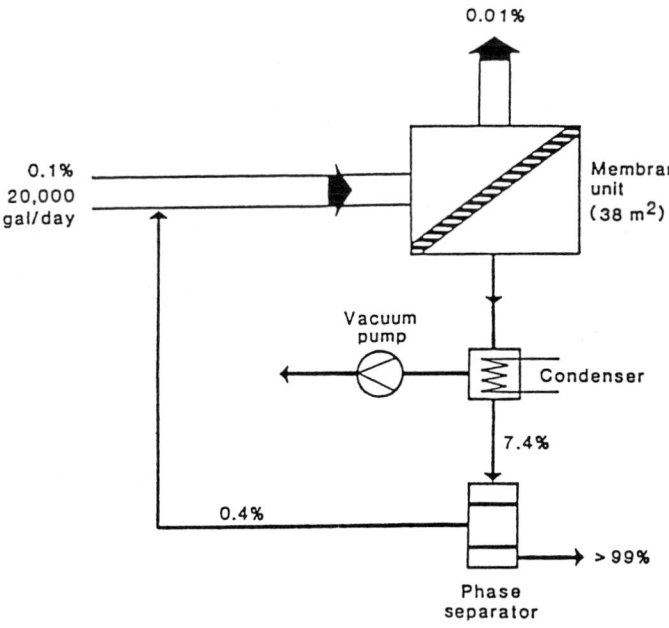

FIGURE 10–29. Removal of 1,1,2-trichloroethane by pervaporation (reprinted from Blume and Baker 1987 with permission).

chemistry depends largely on the type of separation to be made (Cabasso 1983).

Current membranes exhibit only marginal performance for most organic/organic separations of industrial interest. Membranes must be custom-designed for specific process objectives. This factor, combined with the large inertia in the chemical industry toward conventional processing with distillation, has retarded the development of applications for organic/organic separations.

In theory, pervaporation should be a powerful technique for separating organic mixtures. Azeotropes are irrelevant. Small differences in volatility are also irrelevant because the membrane separates on the basis of chemical functionality and interaction rather than volatility. Several obvious examples of applications can be cited:

1. Aromatics/paraffins (i.e., benzene/hexane)
2. Branched hydrocarbons from n-paraffins (i.e., isooctane/hexane)
3. Olefins/paraffins (i.e., pentene/pentane)
4. Isomeric mixtures (i.e., xylenes)
5. Chlorinated hydrocarbons from hydrocarbons (i.e., chloroform from hexane)
6. Purification of dilute streams (i.e., IPA from heptane/hexane).

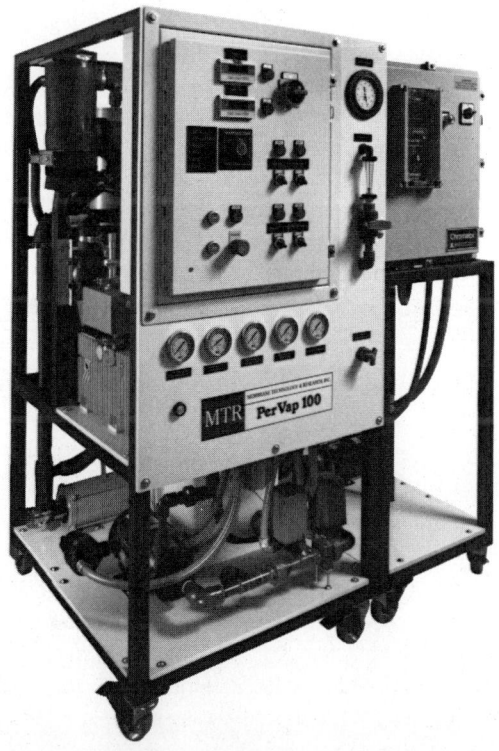

FIGURE 10–30. Pilot plant for removal of organics from aqueous streams. (Courtesy of Membrane Technology and Research, Inc.)

TABLE 10-7. Selected Organic/Organic Pervaporation Membranes.

Membrane Material	Organic Feed Concentration (mol%)	Temperature (°C)	Permeate Pressure (kPa)	Selectivity (α)	Organic Flux (kg/m^2·h)	Reference
Polypropylene	Acetone (50) Butanol (50)	30	1 to 31	—	0–0.025	Ohya et al. 1988
PTFE/PVP	Butanol (10) Cyclohexane (90)	25	<0.1	23.5	0.3	Aptel et al. 1976
PTFE/PVP	Chloroform (65) n-Hexane (35)	25	—	3.9	2.65	Aptel et al. 1976
Polyvinylalcohol composite	IPA (8) n-Hexane (92)	60	—	>900	<0.01	Schneider 1987
Polyvinylalcohol composite	IPA (8) Toluene (92)	60	—	>900	<0.02	Schneider 1987
Polyethylene	IPA (25 to 70) Benzene (30 to 75)	42 to 60	—	—	0.1 to 2	Carter and Jagannadhaswamy 1964

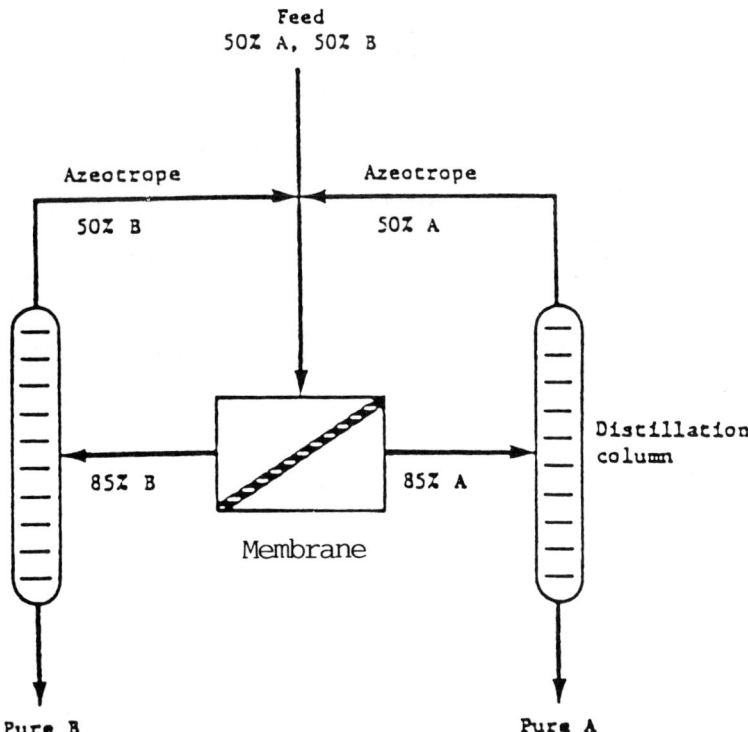

FIGURE 10–31. Integrated pervaporation-rectification process for separation of a 50/50 azeotropic mixture.

Because of the magnitude of these processes in the chemical industry, and the scale of existing distillation processes, it is almost a certainty that industrial pervaporation processes will be integrated, not stand-alone. An example of one such process is shown in Figure 10–31 for separation and recovery of a 50/50 azeotropic mixture. A small pervaporation system is situated between two existing rectification columns. The membrane only enriches the retentate in component B, permeating component A. Both streams can now be economically treated by the existing distillation units, with the top ends of both columns recycled to the pervaporation unit. Not only does such a process allow the capacity of existing distillation trains to be considerably magnified, it requires only marginal selectivities in the pervaporation membranes.

As of 1991, no known commercial plant was utilizing any of the above concepts for organic/organic separations. The first plant for organic/organic separation by pervaporation was reported by Air Products (Chen, Markiewicz, and Venugopal 1989) for the removal of methanol from methyl-tert-butyl ether in the production of octane enhancer for fuel blends. One such scheme for their TRIM™ process is shown in Figure 10–32. The success of this application lies in the high selectivity for the spiral-wound cellulose acetate membrane for methanol over MTBE, and in its ability to utilize MTBE/C_4 mixtures with some methanol remaining in the stream to the debutanizer column. There are several other obvious applications for removal of an organic contaminant from another organic, but few have been reported.

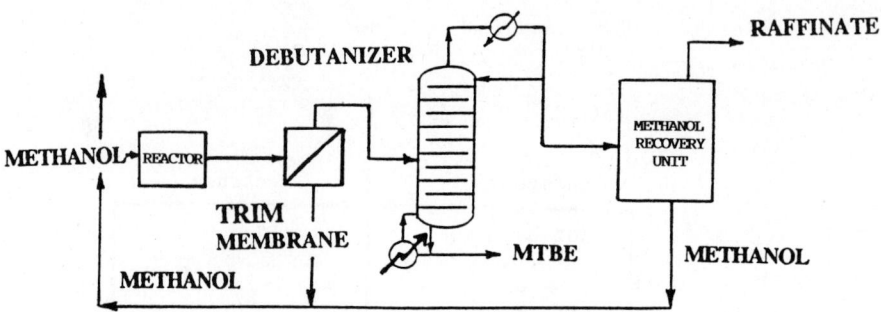

FIGURE 10-32. TRIM™ process for methanol/methyl *t*-butyl ether separation (reprinted from Chen, Markiewicz, and Venugopal 1989 with permission).

VAPOR PERMEATION

In principle, vapor permeation is similar to membrane pervaporation. The feed stream is a mixture of vapors, rather than a liquid. As in pervaporation, the permeate partial pressure is maintained by use of an inert sweep gas or a vacuum. Because the requirements for the membranes are similar to those in pervaporation, membranes currently being employed in vapor permeation are generally the same as those used in pervaporation.

Vapor permeation, as a commercial process, is only now being developed. It is finding particular advantage in the treatment of gaseous streams, where pervaporation would require an additional energy- and capital-intensive compression step. Vapor permeation has other theoretical advantages over pervaporation. First, the feed is a vapor, so that it would be compatible with fractional distillation. Second, large heat inputs would not be required to vaporize the feed, thus eliminating the design of complex modules for maximizing heat transfer, and eliminating the need for interstage heating and its associated plumbing.

Because of the increased volumes to be treated in vapor permeation, and the fact that most of the streams of interest are near their boiling point, resulting in two-phase feeds under certain conditions, considerable work remains to be done before this promising process technology becomes accepted industrial technology. One commercial plant has been installed by Lurgi in West Germany (Sander and Janssen 1989) using GFT plate-and-frame membranes for dehydration of various organics. Two other commercial plants have been reported by MTR in the United States, although details are not available. A review of vapor permeation has been given by Meares (1988). Primary application areas are recovery of organics from solvent-laden vapors and pollution treatment.

An example of an industrial vapor permeation pilot plant is shown in Figure 10-33. This VaporSep unit, manufactured by MTR, removes and recovers volatile organic compounds, such as halogenated hydrocarbons,

FIGURE 10-33. Pilot plant for removal of organics from air streams. (Courtesy of Membrane Technology and Research, Inc.)

from air streams. The unit is said to be operable in several configurations—depending on the removal requirements—on contaminated air streams of up to 100 SCFM (see Chapter 5 for additional information).

CONCLUSIONS

Membrane pervaporation is a rapidly emerging technology for the separation of many organic-aqueous systems. With numerous demonstrated commercial successes in Europe and Asia, it is only now finding commercial application in North America. Because of its great versatility and unique capability, pervaporation is rapidly taking its place among the conventional techniques utilized by the chemical separation community. Organic-organic separation is seen as the most significant long-term potential application for pervaporation. However, considerable membrane and process development remains to be done. In all cases, most successful processes require integration with existing separation unit operations.

NOTATION

General Notation

See the General Notation section at the beginning of this handbook.

Special Notation

a activity, dimensionless
K_s sorption coefficient, mol/L^3 or M/L^3
R gas constant, ML^2/t^2T mol

REFERENCES

Aptel, P., N. Challard, J. Cuny, and J. Neel. 1976. Application of pervaporation process to separate azeotropic mixtures. *J. Membr. Sci.* 1:271–287.

Asada, T. 1987. Future of pervaporation. In *Proceedings of Second International Conference on Pervaporation Processes in the Chemical Industry*, ed. R. A. Bakish, pp. 240–248. Englewood, NJ: Bakish Materials Corporation.

Bartels, C. R., T. G. Dorawala, J. Reale, Jr., and V. Shah. 1988. Plant evaluation of pervaporation process. In *Proceedings of Third International Conference on Evaporation Processes in the Chemical Industry*, ed. R. A. Bakish, pp. 486–493. Englewood, NJ: Bakish Materials Corporation.

Bengston, G., and K. W. Boddeker. 1988. Pervaporation of low volatiles from water. In *Proceedings of Third International Conference on Pervaporation Processes in the Chemical Industry*, ed. R. A. Bakish, pp. 439–448. Englewood, NJ: Bakish Materials Corporation.

Binning, R. C., and F. E. James. 1958a. Now separate by membrane permeation. *Petrol. Refiner.* 37:214.

Binning, R. C., and F. E. James. 1958b. Permeation: a new way to separate mixtures. *Oil Gas J.* 56(21):104–105.

Blume, I., and R. W. Baker. 1987. Separation and concentration of organic solvents from water using pervaporation. In *Proceedings of Second International Conference on Pervaporation Processes in the Chemical Industry*, ed. R. A. Bakish, pp. 111–125. Englewood, NJ: Bakish Materials Corporation.

Blume, I., J. G. Wijmans, and R. W. Baker. 1990. The separation of dissolved organics from water by pervaporation. *J. Membr. Sci.* 49(3):253–286.

Cabasso, I. 1983. Organic liquid mixture separation by permselective polymer membranes. *Ind. Eng. Chem. Process Des. Dev.* 22:313.

Cabasso, I., and Z. Z. Liu. 1985. The permselectivity of ion-exchange membranes for nonelectrolyte liquid mixtures: 1. separation of alcohol/water mixtures with Nafion hollow fibers. *J. Membr. Sci.* 24(1):101–119.

Carter, J. W., and B. Jagannadhaswamy. 1964. Separation of organic liquids by selective permeation through polymeric films. *Brit. Chem. Eng.* 9(8):523–526.

Changlou, Z. 1987. Separation of ethanol-water mixtures by pervaporation-membrane separation process. *Desalination* 62:299–313.

Chen, M. S. K., G. S. Markiewicz, and K. G. Venugopal. 1989. Development of membrane pervaporation TRIM™ process for methanol recovery from CH_3OH/MTBE/C_4 mixtures. *AIChE Symp. Ser.* 85(272):82–88.

Ellinghorst, G., A. Neimoller, H. Scholz, and H. Steinhauser. 1987. Membranes for pervaporation by radiation grafting and curing and by plasma. In *Proceedings of Second International Conference on Pervaporation Processes in the Chemical Industry*, ed. R. A. Bakish, pp. 79–99. Englewood, NJ: Bakish Materials Corporation.

Escudier, J. L., M. Le Bouar, M. Moutounet, C. Jouret, and J. M. Barillere. 1988. Application and evaluation of pervaporation for the production of low alcohol wines. In *Proceedings of Third International Conference on Pervaporation Processes in the Chemical Industry,* ed. R. A. Bakish, pp. 379–386. Englewood, NJ: Bakish Materials Corporation.

Featherstone, W., and T. Cox. 1971. Separation of aqueous-organic mixtures by pervaporation. *Brit. Chem. Eng. Process Technol.* 16(9):817–819.

Fleming, H. L. 1989. Dehydration of organic/aqueous mixtures by membrane pervaporation. In *Proceedings of International Conference on Fuel Alcohols and Chemicals,* ed. W. Kampen. Charlotte, NC: K-Engineering.

Groot, W. J., R. G. M. van der Lans, and K. Ch. A. M. Luyben. 1988. Pervaporation of fermentation products: mass transfer of solutes in silicone membranes and the performance of pervaporation in a fermentation. In *Proceedings of Third International Conference on Pervaporation Processes in the Chemical Industry,* ed. R. A. Bakish, pp. 398–404. Englewood, NJ: Bakish Materials Corporation.

Hauser, J., A. Heintz, G. A. Reinhardt, B. Schmittcker, M. Wesslein, and R. N. Lichtenthaler. 1987. Sorption, diffusion, and pervaporation of water/alcohol mixtures in PVA membranes: experimental results and theoretical treatment. In *Proceedings of Second International Conference on Pervaporation Processes in the Chemical Industry,* ed. R. A. Bakish, pp. 15–34. Englewood, NJ: Bakish Materials Corporation.

Hirotsu, T. 1988. Water-ethanol separation by pervaporation through plasma-graft polymerized membranes of HEMAs. In *Proceedings of Third International Conference on Pervaporation Processes in the Chemical Industry,* ed. R. A. Bakish, pp. 103–109. Englewood, NJ: Bakish Materials Corporation.

Huang, R. Y. M., M. Balakrishnan, and J.-W. Matsuura. 1988. Pervaporation separation of pentane-alcohol mixtures using Nylon 6-polyacrylic acid (PAA) ionically crosslinked membranes: part II. experimental data and theoretical interpretation. In *Proceedings of Third International Conference on Pervaporation Processes in the Chemical Industry,* ed. R. A. Bakish, pp. 212–221. Englewood, NJ: Bakish Materials Corporation.

Karakane, H., M. Tsuyumoto, Y. Maeda, K. Satoh, and Z. Honda. 1988. Separation of water-ethanol by pervaporation through polyelectrolyte complex composite membrane. In *Proceedings of Third International Conference on Pervaporation Processes in the Chemical Industry,* ed. R. A. Bakish, pp. 194–202. Englewood, NJ: Bakish Materials Corporation.

Kaschemekat, J., J. G. Wijmans, R. W. Baker, and I. Blume. 1988. Separation of organics from water using pervaporation. In *Proceedings of Third International Conference on Pervaporation Processes in the Chemical Industry,* ed. R. A. Bakish, pp. 405–412. Englewood, NJ: Bakish Materials Corporation.

Kimura, S., and T. Nomura. 1983. Pervaporation of organic substance water system with silicone rubber membrane. *Maku (Membr.)* 8(3):177–183.

Kraetz, L. 1988. Dehydration of alcohol fuels by pervaporation. *Desalination* 70:481–485.

Kujawski, W., T. Q. Nguyen, and J. Neel. 1988. Pervaporation of water-alcohol mixtures through Nafion 117 and poly(ethylene-co-styrene sulfonate) membranes. In *Proceedings of Third International Conference on Pervaporation Processes in the Chemical Industry,* ed. R. A. Bakish, pp. 355–363. Englewood, NJ: Bakish Materials Corporation.

Meares, P. 1988. The sorption and diffusion of vapours in polymers. In *Proceedings of Third International Conference on Pervaporation Processes in the Chemical Industry,* ed. R. A. Bakish, pp. 12–20. Englewood, NJ: Bakish Materials Corporation.

Niemoller, A. 1988. Radiation-grafted membranes for pervaporation of ethanol/water mixtures. *J. Membr. Sci.* 36:385–404.

Ohya, H., K. Matsumoto, H. Matsumoto, H. Katagiri, Y. Futamura, S. Sata, and Y. Negishi. 1988. Transport of mixed vapors in membrane distillation. In *Proceedings of Third International Conference on Pervaporation Processes in the Chemical Industry,* ed. R. A. Bakish, pp. 501–507. Englewood, NJ: Bakish Materials Corporation.

Rapin, J. L. 1988. The Betheniville pervaporation unit—the first large-scale production plant for the dehydration of ethanol. In *Proceedings of Third International Conference on Pervaporation Processes in the Chemical Industry,* ed. R. A. Bakish, pp. 364–378. Englewood, NJ: Bakish Materials Corporation.

Rautenbach, R., C. Herion, and M. Franke. 1988. Dehydration of multicomponent organic systems by a reverse osmosis/pervaporation hybrid process—module, process design and economics. In *Proceedings of Third International Conference on*

Pervaporation Processes in the Chemical Industry, ed. R. A. Bakish, pp. 274–286. Englewood, NJ: Bakish Materials Corporation.

Reineke, C. E., J. A. Jagodzinski, J. A. Roper, and K. R. Denslow. Highly selective cellulosic polyelectrolyte membranes for the permeation of alcohol-water mixtures. *J. Membr. Sci.* 32(2–3):207.

Sander, U., and H. Janssen. 1989. Industrial applications of vapor permeation. Paper read at the 4th Intl. Conf. on Pervaporation in the Chemical Industry, 4–7 December 1989, Ft. Lauderdale, FL.

Sander, U., and P. Soukup. 1988. Practical experience with pervaporation systems for liquid and vapor separation. In *Proceedings of Third International Conference on Pervaporation Processes in the Chemical Industry,* ed. R. A. Bakish, pp. 508–518. Englewood, NJ: Bakish Materials Corporation.

Schneider, W. H. 1987. Purification of anhydrous organic mixtures by pervaporation. In *Proceedings of Second International Conference on Pervaporation Processes in the Chemical Industry,* ed. R. A. Bakish, pp. 169–175. Englewood, NJ: Bakish Materials Corporation.

Seok, D. R. 1987. Use of pervaporation for separating azeotropic mixtures using two different hollow fiber membranes. *J. Membr. Sci.* 33(1):71.

Wesslein, M., A. Heintz, G. A. Reinhardt, and R. N. Lichtenthaler. 1988. Pervaporation of binary and multicomponent mixtures using PVA membranes: experiments and model calculations. In *Proceedings of Third International Conference on Pervaporation Processes in the Chemical Industry,* ed. R. A. Bakish, pp. 172–180. Englewood, NJ: Bakish Materials Corporation.

IV
Dialysis

11.	Definitions	163
12.	Theory	167
13.	Design	186
14.	Applications	206
15.	Cost Estimates	212

11

Definitions

Stephen B. Kessler
Sepracor Inc.

Elias Klein
University of Louisville

INTRODUCTION AND HISTORY
TERMINOLOGY
GENERAL DESCRIPTION OF
   DIALYSIS
Driving Force for Mass Transfer
Membrane Function and Structure
NOTATION
REFERENCES

INTRODUCTION AND HISTORY

Although nineteenth century scientists attempting to understand the phenomenon of osmosis (e.g., Graham 1861) are generally credited with the discovery of semipermeable membranes, modern exploitation for industrial applications of dialysis must be credited to twentieth century workers. In the first part of this century, two applications of dialysis were widely described. One was the recovery of NaOH from cellulose steeping liquors. These contained both the desired alkali and contaminating hemicelluloses. The steeping liquors, containing 20% NaOH, were dialyzed against water and the NaOH—which permeated the parchment "membranes"—was recovered as a dilute (4 to 5%) solution. The contaminating hemicelluloses remained behind and were used as fuel or partially recycled into the cellulose feed streams. A second early application of dialysis was the recovery of acid from copper leaching solutions. Dilute sulfuric acid was recovered by dialysis because the acid diffused much more rapidly than did the metal salts.

The modern resurgence in applications of dialysis can be traced to development of the artificial kidney by Kolff and others (Kolff and Berk 1944; Quinton, Dillard, and Scribner 1960; Dedrick, Bischoff, and Leonard 1968). During World War II, Kolff was able to demonstrate that cellulose film could be used to reduce the concentration of metabolic endproducts in a patient whose kidneys had failed (Nose, Kjellstrand, and Ivanovich 1986). He managed to use arterial pressure to drive the patient's blood on one side of a membrane while bathing the other side with a saline solution. To reduce the interference from stagnant blood layers—described in Chapter 12 as a concentration boundary layer—he devised a rotating drum to support the narrow width cellulose tubing he used as membrane material. From this very modest beginning, we can trace the development of an entirely new therapy that now maintains more than 200,000 end-stage kidney patients worldwide (Drukker, Parsons, and Maher 1983).

As a consequence of the resurgent interest in dialytic separations—spurred by the growth of the artificial kidney development—a number of sophisticated separations, as well as catalytic processes have been developed. As described above, traditional dialysis is a rate-governed process wherein the membrane separates solutes by size. Both phases consist of aqueous solutions. Processes such as Donnan dialysis and membrane solvent extraction (MSX) are membrane-based equilibrium processes. A further distinction can be made in the case of MSX or catalytic membrane-based reactors (MBR); these operate with both aqueous and organic phases and, once again, the membrane is doing something other than a size separation. In both MSX and MBR, the membrane stabilizes the phase separation. In MBR the membrane also provides a site for catalyst (e.g., enzyme) immobilization. These and other examples of processes that in some way have descended from dialysis will be discussed in detail in later chapters of this handbook.

TERMINOLOGY

In the chapters that follow, a number of terms are used that can be subject to misinterpretation. To reduce this problem as much as possible, some definitions are presented here. These are based principally on the experience of the authors, but "membranologists" are in general agreement about their usage.

Dialysis is a rate-governed membrane process in which a microsolute is driven across semipermeable membrane by means of a concentration gradient. The microsolute diffuses through the membrane at a greater rate than macrosolutes also present in the *feed* solution. If the receiving solution (defined as *dialysate*) is not continuously renewed, the solute concentrations on both sides of the membrane will tend to equalize, negating the driving force for the separation. This is an essential difference between an equilibrium-based process and a rate-governed membrane process.

Common usage designates the fluid phase from which solutes are removed as the *feed* stream. Thus, in hemodialysis, the blood is considered the feed stream since it contains the metabolites that need to be removed. This does not mean, however, that solutes are only transferred out of the stream designated as feed. In fact, during hemodialysis, Ca^{2+} ions, bicarbonate ions, and glucose are transferred from the dialysate stream into the feed stream. Nevertheless, blood is designated as the feed stream because the major—or key—components are delivered to the dialyzer by this stream.

Dialysate is that stream designated to receive the solutes to be removed from the feed stream. It may be similar in composition to the feed, or it may be immiscible with the feed phase. Processes employing two immiscible phases are usually termed membrane solvent extraction. This equilibrium-based process is covered in detail in Chapter 41.

To characterize the transport properties of the dialytic membrane, current practice relies on nomenclature supplied from irreversible thermodynamics. The works of Kedem and Katchalsky (1958) and Katchalsky and Kedem (1962) serve as an excellent source of the nomenclature, as do reviews by Katchalsky and Curran (1965) and Bean (1972). Three terms will be used in the discussions that follow. The first, P_m, describes the diffusive permeability of a specific solute through a given membrane and is given in units of cm/s. If the membrane is homogeneous in cross section, the magnitude of P_m will scale inversely with membrane thickness. When it does scale linearly, it can be interpreted as being proportional to the diffusivity of that solute in the membrane matrix. However, because many membranes now made are asymmetric in structure, such interpretations may be speculative at best.

A second descriptor of membrane properties is the hydraulic permeability coefficient of the membrane, designated L_p. Its units are cm/s-atm, or analogous units. Again, the coefficient may or may not scale inversely with membrane thickness, depending on membrane asymmetry.

The third term used is a description of the relative permeability of solvent and solute i when a solution is forced through the membrane by applied pressure. Two terms are often confused: One is the rejection coefficient R_i and the other is the Staverman coefficient σ (Staverman

1951). The rejection coefficient is an experimental value that describes the fraction of the feed solute concentration that is *not* found in the filtrate. One hundred percent rejection means that none of the solute is found in the filtrate; 40% rejection means that 40% of the solute concentration does not appear in the filtrate (i.e., 60% does appear), and so on. It varies not only with the nature of the solute and the membrane, but also with the transmembrane velocity of the solution. It is not an intrinsic property of the membrane/solute pair. The Staverman coefficient is the limiting value approached by the rejection coefficient as the transmembrane velocity is increased. It is characteristic of a given membrane/solute pair and is defined quantitatively in Chapter 12.

GENERAL DESCRIPTION OF DIALYSIS

Driving Force for Mass Transfer

The use of dialysis for the separation of solutes from a common solvent relies for its driving force on the difference in chemical potential between the *feed* solution and the receiving *dialysate*. The process is diffusive. If the feed and dialysate are composed of the same solvent and differ only slightly in concentration, the assumption is made that the solute concentrations are proportional to their activities a. Thus, the earlier literature describing dialysis from aqueous feeds into aqueous dialysates assumes that concentration gradients describe the driving forces satisfactorily. When the feed and dialysate are composed of differing phases, these approximations no longer hold since the standard states to which the activities are referenced may be different. For exact descriptions of driving forces, we must then resort to estimates of fugacities.

Membrane Function and Structure

The upper limits of pore dimensions in dialysis membranes are purposely chosen to minimize convective transport, which is often an obligatory result of pumping feed and dialysate solutions. Consequently, the solute size that can permeate dialytic membranes has an upper bound determined by the above-mentioned factor and by compromises made during membrane fabrication.

Dialytic membranes are generally fabricated to have pore radii smaller than 60×10^{-8} cm. Pores of this dimension will, to a large extent, retain the common proteins (e.g., albumin, IgG) when only a diffusive gradient exists. Effective pore dimensions are reduced in hydrophobic synthetic polymers by adsorption of proteins, so that membranes with such large pores develop improved protein retention characteristics during use in a dialysis process. The hydrogel membranes (e.g., cellulosics), which are generally fabricated with much smaller pore radii (17×10^{-8} cm), do not adsorb proteins significantly, thus their transport properties are not altered during use.

When the pore dimensions of a dialysis membrane approach those of ultrafiltration membranes, it is important to specify the conditions of use to predict solute transfer. Because dialysis often requires high flow rates, there is frequently a component of convective flow through the membrane. Consequently, a dialytic membrane with large pores acts, in part, as an ultrafilter. If its solute permeability is measured in that mode of operation, quite different results are obtained compared with measurements that are obtained in a purely diffusive mode.

To achieve selective mass transfer in a dialytic process, we rely on a combination of solute and membrane properties. As an initial approximation, two solutes can be separated by dialysis if their diffusion coefficients in the feed differ by at least an order of magnitude. The solute diffusivity in the continuum within the membrane is an indicator of whether or not—and how fast—a solute may permeate the membrane. This diffusivity is modified by the matrix structure of the membrane, usually to a smaller value. In the simplest case of a hydrogel membrane, such as regenerated cellulose, the aqueous diffusion coefficients are reduced by a factor of 10 for small solutes, with higher reductions for larger solutes.

Additional information concerning dialysis membrane materials and structure can be found in Chapter 13.

NOTATION

General Notation

See the General Notation section at the beginning of this handbook.

Special Notation

L_p hydraulic permeability coefficient, L/tp
P_m diffusive permeability of solute in dialysis, L/t
R_i rejection coefficient for species i, dimensionless

REFERENCES

Bean, C. P. 1972. The physics of porous membranes—neutral pores. In *Membranes, A Series of Advances, Vol. 1, Macroscopic Systems and Models,* ed. G. Eisenman, pp. 1–54. New York: M. Dekker.

Dedrick, R. L., K. B. Bischoff, and E. F. Leonard, eds. 1968. *The Artificial Kidney.* AIChE Symposium Series No. 84, Vol. 64. New York: American Institute of Chemical Engineers.

Drukker, W., F. M. Parsons, and J. F. Maher, eds. 1983. *Replacement of Renal Function by Dialysis.* Boston: Martinus Nijhof.

Graham, T. 1861. Liquid diffusion applied to analysis. *Phil. Trans. Roy. Soc. London* 151:183–224.

Katchalsky, A., and P. F. Curran. 1965. *Nonequilibrium Thermodynamics in Biophysics.* Cambridge, Mass.: Harvard University Press.

Katchalsky, A., and O. Kedem. 1962. Thermodynamics of flow processes in biological systems. *Biophys. J.* 2(suppl.):53.

Kedem, O., and A. Katchalsky. 1958. Thermodynamic analysis of the permeability of biological membranes to nonelectrolytes. *Biochim. Biophys. Acta* 27:229.

Kolff, W. J., and H. T. Berk. 1944. The artificial kidney: a dialyzer with a great area. *Acta Med. Scan.* 117:121–134.

Nose, Y., C. Kjellstrand, and P. Ivanovich, eds. 1986. *Progress in Artificial Organs.* Cleveland: ISAO Press.

Quinton, W., D. Dillard, and B. H. Scribner. 1960. Cannulation of blood vessels for prolonged hemodialysis. *Trans. Am. Soc. Artif. Intern. Organs* 6:104.

Staverman, A. J. 1951. Theory of measurement of osmotic pressure. *Rec. Trav. Chim.* 70:344.

12
Theory

Stephen B. Kessler
Sepracor Inc.

Elias Klein
University of Louisville

MASS TRANSFER: MEMBRANE
    Transport in Homogeneous Membranes: Solution-Diffusion Theory
    Transport in Porous Membranes: Continuum Hydrodynamic Theory
    Irreversible Thermodynamics
    Nonideal Effects
MASS TRANSFER: OVERALL
    Mass Transfer in Permeable Ducts with Laminar Flow
    Mass Transfer in Irregular Geometries with or without Turbulent Flow
    Dialysis-Specific Mass Transfer Parameters
    Contribution of Convection to Mass Transfer
NOTATION
REFERENCES

MASS TRANSFER: MEMBRANE

To model effectively membrane transport phenomena that pertain to dialysis, two general types of membrane models can be considered: homogeneous and porous. Homogeneous membranes are thought of as structureless continua, which, in the case of dialysis, consist of polymer-liquid gels. Porous membranes are viewed as an impervious polymer phase, interpenetrated by liquid-filled pores.

The various classes of "real" membranes described in the preceding section can be divided into the categories of homogeneous and porous, although this distinction is not always an obvious one. Since the range of pore sizes in available membranes is relatively continuous, no sharp delineation exists between homogeneous and porous. Bean (1972) has suggested a criterion for making this distinction, based on a comparison of the diffusive and hydraulic permeabilities of a given membrane.

Modeling of transport in homogeneous membranes is well matched by a phenomenological approach and a solution-diffusion theory. This theory assumes convective transport to be negligibly small. The transport properties of porous membranes can be described by continuum hydrodynamic theory, a mechanistic approach that considers the structure of the membrane. Hydrodynamic theory takes both diffusive and convective transport into account. Linear, nonequilibrium (irreversible) thermodynamics constitutes a more general phenomenological approach that can be applied to transport in either class of membrane. This approach is especially useful when multiple solutes and/or driving forces are involved.

Transport in Homogeneous Membranes: Solution-Diffusion Theory

In general, the driving force for mass transfer is a potential gradient across the membrane thickness. Phenomenological models used to describe mass transfer processes in homogeneous membranes are concerned with the relationship between this driving force and the resulting flux. In dialysis, the primary flux is that due to diffusion, which is driven by the concentration gradient component of a gradient in chemical potential. While other driving forces may exist, leading to additional fluxes (e.g., pressure gradient component of chemical potential gradient causing bulk flow), dialysis is fundamentally a diffusive process and is modeled as such by solution-diffusion theory.

The cornerstone of solution-diffusion theory, Fick's first law, is a simple phenomenological model that describes the potential/flux relationship for diffusive mass transfer in binary systems. It states that the diffusive flux is proportional to the gradient in concentration, mole fraction, or chemical potential at constant pressure. In the strictest sense, a system consisting of solute, solvent, and membrane is ternary rather than binary. However, a pseudobinary approach can be taken in which solute A diffuses through component B, where B is defined as both membrane and solvent. The molar flux of solute A can be described by the following general form of Fick's first law:

$$N_{Ay} = x_A(N_{Ay} + N_{By}) - c_t D_{AB} \frac{dx_A}{dy}, \quad (12\text{--}1)$$

where N_{Ay} and N_{By} are the molar fluxes of A and B in the y direction, x_A is the mole fraction of solute A, c_t is the total concentration of A and B, and D_{AB} is the binary diffusion coefficient. The first term on the right represents bulk flow resulting from unequal counter diffusion of A and B, whereas the second term gives the diffusive flux.

In reasonably dilute solutions, the bulk flow term may be neglected and the total molar concentration c_t may be considered constant. Now Eq. (12–1) can be simplified to

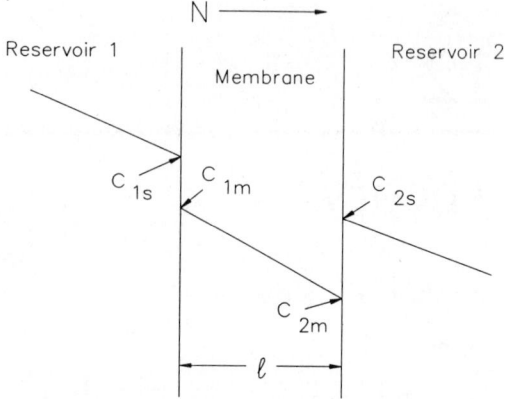

FIGURE 12–1. Typical concentration profile for diffusive transport across a membrane. In this example, the solute is less soluble in the membrane than in the external phases.

$$N_{Ay} = -D_{AB} \frac{dc_A}{dy}. \quad (12\text{--}2)$$

Assuming a linear concentration gradient and integrating Eq. (12–2) across the membrane,

$$N_{Ay} = \frac{D_{AB}(c_{1m} - c_{2m})}{l}, \quad (12\text{--}3)$$

where c_{1m} and c_{2m} are concentrations within the membrane at each of its interfaces and l is the membrane thickness.

Figure 12–1 shows a typical concentration profile for diffusive transport across a membrane. In this particular example, the solute is less soluble in the membrane than in either of the external solutions. Further, c_{1s} and c_{2s} are the concentrations of that solute in the external phases at the membrane/solution interfaces.

A partition coefficient K can be defined as the ratio between membrane and external solution concentrations at equilibrium:

$$K = \frac{c_{1m}}{c_{1s}} = \frac{c_{2m}}{c_{2s}}. \quad (12\text{--}4)$$

(Note: In general $c_{1m}/c_{1s} \neq c_{2m}/c_{2s}$). Combining Eqs. (12–3) and (12–4) and simplifying by eliminating subscripts,

$$N = KD\Delta c/l. \quad (12\text{--}5)$$

Equation (12–5) is a form of Fick's first law that is particularly well suited to membrane problems because external concentrations can be readily measured, while concentrations within the membrane, as in Eq. (12–3), cannot.

Defining a membrane diffusive permeability by

$$P_m = KD/l \qquad (12–6)$$

and a membrane mass transfer resistance by

$$R_m = l/KD = l/P_m, \qquad (12–7)$$

Eq. (12–5) can be written as

$$N = P_m \Delta c = \Delta c / R_m. \qquad (12–8)$$

Fick's second law can be derived either from Fick's first law (Crank 1975) or the equation of continuity (Bird, Stewart, and Lightfoot 1960). It describes one-dimensional diffusion in a liquid film in which there are no chemical changes and no flow taking place:

$$\frac{dc}{dt} = \frac{D \, d^2 c}{dy^2}. \qquad (12–9)$$

Consider a liquid film of thickness $2l$ that contains a solute at concentration c_0. At time $t = 0$, the solute concentration at both surfaces of the film is suddenly increased to c_1. Integration of Eq. (12–9), with appropriate boundary conditions, can provide concentration profiles within the film and fluxes as a function of time.

Colton and Lowrie (1981) illustrate the solution to the unsteady-state diffusion problem posed above in the form of dimensionless concentration profiles. The dimensionless group, Dt/l^2, is the Fourier number. When $Dt/l^2 = 1$, the film is approaching a new steady-state concentration. A characteristic diffusion time, $t_D = l^2/D$, can be defined that estimates the equilibration time for diffusion across a liquid film (or a liquid-filled membrane).

Transport in Porous Membranes: Continuum Hydrodynamic Theory

Yasuda and coworkers (Yasuda, Lamaze, and Ikenberry 1968; Yasuda et al. 1969; Yasuda, Lamaze, and Peterlin 1971) developed a concept of dialytic membranes as homogeneous water-swollen gels in which thermally induced movement of segments of (coiled) polymer molecules leaves an interstitial volume free for solute transport. They concluded that the permeability characteristics of a highly swollen system cannot be described by a single coefficient. Values of solute and solvent permeabilities depend on the conditions of measurement, particularly the magnitude of convective flux relative to diffusive flux (i.e., the transmembrane Peclet number). They concluded that the following generalizations were valid:

1. The permeability of a solute, whose size is small compared to a calculated membrane pore size, is proportional to the degree of hydration of the membrane.
2. Solute permeability coefficients decrease exponentially with increasing molecular size, when the latter is expressed in terms of molecular cross-sectional area.
3. Solute rejection coefficients change markedly when the solute cross section approaches the dimension of the channels.

A different approach was used by Klein, Holland, and Eberle (1979) for a series of dialytic membranes that included not only the homogeneous, swollen gels studied by Yasuda and coworkers, but also glassy polymers having porous structures. In describing transport through either type of structure, the essence of the problem is how to describe the size distribution of the channels through which the solutes must pass, and the interactions between the solute and the channel material. The Yasuda approach was based on the statistically predictable formation and disappearance of void volumes in the hydrogel. The interaction between solute and membrane matrix material is ignored. The approach used by Klein and

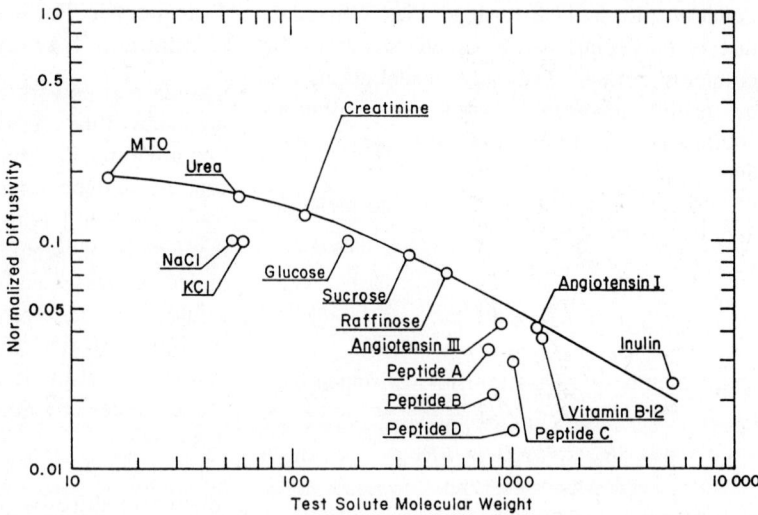

FIGURE 12–2. Plot of normalized diffusivity (D_m/D) as a function of solute molecular weight for a typical dialysis membrane (Cuprophan® 150 PM) (reprinted from Klein 1977 with permission).

coworkers based on earlier single pore channel models (Renkin 1954; Beck and Schultz 1972; Verniory et al. 1973) assumes that the membrane has a pore size distribution that can be represented by a single, hydrodynamically equivalent pore dimension. Since neither the pore fractional cross section A_p nor the actual pore length l_p can be measured independently, an experimental method is used to derive their ratio. From the Hagen-Poiseuille equation, the hydraulic permeability L_p is related to the ratio A_p/l_p and the hydrodynamic pore radius r_p by

$$L_p = \frac{A_p}{l_p} \frac{r_p^2}{8\eta}, \quad (12\text{-}10)$$

where η is the viscosity of the solution. The value of L_p is determined by measuring the volume flux of water across the membrane in response to an applied pressure gradient. Using a diffusional model (Beck and Schultz 1972; Verniory et al. 1973), the ratio A_p/l_p can be related to the diffusive permeability coefficient P_m by

$$P_m = \frac{(A_p/l_p)D(1-q)^2}{K_1} = \frac{D_m}{l_p}, \quad (12\text{-}11)$$

where q is the ratio of solute radius to pore radius, D is the solute diffusion coefficient in the solvent, K_1 is a power series in q, and D_m is the effective solute diffusivity in the membrane.

For a solute such as tritiated water, whose radius is small compared to the pore radius, q is approximately equal to zero, $K_1 = 1$, and the above equations can be solved simultaneously to yield

$$r_p^2 = \frac{8\eta D L_p}{P_m}. \quad (12\text{-}12)$$

Once the single representative membrane "pore" size has been determined in this way, the permeability of any other solute of known dimensions can be calculated for the same membrane. In the absence of any specific solute/membrane interactions, such as ionic or hydrophobic bonding, this model is useful for predicting membrane permeabilities when $q < 0.6$.

On the basis of either model, dialytic transport clearly decreases as solute size increases because of two effects: Increasing molecular size implies lower solute diffusivity and increasing solute sizes produce more interfering collisions with the pore walls. The effect is seen in Figure 12–2 where the ratio of diffusivity in the membrane D_m to the solution diffusivity D (defined as normalized diffusivity) is plotted as a function of solute molecular weight. At the low molecular weight end of the plot, the impedance of the membrane reduces diffusivity to 50 to 80% of the free solution diffusivity. Membrane impedance can be defined quantitatively as (1 − normalized diffusivity). At higher molecular weights—where solute sizes approach the dimensions of the membrane channels—the impedance is more than 99%.

Irreversible Thermodynamics

The simultaneous presence of both concentration and pressure driving forces results in a mixed diffusive-convective transport process. When both diffusive and convective fluxes are of significant magnitudes, both must be accounted for, as well as their interactions. Irreversible thermodynamics provides a framework for analyzing these complex interactions between fluxes and forces. Irreversible thermodynamics has been applied to membranes by Kedem and Katchalsky (1958) and Katchalsky and Kedem (1962) based on the fundamental work of Onsager (1931a, 1931b). In addition to the reviews by Katchalsky and Curran (1965) and Bean (1972), this topic has been discussed from a membrane perspective by Bungay (1986) and Spriggs and Li (1976).

The hydraulic permeability of a membrane, L_p, is expressed in the form of a coefficient of proportionality between volumetric flux (superficial velocity) J_v and transmembrane pressure difference Δp as shown in Eq. (12–13):

$$J_v = L_p \Delta p \quad (\Delta \pi_s = 0). \quad (12\text{–}13)$$

Equation (12–13) applies in the absence of an osmotic pressure difference ($\Delta \pi_s = 0$). If this is not the case, then osmotic pressure differences across the membrane will affect the flux. This situation is described by Eq. (12–14):

$$J_v = L_p \Delta p + L_{pD} \Delta \pi_s, \quad (12\text{–}14)$$

where L_{pD} is the coefficient of osmotic flow. A second phenomenological equation defines the differential rate of transport of solute and solvent, J_D:

$$J_D = L_{Dp} \Delta p + L_D \Delta \pi_s, \quad (12\text{–}15)$$

where L_{Dp} is defined as the ultrafiltration coefficient and L_D is the diffusional mobility per unit osmotic pressure difference.

By Onsager's law, $L_{pD} = L_{Dp}$. Thus, three measurements are required to characterize a given membrane-solvent-solute system. One experiment is suggested by Eq. (12–13): measurement of hydraulic permeability L_p in the absence of an osmotic pressure difference.

In the case of a selective membrane with solute present, a difference in solute and solvent fluxes is observed. This exchange flow is described by Eq. (12–16), and it provides a means of measuring L_{Dp}:

$$J_D = L_{Dp} \Delta p \quad (\Delta \pi_s = 0). \quad (12\text{–}16)$$

A third experiment can be done with no hydrostatic pressure difference, but with solute concentration differing across the membrane ($\Delta \pi_s > 0$). In this case, both an exchange flow and an osmotic flow take place. The exchange flow defines L_D and is described by Eq. (12–17):

$$J_D = L_D \Delta \pi_s \quad (\Delta p = 0). \quad (12\text{–}17)$$

A useful transformation of Eqs. (12–14) and (12–15) is given below. Staverman (1951) defined a reflection coefficient σ:

$$\sigma = -(L_{pD}/L_p). \quad (12\text{–}18)$$

At zero volume flow, the solute flux J_s can be related to the osmotic pressure by a solute permeability coefficient ω:

$$\omega = J_s/\Delta \pi_s \quad (J_v = 0), \quad (12\text{–}19a)$$

where ω is related to the membrane permeability P_m through the van't Hoff equation. Thus,

$$P_m = \omega RT. \quad (12\text{–}19b)$$

With the coefficients defined by Eqs. (12–18), (12–19a), and (12–19b), the transformed phenomenological equations are (Kedem and Katchalsky 1958; Katchalsky and Curran 1965):

$$J_v = L_p \Delta p - \sigma L_p \Delta \pi_s, \quad (12\text{–}20)$$

$$J_s = c_s(1 - \sigma)J_v + \omega\Delta\pi_s. \qquad (12\text{--}21)$$

Equations (12–20) and (12–21) are known as the Kedem-Katchalsky equations.

For systems involving multiple solutes, coupling occurs in the sense that the chemical potential gradient of each solute affects the flux of all solutes. Fick's laws, being limited to binary systems, do not account for coupling. The nonequilibrium thermodynamics approach considers that all fluxes depend on all forces, thus providing an appropriate theoretical framework within which to account for coupling.

Nonideal Effects

The plot of reduced diffusivities as a function of molecular weight can be misleading when applications are projected from such data. On the basis of the reduced diffusivity for the examples shown, almost no transfer of solutes greater than about 5000 dalton would be expected. However, in "real-world" devices mass transfer occurs in the presence of differential transmembrane pressures. The need to flow feed and dialysate solutions past the membrane surfaces creates obligatory pressure drops in the respective channels. When these are unequal, transfer occurs by convective modes as well as by dialytic transfer. Moreover, the presence of nonpermeating species may create osmotic pressures that oppose the applied hydraulic pressure differences. Thus, the computation of mass transfer in a device is often quite a bit more complicated than in a laboratory test cell in which flow and pressure conditions can be controlled more closely.

But even under the more or less ideal circumstances of laboratory test cells, nonideal effects stem from interactions between the feed solution and the membrane. For membranes made from hydrophobic polymers, the best described phenomenon is solute adsorption within the pores. Strong hydrophobic bonding between large solutes, such as peptides and proteins, can alter the effective pore dimensions of the dialysis membranes. This is reflected in reductions of both the hydraulic permeability and the dialytic transfer of large solutes.

MASS TRANSFER: OVERALL

In an overall sense, mass transfer in the dialysis process involves three phases: feed, membrane, and dialysate. In addition to the resistance to solute diffusion encountered in the membrane, diffusional resistances in each of the two liquid phases result in concentration gradients adjacent to the membrane. Within these boundary layers, solute concentration differs from that of the bulk solution. The magnitude of the mass transfer resistance across such a boundary layer is dependent on the thickness of the layer, which depends, in turn, on fluid flow parameters. Mass transfer coefficients are a convenient means of quantifying mass transfer rates in the case of the two moving fluid phases, feed and dialysate, for both laminar and turbulent flow. As shown in Eq. (12–22), the mass transfer coefficient k is defined by an expression of the same form as Fick's first law [see Eqs. (12–5) and (12–8)]. The concentration difference Δc is that between the membrane surface and the bulk liquid:

$$N = k\Delta c, \qquad (12\text{--}22)$$

where k replaces the permeability P_m, which in Fick's first law is based on a constant film thickness. Mass transfer coefficient k is expressed in the same units as P_m, that is, linear velocity, and, like P_m, is the reciprocal of mass transfer resistance, but unlike P_m, is based on a variable film thickness. The combined effects of simultaneous convective and diffusive mass transfer that occur in permeable ducts are difficult to analyze theoretically. Mass transfer coefficients are useful, both as a means of expressing analytical results and as a measurable experimental parameter in cases for which analysis is inadequate.

Consider a dialyzer of unspecified geometry operating countercurrently. The feedside flow is assumed to be laminar, while the dialysate flow is assumed to be turbulent. The velocity profiles for this case are illustrated by Keller (1973), as shown in Figure 12–3(a). Solutes in the feed are convected parallel to the membrane surface, diffuse through the membrane phase, and are then transferred by convection again in

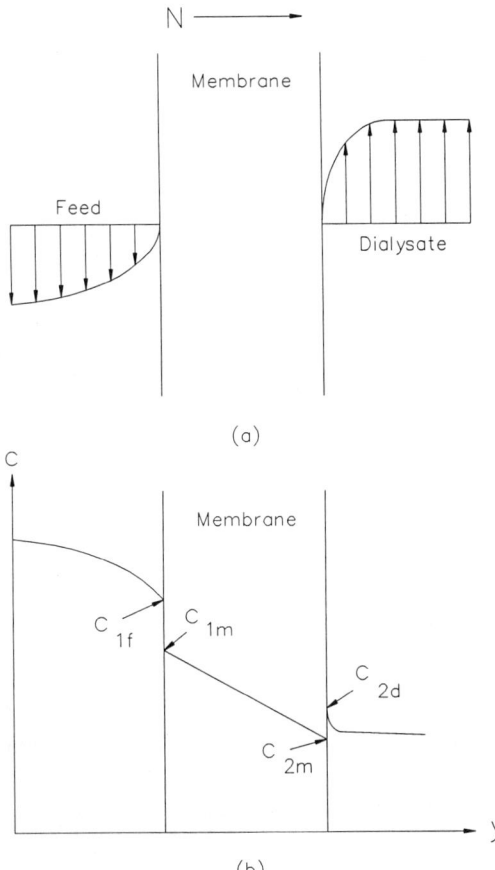

FIGURE 12–3. (a) Velocity profiles for the case of laminar feed flow and turbulent dialysate flow. (b) Concentration profiles corresponding to the velocity profiles in (a) (after Keller 1973).

the dialysate. Figure 12–3(b) shows the resulting (qualitative) concentration profiles at a single point with respect to the length of the dialyzer.

The overall mass transfer resistance R_o can be expressed as the sum of the resistances associated with each phase:

$$R_o = R_f + R_m + R_d, \quad (12\text{–}23)$$

where the subscripts f, m, and d refer to feed, membrane, and dialysate, respectively.

The overall mass transfer coefficient k_o is the reciprocal of R_o:

$$k_o = 1/R_o. \quad (12\text{–}24)$$

Thus,

$$1/k_o = 1/k_f + 1/P_m + 1/k_d. \quad (12\text{–}25)$$

Considering now the length coordinate, both momentum and concentration boundary layers develop along the length of a flow conduit as shown in Figure 12–4. Most practical devices operate such that the momentum boundary layer is fully developed within a negligibly short axial distance, while, in the same devices, concentration boundary layers develop throughout their length. As a result, the local mass transfer coefficient, as defined above, varies throughout the length of the device and is difficult to measure experimentally. A length-averaged, overall mass transfer coefficient is a readily measured parameter. It expresses the relationship between the average molar flux N and a length-averaged concentration difference $\overline{\Delta c}$:

$$N = \overline{k_o} \overline{\Delta c}. \quad (12\text{–}26)$$

Evaluating $\overline{k}_o$ can be done analytically for laminar flow in tubes and rectangular ducts. For irregular geometries and/or turbulent flows, empirical correlations can be used. In either case, the heat transfer literature serves as a basis, and dimensionless groups provide an efficient means of expressing the results.

Mass Transfer in Permeable Ducts with Laminar Flow

In some dialyzer designs, in particular, hemodialyzers for which extracorporeal blood volume must be minimized, the feedside flow is channeled through relatively thin ducts and is laminar in nature. As stated above, this situation lends itself to an analytical approach.

The Sherwood number (analogous to the Nusselt number in heat transfer) is a form of dimensionless mass transfer coefficient. It is defined by

$$\text{Sh} = kL/D, \quad (12\text{–}27)$$

where L is a length characteristic of the duct geometry and D is the diffusion coefficient of the solute species under consideration.

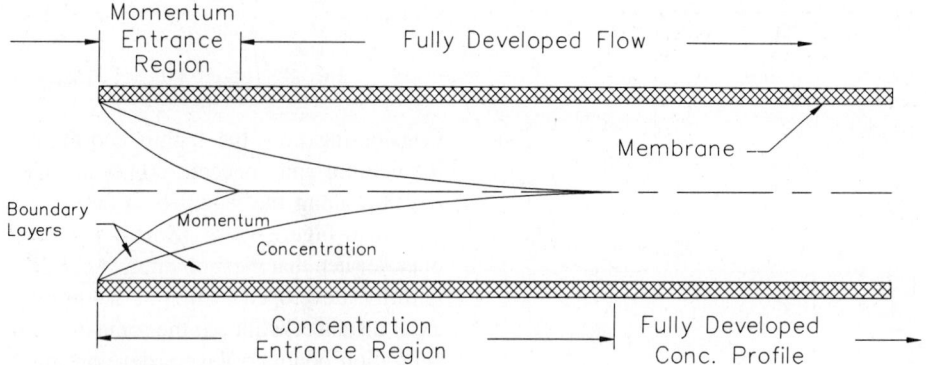

FIGURE 12–4. Flow conduit within a diffusive mass transfer device showing development of momentum and concentration boundary layers (after Colton and Lowrie 1981).

A length-averaged, feedside Sherwood number $\overline{Sh}_f$ can be defined as

$$\overline{Sh}_f = \overline{k}_f L/D. \qquad (12\text{--}28)$$

This $\overline{Sh}_f$ can be related to experimentally measurable concentrations as well as estimated analytically for laminar flow in regularly shaped ducts. Equations (12–29) and (12–30) define a dimensionless concentration c^* and a dimensionless axial length z^* for laminar flow in a porous tube:

$$c^* = (c_b - c_f)/(c_d - c_f), \qquad (12\text{--}29)$$

$$z^* = zD/vd_{ti}^2, \qquad (12\text{--}30)$$

where c_b is the average bulk solute concentration in the feed duct, c_f and c_d are solute concentrations in the feed (as it enters) and dialysate, respectively, z is distance in the axial direction, D is the solute diffusion coefficient, v is average axial velocity, and d_{ti} is the diameter of the tube.

Equation (12–31) defines a wall Sherwood number (Sh_w), the ratio of the concentration gradient in the fluid adjacent to the wall to that in the wall, and Eq. (12–32) defines a length-averaged overall Sherwood number ($\overline{Sh}_o$) in terms of the previously defined feedside and wall Sherwood numbers:

$$Sh_w = k_w d_{ti}/D, \qquad (12\text{--}31)$$

$$1/\overline{Sh}_o = 1/Sh_w + 1/\overline{Sh}_f. \qquad (12\text{--}32)$$

Solution of the convection-diffusion equation with appropriate boundary conditions gives values for $\overline{Sh}_o$, from which values for $\overline{Sh}_f$ may be calculated using Eq. (12–32). Colton and coworkers (Colton and Lowrie 1981; Colton 1969; Colton et al. 1971) have reviewed various solutions for tubes and rectangular ducts. The classical Graetz problem is stated for laminar flow heat transfer in a tube with constant temperature at the wall, analogous to constant wall concentration in the mass transfer problem. Lévêque's solution (Lévêque 1928) assumes a linear velocity profile near the wall and is valid in the concentration entrance region. This solution is a useful approximation in many cases and is given in Eq. (12–33):

$$\overline{Sh}_f = 1.62 z^{*-0.33}. \qquad (12\text{--}33)$$

Colton (1969) solved the case of a constant flux boundary condition with finite wall permeability for rectangular channels, which is more pertinent to a typical dialysis process. Davis and Parkinson (1970) solved the equivalent case for tubes. These solutions are shown graphically by Colton and Lowrie (1981) and Colton (1988). Their graph of the Davis and Parkinson solution for tubes is reproduced in

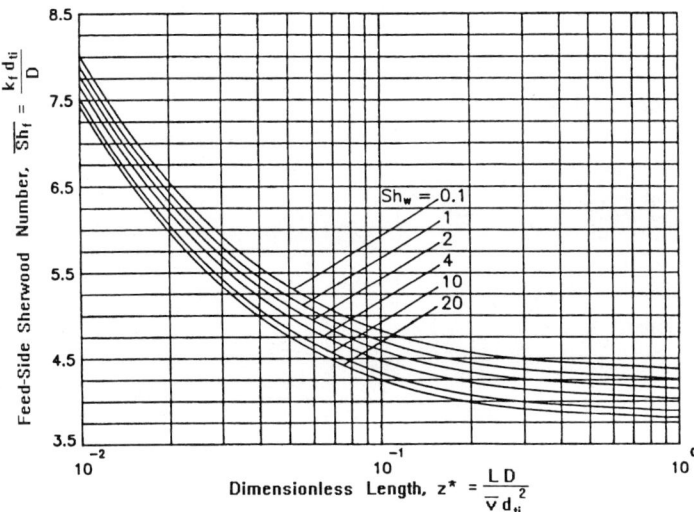

FIGURE 12–5. Davis and Parkinson solution for laminar-flow mass transfer in a tube with permeable walls—constant flux boundary condition. The family of curves represents values of the wall Sherwood number Sh_w (after Colton and Lowrie 1981).

Figure 12–5, which shows a family of curves covering a range of wall Sherwood numbers from $Sh_w = 0.1$ to 20. In the concentration entrance region (dimensionless length $z^* < 10^{-2}$), the Lévêque solution is a good approximation to the Davis and Parkinson solution. At higher values of dimensionless length, corresponding to more fully developed boundary layers, the Lévêque solution tends to underpredict the Sherwood number and does not predict the asymptotic value of $\cong 4$ predicted by both Davis and Parkinson for tubes and Colton for rectangular channels.

Subsequently, Cooney, Davis, and Kim (1974) and Jagannathan and Shettigar (1977) included a variable dialysate-side resistance in reformulating the analysis as a conjugated boundary value problem for parallel-plate and hollow-fiber configurations, respectively. These latter approaches require solution by numerical methods. The analysis of Jagannathan and Shettigar for hollow fibers, which also considers the effects of ultrafiltration, predicts that, for constant membrane area, fiber diameter and length have very little effect on overall mass transfer for low-permeability membranes. However, for high-permeability membranes, overall mass transfer decreases with increasing membrane diameter.

Mass Transfer in Irregular Geometries with or without Turbulent Flow

The dialysate-side flow in many dialyzer designs occurs within an irregular geometry and may be turbulent as well. Analysis of these situations is difficult and the resulting predictions tend to exhibit larger errors than in the above case of laminar flow in well-defined ducts. Instead, correlations from empirical heat and mass transfer studies are usually employed.

One dialysate-side flow configuration that has been treated analytically is the shell side of hollow-fiber dialysis modules. The starting point for these analyses is modeling of laminar flow relative to arrays of cylinders. (Most hollow-fiber dialyzers contain sufficiently small diameter fibers at high enough packing densities that shellside flow is laminar). The "free surface" or "equivalent annulus" model of Happel (1959) applies to flow either parallel or perpendicular to the cylinder's axes and is limited to low packing densities. Sparrow and Loeffler (1959) obtained an analytical solution for the parallel flow case that applies at high as well as at low packing densities, where it agrees with Happel's model.

Hermans (1978), Noda and Gryte (1979), and Gostoli and Gatta (1980) have developed models for hollow-fiber dialyzer performance that incorporate dialysate-side mass transfer resistance. Hermans makes use of an equivalent annulus approximation to model a hollow-fiber dialyzer with shellside flow perpendicular to the fibers. Gostoli and Gatta apply the same approximation to parallel flow in both cocurrent and countercurrent directions. They define a parameter, α, which is the ratio of the inside diameter of the fictitious annulus to the fiber outside diameter. Fiber packing density Φ can be defined by

$$\Phi = Nd_{to}^2/d_S^2, \qquad (12\text{--}34)$$

where N is the number of fibers, d_{to} is the fiber outside diameter, and d_S is the inside diameter of the module shell. The quantity α can also be defined as the square root of Φ. For values of $\alpha < 0.635$ ($\Phi < 0.4$), estimation of Sherwood numbers by the equivalent annulus approximation is accurate to within 5% (Gostoli and Gatta 1980).

Noda and Gryte (1979) applied the complete Sparrow-Loeffler solution to dialysate-side mass transfer without the usual simplifying assumptions. Their model takes into account the interactions between neighboring fibers that occur at higher packing densities. Figure 12–6 shows the values predicted for the various Sherwood numbers by their model. Maxima for both dialysate-side Sherwood number, $\overline{Sh}_d$, and overall Sherwood number, $\overline{Sh}_o$, are predicted to occur at a packing density value of 0.63. Analyses based on the equivalent annulus approximation (Hermans 1978; Gostoli and Gatta 1980) neglect interactions between fibers and predict Sh_o to increase with packing density without reaching a maximum.

All of the above authors assume regular arrays of cylinders, either triangular or square, in their analyses. The authors of the hollow-fiber models (Hermans 1978; Noda and Gryte 1979; Gostoli and Gatta 1980) all recognize that real dialyzers represent randomly distributed arrays whose behavior can deviate significantly from that of ideally distributed models. One form of deviation that can occur is channeling

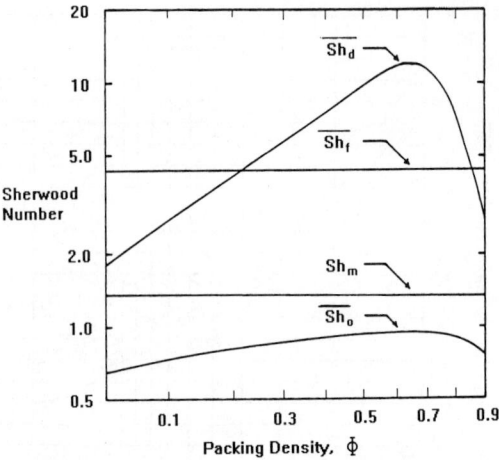

FIGURE 12–6. Sparrow-Loeffler solution for axial-flow shellside mass transfer, including interactions between fibers. Assuming a regular, triangular array, the model predicts maxima in both dialysate-side and overall Sherwood numbers ($\overline{Sh}_d$ and $\overline{Sh}_o$) at a packing density of $\Phi = 0.63$ (after Noda and Gryte 1979).

or maldistribution of flow, which is discussed further in Chapter 13. In general, these deviations compromise the efficiency of real dialyzers and the model predictions should be viewed as an upper limit of achievable performance.

Devices that deviate significantly from model predictions can be treated with correlations obtained empirically for similar devices. This approach can be used in both laminar and turbulent flow and for virtually any geometry. In this type of correlation, the Sherwood number is usually defined in terms of other dimensionless groups whose values can be measured experimentally. The Reynolds number, Re, characterizes fluid flow as a ratio of inertial-to-viscous forces, while the Schmidt number, Sc, expresses the relationship between momentum diffusivity and molecular diffusivity:

$$\text{Re} = \rho v L/\eta, \qquad (12\text{--}35)$$

$$\text{Sc} = \eta/\rho D, \qquad (12\text{--}36)$$

where ρ is the fluid density, v is a characteristic velocity, L is a characteristic length, η is the fluid viscosity, and D is the solute diffusion coefficient.

The general expression for local Sherwood number is given by

$$\text{Sh} = f(\text{Re, Sc, duct geometry}). \quad (12\text{–}37)$$

For a length-averaged Sherwood number, the dependency on duct geometry drops out and correlations of the form of Eq. (12–38) are used:

$$\text{Sh} = a(\text{Re}^b, \text{Sc}^c), \quad (12\text{–}38)$$

where a, b, and c are empirically determined constants.

Spriggs and Li (1976), Yang and Cussler (1986), Klein, Ward, and Lacey (1987), and Prasad and Sirkar (1988) have reviewed mass transfer correlations published for heat and mass transfer studies for flow external to tubes. Values reported in the literature for constant b vary from 0.33 to 0.93. The lower end of this range probably results from flows that are laminar in nature, while the upper end may reflect effects such as movement of flexible tubes (e.g., hollow fibers) in response to the external flow field. A value of 0.67 (⅔) is generally accepted for turbulent flow. Better agreement is seen for the value of c, which ranged from 0.32 to 0.38 and is usually fixed at 0.33 (⅓). Constant a varies considerably and appears to depend on the geometry of the system.

If a given system is similar in its geometry and fluid flow to one for which an established correlation exists, then applying the published values for the constants a, b, and c will probably predict its performance with reasonable accuracy. If this is not the case, then a Wilson plot may be used to correlate data from the new system. In this procedure, first used by Wilson (1915) to analyze heat transfer resistances, the relationship between dialysate mass transfer coefficient $\bar{k}_d$ and dialysate flow is assumed to follow a relationship defined by

$$\bar{k}_d = AQ_d^b, \quad (12\text{–}39)$$

where Q_d is the dialysate flow rate and A and b are constants specific to the flow geometry and

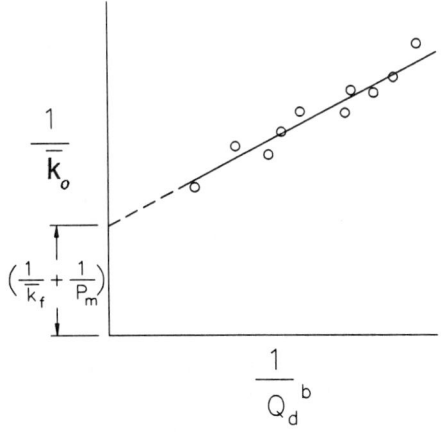

FIGURE 12–7. Wilson plot used to correlate shellside mass transfer data from devices for which a specific correlation has not been established.

solute under study. Then, substituting in Eq. (12–25) gives

$$1/\bar{k}_o = 1/\bar{k}_f + 1/P_m + 1/AQ_d^b. \quad (12\text{–}40)$$

The quantity $\bar{k}_o$ is measured at several values of Q_d and $1/\bar{k}_o$ is plotted versus $1/Q_d^b$ as shown in Figure 12–7. A value for b is chosen that gives the best straight-line fit to the data. The intercept of this line corresponds to infinite dialysate flow rate at which $1/\bar{k}_d$ is zero. Thus, the value of the intercept equals $(1/\bar{k}_f + 1/P_m)$ and $1/\bar{k}_d$ at any Q_d can be determined by subtracting the intercept from the corresponding value of $1/\bar{k}_o$.

Dialysis-Specific Mass Transfer Parameters

In the previous section, mass transfer was discussed for the liquid phases adjacent to the membrane. For practical applications, this analysis must be extended to describe mass transfer for the dialyzer as a whole. Neglecting convective contributions, we can write the following equation for the moles of solute, dQ_i, transferred across an element of membrane of length dz and area dA_m per unit time (Figure 12–8):

$$dQ_i = k_o(c_f - c_d) \, dA_m, \quad (12\text{–}41)$$

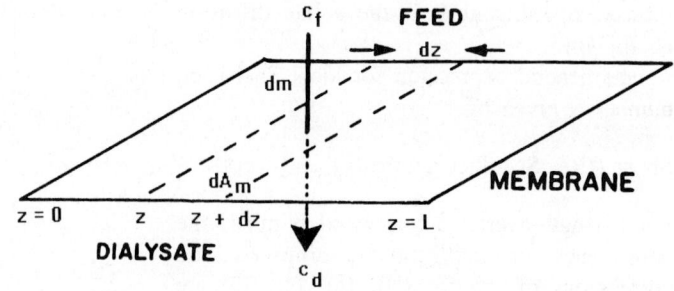

FIGURE 12–8. Diffusive mass transfer across an element of a membrane of length dz and area dA_m (reprinted from Klein, Ward, and Lacey 1987 with permission).

where c_f and c_d are the solute concentrations in the feed and dialysate, respectively, in the element of length dz. Equation (12–41) can be integrated along the length of the membrane to give the following equation for overall solute transfer rate:

$$Q_i = k_o A_m \left(\frac{\Delta c^*_{z=1} - \Delta c^*_{z=0}}{\ln(\Delta c^*_{z=1}/\Delta c^*_{z=0})} \right), \quad (12\text{–}42)$$

where $\Delta c^*_{z=0}$ and $\Delta c^*_{z=1}$ are the transmembrane concentration differences at either end of the device. For countercurrent flow, $\Delta c^*_{z=0} = c_{fi} - c_{do}$ and $\Delta c^*_{z=1} = c_{fo} - c_{di}$. This leads to the following equation for overall mass transfer in a dialyzer with this flow configuration:

$$Q_i = k_o A_m \left(\frac{(c_{fo} - c_{di}) - (c_{fi} - c_{do})}{\ln[(c_{fo} - c_{di})/(c_{di} - c_{do})]} \right). \quad (12\text{–}43)$$

A similar expression can be derived for cocurrent flow by redefining $\Delta c^*_{z=0}$ and $\Delta c^*_{z=1}$.

If ultrafiltration is negligible, inlet and outlet flows are equal for both the feed and dialysate streams. Using this assumption, the following overall mass balances may be written for each stream:

$$Q_i = Q_f(c_{fi} - c_{fo}) = Q_d(c_{do} - c_{di}). \quad (12\text{–}44)$$

The performance of a dialyzer can be described in terms of its *dialysance* D^*, which is defined as the rate of mass transfer divided by the concentration difference between inlet feed and inlet dialysate (Klein, Ward, and Lacey 1987), that is,

$$D^* = \frac{Q_i}{c_{fi} - c_{di}}. \quad (12\text{–}45a)$$

Clearance, defined by Eq. (12–45b), is derived from renal physiology and is similar to dialysance except that dialysate solute concentration is not considered (Colton and Lowrie 1981):

$$C = \frac{Q_i}{c_{fi}}. \quad (12\text{–}45b)$$

If the solute concentration is zero in the dialysate (as in the case of a single-pass dialysate system), then dialysance and clearance have the same value.

Combining Eqs. (12–43), (12–44), and (12–45a), the following expression is obtained for the dialysance of a countercurrent dialyzer in terms of membrane properties and hydrodynamics (k_o, A_m) and operating conditions (Q_f, Q_d):

$$D^* = Q_f \left(\frac{\exp[k_o A_m(1 - Q_f/Q_d)/Q_f] - 1}{\exp[k_o A_m(1 - Q_f/Q_d)/Q_f] - Q_f/Q_d} \right).$$

$$(12\text{–}46)$$

Note that if the feed and dialysate flows are essentially equal ($Q_f \cong Q_d$), then

$$D^* = Q_f \frac{k_o A_m}{Q_f + k_o A_m}. \quad (12\text{–}47)$$

The foregoing assumes that the solute is distributed freely in the solvent phase. This assumption may not always be valid; for example, in protein solutions, small solutes may bind to the protein molecules and exist in equilibrium between the free and bound states. In such circumstances, modifications of Eq. (12–46) must be used to determine dialysance (Farrell et al. 1974).

Equation (12–46) can be used to estimate the degree of separation of two solutes by a given dialyzer. For any solute, the extraction ratio E of solute from the feed stream is given by

$$E = \frac{Q_f c_{fi} - (Q_f - Q_{uf})c_{fo}}{Q_f c_{fi}}. \quad (12\text{–}48)$$

For negligible ultrafiltration, this reduces to

$$E = \frac{c_{fi} - c_{fo}}{c_{fi}}. \quad (12\text{–}49)$$

If the inlet concentration of solute in the dialysate is zero ($c_{di} = 0$), Eq. (12–49) further reduces to

$$E = \frac{D^*}{Q_f}. \quad (12\text{–}50)$$

The extraction ratio represents the fraction of maximum solute concentration change that can be attained under a given set of operating conditions. It is analogous to the effectiveness of a heat exchanger. The terms of Eq. (12–46) can be combined through definition of the following dimensionless parameters:

$$N_t = k_o A_m / Q_f, \quad (12\text{–}51)$$

$$Z = Q_f / Q_d. \quad (12\text{–}52)$$

Here N_t is defined as the number of transfer units and is a measure of the mass transfer size of a dialyzer. Extraction ratio E for countercurrent flow can now be expressed as

$$E = \frac{1 - \exp[-N_t(1 - Z)]}{1 - Z\exp[-N_t(1 - Z)]}. \quad (12\text{–}53)$$

Similar expressions can be derived for other flow configurations including well-mixed dialysate flow (Michaels 1966). For cocurrent flow,

$$E = \frac{1 - \exp[-N_t(1 + Z)]}{1 + Z}. \quad (12\text{–}54)$$

For perpendicular flow,

$$E = \left(\frac{1}{N_t Z}\right) \sum_{n=0}^{\infty} [S_n(N_t) S_n(N_t Z)], \quad (12\text{–}55)$$

where

$$S_n(Y) = 1 - \exp(-Y) \sum_{m=0}^{n} \left(\frac{Y^m}{m!}\right).$$

The term "perpendicular flow" refers to the case wherein the feed and dialysate streams flow at 90° relative to each other. This term was chosen instead of "crossflow" because "crossflow" is used interchangeably with "tangential flow" to describe flow tangential to the membrane surface.

The relationships between E, N_t, and Z are shown graphically for each of the three flow configurations in Figures 12–9, 12–10, and 12–11. The separation factor α_{jk} in a dialytic process can be considered to be the ratio of fractional masses of two solutes removed from their common feed stream, under a given set of operating conditions. By applying Equation (12–48) to each solute, we can see that α_{jk} is equal to the ratio of fractional extractions of the two solutes. Thus, Figures 12–9, 12–10, and 12–11 can be used to estimate the relative separation of two solutes from a knowledge of their overall mass transfer coefficients, membrane area, and feed and dialysate flow rates.

Contribution of Convection to Mass Transfer

The above derivations are based on the assumption that convective mass transfer is negligible; that is, significant ultrafiltration is not occurring. If significant ultrafiltration does occur, mass transfer will be enhanced by the addition of a convective component to the diffusive mass transfer described above. Ultrafiltration will occur from the feed to the dialysate in response to a pressure gradient, which may be either applied, in order to concentrate the feed, or obligatory, as a consequence of the geometry of the device and the desired feed flow rate. Less commonly, if the feed is highly concentrated

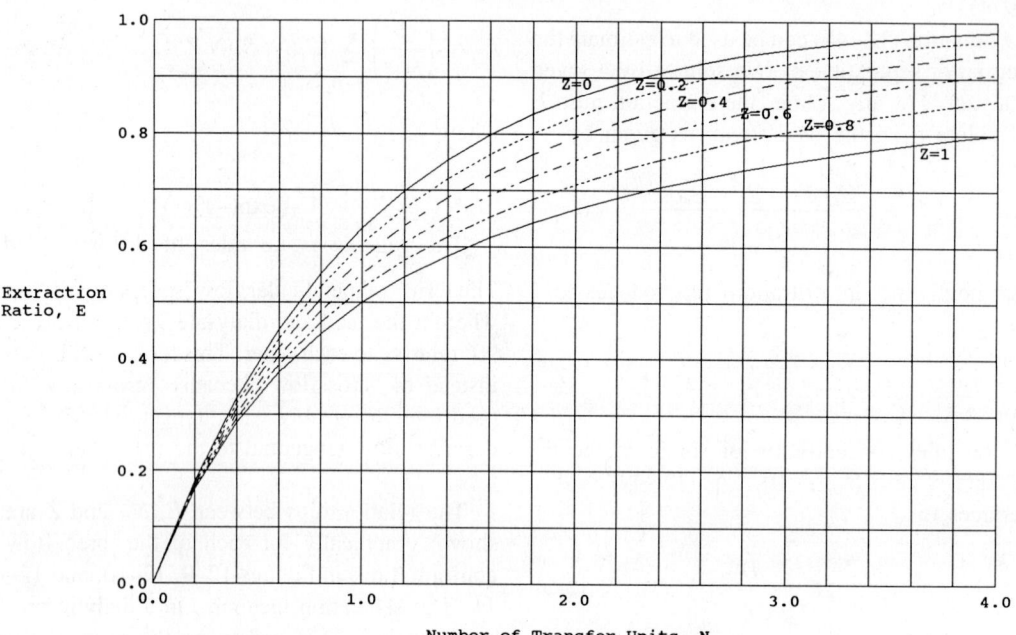

FIGURE 12-9. Extraction ratio E as a function of number of transfer units N_t in a countercurrent flow dialyzer. The graphs of Figures 12-9, 12-10, and 12-11 can be used to estimate the relative separation of two solutes in each of three flow configurations.

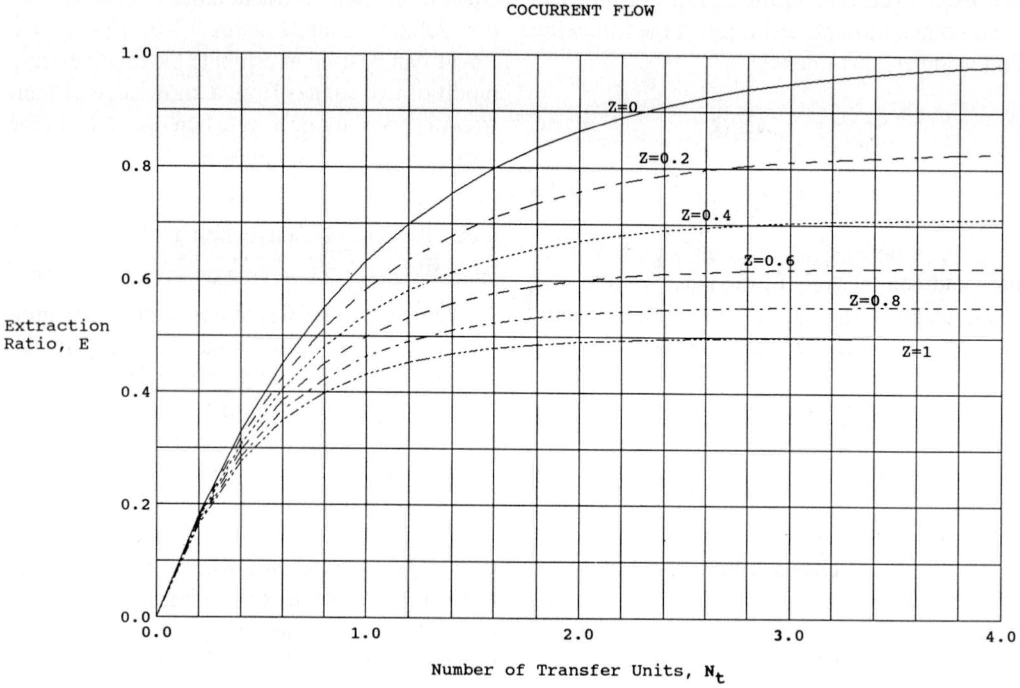

FIGURE 12-10. Extraction ratio E as a function of number of transfer units N_t in a cocurrent flow dialyzer.

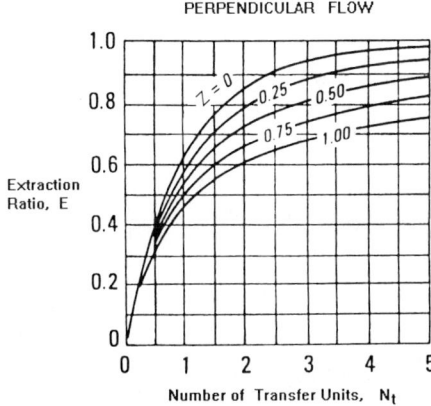

FIGURE 12–11. Extraction ratio E as a function of number of transfer units N_t in a perpendicular flow dialyzer (reprinted from Colton and Lowrie 1981 with permission).

and the dialysate dilute, ultrafiltration may occur from the dialysate to the feed in response to an osmotic gradient. In the first case, ultrafiltration will enhance solute transport while in the second it will impede it.

As discussed earlier, at a steady state the overall solute flux J_s at a point on the membrane can be considered to consist of the sum of a diffusive and convective component:

$$J_s = P_m \Delta c_s + J_v(1 - \sigma)\bar{c}_s, \quad (12\text{-}56)$$

where the first term on the right side, $P_m \Delta c_s$, represents diffusive transfer and the second term, $J_v(1 - \sigma)\bar{c}_s$, convective transfer (Spiegler and Kedem 1966). The average solute concentration within the membrane, $\bar{c}_s$, can be estimated using Eqs. (12–57), (12–58), and (12–59) (Villaroel, Klein, and Holland 1977):

$$\bar{c}_s = c_w - \frac{\Delta c_s}{3}, \quad (12\text{-}57)$$

where c_w is the feedside solute concentration at the membrane wall, which can, in turn, be estimated using concentration polarization theory, of which Eq. (12–58) is an example specific to dialysis:

$$\frac{c_w}{c_f} = \frac{1 + \psi + (1 - \sigma)\xi c_d/c_f}{(1 - \sigma)(1 + \xi) + \psi}, \quad (12\text{-}58)$$

where $\xi = [\exp(\text{Pe}) - 1]^{-1}$ (where Pe is typically <0.1 in dialytic processes) and $\psi = [\exp(\theta) - 1]^{-1}$. For hollow-fiber membranes, θ is given by

$$\theta = 0.709 J_v \left(\frac{zd^3 N}{Q_f D^2}\right)^{0.33}, \quad (12\text{-}59)$$

where z is the axial distance from the hollow-fiber inlet to the point at which c_w is being evaluated.

The relative magnitudes of diffusive and convective transfer are functions of the membrane permeability P_m and the reflection coefficient σ. For species of small molecular size, the resistance to diffusive transfer through a membrane is low, that is, P_m is large, and diffusive transfer always significantly exceeds convective transfer at low Peclet numbers. However, as solute molecular size increases, membrane permeability decreases logarithmically (Farrell and Babb 1973), whereas reflection coefficients increase at a much lesser rate and the relative importance of convection to overall mass transfer increases. This effect is illustrated in the following example.

Figure 12–12 shows the membrane permeability and reflection coefficient as a function of molecular weight for a cellulosic dialysis membrane (Wendt et al. 1979). Consider two solutes, A and B, having molecular weights of 200 and 2000 daltons, respectively, both having feedside concentrations at a point on the membrane surface of 0.1 g/cm³. Assuming negligible dialysate-side concentrations, $\Delta c_s = 0.1$ g/cm³ (referring to both solutes). Using Eqs. (12–57), (12–58), and (12–59), an estimate of the value of $\bar{c}_s$ can be made. For both solutes, Eq. (12–58) estimates the value of c_w/c_f to be unity ($c_w \cong c_f$). Then, Eq. (12–57) gives $\bar{c}_s = 0.067$ g/cm³ for both solutes A and B.

For solute A, $P_m = 3.6 \times 10^{-4}$ cm/s and $1 - \sigma = 0.8$. Substituting these values into Eq. (12–56), the following result is obtained:

$$J_s = (3.6 \times 10^{-5} + 0.054 J_v) \text{ g/s} \cdot \text{cm}^2.$$

$$(12\text{-}60)$$

As the ultrafiltration rate increases from 0 to 0.5 × 10⁻⁴ cm/s, J_s increases from 3.6×10^{-5} g/s·cm² to 3.9×10^{-5} g/s·cm²; that is, convec-

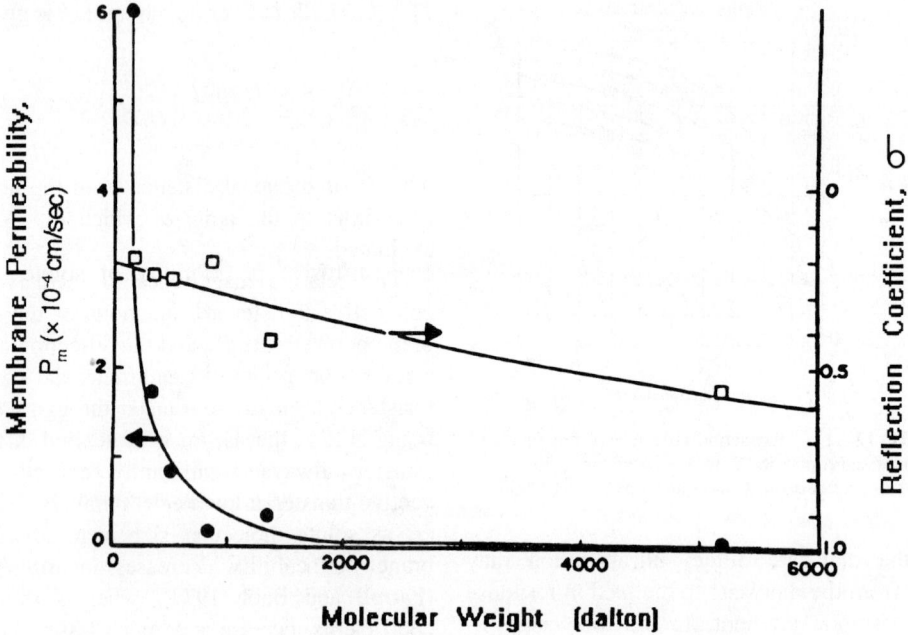

FIGURE 12–12. Membrane permeability and reflection coefficient as a function of solute molecular weight for a typical dialysis membrane (Cuprophan® 150 PM) (after Klein, Ward, and Lacey 1987).

tion increases overall solute transfer by 8%. For solute B, $P_m = 0.40 \times 10^{-4}$ cm/s and $1 - \sigma = 0.6$. Under these conditions, Eq. (12–56) reduces to

$$J_s = (0.40 \times 10^{-5} + 0.04 J_v) \text{ g/s} \cdot \text{cm}^2.$$

(12–61)

Now, for an increase in ultrafiltration rate from 0 to 0.5×10^{-4} cm/s, J_s increases from 0.40×10^{-5} to 0.60×10^{-5} g/s·cm², an increase in overall mass transfer of 50%.

The above considerations deal with a point on the membrane. Application of these concepts to the dialyzer as a whole requires that the feed and dialysate solute concentrations at every point on the membrane be expressed in terms of known concentrations. Such an approach has been developed by Schindhelm, Farrell, and Stewart (1977), who by assuming a zero inlet dialysate concentration and a linear dialysate concentration profile, developed the following expression for the dialysance of solutes greater than 300 dalton molecular weight in a countercurrent dialyzer in the presence of simultaneous diffusion and convection. [Note that the equations for X and Y given in this reference contain misprints; the correct expressions are as given in Eqs. (12–63) and (12–64).]

$$D^* = \frac{X}{1 + Y}, \quad (12\text{–}62)$$

where

$$X = -\frac{k_o A_m + Q_{uf}(1 - \sigma)}{Q_{fi} - Q_{uf}} \frac{Q_{fi}}{Z} [1 - \exp(Z)],$$

(12–63)

$$Y = -[k_o A_m + Q_{uf}(1 - \sigma)] \frac{k_o A_m}{Q_{fi} Q_{do} Z^2}$$

$$\times \left[1 + \left(\frac{Z}{2}\right) + \left(\frac{1}{Z}\right)[1 - \exp(Z)]\right]$$

$$+ \frac{X}{Q_{fi}} + \frac{k_o A_m}{Z Q_{do}}, \quad (12\text{–}64)$$

$$Z = \frac{k_o A_m - Q_{uf} \sigma}{Q_{fi}}. \quad (12\text{–}65)$$

A similar analysis, but one that requires a numerical methods solution, has been described by Jaffrin, Gupta, and Malbranq (1981).

The contribution of convection to overall mass transfer, illustrated previously for a point on the membrane, can now be estimated for the dialyzer as a whole. Referring to the previous example, consider the contribution of convection to dialysance for the two solutes A and B in a dialyzer containing 10^4 cm^2 of cellulosic membrane operating in a countercurrent configuration with feed and dialysate flow rates of 8 and 16 cm^3/s, respectively. For such a membrane, the overall mass transfer coefficients of solutes A and B are on the order of 3.23×10^{-4} and 0.37×10^{-4} cm/s, respectively, while the reflection coefficients are on the order of 0.2 and 0.4, respectively. Using these data in Eqs. (12–62) through (12–65), we can calculate that the dialysance of solute A will increase by 12% (from 2.47 to 2.78 cm^3/s) as the ultrafiltration rate increases from 0 to 0.5 cm^3/s, while the same increase in ultrafiltration rate will result in an increase of 84% (from 0.36 to 0.66 cm^3/s) in the dialysance of solute B. This example further demonstrates the important contribution of convection to overall mass transfer for larger molecular weight species.

NOTATION

General Notation

See the General Notation section at the beginning of this handbook.

Special Notation

A_m membrane area, L^2
A_p total pore opening area/membrane area, dimensionless
C clearance, L^3/t
d_{ti} inside diameter of the tube or hollow fiber, L
d_{to} fiber outside diameter, L
D^* dialysance, L^3/t
E extraction ratio, dimensionless
k mass transfer coefficient, L/t
K distribution coefficient, dimensionless
l_p pore length, L
L_D diffusional mobility per unit osmotic pressure difference, L/tp
L_{Dp} ultrafiltration coefficient, L/tp
L_p hydraulic permeability, L/tp
L_{pD} coefficient of osmotic flow, L/tp
N molar flux, mol/$L^2 t$
N_i molar flux of species i, mol/$L^2 t$
N_t number of transfer units
P_m diffusive permeability of solute in dialysis, L/t
q ratio of solute radius to pore radius, dimensionless
Q_d dialysate flow rate, L^3/t
Q_f feed flow rate, L^3/t
Q_{uf} ultrafiltration rate from the feed to the dialysate, L^3/t
R mass transfer resistance, t/L
y distance, normal direction, L
z distance, axial direction, L
z^* distance, axial direction, dimensionless
Z ratio of feed to dialysate flow rate, Q_f/Q_d, dimensionless

Greek Letters

η viscosity, M/Lt
σ reflection coefficient defined by Eq. (12–18), dimensionless
ω solute permeability coefficient, mol/$L^2 tp$

Subscripts

d dialysate
o overall
s solute

REFERENCES

Bean, C. P. 1972. The physics of porous membranes—neutral pores. In *Membranes, A Series of Advances, Vol. 1, Macroscopic Systems and Models*, ed. G. Eisenman, pp. 1–54. New York: M. Dekker.

Beck, R. E., and J. S. Schultz. 1972. Hindrance of solute diffusion within membranes as measured with microporous membranes of known pore geometry. *Biochim. Biophys. Acta* 255:273–303.

Bird, R., W. E. Stewart, and E. N. Lightfoot. 1960. *Transport Phenomena*. New York: John Wiley and Sons.

Bungay, P. M. 1986. Transport principles—porous membranes. In *Synthetic Membranes: Science, Engineering and Applications*, ed. P. M. Bungay,

H. K. Lonsdale, and M. N. de Pinho. Dordrecht, Holland: D. Reidel Publishing Company.

Colton, C. K. 1969. Permeability and transport studies in batch and flow dialyzers with applications to hemodialysis. Ph.D. diss. Massachusetts Institute of Technology, Cambridge, MA.

Colton, C. K. 1988. Unpublished data.

Colton, C. K., and E. G. Lowrie. 1981. Hemodialysis: physical principles and technical considerations. In *The Kidney,* Vol. II, ed. B. M. Brenner and F. C. Rector, Jr., pp. 2425–2489. New York: W. B. Saunders Company.

Colton, C. K., K. A. Smith, P. Stroeve, and E. W. Merrill. 1971. Laminar flow mass transfer in a flat duct with permeable walls. *AIChE J.* 17: 773.

Cooney, D. O., E. J. Davis, and S.-S. Kim. 1974. Mass transfer in parallel-plate dialyzers—a conjugated boundary value problem. *Chem. Eng. J.* 8:213–222.

Crank, J. 1975. *The Mathematics of Diffusion.* Oxford: Clarendon Press.

Davis, H. R., and G. V. Parkinson. 1970. Mass transfer from small capillaries with wall resistance in the laminar flow regime. *Appl. Sci. Res.* 22:20.

Farrell, P. C., and A. L. Babb. 1973. Estimation of the permeability of cellulosic membranes from solute dimensions and diffusivities. *J. Biomed. Mater. Res.* 7:275–300.

Farrell, P. C., J. W. Esbach, J. E. Vizzo, and A. L. Babb. 1974. Hemodialyzer reuse: estimation of area loss from clearance data. *Kidney Int.* 5:446–450.

Gostoli, C., and A. Gatta. 1980. Mass transfer in a hollow-fiber dialyzer. *J. Membr. Sci.* 6:133–149.

Happel, J. 1959. Viscous flow relative to arrays of cylinders. *AIChE J.* 5:174–177.

Hermans, J. J. 1978. Physical aspects governing the design of hollow-fiber modules. *Desalination* 26:45–62.

Jaffrin, M. Y., B. B. Gupta, and J. M. Malbrancq. 1981. A one-dimensional model of simultaneous hemodialysis and ultrafiltration with highly permeable membranes. *J. Biomech. Eng.* 103:261–266.

Jagannathan, R., and U. R. Shettigar. 1977. Analysis of a tubular hemodialyzer—effect of ultrafiltration and dialysate concentration. *Med. Biol. Eng. Comput.* 15:134–139.

Katchalsky, A., and P. F. Curran. 1965. *Nonequilibrium Thermodynamics in Biophysics.* Cambridge, MA: Harvard University Press.

Katchalsky, A., and O. Kedem. 1962. Thermodynamics of flow processes in biological systems. *Biophys. J.* 2(suppl.):53.

Kedem, O., and A. Katchalsky. 1958. Thermodynamic analysis of the permeability of biological membranes to nonelectrolytes. *Biochim. Biophys. Acta* 27:229.

Keller, K. H. 1973. *Fluid Mechanics and Mass Transfer in Artificial Organs.* Special publication by *Trans. Am. Soc. Artif. Intern. Organs.* Washington, DC: Georgetown University Press.

Klein, E. 1977. Hollow-fiber membrane developments. *J. Appl. Polym. Sci.* 31:361–381.

Klein, E., F. F. Holland, and K. Eberle. 1979. Comparison of experimental and calculated permeability and rejection coefficients for hemodialysis membranes. *J. Membr. Sci.* 5:173–188.

Klein, E., R. A. Ward, and R. E. Lacey. 1987. Membrane processes—dialysis and electrodialysis. In *Handbook of Separation Process Technology,* ed. R. W. Rousseau. New York: John Wiley and Sons.

Lévêque, J. A. 1928. Les lois de la transmission de chaleur par convection. *Ann. Mines* 13:201, 305, 381.

Michaels, A. S. 1966. Operating parameters and performance criteria for hemodialyzers and other membrane-separation devices. *Trans. Am. Soc. Artif. Intern. Organs* 12:387–392.

Noda, I., and C. C. Gryte. 1979. Mass transfer in regular arrays of hollow fibers in countercurrent dialysis. *AIChE J.* 25:113–122.

Onsager, L. 1931a. Reciprocal relations in irreversible processes. I. *Phys. Rev.* 37:405–426.

Onsager, L. 1931b. Reciprocal relations in irreversible processes. II. *Phys. Rev.* 38:2265–2279.

Prasad, R., and K. K. Sirkar. 1988. Dispersion-free solvent extraction with microporous hollow-fiber modules. *AIChE J.* 33:177–188.

Renkin, E. M. 1954. Filtration, diffusion, and molecular sieving through porous cellulose membranes. *J. Gen. Physiol.* 38:225.

Schindhelm, K., P. C. Farrell, and J. H. Stewart. 1977. Convective mass transfer in the artificial kidney. Paper read at 2nd Australian Conference on Heat and Mass Transfer, 16–18 February 1977, University of Sydney.

Sparrow, E. M., and A. L. Loeffler, Jr. 1959. Longitudinal laminar flow between cylinders arranged in regular array. *AIChE J.* 5:325–330.

Spiegler, K. S., and O. Kedem. 1966. Transport

coefficients and salt rejection in uncharged hyperfiltration membranes. *Desalination* 1: 311–326.

Spriggs, H. D., and N. N. Li. 1976. Liquid permeation through polymeric membranes. In *Membrane Separation Processes,* ed. P. Meares. Amsterdam: Elsevier Scientific Publishers.

Staverman, A. J. 1951. Theory of measurement of osmotic pressure. *Rec. Trav. Chim.* 70: 344.

Verniory, A., R. DuBois, P. Decoodt, J. P. Gassee, and P. P. Lambert. 1973. Measurement of the permeability of biological membranes. Application to the glomerular wall. *J. Gen. Physiol.* 62:489–507.

Villaroel, F., E. Klein, and F. Holland. 1977. Solute flux in hemodialysis and hemofiltration membranes. *Trans. Am. Soc. Artif. Intern. Organs* 23:225–233.

Wendt, R. P., E. Klein, E. H. Bressler, F. F. Holland, R. M. Serino, and H. Villa. 1979. Sieving properties of hemodialysis membranes. *J. Membr. Sci.* 5:23–49.

Wilson, E. E. 1915. A basis for rational design of heat transfer apparatus. *Trans. ASME* 37:47.

Yang, M.-C., and E. L. Cussler. 1986. Designing hollow-fiber contactors. *AIChE J.* 32:1910–1916.

Yasuda, H., A. Peterlin, C. K. Colton, K. A. Smith, and E. W. Merrill. 1969. Permeability of solutes through hydrated polymer membranes. III. theoretical background for the selectivity of dialysis membranes. *Die Makromol. Chemie* 126:177–186.

Yasuda, H., C. E. Lamaze, and L. D. Ikenberry. 1968. Permeability of solutes through hydrated polymer membranes. I. diffusion of sodium chloride. *Die Makromol. Chemie* 118:19–35.

Yasuda, H., C. E. Lamaze, and A. Peterlin. 1971. Diffusive and hydraulic permeabilities of water in water-swollen polymer membranes. *J. Polym. Sci.* 9:1117–1131.

13
Design

Stephen B. Kessler
Sepracor Inc.

Elias Klein
University of Louisville

MEMBRANE SELECTION
   Membrane Materials
   Membrane Structures
MEMBRANE MODULES
   Module Types
   Module Design

   Materials Selection
PROCESS AND SYSTEM DESIGN
   Batch versus Continuous Operation
   Staged Operation
NOTATION
REFERENCES

MEMBRANE SELECTION

Membrane Materials

When a dialysis membrane separates two aqueous phases, the problem of material choice is relatively simple. Clearly, the membrane pores must be filled with water during use, so wettability is a primary consideration. As a consequence of this requirement, the predominant materials for dialysis membranes are relatively hydrophilic polymers. At one end of the spectrum are cellulose and poly(ethylene-co-vinyl alcohol) (Eval), and on the other end poly(methylmethacrylate), the latter prewetted by the manufacturer. In between fall cellulose acetate with a degree of substitution (DS) of 2.5, poly(acrylonitrile-co-methallylsulfonic acid), and other acrylonitrile copolymers.

The use of poly(ethersulfone) membranes (PES) for hemodialysis would appear to be a contradiction to the wettability requirement, but many dialyzers made from this material contain a small amount of poly(vinylpyrrolidone) (PVP) as an alloying agent to the PES. The PVP provides the requisite wetting, although there may still be very small diameter pores that are inactive during dialysis because of the exclusion of water.

The membrane in widest use for dialysis is regenerated cellulose. It can be produced by a variety of means, but two processes dominate. The oldest and most widely used process is based on producing a copper amine complex of cellulose in a strongly alkaline solution (i.e., the cuprammonium process). The solution containing 5 to 8% cellulose is coagulated by dissociating the complex and removing excess copper and alkali. The characteristics of membranes produced by this process are very high wet strengths, very thin membranes (as thin as 0.0005 cm), and a uniform cross section. A second process for producing cellulose membranes relies on casting a cellulose acetate membrane from a high-temperature solvent. The resulting cellulose acetate membrane is then hydrolyzed to de-acetylate the polymer to the native cellulose. Because cellulose acetate polymers are generally lower in molecular weight than the cellulose used in the cuprammonium process, these membranes have lower strength and stability to alkali.

The cellulose membranes are atypical in that they contain a high percentage of water, yet do not lose their mechanical integrity. As will be shown later, this high water content correlates with rapid solute diffusivity within the membrane.

When the membrane separates two differing and immiscible phases, the material selection must reflect which phase is to be the pore wetting phase (see Chapter 41 for further details). Thus, if the dialytic process is to remove an organic acid from an organic solvent by extraction with an aqueous base, a lipophilic membrane might be chosen. The organic phase would wet the pores preferentially to the aqueous phase and transfer would occur at the aqueous/membrane interface. The organic layer in the membrane would become depleted and must then be considered as a major contributor to the boundary layer. If, instead, a hydrophilic membrane were selected, the membrane would be wet with the aqueous phase (see Chapter 41 for alternative wetting conditions) and transfer would occur at the organic/membrane interface. Now the salt of the organic acid would accumulate in the membrane structure until it is carried away by the aqueous phase. The boundary layer problem is then reversed. A knowledge of the relative membrane diffusivities, i.e., organic acid in organic solvent in the membrane *vis-à-vis* organic salt in the aqueous membrane phase, would aid in predicting the preferred choice of membrane material.

In either of the above cases, the membrane material would have to be completely resistant to the solvent effects of either phase. Absence of solubility is not a sufficient criteria. Many membrane structures are produced as metastable phases that very slowly undergo compaction of their structure to more stable forms. If the contacting solvent accelerates this process, the membrane life may be shortened such that practical applications are impossible.

Finally, the various fabrication options available with a given material must be considered. Most current dialysis devices are produced in the form of hollow fibers or tubes. Materials formed by sintering are difficult to produce in long lengths such as are required by tubes. Thus, the preponderance of membranes available today is produced by coagulating a melt, or solution, or a suspension via some extrusion process. Because diffusion paths should be kept short, dialytic membranes are generally less than 0.0050 cm thick. Such thin walls coupled with a need to wind, or bend, the tubes dictates that the materials must have relatively high tensile moduli. Consideration of these factors, together with an understanding of the phase-separation mechanisms that lead to semipermeable structures, is what constitutes the art and science of "membranology."

Membrane Structures

Membrane morphology provides important distinctions between dialytic membranes. For simple dialysis applications, such as equilibration of microsolute concentrations in two aqueous phases, the choices are many. Most dialytic membranes have pore radii of $>15 \times 10^{-8}$ cm, which is significantly larger than the Stokes radius of a 500-dalton molecular weight solute. However, when selective separation of solutes is desired, other constraints apply. Dialytic membranes with pore radii in the 75×10^{-8}–100×10^{-8}-cm range are now available. The sieving properties of such membranes allow them to retain macromolecules while selectively removing solutes of "middle" molecular weight. These "high-efficiency" dialysis membranes typically exhibit high levels of protein adsorption, which makes *a priori* prediction of selectivity problematic.

Since device clearance, in the absence of significant ultrafiltration, depends on short diffusion paths within the membrane, the thinner the membrane, the more rapid the mass transfer. So the first choice would be to make a membrane as thin as the material will allow, while still providing the requisite mechanical stability for potting and pumping pressures. In addition to being thin, the membrane must provide the required morphology, i.e., porosity and pore size. These are often conflicting demands. High porosity implies reduced load-bearing elements in the cross section and, thus, limited tensile strengths. Fortunately, the more hydrophilic materials, such as cellulose and copolymers of vinyl alcohol, can be produced as thin, highly efficient hydrogel membranes reinforced by crystalline regions. The more hydrophobic, glassy polymers, such as poly-(sulfones), are produced in asymmetric form to provide a thin solute-resistant skin supported by a more porous substructure. The thick substructure provides the mechanical support needed.

The ability to produce membranes composed of a solute-retentive skin integral with a more porous substructure was a major development in membrane technology. This asymmetric structure was found to permit production of membranes with large L_p values that were still retentive to small molecules. The major benefit of this asymmetry, however, accrues to ultrafiltration and reverse osmosis membranes since the substructure does not produce a stagnant boundary layer in such convective processes. With dialysis membranes, the thickness of the porous substructure can add significantly to the boundary layer problems by producing an unstirred layer of fluid. Such layers create unwanted transport resistances.

Asymmetric membrane structures are produced by the coagulation process used to convert a solution of polymer into a membrane structure. The phase-separation mechanisms that occur as the polymer is precipitated from concentrated solutions include liquid-liquid and liquid-gel transitions. The skin is thought to be caused by a rapid loss of solvent from the polymer solution film into the coagulating bath. The resulting highly concentrated polymer solution precipitates with a morphology different from the underlying polymer solution since the initially coagulated surface acts as a barrier to transfer of the solvent out of—and nonsolvent into—the lower layers of the solution. For the production of asymmetric dialysis membranes, the control of such phase-separation mechanisms becomes rather critical because of the added demand for thin cross sections in dialytic applications.

Table 13–1 lists some commercially available, hollow-fiber dialysis membranes with their basic performance parameters. Both gel-type and asymmetric membrane structures are represented. Unless otherwise noted, UFR (ultrafiltration rate per membrane area) and P_m (diffusive permeability) were measured using saline solutions.

MEMBRANE MODULES

Module Types

Industrial Dialyzers

Early industrial dialyzers comprised three general configurations: the tank type, the plate-and-frame type, and the tube type. Tuwiner (1962) reviews these designs as well as those of early laboratory dialyzers in considerable detail. Of the three types, the plate-and-frame design was most widely used, while the tube type (a bundle of tubular membranes) did not achieve commercial importance.

The Cerini dialyzer, an example of the tank design, was developed in Italy in 1928 to reclaim sodium hydroxide in the rayon industry. It consisted of a $3 \times 1.5 \times 1.2$-m tank, containing 50 membrane bags totaling 300 m^2 of membrane area. Thus, the membrane area per unit volume of this dialyzer was 55 m^{-1}. Each of the cotton bags was supported by an internal metal mesh frame and was individually connected to the water supply. Feed solution flowed into the tank at the bottom and exited at the top by simply overflowing.

The patent literature abounds with examples of plate-and-frame (or filter press) dialyzer designs (Tuwiner 1962). By providing improved membrane support compared to the tank design, these dialyzers allowed the use of much thinner membranes. The lower diffusive resistance of

TABLE 13–1. Performance Parameters of Some Representative Hollow-Fiber Dialysis Membranes.

Description	Material	UFR (mL/h m^2 mm Hg)	P_m/Vitamin B$_{12}$ (10^{-3} cm/min)
Cuprophan® C1 (Enka)	Cellulose	4.0	5.3
Cuprophan® D4 (Enka)	Cellulose	3.0	4.0
Highflux® RC-HP400 (Enka)	Cellulose	30	12
Gambro HF	Cellulose	3.8[a]	4.8
Hospal	PAN-methallyl sulfate	27	9.5
Baxter/Toyoba	Cellulose acetate	4.6	5.7
Fresenius F-6	Polysulfone	4.6	5.9
Fresenius F-60	Polysulfone	33	17
Toray/Hoechst	PMMA	16[a]	10

[a]Values measured from whole blood.

the thin membranes led to a fivefold to tenfold increase in productivity per unit membrane area relative to the Cerini dialyzer. On a membrane area per unit volume basis, plate-and-frame and tank-type dialyzers were comparable.

One of the more recent examples of the plate-and-frame design is the Graver dialyzer. This design uses a poly(vinylchloride) membrane support frame. A "repeat unit" in such a dialyzer consists of two such frames—one perfused with dialysate and one with feed solution—two sheets of membrane, and the necessary gaskets. The Graver dialyzer is comprised of 150 such units. Figure 13–1 illustrates a single-repeat unit in a generic plate-and-frame dialyzer.

In countercurrent operation, the lower density dialysate flows downward and the higher density feed flows upward in adjacent compartments. Thus, the increasing density of the dialysate as solute is transferred to it and the corresponding decrease in density of the feed both serve to facilitate a balanced flow distribution among the cells. Hydrostatic pressure in the dialysate cells is maintained at a higher level than in the feed cells, which ensures that the membranes are pressed against the support structure of the feed frames.

The hollow-fiber dialyzer, developed first for hemodialysis, has been adapted for industrial applications. Enka AG offers industrial dialyzers based on their Cuprophan® regenerated cellulose hollow-fiber membrane. Hollow fibers of 200-μm inside diameter and 16-μm wall thickness are interwoven with thread and grouped into bundles. Three such bundles are contained in the annular space between two stainless steel tubes and encapsulated with polyurethane at each end, which forms tubesheets. The inner tube acts as a conduit feeding the shell side of the module, the flow exiting at one end of the outer tube. Two such submodules of 28-cm effective fiber length are joined in series, comprising a module of 22.5-m^2 membrane area. The membrane area per unit volume of this device is approximately 2200 m^{-1}, which is a fortyfold increase over the tank or plate-and-frame designs.

While the traditional aqueous/aqueous dialysis process is limited in its industrial applications, modern hollow-fiber dialyzers with expanded chemical resistance are finding new application in aqueous/organic solvent extraction processes. An example of an industrial hollow-fiber device of this kind is produced by Sepracor Inc. (more details are available in

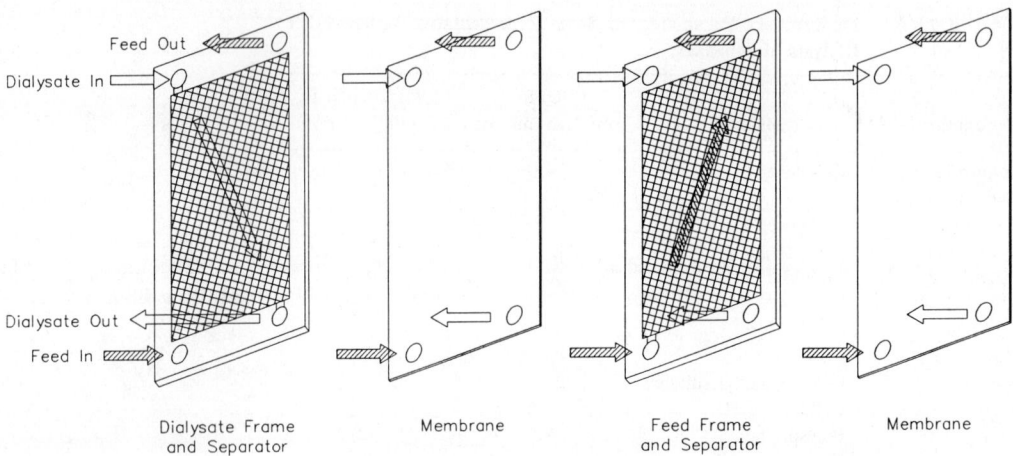

FIGURE 13–1. Single-repeat unit from a typical plate-and-frame dialyzer (with gaskets omitted).

Chapter 41). These modules were developed specifically for applications involving aqueous/nonaqueous extraction processes and are fabricated from materials that are resistant to organic solvents. Available hollow-fiber types include regenerated cellulose and poly(acrylonitrile), both of 200-μm inside diameter. The fibers are encapsulated in epoxy, forming a cartridge that has ends provided with radial O-ring seals. The cartridge is installed in a housing, together comprising a module. Sepracor modules are sized by total available fiber packing volume and range from 1.5 mL to 12 L in size (Figure 13–2). A 12-L module packed with typical regenerated cellulose dialysis fibers has a membrane area of 65 m^2. Based on overall volume occupied, its membrane area per unit volume is comparable to the Enka unit described above.

Hemodialyzers

As described in Chapter 11, Kolff and Berk (1944) demonstrated the feasibility of hemodialysis and applied the process clinically using a "rotating drum" artificial kidney. This device consisted of a drum wrapped with cellulose sausage casing through which blood flowed while the drum was rotated in a bath of dialysate solution.

In 1956 Kolff and coworkers developed the

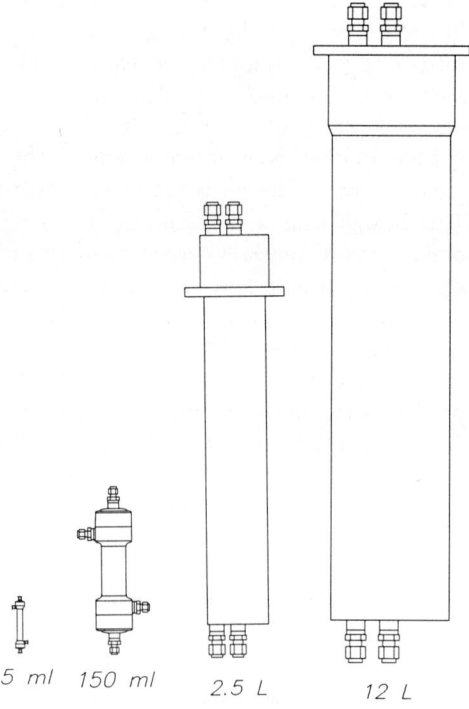

FIGURE 13–2. Hollow-fiber modules for diffusion-based membrane separation processes. The modules are sized by total fiber packing volume and range from lab scale (1.5 mL) to production scale (12 L). (Courtesy of Sepracor Inc.).

coil dialyzer. Like the rotating drum, this design was based on a tubular cellophane membrane. The tubing was flattened and rolled into a coil along with a sheet of plastic mesh. In operation, blood flowed inside of the cellophane tubing around the coil, while dialysate flowed axially through the space formed by the plastic mesh.

Kiil introduced a plate-and-frame hemodialyzer in 1960. Like industrial plate-and-frame modules, it was operated in the countercurrent flow mode. It was a reusable device whose membranes were replaced by disassembly and reassembly of the unit. Due to the amount of labor required to maintain reusable dialyzers, disposable devices of both the coil type and plate-and-frame type were introduced (Figure 13–3). The disposable designs tend to be more compact than their reusable counterparts and are sold presterilized. Hollow-fiber hemodia-

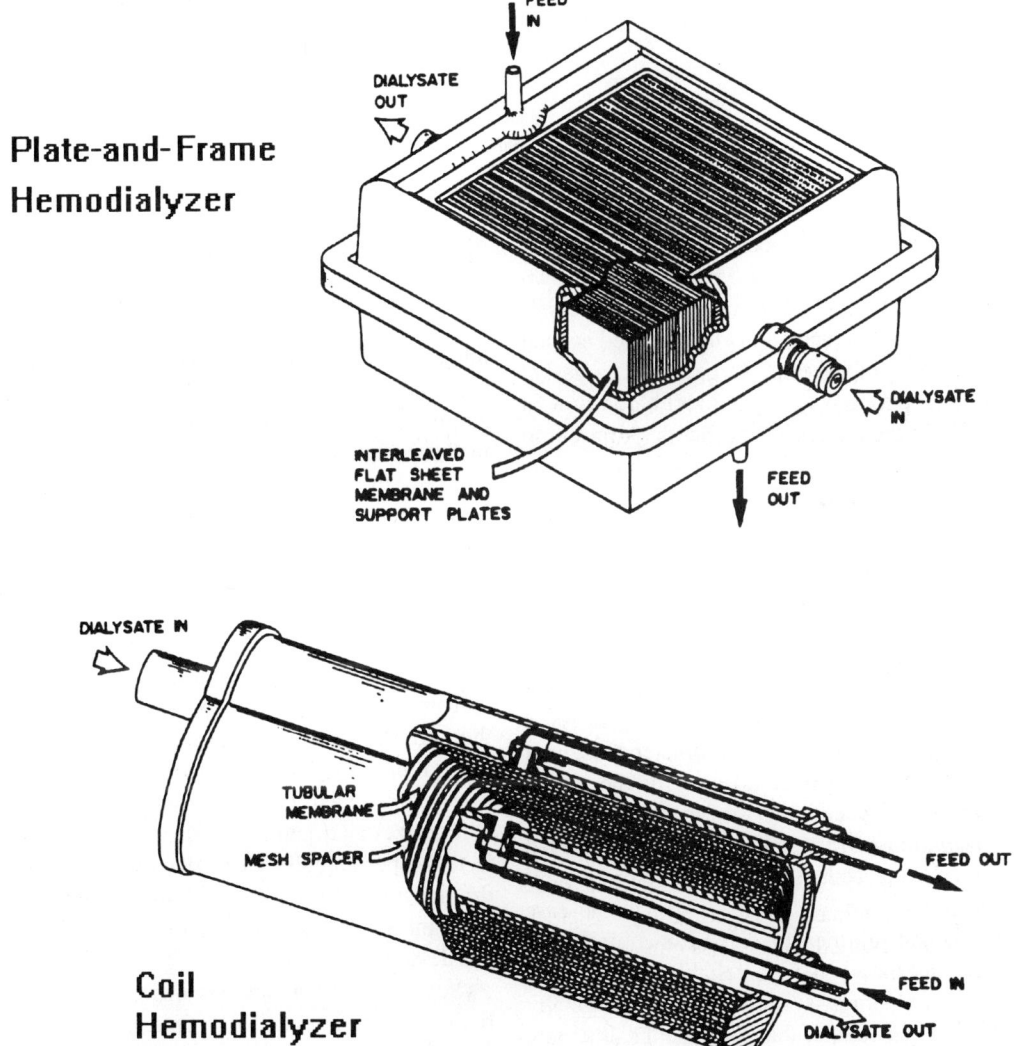

FIGURE 13–3. Hemodialyzers of the coil type and plate-and-frame type (Courtesy of Baxter Healthcare Corp.; Klein, Ward, and Lacey 1987).

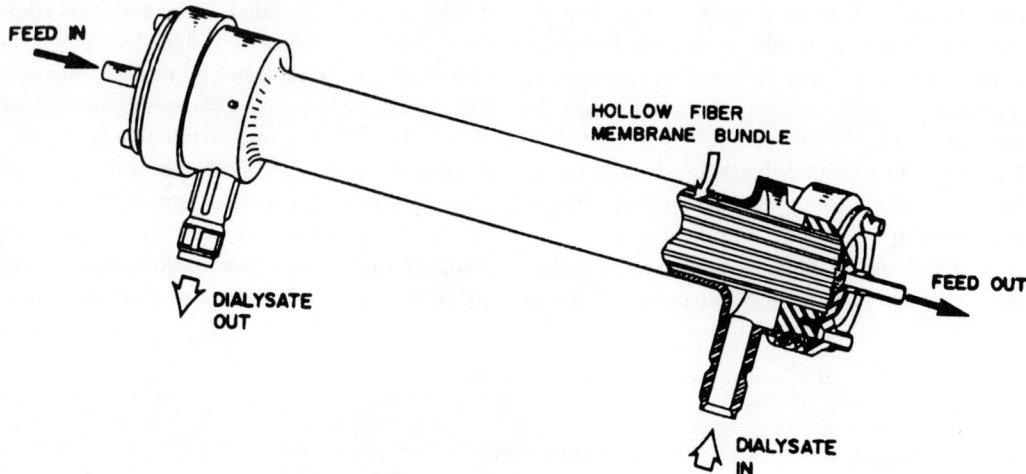

FIGURE 13-4. Typical hollow-fiber hemodialyzer (Courtesy of Baxter Healthcare Corp.; Klein, Ward, and Lacey 1987).

lyzers were first introduced in 1966 (Lipps et al. 1967). The structure of a typical unit is shown in Figure 13-4. A bundle of hollow fibers is contained in a housing and encapsulated at each end forming tubesheets. At each end, a gasket and endcap form headers to direct blood flow in and out of the lumens of the fibers. Adjacent to each tubesheet is a circumferential header, which directs dialysate flow in and out of the shellside space. The device is geometrically similar to a shell-and-tube heat exchanger.

The housing and endcaps of these modules are usually fabricated from a transparent engineering polymer such as polycarbonate or poly(styrene-co-acrylonitrile). The encapsulant that forms the tubesheet is typically a polyurethane, and a low-Durometer elastomer such as silicone rubber is used for the gasket between the endcap and tubesheet.

The hollow-fiber dialyzer has achieved widespread acceptance and has largely supplanted other designs. This is due to the ease with which this design can be manufactured, its compactness, and its reliable performance in extracorporeal systems. While originally intended as disposable items, health care costs have provided incentive to reuse these dialyzers, but only with respect to a single patient.

Module Design

From a design standpoint, the evolution of dialyzer types followed two distinct lines: those intended for industrial applications and those intended for medical applications. Specific examples from both application areas were given in the preceding section.

Many of the design requirements of utmost importance to membrane devices intended for medical use have no relevance to industrial membrane modules (and vice versa). For example, all hemodialyzers are packaged sterile and are intended for single use or reuse only with the same patient. Examples of presterilized industrial membrane devices exist; however, they are the exception and no presterilized industrial dialyzers are known to the authors. The converse is equally true. An industrial dialyzer that utilizes an organic solvent as extractant and is to be periodically cleaned with sodium hydroxide must be constructed from chemically resistant materials, unlike those used for medical devices.

Scale is another obvious difference. Membrane devices intended for medical use are designed to perform mass transfer operations at a scale dictated by human anatomy and physiology. Industrial dialyzers, on the other hand, are scaled according to process economic factors and are typically much larger than their medical counterparts.

Process economics, in addition to affecting scale, also tends to favor long-lived industrial dialyzers, while, as mentioned above, hemodialyzers are intended for single use or limited reuse. The economics of health care as applied to end-stage renal disease (ESRD) is equally important in determining hemodialyzer design—no "overdesign" is tolerated in this cost-conscious and competitive market.

To illustrate the process leading to an optimized module design, a hollow-fiber module will be designed for a hypothetical dialysis process. As described in the preceding sections, hollow-fiber modules have become prevalent in both industrial and medical applications due to their high volumetric efficiency and ease of manufacture.

The hypothetical process for which this hollow-fiber module is being designed is an industrial process in which the product is a compound of 20K-dalton molecular weight ($\bar{M}$) in aqueous solution. It is desired to separate this product from a contaminant whose molecular weight is 200 dalton in a continuous process. The product solute concentration is 5 g/L and required productivity is 50 kg/h. The contaminant concentration is 1 g/L and should be reduced to 0.1 g/L or less by the process. An aqueous dialysate solution is to be used, which is buffered to pH 7.0.

Fiber Dimensions and Number

Yang and Cussler (1986) conclude that the key advantage of hollow-fiber membrane modules is the high ratio of membrane area to unit device volume they provide and that mass transfer coefficients for hollow-fiber devices are, in general, similar to those for other types of equipment. This conclusion provides a clue regarding choice of fiber dimensions—minimization of fiber diameters (inside and outside diameters) maximizes area-to-volume. However, practical lower limits exist for fiber diameters. The internal diameter is limited by the size of any suspended particles that may be present. Wall thickness, which together with inside diameter determines outside diameter, has a lower limit set by fiber strength requirements.

For given diameters, fiber length has an upper limit that is determined by pressure drop and/or mass transfer parameters. In the case of high-flux membranes, excessive translumenal pressure drop (TLP) causes high ultrafiltration rates, the effects of which are discussed in Chapter 12. A further consequence of high TLP is the increased likelihood of Starling's flow or back-filtration, which can cause a further decrease in separation efficiency. If, at the maximum lumenal flow rate allowable within TLP limitations, the desired degree of solute transfer occurs in less than one pass of the feed stream, then fiber length is excessive. At the other extreme, if too short a device is designed, excessively high flow rates are required to achieve a given value of k_o over a given membrane area and recirculation may be required to provide sufficient residence time. Another limit on "shortness" is manufacturing cost. The shorter a device is for a given membrane area, the more potential membrane area is lost to waste and the more expensive the housing becomes.

While not all of these optimization parameters can be included in a single expression, the ones that relate directly to fluid dynamics and mass transfer performance can be dealt with on a quantitative basis. Referring to the hypothetical process described above, based on throughput alone, the minimum feed flow rate required to process 50 kg/h is $Q_f = 10{,}000$ L/h. Assume that the feed solution viscosity is 1.0 cp and that the maximum permissible TLP is 75 kPa (11 psi), relative to the effective fiber length (excluding fiber encapsulated in the tubesheets). Equation (13–1) gives the number of hollow fibers required to process a given feed flow rate within TLP limitations:

$$N = 8\eta L Q_f/(\text{TLP}\,\pi r_{ti}^4), \qquad (13\text{–}1)$$

where η is solution viscosity, L is effective fiber length, Q_f is feed flow rate, and r_{ti} is fiber internal radius.

A further constraint on the number of fibers required is the membrane area needed to provide a sufficiently high rate of transfer of the contaminant. By choosing commercially available dialysis fibers covering a range of dimensions, practical values of both fiber dimensions and permeability can be assumed. Enka AG offers dialysis fibers of 100, 200, and 300 μm inside diameter (dry dimensions), designated C3, C1, and D4, respectively. Based on the experience of the authors (Klein et al. 1976, 1977), these fibers will be assumed in the wet state to have inside diameters of 125, 250, and 375 μm and wall thicknesses of 22, 22, and 32 μm, respectively.

To a first approximation, the diffusive permeabilities of all three fibers chosen are the same for low molecular weight solutes (<200 dalton) (Klein et al. 1976; 1977). It is assumed throughout the following analysis that passage of the 20K-dalton product is negligible through the membranes chosen. A convenient way of estimating required membrane area (for a first iteration of the design) is to refer to clearance values published for hemodialyzers of a given size. Figure 13–5 gives a value of $C = 90$

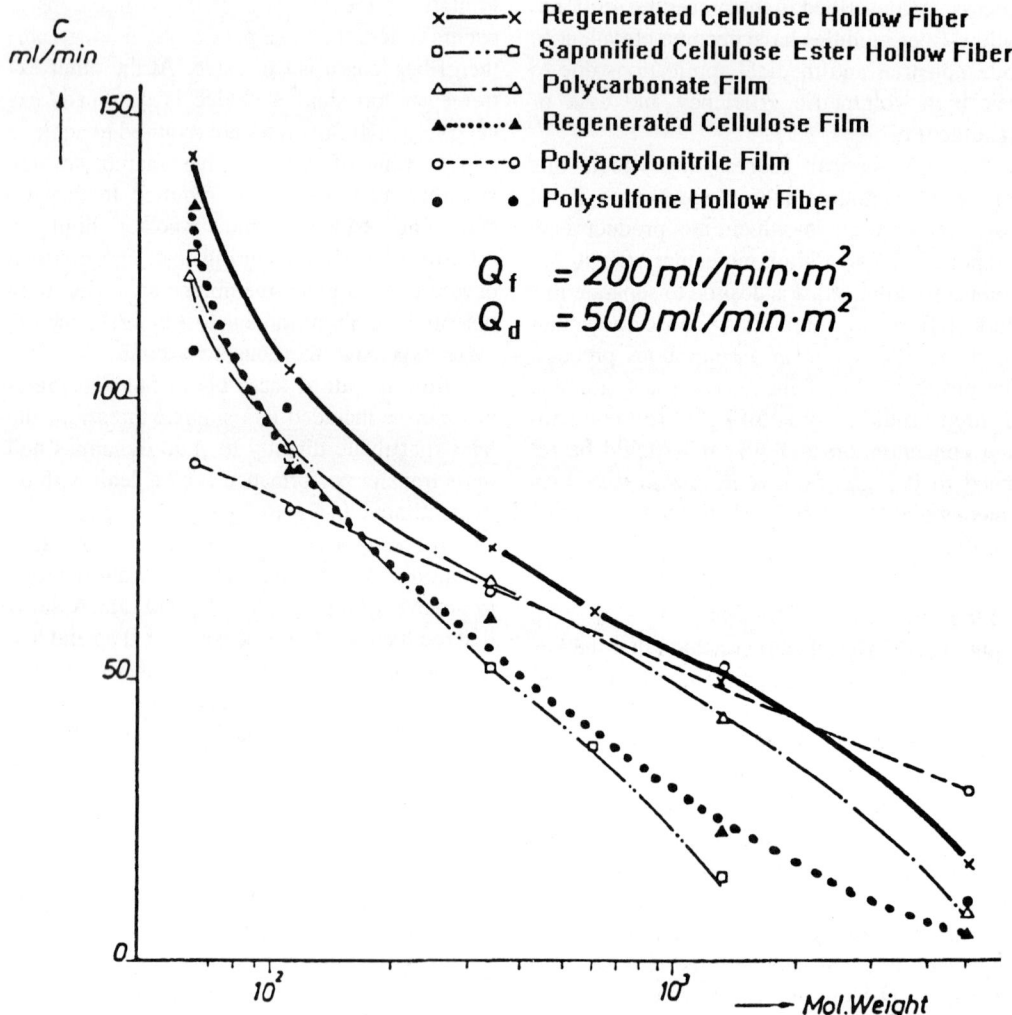

FIGURE 13–5. Clearance of various dialysis membranes as a function of molecular weight (after Enka AG 1982).

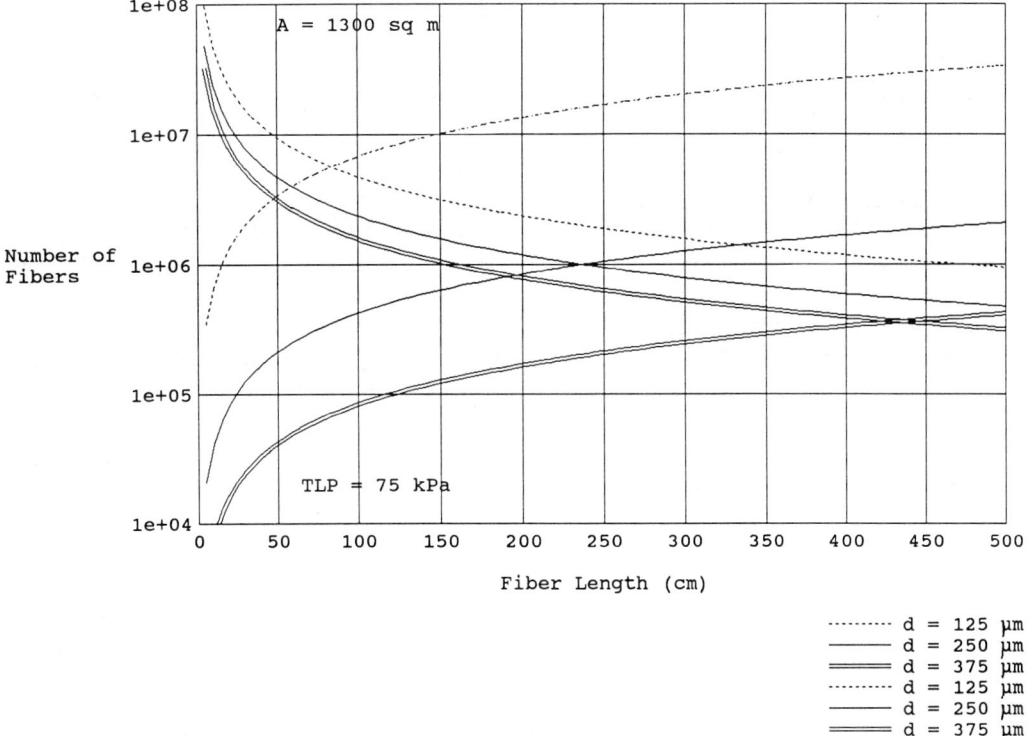

FIGURE 13–6. Number of fibers required as a function of fiber dimensions for constant translumenal pressure drop and area. (Used in designing a hollow-fiber dialysis module for a hypothetical process.)

mL/min for a 1-m^2 dialyzer using one of the hollow fibers chosen above. Dividing Q_f by C gives an estimate of required membrane area, $A_m = 1850$ m^2. Equation (13–2) gives the number of hollow fibers required to provide a given membrane area as a function of fiber dimensions:

$$N = A_m/(2\pi r_{ti} L). \quad (13\text{–}2)$$

Figure 13–6 was plotted to illustrate how the above constraints of TLP and membrane area limit the design space for the fibers chosen. The three curves that originate in the lower left corner of the graph were generated by Eq. (13–1) using the above-stated values for TLP and Q_f — each curve represents a different fiber inside diameter. Values of fiber number and length on or above each of the curves result in acceptable TLP values that are below 75 kPa.

The family of curves that originates in the upper left corner of the graph were generated by Eq. (13–2) with $A_m = 1850$ m^2. Values of fiber number and length on or above these curves result in membrane areas that are sufficient for mass transfer. Thus, for the 125-μm-i.d. fiber, greater than 10^7 fibers are required for most fiber lengths. The minimum number occurs at the intersection point of the two curves: $L = 85$ cm and $N = 5.6 \times 10^6$ fibers. For the 250-μm-i.d. fiber, the minimum number is $N = 1.0 \times 10^6$ at a fiber length of $L = 240$ cm. The corresponding values for the 375-μm-i.d. fiber are $N = 3.6 \times 10^5$ and $L = 440$ cm. At this point in the design process, it would be premature to select from among the three fiber sizes. A detailed evaluation of mass transfer performance may help in the decision.

Flow Configurations

Feed and dialysate flows can be either cocurrent or countercurrent in the case of parallel flows.

Other possibilities discussed in Chapter 12 include well-mixed dialysate and perpendicular flow. Of these possible flow configurations, countercurrent and perpendicular flow offer the best potential mass transfer performance. Cocurrent operations are indicated when minimal convective flow is desired and obligatory pressure drops are encountered. Under cocurrent flow configurations, pressure differences are easier to balance across the membrane along the length of the device. Countercurrent flows are used to maximize concentration differences between the inlet dialysate and the outlet feed, where the latter is most dilute.

Based on Eqs. (12–53) through (12–55), extraction ratio E versus the number of transfer units N_t has been plotted for three of the four flow configurations discussed above. Figures 12–9 through 12–11 of Chapter 12 show these plots of the countercurrent, cocurrent, and perpendicular flow cases. Extraction ratio E is perhaps the most meaningful measure of dialyzer mass transfer performance. In countercurrent flow, at equal values of N_t and Z, somewhat higher values of E are obtained than in the other two flow configurations. This implies that countercurrent flow is the most efficient of the three configurations. Recall, however, that $N_t = \bar{k}_o A_m / Q_f$. Thus, comparison at equal values of N_t implies that (if A_m and Q_f are to be held constant) $\bar{k}_o$ is the same in both countercurrent and perpendicular flow hollow-fiber modules operating at the same Q_d. This is most often not the case and $\bar{k}_o$ is usually higher for the perpendicular flow case.

Yang and Cussler (1986) compared published heat transfer correlations for flow external to tubes in both the parallel and perpendicular flow configurations. They further conducted oxygen transfer experiments with hollow-fiber modules in which external water flow in both flow configurations was tested under conditions where membrane resistance was negligible. Figure 13–7 compares the published correlations as well as the correlations

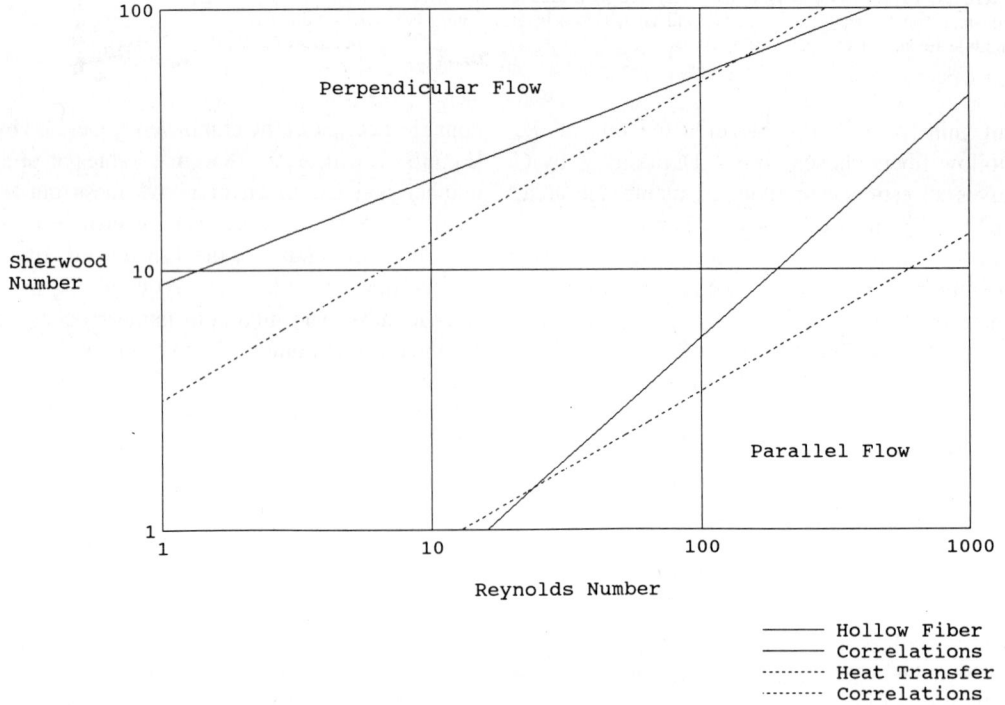

FIGURE 13–7. Mass transfer and heat transfer correlations for flow external to tubes, comparing parallel flow and perpendicular flow.

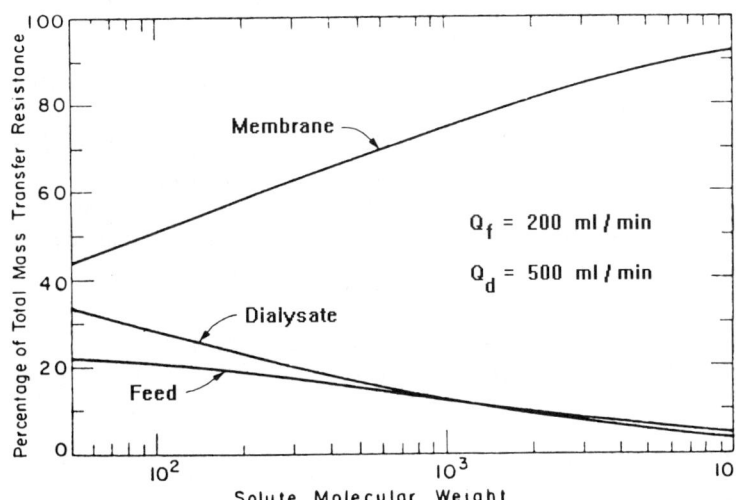

FIGURE 13–8. Relative contributions of feed, membrane, and dialysate mass transfer resistances for a typical hollow-fiber dialyzer operating in countercurrent flow (reprinted from Colton and Lowrie 1981 with permission).

resulting from their experiments. In both cases, the Sherwood number at a given value of Reynolds number is more than an order of magnitude higher in perpendicular flow compared to parallel flow.

Consider the case in which membrane mass transfer resistance is significant, which is more relevant to dialysis. Figure 13–8 shows relative contributions of feed, membrane, and dialysate resistances as a function of solute molecular weight for a typical hollow-fiber dialyzer operating in countercurrent flow. For a solute of 200 dalton $\bar{M}$, R_d contributes about 25% of the overall mass transfer resistance. Applying the results from Yang and Cussler (1986) for perpendicular flow, a dialyzer containing the same fiber as in Figure 13–8, but with dialysate in perpendicular flow, should show a 25% decrease in R_o due to R_d becoming negligible.

By applying the above example to the plots of Figures 12–9 and 12–11 of Chapter 12, $\bar{k}_o$ (perpendicular) is equal to $1.33\bar{k}_o$ (parallel) at the same Reynolds number and comparison should be made between the perpendicular flow case at a value of N_t that is 33% higher than that used in the countercurrent flow case. Compared in this way, for a value of $Z = 0.5$, perpendicular flow attains $E = 0.9$ at $N_t = 5$, while, for countercurrent flow at the same Reynolds number, $N_t = 3.75$ and $E = 0.9$. When enhanced dialysate mass transfer in perpendicular flow is taken into account, per-

pendicular flow and countercurrent flow are predicted to give comparable overall mass transfer performance in a typical dialytic process.

Shellside Pressure Drop

Another consideration in choosing between perpendicular and countercurrent flow is shellside pressure drop. The above mass transfer comparison was done at equal Reynolds numbers, which is an appropriate basis for a pressure drop comparison as well. Happel's (1959) free-surface model predicts pressure drops for laminar flow relative to arrays of cylinders. He states that his model shows good agreement with experimental results (on arrays of cylinders) up to Re = 100. Referring to the Reynolds number definition, Eq. (12–35) of Chapter 12, the Reynolds number for shellside flow is obtained by inserting fiber outside diameter for the characteristic length L in Eq. (12–35) of Chapter 12. The wet outside diameters for the three hollow fibers chosen are 170, 295, and 440 μm. Thus, at Re = 100, average velocities in the shell for the three fiber sizes are $v = 59$, 34, and 23 cm/s, respectively. The free-surface model should be applicable since actual hollow-fiber modules are likely to operate below these velocities.

Application of the free-surface model (Happel 1959) to a moderately packed ($\Phi = 0.5$) triangular array of the 295-μm-o.d. fibers gives resulting values for Darcy's constant of $K_D = 1.18 \times 10^{-6}$ cm^2 for perpendicular flow and $K_D = 3.06 \times 10^{-6}$ cm^2 for parallel flow in the same array [$K_D = \eta v/(\Delta p/l)$]. Since K_D is inversely proportional to pressure drop per unit length, the model predicts a pressure drop that is 2.6 times as large for perpendicular flow as for parallel flow over the same length. However, most practical hollow-fiber module configurations have axial lengths that exceed their transverse flow dimension by factors greater than 2.6:1. Thus, at this level of analysis, comparable pressure drops would be expected for parallel and perpendicular flow.

Keep in mind that the free-surface model applies to rigid, regularly arranged cylinders and that real hollow fibers may deviate significantly. Hermans (1978) compares results predicted by the free-surface model for perpendicular flow with an empirical study using hollow fibers (Dandavati, Doshi, and Gill 1975) and concludes that measured pressure drops exceeded those predicted by the model by a factor of 5:1. Deviations in the direction of higher-than-predicted pressure drops can be attributed to packing of fibers in response to flow, deformation of fibers due to the pressure gradient, nonuniform outside diameter, and "kinkiness." The first two of these causes apply specifically to the perpendicular flow case. Packing of fibers can be controlled by weaving them into a mat or providing local support with supplementary fibers. "Kinkiness," which may increase pressure drop in parallel flow, can be of benefit in perpendicular flow to control packing. In general, the pressure drop behavior in parallel flow is more predictable than in perpendicular flow, and simple fiber bundle configurations (e.g., without weaving or supplementary fibers) give good performance.

For the reasons stated above, the parallel flow regime will be chosen for the module being designed. Predictability, simplicity of design, and module-to-module consistency are good reasons to choose parallel flow for a dialysis process. However, for processes in which one of the fluid boundary layers contributes most of the overall mass transfer resistance, perpendicular flow should be considered. For example, for the experiments conducted by Yang and Cussler (1986), both membrane resistance and gas-phase boundary layer resistance were quite small compared with liquid-phase boundary layer resistance. In this case, perpendicular flow gave overall mass transfer coefficients that exceeded those for parallel flow by more than an order of magnitude, as described in the preceding section.

Prediction of Overall Mass Transfer Performance

Use of Figures 12–9 and 12–11 of Chapter 12 to estimate values for the extraction ratio requires knowledge of a dialyzer's overall mass transfer coefficient $\bar{k}_o$. An experimentally determined value of $\bar{k}_o$ for the solute, membrane, and dialyzer configuration of interest is to be preferred. Estimates of $\bar{k}_o$ can be made based on knowledge of the geometry and fluid dynamics of the dialyzer and the nature of the membrane and solute. Methods have been described in Chapter 12 and this chapter that allow estimation of P_m, $\bar{k}_f$, and $\bar{k}_d$ for a variety of situations. Regardless of whether an experimental measurement or a calculated estimate of $\bar{k}_o$ is used, it is important to make the determination at the projected operating temperature because of the strong temperature dependence of solute diffusivity.

As discussed earlier, the overall mass transfer resistance ($\bar{R}_o = 1/\bar{k}_o$) can be considered the sum of the individual boundary layer and membrane resistances; that is, as shown in Eq. (12–25) of Chapter 12:

$$1/\bar{k}_o = 1/\bar{k}_f + 1/P_m + 1/\bar{k}_d.$$

If the individual mass transfer coefficients can be estimated, then a value for $\bar{k}_o$ can be derived from Eq. (12–25) of Chapter 12. Since three examples of hollow-fiber membranes have been selected, published data can be used as a prediction of P_m. Such studies have been done by Klein et al. (1976, 1977). By interpolation from the results of these studies, P_m for 125- and 250-μm-i.d. fibers, both of which have 22-μm wet wall thicknesses, is estimated to be 4.0 ×

10^{-4} cm/s for a solute of 200 daltons $\tilde{M}$ and the 375-μm-i.d. fiber, which has a wall thickness of 32 μm, has an estimated permeability of 3.1 $\times$ 10^{-4} cm/s.

Before attempting to predict $\overline{k}_d$, a value for packing density should be chosen. Noda and Gryte (1979) predicted $\Phi = 0.63$ as the optimum value for packing density, but they, as well as Yang and Cussler (1986) and Hermans (1978), warn of the effect of channeling on shellside mass transfer. If shellside flow is poorly distributed among the fibers, then relatively stagnant regions exist that severely compromise mass transfer. Yang and Cussler observed this effect in a close-packed module that exhibited flow-independent mass transfer at 1/10 the rate of the same fibers in a loosely packed module. The value of Φ at which channeling becomes a problem is dependent on module design factors, such as fiber bundle configuration (woven or unsupported), shell inside diameter and ratio of fiber length to outside diameter. Of course, it is desirable to approach $\Phi = 0.63$ to maximize both surface-to-volume ratio of the module and mass transfer coefficient per unit flow rate. The authors' experience has been that packing densities in the 50 to 60% range can be achieved without undue channeling provided the fiber length is large in proportion to both fiber and shell diameters. The initially proposed fiber lengths arrived at in Figure 13–6 certainly qualify. Therefore, $\Phi = 0.55$ will be chosen as a reasonable value for the design at hand.

To estimate $\overline{k}_d$, a correlation developed by Wald, Kessler, and Lopez (1990) will be used. The basis for this correlation was a series of experiments using a number of identical hollow-fiber modules with L/d much larger than hemodialyzers, which tends to minimize channeling. Mass transfer performance of this series of modules showed little module-to-module variation. Equation (13–3) gives the correlation

$$\overline{Sh}_d = 0.025 \, Re^{0.94} \, Sc^{0.33}. \quad (13\text{–}3)$$

This correlation is similar to that reported by Yang and Cussler (1986) for their loosely packed hollow-fiber modules. Equation (13–3) predicts higher values for the Sherwood number, especially for large values of L/d, than does the Yang and Cussler correlation. The value of exponent b is similar in both correlations (0.94 and 0.93) and is difficult to explain in relation to laminar flow in bundles of rigid tubes. However, this value is typical of shellside mass transfer in hollow-fiber modules (Colton 1988). The fact that this exponent is much higher than predicted or observed for laminar flow in well-defined ducts and higher even than that observed for turbulent flow can be explained if we consider the freedom that hollow fibers have to move laterally and orient with the flow field. As fluid velocity increases in parallel flow, the resulting forces on the fibers tend to improve the uniformity of fiber distribution, which, in turn, minimizes channeling and improves overall mass transfer.

In estimating $\overline{k}_d$ for the module being designed, a value of 1100 is estimated for the Schmidt number (based on a diffusion coefficient of 9×10^{-6} cm^2/s for a 200-dalton solute). The Reynolds number will be calculated for each of the three "minimum fiber number" points selected in Figure 13–6, assuming a feed flow rate of 10,000 L/h and a dialysate flow rate of 20,000 L/h. Then a value for $\overline{Sh}_d$ is calculated using Eq. (13–3) and $\overline{k}_d$ is calculated using Eq. (12–27) of Chapter 12. The results of these calculations are tabulated in Table 13–2.

Now the fluid-side mass transfer coefficient based on logarithmic mean concentration difference can be estimated. First, inserting the above estimates of membrane permeability and dialysate-side mass transfer coefficient into Eq.

TABLE 13–2. Values of Reynolds Number and Dialysate-Side Sherwood Number and Mass Transfer Coefficient Corresponding to Minimum Fiber Numbers of Figure 13–6.

Fiber Outside Diameter (μm)	Re	$\overline{Sh}_d$	$\overline{k}_d$ (cm/min)
170	9.0	2.0	0.064
295	29	6.0	0.11
440	55	11	0.13

TABLE 13–3. Values of Wall Sherwood Number and Dimensionless Length Used to Determine Feedside Sherwood Numbers.

Fiber Inside Diameter (μm)	Sh_w	z^*
125	0.40	1.2
250	0.91	0.61
375	1.1	0.40

(13–4), values for k_w can be calculated for each of the fiber diameters:

$$1/k_w = 1/P_m + 1/\overline{k}_d. \quad (13\text{–}4)$$

Then, an estimate can be made for the wall Sherwood number, as defined by Eq. (12–31) of Chapter 12, at each fiber inside diameter. Values for dimensionless length z^* can be calculated for the minimum fiber number conditions for each of the three fibers. The above results are tabulated in Table 13–3.

Referring to Figure 12–5 of Chapter 12, values of feedside (lumen-side) Sherwood numbers can be determined from the curves. Feedside mass transfer coefficient $\overline{k}_f$ can then be calculated. Overall mass transfer coefficients can now be determined by adding each of the mass transfer resistances previously determined (reciprocal of each k). These values can be found in Table 13–4.

By comparing the values of membrane permeability and the various mass transfer coefficients as resistances (their reciprocals), the relative contributions to overall resistance can be obtained, as shown in Table 13–5.

To reiterate, the initial design compares the three different fibers at equivalent feed and dialysate flow rates, membrane area, and TLP, operating in countercurrent flow. The membrane area was chosen by approximating overall mass transfer using clearance data for a dialyzer containing one of the fibers chosen. Feed flow rate was chosen as the minimum flow required to provide the required throughput of product, and dialysate flow rate was arbitrarily set at twice the feed flow. From Figure 13–6, conditions were chosen such that the minimum total fiber number for each diameter was selected.

The result, as presented in Table 13–5, is favorable in that for all fibers selected, the sum of feed and dialysate resistances contributes only one-third of overall resistance, the rest being contributed by the membrane.

Further insight into the performance of this design and its optimization can be gained by referring to Figure 12–9 of Chapter 12, the plot of extraction ratio versus the number of transfer units for countercurrent flow. Calculating values of N_t for the three fiber diameters can now be done using the $\overline{k}_o$ values in Table 13–4. These values have been referred to a value of $Z = 0.5$ in Figure 12–9 of Chapter 12 and the results are tabulated in Table 13–6.

These extraction ratio values predict that the initial design conditions will result in between 67 and 75% removal of low molecular weight contaminants, depending on the membrane selected. Since the goal is 90% removal, some adjustments must be made in module design and/or operating conditions to achieve a higher extraction ratio.

At this point, the 375-μm-i.d. fiber will be dropped from consideration due to its lower level of performance (which is due primarily to its thicker wall and resulting lower membrane permeability). With the two remaining fibers, simply increasing dialysate flow rate will in-

TABLE 13–4. Feedside Sherwood Numbers and Mass Transfer Coefficients and Overall Mass Transfer Coefficients.

Fiber Inside Diameter (μm)	$\overline{Sh}_f$	$\overline{k}_f$ (cm/min)	$\overline{k}_o$ (cm/min)
125	4.3	0.19	0.016
250	4.3	0.093	0.016
375	4.3	0.062	0.013

TABLE 13–5. Contributions of Feed, Membrane, and Dialysate Resistances to R_o in Initial Design.

Fiber Inside Diameter (μm)	$\bar{R}_f$ (%)	R_m (%)	$\bar{R}_d$ (%)
125	8	67	25
250	17	68	15
375	21	69	10

TABLE 13–7. Number of Transfer Units N_t and Extraction Ratio E Reflecting a 2× Increase in Q_d and $Z = 0.25$.

Fiber Inside Diameter (μm)	N_t	E
125	2.0	0.82
250	1.9	0.81

crease extraction ratio. An increase from 20,000 to 40,000 L/h will approximately double the values of $\bar{k}_d$ associated with each fiber (since $\bar{k}_d$ depends on $\text{Re}^{0.94}$). Referring again to Figure 12–9 of Chapter 12, a further advantage results from increasing Q_d, reflected in the choice of $Z = 0.25$, which gives higher extraction ratio values for a given value of N_t. The new values are shown in Table 13–7.

While the changes have brought performance closer to the goal of $E = 0.9$, further increases in Q_d will not achieve the goal. Instead, membrane area will be increased by increasing the total number of fibers, while holding fiber length constant. The value of Q_f will remain unchanged, resulting in a decrease in TLP and in $\bar{k}_f$. If the total fiber number is increased by 50%, A_m will increase to 2775 m² and TLP will decrease from 75 to 50 kPa. Since feedside mass transfer is already operating in the asymptotic region of Figure 12–5 of Chapter 12, the decrease in $\bar{k}_f$ will be negligible. Dialysate flow rate will be increased by an additional 50% to maintain the same Reynolds number as in the previous example. Since Q_f has not increased, $Z = 0.17$ in this example. The resulting values of N_t and E (from Figure 12–9 of Chapter 12) are given in Table 13–8.

Physical Design of Membrane Modules

Up to this point, a number of hollow fibers operating in parallel have been considered for design purposes, without regard to modularization. Since the total number of fibers is in excess of 10^6 fibers for the process considered above, multiple modules will be required. For the two fibers still under consideration, the total fiber numbers are now 8.4×10^6 and 1.5×10^6 for the 125- and 250-μm fibers, respectively. Considering the different fiber lengths (85 cm for 125 μm and 240 cm for 250 μm), total module volumes can be calculated as well. At $\Phi = 0.55$, for the 125-μm fiber, an internal module volume of 295 L is required, while the design based on the 250-μm fiber requires 445 L. Assuming a maximum shell inside diameter of 25 cm, seven 85-cm-long modules containing the 125-μm fiber or four 240-cm-long modules containing the 250-μm fiber would meet the requirements. While four of the longer modules occupy more total volume than seven of the shorter modules, manifolding of four modules is simpler. Another practical advantage of the

TABLE 13–6. Number of Transfer Units N_t and Extraction Ratio E for the Initial Design with $Z = 0.5$.

Fiber Inside Diameter (μm)	N_t	E
125	1.8	0.75
250	1.8	0.75
375	1.4	0.67

TABLE 13–8. Number of Transfer Units N_t and Extraction Ratio E Reflecting a 1.5× Increase in A_m, a 3× Increase in Q_d, and $Z = 0.17$.

Fiber Inside Diameter (μm)	N_t	E
125	3.0	0.92
250	2.9	0.91

250-μm-i.d. fiber is that it is more resistant to clogging of the lumen.

Configuration of the modules could be similar to a shell-and-tube heat exchanger or a hollow-fiber hemodialyzer (Figure 13–4). Shellside flow distribution to a 25-cm-diameter bundle of 295-μm-o.d. fibers requires special attention, however. Simply providing a baffled circumferential header at each end of the shell would leave an area of stagnation in the center of the bundle. Instead, flow will be introduced through a distribution tube at one end and in the center of the bundle and will exit circumferentially at the opposite end. The Enka and Sepracor industrial hollow-fiber modules described above use examples of this approach.

Materials Selection

For the process described above, the feed is relatively noncorrosive (pH 7.0). Material selection may be determined more by the method used to clean the modules in this case. For cellulose membranes, a sanitizing solution such as glutaraldehyde or peroxyacetic acid is likely to be used. Either 316 stainless steel or an engineering polymer such as polysulfone can be used to fabricate the module housing. The encapsulant used to form tubesheets at either end of the module could be a polyurethane, silicone rubber, epoxy, or a thermoplastic such as polypropylene.

In the situation where either the feed or dialysate is an organic solvent solution, stainless steel is a good choice for the housing, although more solvent-resistant polymers such as nylon can also be used. The list of possible encapsulants should be shortened to epoxy or polypropylene in this case. Elastomeric seals are limited to elastomers specifically selected for a particular solvent, perfluoro elastomers, or fluoro-polymer-coated seals.

If a membrane is selected that allows steaming-in-place (SIP), then only stainless steel should be considered for a housing material with epoxy or, possibly, a thermoplastic as encapsulant. Ethylene-propylene copolymer elastomers (EPR or EPDM) are a good choice for elastomeric seals in contact with steam.

Cleaning with caustic solutions requires similar materials choices to SIP. If sodium hypochlorite is to be used as a cleaning agent, then stainless steel is not a good choice for a housing material. One of the engineering polymers would be preferred.

PROCESS AND SYSTEM DESIGN

Batch versus Continuous Operation

Dialysis can be used as a unit operation in two basic configurations: as a batch operation or in continuous mode. Figure 13–9 depicts a schematic of batch dialysis. A reservoir of volume V and feed solute concentration c_f is dialyzed continuously with an inlet concentration equal to c_{di}. The feed and dialysate streams enter the dialyzer with flows of Q_f and Q_d, respectively. The feed stream exiting the dialyzer returns to the reservoir while the spent dialysate is discarded to enter another recovery process. Such a configuration could be used for stripping low molecular weight contaminants from a high molecular weight product. It could also be used to recover a low molecular weight product by catalytic cleavage of a high molecular substrate in the feed reactor. In that event the spent dialysate is processed to recover the prod-

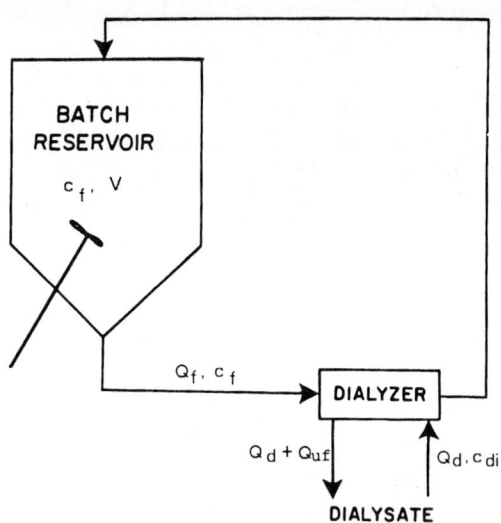

FIGURE 13–9. A batch dialysis process (reprinted from Klein, Ward, and Lacey 1987 with permission).

uct and not discarded. The former represents, in essence, the practice of hemodialysis with the reservoir representing the body pool of water.

A mass balance may be written for the reservoir:

$$\frac{d(Vc_f)}{dt} = -D^*(c_f - c_{di}), \quad (13\text{--}5)$$

where D^* is the dialysance defined previously. A volume balance may also be written for the reservoir:

$$V = V_0 - Q_{uf}t, \quad (13\text{--}6)$$

where V_0 is the initial reservoir volume and Q_{uf} is the rate of ultrafiltration from the feed to the dialysate. Substituting for V in Eq. (13–5) leads to

$$(V_0 - Q_{uf}t)\frac{dc_f}{dt} - Q_{uf}c_f = -D^*(c_f - c_{di}). \quad (13\text{--}7)$$

Rearrangement and integration lead to the following relationship for the reservoir concentration as a function of time:

$$c_f(t) = \frac{D^* c_{di}}{D^* - Q_{uf}}\left(1 - \left(\frac{(V_0 - Q_{uf}t)}{V_0}\right)^b\right)$$

$$+ c_f^0 \left(\frac{(V_0 - Q_{uf}t)}{V_0}\right)^b, \quad (13\text{--}8)$$

where $b = (D^* - Q_{uf})/Q_{uf}$ and c_f^0 is the initial feed solution concentration. If ultrafiltration is negligible, that is, $Q_{uf} = 0$, Eq. (13–5) reduces to

$$V\frac{dc_f}{dt} = -D^*(c_f - c_{di}), \quad (13\text{--}9)$$

and the reservoir solute concentration is given by

$$c_f(t) = c_f^0 \exp\left(\frac{-D^* t}{V}\right) + c_{di}\left[1 - \exp\left(\frac{-D^* t}{V}\right)\right]. \quad (13\text{--}10)$$

Dialysis reactors may also be operated in a continuous mode. This can be represented by adding a feed line to the schematic of Figure 13–9.

The addition of a high molecular weight substrate at a flow rate Q_r to an enzymatic reactor whose products permeate the membrane would exemplify one case. As the substrate is converted to dialyzable product at a rate G, the increasing concentration must be incorporated into the previous equations. Equation (13–5) now becomes

$$\frac{d(Vc_f)}{dt} = -D^*(c_f - c_{di}) + G, \quad (13\text{--}11)$$

and the integrated form incorporates the solute regeneration rate and volume expansion as follows:

$$c_f(t) = \frac{D^* c_{di} + G}{D^* - Q_{uf} + Q_r}\left(1 - \left(\frac{(V_0 - Q_{uf}t)}{V_0}\right)^g\right)$$

$$+ c_f^0 \frac{(V_0 - Q_{uf}t)^g}{V_0}, \quad (13\text{--}12)$$

where

$$g = \frac{D^* - Q_{uf} + Q_r}{Q_{uf} - Q_r}.$$

Staged Operation

To improve the selectivity of a dialytic separation beyond what can be achieved in a single stage, a cascade process consisting of several stages can be used. Rautenbach (1986) has reviewed the cascade process option as it applies to membrane-based processes in general. Noda and Gryte (1981) described a multistaged system with two cross-current dialyzer cascades and a reverse osmosis unit acting as a nonselective solvent stripper. This approach increases the slope of the curves in Figures 12–9, 12–10, and 12–11 of Chapter 12, relating E to N_t, resulting in improved relative separation of two solutes of differing molecular weight. Assuming $Q_f = Q_d$ throughout the system, the extraction ratio for the system is given by Eq. (13–13):

$$E = \frac{N_t^n}{1 + N_t^n}, \quad (13\text{–}13)$$

where n is the number of stages in each cascade.

To illustrate the efficacy of this approach in improving selectivity, assume that two solutes, A and B, are being separated in a countercurrent, single-stage dialysis process with $Q_f = Q_d$. Under these conditions, their extraction ratios are $E_A = 0.33$ and $E_B = 0.60$ with $N_{tA} = 0.5$ and $N_{tB} = 1.5$. The resulting selectivity is the ratio of E_B to E_A, which is 1.82. In a three-stage cascade system, Eq. (13–13) predicts E values of 0.11 and 0.77 for a selectivity of 7.0.

NOTATION

General Notation

See the General Notation section at the beginning of this handbook.

Special Notation

A_m	membrane area, L^2
c_f^0	initial feed solution concentration, mol/L^3
C	clearance, L^3/t
D^*	dialysance, L^3/t
E	extraction ratio, dimensionless
G	conversion rate of the substrate to the dialyzable product, mol/t
k	mass transfer coefficient, L/t
K_D	Darcy's constant, L^2
N	number of hollow fibers
N_t	number of transfer units
P_m	diffusive permeability of solute in dialysis, L/t
Q_d	dialysate flow rate, L^3/t
Q_f	feed flow rate, L^3/t
Q_{uf}	ultrafiltration rate from the feed to the dialysate, L^3/t
r_{ti}	inside radius of hollow fiber, L
R	mass transfer resistance, t/L
TLP	translumenal pressure drop, p
UFR	ultrafiltration rate per membrane area, L/tp
V_0	initial reservoir volume, L^3
y	distance, normal direction, L
z	distance, axial direction, L
z^*	distance, axial direction, dimensionless
Z	ratio of feed to dialysate flow rate, Q_f/Q_d, dimensionless

Greek Letters

η	viscosity, M/Lt
Φ	packing density of a hollow-fiber module, dimensionless

Subscripts

d	dialysate
o	overall

REFERENCES

Colton, C. K. 1988. Unpublished data.

Colton, C. K., and E. G. Lowrie. 1981. Hemodialysis: physical principles and technical considerations. In *The Kidney*, Vol. II, ed. B. M. Brenner and F. C. Rector, Jr., pp. 2425–2489. New York: W.B. Saunders Company.

Dandavati, M. S., M. R. Doshi, and W. N. Gill. 1975. *Chem. Eng. Sci.* 30:877–886.

Enka AG. 1982. Cuprophan: dialyzing tubular and flat membranes. Product brochure.

Happel, J. 1959. Viscous flow relative to arrays of cylinders. *AIChE J.* 5:174–177.

Hermans, J. J. 1978. Physical aspects governing the design of hollow-fiber modules. *Desalination* 26:45–62.

Klein, E., F. F. Holland, A. Donnaud, A. Lebeouf, and K. Eberle. 1977. Diffusive and hydraulic permeabilities of commercially available cellulosic hemodialysis films and hollow fibers. *J. Membr. Sci.* 2:349–364.

Klein, E., F. F. Holland, A. Lebeouf, A. Donnaud, and J. K. Smith. 1976. Transport and mechanical properties of hemodialysis hollow fibers. *J. Membr. Sci.* 1:371–396.

Klein, E., R. A. Ward, and R. E. Lacey. 1987. Membrane processes—dialysis and electrodialysis. In *Handbook of Separation Process Technology*, ed. R.W. Rousseau. New York: John Wiley & Sons.

Kolff, W. J., and H. T. Berk. 1944. The artificial kidney: a dialyzer with a great area. *Acta Med. Scan.* 117:121–134.

Lipps, B. J., R. D. Stewart, H. A. Perkins, G. W. Holmes, E. A. McLain, M. R. Rolfs, and P. P. Oja. 1967. The hollow-fiber artificial kidney. *Trans. Am. Soc. Artif. Intern. Organs* 13:200–207.

Noda, I., and C. C. Gryte. 1979. Mass transfer in regular arrays of hollow fibers in countercurrent dialysis. *AIChE J.* 25:113–122.

Noda, I., and C. C. Gryte. 1981. Multistage membrane separation processes for the continuous fractionation of solutes having similar permeabilities. *AIChE J.* 27:904–912.

Rautenbach, R. 1986. Process design and optimization. In *Synthetic Membranes: Science, Engineering and Applications,* ed. P. M. Bungay, H. K. Lonsdale, and M. N. de Pinho, pp. 457–522. Dordrecht, Holland: D. Reidel Publishing Co.

Tuwiner, S. B. 1962. *Diffusion and Membrane Technology.* New York: Reinhold Publishing Co.

Wald, S. A., S. B. Kessler, and J. L. Lopez. 1990. Characterization of large-scale membrane devices for diffusive transport. Paper read at The International Congress on Membranes and Membrane Processes, 20–24 August 1990, Chicago, IL.

Yang, M.-C., and E. L. Cussler. 1986. Designing hollow-fiber contactors. *AIChE J.* 32:1910–1916.

14
Applications

Stephen B. Kessler
Sepracor Inc.

Elias Klein
University of Louisville

INTRODUCTION
HEMODIALYSIS
   Treatment of End-Stage Renal Disease
   Recent Trends in End-Stage Renal Disease Therapy
   Transport Properties of Blood
   Interaction of Plasma Proteins with Foreign Surfaces
HEMOFILTRATION AND HEMODIAFILTRATION
DONNAN DIALYSIS
ALCOHOL REDUCTION OF BEVERAGES
NOTATION
REFERENCES

INTRODUCTION

The traditional industrial applications of dialysis, such as recovery of sodium hydroxide in rayon processing or separation of nickel sulfate from sulfuric acid in electrolytic copper refining, have been reviewed in detail by Tuwiner (1962). The dialyzers used in these processes were described in the preceding chapter. This chapter is devoted to applications of dialysis that are currently of greater interest.

Hemodialysis is the dominant use of dialysis at this time and is responsible for much of the detailed analysis of the process that has been done in recent years. Donnan dialysis makes use of ion-selective membranes to provide improved selectivity, while alcohol reduction in beer is a recent application that is of interest because it makes use of large-scale hollow-fiber dialyzers. Other related applications are membrane-solvent extraction, in which one of the phases contacting the membrane is an organic solvent, membrane-contained liquid membranes, and enzyme-based membrane reactors. These state-of-the-art new processes are covered in Part X of this handbook.

HEMODIALYSIS

Treatment of End-Stage Renal Disease

Once renal function decreases to less than approximately 5% of normal, life cannot be sustained without recourse to artificial means. Hemodialysis, the dialysis of a patient's blood against a physiological saline solution, replaces kidney function in three principal areas: removal of waste metabolites, removal of excess body water, and restoration of acid-base and electrolyte balances. The waste metabolites include urea, the endproduct of protein catabolism, and creatinine, the endproduct of muscle metabolism. In addition, it is thought that there are larger solutes (up to 13K dalton) that accumulate very slowly in end-stage renal disease (ESRD) patients and may have negative impacts on various metabolic regulatory mechanisms. In the literature these solutes are referred to as "middle molecules" for lack of better identification.

Until recently, chronic therapy was achieved almost exclusively by hemodialysis, either in stand-alone clinics, at home, or in clinics attached to hospitals. In recent years this therapy has been augmented by use of the peritoneal membrane in place of an external synthetic membrane. However, hemodialysis remains the principal technique for maintaining life for patients with ESRD.

Patients with ESRD currently undergo treatment three times per week. The length of the treatment varies with the dialyzer used, the size of the patient, and the quantity of fluid the patient has retained since the preceding treatment. As a general rule, three to five hours of treatment will suffice for all patients. For patients that have regained limited amounts of fluid, the minimum time is determined—as a rough approximation—by the dimensionless ratio Ct/V. In this estimate the value C represents the device clearance, under the conditions in use, t is the length of time of dialysis, and V is the body water pool of the patient. Since clearance has units of mL/min, the ratio is dimensionless. Studies have shown that ratios greater than 1.2 are adequate for maintenance therapy. For patients that have retained significant quantities of fluid, the minimum time predicted by Ct/V may not be long enough to permit loss of the requisite amount of fluid via ultrafiltration.

Hemodialysis membranes have ultrafiltration capacities ranging from 5 to 70 mL/h m^2 mm Hg. Because of the obligatory venous resistance (needle plus vein) ranging from 60 to 120 mm Hg, and an additional blood channel resistance of 20 to 50 mm Hg, the minimum transmembrane pressure difference encountered is between 80 to 170 mm Hg. These values depend, of course, on the rate of blood flow employed.

For low permeability dialyzers containing 1 m^2 of surface area, this predicts that ultrafiltration (in the absence of dialysate-side negative pressure) will produce 400 to 850 mL/h of fluid loss. With the more permeable dialyzers the fluid loss can reach 4 to 8 L/h. Such high rates can produce fluid disequilibrium and are avoided in modern dialysate controllers by mechanical means.

In one approach to the control of ultrafiltration, the dialysate controller provides balanced volumes of dialysate into and out of the shell side of the dialyzer. Introduction of a bias into this balance allows predetermined rates of ultrafiltration. Another series of dialysate controllers uses turbine flow meters to measure the rate of dialysate entering and leaving the device and then balances pressures to achieve predetermined rates of filtration. Dialyzers with high ultrafiltration coefficients are often termed "high-efficiency" dialyzers, although it does not necessarily follow that asymmetric membranes with high ultrafiltration coefficients provide high rates of diffusive transport.

Recent Trends in End-Stage Renal Disease Therapy

Two major alterations in the delivery of hemodialysis therapy are in progress. One is the development of synthesized erythropoietin (EPO), a hormone normally found in kidney tissue that stimulates the formation of red blood cells. The other is continual pressure to reduce the costs of ESRD treatment. The availability of EPO for patients whose usual hematocrit (blood fraction occupied by oxygen carrying red cells) is in the low twenties—compared to a normal value of 40 to 45—has a significant social as well as therapeutic effect. Patients whose hematocrit returns to near normal have better tissue-oxygenating capacity, often resulting in improved appetite and physical well-being. The increased food intake requires higher clearance of metabolites, which, in turn, puts additional demands on clinics that are already strained in meeting current needs.

The hemodialyzer manufacturers are responding to this dilemma by producing "high-efficiency" dialyzers. These devices are derived via two means. One way is merely to increase the surface area and cross section of the blood flow channel. The patient must then be treated with higher blood flow rates. If vascular disease does not mitigate this approach, it can be a simple solution to maintaining the patient in a new equilibrium condition. However, the collateral benefit is that dialysis times can be shortened for patients whose metabolism is not increased by virtue of EPO therapy. Since time on dialysis can translate into costs for the clinic, it is one way of coping with the lowered reimbursement rates. Moreover, increased surface areas in commodity dialyzers are not expensive.

A more costly approach offered by some manufacturers is to introduce new dialyzers with significantly increased ultrafiltration coefficients (UFR). However, higher UFRs do not affect the diffusive permeability of small solutes, such as urea, significantly. Where distinctions can be made is in the transport of large ions, such as inorganic phosphate, and in the range of "middle molecules." This effect is seen in Table 13-1 of Chapter 13 in the diffusive permeability of vitamin B_{12}.

With the use of very high UFR hemodialyzers comes a new problem. The obligatory pressure difference due to pumping blood in excess of 400 mL/min is so high that the resulting UFR, if uncontrolled, would be life threatening to the patient. Consequently, dialysate control machines have been introduced that limit the volume of net ultrafiltration flow from the blood into the dialysate. However, the nature of the pressure drop along the blood flow path, combined with the limited acceptance of fluid into the dialysate compartment, leads to Starling's flow (back-filtration) within the dialyzer. Thus, some of the dialysate can return to the patient's blood. In the past, dialysate has not generally been prepared as a sterile fluid. With the possibility of Starling's flow, concern is raised that bacterial products able to cross the membrane convectively may affect the patient's well-being. This is of concern, especially in view of recent findings that small solutes from bacterial lysis activate neutrophil receptors.

Transport Properties of Blood

Calculation of transport from whole blood requires some modification when the solute-membrane characterization is carried out in saline solution. First, the presence of erythrocytes reduces the available volume in which diffusible solute is distributed. For normal blood approximately 40 vol% of the blood is occupied by the red cells. Although some solutes inside these cells can equilibrate quickly with the fluid plasma (cell free blood) passing through the dialyzer, this is not a general phenomenon. Additionally, the plasma proteins, present to about 7 wt.%, occupy a fraction of the remaining plasma volume. Consequently, the fraction of the feed fluid actually carrying diffusible solute may be as low as 53% of the incoming flow.

Some evidence exists that the presence of red cells increases the diffusivity of solutes near the membrane wall (Collingham 1968). Thus, translation of device performance from saline to whole blood data requires some care.

Interaction of Plasma Proteins with Foreign Surfaces

Whole blood carries within it two humoral defense mechanisms that have been found to be of importance in extracorporeal treatment. The first is the coagulation defense system that protects us from excessive blood loss when the vascular tree is damaged. The second is the complement system that identifies—for subsequent attack and destruction—foreign bodies that have entered the circulation. The latter may be bacteria, viruses, or synthetic surfaces, which can initiate the complement coagulation cascade. Each of these defense mechanisms is complex and they can operate interdependently.

The mechanism of initiation of the coagulation cascade by synthetic surfaces is still a matter of debate. Two observations are important. Very hydrophilic surfaces, such as poly(ethylene oxide) gels or regenerated cellulose, appear to trigger this mechanism only minimally. Very hydrophobic surfaces may or may not initiate the cascade, depending on the protein adsorption that occurs—and whether or not such proteins are denatured following adsorption. As a practical matter, all patients undergoing hemodialysis are given intravascular injections of heparin, a glycosaminoglycan, which catalyzes the destruction of fibrinogen. The latter is a necessary precursor to the formation of thrombin, the clot-forming protein.

The immune protection system is thought to operate via an enzymatic cascade, which eventually assembles an attack complex at the surface of the foreign body. An intermediate product, C5a, signals monocytes and neutrophils to produce stimulants for subsequent cellular attack of the foreign material. Although most synthetic membranes are not subject to enzymatic degradation, they can stimulate the cascade to occur. The by-product of such stimulation is the release of proteins that produce the characteristics of an inflammatory reaction. That is, the presence of the synthetic surface fools the defense mechanism into activating and wasting its enzymatic and oxidative burst defenses.

HEMOFILTRATION AND HEMODIAFILTRATION

The limitations imposed by diffusion rates on dialytic mass transfer led Henderson et al. (1967) to investigate an ultrafiltration (i.e., hemofiltration) procedure for the removal of ESRD metabolites. Because of the limitations on filtration rates produced by high hematocrits, a predilution method was used. In this procedure, the blood is first diluted with saline solution en route to the hemofilter and then ultrafiltered. The diluted plasma water that is ultrafiltered contains the microsolutes that normally cross the dialysis membrane by diffusion. In fact, membranes suitable for hemofiltration are generally more permeable to high molecular weight solutes so that larger species are removed relatively more rapidly than is possible in hemodialysis. However, the removal rate decreases exponentially since the recycled plasma becomes progressively more dilute. Thus, each subsequent increment of ultrafiltered fluid carries with it less and less of the metabolic waste.

The difficulty with this concept is that very large quantities of sterile, endotoxin-free saline solution must be prepared to provide sufficient volume for the diafiltration process. The procedure has found more acceptance in Europe than in the United States.

A combination of the two techniques has been reported at a few centers. To reduce the volume of diafiltration required, the hemofiltration is combined with a purely dialytic step. In this process, termed *hemodiafiltration*, the blood first passes through a dialyzer—with its countercurrent dialysate—where small solutes are removed efficiently. Before it is returned to the patient, the dialyzed blood is diluted with smaller volumes of sterile saline and then passed through a hemofiltration device to remove the dilution fluid. During this step the larger solutes that were not removed efficiently by dialysis are filtered by convection. Alternatively, both dialysis and hemofiltration can be carried out simultaneously using a single membrane device (a hemodiafilter). Hemodiafiltration is more complex than either dialysis or hemofiltration in that it requires both a dialysate fluid and a sterile infusion fluid as well as a more complex fluid management system.

DONNAN DIALYSIS

Donnan dialysis differs from the classical forms of dialysis in that it relies on ion-selective membranes to retard the transfer of either a cation or an anion across the barrier. The charge species that is allowed to cross the membrane will equilibrate across the membrane until the "Donnan equilibrium" conditions (Wallace 1967) are reached. Thus, it can be described as a membrane-based equilibrium process. For the permeable species, the conditions for equilibrium can be written as

$$\frac{(a_{iL})^{1/z_i}}{(a_{iR})^{1/z_i}} = K, \quad (14\text{-}1)$$

where a_i represents the activity of the particular ion of valence z_i, and L and R designate solutions on different sides of the membrane, and K is the Donnan constant.

This equation applies to any ion moving through the membrane. As a result, the constant K describes the ratio of any permeating ionic species and all species of ions of a given charge class will transport to satisfy the equilibrium condition. For example, if tap water, dilute in sodium but high in calcium, is separated from seawater by a cation (only) permeable membrane, the system will attempt to reach the following equilibrium:

$$\frac{(Na)_T}{(Na)_S} = \frac{(Ca)_T^{0.5}}{(Ca)_S^{0.5}} = K, \quad (14\text{-}2)$$

where the parentheses indicate concentrations (ignoring activity coefficients) and the subscripts T and S refer to the tap and seawater phases, respectively. Lake and Melsheimer (1978) showed how this technique might be used to reduce the calcium content of seawater prior to desalination (in order to minimize scaling) using the higher concentrations of the raffinate Na ions to "pump" calcium across a Donnan dialysis barrier.

The mechanism is conceptually simple. A membrane with a high concentration of fixed charges contains an equivalent concentration of mobile counter-ions of opposite charge (gegenions). Only ions with this same charge can displace the counter-ions, and the concentration of all ions of this charge in the membrane is limited to the concentration of fixed charges. Thus, an external cation competes with the mobile membrane cation for association with a fixed negative charge. The total number of such moving ions in the membrane at any time is limited. The analysis usually begins with the assumption that at either side of the membrane a Donnan equilibrium exists between the solution and the membrane face. Within the membrane an ion gradient then exists that will attempt to redistribute toward equilibrium by diffusion.

Two opposing effects are imposed by extremes in membrane structure. If the pore channels are very large, the possibility exists for a co-ion (i.e., one of the same charge as the fixed ion) to diffuse across the membrane without being repelled by fixed charges. This would be a leakage flow and would dissipate the Donnan flux. On the other hand, if the pore channels are very small indeed, so that no leakage flow occurs, the diffusion rates may be very low. If the rates are too low, the membrane appears to be essentially impermeable.

Donnan dialysis has been proposed for a number of applications in which metal ions need to be recovered, using hydrogen ions as the "pumping" ion. The availability of sulfonated perfluoro polymer membranes (Nafion®, Du Pont) has facilitated the use of such techniques for the transfer of cations. However, the transfer of anions remains a more difficult problem because high-efficiency cationic membranes are not available. If membranes of equal utility were available, the use of serial Donnan dialyzers can be foreseen as an efficient means of deionization.

Conceptually, both the cation-exchange and the anion-exchange sections of such a device would operate like continuous ion-exchange columns. Ion-exchange columns are loaded with a particular ion and then, off-line, recharged to their original ion concentration. Donnan dialysis membranes provide this same function continuously. On one side of the membrane the feed solution "loads" the membrane; on the other side of the membrane these ions are off-loaded and the membrane recharged. Thus, a cation- and an anion-exchange membrane operating in series, using strong acid and base as the respective "pump" solutes, would provide continuous mixed bed deionization.

ALCOHOL REDUCTION OF BEVERAGES

Moonen and Niefind (1982) published a report describing a process that produces 1500 L/h of reduced-alcohol beer. In this process, five hollow-fiber modules are used, each having 18 m² of membrane area (90 m² total). Beer flows through the fiber lumina and water is the dialysate. At the rate of 1500 L/h, a 40% reduction in ethanol content is achieved. In his analysis of the process, Jonsson (1986) derives an overall mass transfer coefficient of $k_o = 0.011$ cm/min (which is comparable to the values derived in Chapter 13 for a hypothetical dialysis process).

While superior to distillation, this approach to reducing the alcohol content of beverages is limited by the fact that other low molecular weight constituents of the beer are removed as readily as is ethanol, affecting aroma and taste. Other, more selective, membrane processes used in conjunction with or in place of dialysis can correct this shortcoming.

Another approach, described by a European patent (Tilgner and Schmitz 1981), is the use of a dialysate that consists of either a similar beverage that is unfermented or one that has been previously stripped of alcohol. In either case, the preferred version of the process recirculates the dialysate, while using vacuum distillation to remove ethanol continuously. The presence of flavor compounds in the dialysate is intended to minimize their removal from the beverage being processed.

NOTATION

General Notation

See the General Notation section at the beginning of this handbook.

Special Notation

C clearance, L^3/t
K Donnan constant, dimensionless

REFERENCES

Collingham, R. E. 1968. Mass transfer in flowing suspensions. Ph.D. diss. Univ. of Minnesota.

Henderson, L., A. Besarb, A. Michaels, and L. W. Bluemle. 1967. Blood purification by ultrafiltration and fluid replacement (diafiltration). *Trans. Am. Soc. Artif. Intern. Organs* 16:216–222.

Jonsson, G. 1986. Dialysis. In *Synthetic Membranes: Science, Engineering and Applications*, ed. P. M. Bungay, H. K. Lonsdale, and M. N. de Pinho, pp. 625–646. Dordrecht, Holland: D. Reidel Publishing Company.

Lake, M. A., and S. S. Melsheimer. 1978. Mass transfer characterization of Donnan dialysis. *AIChE J.* 24:130–137.

Moonen, H., and N. J. Niefind. 1982. Alcohol reduction in beer by means of dialysis. *Desalination* 41:327–335.

Tilgner, H. G., and F. J. Schmitz. 1981. European Patent 36175, assigned to Akzo Gmbh.

Tuwiner, S. B. 1962. *Diffusion and Membrane Technology*. New York: Reinhold Publishing Company.

Wallace, R. M. 1967. Concentration and separation of ions by Donnan membrane equilibrium. *Ind. Eng. Chem. Process Des. Dev.* 6:423–431.

15

Cost Estimates

Stephen B. Kessler
Sepracor Inc.

Elias Klein
University of Louisville

INTRODUCTION
COST ESTIMATION FOR A
  COMMODITY PROCESS:
  HEMODIALYSIS
  Materials
  Equipment
  Labor

COST ESTIMATION FOR A
  CUSTOM-DESIGNED PROCESS:
  PURIFICATION OF A
  HYPOTHETICAL 20K-DALTON
  COMPOUND
  Materials
  Equipment
  Labor
REFERENCE

INTRODUCTION

The cost of a process that incorporates a dialysis procedure depends not only on the dialytic device, but also on the ancillary equipment to complete the separation process. Depending on the type of separation considered, the fractional weights of these two cost components may vary over extreme ranges. Unfortunately, dialytic separation processes are not as amenable to cost prediction as are the more classical separations such as distillation. Although the acquisition and replacement costs of the dialytic separator can be predicted with some accuracy, this may represent only a small fraction of the total process costs.

To begin any cost estimate, the contemplated process should be assigned to one of two or three subcategories. These categories may represent widely used dialysis procedures—analogous to commodity production processes—or they may be progressively more specific process categories. The latter can be exemplified by ion recovery as a general dialysis separation, but one with few market applications, or by a highly specific dialysis separation coupled to a particular enzyme conversion process to produce chirally pure products. In this latter process, the dialysis materials and procedures must be specifically tailored to the chemistry involved in the enzymatic conversion. Thus, we can readily distinguish dialytic processes on the basis of commodity or custom-design market size. When the procedure is widely applied, as in hemodialysis, we can expect to find low costs for the dialysis membrane (per unit area) and standardized separation devices and ancillary equipment. However, when the dialytic procedure is part of a highly specific reaction and separation process, both the dialytic device and the ancillary equipment will probably be custom designed. As a consequence, the cost per unit area of the membrane device can be significantly higher. These higher costs are a result not only of the reduced market size; they are higher because the dialysis device contains a significant cost due to proprietary intellectual property.

For the reasons outlined above, this chapter can provide readers only a general guide on how costing of dialytic separations might be approached. Clearly, the analysis for custom-designed processes must be based mainly on projections of costs, while the commodity applications can rely on data from experience. The sections that follow provide examples of both.

COST ESTIMATION FOR A COMMODITY PROCESS: HEMODIALYSIS

Materials

The principal components and materials involved in providing hemodialysis to end-stage renal disease (ESRD) patients are the dialytic device, tubing to convey the blood, saline solution, heparin, and dialysate fluid. All of the above, except the dialysate, must be provided as sterile materials. Even dialysate fluid must be sanitized. Dialyzer costs vary from $10 to $35 for delivered devices containing approximately 1.0 m^2 of surface area. Dialysate fluid can be produced at less than $0.10/L and each treatment requires approximately 150 L (cost per treatment: $15).

Equipment

Capital equipment for hemodialysis clinics can be grouped into three general categories. The most expensive item is the dialysate controller, which dilutes concentrated salt solutions to produce a physiological solution used as dialysate. In addition to the metering function, it provides control of the ultrafiltration rate, infusion of anticoagulants (heparin), and several safety alarm functions. Current (1990) prices vary between $12,000 to $16,000 per machine. Each machine can be used to treat between six and eight patients, depending on the number of shifts operating in any particular clinic.

The second most expensive category of equipment contains safety monitoring and emergency equipment, such as electrocardiographs and resuscitation carts, continuous blood pressure monitors, and central monitoring stations. The third category of equipment costs includes facilities for preparing pure water from municipal supplies. This water must be piped to each treatment station and to any reuse device used for cleaning and sterilizing dialyzers between treatments.

Labor

The single largest item in dialysis center costs is labor. The industry has gradually receded from using graduate nurses for all functions to using such trained personnel only in supervisory roles. Technicians trained for this specific role now perform most of the work in hemodialysis centers in the United States. A typical hemodialysis center treating 100 patients each week will be allowed a reimbursement of $125/treatment by the Health Care Financing Agency (HCFA). This means that total costs—including rents, amortization, and any profits—must be less than $120 to allow for bad debts. Perhaps $45/treatment of this cost is consumed by labor, $30/treatment by supply costs, $10/treatment by plant and machinery maintenance, and $10/treatment for administrative costs. The balance goes to laboratory costs, rental, amortization of capital, and profit, if any. Efforts to reduce staffing levels and introduction of dialyzer reuse have reduced costs continually over the past 10 years. The most cost-effective procedure is clearly home treatment, despite single usage of dialyzers, since it avoids most labor costs.

COST ESTIMATION FOR A CUSTOM-DESIGNED PROCESS: PURIFICATION OF A HYPOTHETICAL 20K-DALTON COMPOUND

Materials

The dialytic process designed in Chapter 13 requires four hollow-fiber membrane modules containing a total of 2800 m² of regenerated-cellulose dialysis fiber. While not described explicitly in the design, it is assumed that the hollow-fiber bundle is contained in a cartridge that is separate from the module housing.

Since the market for hemodialyzers is of a commodity nature, the prices given above for hemodialyzers are misleadingly low for estimating the purchase price of replacement membrane cartridges for custom-designed, process-scale dialyzers. For the four 700-m² modules of the hypothetical process, replacement cartridges will be assumed to cost $70,000 each or $280,000 for a four-module system. These estimates are based on a cost of $100/m² of membrane area as packaged in replacement cartridges.

Frequency of membrane replacement depends on factors such as pH of feed and type and frequency of cleaning regimen. The feed in the current example was specified to be pH 7.0. If it is assumed that membrane fouling occurs slowly and can be reversed with mild cleaning agents, then a 1-yr life expectancy is reasonable. This results in a membrane replacement cost of $280,000 per year (in 1990).

Some of the suppliers of industrial membrane modules designed for use in diffusion-based membrane-separation processes such as dialysis, membrane-solvent extraction, and membrane reactor processes follow:

Enka AG
Division of Akzo, Fibers and Polymers
Business Unit Membrana
Postfach 20 09 16
Ohder Strasse 28
D-5600 Wuppertal 2
Germany

Separations Products Division
Hoechst Celanese Corporation
13800 South Lakes Drive
Charlotte, NC 28217

HPD Incorporated
Department C
HPD Place
P.O. Box 3032
Naperville, IL 60566-7032

Sepracor Incorporated
33 Locke Drive
Marlborough, MA 01752

Equipment

Cost of equipment for the type of process under consideration depends heavily on the degree of automation desired, the extent of inclusion of ancillary equipment and factors such as sanitary design and materials of construction. For example, the feed and dialysate pumps for this system must provide flows of 10,000 L/h and 60,000 L/h, respectively. Pumps of sanitary design that will provide flow rates of this magnitude typically cost five times as much as standard pumps of the same flow capability.

A totally manual, "stripped down" system could be designed to operate a large dialytic process. It is more likely, considering labor expense and process control issues, that some degree of automation would be desired. This factor alone (degree of automation) leads to a broad range of potential systems costs.

A system that would operate the process envisioned in Chapter 13 would cost between $100 and $800/m², depending on the factors outlined above and not including ancillary equipment or membrane modules. Thus, for a 2800-m² system, 1990 costs range from $280,000 to $2,240,000.

Scaling factors for cost estimates of this type of equipment have been reviewed by Remer and Idrovo (1990). They define the relationship between cost and capacity as

$$\frac{\text{cost}_2}{\text{cost}_1} = \left(\frac{\text{capacity}_2}{\text{capacity}_1}\right)^R, \qquad (15\text{--}1)$$

where the exponent R is defined as the scaling factor. If cost estimates for systems of different scales are plotted versus capacity on a log-log plot, R is the slope of the best fit line through the points.

Remer and Idrovo (1990) reviewed reports in the literature of scaling factors determined for a variety of types of equipment used in biopharmaceutical processing. While dialysis systems per se were not covered, scaling factors for both reverse osmosis and ultrafiltration systems were listed for an "unlimited" range of sizes. Published R values for these types of membrane systems ranged from 0.9 to 1.0, indicating a nearly linear relationship between cost and capacity. Dialysis systems should follow the same relationship since the equipment requirements are similar.

When applying such scaling factors, overriding cost factors such as sanitary design and degree of automation must be considered. It goes without saying that the cost of a manually operated pilot system will not scale linearly to that of a fully automated production system.

Labor

Equipment costs for modern, fully automated dialysis systems are much higher than traditional approaches such as plate-and-frame dialysis systems. The advantage of fully automated systems lies in reduced labor costs.

For the type and size of system estimated above, labor costs are typically in the range of 4 to 8 man-hours per 24-h day. These costs do not increase appreciably with scale because the tasks being performed with an automated hollow-fiber dialysis system do not change as the system increases in size.

REFERENCE

Remer, D. S., and J. H. Idrovo. 1990. Cost-estimating factors for biopharmaceutical process equipment. *Biopharm.* 3:36–41.

V

Electrodialysis

16.	Introduction and Definitions	219
17.	Theory	223
18.	Ion-Exchange Membranes	230
19.	Design and Cost Estimates	246
20.	Applications	255

16

Introduction and Definitions

Heinrich Strathmann
Universität Stuttgart

INTRODUCTION
ELECTRODIALYSIS PROCESS
  PRINCIPLES
TYPES OF ELECTRODIALYSIS

FEASIBILITY OF
  ELECTRODIALYSIS
REFERENCES

INTRODUCTION

The history of electrodialysis began with the work of Ostwald (1890). He investigated the properties of semipermeable membranes and was the first to discover that a membrane is impermeable for any electrolyte if it is impermeable either for its cation or its anion. To illustrate this, he postulated the existence of the so-called "membrane potential" at the boundary between the membrane and the solution as a consequence of the difference of concentration. Later, Donnan (1911) confirmed this postulate for the boundary of an ion-exchange membrane and its surrounding solution. Simultaneously, he developed a mathematical equation describing the concentration equilibrium, which resulted in the so-called "Donnan exclusion potential."

Ion-exchange membranes include cation-exchange and anion-exchange membranes. Membranes containing negatively charged groups fixed to the membrane matrix allow the passage of cations but reject anions; they are called *cation-exchange membranes*. On the other hand, *anion-exchange membranes* with positively charged groups fixed to the membrane matrix allow the passage of anions but not cations (see Chapter 18 for further details).

The first basic studies related to ion-selective membranes were carried out by Michaelis and Fujita (1925) with the homogeneous, weak acid collodion membranes. Collodion is a solution of pyroxylin, a low-nitrogen form of nitrocellulose, in a mixture of ethanol and ether. By means of an oxidative treatment, Sollner (1950) obtained membranes with satisfactory electrical properties. However, their mechanical properties were extremely poor. Around 1940, the interest in industrial applications led to the development of synthetic ion-exchange membranes on the basis of phenol-formaldehyde-polycondensation resins (Wassenegger and Jaeger 1940). Later, polystyrene cross-linked with divinylbenzene became almost exclusively the basis of ion-exchange membranes. After the importance of these membranes for electrodialytical purposes was clearly seen, the development increased so fast that in 1950 the first ion-exchange membranes from Juda and McRae (1950) at Ionics Inc. and from Winger, Bodamer, and Kunin (1953) at Rohm and Haas were commercially available. The main use envisaged for the membranes was the desalination of brackish water and seawater. They needed to have high selectivity and low electro-osmotic transfer in contact with very dilute solutions. The electrical resistance, however, was not of the highest priority because it was controlled mainly by the conductivity of the dilute stream. These so-called "heterogeneous" membranes were manufactured by the dispersion of a fine ion-exchange resin powder within the solution of a matrix polymer and by the evaporation of the solvent.

As discussed in Chapters 11 through 15 on dialysis, the concentration gradient of the solute species across the membrane is the driving force for solute transport through the membranes, which are uncharged but not necessarily nonselective. Because of this reason, the extent of solute transfer in dialysis is limited by the condition of zero concentration gradient, at which time dialytic transfer stops. In electrodialysis, the continuous supply of electrical energy prevents the situation of no driving force in most cases. In fact, electrolytes are usually transferred from a less concentrated solution through the ion-exchange membrane to a more concentrated solution because of the electric energy. However, the purification of the solvent in both dialysis and electrodialysis takes place by the removal of the undesirable solute through the membrane, whereas the purification of the solvent in the two other membrane separation processes of reverse osmosis and ultrafiltration takes place by the selective transport of the solvent through the membrane, which rejects the solute.

Electrodialysis is an electrochemical separation process in which electrically charged membranes and an electrical potential difference are used to separate ionic species from an aqueous solution and other uncharged components. Electrodialysis is used widely today for desalination of brackish water, and in some areas of the world it is the main process for the production of potable water. Although of major importance, water desalination and table salt production are by no means the only significant applications. Stimulated by the development of new ion-exchange membranes with better selectivities, lower electrical resistance, and improved thermal, chemical, and mechanical properties, other uses of electrodialysis, especially in the food, drug, and chemical process industries as well as in biotechnology and wastewater treatment, have recently gained a broader interest (Korngold 1984).

Electrodialysis in the classical sense can be utilized to perform several general types of separations, such as the separation and concentration of salts, acids, and bases from aqueous solutions, the separation of monovalent ions from multivalent ions and multiple charged components, and the separation of ionic compounds from uncharged molecules.

Slightly modified electrodialysis is also used today to separate mixtures of amino acids or even proteins. It is also used to produce acids and bases from the corresponding salts when it is combined with electrically forced water dissociation in bipolar membranes. A *bipolar membrane* is prepared by combining a cation-exchange membrane with an anion-exchange membrane in such a manner that one is adjacent to the other. (See Chapters 18 and 20 for further details.)

In many applications, electrodialysis is in direct competition with other separation processes such as distillation, ion exchange, reverse osmosis, and various chromatographic procedures. In other applications, there are very few other economic alternatives to electrodialysis. Although the process has been known in principle for more than 50 years, large-scale industrial utilization began about 20 years ago. Because of the inherent features of the process, it seems quite likely that electrodialysis and related processes will be introduced in new applications in the chemical, food, and pharmaceutical industries (Strathmann 1984).

ELECTRODIALYSIS PROCESS PRINCIPLES

The principle of the process is illustrated in Figure 16–1, which shows a schematic diagram of a typical electrodialysis cell arrangement consisting of a series of anion- and cation-exchange membranes arranged in an alternating pattern between an anode and a cathode to form individual cells. A *cell* consists of a volume with two adjacent membranes. If an ionic solution such as an aqueous salt solution is pumped through these cells and an *electrical potential* established between the anode and cathode, the positively charged cations migrate toward the cathode and the negatively charged anions toward the anode. The cations pass easily through the negatively charged cation-exchange membrane but are retained by the positively charged anion-exchange membrane. Likewise, the negatively charged anions pass through the anion-exchange membrane but are retained by the cation-exchange membrane. The overall result is an increase in the ion concentration in alternate compartments, while the other compartments simultaneously become depleted. The depleted solution is generally referred to as the diluate and the concentrated solution as the *brine* or the *concentrate*. The *driving force* for ion transport in the electrodialysis process is the applied electrical potential between the anode and cathode.

Figure 16–1 shows only two cation-exchange membranes (C) and two anion-exchange membranes (A). An actual electrodialysis stack may have a few hundred of such membranes. The total space occupied by the diluate solution between two contiguous membranes, the concentrated solution between two contiguous membranes next to the diluate chamber, and the two contiguous anion- and cation-exchange membranes make up a *cell pair*. The cell pair is a repeating unit in an electrodialysis stack.

TYPES OF ELECTRODIALYSIS

The schematic described in Figure 16–1 corresponds to one of the most common forms of electrodialysis used in practice for desalination and deionization purposes. Such an arrangement is referred to as the *internally staged process of electrodialysis*. At least five other types of electrodialysis with various arrangements of ion-exchange and neutral membranes exist (Schaffer and Mintz 1966):

1. Single ion-exchange membrane electrodialysis
2. Neutral membrane electrodialysis
3. Cation-neutral membrane electrodialysis
4. Electrogravitation
5. Multicell single-stage electrodialysis.

FEASIBILITY OF ELECTRODIALYSIS

The technical feasibility of electrodialysis as a mass separation process, i.e., its ability to separate certain ions from a given mixture with other molecules, is mainly determined by the ion-exchange membranes used in the system. The economics of the process are determined by the operating costs, which are dominated by the energy consumption and investment costs for a plant of a desired capacity, which again are a

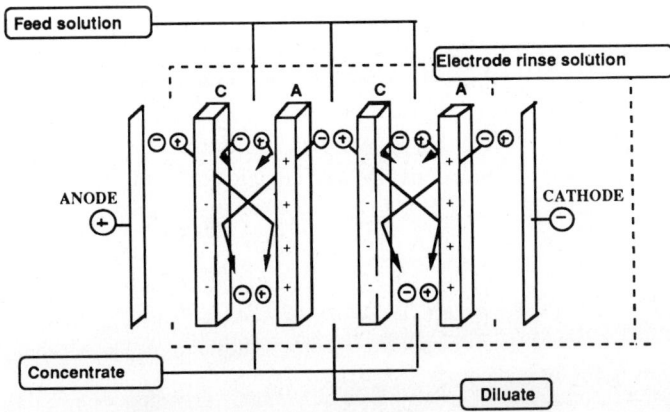

FIGURE 16–1. Schematic diagram of the electrodialysis process.

function of the membrane area and various design parameters such as cell dimensions, flow velocities, and current density (Strathmann 1979). The energy consumption, membranes, and design and cost estimates for the electrodialysis process are covered in Chapters 17, 18, and 19, respectively. In addition, Chapter 20 describes electrodialysis applications.

REFERENCES

Donnan, F. G. 1911. The theory of membrane equilibrium in presence of a non-dialyzable electrolyte. *Z. Electrochem.* 17:572.

Juda, W., and W. A. McRae. 1950. Coherent ion-exchange gels and membranes. *J. Am. Chem. Soc.* 72:1044.

Korngold, E. 1984. Electrodialysis: membranes and mass transport. In *Synthetic Membrane Processes*, ed. G. Belfort, pp. 192–219. New York: Academic Press.

Michaelis, L., and A. Fujita. 1925. The electric phenomenon and ion permeability of membranes: II. permeability of apple peel. *Biochem. Z.* 158:28–37.

Ostwald, W. 1890. Elektrische Eigenschaften Halbdurchlässiger Scheidewande. *Z. Physik. Chemie* 6:71–82.

Schaffer, L. H., and M. S. Mintz. 1966. Electrodialysis. In *Principles of Desalination,* ed. K. S. Spiegler, pp. 3–20. New York: Academic Press.

Sollner, K. 1950. Recent advances in the electrochemistry of membranes of high ionic selectivity. *J. Electrochem. Soc.* 97:139C–151C.

Strathmann, H. 1979. *Trennung von Molekularen Mischungen mit Hilfe Synthetischer Membrane.* Darmstadt: Steinkopff-Verlag.

Strathmann, H. 1984. Ion-exchange membranes in industrial separation processes. *J. Sep. Process Technol.* 5(1):1–13.

Wassenegger, H., and K. Jaeger. 1940. Effecting cation exchange as in removing calcium from hard waters. U.S. Patent 2,204,539.

Winger, A. G., G. W. Bodamer, and R. Kunin. 1953. Some electrochemical properties of new synthetic ion-exchange membranes. *J. Electrochem. Soc.* 100(4):178–184.

17

Theory

Heinrich Strathmann
Universität Stuttgart

ELECTRODIALYSIS ENERGY
  REQUIREMENTS
  Minimum Energy Required for the
    Separation of a Molecular Mixture
  Practical Energy Requirements for
    Ion Transfer
  Pumping Energy Requirements

Energy Consumption in Electrodialysis and Other Separation Processes
ION-EXCHANGE MEMBRANE
  EQUILIBRIA
NOTATION
REFERENCES

In this chapter, the various aspects of energy requirements in electrodialysis are considered first, and then the nature of equilibrium distribution of ions between the solution and the ion exchange membrane is considered.

ELECTRODIALYSIS ENERGY REQUIREMENTS

The energy required in an electrodialysis process is an additive of two terms: (1) the electrical energy to transfer the ionic components from one solution through membranes into another solution and (2) the energy required to pump the solutions through the electrodialysis unit. Depending on various process parameters, particularly the feed solution concentration, either one of the two terms may be dominating, thus determining the overall energy costs. The energy consumption due to electrode reactions can generally be neglected since more than 200 cell pairs are placed between the two electrodes in a modern electrodialysis stack.

Minimum Energy Required for the Separation of a Molecular Mixture

In electrodialysis as in any other separation process, a minimum amount of energy is required for the separation of various components from a mixture. For the removal of salt from a saline solution, this energy is given by

$$\Delta G = RT \ln \left(\frac{a_w^\circ}{a_w} \right), \qquad (17\text{--}1)$$

where ΔG is the Gibbs free-energy change required to remove 1 mol of water from a solution, R the gas constant, T the absolute temperature in degrees K, and a_w° and a_w are the water activities in pure water and the solution, respectively. Expressing water activity in the solution of a monovalent salt by the concentration of the dissolved ionic components, the minimum energy required to remove water from a monovalent salt is given by (Schaffer and Mintz 1966):

$$E_{\text{theo}} = \Delta G$$

$$= 2RT(c_f - c')\left(\frac{\ln(c_f/c'')}{(c_f/c'') - 1} - \frac{\ln(c_f/c')}{(c_f/c') - 1}\right), \quad (17\text{-}2)$$

where ΔG again refers to the Gibbs free-energy change required for the production of 1 L of diluate solution, c is the salt concentration, the subscript f refers to the feed solution, and the superscripts ' and " refer to the diluate and the concentrate, respectively. Furthermore,

$$\Delta G = \sum_i n_i z_i F \Delta \psi \quad \text{for } i = 1, 2, 3, \ldots, n, \quad (17\text{-}3)$$

where F is Faraday's constant (9.652×10^4 amp-s/g-equivalent), z the valence of the ion species i, n the number of moles for species i, and $\Delta\psi$ the potential drop due to the concentration difference in the diluate and concentrate. This potential drop is generally referred to as the concentration potential.

Practical Energy Requirements for Ion Transfer

The total electrical potential drop across an electrodialysis cell consists only partly of the concentration potential. The other part is used to overcome the ohmic resistance of the cell. This ohmic resistance is caused by the friction of the various ions with the membranes and water while being transferred from one solution to another, resulting in an irreversible energy dissipation in the form of heat generation. The potential drop to overcome the ohmic resistance can be significantly higher than the concentration potential; thus, in electrodialysis the energy required in practice generally is significantly higher than the theoretically required minimum energy. Furthermore, energy is required to pump the feed solution, the diluate, and the concentrate through the electrodialysis stack. Depending on various process parameters, particularly the feed solution concentration, either of these three terms may be dominant, thus determining the overall energy costs (Wilson 1960).

The energy necessary to remove salts from a solution is directly proportional to the total current flowing through the stack and the voltage drop between the two electrodes in a stack. The energy consumption in a practical electrodialysis separation procedure can be expressed by (Lacey 1972)

$$E_{\text{prac}} = I^2 n R_e t, \quad (17\text{-}4)$$

where E_{prac} is the energy consumption, I the electric current through the stack, R_e the resistance of a cell pair, n the number of cell pairs in a stack, and t the time.

The electric current needed to desalt a solution is directly proportional to the number of ions transferred through the ion-exchange membranes from the feed stream to the concentrated brine. It is expressed as

$$I = \frac{zFQ_f \Delta c}{\xi}, \quad (17\text{-}5)$$

where Q_f is the volumetric flow rate of the feed solution, Δc the concentration difference between the feed solution and the diluate, ξ the current utilization, and the rest of the symbols have the same definitions given earlier. The current utilization is directly proportional to the number of cell pairs in a stack.

A combination of Eqs. (17-4) and (17-5) gives the energy consumption in electrodialysis as a function of the current applied in the process, the electrical resistance of the stack, i.e., the resistance of the membrane and the electrolyte solution in the cells, the current utilization, and the amount of salt removed from the feed solution:

$$E_{\text{prac}} = \frac{InR_e tzFQ_f \Delta c}{\xi}. \quad (17\text{-}6)$$

Equation (17–6) indicates that the electrical energy required in electrodialysis is therefore directly proportional to the amount of salts that has to be removed from a certain feed volume to achieve the desired product concentration. Energy consumption is also a function of the electrical resistance of a cell pair. Electrical resistance, again, is a function of individual resistances of the membranes and the solutions in the cells. Since the resistance of the solution is inversely proportional to its ion concentration, the overall resistance of a cell will in most cases be determined by the resistance of the diluate solution. The concentration in the diluate cell, however, decreases during the desalting process, and thus its resistance increases accordingly. Under the assumption that the concentration in the diluate is much lower than that in the feed and brine, the energy consumption can be expressed by (Korngold 1984)

$$E_{\text{prac}} = \frac{InbV \log(c_f/c')}{\xi}, \quad (17\text{–}7)$$

where V is the total volume of the diluate solution and b is a constant factor. A typical value for the resistance of an electrodialysis cell pair, i.e., the cation- and anion-exchange membranes plus the diluate and the concentrated solutions, in the desalination of brackish water is within the range of 5 to 500 Ω cm^2. For other applications the electrical resistance of a cell pair might be significantly higher or lower (Lacey 1966). In applications where extremely low diluate concentrations are required, the conductivity in the diluate cell can be improved by using ion conductive spacers.

Pumping Energy Requirements

The operation of an electrodialysis system requires two or three pumps to circulate the diluate, the brine, and eventually the electrode rinse solutions through the stack. The energy required for pumping these solutions is determined by the volumes to be circulated and the pressure drop. It can be expressed by

$$E_p = k_D Q_D \Delta p_D + k_B Q_B \Delta p_B + k_E Q_E \Delta p_E. \quad (17\text{–}8)$$

Here, E_p is the pumping energy, k a constant referring to the efficiency of the pump, Q the volumetric flow rate, Δp the pressure loss, and the subscripts D, B, and E refer to the diluate, brine, and electrode rinse solutions, respectively.

The pressure losses in the various cells are determined by solution flow velocities and the cell design. The energy requirements for circulating the solution through the system may become significant or even dominant when solutions have rather low salt concentrations.

Other energy-consuming processes are the electrochemical reactions at the electrodes. In a stack with a multicell arrangement, however, the energy consumed at the electrodes is generally less than 1% of the total energy used for the ion transfer and can therefore be neglected.

Energy Consumption in Electrodialysis and Other Separation Processes

In many applications electrodialysis is competing with other separation processes. While the theoretically required minimum energy is identical in all processes, significant differences exist as far as the irreversible energy dissipation is concerned. For the desalination of a saline solution, different processes, such as reverse osmosis, ion exchange, and distillation, are used in addition to electrodialysis. All processes require the same theoretical minimum energy. The irreversibly dissipated energy is rather different in the different processes, as can be illustrated by comparing the basic principles of desalination by electrodialysis and reverse osmosis, shown schematically in Figure 17–1.

The basic difference between reverse osmosis and electrodialysis is the following. In reverse osmosis the water passes through the membrane under a driving force of a hydrostatic pressure difference, whereas in electrodialysis the salt passes through the membrane under the driving force of an electrical potential difference. The irreversible energy loss in reverse osmosis is caused by friction experienced by water molecules on their pathways through the membrane matrix. This means that the irreversible energy loss in reverse osmosis is in-

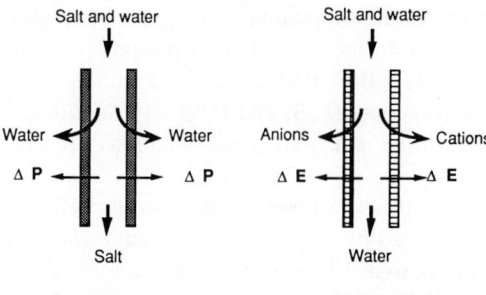

FIGURE 17–1. Schematic diagram illustrating the operating principles of reverse osmosis and electrodialysis.

dependent of the feed water salt concentration. In electrodialysis the irreversible energy loss is caused by the friction of ions on their pathway through the membrane from the diluate to the brine solution. Thus, in electrodialysis their reversible energy loss is directly proportional to the feed salt concentration. For feed solution with low salt concentrations, the energy requirements are therefore generally lower for electrodialysis than for reverse osmosis, and at high feed solution salt concentration, the situation is reversed. This is shown schematically in Figure 17–2 where the irreversible energy consumption is plotted versus the feed solution concentration assuming an identical product water for both cases.

A comparison of mass separation processes with regard to their energy consumptions has to take into account the fact that in electrodialysis the energy required in the form of electricity is a relatively expensive form, but in distillation, a relatively inexpensive form of energy, i.e., heat, can be used. In ion exchange, very little energy is required directly. However, the chemical used for the regeneration of the resin requires a significant amount of energy for its production.

ION-EXCHANGE MEMBRANE EQUILIBRIA

Ion-exchange membranes are ion-exchange resins in film form. Therefore, they consist of highly swollen gels carrying fixed positive or negative charges. There are two different types of ion-exchange membranes:

1. Cation-exchange membranes, which contain negatively charged groups fixed to the polymer matrix
2. Anion-exchange membranes, which contain positively charged groups fixed to the polymer matrix.

In a cation-exchange membrane, the fixed anions are in electrical equilibrium with mobile cations in the interstices of the polymer as indicated in Figure 17–3. This figure shows schematically the matrix of a cation-exchange membrane with fixed anions and mobile cations, the latter referred to as counter-ions. In contrast, the mobile anions, called co-ions, are more or less completely excluded from the polymer matrix because of their electrical charge, which

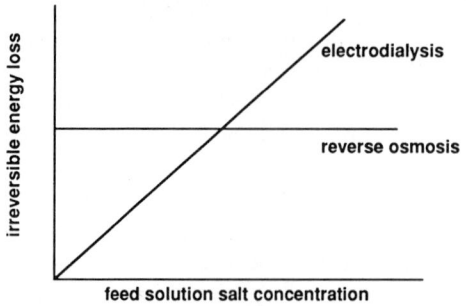

FIGURE 17–2. Schematic diagram showing the irreversible energy losses in electrodialysis and reverse osmosis as a function of feed solution salt concentration.

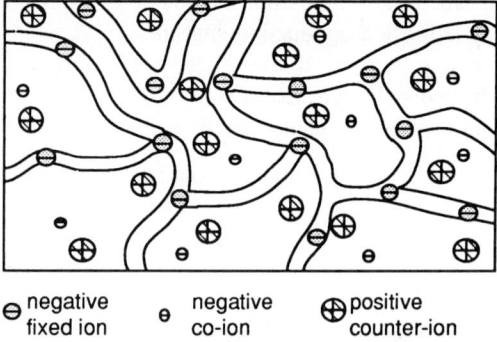

FIGURE 17–3. Schematic diagram of the structure of a cation-exchange membrane showing the polymer matrix with the negative fixed charges, negative co-ions, and positive counter-ions.

is identical to that of the fixed ions. This type of exclusion is called the Donnan exclusion in honor of Donnan (1911) for his pioneering work. Due to the exclusion of the co-ions, a cation-exchange membrane permits transfer of cations only. Anion-exchange membranes carry positive charges fixed on the polymer matrix. Therefore, they exclude all cations and are permeable to anions only.

The selectivity of ion-exchange membranes results from the exclusion of co-ions of the membrane phase. For a cation-exchange membrane in the dilute solution of a strong electrolyte, this can be illustrated using the model of Gregor (1951). Here, the concentration of the cations is higher in the exchanger than in the solution, because the cations are attracted by the negatively charged fixed ions. On the other hand, the concentration of mobile anions in the solution is higher than in the exchanger. An equilibrium in the concentration of electrically neutral charged particles could take place via diffusion. Because electroneutrality is required here as well, the process is limited if charged particles are involved. The passing of cations into the solution and of anions into the cation-exchange membrane leads to a counteracting space charge. Therefore, the so-called Donnan exclusion equilibrium is established between the attempt of diffusion on one side and the establishment of an electrical potential difference on the other. The activity of the counterions in the exchanger is higher than in the solution and the activity of the co-ions is smaller, respectively. The thermodynamical treatment by Donnan and Guggenheim (1932) is based on an equilibrium in the membrane phase (indicated by absence of a superscript) and in the outer phase (indicated by the superscript ') of the electrochemical potential μ_i of all ions that are able to permeate through the membrane:

$$\mu_i = \mu_i'. \qquad (17\text{-}9)$$

In both of the adjacent phases, the concentrations as well as the osmotic pressure and the electrical potential can be different. For the distribution of a special ion, an established electrical potential difference $\psi - \psi'$, the Donnan potential $\Delta\psi_{\text{Don}}$, can be described as a function of the different activities a_i and the swelling pressure π_i:

$$\Delta\psi_{\text{Don}} = \psi - \psi' = \frac{1}{z_i F}\left[RT \ln\left(\frac{a_i}{a_i'}\right) - \tilde{V}_i \pi_i\right],$$

$$(17\text{-}10)$$

where z_i is the valency of ion species i, F Faraday's constant, R the gas constant, T the absolute temperature, $\tilde{V}_i$ the partial molar volume of component i, and π_i the swelling pressure. The numerical value of $\Delta\psi_{\text{Don}}$ is negative for the cation-exchange membrane and positive for the anion-exchange membrane. Unfortunately, the Donnan potential cannot be evaluated by direct measurement; however, it presents the starting point for the calculation of the distribution of the co-ions between the solution and the membrane and hence for the determination of the permselectivity. The swelling pressure, which is directly proportional to the concentration of the fixed ions and inversely proportional to the concentration of the electrolyte, can be substituted, and for 1:1 electrolytes an approximate relation for the concentration of the co-ions in the membrane c_{co} is obtained:

$$c_{\text{co}} = \frac{c'^2}{c_{\text{fixed}}}\left(\frac{\gamma'}{\gamma}\right)^2, \qquad (17\text{-}11)$$

where c' is the electrolyte concentration in the solution, c_{fixed} is the concentration of the fixed ions in the membrane, and γ' and γ are the average activity coefficients of the salt in the solution and the membrane, respectively. This fundamental equation is based on the theory of Teorell (1951) and Meyer and Sievers (1936). However, the more complex structure of modern membranes cannot be adequately described exclusively by this theory. The remaining differences in the observed and expected membrane behavior are primarily a result of the nonuniformity in the distributions of molecular components in the membrane. This results from structural irregularities on a molecular level and from the influence of the electric field. Additionally, the practical application of the thermodynamic relations is rather limited by the difficulties in the experimental measurement of independent interaction, diffusion, resistance, and frictional coefficients.

The Donnan exclusion equilibrium as well as the selectivity increase depend on (1) the concentration of the fixed ions, (2) the increasing valence of the co-ions, (3) the decreasing valence of the counter-ions, (4) the decreasing concentration of the electrolyte solution, and (5) the decreasing affinity of the exchanger with respect to the counter-ions.

Further important parameters for the characterization of ion-exchange membranes are the density of the polymer network, the hydrophobic and hydrophilic properties of the matrix polymer, the distribution of the charge density, and the morphology of the membrane itself. All these parameters do not only determine the mechanical properties, but they also have a considerable influence on the sorption of the electrolytes and the nonelectrolytes and therefore on the swelling.

The most desirable properties for ion-exchange membranes are:

1. *High permselectivity:* An ion-exchange membrane should be highly permeable to counter-ions, but should be impermeable to co-ions.
2. *Low electrical resistance:* The permeability of an ion-exchange membrane for the counter-ions under the driving force of an electrical potential gradient should be as high as possible.
3. *Good mechanical and form stability:* The membrane should be mechanically strong and should have a low degree of swelling or shrinking in transition from dilute to concentrated ionic solutions.
4. *High chemical stability:* The membrane should be stable over a pH range from 0 to 14 and in the presence of oxidizing agents.

Optimization of the properties of ion-exchange membranes is often difficult because the parameters determining the different properties often have opposing effects. For instance, a high degree of cross-linking improves the mechanical strength of the membrane but also increases its electrical resistance. A high concentration of fixed ionic charges in the membrane matrix leads to a low electrical resistance but, in general, causes a high degree of swelling combined with poor mechanical stability. The properties of ion-exchange membranes are determined by two parameters: the basic polymer matrix and the type and concentration of the fixed ionic moiety. The basic polymer matrix determines to a large extent the mechanical, chemical and thermal stability of the membrane. Very often the matrixes of ion-exchange membranes consist of hydrophobic polymers such as polystyrene, polyethylene, and polysulfone. Although these basic polymers are insoluble in water and show a low degree of swelling, they may become water soluble by the introduction of the ionic moieties. Therefore, the polymer matrixes of ion-exchange membranes are very often cross-linked. The degree of cross-linking then determines to a large extent the degree of swelling and the chemical and thermal stability, but it also has a large effect on the electrical resistance and the permselectivity of the membrane.

The type and concentration of the fixed ionic charges determine the permselectivity and electrical resistance of the membrane, but they also have a significant effect on the mechanical properties of the membrane. The degree of swelling, especially, is affected by the concentration of the fixed charges. The following moieties are used as fixed charges in cation-exchange membranes:

$-SO_3^-$ $-COO^-$ $-PO_3^{2-}$ $-HPO_2^-$ $-AsO_3^{2-}$ $-SeO_3^-$.

In anion-exchange membranes, fixed charges may be:

$-NH_3^+$ $-RNH_2^+$ $-R_3N^+$ $=R_2N^+$ $-R_3P^+$ $-R_2S^+$.

These different ionic groups have significant effects on the selectivity and electrical resistance of the ion-exchange membrane. The sulfonic acid group, e.g., $-SO_3^-$, is completely dissociated over nearly the entire pH range, while the carboxylic acid group, $-COO^-$, is virtually undissociated in the pH range < 3. The quaternary ammonium group $-R_3N^+$ again is completely dissociated over the entire pH range, while the primary ammonium group $-NH_3^+$ is only weakly dissociated. Accordingly, ion-exchange membranes are referred to as being weakly or strongly acidic or basic in character. Most commercially available cation-exchange membranes have $-SO_3^-$ or $-COO^-$ groups, and most anion-exchange membranes contain $-R_3N^+$ groups.

NOTATION

General Notation

See the General Notation section at the beginning of this handbook.

Special Notation

a_i activity of ion species i, mol/L^3, g-mol/L or kg-mol/m^3

a_w water activity, mol/L^3, g-mol/L or kg-mol/m^3

E_{prac} energy consumption in a practical electrodialysis separation procedure, E, joule or kWh

I electric current through the electrodialysis stack, A or ampere

n number of cell pairs in an electrodialysis stack

n_i number of moles for ion species i

Q_f volumetric flow rate of the feed solution, L^3/t, L/s or m^3/s

V total volume of the diluate solution, L^3, L, or m^3

z valence of ion species i, dimensionless

Greek Letter

ξ current utilization, dimensionless

Superscripts

$'$ value in the depleted solution (diluate)

$''$ value in the concentrated solution (concentrate)

REFERENCES

Donnan, F. G. 1911. The theory of membrane equilibrium in presence of a non-dialyzable electrolyte. *Z. Elektrochem.* 17:572.

Donnan, F. G., and E. A. Guggenheim. 1932. Exact thermodynamics of membrane equilibrium. *Z. Physik. Chemie* A162:346–360.

Gregor, H. P. 1951. Gibbs-Donnan equilibria in ion-exchange resin systems. *J. Am. Chem. Soc.* 73:642–650.

Korngold, E. 1984. Electrodialysis: membranes and mass transport. In *Synthetic Membrane Processes*, ed. G. Belfort, pp. 192–219. New York: Academic Press.

Lacey, R. E. 1966. *Membrane Processes for Industry*. Birmingham, AL: Southern Research Institute.

Lacey, R. E. 1972. Basis of electromembrane processes. In *Industrial Processing with Membranes*, ed. R. E. Lacey and S. Loeb. New York: John Wiley & Sons.

Meyer, K. H., and J. F. Sievers. 1936. Permeability of membranes. *Helv. Chim. Acta* 19:649–677.

Schaffer, L. H., and M. S. Mintz. 1966. Electrodialysis. In *Principles of Desalination*, ed. K. S. Spiegler, pp. 3–20. New York: Academic Press.

Teorell, T. Z. 1951. Quantitative treatment of membrane permeability. *Z. Elektrochem.* 55:460–469.

Wilson, J. R. 1960. *Demineralization by Electrodialysis*. London: Butterworth.

18

Ion-Exchange Membranes

Heinrich Strathmann
Universität Stuttgart

ION-EXCHANGE MEMBRANES
PREPARATION OF ION-EXCHANGE MEMBRANES
   Preparation Procedure for Homogeneous Ion-Exchange Membranes
   Preparation Procedure for Heterogeneous Ion-Exchange Membranes
SPECIAL-PROPERTY ION-EXCHANGE MEMBRANES
   Monovalent Ion Permselective Membranes
   Proton Permselective Cation-Exchange Membranes
   Anion-Exchange Membranes of High Proton Retention
   Antifouling Anion-Exchange Membranes
   Fluorocarbon-Type Cation-Exchange Membranes
   Alkaline Stable Anion-Exchange Membranes
   Bipolar Membranes
CHARACTERIZATION OF ION-EXCHANGE MEMBRANES
   Mechanical Examination and Swelling Behavior of Membranes
   Long-Term Chemical Stability of Membranes
   Determination of Membrane Ion-Exchange Capacities
   Permselectivity of Ion-Exchange Membranes
   Electrical Resistance of Ion-Exchange Membranes
REFERENCES

ION-EXCHANGE MEMBRANES

The properties and preparation procedures of ion-exchange membranes are closely related to those of ion-exchange resins. As with resins, many types of membranes are possible with different polymer matrixes and different functional groups to confer ion-exchange properties on the product. Although a number of inorganic ion-exchange materials are available, most of them based on zeolites and bentonites (Helfferich 1962), these materials are rather unimportant in ion-exchange membranes and will not be discussed further.

Most commercial ion-exchange membranes can be divided, according to their structure and preparation procedure, into two major categories, either homogeneous or heterogeneous. Homogeneous ion-exchange membranes are produced either by polymerization of functional monomers, for example, by means of polycondensation of phenolsulfonic acid with formaldehyde, or by additional functionalization of a polymer film by sulfonation of a polystyrene film (Friedlander 1968). For the combination of electrical and mechanical properties, the so-called heterogeneous ion-exchange membranes show many more possible variations. The degree of heterogeneity of ion-exchange membranes increases according to the following order as indicated by Molau (1981): (1) homogeneous ion-exchange membranes, (2) interpolymer membranes, (3) microheterogeneous graft- and block-polymer membranes, (4) snake-in-the-cage ion-exchange membranes, and (5) heterogeneous ion-exchange membranes.

From the viewpoint of macromolecular chemistry, all the intermediate forms are considered to be so-called *polymer blends*. As a consequence of the polymer/polymer incompatibility, on one hand, a phase separation of the different polymers is obtained, while on the other hand, a specific aggregation of the hydrophilic and hydrophobic properties of the electrolytes is obtained. A classification of the membrane morphology is then possible, depending on the type and size of the microphase. The membranes are translucent, an indication that inhomogeneities, if any, are smaller than the wavelength of visible light (400 nm). Thus, these membranes are called *interpolymer* or *microheterogeneous* membranes.

Heterogeneous membranes are produced by melting and pressing of a dry ion-exchange resin with a granulated polymer, or by dispersion of the ion-exchange resin in the solution or melting of a matrix polymer (Suryanarayana and Krishnaswamy 1963). In the same manner, the polymer matrix can be polymerized *in situ* directly around the resin particles (Joshi and Ganu 1978). Microheterogeneous membranes, for example, are produced by means of block-copolymerization of ionogenic and nonionogenic monomers, or by graft-copolymerization of functional monomers (Ishigaki et al. 1978). Interpolymer membranes are produced by dissolving compatible functional polymers in one solvent to form a homogeneous, macroscopically transparent solution (Caplan and Sollner 1974), followed by the evaporation of the solvent.

PREPARATION OF ION-EXCHANGE MEMBRANES

As far as their chemical structure is concerned, ion-exchange membranes are very similar to normal ion-exchange resins; from the chemical point of view, these resins would make excellent membranes of high selectivity and conductivity. The difference between membranes and resins arises largely from the mechanical requirements of the membrane process. Unfortunately, ion-exchange resins are mechanically weak, cation resins tend to be brittle, and anion resins soft. They are dimensionally unstable due to the variation in the amount of water imbibed into the gel under different circumstances. Changes in electrolyte concentration, in the ionic form, or in temperature may cause major changes in the water uptake and hence in the volume of the resin. These changes can be tolerated in small spherical beads. But in large sheets that have been cut to fit an apparatus, they are not acceptable. Thus, it is generally not possible to use sheets of material that have been prepared in the same way as a bead resin. The most common solution to this problem is the preparation of a membrane with a backing of a stable reinforcing material that gives the necessary strength and dimensional stability. Preparation procedures for making ion-exchange resins and membranes are described in great detail in the patent literature (Flett 1983).

Preparation Procedure for Homogeneous Ion-Exchange Membranes

The methods of making homogeneous ion-exchange membranes can be summarized in three different categories:

1. polymerization or polycondensation of monomers; at least one of them must contain a moiety that either is or can be made anionic or cationic, respectively

2. introduction of anionic or cationic moieties into a preformed solid film
3. introduction of anionic or cationic moieties into a polymer, such as polysulfone, followed by the dissolving of the polymer and casting it into a film.

Polymerization and Polycondensation of Monomers

One of the first membranes made by polymerization or polycondensation of monomers was prepared from phenol by polycondensation with formaldehyde according to the following reaction scheme (Helfferich 1962):

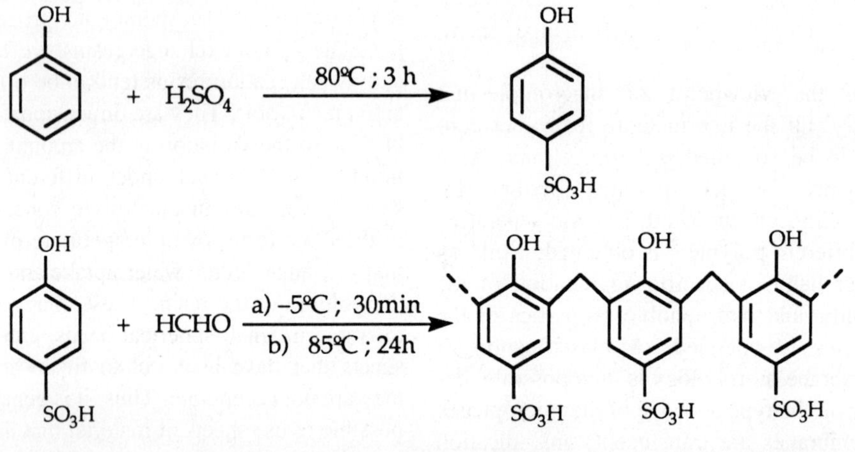

Phenol is treated with concentrated H_2SO_4 at 80°C, which leads to the phenolsulfonic acid in paraform, a brown, crystalline material. This acid is reacted with a 38% solution of formaldehyde in water initially at −5°C for about 30 min and then at 85°C for several hours. The solution is then cast into a film, which forms after cooling to room temperature. Excess monomer is removed by washing the film in water.

The second method of preparing a cation- or anion-exchange membrane, which is very common, is the polymerization of styrene and divinylbenzene and its subsequent sulfonation or amination. The cation-exchange membrane is obtained according to the reaction scheme:

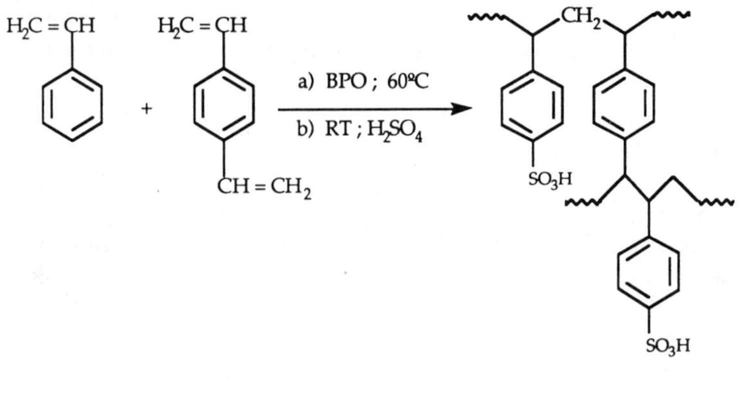

The anion-exchange group is introduced into the polymer by chloromethylation and amination with triamine according to the following reaction scheme:

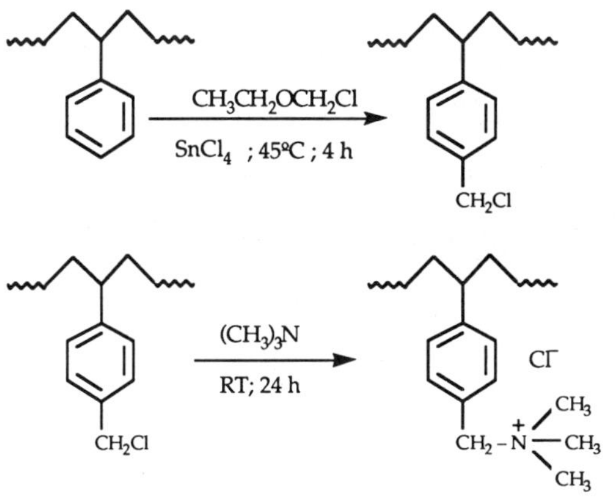

Numerous references in the literature discuss the preparation of ion-exchange membranes by polymerization (Murphy, Paton, and Ansell 1943; McRae and Alexander 1960; Sata, Kuzumoto, and Mizutani 1976). In recent years, membranes based on perfluorocarbon polymers have proved to be very useful in the chlor-alkali industry. They are prepared by copolymerization of tetrafluorethylene with perfluorovinyl ether having a carboxylic or sulfonic ester group at the end of a side chain (Eisenberg and Yeager 1982). Acrylic-based membranes have also become very common in recent years.

Introduction of Anionic or Cationic Moieties into a Solid Preformed Film

Concerning the introduction of anionic or cationic moieties into a preformed film, the monomer may either contain a cross-linking agent such as divinylbenzene, or alternatively it may be grafted onto a film by radiation techniques. Starting with a film makes the membrane preparation rather easy. The starting material may be a hydrophilic polymer, such as cellophane or polyvinylalcohol. More often, however, a hydrophobic polymer such as polyethylene or polystyrene is used.

Ion-exchange membranes made by sulfochlorination and amination of polyethylene sheets, for instance, have low electrical resistance combined with high permselectivity and excellent mechanical strength. The reaction scheme for the preparation of these membranes is given below:

(a) *preparation of the cation-exchange membrane:*

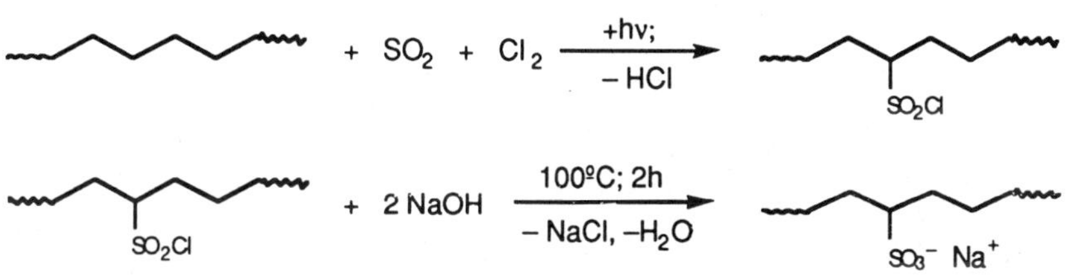

(b) *preparation of anion-exchange membrane:*

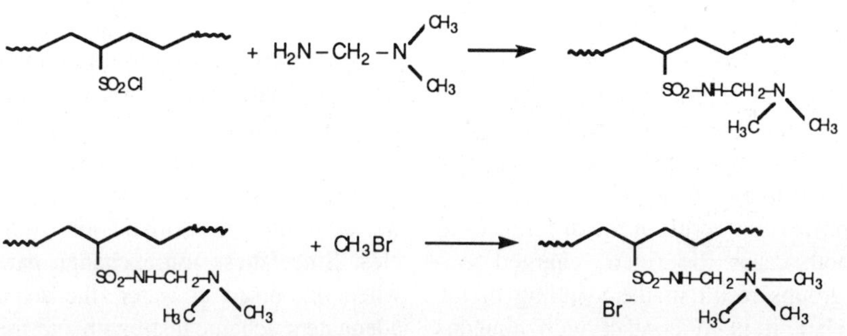

Introduction of Anionic or Cationic Moieties into a Polymer Chain

Membranes can also be prepared by dissolving and casting a functionalized polymer, such as sulfonated polysulfone, into a film. The reaction scheme for the sulfonation of polysulfone is:

The sulfonated polysulfone can be cast as a film on a screen and precipitated after the evaporation of most of the solvent such as DCE (dichloroethane) (Zschocke and Quellmalz 1985). This leads to a reinforced membrane with excellent chemical and mechanical stabilities and good electrochemical properties.

Preparation Procedure for Heterogeneous Ion-Exchange Membranes

These membranes consist of fine colloidal ion-exchange particles embedded in an inert binder such as polyethylene, phenolic resins, or polyvinylchloride. Such membranes can be prepared simply by calendering ion-exchange particles into an inert plastic film. Another procedure is the dry molding of inert film-forming polymers and ion-exchange particles and then milling the mold stock. Also, ion-exchange particles can be dispersed in a solution containing a film-forming binder, and then the solvent is evaporated to give the ion-exchange membrane. Similarly, ion-exchange particles are dispersed in a partially polymerized binder polymer, and then the polymerization is completed.

However, some significant differences between ion-exchange membranes and ion-exchange resins are evident concerning the details of the polymer structures. These differences are primarily a result of the differences in size. In both cases the fixed, charged ion-exchange groups result in the swelling of the polymer when it is in contact with aqueous solutions. The amount of swelling depends to some degree on the ionic strength of the solution. In the case of granular ion-exchange resins, the extent of swelling is limited by cross-linking and by entanglement of the polymers. Typically, the level of cross-linking is about 10%. Due to the spherical symmetry of granular ion-exchange resins and to the fact that they are not physically constrained in use, there is generally no functionally important physical damage to the resins from drying and rewetting or from change in ambient ionic strength.

Dimensional changes that are tolerable during use in the case of granular ion-exchange resins are not acceptable in ion-exchange membranes because of the large sizes of the latter and the fact that they are physically constrained in the electrodialysis stacks in which they are used. As a result, useful ion-exchange capacities (IEC = 1 to 3 meq/g dry membrane) tend to be lower than in the case for granular exchangers (IEC = 3 to 5 meq/g resin), resulting in reduced swelling tendencies. In addition to covalent cross-linking, other strategies are used to limit swelling and changes in the swelling, for example, reinforcing fabrics and including, in the ion-exchange resin polymers, polymer segments that yield microcrystalline regions in the resins or otherwise reduce swelling stresses.

Heterogeneous membranes with useful low electrical resistances contain more than 65 wt.% of the cross-linked ion-exchange particles. Since these ion-exchange particles swell when immersed in water, the achievement of adequate mechanical strength and freedom from distortion combined with low electrical resistance has been difficult. Most heterogeneous membranes that possess adequate mechanical strength generally show poor electrochemical properties. On the other hand, a membrane that contains ion-exchange particles large enough to show desired electrochemical performance exhibits poor mechanical strength.

In general, heterogeneous ion-exchange membranes have relatively high electrical resistances. Homogeneous ion-exchange membranes have a more even distribution of fixed ions and often lower electrical resistances.

SPECIAL-PROPERTY ION-EXCHANGE MEMBRANES

In the literature, numerous methods have been reported for the preparation of ion-exchange membranes with special properties, for instance, to be used for the production of table salt, as battery separators, ion-selective electrodes, or in the chlor-alkali process. Significant effort has also been concentrated on the development of anion-exchange membranes with low fouling tendencies.

Monovalent Ion Permselective Membranes

Since 1972, in Japan table salt has been produced by electrodialytic concentration of sea water. For the specific requirements of this process, ion-exchange membranes that can separate monovalent ions from a mixed solution containing monovalent and multivalent ions have been developed. Tokuyama Soda has commercialized monovalent cation selective membranes (Neosepta® CMS) prepared by forming a thin cationic charged layer on the membrane surface (Sata, Izuo, and Mizutani 1984). Monovalent anion permselective membranes (Neosepta® ACS), which have a thin highly cross-linked layer on the membrane surface, have also been developed. By such means the selectivity of sulfate compared to that of chloride can be reduced from about 0.5 to about 0.01 and of magnesium compared to sodium from about 1.2 to about 0.1.

Proton Permselective Cation-Exchange Membranes

Amphoteric-type ion-exchange membranes are preferentially permeable to hydrogen ions. By coating this membrane with a thin cation charged layer, the selectivity for H^+ versus Na^+ is improved up to $\alpha(H^+/Na^+) = 12$ (Yamane, Izuo, and Mizutani 1965).

Anion-Exchange Membranes of High Proton Retention

By means of traditional membranes, electrodialysis cannot be applied to the recovery of acid in order to reuse the acid because of high proton leakage through the anion-exchange membranes. In general, since protons permeate easily through an anion-exchange membrane, acids cannot be concentrated to more than a certain level by electrodialysis with any degree of high efficiency. Recently developed membranes, such as the Neosepta® ACM (Urano, Ase, and Naito 1984) exhibit low proton permeabilities and enable efficient acid concentration.

Antifouling Anion-Exchange Membranes

Significant improvements have been made in the development of anion-exchange membranes with low fouling tendencies. In conventional electrodialysis plants, the permissible current density in the anion-exchange membrane is smaller than in the cation-exchange membrane, largely due to the risk of precipitates (Korngold et al. 1970). The anion-exchange membrane is more sensitive to fouling. According to some former investigations concerning the permeability of commercial anion-exchange membranes, the upper molecular weight limit for practical electrodialytical separations is in the range of 100 daltons. A molecular weight of 350 daltons is to be considered as a maximum size for any electrotransport through, e.g., the Ionac MA 3475 membrane. The static permselectivity decreases gradually from 98 to 30% when the molecular weight of the solute increases from

59 to 171 daltons (Dohno, Azumi, and Takashima 1975).

Fouling of anion-exchange membranes often occurs when the anion is small enough to penetrate into the membrane structure, but its electromobility is so poor that the membrane is virtually blocked. To overcome this problem, different companies have developed membranes, such as the Neosepta® AMX, which is characterized by a high permeability for large organic anions. In general, the permselectivity of these membranes is lower than that of regular membranes. Since the pore diameter of ion-exchange membranes is about 10 Å, polyelectrolytes of high molecular weight are not harmful to ion-exchange membranes; however, large ionic compounds with molecular weight of several hundred can cause membrane fouling.

One method of improving the permeability of anion-exchange membranes for large organic acids is based on the adjustment of the degree of cross-linking and the chain length of the cross-linker in the polymer network, as indicated by the following structure (Gudernatsch, Krumbholz, and Strathmann 1989):

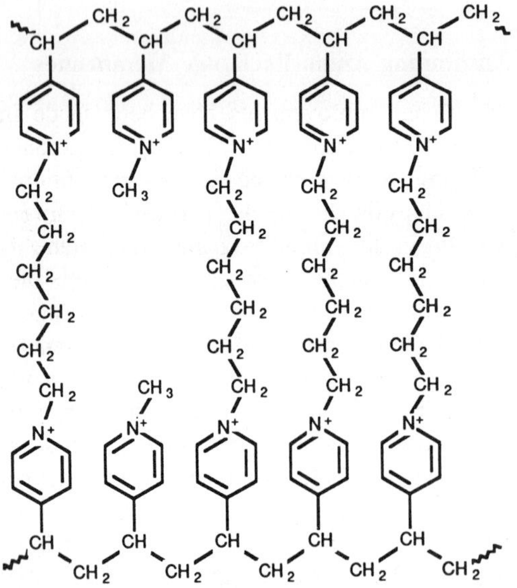

During the membrane formation, poly-(4-vinylpyridine) is quaternized with methyliodide and simultaneously cross-linked together with different dibromoalkanes.

Ionics Inc. produces a macroreticular membrane that is less sensitive to traces of detergents. It is produced by dissolving an organic compound in the membrane-forming system. When the material diffuses from the membrane after the polymerization, large pores are left behind. Alternatively, passive salts such as potassium iodide are added to the solvent of the binder polymer, mostly dimethylformamide. Through these pores, large anionic molecules can penetrate, thus preventing a steep increase in the electrical resistance.

Another type of antifouling anion-exchange membrane is produced by Tokuyama Soda. The membrane is coated with a thin layer of cation-exchange groups causing electrostatic repulsion of organic molecules. Practically, the coating is done by weak sulfonation of the membrane surface, followed by the ordinary chloromethylation and quaternization steps.

Ion-exchange membranes based on aliphatic polymers show reduced organic fouling in natural waters compared to membranes based on aromatic polymers. The aliphatic membranes also allow operation with solutions containing 0.5 ppm of chlorine and for shock chlorination up to 20 ppm of free chlorine.

Fluorocarbon-Type Cation-Exchange Membranes

Most conventional hydrocarbon ion-exchange membranes are degraded by oxidizing agents, especially at elevated temperatures. To adapt ion-exchange membranes to an application in the chlor-alkali industry, a fluorocarbon-type membrane with excellent chemical and thermal stability was developed first by Du Pont as Nafion® (Connolly and Gresham 1966; Grot 1973). The membrane is produced by a several-step procedure, which starts with the synthesis of an ionogenic perfluorovinylether and its copolymerization with tetrafluoroethylene (TFE). From the reaction of TFE with sulfur trioxide in the first step, a cyclic sulfone is formed that rearranges to 2-fluorosulfonyl-difluoroacetylfluoride:

$$\underset{F}{\overset{F}{>}}C=C\underset{F}{\overset{F}{<}} + SO_3 \longrightarrow F_2C\underset{O}{\overset{CF_2}{\diamond}}SO_2 \longrightarrow F-\underset{O}{\overset{O}{\underset{\|}{S}}}-CF_2-C\underset{F}{\overset{O}{\nwarrow}}$$

This intermediate is reacted with hexafluoropropylene oxide to produce a sulfonyl fluoride adduct:

$$F-\underset{O}{\overset{O}{\underset{\|}{S}}}-CF_2-C\underset{F}{\overset{O}{\nwarrow}} + (m+1)\ \underset{F}{\overset{O}{\underset{F}{\triangle}}}_{CF_3} \longrightarrow F-\underset{O}{\overset{O}{\underset{\|}{S}}}-CF_2-CF_2-(OCF-CF_2)_m-OCF-C\underset{F}{\overset{O}{\nwarrow}}$$
$$\qquad\qquad\qquad\qquad\qquad\qquad\qquad\qquad\qquad CF_3 \qquad\quad CF_3$$

By heating with sodium carbonate, sulfonyl fluoride vinyl ether is formed, which is then copolymerized with TFE:

$$F-\underset{O}{\overset{O}{\underset{\|}{S}}}-CF_2-CF_2-(O\underset{CF_3}{\overset{|}{C}}F-CF_2)_m-OCF=CF_2 \quad + \quad \underset{F}{\overset{F}{>}}C=C\underset{F}{\overset{F}{<}}$$

$$\longrightarrow \sim(CF_2-CF_2)_n-\underset{O-(CF_2-CF-O)_m-CF_2-CF_2-\underset{O}{\overset{O}{\underset{\|}{S}}}-F}{\overset{|}{CF}}-CF_2\sim$$
$$\qquad\qquad\qquad\qquad\qquad\qquad\qquad\qquad\qquad CF_3$$

The resulting copolymer is extruded as a film about 120 μm thick. Finally, the ionogenic moiety is converted to membranes, which carry sulfone groups in the bulk of the membrane phase and carboxyl groups on the surface as the charged moieties. Therefore, the $-SO_2F$ groups are reacted with sodium hydroxide to $-SO_3Na$ groups. The $-CF_2SO_2F$ groups are converted to $-COOH$ groups in various ways by the different manufacturers. Du Pont has a separate perfluorocarboxylate copolymer, which is laminated to the perfluorosulfonic acid substrate. Tokuyama Soda again transforms the $-CF_2SO_3H$ groups to $-CF_2SO_2Cl$ and then to $-COOH$ (Onoue et al. 1980). Asahi Chemical reacts the copolymer with hydrazine to form a sulfonic acid followed by oxidation. The membranes of Dow Chemical are produced in a similar way. They have lower electrical resistance and less clustering. Asahi Glass Flemion® membranes are homogeneous or laminar perfluorocarboxylic acids. Their basic copolymer is slightly different from the others. In general, all copolymers are not cross-linked so that the swelling behavior is controlled by the microcrystallinity due to the tetrafluoroethylene content. The $-SO_3H$ groups are attached to the PTFE backbone by a molecular spacer that allows a certain degree of clustering. The economic lifetimes of chlor-alkali membranes under the aggressive conditions are in the range of 3 yr. The lifetime is determined in part by morphological changes and in part by loss of carboxylic acid ion-exchange capacity.

Alkaline Stable Anion-Exchange Membranes

In several technically interesting applications, the economics of the process is affected by the limited stability of currently available anion-exchange membranes in strong alkaline solutions. In the case of cation-exchange membranes, the chemical stability could be improved by perfluorination of the polymer backbone, resulting in membranes such as the Nafion®-, Flemion®-, and Neosepta®-type structures. Comparable attempts with anion-exchange membranes to overcome the poor alkaline stability failed (Matsui et al. 1986). Because the novel fluorocarbon-type anion-exchange membranes cannot overcome the problem of instability under strong basic conditions, it seems reasonable to assume that the alkaline stability of an anion-exchange membrane is determined by the stability of the incorporated positively charged groups against the attack of hydroxyl ions. By determining the disintegration rate of quaternized amines in alkaline solutions, Bauer, Gerner, and Strathmann (1988) developed an anion-exchange membrane with considerably improved alkaline stability. Due to their higher acidity, the cross-linked bis-quaternary structures have been shown to have a much lower alkaline stability than mono-quaternary ammonium groups. The highest alkaline stability was obtained with 1-benzyl-1-azonia-4-aza-bicyclo[2.2.2]-octane hydroxide, the mono-quaternary salt of DABCO.

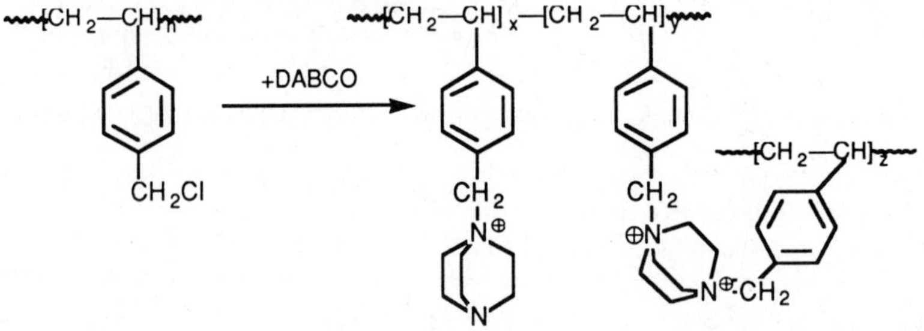

A further advantage arises from the possibility of preparing a well-defined polymer network, because exposure to basic solutions does not diminish the degree of cross-linking but instead converts the fixed ions to a piperazine system.

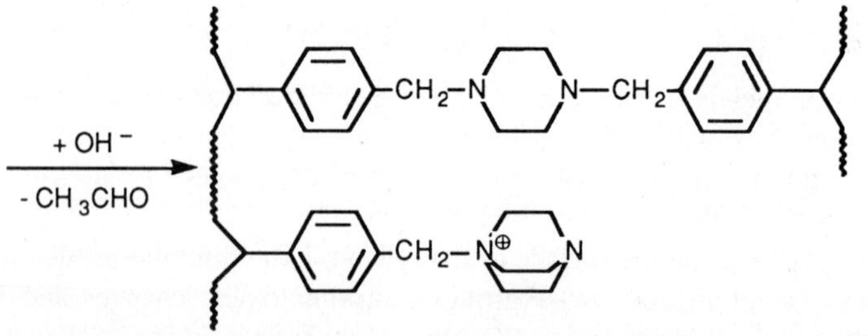

The best electrical properties were obtained with polyvinylidenefluoride (PVDF) as the matrix polymer, but in alkaline solutions the strong basic quaternary groups catalyzed a dehydro-halogenation. Integrated into a polysulfone (PSU) or polyethersulfone (PES) matrix, the anion-exchange membrane did not show any changes in its mechanical or electrical properties.

Bipolar Membranes

Bipolar membranes have recently gained increasing attention as an efficient tool for the production of acids and bases from their corresponding salts by electrically enforced, accelerated water dissociation. The process, which has been known for many years, is economically very attractive and has a multitude of interesting technical applications (Liu, Chlanda, and Nagasubramanian 1977). So far, however, large-scale technical use of bipolar membranes has been rather limited by the availability of efficient membranes.

The principal structure of a bipolar membrane and its function is illustrated in Figure 18–1. This figure shows an anion- and a cation-exchange membrane arranged between two electrodes, similar to conventional electrodialysis. If a NaCl-containing solution is placed between these membranes and an electrical potential gradient is applied, all ionic species are removed from the solution. When no sodium and chloride ions are left in the solution, the transport of the electrical charges through the membranes is accomplished exclusively by protons and hydroxyl ions, which are available even in pure water in a concentration of 10^{-7} mol/L due to the dissociation equilibrium of water. The dissociated water is continuously replenished from the outer phases, and thus an alkaline solution is formed on the anion-

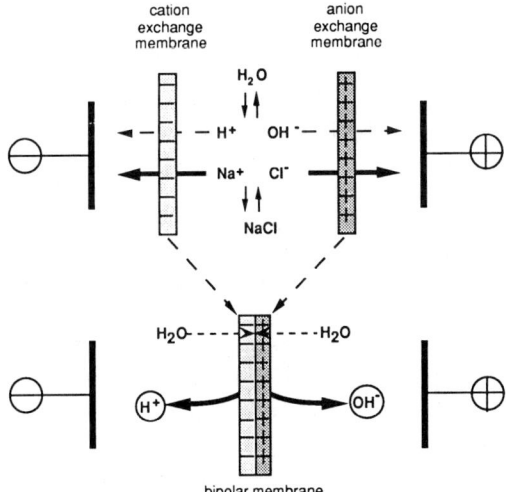

FIGURE 18–1. Schematic diagram showing the configuration and basic function of a bipolar membrane.

exchange side and an acid solution on the cation-exchange side of the bipolar membrane.

Bipolar membranes can be prepared by simply laminating conventional cation- and anion-exchange membranes back to back (Leitz 1972). The total potential drop depends on the applied current density, the resistance of the two membranes, and the resistance of the solution between them. Since the specific resistance of deionized water is very high, the distance between the membranes of opposite polarity should approach zero. Laminated bipolar membranes mostly exhibit an unsatisfactory chemical stability at high pH values and a rather poor water-splitting capability. Single-film bipolar membranes (Chlanda, Lee, and Liu 1976) and multilayer bipolar membranes fulfill most of the practical needs. For example, Bauer, Gerner, and Strathmann (1988) prepared a bipolar membrane by casting a cation selective layer on top of the previously prepared well-defined cross-linked anion-exchange membrane. To minimize the electrical resistance, the thickness of the interphase between the oppositely charged membranes should preferably be less than 5 nm. It could also be shown that in bipolar membranes the chemical stability is determined to a large extent by the properties of the positively charged anion-exchange moieties and by the properties of the matrix polymer. Furthermore, in practical applications, bipolar membranes

should have not only good chemical stabilities but also adequate water-splitting capabilities. From the observed ion fluxes, calculations show that the water dissociation rate in the bipolar region is much faster than in pure water already at current densities in excess of 1 mA cm^{-2}. The reasons for this acceleration are not completely verified. However, the experimental evidence reported in the literature strongly supports the hypothesis that it is caused by a reversible proton transfer reaction between the charged groups of the membrane and the water molecules at the surface.

CHARACTERIZATION OF ION-EXCHANGE MEMBRANES

Ion-exchange membranes used in electrodialysis can be classified in terms of their mechanical and electrical properties, their permselectivity, and their chemical stability. A microscopic examination yields information on whether a membrane is reinforced and the type of reinforcement used. The electrical charge of an ion-exchange membrane can be determined qualitatively by using indicator solutions. A drop of 0.05% solution of methylene blue and methyl orange on a sample stains a golden yellow on top of an anion-exchange membrane and a deep blue on top of a cation-exchange membrane.

Mechanical Examination and Swelling Behavior of Membranes

Detailed mechanical characterization involves the determination of thickness, dimensional stability, tensile strength, and hydraulic permeability. All mechanical measurements should be prepared with pretreated and well-equilibrated membranes. Information related to storage and handling characteristics, the durability, and type of reinforcing material is obtained from the determination of the dimensional changes between the wet and dry states of the membrane. Hydraulic permeability measurements provide information on the diffusive transport of components through a membrane under a hydrostatic pressure driving force. The presence of pinholes in ion-exchange membranes will not only obscure the hydraulic permeability test but will also invalidate any application. Pinholes can be determined by placing a wet membrane sheet on a sheet of white absorbent paper. A 0.2% solution of methylene blue for an anion-exchange membrane or a 0.2% solution of Erythrocein-B for a cation-exchange membrane has to be spread over the entire surface. If no spots of the dye can be observed on the paper, the membrane is free of pinholes and can be tested for its hydraulic permeability. The test is carried out at room temperature using deionized water and a hydrostatic pressure driving force. The permeability can then be calculated from the volumetric flow rate in $L \cdot m^{-2} \cdot h^{-1}$.

The gel water content or swelling capacity determines not only the dimensional stability of the membrane but also affects its selectivity, electrical resistance, and hydraulic permeability. The swelling of a membrane depends on the nature of the polymeric material, the ion-exchange capacity, the cross-linking density, and the homogeneity of the membrane. Usually, the gel water content is expressed by the weight difference between the wet and dry membrane. In a test for the gel water content, a sample is equilibrated for 2 days in deionized water. After removing the surface water from the sample, the wet weight of the swollen membrane W_{wet} is determined. The sample is then dried at 75°C over phosphorus pentoxide under reduced pressure until a constant weight W_{dry} is obtained. The gel water weight percent is obtained from $100(W_{wet} - W_{dry})/W_{dry}$.

Long-Term Chemical Stability of Membranes

The economics of ion-exchange membranes in different applications is determined to a large extent by their chemical stability under process conditions. Membrane deterioration after exposure for certain time periods to various test solutions containing acids, bases, or oxidizing agents is estimated by visual comparison with new, unexposed samples and by determining changes in their mechanical and electrical properties.

Determination of Membrane Ion-Exchange Capacities

The ion-exchange capacity of charged membranes is determined by titrating the fixed ions, e.g., $-SO_3^-$ or $-R_4N^+$ groups with 1 N NaOH or HCl, respectively. For these tests, cation- and anion-exchange membranes are equilibrated for about 24 h in 1 N HCl or NaOH, respectively, and then rinsed free from chloride or sodium for 24 h with deionized water. The ion-exchange capacity of the samples is then determined by back titration with 1 N NaOH or HCl, respectively. Weak base anion-exchange membranes are characterized by equilibration in 1 N sodium chloride and titration with standardized 0.1 N silver nitrate solution. The samples are then dried, and the ion-exchange capacity is calculated for the dry membrane.

Permselectivity of Ion-Exchange Membranes

The permselectivity of an ion-exchange membrane relates the transport of electric charges by specific counter-ions to the total transport of electrical charge through the membrane. An ideal selective cation-exchange membrane would, for example, transmit positively charged ions only. The permselectivity approaches zero when the transport numbers within the membrane $t_\pm$ are the same as in the electrolytic solution $t'_\pm$. The degree of permselectivity depends on the concentration of electrolytes in the membrane and therefore on the ion-exchange capacity and the cross-linking density. When a membrane separates diluate and concentrate solutions, there will be a concentration gradient across the membrane. In this case the permselectivity $\alpha_\pm$ can be calculated from the transfer of counter-ions (dynamic method) according to the following equation (Tuwiner 1962):

$$\alpha_\pm = \frac{t_\pm - t'_\pm}{1 - t'_\pm}. \qquad (18\text{--}1)$$

A faster method for the determination of the apparent permselectivity $\alpha_\pm^a$ is found from a

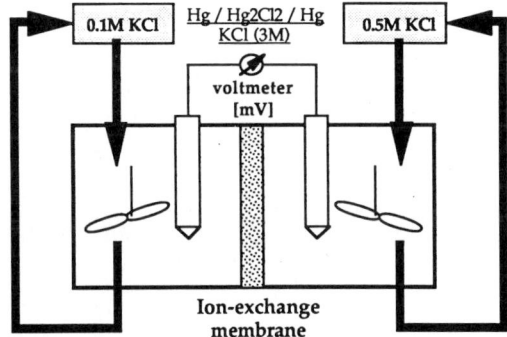

FIGURE 18–2. Experimental setup for determining membrane permselectivities.

potential measurement (static method). The experimental setup of the test procedure is illustrated in Figure 18–2. In this special case, the transport of water through the membrane is not taken into account. The apparent transport number $t_\pm^a$ of an ion is obtained as

$$t_\pm^a = t_\pm - 0.018 t_w c_\pm, \qquad (18\text{-}2)$$

where 0.018 mL $\cdot$ mmol^{-1} means the molar volume of water and t_w the water transfer rate. As an advantage, the determination of the potential between two solutions of different concentrations is approximately independent of any polarization effects at the membrane surface.

The actual test system consists of two cells separated by the membrane sample. The potential difference across the membrane is measured using a set of calomel electrodes. The selectivity is then calculated from the ratio of the experimentally determined to the theoretically calculated potential difference for a 100% permselective membrane.

For a system consisting of standardized aqueous solutions of 0.1 N and 0.5 N KCl at 25°C, for example, this theoretical potential difference amounts to exactly 36.94 mV. It is calculated using the Nernst equation:

$$\Delta\psi_{\text{theo}} = \frac{RT}{zF} \ln\left(\frac{m_1' \, \gamma_1'}{m_2' \, \gamma_2'}\right), \qquad (18\text{-}3)$$

where R is the gas constant, T the absolute temperature, z the electrochemical valence of the ions in the solutions, F the Faraday's constant, m_1' and m_2' the concentrations, and γ_1' and γ_2' the activity coefficients of the two solutions separated by the membrane. The apparent permselectivity $\alpha_\pm^a$ of the membrane is given by

$$\alpha_\pm^a = \frac{\Delta\psi_{\text{exp}}}{\Delta\psi_{\text{theo}}} 100\%, \qquad (18\text{-}4)$$

where $\Delta\psi_{\text{exp}}$ is the measured potential difference between the two electrolyte solutions. The absolute value of $\Delta\psi_{\text{exp}}$ is positive for a cation-exchange membrane and negative for an anion-exchange membrane (Korngold 1984).

Electrical Resistance of Ion-Exchange Membranes

The electrical resistance of ion-exchange membranes is one of the factors that determines the energy requirements of electrodialysis processes. It is, however, generally considerably lower than the resistance of the dilute solutions surrounding the membrane since the ion concentration in the membrane is very high. The specific membrane resistance is usually reported in ohm $\cdot$ cm. From the engineering point of view, the membrane area resistance in units of ohm $\cdot$ cm^2 is more useful. The area resistance of ion-exchange membranes is determined by conductivity measurements in a cell that consists of two well-stirred chambers separated by the membrane as shown in Figure 18–3.

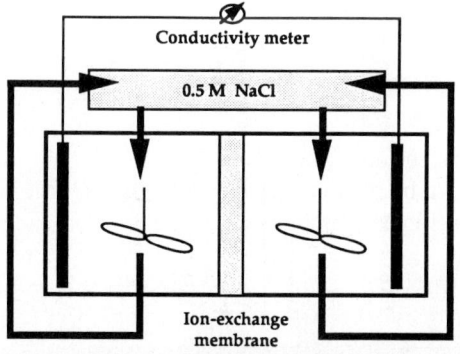

FIGURE 18–3. Experimental setup for determining the electrical resistance of ion-exchange membranes.

TABLE 18–1. Properties of Commercial Ion-Exchange Membranes.

Membrane	Type	Structure Properties	Backing	IEC (meq/g)	Thickness (mm)	Gel Water (%)	Area Resistance 0.5 N NaCl, 25°C ($\Omega \cdot cm^2$)	Permselectivity 1.0/0.5 N KCl (%)
Asahi Chemical Industry Company Ltd., Chiyoda-ku, Tokyo, Japan								
K 101	Cation	Styrene/DVB	Yes	1.4	0.24	24	2.1	91
A 111	Anion	Styrene/DVB	Yes	1.2	0.21	31	2–3	45
Asahi Glass Company Ltd., Chiyoda-ku, Tokyo, Japan								
CMV	Cation	Styrene	PVC	2.4	0.15	25	2.9	95
AMV	Anion	Butadiene	PVC	1.9	0.14	19	2–4.5	92
ASV	Anion	Univalent		2.1	0.15	24	2.1	91
DMV	Cation	Dialysis			0.15		—	—
Flemion®	Cation	Perfluorinated						
Ionac Chemical Company, Sybron Corporation, Birmingham, NJ 08011, U.S.A.								
MC 3470	Cation		Tergal	1.5	0.6	35	6–10	68
MA 3475	Anion		Tergal	1.4	0.6	31	5–13	70
MC 3142	Cation			1.1	0.8		5–10	—
MA 3148	Anion		Tergal	0.8	0.8	18	12–70	85
Ionics Inc., Watertown, MA 02172, U.S.A.								
61AZL386	Cation		Modacrylic	2.3	0.5	46	~6	—
61AZL389	Cation		Modacrylic	2.6	1.2	48	—	—
61CZL386	Cation		Modacrylic	2.7	0.6	40	~9	—
103QZL386	Anion		Modacrylic	2.1	0.63	36	~6	—
103PZL386	Anion		Modacrylic	1.6	1.4	43	~21	—
204PZL386	Anion		Modacrylic	1.9	0.57	46	~8	—
204SXZL386	Anion		Modacrylic	2.2	0.5	46	~7	—
204U386	Anion		Modacrylic	2.8	0.57	36	~4	—

Membrane	Type							
Du Pont Company, Wilmington, DE 19898, U.S.A.								
N 117	Cation	Perfluorinated	0.9	No	0.2	16	1.5	—
N 901	Cation	Perfluorinated	1.1	PTFE	0.4	5	3.8	96
Pall RAI, Inc., Hauppauge, NY 11788, U.S.A.								
R-5010-L	Cation		1.5	LDPE	0.24	40	2–4	85
R-5010-H	Cation		0.9	LDPE	0.24	20	8–12	95
R-5030-L	Anion		1.0	LDPE	0.24	30	4–7	83
R-5030-H	Anion		0.8	LDPE	0.24	20	11–16	87
R-1010	Cation	Perfluorinated	1.2	No	0.1	20	0.2–0.4	86
R-1030	Anion	Perfluorinated	1.0	No	0.1	10	0.7–1.5	81
Rhone-Poulenc Chemie GmbH, Frankfurt, Germany								
CRP	Cation		2.6	Tergal	0.6	40	6.3	65
ARP	Anion		1.8	Tergal	0.5	34	6.9	79
Tokuyama Soda Company Ltd., Nishi-Shimbashi, Minato-ku, Tokyo 105, Japan								
CL-25T	Cation		2.0	PVC	0.18	31	2.9	81
ACH-45T	Anion		1.4	PVC	0.15	24	2.4	90
ACM	Anion	Low H$^+$-transport	1.5	PVC	0.12	15	4–5	—
AMH	Anion	Chemical resistant	1.4	—	0.27	19	11–13	—
CMS	Cation	Univalent	>2.0	PVC	0.15	38	1.5–2.5	—
ACS	Anion	Univalent	>1.4	PVC	0.18	25	2–2.5	—
AFN	Anion	Antifouling	<3.5	PVC	0.15	45	0.4–1.5	—
AFX	Anion	Dialysis	1.5	PVC	0.14	25	1–1.5	—
Neosepta®-F		Perfluorinated						

The cell is filled with a 0.5 N solution of NaCl. The electrical resistance of the cell is measured with and without a membrane separating the two chambers. The membrane area resistance is then calculated from the difference in the conductivity of the two measurements using the equation

$$R_A = A\left(\frac{1}{k_T} - \frac{1}{k'}\right), \quad (18\text{-}5)$$

where R_A is the area resistance of the membrane per unit area, k_T the conductivity of the cell and the membrane, k' the mean conductivity of the cell, and A the cell constant.

Properties of commercially available membranes are listed in Table 18–1. This table, however, is not exhaustive; more special property membranes are available on the market today.

REFERENCES

Bauer, B., F.-J. Gerner, and H. Strathmann. 1988. Development of bipolar membranes. *Desalination* 68:279–292.

Caplan, S. R., and K. Sollner. 1974. Influence of the characteristics of the activating polyelectrolyte in the preparation and on the properties of interpolymer ion-exchange membranes: rational principles of membrane preparation and their experimental test. *J. Colloid Interface Sci.* 40(1):46–66.

Chlanda, F. P., L. T. C. Lee, and K. J. Liu. 1976. Bipolar membranes and method of making same. U.S. Patent 4,116,889.

Connolly, D. J., and W. F. Gresham. 1966. Fluorocarbon vinyl ether polymers. U.S. Patent 3,282,875.

Dohno, R., T. Azumi, and S. Takashima. 1975. Permeability of monocarboxylate ions across an anion-exchange membrane. *Desalination* 16(1):55–64.

Eisenberg, A., and H. L. Yeager. 1982. *Perfluorinated Ionomer Membranes*. ACS Symp. Series No. 180. Washington, DC: American Chemical Society.

Flett, D. S. 1983. *Ion Exchange Membranes*. Chichester, U.K.: E. Horwood Ltd.

Friedlander, H. Z. 1968. Membranes. *Encycl. Polym. Sci. Technol.* 8:620–638.

Grot, W. G. 1973. Laminates of support material and fluorinated polymer containing pendant side chains containing sulfonyl groups. U.S. Patent 3,770,567.

Gudernatsch, W., Ch. Krumbholz, and H. Strathmann. 1989. Development of an anion-exchange membrane with improved permeability for organic acid of high molecular weight. In *Proc. 6th Intl. Symp. on Synthetic Membranes in Science and Technology*, 4–8 September 1989, Tübingen, Germany, pp. 223–226.

Helfferich, F. 1962. *Ion Exchange*. New York: McGraw-Hill Book Company.

Ishigaki, I., N. Kamiya, T. Sugo, and S. Machi. 1978. Synthesis of an ion-exchange membrane by radiation-induced grafting of acrylic acid onto poly(tetrafluoroethylene). *Polym. J.* 10(5):513–519.

Joshi, K. M., and G. M. Ganu. 1978. Measurement of ionic activities in aqueous solutions using an epoxy-based ion-exchange membrane electrode. *Indian J. Technol.* 16(2):53–55.

Korngold, E. 1984. Electrodialysis: membranes and mass transport. In *Synthetic Membrane Processes*, ed. G. Belfort, pp. 192–219. New York: Academic Press.

Korngold, E., F. De Körösy, R. Rahav, and M. F. Taboch. 1970. Fouling of anion selective membranes in electrodialysis. *Desalination* 8(2):195–220.

Leitz, F. B. 1972. Apparatus for electrodialysis of electrolytes employing bilaminar ion-exchange membranes. U.S. Patent 3,654,125.

Liu, K. J., F. P. Chlanda, and K. J. Nagasubramanian. 1977. Use of bipolar membranes for generation of acid and base: an engineering and economic analysis. *J. Membr. Sci.* 2:109.

Matsui, K., E. Tobita, K. Sugimoto, K. Kondo, T. Seita, and A. Akimoto. 1986. Novel anion exchange membranes having fluorocarbon backbone: preparation and stability. *J. Appl. Polym. Sci.* 32(3):4137–4143.

McRae, W. A., and S. S. Alexander. 1960. Sulfonating reagent and its use in preparing cation-exchange membranes. U.S. Patent 2,962,454.

Molau, G. E. 1981. Heterogeneous ion-exchange membranes. *J. Membr. Sci.* 8(3):309–330.

Murphy, E. A., F. J. Paton, and J. Ansell. 1943. Apparatus for the electrical treatment of colloidal dispersions. U.S. Patent 2,331,494.

Onoue, Y., T. Sata, A. Nakahara, and J. Itoh. 1980. Process for preparing fluorine-containing polymers having carboxyl groups. U.S. Patent 4,200,711.

Sata, T., R. Izuo, and Y. Mizutani. 1984. Study of

membrane for selective permeation of specific ions. *Soda to Enso* 35(415):313–336.

Sata, T., K. Kuzumoto, and Y. Mizutani. 1976. Modification of anion-exchange membranes with polystyrene sulfonic acid. *Polym. J.* 8:225–226.

Suryanarayana, N. P., and N. Krishnaswamy. 1963. Ion-exchange membranes based on rubber. *J. Polym. Sci. Part B*. 1(9):491–495.

Tuwiner, S. B. 1962. *Diffusion and Membrane Technology*. New York: Reinhold Publishing Co.

Urano, K., T. Ase, and Y. Naito. 1984. Recovery of acid from wastewater by electrodialysis. *Desalination* 51(2):213–226.

Yamane, R., T. Izuo, and Y. Mizutani. 1965. Ion-exchange membranes XXIV: permeability of the amphoteric ion-exchange membranes. *Denki Kagaku* 33(8):589–593.

Zschocke, P., and D. Quellmalz. 1985. Novel ion exchange membranes based on an aromatic polyethersulfone. *J. Membr. Sci.* 22(2&3):325–332.

19

Design and Cost Estimates

Heinrich Strathmann
Universität Stuttgart

LIMITING CURRENT DENSITY
   AND CURRENT UTILIZATION
ELECTRODIALYSIS STACK DESIGN
DESIGN AND ECONOMICS

Capital Costs
Operating Costs
NOTATION
REFERENCES

The performance of electrodialysis as a unit operation is determined by several process and equipment design parameters, such as feed flow velocities, cell and spacer construction, stack design, etc. These parameters affect directly the costs of the process as well as indirectly by means of the limiting current density and the current utilization.

LIMITING CURRENT DENSITY AND CURRENT UTILIZATION

The limiting current density is the maximum current that may pass through a given membrane area without creating adverse effects, i.e., higher electrical resistance or lower current utilization. The limiting current density is determined by the ion concentration in the diluate flow stream and by concentration polarization effects as indicated in Figure 19–1. This figure shows the concentration gradients of cations in the boundary layers at the surfaces on the two sides of a cation-exchange membrane during an electrodialysis desalting process. Similar anion

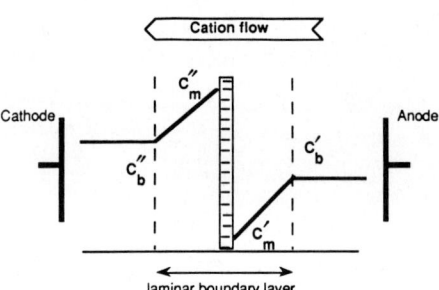

FIGURE 19–1. Schematic diagram of the concentration profiles of the cations in the laminar boundary layers at both surfaces of a cation-exchange membrane during electrodialysis, where c is the cation concentration, the subscripts b and m refer to the bulk solution and the membrane surface, and the superscripts $'$ and $''$ refer to the diluate and the concentrate.

concentration profiles are obtained at the surfaces of an anion-exchange membrane.

The transport of charged species to the anode or cathode through a set of ion-exchange membranes leads to a concentration decrease of counter-ions in the laminar boundary layer at the membrane surface facing the diluate cell and an increase at the surface facing the brine cell. The effect of concentration polarization due to a concentration increase in the brine is less severe. But the decrease in the concentration of counter-ions in the diluate directly affects the limiting current density and increases the electrical resistance of the solution in the boundary layer. The limiting current density is the current density at which the ion concentration at the surfaces of the cation- and/or anion-exchange membranes in the cells with the depleted solution will become zero. The limiting current density can be described by

$$i_{\lim} = \frac{c'_b D z_+ F}{Y_b(t_+ - t'_+)} = \frac{c'_b z_+ F k}{t_+ - t'_+}, \qquad (19\text{--}1)$$

where $i_{\lim}$ is the limiting current density, c'_b the bulk solution concentration in the cell with the depleted solution, D and z the diffusion coefficient and the electrochemical valence of the ions in the solution, F Faraday's constant, Y_b the boundary layer thickness, t and t' are the ion transport numbers in the membrane and the solution, respectively, and the subscript + refers to cations (on the other hand, the subscript − refers to anions). The constant k is the mass transfer coefficient, taking into account the influence of the hydrodynamics, flow channel geometry, spacer design, etc. (Mintz 1963).

According to Eq. (19–1), the limiting current density is proportional to the ion concentration in the diluate and the mass transfer coefficient, which is determined mainly by the cell geometry and the feed solution flow velocity. If in electrodialysis the limiting current density is exceeded, the process efficiency will be drastically diminished because of the increasing electrical resistance of the solution and because of water splitting, which leads to both pH changes and additional operational problems. The mass transfer coefficient can be related by the well-known relations for mass transfer to the Sherwood number, which again is a function of the Schmidt and Reynolds numbers. Introducing the proper relations into Eq. (19–1) leads to an expression that describes the limiting current density as a function of the feed flow velocity in the electrodialysis stack:

$$i_{\lim} = c'_b a v^b, \qquad (19\text{--}2)$$

where c'_b is the concentration in the diluate cell, v is the linear flow velocity of the solution through the electrodialysis cells, and a and b are constants whose values are determined by a series of parameters such as cell and spacer geometry, solution viscosity, ion transport numbers in the membrane and the solution, etc.

The constants a and b and therefore the limiting current density are a function of the electrodialysis stack design and must be determined experimentally. The limiting current density can be determined by measuring the total resistance of a cell pair and the pH value in the diluate cell as a function of the current density i. When the pH value is plotted versus $1/i$, a sharp decrease in the pH value is noted when the limiting current density is exceeded. Likewise, when the total resistance of a cell pair is plotted versus $1/i$, a minimum is obtained at the limiting current density. This is shown schematically in Figures 19–2(a) and 19–2(b).

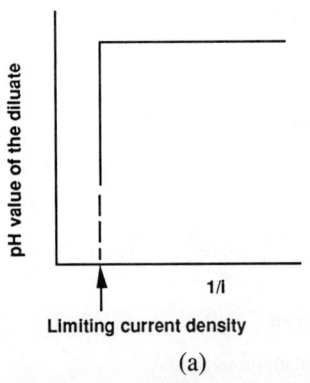
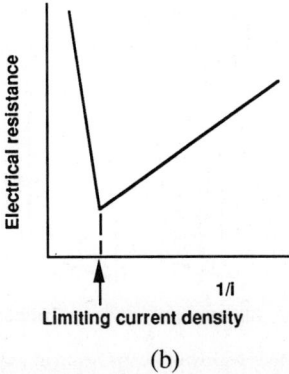

FIGURE 19–2. Schematic diagrams illustrating the determination of the limiting current density by plotting: (a) the pH value of the diluate cell versus $1/i$ and (b) the total resistance of a cell pair versus $1/i$.

The limiting current density determines the minimum membrane area required to achieve a certain desalting effect. Another very important parameter for the overall performance of the electrodialysis process is the current utilization. The current utilization determines the portion of the total current that passes through an electrodialysis stack that is actually used to transfer ions from a feed solution. The current utilization, which is always less than 100%, is affected by three factors: (1) membrane selectivity, (2) osmotic and ion-bound water transport, and (3) current passing through the stack manifold.

The water transferred due to osmosis and ion hydration can be significant at higher brine salt concentrations. The membrane selectivity also depends on the salt concentration, due to a phenomenon referred to as the *Donnan equilibrium relationship* [see Eq. (17–11) of Chapter 17]. For a diluted feed or brine solution, i.e., the feed solution or brine concentration is much lower than the concentration of the fixed charges in the membrane, an ion-exchange membrane is more or less strictly semipermeable, meaning that the membrane is permeable to counter-ions only. However, when the ion concentration in the feed solution is of the same order as that of the fixed charges in the membrane, co-ions may also enter the membrane, and the selectivity of which will then be decreased, with the consequence that the current efficiency will be decreased also.

ELECTRODIALYSIS STACK DESIGN

A typical electrodialysis stack design is shown in Figure 19–3. An electrodialysis stack is essentially a device to hold an array of membranes between electrodes in such a way that the streams being processed remain separated. The gaskets not only separate the membranes but also contain manifolds to distribute the process fluids in the different compartments. The supply ducts for the diluate and the brine are formed by matching holes in the gaskets, the membranes, and the electrode cells. The distance between the membrane sheets, i.e., the cell thickness, should be as small as possible since water, even with salt in it, has a relatively high electrical resistance.

In industrial-size electrodialysis stacks, membrane distances are typically between 0.5 and 2 mm (Mintz 1963). A spacer is introduced between the individual membrane sheets both to support the membrane and to help control the feed solution flow distribution. The most

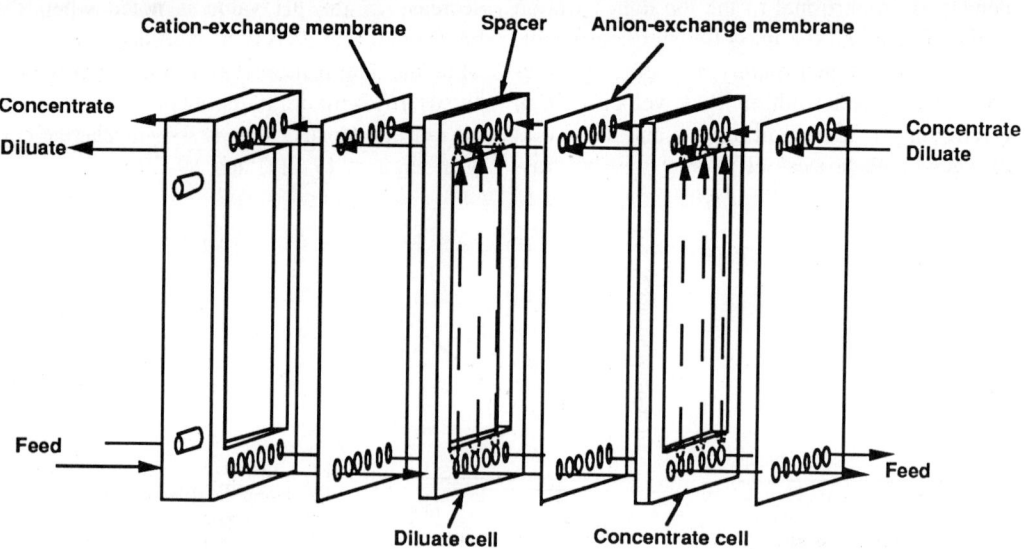

FIGURE 19–3. Exploded view of components in an electrodialysis stack.

serious design problem for an electrodialysis stack is that of assuring uniform flow distribution in the various compartments. In a practical electrodialysis system, 200 to 1000 cation- and anion-exchange membranes are installed in parallel to form an electrodialysis stack with 100 to 500 cell pairs (Lacey 1966).

As in other membrane separation processes, electrodialysis is affected by concentration polarization and membrane fouling. The magnitude of concentration polarization is largely determined by the electrical current density, by the cell (particularly the spacer design), and by the flow velocities of the diluate and brine solutions. Concentration polarization effects in electrodialysis lead to the depletion of ions in the laminar boundary layer at the membrane surfaces in the cell containing the diluate flow stream and to the increase of ions in the laminar boundary layer at the membrane surfaces in the cell containing the brine solution. Concentration polarization affects the separation efficiency by decreasing the limiting current density.

More difficult to control and in its consequences much more severe are membrane fouling effects due to adsorption of polyelectrolytes, such as humic acids, surfactants, proteins, etc. Because of their size, these components often penetrate the membrane only partially, thus resulting in severely reduced ion permeability of the membrane. Surface fouling can also increase concentration polarization.

In designing an electrodialysis stack, several general criteria concerning mechanical, hydrodynamic, and electrical properties have to be considered. Since some of the criteria have opposing tendencies, the final stack construction is generally a compromise between several conflicting parameters.

A proper electrodialysis stack design should provide a maximum effective membrane area per unit stack volume. The solution-distribution design should ensure equal and uniform flow distribution through each compartment. Any leakage between the diluate, concentrate, and electrode cells has to be prevented. The spacer screen should provide a maximum of mixing of the solutions at the membrane surfaces and cause a minimum in pressure loss.

Most stack designs used in current large-scale electrodialysis plants are one of the two basic types, tortuous path and sheet flow (Lacey 1969). These designations refer to the type of the solution flow path in the compartments of the stack. In the tortuous-path stack, the membrane spacer and gasket have a long serpentine cutout, which defines a long narrow channel for the fluid path. The objective is to provide a long residence time for the solution in each cell in spite of the high linear velocity that is required to limit polarization effects. A tortuous-path spacer gasket is shown schematically in Figure 19–4(a).

In stack designs employing the sheet-flow principle, a peripheral gasket provides the outer seal, and the solution flow is approximately in a straight path from the entrance to the exit ports, which are located on opposite sides in the gasket. This is illustrated in Figure 19–4(b), which shows the schematic diagram of a sheet-flow spacer of an electrodialysis stack.

Solution flow velocities in sheet-flow stacks lie typically between 5 and 10 cm/s, whereas in tortuous-path stacks solution flow velocities of 15 to 50 cm/s are used. Because of higher flow velocities and longer flow paths, higher pressure drops of the order of 2 to 3 bars are obtained in tortuous-path stacks than in sheet-flow systems where pressure drops of 1 to 2 bars occur. However, higher velocities help to reduce the deposition of suspended solids and biological materials. Several other stack concepts are described in the literature; most of them provide three or four independent solution flow cells, which are used, for example, in combination with bipolar membranes.

DESIGN AND ECONOMICS

In addition to the actual stack, an electrodialysis plant consists of several components essential for proper operation. A flow diagram of a typical unidirectional electrodialysis plant is shown in Figure 19–5. After proper pretreatment, the feed solution is pumped through the actual electrodialysis unit, which generally consists of one or more stacks in series or parallel. A deionized solution and a concentrated brine are obtained.

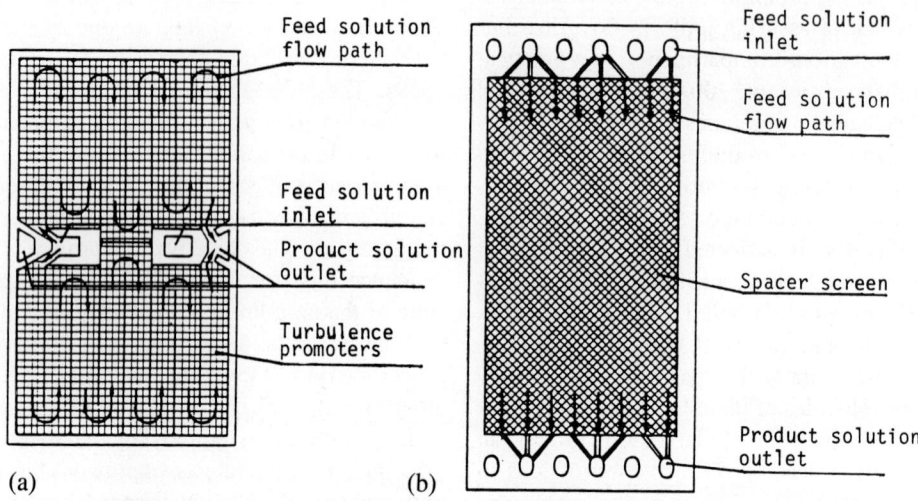

FIGURE 19-4. Schematic diagrams of (a) a tortuous-path electrodialysis spacer gasket and (b) a sheet-flow electrodialysis spacer gasket.

The concentrated and depleted process streams leaving the last stack are collected in storage tanks when the desired degree of concentration or depletion is achieved, or they are recycled if further concentration or depletion is desired. Sometimes acid is added to the concentrated stream to prevent scaling of carbonates and hydroxides. To prevent the formation of free chlorine by anodic oxidation, the electrode cells are sometimes rinsed with a separate solution that does not contain any chloride ions. In many cases, however, the feed or brine solution is also used in the electrode cells.

In addition to the unidirectional electrodialysis, an operating mode referred to as *electrodialysis reversal* is often used. In this operating mode, the polarity of the current is changed at specific time intervals ranging from a few min-

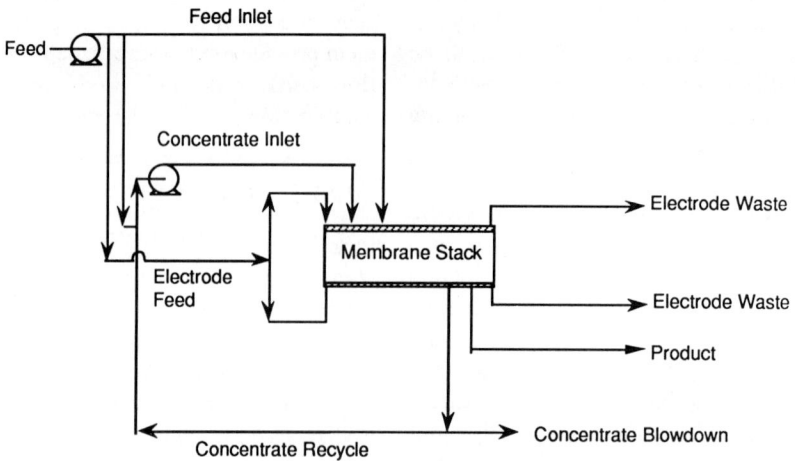

FIGURE 19-5. Flow diagram of a typical unidirectional electrodialysis desalination plant.

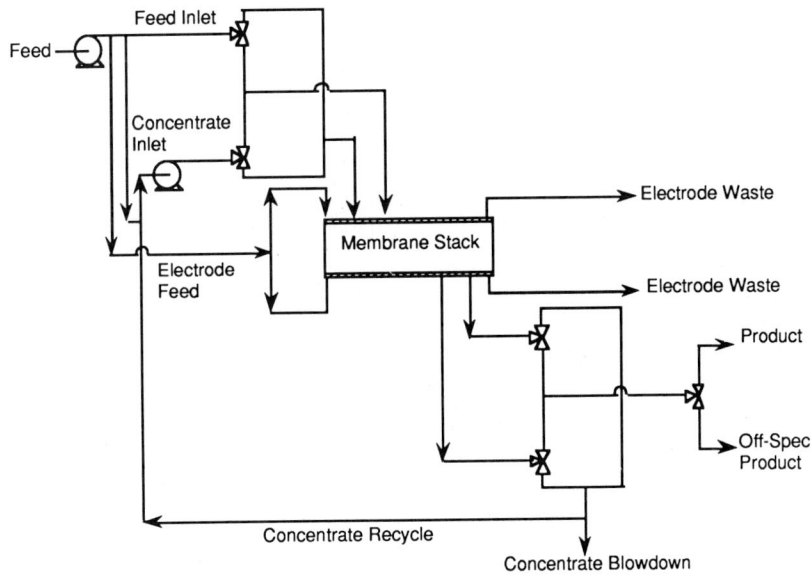

FIGURE 19–6.   Typical electrodialysis reversal plant.

utes to several hours. In the reverse polarity operating mode, the hydraulic flow streams are reversed simultaneously, i.e., the diluate cell will become the brine cell and vice versa. The advantage of the reverse polarity operating mode is that precipitations in the brine cells are prevented to a large extent. Or, if some precipitations do occur, they will be redissolved when the brine cell becomes the diluate cell in the reverse operating mode (Katz 1979). The flow scheme of a typical electrodialysis reversal plant is shown in Figure 19–6.

Process design and economics are closely related in electrodialysis. As in most other membrane processes, the total cost is the sum of fixed charges associated with amortization of the plant capital cost and operating costs, such as energy and labor costs. Membrane replacement costs are often regarded as a separate item because of the relatively short life of the membrane.

Capital Costs

Capital costs include depreciable items, such as the electrodialysis stacks, pumps, electrical equipment, membranes, etc., and nondepreciable items, such as land and working capital.

The capital costs of an electrodialysis plant depend strongly on the total membrane area required for a certain plant capacity. The required membrane area, however, is proportional to the amount of ionic species removed from a given feed solution, as indicated by the following equation (Lacey 1969):

$$A = \frac{zFQ\Delta cn}{i\xi}, \qquad (19\text{--}3)$$

where A is the effective cell pair area, Q the volumetric flow rate of the produced potable water, Δc the difference in the salinity of the feed and product water, n the number of cells in a stack, i the current density, which should be about 80% of the limiting current density, ξ the current utilization, and the rest of the symbols have the same definitions given earlier. Note that the actual membrane area is often significantly higher than the effective membrane area calculated by Eq. (19–3) because of the so-called "shadow effect" of spacers. The limiting current density is a function of the diluate concentration, which is changing during the desalting process from the concentration of the original feed to the product solution concentration. The calculation of the minimum membrane area required for a given desalting capacity is based on an average limiting current density, which is a function of the average diluate concentration given by

$$\bar{i}_{\lim} = a\bar{c}' = a\frac{c_f - c'}{\log(c_f/c')}, \quad (19\text{–}4)$$

where $\bar{i}_{\lim}$ is an average limiting current density, a is a constant factor that depends on the cell and spacer geometry and feed flow velocity, $\bar{c}'$ is the average diluate concentration, and c_f and c' are the feed and diluate concentrations, respectively. Substitution of Eq. (19–4) into (19–3) leads to

$$A_{\min} = a'zFQ \log(c_f/c'), \quad (19\text{–}5)$$

where $A_{\min}$ is the minimum membrane area required for a certain plant capacity and given feed and product solution concentrations, a' is a constant for a given plant design and operating mode, and the rest of the symbols have the same definitions given earlier.

For a certain plant capacity, the required membrane area is directly related to the feed water concentration assuming the same product water concentration. This is illustrated in Figure 19–7.

For typical brackish water of approximately 3000 ppm total dissolved solids (TDS) and an average current density of 12 mA/cm², the required membrane area for a plant capacity of 1 m³ of product per day is about 0.4 m² of cation-exchange membranes and 0.4 m² of anion-exchange membranes. Other items such as pumps, piping, and tanks, etc., do not depend on feed water salinity but on plant size. For desalination of brackish water with a salinity of approximately 3000 ppm, the total capital cost for a plant with a capacity of 1000 m³/day will be in the range of US$200 to 300 per cubic meter per day capacity. The cost of the actual membrane is less than 30% of the total capital costs. Assuming a useful life of 5 yr for the membranes (up to 7 yr is common for many brackish water applications) and 10 yr for the rest of the equipment, a feed water salinity of 3000 ppm, and a 24-h operating day, the total amortization of the investment is about US$0.10 to 0.15 per cubic meter of potable water with a salinity of less than 500 ppm. The operating costs are mainly determined by the required energy. For some applications, which have high brine concentrations, the current density may be limited by the manifold shorting, resulting in a requirement for more membrane area to achieve equivalent desalination.

Operating Costs

As pointed out before, the energy in electrodialysis is determined by the electrical energy required for the actual desalting process and the energy necessary for pumping the solution through the stack. The energy for the actual desalting process, i.e., the ion transfer from the feed solution to the brine, is directly proportional to the amount of ionic species to be removed, as indicated in Eq. (17–6) of Chapter 17. The energy requirement for the production of potable water as a function of the feed water concentration is shown in Figure 19–8. The case considered is a NaCl feed solution with the product having a salt concentration of less than 500 ppm.

The pumping energy is independent of the feed solution salinity. But it does depend on the feed water recovery rate and temperature. Assuming a pressure drop in the unit of approximately 400 kPa (4 bars), a pump efficiency of

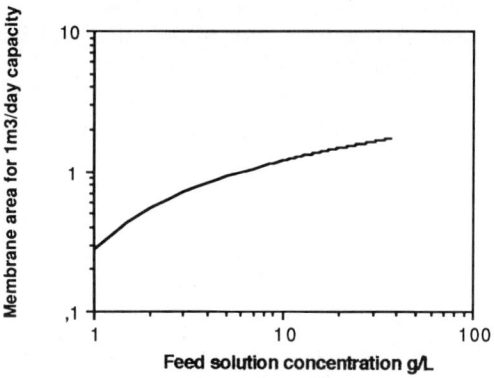

FIGURE 19–7. Schematic diagram of the required membrane area in electrodialysis desalination as a function of the feed water concentration at constant current density and plant capacity.

Design and Cost Estimates 253

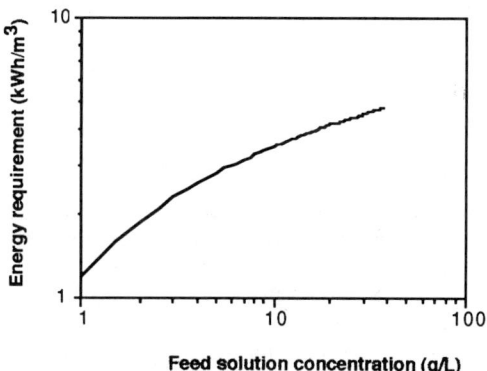

FIGURE 19-8. Energy requirements for the production of potable water with a solid content of 500 ppm as a function of the feed solution concentration ($\Delta\psi$ per cell pair = 0.8 V).

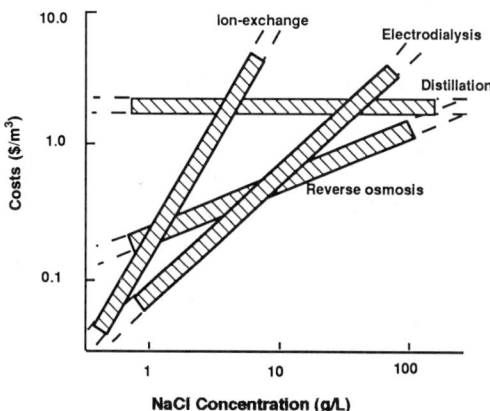

FIGURE 19-10. Water desalination costs as a function of the feed solution concentration for ion exchange, electrodialysis, reverse osmosis, and distillation.

70%, and 50% recovery, the total pumping energy will be about 0.4 kWh/m³ product water. This indicates that at low feed water salt concentrations, the cost for pumping the solution through the unit might become quite significant.

According to Eqs. (17-6) and (17-7) of Chapter 17, the energy cost increases with increasing current density while the required membrane area decreases with increasing current density. Thus, the total desalination cost, which is the sum of capital, energy, and operating costs, will reach a minimum at a certain current density as illustrated in Figure 19-9, where the total cost is shown as a function of the applied current density for a given feed solution.

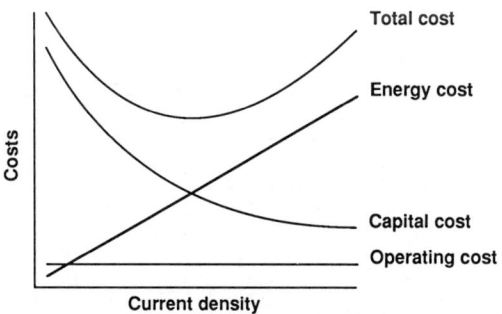

FIGURE 19-9. Schematic diagram of electrodialysis process costs as a function of the applied current density.

Furthermore, quite interesting is a comparison of the costs of desalination by various processes (Strathmann 1984) as a function of the feed water salinity, as shown in Figure 19-10. This figure indicates that at very low feed solution salt concentrations, ion exchange is the most economical process. But its costs increase sharply with feed solution salinity. At about 500 ppm, electrodialysis becomes the more economical process, while at around 5000 ppm, reverse osmosis is the less costly process. At very high feed solution salt concentrations, in excess of 100,000 ppm, multistage flash evaporation becomes the most economical process. The costs of potable water produced from brackish water sources are in the range of US$0.20 to 0.50 per cubic meter.

NOTATION

General Notation

See the General Notation section at the beginning of this handbook.

Special Notation

A effective cell pair area, L^2 or m²

c' concentration in the depleted solution (diluate), M/L^3, g/L, or kg/m³

Δc difference in the salinity of the feed and product water, M/L^3, g/L, or kg/m^3
k mass transfer coefficient, L/t or m/s
n number of cells in an electrodialysis stack
Q volumetric flow rate of the produced potable water, L^3/t or m^3/s
Y_b boundary layer thickness, L or m
z electrochemical valence of ions in the solution, dimensionless

Greek Letter
ξ current utilization, dimensionless

REFERENCES

Katz, W. E. 1979. The electrodialysis reversal (EDR) process. *Desalination* 28:31–40.

Lacey, R. E. 1966. *Membrane Processes for Industry*. Birmingham, AL: Southern Research Institute.

Lacey, R. E. 1969. Electrosorption and desorption process for demineralization. Report 398. U.S. Office Saline Water. Res. Dev. Program.

Mintz, M. S. 1963. Electrodialysis: principles of process design. *Ind. Eng. Chem.* 55(6):18–28.

Strathmann, H. 1984. Ion-exchange membranes in industrial separation processes. *J. Sep. Process Technol.* 5(1):1–13.

20

Applications

Heinrich Strathmann
Universität Stuttgart

CURRENT APPLICATIONS OF
 ELECTRODIALYSIS
  Desalination of Brackish Water by
   Electrodialysis
  Production of Table Salt
  Electrodialysis in Wastewater Treatment
  Concentration of Reverse Osmosis
   Brines
  Electrodialysis in the Chemical,
   Food, and Drug Industries

Production of Ultrapure Water
OTHER ELECTRICALLY DRIVEN
 MEMBRANE PROCESSES AND
 APPLICATIONS
  Diffusion Dialysis
  Chlorine-Alkaline Electrolysis
  Acid and Base Production by
   Electrodialytic Water Dissociation
NOTATION
REFERENCES

CURRENT APPLICATIONS OF ELECTRODIALYSIS

Electrodialysis was first developed for the desalination of saline solutions, in particular, brackish water. The production of potable water is still currently the most important industrial application of electrodialysis. But other applications, such as the treatment of industrial effluents, the production of boiler feed water, the demineralization of whey, the deacidification of fruit juices, etc., are gaining increasing importance with large-scale industrial installations (Ahlgren 1972; Korngold, Kock, and Strathmann 1978). Another application of electrodialysis, which is limited regionally to Japan and Kuwait, has gained considerable commercial importance: the production of table salt from seawater. Diffusion dialysis and the use of bipolar membranes have significantly expanded the application of electrodialysis in recent years (Strathmann 1985). Some of the more important industrial applications of electrodialysis and related processes are listed in Table 20–1.

Desalination of Brackish Water by Electrodialysis

In terms of the number of installations, the most important large-scale application of electrodialysis is the production of potable water from brackish water. Here, electrodialysis competes directly with reverse osmosis and multistage flash evaporation. For water with relatively low

TABLE 20–1. Industrial Applications of Electrodialysis and Related Processes.

Application	Membranes and Stack Design	Status of the Art	Key Problems
Brackish water desalination	Anion- and cation-exchange membranes, tortuous-path and sheet-flow stacks	Commercial	—
Boiler feed water, industrial process water	Anion- and cation-exchange membranes, tortuous-path stack	Commercial	Scaling costs
Production of table salt	Anion- and cation-exchange membranes, sheet-flow stack	Commercial	Costs
Industrial effluent	Anion- and cation-exchange membranes, tortuous-path and sheet-flow stacks	Commercial	Costs
Nontoxic treatment	Sheet-flow and tortuous-path stacks	—	Removal
Food and pharmaceutical industries	Anion- and cation-exchange membranes, sheet-flow and tortuous-path stacks	Commercial	Membrane fouling, product loss
Diffusion dialysis of acids	Cation-exchange membranes, sheet-flow stack	Commercial	Costs
Ultrapure water	Filled-cell stack	Commercial	Process reliability
Water dissociation	Bipolar membranes, three-cell stack	Commercial	Membrane performance

salt concentration (less than 5000 ppm), electrodialysis is generally the most economic process, as indicated in Chapter 19. One significant feature of electrodialysis is that the salts can be concentrated to comparatively high values (in excess of 18 to 20 wt.%) without affecting the economics of the process severely. Most modern electrodialysis units operate with so-called *electrodialysis reversal* (EDR), i.e., the anode and cathode, and with that the diluate and concentrate cell systems are exchanged periodically, preventing a scaling due to concentration polarization effects (Ionics 1988) (see Chapter 19). In brackish water desalination, more than 2000 plants with a total capacity of more than 1,000,000 m^3 of product water per day are installed, requiring a membrane area in excess of 1.5 million square meters. Installations in Russia and China for the production of potable water are estimated as being of the same order of magnitude. Exact data, however, are difficult to obtain.

In many cases now, EDR is used in combination with reverse osmosis (RO) and/or ultrafiltration (UF). UF/EDR provides an excellent pretreatment for the RO. A large number of trailer-mounted units with these three membrane systems are used in the power and semiconductor industries.

Production of Table Salt

The production of table salt from seawater by the use of electrodialysis to concentrate sodium chloride up to 200 g/L prior to evaporation is a technique developed and used nearly exclusively in Japan. More than 350,000 tons of table salt are produced annually by this technique requiring more than 500,000 m^2 of installed ion-exchange membranes (Tokuyama

Soda 1988). The key to the success of this technology has been the low-cost, highly conductive membranes with a preferred permeability of monovalent ions. However, note that in Japan this salt production procedure is highly subsidized.

Electrodialysis in Wastewater Treatment

The main application of electrodialysis in wastewater treatment systems is in processing rinse waters from the electroplating industry. Here, the complete recycle of the water and metal ions is achieved by electrodialysis. Compared to reverse osmosis, electrodialysis has the advantage of being able to utilize more thermally and chemically stable membranes, so that processes can be run at elevated temperatures and in solutions of very low or high pH values. Furthermore, the concentrations that can be achieved in the brine can be significantly higher. The disadvantage of electrodialysis is that only ionic components can be removed and additives usually present in a galvanic bath cannot be recovered.

An application that has been studied in a pilot plant stage is the regeneration of chemical copper plating baths. In the production of printed circuits, a chemical process is often used for copper plating. The components to be plated are immersed into a bath containing, besides the copper ions, a strong complexing agent, for example, ethylenediaminetetraacetic acid (EDTA), and a reducing agent such as formaldehyde. Since all constituents are used in relatively low concentrations, the copper content of the bath is soon exhausted, and $CuSO_4$ has to be added. During the plating process, formaldehyde is oxidized to formate. After prolonged use, the bath becomes enriched with Na_2SO_4 and formate and consequently loses its useful properties. By applying electrodialysis in a continuous mode, the Na_2SO_4 and formate can be selectively removed from the solution, without affecting the concentrations of formaldehyde and the EDTA complex. Therefore, the useful life of the plating solution is significantly extended (Korngold, Kock, and Strathmann 1978).

Several other potential applications of electrodialysis in wastewater treatment systems have been studied on a laboratory scale and are reported in the literature. While in most of these applications the average plant capacity is considerably lower than that for brackish water desalination or table salt production, there are also a significant number of large plants installed for the treatment of refinery effluents and cooling tower waste streams.

Concentration of Reverse Osmosis Brines

A further application of electrodialysis is concentration of reverse osmosis brines. Because of limiting membrane selectivity and the osmotic pressure of concentrated salt solutions, the concentration of brine in reverse osmosis desalination plants cannot exceed certain values. Often the disposal of large volumes of brine is difficult, and further concentration is desirable. This further concentration may be achieved at reasonable costs by means of electrodialysis.

Electrodialysis in the Chemical, Food, and Drug Industries

The use of electrodialysis in the food, drug, and chemical industries has been studied quite extensively in recent years. Several applications have considerable economic significance and are already well established. One is the demineralization of cheese whey. Normal cheese whey contains between 5.5 and 6.5% of dissolved solids in water. The primary constituents in whey are lactose, protein, minerals, fat, and lactic acid. Whey provides an excellent source of protein, lactose, vitamins, and minerals, but in its normal form it is not considered a proper food material because of its high salt content. With the ionized salts substantially removed, whey approaches the composition of human milk, and therefore provides an excellent source for the production of baby food. The partial demineralization of whey can be carried out quite efficiently by electrodialysis. The process is used extensively and described in detail in the literature.

The removal of tartaric acid from wine is another possible application. Particularly in the production of bottled champagne, the formation of crystalline tartar in wine must be avoided; therefore, the concentration of tartaric acid must be reduced to a value that does not exceed the solubility limit. This again can be done efficiently by electrodialysis.

Desalting of dextran solutions, another application for electrodialysis, is of technical significance as a large-scale industrial process.

Several other applications of electrodialysis in the pharmaceutical industry have been studied on a laboratory scale. Most of these applications are concerned with desalting solutions containing active agents that have to be separated, purified, or isolated from certain substrates. Here, electrodialysis is often in competition with other separation procedures such as dialysis, solvent extraction, etc. In many cases, electrodialysis is the superior process as far as economics and the quality of the product are concerned. In particular, the separation of amino acids and other organic acids by electrodialysis seems to be of interest to the pharmaceutical and chemical industries. However, the deionization of cheese whey, with an installed capacity of more than 35,000 m^2 of membrane area for the production of more than 150,000 tons of desalted lactose per year, is economically by far the most important application of electrodialysis in the food industry today.

Production of Ultrapure Water

More recently, electrodialysis has been used for the production of ultrapure water for the semiconductor industry. A combination with mixed bed ion-exchange resins, as shown in Figure 20–1, seems attractive because completely deionized water is obtained without the chemical regeneration of the ion-exchange resin (Kedem and Maoz 1976). This process has been commercialized recently.

The electrodialysis industry has experienced a steady growth since it made its appearance as an industrial-scale separation process about 15 years ago. Currently, the desalination of brackish water and the production of table salt are still the dominant applications, but new areas of application in the food and chemical process industries are gaining interest rapidly.

OTHER ELECTRICALLY DRIVEN MEMBRANE PROCESSES AND APPLICATIONS

Electrodialysis is today by far the most important industrial membrane separation process using ion-exchange membranes and the driving force of an electrical potential gradient. However, several other processes are rapidly achieving technical relevance, such as diffusion dialysis, regular electrolysis used for the production of chlorine and caustic soda, and the electrodialysis with bipolar membranes used for the production of acids and bases from the corresponding salts. Most of these processes have been de-

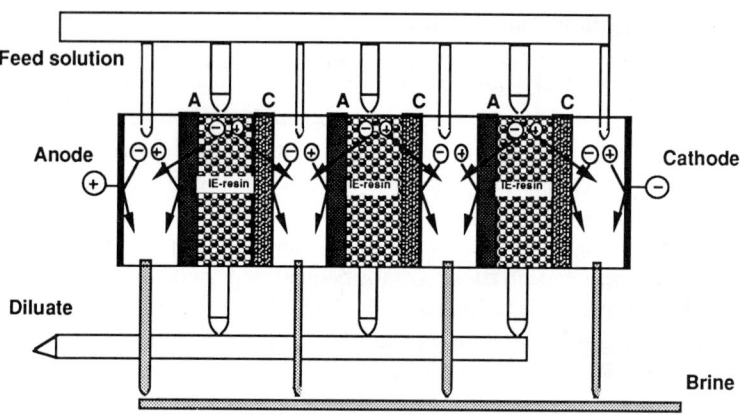

FIGURE 20–1. Principle of electrodialytic regeneration of mixed bed ion-exchange resins for the production of deionized water.

veloped only recently, and their large-scale industrial utilization is still in the beginning stages.

Diffusion Dialysis

A process in which ion-exchange membranes in an electrodialysis stack cell arrangement are used without any applied electrical potential gradient is referred to as *diffusion dialysis*. In this process, a concentration difference in two solutions separated by an ion-exchange membrane is used as the driving force for the transport of ions. The principle of the process is shown schematically in Figure 20–2. Additional material on this process is provided in Chapters 11 through 15.

Figure 20–2 shows a $CuSO_4$ solution and 1 N H_2SO_4 separated by a cation-exchange membrane. Since the H^+ ion concentration in acid solution I is significantly higher (pH = 1) than the H^+ ion concentration in the copper sulfate solution II (pH = 7), there will be a driving force for the transport of H^+ ions from solution I into solution II. Since the membrane is permeable to cations only, an electrical potential buildup results, which will counterbalance the concentration-difference driving force of the H^+ ions. This electrical potential difference, which is referred to as the *diffusion potential*, will cause a flux of Cu^{++} ions against their concentration gradient from solution II into solution I. As long as the H^+ ion concentration difference between the two phases separated by the cation-exchange membrane is maintained, Cu^{++} ions will be transported until their concentration difference is of the same order of magnitude as the H^+ ion concentration difference.

The process can be carried out accordingly with anions through anion-exchange membranes. An example of anion diffusion dialysis is the sweetening of citrus juices. In this process, hydroxyl ions furnished by a caustic solution replace the citrate ions in the juice.

Chlorine-Alkaline Electrolysis

The electrolytic production of chlorine and caustic soda by the use of a cation-exchange membrane as a separation medium has gained significant technical and commercial importance. The principle of the process is illustrated in the schematic drawing of Figure 20–3, which shows an electrolysis cell arrangement consisting of two chambers separated by a cation-exchange membrane.

One compartment contains an anode and a sodium chloride feed solution. The other compartment contains the cathode and water at the beginning of the process. When an electrical potential difference between the two electrodes is applied, the positively charged sodium ions will migrate toward the cathode to produce hydrogen and hydroxyl ions by an electrochemical reaction at the cathode. The negatively charged

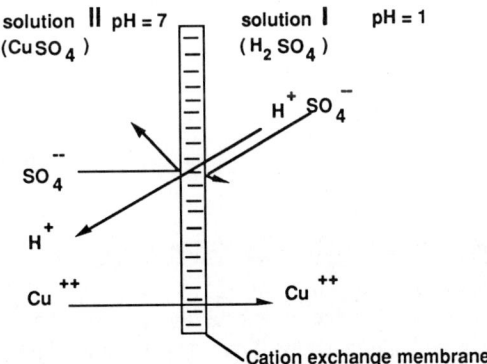

FIGURE 20–2. The principle of diffusion dialysis illustrating the transport of Cu^{++} ions through a cation-exchange membrane utilizing a diffusion potential built up by the flux of H^+ ions.

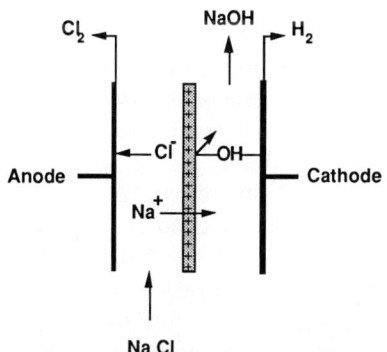

FIGURE 20–3. Schematic of the chlorine-alkaline production process.

chloride ions move toward the anode and are oxidized to form chlorine. Thus, sodium chloride is electrochemically converted into chlorine, caustic soda, and hydrogen. Migration of the hydroxyl ions is prevented by the cation-exchange membrane. Therefore, the current utilization in the electrolytic chlorine and caustic soda production is close to 100%. The compartment containing the produced sodium hydroxide is usually operated in a feed-and-bleed mode, and its sodium hydroxide concentration is kept as high as possible. In industrial production processes, sodium hydroxide concentrations in excess of 10 wt.% are obtained. Since the sodium chloride concentration is also kept rather high, the electrical resistance of the solutions is comparatively low, and the cell system can be operated at relatively high current densities up to a few thousand amps per square meter. The main problem in the electrolytic production of chlorine and caustic soda is the stability of the cation-exchange membrane, which faces a strong caustic environment on one side and the solution containing free chlorine on the other side. Today, membranes based on fluorinated hydrocarbon polymers, such as the Nafion® membrane, have demonstrated useful lifetimes of several years in operation, even at elevated temperatures.

Acid and Base Production by Electrodialytic Water Dissociation

As indicated earlier, the electrodialytic water dissociation with bipolar membranes is an efficient tool for the production of acids and bases from the corresponding salts. So far, however, the large-scale use of the process has been rather limited because of the shortcomings of current bipolar membranes, which have to meet certain requirements as far as their water-splitting capability, electrical properties, and chemical stability are concerned. But recent progress in the development of efficient bipolar membranes has increased the technical and industrial importance of this process (Chlanda, Lee, and Liu 1976; Bauer, Effenberger, and Strathmann 1989). The function of a bipolar membrane and the basic principle of the process were illustrated in Chapter 18 (Figure 18–1).

The electrodialytic production of acids and bases with bipolar membranes has the advantage of being very energy efficient. The theoretical energy required for the process is that needed to establish the desired concentration of H^+ and OH^- ions in the outer phases of the membrane from their concentration in the membrane, which is approximately 10^{-7} mol/L. The free-energy change ΔG of this process is

$$\Delta G = RT \ln \left(\frac{a_{H^+}^i \, a_{OH^-}^i}{a_{H^+}^o \, a_{OH^-}^o} \right), \quad (20\text{–}1)$$

where R is the gas constant, T the absolute temperature, and a the activity. The superscripts o and i refer to the outside solutions and the interphase between cation- and anion-exchange membranes, respectively.

For the generation of one molar solutions of H^+ and OH^- ions, i.e., $a_{H^+}^o = 1$ and $a_{OH^+}^o = 1$, Eq. (20–1) reduces to

$$\Delta G = RT \ln(a_{H^+}^i \, a_{OH^-}^i) = RT \ln K_w, \quad (20\text{–}2)$$

where K_w is the ionic product of water.

For electrolyte solutions, the free-energy change can be related to the electromotive force by

$$\Delta G = -zF\Delta\psi, \quad (20\text{–}3)$$

where $\Delta\psi$ is the reversible electromotive force. Combining Eqs. (20–2) and (20–3) for $z = 1$ leads to

$$\Delta\psi = -\frac{RT}{F} \ln K_w. \quad (20\text{–}4)$$

The free-energy change and the electromotive force for the generation of H^+ and OH^- from water in a perfectly semipermeable membrane can be calculated by Eqs. (20–3) and (20–4), respectively.

The free-energy change for the dissociation of one mole water and thus the production of

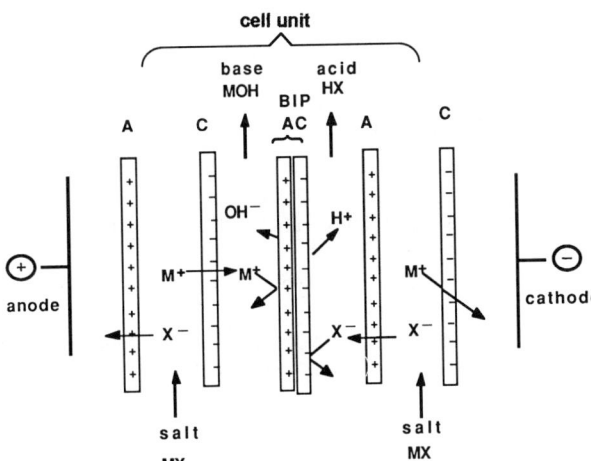

FIGURE 20-4. Electrodialysis cell arrangement with a bipolar membrane for the production of an acid and a base from the corresponding salt.

one molar acid at 25°C is

$$\Delta G = -0.0222 \text{ kWh}.$$

The reversible electromotive force for the process is

$$\Delta \psi = 0.828 \text{ V}.$$

The actual potential drop across the bipolar membrane is always higher because of irreversible effects due to the electrical resistance of the bipolar membrane. To minimize the irreversible energy losses in the bipolar membrane, its electrical resistance should be as low as possible.

A typical arrangement of an electrodialysis stack with a bipolar membrane as used for the production of an acid and a base is illustrated in Figure 20-4, which shows the production of an acid HX and a base MOH from a salt MX. A repeating cell unit in this arrangement consists of three individual cells. One hundred to 200 repeating cell units may be placed between two electrodes in industrial-size stacks.

The generation of acids and bases via an electrolysis process, however, requires considerably more energy. This is evident from the very nature of the process, which entails the coproduction of H_2 and chlorine or O_2 as indicated in Figure 20-3. This requires some additional energy input. The theoretical energy in electrolysis varies slightly, depending on the particular salt being processed, the concentration of the acid and base generated, and the temperature of operation. For production of 1 N acids and bases at 25°C, the theoretical free-energy change varies between 0.056 and 0.058 kWh. The electromotive force thus varies between 2.1 and 2.2 V at 25°C. For a practical electrolysis process, however, significantly larger amounts of energy must be provided to overcome the overvoltage for gas release at the electrodes.

The major problem with the use of bipolar membranes for the production of acids and bases from the corresponding salts is obtaining high-purity high-concentration products. However, electrolysis is very good in this respect.

NOTATION

General Notation

See the General Notation section at the beginning of this handbook.

Special Notation

a activity, mol/L^3, g-mol/L, or kg-mol/m^3
K_w ionic product of water, (mol/L^3)^2, (g-mol/L)^2, or (kg-mol/m^3)^2
z electrochemical valence of ions in the solution, dimensionless

Greek Letter

$\Delta\psi$ reversible electromotive force, V

Superscripts

i value in the interphase between cation- and anion-exchange membranes

o values in the solutions outside the membrane

REFERENCES

Ahlgren, R. M. 1972. Electro membrane processes for recovery of constituents from pulping liquors. In *Industrial Processing with Membranes,* ed. R. E. Lacey and S. Loeb, pp. 71–81. New York: John Wiley & Sons.

Bauer, B., F. Effenberger, and H. Strathmann. 1989. Anion-exchange membranes with improved alkaline stability. In *Proc. 6th Intl. Symp. on Synthetic Membranes in Science and Technology,* 4–8 September 1989, Tübingen, Germany, pp. 115–118.

Chlanda, F. P., L. T. C. Lee, and K. J. Liu. 1976. Bipolar membranes and method of making same. U.S. Patent 4,116,889.

Ionics Inc. 1988. Product bulletin. Watertown, MA.

Kedem, O., and Y. Maoz. 1976. Ion conducting spacer for improved electrodialysis. *Desalination* 19(1–3):465–470.

Korngold, E., K. Kock, and H. Strathmann. 1978. Electrodialysis in advanced waste water treatment. *Desalination* 24(1–3):129–139.

Strathmann, H. 1985. Electrodialysis and its application in the chemical process industry. *Sep. Purif. Methods* 14(1):41–66.

Tokuyama Soda. 1988. Product bulletin. Tokyo, Japan.

VI

Reverse Osmosis

21.	Introduction and Definitions	265
22.	Theory	269
23.	Design	281
24.	Selected Applications	312
25.	Cost Estimates	355

21

Introduction and Definitions

Dibakar Bhattacharyya and Michael E. Williams
University of Kentucky

INTRODUCTION
MEMBRANE PROCESS
  DESCRIPTIONS AND
  TERMINOLOGY

DRIVING FORCE
NOTATION
REFERENCES

INTRODUCTION

Reverse osmosis (RO) technology has grown extensively in recent years; many new types of membranes are now available, leading to an increase in various applications. Improvements have been made in membrane materials making them more pH, temperature, and chlorine resistant than the traditional cellulose acetate membranes. The ability of membranes to separate simultaneously, or selectively, organic and inorganic solutes from aqueous systems without phase change offers substantial energy savings and flexibility in the design of separation processes. The industrial development of noncellulosic, thin-film composite (TFC) membranes has provided better flux performance and enhanced separations of organics under lower operating pressures than those obtained with cellulosic membranes.

The ideal reverse osmosis membrane should be resistant to chemical and microbial attack, and the separation and mechanical characteristics should not change after long-term operation. Commercial RO membranes are either asymmetric (one polymer) or thin-film composite structures. Applications of RO membranes include treatment of water and hazardous wastes, separation processes in the food, beverage, and pulp and paper industries, recovery of organic and inorganic materials from chemical processes, etc. Extensive reviews of the development of membrane and RO technology, membrane materials, transport fundamentals, and applications can be found in Parekh (1988), Rautenbach and Albrecht (1989), Sirkar and Lloyd (1988), Sourirajan and Matsuura (1985), Lee (1987), Lloyd (1985), Belfort (1984), Kesting (1985), Drioli and Nakagaki (1986), Cecille and Toussaint (1989), Koros et al. (1988), and Riley (1990).

MEMBRANE PROCESS DESCRIPTIONS AND TERMINOLOGY

A reverse osmosis membrane acts as a barrier to flow, allowing selective passage of a particular species (solvent) while other species (solutes) are retained partially or completely. Solute separation and permeate water (solvent) flux characteristics of membranes depend on the membrane material selection, the preparation procedures, and the structure of the membrane barrier layer (Lloyd 1985; Sourirajan and Matsuura 1985). Primary separation of solutes occurs at the thin-film (skin) barrier layer. Two situations may arise depending on the extent of solvent-membrane affinity and solute-membrane affinity: solvent preferential sorption and solute preferential sorption.

Osmosis is a natural phenomenon in which water passes through a semipermeable (no solute flow) membrane from the side with lower solute concentration to the higher solute concentration side until equilibrium of solvent (water) chemical potential is restored. At equilibrium the pressure difference between the two sides of the membrane is equal to the osmotic pressure difference. To reverse the flow of water, a pressure difference greater than the osmotic pressure difference is applied; as a result, separation of water from solutions becomes possible. This phenomenon is termed *reverse osmosis*. It has also been termed *hyperfiltration*.

Reverse osmosis processes can be classified into three types: high-pressure RO (5.6 to 10.5 MPa, such as seawater desalination), low-pressure RO (1.4 to 4.2 MPa, such as brackish water desalination), and nanofiltration or "loose RO" (0.3 to 1.4 MPa, such as partial demineralization or 0 to 20% NaCl rejection). High-pressure and low-pressure RO are typically used for very high rejection of inorganics (95 to 99.9% NaCl rejection) and for moderate to high rejections of low molecular weight organics, respectively. Organic rejection depends on membrane polymer types and structures and membrane/solute interactions.

The terminology used for pressure-driven membrane processes (such as RO) was recently reviewed by Gekas (1988). The important operating variables for RO are feed flow rate and concentrations of dissolved solutes, types of solutes, transmembrane pressure (Δp), temperature (T), pH, and concentration of suspended solids (if any). Any membrane process produces two streams: the permeate (portion of the feed passing through the membrane) and the retentate or concentrate (portion of the feed not passing through the membrane). A schematic of an RO process is shown in Figure 21–1. The process is generally evaluated in terms of three parameters: observed solute rejection R, water flux J_w, and water recovery r. The definitions of these and other parameters are given below:

$$R = \text{observed solute rejection} = 1 - \frac{c''}{c'}. \quad (21\text{--}1)$$

With significant water recovery, R can also be defined as

$$R = 1 - \frac{c''}{c'_b} \quad \text{or} \quad 1 - \frac{c''}{c'_c}, \quad (21\text{--}2)$$

R' = intrinsic membrane solute rejection

$$= 1 - \frac{c''}{c'_W}, \quad (21\text{--}3)$$

J_w = water flux

$$= \frac{\text{volumetric, mass, or molar permeation rate}}{(\text{membrane area})}, \quad (21\text{--}4)$$

$r = \dfrac{\text{solvent or water recovery}}{\text{for batch system}} = \dfrac{\Sigma J_w A_x \Delta t}{V_o}, \quad (21\text{--}5)$

$r = \dfrac{\text{solvent or water recovery}}{\text{for continuous system}} = \dfrac{J_w A_x}{Q_i}. \quad (21\text{--}6)$

DRIVING FORCE

The thermodynamic driving forces involved in solute and solvent transport through RO membranes are the respective chemical potential gradients across the membrane: $-\nabla \mu_s$ (solute chemical potential gradient) and $-\nabla \mu_w$ (water chemical potential gradient). Since these are not measurable quantities, driving forces for RO

Introduction and Definitions

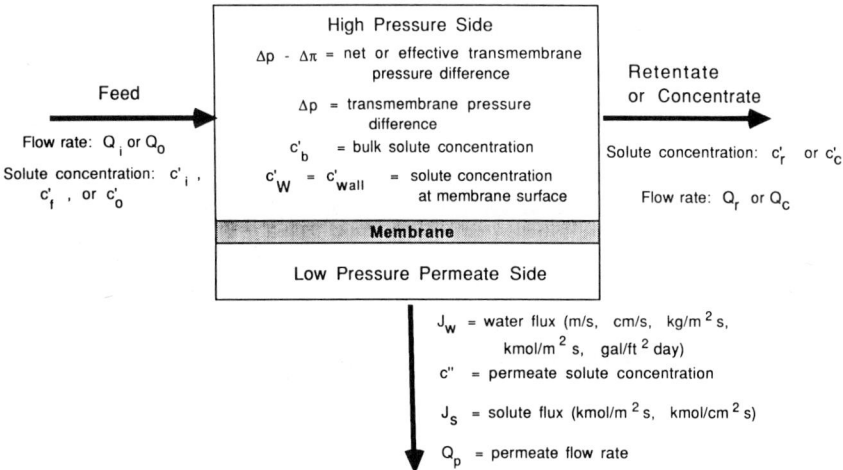

FIGURE 21-1. Reverse osmosis process schematic.

are generally related to differences in concentration (chemical activity) and pressure across the membrane. When the membrane polymer contains fixed charged groups (such as sulfonic acid, carboxylic acid, etc.), high salt separations can be achieved, even with porous membranes, by means of Donnan exclusion. For example, with negatively charged membranes, the rejection follows the order: $Na_2PO_4 > Na_2SO_4 > NaCl$. Commercially available new-generation nanofiltration (low-pressure) membranes usually have negatively charged groups in polymers and are being used for selective separations and/or partial water demineralization. Salt rejection by the Donnan exclusion mechanism is always decreased by increasing salt concentration.

For high-rejection membranes, the influence of pressure driving force on solute flux is very small and is often neglected. For imperfect (rejection R less than 1.0) membranes, a coupling coefficient (the Staverman reflection coefficient, σ) is often incorporated in the transport equations. The reflection coefficient is defined as: $\sigma = -L_{\pi p}/L_p$, where L_p = hydraulic permeability of membrane and $L_{\pi p}$ = coupling coefficient (Pusch 1986; Kedem 1972). The reflection coefficient can be calculated by measuring the pressure difference at which the volumetric permeation flux (usually called volumetric flux) is zero for a given osmotic pressure difference across the membrane:

$$\sigma = \frac{\Delta p}{\Delta \pi} \text{ at volumetric } J_w = 0. \quad (21\text{-}7)$$

The magnitude of σ is ≤ 1, and is usually positive. A value of $\sigma = 1$ indicates that the membrane is completely impermeable to solute. The limiting salt rejection (often defined as R_∞) for a given membrane material is equal to σ.

NOTATION

General Notation

See the General Notation section at the beginning of this handbook.

Special Notation

c'_c	solute concentration in the concentrate, M/L^3
c'_r	solute concentration in the retentate, M/L^3
c'_W	feed solute concentration at the membrane surface, M/L^3
J_w	water flux, $M/L^2 t$, $mol/L^2 t$, or L/t
L_p	hydraulic permeability of membrane, L/tp

$L_{\pi p}$ coupling coefficient, L/tp
R observed solute rejection, dimensionless
R' intrinsic membrane solute rejection, dimensionless

Greek Letter

σ Staverman reflection coefficient, dimensionless

REFERENCES

Belfort, G. 1984. *Synthetic Membrane Processes: Fundamentals and Water Applications*. New York: Academic Press.

Cecille, L., and J. Toussaint. 1989. *Future Industrial Prospects of Membrane Processes*. New York: Elsevier.

Drioli, E., and M. Nakagaki. 1986. *Membrane and Membrane Processes*. New York: Plenum Press.

Gekas, V. 1988. Terminology for pressure-driven membrane operations. *Desalination* 68:77.

Kedem, O. 1972. Water and salt transport in hyperfiltration. Chap. 1 in *Reverse Osmosis Membrane Research*, ed. H. Lonsdale and H. Podall. New York: Plenum Press.

Kesting, R. 1985. *Synthetic Polymeric Membranes: A Structural Perspective*. New York: Wiley-Interscience.

Koros, W., G. Fleming, S. Jordan, T. Kim, and H. Hoehn. 1988. Polymeric membrane materials for solution-diffusion based permeation separations. *Prog. Polym. Sci.* 13:339.

Lee, E. 1987. Membranes, synthetic, applications. *Encycl. Phys. Sci. Technol.* 8:20.

Lloyd, D. 1985. Material science of synthetic membranes. *ACS Symp. Series No. 269*. Washington, DC: American Chemical Society.

Parekh, B. 1988. *Reverse Osmosis Technology*. New York: Marcel Dekker.

Pusch, W. 1986. Measurement techniques of transport through membranes. *Desalination.* 59:105.

Rautenbach, R., and R. Albrecht. 1989. *Membrane Processes*. New York: John Wiley & Sons.

Riley, R. 1990. Reverse osmosis. In *Membrane Separation Systems*. Report DOE/ER/30133-H1, P5-1, U.S. Department of Energy.

Sirkar, K., and D. Lloyd. 1988. New membrane materials and processes for separation. *AIChE Symp. Series No. 261*. New York: American Institute of Chemical Engineers.

Sourirajan, S., and T. Matsuura. 1985. *Reverse Osmosis/Ultrafiltration Principles*. Ottawa, Canada: National Research Council of Canada.

22
Theory

Dibakar Bhattacharyya and Michael E. Williams
University of Kentucky

SOLUTION-DIFFUSION MODEL
SOLUTION-DIFFUSION-
  IMPERFECTION MODEL
PREFERENTIAL
  SORPTION–CAPILLARY FLOW
  MODEL
SURFACE FORCE–PORE FLOW
  MODEL
IRREVERSIBLE
  THERMODYNAMICS MODEL

DONNAN EQUILIBRIUM MODEL
EXTENDED NERNST-PLANCK
  MODEL
EFFECT OF OPERATING
  VARIABLES
CONCENTRATION POLARIZATION
NOTATION
REFERENCES

Various reviews (Pusch 1986; Soltanieh and Gill 1981; Mazid 1984; Dickson 1988; Dresner and Johnson 1980; Jonsson 1980; Rautenbach and Albrecht 1989; and Sourirajan and Matsuura 1985) on reverse osmosis (RO) transport mechanisms and models have been reported in the literature. Several models have been developed to describe solute and solvent fluxes through RO membranes. The transport models can be divided into three types: nonporous or homogeneous membrane models (solution-diffusion, extended solution-diffusion, and solution-diffusion-imperfection models), pore-based models (preferential sorption–capillary flow, finely porous, and surface force–pore flow models), and irreversible thermodynamics phenomenological models (such as Kedem-Katchalsky and Spiegler-Kedem models). Most models for reverse osmosis membranes assume diffusion or pore flow through the membrane while charged membrane theories include electrostatic effects. A model that includes prediction of reverse osmosis rejection to multi-ionic salt solutions was reported by Brusilovsky and Hasson (1989). For nanofiltration membranes, which are often negatively charged, Donnan exclusion models and the extended Nernst-Planck model can be used to determine solute fluxes.

Reverse osmosis membranes, in practice, have either an asymmetric or a thin-film composite structure in which a thin skin at the top

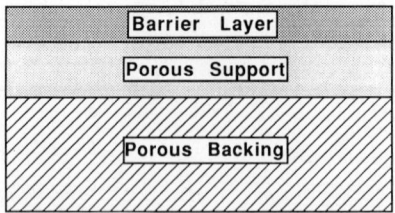

FIGURE 22–1. Thin-film composite membrane structure.

facing the feed solution acts as the selective layer. The pressure drop in the porous backing of such membranes is usually very small; further, this backing has almost no influence on the solute rejection properties of the membrane. Transport and rejection models of RO membranes, therefore, focus only on the thin membrane skin or surface layer. Figure 22–1 provides a schematic of such a membrane structure.

SOLUTION-DIFFUSION MODEL

The solution-diffusion model was originally developed by Lonsdale, Merten, and Riley (1965). This model assumes that both the solute and solvent dissolve in the nonporous and homogeneous surface layers of the membrane and then each diffuses across it in an uncoupled manner due to its own chemical potential gradient; this gradient is the result of concentration and pressure differences across the membrane. Differences in the solubilities (partition coefficients) and diffusivities of the solute and solvent in the membrane phase are important in this model since these differences strongly influence the fluxes through the membrane.

According to this model, based on pure diffusion, water (solvent) flux through the membrane is given by

$$J_w = \frac{P_w}{l}(\Delta p - \Delta \pi), \quad (22\text{–}1)$$

where $P_w = D_w c_{\text{water}} \bar{V}_w / R_g T$, Δp is the pressure difference across the membrane, and $\Delta \pi$ is the osmotic pressure difference between the feed and permeate at the membrane surfaces. The solute flux J_s according to this model is given by

$$J_s = \frac{P_s}{l}(c'_W - c''), \quad (22\text{–}2)$$

where $P_s = D_{sm} K_s$, c'_W is the concentration of solute in the feed solution at the membrane surface, and c'' is the concentration of solute in the permeate solution. In these equations, D_w and D_{sm} represent diffusivities of water and solute, respectively, c_{water} is the concentration of water in the membrane, $\bar{V}_w$ is the partial molar volume of water, R_g is the gas constant, T is the absolute temperature, K_s is the solute distribution coefficient, and l is the membrane thickness. In Eq. (22–1) the quantity P_w/l is frequently referred to as A, the pure water permeability constant; the quantity P_s/l in Eq. (22–2) is also often referred to as B, the solute transport parameter. These equations predict that the solute flow through the membrane is independent of water flow; rejection by the membrane depends on the solute distribution coefficient between the solution and membrane phase and the solute diffusivity in the membrane phase. The equations also predict that feeds with higher osmotic pressures result in less flux through the membrane. Using Eqs. (22–1) and (22–2), intrinsic membrane solute rejection as a function of pressure can be expressed as

$$R' = \left(1 + \frac{P_s/l}{J_w}\right)^{-1} = \left[1 + \left(\frac{P_s}{P_w}\right)\left(\frac{1}{\Delta p - \Delta \pi}\right)\right]^{-1}.$$
$$(22\text{–}3)$$

This model is applicable to membranes with relatively small water content.

The regular solution-diffusion model presented above shows that as $\Delta p \to \infty$, rejection is equal to unity. Jonsson (1980) modified the solution-diffusion model to include the effect of pressure on solute transport. Also, the model cannot characterize negative solute rejections

($R < 0$). Pusch (1986) developed an extended solution-diffusion model to predict negative rejections for systems such as phenol separation with cellulose acetate membranes. The extended model includes two transport parameters.

SOLUTION-DIFFUSION-IMPERFECTION MODEL

The solution-diffusion-imperfection model (Sherwood, Brian, and Fisher 1967) is an extension of the solution-diffusion model to include pore flow as well as diffusion of solute and solvent through the membrane. This model recognizes that imperfections (pores) exist on the membrane surface through which solute and solvent can flow. The water flux through the membrane is given by

$$J_w = \frac{P_w}{l}(\Delta p - \Delta \pi) + \frac{P_3}{l}\Delta p, \quad (22\text{-}4)$$

where P_3/l is a coupling coefficient. The first term in Eq. (22–4) accounts for diffusion as in the solution-diffusion model, while the second term is the pore flow contribution to the water flux. The solute flux is given by

$$J_s = \frac{P_2}{l}(c_W' - c'') + \frac{P_3}{l}\Delta p c_W', \quad (22\text{-}5)$$

where P_2/l is a solute permeability coefficient equivalent to P_s/l as in Eq. (22–2); the second term in Eq. (22–5) accounts for pore flow of the solute through the membrane. Combining Eqs. (22–4) and (22–5) results in the following for rejection:

$$R' = \left[1 + \left(\frac{P_2}{P_w}\right)\left(\frac{1}{\Delta p - \Delta \pi}\right) + \left(\frac{P_3}{P_w}\right)\left(\frac{\Delta p}{\Delta p - \Delta \pi}\right)\right]^{-1}. \quad (22\text{-}6)$$

The concentration and pressure dependence of this three-coefficient model is a serious limitation to its application to design estimation (Soltanieh and Gill 1981).

PREFERENTIAL SORPTION–CAPILLARY FLOW MODEL

The preferential sorption–capillary flow (PSCF) model proposed by Sourirajan (Sourirajan 1970; Sourirajan and Matsuura 1985) assumes that the mechanism of separation is determined by both surface phenomena and fluid transport through the pores. In contrast to the solution-diffusion model, the membrane is assumed to be microporous. The model states that the membrane barrier layer has chemical properties such that is has a preferential sorption for the solvent or preferential repulsion for the solutes of the feed solution. As a result, a layer of almost pure solvent is preferentially sorbed on the surface and in the pores of the membrane. Solvent from this layer is then forced through the membrane capillary pores under pressure.

The water flux according to this model is given by

$$N_w = A\{\Delta p - [\pi(x_s') - \pi(x_s'')]\}, \quad (22\text{-}7)$$

where A is the pure water permeability constant, Δp is the applied pressure difference, $\pi(x_s)$ is the osmotic pressure of a solution with solute mole fraction x_s, and x_s' and x_s'' are the mole fractions of solute in the feed and permeate solutions, respectively. The solute flux is expressed by

$$N_s = \frac{c_T K_s D_{sm}}{l}(x_s' - x_s''), \quad (22\text{-}8)$$

where c_T is the total molar concentration, K_s is the solute distribution coefficient, D_{sm} is the diffusion coefficient of the solute in the membrane, and l is the membrane thickness.

SURFACE FORCE–PORE FLOW MODEL

The surface force–pore flow (SFPF) model (Matsuura and Sourirajan 1981; Sourirajan and

Matsuura 1985), which is a quantitative expression of the PSCF model, allows characterization and specification of a membrane precisely as a function of pore size distribution (or average pore size) along with a quantitative measure of surface forces arising between solute-solvent and the membrane wall inside the transport corridor. In this model, the following assumptions are made:

1. The transport of solute and solvent through the membrane pore is governed by interaction forces, friction forces, and driving forces (i.e., diffusive force) due to gradients in chemical potential of the solute and the solvent.
2. The pores of the membrane are assumed to be cylindrical.
3. A molecule thick layer of pure water is considered preferentially sorbed in the region immediately adjacent to the membrane wall.
4. A solute potential field, which controls the radial distribution of solute, exists within the pore.

A brief description of the mathematical formulation is discussed below. The solvent velocity profile equation in the pore written in dimensionless form is

$$\frac{d^2 v_p^+(r_p^+)}{d(r_p^+)^2} + \frac{1}{r_p^+} \frac{dv_p^+(r_p^+)}{dr_p^+} + \frac{\beta_2}{\beta_1}$$
$$+ \frac{1}{\beta_1}[c_s(r_p^+) - 1]\{1 - \exp[\bar{\Phi}(r_p^+)]\}$$
$$- \frac{[b(r_p^+) - 1]c_s(r_p^+)v_p^+(r_p^+)}{\beta_1} = 0,$$

(22–9)

with the following boundary conditions:

$$\frac{dv_p^+(r_p^+)}{dr_p^+} = 0 \quad \text{at } r_p^+ = 0, \quad (22\text{–}10)$$

$$v_p^+(r_p^+) = 0 \quad \text{at } r_p^+ = 1, \quad (22\text{–}11)$$

where $v_p^+(r_p^+)$ is the dimensionless velocity and $c_s(r_p^+)$ is defined as

$$c_s(r_p^+) = \frac{\exp[v_p^+(r_p^+)]}{1 + (b(r_p^+)/\exp[-\Phi(r_p^+)])\{\exp[v_p^+(r_p^+)] - 1\}}.$$

(22–12)

The intrinsic membrane solute rejection R' is given by

$$R' = 1 - \frac{\int_0^1 c_s(r_p^+)v_p^+(r_p^+)r_p^+ \, dr_p^+}{\int_0^1 v_p^+(r_p^+)r_p^+ \, dr_p^+}.$$

(22–13)

The above model assumes that the transport of water and solute takes place through cylindrical pores of radius r_p running across the active skin layer of thickness l. In the SFPF model, it is also assumed that there is an area into which the center of the water molecule cannot enter due to its collision with the membrane pore wall; thus two radius values, r_a and r_p, are defined:

$$r_p = r_a + d_w, \quad (22\text{–}14)$$

where $d_w = 0.87$ Å.

Expressions for the potential function accounting for the force exerted on the solute molecule by the pore wall are given in dimensionless form for an ionized solute by

$$\bar{\Phi}(r_p^+) = \frac{\bar{A}/r_a}{(r_p/r_a) - r_p^+}, \quad (22\text{–}15)$$

where $\bar{A}$ is a measure of the resultant electrostatic force; for a nonionized organic solute

$$\bar{\Phi}(r_p^+) = -\frac{\bar{B}/r_a^3}{[(r_p/r_a) - r_p^+]^3}, \quad (22\text{–}16)$$

where $\bar{B}$ is a measure of the resultant short-range van der Waals force. The friction parameter, b, in Eq. (22–9) is defined as the ratio of the frictional force acting on the solute moving in the membrane pore to the frictional force of the bulk solution. The parameter b is a function of $\bar{d}/r_p$, where $\bar{d}$ is the distance characteristic of the steric hindrance (Sourirajan and Matsuura 1985); $\bar{d}$ can be approximated as the Stokes radius of the solute.

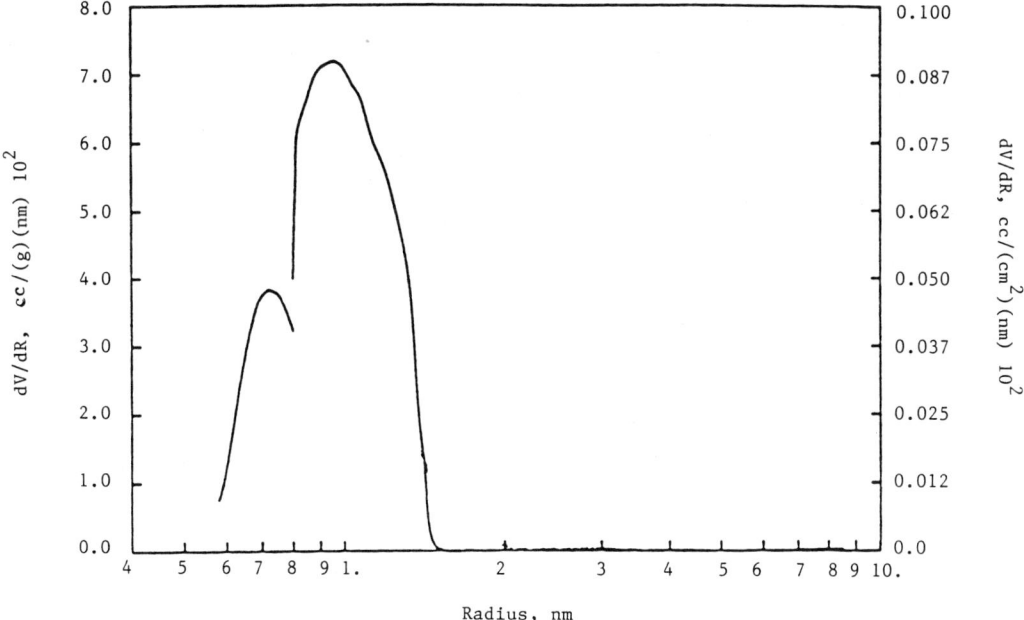

FIGURE 22–2. Pore volume distribution of a low-pressure RO membrane (FT30); here R stands for pore radius.

A more realistic procedure is to express the transport of material through a distribution of pores r_{p_i}, each having a frequency of distribution $Y_i(r_p)$. More generally, the rejection R' can be written in terms of pore size distribution as

$$R' = 1 - \frac{\sum_{i=1}^{m} Y_i(r_p)\left[\left(\int_0^1 c_s(r_p^+)v_p^+(r_p^+)r_p^+ \, dr_p^+\Big|_{r_{p_i}}\right)\Delta r_{p_i}\right]}{\sum_{i=1}^{m} Y_i(r_p)\left[\left(\int_0^1 v_p^+(r_p^+)r_p^+ \, dr_p^+\Big|_{r_{p_i}}\right)\Delta r_{p_i}\right]}$$

(22–17)

Bhattacharyya et al. (1986) have reported a numerical solution technique for the above equations by using a spline collocation package. The solution technique required one piece of experimental RO data for a particular solute and average pore diameter or pore size distribution data. Typical pore size distribution data of a FilmTec FT30 RO membrane obtained by N_2 and CO_2 adsorption-desorption data are shown in Figure 22–2. The numerical results for various nonionized organics (dilute solutions) agreed very well with the experimental RO results. Figure 22–3 shows some experimental results of rejection versus pressure for selected phenolics along with the numerical solutions of the SFPF model.

Mehdizadeh and Dickson (1989) developed an improved version of the original SFPF model and provided numerical results obtained by orthogonal collocation. This version included modified forms of boundary conditions, potential function term, material balance on the solute in the pore, and osmotic effect corrections.

IRREVERSIBLE THERMODYNAMICS MODEL

The thermodynamics of irreversible processes indicate that the flow of each component in a solution is related to the flows of other components. Kedem and Katchalsky (1958) developed relationships for flow in membranes using phenomenological equations of transport. Pusch (1986) and Soltanieh and Gill (1981) have also provided excellent reviews of these models.

For dilute solutions the transport equations

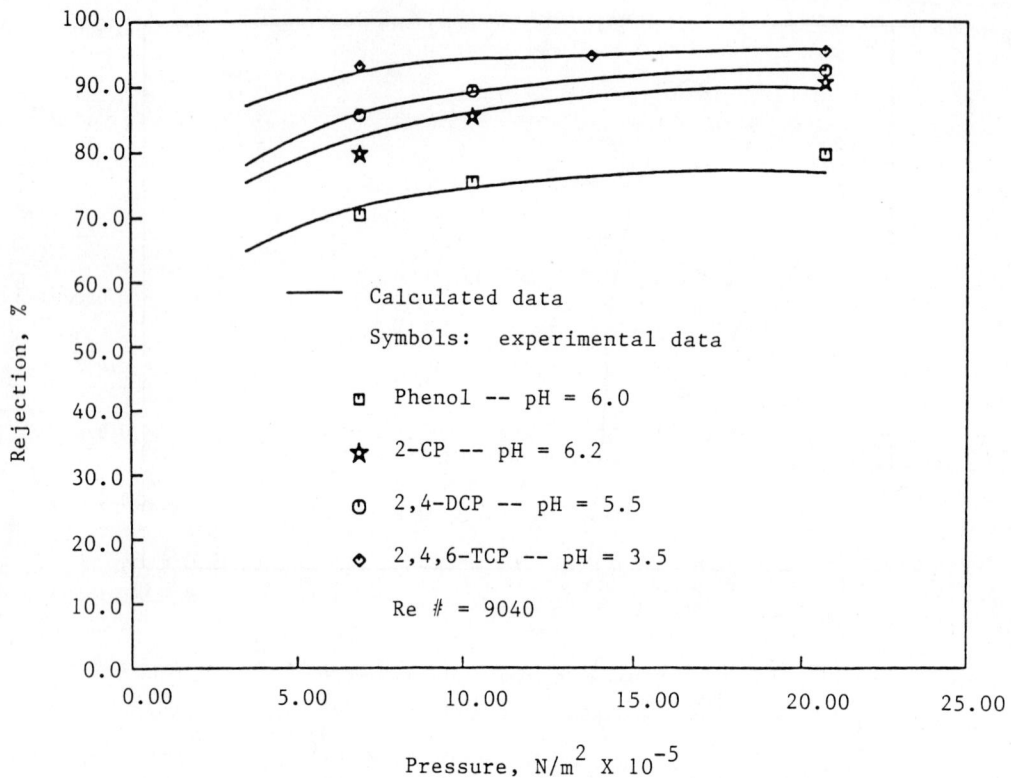

FIGURE 22–3. Effect of pressure on experimental and calculated rejections of single-component solutions of phenol, 2-chlorophenol (2-CP), 2,4-dichlorophenol (2,4-DCP), and 2,4,6-trichlorophenol (2,4,6-TCP).

can be represented by

$$J_w = L_p(\Delta p - \sigma \Delta \pi), \quad (22\text{--}18)$$

$$J_s = (c_m)_{av}(1 - \sigma)J_w + \frac{P_s}{l}(c'_W - c''). \quad (22\text{--}19)$$

In Eq. (22–19), $(c_m)_{av}$ is represented by

$$(c_m)_{av} = (c'_W - c'')/\ln(c'_W/c''). \quad (22\text{--}20)$$

For the case of $\sigma = 1$, that is, in the absence of solute and solvent flow coupling, Eqs. (22–18) and (22–19) reduce to the solution-diffusion model transport equations.

DONNAN EQUILIBRIUM MODEL

A dynamic equilibrium occurs when a charged membrane is placed in a salt solution. The counter-ion (opposite in charge of the fixed charge in the membrane) concentration is higher while the co-ion (same sign charge as the fixed membrane charge) concentration is lower in the membrane phase than in the bulk solution, creating a Donnan potential. This potential prevents the diffusion of the counter-ion from the membrane phase to the bulk solution and the diffusion of the co-ion from the bulk solution to the membrane phase. A potential also occurs when an applied pressure gradient forces water flow through the membrane. The effect of the Donnan potential is to repel the co-ion from the membrane, and because of electroneutrality requirements, the counter-ion is also rejected. The fixed charged groups in most charged RO membranes are carboxylic and sulfonic groups.

For the salt $M_{z_y}Y_{z_m}$, which ionizes to M^{z_m+} and Y^{z_y-}, at equilibrium the salt distribution coefficient K^* is given by

$$K^* = \left[\frac{c_{y(m)}}{c_y}\right] = \left[z_y^{z_y}\left(\frac{c_y}{c_m^*}\right)^{z_y}\left(\frac{\gamma}{\gamma_m}\right)^{z_y+z_m}\right]^{1/z_m}, \quad (22\text{-}21)$$

where z_i represents the charge of species i, c_y and $c_{y(m)}$ are the concentrations of co-ion Y in the bulk solution and membrane phase, respectively, γ and γ_m are activity coefficients, and c_m^* is the charge capacity of the membrane. This equation applies to negatively charged membranes useful for nanofiltration. The rejection is then approximated as

$$R' = 1 - K^*. \quad (22\text{-}22)$$

The model (Bhattacharyya and Cheng 1986) predicts that the rejection is a function of the membrane charge capacity, the feed solute concentration, and the charge of the ions. While it does provide a qualitative description of solute rejection, this model does not take into account diffusive and convective fluxes; these are also important in the charged membrane process.

EXTENDED NERNST-PLANCK MODEL

The extended Nernst-Planck equation has been used by Lakshminarayanaiah (1969) and Dresner (1972) to describe the flux of ions through a charged membrane. The equation is

$$J_j = J_w c_{j(m)} + z_j c_{j(m)} \frac{FE}{R_g T} - D_{j(m)} \frac{dc_{j(m)}}{dx}$$

$$- c_{j(m)} D_{j(m)} \frac{d(\ln \gamma_{j(m)})}{dx}, \quad (22\text{-}23)$$

where J_j is the flux of ion j, $c_{j(m)}$ is the concentration of ion j in the membrane, $D_{j(m)}$ is the diffusivity of ion j in the membrane, z_j is the charge of ion j, $\gamma_{j(m)}$ is the activity coefficient of ion j in the membrane, and E and F are the Donnan potential and Faraday's constant, respectively. In Eq. (22-23) the first term represents solute flux due to convection, the second term accounts for flux due to the Donnan potential, and the last two terms describe the salt flux due to diffusion. As with the Donnan equilibrium model, this model predicts that the solute rejection is a function of feed concentration and charge of the ion. However, the Nernst-Planck equation includes the effects of convective and diffusional fluxes, which can be important.

EFFECT OF OPERATING VARIABLES

Intrinsic membrane solute rejections (and hence product quality) and water flux are functions of concentration, pressure, temperature, and pH (for ionizable organic solutes). Selected generalized curves (DOE 1990) illustrating the effects of some of these variables are shown in Figure 22-4. Water flux, for example, is shown to increase linearly with applied pressure; this behavior is predicted by several of the transport models. Temperature also affects water flux because increases in temperature result in in-

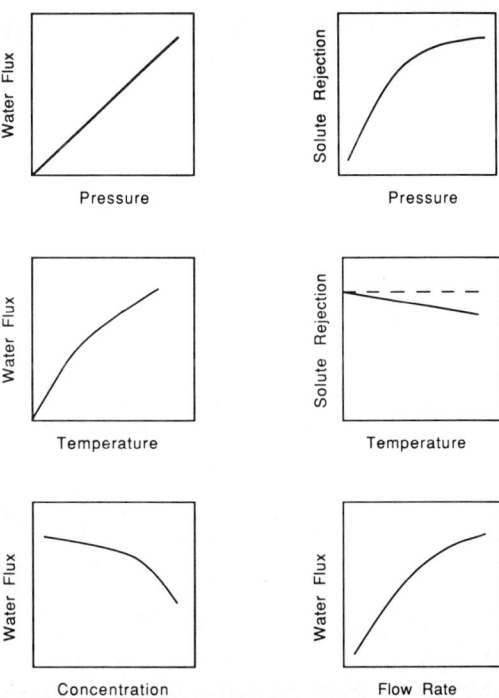

FIGURE 22-4. Curves showing effect of variables on RO membranes (DOE 1990).

creases in osmotic pressure and solute and solvent permeability; the increase in solvent permeability results in an increase in water flux. This water flux increase can often be described by an Arrhenius temperature dependence (Dickson 1988; Mehdizadeh, Dickson, and Eriksson 1989) on pure water permeability constant. Pure water flux change with temperature can also be predicted by water viscosity changes. The water flux dependence on solute concentration and feed water flow rate is also shown in the figure. Higher concentration solutions have lower water fluxes because of the higher osmotic pressure of these solutions; this behavior is predicted by several of the models. Water flux is greater at high flow rates (i.e., high water velocity over membrane) since this minimizes concentration polarization (see the Concentration Polarization section).

Solute rejection generally increases with pressure up to an asymptotic value [i.e., as predicted by Eq. (22–6)] as shown in Figure 22–4. For organic solutes that have strong interaction with membrane polymers, rejection often decreases with pressure, which can be adequately predicted by the SFPF model. Rejection of ionizable organic solutes is a strong function of pH. For example, Figure 22–5 shows typical rejection of propionic acid (pK_a = 4.9) with a low-pressure RO membrane. The dramatic effect of ionization is demonstrated in this figure. Solute rejection generally decreases with increasing temperature; however, rejection can remain virtually constant over a wide temperature range for some membranes and solutes.

Some organic solutes have been found to adsorb on the membrane surface and in the pores. The adsorbed solute may cause large drops in water flux due to a reduction in the path available for water transport through the membrane. The drop in flux caused by organic adsorption has been modeled by Williams (1989) by including an adsorption resistance term in the expression for J_w:

$$J_w = \frac{1}{R_m + R_{ads}} (\Delta p - \Delta \pi), \quad (22\text{–}24)$$

where R_m is the membrane resistance ($R_m = 1/A$) and R_{ads} is the resistance due to solute adsorption on the membrane. For the case of no solute adsorption, Eq. (22–24) reduces to Eq. (22–1).

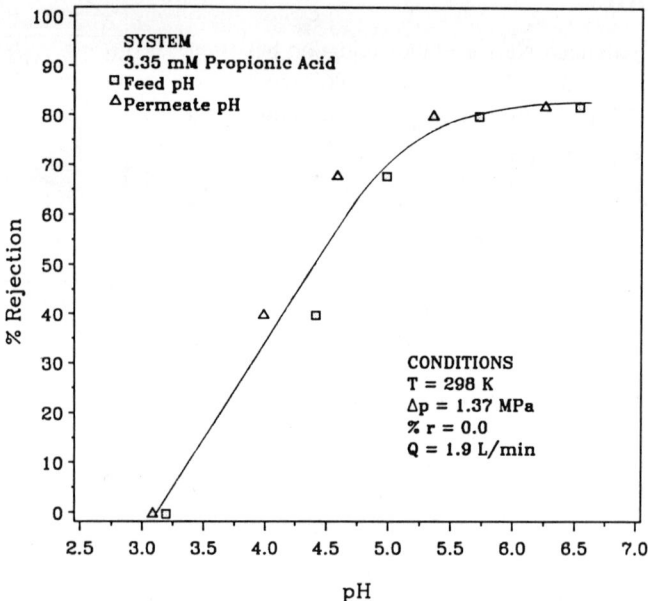

FIGURE 22–5. Effect of pH on rejection of propionic acid by a nanofiltration membrane (NF40 spiral-wound module).

CONCENTRATION POLARIZATION

The term *concentration polarization* is used to describe the accumulation of membrane rejected solute at the membrane surface where the solute concentration is much higher than that of the bulk feed solution (Figure 22–6). As water passes through the membrane, the convective flow of solute to the membrane surface is much larger than the diffusion of the solute back to the bulk feed solution; as a result, the concentration of the solute at the membrane surface increases. Possible negative effects of concentration polarization include:

1. Decrease in water flux due to an increase in osmotic pressure at the membrane surface
2. Increase in solute flux through the membrane
3. Precipitation of the solute if the surface concentration exceeds its solubility limit, leading to plugging of the membrane pores and hence reduced water flux
4. Changes in separation properties of the membrane
5. Fouling (which is the accumulation of material at the surface that plugs the membrane pores and so reduces water flux) could occur.

The extent of concentration polarization can be reduced by promoting good mixing of the bulk feed solution with the solution near the membrane surface. This can be done by modifying the membrane module to enhance the mixing, by including turbulence promoters in the feed channel, or by increasing the feed flow rate to increase the axial velocity and so promote turbulent flow. Reviews of concentration polarization and fouling have been reported by Matthiasson and Sivik (1980) and Gekas and Hallstrom (1987).

Concentration polarization greatly complicates the modeling of membrane systems because experimental determination of the wall concentration is very difficult. For very high flow rates, enough mixing occurs, and the wall concentration can be assumed to equal the bulk concentration; but at lower flow rates this assumption could cause substantial error. To calculate the wall concentration exactly, the Navier-Stokes diffusion-convection equation must be solved. For flow between parallel plate membranes, the equation becomes

$$u \frac{\partial c'}{\partial z} + v \frac{\partial c'}{\partial y} - D_{sw}\left(\frac{\partial^2 c'}{\partial z^2} + \frac{\partial^2 c'}{\partial y^2}\right) = 0, \quad (22\text{--}25)$$

with the boundary conditions:

$$c'(0,y) = c'_0, \quad (22\text{--}26)$$

$$\frac{\partial c'(z,0)}{\partial y} = 0, \quad (22\text{--}27)$$

$$D_{sw}\frac{\partial c'[z,\delta(z)]}{\partial y} = v_w\{c'[z,\delta(z)] - c''\}, \quad (22\text{--}28)$$

$$c'[z,\delta(z)] = c'_w = \frac{c''}{(1 - R')}. \quad (22\text{--}29)$$

No exact analytical solutions exist to this equation, but several approximate solutions have been developed.

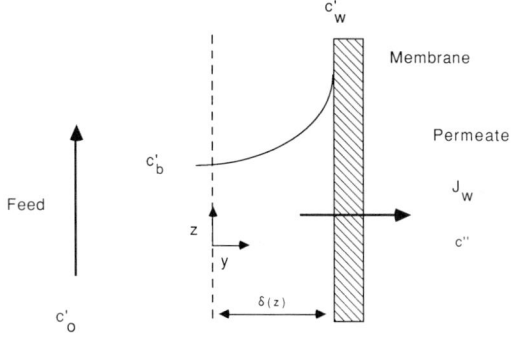

FIGURE 22–6. Concentration profile.

c'_0 = initial concentration
c'_b = bulk concentration
c'_w = wall concentration

The film theory developed by Brian (1966) is the most well-known technique. Brian assumed that the boundary layer is stagnant and its thickness does not change with channel length. Eq. (22–25) then becomes

$$v_w \frac{dc'}{dy} = D_{sw} \frac{d^2c'}{dy^2}, \quad (22\text{–}30)$$

which can be integrated to obtain

$$\frac{c'_W(z) - c''}{c'_0 - c''} = \exp\left(\frac{v_w \delta}{D_{sw}}\right), \quad (22\text{–}31)$$

where δ is the film thickness and can be calculated using the relation

$$\delta = \frac{d_h}{\text{Sh}}. \quad (22\text{–}32)$$

Gekas and Hallstrom (1987) have reviewed correlations of Sherwood numbers in membrane modules. For turbulent flow, the correlation

$$\text{Sh} = 0.023 \, \text{Re}^{0.8} \, \text{Sc}^{0.33} \quad (22\text{–}33)$$

is most often used. For laminar flow the Sherwood number may be calculated from

$$\text{Sh} = 1.86 \, (\text{Re} \, \text{Sc} \, d_h/L)^{0.33}. \quad (22\text{–}34)$$

Sherwood et al. (1965) obtained an asymptotic solution, valid in the turbulent regime only, by assuming that the film layer is laminar:

$$\frac{c'_W}{c'_0} = \frac{\exp(2v_w \text{Sc}^{2/3}/fu_0)}{R' + (1 - R')\exp(2v_w \, \text{Sc}^{2/3}/fu_0)}, \quad (22\text{–}35)$$

where f is the Fanning friction factor calculated from

$$f = 0.08 \, \text{Re}^{-1/4}. \quad (22\text{–}36)$$

Sherwood et al. (1965) also provide concentration polarization equations for other conditions. For laminar flow between two flat membranes, the following equations are used:

$$\frac{c'_W}{c'_b} = 1.536 \, (\epsilon_0)^{1/3} + 1 \quad \text{for } \epsilon_0 \ll 1, \quad (22\text{–}37)$$

$$\frac{c'_W}{c'_b} = \epsilon_0 + 5\left\{1 - \exp\left[-\left(\frac{\epsilon_0}{3}\right)^{1/2}\right]\right\}$$
$$+ 1 \quad \text{near channel entrance}, \quad (22\text{–}38)$$

$$\frac{c'_W}{c'_b} = \frac{1}{3\sigma_0^2} + 1 \quad \text{far downstream}, \quad (22\text{–}39)$$

where

$$\sigma_0 = \frac{D_{sw}}{v_w h}, \quad (22\text{–}40)$$

$$\epsilon_0 = \frac{v_w L}{3u_0 \sigma_0}. \quad (22\text{–}41)$$

These equations assume 100% solute rejection and a constant solvent flux throughout the membrane module.

NOTATION

General Notation

See the General Notation section at the beginning section of this handbook.

Special Notation

A	pure water permeability constant, L/tp
$\underline{A}$	constant characterizing electrostatic repulsion force, L
b	frictional function
B	solute transport parameter, $M/L^2 tp$
$\underline{B}$	constant characterizing van der Waals interaction force, L^3
c_m^*	charge capacity of membrane, mol/L^3
c_s	dimensionless solute concentration normalized with respect to the feed solute concentration at the membrane surface
c'_W	feed solute concentration at the membrane surface M/L^3
d_w	diameter of water molecule, L
$\underline{d}$	constant characterizing steric repulsion at membrane interface, L
E	membrane Donnan potential, V/L

F	Faraday constant, at/g-equivalent
h	membrane half channel height, L
K^*	solute distribution coefficient in charged membrane, dimensionless
L_p	hydraulic permeability of membrane, L/tp
P_s	solute permeability, L^2/tp
P_s/l	membrane constant for solute transport by diffusion, L/tp
P_w	water permeability, L^2/tp
P_w/l	membrane constant for water transport by diffusion, L/tp
P_2/l	membrane constant for transport by diffusion, L/tp
P_3/l	membrane constant for solute transport by pore flow, L/t
r_p^+	dimensionless pore radius, r/r_p
R_{ads}	membrane adsorption resistance, pt/L
R_g	gas constant, ML^2/t^2T mol
R_m	membrane resistance, pt/L
R'	intrinsic membrane solute rejection, dimensionless
u	axial velocity, L/t
v	transverse velocity, L/t
v_w	water permeation velocity, L/t
v_p	velocity of solvent in membrane pore, L/t
v_p^+	dimensionless solvent velocity in membrane pore, $v_p l/D_{sm}$
Y_i	pore radius distribution function

Greek Letters

β_1	dimensionless solution viscosity, $\eta D_s/R_g T r_a^2 c_W^1$
β_2	dimensionless operating pressure, $p/R_g T c_W^1$
δ	concentration boundary layer thickness, L
$\bar{\Phi}$	dimensionless potential function of force exerted on solute molecule by membrane pore wall

Subscripts

s	solute
w	water

REFERENCES

Bhattacharyya, D., and C. Cheng. 1986. Separation of metal chelates by charged composite membranes. In *Recent Developments in Separation Science*, ed. N. N. Li, Vol. 9, p. 707, Boca Raton, FL: CRC Press.

Bhattacharyya, D., M. Jevtitch, J. Schrodt, and G. Fairweather. 1986. Prediction of membrane separation characteristics by pore distribution measurements and surface force-pore flow model. *Chem. Eng. Comm.* 42:111.

Brian, P. 1966. Mass transport in reverse osmosis. In *Desalination by Reverse Osmosis*, ed. U. Merten. Cambridge, MA: MIT Press.

Brusilovsky, M., and D. Hasson. 1989. Prediction of reverse osmosis membrane salt rejection in multi-ionic solutions from single salt data. *Desalination* 71:355.

Dickson, J. 1988. Fundamental aspects of reverse osmosis. Chap. 1 in *Reverse Osmosis Technology*, ed. B. Parekh. New York: Marcel Dekker.

DOE. 1990. *Membrane Separation Systems*. Report. DOE/ER/30133-H1, P5-1, U.S. Department of Energy.

Dresner, L. 1972. Some remarks on the integration of the extended Nernst-Planck equations in the hyperfiltration of multicomponent solutions. *Desalination* 10:27.

Dresner, L., and J. Johnson. 1980. Hyperfiltration (reverse osmosis). Chap. 8 in *Principles of Desalination*. ed. K. Spiegler and A. Laird. New York: Academic Press.

Gekas, V., and B. Hallstrom. 1987. Mass transfer in the membrane concentration polarization layer under turbulent cross flow. *J. Membr. Sci.* 30:153.

Jonsson, G. 1980. Overview of theories for water and solute transport in UF/RO membranes. *Desalination* 35:21.

Kedem, O., and A. Katchalsky. 1958. Thermodynamic analysis of the permeability of biological membranes to non-electrolytes. *Biochim. Biophys. Acta* 27:229.

Lakshminarayanaiah, N. 1969. *Transport Phenomena in Membranes*. New York: Academic Press.

Lonsdale, H., U. Merten, and R. Riley. 1965. Transport properties of cellulose acetate osmotic membranes. *J. Appl. Polym. Sci.* 9:1341.

Matsuura, T., and S. Sourirajan. 1981. Reverse osmosis transport through capillary pores under the influence of surface forces. *Ind. Eng. Chem. Process Des. Dev.* 20:273.

Matthiasson, E., and B. Sivik. 1980. Concentration polarization and fouling. *Desalination* 35:59.

Mazid, M. 1984. Mechanisms of transport through reverse osmosis membranes. *Sep. Sci. Technol.* 19:357.

Mehdizadeh, H., J. Dickson, and P. Eriksson. 1989. Temperature effects on the performance of thin-film composite aromatic polyamide membranes. *Ind. Eng. Chem. Res.* 28:814.

Mehdizadeh, H., and J. Dickson. 1989. Theoretical modification of the surface force-pore flow model for reverse osmosis transport. *J. Membr. Sci.* 42:119.

Pusch, W. 1986. Measurement techniques of transport through membranes. *Desalination* 59:105.

Rautenbach, R., and R. Albrecht. 1989. *Membrane Processes*. New York: John Wiley & Sons.

Sherwood, T., P. Brian, R. Fisher, and L. Dresner. 1965. Salt concentration at phase boundaries in desalination by reverse osmosis. *Ind. Eng. Chem. Fundam.* 4:113.

Sherwood, T., P. Brian, and R. Fisher. 1967. Desalination by reverse osmosis. *Ind. Eng. Chem. Fundam.* 6:2.

Soltanieh, M., and W. Gill. 1981. Review of reverse osmosis membranes and transport models. *Chem. Eng. Comm.* 12:279.

Sourirajan, S. 1970. *Reverse Osmosis*. New York: Academic Press.

Sourirajan, S., and T. Matsuura. 1985. *Reverse Osmosis/Ultrafiltration Principles*. Ottawa, Canada: National Research Council of Canada.

Williams, M., 1989. Separation and purification of dilute hazardous organics by ozonation-low pressure composite membrane process. M.S. thesis, Department of Chemical Engineering, Univ. of Kentucky, Lexington.

23

Design

Dibakar Bhattacharyya and Michael E. Williams
University of Kentucky

Roderick J. Ray and Scott B. McCray
Bend Research, Inc.

GENERAL DESIGN
  CONSIDERATIONS
MEMBRANES
REVERSE-OSMOSIS MEMBRANE
  MODULES
    Modules
    Discussion and Conclusions
THEORETICAL ASPECTS OF
  MODULE DESIGN
OTHER DESIGN CONSIDERATIONS
FEED PRETREATMENT

Pretreatment to Prevent Chemical
  Damage to the Membrane
Pretreatment to Prevent Fouling
Suspended Solids and Particulates
Colloids
Scale-Forming Salts
Metal Oxides
Biological Foulants
Organic Foulants
NOTATION
REFERENCES

GENERAL DESIGN CONSIDERATIONS

Membrane performance prediction in reverse osmosis (RO) processes is important both in designing membrane systems and module arrangements and in choosing appropriate operating conditions to obtain a desired separation with maximum water flux (Sirkar, Dang, and Rao 1982; Belfort 1984; Rautenbach and Albrecht 1989; Bhattacharyya et al. 1990; DOE 1990). Some of the limiting factors in RO operations are concentration polarization, particulate fouling, bacterial fouling, and organics adsorption (Gilron and Hasson 1987; Siler 1987; Potts, Ahlert, and Wang 1981; Ridgway, Rigby, and Argo 1985; Williams, Deshmukh, and Bhattacharyya 1990).

The design variables include module hydrodynamics and flow rates, water characteristics (solute concentrations and diffusivities, osmotic pressure, viscosity, etc.), operating pressure, and fouling potential as a function of water recovery. The design should include the following considerations:

1. *Solution variables:*
   - suspended solids
   - dissolved inorganics
   - microorganisms
   - dissolved organics
   - sparingly soluble materials
   - organic solvents, oxidizing chemicals
   - temperature, pH
2. *Minimum pretreatment requirements*

3. *Membrane variables:*
   - polymer type and module geometry
   - pressure, flow rate, pressure loss/module
   - water recovery and concentration levels
   - minimum tolerable flux and desired flux level
   - module arrangements
   - cleaning requirements
4. *Membrane integration with other processes.*

For well-characterized feed solutions, membrane module performance can be predicted as a function of operating variables. Many approximate calculation methods (Prasad and Sirkar 1985; Evangelista and Jonsson 1988b; Soltanieh and Gill 1984) exist for special conditions, but there is no exact analytical solution that can handle all conditions. Numerical solutions, on the other hand, can be effectively utilized to predict membrane performance. Complex feed solutions with high fouling potential often require pilot plant experiments to obtain design information for full-scale plants. The fouling problems depend on the extent of pretreatment and module types. The flux loss due to particulates or bacterial adhesion generally increases in the following order: tubular < spiral-wound < hollow-fiber.

MEMBRANES

Membrane separation is governed by both the chemical nature of the membrane polymer and the physical structure of the membrane. The desired separation attainable with a particular membrane depends on the relative permeability of the membrane for the solution components. Excellent reviews of RO-type membrane materials have been reported by Lloyd and Meluch (1985), Kesting (1985), Koros et al. (1988), Cadotte et al. (1981), Sourirajan and Matsuura (1985), Lonsdale (1987), and DOE (1990).

Ever since the development of the first asymmetric, cellulose acetate membranes by Loeb and Sourirajan (1963), significant progress has been made in the field of noncellulosic RO membrane materials with broad pH and temperature tolerances. The cellulosic polymers, linear and cross-linked aromatic polyamides, cross-linked polyetherurea, and various other thin-film composite membranes are by far the most important RO membrane materials. An excellent review of various preparation methods is available in a recent DOE report (1990). For applications of RO membranes in the food processing area, the following membranes have received U.S. Food and Drug Administration (FDA) approval: linear aromatic polyamide hollow-fiber membranes and thin-film composite membranes (spiral-wound) of the type cross-linked fully aromatic polyamide, polypiperazinamide, and aryl-alkyl polyetherurea.

According to Sourirajan and Matsuura (1985), the material science of RO membranes has two major aspects: the physicochemical basis for the choice of materials for membranes, and the physicochemical factors related to membrane preparation for different performance characteristics. Current industrial practice is to classify RO membranes into three categories: high-pressure seawater desalination RO, low-pressure brackish water RO, and ultralow-pressure RO or nanofiltration for water softening and selective separations. As the use of membrane separations extends beyond desalination/demineralization to industrial solutions/wastes containing organics, knowledge of organic-polymer interactions (Sourirajan and Matsuura 1985; Williams, Deshmukh, and Bhattacharyya 1990; Bhattacharyya et al. 1986; Pusch, Yu, and Zheng 1989; and Jiang et al. 1989) is essential for optimum membrane selection and design. High-performance liquid chromatography (HPLC) experiments with appropriate membrane polymer packing materials or the surface force–pore flow (SFPF) model presented in Chapter 22 provide a way to determine organic/membrane interaction parameters.

Performance characteristics of various commercially available RO and nanofiltration membranes in terms of salt and selected organic rejections are summarized in Tables 23–1 and 23–2.

TABLE 23-1. Characteristics of Selected Reverse Osmosis Membranes.

Membrane Material and Commercial Name	Manufacturer	Test Module[a]	Solute	Test Conditions	Flux $\times 10^4$ cm^3/cm^2 s (gfd)	% Rejection
Cellulose acetate		F	NaCl	50,000 ppm, 8 MPa	9.17 (19.4)	98
			Methanol	1.7 MPa		7
			Ethanol	23–138 ppm, 1.7 MPa		10
			Phenol	1.7 MPa		0
	UOP	T	NaCl	5000 ppm, 25°C, 4.1 MPa, $r = 0\%$	4.8 (10.2)	98
			Methanol	1000 ppm, 25°C, 4.1 MPa, $r = 0\%$		<0
			Ethanol	"		2
			Urea	"		26
			Phenol	"		17
815 PR	Osmonics	S–	NaCl	2000 ppm, 25°C,		90
815 SR		40 × 7.9 in.	"	2.9 MPa, $r = 10\%$		95
815 HR			"	pH 5.0–6.0		97.5
CA 990	DDS		NaCl	2500 ppm, 25°C, pH 7, 4 MPa	19.5 (41.3)[b]	90–92
CA 995			NaCl	"	13.9 (29.5)[b]	95–97
CA 999			NaCl	"	5.57 (11.8)[b]	98–99.5
SC-3000	Toray		NaCl	1500 ppm, 25°C, 1.5 MPa	3.47 (7.37)	96
			Methanol	1000 ppm, 25°C, 1.5 MPa		5
			Ethanol	"		9
			Urea	"		26
			Phenol	"		0
Cellulose diacetate and triacetate						
CA 865PP	DDS		NaCl	2500 ppm, 20°C, 3 MPa	23.7 (50.2)	26–34
CA 930			NaCl	2500 ppm, 20°C, 1 MPa	16.7 (35.4)	26–34
CA 960PP			NaCl	2500 ppm, 20°C, 3 MPa	19.5 (41.3)	55–65
CA 990PP			NaCl	2500 ppm, 20°C, 4 MPa	16.7 (35.4)	>85
CA 992PP			NaCl	"	16.7 (35.4)	>90
CA 995PP			NaCl	"	11.1 (23.6)	>94

[0.456 L/s (10,400 gpd)][b]
[0.355 L/s (8100 gpd)][b]
[0.280 L/s (6400 gpd)][b]

TABLE 23-1. (*Continued*)

Membrane Material and Commercial Name	Manufacturer	Test Module[a]	Solute	Test Conditions	Flux $\times 10^4$ cm^3/cm^2 s (gfd)	% Rejection
Cellulose triacetate	Envirogenics	F	NaCl	5000 ppm, 25°C, 4.1 MPa, r = 0%	2.31 (4.9)	98
			Methanol	1000 ppm, 25°C, 4.1 MPa, r = 0%		<0
			Ethanol	"		23
			Urea	"		38
			Phenol	"		<0
Cellulose acetate butyrate	Universal Water	F	NaCl	5000 ppm, 25°C, 4.1 MPa, r = 0%	0.65 (1.4)	>99
			Methanol	1000 ppm, 25°C, 4.1 MPa, r = 0%		<0
			Ethanol	"		1
			Urea	"		8
			Phenol	"		10
Aromatic polyamide						
B9	Du Pont	HF – 2 ft × 5 in.	NaCl	5000 ppm, 25°C, 2.8 MPa, r = 75%	[0.055 L/s (1250 gpd)]	93
			Methanol	1000 ppm, 25°C, 2.8 MPa, r = 75%		28
			Ethanol	"		36
			Urea	"		33
			Phenol	"		43
B10	Du Pont	HF – 4 ft × 5 in.	NaCl	5000 ppm, 25°C, 5.2 MPa, r = 75%	[0.197 L/s (4500 gpd)]	99
			Methanol	1000 ppm, 25°C, 5.2 MPa, r = 75%		10
			Ethanol	"		15
			Urea	"		41
			Phenol	"		64
Cross-linked aromatic polyamide						
FT30-SW30HR-8040	FilmTec	S– 40 in. × 7.9 in.	NaCl	35,000 ppm, 25°C, pH 8. 5.5 MPa, r = 10%	5.71 (12.1) [0.175 L/s (4000 gpd)]	99.5

Membrane	Manufacturer	Module	Solute	Conditions	Flux	Rejection (%)
FT30-SW30-8040		S–40 in. × 7.9 in.	NaCl	35,000 ppm, 25°C, pH 8.8, 5.5 MPa, $r = 10\%$	8.58 (18.2) [0.263 L/s (6000 gpd)]	99.1
FT30-BW30-8040		S–40 in. × 7.9 in.	NaCl	2000 ppm, 25°C, pH 8, 1.6 MPa, $r = 15\%$	10.7 (22.7) [0.329 L/s (7500 gpd)]	98
FT30-BW			Methanol	2000 ppm, 25°C, pH 7, 1.6 MPa		25
			Ethanol	"		70
			Urea	"		70
		F	Phenol	51 ppm, pH 7.4, $r = 83\%$, 2.1 MPa		90
		F	Phenol	85 ppm, pH 11.4, $r = 39\%$, 2.1 MPa		99
FT30-TW30-440		S–40 in. × 3.9 in.	NaCl	2000 ppm, 25°C, pH 8, 1.6 MPa, $r = 15\%$	10.6 (22.5) [0.079 L/s (1800 gpd)]	98
FT30-SGHR-8040		S–40 in. × 7.9 in.	NaCl	32,300 ppm, 25°C, pH 8.8, 5.5 MPa, $r = 10\%$	9.72 (20.6) [0.298 L/s (6800 gpd)]	98.4
HR 95	DDS		NaCl	2500 ppm, 25°C, 4 MPa	20.9 (44.3)	95–97
			Methanol	1000 ppm, 25°C		42
			Ethanol	9000–26,000 ppm, 25°C, 4 MPa		70–75
			Urea	1000–20,000 ppm, 25°C, 4 MPa		70
			Phenol	30–1000 ppm		80–90
HR 98	DDS		NaCl	2500 ppm, 25°C, 4 MPa	20.9 (44.3)	98.5–99.5
			Methanol	1000 ppm, 25°C, 7.1 MPa		43
			Ethanol	"		90
411 HR (PA)	Osmonics	S–40 in. × 3.9 in.	NaCl	2000 ppm, 25°C, pH 8, 1.6 MPa, $r = 10\%$	[0.079 L/s (1800 gpd)]	98
815 HR (PA)		S–40 in. × 7.9 in.	NaCl	"	[0.329 L/s (7500 gpd)]	98
Polyvinylalcohol (TFC) NTR-729HF	Nitto		NaCl	1500 ppm, 25°C, pH 6–7, 0.99 MPa	15.1 (32)[b]	92

TABLE 23-1. (*Continued*)

Membrane Material and Commercial Name	Manufacturer	Test Module[a]	Solute	Test Conditions	Flux × 10⁴ cm³/cm² s (gfd)	% Rejection
NTR-739HF			Ethanol	"		25
			NaCl	"		95
			Ethanol	"		30
Aryl-alkyl polyamide/ polyurea						
NTR-7197	Nitto	S	NaCl	5000 ppm, 25°C, 4.1 MPa	26.7 (56.5)[b]	98.5
NTR-7199		S	NaCl	1500 ppm, 25°C, 1.5 MPa	11.6 (24.6)	99.4
UTC-40HF	Toray	S	NaCl		29.0 (61.5)	90
UTC-40HR			NaCl	1000 ppm, 25°C, 1.5 MPa	11.6 (24.5)	98–99
			Methanol	"		9
			Ethanol	"		34
			Urea	"		28
			Phenol	"		26
PEC 1000 (polyfuran)			NaCl	35,000 ppm, 25°C, pH 7, 5.5 MPa	4.05 (8.59)	99.92
			Methanol	55,000 ppm, 25°C, pH 6.9, 5.5 MPa	4.40 (9.33)	41
			Ethanol	60,000 ppm, 25°C, pH 6.9, 5.5 MPa	2.67 (5.65)	97
			Urea	10,000 ppm, 25°C, pH 6.9, 5.5 MPa	6.46 (13.7)	85
			Phenol	10,000 ppm, 25°C, pH 5.2, 5.5 MPa	2.78 (5.89)	99.0
Cross-linked polyethylenimine						
NS100	UOP	F	NaCl	5000 ppm, 25°C, 4.1 MPa, $r = 0\%$	4.58 (9.71)	99
			Methanol	1000 ppm, 25°C, 4.1 MPa, $r = 0\%$		17
			Ethanol	"		61
			Urea	"		79
			Phenol	"		64
NS200	North Star	F	NaCl	5000 ppm, 25°C, 4.1 MPa, $r = 0\%$	3.16 (6.69)	97

Membrane	Mfr.	Module	Solute	Conditions	Flux gfd (F)	r(%)
			Methanol	1000 ppm, 25°C, 4.1 MPa, $r = 0\%$		36
			Ethanol	"		84
			Urea	"		78
			Phenol	"		85
Other Membranes						
TFC-LP	UOP	S	NaCl	BW, 1.4 MPa net	10.3 (21.8)	97
TFC-HF		S	NaCl	SW, 2.8 MPa net	9.48 (20.1)	99
TFC-SS		S	NaCl	SW, 2.8 MPa net	6.13 (13.0)	99.5
TFCL-HP		S	NaCl	SW, 2.8 MPa net	>9.43 (>20)	>99
TFCL-LP		S	NaCl	BW, 1.4 MPa net	>9.43 (>20)	>97
TFC-801		S	NaCl	35,000 ppm, 25°C, pH 5.5, 5.5 MPa	5.38–7.74 (11.4–16.4)	99.14–99.66
TFC-402		S	NaCl	10,000 ppm, 25°C, pH 5.5, 2.8 MPa	10.8–12.6 (22.8–26.8)	99.17–99.41
TFC-202		S	NaCl	5000 ppm, 25°C, pH 5.5, 1.4 MPa	8.40–10.6 (17.8–22.4)	98.7–99.17
HT-RO		S	NaCl	1000 ppm, 25°C, pH 6, 1.7 MPa	11.3 (24)	99.2
PA-300			NaCl	5530 ppm, 2.8 MPa	9.43 (20)	98
			NaCl	35,000 ppm, 6.9 MPa	9.43–11.8 (20–25)	99.4
			Ethanol	700 ppm, 25°C, pH 4.7, 6.9 MPa		90
			Urea	1250 ppm, 25°C, pH 4.9, 6.9 MPa		80–85
			Phenol	100 ppm, 25°C, pH 4.9, 6.9 MPa		93
			Phenol	100 ppm, 25°C, pH 12, 6.9 MPa		>99

Notes: [ ] = Quantities in brackets represent module production of permeate water. SW = Seawater. BW = Brackish water. gfd = gallons/ft^2 day. gpd = gallons/day. r(%) = % recovery. ppm = parts per million by weight.

[a] F—flat-sheet membrane; S—spiral-wound module; T—tubular membrane; HF—hollow-fiber module.
[b] Pure water flux.

TABLE 23–2. Characteristics of Selected Nanofiltration Membranes.

Commercial Name	Manufacturer	Test Module[a]	Solute	Test Conditions	Flux $\times 10^4$ cm^3/cm^2 s (gfd)	% Rejection
NF40	FilmTec		NaCl	2000 ppm, 25°C, 1.6 MPa	12.0 (25.4)	45
			Glucose	"		90
NF40-8040			MgSO$_4$	2000 ppm, 25°C, pH 8, 1.6 MPa	10.0 (21.2)	95 min.
NF40HF			NaCl	2000 ppm, 25°C, 0.9 MPa	12.0 (25.4)	40
			MgSO$_4$	"		95
			Glucose	"		90
NF50			NaCl	2000 ppm, 25°C, 0.4 MPa		50
			MgSO$_4$	"		90
			Glucose	"		90
NF70			NaCl	2000 ppm, 25°C, pH 7, 0.48 MPa	10.0 (21.2)	80
			MgSO$_4$	2000 ppm, 25°C, pH 8, 0.48 MPa	[0.307 L/s (7000 gpd)]	95 min.
			Glucose	"		98
XP20			NaCl	1.0 MPa	10 (21.2)	20
			MgSO$_4$	"		85
			Glucose	"		55
XP45			NaCl	0.7 MPa	10 (21.2)	50
			MgSO$_4$	"		97.5
NTR-7250	Nitto		NaCl	2000 ppm, 2.0 MPa	35.4 (75)[b]	50
			MgSO$_4$	"	23.6 (50)	98
NTR-7410			NaCl	5000 ppm, 25°C, 0.99 MPa	138 (292)[b]	15
			Na$_2$SO$_4$	"		55
			MgSO$_4$	"		9
NTR-7450			NaCl	25°C, 1 MPa	25.9 (55)[b]	51
			Na$_2$SO$_4$			92
			MgSO$_4$			32
			Glucose			93

UTC-20HF	Toray	S	NaCl	1500 ppm, 25°C, 1.5 MPa	40.6 (86)	50
			MgSO$_4$	2000 ppm, 25°C, 1.5 MPa		99.8
UTC-20HR		S	NaCl	1500 ppm, 25°C, 1.5 MPa	28.3 (60)	70–80
			MgSO$_4$	2000 ppm, 25°C, 1.5 MPa	11.1 (23.6)	>94
			Glucose	1000 ppm, 25°C, 1.5 MPa		85
MPT-20	Membrane Products Kiryat Weizmann		NaCl + low MW organics	35,000 ppm (5% organics), 45°C, 2.5 MPa	11.8 (25)	
			NaCl	50,000 ppm, 25°C, 2.5 MPa		0
			Glucose	10,000 ppm, 25°C, 2.5 MPa		75
MPT-30			NaCl + low MW organics	100,000 ppm (5% organics), 50°C, 2.5 MPa	9.43 (20)	
			NaCl	50,000 ppm, 25°C, 2.5 MPa		10
			Glucose	10,000 ppm, 25°C, 2.5 MPa		70
Desal-5	Desalination		NaCl	1000 ppm, 1 MPa	12.8 (27.1)	47
			Glucose	25°C, 1 MPa		83
DRC-1000	Celfa		NaCl	3500 ppm, 1.0 MPa	13.9 (29.5)	10
HC50	DDS		NaCl	2500 ppm, 4.0 MPa	22.3 (47.2)	60
NF-PES-10/PP 60	Kalle		NaCl	5000 ppm, 4.0 MPa	111.3 (236)	15
NF-CA-50/PET 100			NaCl	5000 ppm, 4.0 MPa	33.4 (70.8)	55
SU200HF	Toray		NaCl	1500 ppm, 1.5 MPa	41.7 (88.5)	50
SU600			NaCl	500 ppm, 0.35 MPa	7.79 (16.5)	55
			Glucose	25°C, 1 MPa		93
SU700			Glucose	"		99

Notes: [ ] = Quantities in brackets represent module production of permeate water. gfd = gallons/ft^2 day. gpd = gallons/day. r(%) = % recovery. ppm = parts per million by weight.
[a]F—flat-sheet membrane; S—spiral-wound module; T—tubular membrane; HF—hollow-fiber module.
[b]Pure water flux.

289

REVERSE-OSMOSIS MEMBRANE MODULES

The economics and utility of RO processes depend as much on the inexpensive and efficient packaging of RO membranes (i.e., modularization) as on the membrane technology itself. The first modules were of the plate-and-frame or tube-in-shell configurations, and these module types are still commercially available for specialty applications (Lonsdale 1982). However, advances in module technology have produced new module types—spiral-wound and hollow-fiber modules—that are less expensive and have superior performance.

An example of the state of the art today is a membrane module that can desalt seawater in a single pass, is guaranteed for 5 years of normal operation, and has a retail cost as low as $2/ft^2 ($19/m^2) (Riley 1989). To produce modules that meet these specifications, developers of today's membrane modules have had to satisfy a myriad of performance and economic requirements. Briefly, these include:

1. *Mechanical:* The membrane module must support the membrane at pressures of up to 1200 psi (85 kg/cm^2), even with periodic depressurization and cleaning of the membrane. This can be difficult because the bulk membrane—especially the separating layer or "skin"—is relatively fragile. Furthermore, the module design must minimize pressure drops in the feed and product streams.
2. *Hydrodynamics:* Module design must minimize membrane fouling and the effects of concentration polarization. A module susceptible to fouling will require more expensive and complicated pretreatment. Significant concentration polarization will result in lower quality permeate and lower flux.
3. *Economics:* To be economical, modules must have long lifetimes, be inexpensive to fabricate, and be easy to replace.

Commercially available membrane modules can be classified into four basic types: (1) spiral-wound, (2) hollow-fiber, (3) tubular, and (4) plate-and-frame. Table 23–3 is a summary of the general characteristics of the four basic membrane-module types.

Others have compiled detailed lists of membrane-module suppliers and their available products (Allegrezza 1988). Due to the rapid pace of innovation in the membrane field, the information in such lists becomes outdated quickly. Since the purpose of this chapter is to provide sufficient information to design and estimate costs for RO systems, we will include only general performance and cost information for each type of membrane module. The reader can use this information to determine the type

TABLE 23–3. General Characteristics of RO Membrane Modules.

Characteristic	Module Type			
	Spiral-Wound	Hollow-Fiber	Tubular	Plate-and-Frame
Typical packing density (ft^2/ft^3)	245	1830	21	150
[m^2/m^3]	[800]	[6000]	[70]	[500]
Required feed flow rate (ft^3/ft^2-s)	0.8–1.6	~0.016	3–15	0.8–1.6
[m^3/m^2-s]	[0.25–0.5]	[~0.005]	[1–5]	[0.25–0.5]
Feedside pressure drop (psi)	43–85	1.4–4.3	28–43	43–85
[kg/cm^2]	[3–6]	[0.1–0.3]	[2–3]	[3–6]
Membrane fouling propensity	High	High	Low	Moderate
Ease of cleaning	Poor to good	Poor	Excellent	Good
Typical feed stream filtration requirements	10- to 25-μm filtration	5- to 10-μm filtration	Not required	10- to 25-μm filtration
Relative expense	Low	Low	High	High

of membrane module best suited for the application of interest. By contacting suppliers directly, up-to-date cost information can be obtained. A list of suppliers of membrane modules and their addresses is given in Table 23–4. Independent cost analyses may be undertaken using the information provided in Chapter 25.

TABLE 23–4. Suppliers of RO Membrane Modules.

Module Type	Application		
	Seawater Desalination	Brackish Water Treatment	LPRO/MS
Spiral-wound	FilmTec Corp. Div. of Dow 7200 Ohms Lane Minneapolis, MN 55435-3482	Desalination Systems Inc. 1238 Simpson Way Escondido, CA 92025	Desalination Systems Inc. 1238 Simpson Way Escondido, CA 92025
	Fluid Systems Div. UOP Inc. 10124 Old Grove Rd. San Diego, CA 92151	Du Pont Company "Permasep" Products Glasgow Site Wilmington, DE 19898	FilmTec Corp. Div. of Dow 7200 Ohms Lane Minneapolis, MN 55435-3482
	Nitto Denko America Inc. (formerly Hydranautics) P.O. Box 3690 Santa Barbara, CA 93130	FilmTec Div. of Dow 7200 Ohms Lane Minneapolis, MN 55435-3482	Nitto Denko America Inc. 2800 Corvin Dr. Santa Clara, CA 95051
	Nitto Denko America Inc. 2800 Corvin Dr. Santa Clara, CA 95051	Fluid Systems Div. UOP Inc. 10124 Old Grove Rd. San Diego, CA 92151	SEPAREM S.P.A. Via per Oropa 118 13051 Biella, Italy
	Toray Industries Inc. Membrane Products Dept. Toray Bldg. 2-2 Nihonbashi-muromachi Chuo-ku, Tokyo 103, Japan	Nitto Denko America Inc. (formerly Hydranautics) P.O. Box 3690 Santa Barbara, CA 93130	Sumitomo Chemical Co. Ltd. 2-Chome, Tsukahara Takatsuki City Osaka 569, Japan
		Millipore Corp. 80 Ashby Rd. Bedford, MA 01730	Toray Industries Inc. Membrane Products Dept. Toray Bldg. 2-2 Nihonbashi-muromachi Chuo-ku, Tokyo 103, Japan
		Osmonics Inc. 5951 Clearwater Dr. Minnetonka, MN 55343	
		SEPAREM S.P.A. Via per Oropa 118 13051 Biella, Italy	
		Sumitomo Chemical Co. Ltd. 2-Chome, Tsukahara Takatsuki City Osaka 569, Japan	

TABLE 23-4. (*Continued*)

Module Type	Application		
	Seawater Desalination	Brackish Water Treatment	LPRO/MS
Hollow-Fiber	Du Pont Company "Permasep" Products Glasgow Site Wilmington, DE 19898	Du Pont Company "Permasep" Products Glasgow Site Wilmington, DE 19898	FilmTec Corp. Div. of Dow 7200 Ohms Lane Minneapolis, MN 55435-3482
	Toyobo Company Ltd. Membrane Division 2-8 Dojima Jama 2-Chome Kita-ku, Osaka 530, Japan	FilmTec Corp. Div. of Dow 7200 Ohms Lane Minneapolis, MN 55435-3482	
		Toyobo Company Ltd. Membrane Division 2-8 Dojima Jama 2-Chome Kita-ku, Osaka 530, Japan	
Tubular		Koch Membrane Systems 850 Main St. Wilmington, MA 01887	
		Nitto Denko America Inc. 2800 Corvin Dr. Santa Clara, CA 95051	
		Paterson Candy Int. Ltd. Laverstock Mill Witchurch, Hampshire RG287NR United Kingdom	
		Stork Friesland BV P.O. Box 13 8400 AA Gorredijk The Netherlands	
		Sumitomo Chemical Co. Ltd. 2-Chome, Tsukahara Takatsuki City Osaka, 569, Japan	
		Teijin America Inc. 10 Rockefeller Plaza New York, NY 10020	
Plate-and-Frame		Dow Danmark A/S (formerly De Danske Sukkerfabrikker) DK-4900 Naksov, Denmark	

Modules

Spiral-Wound Modules

Spiral-wound modules allow the efficient packaging of flat-sheet membrane in a convenient cylindrical form. Figure 23–1 is a schematic of a simple single-leaf spiral-wound module. In this module type, a membrane "sandwich" is rolled around a product water collection tube. The membrane sandwich consists of two flat sheets of membrane with a product water channel spacer in between and a feed channel spacer. Three sides of this sandwich are glued together and the fourth side is glued into the product water collection tube.

In spiral-wound modules, the feed stream is passed into one end (or "face") of the cylindrical module and along one side of the membrane sandwich. Water permeates the membrane and passes into the product water channel, where it travels in a spiral until it reaches the center of the module. There, it flows through small holes in the product water collection tube and exits the module through the product water outlet. The brine that does not permeate the membrane exits the module through the brine outlet at the opposite end of the module.

Typically, the feed channel spacer is made of a polypropylene mesh and its design depends on the application. For example, for seawater desalination, the mesh is designed to maximize turbulence, minimize the pressure drop, and promote a high packing density. For applications in which the feed streams contain species that tend to foul membranes, the mesh is designed to promote uninterrupted flow on the feed side of the module, instead of promoting turbulent flow.

The product water channel spacer, which is placed between the two sheets of membrane, is usually made of a finer mesh than that used for the feed channel spacer. It must be fine enough to provide the support necessary for the membrane to withstand high feed pressures, but open enough to provide a low-resistance pathway for the product water to travel to the product water collection tube. To prevent the components of the spiral-wound module from being pushed out of alignment axially by the pressure of the feed stream, an antitelescoping support may be used on the end of the module.

To make manufacturing easier and to avoid the high pressure drops associated with longer permeate channels, most commercial spiral-wound modules are of the multileaf type. Instead of one long membrane sandwich being wound around the product water collection tube, multileaf modules contain multiple-membrane sandwiches, each of which is attached to the product water collection tube. Figure 23–2 shows a cutaway view of a multileaf spiral-wound module. For the past several years, the industry standard for multileaf spiral-wound modules has been an 8-in.-diameter (203-mm), 40-in.-long (1-m) module containing about 365 ft^2 (33.9 m^2) of membrane (Riley 1989). Recently, however, larger-diameter modules have been fabricated. Multileaf mod-

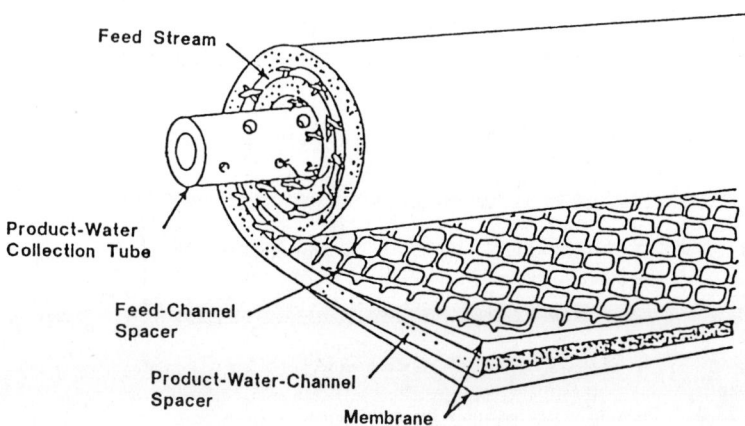

FIGURE 23–1. Single-leaf spiral-wound module.

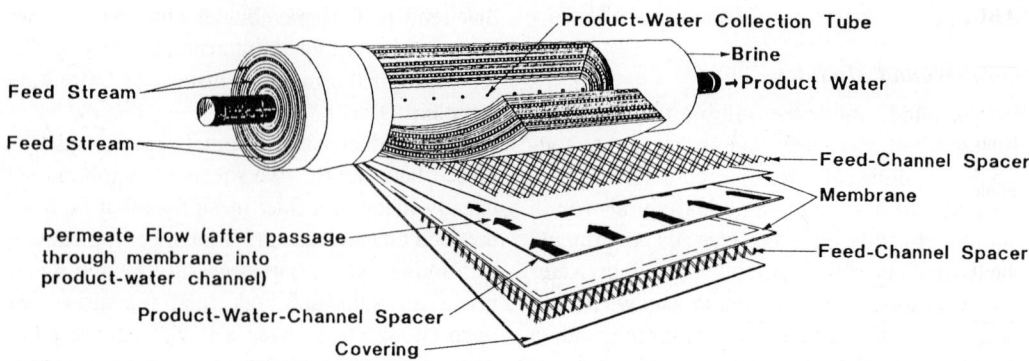

FIGURE 23–2. Cutaway view of a multileaf spiral-wound module.

ules are available that are 11 in. (279 mm) in diameter and that contain about 700 ft^2 (65 m^2) of membrane. Longer modules are planned as well—60-in.-long (1.5-m) spiral-wound modules are now commercially available.

To be used, the spiral-wound module must be contained within a pressure vessel assembly. As shown in Figure 23–3, the pressure vessel assembly consists of the cylindrical housing for the module(s), the plumbing to connect the modules together in series (if appropriate), and the plumbing to connect the feed inlet, product water outlet, and brine outlet ports to the RO system.

A pressure vessel assembly that contains six 8-in.-diameter (203-mm) spiral-wound modules would cost about $4500, or 40% of the total module cost for a high-pressure (seawater desalination) application. The lifetime of this pressure vessel assembly may be as long as 20 years. (Most RO plants have expected lifetimes of 10 to 20 years; pressure vessels are usually warranted for 3 years.)

Table 23–5 summarizes productivity and cost data for spiral-wound modules for seawater desalination. Shown in the table are typical productivities for modules that contain the common membrane types used in spiral-wound modules today: cellulose acetate or thin-film composite membranes. Table 23–6 summarizes the data for spiral-wound modules for brackish water treatment, and Table 23–7 summarizes the data for spiral-wound modules for low-pressure-RO/membrane-softening (LPRO-MS) applications. Typical operating conditions and limits for spiral-wound modules are given in Table 23–8.

Hollow-Fiber Modules

Hollow-fiber modules are based on a much different approach to packaging large quantities of membrane area into a small volume. Figure 23–4 shows the hollow-fiber module configuration. The hollow-fiber membranes used in this configuration are inherently less water permeable than the flat-sheet thin-film composite or asymmetric membranes used in the spiral-wound modules. However, hollow-fiber mem-

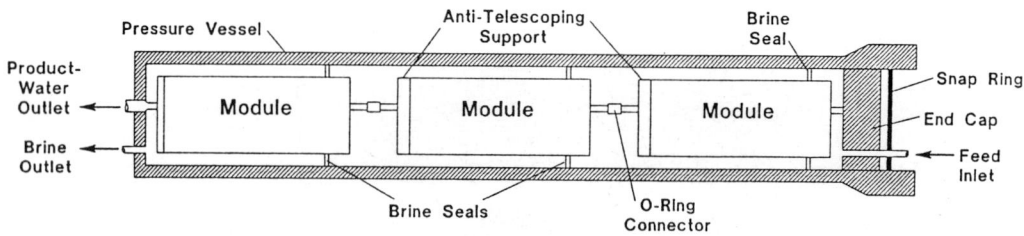

FIGURE 23–3. Pressure vessel assembly for spiral-wound modules.

TABLE 23–5. **Summary of Typical Productivities and Costs for Large Spiral-Wound Modules for Seawater Desalination.**

Membrane Type	Module Diameter[a] (in.)	[mm]	Module Productivity (gpd)	[m³ pd]	NaCl Rejection (%)	Cost[b] ($/gpd)	[$/m³ pd]
Cellulose acetate	4	[102]	580	[2.2]	96	c	c
	8	[203]	2300	[87]	96	c	c
Thin-film composite	4	[102]	1000	[3.8]	99.4	0.66	[173.70]
	8	[203]	4000	[15.1]	99.4	0.38	[101.30]

[a] All modules are ~40 in. (1016 mm) long unless otherwise noted.
[b] Module costs obtained from manufacturers, based on single-module costs. Cost does not include pressure vessel.
[c] Cost information is unavailable.

TABLE 23–6. **Summary of Typical Productivities and Costs for Large Spiral-Wound Modules for Brackish Water Treatment.**

Membrane Type	Module Diameter[a] (in.)	[mm]	Module Productivity (gpd)	[m³ pd]	NaCl Rejection (%)	Cost[b] ($/gpd)	[$/m³ pd]
Cellulose acetate	4	[102]	2000	[7.6]	95	0.15	[39.60]
	8	[203]	8000	[30.3]	95	0.12	[31.70]
Thin-film composite	4	[102]	1800	[6.8]	98	0.27	[71.30]
	8	[203]	7500	[28.4]	98	0.16	[42.30]

[a] All modules are ~40 in. (1016 mm) long unless otherwise noted.
[b] Module costs obtained from manufacturers, based on single-module costs. Cost does not include pressure vessel.

TABLE 23–7. **Summary of Typical Productivities and Costs for Large Spiral-Wound Modules for LPRO/MS Applications.**

Membrane Type	Module Diameter[a] (in.)	[mm]	Module Productivity (gpd)	[m³ pd]	NaCl Rejection (%)	Cost[b] ($/gpd)	[$/m³ pd]
Cellulose acetate	4	[102]	1640	[6.2]	85	c	c
			2200	[8.3]	75	0.15	[39.60]
	8	[203]	6560	[24.8]	85	c	c
			8000	[30.3]	75	0.10	[26.40]
Thin-film composite	4	[102]	1800	[6.8]	96	0.18	[47.50]
			2100	[7.9]	92	0.24	[63.40]
			1700	[6.4]	80	c	c
			1800	[6.8]	40	c	c
	8	[203]	7500	[28.4]	96	0.13	[34.30]
			9500	[36.0]	92	0.16	[42.30]
			7000	[26.5]	80	c	c
			7500	[28.4]	40	c	c

[a] All modules are ~40 in. (1016 mm) long unless otherwise noted.
[b] Module costs obtained from manufacturers, based on single-module costs. Cost does not include pressure vessel.
[c] Cost information is unavailable.

TABLE 23-8. Typical Operating Conditions and Limits of Large Spiral-Wound Modules for Various Applications.

Application	Membrane Type	Maximum Feed Temperature		Typical pH Range	Allowable Chlorine Concentration (ppm)	Pressure (psi)
		(°F)	[°C]			
Seawater desalination	Cellulose acetate	113	[45]	4–7	0.2–2	800
	Thin-film composite	113	[45]	2–11	<0.1	800
Brackish water treatment	Cellulose acetate	104	[40]	3–7	0.2–2	400
	Thin-film composite	113	[45]	2–11	<0.1	400
LPRO/MS applications	Cellulose acetate	104	[40]	3–7	0.2–2	200
	Thin-film composite	104	[40]	3–10	<1.0	150

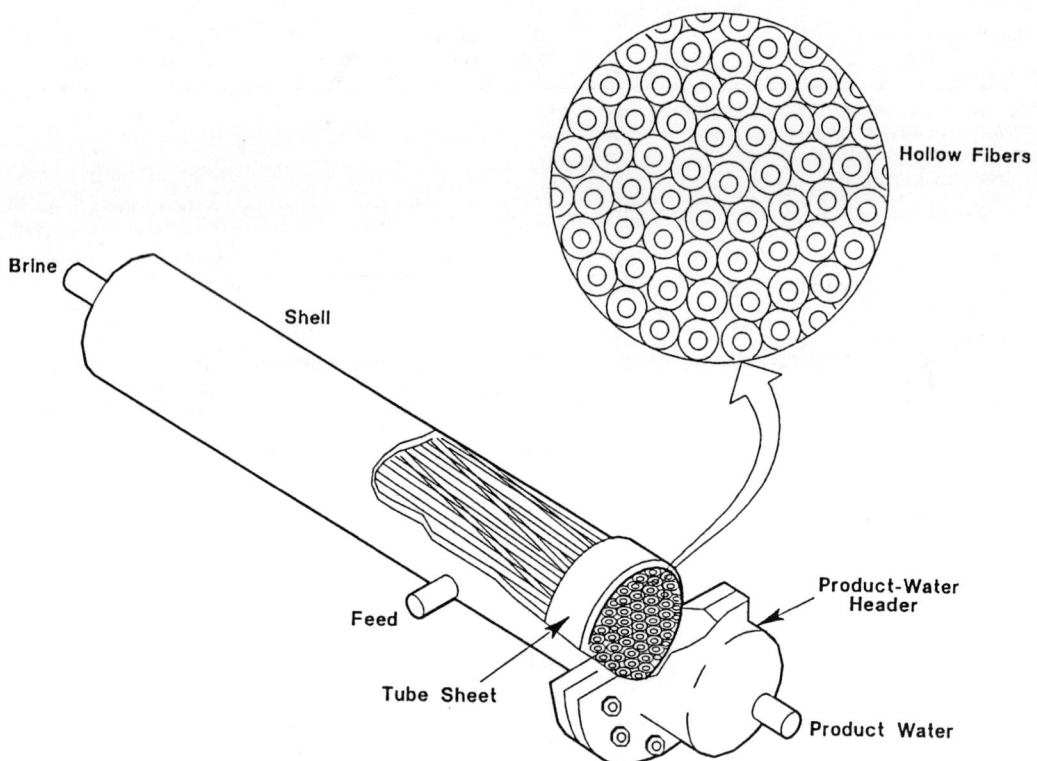

FIGURE 23-4. Cutaway view of hollow-fiber membrane module.

TABLE 23-9. Summary of Typical Productivities and Costs for Large Hollow-Fiber Modules for Seawater Desalination and Brackish Water Treatment.

Application	Membrane Type	Module Diameter (in.)	[mm]	Module Productivity (gpd)	[m³ pd]	NaCl Rejection (%)	Cost[a] ($/gpd)	[$/m³ pd]
Seawater desalination	Cellulose acetate	5.7	[144]	325	[1.2]	99.4	b	b
		8.3	[210]	1,300	[5.0]	99.4	b	b
		11.7	[298]	7,300	[27.5]	99.4	b	b
		14.2	[360]	9,300	[35.0]	99.4	b	b
	Aramid[c]	4	[102]	1,000	[3.8]	99	2.17	[573.30]
		8	[203]	5,000	[18.9]	99	1.22	[322.90]
Brackish water treatment	Cellulose acetate	5.9	[150]	6,500	[24.6]	94	b	b
		9.5	[241]	16,000	[60.6]	96	b	b
		10.75	[273]	28,000	[106]	96	b	b
	Aramid	4	[102]	4,200	[15.9]	90[d]	0.31	[81.10]
		8	[203]	16,000	[60.6]	90[d]	0.24	[63.70]
		10	[254]	25,000	[94.6]	90[d]	0.24	[63.70]

[a]Module costs obtained from manufacturers, based on single-module costs. Cost does not include pressure vessel.
[b]Cost information is unavailable.
[c]Du Pont's linear aromatic polyamide membrane.
[d]Minimum salt rejection.

branes have higher salt rejections, can be operated often at higher pressures, and make higher module packing densities possible. Because of these trade-offs, the costs per unit of water produced are comparable to those of spiral-wound modules.

In the hollow-fiber module configuration, the fibers are pressurized from the outside, and the product water passes into the interior (lumen) of the fibers. The product water flows down the lumen through a tubesheet and into a product water header. Because the fibers are pressurized from the outside, fibers with less mechanical strength can be used than if fibers were pressurized from the inside. Furthermore, the pressure drop down the lumen of the fibers is reduced because the permeate stream has a lower flow rate than the feed stream.

Table 23-9 summarizes the productivity and cost data for hollow-fiber membrane modules for seawater desalination and brackish water treatment based on two common membrane types. Data for LPRO/MS applications are not included because hollow-fiber modules are not used for this application.

Tubular Modules

Tubular membrane modules, shown in Figure 23-5, can contain up to 30 tubes and can be up to 20 ft (6 m) in length. The membranes are normally supported within stainless steel tubes. The tubes are connected in series in most designs; therefore, one module can operate in a multipass configuration to maximize recovery, given the large diameter of the feed channels. With this module design, the feed channels and, most importantly, the permeate channels can be easily cleaned, making the modules appropriate for food and dairy applications in which frequent cleaning is necessary.

Besides ease of cleaning, a major advantage to the tubular configuration is that the tube

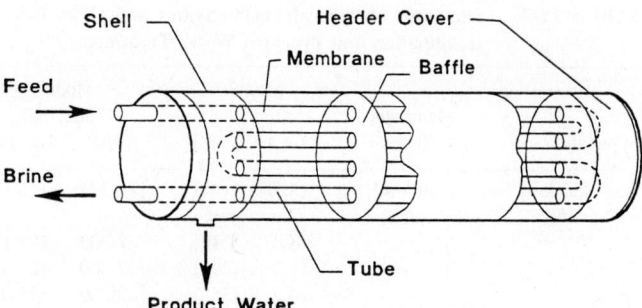

FIGURE 23-5. Tubular membrane module.

diameter (typically 0.5 in. for RO applications) is large enough to promote turbulent flow under most conditions without an excessive pressure drop. This feature makes the modules very resistant to fouling. However, this module type has two major disadvantages: (1) high energy costs due to the large feed channels (and the corresponding need to pump a relatively large volume of water through the modules) and (2) high capital costs, primarily due to the modules' low packing density.

Table 23-10 lists performance and cost data for commercial tubular modules. As indicated in Table 23-10, tubular modules that exhibit a

TABLE 23-10. Summary of Productivities and Costs for Large Tubular Modules.

Membrane Type	Manufacturer	Membrane Area[a] (ft^2)	[m^2]	Module Productivity (gpd)	[m^3 pd]	NaCl Rejection (%)	Cost[b] ($/gpd)	[$/m^3$ pd]
Cellulose acetate	Koch	40.7	[3.8]	410	[1.5]	86	c	c
				810	[3.0]	97	c	c
	Nitto Denko	17.4	[1.6]	350	[1.3]	95	c	c
				525	[2.0]	50	c	c
	Paterson Candy Int. Ltd.	28.0	[2.6]	—		80–98	~0.5	[132.10]
	Stork	55.5	[5.2]	1000	[3.8]	97	0.20	[54.10]
				1400	[5.3]	95		
Thin-film composite	Paterson Candy Int. Ltd.	28.0	[2.6]	1000	[3.8]	99	0.54	[142.70]
	Stork	55.5	[5.2]	1600	[6.1]	98	0.30	[78.40]
	Sumitomo	11.8	[1.1]	210	[0.8]	98	c	c
	Teijin	13.6	[1.3]	400	[1.5]	98	c	c
				500	[1.9]	95		
				240	[0.9]	85[d]	c	c
				400	[1.5]	30[d]	c	c
				500	[1.9]	15[d]	c	c

[a]Some manufacturers offer more than one size module.
[b]Module costs obtained from manufacturers, based on single-module costs. Cost does not include pressure vessel.
[c]Cost information is unavailable.
[d]Sucrose rejection.

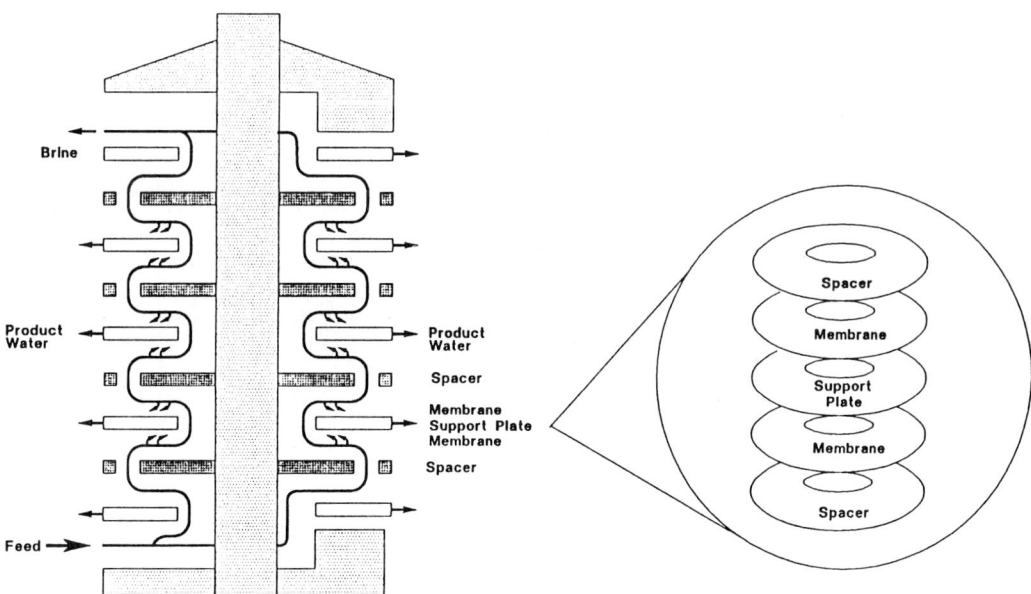

FIGURE 23–6. Schematic of a plate-and-frame module.

range of NaCl rejections and productivities are available for RO applications.

Plate-and-Frame Modules

Plate-and-frame modules make use of flat-sheet membrane, as do spiral-wound modules. Figure 23–6 is a schematic of a plate-and-frame module that shows the flow path used for the feed stream. As with spiral-wound modules, plate-and-frame modules use a membrane sandwich—i.e., two flat-sheet membranes separated by a support plate that also acts as a product water channel.

In some designs, such as the one shown in Figure 23–6, the membrane sandwiches are in the form of disks. The membrane disks are separated by spacers that allow the feed solution to flow radially inward on one side of the membrane disk and radially outward on the other side. This module design produces long feed channels, making high recoveries possible. These modules have low packing densities and are correspondingly expensive. They are primarily used to treat fouling-prone feed streams and to produce potable water in small-scale applications.

As of 1991, Dow Danmark A/S (formerly De Danske Sukkerfabrikker) of Naksov, Denmark, was the only manufacturer of plate-and-frame modules.

Discussion and Conclusions

Current sales of spiral-wound modules account for about 74% of the total RO market, and sales of hollow-fiber modules total about 26% of the market (Riley 1989). Sales of tubular and plate-and-frame modules are relatively small compared with those of spiral-wound and hollow-fiber modules. Table 23–4 lists the main suppliers of RO membrane modules and the types of modules available from each. Other types of modules available or under development (such as tubeside feed inside-skinned hollow fibers) are not discussed here.

At present, the cost for spiral-wound and hollow-fiber modules represents only a small fraction of the total operating costs (see Chapter 25); further reductions in module costs will probably not have a significant impact on the RO market. Future innovations are likely to be focused on the design of fouling-resistant modules and/or chemically resistant membranes. The major cost savings in the future will result

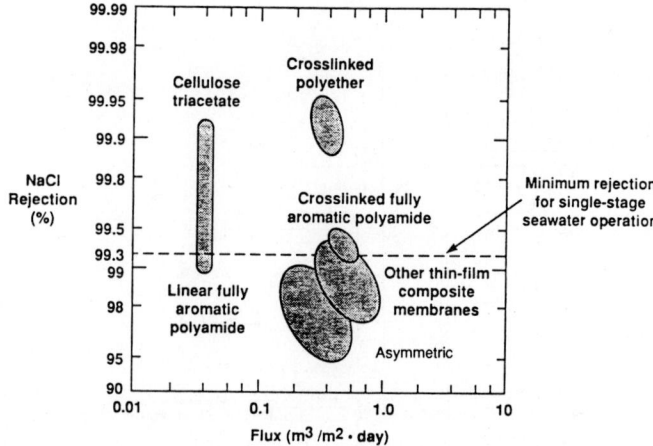

FIGURE 23-7. Desalination performance characteristics of high-pressure membranes operating at 5.6 MPa and 25° C (DOE 1990).

from reducing the need for pretreatment, extending the life of the membranes, and increasing the range of applications for which RO is suitable.

THEORETICAL ASPECTS OF MODULE DESIGN

Spiral-wound modules are the most widely used configuration in the RO industry, principally because they are less susceptible to fouling (DOE 1990). The nature and characteristics of spacer materials is also an important aspect of RO technology. Optimization of spacers to minimize pressure drop (particularly in nanofiltration modules) is an important design consideration. Membrane water flux behavior determines water recovery per module. Selected water flux and salt rejection behaviors (DOE 1990) of some commercially available membranes are shown in Figure 23-7 (high-pressure RO) and Figure 23-8 (nanofiltration or loose RO).

The numerical predictions of concentration polarization (c'_W/c'_b) and flux behavior of simple configurations (two parallel plate membranes without spacers) are discussed first. Bhattacharyya et al. (1990) and Back (1987) adopted a finite element numerical technique to solve Eqs. (22-25) to (22-29) by using appropriate velocity profile equations for laminar and turbulent flow. Figure 23-9 shows a typical boundary layer concentration profile (as c'/c'_b

dimensionless concentration) as a function of axial and transverse distance for 5000 mg/L NaCl and at $\Delta p = 3.04$ MPa and Re = 1333. The effects of channel height, Reynolds number, temperature, and feed concentration on water flux are shown in Figures 23-10 and 23-11. The significant flux drop (Figure 23-10) at very high Reynolds number and low channel height is due to high pressure loss. Temperature effects are also interesting because of two opposing factors: flux decrease due to osmotic pressure increase, and flux increase due to an increase of water permeability coefficient with temperature.

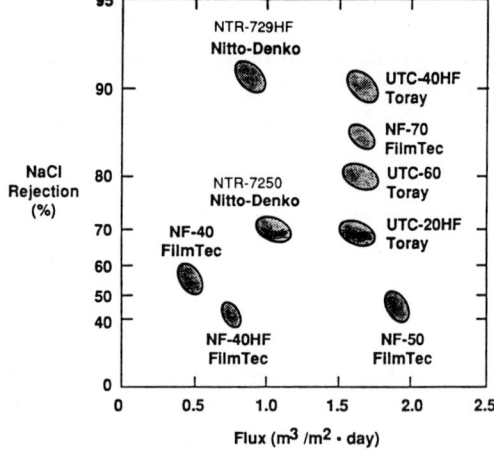

FIGURE 23-8. Desalination performance of several commercial nanofiltration membranes operating on 500 mg/L NaCl feed at 0.8 MPa and 25°C (DOE 1990).

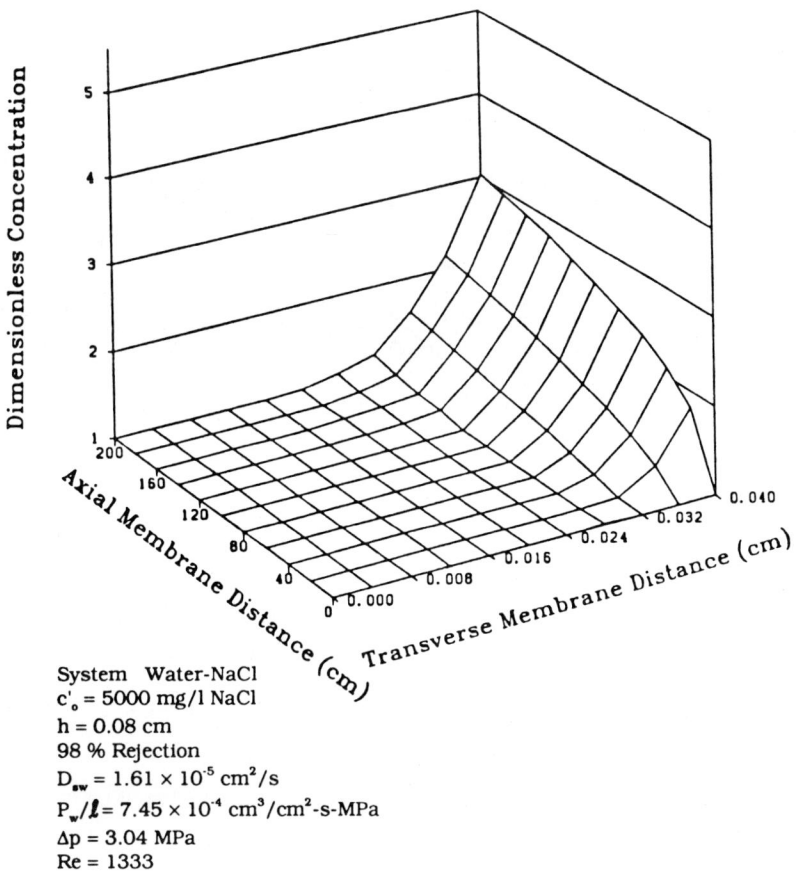

FIGURE 23–9. Concentration profile of a 5000 mg/L NaCl solution between parallel plates.

Various analytical and numerical models of differing complexity have been reported in the literature to predict and optimize spiral-wound module performance (Evangelista and Jonsson 1988a; Evangelista and Jonsson 1988b; Belfort 1988; Prasad and Sirkar 1985; Bhattacharyya et al., 1990; Schock and Miquel 1987; Wiley, Fell, and Fane 1985; Rautenbach and Dahm 1987). The packaging density and, thus, optimum permeation rate per unit volume is mainly determined by feed and permeate channel spacers, compatible with acceptable pressure drop. Belfort (1988) has provided an excellent review on the comparisons of various modules based on fluid mechanics.

The performance characteristics of spiral-wound modules in terms of pressure drop and mass transfer correlations have been reported by Schock and Miquel (1987). For pressure drop and mass transfer calculations, the Reynolds number is often defined in terms of hydraulic diameter d_h:

$$d_h = \frac{4(\text{total volume} - \text{spacer volume})}{(\text{wetted surface of flat channel} + \text{wetted surface of spacer})}.$$

(23–1)

Schock and Miquel (1987) correlated the data for various commercial modules and found the following mass transfer correlation to be valid:

$$Sh = 0.065 \, Re^{0.875} \, Sc^{0.25}. \quad (23\text{–}2)$$

Several excellent analytical solutions (with various assumptions) have been reported for the

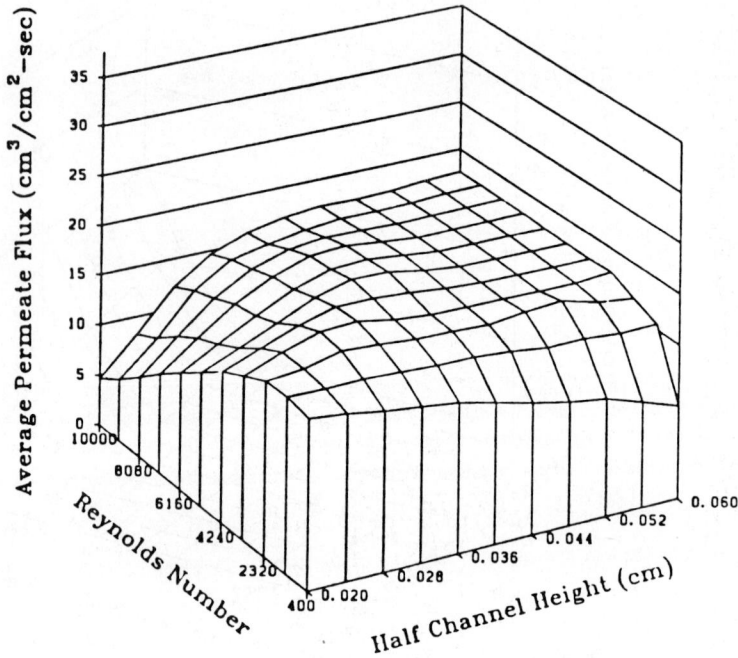

FIGURE 23-10. Effect of Reynolds number and channel height on permeate flux.

System Water-NaCl
c'_o = 5000 mg/l NaCl
h = 0.08 cm
L = 100 cm
98 % Rejection
D_{sw} = 1.61 × 10^{-5} cm^2/s
P_w/ℓ = 7.45 × 10^{-4} cm^3/cm^2-s-MPa
Δp = 5.07 MPa

determination of productivity (Q_p/V_m), membrane area, and average permeate quality (Evangelista and Jonsson 1988b; Sirkar, Dang, and Rao 1982; Prasad and Sirkar 1985). For the case of low water recovery per module, Evangelista and Jonsson (1988b) have reported the following simplified expression relating productivity (specific permeation rate) to operating variables and membrane geometry for spiral-wound modules

$$\frac{Q_p}{V_m} = \frac{L_p \tanh(\alpha_p W)(\Delta p_e - A_b \eta u_F L/16 h_b^2)}{\alpha_p W(h_b + h_p + h_m)},$$

(23-3)

where Δp_e is effective pressure (applied pressure – pressure loss – osmotic pressure). Simple analytical expressions for single solute (Sirkar, Dang, and Rao 1982) and multisolute (Prasad and Sirkar 1985) systems also have been reported for the prediction of spiral-wound module length and average permeate solute concentration. For highly rejected, dilute solutions, their analytical expressions agreed very well with numerical solutions involving various design examples. These models provide valuable tools for membrane performance prediction both under laminar and turbulent flow conditions.

The efficiency of spiral-wound elements can be enhanced significantly by changing the spacer dimensions. Thinner feed and permeate spacers will produce modules with a high packing density at the expense of larger pressure drop and possibly enhanced fouling. Evangelista and Jonsson (1988a) have reported an elegant analytical method to calculate specific permeation

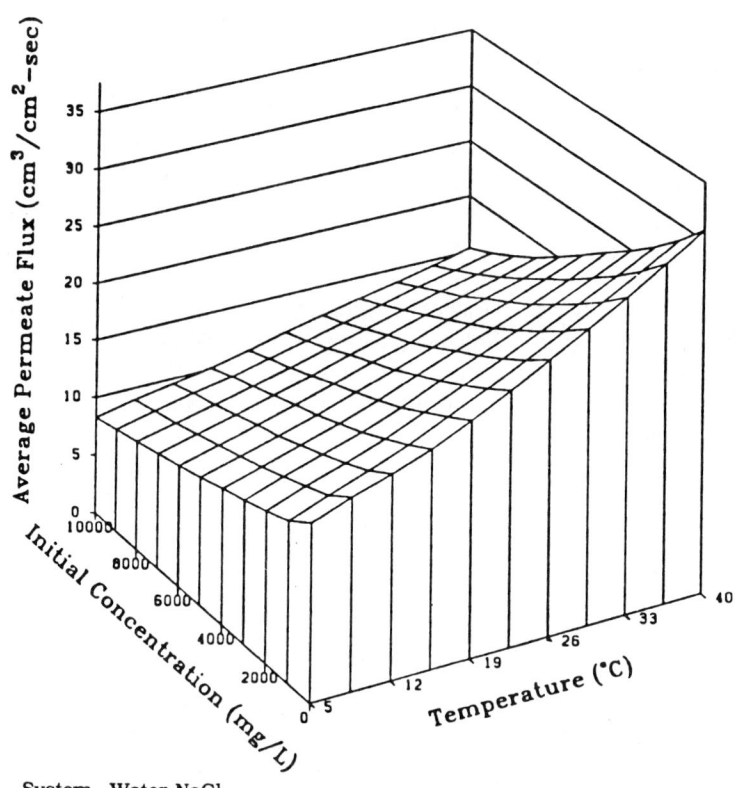

FIGURE 23-11. Effect of feed concentration and temperature on average permeate flux.

System Water-NaCl
h = 0.08 cm
L = 100 cm
98 % Rejection
D_{sw} (25°C) = 1.61 × 10^{-5} cm^2/s
P_w/ℓ (25°C) = 7.45 × 10^{-4} cm^3/cm^2-s-MPa
u_o = 75 cm/s
Δp = 3.04 MPa

rate and optimum feed and permeate spacer thickness. For a module with low recovery, they have reported simplified equations for specific permeation rate and optimum thickness for feed and permeate channels. At a fixed membrane width W and module length L, the following expressions have been reported for optimum permeate channel thickness (h_p^{opt}) and feed (brine) channel thickness (h_b^{opt}):

$$h_p^{opt} = \left(\frac{3}{32} A_p \mu L_p (h_b + h_m)\right)^{1/4} W^{1/2}, \tag{23-4}$$

$$h_b^{opt} = \left(\frac{5}{16} \frac{A_b \mu u_F L (h_p + h_m)}{\Delta p}\right)^{1/3}. \tag{23-5}$$

As can be seen, for example, an optimum brine channel thickness required increases in the spacer friction parameter (A_b), solution viscosity (μ), feed velocity (u_F), permeate channel height (h_p), and module length (L). In the above equations L_p is the hydraulic permeability of the membrane and h_m is the thickness of the membrane plus the support. Evangelista and Jonsson (1988b) have also determined optimum geometrical characteristics of modules by a numerical method using a three-parameter model.

Another way of evaluating the effects of spacers on module performance is in terms of the "eddy constant" as introduced by Miyoshi, Fukuumoto, and Kataoka (1987). Back (1987) used Miyoshi's velocity profile equation containing an eddy constant (for spacer) term m and a finite element technique to solve Eqs. (22–25) to (22–29). The eddy constant $[m = (\text{eddy viscosity}) \cdot h/(yv)]$ was approximated by

$$m = 2.1 \times 10^5 \, [n\,(t-d)]^{2.4} \left(\frac{1-\varepsilon_s}{\varepsilon_s^3}\right)^{0.8},$$

(23–6)

where n is the number of spacers, d is the diameter of an individual spacer fiber, t is spacer thickness, and ε_s is the void fraction of spacer. As m increases, the velocity profile becomes flatter, with larger velocity gradients close to the wall, which enhances the mass transfer. Computer simulations provided water flux, concentration polarization, and permeate concentration as functions of eddy constant. A typical simulation plot of the effects of m and channel height on wall concentration is shown in Figure 23–12. Best separation occurs at the highest eddy constant and the thinnest channel. However, for thin channels and high m values the module pressure loss would be significant.

Productivity calculations for hollow-fiber modules have been reported by various authors (Rautenbach and Albrecht 1989; Doshi 1988; Soltanieh and Gill 1982; Soltanieh and Gill 1983). Maximization of hollow-fiber productivity involves optimizing fiber diameter and length, minimizing friction losses, and maximizing surface area per unit volume. Doshi, Gill, and Kabadi (1977) have derived a simplified equation connecting productivity (for

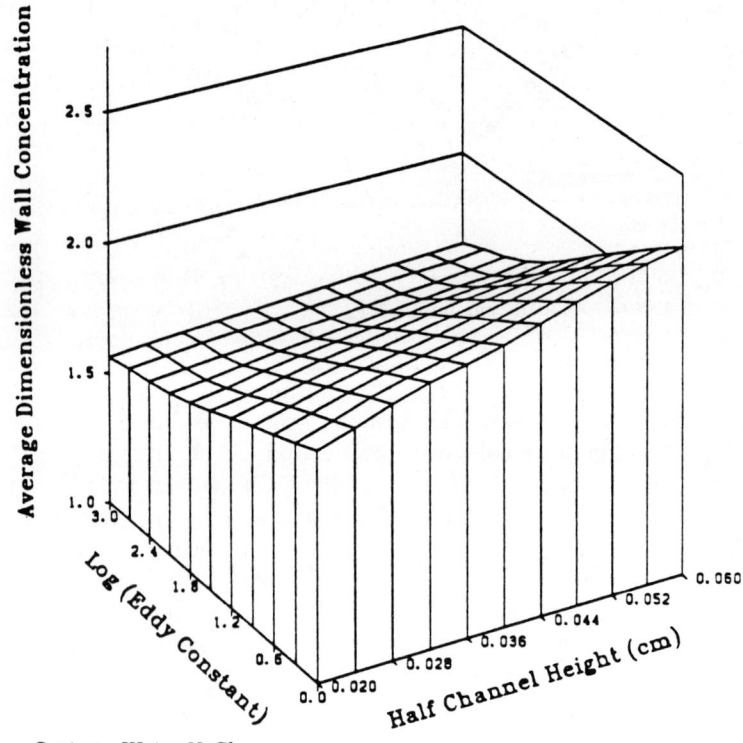

FIGURE 23–12. Effect of log eddy constant and half-channel height on average dimensionless wall concentration.

System Water-NaCl
$c'_o = 5000$ mg/l NaCl
$L = 100$ cm
98 % Rejection
$D_{sw} = 1.61 \times 10^{-5}$ cm^2/s
$P_w/\ell = 7.45 \times 10^{-4}$ cm^3/cm^2-s-MPa
$u_o = 75$ cm/s
$\Delta p = 3.04$ MPa

pure water or very dilute solutions) with variables such as fiber diameter and length, void fraction, permeability, and applied pressure:

$$\frac{Q_p}{V_m} = \frac{P_w}{l}\left(\frac{2(1-\varepsilon)}{r_o\beta_m[\coth(\beta_m) + \beta_m L_s/L]}\right)\Delta p.$$

(23-7)

Concentration polarization and osmotic pressure effects in hollow-fiber modules have also been reported by Hermans (1978) and Soltanieh and Gill (1982).

OTHER DESIGN CONSIDERATIONS

Many variables can influence reverse osmosis process design. Although manufacturers report separation characteristics of new membranes, the performance behavior may change with operating time, particularly for long-term operation. For example, Klinko, Light, and Cummings (1985) reported an equation to compute rejection drop with time for a UOP spiral-wound module (2021HF) as

$$R = 1 - (1 - R_0)\left(\frac{\theta}{\theta_o}\right)^{0.0001p - 0.02},$$

(23-8)

where p is the applied pressure in psi, θ is the time in days, and θ_o is the starting time in days.

Key factors for selecting the best membrane for a particular application may include high flux and rejection, chlorine tolerance, low fouling rate, low bacteria adhesion, long life, and stability to extremes in pH, temperature, and chemical concentrations (Light, Chu, and Haddock 1988). For example, high-temperature (up to 90°C) operations of UOP-TFC membranes (Chu, Campbell, and Light 1988) and sulfonated polysulfone membranes (Coplan, Kopylenko, and Kozminiski 1984) have been reported in the literature. The high-temperature RO membrane modules are required for a variety of dewatering, concentrating, and component separation applications in food and beverage processes that operate at elevated temperatures.

FEED PRETREATMENT

The RO membranes currently available are generally not robust enough to operate directly on typical feed water streams. Feed waters usually contain components that can adversely affect the performance and lifetime of the RO membrane system. Therefore, the performance of an RO system will only be as good as the system used to pretreat the water before it enters the system. Virtually every RO system includes some level of feed pretreatment designed to (1) extend the lifetime of the membranes, (2) prevent fouling of the membranes, and (3) maintain the performance (i.e., rejection and recovery) of the system. As pointed out in Chapter 25, the capital and operating costs associated with the pretreatment subsystem account for up to 50% of the total production cost.

The extent (and therefore the cost) of pretreatment will depend on several factors, including (1) the type of membrane modules used, (2) the composition of the feed stream, and (3) the desired performance of the system. In some cases, simple filtration is all that is required; in other cases, several treatment steps in series are necessary. Most RO systems, and expecially for large systems [i.e., those that treat > 100,000 gallons per day (gpd)], the pretreatment system must be designed specifically for the individual application. To design the pretreatment system, detailed information about the composition of the feed water (including any seasonal variations) and the desired performance of the RO system (i.e., the recovery and permeate quality) must be known.

In this section, the pretreatment methods have been divided into two broad categories: pretreatment to prevent chemical damage to the membrane and pretreatment to prevent fouling. Although the pretreatment methods used in each of these categories are sometimes the same, the reasons for the pretreatment are dramatically different. Details of these pretreatment methods are discussed below.

Pretreatment to Prevent Chemical Damage to the Membrane

The concentration of chlorine and the pH of the feed water are the two most common factors that can result in chemical damage to RO membranes. Table 23-11 gives the allowable concentration of chlorine and the acceptable pH range for different types of typical RO membranes. Chlorine is often added to feed waters to control microbial growth (see below). Unfortunately, many RO membranes, especially polyamide membranes, are damaged by even low concentrations of chlorine. Therefore, the feed water must be dechlorinated before it enters the membrane system.

Several methods are used for dechlorination. The most popular are (1) treatment with sodium bisulfite ($NaHSO_3$), (2) carbon filtration, and (3) treatment with gaseous sulfur dioxide (SO_2) (Bates, Coulter, and Thomas 1988; Al-Borno and Abdel-Jawad 1989; Parise, Parekh, and Smith 1988; Ko and Guy 1988). Each of these pretreatment methods is effective; the choice of method will depend on the chemical costs and other feed water pretreatment requirements.

While dechlorination is vital for most polyamide membranes, control of pH is particularly important for membranes based on cellulose acetate. These membranes undergo rapid hydrolysis below pH 4 and above pH 7. The rate of hydrolysis is rapid under acidic conditions and is even faster under basic conditions (Vos, Burris, and Riley 1966). Therefore, tight control of pH is essential. In most applications, especially for seawater desalination, the pH of the feed must be lowered by adding hydrochloric or sulfuric acid. In situations where the pH must be raised, caustic (NaOH) is usually used.

Pretreatment to Prevent Fouling

Most feed waters contain contaminants (foulants) that reduce membrane productivity (flux) over time. The extent to which a membrane fouls will depend on the module configuration and on the types and concentrations of contaminants in the feed water. As discussed in this chapter, module configuration can affect the propensity of a membrane to foul. For example, membranes in spiral-wound and shellside feed hollow-fiber modules foul readily and therefore require an extensive pretreatment system. On the other hand, tubular modules are more resistant to fouling due to the hydrodynamics within them; therefore, only minimal pretreatment is required. These factors must be considered when designing a pretreatment system.

The types of contaminants that foul RO membranes can be divided into six categories: (1) suspended solids and particulates, (2) colloids, (3) scale-forming salts, (4) metal oxides, (5) biological foulants, and (6) organic foulants (Bates, Coulter, and Thomas 1988; Parise, Parekh, and Smith 1988; Osta and Bakheet 1987). Each of these foulants can dramatically reduce the performance of a membrane. Pretreatment methods used to prevent fouling from these contaminants are summarized in Table 23-12 and are discussed below.

Suspended Solids and Particulates

Suspended solids and other large particulates can foul membrane surfaces, plug feed water passages, and damage pumps and instrumentation (Bates, Coulter, and Thomas 1988; Parise, Parekh, and Smith 1988; Osta and Bakheet 1987; Ko and Guy 1988). Commonly, sus-

TABLE 23-11. Allowable Concentrations of Chlorine and pH Operating Ranges for Typical RO Membranes.

RO Membrane Type	Allowable Chlorine Concentration (ppm)	Allowable pH Range
Cellulose acetate-based	0.3–1.0	4–6
Polyamide[a]	<0.05	4–11
TFC—not resistant[b]	0[c]	3–11
TFC—minimal resistance[d]	0.05	3–11
TFC—chlorine resistant[e]	1.0	3–11

[a]Linear polyamide such as Du Pont B-9.
[b]Polyamide or polyurea.
[c]Feed must be dechlorinated.
[d]Aromatic polyamide.
[e]Sulfonated polysulfone.

TABLE 23-12. Pretreatment Methods Used for Foulants Typically Encountered in RO Systems.

Foulant Type	Pretreatment Methods
Suspended solids and particulates	Coarse screening Use of hydrocyclones Use of cartridge filters Use of multimedia filters
Colloids	Coagulation/flocculation followed by filtration Ultrafiltration
Scale-forming salts	Acidification Water softening using lime and lime soda Use of antiscale agents
Metal oxides	Acid cleaning Selection of proper materials
Biological foulants	Chlorination Ozonation Exposure to ultraviolet light Use of concentrated sodium bisulfite Use of copper sulfate
Organic foulants	Coagulation followed by filtration Carbon adsorption Chemical oxidation Ultrafiltration Microfiltration

pended solids are removed using a series of filters. Coarse screening or hydrocyclones are used to remove large particulates, and then a cartridge filter (Lepore and Ahlert 1988) or a multimedia filter containing sand, garnet, and anthracite is used to remove finer particles (Bates, Coulter, and Thomas 1988; Parise, Parekh, and Smith 1988; Lepore and Ahlert 1988). As a rule of thumb, particles larger than one-fifth the size of the smallest channel within the module should be removed (Parise, Parekh, and Smith 1988). Typically, feed water to a spiral-wound module should be treated with a 20- to 50-μm filter, while shellside feed hollow-fiber modules require a 5-μm filter (Ko and Guy 1988).

Colloids

Colloids are usually charged particles smaller than 1 μm in diameter. They are common in surface waters and if not removed from feed waters can drastically reduce the productivity of a membrane.

Several techniques can be used to remove colloids, the most common of which is coagulation/flocculation followed by conventional filtration. Typical coagulants used are alum ($Al_2[SO_4]_3$), ferric chloride ($FeCl_3$), and polymer or polyelectrolyte materials (Parise, Parekh, and Smith 1988; Osta and Bakheet 1987). The same types of filters used to remove suspended solids are used to remove the coagulated particles from the treated feed water.

Another common pretreatment technique that is used for the production of ultrapure water is ultrafiltration (UF) (Parise, Parekh, and Smith 1988). The UF membrane acts as an extremely fine filter, capable of removing small colloids. While UF can be used to produce a high-quality feed, the UF membranes themselves can foul. Therefore, pretreatment systems employing UF membranes are designed to have regular cleaning cycles to maintain the performance of the UF membranes.

Scale-Forming Salts

In RO systems in which high recoveries are required, the solubility limits of many salts and other materials can be exceeded, causing the material to precipitate on the membrane surface. Compounds most frequently encountered are calcium carbonate; sulfate salts of calcium, barium, and strontium; calcium fluoride; and silica (Ko and Guy 1988; Osta and Bakheet 1987; Al-Borno and Abdel-Jawad 1989; Lepore and Ahlert 1988; Parise, Parekh, and Smith 1988).

Several methods are used to minimize or eliminate the formation of these scale deposits. The first is acidification by acid injection. Acid injection converts bicarbonate alkalinity to CO_2, eliminating the formation of $CaCO_3$ scale.

The second method used is water softening using lime or lime soda. In this process, hydrated lime or soda ash is added to soften the water. Calcium and magnesium hydroxides are then removed as precipitates. This process can also remove some of the silica present in the feed water by absorption into the magnesium hydroxide precipitate (Ko and Guy 1988). Coagulants are often added to the treated water to aid in removing the precipitate.

The third method used is the addition of antiscale agents or so-called "threshold" agents. These compounds reduce the rate at which scale forms, allowing the system to operate with concentrations above the solubility limit. One of the most common threshold agents used to control calcium sulfate formation is sodium hexametaphosphate (SHMP). Many other threshold agents are available that control most of the scale-forming compounds commonly encountered in RO systems (Amjad 1987, 1988).

Metal Oxides

Metal oxides are often present in feed waters to RO systems due to corrosion in piping and in the system. Metal oxide deposits on membrane surfaces reduce flux. If they are discovered early enough, metal oxide deposits can be cleaned from the membrane surface using acids. However, the best method to prevent fouling due to metal oxides is to select the proper materials for the system so corrosion does not occur (Osta and Bakheet 1987; Lepore and Ahlert 1988; Bates, Coulter, and Thomas 1988).

Biological Foulants

The formation of biological foulants or slimes on the membrane surface is undesirable because it reduces the flux through the membrane and reduces the effective salt rejection of the membrane. To prevent slime formation, the feed water is often disinfected before it enters the RO system. Chlorination to 0.5 ppm by injection of chlorine gas or addition of hypochlorite is the most common method used. However, as discussed above, many RO membranes are damaged by chlorine. Therefore, the feed must then be dechlorinated, usually with sodium bisulfite, before it enters the system. Other disinfectants that have been used include ozone, ultraviolet light, formaldehyde, concentrated sodium bisulfite, and copper sulfate (Lepore and Ahlert 1988; Bates, Coulter, and Thomas 1988; Osta and Bakheet 1987; Al-Borno and Abdel-Jawad 1989; Ko and Guy 1988; Parise, Parekh, and Smith 1988).

Organic Foulants

When feed waters contain many contaminants, organic materials may deposit on the membrane

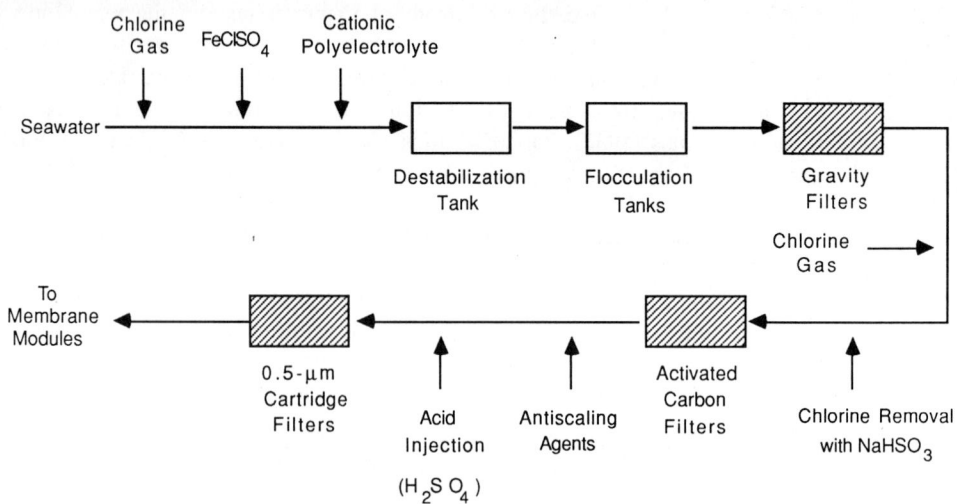

FIGURE 23-13. Pretreatment train at Doha reverse osmosis plant (DROP) in Kuwait (adapted from Al-Borno and Abdel-Jawad 1989).

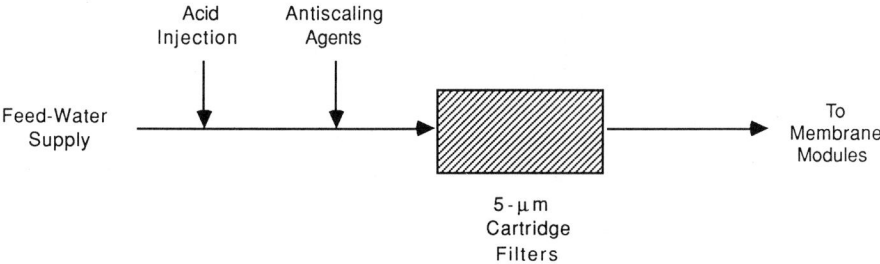

FIGURE 23–14. Pretreatment train for membrane-softening application (adapted from Watson and Hornburg 1989).

surface, decreasing membrane performance. Several methods can be used to remove these organic materials, including coagulation and filtration, carbon adsorption, chemical oxidation, and UF or microfiltration (Bates, Coulter, and Thomas 1988).

As stated above, most pretreatment systems must be designed specifically for the specific application. For example, Figure 23–13 depicts the pretreatment train used at the Doha RO plant (DROP) in Kuwait (Al-Borno and Abdel-Jawad 1989). This seawater-desalination plant uses a variety of pretreatment techniques to ensure that the lifetime of the RO system is long. However, in other applications, especially when the feed water is relatively clean (e.g., well water), simpler pretreatment trains can be employed. Figure 23–14 shows the pretreatment train used for an LPRO/MS application (Watson and Hornburg 1989). Here, acid injection, addition of antiscale agents, and simple cartridge filtration are used for pretreatment.

Although most pretreatment systems must be designed specifically for each application and although feed waters vary considerably from site to site, general RO pretreatment schemes have been developed over the years. These general schemes must be modified for each RO system, depending on the nature of the feed water and the desired performance of the RO system. With a properly designed pretreatment train, a reverse osmosis system lifetime of several years can be expected under normal operating conditions.

NOTATION

General Notation

See the General Notation section at the beginning of this handbook.

Special Notation

A_b feed spacer friction parameter, dimensionless

A_p permeate spacer friction parameter, dimensionless

c'_w concentration of feed solute at the membrane surface, M/L^3

d diameter of individual spacer fiber in spiral-wound module, L

h membrane half-channel height, L

h_b feed channel thickness for spiral-wound module, L

h_m thickness of membrane plus support, L

h_p permeate channel thickness for spiral-wound module, L

L_p hydraulic permeability of membrane, L/tp

L_s fiber seal length, L

m eddy spacer constant, dimensionless

n number of spacers, dimensionless

P_w water permeability, L^2/tp

R_0 initial rejection, dimensionless

t spacer thickness in spiral-wound module, L

u_F feed velocity, L/t

V_m volume of membrane module, L^3

W membrane width, L

Greek Letters

α_p $[A_p \eta L_p/(8h_p^3)]^{1/2}$

β_m $[16(P_w/l)\eta r_o L^2/(r_i^4)]^{1/2}$

Δp_e effective feedside pressure, p
ϵ_s void fraction of spacer in spiral-wound module, dimensionless
θ time, t
θ_o starting time, t

Subscripts
s solute
w water

REFERENCES

Al-Borno, A., and M. Abdel-Jawad. 1989. Conventional pretreatment of surface seawater for reverse osmosis application, state of the art. *Desalination* 74:3.

Allegrezza, A. 1988. Commercial reverse osmosis membranes and modules. Chap. 2 in *Reverse Osmosis Technology*. ed. B. Parekh. New York: Marcel Dekker.

Amjad, Z. 1987. Advances in scaling and deposit control for RO systems. *Ultrapure Water* 4:34.

Amjad, Z. 1988. State of the art deposit control agents for reverse osmosis applications. Paper read at 6th Membrane Technology/Planning Conference, 1–3 November 1988, at Cambridge, MA.

Back, S. 1987. Prediction of concentration polarization and flux behavior of reverse osmosis membrane systems by numerical analysis. M.S. thesis, Department of Chemical Engineering, Univ. of Kentucky, Lexington.

Bates, W., B. Coulter, and D. Thomas. 1988. Practical application and guidelines for the use of pretreatment techniques for reverse osmosis systems utilized for the production of high purity water. Paper read at 2nd Annual High Purity Water Conference and Exposition, 28–30 March 1988, at Philadelphia, PA.

Belfort, G. 1984. *Synthetic Membrane Processes: Fundamentals and Water Applications*. New York: Academic Press.

Belfort, G. 1988. Membrane modules: comparison of different configurations using fluid mechanics. *J. Membr. Sci.* 35:245.

Bhattacharyya, D., S. Back, R. Kermode, and M. Roco. 1990. Prediction of concentration polarization and flux behavior in reverse osmosis by numerical analysis. *J. Membr. Sci.* 48:231.

Bhattacharyya, D., M. Jevtitch, J. Schrodt, and G. Fairweather. 1986. Prediction of membrane separation characteristics by pore distribution measurements and surface force-pore flow model. *Chem. Eng. Comm.* 42:111.

Cadotte, J., R. King, R. Majerle, and R. Petersen. 1981. Interfacial synthesis in the preparation of reverse osmosis membranes. *J. Macromol. Sci. Chem.* A15:727.

Chu, H., J. Campbell, and W. Light. 1988. High-temperature reverse osmosis membrane element. *Desalination* 70:65.

Coplan, M., M. Kopylenko, and J. Kozminiski. 1984. Implications to the CPI of a new high performance reverse osmosis membrane. Reprint of paper read at Annual AIChE Meeting, San Francisco, CA.

DOE. 1990. *Membrane Separation Systems*. Report. DOE/ER/30133-H1, U.S. Department of Energy.

Doshi, M. 1988. Modeling of reverse osmosis membrane devices. Chap. 3 in *Reverse Osmosis Technology*. ed. B. Parekh. New York: Marcel Dekker.

Doshi, M., W. Gill, and V. Kabadi. 1977. Optimal design of hollow fiber modules. *AIChE J.* 23:765.

Evangelista, F., and G. Jonsson. 1988a. Optimal design and performance of spiral wound modules—numerical method. *Chem. Eng. Comm.* 72:69.

Evangelista, F., and G. Jonsson. 1988b. Optimal design and performance of spiral wound modules—analytical method. *Chem. Eng. Comm.* 72:83.

Gilron, J., and D. Hasson. 1987. Calcium sulphate fouling of reverse osmosis membranes: flux decline mechanism. *Chem. Eng. Sci.* 42:2351.

Hermans, J. 1978. Physical aspects governing the design of hollow fiber modules. *Desalination* 26:45.

Jiang, J., S. Mingji, F. Minling, and C. Jiyan. 1989. Study on the interaction between membranes and organic solutes by HPLC method. *Desalination* 17:107.

Kesting, R. 1985. *Synthetic Polymeric Membranes: A Structural Perspective*. New York: Wiley-Interscience.

Klinko, K., W. Light, and C. Cummings. 1985. Factors influencing optimum seawater reverse osmosis system designs. *Desalination* 54:3.

Ko, A., and D. Guy. 1988. Brackish and seawater desalting. Chap. 5 in *Reverse Osmosis Technology*. ed. B. Parekh. New York: Marcel Dekker.

Koros, W., G. Fleming, S. Jordan, T. Kim, and H. Hoehn. 1988. Polymeric membrane materials for solution-diffusion based permeation separations. *Prog. Polym. Sci.* 13:339.

Lepore, J., and R. Ahlert. 1988. Fouling in membrane processes. Chap. 4 in *Reverse Osmosis*

Technology, ed. B. Parekh. New York: Marcel Dekker.

Light, W., H. Chu, and K. Haddock. 1988. Reverse osmosis elements for dechlorinated and nonchlorinated feed water processing. *AIChE Symp. Series No. 261*. New York: American Institute of Chemical Engineers.

Lloyd, D., and T. Meluch. 1985. Selection and evaluation of membrane materials for liquid separations. *ACS Symp. Series No. 269*. Washington, DC: American Chemical Society.

Loeb, S., and S. Sourirajan. 1963. Sea water demineralization by means of an osmotic membrane. *Adv. Chem. Ser.* 38:117.

Lonsdale, H. 1982. The growth of membrane technology. *J. Membr. Sci.* 10:81.

Lonsdale, H. 1987. The evolution of ultrathin synthetic membranes. *J. Membr. Sci.* 33:121.

Miyoshi, H., T. Fukuumoto, and T. Kataoka. 1987. A consideration on flow distribution in an ion exchange compartment with spacer. *Desalination* 42:47.

Osta, T., and L. Bakheet. 1987. Pretreatment system in reverse osmosis plants. *Desalination* 63:71.

Parise, P., B. Parekh, and R. Smith. 1988. Reverse osmosis for producing pharmaceutical-grade waters. Chap. 8 in *Reverse Osmosis Technology*. ed. B. Parekh. New York: Marcel Dekker.

Potts, D., R. Ahlert, and S. Wang. 1981. A critical review of fouling of RO membranes. *Desalination* 36:235.

Prasad, R., and K. Sirkar. 1985. Analytical design equations for multicomponent reverse osmosis processes by spiral wound modules. *Ind. Eng. Chem. Process Des. Dev.* 24:350.

Pusch, W., Y. Yu, and L. Zheng. 1989. Solute-solute and solute-membrane interactions in hyperfiltration of binary and ternary aqueous organic feed solutions. *Desalination* 75:3.

Rautenbach, R., and R. Albrecht. 1989. *Membrane Processes*. New York: John Wiley & Sons.

Rautenbach, R., and W. Dahm. 1987. Design and optimization of spiral-wound and hollow fiber RO modules. *Desalination* 65:259.

Ridgway, H., M. Rigby, and D. Argo. 1985. Bacterial adhesion and fouling of reverse osmosis membranes. *JAWWA* 77:97.

Riley, R. 1989. Separation Systems, International, La Jolla, CA. Personal communication, September 20.

Schock, G., and A. Miquel. 1987. Mass transfer and pressure loss in spiral-wound modules. *Desalination* 64:339.

Siler, J. 1987. Reverse osmosis membranes concentration polarization and surface fouling: predictive models and experimental verifications. Ph.D. diss. Department of Chemical Engineering, Univ. of Kentucky, Lexington.

Sirkar, K., P. Dang, and G. Rao. 1982. Approximate design equations for reverse osmosis desalination by spiral-wound modules. *Ind. Eng. Chem. Process Des. Dev.* 21:517.

Soltanieh, M., and W. Gill. 1982. Analysis and design of hollow fiber reverse osmosis systems. *Chem. Eng. Comm.* 18:311.

Soltanieh, M., and W. Gill. 1983. A note on the effect of fiber length on the productivity of hollow fiber modules. *Chem. Eng. Comm.* 22:109.

Soltanieh, M., and W. Gill. 1984. An experimental study of the complete mixing model for radical flow hollow fiber RO systems. *Desalination* 49:57.

Sourirajan, S., and T. Matsuura. 1985. *Reverse Osmosis/Ultrafiltration Principles*. Ottawa, Canada: National Research Council of Canada.

Vos, K., F. Burris, Jr., and R. Riley. 1966. Kinetic study of the hydrolysis of cellulose acetate in the pH range of 2–10. *J. Appl. Polym. Sci.* 5:211.

Watson, B., and C. Hornburg. 1989. Low-energy membrane nanofiltration for removal of color, organics, and hardness from drinking water supplies. *Desalination* 72:11.

Wiley, D., C. Fell, and A. Fane. 1985. Optimization of membrane module design for brackish water desalination. *Desalination* 52:249.

Williams, M., R. Deshmukh, and D. Bhattacharyya. 1990. Separation of hazardous organics by reverse osmosis membranes. *Environmental Prog.* 9:118.

24

Selected Applications

Michael E. Williams and Dibakar Bhattacharyya
University of Kentucky

Roderick J. Ray and Scott B. McCray
Bend Research, Inc.

INTRODUCTION
SEAWATER AND BRACKISH
  WATER DESALINATION
    Effect of Feed Water Composition
      and Characteristics
    Feed Water Intake Systems
    Feed Water Pretreatment
    Membranes and Membrane Configurations
    Post-Treatment
    Concentrate Disposal
    Power Recovery Devices
    Corrosion
    Process Control and Measurements
    Operation and Maintenance
    Hybrid Processes
    Computer-Aided Process Design
    Case Study: Jeddah 1 RO Desalination Plant
    Case Study: RA's Abu Jarjur Desalination Plant
WASTEWATER AND HAZARDOUS
  WASTE TREATMENT
    Industrial and Hazardous Wastewater
    Municipal Wastewater
APPLICATIONS OF RO IN THE
  CHEMICAL PROCESSING
  INDUSTRY
    Applications in the Electroplating
      and Metal-Finishing Industry
    Applications in the Pulp and Paper
      Industry
    Applications in the Textile Industry
    Applications in the Petroleum Industry
    Applications in the Power-Generation
      Industry
    Other Wastewater Applications
NANOFILTRATION APPLICATIONS
  IN WASTEWATER TREATMENT
  AND CPI INDUSTRIES
SURFACE WATER AND
  GROUNDWATER TREATMENT BY
  MEMBRANES
    Reverse Osmosis Membranes
    Nanofiltration Membranes
APPLICATIONS IN THE FOOD
  PROCESSING INDUSTRY
    Water Treatment Applications
    Product and Chemical Recovery Applications
    Concentration/Dewatering Applications
    Fractionation Applications
    Conclusions
APPLICATIONS OF HYBRID
  PROCESSES
REFERENCES

INTRODUCTION

The development of new-generation membranes that can tolerate wide pH ranges, higher temperatures, and harsh chemical environments and that have improved water flux and solute rejection characteristics has resulted in many applications for the reverse osmosis (RO) process. In addition to the traditional seawater and brackish water desalination processes, RO membranes have found applications in wastewater treatment, production of ultrapure water, water softening, food processing, and many other applications. Table 24-1 shows some applications of RO along with selected references. Membrane processes for these applications have several advantages over many of the traditional separation techniques such as distillation, extraction, ion exchange, and adsorption. No energy-intensive phase changes or potentially expensive solvents or adsorbents are needed for membrane separations, and simultaneous separation and concentration of both inorganic and organic compounds is possible with the RO process. Also, the RO process is inherently simple to design and operate compared to many traditional separation processes. Reverse osmosis can also be combined with ultrafiltration, pervaporation, distillation, and other separation techniques to produce hybrid processes that result in highly efficient and selective separations.

In addition to traditional RO membranes, nanofiltration ("loose RO") membranes, which have low salt rejection and high water fluxes at low pressures, are finding increasing numbers

TABLE 24-1. Selected Applications of Reverse Osmosis.

Application	Selected References
Seawater and brackish water desalination	Spiegler and Laird 1980; Ko and Guy 1988
Wastewater treatment	
General	Slater, Ahlert, and Uchrin 1983a; Ghabris, Abdel-Jawad, and Aly 1989
Industrial and municipal wastewater	Bhattacharyya et al. 1987; Bhattacharyya and Williams 1990; Nusbaum and Argo 1984
Electroplating wastewater	Sato et al. 1977; Spatz 1979; Imasu 1985
Pulp and paper wastewater	Olsen 1980; Paulson and Spatz 1983; Chakravorty and Srivastava 1987
Textile wastewater	Brandon et al. 1981; Porter and Goodman 1984; Slater, Ahlert, and Uchrin 1987
Nanofiltration applications in wastewater treatment	
Textile	Simpson, Kerr, and Buckley 1987
Pulp and paper	Bindoff et al. 1987
Electroplating	Cadotte et al. 1988
Dye manufacturing	Perry and Linder 1989
Food processing	Ikeda et al. 1988; Cadotte et al. 1988
Hazardous wastewater	Bhattacharyya, Adams, and Williams 1989; Bhattacharyya and Williams 1990
Surface water and groundwater treatment	
Reverse osmosis	Wilson and Duran 1982; Eisenberg and Middlebrooks 1986; Baier et al. 1987
Nanofiltration	Conlon 1985; Simpson, Kerr, and Buckley 1987; Cadotte et al. 1988; Taylor et al. 1989

of applications. Nanofiltration membranes are often negatively charged so anion repulsion mainly determines salt rejection; as a result, monovalent ions (i.e., Cl^-) are rejected less than highly charged ions (i.e., SO_4^{2-}), and as monovalent salt concentrations increase, rejections of these monovalent ions decrease sharply. Because of this, solutions with high osmotic pressures due to high concentrations of monovalent salts will pass through the membrane even at low pressures while highly charged ions and high molecular weight (>200) organics will be rejected. These characteristics make nanofiltration membranes ideal for many applications, including water softening, desalting, food processing, and wastewater treatment.

SEAWATER AND BRACKISH WATER DESALINATION

The desalination of seawater and brackish water to produce potable water is the most mature application of the RO process; this process has been used successfully over the last two decades in virtually every part of the world. Recently, RO processes were estimated to represent 23.4% of the 11.5 million m^3/day (3.0 billion gal/day) of potable water produced worldwide by all desalination processes (Wangnick 1986). Competing desalination processes include multistage flash (MSF) evaporation, vapor compression (VC), multieffect distillation (MED), solar evaporation, freezing processes, and electrodialysis processes (Spiegler and Laird 1980). Many comparisons between these different processes have been made, and while MSF is the major producer of potable water, many agree that RO will account for an increasingly larger fraction of desalinated water in the future because of the advantages of the RO process. These advantages include lower energy requirements, lower capital and operating costs, low land and space requirements, reduced equipment corrosion problems, ease of construction, operation, and maintenance due to the modular nature of the process, and short construction times (Sackinger 1982; Glueck-

TABLE 24–2. Selected Desalination Plant Locations and Capacities.

Plant Location	Capacity, m^3/day (mgd)	Reference
Seawater Plants		
Jeddah, Saudi Arabia (SA)	57,000 (15)	Muhurji et al. 1989
Al-Birk, SA	2,300 (0.6)	Hassan et al. 1989
Umm Lujj, SA	4,400 (1.2)	Light et al. 1988
Doha, Kuwait	3,000 (0.8)	Malik et al. 1987
Yanbu, SA	5,000 (1.3)	Wojcik 1983
Jeddah, SA	2,300 (0.6)	Wojcik 1983
Jeddah, SA	12,000 (3.2)	Muirhead, Beardsley, and Aboudiwan 1982
Key West, Florida	11,400 (3.0)	Boesch 1981
Brackish Water Plants		
RA's Abu Jarjur, Bahrain	46,000 (12.2)	Al-Arrayedh, Ericsson, and Amin-Saad 1987
Manfouha I, SA	27,300 (7.2)	Wojcik 1983
Manfouha II, SA	36,400 (9.6)	Wojcik 1983
Malez, SA	18,200 (4.8)	Wojcik 1983
Shemessy, SA	27,300 (7.2)	Wojcik 1983
Salbukh, SA	38,400 (10.1)	Wojcik 1983
Buwayb, SA	45,000 (11.9)	Wojcik 1983
Jubail, SA	15,000 (4.0)	Wojcik 1983
Shedgum, SA	5,300 (1.4)	Wojcik 1983
Dhahran, SA	3,500 (0.9)	Wojcik 1983
Berri, SA	6,800 (1.8)	Wojcik 1983
Riyadh, SA	4,500 (1.2)	Wojcik 1983
Majmaah, SA	3,800 (1.0)	Wojcik 1983
Meccah, SA	15,000 (4.0)	Wojcik 1983

stern and Kantor 1983; Soo-Hoo et al. 1983; Akashah et al., 1987; Darwish 1987).

Numerous RO desalination plants with widely varying capacities and feed waters have been installed in various locations around the world. These plants use various pretreatment schemes, membranes, and membrane configurations. Table 24–2 lists locations of some of these plants and capacities.

Many factors affect the design of a RO desalination plant. These include site-specific and designer-specified variables. Site-specific vari-

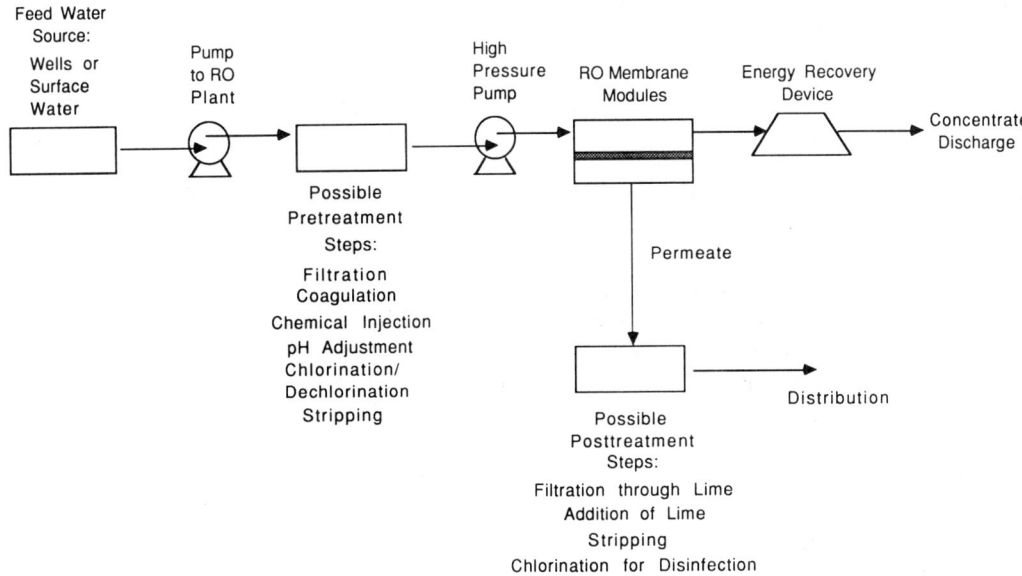

FIGURE 24-1. Typical desalination process.

ables are those that depend on the location of the desalination plant such as feed water characteristics and composition and energy costs. Designer-specified variables include the type and configurations of membranes and pretreatment methods that are used, the product water quality and quantity required, the operating pressure, and the percent recovery of the feed water. Other factors must also be considered because of the nature of the desalination process. These include the use of corrosion-resistant materials because of the corrosive nature of brackish water and seawater and the use of construction materials that can withstand the high pressures of the RO process for long periods of time.

Many different designs are used in desalination plants around the world. These plants have various degrees of complexity and sophistication. The larger and more expensive plants use several pretreatment stages, highly efficient membranes and membrane configurations, high-efficiency pumps, energy recovery devices, and sophisticated control systems. In contrast, simple RO desalination plants have been successful with only minimum pretreatment, simple pumping systems, and inexpensive membranes (Glueckstern and Kantor 1983). However, most desalination plants have several basic components in common. These consist of the feed water intake system, various pretreatment steps, RO membranes, product water post-treatment, and power recovery systems (for larger seawater plants). These components are shown in Figure 24-1.

Effect of Feed Water Composition and Characteristics

The composition and characteristics of the feed water used in RO desalination plants directly affect the RO process. The total dissolved solids (TDS) concentration and to a lesser extent the temperature determine the osmotic pressure of the feed, which affects the water flux through RO membranes [since $J_w = (P_w/l)(\Delta p - \Delta \pi)$]. Higher osmotic pressures of more concentrated and higher temperature feed waters result in less water flux through the RO membranes and so make higher operating pressures or more membrane area necessary to produce a given quantity of permeate. The osmotic pressure of seawater can be approximated as 10 psi/1000 ppm of TDS (Weber

1972) or from the relation (Ko and Guy 1988):

$$\pi = 1.19(T + 273)\sum M_i, \quad (24\text{-}1)$$

where M_i is the molar concentration (mol/L) of the individual ions and T is the temperature in degrees centigrade of the feed water. Other properties of seawater (viscosity, vapor pressure, etc.) are discussed in Spiegler and Laird (1980).

The composition of brackish water and seawater varies considerably depending on the location of its source. While seawater contains approximately 35,000 ppm TDS in regions around the United States, Europe, and Japan, seawater in the Middle East can range between 40,000 and 50,000 ppm TDS (Kurihara et al. 1983). A typical analysis of seawater from the Arabian Gulf is shown in Table 24–3 along with the temperature range (Ericsson, Hallmans, and Vinberg 1987). The temperature of the feed water must be considered since this can affect water flux and salt passage of the membranes used in the process; each type of membrane has a temperature above which the product water specifications are not met (Soleman, de Graauw and van Putten 1983). This can be an important consideration for feed waters that have large seasonal variations in temperature or in Middle Eastern brackish waters and seawaters that have extremely high temperatures (Kurihara et al. 1983). Some feeds (particularly seawater) will also contain other components such as suspended solids, colloids, organics such as humic materials, and microorganisms. These can severely reduce the operating life and performance of RO membranes and so must be removed by pretreatment methods. The extent of these compounds present in the feed water determines the types of pretreatment necessary (Allard, Rovel, and Treille 1976).

Feed Water Intake Systems

Two types of intake systems have been widely used to obtain feed water for the RO process. For seawater the preferred intake system is through seawells that are drilled near the coast; these offer several advantages over the direct intake of feed from open waters because of the excellent mechanical cleaning of the seawater by subterranean filtration. As a result, feeds from seawells typically have fewer suspended solids and organic materials present and lower dissolved oxygen, which can damage the membranes used in the process and cause increased corrosion problems. Also, seawells typically contain fewer biologically active waters, and seasonal temperature fluctuations of the feed water are reduced or avoided with seawells (Soleman, de Graauw, and van Putten 1983; Heyden 1985). These characteristics can result in substantially less stringent pretreatment requirements (Pepper 1981). However, if these wells contain much higher TDS than the open seawater or do not provide a steady feed source, then the feed water must be taken from the surface water near the coast.

With open water intake systems special attention must be given to seasonal temperature fluctuations, varying feed water composition at low and high tides, and the possible uptake of dirt, sand, seaweed, fish, and algae, which can plug the intake system (Soleman, de Graauw, and van Putten 1983). Problems with surface seawater intake systems have been extensively studied and several designs (such as the use of intake screens) are available that help offset these problems (Kreshman 1985; Ericsson, Hallmans, and Vinberg 1987; Nada, Zahrani, and Ericsson 1987; Muhurji et al. 1989). Brackish water feed is usually taken from wells,

TABLE 24–3. Analysis of Arabian Gulf Seawater.

Temperature range (°C)	11–35
pH	8.1
Dissolved oxygen	5.5 mg/L
Total dissolved solids	43,800 mg/L
Total alkalinity (as calcium carbonate)	144 mg/L
Carbon dioxide	2 mg/L
Total hardness (as calcium carbonate)	8010 mg/L
Calcium	508 mg/L
Magnesium	1618 mg/L
Sodium	13,440 mg/L
Potassium	483 mg/L
Strontium	17 mg/L
Bicarbonate	176 mg/L
Chloride	24,090 mg/L
Sulfate	3384 mg/L
Bromide	83 mg/L
Fluoride	1 mg/L

although brackish rivers have also been used as feed sources. In general, the unique characteristics of each plant site must be taken into consideration when choosing the best feed water intake system for a particular plant site. This could include drilling test wells or determining the best location along the coastline to place a desalination plant so that problems with surface water intake systems are minimized.

Feed Water Pretreatment

Feed water pretreatment is essential to prevent membrane fouling or damage by harmful brackish water and seawater components or oxidizing agents. The degree of pretreatment that is needed is determined by both the quality of the intake feed water and the type of membrane used in the plant; different manufacturers require varying degrees of quality for the membrane feed water (Soleman, de Graauw, and van Putten 1983; Jawad 1989). The raw feed water can contain materials such as silt and other suspended solids, colloidal material, microbiological organisms, and dissolved solids. The suspended solids and colloidal materials must be removed to prevent fouling of the membrane; this fouling can result in substantial reductions in water flux through the membrane and permeate water quality. Microorganisms must be controlled because these can also foul the membrane by forming slime layers on the membrane surface, greatly reducing flux. Also, many membranes are attacked and degraded by microorganisms leading to loss of quality and bacterial contamination of the product water (Allard, Rovel, and Treille 1976; Pepper 1981; Osta and Bakheet 1987; Al-Borno and Abdel-Jawad 1989). In addition, some of the salts present in seawater are near saturation and so when these are concentrated by the membrane there is potential for precipitation or scale formation on the membrane surface which can greatly reduce the water flux. The salts of greatest concern are calcium carbonate and calcium sulfate; these must be removed or stabilized before the feed is passed to the membrane (Pepper 1981; Al-Borno and Abdel-Jawad 1989).

The species of most concern and pretreatment methods typically used to remove them are shown in Table 24-4. Generally, a larger number of pretreatment methods results in a better quality feed water to the membrane. However, since the cost of pretreatment is usually the major operating cost in the RO process, in most cases the cost of additional pretreatment must be balanced against the cost of membrane replacement that would be necessary if fewer pretreatment steps were applied (Pepper 1981; Jawad 1989). The minimum amount of pretreatment that must be provided can be determined for a particular water recovery (Ebrahim and Darwish 1989).

Suspended Solids

Suspended solids present in the feed water must be removed to prevent clogging of the membrane. Sand or multimedia filtration (gravel, sand, and anthracite) followed by 5- to 25-μm cartridge filtration is usually used to remove suspended solids (Allard, Rovel, and Treille 1976; Kawana et al. 1987; Osta and Bakheet 1987). Coagulation followed by settling and sand filtration and in-line coagulation and direct filtration have also been used (Tsuge, Suda, and Matsumura 1981; Al-Borno and Abdel-Jawad 1989). Ultrafiltration pretreatment can also remove suspended solids (Heyden 1985).

Precipitation and Scale Formation

As the feed water is concentrated by the RO membrane, the solubility limits of some of the components in the concentrate may be exceeded. This can cause formation of a precipitate or scale on the surface of the membrane, causing large flux drops. The components of concern include $CaCO_3$, $MgCO_3$, $CaSO_4$, silica (SiO_2), $BaSO_4$, $SrSO_4$, CaF_2, and iron and other metal oxide precipitates (from corrosion). One method of preventing precipitation is to operate at low water recoveries so that the solubility limits are not exceeded and no precipitation occurs (Nada, Zahrani, and Ericsson 1987). Pretreatment to avoid scaling or precipitation may consist of injection of acid or chelating agents, ion-exchange softening, or chemical precipitation (Allard, Rovel, and Treille 1976). Acid is

TABLE 24-4. Pretreatment Methods Used in Seawater and Brackish Water Desalination.

Species	Problem	Pretreatment Method
Suspended solids	Fouling of membrane by particles causes reduced flux	Sand filtration Multimedia filtration Coagulation filtration Cartridge filtration Ultrafiltration
Precipitation/scale formation by $CaCO_3$, $MgCO_3$, $CaSO_4$, SiO_2, $BaSO_4$, $SrSO_4$, CaF_2, metal (i.e., Fe) hydroxide	Fouling of membrane by precipitate or scale causes reduced flux	Operate at low water recovery (so solubility not exceeded) Acid or chelating agent addition to prevent precipitation Chemical precipitation (i.e., lime-softening) of sparingly soluble components Addition of antiscalants (i.e., SHMP) Sand filtration to remove SiO_2
Colloids (clays, iron colloids, $Al(OH)_3$)	Fouling of membrane by colloids causes reduced flux	Coagulation followed by filtration Ultrafiltration
Microorganisms	Slime layers on membranes cause reduced flux; some membranes (i.e., cellulose acetate) degraded by microorganisms	Chlorination (shock treatment or continuous addition) Sodium bisulfite addition (shock treatment or continuous addition) UV light treatment Ozonation Copper sulfate addition (for algae, plankton) Chloramine addition
Chlorine	Chlorine added for disinfection will damage most membranes	Sodium bisulfite addition Activated carbon filters
Hydrogen sulfide	Not removed by membrane	Oxidation to sulfate Air stripping
Organics	Adsorption on membrane can cause loss of water flux over time; some high MW organics can coagulate to form colloids	Activated carbon Replace use of cationic polymers (coagulant), which cause formation of organic colloids
Dissolved oxygen	Oxygen can damage some types of membranes; oxygen can cause increased corrosion problems	Sodium bisulfite addition Vacuum deaeration
pH	Should be in acceptable operating range of membrane	Adjust with acid (HCl, H_2SO_4) or base (lime, NaOH)

usually added to cause the formation of CO_2 from $CaCO_3$ and $MgCO_3$ and thus eliminate carbonate scaling (Allard, Rovel, and Treille 1976; Pepper 1981; Osta and Bakheet 1987; Nada, Zahrani, and Ericsson 1987). Lime-softening can also be used to prevent $CaCO_3$, $MgCO_3$, and some silica scaling (Allard, Rovel, and Treille 1976). Antiscalants are typically added to prevent $CaSO_4$ precipitation but can also prevent precipitation of $BaSO_4$, $SrSO_4$, $CaCO_3$, and CaF_2. These agents bind the calcium and slow down the rate of precipitation. Sodium hexametaphosphate (SHMP) is the antiscalant most commonly used; however, polymeric compounds (such as polyelectrolytes and polyacrylates) are gaining wide acceptance (Allard, Rovel, and Treille 1976; Pepper 1981; Amjad 1985; Osta and Bakheet 1987; Nada, Zahrani, and Ericsson 1987). Silica scale can be reduced by decreasing recovery, chemical pretreatment, or sand filtration (Pepper 1981; Nada, Zahrani, and Ericsson 1987). The use of corrosion-resistant materials in the desalination plant can reduce precipitates of metal oxides.

Colloids

Colloidal particles inside the membrane modules can, under favorable conditions, coagulate and foul the membrane, causing a large drop in flux. This is particularly a problem for surface seawater treatment. Colloids include small particles of aluminum silicate (clays), iron colloids from corrosion, and aluminum hydroxide. Colloids are usually removed by coagulation followed by filtration (Allard, Rovel, and Treille 1976; Pepper 1981; Arrayedhy 1987). Coagulants that are used include alum ($Al_2[SO_4]_3$), $FeCl_3$, $FeClSO_4$, or coagulating polymers or polyelectrolytes. Ultrafiltration is also effective in removing colloids (Heyden 1985).

Microorganisms

Microorganisms can form slime layers on membranes and cause large flux drops. Also, some membranes (particularly cellulose acetate) are degraded by microorganisms. As a result microorganisms must be removed before the membranes. The most common method of pretreatment is by chlorination. Chlorination can be performed either by shock treatment or continuous addition. Shock treatment typically consists of injecting 5 to 10 mg/L chlorine for 15 minutes every 8 hours; continuous addition involves continuously injecting 0.5 to 2 mg/L to the feed water. Treatment with formaldehyde or sodium bisulfite will also limit the extent of biological growth; sodium bisulfite shock treatment is frequently used (Osta and Bakheet 1987; Arrayedhy 1987). Chlorination followed by dechlorination with periodic sodium bisulfite shock treatment is also very effective in controlling the growth of microorganisms (Applegate and Erkenbrecher 1987); sodium metasulfite and sulfur dioxide also have been used. Ultraviolet light, ozone, and copper sulfate have also been used for seawater disinfection (Al-Borno and Abdel-Jawad 1989). Copper sulfate can be added to inhibit growth of several microorganisms (such as algae and plankton); quantities of <0.5 mg/L of $CuSO_4 \cdot 5\ H_2O$ are sufficient in most cases (Light et al. 1988). The use of chloramines has also recently been reported for the disinfection of membrane feed water (Applegate, Erkenbrecher, and Winters 1989). The chloramines successfully prevented biological growth but did not react with humic materials present to form trihalomethane compounds, which are sometimes a problem in the chlorination process.

Chlorine (Dechlorination)

Chlorination is used in most desalination processes to prevent biological growth on membranes. However, most membranes cannot tolerate residual chlorine concentrations so it is necessary to dechlorinate the feed water before it is passed to the membrane. Sodium bisulfite is added to remove excess chlorine in almost all desalination plants (Al-Borno and Abdel-Jawad 1989). From 1.5 to 6 mg/L of sodium bisulfite is added to react with residual chlorine concentrations of 1 mg/L. Activated carbon filters also can serve to remove residual chlorine and are sometimes used after sodium bisulfite addition to provide extra protection of the membranes against chlorine damage (Ebrahim and Malik 1987a, 1987b).

Hydrogen Sulfide

Most hydrogen sulfide can be removed by placement of a degassing tower before or after the membrane modules. When degassing is done before the membranes, calcium hypochlorite or chlorine is added to oxidize residual hydrogen sulfide and colloidal sulfur formed to sulfate (Pepper 1981). To avoid the possibility of sulfur generation and formation of colloids that can foul the membranes, air stripping of the hydrogen sulfide from the permeate can be used (Arrayedhy 1987).

Organics

Organic materials such as humic and fulvic acids and purgeable organic compounds such as trihalomethanes (products of chlorination reactions with natural organic materials in untreated feed water) can result in an increased drop in membrane water flux over time (Moody et al. 1983; Bergman, Harlow, and Laverty 1985). Organic fouling components can affect water flux and salt rejection and are difficult to detect since these cannot be determined by the silt density index (SDI) or fouling index analysis. Most dissolved organics in seawater range from 1 mg/L in open ocean water to 80 to 100 mg/L in coastal waters and resemble humic acids. These compounds can coagulate and cause fouling if membranes with high water fluxes are used or cationic polymers (as coagulant) are used in pretreatment. Replacement of the cationic polymers in pretreatment can reduce this type of organic fouling (Winters 1987). In addition, activated carbon can be used to remove organics from the feed water. This is sometimes used when the feed waters from wells contain low concentrations of dissolved organics caused by petroleum contamination (Arrayedhy 1987).

Dissolved Oxygen

Dissolved oxygen must be removed in some cases to prevent membrane damage and reduce corrosion problems. Addition of sodium bisulfite or vacuum deaeration can be used to remove dissolved oxygen from the feed water.

pH

The pH of the feed water sometimes must be adjusted before it is fed to the membranes. For cellulose acetate membranes, the feed pH should be in the range of 4 to 7 to minimize membrane hydrolysis. Also, many times membrane rejection can be increased by operating at an optimum feed pH.

Membranes and Membrane Configurations

Many desalination membranes are commercially available. The type of membrane used is determined by the water flux and product water quality requirements, stability of the membrane under the operating conditions of the desalination plant, and membrane costs. The most widely used membrane module configurations for desalination are the spiral-wound or hollow-fiber membrane. These allow large membrane surface area per module and so minimize space requirements.

Desalination can be carried out by a single-pass or double-pass RO arrangement. In the single-pass configuration, highly rejecting membranes are utilized to give permeate water that can be directly utilized; this configuration is shown in Figure 24–2. In double-pass RO, the permeate of a set of seawater membranes serves as the feed to a set of brackish water membranes as shown in Figure 24–3. Part of the permeate from the seawater membranes bypasses the brackish water membranes and is blended with the high-quality permeate of the brackish water membranes to give a product water of acceptable quality. This arrangement minimizes the number of brackish water mem-

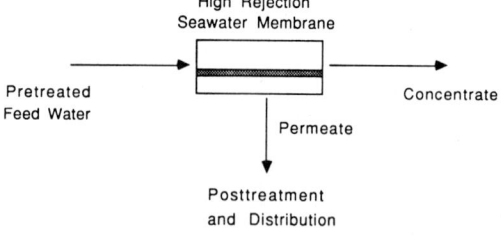

FIGURE 24–2. Single-pass RO process.

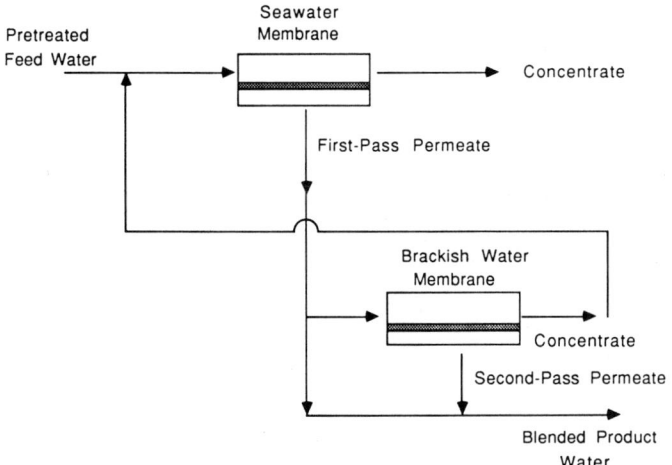

FIGURE 24–3. Double-pass RO process.

branes needed and thus reduces capital and operating costs. The concentrate streams from the brackish water membranes are mixed with the feed water to the seawater membranes since it has low TDS compared to seawater and has already been pretreated. The type of configuration used depends on the type of membrane used and the salinity of the feed water.

Double-pass operation usually requires less energy than single-pass operation for membrane salt rejections up to 99.2%; for membranes with higher salt rejections, single-pass systems are more efficient (Kaschemekat et al. 1983). However, single-pass membranes with high rejections usually have lower flux, and double-pass RO desalination is usually more reliable (Soleman, de Graauw, and van Putten 1983). For seawater desalination feeds with salinity of 35,000 to 38,000 mg/L, the single-pass process is favored because of the simplicity in design and operation and no interstage pumps are required to provide the feed pressure to the second-pass membranes. Above 38,000 mg/L a double-pass arrangement is usually favored (Klinko, Light, and Cummings 1985); however, single-pass RO desalination is possible for high-salinity feed waters (Kurihara et al. 1983).

Recoveries for seawater feeds are typically in the range of 25 to 35%. High product water recovery rates are favored because these result in lower energy consumption and savings in feed water pretreatment costs since less feed is required to produce a specified quantity of product water (Kaschemekat et al. 1983; Sackinger 1985; Winters 1987). Higher recoveries can be achieved in two ways: by increasing the membrane area or by increasing the operating pressure to increase permeate flux. More membrane area can be achieved by placing the membrane modules in stages as shown in Figure 24–4. The concentrate from the first stage serves as the feed to the second stage and so on. However, savings in energy and pretreatment costs must be balanced with the additional costs of the extra membrane modules required (Kaschemekat et al. 1983; Wade 1987; Sulpizio, Light, and Perlman 1989). Higher operating pressures can also increase recoveries. High-pressure operations up to 8.3 MPa (1200 psi) with 50% recovery have been reported (Sackinger 1985; Winters 1987).

Post-Treatment

The permeate product water from the membranes is usually treated to remove carbon dioxide, reduce its corrosive nature, and provide disinfection. Carbon dioxide formed when acid is added in pretreatment passes through the membranes. A decarbonator is sometimes used to remove CO_2 followed by blending with oxygen to improve the taste of the product water (Nada, Zahrani, and Ericsson 1987). More frequently the permeate is passed through a lime-

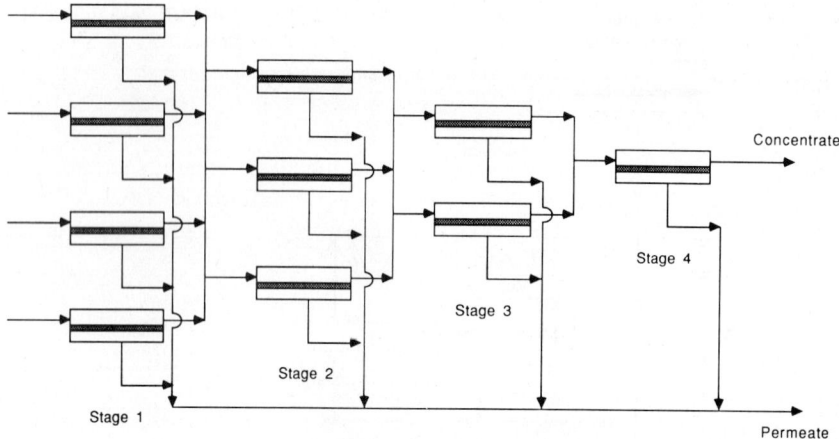

FIGURE 24-4. RO stages to provide high water recoveries.

stone ($CaCO_3$) bed, lime [$Ca(OH)_2$] is added, or caustic soda (NaOH) is added to cause the formation of calcium bicarbonate ($CaHCO_3$) (Heyden 1985; Al-Awadi and Abdel-Jawad 1987; Nada, Zahrani, and Ericsson 1987; Malik et al. 1987; Hassan et al. 1989). This helps reduce corrosion problems in the water distribution system. Also, disinfection by chlorination (addition of chlorine gas or sodium or calcium hypochlorite) is usually performed to protect against contamination with microorganisms in the distribution system and to remove any microorganisms that may have passed the membranes into the permeate.

Concentrate Disposal

For desalination plants located near coastal areas, the concentrate stream from RO is discharged directly to the open ocean. The discharge point is usually placed a distance away from feed water intake systems to avoid the possibility of increasing the salinity of the feed water. When the concentrate cannot be directly discharged, such as could be the case in some brackish water desalination plants, it is frequently placed in evaporation ponds (Finlay and Ferguson 1977); however, attention should be given to the possibility of salt-contaminated water leaching from evaporation ponds and contaminating groundwater. Distillation following RO for brackish water has been suggested as one method of reducing brackish water concentrate disposal problems (Leitner 1983). It has also been suggested that valuable by-products such as NaCl, chlorine (by electrolysis process), and trace metals can be obtained from the concentrate (Al-Mutaz 1987).

Power Recovery Devices

In seawater desalination the concentrate from the RO stages is at high pressure. Energy recovery devices are used to recover work from the concentrate stream. The addition of energy recovery equipment typically reduces net power consumption by 25 to 40% (Soo-Hoo et al. 1983). Pelton wheels and reverse-running pumps are the most frequently used energy recovery devices although others such as a hydraulic turbocharger (22% reduction in energy requirements) (Lozier, Oklejas, and Silbernagel 1989) and a pressure exchanger (recovery of up to 60% of pumping energy) (Darwish, Abdel-Jawad, and Hauge 1989) have been proposed. In Pelton wheel turbines the concentrate passed through a turbine rotates the shaft of the turbine; the shaft work produced along with a motor to provide extra power can be used to drive the feed pump to the RO membranes (Hoeting 1982; Wilson, Gruendisch, and Calder-Potts 1987; Darwish, Abdel-Jawad, and Hauge 1989). Reversed centrifugal pumps used as turbines can be used to drive feed pumps directly; the feed pressure can be raised by subsequent booster pumps.

Corrosion

Brackish water and seawater are highly corrosive; these are even more corrosive when pressurized in the RO process and if acidified as is done in pretreatment in many desalination plants. Corrosion can result in equipment and piping failures and produce metal oxides that can foul RO membranes. To reduce corrosion in the RO desalination plant, contact with metallic surfaces should be minimized by using plastic materials such as PVC or polyethylene in the low-pressure (permeate) sections of the plant. Stainless steel should be used in the high-pressure parts of the plant. In brackish water desalination plants, AISI 316L stainless steel can be used, but in more saline waters such as seawater higher alloy stainless steels and nickel-base alloys are needed to avoid pitting and crevice corrosion. A number of stainless and alloy steel materials have been used in desalination plants and have shown good resistance to corrosion (Oldfield and Todd 1985; Nordin, Ericsson, and Wallen 1985; Hassan and Malik 1989; Carew et al. 1989). Oldfield and Todd (1985) discuss several materials that are suitable for use in an RO desalination plant.

Process Control and Measurements

Several parameters must be measured and controlled in the RO desalination plant. These include the following: membrane feed pH to avoid alkaline ($CaCO_3$) scale formation, prevent membrane damage (particularly important for cellulose acetate membranes), and optimize salt rejection; temperature to protect against membrane damage from excessive temperatures and enhance productivity for low-temperature feed waters; pressure to provide adequate membrane flux and protect against pressure surges, which could damage the membrane; flow to assure adequate membrane flux and prevent concentration polarization; and conductivity to assure product water quality and detect fouling,

leaking, or damage to the RO modules (Mindler and Epstein 1986). Control systems and monitors can also be used for pre- and post-treatment processes. Control systems can improve data acquisition, improve reliability, allow optimization of operating parameters, and provide savings in manpower needed for operation (Glueckstern et al. 1985; Fredkin and Banks 1989).

Operation and Maintenance

Operation and maintenance in a RO desalination plant consists of operation and repair of equipment (pumps, piping), maintenance of pre- and post-treatment systems (including filter changes, maintenance of proper chemical levels, and backflushing sand filters), and monitoring of feed and product water quality to ensure these are of sufficient quality. Individual membrane modules should be checked regularly to determine if they need to be replaced or cleaned. Also, since most desalination plants are placed in harsh environments (such as near coastal areas), process control and other instruments should be checked regularly to monitor for corrosion problems.

Hybrid Processes

Hybrid processes involving the integration of RO with other processes have also been used for desalination. Figure 24-5 shows a combination of a MSF distillation process and RO. Advantages of the RO/MSF process include lower capital costs and energy requirements for a combined RO/MSF process compared to that of a MSF plant with the same capacity; smaller feed water requirements since the concentrate of the RO is used as the feed to the MSF unit; the permeate water of RO with a relatively high TDS can be blended with the high purity water of the MSF units to obtain suitable quality product water, eliminating the need to use double-pass RO desalination; mixing with the RO permeate reduces the temperature of the MSF product water; and electric power generation from MSF can be utilized in the RO plant (Dar-

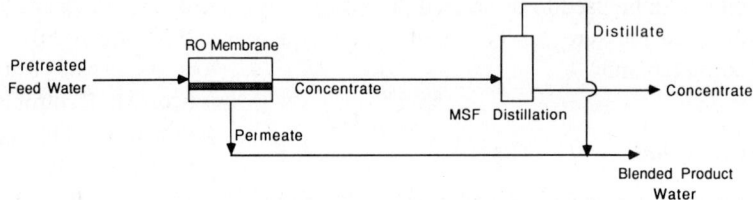
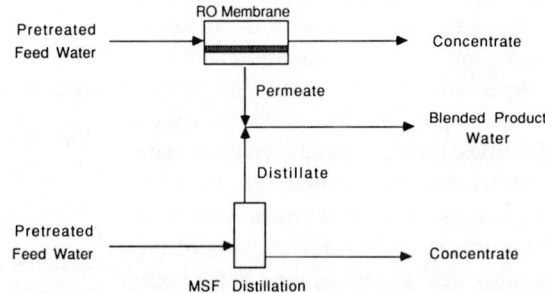

FIGURE 24-5. Hybrid RO/MSF desalination processes.

wish 1987; Al-Mutaz, Soliman, and Daghthem 1989; Awerbuch et al. 1989).

Computer-Aided Process Design

The design of a RO desalination plant requires consideration of the different variables that have been discussed. Computer programs are available to simplify greatly the design procedure, and these are used by most membrane system manufacturers (Ko and Guy 1988).

Case Study: Jeddah 1 RO Desalination Plant

The Jeddah 1 RO plant in Jeddah, Saudi Arabia, is a 15 mgd RO seawater desalination plant that has been operating since April 1989 (Muhurji et al. 1989). A simplified schematic of the desalination process is shown in Figure 24-6. Seawater (43,000 ppm) is taken from the Red Sea and pumped to the RO plant. Pretreatment consists of addition of sodium hypochlorite to control bacteria and algae growth, addition of ferric chloride as a coagulant, and multimedia (gravel, sand, and anthracite) filtration. Sulfuric acid is added to prevent scale formation and membrane hydrolysis, and the feed is passed through micron cartridge filters. The plant uses Toyobo Hollosep hollow-fiber membrane modules made of cellulose triacetate at an operating pressure of 5.9 MPa (850 psi). The product water has a TDS of 145 mg/L, and the plant has exceeded or met all of its design requirements.

Case Study: RA's Abu Jarjur Desalination Plant

The RA's Abu Jarjur desalination plant in Bahrain has a capacity of 10 mgd (Al-Arrayedh, Ericsson, and Ohtani 1985; Al-Arrayedh, Ericsson, and Amin-Saad 1987). A schematic of the plant is given in Figure 24-7. The feed water is taken from brackish wells. Pretreatment consists of addition of coagulant (polyelectrolyte) followed by multimedia (sand and anthracite) filtration. Activated carbon (which is regenerated) is used to remove hydrocarbons contained in the brackish water. Sulfuric acid to control $CaCO_3$ scale, SHMP to control $CaSO_4$ scale, and $NaHSO_3$ for shock treatment disinfection are also added before microfiltration. The membranes used consist of Du Pont Permasep B-10 hollow-fiber modules operating at 6.5 MPa (940 psi). Post-treatment includes stripping of H_2S that passed through the membrane and addition of chlorine and lime

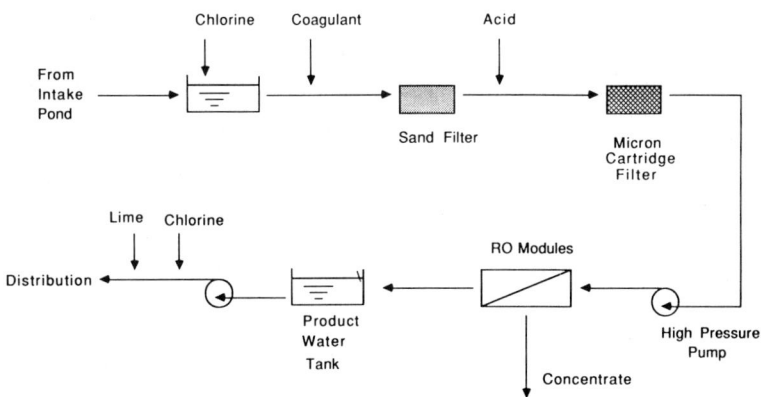

FIGURE 24–6. Schematic of Jeddah RO desalination plant.

for disinfection and corrosion protection. The plant produces water with TDS of 190 mg/L from feed water containing 13,000 mg/L TDS.

WASTEWATER AND HAZARDOUS WASTE TREATMENT

Membrane processes have received considerable attention for the separation and concentration of inorganics and organics from various wastewaters. The development of low-pressure processes has made RO an attractive alternative for the treatment of dilute aqueous wastes since these offer high fluxes and solute separations and can operate over wide temperature and pH ranges. RO processes for wastewater treatment have been applied to the chemical, textile, petrochemical, electrochemical, pulp and paper, and food industries as well as for the treatment of municipal wastewater and landfill leachates (Slater, Ahlert, and Uchrin 1983a; Ghabris, Abdel-Jawad, and Aly 1989). Membrane pro-

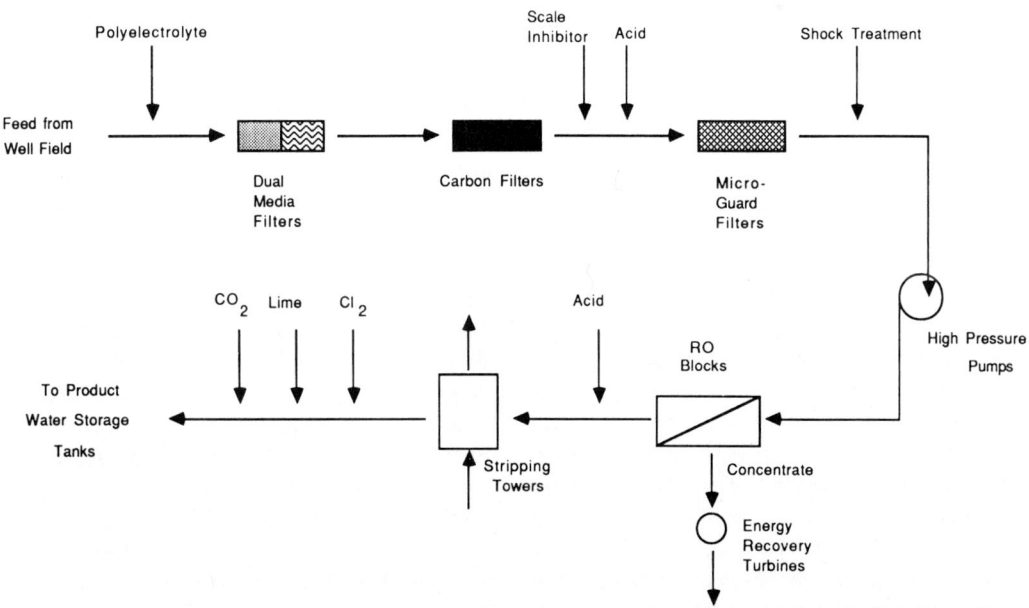

FIGURE 24–7. Schematic of RA's Abu Jarjur RO desalination plant (adapted from Al-Arrayedh, Ericsson, and Amin-Saad 1987).

cesses have been combined with or substituted for traditional advanced treatment technologies such as biological treatment, adsorption, stripping, oxidation, and incineration. The membrane processes can be used to concentrate and purify simultaneously wastewater containing both inorganics and organics and produce a 20- to 50-fold decrease in waste volumes that must be treated with other processes such as incineration. In addition, charged (nanofiltration) membranes allow for the possibility of selective separation of hazardous components.

Factors affecting solute removal from aqueous wastes by RO membranes are as follows: the type of solute, feed concentration, extent of water recovery, feed pH, temperature, operating pressure, and type of membrane (Slater, Ahlert, and Uchrin 1983a; Ghabris, Abdel-Jawad, and Aly 1989). Pretreatment of the waste streams to remove particulates, oils and greases, suspended solids, colloids, and oxidizing agents is usually necessary to prevent fouling or damage to the membrane. The pretreatments most frequently used include filtration, coagulation, pH adjustment, and adsorption; however, the exact nature of the pretreatment processes that are required depends on the characteristics of the feed. As in the desalination process, in many cases the costs of pretreatment must be balanced against those of membrane replacement.

Industrial and Hazardous Wastewater

The interest in applications of membrane processes to treatment of wastewater containing hazardous organics and inorganics has generated a considerable amount of solute separation data in the literature for many types of RO membranes. Early studies involved the use of cellulose acetate membranes. Many organic and inorganic solute rejections from earlier work are compiled in Sourirajan (1970) and Sourirajan and Matsuura (1985). Edwards and Schubert (1974) described the use of cellulose acetate membranes for the separation of dilute solutions of salts of the herbicide 2,4-D; separations as high as 50% were obtained. Lonsdale et al. (1969) used cellulose acetate membranes to achieve 92.8% retention of the acid of 2,4-D, and others reported that the isopropyl ester of 2,4-D was retained at levels of more than 99.7% (Hinden, Bennett, and Narayanan 1969). Hinden, Bennett, and Narayanan (1969) also reported some success in the separation of several other compounds. Shuckrow, Pajak, and Osheka (1981) compiled studies of the treatment of industrial wastewaters with a wide range of contaminants using cellulose acetate membranes. For typical industrial wastewater, 95% removals of total organic carbon (TOC) were obtained, and chemical oxygen demand (COD) removals of more than 90% were achieved. Rejections of organic compounds were as follows: benzene, 59%; bis(2-ethylhexyl)phthalate, 67%; acenapthene, 73%; phenol, 25%; and methylene chloride, 10%. Removal of inorganic compounds included >42% for cyanide, 67% for chromium, >92% for arsenic, and 97% for zinc.

Chian, Bruce, and Fang (1975) compared cellulose acetate and composite (NS-100) membranes for the concentration of 15 major pesticides, which included seven chlorinated hydrocarbons and four organophosphorous compounds. Rejections of more than 99% were obtained with both membranes for pesticide concentrations ranging from 0.04 to 1.5 mg/L. A higher flux was obtained with the NS-100 composite membrane under the same operating conditions, but both membranes exhibited significant adsorption (up to 100% adsorption) of the pesticides on the membrane. Fang and Chian (1976) also compared rejections of polar organic compounds from aqueous solutions using cellulose acetate, composite (aromatic polyamide), and several other types of membranes. The 13 low molecular weight organics studied contained different functional groups, including acid, aldehyde, amide, amine, ester, ether, ketone, phenol, and alcohol; feed concentrations were 1000 mg/L. The aromatic polyamide membranes had rejections ranging from 10 to 95% while the cellulose acetate membranes had rejections ranging from 0 to 80%.

Scherm and Lawson (1977) used spiral-wound polyamide membranes to reduce COD in dilute wastewater from an organic chemical manufacturing plant after it had been treated by an activated sludge process, activated carbon, and multimedia filtration. Chloride rejections of 78% and organic rejections of >90% were maintained over seven weeks of operation at 80% water recovery. Removal of carcinogenic compounds from industrial waste streams using spiral-wound and hollow-fiber polyamide membranes was also reported (Light 1981). TOC rejections of 92.5% were obtained with both membranes for feeds containing 70 mg/L. Studies with dilute synthetic wastewaters containing carcinogenic species showed high removals (greater than 99%) of polynuclear aromatic hydrocarbons (PAH), aromatic amines, and nitrosamines could be achieved with little fouling problems.

Kurihara et al. (1981) compiled rejections for several organic and inorganic compounds for a composite membrane. The PEC-1000 membrane they studied had rejections of 97% for ethanol, 94% for ethylene glycol, 99% for phenol, 97% for acetone, and 99.8% for tetrahydrofuran. Rejections of copper, cobalt, iron, nickel ions, and NaCl were greater than 99%. Lynch et al. (1984) compared the performances of cellulose acetate membranes to the FT30 composite membrane. The FT30 membrane had greater rejection (>90%) of most organic compounds compared to the cellulose acetate membranes; however, significant adsorption of 2,4-dichlorophenol and other organics on the FilmTec FT30 membrane was found. Rickabaugh et al. (1986) also reported polyamide membranes gave much higher separations of chlorinated hydrocarbons and pesticides (in the range of 95.2 to 99.8%) than cellulose acetate membranes.

Bhattacharyya et al. (1984) have used two types of composite membranes for the separation of biotreated coal-liquefaction wastewater. These were the FT30 aromatic polyamide membrane (in a spiral-wound module) from FilmTec and an asymmetric aliphatic-polyamide membrane (hollow-fiber module) from Du Pont. Feed pretreatment consisted of sand filtration followed by 5-μm cartridge filtration. At 90% water recovery, both membranes removed 94 to 98% of the organics and 100% of the wastewater color. The hollow-fiber module rejected 77% of TDS and conductivity and 80% of chloride; the spiral-wound module rejections were slightly higher at 84% removal of TDS and conductivity and 91% of chloride. Fluxes with the spiral-wound module were also higher.

Siler and Bhattacharyya (1985) studied membrane separation and concentration of oil shale retorting wastewaters using the FT30 membrane. This raw wastewater had high concentrations of organics (aliphatic acids, phenolics), inorganics (NH_3, Cl^-, S^{2-}), color, and odor along with oils and fine suspended solids. Rejections by the membrane (with and without various pretreatments) were 60 to 94% for conductivity, 15 to 90% for NH_3, and 75 to 88% for TOC (with phenol rejections ranging from 75 to 94%). Flux drops caused by the wastewaters and solute rejections were affected by pretreatment with activated carbon, stripping, and pH adjustment.

Bhattacharyya et al. (1987) and Bhattacharyya and Madadi (1988) have done extensive work on the separation of various priority pollutants in dilute solution with the FT30 membrane. Experiments showed that rejections of >98% for PAH compounds (napthalene, anthracene, phenanthrene) were possible with little decrease in permeate water flux. For ionizable organics such as phenol, chlorophenols, and nitrophenols, they found that rejections and flux drops were highly dependent on operating pH values; flux drops were lower and rejections were higher at higher feed pH values at which the compounds were ionized. Flux drop and rejection behaviors as a function of feed pH are shown in Figures 24–8 and 24–9 for some of the organics studied. Membrane rejections (at pH 11) were 99.5 to 99.8% for phenol, 2-chlorophenol, 2,4-dichlorophenol, and 2,4,6-trichlorophenol and 98.9 to 99.9% for 2-nitrophenol, 4-nitrophenol, and 2,4-dinitrophenol. In these studies they also found substantial water flux drop for the nonionized chloro and nitrophenols due to adsorption of these on the membrane. Pusch, Yu, and Zheng (1989) studied rejections of four composite membranes (HR-95, HR-98, FT30, PEC-1000) and two asymmetric mem-

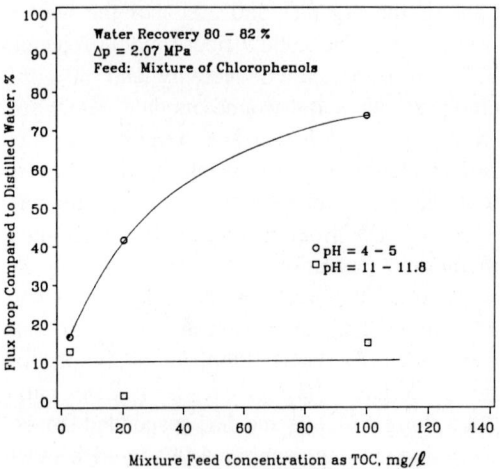

FIGURE 24–8. Permeate flux drop as a function of feed concentration for low- and high-pH conditions (Bhattacharyya et al. 1987).

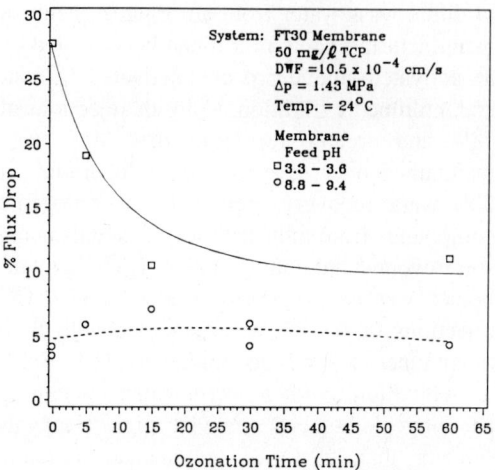

FIGURE 24–10. Effect of preozonation time on membrane flux drop for 2,4,6-trichlorophenol (TCP) solutions (Bhattacharyya and Williams 1990).

branes (SOLROX SC-200 and SC-1000) containing a variety of organic compounds, including methanol, ethanol, acetonitrile, formamide, benzylalcohol, phenol, benzaldehyde, benzoic acid, and aniline in single and multicomponent systems. Rejections ranged from only 25% to as high as 99.9% depending on the membrane and solute type; the composite membranes gave higher rejections. Williams, Deshmukh, and Bhattacharyya (1990) and Bhattacharyya and Williams (1990) showed that separations of selected chlorophenols and chloroethanes could be achieved using an ozonation-membrane process. Partial ozonation was used to convert the hazardous organics into organic acid intermediate compounds; the intermediates caused substantially less water flux drop than the chlorophenol and chloroethane compounds for the FT30 membrane studied. Figure 24–10 illustrates the improvements in flux drop resulting from ozonation of the hazardous organic trichlorophenol. The TOC rejections of the intermediates were in the range of 80 to 96%. Overall removal of >99.8% of a model hazardous organic compound (trichlorophenol) was shown to be possible as illustrated in Figure 24–11.

Municipal Wastewater

The RO process has also been applied to the treatment of municipal wastewater. Most of the attention given in municipal wastewater treatment by RO has focused on the removal of dissolved solids since these are not removed by conventional municipal treatment processes. However, RO processes can also remove organics, color, and nitrate as well as lower TDS concentrations. Most studies of municipal wastewater treatment have indicated that pretreatment and cleaning were necessary to main-

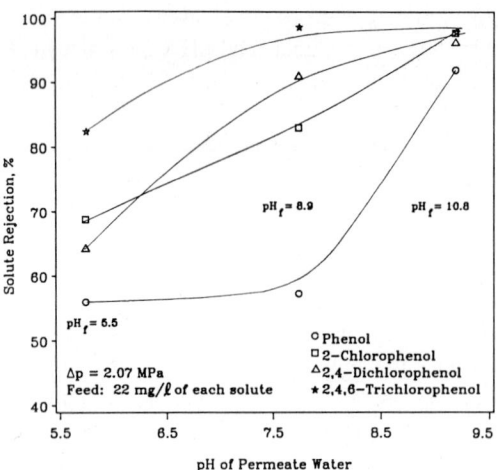

FIGURE 24–9. Effect of pH on rejections of chlorophenols at 2.1 MPa (Bhattacharyya et al. 1987).

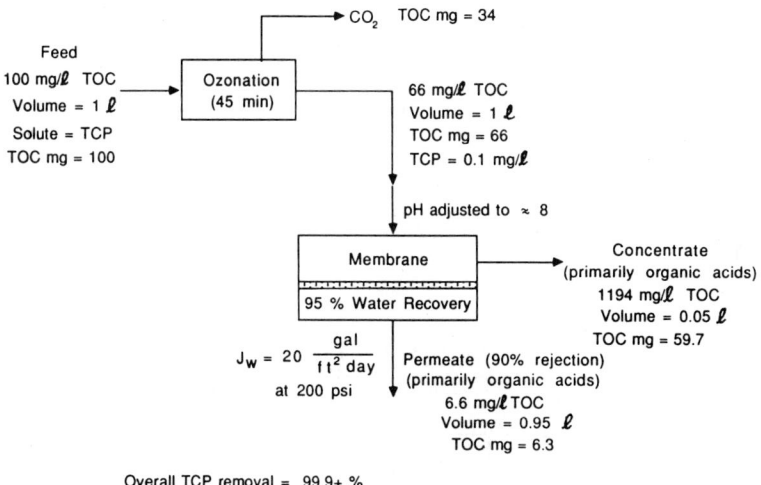

FIGURE 24-11.
Ozonation membrane process for removal of TOC and trichlorophenol.

tain high membrane fluxes and high permeate water quality.

Cruver (1975) reported results from pilot plant studies of municipal wastewater treatment by RO using various pretreatment techniques (clarification, media filtration, activated carbon). The TDS in the primary effluent feed water was reduced by >93%, turbidity, color, and suspended solids by 100%, nitrate by >75%, and COD by >65%. Membrane fluxes were maintained with regular chemical cleaning. Fang and Chian (1976) reported separation results of municipal sewage by aromatic polyamide membranes. TOC rejection of raw sewage was 70%; biological pretreatment of the sewage increased TOC rejection to as high as 94%. Lim and Johnston (1976) used cellulose acetate membranes to separate nutrients from raw and secondary municipal wastewaters. Phosphate removals of 100% were obtained, and rejections of ammonia and nitrate were 85 and 86%, respectively. Calcium removal was 92%, while that of chloride was 83%. Aluminum, iron, and sulfate ions (used in coagulation of the wastewater) were rejected 97%.

Tsuge and Mori (1977) used tubular reverse osmosis membranes for the reclamation of secondary effluent from a municipal sewage treatment plant. Pretreatment systems included coagulation-sedimentation, double media sand filtration, and activated carbon adsorption in various combinations. Rejections were greater than 93% for conductivity, >91% for TOC, and 100% for turbidity; the product water was within World Health Organization (WHO) drinking water standards for almost all species present. High fluxes were maintained by periodic water flushing and chemical cleaning. Van den Huvel, Zoetemeyer, and Beolhouver (1981) proposed using RO to concentrate wastewater prior to anaerobic digestion. They used tubular cellulose acetate membranes to achieve greater than 98% COD rejection. Stenstrom et al. (1982) used a pilot plant to study municipal wastewater reclamation by tubular cellulose acetate membranes over a 3-yr period. Effluent from a trickling filter biological treatment process was fed to the RO system after a variety of pretreatments, including multimedia filtration, filtration with cationic organic polymer, clarification, and chemical addition (chlorine, acid). Membrane fluxes were improved with regular citric acid cleaning to remove insoluble salts on the membrane surface. Chlorination, coagulation, and clarification of the membrane feed water also resulted in significant improvements in membrane flux. The TDS rejections were 81% (for feed TDS up to 1090 mg/L), and TOC rejections were >94% (for feed TOC up to 26.5 mg/L), making the permeate water suitable for reuse.

Richardson and Argo (1977), Allen and Elser (1979), Argo and Montes (1979), and Nusbaum and Argo (1984) have reported on the

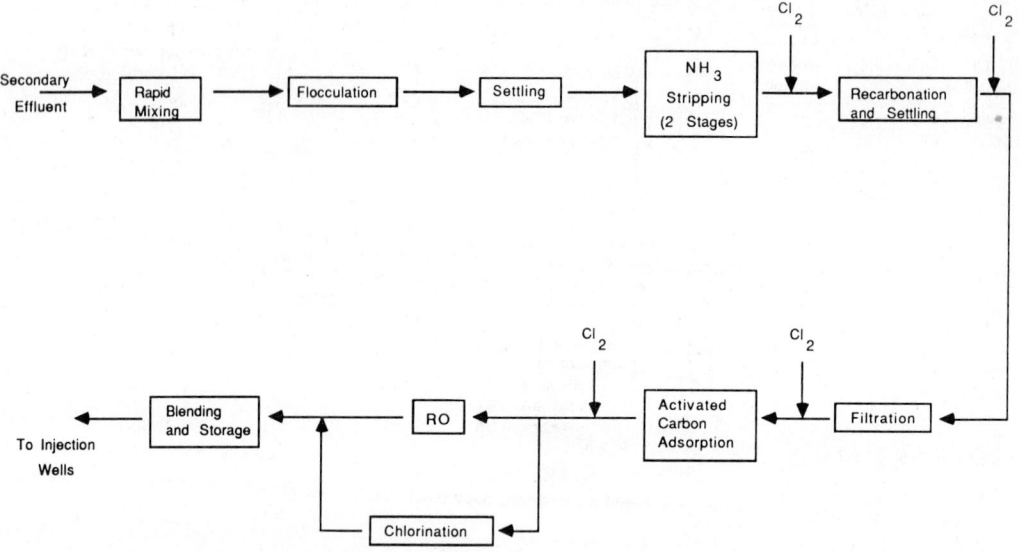

FIGURE 24–12. Schematic of Water Factory 21 (adapted from Nusbaum and Argo 1984).

success of the large-scale use of RO for municipal wastewater treatment at Water Factory 21 in Orange County, California. The plant influent consists of secondary treated wastewater; the capacity of the plant is 15 mgd. Processes at Water Factory 21 include chemical clarification and settling, ammonia stripping, recarbonation, multimedia filtration, activated carbon adsorption, reverse osmosis, and chlorination; these are shown in Figure 24–12. The plant effluent is mixed with groundwater and injected into aquifers that supply water for a variety of uses. The RO system in the plant has a capacity of 5 mgd and is used primarily to reduce TDS in the effluent, which ranges from 1100 to 1400 mg/L. For cellulose acetate membranes at operating pressures up to 4.1 MPa (600 psi) TDS rejections were 85 to 88% and average COD removal was 89%. Water Factory 21 has shown that large-scale municipal wastewater reclamation is possible and has identified some of the problems that occur.

APPLICATIONS OF RO IN THE CHEMICAL PROCESSING INDUSTRY

Reverse osmosis technologies have characteristics that make their use advantageous for many applications in the chemical processing industry (CPI). For certain CPI applications, RO offers advantages over competing technologies, including (1) low energy requirements, (2) low processing temperatures (and attendant minimization of thermal damage to chemicals during processing), (3) continuous rather than batch operation, (4) modular construction, and (5) simple system designs. For many applications, these advantages can reduce capital and operating costs if RO is used instead of a competing technology.

The main use of RO by the CPI is for treatment of wastewater to reduce the discharge of potentially hazardous wastes. The CPI discharges about 18 billion gallons of wastewater daily (U.S. Congress 1988) and spends more than $3.4 billion each year to treat this wastewater (Benn 1986). As the list of chemicals regulated by the U.S. Environmental Protection Agency (EPA) grows (Hanson 1990), so will the need for efficient wastewater treatment processes.

Reverse osmosis is also used by the CPI for the recovery and reuse of feedstocks and/or products. Because RO has low energy requirements and makes possible low-temperature operation, RO is ideal for this application. RO is used by many types of industries within the

CPI, including (1) the electroplating and metal-finishing industry, (2) the pulp and paper industry, (3) the textile industry, (4) the petroleum industry, and (5) the power-generation industry. Details of the use of RO in these industries are presented below.

Applications in the Electroplating and Metal-Finishing Industry

Many of the wastewaters and process waters in the electroplating and metal-finishing industries contain heavy metals. These metals must be removed, recovered, or recycled for environmental and economic reasons. In many situations, RO provides an attractive alternative to conventional technologies (e.g., precipitation, ion exchange, complexation, adsorption, evaporation, electrochemical operations) (Peters, Ku, and Bhattacharyya 1985). As a result, the membrane processes have been increasingly used for recovery of plating chemicals in the electroplating industry; these processes allow for the concentration of valuable materials that can be reused in the electroplating process and the elimination of costly treatment processes that were previously used to dispose of the electroplating waste.

Reverse osmosis is an accepted process for rinse water recovery so that it can be recycled. Recoveries up to 95% have been obtained with high separations of most metal species present in wash water feeds (Goldsmith 1975; Skovronek and Stinson 1977; Kremen, Hayes, and Dubos 1977). McNulty, Goldsmith, and Gollan (1977) reported the treatment of rinse water from nickel electroplating baths using hollow-fiber membranes and found that rejections of nickel, total solids, and conductivity were high. Figure 24–13 shows a schematic of a typical RO system for a metal-plating shop.

Schrantz (1975) reported the recovery of copper from plating wastewater that allowed one-third less copper use in the plating operation. Sato et al. (1977) described the use of ROGA spiral-wound membranes for the reclamation of metal-plating wastewater containing chromium and cyanide. Figure 24–14 shows a simplified schematic of the wastewater

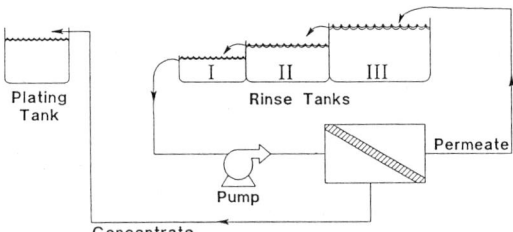

FIGURE 24–13. RO process for a metal-plating shop.

treatment system. The product water was of sufficient quality to be reused in the plating process. Koga and Ushikoshi (1977) reported the use of RO for the recovery of aluminum electrodeposition paints using a closed recovery system (ALCLOSE RO), which utilized Du Pont aromatic polyamide membranes. It was shown that the plating solution could be concentrated and sent back to the electrodeposition tanks while the permeate could be used as rinse water even for recoveries up to 99%. It was estimated that the paint recovery process could result in substantial savings in operating costs.

Kamizawa et al. (1978) reported success for the recovery of gold-plating rinse wastewater using aromatic polyamide membranes. Spatz (1979) reported the use of RO for nickel recovery from the plating bath. Water recoveries of 97% were achieved. The permeate water was used in rinsing operations, and the concentrate was recycled back to the plating bath, resulting in significant cost savings because of reduction in nickel consumption and waste treatment costs. Kosarek (1981) compiled results of rejections of inorganic compounds from electroplating wastewaters. Antimony, arsenic, beryllium, cadmium, chromium, copper, lead, mercury, nickel, selenium, silver, thallium, zinc, and cyanide ions were more than 90% removed by RO processes. Robinson (1983) described the use of RO for recovery of nickel from plating baths that resulted in substantial costs savings due to recovery of the nickel for reuse in the operation.

Thorsen (1985) reported investigating the use of DDS membranes (HR-98) to recover phosphoric acid from the polishing step of an electrolytic process for aluminum products.

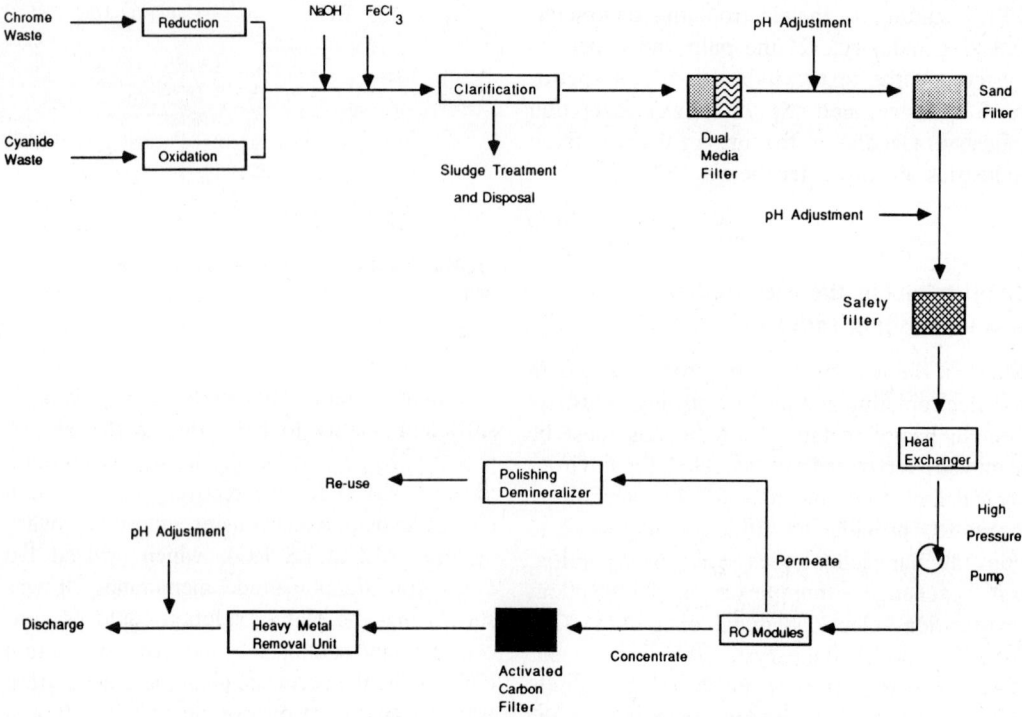

FIGURE 24-14. Schematic of system for plating wastewater reclamation.

Recoveries of 96 to 98% could be achieved, and the membranes appeared to be stable even for the pHs of 0.9 to 1.0 found at the end of the recovery runs. Imasu (1985) discussed the use of RO to recover plating chemicals using cellulose acetate and FT30 membranes at three plating shops. At 80% water recovery, rejections were suitable to allow reuse of the permeate water in the plating processes. The RO process was cost-effective in both treating the plating wastes and producing water for use in the plating process. Slater, Ferrari, and Wisniewski (1987) used RO to remove cadmium from metal processing wastewaters. The thin-film composite membranes they used had rejections of >99.9% for feeds containing 165 mg/L. Other metals present and conductivity were rejected in excess of 98%; the product water was of high quality.

Applications in the Pulp and Paper Industry

Large quantities of water are used in pulp and paper processing, thus generating large quantities of wastewater. Heightened awareness of environmental problems, increased incentives for water conservation and reuse, and increased incentives for the recovery and recycle of valuable chemicals have led the pulp and paper industry to examine technologies for treatment of these wastewaters.

The use of RO has been studied for this application because of the potential economic and technological advantages RO offers. However, to date, few commercially successful RO installations exist for pulp and paper applications. This is partly due to the harsh conditions under which the RO membranes would sometimes need to be operated (temperatures above 60°C and pH values greater than 9) and partly due to the fouling nature of the wastewaters. The high concentrations of suspended solids and the high biological and chemical oxygen demand (BOD and COD) of the wastewaters can lead to rapid declines in flux if pretreatment is not extensive (Buckley et al. 1985). In most cases, plate-and-frame or tubular modules are used for the treatment of pulp and paper wastewater because they are easy to clean and resist fouling (Chakravorty and Srivastava 1987; Jonsson and Wimmerstedt 1985).

The applications of RO in the pulp and paper industry have been mostly for the removal of dissolved solids, color, and organics from wastewaters. The RO process has primarily been used to reduce the volumes of wastewaters so that these could be further concentrated by evaporation. Morris, Nelson, and Walraver (1972) used RO to separate wastewaters from pulp and paper mills. Recoveries of 99% were achieved for feeds consisting of 10,000 mg/L dissolved solids; the process produced a high-quality permeate. Wiley et al. (1978) used RO membranes to remove color and reduce COD by 95%.

Olsen (1980) reported the use of RO to concentrate spent sulfite liquor (SSL). The SSL wastes contain lignosulphates, mono- and polysaccharides, and other organic compounds and various inorganic compounds. Reverse osmosis is used in a full-scale pulp and paper facility to concentrate the SSL wastes from 6 to 12% solids so that it can be further concentrated by evaporation. Paulson and Spatz (1983) also have indicated that RO was used to concentrate solids from less than 2% to more than 10% in SSL wastes prior to further concentration by evaporation; this RO process greatly reduced evaporation costs by reducing the volume to be evaporated. They reported results from short-term tests that showed that rejections of 98.9% of solids, 90.6% of BOD, and 97.0% of COD for feed concentrations of 30,150 mg/L solids, 7,020 mg/L BOD, and 30,560 mg/L COD could be achieved with the RO process. They also reported results from a process utilizing ultrafiltration ahead of a high-pressure (2.76- to 4.14-MPa) RO membrane. The ultrafiltration membrane removed most of the color-causing compounds while the RO membrane concentrated most of the conductive solids and all the remaining refractive compounds; this process produced an excellent quality permeate. A schematic of the system is shown in Figure 24–15.

Jonsson and Wimmerstedt (1985) reported results for the treatment of waste paper white water. The thin-film composite membranes (PCI ZF99) used had 99.4% TDS rejection and >99.8% COD rejection for water recoveries of 80%. Simpson and Groves (1983) used RO to treat highly colored bleach effluent from a kraft mill. Rejections of both inorganic and organic compounds were above 90% for solid concentrations in the range of 5,000 to 30,000 mg/L. Chakravorty and Srivastava (1987) and Chakravorty (1989) reported the use of RO to recover water from pulp and paper mill effluents. Rejections of 90 to 95% for color, TDS, organics, inorganics, COD, and BOD were obtained for RO feeds consisting of the permeate from ultrafiltration membranes. No drop in RO membrane flux was observed at a pressure of 2.7 MPa (400 psi) during the studies. Feed pretreatment to the ultrafiltration membrane consisted of settling, pH adjustment, and microfiltration. Basu and Sakpal (1987)

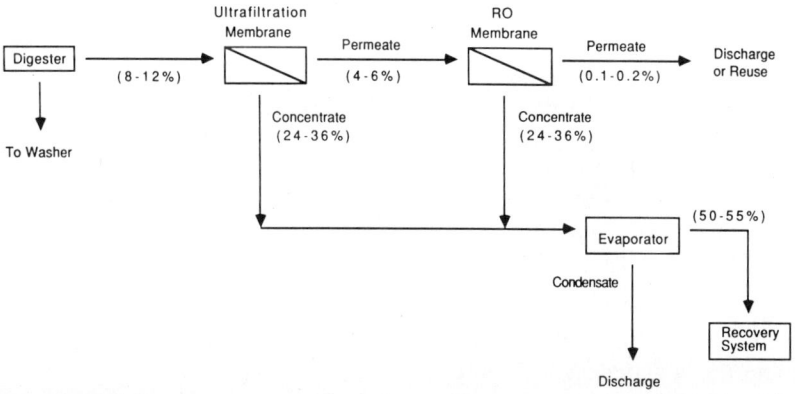

FIGURE 24–15. Use of UF and RO for preconcentration of digester effluent prior to evaporation.

used RO membranes to treat black liquor from an alkaline pulping process. Cellulose diacetate membranes in spiral-wound, tubular, and plate-and-frame modules were used. Pretreatment consisted of primary sedimentation, pH adjustment, clarification, multimedia sand filtration, and cartridge filtration. Color rejection of over 95% and chloride rejections of over 80% were obtained for feed concentrations from 2 to 12% TDS.

Applications in the Textile Industry

Large volumes of water are used in the textile industry for many purposes, including washing, scouring, dyeing, printing, bleaching, and finishing cloth (El-Nashar 1980). The wastewaters generated in these processes contain a wide range of contaminants including salts, dyestuffs (soluble and colloidal), fatty acids, surfactants, scouring agents, oils, greases, and oxidizing and reducing agents (Treffry-Goatley, Buckley, and Groves 1983). The temperatures of these wastewaters often vary widely (30 to 90°C), as do the pH levels (4 to 12) (Porter and Goodman 1984; Brandon et al. 1981).

Several investigators have examined the use of RO to treat textile wastewaters. RO could be used to (1) reduce the volume of wastewater generated, because the purified water could be reused; (2) recover and recycle valuable components from the waste streams (e.g., dyestuffs); and/or (3) recover the thermal energy in hot wastewaters. Many investigators consider RO an excellent treatment option for textile wastewaters because RO systems can treat feed solutions of variable composition and because RO systems have low energy requirements.

The majority of the applications investigated for RO have been for the treatment of textile dyehouse effluents (El-Nashar 1980; Treffry-Goatley, Buckley, and Groves 1983; Porter and Goodman 1984; Brandon et al. 1981; Buckley et al. 1985; Slater, Ahlert, and Uchrin 1987). Figure 24–16 shows a schematic of a typical system for this application (Brandon et al. 1981; Porter and Goodman 1984). Here, the wastewater from the various washing and rinsing steps is treated by the RO membrane. Permeate is reused as washwater, and the concentrate can be either reused or discarded.

Brandon et al. (1981) and Porter and Goodman (1984) used dynamically formed zirconium oxide-polyacrylate membranes in tubular form to treat textile dyehouse effluents. They found that more than 97% of the dye could be removed at a recovery of 95%. These membranes were also able to operate at temperatures of 55 to 85°C, thus recovering a large part of the energy needed to heat the hot (85 to 95°C) washwater. Similar performance was obtained using spiral-wound RO modules (Treffry-Goatley, Buckley, and Groves 1983). For both the tubular and spiral-wound modules, flux declined slightly due to fouling. However, with routine cleaning, the performance of the membranes could be maintained for extended periods.

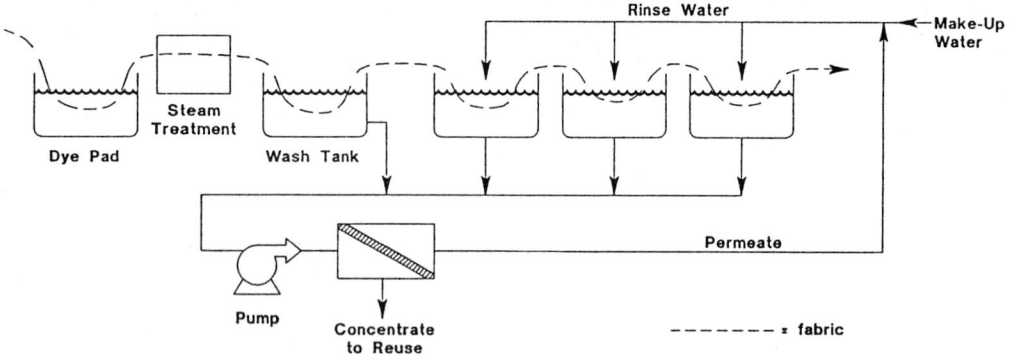

FIGURE 24–16. Schematic of a RO process approved for treating textile dyehouse effluents.

El-Nashar (1980) examined the economic feasibility of using RO to treat dyehouse effluents. His studies showed that RO offered economic advantages because (1) the system used less water; (2) less wastewater was generated; and (3) less energy was required (because energy was recovered through reuse of hot water). In fact, the cost savings exceeded the cost of the RO system when the system was operated at recoveries of more than 90%. Thus, using RO to treat textile wastewater is technically and economically feasible.

Applications in the Petroleum Industry

Reverse osmosis has also been used to treat many process streams in the petroleum industry. Because many of the separations performed in the petroleum industry involve chemicals that damage or degrade membranes, RO has been used almost exclusively to treat wastewaters. The composition of wastewaters generated by the petroleum industry is almost as diverse as the range of products produced. In the past, many of these wastewaters have been treated using biological processes (e.g., activated sludge). However, these wastewaters contain components that can inhibit the action of the organisms in the sludge, rendering the process ineffective. An alternative treatment technique, air flotation, also has limitations. Air flotation is used to remove most of the suspended or "free" oils from wastewaters, but it cannot be used to remove water-soluble low molecular weight organics. In some cases, the concentration of oils in the treated wastewater remains too high to allow discharge into the environment. As a result, other treatment options are employed, such as solvent extraction, wet oxidation, carbon adsorption, and RO.

RO-based systems for the recovery of oily wastewaters are now on the market or in the pilot plant stage. However, extensive pretreatment is often required to maintain high performance because of the composition of the wastewater. For example, in one system, the oily wastewater is passed through a cyclone separator, a polishing filter, and a bank of charcoal to remove suspended and some dissolved organic solids. Finally, the feed is passed through an RO unit containing spiral-wound modules (Anonymous 1984).

Extensive pretreatment is also required in another RO-based system that was designed to treat wastewater from an *in situ* oil-shale retort operation (Sareen, Dickehuth, and Van Sittert 1982). In this case, the pretreatment train consists of an oil/water separator, an air flotation unit, a clarifier, a sand filter, an activated-carbon filter, and an ion-exchange unit.

Despite the need for these complicated pretreatment trains, RO is still considered a viable treatment method for oily wastewaters produced by the petroleum industry and for oily wastewaters produced by many other industries. For example, RO has been investigated to treat oily wastewater at a facility that manufactures diesel engines (Spatz 1981) and to treat oily wastewaters generated in the metal-cutting and metal-forming industry (Cartwright 1989).

Applications in the Power-Generation Industry

The primary uses of RO by the power-generation industry have been to produce boiler-quality water for steam generation (Patra et al. 1987; Prabhakar, Misra, and Ramani 1987; D'Auria, Itteilag, and Pastrick 1987) and to reduce wastewater effluents (Patra et al. 1987; Schutte et al. 1987; Bryant et al. 1987; Hess et al. 1988). The treatment of wastewater effluents (primarily from cooling tower blowdown) is similar to the treatment of brackish water.

Other Wastewater Applications

Reverse osmosis processes have also been used for other applications in waste treatment. Chian and De Walle (1977) used cellulose acetate and a composite membrane to treat sanitary landfill leachate; TOC rejections of >91% for both membranes were obtained. Slater, Ahlert, and Uchrin (1983a, 1983b) reported the use of tubular cellulose acetate membranes to treat industrial landfill leachate that had been pretreated by lime coagulation; TDS removal was 98%, and COD was reduced by 68%. No signs of significant fouling or concentration polarization were found in the studies even at 75% recovery. Dagon (1978) reported the use of RO for the recovery of valuable materials from photographic processing effluents; it has been used for silver recovery, fixer reuse, and bleach and color developer regeneration. Daignault (1977) showed RO processes could remove 99% of silver and 98% of iron and substantially lower BOD and COD from a waste stream of a photographic processing plant.

Muratova et al. (1979) reported the use of RO for the treatment of petrochemical wastewater. The TDS of the feed (1200 mg/L) was reduced by 70%, and COD rejection was 51% for a feed stream containing 49 mg/L. Ghassemi, Yu, and Quinlivan (1981) reported results for the separation of coolant oils in wastewater; TDS rejections were 97% for feeds of 1400 mg/L, and TOC was reduced by 97.5% for feeds containing 2000 mg/L. Terril and Neufeld (1983) used cellulose acetate membranes to treat blast-furnace scrubber water. Rejections of >99% were obtained for calcium, magnesium, zinc, and sulfate, 93% for ammonia, 91% for fluoride, 94% for chloride, and 97% for TDS. The permeate was of sufficient quality for use in the recycle system for water recoveries up to 70%.

Davis et al (1987) reported on the use of RO to treat wastewater contaminated with heavy metals, including arsenic, cadmium, chromium, copper, molybdenum, nickel, vanadium, and tungsten; RO was indicated as a cost-effective alternative for treatment of these wastewaters. Hays et al. (1988) used composite membranes to concentrate and recover ammonium nitrate from condensate streams from a manufacturing plant. Rejections of over 90% at 85% recovery were obtained for feeds containing 1709 mg/L nitrate and 325 mg/L ammonium ions. Hsiue et al. (1989) have studied the separation of uranium conversion process effluent compounds from radioactive wastewaters. By means of FilmTec's FT30 membrane, uranium rejections of greater than 98% were achieved, and it was shown that the radioactivity of the permeate water could be reduced to the lower limit detection value.

NANOFILTRATION APPLICATIONS IN WASTEWATER TREATMENT AND CPI INDUSTRIES

Nanofiltration membranes such as the NF40, NF70, XP45, MPT-20, MPT-30, NTR-7410, and NTR-7450 have also received increasing attention for wastewater treatment. These membranes are often negatively charged so it is primarily the anion repulsion that determines salt rejection: the higher the charge on the anion the greater the salt rejection. In addition, these membranes also reject organic compounds with molecular weights above 200 to 500 (Eriksson 1988; Cadotte et al. 1988). Therefore, the unique properties of these membranes allow for the possibilities of selective separations, including separation of solutes with charge differences and separation of high molecular weight organics from high-concentration monovalent salt solutions. In addition these membranes have high fluxes at low pressures. These characteristics have been exploited for several wastewater treatment applications.

Simpson, Kerr, and Buckley (1987) reported the use of nanofiltration for the removal of hardness and organic impurities in one stage of the recovery of pollutants from a textile mill. The removal was necessary to prevent fouling of a cation-exchange membrane by calcium and magnesium hydroxide in the final stages of the process. Rejections of the membrane included 29% of conductivity, 33% of sodium, 48% of calcium, 67% of magnesium, and 47% of soluble organic carbon in the waste streams. Bindoff et al. (1987) used nanofiltration membranes to remove color-causing compounds from the effluent of the caustic extraction stage of a wood pulping process. The effluents contained lignin and chlorinated lignin derivatives, many of which were negatively charged, and monovalent cations such as sodium whose removal was not required. The nanofiltration membrane removed color >98% while allowing the inorganics to pass through so that the effluent could be reused in the extraction process. The use of nanofiltration membranes resulted in lower operating costs because of lower operating pressures than reverse osmosis membranes since the monovalent ions were not removed.

Perry and Linder (1989) described the successful use of the MPT-20 and MPT-30 nanofiltration membranes in industrial dye manufacturing waste treatment. These negatively charged membranes have high rejections of many low molecular weight dyes while having low rejections of sodium chloride (<10%). Negative NaCl rejections have also been observed for feeds consisting of high NaCl concentrations mixed with low concentrations of organics. Cadotte et al. (1988) have listed several applications for nanofiltration membranes. The XP20 membrane was used to recover copper EDTA salt at pH 12 from copper-plating baths. The XP45 membrane has been utilized to remove sulfates (98%) from seawater while allowing sodium chloride to pass through; the resulting sulfate-free seawater could be used in offshore oil well applications to enhance oil recovery. Another application of the XP45 membrane involves the separation of sodium chloride from cheese whey; the cheese whey is rejected by the membrane while the salt is not. The NF70 membrane was used to treat olive processing wastewater, which contained organics and salts. Treated water was of sufficient quality for recycle to the process.

Ikeda et al. (1988) reported the use of negatively charged nanofiltration membranes to decolor soy sauce solutions. The NTR-7410 spiral-wound module removed 80% of color while rejecting only 6% of the NaCl (feed concentration of 14.1%) in the feed. Paper pulping wastewaters containing color-causing compounds such as lignin sulphonates were also separated using the NTR-7410 membrane; COD rejections of 90% were obtained with high fluxes because the negatively charged lignin sulphonates did not foul the negatively charged membrane.

Bhattacharyya, Adams, and Williams (1989) have investigated the separation of selected inorganic and organic compounds using the NF40 membrane. It was shown that solutions of sucrose could be separated effectively from sodium chloride solutions using the nanofiltration membrane; the sucrose rejection was over 90% because of its size while the salt rejection was less than 20%. It was also shown that mixtures of cadmium and nickel could be selectively separated by adding sodium chloride to the solution. The cadmium formed the neutral complex $[CdCl_2]^0$ with the chloride and so passed through the membrane; the nickel did not complex and so was rejected. For 0.5 M NaCl concentrations, nickel rejection was 87% while the cadmium (in complexed form) rejection was only 16%. Experiments with the phenol system showed that extent of ionization greatly affected solute rejection. Rejection was less than 5% when the phenol was not ionized at pH values below 8 but increased to 71% when the solute was ionized at feed pH 11.

Williams, Deshmukh, and Bhattacharyya (1990) and Bhattacharyya and Williams (1990) used the NF40 membrane for the separation of hazardous organics (trichloroethane, tetrachloroethane, chlorophenol, dichlorophenol, and trichlorophenol). Flux drops and rejections were highly dependent on feed pH values for the ionizable compounds studied; when the compounds were ionized, they became negatively charged and so did not interact with the membrane as much, causing less flux drop. Partial ozonation of the membrane feed resulted in decreased flux drop and increased rejection; ozonation converted the hazardous organics into organic acid intermediates. The ionized organic acids did not cause as much flux drop and were rejected better at lower pH values since these have low pK_a values. Figure 24–17 shows the improvements in TOC rejection after feed preozonation for a mixture of hazardous organics.

SURFACE WATER AND GROUNDWATER TREATMENT BY MEMBRANES

The incidences of surface water and groundwater contamination by toxic inorganic and organic compounds are becoming more and more frequent. A number of recent studies in the

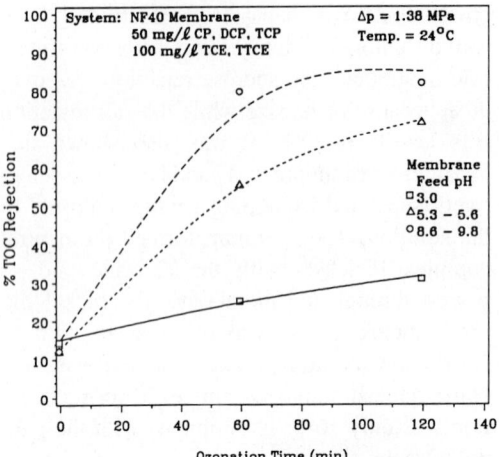

FIGURE 24-17. Effect of ozonation time on TOC rejection by an NF40 membrane of mixture containing 2-chlorophenol (CP), 2,4-dichlorophenol (DCP), 2,5,6-trichlorophenol (TCP), 1,1,2-trichloroethane (TCE), and 1,1,2,2-tetrachloroethane (TTCE).

United States have shown the presence of agricultural chemicals (pesticides, herbicides, fungicides, fertilizers, etc.) and other organic compounds in water supplies. The problem of groundwater contamination is particularly important since it is the primary source for 81% of the public's drinking water (Stacha and Pontius 1984). In addition, increased potable water demand has forced many communities to consider alternative sources of drinking water that are of lower quality. These new sources require more treatment to remove color, organics, odor, hardness, and other dissolved materials.

Five types of contaminants of public health importance have been identified by the EPA. These contaminants are included in the National Interim Primary Drinking Water Regulations (NIPDWR) (*Federal Register* 1975) and include: (1) inorganic contaminants, including arsenic, lead, fluoride, nitrate, and others; (2) organic contaminants, which include various pesticides and herbicides; (3) microbiological contaminants such as bacteria and viruses; (4) turbidity; and (5) radiological contaminants. Many other countries and the World Health Organization (WHO) have also set drinking water standards; maximum contaminant levels (MCLs) for several compounds are given in Table 24-5 (Montgomery 1985; Rice 1985; JAWWA 1985).

The inorganic contaminants in drinking water sources are associated with both natural processes of chemical weathering and soil leaching and human activities such as wastewater from manufacturing and mining and leachates from hazardous waste landfills. Organic compounds that are a cause for concern in drinking water result from sources such as industrial and municipal discharges, urban, rural, and agricultural runoff, hazardous landfill leachates, and naturally occurring materials. These contaminants include pesticides, herbicides, and other agricultural chemicals, volatile and synthetic organic carbons (VOCs and SOCs), priority pollutant organics, and naturally occurring materials with high molecular weights such as humic and fulvic acids. Trihalomethanes (THMs) such as chloroform, bromodichloromethane, and tribromomethane have also been identified as a problem in finished drinking water from many treatment plants. Trihalomethanes are formed as by-products of disinfection (chlorination) when naturally occurring organics such as humic and fulvic acids are present; THMs are considered carcinogenic.

New standards and the necessity of using lower quality waters will require the use of more treatment processes to produce drinking water. Treatment processes that can be used include coagulation, lime softening, ion exchange, oxidation (aeration, chlorination, ozonation), and activated carbon adsorption. Membrane processes can also be used and have advantages over other treatment processes. Reverse osmosis membranes can simultaneously remove hardness, color, many kinds of bacteria and viruses, and many organic compounds such as pesticides and herbicides as well as other naturally occurring organic materials. Nanofiltration membranes can also remove hardness and many high molecular weight organic compounds. As a result many applications of membrane processes have been reported for treatment to produce drinking water from surface and groundwater.

TABLE 24-5. Selected Drinking Water Standards Identified by the U.S. EPA and WHO.

	Maximum Contaminant Level, U.S. EPA[a]	Maximum Permissible Level, WHO[a]
Arsenic	0.05	0.05
Barium	1	—
Cadmium	0.010	—
Chloride[b]	250	600
Chromium	0.05	—
Copper[b]	1	1.5
Fluoride	1.4–2.4[c]	0.6–1.7[c]
Iron[b]	0.3	1.0
Lead	0.05	0.1
Magnesium	—	150
Manganese[b]	0.05	0.5
Mercury	0.002	0.001
Nitrate	10	—
Selenium	0.01	0.01
Sulfate[b]	250	400
Silver	0.05	—
Zinc[b]	5	15
TDS[b]	500	1500
Total Hardness (as $CaCO_3$)	—	500
Corrosivity[b]	Noncorrosive	—
Color[b] (color units)	15	50
Odor[b] (threshold odor number)	3	—
Turbidity (TU)	1	25
pH[b]	6.5–8.5	6.5–9.2
Coliform (organisms/100 ml)	1	1
Foaming agents[b]	0.5	1.0
Organics		
Trihalomethanes ($\mu g/L$)	100	—
Trichloroethylene	0.005	—
Carbon tetrachloride	0.005	—
Vinyl chloride	0.002	—
1,2-Dichloroethane	0.005	—
Benzene	0.005	—
1,1-Dichloroethylene	0.007	—
1,1,1-Trichloroethane	0.02	—
p-Dichlorobenzene	0.75	—
Endrin	0.0002	—
Lindane	0.004	—
Methoxychlor	0.1	—
Toxaphene	0.005	—
2,4-D	0.1	—
2,4,5-TP Silvex	0.01	—
Radioactivity[d]		
Gross alpha (pCi/L)	15	—
Radium-226 and 228 (pCi/L)	5	—
Tritium (pCi/L)	20,000	—
Strontium-90 (pCi/L)	8	—

[a]All units in mg/L unless indicated otherwise.
[b]EPA National Secondary Drinking Water Regulations. All others EPA National Primary Drinking Water Regulations.
[c]Depends on temperature.
[d]pCi = picoCurie.

Reverse Osmosis Membranes

Large amounts of data have been generated for the use of RO membranes to remove various contaminants that are found in drinking water supplies. Sourirajan (1970) and Sourirajan and Matsuura (1985) list separation results for many different compounds. In addition, much of the information concerning wastewater treatment also can be related to the treatment of drinking water. A recent book (Eisenberg and Middlebrooks 1986) provided an in-depth review of RO treatment of drinking water. In many cases it was indicated that RO could remove a wide range of contaminants and meet MCLs.

Several studies on the use of RO for drinking water treatment are of particular importance. Chian, Bruce, and Fang (1975) investigated the separation of a variety of pesticides and herbicides (aldrin, lindane, heptachlor, heptachlor epoxide, DDE, DDT, dieldrin, diazinon, methylparathion, malathion, parathion, randox, trifluralin, atrazine, and captan). Rejections were high in most cases (>98%) but adsorption of these on the cellulose acetate and NS-100 composite membranes was significant for most of the compounds studied. They concluded that adsorption played a large role in the separation. Johnston and Lim (1978) also studied removal of several chlorinated hydrocarbons and organic phosphate pesticides with cellulose acetate membranes. Rejections of chlordane, endrin, parathion, aldrin, and DDT were 100% and that of malathion was 94%. However, significant adsorption of the pesticides on the membranes used was observed. The adsorbed solutes could eventually pass through the membrane and contaminate the permeate. These results indicated that careful attention must be given to adsorption of organics onto membranes used so that these contaminants do not permeate the membrane and contaminate water supplies.

Nusbaum and Riedinger (1980) indicated many low molecular weight organic compounds were poorly rejected by the RO membranes they studied but that large molecular weight compounds such as lignins and humic and fulvic acids were highly removed. These materials are among those that result in the formation of trihalomethanes upon chlorination. Odegaard and Koottatep (1982) also reported high separations of humic substances by RO. Bhattacharyya and Williams (1990) reported that high separations of humic acids could be achieved by the FilmTec FT30-BW membrane. In addition, the presence of humic acids did not significantly affect flux or rejection of other hazardous organics (i.e., trichlorophenol) for feeds with high pH (>9).

Huxstep (1981) used pilot-scale plants to treat drinking water with two different membranes. Each plant had a capacity of 45,000 gal/day. The low-pressure membrane [operating at 1.27 MPa (184 psi)] studied had TDS rejection of 49% for feeds of 1214 mg/L, 64% rejection of hardness (as $CaCO_3$) for feeds of 423 mg/L, 64% rejection for magnesium for feeds of 234 mg/L, 42% chloride rejection for feeds of 452 mg/L, 84% rejection of sulfate for feeds of 143 mg/L, 35% rejection of sodium for feeds of 201 mg/L, and 13% color rejection and 25% TOC rejection for feeds of 28 mg/L and 16 mg/L, respectively. Rejections of the RO system operating at 339 psi were higher and included 95% for TDS, 97% for hardness, 98% for magnesium, 96% for chloride, 99% for sulphate, 92% for sodium, 75% of color, and 56% of TOC for feeds with the same concentration as in the lower pressure RO system.

Regunathan, Beauman, and Kreusch (1983) presented results for the removal of contaminants by a RO-carbon adsorption system. Data for the effluent showed 95 to 97% removal of THM for 1300 gal of treated water. The pesticides endrin and methoxychlor were highly removed by the RO membrane. Inorganics removals ranged from 40% for nitrate to 98% for sulfate.

Wilson and Duran (1982) studied RO (Envirogenic spiral-wound cellulose acetate and Du Pont hollow-fiber) membranes for upgrading drinking water supplies for small Mexico communities to meet drinking water standards. A mobile demonstration unit was operated for 500 to 2000 hours in nine communities using existing drinking water supplies as feeds. In all cases the permeate water from the demonstration unit met drinking water standards. Remov-

TABLE 24-6. Quality of Groundwater Before and After Treatment by RO (Sorg and Love 1984).

Parameter	Influent	Cellulose Acetate		Polyamide		Thin-Film Composite	
		Permeate	Reject	Permeate	Reject	Permeate	Reject
Total dissolved solids	252	12	281	20	304	2	272
Chloride	20.2	2.36	19.0	2.08	21.7	<0.1	17.5
Sulfate	28	<15.0	31.0	<15.0	36.0	<15.0	33.0
Sodium	10.2	1.5	11.6	3.2	12.8	1.6	12.0
Barium	<0.2	—	—	—	—	0.3	—
Arsenic	<0.005	—	—	—	—	—	—
Selenium	<0.005	—	—	—	—	—	—
Fluoride	<0.1	—	0.2	—	0.1	—	—
Hardness, as $CaCO_3$	258	11	282	15	328	1	295
Alkalinity, as $CaCO_3$	173	10	184	18	191	4	189
Conductance, μmhos/cm	448	24	454	42	652	5	464
Chromium	<0.005	—	—	—	—	—	—
Silver	<0.03	—	—	—	—	—	—
Copper	<0.02	—	—	—	—	—	—
Manganese	0.12	<0.03	0.14	<0.03	0.16	<0.03	0.14
Lead	<0.005	—	—	—	—	—	—
Iron	<0.1	—	—	0.13	—	—	0.10
Cadmium	<0.002	—	—	—	—	—	—
Zinc	<0.02	—	—	—	—	—	—
Mercury	<0.0005	—	—	—	—	—	—
Turbidity, NTU	0.46	0.06	0.38	0.07	0.24	0.07	0.46
Color, units	4	3	3	4	4	2	5
pH, units	7.70	6.10	7.60	6.60	7.70	5.50	7.80

Notes: All units mg/L except as noted.
— Indicates below detection.

al of radium-226 by RO was also found to be effective in a Florida study conducted by Sorg, Forbes, and Chambers (1980). The permeate water met drinking water requirements of 5 pCi/L.

Sorg and Love (1984) conducted studies with actual groundwaters in which only a few of the contaminants being studied were spiked. The membranes studied were cellulose acetate, cellulose triacetate, thin-film composite (aromatic polyamide), and polyamide membranes; the manufacturers were Toray, FilmTec, Dow, and Du Pont. Most of the inorganic contaminants were highly rejected (>90%) although some such as fluoride, nitrate, and arsenic (III) were only moderately rejected. Table 24-6 shows selected separation results. For the VOCs studied, cellulose acetate and Du Pont polyamide membranes showed less than 25% rejections. FilmTec composite (FT30) membranes showed encouraging results for short durations, with rejections ranging from 80 to 88% for TOC and total organic halogens. However, rejections of some of the compounds decreased with operating time.

Baier et al. (1987) used RO to remove several agricultural chemicals from groundwater using cellulose acetate, thin-film composite spiral-wound, and hollow-fiber polyamide membranes. The compounds studied along with feed and permeate concentrations are shown in Table 24-7. The thin-film composite and polyamide hollow-fiber membranes gave permeates with lower contaminant concentrations. Pilot plant data were also presented for a polyamide hollow-fiber membrane unit. Groundwater containing ~20 μg/L aldicarb

TABLE 24-7. Separation Results for Groundwater Containing Agricultural Chemicals.

Membrane[a]	Aldicarb Sulfoxide		Aldicarb Sulfone		1,2-Dichloropropane		Carbofuran		Nitrate	
	Feed	Permeate	Feed	Permeate	Feed	Permeate	Feed	Permeate	Feed[b]	Permeate[b]
A	39	<1.0	47	3.0	24	23	14	2	9.6	2.5
B	39	<1.0	47	1.0	24	12	14	<1	9.6	1.0
C	39	<1.0	47	2.0	24	6	14	<1	9.6	1.0
D	39	<1.0	47	<1.0	24	5	14	<1	9.6	0.6
E	39	1.0	47	2.0	24	15	14	<1	9.6	0.4
F	39	2.0	47	3.0	24	15	14	<1	9.6	0.5

[a] A: Culligan; 2-in.-diameter tubular cellulose acetate membrane with feed flow rate of 2.2 L/min and 6% recovery.
B: Du Pont; 2.5-in. thin-film composite polyamide spiral-wound membrane with feed flow rate of 1.0 L/min and 10% recovery.
C: Du Pont; 4.0-in. polyamide hollow-fiber membrane with feed flow rate of 2.6 L/min and 5% recovery.
D: FilmTec; 2.5-in. thin-film composite polyamide spiral-wound membrane with feed flow rate of 2.2 L/min and 5% recovery.
E: Hydranautics; 2.5-in. thin-film composite polyamide spiral-wound membrane with feed flow rate of 1.3 L/min and 13% recovery.
F: Fluid Systems; 2.5-in. thin-film composite polyamide spiral-wound membrane with feed flow rate of 1.2 L/min and 16% recovery.
[b] Units in mg/L. All other units in µg/L.

sulfone, ~15 µg/L aldicarb sulfoxide, and ~20 µg/L 1,2-dichloropropane passed through 5-µm cartridge filters and then through the membranes [operated at 2.8 MPa (400 psi) and 5 gal/min feed rate]. The product water contained <2 µg/L aldicarb sulfone, <1 µg/L aldicarb sulfoxide, and <7 µg/L 1,2-dichloropropane. Rejections did not significantly change over the course of the nine-month study. Membrane fluxes were maintained by daily water rinses and monthly cleanings.

Nanofiltration Membranes

Nanofiltration membranes also have been used for drinking water treatment. These membranes can serve as water softeners, reduce TDS concentrations, remove color, and remove high molecular weight organics. These membranes have high fluxes at low pressures and so in many cases processes with nanofiltration membranes have lower capital and operating costs than RO processes.

Conlon (1985) has discussed several applications of nanofiltration for water softening and removal of organics. FilmTec NF50 membranes were used to replace coagulation, filtration, and ammonia treatment processes in a water treatment plant. Color removal by the membrane was 96%, TOC levels were reduced by an average of 84%, and trihalomethane forming potential was reduced to levels meeting regulations. Test results at another water treatment plant indicated that the NF50 membrane permeate was substantially better than the feed in terms of hardness and TDS.

Simpson, Kerr, and Buckley (1987) studied the removal of sodium carbonate by the NF40 membrane over a feed pH range of 6.6 to 11.0. At high pH carbonate ions predominated and were highly rejected by the negatively charged membrane, while at lower pH values bicarbonate ions were more prevalent and were rejected less by the membrane because of the bicarbonate ion's lower charge. Sodium ions were rejected to an extent that balanced electroneutrality requirements. Sodium rejections were as high as 95% at feed pH > 11. The results indicate the membrane could be used for water softening.

Eriksson (1988) reported that nanofiltration membranes such as the NF50, NF70, NF40, and NF40HF could be used for partial water desalination. Partial desalination or nanofiltration treatment to remove primarily one or only a few species has advantages over reverse osmosis membranes because of lower energy requirements from lower operating pressure requirements. Results were reported for surface water treatment in Florida using the NF50 membrane. At 0.45 MPa, 94% TOC rejection, 96% color rejection, 76% alkalinity (as $CaCO_3$) rejection, 77% hardness (as $CaCO_3$) rejection, 52% chloride rejection, and 84% rejection of TDS were obtained. The NF70 membranes were tested on well water in Iowa. Calcium and magnesium rejections were 88%, and strontium and barium rejections were 90 and 100%, respectively. Chloride, alkalinity (as $CaCO_3$), fluoride, sulfate, iron, silicon, phosphorus, and boron rejections were 45, 59, 29, 98, 100, 28, 60, and 4%, respectively. The high rejections of hardness by the NF70 makes this membrane suitable for water softening.

Cadotte et al. (1988) and FilmTec (1988) reported the use of the NF70 thin-film composite membrane to treat drinking water in Palm Beach County, Florida. The feed water from a shallow well contained more than 500 mg/L TDS and organic contaminants from agricultural runoff and natural sources. The organic contaminants caused color and were identified as trihalomethane precursors. The 42,000 gal/day treatment system that was installed consisted of chemical pretreatment, nanofiltration, degasification to remove H_2S, and chemical post-treatment. Table 24–8 shows the feed and permeate water for the system operating at 0.69 MPa (100 psi). The system removed over 97% of the total organic halogens, more than 90% of the TOC, and reduced trihalomethane (THM) formation potential, which results from chlorination of the permeate. The membrane also removed 92% of the total hardness.

Cadotte et al. (1988) also reported that the NF70 membrane can serve as a "water softener"

TABLE 24-8. Feed and Permeate Water of NF70 Membrane at Florida Site.

	Feed Water	Permeate
Alkalinity, mg/L $CaCO_3$	283	85
Chloride, mg/L	64	22
Color, CPU	38	2
Calcium hardness, mg/L $CaCO_3$	284	22
Total hardness, mg/L $CaCO_3$	316	24
Sulfate, mg/L	20	6
Total dissolved solids, mg/L	396	134
Total organic carbon, mg/L	15.4	1.5
Total organic halogen, μg/L (seven-day formation potential)	2000	51
Trihalomethanes, μg/L (seven-day formation potential)	630	56

and also remove organics. Nanofiltration can lower TDS and hardness, reduce organics that impart taste and odor, and remove radium ions. The low-pressure RO system can remove these contaminants in a single process, replacing multiple processes such as lime softening followed by activated carbon adsorption. The NF70 membrane was also used to remove fluoride from water supplies; fluoride rejection was 90%, reducing feed levels from 4 to 0.4 mg/L.

Dykes and Conlon (1989) reported on the use of nanofiltration for water softening and THM precursor removal. Membrane treatment for organics (THM precursors) was identified as a reliable method of meeting existing and future limits. Pilot tests indicated that requirements could be met using the low-pressure [0.69-MPa (100-psi)] RO process. Watson and Hornburg (1989) also indicated that nanofiltration could be used to remove color, organics, and hardness from drinking water cost effectively.

Taylor et al. (1989) have reported on the rejection of synthetic organic compounds (SOCs) from groundwater by nanofiltration. The SOCs were ethylene dibromide (EDB), dibromochloropropane (DBCP), chlordane, heptachlor, methoxychlor, and alachor; the membrane used was the FilmTec NF70 membrane. The rejection varied from 0% for EDB to 100% for chlordane, methoxychlor, heptachlor, and alachor. Average rejection of DBCP was 35%. Adsorption of methoxychlor, heptachlor, and chlordane on the membrane was observed. Trihalomethane-forming compounds, total organic halide-forming compounds, dissolved organic compounds, and color were reduced by 95, 93, 95, and 91%, respectively. Inorganic rejections included 85% for TDS, 100% for SO_4, 64% for sodium, 87% for iron, 89% for hardness (as $CaCO_3$), and 78% for alkalinity. Water fluxes varied from 9.4×10^{-4} to 11.8×10^{-4} cm/s [20 to 25 gal/ft^2/day, (gfd)] at pressures near 0.69 MPa (100 psi).

Rautenbach and Groschl (1990) suggested another application of nanofiltration: the separation of nitrate from well water. Nitrate rejections of the NF40 membrane are low for feeds containing sulfate. Sulfate added to well water containing nitrate would result in nitrate passing through the membrane while sulfate and other highly charged species would be rejected. The permeate then could be treated with ion exchange to remove the nitrate. Bhattacharyya and Williams (1990) also indicated that the NF40 membrane gave high rejections of humic acids (THM precursors) even at high water recoveries (80%) with little drop in permeate water flux.

APPLICATIONS IN THE FOOD PROCESSING INDUSTRY

Food processing applications such as dewatering and the concentration of foodstuffs were among the first uses of RO technology. Food processing applications were a logical use of RO because, where practical, the technology offers many potential advantages over conventional technologies used by the food processing industry. These advantages include (1) low relative energy requirements and low costs; (2) lower processing temperatures and reduction of thermal damage to food during processing; and (3) simpler system designs. The

TABLE 24-9. Types of Separations for which RO is Used in the Food Processing Industry.

Type of Separation	Example
Water treatment	Pretreatment of boiler water Water softening Recycle of hot process waters
Product and chemical recovery	Recovery of sugars and acids from rinse waters from fruit cocktail dicer Recovery of peach by-products Recovery of caustic for peeling operations Regeneration of cleaning solutions and sanitizers Recovery of sweet potato stillage
Concentration/ denaturing	Juices (e.g., tomato, orange, apple, grape) Water from processing fish Concentration of milk or whey for cheese production or for transport Maple syrup
Fractionation	Fruit juice clarification Recovery of flavors, fragrances, pectins, and proteins Removal of limonin from orange juice Separation of sugars from proteins in tomato serum Removal of alcohol from wine Removal of citric acid from pasta-blancher water

overall result of these factors is reduced capital and operating costs. Table 24-9 lists examples of the main types of RO applications in food processing and examples for each type of separation.

Currently, the biggest user of RO for food processing applications is the dairy industry. However, because of recent advances, the use of RO membranes for food processing applications has expanded into new areas. Improved membranes better suited for use in food processing have been developed, and modules and systems have been developed that are easy to clean and sanitize. Reverse osmosis technology is now being used in meat, poultry, and seafood applications, in fruit and vegetable processing, and in the beer and wine industries.

Four main types of application for RO technology in food processing are considered below: (1) water treatment, (2) product and chemical recovery, (3) concentration/denaturing, and (4) fractionation.

Water Treatment Applications

RO membranes are currently used for several water treatment applications in the food industry (Pedersen 1986): demineralization of boiler feed waters, water softening to protect coolers and to prevent spotting and discoloration, and recycling of process waters. Of these applications, recycling of process waters is the most common use because of the high costs of producing the process water, the large quantities of water used, and the costs associated with wastewater disposal (Merlo, Pedersen, and Rose 1985). In many cases, reducing the volume of waste water can save enough in disposal costs to pay for the RO system used to recycle the water. Reverse osmosis systems can also be used to recycle warm ($\sim 70°C$) process water, thus reducing energy costs.

Specific examples of using RO for water treatment are recycling hot waters used to wash empty containers (Merlo, Pedersen, and Rose 1985); removal of olive particles from flume water used as boiler feed (Pedersen 1986); and removal of oils, fats, and bacteria from meat processing wastewaters (Gekas, Hallstrom, and Tragardh 1985). Spiral-wound and hollow-fiber modules are typically used for this application.

Product and Chemical Recovery Applications

These applications are closely associated with the treatment of wastewaters by RO membranes. In many food processing applications, products (e.g., sugars, proteins) are lost during washing, blanching, peeling, or other operations as the food is processed. This loss of product to a waste stream (and the associated lost revenue and increased costs for disposal) makes the recovery of the product attractive.

Specific examples of the use of RO membranes for this application include recovery of sugars and acids from the rinse water from fruit-cocktail dicers for use as filler in canned products (Merlo, Pedersen, and Rose 1985), recovery of peach by-products, and recovery of sweet potato stillage (Wu 1988a). Because of the fouling-prone nature of the feed stream, tubular and plate-and-frame modules are usually used for this application. Depending on the application, asymmetric cellulose acetate and thin-film composite membranes are used.

In addition to the recovery of food products, RO is also used to recover and recycle the chemicals required in food processing. For example, caustic solution is often used in peeling operations; RO can be used to recycle this caustic solution, reducing the costs for chemicals (Merlo, Pedersen, and Rose 1985). Additionally, cleaning solutions and sanitizers can be regenerated using RO (Pedersen 1986).

Concentration/Dewatering Applications

Many foods contain high concentrations of water when they are harvested. Removal of some of this water can reduce processing, storage, or transportation costs for these products. Reverse osmosis membranes are often well suited for this application for several reasons: (1) operation at low temperatures minimizes thermal damage to the product, (2) energy costs are reduced, and (3) system designs are simpler so capital costs are lowered (Gregor 1985; Merlo, Pedersen, and Rose 1985; Ray 1987; Parkinson 1983; Hurly 1987; Wu 1988a; Sheu and Wiley 1983). However, for RO membranes to be effective, they must be capable of operating on feed streams containing high osmotic pressures and high concentrations of suspended solids. For this reason, hybrid systems are often used: RO is used to dewater partially and concentrate the product, and a conventional process (such as evaporation) is used to complete the concentration process (Ray 1987).

Because of the high osmotic pressures and high concentrations of suspended solids encountered in dewatering applications, it is not surprising that the biggest problem associated with this application is fouling (Kulozik and Kessler 1988; Merlo, Pedersen, and Rose 1985). Fouling-resistant modules (e.g., tubular modules) are typically used for this application—with considerable success (Parkinson 1983). A list of available tubular modules and membrane types is given in Chapter 23.

The most common use of this application is in the dairy industry where RO has been used to treat whole whey and milk and to treat permeate from ultrafiltration of whey and milk (Gregor 1985; Horton 1989; Ray 1987; Pepper, Orchard, and Merry 1985; Ryder 1985). An estimated 750,000 ft^2 of RO membrane (primarily in tubular modules) has been installed for dairy applications at a cost of more than $60 million (Horton 1989).

Water softening by membranes is also used by the dairy industry for many applications, including conversion of "salt whey" (the whey from cheddar and similar cheeses) to normal sweet whey, conversion of hydrochloric acid casein whey to sweet whey, and demineralization and concentration of whey and permeate from the ultrafiltration of milk for the manufacture of lactose and lactose derivatives (Horton 1989).

Nondairy applications of RO for concentration and dewatering of food products include concentration of fruit juices (e.g., tomato, orange, apple, grape) (Merlo, Pedersen, and Rose 1985; Parkinson 1983; Hurly 1987; Sheu and Wiley 1983; Gekas, Hallstrom, and Tragardh 1985; Pepper, Orchard, and Merry 1985); concentration of water from fish processing (Merlo, Pedersen, and Rose 1985); concentration of maple syrup (Gekas, Hallstrom, and Tragardh 1985); and concentration of corn light steepwater (Ray 1987; Wu 1988b).

Fractionation Applications

One of the areas with the largest growth potential for RO applications is the fractionation of food products. Membranes will play an increasing role in food processing because of the current emphasis on natural products and low-calorie, low-cholesterol, low-caffeine, and high-fiber foods. Examples of fractionation applications include clarification of juices (Pedersen 1986; Pepper, Orchard, and Merry 1985); recovery of flavors and fragrances from natural products (Pedersen 1986); removal of limonin from orange juice (Merlo, Pedersen, and Rose 1985); separation of sugars from proteins in tomato serum (Merlo, Pedersen, and Rose 1985); removal of alcohol from beer or wine (Merlo, Pedersen, and Rose 1985; Bui et al. 1986; Mooney and Light 1985); and removal of citric acid from pasta-blancher water (Merlo, Pedersen, and Rose 1985). Tubular modules are typically used for this application.

Conclusions

RO has excellent potential for expanded use in food processing. RO membranes have been used with commercial success for many applications, such as water treatment, product recovery, product concentration, and product fractionation. As membranes become more universally accepted for these applications over the next decade, the use of membranes is likely to grow steadily.

Several developments in RO membrane technology will speed the acceptance of RO for food processing. These developments include (1) modules that are more resistant to fouling, (2) membranes and modules that can withstand repeated steam sterilization, (3) membranes that resist damage from chemicals such as chlorine and caustics, and (4) membranes and modules that can operate continuously at temperatures above 70°C.

APPLICATIONS OF HYBRID PROCESSES

For many separations it is neither desirable, nor at times even possible, to use RO to carry out the entire separation desired. For example, in treating wastewater streams, the removal of high percentages of the water can result in osmotic pressures, viscosities, or concentrations of suspended solids that are too high to make treatment by RO practical. Engineers have therefore begun designing hybrid systems in which the RO process is combined with another process, in series or in a recycle stream.

The advantages of hybrid processes include the following: (1) the separation can be effected over a broader range of feed concentrations than is possible with either RO or the other unit operation; (2) the process can be optimized more easily because the flow rates can be adjusted specifically for each unit operation; (3) the system can be adjusted to accommodate changes in the feed stream; (4) the lifetime of the RO membrane, a key issue in process economics, can often be extended by continuously adjusting the system to operate within the membrane's optimal range; and (5) the overall production cost associated with a separation is often lower in a hybrid process than for either unit operation alone.

An example of a hybrid process is the combination of RO with pervaporation for the removal of organics from water. Figure 24–18 shows a schematic of this hybrid process. In this system, an aqueous feed stream contaminated with an organic is fed to a pervaporation module. The pervaporation module is designed to remove selectively the organic from water, producing an organic-rich permeate. The retentate from the pervaporation module, now with a reduced concentration of the organic, is sent to a RO module. The RO module rejects the organic, producing a high-quality permeate that is suitable for discharge. The retentate from the RO module, concentrated in organic, is then recycled back to the pervaporation module. Thus, the hybrid system produces two streams: (1) a high-quality stream that meets discharge regulations for the organic and (2) a stream concentrated in the organic that can be recycled, or disposed of in an environmentally acceptable manner. With this system design, each unit operates under conditions as close to optimal as possible. The two units work in synergy for a separation system that is efficient and flexible, since the system can be adjusted to accommodate changes in the feed stream or changes that occur in the membrane over time.

Other synergistic hybrid systems can be designed by combining RO with other unit operations. Additional examples include:

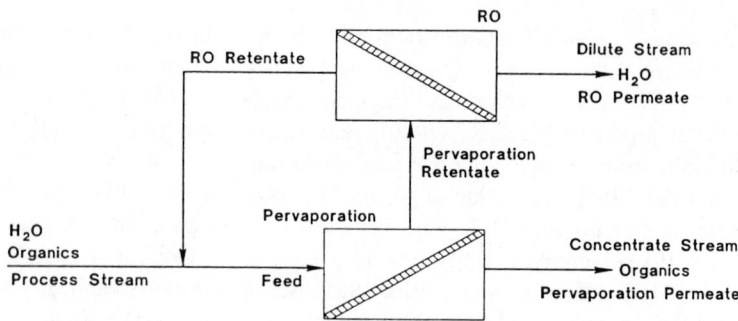

FIGURE 24–18. Schematic of an RO/pervaporation hybrid system.

1. Combining RO with coupled or facilitated transport to remove selectively heavy metal ions from water. Here, RO would be used to concentrate heavy metal ions in a stream that is sent to the coupled- or facilitated-transport membrane, resulting in a high flux of metal ions through the coupled- or facilitated-transport membrane.
2. Using RO combined with electrodialysis (ED) to treat streams with high salinity. Rather than using ED to produce water of the desired quality, a partially desalinated stream from the ED process is sent to an RO membrane to produce water of the desired quality. This design allows the ED process to operate in an optimal range where resistances through the ED stack are sufficiently low for the process to be economical.
3. Using RO combined with crystallization for the recovery of valuable products (e.g., proteins) from solution. RO is used to concentrate the dissolved species and temperature change (or another means of promoting crystallization) is used to recover the valuable products.

As RO is applied to more demanding separations, the use of hybrid systems to accomplish these separations is likely to increase. The advantages of hybrid systems over single-unit operations make them attractive for many applications.

REFERENCES

Akashah, S., M. Abdel-Jawad, M. Abdelhalim, and J. Dahdah. 1987. Cost and economic analysis of Doha reverse osmosis plant (Kuwait). *Desalination* 64:65.

Al-Arrayedh, M., B. Ericsson, M. Amin-Saad. 1987. Reverse osmosis desalination plant, RA's Abu Jarjur, State of Bahrain—two years operational experience for the 46,000 m^3/day RO plant. *Desalination* 65:197.

Al-Arrayedh, M., B. Ericsson, and M. Ohtani. 1985. Construction and operation of 46,000 m^3/day reverse osmosis desalination plant, RA's Abu Jarjur, Bahrain. *Desalination* 55:319.

Al-Awadi, F., and M. Abdel-Jawad. 1987. Evaluation of the three post-treatment systems at Doha seawater reverse osmosis plant—Kuwait. *Desalination* 63:109.

Al-Borno, A., and M. Abdel-Jawad. 1989. Conventional pretreatment of surface seawater for reverse osmosis application, state of the art. *Desalination* 74:3.

Allard, J., J. Rovel, and P. Treille. 1976. Importance of pretreatment in RO plant design and its incidence on O and M costs. *Desalination* 19:169.

Allen, P., and G. Elser. 1979. They said it couldn't be done—the Orange County, California experience. *Desalination* 30:23.

Al-Mutaz, I. 1987. By-product recovery from Saudi desalination plants. *Desalination* 64:97.

Al-Mutaz, I., M. Soliman, and A. Daghthem. 1989. Optimum design for a hybrid desalting plant. *Desalination* 76:177.

Amjad, Z. 1985. Applications of antiscalants to control calcium sulfate scaling in reverse osmosis systems. *Desalination* 54:263.

Anonymous. 1984. *Membrane and Separation Technology News.* 2:1.

Applegate, L., and C. Erkenbrecher, Jr. 1987. Monitoring and control of biological activity in Permasep seawater RO plants. *Desalination* 65:331.

Applegate, L., C. Erkenbrecher, and H. Winters. 1989. New chloramine process to control aftergrowth and biofouling in Permasep B-10

RO surface seawater plants. *Desalination* 74: 51.

Argo, D., and J. Montes. 1979. Wastewater reclamation by reverse osmosis. *JWPCF* 51:590.

Arrayedhy, M. 1987. Pre- and post-treatment at the RO plant at RA's Abu Jarjur, Bahrain. *Desalination* 63:81.

Awerbuch, L., S. May, R. Soo-Hoo, and V. Van Der Mast. 1989. Hybrid desalting systems. *Desalination* 76:189.

Baier, J., B. Lykins, C. Fronk, and S. Kramer. 1987. Using reverse osmosis to remove agricultural chemicals from groundwater. *JAWWA* August:55.

Basu, S., and V. Sakpal. 1987. Membrane separation technique in simplification of soda recovery process in pulp and paper industry. *Desalination* 67:371.

Benn, J. 1986. Membranes for industrial waste applications. Paper read at the 1986 4th Annual Membrane Technology/Planning Conference, 5–7 November 1986, Cambridge, MA.

Bergman, R., H. Harlow, and P. Laverty. 1985. Resolution of RO flux decline problems at Englewood, Florida. *Desalination* 54:55.

Bhattacharyya, D., R. Adams, and M. Williams. 1989. Separation of selected organics and inorganic solutes by low pressure reverse osmosis membranes. In *Biological and Synthetic Membranes*. ed. D. Butterfield. New York: Alan R. Liss.

Bhattacharyya, D., T. Barranger, M. Jevtitch, and S. Greenleaf. 1987. Separation of dilute hazardous organics by low pressure composite membranes. Report EPA/600/87/053. U.S. Environmental Protection Agency.

Bhattacharyya, D., M. Jevtitch, J. Ghosal, and J. Kozminski. 1984. Reverse-osmosis membrane for treating coal-liquefaction wastewater. *Environmental Prog.* 3:95.

Bhattacharyya, D., and M. Madadi. 1988. Separation of phenolic compounds by low pressure composite membranes: mathematical model and experimental results. *AIChE Symp. Series No. 261*. New York: American Institute of Chemical Engineers.

Bhattacharyya, D., and M. Williams. Submitted 1990. Separation of hazardous organics by low pressure reverse osmosis membranes—phase II. U.S. Environmental Protection Agency.

Bindoff, A., C. Davies, C. Kerr, and C. Buckley. 1987. The nanofiltration and reuse of effluent from the caustic extraction stage of wood pulping. *Desalination* 67:453.

Boesch, W. 1981. Building and commissioning the 3.0 mgd seawater RO plant for Key West Florida. *Desalination* 38:485.

Brandon, C., D. Jernigan, J. Gaddis, and H. Spencer. 1981. Closed cycle textile dyeing: full scale renovation of hot wash water by hyperfiltration. *Desalination* 39:301.

Bryant, T., J. Stuart, I. Fergus, and R. Lesan. 1987. The use of reverse osmosis as a 35,600 m^3/day concentrator in the waste water management scheme at 4640 MW Bayswater/Liddel power station complex—Australia. *Desalination* 67:327.

Buckley, C., K. Treffry-Goatley, M. Simpson, A. Bindoff, and G. Groves. 1985. Pretreatment, fouling, and cleaning in the membrane processing of industrial effluents. *ACS Symp. Series No. 261*. Washington, DC: American Chemical Society.

Bui, K., R. Dick, G. Moulin, and P. Galzy. 1986. A reverse osmosis process for the production of low ethanol content wine. *Am. J. Enol. Vitic.* 37:297.

Cadotte, J., R. Forester, M. Kim, R. Petersen, and T. Stocker. 1988. Nanofiltration membranes broaden the use of membrane separation technology. *Desalination* 70:77.

Carew, J., M. Abdel-Jawad, A. Julka, and Y. Al-Wazzan. 1989. Performance of materials used in seawater reverse osmosis plants. *Desalination* 74:85.

Cartwright, P. 1989. The application of membrane separation technologies to industrial processes. In *43rd Purdue Industrial Waste Conference Proc.* Chelsea, Mich.: Lewis Publishers.

Chakravorty, B. 1989. Effluent treatment by membrane technology—opportunities and challenges. In *Biological and Synthetic Membranes*. ed. D. Butterfield, New York: Alan R. Liss.

Chakravorty, B., and A. Srivastava. 1987. Application of membrane technologies for recovery of water from pulp and paper mill effluents. *Desalination* 67:363.

Chian, E., W. Bruce, and H. Fang. 1975. Removal of pesticides by reverse osmosis. *Environmental Sci. Technol.* 9:364.

Chian, E., and F. De Walle. 1977. Evaluation of leachate treatment: Volume II, biological and physical-chemical processes. Report EPA-600/2-77-186b. U.S. Environmental Protection Agency.

Conlon, W. 1985. Pilot field test data for prototype ultra low pressure reverse osmosis elements. *Desalination* 56:203.

Cruver, J. 1975. Waste-treatment applications of reverse osmosis. *Trans. ASME.* February:246.

Daignault, L. 1977. Pollution control in the

photoprocessing industry through regeneration and reuse. *J. Appl. Photo. Eng.* 3:94.

Dagon, T. 1978. Photographic processing effluent control. *J. Appl. Photo. Eng.* 4:62.

D'Auria, G., T. Itteilag, and R. Pastrick. 1987. The impact of reverse osmosis on makeup water chemistry at Millstone Unit Two Nuclear Power Station. *Ultrapure Water* 4:22.

Darwish, M. 1987. Critical comparison between energy consumption in large capacity reverse osmosis (RO) and multistage flash (MSF) seawater desalting plants. *Desalination* 63:143.

Darwish, M., M. Abdel-Jawad, and L. Hauge. 1989. A new dual-function device for optimal energy recovery and pumping for all capacities of RO systems. *Desalination* 75:25.

Davis, G., D. Paulson, R. Gosik, and G. Van Riper. 1987. Heavy metals contaminated waste water treatment with reverse osmosis—two case histories. Paper read at AIChE Annual Meeting, 15–20 November 1987, New York.

Dykes, G., and W. Conlon. 1989. Use of membrane technology in Florida. *JAWWA* 81:43.

Ebrahim, S., and B. Darwish. 1989. Scaling potential of Kuwait seawater for reverse osmosis desalination. *Desalination* 74:141.

Ebrahim, S., and A. Malik. 1987a. Pretreatment of surface seawater feed at DROP. *Desalination* 63:95.

Ebrahim, S., and A. Malik. 1987b. Membrane fouling and cleaning at DROP. *Desalination* 66:201.

Edwards, V., and P. Schubert. 1974. Removal of 2,4-D and other persistent organic molecules from water supplies by reverse osmosis. *JAWWA* Oct.:610.

Eisenberg, T., and E. Middlebrooks. 1986. *Reverse Osmosis Treatment of Drinking Water*. Boston: Butterworth.

El-Nashar, A. 1980. Energy and water conservation through recycle of dyeing wastewater using dynamic Zr(IV)-PAA membranes. *Desalination* 33:21.

Ericsson, B., B. Hallmans, and P. Vinberg. 1987. Optimization for design of large RO seawater desalination plants. *Desalination* 64:459.

Eriksson, P. 1988. Nanofiltration extends the range of membrane filtration. *Environmental Prog.* 7:58.

Fang, H., and E. Chian. 1976. Reverse osmosis separation of polar organic compounds in aqueous solution. *Environmental Sci. Technol.* 10:364.

Federal Register. 1975. 40:59566.

FilmTec Corporation. 1988. Field data.

Finlay, W., and P. Ferguson. 1977. Design and operation of a turnkey 17,000,000 GPD reverse osmosis water treatment plant—Buwayb, Riyadh V, Saudi Arabia. *Desalination* 23:389.

Fredkin, E., and R. Banks. 1989. Computerized instrumentation and control for reverse osmosis systems. *Desalination* 75:141.

Gekas, V., B. Hallstrom, and G. Tragardh. 1985. Food and dairy applications: the state of the art. *Desalination* 53:95.

Ghabris, A., M. Abdel-Jawad, and G. Aly. 1989. Municipal wastewater renovation by reverse osmosis, state of the art. *Desalination* 75:213.

Ghassemi, M., K. Yu, and S. Quinlivan. 1981. Feasibility of commercial water treatment techniques for concentrated waste spills. Report EPA-600/2-81-213. U.S. Environmental Protection Agency.

Glueckstern, P., and Y. Kantor. 1983. Seawater vs. brackish water desalting: technology, operating problems and overall economics. *Desalination* 44:51.

Glueckstern, P., M. Wilf, J. Etgar, and J. Ricklis. 1985. Use of microprocessors for control. *Desalination* 55:469.

Goldsmith, R. 1975. Membrane processing in the metal finishing industry. *J. Eng. Indus.* 97:238.

Gregor, H. 1985. Energy efficient membrane separation processes for the corn wet milling industry. Report DE-AC07-83ID12480. U.S. Department of Energy.

Hanson, D. 1990. Hazardous wastes: EPA adds 25 organics to RCRA list. *Chem. Eng. News* 68:11.

Hassan, A., S. Al-Jarrah, T. Al-Lohibi, A. Al-Hamdan, L. Bakheet, and M. Al-Amri. 1989. Performance evaluation of SWCC SWRO plants. *Desalination* 74:37.

Hassan, A., and A. Malik. 1989. Corrosion resistant materials for seawater RO plants. *Desalination* 74:157.

Hays, D., M. Miget, M. Motilall, and G. Davis. 1988. Recovery of ammonium nitrate by reverse osmosis. Report prepared by Osmonics, Inc., Minnetonka, MN.

Hess, M., G. Jones, W. Micheletti, and J. Tomlinson. 1988. Wastewater concentration by seeded reverse osmosis: a field demonstration in the electric power industry. *Environmental Prog.* 7:7.

Heyden, W. 1985. Seawater desalination by reverse osmosis—plant design, performance data, operation, and maintenance (Tamajib, Arabian Gulf Coast). *Desalination* 52:187.

Hinden, E., P. Bennett, and S. Narayanan. 1969. Organic compounds removal by reverse osmosis. *Water and Sewage Works* 116:66.

Hoeting, W. 1982. Seawater reverse osmosis with energy recovery. *Desalination* 40:357.

Horton, B. 1989. Horton International, Inc. Cambridge, MA. Personal communication.

Hsiue, G., L. Pung, M. Chu, and M. Shieh. 1989. Treatment of uranium effluent by reverse osmosis membrane. *Desalination* 71:35.

Hurly, P. 1987. New filters clean up in new markets. *High Technol.* August:21.

Huxstep, M. 1981. Inorganic contaminant removal from drinking water by reverse osmosis. Report EPA-600/2-81. U.S. Environmental Protection Agency.

Ikeda, K., T. Nakano, H. Ito, T. Kubota, and S. Yamamoto. 1988. New composite charged reverse osmosis membrane. *Desalination* 68:109.

Imasu, K. 1985. Wastewater recycle in the plating industry using brackish water reverse osmosis elements. *Desalination* 56:137.

Jawad, M. 1989. Future for desalination by reverse osmosis. *Desalination* 72:23.

JAWWA (December 1985).

Johnston, H., and H. Lim. 1978. Removal of persistent contaminants from municipal effluents by reverse osmosis. Report 85. Toronto, Ontario: Ontario Min. Environ.

Jonsson, A., and R. Wimmerstedt. 1985. The application of membrane technology in the pulp and paper industry. *Desalination* 53:181.

Kamizawa, C., H. Masuda, M. Matsuda, T. Nakane, and H. Akami. 1978. Studies on the treatment of gold plating rinse by reverse osmosis. *Desalination* 27:261.

Kaschemekat, J., W. Hilgendorff, K. Boeddecker, A. Hassan, and A. Malik. 1983. Two-stage reverse osmosis desalination. *Desalination* 46:151.

Kawana, S., T. Yoshioka, H. Harashina, and T. Mikuni. 1987. Seawater pretreatment by continuous sand filter for seawater RO (reverse osmosis) desalination plant. *Desalination* 66:339.

Klinko, K., W. Light, and C. Cummings. 1985. Factors influencing optimum seawater reverse osmosis system designs. *Desalination* 54:3.

Ko, A., and D. Guy. 1988. Brackish and seawater desalting. Chap. 5 in *Reverse Osmosis Technology*. ed. B. Parekh. New York: Marcel Dekker.

Koga, S., and K. Ushikoshi. 1977. ALCLOSE reverse osmosis system. *Desalination* 23:105.

Kosarek, L. 1981. Removal of various toxic heavy metals and cyanide from water by membrane processes. In *Chemistry in Water Reuse*, ed. W. J. Cooper. Vol. 1. Ann Arbor, MI: Ann Arbor Science.

Kremen, S., C. Hayes, and M. Dubos. 1977. Large-scale reverse osmosis processing of metal finishing rinse waters. *Desalination* 20:71.

Kreshman, S. 1985. Seawater intakes for desalination plants in Libya. *Desalination* 55:493.

Kulozik, U., and H. Kessler. 1988. Permeation rate during reverse osmosis of milk influenced by osmotic pressure and deposit formation. *J. Food Sci.* 53:1377.

Kurihara, M., N. Harumiya, N. Kanamaru, T. Tonomura, and M. Nakasatomi. 1981. Development of the PEC-1000 composite membrane for single-stage seawater desalination and the concentration of dilute aqueous solutions containing valuable materials. *Desalination* 38:449.

Kurihara, M., Y. Nakagawa, H. Takeuchi, N. Kanamaru, and T. Tonomura. 1983. Single-stage seawater desalination at high temperature and salinity as present in the Middle East using PEC-1000 membrane modules. *Desalination* 46:101.

Leitner, G. 1983. Waste concentration vs evaporation ponds for reverse osmosis reject water—a case study. *Desalination* 44:31.

Light, W. 1981. Removal of chemical carcinogens from water/wastewater by reverse osmosis. In *Chemistry in Water Reuse*, ed. W. J. Cooper. Vol. 1. Ann Arbor, MI: Ann Arbor Science.

Light, W., J. Perlman, A. Riedinger, and D. Needham. 1988. Desalination of non-chlorinated surface seawater using TFC® membrane elements. *Desalination* 70:47.

Lim, M., and H. Johnston. 1976. Reverse osmosis as an advanced treatment process. *JWPCF* 48:1820.

Lonsdale, H., C. Milstead, B. Cross, and F. Graber. 1969. Study of rejection of various solutes by reverse osmosis membranes. Report 447. U.S. Office of Saline Water R&D Program.

Lozier, J., E. Oklejas and M. Silbernagel. 1989. The hydraulic turbocharger: a new type of device for the reduction of feed pump energy consumption in reverse osmosis systems. *Desalination* 75:71.

Lynch, S., J. Smith, L. Rando, and W. Yauger. 1984. Isolation or concentration of organic substances from water—an evaluation of reverse osmosis concentration. Report EPA-600/1-84-018. U.S. Environmental Protection Agency.

Malik, A., K. Mousa, N. Younan, and B. Rao. 1987. Performance evaluation of three different seawater RO membranes at DROP in Kuwait. *Desalination* 63:163.

McNulty, K., R. Goldsmith, and A. Gollan. 1977. Reverse osmosis field test: treatment of watts nickel rinse waters. Report EPA-600/2-77-039. U.S. Environmental Protection Agency.

Merlo, C., L. Pedersen, and W. Rose. 1985. Hyperfiltration/reverse osmosis: a handbook on membrane filtration for the food industry. Report DE-FG01-84CE40691. U.S. Department of Energy.

Mindler, A., and A. Epstein. 1986. Measurements and control in reverse osmosis desalination. *Desalination* 59:343.

Montgomery, J. 1985. *Water Treatment Principles and Design*. New York: John Wiley & Sons.

Moody, C., J. Kaakinen, J. Lozier, and P. Laverty. 1983. Yuma desalting test facility: foulant component study. *Desalination* 47:239.

Mooney, L., and W. Light. 1985. Food applications for sanitary design spiral-wound membrane elements. Paper read at the International Membrane Technology Symposium, 28–30 May 1985. Tylosand, Sweden.

Morris, D., W. Nelson, and G. Walraver. 1972. Recycle of papermill waste waters and application of reverse osmosis. Report EPA-12040 FUB 01/72. U.S. Environmental Protection Agency.

Muhurji, M., G. Faggard, V. Van Der Mast, and H. Iamai. 1989. Jeddah 1 RO plant—phase I 15 MGD reverse osmosis plant. *Desalination* 76:75.

Muirhead, A., S. Beardsley, and J. Aboudiwan. 1982. Performance of the 12,000 m^3/D seawater reverse osmosis desalination plant at Jeddah, Saudi Arabia January 1979 through January 1981. *Desalination* 42:115.

Muratova, N., V. Shevchenko, G. Stychinskii, A. Dolganova, and I. Smirnov. 1979. Study of wastewater treatment by reverse osmosis. *Int. Chem. Eng.* 19:350.

Nada, N., A. Zahrani, and B. Ericsson. 1987. Experience on pre- and post-treatment from seawater desalination plants in Saudi Arabia. *Desalination* 66:303.

Nordin, S., B. Ericsson, and B. Wallen. 1985. Optimization of high pressure piping in reverse osmosis plants. *Desalination* 55:247.

Nusbaum, I., and D. Argo. 1984. Design, operation, and maintenance of a 5-mgd wastewater reclamation reverse osmosis plant. In *Synthetic Membrane Processes: Fundamentals and Water Applications*. ed. G. Belfort. New York: Academic Press.

Nusbaum, I., and A. Riedinger. 1980. Water quality improvement by reverse osmosis. In *Water Treatment Plant Design*. ed. R. Sanks. Ann Arbor, MI: Ann Arbor Science Publications, Inc.

Odegaard, H., and S. Koottatep. 1982. Removal of humic substances from natural waters by reverse osmosis. *Water Res.* 16:613.

Oldfield, J., and B. Todd. 1985. The use of stainless steels and related alloys in reverse osmosis desalination plants. *Desalination* 55:261.

Olsen, O. 1980. Membrane technology in the pulp and paper industry. *Desalination* 35:291.

Osta, T., and L. Bakheet. 1987. Pretreatment system in reverse osmosis plants. *Desalination* 63:71.

Parkinson, G. 1983. Reverse osmosis: trying for wider applications. *Chem. Eng.* May:26.

Patra, R., S. Prabhakar, B. Misra, and M. Ramani. 1987. Water and effluent water treatment using reverse osmosis. *Desalination* 67:507.

Paulson, D., and D. Spatz. 1983. Reverse osmosis/ultrafiltration membrane applied to the pulp and paper industry. In *Proc. Technical Association of the Pulp and Paper Industry, 1983 International Dissolving and Specialty Pulps Conferences*. Atlanta, GA: TAPPI Press.

Pedersen, L. 1986. Membrane in food applications: new developments. Paper read at the 1986 4th Annual Membrane Technology/Planning Conference, 5–7 November 1986, Cambridge, MA.

Pepper, D. 1981. Pretreatment and operating experience on reverse osmosis plant in the Middle East. *Desalination* 38:403.

Pepper, D., A. Orchard, and A. Merry. 1985. Concentration of tomato juice and other fruit juices by reverse osmosis. *Desalination* 53:157.

Perry, M., and C. Linder. 1989. Intermediate reverse osmosis ultrafiltration (RO UF) membranes for concentration and desalting of low molecular weight organic solutes. *Desalination* 71:233.

Peters, R., Y. Ku, and D. Bhattacharyya. 1985. Evaluation of recent treatment techniques for removal of heavy metals from industrial wastewaters. *AIChE Symp. Series No. 243*. New York: American Institute of Chemical Engineers.

Porter, J., and G. Goodman. 1984. Recovery of hot water, dyes, and auxiliary chemicals from textile wastestreams. *Desalination* 49:185.

Prabhakar, S., B. Misra, and M. Ramani. 1987. Reverse osmosis for the production of boiler feed water to power plants. *Indian J. Technol.* 25:405.

Pusch, W., Y. Yu, and L. Zheng. 1989. Solute-solute and solute-membrane interactions in hyperfiltration of binary and ternary aqueous organic feed solutions. *Desalination* 75:3.

Rautenbach, R., and A. Groschl. 1990. Separation

potential of nanofiltration membranes. *Desalination* 77:73.

Ray, R. 1987. The use of membranes in hybrid industrial separation systems. Final Report EPRI EM-5231, Research Project 2662-2. Electric Power Research Institute.

Regunathan, P., W. Beauman, and E. Kreusch. 1983. Efficiency of point-of-use devices. *JAWWA* 75:42.

Rice, R. 1985. *Safe Drinking Water*. New York: Lewis Publishers.

Richardson, N., and D. Argo. 1977. Orange County's 5 MGD reverse osmosis plant. *Desalination* 23:563.

Rickabaugh, J., S. Clement, J. Martin, and M. Sunderhaus. 1986. Chemical and microbial stabilization techniques for remedial action sites. *Proc. 12th Annual Hazardous Wastes Symposium,* U.S. EPA, Cincinnati, OH.

Robinson, G. 1983. Recovery pays off for Chicago job shop plater. *Products Finishing* (June).

Ryder, D. 1985. Use of membranes in the manufacture of hard and semi hard cheeses. *Desalination* 53:129.

Sackinger, C. 1982. Energy advantages of reverse osmosis in seawater desalination. *Desalination* 40:271.

Sackinger, C. 1985. Seawater reverse osmosis at high pressures. *Desalination* 54:19.

Sareen, S., D. Dickehuth, and T. Van Sittert. 1982. Wastewater treatment study: MIS retorts. In *Proc. 15th Oil Shale Symposium.* ed. J. Gary. Golden, CO: Colorado School of Mines.

Sato, T., M. Imaizumi, O. Kato, and Y. Taniguchi. 1977. RO applications in wastewater reclamation for re-use. *Desalination* 23:65.

Scherm, M., and C. Lawson. 1977. Pilot demonstration of renovation and reuse of wastewaters from organic chemical manufacturing. *Ind. Water Eng.* 14:16.

Schrantz, J. 1975. Big savings with reverse osmosis and acid copper. *Ind. Fin.* 51:30.

Schutte, C., T. Spencer, J. Aspden, and D. Hanekom. 1987. Desalination and reuse of power plant effluents: from pilot plant to full scale application. *Desalination* 67:255.

Sheu, M., and R. Wiley. 1983. Preconcentration of apple juice by reverse osmosis. *J. Food Sci.* 48:422.

Shuckrow, A., A. Pajak, and J. Osheka. 1981. Concentration technologies for hazardous aqueous waste treatment. Report EPA-600/2-81-019. U.S. Environmental Protection Agency.

Siler, J., and D. Bhattacharyya. 1985. Low pressure reverse osmosis membrane: concentration and treatment of hazardous wastes. *Hazardous Waste Hazardous Mater.* 2:45.

Simpson, M., and G. Groves. 1983. Treatment of pulp/paper bleach effluents by reverse osmosis. *Desalination* 47:327.

Simpson, J., C. Kerr, and C. Buckley. 1987. The effect of pH on the nanofiltration of the carbonate system in solution. *Desalination* 64:305.

Skovronek, H., and M. Stinson. 1977. Advanced treatment approaches for metal finishing wastewater. Part 1. Report EPA-600/J-77-056a. U.S. Environmental Protection Agency.

Slater, C., R. Ahlert, and C. Uchrin. 1983a. Applications of reverse osmosis to complex industrial wastewater treatment. *Desalination* 48:171.

Slater, C., R. Ahlert, and C. Uchrin. 1983b. Treatment of landfill leachates by reverse osmosis. *Environmental Prog.* 2:251.

Slater, C., R. Ahlert, and C. Uchrin. 1987. Reverse osmosis processes for the renovation and reuse of hazardous industrial wastewaters. *Current Prac. Environmental Sci.* 3:1.

Slater, C., A. Ferrari, and P. Wisniewski. 1987. Removal of cadmium from metal processing wastewaters by reverse osmosis. *J. Environmental Sci. Health.* A22:707.

Soleman, M., J. de Graauw, and W. van Putten. 1983. Large seawater reverse osmosis plants: design considerations. *Desalination* 46:163.

Soo-Hoo, R., L. Awerbuch, S. May, M. Mattson, and S. Kremen. 1983. Parametric study on seawater reverse osmosis desalting plants. *Desalination* 46:3.

Sorg, T., R. Forbes, and D. Chambers. 1980. Removal of radium 226 from Sarasota County, FL, drinking water by reverse osmosis. *JAWWA* 72:230.

Sorg, T., and O. Love. 1984. Reverse osmosis treatment to control inorganic and volatile organic contaminants. Report EPA-600/D-84-198. U.S. Environmental Protection Agency.

Sourirajan, S. 1970. *Reverse Osmosis*. New York: Academic Press.

Sourirajan, S., and T. Matsuura. 1985. *Reverse Osmosis/Ultrafiltration Principles*. Ottawa, Canada: National Research Council of Canada.

Spatz, D. 1979. A case history of reverse osmosis used for nickel recovery in bumper recycling. *Plating Surface Finishing* (July).

Spatz, D. 1981. Multiyear experience with oily and organic chemical waste treatment using reverse osmosis. *ACS Symp. Series No. 154*. Washington, DC: American Chemical Society.

Spiegler, K., and A. Laird, eds. 1980. *Principles of Desalination,* 2nd ed. New York: Academic Press.

Stacha, J., and F. Pontius. 1984. An overview of water treatment practices in the United States. *JAWWA* 76:75.

Stenstrom, M., J. Davis, J. Lopez, and J. McCutchan. 1982. Municipal wastewater reclamation by reverse osmosis—a 3-year case study. *JWPCF* 54:43.

Sulpizio, T., W. Light, and J. Perlman. 1989. Economic and technical factors affecting seawater reverse osmosis process design. *Desalination* 74:187.

Taylor, J., S. Durancean, L. Mulford, D. Smith, and W. Barrett. 1989. SOC rejection by nanofiltration. Report EPA/600/2-89/023. U.S. Environmental Protection Agency.

Terril, M., and R. Neufeld. 1983. Reverse osmosis of blast—furnace scrubber water. *Environmental Prog.* 2:121.

Thorsen, T. 1985. Recovery of phosphoric acid with RO. *Desalination* 53:217.

Treffry-Goatley, K., C. Buckley, and G. Groves. 1983. Reverse osmosis treatment and reuse of textile dyehouse effluents. *Desalination* 47: 313.

Tsuge, H., and K. Mori. 1977. Reclamation of municipal sewage by reverse osmosis. *Desalination* 23:123.

Tsuge, H., S. Suda, and T. Matsumura. 1981. Pretreatment of seawater RO feed by means of coagulation-filtration. *Desalination* 38:425.

U.S. Congress Office of Technology Assessment. 1988. Using desalination technologies for water treatment. Report OTA-BP-O-46.

Van den Huvel, J., R. Zoetemeyer, and C. Beolhouver. 1981. Purification of municipal wastewater by subsequent reverse osmosis and anaerobic digestion. *Biotech. Bioeng.* 23:2001.

Wade, N. 1987. R.O. design optimization. *Desalination* 64:3.

Wangnick, K. 1986. World market of desalting plants. *IDA Magazine* 1:53.

Watson, B., and C. Hornburg. 1989. Low-energy membrane nanofiltration for removal of color, organics and hardness from drinking water supplies. *Desalination* 72:11.

Weber, W. 1972. *Physicochemical Processes for Water Quality Control*. New York: John Wiley & Sons.

Wiley, A., L. Dambruch, P. Parker, and H. Dugal. 1978. Treatment of bleach plant effluents: a combined reverse osmosis/freeze concentration process. *TAPPI* 61:77.

Williams, M., R. Deshmukh, and D. Bhattacharyya. 1990. Separation of hazardous organics by reverse osmosis membranes. *Environmental Prog.* 9:118.

Wilson, D., and R. Duran. 1982. Water treatment for small public supplies. *Water Resources Research Report No. 141*. New Mexico State University.

Wilson, W., A. Gruendisch, and I. Calder-Potts. 1987. The use of Pelton wheel turbines for energy recovery in reverse osmosis systems. *Desalination* 65:231.

Winters, H. 1987. Control of organic fouling at two seawater reverse osmosis plants. *Desalination* 66:319.

Wojcik, C. 1983. Desalination of water in Saudi Arabia by reverse osmosis performance study. *Desalination* 46:17.

Wu, Y. 1988a. Characterization of sweet potato stillage and recovery of stillage solubles by ultrafiltration and reverse osmosis. *J. Agric. Food Chem.* 36:252.

Wu, Y. 1988b. Reverse osmosis and ultrafiltration of corn light steep-water solubles. *Cereal Chem.* 65:105.

25
Cost Estimates

Roderick J. Ray
Bend Research, Inc.

INTRODUCTION
BASIS AND DESCRIPTIONS OF
  COST CATEGORIES
  Capital Costs: Direct
  Capital Costs: Indirect
  Operating Costs
  Capital Recovery Costs
METHODOLOGY FOR ESTIMATING
  RO SYSTEM COSTS
SENSITIVITY STUDIES
  Sensitivity Studies for Capital Costs
  Sensitivity Studies for Operating
    Costs
SMALL-SCALE AND SPECIALTY
  SYSTEMS
  Small Systems
  Specialty Systems
CONCLUSIONS
APPENDIX: WORKSHEETS
ACKNOWLEDGMENTS
NOTATION
REFERENCES

INTRODUCTION

The key factor in determining the cost of water treatment by reverse osmosis (RO) is the inherent variation in capital costs and operating costs for RO systems. Capital costs for RO systems put out for bid can easily vary by up to a factor of 2 (Birkett 1988; Glueckstern and Arad 1984). Operating costs also vary over a wide range, even for the same type of application (Birkett 1988; Applegate 1984). Several reasons cause these cost variations, including (1) widely different feed stream compositions; (2) widely different capabilities in RO technology, including membranes, membrane modules, equipment, and systems design; (3) the competitive commercial environment of the RO industry and associated technological breakthroughs; (4) system size; and (5) desired product purity.

These cost variations make it difficult to derive cost equations that allow the reader to predict accurately the costs of an RO system "from scratch." Many excellent analyses have been done of the costs of RO systems for applications ranging from seawater desalination to producing ultrapure water. (See, for example, Birkett 1988; Hornberg and Morin 1986; Taylor, Smith, and Goigel 1989; Applegate 1984; Andrews and Bergman 1986; Al-Marafie and Darwish 1989; Ericsson and Hallmans 1985). The reader is referred to these references for the details of specific RO applications. Furthermore, the major manufacturers of membrane modules and systems (see Chapter 24) have developed sophisticated computer programs to help them prepare bids on RO systems given certain detailed input parameters, such as exact feed composition, power costs, and water quality requirements.

The approach adopted here is to provide a methodology for estimating the cost of an RO system for a given application with sufficient accuracy that the economic feasibility can be established without resorting to a full-scale design and bid process. To this end, three applications areas have been considered here: seawater desalination, brackish water treatment, and low-pressure reverse osmosis/membrane-softening (LPRO/MS) treatment. The methodology is applied to both large systems (those with capacities of $\geq$50,000 gal/day or $\geq$190 m^3/day) and, in a separate section, to specialty systems and systems smaller than 50,000 gal/day (190 m^3/day).

In each of these areas, a reasonable average or "most representative" set of costs for the capital and operating costs associated with RO has been determined. The reader can use these representative costs as a good estimate of the costs to be expected for a given application, or the reader can "correct" these representative costs using the appropriate equation to increase or decrease them as warranted for a given application. In this way, the reader can make use of whatever specific information is available for a given application, but specific information for every cost category is not required to estimate the RO costs. This methodology, therefore, can be used to estimate the costs of RO treatment for a given application by making an already reasonable estimate more accurate—rather than building those costs from "first principles."

In this chapter, each cost category for capital costs and operating costs is discussed; the basis used for the representative cost in each category is reviewed. The methodology for estimating the costs of RO systems, based on the representative values for each cost category, is then explained. Sensitivity studies of certain cost categories are reviewed, and the application of the costing methodology to small-scale and specialty systems is discussed. Finally, conclusions about the methodology are offered.

BASIS AND DESCRIPTIONS OF COST CATEGORIES

The cost categories for capital and operating costs used in this costing method are shown in

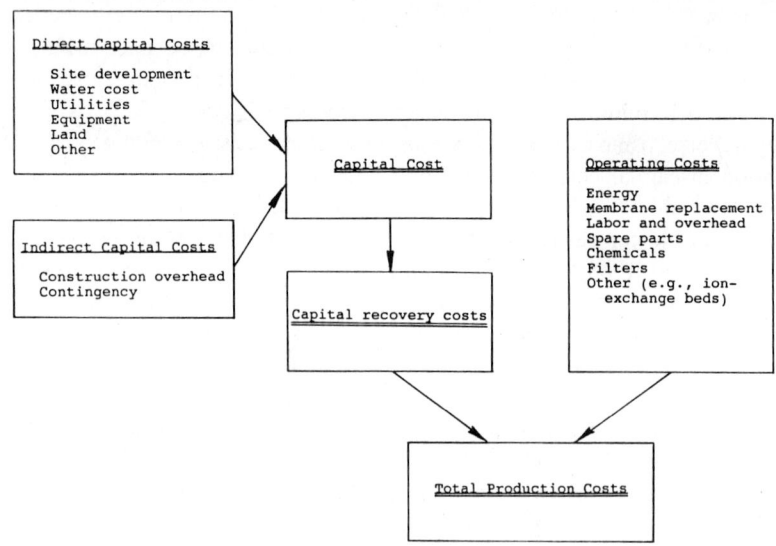

FIGURE 25–1. Reverse osmosis cost categories.

Figure 25–1. These cost categories cover costs that can be expected in most RO applications and allow standardization of the cost information found in the literature. This section provides a brief discussion of each cost category, presenting the average and/or most representative cost for each and the equation to be used to correct the representative cost if more specific information is available. The average or most representative costs presented here are based on a plant capacity of 1,000,000 gal/day (3800 m^3/day). Using the size-correction factor discussed later in this section, however, these costs can be corrected for system sizes from about 100,000 gal/day (380 m^3/day) to 50,000,000 gal/day (190,000 m^3/day).

Capital Costs: Direct

Site Development

Site development costs include the costs of buildings, roads, fences, and any other site modifications that must be made for the equipment to be installed. The costs of the feed water supply and the brine disposal systems are not included, nor are the capital costs associated with installation of utilities.

The representative costs of site development are assumed to be $0.10/gal/day ($26.42/m^3/day) for seawater, brackish water, and LPRO/MS systems. However, if an RO plant is being retrofitted into an existing water treatment facility, site development costs can be essentially zero. Conversely, in remote or rugged areas or in other special situations, site development costs can be quite high. In the literature, these costs range between $0 and $5.50/gal/day ($1450/m^3/day) (Al-Marafie and Darwish 1989; Taylor et al. 1989b; Hornberg and Morin 1986; Ericsson and Hallmans 1985; Wade, Heaton, and Boulter 1985).

Water Costs

Water costs are the costs associated with the feed water supply and brine disposal. Representative water costs and ranges reported in the literature are listed in Table 25–1 for the seawater desalination, brackish water treatment, and LPRO/MS applications. Although the representative costs are close for the three applications areas, cost components within this category can vary widely. Comprehensive discussions on the topic of water costs are contained in publications by Taylor et al. (1989b), the Office of Water Technology (Anonymous, 1989), and Rogers (1984).

Costs associated with the feed water supply are affected by three main factors: (1) the system used to supply the feed water, (2) the need for storage tanks, and (3) the percentage of feed water that is recovered as permeate.

The system used to supply the feed water is highly application dependent. For seawater desalination, sophisticated sea wells or beach wells with complex pumping and plumbing systems are often needed. For brackish water treatment, feed water is supplied from a well, which may need to be quite deep. LPRO/MS applications typically involve the treatment of groundwater or water from shallow wells, but in some cases deep wells are also required.

Storage tanks can also add to the costs of the feed water supply. These are often needed for smaller applications and can provide a backup if the feed water supply is interrupted or if a total system failure occurs. The tanks can also be used as pretreatment tanks or product water tanks, as appropriate.

TABLE 25–1. Representative Costs and Ranges Reported in the Literature for Water.

Application	Representative Cost		Range	
	($/gal/day)	[$/m^3/day]	($/gal/day)	[$/m^3/day]
Seawater	0.09	[23.78]	0.05–1.00	[13.20–264.20]
Brackish water	0.09	[23.78]	0.05–0.50	[13.20–132.10]
LPRO/MS	0.20	[52.84]	0.05–0.45	[13.20–118.90]

TABLE 25–2. Representative Costs and Ranges Reported in the Literature for Utilities.

Application	Representative Cost		Range	
	($/gal/day)	[$/m³/day]	($/gal/day)	[$/m³/day]
Seawater	0.16	[42.27]	0.10–0.45	[26.42–118.90]
Brackish water	0.11	[29.06]	0.05–0.30	[13.20–79.26]
LPRO/MS	0.06	[15.85]	0.03–0.10	[7.93–26.42]

The percentage of feed water that must be recovered as permeate determines the volume of feed water that must be supplied and the volume of brine produced as waste; thus, it affects feed water supply costs and brine disposal costs.

Capital costs for brine disposal can also be significant and will vary depending on the type of disposal system used, environmental regulations affecting the site, and the volume of brine that must be disposed. The five typical brine disposal methods are (1) disposal in the ocean; (2) "land-spreading," in which the brine is pumped over land; (3) disposal in a sewer (an option only in small-scale systems); (4) injection wells; and (5) evaporation/crystallization ponds.

Utilities

Utilities include power supply systems for electricity and high-voltage alternating current, and external plumbing. Representative capital costs for utilities and the ranges for those costs are shown in Table 25–2.

Equipment

Equipment is, of course, the dominant category of the direct capital costs. Equipment costs include (1) the feed water pretreatment system, (2) the membrane modules, (3) the RO system, (4) shipping and installation of the system, and (5) the engineering costs associated with the equipment design. Table 25–3 shows representative costs for each application area and the ranges for these costs reported in the literature. Table 25–4 shows the typical percentage of the total equipment costs associated with each equipment component.

The pretreatment system can range from a small fraction of the equipment costs (e.g., for a system operating on municipal water or relatively clean process waters) to more than one-half of the equipment costs in certain applications for seawater desalination or surface water treatment.

Chapter 23 details the costs of membrane modules. Normally, the cost of membrane modules is 20 to 35% of the RO system costs. This can vary significantly, depending on the application and membrane type chosen. As a rule, the module replacement costs account for a greater percentage of total production costs than does the cost of the initial membrane modules.

Costs for the RO system includes the pumps, controls, plumbing, electrical subsystems, membrane module pressure vessels, and skids (for certain plants). Also included in many larg-

TABLE 25–3. Representative Costs and Ranges Reported in the Literature for Equipment.

Application	Representative Cost		Range	
	($/gal/day)	[$/m³/day]	($/gal/day)	[$/m³/day]
Seawater	3.34	[882.43]	2.50–4.50	[660.50–1188.90]
Brackish water	1.55	[409.51]	0.75–2.00	[198.15–528.40]
LPRO/MS	0.75	[198.15]	0.40–1.10	[105.68–290.62]

TABLE 25-4. Representative Percentages of Equipment Costs for Each Component.

Component	Application		
	Seawater (%)	Brackish Water (%)	LPRO/MS (%)
Pretreatment	15	20	15
Membrane modules	15	15	20
RO system	60	55	55
Shipping and installation	5	5	5
Equipment-related engineering	5	5	5

er, high-operating-pressure systems is power recovery equipment, which can account for a significant percentage of the total equipment costs.

Shipping costs are normally only a small percentage of the total equipment costs, but can become significant for remote locations and for very large systems. Another factor of importance is shipping time, which can alter total construction time significantly and affect indirect capital costs.

Finally, the engineering costs associated with the equipment design are a small percentage of many RO systems—especially for conventional applications. However, in certain special situations, these costs can be very high. For example, extensive engineering work on the pretreatment system may be required in cases wherein feed waters contain unusual compositions of certain scale-forming compounds.

Land

Every RO system must be located somewhere, and the land required will normally have a price. A cost for land has not been included in the costing method presented here, however, because many of the RO systems costs reported in the literature show land-related costs to be negligible or nonexistent. Rather, land is included as a category to remind the reader to investigate this cost and to include it if it is known or if it is a relatively large value (Birkett 1988; Hornberg and Morin 1986; Larson and Leitner 1979; Wade, Heaton, and Boulter 1985).

Other

For certain applications, capital costs must be considered that are not typically found in most systems. For example, in ultrapure water production, a sophisticated post-treatment subsystem that might include ion-exchange beds or other high-cost equipment must be designed, fabricated, and installed.

System Size Correction

Economies of scale affect the costs of RO plants, just as they affect other unit operations in the chemical process industry. The equation normally used to estimate the cost of a system, given a known cost for a system of a different size, is

$$\begin{pmatrix} \text{size-corrected} \\ \text{system} \\ \text{direct capital cost} \end{pmatrix} = \begin{pmatrix} \text{base system} \\ \text{direct capital} \\ \text{cost} \end{pmatrix} \times \left(\frac{\text{actual system size}}{\text{base system size}} \right)^n,$$

(25-1)

where the base system size is 1,000,000 gal/day (or 3800 m^3/day) and the exponent n, known as the "scale factor," is determined from actual cost data. Many conventional unit operations, such as distillation, follow the "0.6 scaling rule"; that is, the scale factor, n, is 0.6.

For membrane systems, the scale factor is normally larger, with n being in the range of 0.75 to 1.00 (Birkett 1988; Leitner 1987). This higher value for n implies that membrane systems reach a point at which increases in size do not result in significant savings—from a capital cost standpoint they do not scale up as well as other unit operations do. Conversely, however, this larger value for n implies that membrane systems do "scale down" well—to a point at which scale factor n takes on a smaller value. For ready-made plants this point is about 100,000 gal/day (380 m^3/day), whereas for tailor-made plants this point is about 50,000

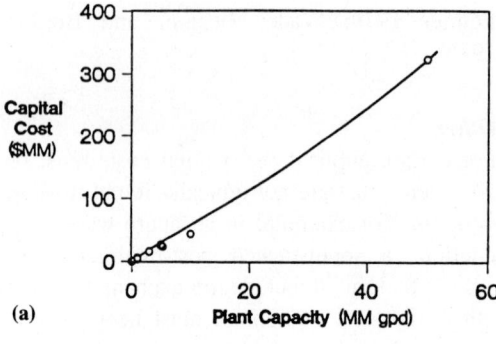

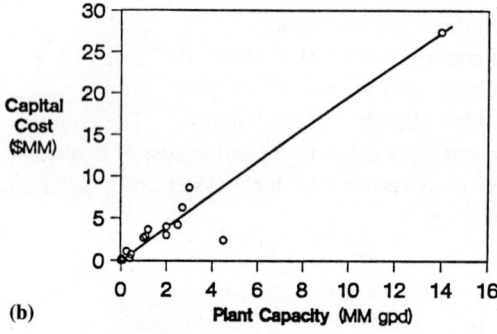

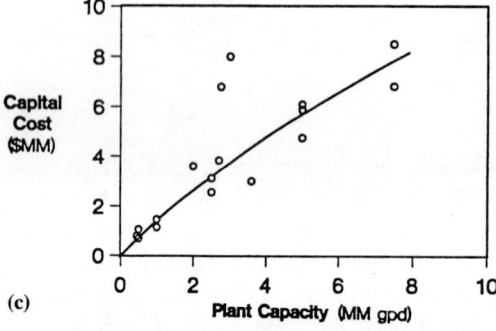

FIGURE 25–2. Total capital costs of RO systems as a function of plant capacity for (a) seawater, (b) brackish water, and (c) LPRO/MS applications ($MM = $1,000,000; MM gpd = 1,000,000 gal/day).

termined from Figure 25–2 for each application, are shown in Table 25–5.

Equation (25–1) corrects the direct capital cost of the system for size. However, capital costs for RO systems are usually given in $/gal/day. To correct these values, Eq. (25–1) must be modified as follows:

$$\begin{pmatrix} \text{size-corrected} \\ \text{system} \\ \text{direct capital cost} \\ (\$/\text{gal}/\text{day}) \end{pmatrix} = \begin{pmatrix} \text{base system} \\ \text{direct capital} \\ \text{costs} \\ (\$/\text{gal}/\text{day}) \end{pmatrix}$$

$$\times \left(\frac{\text{actual system size}}{\text{base system size}} \right)^{n-1}.$$

(25–2)

This size correction method can be used in several ways. For instance, if specific cost data are unavailable for any of the direct capital cost categories for a given application, then the total direct capital costs can be corrected for system size using Eq. (25–2) and the appropriate value for n from Table 25–5. If cost data are known for one or more of the direct capital cost categories, this size correction factor can be used for the categories for which data are not available; then the costs can be totaled to yield a total corrected capital cost. In this way, a size-corrected direct capital cost total can be determined.

Capital Costs: Indirect

Indirect capital costs are subsidiary costs associated with construction of the RO plant and with the uncertainty in the time frame of construction and in the direct costs. As with the direct capital cost categories listed earlier, a

gal/day (190 m³/day). Figure 25–2 shows a plot of the capital costs of RO systems for the seawater desalination, brackish water treatment, and LPRO/MS applications as a function of plant capacity. These costs are based on reports in the literature and personal communication with personnel at firms that supply commercial RO systems. The scale factors, de-

TABLE 25–5. Scale Factors for System Size Correction.

Application	Scale Factor
Seawater	0.95
Brackish water	0.87
LPRO/MS	0.85

representative value for these costs is presented here, but these costs can be significantly smaller or larger than the representative values, given special circumstances.

Construction Overhead

Construction overhead includes the following cost components: wages and fringe benefits, field supervision, temporary facilities, construction equipment, contractors' fees, and the engineering costs associated with the overall site construction and system installation. Construction overhead will depend on whether the system is custom built or is a "standard" unit. The representative value for this category is 12% of the total direct capital costs.

Contingency

A representative cost for the contingency fund for an RO plant is 10% of the total direct capital costs. This value can be significantly larger depending on location. For example, a plant being constructed in a politically unstable country may require a larger contingency fund than normal.

Operating Costs

Energy

In many RO systems, energy costs are the largest single component in the total cost of the water produced. Energy costs include costs for pumps for feed water wells, high-pressure pumps, pretreatment systems, and instrumentation. By far the dominant energy cost in most installations is for the high-pressure pumps. Significant work has been directed at developing power recovery technology, which now can recover up to 50% of the power used by high-pressure pumps in some seawater desalination installations. To produce water of the desired quality, some seawater desalination installations have two stages. In these plants, the permeate from the first stage is used as feed for the second stage. Multiple-stage systems are described in Chapter 24.

Table 25-6 shows the representative energy costs for the three application areas. Also included in this table are the ranges of energy costs reported in the literature for each application. Table 25-7 lists the bases used for the representative costs.

TABLE 25-6. Representative Costs and Ranges Reported in the Literature for Energy.

Application	Representative Cost		Range	
	($/1000 gal)	[$/m^3]	($/1000 gal)	[$/m^3]
Seawater	1.60	[0.42]	0.40–2.85	[0.11–0.75]
Brackish water	0.43	[0.11]	0.30–0.95	[0.08–0.25]
LPRO/MS	0.17	[0.05]	0.15–0.30	[0.04–0.08]

TABLE 25-7. Bases for Representative Energy Costs.

Item	Seawater	Brackish Water	LPRO/MS	Reference
Operating pressure (psig)	1000	450	200	—
Power cost ($/kWh)	0.075	0.075	0.075	Birkett 1988
Recovery (%)	40	60	75	Larson and Leitner 1979; Birkett 1985; Taylor et al. 1989a, 1989b
Pump efficiency (%)	85	85	85	Calder Ltd. 1989
Energy recovery (%)				
Without energy-recovery equipment	0	0	0	—
With energy-recovery equipment[a]	42	28	NA	

[a] Energy recovery = 0.7 × (100-recovery).

These representative energy costs can be corrected for a single-stage system using the equation

$$\begin{pmatrix} \text{energy} \\ \text{operating} \\ \text{cost (\$/1000 gal)} \end{pmatrix} = 0.724 \times p_r \times E_c \times F_{R1}^{-1} \times P_e^{-1} \times (100 - E_R) \qquad (25\text{-}3\text{a})$$

or

$$\begin{pmatrix} \text{energy operating} \\ \text{cost} \\ (\$/\text{m}^3/\text{day}) \end{pmatrix} = 0.191 \times p_r \times E_c \times F_{R1}^{-1} \times P_e^{-1} \times (100 - E_R), \qquad (25\text{-}3\text{b})$$

where

p_r = operating pressure (psig)
E_c = cost of electricity (\$/kWh)
F_{R1} = percent of first-stage feed recovered as permeate (%)
P_e = combined efficiency of pump and motor (%)
E_R = percent of feed pump energy recovered by energy recovery equipment (%).

For a two-stage system, substitute the equation

$$\begin{pmatrix} \text{energy operating} \\ \text{costs} \\ (\$/1000 \text{ gal}) \end{pmatrix} = 0.724 \times p_r \times E_c \times \frac{100}{F_{R1} \times F_{R2}} \times P_e^{-1} \times (100 - E_R), \qquad (25\text{-}3\text{c})$$

where F_{R2} is the percent of first-stage permeate recovered as second-stage permeate. The 0.724 multiplier has units of kWh/psig-1000 gal and is derived from

$$\frac{0.724 \text{ kWh}}{\text{psig-1000 gal}}$$

$$= \frac{1000 \text{ gal}}{1000 \text{ gal}} \times \frac{8.34 \text{ lb}}{\text{gal}} \times \frac{1 \text{ atm}}{14.7 \text{ psig}} \times \frac{33.9 \text{ ft H}_2\text{O}}{\text{atm}} \times \frac{\text{kW-min}}{44260 \text{ ft lb}} \times \frac{\text{h}}{60 \text{ min}} \times 100.$$

If site-specific information is available for any of the categories in Table 25–7, Eq. (25–3a), (25–3b), or (25–3c) can be used to calculate a corrected energy cost. For example, if the power cost is known to be \$0.09/kWh instead of \$0.075, but no other specific information is available, the reader can still calculate a more accurate energy cost using the actual cost of energy and the average values for the other variables affecting the energy costs. This equation can also be used to determine whether to invest in energy recovery equipment, using the information given on equipment capital costs and capital recovery cost calculations in the last sections.

Membrane Replacement

The replacement of membrane modules is another key operating cost. This cost can quickly dominate total production costs if an operator error or a sudden change in the feed stream composition occurs and membrane modules are damaged. The cost of the replacement modules is, of course, an important variable. Even more important, however, is membrane lifetime. Manufacturers typically guarantee between 3

Cost Estimates

TABLE 25–8. Representative Costs and Ranges Reported in the Literature for Membrane Replacement.

Application	Representative Cost ($/1000 gal)	[$/m³]	Range ($/1000 gal)	[$/m³]
Seawater	0.41	[0.11]	0.05–1.70	[0.01–0.45]
Brackish water	0.18	[0.05]	0.05–1.30	[0.01–0.34]
LPRO/MS	0.18	[0.05]	0.05–0.50	[0.01–0.13]

and 5 years, and there are examples in the literature of membrane lifetimes of more than 7 years. For a representative value, a conservative membrane lifetime estimate of 3 years has been used here.

The cost per module does not vary much, even for larger applications. Normally, a small cost reduction is given if more than 10 membrane modules are purchased, and this reduction increases slightly as the number of modules purchased increases. However, at about 100 modules, no further price reduction is usually offered, so for most large-scale applications, the cost per module remains constant.

Table 25–8 shows the representative costs and the range for seawater, brackish water, and LPRO/MS applications. These representative values are corrected using Eq. (25-4a) or (25-4b) or Eq. (25-5a) or (25-5b). Equations (25-4a) and (25-4b) are based on membrane cost:

$$\left(\begin{array}{c}\text{unit replacement cost for membranes (URC}_m\text{)}\\(\$/1000\text{ gal})\end{array}\right) = 2.74 \times M_c \times M_P^{-1} \times M_l^{-1} \quad (25\text{-}4a)$$

or, for a two-stage system,

$$\text{URC}_m = 2.74 \times \frac{100}{F_{R1}} \times M_c \times M_P^{-1} \times M_l^{-1}, \quad (25\text{-}4b)$$

where M_c is the membrane cost ($/ft²), M_p is the membrane production rate gal/ft²/day, and M_l is the membrane life (yr).

Equations (25–5a) and (25–5b) are based on module cost:

$$\left(\begin{array}{c}\text{unit replacement cost for modules (URC}_M\text{)}\\(\$/1000\text{ gal})\end{array}\right) = 2.74 \times M_C \times M_P^{-1} \times M_L^{-1} \quad (25\text{-}5a)$$

or, for a two-stage system,

$$\text{URC}_M = 2.74 \times \frac{100}{F_{R1}} \times M_C \times M_P^{-1} \times M_L^{-1}, \quad (25\text{-}5b)$$

where M_C is the module cost ($/module), M_P is the module production rate (gal/day), and M_L is the module life (yr). The 2.74 multiplier has units of yr/1000 day and is derived from

$$\frac{2.74\text{ yr}}{1000\text{ day}} = \frac{1000\text{ day}}{1000\text{ day}} \times \frac{1\text{ yr}}{365\text{ day}}.$$

Equations (25–4a) and (25–4b) are based on membrane productivity data (in gal/ft²/day) and costs per square foot of membrane. Equations (25–5a) and (25–5b) are based on productivity per module (in gal/day/module) and costs per module. This allows the use of laboratory data [Eqs. 25–4a and 25–4b] or manufacturer data [Eqs. 25–5a and 25–5b]. Table 25–9 shows

TABLE 25–9. Bases for Representative Costs for Membrane Replacement in Table 25–8.

Item	Seawater	Brackish Water	LPRO/MS
Water productivity			
Flux (gal/ft²/day)	15	25	25
[m³/m²/day]	[0.6]	[1.0]	[1.0]
By module (gal/day/module)	4000	7500	7000
[m³/day/module]	[15]	[28]	[26]
Cost			
$/ft²	7	5	5
$/m²	73	54	54
$/module	1800	1500	1400
Lifetime (yr)	3	3	3

the bases used for the representative values given in Table 25–8.

Labor and Overhead

Labor is another key contributor to operating costs. Table 25–10 shows the representative labor costs for the three applications and the observed ranges. Labor costs can vary considerably for two major reasons: (1) the direct cost of labor varies over a wide range in the United States and in other parts of the world and (2) the number of operators required to operate a given RO plant varies. These average costs can be corrected using

$$\begin{pmatrix} \text{unit cost of labor (UCL)} \\ (\$/1000 \text{ gal}) \end{pmatrix} = 0.08 \times Lh \times S \times Ws \times PL_C^{-1} \times (L_{OH} + 100), \quad (25\text{–}6)$$

where

Lh = labor cost per hour ($/h)
S = shifts per day (number/day)
Ws = workers per shift (number/shift)
PL_C = plant capacity [1000 gal/day (3.8 m³/day)]
L_{OH} = labor overhead (%).

The 0.08 multiplier has the units of hour and is derived from

0.08 h = 8 h/100.

Table 25–11 shows the bases used to obtain the representative costs in Table 25–10.

For smaller RO plants, an important question is whether operators can be "shared" with other unit operations on a site or in a plant. For many smaller RO systems, very little operator time is required, and therefore the labor costs can be very high per unit of water produced if the

TABLE 25–10. Representative Costs and Ranges Reported in the Literature for Labor.

	Representative Cost		Range	
Application	($/1000 gal)	[$/m³]	($/1000 gal)	[$/m³]
Seawater	0.30	[0.08]	0.15–0.75	[0.04–0.20]
Brackish water	0.30	[0.08]	0.15–0.75	[0.04–0.20]
LPRO/MS	0.20	[0.05]	0.10–0.45	[0.03–0.12]

TABLE 25-11. Bases for Representative Labor Costs in Table 25-10.

Item	Seawater	Brackish Water	LPRO/MS
Number of shifts/day	1	1	1.3
Number of workers/shift	2	2	1
Labor cost/h ($/h)	14.5	14.5	14.5
Labor overhead (%)	30	30	30

operator cannot be shared. For larger plants [e.g., about 500,000 gal/day (2000 m³/day) for seawater desalination applications (Buros 1979)], the cost of a full-time operator does not account for as large a percentage of total production costs. The amount and sophistication of the pretreatment or post-treatment systems is also a factor in the amount of labor required to operate a given system.

Spare Parts

Spare parts refers here to replacement parts needed to maintain the system such as pump parts, valves, control system components, and miscellaneous parts. This category does not include consumable chemicals, filters, or membrane modules. The representative value for spare parts is $0.09/1000 gal ($0.02/m³) for seawater desalination and brackish water treatment, and $0.02/1000 gal ($0.01/m³) for LPRO/MS applications.

Chemicals

For many RO applications, chemical costs can be predicted with relative precision. The chemicals needed for pretreatment account for the bulk of the chemicals used, but significant amounts of chemicals can also be needed for post-treatment (e.g., for the chlorination of product water). The representative chemical costs and reported ranges are shown in Table 25-12. Table 25-13 lists commonly used pretreatment chemicals, their cost per pound (and per kilogram), and a typical cost per 1000 gal (and per m³) water produced (Leitner 1987; Birkett 1988; Saltech 1981).

Because chemical costs vary so widely among applications and because they can constitute a major portion of operation costs, chemical costs must be estimated accurately. In some cases, chemical costs are so high that water treatment by RO is prohibitive. (The ranges shown in Table 25-12 apply to systems actually built and operated.) Certain feed waters, for instance, could require the addition of millions of pounds of Na_2SO_3 or SO_2 per year. This situation would result not only in high chemical costs, but in high labor costs and increased capital costs for the additional pretreatment equipment needed.

Due to these considerations, a full analysis of the feed water, preferably over the course of an entire year, is needed. The results of this analysis should be supplied to prospective

TABLE 25-12. Representative Costs and Ranges Reported in the Literature for Chemicals.*

	Representative Cost		Range	
Application	($/1000 gal)	[$/m³]	($/1000 gal)	[$/m³]
Seawater	0.16	[0.04]	0.03–0.60	[0.01–0.16]
Brackish water	0.15	[0.04]	0.03–0.55	[0.01–0.14]
LPRO/MS	0.09	[0.02]	0.03–0.20	[0.01–0.05]

*From Birkett 1988; Hornberg and Morin 1986; Andrews and Bergman 1986; Al-Marafie and Darwish 1989; Ericsson and Hallmans 1985; Taylor et al. 1989a, 1989b; Wade, Heaton, and Boulter 1985; Birkett 1985; Smith and Swinton 1988; Buros 1979; Hydranautics Water Systems 1983, 1986.

TABLE 25-13. Costs of Common Pretreatment Chemicals Used for Seawater and Brackish Water Applications (Leitner 1987; Hydranautics Water Systems 1983, 1986).

Chemical	Cost			
	($/lb)	[$/kg]	($/1000 gal)	[$/m^3]
Antifoam	1.15	[2.54]	0.01	[0.003]
Sulfuric acid (100%)	0.26	[0.58]	0.04	[0.01]
Antiscalant	2.00	[4.38]	0.04	[0.01]
Sodium hexametaphosphate	0.35	[0.77]	0.02	[0.005]
Caustic (NaOH)	0.23	[0.51]	0.02	[0.005]
Sodium sulfite	0.07	[0.13]	0.02	[0.005]
Chlorine	0.15	[0.33]	0.01	[0.003]

TABLE 25-14. Representative Costs and Ranges Reported in the Literature for Filters.

	Representative Cost		Range	
Application	($/1000 gal)	[$/m^3]	($/1000 gal)	[$/m^3]
Seawater	0.05	[0.01]	0.04–0.55	[0.01–0.14]
Brackish water	0.05	[0.01]	0.02–0.20	[0.005–0.05]
LPRO/MS	0.02	[0.005]	0.01–0.10	[0.003–0.03]

membrane element suppliers so that their guarantees of membrane lifetime are realistic given system operating conditions.

Filters

The representative costs for filters and reported ranges are shown in Table 25–14. Filters are normally rated at 5 to 25 μm and have widely varying lifetimes, depending on feed water quality. In many cases, the filter lifetime can vary significantly with the seasons as well, especially when treating surface waters or water from shallow wells.

Other

This category is included to account for any unusual costs specific to an application (for example, the production of ultrapure water, which requires ion-exchange beds for post-treatment, or applications that require ozonation or decarbonization of the product water).

Capital Recovery Costs

The total capital recovery costs often are a key in determining the feasibility of a project. In fact, in cases where financing may be difficult to obtain, a go/no-go decision may be based more on total capital costs than on the total production costs, which are discussed below.

To determine the ultimate cost of the product water produced by an RO system, it is necessary to determine what amount of the total production costs is due to the capital costs (see Figure 25–1). The capital recovery costs are calculated based on interest rates and equipment lifetime. The capital recovery cost is then calculated using

$$\begin{pmatrix} \text{capital recovery} \\ \text{costs} \\ (\$/1000 \text{ gal}) \end{pmatrix} = \frac{(\text{total capital cost}) \times 1000 \times i \times [1 + (i/100)]^r}{365 \times (100 - Dt) \times [(1 + i/100)^r - 1]}, \quad (25\text{–}7)$$

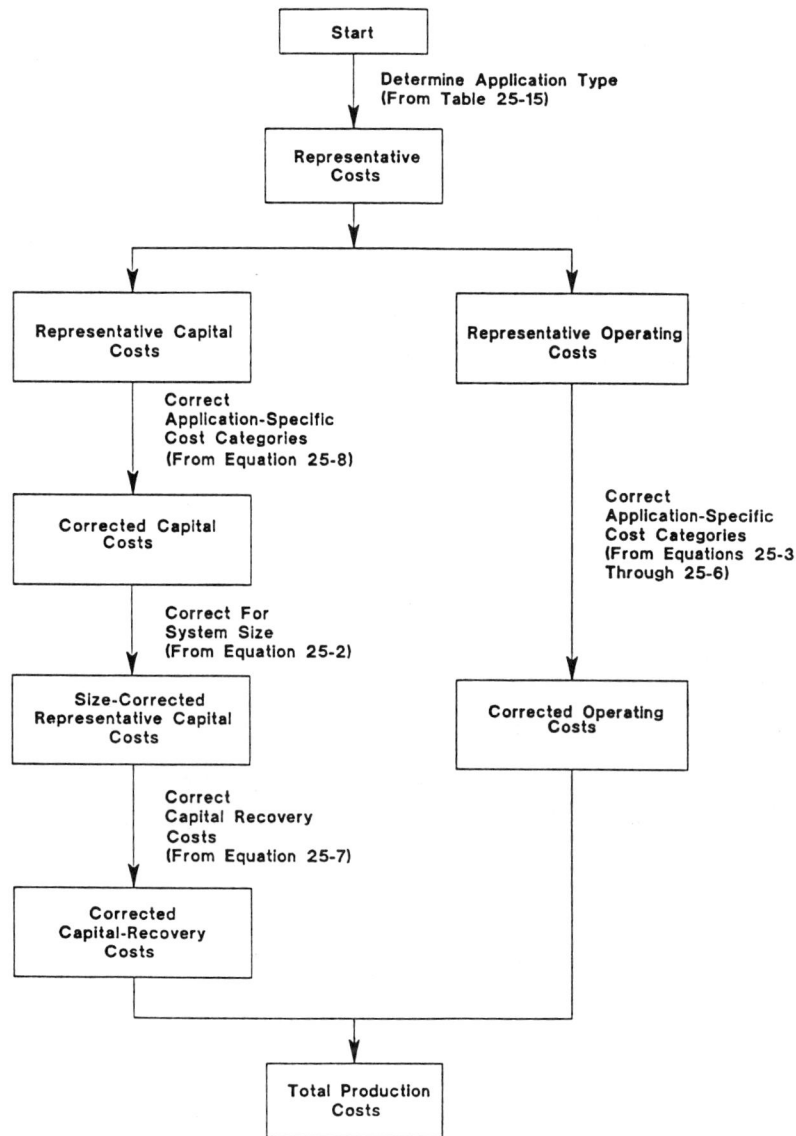

FIGURE 25–3. Flow chart for RO cost estimation.

where r is system lifetime (yr), i is the annual interest rate (%), and Dt represents downtime (%). A system lifetime (exclusive of membrane replacement) of 15 years, an interest rate of 12%, and a downtime percentage of 15% are used as representative values.

METHODOLOGY FOR ESTIMATING RO SYSTEM COSTS

Figure 25–3 shows in detail the method used here for cost estimation. As mentioned earlier, this costing method is based on "correcting" representative capital and operating costs for each application using the correction equations from the previous section. In this way, the reader can use any specific information available for a given application, even if it is incomplete, to obtain the most accurate cost estimate possible. For variables for which no specific information is known, the sensitivity studies can easily be performed to determine how a change in the variable affects the total production cost.

The representative cost data for each applica-

TABLE 25-15. Summary of Representative Costs for RO Treatment.

Cost Category	Application Type					
	Seawater		Brackish Water		LPRO/MS	
CAPITAL COSTS ($/gal/day [$/m^3/day])						
Direct						
Site development	0.10	[26.42]	0.10	[26.42]	0.10	[26.42]
Water	0.09	[23.78]	0.09	[23.78]	0.20	[52.84]
Utilities	0.16	[42.27]	0.11	[29.06]	0.06	[15.85]
Equipment	3.34	[882.43]	1.55	[409.51]	0.75	[198.15]
Land	—	—	—	—	—	—
Other	—	—	—	—	—	—
Total Direct Capital Costs	3.69	[974.90]	1.85	[488.77]	1.11	[293.26]
Indirect						
Construction overhead	0.44	[116.98]	0.22	[58.65]	0.13	[35.19]
Contingency	0.37	[97.49]	0.19	[48.88]	0.11	[29.33]
Other	—		—		—	
Total Indirect Capital Costs	0.81	[214.47]	0.41	[107.53]	0.24	[64.52]
Total Capital Costs	4.50	[1189.37]	2.26	[596.30]	1.35	[357.78]
OPERATING COSTS ($/1000 gal [$/m^3])						
Energy	1.60	[0.42]	0.43	[0.11]	0.17	[0.04]
Membrane replacement	0.41	[0.11]	0.18	[0.05]	0.18	[0.05]
Labor and overhead	0.30	[0.08]	0.30	[0.08]	0.20	[0.05]
Spare parts	0.09	[0.02]	0.09	[0.02]	0.02	[0.01]
Chemicals	0.16	[0.04]	0.15	[0.04]	0.09	[0.02]
Filters	0.05	[0.01]	0.05	[0.01]	0.02	[0.01]
Other (ion-exchange beds)	—		—		—	
Total Operating Costs	2.61	[0.68]	1.20	[0.31]	0.68	[0.18]
Capital Recovery Costs	2.13	[0.56]	1.07	[0.28]	0.64	[0.17]
Total Production Costs	4.74	[1.24]	2.27	[0.59]	1.32	[0.35]

tion area are summarized in Table 25-15. The reader begins the cost estimation by choosing the application type from Table 25-15 that best fits the application of interest. Capital costs are then corrected, as indicated in Figure 25-3. The categories of capital costs for which actual costs are unknown are corrected for the system size (desired water production rate) using Eq. (25-2) and the appropriate scale factor (n). Then, the capital cost is corrected using any specific available information on the given application. This is done by correcting the representative costs in Tables 25-1 through 25-3 with actual cost information, using

$$\text{corrected capital cost (\$/gal/day)} = LD + SD + WS + U + EQ, \quad (25\text{-}8)$$

where

LD = cost of land ($/gal/day)
SD = cost of site development ($/gal/day)
WS = cost of water supply ($/gal/day)
U = cost of utilities ($/gal/day)
EQ = cost of equipment ($/gal/day).

(To convert values from gallons per day to cubic meters per day, use costs for each based on dollars per cubic meter per day.) Indirect costs are then calculated using specific available information or the representative percentages given earlier (construction overhead = 12% of total direct capital costs, and contingency = 10% of total direct capital costs). Finally, the corrected capital costs are calculated using Eq. (25-2).

To correct operating costs, the representative costs in Table 25-15 are corrected using any specific available information. Equations (25-3) through (25-6) can be used. The corrected operating costs are then totaled and added to the corrected capital recovery costs to obtain the total production costs.

The appendix at the end of this chapter contains worksheets for each application plus an example to illustrate how the costing methodology works.

SENSITIVITY STUDIES

In this section, the results of sensitivity studies of certain cost categories are reported for the three application areas studied (i.e., seawater, brackish water, and LPRO/MS). Cost parameters were chosen for sensitivity studies because either their contribution to the total production costs is important or their range in the literature is particularly wide. Of course, the equations presented earlier can be used to generate other sensitivity studies for a specific application. The literature provides similar sensitivity studies that show the same general trends as discussed in this section (Lykins, Clark, and Fronk 1988).

The figures in this section show the relative effects of varying operational parameters on total production costs. The ranges chosen for parameters are consistent with the ranges listed in this text or are within the normal operating ranges for the different applications.

Sensitivity Studies for Capital Costs

Figure 25-4 shows the variation in total production costs with a variation in total capital costs.

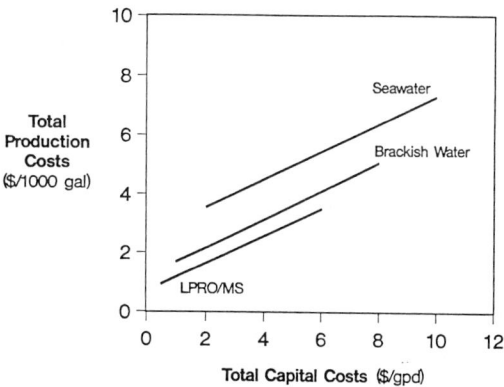

FIGURE 25-4. Total production costs as a function of total capital costs ($/gpd = $/gal/day).

This graph will allow the reader to determine the effect of any significant variation in any capital cost category from the representative values shown in Table 25-15. As the graph shows, if the total capital cost varies by an equal dollar amount for each application, the percentage change will be larger for the brackish water and LPRO/MS applications than for the seawater desalination application.

Sensitivity Studies for Operating Costs

Energy costs are a key component of the operating costs in most RO applications, with their relative importance being inversely proportional to operating pressure (see Table 25-7). Total energy costs are a strong function of electricity costs, operating pressure, and the percentage of the feed water recovered as product. The effect of the cost of electricity is shown in Figure 25-5 for the three applications areas. As expected, the total production costs are the most sensitive to the cost of electricity for seawater desalination, less sensitive for brackish water, and least sensitive for LPRO/MS installations. This sensitivity to the cost of electricity has resulted in innovations such as combined cogeneration/seawater desalination plants, and energy recovery technology (Al-Marafie and Darwish 1989; Glueckstern and Arad 1984).

The effect of operating pressure on total production costs is shown in Figure 25-6. As can be seen, changing the operating pressure has the

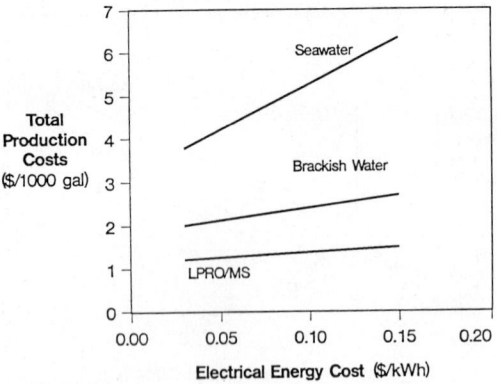

FIGURE 25–5. Total production costs as a function of electrical energy costs.

greatest effect on total production costs of a seawater desalination system, less effect on a brackish water system, and the least effect on an LPRO/MS application. This is because the energy costs for seawater desalination systems constitute a higher percentage of the total production costs than for brackish water or LPRO/MS applications; since energy costs are proportional to operating pressure, changing the operating pressure for a seawater desalination system will affect energy costs more and, therefore, have a greater effect on total production costs. Also the percent recovery is lower for seawater desalination, so a much greater volume of water is being pumped at a given operating pressure.

The effect of energy recovery technology on total production costs is shown in Table 25–16.

Considerable innovation has occurred in the area of energy recovery technology, and economical units are now available for plants with capacities as small as 10,000 gal/day.

As expected, energy recovery technology has the greatest impact on total production costs for seawater desalination applications. From the bases in Tables 25–7 and 25–16, total production costs are seen to be reduced by 20% if energy recovery technology is used. The use of energy recovery technology has less of an impact on total production costs for brackish water and LPRO/MS applications—often, it is not worth the added capital investment.

Another key variable affecting energy costs and total production costs is the percentage of water recovered from the feed stream by a given RO system. The sensitivity of the total production costs to recovery is quite complex and depends on a number of factors. Figure 25–7 shows the basic dependence of total production costs on percent recovery. At very low recoveries, total production costs will be high due to energy wasted on unrecovered water and the need for a larger system (especially the feed stream side of the system) to produce a given amount of water.

At very high recoveries, other factors—primarily high average feed stream salinities—will drive up the costs. Factors driving up the total production costs at high recoveries include (1) low fluxes due to high osmotic pressures—resulting in more membrane area being required, (2) the need for "stages" to produce

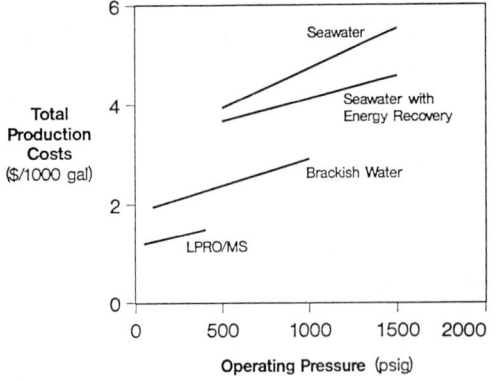

FIGURE 25–6. Total production costs as a function of operating pressure.

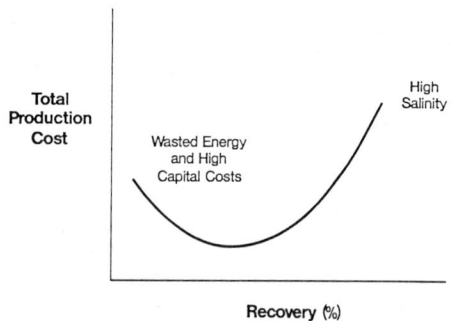

FIGURE 25–7. Total production costs as a function of percent recovery.

TABLE 25–16. Effect of Energy Recovery Technology on Capital and Production Costs (Glueckstern and Arad 1984).

	Capital Cost		Production Cost (including capital cost)	
	without ($/gal/day) [$/m³/day]	with ($/gal/day) [$/m³/day]	without ($/1000 gal) [$/m³]	with ($/1000 gal) [$/m³]
Seawater (35% recovery)	4.50 [1188.90]	4.64 [1225.0]	4.74 [1.25]	4.13 [1.09]
Brackish water (60% recovery)	2.26 [597.10]	2.33 [615.60]	2.27 [0.60]	2.15 [0.57]
LPRO/MS (75% recovery)	1.35 [356.70]	—	1.32 [0.35]	

water that meets water-quality specifications, (3) increased pretreatment costs to avoid the formation of scale at high concentrations of dissolved solids, and (4) increased energy costs due to the high required operating pressures associated with high osmotic pressures.

With the costs high at low and high recoveries, there is a recovery point at which the costs will be at a minimum value, as shown in Figure 25–7. Due to the myriad of factors affecting these costs, the percent recovery corresponding to this minimum will depend strongly on the specific application. (Table 25–7 lists the typical recoveries used for the three application areas.) In Figure 25–8, the effect of percent recovery on total production costs is shown for a single-stage system. In generating the data used in this figure, it was assumed that the plant is operating at a recovery that corresponds to the minimum cost (as represented in Figure 25–7). Therefore, Figure 25–8 illustrates the effect that percent recovery (and, thus, energy costs) has on total production costs for a plant operating under optimal conditions.

Figures 25–9 and 25–10 show the effect membrane performance has on total production costs. The cost of the membrane modules (assuming here 8-in.-diameter spiral-wound composite-membrane modules), shown in Figure 25–7, affects the total production costs through the capital costs (the initial set of modules installed) and through the operating costs

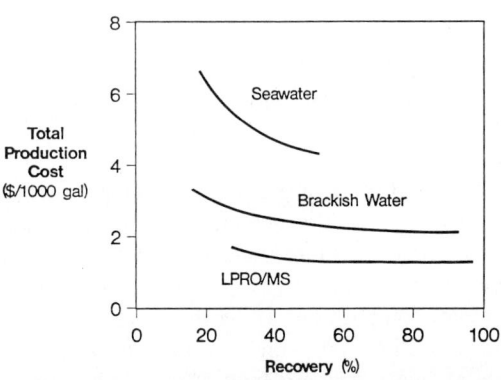

FIGURE 25–8. Total production costs as a function of percent recovery for single-stage systems.

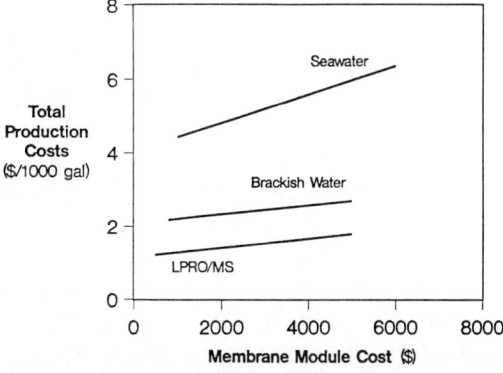

FIGURE 25–9. Total production costs as a function of membrane module cost assuming module productivity given in Table 25–9.

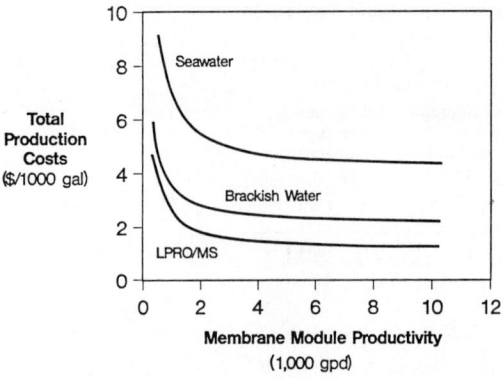

FIGURE 25–10. Total production costs as a function of membrane module productivity assuming module costs given in Table 25–9 (gpd = gal/day).

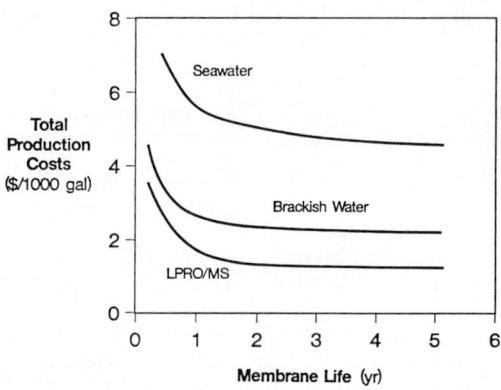

FIGURE 25–11. Total production costs as a function of membrane life assuming operation at conditions given in Table 25–9.

(membrane replacement costs). The total production costs are not a relatively strong function of membrane costs at average lifetimes of 3 years. Membrane module costs have dropped considerably over the last few years, decreasing their impact on total production costs. Note also from Figure 25–9 that a further drop in membrane module costs will not have a large effect on the total production costs.

As can be seen in Figure 25–10, beyond a certain point, increases in productivity do not result in any significant savings in total production costs. However, decreases in module productivity will definitely increase total production costs. Module productivity will decrease if membrane fouling is a problem, but a good pretreatment system can be used to prevent a decrease in module productivity.

A more interesting cost effect—that of membrane lifetime—is shown in Figure 25–11. As this figure shows, beyond a certain point, increases in membrane lifetime will not affect the total production costs significantly. However, a decrease in membrane lifetime does have a significant effect. Upsets in the system, or operator error, obviously can have a large effect on the total production costs if membrane modules need to be replaced.

Labor costs can affect total production costs significantly. A key question is whether a part-time, rather than a full-time, operator must be used. As discussed previously, many small-scale plants do not require a full-time operator, but must pay for one because it is not possible to hire a part-time operator or to split labor costs with an adjacent facility. In such cases, the cost of a full-time operator can drive up total production costs significantly. The impact is illustrated in Figure 25–12(a), which shows total production costs as a function of plant capacity, assuming a full-time operator is required. If a part-time operator can be hired or a full-time worker's time can be shared with an adjacent facility, the overall impact of labor costs on the total production costs is lowered substantially, as shown in Figure 25–12(b). As in other cost categories, because the total production costs are lower for brackish water and LPRO/MS applications, a given increase in labor costs will result in a larger percentage change for those applications than for seawater desalination applications.

Labor costs may be further reduced by improved system design and process control, which may allow more complete automation of plants. For larger plants, this may make it possible to use a part-time rather than a full-time operator or, for smaller plants, reduce the time required of a part-time operator. For instance, improved pretreatment techniques and breakthroughs in membrane technology have decreased the sensitivity of RO systems to feed stream composition. These improvements may make further reduction of labor costs possible, by reducing operator time.

Finally, in some applications, the chemical

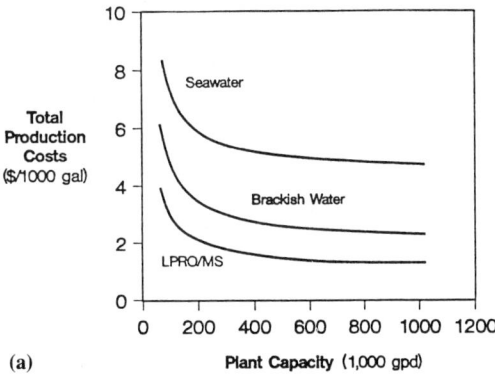

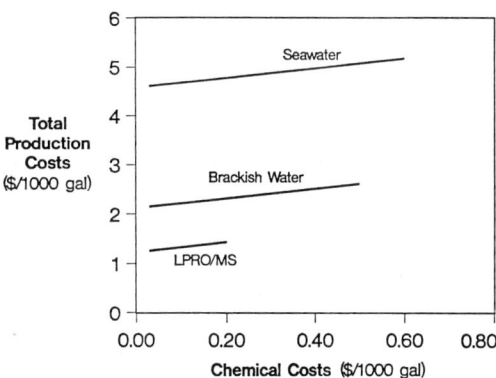

FIGURE 25–13. Total production costs as a function of chemical costs.

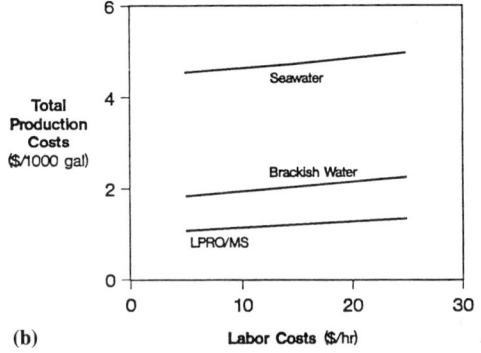

FIGURE 25–12. Total production costs as a function of plant capacity assuming a full-time operator (a), and total production costs as a function of labor costs assuming a part-time operator (b).

costs can be several times the representative value used in Table 25–12. The sensitivity of total production costs to the costs of pretreatment chemicals is shown in Figure 25–13. These costs can be significant, especially in remote areas, where transportation and handling costs can be high. The effect of changes in the cost of filters or spare parts or other operating costs will follow a pattern such as that shown in Figure 25–13 and have a similar impact on total production costs.

SMALL-SCALE AND SPECIALTY SYSTEMS

The methodology described thus far has focused primarily on estimating RO system costs for systems with production capacities of 50,000 gal/day (200 m³/day) or more. In this section, information is presented to aid in estimating costs for smaller RO plants. As the system size decreases, the percentage of variation in the capital, operating, and total production costs can be considerable. Therefore, it is important (and usually easier than for large systems) to secure quoted price information from vendors.

Small Systems

Table 25–17 lists examples (and size ranges) of applications in which smaller RO systems are common. Such systems range from small municipal systems producing potable water from seawater to so-called "point-of-use" (POU) systems operating on line pressure. In smaller systems, many of the cost categories (see Table 25–15) are quite different from those for larger systems. Many of these systems are skid-mounted and are often part of larger water treatment and delivery systems. As such, costs in many of the cost categories discussed above for large systems (see Table 25–15) can be quite different.

Land costs can be negligible, because skid-mounted systems often do not take up significant space. For example, the size of one 100,000 gal/day (380 m³/day) plant is 13 × 25 × 8 ft tall. Site development costs often are limited to providing power and water to the system and plumbing for the return of brine to a convenient sink. Water supply costs can vary

TABLE 25-17. Small-Scale RO Systems.

Application	Example Manufacturer	Production Capacity	
		(gal/day)	(m³/day)
Ultrapure water	Osmonics	150–1000	0.06–3.8
	Culligan	50–22,000	0.2–83.3
Home use	Calpure	3–10	0.01–0.04
Industrial	Culligan	26,000–100,000	98.4–380.0

widely for small systems depending on the feed water source.

Figure 25–14 is a plot of all types of small-scale plants with plant capacities between 5 and 100,000 gal/day (0.02 and 380 m³/day). Note that these costs correlate well without breaking them out by application area (seawater, brackish water, or LPRO/MS). These equipment costs vary significantly, but correlate (with a correlation coefficient of 0.97) between 5 and 100,000 gal/day (0.02 and 380 m³/day) using Eq. (25–1) with a scaling factor of 0.52. Note that from this information, at a capacity of 1500 gal/day (5.7 m³/day), the equipment costs are already about \$4.50/gal/day (\$1190/m³/day), or only 50% higher than average equipment costs for a 1,000,000 gal/day (3800 m³/day) plant.

Included in the equipment costs is the cost for instrumentation. The instrumentation required to operate a small RO system is basically the same as that for a large RO system, and so equipment costs are fixed costs. For the large RO systems, these fixed costs become a very small part of the equipment cost, but the same fixed costs become very significant on a small RO system. This is one reason the total production costs for small RO systems are higher.

Operating costs can also vary significantly from those of larger-scale RO plants. Energy costs have often been high enough in smaller RO plants to cause potential users to choose another technology. There are two primary reasons for this: (1) many small-scale systems are installed in remote locations where energy costs are higher than those found in industrialized areas and (2) until recently, energy recovery technology has not been available for smaller high-pressure pumps. As for larger-scale plants, the specific energy costs of a given small-scale application must be determined to estimate operation costs accurately.

Many small-scale plants still make use of the largest available RO modules [e.g., 8-in.- or 11-in.-diameter (203- or 279-mm) spiral-wound modules]. Therefore, membrane replacement costs are often not significantly higher than those for larger plants. However, if the plant is so small that the use of smaller modules is required, membrane replacement costs can rise. For example, a 4-in.-diameter spiral-wound RO element for brackish water treatment costs about \$0.27/gal/day (\$71/m³/day) (FilmTec 1988a, 1988b, 1989), whereas an 8-in.-diameter module averages about \$0.18/gal/day (\$49/m³/day). Another factor can be pretreatment. In some small-scale RO plants, lack of full-time operator attention or other factors can reduce pretreatment effectiveness and, thus, reduce membrane lifetime.

Labor costs are perhaps the most variable cost as system size is scaled down. In many instances, the small-scale RO plant is a small

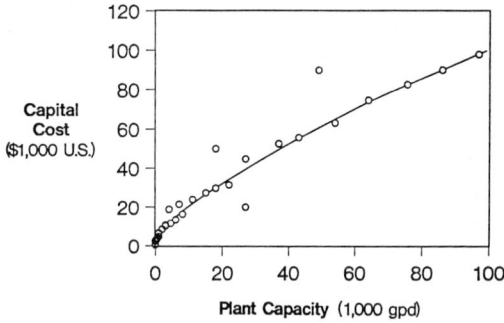

FIGURE 25-14. Capital costs as a function of plant capacity for small ready-made RO systems.

part of a maintenance person's responsibilities; therefore, labor costs are minimal. In one study of seawater desalination plants with capacities of between 5000 and 75,000 gal/day (19 and 284 m^3/day), labor costs were not even mentioned (Matz and Fisher 1981). Where the feed conditions are particularly variable and/or where the plant's operation is critical, an operator may be assigned part or full time to the plant, and the labor costs per unit water produced can be quite high.

The best approach for estimating the costs of small-scale systems is to obtain a quote from a vendor. Failing this, then, the costing method presented in the previous section, corrected using the specific information available and the information above, will give reasonable cost ranges.

The POU systems are a special case. These systems can operate on available line pressure and can be installed under the sink or on skids with pretreatment and post-treatment subsystems and high-pressure pumps. Costs for these systems can vary from $300 to $1000 for LPRO/MS systems and from $80 to $800 for RO systems for the under-the-sink variety (Anonymous 1989). Small skid-mounted units, such as those used on sailboats, can cost up to

TABLE 25–18. Summary of Representative Costs for Tubular RO Treatment (Pain 1989).

Cost Category	Representative Value	
CAPITAL COSTS	($/gal/day	[$/m^3/day])
Direct		
Site development	0.10	[26.42]
Water	0.20	[52.84]
Utilities	0.06	[15.85]
Equipment	7.10	[1875.82]
Land	—	
Other	—	
Total Direct Capital Costs	7.46	[1970.93]
Indirect		
Construction overhead	0.17	[44.91]
Contingency	0.15	[39.63]
Other	—	
Total Indirect Capital Costs	0.32	[84.54]
Total Capital Costs	7.78	[2055.47]
OPERATING COSTS	($/1000 gal	[$/m^3])
Energy	3.10	[0.82]
Membrane replacement	1.79	[0.47]
Labor and overhead	1.54	[0.41]
Spare parts	0.20	[0.05]
Chemicals	0.16	[0.04]
Filters	0.05	[0.01]
Other (ion-exchange beds)	—	
Total Operating Costs	6.84	[1.81]
Capital Recovery Costs	3.68	[0.97]
Total Production Costs	10.53	[2.78]

$10,000, depending on their size and sophistication (Spencer 1990).

Total production costs for under-the-sink units typically range from $0.06 to $0.25/gal ($60 to $250 per 1000 gallons or $16 to $66 per cubic meter). Membrane replacement, pretreatment cartridges and filters, and feed water costs all contribute to these costs. These systems typically recover between 10 and 20% of the feed water. This is a rapidly growing market and, because of competition and technological breakthroughs, the costs are dropping rapidly.

Specialty Systems

Many RO systems are designed and built for specialty applications. For example, RO specialty systems are increasingly being installed as subsystems in larger processing systems. The costs of such installations depend, of course, on the specific situation, but the information and methods presented here can often be used to gain at least a rough cost estimate.

Other specialty systems include those in which specialty RO modules, such as tubular membranes or the DDS plate-and-frame modules, are used. Table 25–18 gives an example of the cost of a tubular RO system for an application involving the processing of about 100,000 gal/day (380 m^3/day) of an aqueous emulsion. This system was part of a larger system and was to be operated in a "feed-and-bleed" recycle mode. The feed stream had a very high loading of a colloidal organic material and was therefore likely to cause fouling. Although fabricating and operating this system was much more expensive than a more conventional RO system, it was far less expensive than the competing process, which was labor-intensive.

CONCLUSIONS

The costs of RO are difficult to estimate accurately without specific information. As Birkett (1988) said, "The number of variables is so great, and the changes in technology sufficiently rapid, that it is impossible to talk in absolutes." Guided by this advice, this chapter provides a costing methodology that gives the reader a representative set of costs as a basis. The methodology is then provided for the correction of these representative costs to adjust for the size and specific characteristics of the system of interest for which information is available.

APPENDIX: WORKSHEETS

Cost Estimates for RO Systems
*Worksheet No. 1**

Application: *Seawater*

Variable	Equation	Representative Value	Value Used
CAPITAL COSTS			
Base system size (1000 gal/day)	25–2	1000	_____
Scale factor, n	25–2	0.95	_____
Construction overhead (%)	—	12	_____
Contingency (%)	—	10	_____
ENERGY OPERATING COSTS			
Operating pressure, p_r (psig)	25–3	1000	_____
Cost of electricity, E_c ($/kWh)	25–3	0.075	_____
First-stage recovery, F_{R1} (%)	25–3	40	_____
Second-stage recovery, F_{R2} (%)	25–3	—	_____
Pump/motor efficiency, P_e (%)	25–3	85	_____
Energy recovery, E_R (%)	25–3	0	_____
MEMBRANE REPLACEMENT			
Membrane cost, M_c ($/ft^2)	25–4	7	_____
Membrane production rate, M_p (gal/ft^2/day)	25–4	15	_____
Membrane life, M_l (yr)	25–4	3	_____
Module cost, M_C ($/module)	25–5	1800	_____
Module production rate, M_P (gal/day)	25–5	4000	_____
Module life, M_L (yr)	25–5	3	_____
LABOR AND OVERHEAD			
Labor cost, Lh ($/h)	25–6	14.5	_____
Shifts per day, S (number/day)	25–6	1	_____
Workers per shift, Ws (number/shift)	25–6	2	_____
Plant capacity, PL_C (1000 gal/day)	25–6	(see above)	_____
Labor overhead, L_{OH} (%)	25–6	30	_____
CAPITAL RECOVERY			
System lifetime, r (yr)	25–7	15	_____
Annual interest rate, i (%)	25–7	12	_____
Downtime, Dt (%)	25–7	15	_____

*This worksheet is designed to identify the variables used to correct the costs using equations given in the text.

Cost Estimates for RO Systems
*Worksheet No. 2**

Application: *Seawater*
Plant Size (1000 gal/day) _____

Capital Costs ($/gal/day)	Column 1 Uncorrected Representative Costs	Column 2 Corrected Representative Costs	Column 3 Value Used
DIRECT			
Site development, SD	0.10	_____	_____
Water, WS	0.09	_____	_____
Utilities, U	0.16	_____	_____
Equipment, EQ	3.34	_____	_____
Land	—	—	_____
Other	—	—	_____
Total Direct Capital Costs			_____
INDIRECT			
Construction overhead			_____
Contingency			_____
Other			_____
Total Indirect Capital Costs			_____
Total Capital Costs			_____

PROCEDURE:
1. Using the scale factor from Worksheet No. 1, convert all representative direct capital costs (column 1) to the plant size specified using Eq. (25–2) and enter in column 2.
2. If specific cost information is known, enter in column 3; otherwise, enter values from column 2 into column 3.
3. Add values in column 3 to obtain total direct capital costs.
4. Calculate construction and overhead costs by multiplying total direct costs by the percentage given in Worksheet No. 1. Enter value in column 3.
5. Repeat step 4 for contingency. Enter value in column 3.
6. Enter other indirect capital costs in column 3.
7. Add values to obtain total indirect capital costs.
8. Add total direct capital costs and total indirect capital costs to obtain total capital costs.

*This worksheet is used to estimate the capital costs of an RO system.

Cost Estimates for RO Systems
Worksheet No. 3*

Application: *Seawater*
Plant Size (1000 gal/day): ──────

Operating Costs ($/1000 gal)	Column 1 Uncorrected[a] Representative Cost	Column 2 Value Used
Energy	1.60	──────
Membrane replacement	0.41	──────
Labor and overhead	0.30	──────
Spare parts	0.09	──────
Chemicals	0.16	──────
Filters	0.05	──────
Other	—	──────
Total Operating Costs		──────
Capital Recovery Costs		──────
Total Production Costs		──────

PROCEDURE:
1. Correct the energy costs using the information from Worksheet No. 1 and Eq. (25–3). Enter results in column 2.
2. Correct the membrane replacement costs using the information from Worksheet No. 1 and either Eq. (25–4) or (25–5). Enter result in column 2.
3. Correct the labor and overhead costs using the information from Worksheet No. 1 and Eq. (25–6). Enter result in column 2.
4. Enter either representative costs or specific known costs for spare parts, chemicals, filters, or other items in column 2.
5. Add column 2 to obtain total operating costs.
6. Calculate capital recovery costs using the information in Worksheet No. 1, the total capital costs from Worksheet No. 2, and Eq. (25–7). Enter result in column 2.
7. Add total operating costs and capital recovery costs to obtain total production costs.

*This worksheet is used to estimate the total production costs for an RO system.
[a]If no specific information is known, enter representative value in column 2.

Example: Diablo Canyon Seawater RO Plant

This is an example of how to estimate costs for a RO plant located at the Diablo Canyon power plant using the methodology explained in the text. Information about the Diablo Canyon power plant, taken from published literature (Hydranautics Water Systems 1986), was used to complete Worksheet Nos. 1, 2, and 3 to estimate capital and operating costs.

Worksheet No. 1 was filled in using data found in the literature and representative values were used for the variables for which specific data were not available. Worksheet No. 2 was filled in following the procedure outlined on the worksheet. Worksheet No. 3 was filled in using the plant design data and the chemical costs found in the literature. The total operating cost estimated using Worksheet No. 3 was $3.49/1000 gal/day, compared with $3.34/1000 gal/day, the actual plant operating cost. The capital recovery costs estimated from Worksheet No. 3 were $2.19/1000 gal/day, compared with actual costs of $2.00/1000 gal/day.

These comparisons show that, using the available plant design and plant capacity data along with the costing methodology, the capital and operating cost estimates are within 10% of the actual reported costs for the Diablo Canyon power plant.

**Example: Diablo Canyon
Cost Estimates for RO Systems
Worksheet No. 1***

Application: *Seawater*

Variable	Equation	Representative Value	Value Used
CAPITAL COSTS			
Base system size (1000 gal/day)	25–2	1000	576
Scale factor, n	25–2	0.95	0.95
Construction overhead (%)	—	12	12
Contingency (%)	—	10	10
ENERGY OPERATING COSTS			
Operating pressure, p_r (psig)	25–3	1000	1,000
Cost of electricity, E_c ($/kWh)	25–3	0.075	0.075
First-stage recovery, F_{R1} (%)	25–3	40	45
Second-stage recovery, F_{R2} (%)	25–3	—	85
Pump/motor efficiency, P_e (%)	25–3	85	85
Energy recovery, E_R (%)	25–3	0	0
MEMBRANE REPLACEMENT			
Membrane cost, M_c ($/ft²)	25–4	7	
Membrane production rate, M_p (gal/ft²/day)	25–4	15	
Membrane life, M_l (yr)	25–4	3	
Module cost, M_C ($/module)	25–5	1800	1,800
Module production rate, M_P (gal/day)	25–5	4000	4,000
Module life, M_L (yr)	25–5	3	3
LABOR AND OVERHEAD			
Labor cost, Lh ($/h)	25–6	14.5	14.5
Shifts per day, S (number/day)	25–6	1	1
Workers per shift, Ws (number/shift)	25–6	2	2
Plant capacity, PL_C (1000 gal/day)	25–6	(see above)	576
Labor overhead, L_{OH} (%)	25–6	30	30
CAPITAL RECOVERY			
System lifetime, r (yr)	25–7	15	15
Annual interest rate, i (%)	25–7	12	12
Downtime, Dt (%)	25–7	15	15

*This worksheet is designed to identify the variables used to correct the costs using equations given in the text.

Example: Diablo Canyon
Cost Estimates for RO Systems
*Worksheet No. 2**

Application: *Seawater* Plant Size (1000 gal/day) __576__ Capital Costs ($/gal/day)	Column 1 Uncorrected Representative Costs	Column 2 Corrected Representative Costs	Column 3 Value Used
DIRECT			
Site development, SD	0.10	0.103	0.103
Water, WS	0.09	0.092	0.092
Utilities, U	0.16	0.164	0.164
Equipment, EQ	3.34	3.433	3.433
Land	—	—	—
Other	—	—	—
Total Direct Capital Costs			3.79
INDIRECT			
Construction overhead			0.46
Contingency			0.38
Other			—
Total Indirect Capital Costs			0.84
Total Capital Costs			4.63

PROCEDURE:
1. Using the scale factor from Worksheet No. 1, convert all representative direct capital costs (column 1) to the plant size specified using Eq. (25–2) and enter in column 2.
2. If specific cost information is known, enter in column 3; otherwise, enter values from column 2 into column 3.
3. Add values in column 3 to obtain total direct capital costs.
4. Calculate construction and overhead costs by multiplying total direct costs by the percentage given in Worksheet No. 1. Enter value in column 3.
5. Repeat step 4 for contingency. Enter value in column 3.
6. Enter other indirect capital costs in column 3.
7. Add values to obtain total indirect capital costs.
8. Add total direct capital costs and total indirect capital costs to obtain total capital costs.

*This worksheet is used to estimate the capital costs of an RO system.

Example: Diablo Canyon
Cost Estimates for RO Systems
*Worksheet No. 3**

Application: *Seawater* Plant Size (1000 gal/day): __576__ Operating Costs ($/1000 gal)	Column 1 Uncorrected[a] Representative Cost	Column 2 Value Used
Energy	1.60	1.67
Membrane replacement	0.41	0.91
Labor and overhead	0.30	0.52
Spare parts	0.09	0.09
Chemicals	0.16	0.25
Filters	0.05	0.05
Other	—	–
Total Operating Costs		3.49
Capital Recovery Costs		2.19
Total Production Costs		5.68

PROCEDURE:
1. Correct the energy costs using the information from Worksheet No. 1 and Eq. (25–3). Enter results in column 2.
2. Correct the membrane replacement costs using the information from Worksheet No. 1 and either Eq. (25–4) or (25–5). Enter result in column 2.
3. Correct the labor and overhead costs using the information from Worksheet No. 1 and Eq. (25–6). Enter result in column 2.
4. Enter either representative costs or specific known costs for spare parts, chemicals, filters, or other items in column 2.
5. Add column 2 to obtain total operating costs.
6. Calculate capital recovery costs using the information in Worksheet No. 1, the total capital costs from Worksheet No. 2, and Eq. (25–7). Enter result in column 2.
7. Add total operating costs and capital recovery costs to obtain total production costs.

*This worksheet is used to estimate the total production costs for an RO system.
[a] If no specific information is known, enter representative value in column 2.

Cost Estimates for RO Systems
*Worksheet No. 1**

Application: *Brackish Water*

Variable	Equation	Representative Value	Value Used
CAPITAL COSTS			
Base system size (1000 gal/day)	25–2	1000	_____
Scale factor, n	25–2	0.87	_____
Construction overhead (%)	—	12	_____
Contingency (%)	—	10	_____
ENERGY OPERATING COSTS			
Operating pressure, p_r (psig)	25–3	450	_____
Cost of electricity, E_c ($/kWh)	25–3	0.075	_____
First-stage recovery, F_{R1} (%)	25–3	60	_____
Second-stage recovery, F_{R2} (%)	25–3	—	_____
Pump/motor efficiency, P_e (%)	25–3	85	_____
Energy recovery, E_R (%)	25–3	0	_____
MEMBRANE REPLACEMENT			
Membrane cost, M_c ($/ft^2)	25–4	5	_____
Membrane production rate, M_p (gal/ft^2/day)	25–4	25	_____
Membrane life, M_l (yr)	25–4	3	_____
Module cost, M_C ($/module)	25–5	1500	_____
Module production rate, M_P (gal/day)	25–5	7500	_____
Module life, M_L (yr)	25–5	3	_____
LABOR AND OVERHEAD			
Labor cost, Lh ($/h)	25–6	14.5	_____
Shifts per day, S (number/day)	25–6	1	_____
Workers per shift, Ws (number/shift)	25–6	2	_____
Plant capacity, PL_C (1000 gal/day)	25–6	(see above)	_____
Labor overhead, L_{OH} (%)	25–6	30	_____
CAPITAL RECOVERY			
System lifetime, r (yr)	25–7	15	_____
Annual interest rate, i (%)	25–7	12	_____
Downtime, Dt (%)	25–7	15	_____

*This worksheet is designed to identify the variables used to correct the costs using equations given in the text.

Cost Estimates for RO Systems
*Worksheet No. 2**

Application: *Brackish Water*
Plant Size (1000 gal/day): _____

Capital Costs ($/gal/day)	Column 1 Uncorrected Representative Costs	Column 2 Corrected Representative Costs	Column 3 Value Used
DIRECT			
Site development, SD	0.10	_____	_____
Water, WS	0.09	_____	_____
Utilities, U	0.11	_____	_____
Equipment, EQ	1.55	_____	_____
Land	—	—	_____
Other	—	—	_____
Total Direct Capital Costs			_____
INDIRECT			
Construction overhead			_____
Contingency			_____
Other			_____
Total Indirect Capital Costs			_____
Total Capital Costs			_____

PROCEDURE:
1. Using the scale factor from Worksheet No. 1, convert all representative direct capital costs (column 1) to the plant size specified using Eq. (25–2) and enter in column 2.
2. If specific cost information is known, enter in column 3; otherwise, enter values from column 2 into column 3.
3. Add values in column 3 to obtain total direct capital costs.
4. Calculate construction and overhead costs by multiplying total direct costs by the percentage given in Worksheet No. 1. Enter value in column 3.
5. Repeat step 4 for contingency. Enter value in column 3.
6. Enter other indirect capital costs in column 3.
7. Add values to obtain total indirect capital costs.
8. Add total direct capital costs and total indirect capital costs to obtain total capital costs.

*This worksheet is used to estimate the capital costs of an RO system.

Cost Estimates for RO Systems
Worksheet No. 3*

Application: *Brackish Water*
Plant Size (1000 gal/day): _____

Operating Costs ($/1000 gal)	Column 1 Uncorrected[a] Representative Cost	Column 2 Value Used
Energy	0.43	_____
Membrane replacement	0.18	_____
Labor and overhead	0.30	_____
Spare parts	0.09	_____
Chemicals	0.15	_____
Filters	0.05	_____
Other	—	_____
Total Operating Costs		_____
Capital Recovery Costs		_____
Total Production Costs		_____

PROCEDURE:
1. Correct the energy costs using the information from Worksheet No. 1 and Eq. (25–3). Enter results in column 2.
2. Correct the membrane replacement costs using the information from Worksheet No. 1 and either Eq. (25–4) or (25–5). Enter result in column 2.
3. Correct the labor and overhead costs using the information from Worksheet No. 1 and Eq. (25–6). Enter result in column 2.
4. Enter either representative costs or specific known costs for spare parts, chemicals, filters, or other items in column 2.
5. Add column 2 to obtain total operating costs.
6. Calculate capital recovery costs using the information in Worksheet No. 1, the total capital costs from Worksheet No. 2, and Eq. (25–7). Enter result in column 2.
7. Add total operating costs and capital recovery costs to obtain total production costs.

*This worksheet is used to estimate the total production costs for an RO system.
[a]If no specific information is known, enter representative value in column 2.

Cost Estimates for RO Systems
*Worksheet No. 1**

Application: *LPRO/MS*

Variable	Equation	Representative Value	Value Used
CAPITAL COSTS			
Base system size (1000 gal/day)	25–2	1000	_____
Scale factor, n	25–2	0.85	_____
Construction overhead (%)	—	12	_____
Contingency (%)	—	10	_____
ENERGY OPERATING COSTS			
Operating pressure, p_r (psig)	25–3	200	_____
Cost of electricity, E_c ($/kWh)	25–3	0.075	_____
First-stage recovery, F_{R1} (%)	25–3	75	_____
Second-stage recovery, F_{R2} (%)	25–3	—	_____
Pump/motor efficiency, P_e (%)	25–3	85	_____
Energy recovery, E_R (%)	25–3	0	_____
MEMBRANE REPLACEMENT			
Membrane cost, M_c ($/ft^2)	25–4	5	_____
Membrane production rate, M_p (gal/ft^2/day)	25–4	25	_____
Membrane life, M_l (yr)	25–4	3	_____
Module cost, M_C ($/module)	25–5	1400	_____
Module production rate, M_P (gal/day)	25–5	7000	_____
Module life, M_L (yr)	25–5	3	_____
LABOR AND OVERHEAD			
Labor cost, Lh ($/h)	25–6	14.5	_____
Shifts per day, S (number/day)	25–6	1.3	_____
Workers per shift, Ws (number/shift)	25–6	1	_____
Plant capacity, PL_C (1000 gal/day)	25–6	(see above)	_____
Labor overhead, L_{OH} (%)	25–6	30	_____
CAPITAL RECOVERY			
System lifetime, r (yr)	25–7	15	_____
Annual interest rate, i (%)	25–7	12	_____
Downtime, Dt (%)	25–7	15	_____

*This worksheet is designed to identify the variables used to correct the costs using equations given in the text.

Cost Estimates for RO Systems
*Worksheet No. 2**

Application: *LPRO/MS*
Plant Size (1000 gal/day) _____

Capital Costs ($/gal/day)	Column 1 Uncorrected Representative Costs	Column 2 Corrected Representative Costs	Column 3 Value Used
DIRECT			
Site development, SD	0.10	_____	_____
Water, W	0.20	_____	_____
Utilities, U	0.06	_____	_____
Equipment, EQ	0.75	_____	_____
Land	—	—	_____
Other	—	—	_____
Total Direct Capital Costs			_____
INDIRECT			
Construction overhead			_____
Contingency			_____
Other			_____
Total Indirect Capital Costs			_____
Total Capital Costs			_____

PROCEDURE:
1. Using the scale factor from Worksheet No. 1, convert all representative direct capital costs (column 1) to the plant size specified using Eq. (25–2) and enter in column 2.
2. If specific cost information is known, enter in column 3; otherwise, enter values from column 2 into column 3.
3. Add values in column 3 to obtain total direct capital costs.
4. Calculate construction and overhead costs by multiplying total direct costs by the percentage given in Worksheet No. 1. Enter value in column 3.
5. Repeat step 4 for contingency. Enter value in column 3.
6. Enter other indirect capital costs in column 3.
7. Add values to obtain total indirect capital costs.
8. Add total direct capital costs and total indirect capital costs to obtain total capital costs.

*This worksheet is used to estimate the capital costs of an RO system.

Cost Estimates for RO Systems
*Worksheet No. 3**

Application: *LPRO/MS*
Plant Size (1000 gal/day): _____

Operating Costs ($/1000 gal)	Column 1 Uncorrected[a] Representative Cost	Column 2 Value Used
Energy	0.17	_____
Membrane replacement	0.18	_____
Labor and overhead	0.20	_____
Spare parts	0.02	_____
Chemicals	0.09	_____
Filters	0.02	_____
Other	—	_____
Total Operating Costs		_____
Capital Recovery Costs		_____
Total Production Costs		_____

PROCEDURE:
1. Correct the energy costs using the information from Worksheet No. 1 and Eq. (25–3). Enter results in column 2.
2. Correct the membrane replacement costs using the information from Worksheet No. 1 and either Eq. (25–4) or (25–5). Enter result in column 2.
3. Correct the labor and overhead costs using the information from Worksheet No. 1 and Eq. (25–6). Enter result in column 2.
4. Enter either representative costs or specific known costs for spare parts, chemicals, filters, or other items in column 2.
5. Add column 2 to obtain total operating costs.
6. Calculate capital recovery costs using the information in Worksheet No. 1, the total capital costs from Worksheet No. 2, and Eq. (25–7). Enter result in column 2.
7. Add total operating costs and capital recovery costs to obtain total production costs.

*This worksheet is used to estimate the total production costs for an RO system.
[a]If no specific information is known, enter representative value in column 2.

ACKNOWLEDGMENTS

The University of Kentucky contributors would like to acknowledge the National Science Foundation (EPSCoR grant) and the U.S. Environmental Protection Agency for partial support of the research results presented in this part of the handbook.

The Bend Research contributors would like to acknowledge David D. Newbold, Randi Wright-Wytcherley, and Ann C. Malkin for their help in preparing this part of the handbook.

NOTATION

General Notation

See the General Notation section at the beginning of this handbook.

Special Notation

Dt	downtime, %
E_c	cost of electricity, $/kWh
EQ	cost of equipment, $/gal/day
E_R	percent of feed-pump energy recovered by energy recovery equipment, %
F_{R1}	percent of first-stage feed recovered as permeate, %
F_{R2}	percent of first-stage permeate recovered as second-stage permeate, %
i	annual interest rate, %
LD	cost of land, $/gal/day
Lh	labor cost per hour, $/h
L_{OH}	labor overhead, %
M_c	membrane cost, $/ft^2
M_C	module cost, $/module
M_l	membrane life, yr

M_L module life, yr
M_p membrane production rate, gal/ft^2/day
M_P module production rate, gal/day
n scale factor for system size correction, dimensionless
P_e combined efficiency of pump and motor, %
PL_C plant capacity, 1000 gal/day
p_r operating pressure, psig
r system lifetime, yr
S shifts per day, number/day
SD cost of site development, $/gal/day
U cost of utilities, $/gal/day
UCL unit cost of labor, $/1000 gal
URC_m unit replacement cost for membranes, $/1000 gal
URC_M unit replacement cost for modules, $/module
WS cost of water supply, $/gal/day
W_s workers per shift, number/shift

REFERENCES

Al-Marafie, A. M. R., and M. A. Darwish. 1989. Water production in Kuwait—its management and economics. *Desalination* 71:45–55.

Andrews, W. T., and R. A. Bergman. 1986. The Malta seawater RO facility. *Desalination* 60:135–144.

Anonymous. 1989. *Using Desalination Technologies for Water Treatment*, Washington, DC: Office of Water Technology.

Applegate, L. E. 1984. Membrane separation processes. *Chem. Eng.* 91(6)64–89.

Birkett, J. D. 1985. Economic factors in reverse osmosis. Paper read at the 3rd Annual Conference on Membrane Technology and Planning, 30–31 October 1985, Cambridge, MA.

Birkett, J. D. 1988. Factors influencing the economics of reverse osmosis. Chap. 11 in *Reverse Osmosis Technology*, ed. B. S. Parekh. New York: Marcel Dekker.

Buros, O. K. 1979. Economic aspects of membrane processes. *Desalination* 30:595–603.

Calder Ltd. 1989. *Calder Pressure Systems*, Publication 9003/1989, Worcester, England.

Ericsson, B., and B. Hallmans. 1985. A comparative study of the economics of RO and MSC in the Middle East. *Desalination* 55:441–459.

FilmTec Corporation. 1988a. Technical bulletin 1001B, Minneapolis, MN.

FilmTec Corporation. 1988b. Technical bulletin 1009B, Minneapolis, MN.

FilmTec Corporation. 1989. Technical sales information 6009, Minneapolis, MN.

Glueckstern, P., and N. Arad. 1984. Economics of the application of membrane processes. part 1: desalting brackish and sea waters. Chap. 12 in *Synthetic Membrane Processes*. Orlando, FL: Academic Press.

Hornberg, D. C., and O. J. Morin. 1986. A cost study of membrane softening and low pressure reverse osmosis systems. *IDA Magazine* 1(3):17–26.

Hydranautics Water Systems. 1983. Applications report, Vol. VI, No. 1, Santa Barbara, CA.

Hydranautics Water Systems. 1986. Applications report, Vol. XV, No. 2, Santa Barbara, CA.

Larson, T. J., and G. F. Leitner. 1979. Desalting seawater and brackish water: a cost update. *Desalination* 30:525–539.

Leitner, G. F. 1987. Economic feasibility of the reverse osmosis process for seawater desalination. *Desalination* 63:135–142.

Lykins, Jr., B. W., R. M. Clark, and C. A. Fronk. 1988. Reverse osmosis for removing synthetic organics from drinking water: a cost and performance evaluation. NTIS Publication PB88-225016. Report to the Water Engineering Laboratory, Office of Research and Development, U.S. Environmental Protection Agency.

Matz, R., and U. Fisher. 1981. A comparison of the relative economics of seawater desalination by vapour compression and reverse osmosis for small to medium capacity plants. *Desalination* 36:137–151.

Pain, L. 1989. Paterson Candy International Membrane Systems Ltd., Witchurch, Hampshire, United Kingdom. Personal communication. March 29, 1989.

Rogers, A. N. 1984. Economics of the application of membrane processes. Part 2: wastewater treatment. Chap. 13 in *Synthetic Membrane Processes*. Orlando, FL: Academic Press.

Saltech Corporation. 1981. Company literature, El Paso, TX.

Smith, B. R., and E. A. Swinton. 1983. Desalination costs in Australia: a survey of operating plants. *Desalination* 70:3–15.

Spencer, P. 1990. Village Marine Technology, Gardena, CA. Phone communication. February 6, 1990.

Taylor, J. S., S. J. Duranceau, W. M. Barrett, and J. F. Goigel. 1989a. Assessment of potable water membrane applications and research needs. Report prepared for the AWWA Research Foundation, October 1989.

Taylor, J. S., L. A. Mulford, S. J. Duranceau, and W. M. Barrett. 1989b. Cost and performance of a membrane pilot plant. *AWWA* 81(11):52–60.

Taylor, J. S., D. K. Smith, and J. F. Goigel. 1989. Cost and performance of membrane plants. Orlando, FL: Civil Engineering Environmental Science Department, Univ. of Central Florida.

Wade, N. M., R. Heaton, and D. G. Boulter. 1985. Desalination and water reuse comparison of MSF and RO in dual purpose power and water plants. *Desalination* 55:373–386.

VII

Ultrafiltration

26. Introduction and Definitions   393
27. Theory and Mechanistic Concepts   398
28. Membranes   408
29. Module and Process Configuration   432
30. Applications and Economics   446

26

Introduction and Definitions

Sudhir S. Kulkarni
National Chemical Laboratory

Edward W. Funk and Norman N. Li
Allied Signal

OVERVIEW
ULTRAFILTRATION MEMBRANE
  PROPERTIES
DEFINITIONS
  Flux
  Rejection
Molecular Weight Cutoff Profile
Concentration Polarization
Fouling
Diafiltration
NOTATION
REFERENCES

OVERVIEW

Ultrafiltration (UF) is primarily a size-exclusion-based pressure-driven membrane separation process. UF membranes typically have pore sizes in the range from 10 to 1000 Å and are capable of retaining species in the molecular weight range of 300 to 500,000 daltons. Typical rejected species include sugars, biomolecules, polymers, and colloidal particles. Most UF membranes are described by their nominal molecular weight cutoff (MWCO), which is usually defined as the smallest molecular weight species for which the membrane has more than 90% rejection.

The view that separation by UF is based on relative molecular sizes is true only as a first approximation. In many instances, the chemistry of the solute/membrane interaction is also important. The MWCO of any given membrane can vary with changing feed chemistries as well as with factors such as molecular orientation, molecular configuration, operating conditions, etc. From the viewpoint of transport fundamentals, the distinction between reverse osmosis (RO) and UF is purely artificial (Sourirajan 1977). However, the nature of the larger molecules that are usually separated by UF leads to significant practical differences between UF and RO processes. As a consequence of the higher molecular weight of species separated in a UF process, osmotic pressure differentials are smaller. Simultaneously, the liquid phase diffusivity of these species is also lower; hence, membrane fouling and concentration polarization problems are more significant in UF.

The driving force for transport across the membrane in both UF and RO processes is a pressure differential. Because UF membranes do not typically reject salts, osmotic pressure differentials are small compared to reverse osmosis. UF processes operate at 2 to 10 bars, although in some cases up to 25 to 30 bars have been used.

Feed (liquid) phase mass transfer resistance

and resistance due to gel layer formation on the membrane surface are extremely important effects in UF processing. Consequently, system design and operating protocol are as important as membrane selection. Membrane selection is aimed at decreasing the fouling tendencies of the membrane surface. The base polymer surface chemistry can be modified in order to increase hydrophilicity, which increases flux and reduces fouling in most aqueous applications. In commercial modules, concentration polarization is decreased by increasing the fluid shear at the membrane surface or by turbulence inducers such as channel spacers. New module designs have been demonstrated at the laboratory level.

UF processes are perhaps the most widely used membrane process next to dialysis and microfiltration. UF can be thought of as performing one or more of the following functions:

1. feed clarification
2. concentration of rejected solutes
3. fractionation of solutes.

Separation is efficient when there is at least a tenfold difference in the sizes of the species. This applies to both solvent:solute concentration processes as well as solute:solute fractionation.

UF membrane processes are generally used in the food, beverage, and dairy industries, for effluent treatment, and for biotechnology and medical applications. A 1982 estimate (Lonsdale 1982) placed UF as a US$50 million industry worldwide. Seventy-five percent of this amount was industrial separations; the remainder was laboratory applications. A recent publication (Riedinger and Faul 1988) seems to agree with the previous estimate. The sale of UF membranes worldwide appeared to be 150 million DM/a (German marks/annum) in 1982, ~250 million DM/a in 1985, and was expected to reach 350 million DM/a in 1990. Several excellent reviews of ultrafiltration have been written (Michaels 1968; Porter 1979; Porter and Michaels 1971, 1972; Beaton 1984; Cooper 1980; Madsen 1977; Torrey 1984; Cheryan 1986).

Chapters 27 through 30 cover various aspects of UF. Theoretical aspects are discussed in Chapter 27. Chapter 28 covers the various UF membranes and their characterization, while membrane modules and processes are reviewed in Chapter 29. Various UF applications are discussed in Chapter 30.

ULTRAFILTRATION MEMBRANE PROPERTIES

Several membrane characteristics are important in determining a membrane's suitability for separation applications. These overall characteristics can be summed up as (1) porosity, (2) morphology, (3) surface properties, (4) mechanical strength, and (5) chemical resistance. These characteristics depend on the membrane material as well as the fabrication technique. To a great extent, these properties are interrelated; for example, a highly porous membrane structure can be maintained only if the polymer has adequate mechanical strength. Properties such as resistance to compaction under pressure, cleaning chemicals, bacterial degradation, and temperature are important for industrial use. Surface properties and pore morphology have a bearing on fouling properties, flux through the membrane, and solute separation.

The most important membrane properties are obviously the membrane productivity (flux) and the extent of separation (rejections of various feed components). Because of the relatively large size of molecules rejected by the membrane as well as the high fluxes of most UF membranes, the phenomena of concentration polarization and fouling are more significant when compared to reverse osmosis systems. These important parameters are defined below. In actual practice the achieved separation is a function of intrinsic membrane properties, process operating conditions, and module geometry.

DEFINITIONS

Flux

The volumetric flux (permeate volume/membrane area, time) is given by

$$J_v = P(\Delta p - \Delta \pi) / l, \qquad (26\text{--}1)$$

where P is a permeability coefficient, Δp is the hydrostatic pressure difference, $\Delta \pi$ is the osmotic pressure difference between feed and permeate phases, and l is the membrane thickness. Osmotic effects are usually smaller in UF than in RO. Cases occur in which the osmotic pressure term in Eq. (26–1) can be neglected; however, this should be checked for each application since it can lead to significant error even in the case of large molecules such as polyethyleneoxide.

Rejection

The observed solute rejection R_i for a given species i is given by

$$R_i = 1 - c_{ip} / c_{ir}, \qquad (26\text{--}2)$$

where c_{ip} is the concentration in the permeate while c_{ir} is the corresponding retentate side value.

When UF is performed in a batch process, the rejection can be calculated from either the permeate or retentate concentrations and the corresponding volume ratios. If V_0 and c_f are the initial feed volume and concentration, respectively, while V_p is the total permeate volume and V_r the remaining retentate volume, the rejection can be calculated as (Blatt 1976)

$$R_i = \frac{\ln(c_r/c_f)}{\ln(\text{VCF})} \qquad (26\text{--}3)$$

or as

$$R_i = \frac{\ln[\text{VCF} - (c_p/c_f)(\text{VCF}-1)]}{\ln(\text{VCF})}. \qquad (26\text{--}4)$$

The volume concentration factor (VCF) (V_0/V_r) is also an important measure of the achievable separation.

Molecular Weight Cutoff Profile

The rejection characteristics of UF membranes are usually expressed as a nominal molecular weight cutoff (MWCO). This number refers to the molecular weight (daltons) of a species, which would be expected to have an R_i value of at least 0.9 [see Eq. (26–2)]. A MWCO profile or retention curve is constructed by measuring the R_i value of chemically similar compounds of varying molecular weights. Such a profile can be characterized by an average MWCO and a dispersion coefficient Δ, which is a measure of the sharpness of the separation possible (Cherkasov 1990). Cherkasov has also suggested a parameter N_m, which is the number of theoretical stages of chromatographic separation equivalent to the membrane separation:

$$N_m = 30/\ln(\Delta^2 + 1). \qquad (26\text{--}5)$$

Membranes with low Δ values will be able to fractionate solutes more efficiently. The Δ values of conventional membranes are found to lie in the range of 0.7 to 2.8 (N_m = 100 to 20) while an isoporous membrane had a Δ value of 0.25 (N_m = 500).

Concentration Polarization

Concentration polarization is a common feature of all pressure-driven membrane processes such as UF. Solute rejected by the membrane builds up at its surface to a concentration c_W. The value of c_W is determined by the balance between solute brought to the membrane surface by convective flow of the solvent and that which back-diffuses to the bulk. At times, however, c_W reaches its solubility limit, which is lower than what the hydrodynamics would predict. As a result of this effect at the surface, the membrane effectively experiences a higher feedside concentration resulting in reduced flux as well as reduced apparent rejection.

The severity of concentration polarization can be lessened by operating conditions. It is particularly significant with the high-flux membranes used in ultrafiltration. Concentration polarization leads to smaller incremental increases in flux as pressure is increased until a gel layer is formed, at which point the flux shows no further increase with pressure. The flux at this point is called the *limiting flux*.

Fouling

Fouling refers to the deposition of some feed components on the membrane surface or within the membrane pores. For example, if concentration c_W reaches the point where solute precipitates or forms a thixotropic gel, this gel layer can provide an additional resistance in series with the membrane itself (Vilker, Colton, and Smith 1981):

$$J_v = (\Delta p - \Delta \pi)/(R_m + R_c), \quad (26\text{-}6)$$

where R_m ($= l/P$) is the membrane resistance to mass transport and R_c is the corresponding resistance of the gel layer or cake. The value of R_c can be so much higher than R_m that flux becomes independent of membrane permeability. Increased applied pressure may only lead to an increase or densification of the gel layer, thus canceling the expected flux increase. Alternatively, solute may be deposited within the membrane pores as a consequence of factors such as pore geometry/tortuosity or solute/pore wall interactions. The pores may then be completely blocked or be effectively reduced in diameter. As a result of either of these mechanisms, the flux through the membrane is reduced while the rejection may be either constant or may increase.

Fouling may be reversed by membrane cleaning; however, some irreversible fouling may also occur, which over time necessitates membrane replacement. The effect of fouling species on the module also needs to be considered.

While both concentration polarization and fouling reduce flux, they have opposing effects on the observed rejection. Another way to distinguish the two phenomena is through their time dependence. Concentration polarization is dependent on operating parameters such as pressure, temperature, feed concentration, and velocity but is not a function of time. Fouling is partially dependent on these variables, particularly feed concentration, but is also time dependent.

When distinguishing between concentration polarization and fouling in actual operation, note that other mechanisms of flux decline also exist, namely, membrane compaction, changes in the feed composition over time, etc.

Diafiltration

Recovery of membrane-permeable solutes may be limited by factors such as reduced flux, high feed viscosity, solubility limits of nonpermeating solutes, etc. The term *diafiltration* refers to dilution of the concentrate by solvent (water) and continuation of ultrafiltration until satisfactory removal of the permeable species is reached. The diluent may be added continuously to make up the volume lost through permeation or discontinuously by diluting back the concentrate obtained after a batch ultrafiltration operation.

NOTATION

General Notation

See the General Notation section at the beginning of this handbook.

Special Notation

c_{ip}	concentration in the permeate, mol/L^3 or M/L^3
c_{ir}	concentration in the retentate, mol/L^3 or M/L^3
c_W	feed solute concentration at the membrane surface, mol/L^3 or M/L^3
P	permeability coefficient, $L^3 t/M$ or L^2/pt
R_i	observed solute rejection, dimensionless
VCF	volume concentration factor, ratio of initial feed volume to retentate, dimensionless

Greek Letter

Δ	dispersion coefficient for measuring the sharpness of separation, dimensionless

REFERENCES

Beaton, N. C. 1984. Industrial ultrafiltration. In *Recent Developments in Separation Science,* ed. N. N. Li, Vol. 7, p. 1. Boca Raton, FL: CRC Press.

Blatt, W. F. 1976. Principles and practice of ultrafiltration. In *Membrane Separation Processes,* ed. P. Meares, Chap. 3, pp. 81–120. Amsterdam: Elsevier Scientific Publishing Co.

Cherkasov, A. N. 1990. Selective ultrafiltration. *J. Membr. Sci.* 50:109–130.

Cheryan, M. 1986. *Ultrafiltration Handbook.* Lancaster, PA: Technomic Publishing Co.

Cooper, A. R. Ed. 1980. *Ultrafiltration Membranes and Applications.* New York: Plenum Press.

Goldsmith, R. L. 1971. Macromolecule ultrafiltration with microporous membranes. *Ind. Eng. Chem. Fundam.* 10(1):113.

Lonsdale, H. K. 1982. The growth of membrane technology. *J. Membr. Sci.* 10:81–181.

Madsen, R. F. 1977. *Hyperfiltration and Ultrafiltration in Plate-and-Frame System.* Amsterdam: Elsevier Scientific Publishing Co.

Michaels, A. S. 1968. Ultrafiltration. In *Progress in Separation & Purification,* ed. E. S. Perry, Vol. I. New York: Wiley-Interscience.

Porter, M. C. 1979. Membrane filtration. In *Handbook of Separation Techniques for Chemical Engineers,* ed. P. A. Schweitzer. New York: McGraw-Hill Book Co.

Porter, M. C., and A. S. Michaels. 1971. Membrane ultrafiltration: applications in food processing, parts 1–4. *Chem. Tech.:* 56–63, 248–254, 440–445, 633–637.

Porter, M. C., and A. S. Michaels. 1972. Membrane ultrafiltration: applications in food processing, part 5. *Chem. Tech.:* 56–61.

Riedinger, H., and W. Faul. 1988. The focusing of membrane R & D on areas of commercial importance. *J. Membr. Sci.* 36:5–18.

Sourirajan, S. 1977. Reverse osmosis—a general separation technique. In *Reverse Osmosis and Synthetic Membranes: Theory-Technology-Engineering,* ed. S. Sourirajan, p. 3. Ottawa, Canada: National Research Council Canada.

Torrey, S. Ed. 1984. *Membrane and Ultrafiltration Technology Developments Since 1981.* Park Ridge, NJ: Noyes Data Corp.

Vilker, V. L., C. K. Colton, and K. A. Smith. 1981. Concentration polarization in protein ultrafiltration: part II. theoretical and experimental study of albumin ultrafiltered in unstirred cell. *AIChE J.* 27(4):637–645.

27

Theory and Mechanistic Concepts

Sudhir S. Kulkarni
National Chemical Laboratory

Edward W. Funk and Norman N. Li
Allied Signal

FUNDAMENTALS
   Basic Models
   Surface Force–Pore Flow Model
   Hindered Transport Models
CONCENTRATION POLARIZATION/FOULING

Fundamentals
   Mechanistic Interpretation of Flux Decline
NOTATION
REFERENCES

FUNDAMENTALS

It is sometimes useful to think of ultrafiltration (UF) as simply flow through pores in which separation is a sieving process based on relative molecular size. However, starting from this simple concept, more detailed models have evolved.

Transport through UF membranes can be thought of as flow through pores where the flow is modified by factors such as pore tortuosity and distribution, interaction of the feed components with the membrane, and interaction of the feed components (e.g., solute:solvent) among themselves.

Basic Models

In the simplest view, UF membranes can be considered to reject larger molecules through a sieving mechanism with solvent flowing in pores and solute molecules having sizes less than the pore diameter being carried convectively (Blatt et al. 1970). Assuming a uniform pore size and laminar flow through these pores, the volumetric flux J_v, defined in Chapter 26, Eq. (26–1), can be expressed in terms of the Hagen-Poiseuille equation. If the membrane is considered as an aggregate of cylindrical pores per unit area, n_p, with diameter d_p and length l, the flux of permeate having viscosity η can be expressed as

$$J_v = \left(\frac{n_p d_p^4}{128\eta}\right) \frac{\Delta p}{l}. \quad (27\text{--}1)$$

This equation predicts that the flux will be proportional to the fourth power of the pore diameter.

The observed solute rejection R_i was defined in Chapter 26, Eq. (26–2). When UF is conceptualized as simply a molecular sieving process, R_i can be related to the ratio of the molecular diameter of the species and the pore diameter by the Ferry (1936) equation:

$$R_i = [\lambda(2 - \lambda)]^2 \quad \text{for } \lambda < 1, \quad (27\text{--}2)$$

$$R_i = 1 \quad \text{for } \lambda \geq 1, \quad (27\text{--}3)$$

where λ is the ratio of the species diameter d_i to the pore diameter d_p.

Zeman and Wales (1981) have modified this equation to also take into account the solute velocity lag (solute moving slower than the solvent):

$$R_i = 1 - \{1 - [\lambda(2 - \lambda)]^2\} \exp(-0.7146\lambda^2).$$

$$(27\text{--}4)$$

This correction is significant for λ values greater than 0.5. Within the same framework (Zeman and Wales 1981; Zeman 1983), consideration has been given to the effects of other variables such as polydisperse solutes, membrane pore size distribution, and van der Waals interaction forces on the rejection coefficient.

A number of more sophisticated models are available for the flow of solutes through pores. A recent review by Deen (1987) summarizes these models. Smith and Deen (1980) include colloidal forces (electrostatic and dispersion) between solutes. Glandt (1980) and Mitchell and Deen (1986) discuss the case of finite concentrations where solute/solute interactions are taken into account.

Expressing the flux through a pore simply in terms of the Hagen-Poiseuille equation as in Eq. (27-1) does not incorporate any membrane characteristic other than the pore diameter. Spiegler (1958) proposed that the transport through the membrane was controlled by the balance of applied and frictional forces. This led to Merten's proposal of a useful "finely porous model" (Merten 1966), the derivation of which has been given by Jonsson and Boessen (1975) and also by Soltanieh and Gill (1981). A newer model (Matsuura and Sourirajan 1981) combines all these features of the solute size, membrane pore diameter, solute/membrane friction, and solute/membrane interaction. This model is discussed below.

Surface Force–Pore Flow Model

Matsuura and Sourirajan (1981) have developed an extensive model for flow through cylindrical circular pores under the influence of surface forces such as electrostatic repulsion, van der Waals dispersion, and friction. This model is particularly effective for studying membranes with pore sizes less than 30 Å or where other methods are more tedious. The solute rejection is

$$R_i = 1 - \frac{1}{c_f} \frac{\int_{-\infty}^{\infty} Y(r) \left[\int_0^r c_p(r')v(r')r' \, dr' \right] dr}{\int_{-\infty}^{\infty} Y(r) \left[\int_0^r v(r')r' \, dr' \right] dr},$$

$$(27\text{--}5)$$

where $Y(r)$ is the pore size distribution, c_f and c_p are the feed and permeate solute concentrations, respectively, and v is the solvent velocity. The $Y(r)$ term can be expressed as a normal distribution with an average pore radius r_p and a standard deviation σ_d. Solvent velocity v is calculated from the differential equation:

$$\frac{d^2v}{dr^2} + \frac{1}{r}\frac{dv}{dr} + \frac{\Delta p}{\eta l}$$

$$+ \frac{RT}{\eta l}(c_p - c_f)\left[1 - \exp\left(-\frac{\Phi}{RT}\right)\right]$$

$$- \left(\frac{(b-1)X_{AB}c_p v}{\eta}\right) = 0, \quad (27\text{--}6)$$

where at $r' = 0$, $dv/dr = 0$ and at $r' = r$, $v = 0$. Here η is the viscosity, X_{AB} is a constant relating the frictional force between solute and solvent to their relative velocity, b is the ratio of the frictional force of the moving solute to the frictional force of the bulk solution, and Φ is the potential function for the force exerted on the solute by the pore wall. Parameter b has been expressed as a function of the ratio of the steric hindrance distance $\overline{D}$ to the pore radius r_p. The electrostatic and van der Waals interaction forces are expressed in terms of constants $\overline{A}$ and $\overline{B}$, e.g., the van der Waals interaction force is modeled as

$$\Phi(r) = -\frac{(\bar{B}/r_a^3)}{[(r_p/r_a) - r]^3}, \quad (27\text{-}7)$$

where r_a is an effective reduced radius:

$$r_a = r_p - d_w. \quad (27\text{-}8)$$

Matsuura and Sourirajan (1981) described procedures to determine the interaction parameters $\bar{A}$, $\bar{B}$, and $\bar{D}$ by HPLC measurements. Values of r_p and d_w can then be chosen to best fit the rejection data according to Eq. (27-5).

This model has been used mainly for characterizing reverse osmosis membranes by water and salt transport measurements (see Chapter 22), although it has clear relevance to UF membranes. A variation on this experimental procedure using a gas adsorption technique was shown by Bhattacharya, Jevtitch, and Schrodt (1986). They also developed the pore size distribution for an ultrafiltration membrane as well.

Hindered Transport Models

These theories describe the transport of solute molecules in terms of hindrance factors for diffusive and convective transport in the pores of a membrane compared to bulk solution (Deen 1987; Anderson and Quinn 1974). The steady-state isothermal transport of a solute of radius r_i in a cylindrical pore of radius r_p filled with solvent (treated as a continuum) is given by

$$-k°T \frac{d \ln c_i}{dz} = 6\pi\eta r_i K_W(v_i - K_L v), \quad (27\text{-}9)$$

where v_i and c_i are the solute velocity and concentration, v_w is the solvent velocity, η the viscosity, z a coordinate in the direction of the pore axis, T the absolute temperature, and $k°$ Boltzmann's constant. The hydrodynamic coefficient K_W represents the drag on the solute from the pore walls and K_L represents the lag of the solute velocity behind the solvent velocity. Deen (1987) has derived a simple expression for the solute flux ($J_i = v_i c_i$) by averaging across the pore cross section and by assuming that the axial and radial dependencies of c are mathematically separable, i.e.,

$$c = g(z) \exp\left(-\frac{E(\beta)}{kT}\right), \quad (27\text{-}10)$$

where $E(\beta)$ represents the potential function for long-range interactions (electrostatic, dispersion forces, etc.) as a function of radial position in the pore ($\beta = r/r_p$). The overall solute flux is given by

$$\bar{J}_i = K_C \bar{v} c_f \frac{1 - (c_p/c_f) \exp(-Pe)}{1 - \exp(-Pe)}, \quad (27\text{-}11a)$$

$$Pe = \frac{\bar{v}l}{D_\infty} \left(\frac{K_C}{K_D}\right), \quad (27\text{-}11b)$$

where K_D (K_W, λ, E) is the hindrance factor for diffusion, K_C(K_L, λ, E) is the corresponding hindrance factor for convective filtration, λ is the dimensionless ratio of the solute size to pore size ($\lambda = r_i/r_p$ for cylindrical pores), l is the pore length, $\bar{v}$ is the average solvent velocity in the pore, and D_∞ is the solute diffusivity in the bulk solution. Expressions for K_D and K_C for neutral solutes ($E = 0$) and various ranges of λ values have been given by Bungay and Brenner (1973) (for $0 \leq \lambda < 1$), by Anderson and Quinn (1974) (for $0 \leq \lambda < 0.4$), by Brenner and Gaydos (1977) for $\lambda < 0.1$, and by Mavrovouniotis and Brenner (1986) for $\lambda > 0.9$. Formulations for K_D and K_C in the case of nonspherical solutes are given by Anderson and Quinn (1974). Values for these functions are also given as a function of λ for slit-shaped pores ($\lambda = r_i$/half-width of slit) (Weinbaum 1981). The effects of electrostatic interactions have been described by Smith and Deen (1983) and Malone and Anderson (1978). The transport of macromolecules, including charge effects, is described by Mitchell and Deen (1984). These theories have been extensively tested with isoporous, track-etched mica, and polycarbonate membranes.

CONCENTRATION POLARIZATION/FOULING

Fundamentals

It was observed very early that the flux in ultrafiltration processes does not increase linearly

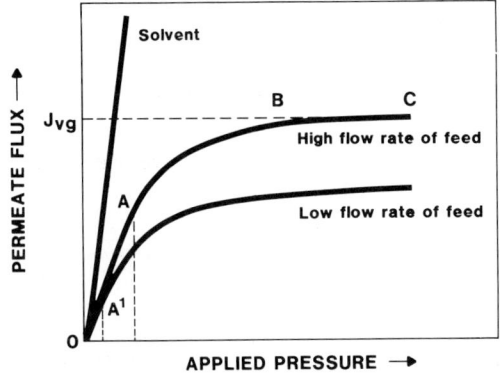

FIGURE 27–1. Schematic flux versus applied pressure plot comparing typical UF behavior with pure solvent flux.

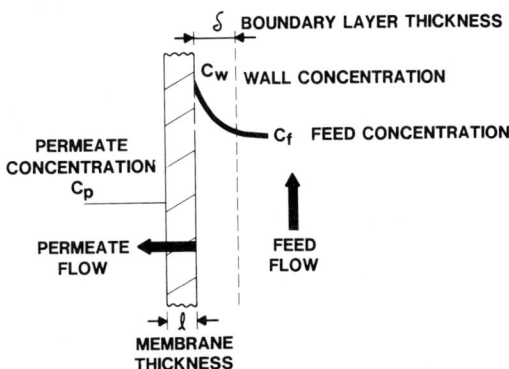

FIGURE 27–2. Boundary layer model for crossflow filtration.

with pressure, after a critical pressure has been reached. Also the flux of even moderately concentrated solutions (e.g., 1% dextran) is far lower than that of pure solvent. This is illustrated schematically in Figure 27–1. Initially, the UF flux increases linearly with increased applied pressure (section O-A, Figure 27–1). After a certain pressure, no increase in flux can be obtained (section B-C, Figure 27–1). These observations indicated the importance of fluid phase mass transfer effects. The buildup of rejected species at the membrane surface is known as *concentration polarization*.

Flux reduction in UF is usually ascribed either to an increase in osmotic pressure at the membrane surface or a decrease in hydraulic permeability due to formation of a gel layer. Both modes of flux reduction have been experimentally verified. As mentioned in Chapter 26, the flux can be written as (Vilker, Colton, and Smith 1981)

$$J_v = (\Delta p - \Delta \pi)/(R_m + R_c), \qquad (27\text{–}12)$$

where R_m is the resistance due to the membrane, R_c is the resistance offered by the deposited cake, and $\Delta \pi$ is the osmotic pressure difference that builds up as a result of solute rejection.

Resistances in Series Model

Flux decline has been blamed on the additional resistance offered by two layers on the membrane feedside surface: the boundary layer and a gel or cake layer. The flux reduction phenomenon has been analyzed by a boundary layer model (Michaels 1968; Brian 1965; Kozinski and Lightfoot 1972) illustrated in Figure 27–2 where δ is the distance over which the concentration changes from c_f (bulk feed concentration) to c_W (concentration at the membrane surface). The distance (δ) is controlled by the module flow and diffusion conditions. Solute is transported to the membrane surface by the convective flow of the permeant; this is balanced by diffusion back to the bulk (Belfort and Nagata 1985):

$$J_v(c - c_p) = -D \left(\frac{dc}{dx} \right). \qquad (27\text{–}13)$$

Integrating this equation across the boundary layer, the flux can be expressed as

$$J_v = k \ln[(c_W - c_p)/(c_f - c_p)], \qquad (27\text{–}14)$$

where $k (= D/\delta)$ is the mass transfer coefficient. As the flux increases with increasing applied pressure, the value of c_W relative to c_f also increases. As c_W reaches the solubility limit, further increases will cause a precipitate or thixotropic gel to be deposited on the membrane surface. Further increases of pressure will not give any more improvement in flux. When c_p is assumed to be small, this limiting flux can be approximated as (Porter 1972)

$$J_{vg} = k \ln(c_g/c_f). \qquad (27\text{–}15)$$

Equation (27–15) can be used to predict the limiting flux in various UF systems if the gel concentration c_g is known and the appropriate correlations for the mass transfer coefficient k are used. The gel concentration is mainly a function of the solute-solvent system and the operating temperature but is independent of the membrane characteristics, feed concentration, flow conditions, and operating pressure. Values of c_g have been found to be around an average value of 25% (weight basis) for macromolecular solutes (with a range of 5 to 50%) and an average of 65% for colloidal dispersions (range 50 to 75%). Li and Ho (1984) have also tabulated the experimentally determined gel concentrations for various common solutes and colloidal dispersions.

The mass transfer coefficient k can be estimated from correlations of the Sherwood number (Sh = kd_h/D) in terms of the Reynolds number (Re = $d_h v_f/\nu$) and the Schmidt number (Sc = ν/D) as discussed by Brian (1965). The symbol ν refers to the kinematic viscosity, d_h is the hydraulic diameter of the device through which the feed flows in contact with the membrane, and v_f is the velocity of the feed flow. Detailed correlations have been summarized by Hwang and Kammermeyer (1975). The correlation for turbulent flow is

$$\text{Sh} = 0.023 \text{ Re}^{0.8} \text{ Sc}^{0.33}. \quad (27\text{-}16)$$

Those for fully developed [Eq. (27–17)] and for developing [Eq. (27–18)] laminar flow (Porter 1972) are

$$\text{Sh} = 1.62 \, [\text{Re Sc } (d_h/L)]^{0.33}, \quad (27\text{-}17)$$

$$\text{Sh} = 0.664 \text{ Re}^{0.5} \text{ Sc}^{0.33} (d_h/L)^{0.5}. \quad (27\text{-}18)$$

Equation (27–18) is valid only if the device channel length is of similar magnitude as the length L^* required for flow profile to be developed ($L^* = 0.029$ Re d_h).

Correlations for k as a function of the system variables have been given by Porter (1972). A dimensional analysis of Eqs. (27–16) to (27–18) shows that k depends on the various operating variables:

$$k \propto D^{2/3} v^x \nu^{0.33-x} d_h^{-(1+x-y)} L^{-y}, \quad (27\text{-}19)$$

where the exponents x and y are

$x = 0.8, y = 0$ for turbulent flow, $\quad (27\text{-}20)$

$x = 0.33, y = 0.33$ for fully developed laminar flow, $\quad (27\text{-}21)$

$x = 0.50, y = 0.5$ for developing laminar flow. $\quad (27\text{-}22)$

Equation (27–19) illustrates why concentration polarization is more of a problem in UF compared to reverse osmosis. The solute diffusivities D of the macromolecular species typically involved in UF are two orders of magnitude less than those of salts. Correspondingly, values of k in UF processes ($k \propto D^{2/3}$) are approximately 5% of similar values in reverse osmosis.

A critical review of the various Sherwood number correlations in the turbulent regime has been made by Gekas and Hallstrom (1987). For the case of Re $> 10^4$ and Sc $> 10^3$, the correlations have the asymptotic form

$$\text{Sh} \sim f^m \text{ Re Sc}^{1/3}, \quad (27\text{-}23)$$

where the friction factor f is also a function of Re. The correlations in Eqs. (27–16) to (27–18) are based on flow in smooth, nonporous tubes or ducts, while in the case of membrane transport, the surfaces are porous and relatively rough. Moreover, bulk properties (viscosity, diffusivity) used to calculate the Re and Sc numbers are not the same as at the higher solute concentrations accumulating at the membrane wall through concentration polarization. Belfort and Nagata (1985) have estimated the change in the friction factor on going from a nonporous to a porous tube in terms of a wall Reynolds number based on the permeate flux. Gekas and Hallstrom (1987) have suggested that a correction factor of $(\text{Sc}/\text{Sc}_W)^{0.11}$ be used to modify the Sh calculated from Eq. (27–23) where Re and Sc are calculated from the bulk properties and Sc_W is calculated from the fluid properties at the membrane surface.

Porter (1972) has given an alternative correlation for k in the fully developed laminar flow mode by relating it to the shear rate at the membrane surface:

$$k = 0.816 \, \dot{\gamma}^{0.33} D^{0.67} L^{-0.33}. \quad (27\text{--}24)$$

The shear rate in laminar flow depends on the channel geometry and the feed velocity according to the relations:

$$\dot{\gamma} = 8v/d_t \text{ for tubes with diameter } d_t, \quad (27\text{--}25)$$

$$\dot{\gamma} = 6v/b \text{ for rectangular channels of height } b. \quad (27\text{--}26)$$

The basic boundary layer model has been elaborated using a more rigorous unsteady-state analysis (Van der Berg and Smolders 1989).

Osmotic Pressure Model

The osmotic pressure model explains the flux reduction by the increased osmotic pressure difference from the feed/membrane surface to the permeate phase that has to be overcome. Goldsmith (1971) has shown that even in the case of polymeric solutes a significant osmotic pressure difference can be developed. Vilker, Colton, and Smith (1981) have shown that UF flux decline of albumin solutions can be modeled by the osmotic pressure approach. In the case of dilute solutions, the osmotic pressure π can be related to the solute concentration by the van't Hoff-type relation:

$$\pi = RTc^n, \quad (27\text{--}27a)$$

$$\pi = RT \sum_{j=1}^{N} a_j c^j, \quad (27\text{--}27b)$$

where T is the absolute temperature and R is the gas constant. Exponent n is usually 1 for low molecular weight species and dilute solutions. However, for large molecules, the relation between solute concentration and osmotic pressure is not linear and n is not equal to 1 or π can be expressed as a power series [Eq. (27–27b)] in c. Unfortunately, osmotic pressure correlations are usually not available for the large molecules, which are the major foulants in ultrafiltration processes. Usually because of the low molarity differences of the solutions involved, $\Delta\pi$ can be estimated to be small. However, in some cases, as the large molecular weight species accumulate at the membrane surface with back-diffusion into the bulk, the actual osmotic pressure difference $[\pi_W(c_W) - \pi_p(c_p)]$ may be significant (Goldsmith 1971).

The basic flux equation [Eq. (26–1)] can be written in the form

$$J_v = (P/l)[\Delta p - (\pi_W - \pi_p)] \quad (27\text{--}28)$$

to account for the fact that the actual osmotic pressure difference experienced by the membrane is higher than $(\pi_f - \pi_p)$. The osmotic pressure at the membrane surface π_W can be estimated from Eqs. (27–27a) and (27–27b) and the boundary layer model [Eq. (27–14) or (27–15)] if the osmotic pressure at the gel conditions π_g is calculated from Eqs. (27–27a) and (27–27b) and the known values of c_g; π_g can be substituted for π_W in Eq. (27–28) in order to calculate the limiting flux.

Mechanistic Interpretation of Flux Decline

Fane (1986) has discussed initial ultrafiltration flux decline in terms of the boundary layer theory, adsorption, and pore plugging. Long-term UF flux decline was attributed to deposition/adsorption and cake consolidation. A similar analysis by Howell and Velicangil (1987) divides the UF process into three time intervals:

1. *First few seconds*: A quasi-steady-state concentration polarization layer is set up.
2. *1 to 10 minutes*: Solute adsorption.
3. *Long-term*: Gelation.

A fundamental understanding of fouling may be key to UF process development. The relative importance of pore plugging and adsorption has been studied. Choe et al. (1986) concluded that both effects could be seen using the IRIS 3042 and YM5 membranes with polyacrylic acid (PA) and bovine serum albumin (BSA) solutions. Both solutes were of similar molecular weights, 90,000 daltons for PA and 68,000 for the BSA. Flux reduction was indicated by the BSA to be primarily through fouling of the membrane surface through hydrophobic interactions, while the more flexible PA caused flux reduction through membrane pore plugging. That BSA causes flux reduction through surface adsorption rather than pore plugging has also been confirmed by Jonsson and Johansen (1989). Jonsson and Johansen constructed a crossflow device with flow on either side of the membrane. Under zero flux or reverse flux conditions, adsorption is the only fouling mechanism possible, while pore plugging can occur during normal flux conditions. Flux reduction with BSA was demonstrated after membrane exposure to all three flow conditions.

A similar earlier study by Sirkar and Prasad (1986) dealt with various chemistry-related and operational UF parameters for the test case of BSA. Flux reduction by protein adsorption was clearly shown by exposing membranes to BSA solutions for various times before starting the UF experiment. The effect of other feed variables such as pH or ionic strength could be rationalized through their effect on BSA aggregation.

Hanemaaijer et al. (1989) have shown that in the case of whey UF, flux decline can be attributed to precipitation of poorly soluble salts ($CaPO_4$) and protein adsorption within the membrane pores. These results were confirmed by Dejmek and Nilsson (1989) who analyzed whey UF data in terms of two competing models describing the total resistance to transport R_t by

$$R_t = R_m + R_c, \qquad (27\text{–}29)$$

where R_c is an additional cake resistance as in a layer of fouling materials on top of the membrane resistance R_m. On the other hand, internal fouling through precipitation or adsorption in the pores can be modeled as an interacting series resistance:

$$R_t = R_m(1 - \Delta/r_p)^{-4}, \qquad (27\text{–}30)$$

where Δ is the thickness of the internal fouling layer within the pores of radius r_p. A regression analysis showed a better fit to the interacting series resistance model.

The effect of concentration polarization is to reduce the observed solute rejection, while fouling typically increases the rejection. If R_i is the observed solute rejection ($R_i = 1 - c_p/c_f$), while the intrinsic (true) membrane solute rejection is given by

$$R_i' = 1 - (c_p/c_w), \qquad (27\text{–}31)$$

the boundary layer analysis shows that R_i will always be less than R_i'. They are related by

$$R_i = \frac{R_i' \exp(-J_v/k)}{R_i' \exp(-J_v/k) + 1 - R_i'}. \qquad (27\text{–}32)$$

The observed rejection would increase with increasing solute diffusivity and decreasing flux. Jonsson (1985) has analyzed the effect of pressure/flux on rejection profiles.

The increase in rejection with fouling is usually ascribed either to sieving through the gel layer on the membrane surface or to a reduction in pore size and the use of the Ferry equation [Eq. (27–2)]. The effect of increasing pressure here is usually to increase rejection, either by overcoming osmotic pressure or gel layer compression.

The effect of pH on protein adsorption and UF flux decline has been studied (Swaminathan, Chaudhuri, and Sirkar 1981). Flux reduction is most damaging at the protein isoelectric point.

NOTATION

General Notation

See the General Notation section at the beginning of this handbook.

Special Notation

$\overline{A}$	constant characterizing the electrostatic repulsion force, L
b	ratio of frictional force of solute to the bulk solution, dimensionless

$\bar{B}$	constant characterizing the van der Waals interaction force, L^3
c_g	solute gel concentration, mol/L^3 or wt.%
c_i	solute concentration, mol/L^3
c_p	solute concentration in the permeate, mol/L^3
c_w	solute concentration at the membrane surface, mol/L^3
d_w	diameter of water or solvent molecule, L
D_∞	solute diffusivity in the bulk solution, L^2/t
$\bar{D}$	steric hindrance distance, L
E	potential function for long-range interactions, ML^2/t^2 or E
k	mass transfer coefficient, L/t
$k°$	Boltzmann's constant, 1.38×10^{-16} g·cm^2/s^2·°K
K_C	hindrance factor for convective filtration, dimensionless
K_D	hindrance factor for diffusion, dimensionless
K_L	hydrodynamic coefficient representing the lag of the solute velocity behind the solvent velocity, dimensionless
K_W	hydrodynamic coefficient representing the drag on the solute from the pore walls, dimensionless
n_p	number of cylindrical pores per unit area
P	permeability coefficient
r_i	solute radius, L
R	gas constant, ML^2/t^2T mol
R_i	observed solute rejection, dimensionless
R_i'	intrinsic (true) membrane solute rejection, dimensionless
v	solvent velocity, L/t
v_i	solute velocity, L/t
X_{AB}	constant relating friction force between solute and solvent, Et/L^2 mol
$Y(r)$	pore size distribution, dimensionless

Greek Letters

η	viscosity, M/Lt
λ	ratio of species diameter to pore diameter, dimensionless
π_g	gel osmotic pressure, M/L^2t or p
π_p	permeate osmotic pressure, M/L^2t or p
π_w	osmotic pressure at the membrane surface, M/L^2t or p
Φ	potential function for force exerted on the solute by pore wall, E/mol

REFERENCES

Anderson, J. L., and J. A. Quinn. 1974. Restricted transport in small pores: model for steric exclusion and hindered particle motion. *Biophys. J.* 14(2):130–150.

Belfort, G., and N. Nagata. 1985. Fluid mechanics and crossflow filtration: some thoughts. *Desalination* 53:57–59.

Bhattacharyya, D., M. Jevtitch, and J. T. Schrodt. 1986. Prediction of membrane separation characterization by pore size distribution measurement and surface force-pore flow model. *Chem. Eng. Commun.* 42:111–128.

Blatt, W. F., A. Dravid, A. S. Michaels, and L. Nelsen. 1970. Solute polarization and cake formation in membrane ultrafiltration: causes, consequences and control techniques. In *Membrane Science and Technology,* ed. J. E. Flynn. New York: Plenum Press.

Brenner, H., and L. J. Gaydos. 1977. The constrained Brownian movement of spherical particles in cylindrical pores of comparable radius. *J. Colloid Interface Sci.* 58:312–356.

Brian, P. L. T. 1965. Concentration polarization in RO desalination with variable flux and incomplete rejection. *Ind. Eng. Chem. Fundam.* 4:439–445.

Bungay, P. M., and H. Brenner. 1973. The motion of a closely fitting sphere in a fluid-filled tube. *Int. J. Multiphase Flow* 1:25–26.

Chan, K., T. Matsuura, and S. Sourirajan. 1982. Interfacial forces, average pore size, and pore size distribution of ultrafiltration membranes. *Ind. Eng. Chem. Prod. Res. Dev.* 21:605–612.

Choe, T. B., P. Masse, A. Verdier, and M. J. Clifton. 1986. Membrane fouling in ultrafiltration of polyelectrolyte solutions: polyacrylic acid and bovine serum albumin. *J. Membr. Sci.* 26:17–30.

Deen, W. M. 1987. Hindered transport of large molecules in liquid-filled pores. *AIChE J.* 33:1409–1425.

Dejmek, P., and J. L. Nilsson. 1989. Flux-based measures of adsorption to ultrafiltration membranes. *J. Membr. Sci.* 40:189–197.

Fane, A. G. 1986. Ultrafiltration: factors influencing flux and rejection. In *Progress in Filtration and Separation,* ed. R. J. Wakeman, Vol. 4. Amsterdam: Elsevier Scientific Publishing Co.

Ferry, J. D. 1936. Ultrafilter membranes and ultrafiltration. *Chem. Rev.* 18:373–455.

Gekas, V., and B. Hallstrom. 1987. Mass transfer in the membrane concentration polarization layer under turbulent cross flow: I. critical literature review and adaptation of existing Sherwood correlations to membrane operations. *J. Membr. Sci.* 30:153–170.

Glandt, E. D. 1980. Density distribution of hard spherical molecules inside small pores of various shapes. *J. Colloid Interface Sci.* 77:512–524.

Goldsmith, R. L. 1971. Macromolecule ultrafiltration with microporous membranes. *Ind. Eng. Chem. Fundam.* 10(1):113.

Hanemaaijer, J. H., T. Robbersten, Th. Van der Boomgaard, and J. W. Gunnink. 1989. Fouling of ultrafiltration membranes: role of protein adsorption and salt precipitation. *J. Membr. Sci.* 40:199–217.

Howell, J. A., and O. Velicangil. 1987. Theoretical considerations of membrane fouling and its treatment for protein ultrafiltration. *J. Appl. Polym. Sci.* 27:21–32.

Hwang, S. T., and K. Kammermeyer. 1975. *Membranes in Separations.* New York: John Wiley and Sons.

Jonsson, G. 1985. Molecular weight cut-off curves for ultrafiltration membranes of varying pore sizes. *Desalination* 53:3–10.

Jonsson, G., and C. E. Boessen. 1975. Water and solute transport through cellulose acetate reverse osmosis membranes. *Desalination* 17:145–165.

Jonsson, G., and P. L. Johansen. 1989. Selectivity of ultrafiltration membranes. In *Indo-ECC Membrane Workshop.* New Delhi: Department of Science and Technology.

Kozinski, A. A., and E. N. Lightfoot. 1972. Protein ultrafiltration: a general example of boundary layer filtration. *AIChE J.* 18(5):1030.

Li, N. N., and W. S. Ho. 1984. Membrane processes. In *Perry's Chemical Engineers' Handbook,* ed. R. H. Perry and D. Green, 6th ed., pp. 17-14–17-35. New York: McGraw-Hill Book Co.

Malone, D. M., and J. L. Anderson. 1978. Hindered diffusion of particles through small pores. *Chem. Eng. Sci.* 33:1429–1440.

Matsuura, T., and S. Sourirajan. 1981. Interfacial parameters governing RO for different polymer material-solution systems through gas and liquid chromatography data. *Ind. Eng. Chem. Process Des. Dev.* 20:273. (See also 1983. *J. Colloid Interface Sci.* 95:10.)

Mavrovouniotis, G. M., and H. Brenner. 1986. Hindered sedimentation and dispersion coefficients for rigid, closely fitting Brownian spheres in circular cylindrical pores containing quiescent fluids. Paper read at AIChE Ann. Mtg., 2–7 November 1986, Miami Beach, FL, Paper No. 85b.

Merten, U. 1966. Transport properties of osmotic membranes. In *Desalination by Reverse Osmosis,* ed. U. Merten, pp. 15–54. Cambridge, MA: MIT Press.

Michaels, A. S. 1968. Ultrafiltration. In *Progress in Separation and Purification,* ed. E. S. Perry, Vol. I, New York: Wiley-Interscience.

Mitchell, B. D., and W. M. Deen. 1984. Theoretical effects of macromolecule concentration and charge on membrane rejection coefficients. *J. Membr. Sci.* 19:75–100.

Mitchell, B. D., and W. M. Deen. 1986. Effect of concentration on the rejection coefficients of rigid macromolecules in track-etch membranes. *J. Colloid Interface Sci.* 113:132–142.

Porter, M. C. 1972. Concentration polarization with membrane ultrafiltration. *Ind. Eng. Chem. Prod. Res. Dev.* 11:234.

Rautenbach, R., and R. Albrecht. 1989. *Membrane Processes.* New York: John Wiley and Sons.

Sirkar, K. K., and R. Prasad. 1986. Protein ultrafiltration—some neglected considerations. In *Membrane Separations in Biotechnology,* ed. W. C. McGregor, pp. 37–59. New York: Marcel Dekker.

Smith, F. G., III, and W. M. Deen. 1980. Electrostatic double layer interactions for spherical colloids in cylindrical pores. *J. Colloid Interface Sci.* 78:44–52.

Smith, F. G., III, and W. M. Deen. 1983. Electrostatic effects on the partitioning of spherical colloids between dilute bulk solution and cylindrical pores. *J. Colloid Interface Sci.* 91:571–590.

Soltanieh, M., and W. N. Gill. 1981. Review of RO membranes and transport models. *Chem. Eng. Commun.* 12:279–363.

Spiegler, K. S. 1958. Transport processes in ionic membranes. *Trans. Faraday Soc.* 54:1408–1428.

Swaminathan, T., M. Chaudhuri, and K. K. Sirkar. 1981. Effect of pH on solvent flux during stirred ultrafiltration of proteins. *Biotech. Bioeng.* 23:1873–1880.

Van der Berg, G. B., and C. A. Smolders. 1989. Boundary layer resistance model for unstirred ultrafiltration: a new approach. *J. Membr. Sci.* 40:149.

Vilker, V. L., C. K. Colton, and K. A. Smith. 1981. Concentration polarization in protein ultrafiltration: part II. theoretical and experimental study of albumin ultrafiltered in unstirred cell. *AIChE J.* 27(4):637–645.

Weinbaum, S. 1981. Strong interaction theory of particle motion through pores and near boundaries in biological flows at low Reynolds number. *Lect. Math. Life Sci.* 14:119.

Zeman, L. 1983. Adsorption effects in rejection of macromolecules by ultrafiltration membranes. *J. Membr. Sci.* 15:213–230.

Zeman, L., and M. Wales. 1981. Polymer solute rejection by ultrafiltration membranes. In *Synthetic Membranes: Vol. II. Hyperfiltration and Ultrafiltration Uses,* ed. A. F. Turbak, ACS Symp. Ser. No. 154, pp. 412–434. Washington, DC: American Chemical Society.

28

Membranes

Sudhir S. Kulkarni
National Chemical Laboratory

Edward W. Funk and Norman N. Li
Allied Signal

POLYMER MEMBRANES
   Membrane Fabrication
   Membrane Modification
   Characteristics of Commercially
     Available Membranes
INORGANIC MEMBRANES

MEMBRANE CHARACTERIZATION
   Membrane Porosity and Morphology
   Surface Properties
ACKNOWLEDGMENTS
NOTATION
REFERENCES

Most ultrafiltration (UF) membranes are polymeric in nature, although recently inorganic membranes have also become available. The formation of an asymmetric membrane structure, i.e., an upper skin that is permselective and a more porous substructure for mechanical support, is an important element in the success of UF membranes. While many polymers, in particular, have been examined for use as membrane materials (Lloyd and Meluch 1985), only a few are widely used. Table 28–1 lists various polymeric and inorganic materials for UF membrane manufacture. Various polymer membranes have been reviewed by Pusch and Walch (1982), Kesting (1971), and Cabasso (1980a). An excellent review of inorganic membranes has been given by Hsieh (1988). An exhaustive list of membrane types made by various manufacturers has been compiled by Cheryan (1986). Table 28–2 lists product information from various UF membrane manufacturers. A list of manufacturers of UF membranes is given in Table 28–3.

TABLE 28–1. Typical Ultrafiltration Membrane Materials.

Polymeric
Polysulfone
Polyethersulfone
Cellulose acetate
Regenerated cellulose
Polyamides
Polyvinylidenefluoride
Polyacrylonitrile
Inorganic
γ-Alumina/α-Alumina
Borosilicate glass
Pyrolyzed carbon
Zirconia/Stainless steel or Zirconia/Carbon

POLYMER MEMBRANES

Membrane Fabrication

The most common UF membranes are based on polysulfone, cellulose acetate, polyamides, and various fluoropolymers. These are typically made by the phase inversion method in which

TABLE 28–2. Performance Characteristics of Selected Commercially Available Ultrafiltration Membranes.

Supplier/Trade Name	Module Type	Membrane Type	Separation Characteristics MWCO or Pore Size	Flux Characteristics Water Flux	Maximum Operating Conditions Pressure (bar)	Temperature (°C)	pH Range
ALCOA/ Membralox	4- or 6-mm-i.d. tubes in multi-channel elements with 1–19 tubes/element; 0.2–3.6 m² area/element. Single tubes 7 mm and 15 mm i.d. also available	γ-alumina on α-alumina support	4–100 nm	10 l/m² · h · bar at 20°C for 4-nm membrane	8–25 (may be limited by module construction)	300	
ALCAN/Anopore	Disk	Al_2O_3	25 nm				
AMICON (WR Grace)	Flat sheet disks. Hollow-fiber modules. Fiber i.d. = 0.2, 0.5, 1.1 mm; length = 20, 64 cm. Module area = 0.3–9.5 ft²	Polysulfone		F = flux mL/ min. cm² of sheet at 55 psi H = 0.3 ft² module, 1.1 mm i.d. typical flux mL/min			
		P 10	10K	2.5–4.0 (F) 40–100 (H)	25 psi for most modules	50	
		P 30	30K	6–10 (F) 40–100 (H)	16 psi for 0.2-mm-i.d. modules		
	Spiral-wound modules 0.9–10 ft² with 0.8-mm feed channel spacers. Available only with YM membranes	P 100	100K	170–220 (H)			

TABLE 28-2. (continued)

Supplier/Trade Name	Module Type	Membrane Type	Separation Characteristics MWCO or Pore Size	Flux Characteristics Water Flux	Maximum Operating Conditions		
					Pressure (bar)	Temperature (°C)	pH Range
ASAHI	0.8- or 1.4-mm-i.d. hollow-fiber modules; 35 or 112 cm long. Areas: 0.2–4.7 m²	Hydrophilic Polymer YM 2	1K	0.02–0.035 (F)		75	
		YM 5	5K	0.06–0.12 (F)			
		Acrylic vinyl polymer X 50	50K	1.0–1.8 (F)		75	
		X 100	100K	0.4–2.0 (F)			
		Polyacrylonitrile		l/m² · h at 1 atm, 25°C for 4.7 m², 0.8-mm-i.d. module			
		AIV-3010	6K	36	3	50	2–10
		ACV-3010	7K	170	3	50	2–10
BERGHOF	BTU series: 11.5-mm tubular modules with 0.2–3.5 m² area/module	Noncellulosic polymers		PEG soln. flux L/m² · h at pressure (bar)			
		BTU-1020	20K	80/4	8	80	2–12
		-2020	20K	50/2	10	40	2–12
		-20100	100K	83/2	10	40	2–12
		-3508	8K	40/2	10	60	2–13
		-3020	20K	50/2	10	60	2–13
		-35100	100K	124/2	10	60	2–13
		-4208	8K	40/2	10	40	2–8
		-4220	20K	53/2	10	40	2–8
		-6050	50K	120/4	8	80	2–12
	BMK series: 0.6-, 1.1-, or 1.5-mm-i.d. hollow-fiber modules with 0.2–4.0 m² area/module	Polyamide or polysulfone	2K, 10K, 30K, 50K, 100K	—	2	50–60	—

Manufacturer	Module description	Membrane / Model	MWCO	Flux (L/m²·h·atm; F: flat sheet, T: tubular)	Max pressure	Max temp (°C)	pH range
DAICEL	14.5-mm-i.d. flat sheet and tubular modules	Polyacrylonitrile					
		-HH	5K	13(F), 21(T)	10	45	—
		-H	10K	25(F), 42(T)			
		-M	20K	50(F), 67(T)			
		-L	40K	83(F), 108(T)			
		Polyethersulfone					
		-40	40K	4(T)	10	90	—
DDS	Plate-and-frame modules with 0.5- to 1.0-mm feed channel spacers	Cellulose Acetate		$1/m^2 \cdot h$ at 5 atm, 20°C			
		800	6K	80	15	50	2–8
		600	20K	150			
		500	65K	350			
		Polysulfone					
		GR81P	10K	200	15	80	0–14
		GR61P	20K	200			
		TFC					
		FS 60P	30K	200	15	80	0–14
DESALINATION SYSTEMS INC/DSI	2-, 4-, and 8-in.-diameter spiral-wound modules, 12–40 in. long. Test modules: 6–20 ft², production modules: 60–350 ft², feed channel spacers: 30, 50, 90 mils. Available in tape or FRP wrap	Polysulfone type (E series)	Flat sheet % Rej. of 1% dextran at 7 psi	Test results water perm. $L/m^2 \, h$ atm	2–6 (limited by module)	100 (membrane) 50–60 (module)	1–13
		E-100	95% of 35K dextran	36–108			
		E-500	96% of 500K dextran	180–360			
			4-in. × 40-in. spiral test results with 1000 ppm NaCl at 25°C and various test pressures				
		TFC type (G series)	NaCl % rej.	gal/day	40 (module)	50	2–10
		G-5	85	1400 (310 psi)			
		G-10	60	1680 (150 psi)			
		G-20	25	1550 (75 psi)			
		G-50	8	1600 (40 psi)			

TABLE 28-2. (continued)

Supplier/Trade Name	Module Type	Membrane Type	Separation Characteristics MWCO or Pore Size	Flux Characteristics Water Flux	Maximum Operating Conditions		
					Pressure (bar)	Temperature (°C)	pH Range
DORR-OLIVER (WR Grace)/ Ioplate.	Plate-and-frame modules with 1.0- or 2.5-mm feed spacers for industrial processes	Cellulosic C series	1K, 5K, 10K 30K, 100K	at 2 atm, 50°C L/m². h 210 (10K) 500 (100K)	75	3.5-10	
		Dynel D series	50K, 100K	500 (50K) 850 (100K)	60	2-12	
		Polysulfone F, K, S series	8K, 10K, 20K, 30K, 50K, 75K, 100K, 200K				
		S-10	10K	600			
		S-30	30K	700			
		Polyamide MFA series	20K, 200K	425	60	2-12	
FILTRON	Flat sheet disks & cassettes (individual or pre-assembled) (0.14-25.7 m² area) using either screen channel or open channel feed separators	NOVA series (polyethersulfone)	Nominal MWCO 1K, 3K, 5K, 8K, 10K, 30K, 50K, 100K.	Expected DI water flux in stirred cells at 25°C & 55 psi (mL/min/cm²) for various nominal MWCO	Cassette operation 5	60	1-14
		OMEGA series (PES, modified to resist fouling by antifoam agents)	Same as NOVA series plus 300K, 1000K.	0.05-0.10 (1K) 0.10-0.25 (3K) 0.25-0.50 (5K) 0.6-1.2 (8K) 0.9-1.9 (10K)	5	60	1-12

Manufacturer	Series	Description	MWCO / Performance notes	Flux / Performance	Max pressure bars (°C)	Max temp °C	pH range (°C)
	ALPHA series (PES modified to resist fouling by antifoam agents)		3K, 10K, 30K, 100K	1.0–4.0 (30K), 2.3–4.5 (50K), 4.5–10.5 (100K)	5	60	3–12
	SIGMA series		Same as for ALPHA series plus 300K, 1000K	8–20 (300K), 25–500 (1000K)	5	121 (autoclavable)	3–12
					bars at °C		Range at °C
FLUID SYSTEMS	Polyethersulfone	3.8- to 4.3-in.-diameter spiral-wound modules 33–39 in. length. Designed to 3A standards for dairy applications. Feedchannel spacers: 30, 41, 80 mils.	6–10K	RO water flux 65 gal/ft² · day 50 psi, 20°C in modules	5.5 (75), 8 (50), 10 (25)	75	2–9 (75), 1–12 (50), 1–13 (25)
GORE-TEX®	PTFE	Disks, 3-mil thick	0.02 μm (350 psi min. H₂O entry pressure)	1.2 mL/min · cm² MeOH at 27.5 in. Hg pressure drop & 21°C		−230 to 315°C	
NADIR (HOECHST)		Flat-sheet membrane in either sheet stock, plate-and-frame, or spiral-wound module form	Results of flat sheet tests in stirred cells with aq. solutes at 43 psi and 20°C		Maximum conditions refer to membrane only		
			MWCO/Avg. % Rejection of solute	Aq. solution flux L/m² · h			
	PES (hydrophilic)						
	UF-PES-4/PP 60		4K/75% of 5K inulin.	30–50 (1% inulin)	40	90	1–14
	UF-PES-8/PP 100		8K/90% of 10K dextran	70–120 (1% dextran)	25	90	1–14
	UF-PES-P5/PP 60		15K/80% of 10K dextran	120–250 (2% PVP 49K)	15	90	1–14
	UF-PES-20/PP 100		20K/85% of 10K dextran	100–150 (2% PVP 49K)	15	90	1–14
	UF-PES-25/PET 100		25K/92% of 49K PVP	140–220 (2% PVP)	15	90	2–10

TABLE 28–2. (continued)

Supplier/Trade Name	Module Type	Membrane Type	Separation Characteristics MWCO or Pore Size	Flux Characteristics Water Flux	Maximum Operating Conditions			
					Pressure (bar)	Temperature (°C)	pH Range
		UF-PES-50/PP 100	50K/75% of 49K PVP	150–250 (2% PVP)	15	90	1–14
		PES (hydrophobic) UF-PES-25N	25K/92% of 49K PVP	150–250 (2% PVP)	10	90	1–14
		Polysulfone (Hydrophilic) UF-PS-100M/PP 100	100K/85% of 2000K dextran	300–500 (1% 2000K dextran)	10	90	1–14
		Cellulose Acetate UF-CA-1/PP 100	1K/98% of 5K inulin	20–25 (1% inulin)	20	40	2–8
		UF-CA-5/PP 100	5K/95% of 5K inulin	30–40 (1% inulin)	20	40	2–8
		UF-CA-30/PAP 25	30K/80% of 49K PVP	150–300 (2% PVP)	15	40	2–8
		UF-CA-100/PET 100	100K/88% of 110K dextran	250–400 (1% dextran)	10	40	2–8
		Regen. cellulose UF-C-30/PP 100	30K/72% of 49K PVP	400–900 (2% PVP)	10	60	1–12
		Aromatic Polyamide UF-PA-20/PP 100	20K/95% of 49K PVP	150–250 (2% PVP)	10	80	2–12
		Polyvinylidene-fluoride UF-PVDF-30N	30K/75% of 49K PVP	250–400 (2% PVP)	15	80	1–10

Company	Configuration	Membrane	Nominal MWCO	Properties			pH range
	1-in. tubular membrane on PET support	Polyacrylonitrile UF-PAV-30	30K/80% of 49K PVP	500–1000 (2% PVP)	10	60	1–10
		Polysulfone types					
		UF-PES-30	30K/87% of 49K PVP	200–300 (2% PVP)	15	90	2–10
		UF-PS-100M	100K/92% of 2000K dextran	300–500 (1% dextran)	10	90	2–10
		Cellulose Acetate					
		UF-CA-10	10K/70% of 10K dextran	40–70 (1% dextran)	20	40	2–8
		UF-CA-30	30K/70% of 49K PVP	50–90 (2% PVP)	15	40	2–8
		UF-CA-100	100K/90% of 500K dextran	250–450 (1% dextran)	10	40	2–8
KALLE	Tubular and flat sheet	Cellulose acetate	2K–100K		20	10	2–8
		Polyamide	20K–100K		10	60	2–12
		Polysulfone	8K–25K		10	90	1–14
MILLIPORE	13- to 150-mm disks. Cassettes (0.5–5, 15, 25 ft² area) with screen (20, 50 mesh) or linear path feed spacers. Prostak: 0.7-mm open rectangular channel. Pellicon: 0.3-mm turbulence promoted channel. Spiral-wound modules (15, 50 ft²) 0.7-mm feed channel spacer.		Nominal MWCO	Water flux gal/min at 50 psi, 25°C for 50 ft² spiral(2) (1)	For cassette operation		
		Polyethersulfone				50	2–14
		PTGC	10K	2.0	80–100 psi		2–14
		PTTK	30K	3.2	(May be less than		2–14
		PTHK	100K	5.3	depending on tubings)		
				(1) Pellicon			
				(2) Prostak			
		Polysulfone					
		PTMK	300K				2–14
		Mixed cellulosic esters					
		PCAC	1K	0.45			4–8

TABLE 28-2. (continued)

Supplier/Trade Name	Module Type	Membrane Type	Separation Characteristics MWCO or Pore Size	Flux Characteristics Water Flux	Maximum Operating Conditions Pressure (bar)	Maximum Operating Conditions Temperature (°C)	Maximum Operating Conditions pH Range
		Cellulose PLGC	10K	2.0			2–13
		PLMK	300K				
		Polyvinylidene-fluoride PKMK	300K				2–14 (pH range is restricted to 3–11 in spirals).
KOCH MEMBRANE SYSTEMS	2-, 4-, and 8-in.-diameter spiral-wound modules with feed channel spacers from 20–80 mils. 0.5- and 1-in.-i.d. tubular modules available in various designs. Series-Cor: 15.5-in. tubes in 3-in. shell. Ultra-cor: 7.5-in. tubes. Super-cor: 19.5-in. tubes. Most standard membranes are available in both spiral and tubular form with the exceptions marked * available only as spirals and + only as tubular	Polyethersulfone HFK-328*	1–10K	gfd at 50 psi, 40°C S: 4-in. spiral flux F: Flat sheet flux 30–50(S), 80–100(F)	Membrane limits at pH6 at 25°C** 10	90	1–13
		HFK-131*	6–20K	35–100(S), 200–500(F)	10	90	1–13

	PVDF HFM-100	10–30K	50–90(S), 300–400(F)	10	90	1–11
	HFM-180	65–300K	100–200(S), 500–700(F)	10	90	1–11
	HFM-251	65–300K	—	10	50	1–11
			$1/m^2 \cdot h$ at 3.3 atm and 25°C			
	PVDF (cationic) HFM-163	20–80K	~170	10	60	2–11
	HFM-183	20–80K				
	PVDF (anionic) HFM-276$^+$	110–600K	~1000	10	90	1–13
			**pressure in module may be restricted to 100–125 psi. Temperature may be restricted to 54°C (tubular) or 80°C (spiral)			
NITTO/ Hydranautics	4-in.-diameter spiral-wound modules	50K (Based on PEG)	386 gal/h in 4- × 40-in. spiral with RO water at 29 psi, 25°C	10	40 (limited by module)	2–11
	11.5-mm-i.d. tubular modules with 4 or 18 tubes/shell of length 1.3–2.9 m long	Hydrophilic Polyolefin 2120 20K				
		Hydrophilic polyolefin 2000 series 20K, 100K	—	4–10	40	1–13 (25°C) 3–10 (40°C)
	11.5-mm-i.d., 18 tubes, 1.3 m long shell	Polyimide 4200 series 8K, 20K				
		Polysulfone 3000 series 8K, 20K, 100K.	—	10	40	2–8
	Hollow-fiber module i.d./o.d. = 0.55/1.0 mm	Polysulfone type 3050 20K	Water flux at 14.2 psi and 25°C >0.44 gpm for 4.3 ft^2 module			

TABLE 28-2. (*continued*)

Supplier/Trade Name	Module Type	Membrane Type	Separation Characteristics MWCO or Pore Size	Flux Characteristics Water Flux	Maximum Operating Conditions Pressure (bar)	Temperature (°C)	pH Range
	i.d./o.d. = 1.1/1.9 mm. shell 3.5- to 11.4-cm diameter by 0.5–1 m. long	3250	20K	— do —			
OSMONICS/ SEPA	Spiral-wound modules 5, 10, 21 cm diameter and 66, 102 cm long. Feed spacers: 24, 34, 45 mils	Cellulose acetate		L/m² · hr, at 35 atm, 25°C			
			1K	85	14	40	2–8.5
			20K	210			
			50K	600			
		Polysulfone	2K	130	14	100	1–13
			20K	350			
			50K	1000			
			100K	800			
		Fluoro-polymer	2K	130	14	90	1–12
			20K	350			
RHONE-POULENC/Iris	Flat sheet used in plate frame modules 0.5- and 1.5-mm feed spacers. Areas from 0.11–10 m²	PAN copolymer		L/m² · h, at 2 atm, *RT*			
		3038	25K	830	—	50	3–10
		3038 TS	10K	167	—	50	3–10
		3042	20K	500	—	40	3–10
		PVDF 3065	20K	1670	—	80	1–10
		Sulfonated Polysulfone 3026	15K	500	—	80	3–14

Manufacturer	Description	Membrane	MWCO	$L/m^2 \cdot h$ in tubular modules at 4 atm, 25°C	Temp (°C)	pH
Carbosep (SFEC)	6-mm-i.d. tubular membranes up to 1.2 m long. Available in areas from 0.1–5.7 m^2	ZrO_2/carbon composite			300°C may be limited to 150°C by system	0–14
		M1	60–80K			
		M4	20K			
		M5	10K			
		M9	300K			
PCI Membrane	12.5-mm-i.d. tubular modules, 2 × 9 tubes in series, 3.66 m long					
		Polyethersulfone				
		ES 404	4K	30	70	2–12
		ES 209	9K	30	70	2–12
		ES 625	25K	15 900	70	2–12
		Polysulfone				
		PU 608	8K	1 800	70	2–12
		PU 120	20K	15 800	70	2–12
		Polyacrylonitrile				
		AN 620	25K	10 180	60	2–10
		PVDF				
		FP 100	100K	10 4000	70	2–12
		FP 200	200K	10	70	2–12
		Cellulose Acetate				
		CA 202	2K	25 25	30	3–6
		CA 407	7K	20 100	30	3–6
ROMICON	0.5- and 1.1-mm-i.d. hollow-fiber modules 31, 63, 109 cm long. Membrane areas from 0.17–4.9 m^2	Polysulfone	1K, 2K, 5K, 10K, 30K, 50K, 100K			
SARTORIUS	Flat-sheet membranes for stirred cell, plate-and-frame or centrifugal devices	Cellulose triacetate	5K	Water flux at 1 bar ml/min. cm^2 0.02		

TABLE 28-2. (continued)

Supplier/Trade Name	Module Type	Membrane Type	Separation Characteristics MWCO or Pore Size	Flux Characteristics Water Flux	Maximum Operating Conditions		
					Pressure (bar)	Temperature (°C)	pH Range
		Polysulfone	10K	0.05			
		Cell. Nitrate	20K	0.4			4–8
		Cell. Acetate	100K	1.6			1–14
		Regen. Cellulose	10K, 50K				
			20K, 70K, 160K				
			20K, 70K, 160K				
SCHOTT/Bioran	Hollow-fiber 0.3-mm-i.d. 0.05-m² test module	Glass (>96% SiO₂) can be modified to be more hydrophilic	10, 13.4, 19, 27, 44, 90 nm pore diameter	10 L/m² · h · bar for 44-nm pore diameter			
TEIJIN	Tubular	Polysulfone Tu series			10	90	
WAFILIN	14-mm tubular modules	Non-cellulosic polymers		l/m² · h at 1 atm, 25°C			
		WFS series					
		8010	20K	30–50	10		
		6010	35K	70–90	6.3		
		5010	100K	150–200	3	85	2–12
		WFA series					
		7010	120K	15–25			
		5010	150K	30–50	9		
		4010	200K	70–90	5	55	2–12
		3010	360K	150–200			
ZENON	Tubular Modules: i.d./area						
	24 mm/2.2 ft²	ZM-1	Low-medium	Waste water 100–200 gfd	4	40	2–12
	12 mm/7.0 ft²	ZM-7	Low-medium		4	40	2–12
	21 mm/10.5 ft²	ZM-8	Low-medium	Waste water 100–200 gfd	5.5	65	2–12

TABLE 28–3. Ultrafiltration Membranes and Membrane System Manufacturers.

SCT, Alcoa Separations Technology
Ceramic Membranes Dept., PO Box 113,
65001 Tarbes, France

A/G Technology Corp.
34 Wexford St.
Needham, MA 02194, USA

Amicon Corp.,
17, Cherry Hill Drive
Danvers, MA 01923, USA

Asahi-Kasei
Hibiya Mitsui Bldg.,
1-2 Yurakucho, 1-Chome
Chiyoda-Ku, Tokyo, Japan

Berghof Membranetechnik GmbH
Harreststrasse 1, D-7412 Eningen
U.A., Germany

Brunswick Technetics
Membrane Filter Products,
4116 Sorrento Valley Blvd.,
San Diego, CA 92121, USA

Carre Inc.,
109 Debra St., PO Box 1555
Seneca, SC 29678, USA

Daicel
3-8-1 Toranomon Bldg, Kasumigasaki
Chiyoda-Ku, Tokyo, Japan

De Danske Sukker Fabrikker
6 Tietgensvej, PO Box 149, DK-4900
Nakskov, Denmark

Desalination Systems Inc.
1107 West Mission Ave,
Escondido, CA 92025, USA

Dorr-Oliver Inc.
77 Havenmeyer Lane
Stanford, CT 06904, USA

Filtron Technology Corp.
500 Main St., PO Box 119
Clinton, MA 01510, USA

Fluid Systems
10054 Old Grove Rd.
San Diego, CA 92131, USA

W. L. Gore & Associates Inc.
Membrane Products Div.
101 Lewisville Rd, PO Box 1550
Elkton, MD 21921, USA

Hoechst Aktiengesellschaft
Werk Kalle, D-2600 Wiesbaden 1
Germany

Koch Membrane Systems
850 Main Street
Wilmington, MA 01887, USA

Kuraray Company Ltd
12–39, 1-Chome, Umeda, Kita-Ku,
Osaka 530, Japan

Millipore Corporation
80 Ashby Road
Bedford, MA 10730, USA

Nitto Denko Corp.
Membrane Division, Mori Bldg,
3rd Fl., 31, 5-7-2 Kojimachi,
Chiyoda-Ku, Tokyo 102, Japan

Hydronautics
8444 Miralani Dr.
San Diego, CA 92126, USA

Osmonics Inc.
5951 Clearwater Dr.
Minnetonka, MN 55343, USA

PCI Membrane Systems Inc.
Laverstoke Mill, Whitchurch
Hampshire RG28 7NR, England

Rhone Poulenc Tech-Sep
Rue Penberton, Saint-Maurice-de-
Beynost, BP 347-FO1703 Miribel
Cedex, France

Romicon Inc.
100 Cummings Park
Woburn, MA 01801, USA

Sartorius GmbH
Postfach 3243, Weender Landstrasse
94-108, D-3400 Goettingen, Germany

Schleicher & Schuell GmbH
D-3354, Dassel Kr.,
Einbeck, Germany

Schott Glaswerke
Geschaftsbereich Chemie
Produktgruppe Apparate-und
Anlagenbau, Postfach 2480,
Hattenbergstrasse 10, D-6500 Mainz 1
Germany

Wafilin BV
Bruchterweg 88, Post Box 5
7700 AA Hardenberg,
The Netherlands

Western Dynetics Inc.
1152 Tourmaline Dr.
Newbury Park, CA 91320, USA

the homogeneous polymer solution is converted to a porous polymer framework through exchange of the solvent with the precipitating nonsolvent. The asymmetric structure, which is responsible for the high fluxes, is composed of the thin, dense skin and a porous substructure. Wijmans et al. (1985) account for the formation of these two layers by two different phase separation phenomena:

1. Gelation for formation of the skin layer
2. Liquid-liquid demixing followed by gelation for the porous layer.

Gelation is induced by the concentration changes occurring on exchange of solvent for the precipitating liquid. The gelation of the bulk solution will be affected by the skin layer formation, which acts as a diffusion barrier (Strathmann and Kock 1977). The ternary phase diagram (polymer-solvent-nonsolvent) is a useful tool for understanding polymer membrane structure as a function of casting solution composition (Lonsdale and Kock 1977). Such phase diagrams have been obtained for casting systems of cellulose acetate, polysulfone (Wijmans et al. 1985), and polyphenylene oxide (Wijmans, Rutten, and Smolders 1985).

Membrane Modification

Hydrophobic polymers such as polysulfone or polyvinylidenefluoride may need to be modified to obtain higher fluxes, less fouling, etc. Some techniques are discussed by Cabasso (1980b). Methods of altering the surface chemistries are as follows:

1. *Reacting base polymer with hydrophilic pendent groups and then casting the membrane:* A common example is sulfonation of polysulfone with chlorosulfonic acid (Quentin 1973).
2. *Surface grafting of hydrophilic species on a previously made membrane:* This is difficult to do without damaging the base membrane by most chemical methods. Cabasso (1980b) describes sulfonation of a polysulfone hollow-fiber membrane using sulfuric acid. Polyvinylidenefluoride (PVDF) can be reacted with compounds such as cellulose ether or inositol in alkaline medium (Madsen 1989). Polysulfone treatment with an NH_3 plasma and further reactions with chlorinated dyes in alkaline medium have been described (Wolff, Steinhauser, and Ellinghorst 1988). Polyvinylchloride UF membranes have been made more hydrophilic by rf discharge treatment (Vigo, Nicchia, and Uliana 1988).

A simpler technique is to coat the base polymer membrane; for example, PVDF membranes can be coated with vinyl alcohol–vinyl acetate copolymer (Kasai and Koyama 1980). There are also reports of polyvinylpyrrolidone being entrapped in a polyetherimide matrix during casting; this increases the membrane hydrophilicity (Smolders 1989).
3. *Polymer blends:* Examples are blends of cellulose acetate with polystyrene and polyphenylene oxide phosphonate esters (Cabasso 1980b) and also a blend of poly-(vinylidenefluoride)/cationic polyelectrolyte (Mir 1983). Membranes made by mixing inorganics and polymers have also been demonstrated and show improved pressure resistance (Kulprathipanja et al. 1988).

Characteristics of Commercially Available Membranes

Typical data with commercially available polysulfone and cellulose acetate (CA) membranes (Nadir/Hoechst) are shown in Table 28–4. The same polymer can be made in a wide range of surface porosities as illustrated for polysulfone in Figure 28–1. All these membranes were made with the same concentration of polysulfone in the casting solution; the solvents and casting conditions have been varied. Macrovoids can be induced in the porous substructure as shown in Figure 28–2 for the case of a polysulfone hollow-fiber membrane. Macrovoids generally lead to less mass transport resistance but limit operational pressure and may lead to defects under compression.

Table 28–4 shows that CA membranes typi-

TABLE 28–4. Data for Polyethersulfone (PES), Polysulfone (PS), and Cellulose Acetate (CA) Membranes Tested with 1% Dextran (10K/MW) [reprinted from Nadir Separations (Hoechst), Product Brochure April 1988, with permission].

Membrane/Backing	Flux (L/m² · h)	Rejection (%)
PES-4/PP60	30–50	92–94
PES-8	70–12	87–92
PS-100/PP-100	300–500	80–90
PS-200 M/PP 100	500–1000	0–10
CA-5/PP 60	30–40	97–98
CA-100/PET 100	250–400	86–92

Note: PP = polypropylene, PET = polyester.

cally have better flux at equivalent rejections than polysulfone. However, polysulfone is necessary for many separations because of its higher stability (see Table 28–5). Polyethersulfone is used in place of polysulfone especially for food industry applications involving steam sterilization. The range of operational conditions in Table 28–5 may be exceeded during brief cleaning cycles. More extreme conditions may also be used in specific cases; for example, Koch (1988) recommends a pH range for polysulfone of 1 to 13 at 25°C and temperatures up to 90°C at a pH of 6 (Koch 1988). Compatibility of various membrane materials with different inorganic and organic media is shown in Table 28–6. This table is a compilation of the information from various membrane manufacturers. It is to be used only as a general guide and does not include the effects of other membrane properties such as morphology that also have an effect on stability at given conditions of pressure, temperature, solvent, etc. While most UF applications are aqueous, Table 28–7 indicates the feasibility of newer organic phase separations.

Flux and rejection data for other typical polymeric membranes in relation to polyethersulfone are shown in Table 28–8. Other sources of similar data are Cheryan (1986) and Olsen (1987).

Thin-film-composite (TFC) membranes of the type used in reverse osmosis or nanofiltration are not as common in true UF. Some examples of TFC UF membranes are the G series membranes from Desalination Systems Inc. consisting of cross-linked noncellulosic films on a polysulfone support.

Hvid et al. (1990) have discussed the preparation of polyurea/polyurethane TFC membranes with molecular weight cutoffs (based on dextrans) in the range from ~3,000 to 500,000 daltons. DiLeo and coworkers (DiLeo and Allegrezza 1991; DiLeo, Allegrezza, and Burke 1991) have developed an asymmetric compos-

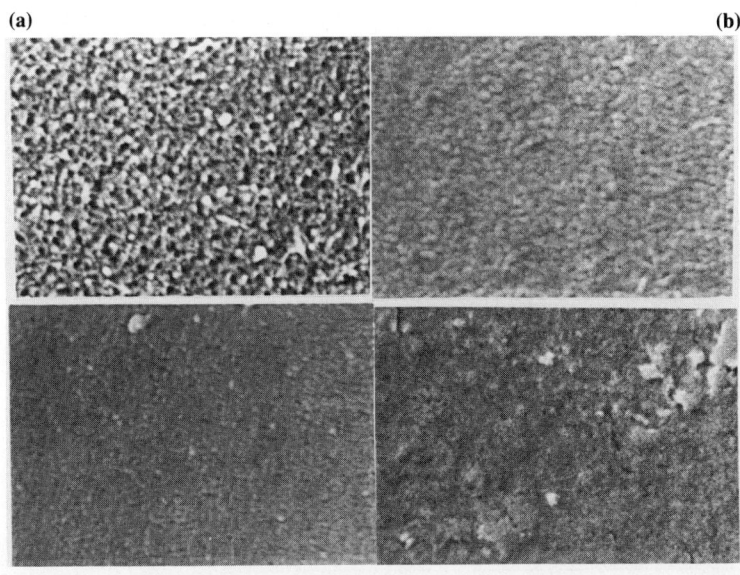

FIGURE 28–1. SEM micrographs of laboratory cast polysulfone membrane surfaces: (a) 17% PS in dimethylformamide, (b) 17% PS in *n*-methyl pyrollidone, (c) 17% PS in dimethyl acetamide, (d) 17% PS in *n*-MP + 2-pyrollidone, and (d) commercial membrane (Fluid Systems, UOP). Magnification 40,000×.

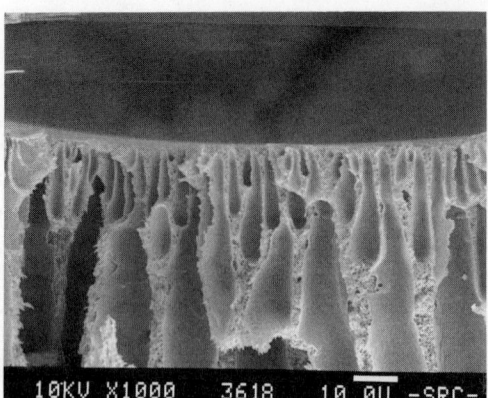

FIGURE 28–2. Inside edge cross section of polysulfone hollow fiber showing macrovoids. Magnification 1000×.

ite membrane, which has a surface skin with ultrafiltration separation properties and an intermediate zone supported by a porous membrane substrate (Durapore®). The intermediate zone has an average pore size smaller than the porous substrate (of about 0.2 μm). This membrane is particularly useful for selectively retaining virus from protein-containing solutions.

INORGANIC MEMBRANES

Alumina membranes with pore sizes in the UF range are made by Alcoa (~40- to 1000-Å pore size; monolith or tubular construction) and by Anotec (~250-Å disks). The former membranes are made by a "slip-casting" method in which repeated layers of uniform particles with decreasing sizes are coated on a porous α-Al_2O_3 support (Hsieh 1988; Hsieh, Bhave, and Fleming 1988; Bhave 1991). The porous support has 15-μm pores; the tube inside diameter is 4 or 6 mm (Munari et al. 1989). The water permeability of an ~40-Å pore size membrane is ~ 10 L/m^2 · h · bar. Due to their γ-Al_2O_3 skin layer, these UF membranes have less stability than their microfiltration (MF) counterparts in which both support and coating are of α-Al_2O_3. Due to formation of a secondary film layer through gelation, MF membranes have been used successfully for ultrafiltration applications such as protein recovery from skim or whole milk (Gillot and Garcera 1985).

Slip-casting techniques have also been used to coat specially designed honeycomb cordierite monoliths. UF membranes with pore diameters of 50 Å are being developed. This approach is important because of its potential for high membrane area packing density, which may bring the per area cost of inorganic membranes closer to those of polymeric materials.

Isotropic microporous glass membranes based on phase separation and acid leaching of the Na_2O-B_2O_3-SiO_2 system have been described for use in reverse osmosis (McMillan and Matthews 1976). Schnable, Langer, and

TABLE 28–5. Ranges of Operating Variables for Various Polymeric Membranes (reprinted from PCI Membrane Systems, Publication No. TPRO 78.2, with permission).

Polymer	pH Range	Nominal MWCO × 10^3	Maximum Pressure (bars)	Maximum Temperature (°C)
Polyethersulfone	2–12	4–9	30	70
		25	15	
Polysulfone	2–12	8–20	15	70
Polyacrylonitrile	2–10	25	10	60
Polyvinylidenefluoride	2–12	100, 200	10	70
Cellulose acetate[a]	3–6	2	25	30
		7	20	

[a]The bacterial resistance of CA is also worse than the other polymers.

TABLE 28–6. Chemical Compatibility of Polymeric Membranes.

Chemical	Membrane Based on		
	Cellulose Triacetate	Polysulfone	Polyvinylidene-fluoride
Hydrochloric acid, conc	N	R	R
Sodium hydroxide, 3N	N	R	R
Amyl alcohol	R	R	R
Methanol	L	R	R
Isopropanol	R	R	R
Ethyl acetate	N	N	L
Diethyl ether	R	R	R
Dioxane	N	N	N
Tetrahydrofuran	N	N	R
Glycerine	R	R	R
Ethylene glycol	L	R	R
Aromatics (benzene, toluene, xylene)	R	N	R
Hexane	R	R	R
Carbon tetrachloride	L	N	R
Chloroform	N	N	R
Freon TF	R	R	R
Acetone	N	N	N
Formaldehyde, 37%	R	R	R

Notes: R = resistant in normal usage. N = not resistant; may shrivel, swell, or dissolve. L = limited exposure in noncritical applications.

Breitenbach (1988) have described glass membranes with pore sizes in the range of 57 to 113 Å used to fractionate proteins. UF membranes based on SiO_2 are also made by Schott Glass and Fuji Filters.

Zirconia membrane systems have been commercialized by several vendors; some of these are "dynamically" formed membranes. A variety of supports for the ZrO_2 have been used: carbon (Gaston County Dyeing Machine Company and Societe de Fabricaton d'Elements Catalytiques), stainless steel (Carre/Du Pont), and alumina (Dynaceram/TDK). The last two are dynamically formed. Carbon-based membranes have been recently commercialized by GFT.

TABLE 28–7. Hydrocarbon Rejections (GPC Measurement) by Polysulfone Membrane as a Function of Feed Solution.

Feed	Molecular Weight	Rejection (%)
1% n-C_{24}	338	0
15% light crude	400	2
15% heavy crude	400	60
	1,000	95
	20,000	99
15% light crude + 0.1% heavy crude	400	40

Note: The conditions were: stirred cell; feed solution in n-pentane and 150 psig with permeate at 30 psig.

MEMBRANE CHARACTERIZATION

In addition to the membrane material itself, the membrane porosity/morphology and its surface properties are important parameters determining its use in separation. Techniques to characterize these two properties are discussed below.

Membrane Porosity and Morphology

Most UF membrane manufacturers characterize their products by a single pore size or molecular weight cutoff value. These values are usually obtained by measuring the rejection of various macromolecules of increasing hydrodynamic diameter or molecular weight. It is understood that this single value is not an adequate measure

TABLE 28–8. Data for Various Membranes with 2% Polyvinylpyrrolidone [49K MW, All Membranes Have the Same Polypropylene Backing (PP 100)] [reprinted from Nadir Separations (Hoechst), Product Brochure April 1988, with permission].

Polymer/Membrane	Flux (l/m² · h)	Rejection (%)
Polyethersulfone		
PES-20	100–150	94–96
PES-50	150–250	70–80
Aromatic polyamide		
PA-20	150–250	94–96
Polyvinylidenefluoride (hydrophobic)		
PVDF-30N	250–400	70–80
Polyacrylonitrile		
PAN-30	500–1000	75–85

of the membrane; however, it is useful for preliminary selection. The relation between membrane flux and rejection characteristics and its average pore size was discussed in Chapter 27 [Eqs. (27–2) and (27–3)].

Most UF membranes have a pore size distribution; however, the effect of this heteroporosity on rejection profiles may be relatively small in actual application. The pore size distribution may be a reason for the varying mean pore sizes that are measured by different methods. Membranes with uniform pore sizes have been designed; examples are the track-etched Nuclepore membrane or the crystalline bacterial-cell-envelope layer membrane (Sara et al. 1988).

The major techniques available to determine membrane pore size are:

1. Bubble-pressure breakthrough (BP)
2. Mercury porosimetry (MP)
3. Solute retention challenge (SR)
4. Electron microscopy (SEM or TEM)
5. Adsorption-based methods
6. NMR measurements.

Methods such as BP, MP, or SEM work best for pore sizes greater than 100 Å (0.01 μm), although with special techniques, pores as small as 10 Å can be measured.

Bubble-Pressure Breakthrough

In general terms, this method measures the pressure required to force one immiscible fluid through the pores of a membrane previously filled with a second immiscible fluid. This was originally done by placing a water-filled membrane with air impingement from below. Bubbles of air penetrate the membrane into an overlaying water layer. The largest pores open at the lowest pressure; thus, by slowly increasing the air pressure (1 bar/min) and monitoring air passage, a pore size distribution can be estimated. When all the pores are open, further flow increases are proportional to pressure increase, at which time, the measurement can be terminated (see Figure 28–3).

Practical problems exist in that the high water/air surface tension (73 dynes/cm) requires relatively high pressures for pore sizes in the UF range. The pressure required can be reduced by using alcohols or a water-alcohol mixture as the imbibed fluid instead, isopropanol or isobutanol being common. The bubble point values also depend on parameters such as the rate of pressure increase, polymer material, capillary length, and temperature (through its effect on viscosity and interfacial tension). Bechhold et al. (1931) have developed the following equation to account for these parameters:

$$d_p = \frac{4\gamma}{p}\left[1 + \frac{2l}{\gamma}\left(\frac{dp}{dt}\frac{\eta_1 + \eta_2}{2}\right)^{1/2}\right], \quad (28\text{–}1)$$

where γ is the interfacial tension, l is the capillary length, and η_i refers to the viscosities of the two immiscible fluids. Microprocessor-controlled instruments are currently available

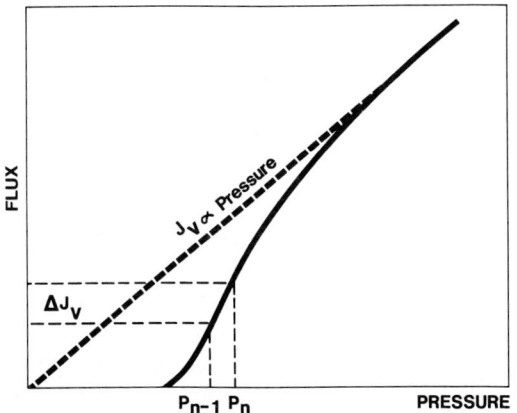

FIGURE 28–3. Schematic flux versus pressure plot for bubble pressure/solvent permeability measurement.

that run air bubble pressure tests as part of a membrane integrity check. Methods for determining pore sizes by bubble pressure techniques are described in ASTM F316-80 and British Standard 6410.

Pore size distribution can also be estimated by using an immiscible liquid to expel imbibed water. This technique is described (Capannelli, Vigo, and Munari 1983; Munari et al. 1989) for situations in which the liquid pair (e.g., isobutanol-water) is chosen to minimize the interfacial tension. Pore sizes as low as 10 Å and as high as 700 Å could be estimated.

A step-by-step procedure to generate the pore size distribution using the flux data generated by a bubble pressure test is described by Kesting (1971). If ΔJ_n is the incremental flux generated by a pressure increase ($\Delta p_n = p_n - p_{n-1}$), the number of pores n_n is:

$$n_n = \frac{\Delta J_n [(p_n + p_{n-1})/2]^3 \eta l}{2\pi\gamma^4} \quad (28\text{–}2)$$

corresponding to a pore diameter of

$$d_p = \frac{4\gamma}{(p_n + p_{n-1})/2}. \quad (28\text{–}3)$$

Several caveats should be mentioned concerning the use of the bubble pressure test. First, it is only strictly applicable when the pore is much larger than the solvent molecule. Second, the method assumes complete wetting of the pore with the liquid. Third, it is based on the assumption that the pores are cylinders aligned perpendicular to the surface; it is not known what the effect is of geometry and shape factors. Finally, for small pores, the high pressures required may compact the membrane, changing the pore size and structure.

Mercury Porosimetry

This method is a variation of the bubble pressure method where mercury is used to fill a dry membrane. The pore size is given by

$$d_p = -4\gamma \cos\theta/p, \quad (28\text{–}4)$$

where θ is the mercury-polymer contact angle. The surface tension of mercury is 485 dynes/cm; hence, high pressures need to be used that may compress the membrane and distort the pore mouth opening (Liabrastre and Orr 1978). A further disadvantage is that the membrane is tested dry.

Solute Retention

The most common method of characterizing the pore size of a given UF membrane is by measuring the rejection of various solutes of increasing molecular weights or hydrodynamic sizes. Most manufacturers designate a molecular weight cutoff (MWCO), which is the molecular weight above which rejections are expected to be higher than 90%.

Standard testing procedures are not yet defined. Various solutes have been used for rejection testing: salts, sugars, purified proteins (albumin, gamma globulin), dextrans, and polyethyleneglycols. By correlating molecular weight (MW) versus solute size as indicated in Figure 28–4, it is possible to estimate the pore size by determining the rejections for various solutes. The construction of such a MWCO versus membrane pore size plot has been discussed by Sarbolouki (1982). Such plots have to be used cautiously since the rejection depends greatly on molecular conformation. Globular proteins are rejected easier than branched polysaccharides or flexible polymers (Porter 1979; Cheryan 1986); for example, a

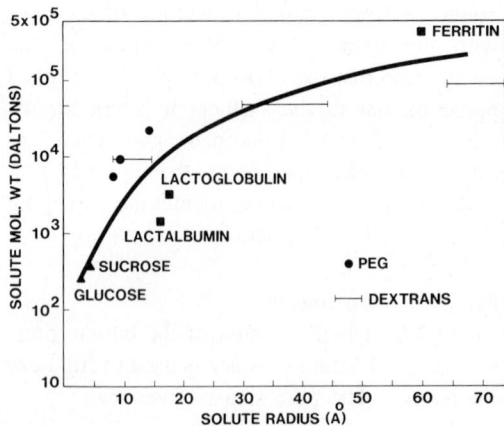

FIGURE 28–4. Molecular weight of various solutes versus the reported solute radius.

membrane that rejects insulin (MW 5700) allows passage of a dextran with 10,000 MW.

Membranes should be tested in conditions where gel formation/concentration polarization is minimized. This can be done by testing at low solute concentrations (~0.1wt.%), high agitation or feed flow rates, and low pressures. Membrane samples can be tested in sequence, each with a different solute. A simpler procedure is described by Cooper and Van Derveer (1979) and Schock, Miguel, and Birkenberger (1989) where the rejection of dextran was measured using GPC. A similar method using maltodextrins was used by Klein, Feldhoff, and Turnham (1983). They also described the correction of the solute rejection to account for boundary layer effects. This correction is based on Eq. (27–32) and the appropriate correlations for the mass transfer coefficient [Eqs. (27–16) to (27–26)]. The danger in this approach is that rejected high molecular weight species may cause higher rejections to be observed for smaller species due to formation of a "dynamic" membrane. This phenomenon is illustrated in Table 28–7.

Zeman (1983) has discussed a similar case in which dextran rejections were higher after the membrane was used in the ultrafiltration of whey. Zeman (1983) also discusses membrane adsorption effects on the observed rejection of proteins.

Electron Microscopy

Scanning electron microscopy (SEM) can be used directly only with dry membranes to measure total surface porosity and pore sizes (see Figure 28–1). This method is limited in resolution power to pore sizes of greater than 50 Å. Transmission electron microscopy (TEM) or scanning transmission electron microscopy (STEM) has better resolution but requires more sample preparation, embedding, and microtoming. SEM combined with image analyses has been used to quantify surface porosity (Martinez-Villa, Arribas, and Tejerina 1988; Vivier, Pons, and Portala 1989).

Adsorption-Based Methods

Analysis of pore size and distribution by gas adsorption/desorption is based on the Kelvin equation, which relates the reduced vapor pressure p of a liquid from a curved surface to the equilibrium vapor pressure p^0 of the same liquid from a plane surface. The curvature of the concave meniscus for vapor condensed in a pore depends on its pore size r_p, so that at a given pressure, liquid condenses in smaller pores before larger ones. The Kelvin equation is given by

$$RT \ln \frac{p}{p^0} = \frac{-2\gamma \tilde{V}_l}{r_p - \Delta} \cos \theta, \quad (28\text{–}5)$$

where R is the gas constant, T is the absolute temperature, γ, θ, and $\tilde{V}_l$ are the surface tension, contact angle, and molar volume of the condensed (liquid) vapor, respectively, and Δ is the thickness of the adsorbed layer of vapor in the pores. The calculation procedure has been described by Dollimore and Heal (1964). The radius can also be estimated from BET measurements of the total surface area and pore volume by

$$r_p = 2 \cdot \text{pore volume/surface area.} \quad (28\text{–}6)$$

The adsorption/desorption method can be used to characterize pores in the size range of 7 to 100 Å. Disadvantages are that the sample must be dried before examination and it is not possible to exclude the contribution from "dead-end" pores that do not contribute to transport characteristics.

Mey-Marom and Katz (1986) have described a modified permeability/adsorption method that measures only the active pores. A water-saturated membrane was gradually dried by lowering the humidity of the surrounding air, with the largest pores drying first. At each equilibrium humidity level, low-pressure permeability measurements were used to estimate the volume of the desired pores. This information was used to estimate the pore size distribution in the membrane.

Thermoporometry is another new method suggested by Smolders and Vugteen (1985) based on the fact that the solidification point of condensed vapor in the pore is also a function of the interface curvature and thereby can be related to the pore size. The degree of undercooling ΔT (°K), observed on heating a frozen sample of membrane saturated with water, was related to the pore radius (Å) by

$$r_p = 6.8 - (323.3/\Delta T). \quad (28\text{--}7)$$

This method can be used for pore sizes in the range from 20 to 200 Å.

NMR Measurements

Glaves and Smith (1989) have demonstrated the determination of pore sizes in water-saturated membranes using nuclear magnetic resonance (NMR), spin-lattice relaxation measurements. The spin-lattice relaxation in a magnetic field is faster for liquid near the pore wall compared to liquid away from the wall. Combined with fast diffusional exchange between these two populations, an average relaxation time constant T_1 is obtained for every pore size present:

$$1/T_1 = \alpha + \beta/r_p. \quad (28\text{--}8)$$

The distribution of T_1 values is obtained from the measured fluid magnetization vectors by deconvolution techniques. The major difficulty in this technique is the determination of the constants α and β in Eq. (28–8) to relate T_1 to the pore radius r_p. This can be done by either calibrating the NMR signal using membranes of the same material with known pore size distribution or by surface area and pore volume measurements to estimate the fraction of liquid on the pore surface and that in the bulk pore.

This technique can be used to characterize pore sizes in the range of 50 Å to 5 μm without the necessity of assuming a cylindrical geometry.

Surface Properties

Conventional polymer techniques such as electron spectroscopy for chemical analyses (ESCA), attenuated total reflection infrared spectroscopy (ATR-IR), and contact angle measurements are used to probe membrane surfaces (Fonteyn, Bijsterbosch, and Van't Riet 1987; Oldani and Schock 1989). By varying the incident or takeoff angle, surfaces can be probed to varying penetration depths by both ATR-IR or ESCA. Contact angles can be measured by various techniques:

1. Direct observation of a liquid drop on a flat-sheet membrane (accuracy is 2 to 5 deg).
2. Tilt-rolloff method for flat sheets (±2.5 deg).
3. Critical wetting tension using a calibrated set of liquids of varying surface tensions.
4. Wilhelmy plate method or modified method for flat-sheet or tubular membranes, respectively (±0.5 deg).

Surface properties are measured to obtain intrinsic membrane properties (after washing away plasticizers, etc.), to monitor effects of membrane modification by chemical treatment or coatings, or to detect foulants. These properties are further discussed in Chapter 29.

ACKNOWLEDGMENTS

The generous assistance provided by J. L. Short (Koch Membrane Systems), Fritz-Feo Grabley (Nadir Separations/Hoechst), D. R. Pearson (PCI Membrane Systems), and B. S. Parekh (Millipore) in collecting information is deeply appreciated.

NOTATION

General Notation

See the General Notation section at the beginning of this handbook.

Special Notation

l capillary length, L
R gas constant, ML^2/t^2T mol

Greek Letters

γ interfacial tension, M/t^2 or E/L^2
Δ thickness of the adsorbed layer of vapor in the pores, L
η viscosity, M/Lt
θ contact angle for polymer with mercury

REFERENCES

Bechhold, H., M. Schlesinger, K. Silbereisen, L. Maier, and W. Nurnberger. 1931. Pore diameters of ultrafilters. *Kolloid Z.* 55:172–198.

Bennasar, M., B. Tarodo de la Fuente, J. Gillot, and D. Garcera. 1984. *Filtra 84*, Conf. Societe Francaise de Filtration, Paris.

Bhave, R. R. 1991. Inorganic Membranes: Synthesis, Characteristics, and Applications. New York: Van Nostrand Reinhold.

Cabasso, I. 1980a. Hollow-fiber membranes. In *Kirk-Othmer Encyclopedia of Chemical Technology*, 3rd ed., Vol. 12, pp. 492–517. New York: John Wiley & Sons.

Cabasso, I. 1980b. Practical aspects in the development of a polymer matrix for ultrafiltration. In *Ultrafiltration Membranes and Applications*, ed. A. R. Cooper. New York: Plenum Press.

Capannelli, G., F. Vigo, and S. Munari. 1983. Ultrafiltration membranes—characterization methods. *J. Membr. Sci.* 15:289–313.

Cheryan, M. 1986. *Ultrafiltration Handbook*. Lancaster, PA: Technomic Publishing Co.

Cooper, A. R., and D. S. Van Derveer. 1979. Characterization of ultrafiltration membranes by polymer transport measurements. *Sep. Sci. Technol.* 14:551–556.

DiLeo, A. J., and A. E. Allegrezza, Jr., 1991. Validatable virus removal from protein solutions. *Nature* 351:420–421.

DiLeo, A. J., A. E. Allegrezza, Jr., and E. T. Burke. 1991. Membrane, process and system for isolating virus from solution. U.S. Patent 5, 017, 292.

Dollimore, D., and G. R. Heal. 1964. An improved method for the calculation of pore size distribution from adsorption data. *J. Appl. Chem.* 14:109–114.

Fonteyn, M., B. H. Bijsterbosch, and K. Van't Riet. 1987. Chemical characterization of ultrafiltration membranes by spectroscopic techniques. *J. Membr. Sci.* 36:141–145.

Gillot, J., and D. Garcera. 1985. New ceramic filtration media for tangential microfiltration and for ultrafiltration. *Informations Chemie* 261:193–197.

Glaves, C. L., and D. M. Smith. 1989. Membrane pore structure analysis via NMR spin-lattice relaxation measurements. *J. Membr. Sci.* 46:167–184.

Hsieh, H. P. 1988. Inorganic membranes. *AIChE Symp. Ser.* 84 (261):1–18.

Hsieh, H. P., R. R. Bhave, and H. L. Fleming. 1988. Microporous alumina membranes. *J. Membr. Sci.* 39:221–241.

Hvid, K. B., P. S. Nielsen, and F. F. Stengaard. 1990. Preparation and characterization of a new ultrafiltration membrane. *J. Membr. Sci.* 53:189–202.

Kasai, M., and N. Koyama. 1986. Hydrophilic porous membranes. Japanese Patent 61,161,103 A2.

Kesting, R. E. 1971. *Synthetic Polymeric Membranes*. New York: McGraw-Hill Book Co.

Klein, E., P. Feldhoff, and T. Turnham. 1983. Molecular weight spectra of ultrafilter rejection: I. a simple HPLC method using homologous maltodextrin polymers. *J. Membr. Sci.* 15:15–26.

Koch Membrane Systems. 1988. *The Koch Difference* (5/88).

Kulprathipanja, S., E. W. Funk, S. S. Kulkarni, and Y. A. Chang. 1988. Separation of a monosaccharide with mixed-matrix membranes. U.S. Patent 4,735,193.

Liabrastre, A. A., and C. Orr. 1978. An evaluation of pore structure by mercury penetration. *J. Colloid Interface Sci.* 64:1–18.

Lloyd, D. R., and T. B. Meluch. 1985. Selection and evaluation of membrane materials for liquid separations. In *Materials Science of Synthetic Membranes*, ed. D. R. Lloyd, ACS Symp. Ser. No. 269. Washington, DC: American Chemical Society.

Lonsdale, H. K., and A. J. Kock. 1977. The formation mechanism of phase inversion membranes. *Polym. Eng. Sci.* 25(17):1074.

Madsen, R. F. 1989. Applications of membrane filtration in the food and biochemical industry. In *Indo-EEC Membrane Workshop*. New Delhi: Department of Science and Technology.

Martinez-Villa, F., J. I. Arribas, and F. Tejerina. 1988. Quantitative microscopic study of surface characteristics of track-etched membranes. *J. Membr. Sci.* 36:19–30.

McMillan, P. W., and C. E. Matthews. 1976. Microporous glasses for reverse osmosis. *J. Mater. Sci.* 11:1187–1199.

Mey-Marom, A., and M. G. Katz. 1986. Measurement of active pore size distribution of microporous membranes: a new approach. *J. Membr. Sci.* 27(2):119–130.

Mir, L. 1983. Positive-charged ultrafiltration membrane for the separation of cathodic/electrodeposition paint compositions. U.S. Patent 4,412,922.

Munari, S., A. Bottino, P. Moretti, G. Capannelli, and I. Becchi. 1989. Permoporometric study on ultrafiltration membrane. *J. Membr. Sci.* 41:69–86.

Oldani, M., and G. Schock. 1989. Characterization of ultrafiltration membranes by IR, ESCA and contact angle measurements. *J. Membr. Sci.* 43:243–258.

Olsen, O. J. 1987. Membrane filtration as a tool in biotechnology downstream processing. *Desalination* 62:329–339.

Porter, M. C. 1979. Membrane filtration. In *Handbook of Separation Techniques for Chemical Engineers*, ed. P. A. Schweitzer. New York: McGraw-Hill Book Co.

Pusch, W., and A. Walch. 1982. Synthetic membranes—preparation, structure and application. *Angew. Chem. Int. Ed. Engl.* 21:660–685.

Quentin, J. P. 1973. Sulfonated polyarylether sulfones. U.S. Patent 3,709,841.

Ronner, J. A., S. Groot-Wassink, and C. A. Smolders. 1989. Investigation of liquid-liquid demixing and aggregate formation in a membrane system by means of PICS. *J. Membr. Sci.* 42:27–45.

Sara, M., C. Manigley, G. Wolf, and U. B. Sleytr. 1988. Isoporous ultrafiltration membranes from bacterial cell envelope layers. *J. Membr. Sci.* 36:179–186.

Sarbolouki, M. 1982. Properties of asymmetric polyimide ultrafiltration membranes: pore size and morphology characterization. *Sep. Sci. Technol.* 17:381. (See also 1984. *J. Appl. Polym. Sci.* 29:743–753.)

Schnable, R., P. Langer, and S. Breitenbach. 1988. Separation of protein mixtures by Bioran porous glass membranes. *J. Membr. Sci.* 36:55–66.

Schock, G., A. Miguel, and R. Birkenberger. 1989. Characterization of ultrafiltration membrane cutoff determination by gel permeation. *J. Membr. Sci.* 41:55–67.

Smolders, C. A. 1989. Polymeric materials for membrane separation processes: a rationalized approach. In *Indo-ECC Membrane Workshop*. New Delhi: Department of Science and Technology.

Smolders, C. A., and E. Vugteen. 1985. New characterization methods for asymmetric ultrafiltration membranes. In *Materials Science of Synthetic Membranes*, ed. D. R. Lloyd, *ACS Symp. Ser. No. 269*, pp. 327–338. Washington, DC: American Chemical Society.

Strathmann, H., and K. Kock. 1977. The formation mechanism of phase inversion membranes. *Desalination* 21:241–255.

Vigo, F., M. Nicchia, and C. Uliana. 1988. PVC ultrafiltration membranes modification by high frequency discharge treatment. *J. Membr. Sci.* 36:187–199.

Vivier, H., M. H. N. Pons, and J. A. F. Portala. 1989. Study of microporous membrane structure by image analysis. *J. Membr. Sci.* 46:81–91.

Wijmans, J. G., J. Kant, M. H. V. Mulder, and C. A. Smolders. 1985. Phase separation phenomena in solutions of polysulfone in mixtures of a solvent and a nonsolvent: relationship with membrane formation. *Polymer* 26:1539–1545.

Wijmans, J. G., H. J. Rutten, and C. A. Smolders. 1985. Phase separation phenomena in solutions of poly (2,6 dimethyl-1,4 phenylene oxide) in mixtures of trichloroethylene, 1-octanol and methanol. *J. Polym. Sci., Polym. Phys.* 23:1941–1955.

Wolff, J., H. Steinhauser, and G. Ellinghorst. 1988. Tailoring of ultrafiltration membranes by plasma treatment and their application for the desalination and concentration of water solutions of organic substances. *J. Membr. Sci.* 36:207–214.

Zeman, L. J. 1983. Adsorption effects in rejection of macromolecules by ultrafiltration membranes. *J. Membr. Sci.* 15:213–230.

29

Module and Process Configuration

Sudhir S. Kulkarni
National Chemical Laboratory

Edward W. Funk and Norman N. Li
Allied Signal

MODULES
   Spiral-Wound Modules
   Plate-Frame Modules
   Tubular and Hollow-Fiber Membranes
   Monolithic Tubular Modules
CONTROL OF MEMBRANE FOULING
   Identification of Foulants and Feed Pretreatment
   Polarization and Fouling Control Strategies
PROCESS CONFIGURATIONS
   Batch Concentration
   Feed and Bleed
   Diafiltration
NOTATION
REFERENCES

MODULES

Several membrane configurations are available both commercially (see Figure 29–1) and at the laboratory level. Polymeric membranes can be cast or extruded, as either flat sheets or in cylindrical geometry. Flat-sheet membranes are used in spiral-wound or plate-and-frame (or "plate-frame") modules while other modules are of the tubular or hollow-fiber type. Inorganic membranes are used commercially only in tubular or monolith form; flat disks are available for laboratory use.

A brief comparison of various modules is given in Table 29–1. The choice of membrane module to be used depends on the ability to handle the fouling or plugging (suspended solids) characteristics of the retentate stream.

Other criteria are the holdup volume, which is particularly important for concentration applications, and the cost of the required membrane area.

More polymers are available as flat sheets than in the tubular geometries. [The most used membranes—polysulfone (PS), polyethersulfone (PES), polyvinylidenefluoride (PVDF), and cellulose acetate (CA)—are available in both.] The flux, the packing density (membrane area/volume), and the module preparation cost are the other important parameters.

Modules (see Table 29–1) can be classified as turbulent-flow wide-bore tubular membranes or as laminar-flow thin-channel devices (spiral-wound, hollow-fiber, plate-frame). The feed channel clearance has to be sufficient to resist plugging by suspended solids.

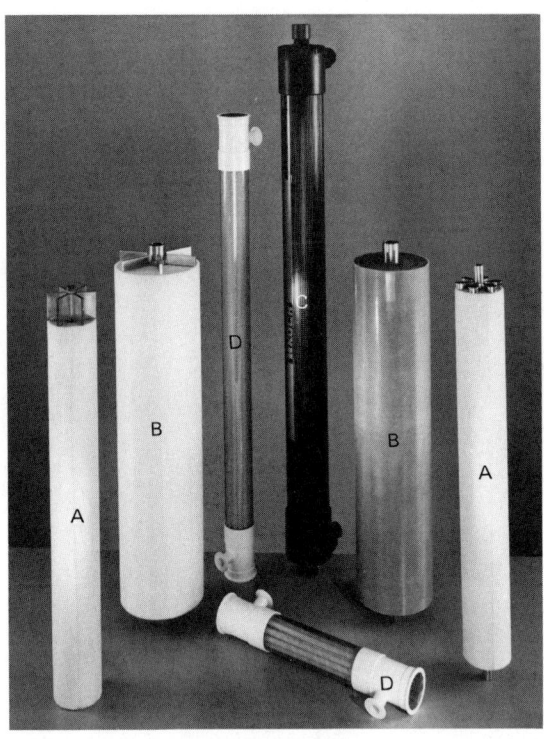

FIGURE 29-1. Commercially available tubular and spiral-wound UF modules: (a) 4-in. spiral-wound membranes (sanitary design); (b) 8-in. spiral-wound membranes; (c) spiral module; and (d) tubular modules containing 0.5-in. tubes of different lengths. (Courtesy of J. L. Short, Koch Membrane Systems.)

Theoretical models (Belfort 1989) have been developed for estimating flow profiles, pressure profiles, and wall shear for rectangular slits (Doughty and Perkins 1970), which approximate plate-frame and tubular (Terrill 1983) geometries. Permeate flow through the porous membrane is usually modeled as uniform suction through the walls, though a variable suction velocity profile has also been modeled (Terrill 1983). Spiral-wound geometry has been modeled as flow through the annulus between concentric cylinders (Chatterjee and Belfort 1986). The effect of turbulence promoters such as the feed spacer has been studied by Levy and Earle (1990).

Other conditions being equal, membranes in tubular modules are more susceptible to particulate capture than those in rectangular slit geometry. In principle, particle trajectory calculations can help design the optimum channel height-to-module length ratio in order to avoid fouling. Another success story is the various theories (Belfort 1989) that explain the different fouling characteristics of colloids versus macromolecules or low molecular weight solutes. Also, a predictive model is available for steady-state laminar flow through porous ducts, valid for any value of Reynolds number (Belfort 1988, 1989). Unfortunately, currently available modules have not been optimized based on theoretical models. However, similar models have been useful for the development of experimental modules based on pulsatile flows, rotating devices, etc. The new experimental designs are reviewed later in this chapter. The commercially available modules are discussed below. Table 29–1 lists some suppliers of various module configurations.

Spiral-Wound Modules

This is the best configuration when feed conditions permit its use. The basic design consists of a membrane sandwich around a permeate carrier, wound as a spiral on a permeate tube along with a feed channel spacer. This module requires low capital cost but is relatively easily plugged. Standard feed spacers are only 0.25 to

TABLE 29-1. Comparison of UF Modules.

Type	Feed Channel Height (mm)	Typical Area Packing Density (m²/m³)	Feed Velocity (m/s)	Reynolds Number[a]	Membrane Cost ($/ft²)	Holdup Volume	Typical Suppliers
Hollow-fiber	1–2.5	1200	0.5–3.5	10–1000	20–30	Very low	Amicon Romicon
Narrow-tube/monolith[b]	3–8	200	—	—	—	Moderate	Alcoa
Wide-tube	10–25	60	2–6	10000–30000	8–70	High	Koch PCI Wafilin
Plate-frame	0.3–1	300	0.7–2.0	100–6000	8–20	Moderate	DDS Dorr-Oliver Millipore
Spiral-wound	0.5–1	600	0.2–1.0	100–1000[c]	4–20	Low	Fluid Systems Koch Osmonics

[a]Transition from laminar to turbulent flow occurs at Re = 4000 for porous tubes and slits compared to 2100 for nonporous tubes. Thus, tubular modules are operated in the turbulent regime, while the others are normally in laminar flow.
[b]Honeycomb monolith modules are being developed that will have feed channel heights of 2 to 3 mm.
[c]Although Re is low, turbulence is induced by the feed space mesh.

0.50 mm thick and require feed turbidities to be less than 1 NTU (Nephelos turbidity unit, based on the intensity of scattered light from the solution, as described by Keily and Rogers 1955). Various manufacturers offer modules for more turbid feeds based on increasing the effective feed channel height by two techniques:

1. *Loose-wrap module*: This module does not use either the usual outer wrap or concentrate seal. As a result, the spiral unwinds slightly and expands to fit the pressure housing when exposed to the feed. This results in higher effective channel height. Another advantage of this module is that eliminating the concentrate or brine seal eliminates dead flow areas resulting in a "sanitary" design (Mooney and Light 1985).
2. *Clear-channel spacer*: Standard spirals use a diamond-mesh polypropylene netting (10 to 20 mils thick) as the feed spacer. This design is expected to promote local turbulence but suffers from plugging by colloidal particles and fouling. Thicker spacers can be used where the effective channel height may be 1 mm. Also a "clear channel" (Paulson, Phelps, and Gach 1987) spacer may be used that allows higher flow rates at normal pressure drop.

Plate-Frame Modules

Plate-frame modules have been extensively discussed by Madsen (1977). The heart of the plate-frame module is the plastic support plate that is sandwiched between two flat-sheet membranes. The membranes are sealed to the plate by either gaskets with locking devices, glue, or are directly bonded. The plate is internally porous and provides a flow channel for the permeate, which is collected from a tube on the side of the plate. Ribs or grooves on the face of the plate provide a feedside flow channel. The feed channel can be a clear path with channel heights from 0.3 to 0.75 mm. The higher channel heights are necessary for high-viscosity feeds; reduction in power consumption of 20 to 40% can be achieved by using a 0.6-mm channel compared to a 0.3-mm channel (Madsen 1989). Alternatively, retentate separator screens (20 or 50 mesh, polypropylene) can be used as in the spiral-wound modules.

Commercial plate-frame units are usually horizontal with the membrane plates mounted vertically. They can be run with each plate in parallel or in combinations of sets of parallel plates in two or three series. Laboratory modules (e.g., from Millipore Corporation) are also available as preformed stacks of up to 10 plates. UF modules are usually restricted to feedside pressures of 5 to 20 bars.

Tubular and Hollow-Fiber Membranes

Membranes with cylindrical geometry can be further differentiated based on the bore diameter:

1. Hollow fibers, 0.5 to 2.5 mm
2. Narrow-bore tubes, 3 to 8 mm
3. Wide-bore tubes, 10 to 25 mm.

Hollow-fiber membranes are spun separately, bundled, and potted into cartridge housings. An excellent review of hollow-fiber membrane technology has been given by Cabasso (1980). Tubular membranes are usually cast in place within supporting tubes or, in the case of the 25-mm tubes made by PCI Membrane Systems, can also be inserted into individual tubes. Examples of various tubular membranes can be seen in Figure 29-2. Feed normally flows through the bore of the tube with a few exceptions such as the Ultradyne (Western Dynetics) design in which the membrane is coated on the porous tube exterior and feed flows through the annulus between membrane and external holder. The bore size required is determined by the size of the largest particulate in the feed; the tube inside diameter should be at least 10 times the particle diameter.

Tubular modules have several disadvantages: low surface area/volume ratio, high flow rates requiring large energy consumption, and high holdup volume. However, the ease of cleanup and ability to withstand plugging can overcome these drawbacks.

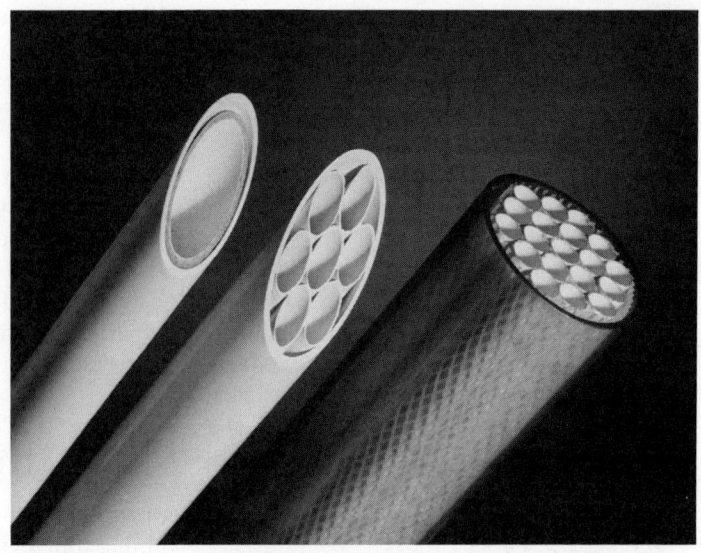

FIGURE 29–2. Cutaway view of various tubular systems. Tube diameters are 0.5 to 1 in. Membranes are cast into support tubes within PVC housings. (Courtesy of J. L. Short, Koch Membrane Systems.)

Monolithic Tubular Modules

Inorganic ultrafiltration membranes (pore size < 0.1 μm) are available in disk, tubular, or honeycomb monolithic form (Hsieh 1988). The monolithic multitube configuration is illustrated in Figure 29–3. The monolith provides higher mechanical strength and area packing density compared to tubular modules; this partially addresses the current problems of alumina membranes; namely, the requirement of careful handling to prevent breakage and high cost/membrane area. A new honeycomb device with very high packing density (approaching that of hollow-fiber modules) is understood to be under development.

CONTROL OF MEMBRANE FOULING

The choice of membrane, module, and process configuration are all important to achieve a high degree of separation without the membrane productivity being unduly hampered by fouling. The fundamental aspects of the fouling phenomena were covered in Chapter 27 where flux and rejection-based experiments to characterize fouling were also described. In addition, surface science and microscopy techniques that have been commonly used for identification of possible foulants were briefly covered. In this section, some operational considerations *vis-à-vis* fouling are reviewed.

Identification of Foulants and Feed Pretreatment

SEM can usually confirm the gel layer hypotheses visually. Fouling can be seen with deposited proteins, crystallized inorganic materials, polysaccharides, or heavy hydrocarbons. The thickness or density of the foulant cake on the membrane surface can be estimated by electron microscopy.

ATR-IR, ESCA, and SEM/EDX have been routinely used to characterize foulants on membrane surfaces. Oldani and Schock (1989) have described the use of ATR-FTIR, ESCA, and contact angle measurements in studying the fouling of BSA and dextran on bisphenol-A polysulfone and polyethersulfone membranes. Hanemaaijer et al. (1989) have used radioactive Ca^{45} and P^{32} to identify fouling mechanisms.

It is useful to analyze foulants after some initial separation steps. Figures 29–4(a) and 29–4(b) show the water soluble and water insoluble fractions of the foulants from UF of hydrolyzed starch. The water soluble fraction is mostly saccharides while in the insoluble fraction it is now possible to detect problem species such as

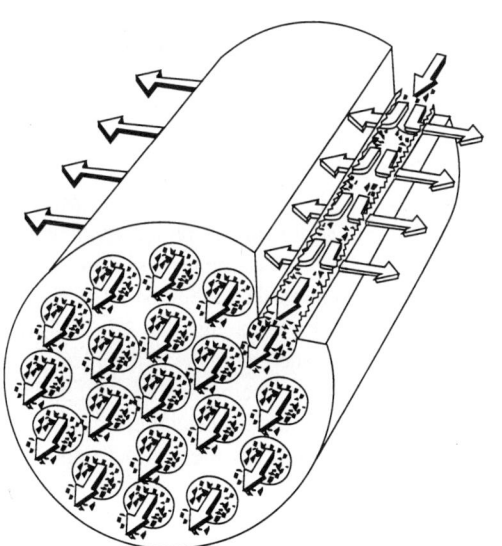

FIGURE 29-3. Ceraflo (Norton) ceramic filter element containing nineteen 3-mm-i.d. tubes. Feed flows through the bore while permeate travels through the microporous walls of the element into the surrounding shell for withdrawal (reprinted from Millipore Corporation, Lit. No. SD117 May 1989, with permission).

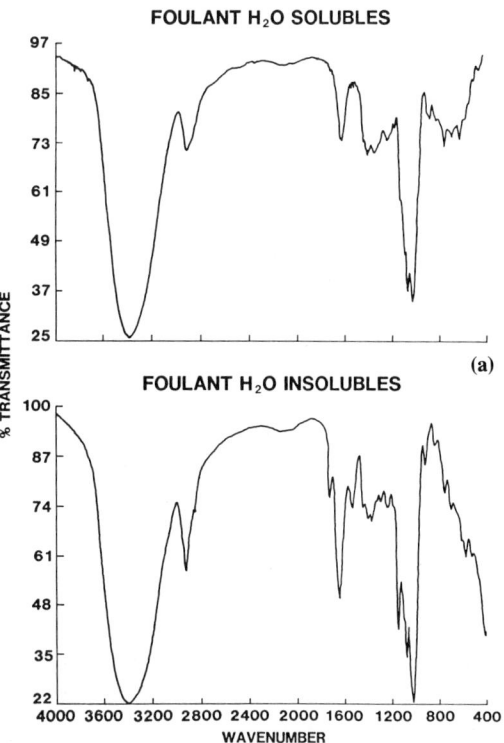

FIGURE 29-4. FTIR analysis of surface layer of hydrolyzed starch fouled membrane: (a) soluble foulants and (b) insoluble species.

proteins and fatty acid esters in addition to saccharides.

Identification of the fouling species is a prerequisite for designing appropriate feed pretreatment systems. In some cases it may be economically possible to adjust the feed pH so that protein foulants will be far from their isoelectric point; this will decrease their tendency to form gel layers on the membrane surface. Antiscalants are used to increase the concentration of mineral species that can be held in solution before precipitation. Prefiltration or centrifugation is used to remove suspended solids, which may "blind" the membrane surface or plug the module. Heat pretreatment (65°C) followed by settling (1 h) can be used to remove immunoglobulins and fats from cheese whey before ultrafiltration (Aptel and Clifton 1986). In some cases the interaction between organic foulants, e.g., proteins and ionic species, can be important. Divalent ions such as Ca^{2+} or Mg^{2+} may cause precipitation through bridging across macromolecular chains; hence, the selective removal of these ions (ion exchange) may be beneficial. Monovalent ions (Na^+, K^+) may actually prevent fouling and precipitation; for example, the effect of the ionic strength on the solubility product of $CaSO_4$ is well known.

Polarization and Fouling Control Strategies

Strategies to control concentration polarization and fouling can be classified as:

1. Membrane material
2. Flow manipulation
3. Additional force field
4. Cleaning procedures.

Membrane Material

The influence of the membrane material on the fouling characteristics is interpreted in terms of

TABLE 29–2. Protein (IgG) Adsorption on Various Polymeric Membranes (reprinted from Millipore Corporation, Tech. Brief TB 032, with permission).

Polymer	Protein Adsorbed (g/m^2)
Polyethersulfone/polysulfone	0.5–0.7
Regenerated cellulose	0.1–0.2
Polyvinylidenefluoride (modified)	0.04

its adsorption of the solute. Matthiasson (1983) showed the relation between concentration of BSA in solution, the amount adsorbed in static adsorption on to the membrane, and the corresponding flux decline. In the case of cellulose acetate membranes, only monolayer adsorption was indicated; however, for polysulfone or polyamide membranes, increasing the solute concentration increased adsorption and decreased permeability. Protein adsorption data for membranes made from various polymers are shown in Table 29–2. In addition to fouling, adsorption is important in small systems because it can affect recovery.

Adsorption effects are usually justified in terms of van der Waals interactions, hydrogen bonding, electrostatic effect, charge transfer effects, dipole moments, etc. Sugars and polysaccharides bind to membranes through hydrophilic interactions, while proteins, surfactants, etc., may have both hydrophilic as well as hydrophobic interactions. The use of positively charged membranes for cathodic paint recycle illustrates the importance of electrostatic effects (Mir 1983). Similarly, hydrophilic polymers are generally less fouling in aqueous environments. Gregor (1977) has claimed low flux declines while using highly sulfonated polystyrene-based membranes for a variety of fouling streams: paper mill wastes, whey recycle, etc. Some methods for increasing the hydrophilicity of polymeric membranes were discussed in Chapter 28.

Surfactant treatment can be used to modify the surface properties of membranes. Fane et al. (1985) showed 20% higher fluxes and 10% lower flux declines with passive adsorption of nonionic surfactants on polysulfone membranes used for albumin UF. The effects of surfactants used need not be uniformly beneficial (Swamikannu et al. 1988). Kim et al. (1989) have demonstrated 30% higher fluxes with a Langmuir-Blodgett (L-B) layer film composed of nonionic surfactants (nonyl phenols, alcohols) supported on a polysulfone membrane base. The L-B film decreases water flux by 10% but gives a higher UF flux using BSA. In addition to modifying surface chemistry, film formation may also help by reducing surface roughness; this is indicated by the beneficial effect of stearic acid (which increases the contact angle) as described by Kim et al. (1989). Another possibility is the increased wetting of smaller pores that were hitherto inaccessible to the solvent.

Flow Manipulation

Commercial membrane modules are designed to control concentration polarization by one of two methods: increased shear at the membrane surface (thin channel devices, high flow rates) or turbulence inducers (feed spacer design, static mixers). Some of the design choices in commercial modules were described earlier in this chapter, while the basic design equations were reviewed in Chapter 27. The role of fluid mechanics in UF performance is crucially important. Two excellent reviews by Belfort (1987; 1989) are available on the theoretical background of various techniques to control concentration polarization. Much innovative work has been done; however, it has not yet been used on a large scale. Some of the new innovations are described as follows:

1. *Pulsatile flow*: Bauser et al. (1982) reported increases of 40% in the flux of whey and whole blood after 2 to 3 hours at 1 Hz using microporous membranes.
2. *Vortice formation by surface corrugation*: A device was constructed in which the membrane covers a series of porous, furrowed channels. Pulsatile flow causes vortices to be formed above and below the center flow line; when the flow direction is reversed,

these vortices are ejected. Jeffree et al. (1981) has demonstrated up to sixfold increase in the UF flux for bovine whole blood using this principle. Similar results have been shown by van der Waal, Stevanovic, and Racz (1989).

3. *Taylor vortices*: Laboratory devices have been constructed with membranes covering the annulus between two concentric cylinders through which the solution flows. On rotating the inner cylinder at a speed high enough to induce Taylor vortices to be formed as a result of flow instabilities, a cleaning mechanism is generated for sweeping the polarization layer from the membrane surface. Kroner and Nissinen (1988) showed that such a device gave higher fluxes and lower enzyme rejections as compared to turbulent flow conditions for enzyme recovery from cell debris. These devices are commercially available from various companies (Sulzer, Membrex, and Hemasciences/Baxter Healthcare).

Vigo, Uliana, and Ravina (1990) have demonstrated flux enhancement by using feed oscillations (0 to 50 Hz) in combination with the rotating concentric cylinder device. A critical frequency is observed, above which the flux begins to increase and reaches a plateau value that is dependent on the pressure. The critical frequency value is dependent on the amplitude of the oscillations.

4. *Other novel module designs*: An example is a novel tubular design (Schubert and Todd 1980) that suggests using a "scoop" to remove boundary layer buildup at various points along the tube length.

Additional Force Field

Use of an additional field to supplement feed flow or convection as a way of controlling concentration polarization is an attractive possibility. This concept has been demonstrated on a laboratory level in the case of centripetal and electrical force fields.

Electrical Field

Murlidhara and Huffman (1988) have discussed a device using a polymeric membrane with Pt electrodes in the feed and permeate chambers connected to a dc power supply. Foulants deposited on the membrane (usually negatively charged) are pulled away on the basis of their electrophoretic mobility. Results show an increase in flux of 20 to 60% for this electrofiltration compared to conventional UF (see Table 29–3). Feed conductivity for successful electrofiltration is in the range of 0.01 to 1 Ω^{-1}.

A variation on this continuous electrical field enhanced filtration is the use of short, intermittent electric pulses to clear the membrane surface. This has been demonstrated for microfiltration using stainless steel membranes.

Inorganic electroconductive membranes for electroultrafiltration have been demonstrated by Guizard et al. (1989). These are tubular membranes with a coaxial Pt electrode in the tube. The membranes have been made by coating 0.2-μm α-alumina membranes with Ni alloy or from sol-gel slip casting using RuO_2-TiO_2.

Rios, Rakotoarisoa, and De la Fuente (1988)

TABLE 29–3. Comparison of Conventional and Electrofiltration Fluxes (reprinted from Murlidhara and Huffman 1988 with permission).

Feed	Membrane	Run Time (h)	Flux Conventional	Electrofiltration (gfd)
Tap water	0.2 μm (MF)	18	329	548
		23	289	508
1% sewage sludge	0.2 μm (MF)	8	0.1 (blinded)	5.2
2% skim milk	10K MWCO	8	3.4	4.2

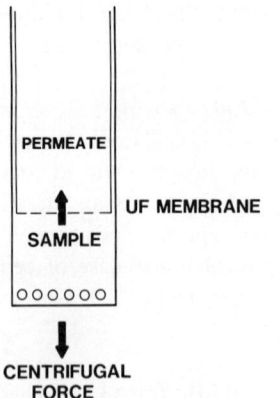

FIGURE 29–5. Schematic of Centrisart I (Sartorius). Centrifugal force in opposing direction to permeate flow pulls away fouling particulates.

have made a comparison between the flux of a tubular membrane with and without the presence of an electric field. A comparison was also made with a case in which stainless steel balls were fluidized within the tube bore in order to provide a sweeping action across the membrane surface (Rios, Rakotoarisoa, and De la Fuente 1987). The electrofiltration fluxes were better than the empty tube alone and comparable to the fluidized balls case.

Centripetal Force

Figure 29–5 shows a laboratory device which is spun in a centrifuge during separation (Sartorius, technical bulletin). Permeate is forced from the feed chamber through the membrane in a direction opposite to the centrifugal force which pulls fouling particles away.

Cleaning Procedures

Fouled membranes are cleaned by clean-in-place (CIP) procedures that minimize downtime. Not all cleaning protocols involve external chemicals. For example, periodic reversal in flow direction may help to prevent particulates from plugging up the module inlet. In another case, periodic reductions in feed pressure along with continued flow helped control gel layer growth and enabled stable fluxes to be maintained in a spiral module (see Figure 29–6). Back-flushing from the permeate side is possible with some tubular membranes; this is a powerful method of removing gross foulants.

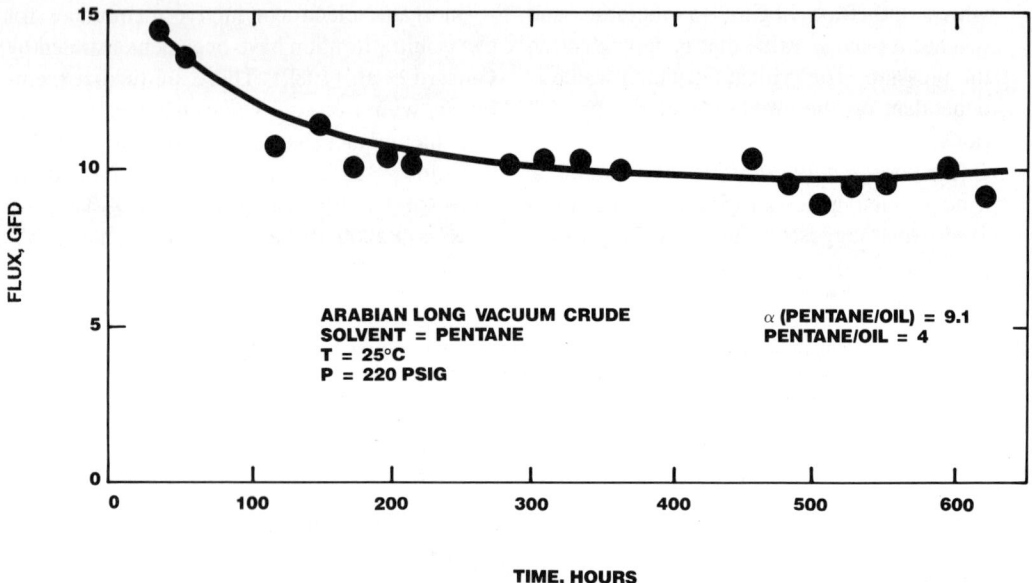

FIGURE 29–6. Heavy crude UF with pressure reduction cleaning cycles (reprinted from Kulkarni, Funk, and Li 1986 with permission).

Tubular membranes of 25 mm diameter are also physically cleaned by the "spongeball" technique.

Proprietary cleaning solutions are available from a variety of sources such as Pfizer (Flocon™), BF Goodrich (MT series), Monarch, and King-Lee (KL series, Diamite series). Cleaners based on mineral acids, sodium hexametaphosphate, polyacrylates, and ethylenediaminetetraacetic acid (EDTA) are used for salt precipitates and mineral scalants. Sodium hydroxide-based cleaners, sometimes supplemented with hypochlorite, are useful for solubilizing fats and proteins. Enzyme cleaners based on either proteases, amylases, glucanases, etc., are used in specific instances under neutral pH conditions. The characterization of foulants discussed earlier in this chapter is the key to cleaning solution selection. In addition, the choice is dictated by the compatibility of the membrane to the chemicals at the cleaning temperature. Cleaning solutions are usually circulated without pressure to prevent deeper penetration of the foulants into the membrane.

Cleaning solutions may have temporary effects on the membrane rejection in addition to the anticipated changes caused by removal of the gel layer or pore opening. Jonsson and Johansen (1989) have discussed a case in which this decreased rejection was attributed to swelling of the polysulfone membrane surface layer as a result of exposure to the cleaning solution (Ultrasil-10).

Since many installations involve several modules in series, the CIP procedures have to be designed so that the foulants removed from the first module do not just get subsequently deposited on the later modules.

PROCESS CONFIGURATIONS

The most common UF process configurations are

1. Batch concentration
2. Feed and bleed
3. Diafiltration.

Examples of these and other configurations are given by Cheryan (1986). While process design also involves specification of pumps, piping, control systems, etc., this is beyond the scope of the present chapter.

In some cases (e.g., wastewater processing) this type of design can be done with reference to standard handbooks on fluid systems, materials corrosion, and equipment manufacturer's literature. However, UF processes for food and biotechnology application generally have special requirements.

Sanitary processing is a must in all food processing and most biotechnology applications. The standard applicable is that of the U.S. Public Health and Dairy Industry Committee (1966/67). Industry norm is for all piping, vessels, etc., to be made of Type 316 stainless steel. The minimum surface finish required is No. 4 mill finish, though in pharmaceutical processing electropolishing is used (Parise, Parekh, and Smith 1988). Sanitary fittings are quick release clamps, e.g., triclover fittings. Diaphragm-type transducers are used for pressure sensors so that the stagnant volumes present in Bourdon-type gauges are eliminated. Other sensors (flow, temperature, etc.) have to be positioned so that stagnant areas are not caused by the insertion of the sensor in the flowing stream. Materials for gaskets, module components, etc., may have to conform to the Code of Federal Regulations (1982). Stainless steel rotary lobe pumps are a good choice for some biotechnology applications; these pumps are easily sterilized and cause little shear-denaturing of sensitive biological molecules. For high pressures, multistage centrifugal pumps are available in which a small portion of the discharge fluid is recirculated to flush the annular space between the pump stages and the casing (Parise, Parekh, and Smith 1988).

Batch Concentration

Batch operation [see Figures 29–7(a) and 29–7(b)] is the simplest configuration and requires the least membrane area to achieve a given separation/unit time. The configuration in Figure 29–7(a) requires a high pumping cost; this can be reduced by the two pump (feed and

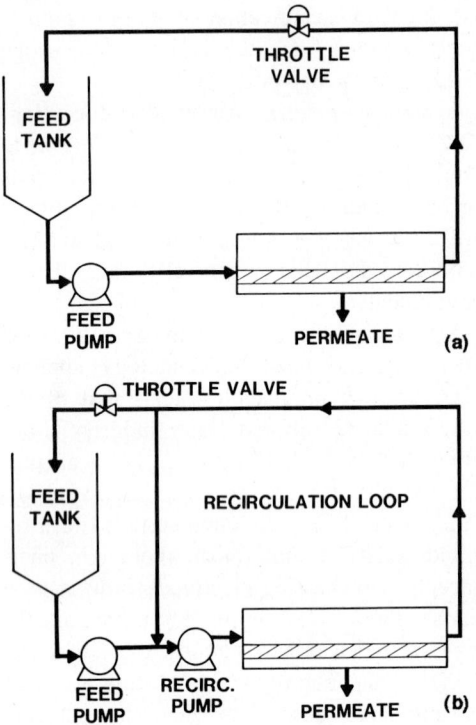

FIGURE 29-7. Batch concentration schematic: (a) total recycle and (b) partial recycle.

recirculating) arrangement in Figure 29–7(b). Batch operation needs buffer storage capacity to be installed in order to be coupled to a continuous process. The equations describing batch operation are analogous to those used for plug flow reactors:

$$-\frac{dV}{dt} = J_v(c)A, \quad (29\text{--}1)$$

$$V_0 c_0 = V_t c_t + A \int_0^T J_v(t) c_p(t)\, dt. \quad (29\text{--}2)$$

In these equations, V is the feed volume remaining, J_v the permeate flux, and c the feed concentration at any time t. The subscripts 0 and t refer to initial conditions and those at any arbitrary time t. Concentration c_p refers to the solute concentration in the permeate.

The relation between flux and concentration at the various levels of operating variables (pressure, temperature, flow rates) may be available from the module manufacturer based on past experience or it will have to be generated by the process developer. The basic equations for flux as a function of concentration and operational variables were discussed earlier [Eqs. (27–13) to (27–19)]. An estimate of the area required can also be made using Eq. (27–15) and estimates for the gel concentration c_g. In general, flux will decrease with increasing solute concentration as shown in Figure 29–8 for the case of apple juice concentration (Blanck and Eykamp 1986).

For the case where rejection is independent of feed solute concentration, the retentate concentration, c_r, can be related to the original concentration c_0 by means of the rejection R_i and volume concentration factor VCF (Cheryan 1986):

$$R_i = 1 - c_p/c_r, \quad (29\text{--}3a)$$

$$\text{VCF} = V_0/V_r, \quad (29\text{--}3b)$$

$$c_r = c_0\, (\text{VCF})^{R_i}. \quad (29\text{--}3c)$$

The rejection is expected to increase as the gel layer establishes itself on the membrane surface and starts acting as a secondary dynamic mem-

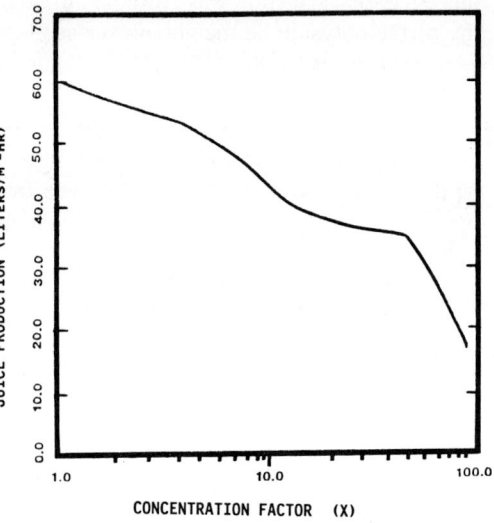

FIGURE 29-8. Flux reduction with increasing concentration of apple juice (reprinted from Blanck and Eykamp 1986 with permission).

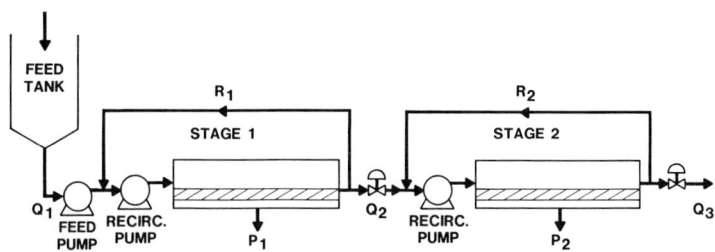

FIGURE 29–9. Continuous operation schematic with two feed-and-bleed stages.

brane. In many cases where the gel layer is established immediately at very low concentration factors the above assumption of constant rejection is adequate for calculations.

Feed and Bleed

The single-pass continuous flow configuration popular in reverse osmosis is rarely used in UF because of material costs. The feed-and-bleed configuration is commonly used for continuous operation (see Figure 29–9) especially in the food industries. Increasing the number of stages increases efficiency so that more than four stages approaches that of batch operation, at the cost of increasing the number of recirculating pumps. The equations describing each individual stage are analogous to those describing the continuous stirred tank reactor (CSTR) in reaction engineering:

$$Q_{n-1} = A_{n-1} J_v(c_n) + Q_n \quad (29\text{-}4)$$

and

$$Q_{n-1} c_{n-1} = A_{n-1} J_v(c_n) c_{np} + Q_n c_n, \quad (29\text{-}5)$$

where Q_n represents the feed flow rate to the nth stage with area A_n. If recovery proceeds to a high extent, the feed flow rate to the later stages will correspondingly decrease. The minimum flow rate required to operate the module so as to avoid fouling problems has to be maintained by either using a tapered cascade-type module arrangement (decreasing number of modules in parallel for each stage) or by changing the module geometry/configuration (e.g., smaller diameter spiral-wound modules in the later stages). The feed-and-bleed configuration operates at the retentate concentration; hence, membrane productivity is lowest for a single stage that operates at the fluxes corresponding to the final concentration. Adding more stages gives productivities closer to those of the batch system. In practice, three to four stages usually give a good approximation to the batch conditions. The feedside (retentate) concentration at any given stage n can be expressed by

$$c_n = c_{n-1} \left(\frac{\text{VCF}_n}{\text{VCF}_n - R_i(\text{VCF}_n - 1)} \right), \quad (29\text{-}6)$$

where $\text{VCF}_n \, (= Q_{n-1}/Q_n)$ is the volume concentration factor (retentate/feed flow rate) in that stage and c_{n-1} is the incoming stream concentration.

Diafiltration

Diafiltration involves the addition of water to the retentate and continuing UF in order to overcome low fluxes at high concentrations or to get better removal of permeable species. This makes sense especially when the permeate is the desired product. Diafiltration may be either sequential or continuous.

Optimization of the permeate/bleed ratios or of the rate of addition of water during diafiltration can be done numerically. The area in each stage of a multistage operation cannot be varied continuously because of module constraints. Kovasin, Hughes, and Hill (1986) discuss the relative merits of dynamic programming and complex search methods for optimizing spent sulfite pulp liquor UF.

Diafiltration can be done in sequential fashion with concentration followed by dilution in several stages. When done in this sequential concentration/dilution mode, the retentate concentration of any species can be expressed as

$$c_r = c_0(\text{VCF})^{1+n(R_i-1)}, \quad (29\text{-}7)$$

where n is the number of sequential diafiltration stages. When diafiltration is done by continuously adding makeup diluent (e.g., water) to balance the permeate being lost, the retentate concentration can be expressed in terms of a dilution factor (DF) with respect to the original feed volume:

$$\text{DF} = \frac{\text{volume of liquid added as diluent}}{\text{initial feed volume}} \quad (29\text{-}8)$$

and

$$c_r = c_0 \exp[-\text{DF}(1 - R_i)]. \quad (29\text{-}9)$$

An important aspect of the diafiltration process is the large amount of convective flow of solvent through the membrane in addition to the diffusive solute transport. In discontinuous diafiltration, the problem of concentration polarization and gel formation at the surface is more severe due to the concentration of feedstock in each cycle. The large convective flow also puts a greater demand on the membrane since the solutes that should be retained must have reasonably high rejection coefficients or they will essentially be "washed out" in the diafiltration process.

NOTATION

General Notation

See the General Notation section at the beginning of this handbook.

Special Notation

A_n membrane area for stage n, L^2
c_g solute gel concentration, mol/L^3 or wt.%
c_p solute concentration in the permeate, mol/L^3
c_t feed solute concentration at time t, mol/L^3
DF   dilution factor, dimensionless
Q_n feed flow rate for stage n, L^3/t
R_i observed solute rejection, dimensionless
V_t feed volume remaining at time t, L^3
VCF   volume concentration factor, dimensionless

REFERENCES

Aptel, P., and M. Clifton. 1986. Ultrafiltration. In *Synthetic Membrane: Science, Engineering and Applications,* ed. P. M. Bungay, H. K. Lonsdale, and M. N. de Pinho, pp. 249–305. Dordrecht, Holland: D. Reidel Publishing Co.

Bauser, H., H. Chmiel, N. Stroh, and E. Walitza. 1982. Interfacial effect with microfiltration membranes. *J. Membr. Sci.* 11:321.

Belfort, G. 1987. Transport properties of osmotic membranes. In *Advanced Biochemical Engineering,* ed. H. R. Bungay and G. Belfort, pp. 239–297. New York: John Wiley & Sons.

Belfort, G. 1988. Membrane modules: comparison of different configurations using fluid mechanics. *J. Membr. Sci.* 35:245–270.

Belfort, G. 1989. Fluid mechanics in membrane filtration recent developments. *J. Membr. Sci.* 40:123–147.

Blanck, R. G., and W. Eykamp. 1986. Fruit juice ultrafiltration. *AIChE Symp. Ser. No.* 82(250): 59–64.

Cabasso, I. 1980. Hollow-fiber membranes. In *Kirk-Othmer Encyclopedia of Chemical Technology,* 3rd ed., Vol. 12, pp. 492–517. New York: John Wiley and Sons.

Chatterjee, S. G., and G. Belfort. 1986. Fluid flow in idealized spiral-wound membrane module. *J. Membr. Sci.* 28:191–208.

Cheryan, M. 1986. *Ultrafiltration Handbook.* Lancaster, PA: Technomic Publishing Co.

Code of Federal Regulations. 1982. *Title 21—Food and Drugs. Parts 175.300, 177.2600,* April 1982.

Doughty, J. R., and H. C. Perkins. 1970. Hydrodynamic entry length for laminar flow between parallel porous plates. *J. Appl. Mech.* 37:548–550.

Fane, A. G., C. J. D. Fell, and K. J. Kim. 1985. The effect of surfactant pretreatment on the ultrafiltration of proteins. *Desalination* 53:37–56.

Gregor, H. P. 1977. Crosslinked interpolymer fixed-charge membranes. U.S. Patent 4,012,324.

Guizard, C., F. Legault, N. Idrissi, A. Larbot, L. Cot, and C. Gavach. 1989. Electronically conductive mineral membranes designed for electroultrafiltration. *J. Membr. Sci.* 41:127–142.

Hanemaaijer, J. H., T. Robbertsen, Th. Van der Boomgaard, and J. W. Gunnink. 1989. Fouling of ultrafiltration membranes role of protein adsorption and salt precipitation, *J. Membr. Sci.* 40:199–217.

Hsieh, H. P. 1988. Inorganic membranes. *AIChE Symp. Ser.* 84 (261):1–18.

Jeffree, M. A., J. A. Peacock, I. J. Sobey, and B. J. Bellhouse. 1981. Gel layer limited hemofiltration rates can be increased by vortex mixing. *Clin. Exp. Dial. Aspheresis* 5:373.

Jonsson, G., and P. L. Johansen. 1989. Selectivity of ultrafiltration membranes. In *Indo-ECC Membrane Workshop.* New Delhi: Department of Science and Technology.

Keily, H. J., and L. B. Rogers. 1955. Instrumental variability of a model 7 Coleman Photonephelometer. *Analy. Chem.* 27(3):459–461.

Kim, K. J., A. G. Fane, and C. J. D. Fell. 1989. The performance of ultrafiltration membranes precipitated by polymers. *J. Membr. Sci.* 43:187–204.

Kovasin, K. K., R. R. Hughes, and C. G. Hill, Jr. 1986. Optimization of an ultrafiltration-diafiltration process using dynamic programming. *Comput. Chem. Eng.* 10(2):107–114.

Kroner, K. H., and V. Nissinen. 1988. Dynamic filtration of microbial suspensions using an axially rotating filter. *J. Membr. Sci.* 36:85–100.

Kulkarni, S. S., E. W. Funk, and N. N. Li. 1986. Hydrocarbon separations by polymeric membranes. *AIChE Symp. Ser.* 82 (250):78–84.

Levy, P. F., and R. S. Earle. 1990. The effect of channel spacers and channel height on flux, energy requirements and process economics in crossflow ultrafiltration. In *Proc. Int. Congress on Membranes and Membrane Processes,* Chicago, IL, pp. 1014–1016, Paper 54-5.

Madsen, R. F. 1977. *Hyperfiltration and Ultrafiltration in Plate-and-Frame Systems.* Amsterdam: Elsevier Scientific Publishing Co.

Madsen, R. F. 1989. Applications of membrane filtration in the food and biochemical industry. In *Indo-ECC Membrane Workshop.* New Delhi: Department of Science and Technology.

Matthiasson, E. 1983. Role of macromolecular adsorption in fouling of ultrafiltration membranes. *J. Membr. Sci.* 16:23–26.

Mir, L. 1983. Positive-charged ultrafiltration membrane for the separation of cathodic/electrodeposition paint compositions. U.S. Patent 4,412,922.

Mooney, L. A., and W. G. Light. 1985. Food applications for sanitary design spiral-wound membrane elements. In *Int. Membrane Technology Symp.,* Tylosand, Sweden.

Murlidhara, H. S., and W. S. Huffman. 1988. Electromembrane technology: a novel approach for antifouling. In *Proceedings of 6th Ann. Membrane Technology/Planning Conf.,* Cambridge, MA. Norwalk, CT: Business Communications Co. Inc.

Oldani, M., and G. Schock. 1989. Characterization of ultrafiltration membranes by IR, ESCA and contact angle measurements. *J. Membr. Sci.* 43:243–258.

Parise, P. L., B. S. Parekh, and R. T. Smith. 1988. RO for producing pharmaceutical-grade waters. In *Reverse-Osmosis Technical Applications for High Purity Water Production,* ed. B. S. Parekh, Chap. 8, p. 355. New York: Marcel Dekker.

Paulson, D. J., B. W. Phelps, and G. J. Gach. 1987. Design innovations for processing high fouling solutions with spiral-wound membrane elements. Osmonics Corp.

Rios, G. M., H. Rakotoarisoa, and T. De la Fuente. 1987. Basic transport mechanisms of ultrafiltration in the presence of fluidized particles. *J. Membr. Sci.* 34:331–343.

Rios, G. M., H. Rakotoarisoa, and T. De la Fuente. 1988. Basic transport mechanisms of ultrafiltration in the presence of an electric field. *J. Membr. Sci.* 38:147–160.

Sartorius GmbH. Small volume filtration in a centrifuge. In *Laboratory Filtration,* Publication FL 0001e, p. 134.

Schubert, J. P., and D. K. Todd. 1980. Hyperfiltration scoop apparatus and method. U.S. Patent 4,218,314.

Swamikannu, A. X., S. S. Kulkarni, E. W. Funk, and R. A. Madsen. 1988. Recovery of space station hygiene water by membrane technology. In *18th Intersociety Conference on Environmental Systems (SAE),* San Francisco, CA.

Terrill, R. 1983. Laminar flow in a porous tube. *Trans. ASME J. Fluids Eng.* 105:303–307.

U.S. Public Health Service and the Dairy Industry Committee, International Association of Milk, Food and Environmental Sanitarians. 1966/67. 3-A, Accepted Practice for Permanently Installed Sanitary Product-Pipelines and Cleaning Systems, 605–02, 1966, revised 1967.

Van der Waal, M. J., S. Stevanovic, and I. Racz. 1989. Mass transfer in corrugated-plate membrane modules: II. ultrafiltration experiments. *J. Membr. Sci.* 40:261.

Vigo, F., C. Uliana, and E. Ravina. 1990. The vibrating ultrafiltration module performance in the low frequency region. *Sep. Sci. Technol.* 25:63–82.

30

Applications and Economics

Sudhir S. Kulkarni
National Chemical Laboratory

Edward W. Funk and Norman N. Li
Allied Signal

APPLICATIONS
   Established Applications
   Emerging Applications

PROCESS ECONOMICS
REFERENCES

APPLICATIONS

Many ultrafiltration (UF) applications are already being practiced in industry and more are being studied. Porter (1977) listed 34 UF applications in 1977; these included applications in the food industry, pharmaceuticals and biotechnology, water purification, and waste treatment in the chemical and paper industries. Various applications are discussed in detail by Cheryan (1986) and by Aptel and Clifton (1986).

Established Applications

Several important UF processes are tabulated in Table 30–1. The most prominent use of UF processes is in the food industry, where UF is used, for example, to recover lactalbumin and lactoglobulin from cheese whey (Cheryan 1986; Aptel and Clifton 1986; Michaels 1968), to preconcentrate milk before cheese making (Cheryan 1986; Aptel and Clifton 1986), and for fruit juice clarification (Blanck and Eykamp 1986). These applications use polysulfone membranes, the first two mainly in the spiral-wound form, the third with tubular modules.

Other important applications are recovery of electrocoat paint (Breslau et al. 1980) and the purification of water for the pharmaceutical industry (Marcus and Postrick 1988). Newer applications in juice clarification, vinegar brewing, and gelatin concentration have been recently described (Short 1988). A few applications are reviewed.

Dairy Applications

Cheese whey, which is the supernatant stream produced in cheese or casein processes, comprises 90% of the original milk volume. Typical whey from cow's milk contains 10% of valuable proteins (based on dry solids content), the remainder being mostly lactose in addition to inorganics, ash, etc. By increasing the relative protein concentration to 35% and then spray drying, the protein-enriched dry solids have an increased value as a livestock feed supplement. By using diafiltration, the protein concentration can be enriched to as high as 80%; this is a premium whey product that is used in the food industry for gelation, emulsification, etc.

UF is also being used for preconcentrating the protein content in milk prior to cheese-

TABLE 30–1. List of Some Ultrafiltration Applications.

Process	Separation
Electrophoretic paint	Process rinse water, recycle paint to dip tank, allow reuse of rinse water
Cheese whey	Concentrate/fractionate proteins from lactose and inorganics
Juice clarification	Remove haze components from apple juice
Textile sizing agents	Recover polyvinyl alcohol after scouring of woven goods
Wine clarification	Remove haze components from red and white wines
Oil/water emulsions	Metal cutting oils (lubricants) concentrated from wastewater for incineration
Polymer latex	Latex emulsions concentrated from wastewater
Dewaxing	Separation of wax components from lower paraffins
Deasphalting	Solvent recovery/recycle for deasphalting of heavy crudes
Egg-white preconcentration	Partial dewatering before spray drying
Fermentation broth	Separate low molecular weight organics/therapeutic agents from cells or cell debris
Kaolin concentration	Partial dewatering of clay slurry before centrifugation
Water treatment	Concentration before sludge dewatering
Affinity membranes	Retain ligand complex from noncomplexed proteins
Reverse osmosis pretreatment	Retain colloidal silica, bacteria

making. This is primarily used for soft cheeses, such as feta cheese in Denmark. The process advantages are that less protein is lost in the whey and smaller cheese-making equipment can be used. These developments have been reviewed by Aptel and Clifton (1986).

The largest use of ultrafiltration is in the production of whey protein concentrate. This operation is described by Robichaux and Ellis (1982). Pasteurized whey is first clarified to remove suspended particulates (curd fines) and centrifuged at high speed to remove fat. Pasteurization also decreases membrane fouling by increasing the binding of calcium phosphate to proteins (Mohr et al. 1990). Adjustment of pH to values away from the average isoelectric point of whey proteins (4.5) gives increased fluxes (DeBoer and Hiddink 1980; Mohr et al. 1990).

The polysulfone membranes used in this application have nominal MWCOs of 10 to 15K daltons. Whey typically contains 6 to 7% (dry basis) total solids (TS) of which the protein content is ~10% (dry basis), the remainder being mostly lactose and salts. Typical UF concentrates contain ~10% TS with a protein content of 35% while the permeate stream contains lactose and salts. Operating conditions are 50 to 55°C at 25- to 75-psig feed pressures. Cleaning (caustic/chlorine-based cleaning agents) is required at least once a day.

Diafiltration can be used to produce concentrates of ~12% TS with a protein content up to 80% (Kosikowski 1986). The diafiltration ratio required (water:original UF concentrate) is 3:1.

UF has also been used to concentrate both whole and skim milk and decrease its lactose content. The operational conditions are similar to those used for whey processing. The degree of concentration is limited primarily by the protein content (maximum <18%) and fats. The

TABLE 30–2. Feed, Permeate, and Concentrate Compositions for the Fourfold Concentration of Whole Milk (Balance Is Water) (reprinted from Kosikowski 1986 with permission).

	Composition (wt.%)		
	Feed	Permeate	Concentrate
Protein	3.2	0.2	12.3
Fats	3.4	0	14.0
Lactose	4.4	4.9	2.8
Ash/salts	0.7	0.5	1.4
Acids	0.14	0.08	0.33
Total solids	11.8	4.6	30.8

composition of the UF concentrate and permeate for the fourfold concentration of whole milk (Kosikowski 1986) is shown in Table 30–2.

Fruit Juice Clarification

Fruit juice clarification by membranes is now an established application (Blanck and Eykamp 1986; Hacket and Swientek 1986; Garrison 1986). Plants with capacities of 25,000 gal/day (Hacket and Swientek 1986) for apple juice clarification are operational. Conventional clarification requires several unit operations such as centrifugation, treatment with pectinases, and rotary vacuum filtration with diatomaceous earth. By comparison, membrane clarification provides simpler operation, better separation, and higher juice yield. UF processes have also been developed for other fruit juices such as grape, pineapple, pear, and cranberry (Blanck and Eykamp 1986; Garrison 1986; Paulson, Wilson, and Spatz 1985).

Both tubular modules and spiral-wound modules with clear channel spacers have been used for juice processing by UF. Depectinized juices such as apple or pineapple, which are processed at ~50°C, have higher fluxes (~100 L/m·h) than grape juice (~10 L/m·h), which is processed at 7°C (Blanck and Eykamp 1986).

The yield (1 − 1/volume concentration factor) of recovered juice is a crucial item in the process economics. The VCF is limited by the flux reduction caused by the increase in viscosity and fouling from the increased concentrations of suspended solids, macromolecules, etc., in the feed. The change in flux for apple juice clarification as a function of the VCF was shown in Figure 29–8. Other related applications that have been extensively tested are UF processes in the grape/wine industry and in tomato processing (Mohr et al. 1990).

Downstream Processing

Recovery of vaccines and antibiotics from fermentation broth is an area of increasing importance. UF processing is simply one step along a detailed purification procedure for biologicals such as interferon. A combined process (Kalyanpur, Skea, and Siwak 1984) for isolation of Cephalosporin C from a batch fermentation broth is shown in Figure 30–1. Cells are first removed by 0.2-μm microfiltration and then proteins and polysaccharides by UF/diafiltration using a 10K MWCO membrane. The UF permeate is concentrated by RO and the antibiotic is finally purified by HPLC.

Effluent Treatment

Various difficult waste streams are being successfully recycled. Contaminated caustic from textile scouring or mercerizing operations is recycled after permeating through in-place coated stainless steel tubular membranes. Oil and grease from a can washing operation has been recovered using polyvinylidenefluoride tubular membranes (McTeer and Wett 1989). The waste stream contains, among other impurities, 2000 to 3000 ppm oil and grease. This is separated into a permeate with less than 50 ppm grease, and a concentrate with 5 to 8% grease; this concentrated grease is then recovered (see Figure 30–2).

The paper and pulp industry has several effluent control and product recovery applications. Lignosulfonate separation from spent sulfite liquor and lignin and color removal from Kraft black liquor are evolving applications with some limited commercial use. A 1400-ton/day calcium sulfite treatment plant is in operation (Patterson Candy Bulletin). Effluent treatment by UF could include bleach effluent

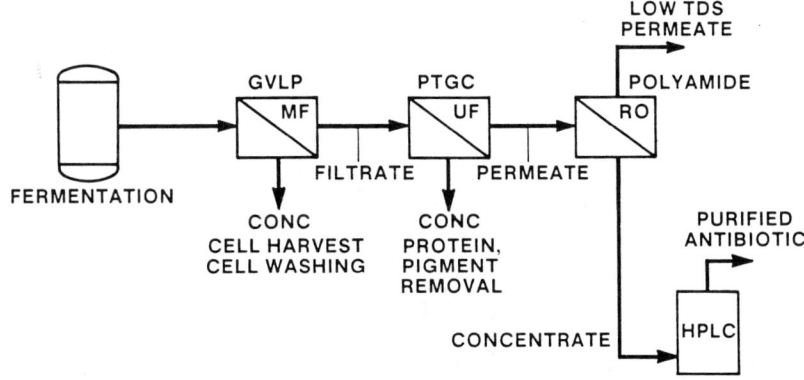

FIGURE 30-1. Schematic process for recovery of Cephalosporin C (reprinted from Kalyanpur, Skea, and Siwak 1984 with permission).

treatment, deresining of pulp wash water, whitewater processing for fiber recovery and de-inking effluent. Bedwell, Yates, and Brubaker (1988) have developed a cross-flow tubular membrane that can be tailored to treat a wide variety of difficult aqueous feeds by controlling the dynamic layer on the membrane surface.

Emerging Applications

Ultrafiltration has also been investigated for organic (nonaqueous) applications that could be of potential commercial importance in the petrochemical and food industries for the recovery of solvent from extraction processes (Kulkarni, Funk, and Li 1986). The processing scheme for separating low molecular weight hydrocarbons (primarily solvent) from heavy hydrocarbons (shown in Figure 30-3) has been successfully run on a pilot plant scale with the use of membrane cleaning techniques to control surface fouling by high molecular weight hydrocarbons.

A potentially useful processing route for nat-

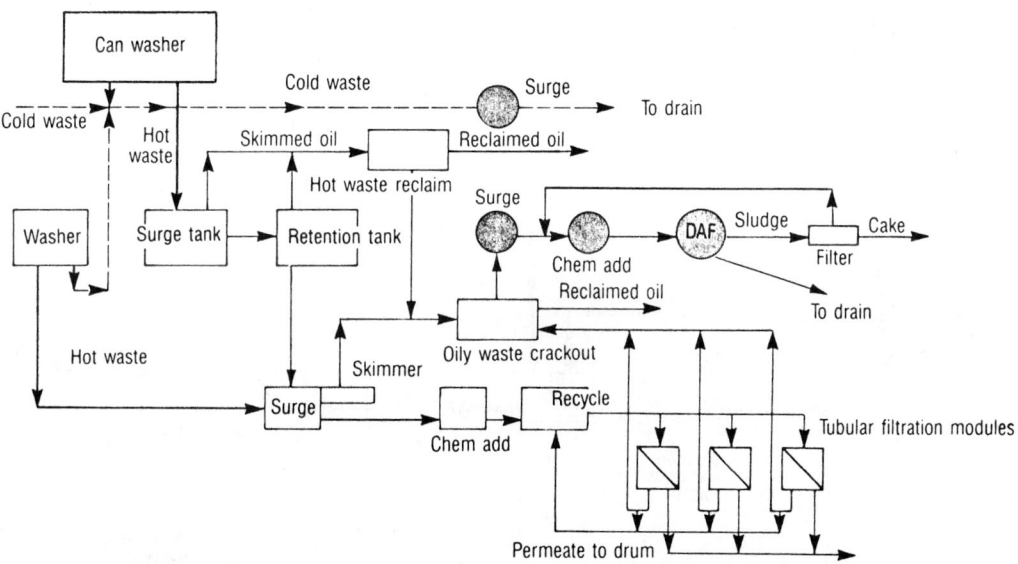

FIGURE 30-2. Schematic process for oil reclaiming from wash water (reprinted from McTeer and Wett 1989 with permission).

FIGURE 30–3. Deasphalting process with membranes for solvent recovery (reprinted from Kulkarni, Funk, and Li 1986 with permission).

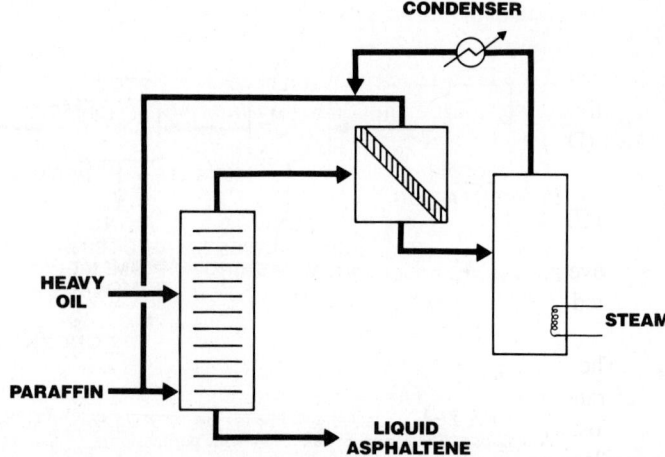

ural products and pharmaceuticals is to combine the product formation with membrane separation in a continuous integrated process. An example is to combine ultrafiltration with enzyme saccharification as illustrated in Figure 30–4 to produce glucose from liquefied cornstarch (Knapik and Bozzano 1986).

Chemical complexation combined with separation can be used in the production of biologically active materials. Matthiasson and Ling (1986) have discussed the use of affinity membranes to separate biologically active Concanavalin A from the inactive form. Active Con A complexes with sugar residues on the surface of heat-treated *Saccharomyces cerevisiae* cells; the complex is retained while the uncomplexed inactive form permeates UF membranes with MWCO in the range of 300 to 1000K. A 0.5 M glucose solution is then used to disassociate the active Con A, which permeates through the membrane in a second step.

Changing the state of aggregation or complexation of the key components in the feedstock is a possible route to new process applications by achieving finer separations than would be expected from pore size considerations alone. Scamehorn and Harwell (1987) discuss the use of surfactants, which form micelles encapsulating the organic components; the rejection of organics in the micellar phase is much higher than that of free dissolved organic. This idea has also been used for increasing the rejection of dissolved metals (Nguyen, Aptel, and Neel 1980; Christian et al. 1988). Criteria

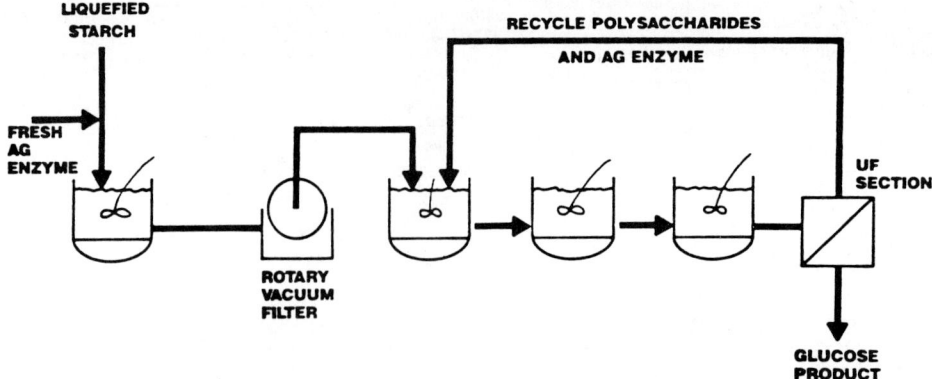

FIGURE 30–4. Process for combined saccharification/ultrafiltration for the production of glucose (Knapik and Bozzano 1986).

for the choice of surfactants have been discussed by Kandori and Schecter (1990). A similar idea is the use of chelating agents to aid in selectively separating multicomponent solutions (Ditnerskii, Volcheck, and Zhilin 1988).

PROCESS ECONOMICS

The overall process economics will be strongly dictated by the following major factors:

1. The initial capital investment for the membrane modules and associated equipment
2. Cost of replacing membrane modules
3. Cleaning chemicals
4. Energy/power usage
5. Labor
6. Depreciation.

The first two are considered the dominant factors while other costs, such as labor, depreciation, etc., are typical of conventional industrial processes. These costs are balanced by the various possible economic benefits of the process such as recovery of product, material and energy savings from recycling of process streams, improvement in performance of downstream processing units, improvement of product quality, and pollution control.

Installed capital costs are generally in the range of US$600 to 1200/m² of membrane area. The cost of polymeric UF membranes lies in the range of US$100 to 500/m²; the costs for membranes in the spiral-wound form usually lie between $100 to 200/m² while in the tubular form the cost may be greater than $250/m². Cheryan (1986) provides a summary of the expected costs for replacing membranes and typical costs for cleaning chemicals. Note that routine use of the "expected" costs without sufficient pilot plant data can be dangerous. It often occurs that fouling and membrane deterioration with real commercial feedstocks are quite different from synthetic or aged samples (e.g., the real foulants may not be present or are precipitated in aged samples).

There is no widely accepted and fundamental approach to estimate membrane life. A commonly used procedure is to plot the logarithm of the flux (at a defined point just after a cleaning cycle) versus logarithm of time. This linear plot is used to estimate a point in time (life) at which the flux is deemed unacceptable. It is also possible for a membrane to fail via loss in rejection. For many UF process applications, a minimum 1-yr life is required for favorable process economics. Most membrane manufacturers guarantee membrane durability for a specified period, generally 1 to 1.5 years for modules being used in conventional processes. The guarantee specifies ranges of flow conditions, and limits on pressures, temperatures, presence of oxidants (e.g., chlorine), organic solvents, and pH for the warranty to be valid. Actual life expectancy is expected to be longer than the minimum guaranteed time.

The energy use for heating/cooling solutions, pumping the feed and recycle streams can be estimated from engineering handbooks. Typically, these costs are not large and amount to only 10 to 15% of the total operating cost. Typical energy consumption lies in the range of 3 to 10 kWh/m³ of permeate. Eykamp (1990a) has observed that in some cases energy consumption in UF processes has now been reduced to 0.3 to 1.5 kWh/m³ of permeate or a design value of 10 to 150 W/m² of installed membrane area.

Beaton and Steadly (1982) have provided a breakdown of the various costs for a typical industrial ultrafiltration plant. This breakdown, shown in Table 30-3 shows that the capital cost and cost of membrane replacement are the major items. Both of these costs are related strongly to the membrane flux and life and to the pricing of the modules.

TABLE 30–3. **Relative Component Costs in the Total Operational Cost of an Ultrafiltration Plant (Beaton and Steadly 1982, with permission).**

	% of Total
Capital cost	38
Membrane replacement	27
Energy/power	16
Labor	10
Cleaning solutions	5
Maintenance	4

TABLE 30-4. Cost of Producing 35% Whey Protein Concentrate by Ultrafiltration (Eykamp 1990a).

	Cost/kg product (US$/kg)	% of Total Cost
Capital cost	0.032	31
Cleaning[a]	0.025	24
Membrane replacement	0.024	23
Energy/power	0.012	11
Labor and maintenance	0.012	11
TOTAL	0.105	100

[a]Also includes value of protein lost during cleaning.

The costs for specific cases may vary significantly from the typical case. Table 30-4 shows the total cost and cost breakdown into the various components for producing 35% whey protein concentrate (Eykamp 1990a, 1990b). In this case the cost of cleaning the modules (cleaning time is 4 h in every 24-h cycle) has also become as important a cost element as the capital and membrane costs. The cleaning cost in Table 30-4 also includes the value of proteins adsorbed or precipitated in the membrane system that are lost during cleaning.

Cost estimates are available for several UF processes of industrial importance. Bemberis and Neely (1986) have listed costs for Kaolin clay dewatering, fermentation broth clarification, apple juice clarification, and egg-white preconcentration.

In discussing these process economics, remember that ultrafiltration is a relatively new process and that the economics can be very different for different processes. Two particular areas can show very large potential differences: membrane replacement and labor. Often favorable estimates of membrane life are made using short pilot plant runs and real-world experience gives a much lower membrane life; this can be due to a host of factors including the nature of real feeds, unexpected contaminants, operator error, and upsets in the plant operation. The labor component in Table 30-3 is estimated to be only 10%; however, this can grow dramatically if more frequent or intensive cleaning is required or if significant operator time is required to adjust process variables.

REFERENCES

Aptel, P., and M. Clifton. 1986. Ultrafiltration. In *Synthetic Membranes: Science, Engineering and Applications,* ed. P. M. Bungay, H. K. Lonsdale, and M. N. de Pinho, pp. 249–305. Dordrecht, Holland: D. Reidel Publishing Co.

Beaton, N. C., and H. Steadly. 1982. Industrial ultrafiltration. In *Recent Developments in Separation Science,* ed. N. N. Li, Vol. 7, pp. 1–29. Boca Raton, FL: CRC Press.

Bedwell, W. B., S. F. Yates, and I. M. Brubaker. 1988. Crossflow microfiltration—fouling mechanism studies. *Sep. Sci. Technol.* 23:531.

Bemberis, I., and K. Neely. 1986. Ultrafiltration as a competitive unit process. *AIChE Symp. Ser.* 82 (250):65–77.

Blanck, R. G., and W. Eykamp. 1986. Fruit juice ultrafiltration. *AIChE Symp. Ser.* 82 (250):59–64.

Breslau, B. R., A. J. Testa, B. A. Milnes, and G. Medjanis. 1980. Advances in hollow fiber ultrafiltration technology. *Polym. Sci. Technol.* 13:109–127.

Cheryan, M. 1986. *Ultrafiltration Handbook.* Lancaster, PA: Technomic Publishing Co.

Christian, S. D., S. N. Bhat, E. E. Tudcer, J. F. Scamehorn, and D. A. El-Sayed. 1988. Micellar enhanced ultrafiltration of chromate anion from aqueous streams. *AIChE J.* 34:189.

DeBoer, R., and J. Hiddink. 1980. Membrane processes in the dairy industry. *Desalination* 35:169–192.

Ditnerskii, Y., K. Volchek, and Y. Zhilin. 1988. Ultrafiltration combined with selective complex formation as an effective method for separation of multicomponent solutions. *Khim. Ind. (Sofia)* 60(3):130–139.

Eykamp, W. 1990a. Ultrafiltration. In *Membrane Separation Systems: A Research Needs Assessment,* pp. 7-1–7-35. U.S. Department of Energy, Office of Energy Research, Office of Program Analysis.

Eykamp, W. 1990b. Ultrafiltration and microfiltration. Plenary Lecture, Session 64, Int. Congress on Membranes and Membrane Processes, 20–24 August 1990, Chicago, IL.

Garrison, J. B. 1986. Fresh juice forecast: Clear and uncloudy. *Beverage World* Sep.:197.

Hacket, R., and R. J. Swientek. 1986. UF system clarifies 25000 gallons per day of apple juice. *Food Process.* Jan.:80.

Honer, C. 1990. *Making History: Dairy Foods.* Tempe, AZ: Gorman Publishing Co.

Kalyanpur, M., W. Skea, and M. Siwak. 1984. Isolation of Cephalosporin C from fermentation broths. Millipore Technical Bulletin TB 007.

Kandori, K., and R. S. Schechter. 1990. Selection of surfactants for micellar-enhanced ultrafiltration. *Sep. Sci. Technol.* 25:83–108.

Knapik, P., and U. G. Bozzano. 1986. Process modelling of HFCS production: advantages of saccharification/ultrafiltration. Paper read at Corn Refiners Association, 21–24 September 1986, Arlington, VA.

Kosikowski, F. V. 1986. Membrane separations in food processing. In *Membrane Separations in Biotechnology,* ed. W. C. McGregor, pp. 201–254. New York: Marcel Dekker.

Kulkarni, S. S., E. W. Funk, and N. N. Li. 1986. Hydrocarbon separations with polymer membranes. *AIChE Symp. Ser.* 82 (250):78–84.

Marcus, D. L., and R. Postrick. 1988. Ultrafiltration—its role in today's water purification systems. *Ultrapure Water* Apr.:40–45.

Matthiasson, B., and T. G. I. Ling. 1986. Ultrafiltration affinity purification: a process for large scale biospecific separation. In *Membrane Separations in Biotechnology,* ed. W. C. McGregor. New York: Marcel Dekker.

McTeer, P., and T. Wett. 1989. Membranes KO oily wastes. *Chem. Process.* Sep.:52(10):106.

Michaels, A. S. 1968. Ultrafiltration. In *Progress in Separation and Purification,* ed. E. S. Perry. Vol. I, pp. 297–334. New York: Wiley-Interscience.

Mohr, C. M., S. A. Leeper, D. E. Engelgau, and B. L. Charboneau. 1990. *Membrane Applications and Research in Food Processing.* Park Ridge, NJ: Noyes Data Corp.

Nguyen, Q. T., P. Aptel, and J. Neel. 1980. Application of ultrafiltration to the concentration and separation of solutes of low molecular weight. *J. Membr. Sci.* 6:71–82.

Patterson-Candy Int. Membrane Systems. Ultrafiltration and RO for the pulp and paper industry. TPRO 44.2.

Paulson, D. J., R. L. Wilson, and D. D. Spatz. 1985. Reverse osmosis and ultrafiltration applied to the processing of fruit juices. In *Reverse Osmosis and Ultrafiltration,* ed. S. Sourirajan and T. Matsuura. ACS Symp. Ser. No. 281, pp. 325–344.

Porter, M. C. 1977. What, when and why of membranes MF, UF and RO. *AIChE Symp. Ser.* 73 (171):83–103.

Robichaux, W., and R. F. Ellis. 1982. Ultrafiltration plant recovers 35% protein concentrate from whey. *Food Process.* Jan.:102–103.

Scamehorn, J. F., and J. H. Harwell. 1987. An overview of surfactant based separation processes. *Surface Sci. Ser.* 26:169–185.

Short, J. L. 1988. New applications for crossflow membrane filtration. *Desalination* 70:341–352.

VIII

Microfiltration

31. Definitions	457
32. Theory for Deadend Microfiltration	461
33. Theory for Crossflow Microfiltration	480
34. Deadend Microfiltration: Applications, Design, and Cost	506
35. Crossflow Microfiltration: Applications, Design, and Cost	571

31
Definitions

Robert H. Davis
University of Colorado

GENERAL DESCRIPTION
CROSSFLOW VERSUS DEADEND
  MICROFILTRATION
MEMBRANE FOULING AND FLUX
  DECLINE

SUMMARY OF CHAPTERS ON
  MICROFILTRATION
NOTATION

GENERAL DESCRIPTION

When pressure-driven flow through a membrane or other filter medium is used to separate micron-sized particles from fluids, the process is called *microfiltration*. Although the exact size range is a matter of debate, microfiltration is generally defined to be the filtering of a suspension containing colloidal or fine particles with linear dimensions in the approximate range of 0.02 to 10 µm. This size range encompasses a wide variety of natural and industrial particles, as shown in Figure 31–1. These particles are generally larger than the solutes that are separated by reverse osmosis and ultrafiltration. Consequently, the osmotic pressure for microfiltration is negligible, and the transmembrane pressure drop, which drives the microfiltration process, is relatively small (1 to 50 psi, typically). Also, the membrane pore size and permeate flux are typically larger for microfiltration than for ultrafiltration and reverse osmosis.

During microfiltration, the imposed pressure drop, which is the driving force for the process, causes the suspending fluid and small solute species to pass through the membrane or other filter medium and be collected as permeate. The particles are retained by the filter medium and collected as concentrated retentate. The mechanism by which the particles are retained depends on the type of filter medium and the nature of its interactions with the particles being filtered. When a membrane with pores that are smaller than the particles is used as the filter medium, then a sieving mechanism applies. As shown in Figure 31–2(a), the particles are physically blocked from passing into or through the membrane. This sieving mechanism is sometimes called *surface filtration*.

A different mechanism applies if the membrane has pores that are larger than the particles or if the filter medium is a fibrous or granular material with interstices that are large compared to the particles being filtered. In these cases, particles can enter the interior of the filter medium. They are removed from the suspend-

FIGURE 31–1. Sizes of several natural and industrial particles (mfp = mean free path; MW = molecular weight).

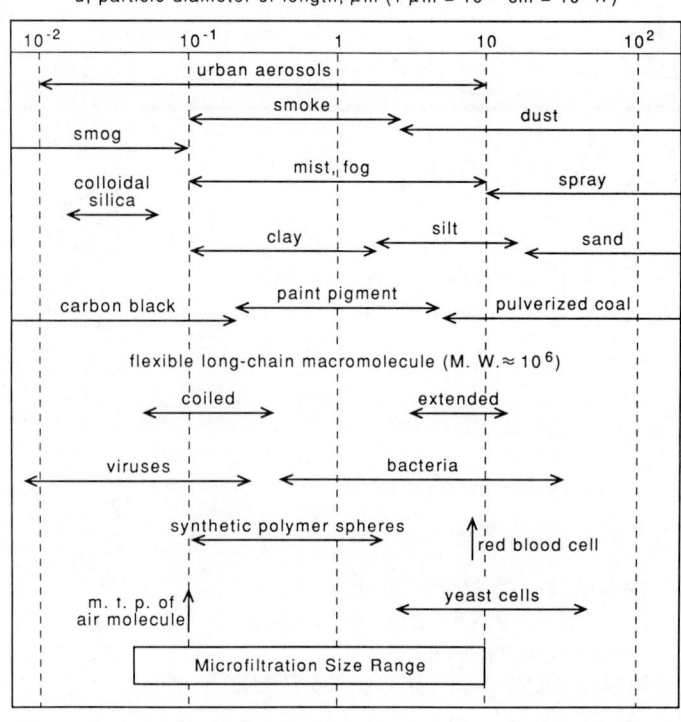

ing fluid if they come in contact with and adhere to the collecting surfaces of the filter medium, as shown in Figure 31–2(b). This type of filtration mechanism is called *depth filtration* since it can occur throughout the depth of the filter medium. Deep-bed filters employing depth filtration are common in high-volume applications such as the filtration of cleanroom and household air. Depth filtration is also employed on both the laboratory and industrial scale for biological and other separations using woven paper and cloth filters, or using membranes with pore sizes larger than the particles being filtered.

CROSSFLOW VERSUS DEADEND MICROFILTRATION

The microfiltration process is carried out in two types of configurations: deadend and crossflow. Most of us have performed deadend filtration in the laboratory using a circular piece of filter paper and a funnel. As shown in Figure 31–3(a), clarified fluid (permeate) is forced perpendicular through the filter, while all or most of the particles are retained. The driving force for the flow of permeate is the pressure drop across the membrane that results from the hydrostatic pressure of the feed suspension and

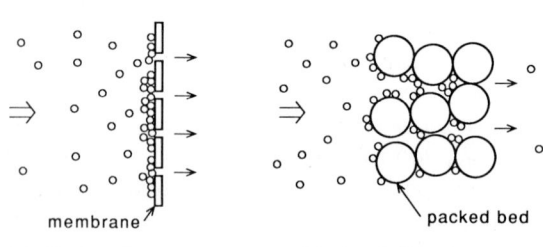

FIGURE 31–2. Schematics of surface filtration and depth filtration.

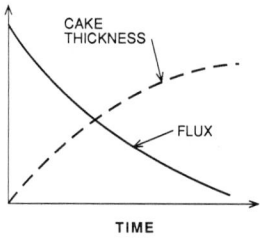
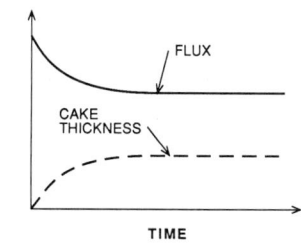
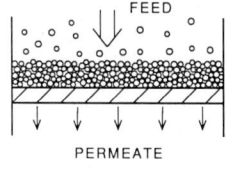
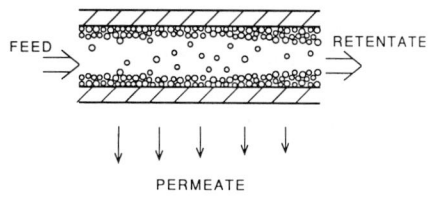

FIGURE 31-3. Schematics of deadend filtration and crossflow filtration.

from applying either suction to the permeate side or pressure to the feed side, or both.

In deadend filtration, the retained particles build up with time as a cake layer if a membrane, screen, or other surface filter medium is used. If a depth filter medium is used, then the retained particles build up in the void spaces. In either case, the particle buildup results in an increased resistance to filtration and causes the permeate flux rate to decline if the pressure drop is held constant, or causes the pressure drop to increase if the flux rate is held constant. As a result, the deadend filtration process must be stopped periodically in order to remove the particles or to replace the filter medium. As such, deadend filtration is by nature a batch process. An exception is rotary drum vacuum filtration, for which cake buildup occurs for only a portion of the rotation cycle. The cake is then discharged by a knife blade or other device at the end of the cycle.

During the past two decades, the crossflow configuration has been increasingly used as an attractive alternative to the deadend configuration. For crossflow microfiltration, sometimes referred to as *tangential-pass filtration*, the sieving mechanism of surface filtration is dominant. The filter operation is similar to that of ultrafiltration and reverse osmosis in that the bulk suspension is made to flow tangential to the surface of a membrane. Although this can be accomplished on a small scale using a batch stir-cell, the common mode of operation is to pump the suspension to be filtered through a narrow tube or channel having microporous membrane walls. The imposed transmembrane pressure drop causes a crossflow of permeate through the membrane to occur. As shown in Figure 31-3(b), the permeate flow carries particles to the membrane surface, where they are rejected and made to form a thin cake layer, which is analogous to the gel layer in ultrafiltration. Unlike deadend filtration, this cake layer does not build up indefinitely. Instead, the high shear exerted by the suspension flowing tangential to the membrane surface sweeps the deposited particles toward the filter exit so that the cake layer remains relatively thin.

As in ultrafiltration, the crossflow configuration for microfiltration is effective in controlling concentration-polarization and the associated cake buildup. Because of this, relatively high fluxes may be maintained over prolonged time periods. As shown in Figure 31-3(b), a steady or quasi-steady flux is achieved once the cake layer has reached its steady-state thickness. In practice, long-term flux decline is sometimes observed even after cake buildup has stopped. This may be the result of cake or membrane compaction or of membrane fouling.

MEMBRANE FOULING AND FLUX DECLINE

Crossflow or tangential-pass microfiltration uses a membrane in order to separate the particles from the suspending fluid. Deadend microfiltration also employs a membrane for many separations. If membrane separations are to be economical, high fluxes are required. Unfortunately, most membrane separations exhibit flux decline as a result of fouling. Fouling may be defined as the deposition of matter on or in the membrane such that membrane performance is altered. Fouling is poorly understood and is one of the major obstacles to more widespread use of crossflow microfiltration in solid-liquid separations. Although fouling is very complicated, we find that considerable understanding may be gained by defining two classes of fouling:

1. *Internal membrane fouling:* The attachment of material within the internal pore structure of the membrane or directly to the membrane surface due to adsorption, precipitation, pore plugging, particulate adhesion, etc.
2. *External cake fouling:* The formation of a stagnant cake layer on the membrane surface due to concentration-polarization as the material being filtered is carried to the membrane by the permeate flow and is then rejected by the membrane.

The external cake and the fouled membrane represent two resistances to filtration in series. Each of these resistances generally increases with time, resulting in a decline in the permeate flux with time. The external cake resistance increases due to cake buildup and compaction. The membrane resistance increases due to membrane fouling and compaction. The cake resistance may be reduced by using crossflow or tangential-pass filters, in which the high shear exerted by the suspension flowing tangential to the membrane sweeps the deposited particles toward the filter exit so that the cake layer remains thin. The membrane resistance may be reduced by using asymmetric membranes with only thin regions controlling the pore size and by choosing membrane materials that do not foul easily. Backflushing and a variety of other techniques have also been proposed for controlling flux decline during microfiltration. The proper approach for minimizing fouling and flux decline in a given situation clearly depends on whether the membrane or the cake layer provides the dominant resistance.

SUMMARY OF CHAPTERS ON MICROFILTRATION

As with the other parts of this handbook, the microfiltration part covers definitions, theory, design, applications, and cost estimates. This brief chapter on definitions has provided a general description of both deadend and crossflow microfiltration processes. Chapter 32 describes theories of deadend microfiltration, including the various depth filtration mechanisms and the surface filtration mechanism leading to cake buildup and flux decline. In Chapter 33, recent theoretical developments to describe crossflow microfiltration processes are presented. The focus is on predictive models for transient and steady-state permeate flux. The various effects of material properties and operating conditions on the permeate flux are described. Chapter 34 presents design and economic considerations for various deadend microfiltration applications, whereas Chapter 35 presents similar information for crossflow microfiltration applications. Available membranes and membrane modules are also tabulated and discussed in Chapters 34 and 35. The applications described include analytical uses, sterilization and purification of beverages, small-volume parenteral solutions and process fluids, cell culture perfusion, harvesting of cells from fermentation broths, antibiotics processing, and others.

NOTATION

General Notation

See the General Notation section at the beginning of this handbook.

Special Notation

mfp   mean free path, L
MW   molecular weight, M/mol

32

Theory for Deadend Microfiltration

Robert H. Davis
University of Colorado

Donald C. Grant
FSI International

DARCY'S LAW
    Membrane Resistance
    Cake Resistance
TRANSIENT OPERATION OF DEADEND MICROFILTERS
    Deadend Microfiltration with Constant Flux
    Deadend Microfiltration with Constant Pressure Drop
STEADY OPERATION OF CONTINUOUS ROTARY DRUM FILTERS
PARTICLE CAPTURE IN GASES
    Capture Mechanisms
    Definitions of Particle Capture Efficiency
    Theory for Single Collector Efficiency
    Relationship between Single and Overall Efficiency
    Results and Discussion for Airborne Particle Capture
PARTICLE CAPTURE IN LIQUIDS
    DLVO Theory for Particle Attraction and Repulsion
    Results and Discussion for Liquid-borne Particle Capture
NOTATION
REFERENCES

In this chapter, theories for deadend microfiltration are presented. The first part of the chapter focuses on the sieving or surface filtration mechanism, which is dominant when the particles are physically too large to pass through the pores of the filter medium. The primary goal is to predict the flux decline due to the buildup of the rejected particles on the membrane surface. We start with Darcy's law for the relationship between flux and pressure drop across a cake layer and membrane in series. This relationship is used to describe transient cake buildup and flux decline for batch operation of deadend microfilters. The analysis is then extended to continuous operation of rotary drum vacuum filters.

The second part of this chapter focuses on the various depth filtration mechanisms that occur when smaller particles are able to enter the interior of the membrane pores. The primary goal is to predict the efficiency of a given filter in removing particles of a given size. Particle capture by depth filtration from both gases and liquids is reviewed. In general, particles are more easily removed by depth filtration mechanisms from gases than from liquids.

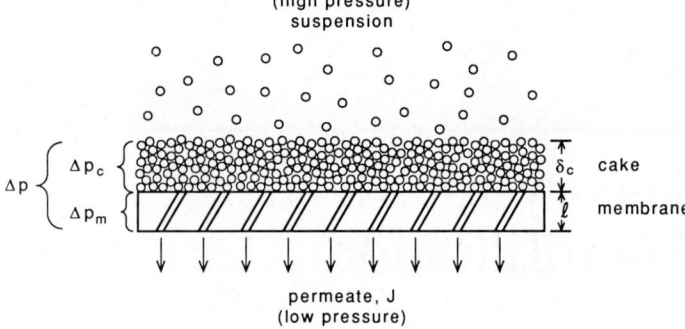

FIGURE 32–1. Schematic of filter cake and membrane in series.

DARCY'S LAW

When the sieving mechanism of microfiltration is dominant, a cake layer of rejected particles usually forms on the membrane surface, as shown in Figure 32–1. The pressure-driven permeate flux through this cake layer and the membrane may be described by Darcy's law:

$$J \equiv \frac{1}{A}\frac{dV}{dt} = \frac{\Delta p}{\eta_0(R_m + R_c)}, \quad (32\text{--}1)$$

where J is the permeate or volumetric flux and denoted commonly as J_v in the membrane literature, V is the total volume of permeate, A is the membrane area, t is the filtration time, Δp is the pressure drop imposed across the cake and membrane, η_0 is the viscosity of the suspending fluid, R_m is the membrane resistance (which can increase with time due to membrane fouling and compaction), and R_c is the cake resistance (which can increase with time due to cake buildup and compression). If this equation is to be useful, knowledge regarding the membrane and cake resistances is needed. Although such knowledge is best gained from experimental measurements, we present semi-empirical formulas that may be used to estimate R_m and R_c.

Membrane Resistance

Membrane resistance clearly depends on the membrane thickness, its nominal pore size, and various morphological features such as the tortuosity, porosity, and pore size distribution. For a membrane whose pores consist of cylindrical capillaries of uniform radius perpendicular to the face of the membrane, the resistance can easily be calculated. Using the Hagen-Poiseuille equation, the flux through such a membrane is

$$J = \frac{n_p \pi r_p^4 \Delta p_m}{8\eta_0 l}, \quad (32\text{--}2)$$

where n_p is the number of pores per unit area, r_p is the pore radius, l is the membrane thickness, and Δp_m is the transmembrane pressure drop. From this, the membrane resistance is given by

$$R_m \equiv \frac{\Delta p_m}{\eta_0 J} = \frac{8l}{n_p \pi r_p^4}, \quad (32\text{--}3)$$

indicating that the membrane resistance increases with increasing membrane thickness and decreases with increasing pore size and number density. It may also increase with time if fouling or particle capture in the membrane interior occurs.

It is often convenient to define the porosity, $\epsilon_m \equiv$ membrane void volume/total volume, and the specific surface area, $S_m \equiv$ pore surface area/solids volume. For a membrane with uniform cylindrical pores, it is easy to show that $\epsilon_m = n_p \pi r_p^2$ and $S_m = 2\pi n_p r_p/(1 - \epsilon_m)$. Using these parameters, Eq. (32–3) becomes

$$R_m = \frac{K(1 - \epsilon_m)^2 S_m^2 l}{\epsilon_m^3}, \quad (32\text{--}4)$$

where $K = 2$ for membranes with uniform cylindrical pores. For other membranes, Eq. (32–4) may still be used, but with the value of the constant K varying with the membrane morphology and pore structures (Gutman 1987).

Cake Resistance

When a cake is incompressible, its porosity and resistance are independent of the imposed pressure drop. The cake resistance is then often estimated by the Carman-Kozeny equation (Carman 1938), which is of the same form as Eq. (32–4):

$$R_c = \frac{K(1 - \epsilon_c)^2 S_c^2 \delta_c}{\epsilon_c^3}, \quad (32\text{--}5)$$

where δ_c is the cake thickness, ϵ_c is the void fraction of the cake, and S_c is the solids surface area per unit volume of solids in the cake. For rigid spherical particles of radius r, the specific surface area is $S_c = 3/r$, the void fraction of a randomly packed cake is $\epsilon_c \approx 0.4$, and the constant K is reported by Grace (1953) to have a value of 5.0.

Noting that the cake resistance is proportional to its thickness, a specific cake resistance per unit thickness is defined as

$$\hat{R}_c \equiv R_c/\delta_c. \quad (32\text{--}6)$$

Often, cake permeability K_c is reported instead, where $K_c = \hat{R}_c^{-1}$. An alternative quantity is the specific cake resistance on a mass basis:

$$R'_c \equiv R_c/w, \quad (32\text{--}7)$$

where w is the mass of cake deposited per unit area of membrane. Of course, these quantities are easily related:

$$w = \rho_s(1 - \epsilon_c)\delta_c, \quad (32\text{--}8)$$

$$\hat{R}_c = \rho_s(1 - \epsilon_c)R'_c, \quad (32\text{--}9)$$

where ρ_s is the mass density of the solids comprising the cake.

Many cake materials, such as flocculated clays and microbial cells, are highly compressible. Compressible cakes exhibit a decrease in void volume and an increase in the specific resistance as the compressive pressure is increased. Porter (1977) and Belter, Cussler, and Hu (1988) suggest that the effects of cake compressibility may be estimated by assuming that the specific cake resistance is a power law function of the imposed pressure drop:

$$R'_c = \alpha_0(\Delta p)^s, \quad (32\text{--}10)$$

where α_0 is a constant related primarily to the size and shape of the particles forming the cake, and s is the cake compressibility, which varies from zero for an incompressible cake to a value near unity for a highly compressible cake. These quantities are easily estimated by measuring the specific cake resistance at various pressure drops and then plotting the logarithm of R'_c versus the logarithm of Δp.

Although useful in practice, Eq. (32–10) represents an oversimplification of the behavior of compressible filter cakes. For example, it assumes that the cake compression depends on the pressure drop across both the membrane and the cake layer and not that across just the cake. Also, the porosity and specific cake resistance vary throughout the cake height. A more thorough analysis of cake compression is presented by Ward (1987) among others.

TRANSIENT OPERATION OF DEADEND MICROFILTERS

In this section, we briefly review the theory for the transient cake buildup, and the associated flux decline for a conventional deadend microfilter, employing a membrane or other filter medium that removes the suspended particles by the sieving mechanism of surface filtration. We consider a batch filtration process that starts with a clean membrane. A cake layer then builds up with time as the filtration proceeds.

The flux at any given time is governed by Darcy's law in a form derived by combining Eqs. (32–1) and (32–6):

$$J = \frac{\Delta p}{\eta_0(R_m + \hat{R}_c \delta_c)}. \quad (32\text{--}11)$$

This expression requires the cake layer thickness $\delta_c(t)$ to be known. This may be determined with the aid of a particle mass balance at the edge of the growing cake layer:

$$\left(J + \frac{d\delta_c}{dt}\right)\phi_s = \phi_c \frac{d\delta_c}{dt}, \quad (32\text{--}12)$$

where $\phi_s = 1 - \epsilon_s$ is the solids volume fraction in the suspension being filtered, and $\phi_c = 1 - \epsilon_c$ is the solids volume fraction in the cake, just below its top surface. The left-hand side of Eq. (32–12) represents the flux of particles into the surface of the cake, and it takes into account that this flux is due to the relative motion of the downward flow of suspension and the upward growth of the cake. As discussed by Doshi and Trettin (1981), the contribution to this flux due to back-diffusion of colloidal and fine particles in unstirred cells is negligible. The right-hand side of Eq. (32–12) represents the buildup of particles in the cake. Combining Eqs. (32–11) and (32–12) yields a first-order ordinary differential equation for the cake layer thickness:

$$\frac{d\delta_c}{dt} = \frac{\phi_s J}{\phi_c - \phi_s} = \frac{\phi_s \Delta p}{(\phi_c - \phi_s)\eta_0(R_m + \hat{R}_c \delta_c)}. \quad (32\text{--}13)$$

This equation is subject to the initial condition, $\delta_c = 0$ at $t = 0$. The form of its solution depends on whether or not the pressure drop is constant during the batch filtration.

Deadend Microfiltration with Constant Flux

If a positive-displacement pump is used, then the permeate flux through the filter remains constant at its initial value, $J = J_0$. The total volume of permeate V then increases linearly with time:

$$V(t) = AJ_0 t. \quad (32\text{--}14)$$

The thickness of an incompressible cake is given by

$$\delta_c(t) = \frac{\phi_s J_0}{\phi_c - \phi_s} t. \quad (32\text{--}15)$$

To maintain the constant permeate flux, the pressure drop must increase with time. From Eq. (32–11), this increase is

$$\Delta p(t) = J_0 \eta_0 \left(R_m + \frac{\hat{R}_c \phi_s J_0}{\phi_c - \phi_s} t\right). \quad (32\text{--}16)$$

For compressible cakes, the analysis is more complicated in that the increase in pressure drop with time causes changes in the cake porosity and in the specific cake resistance. In general, the process must be stopped when the pressure drop rises to the point where the pump is unable to deliver the specified flow rate.

Deadend Microfiltration with Constant Pressure Drop

Deadend batch filtration is often carried out with a constant imposed pressure drop. In this case, the permeate flux decreases with time due to cake buildup. Equation (32–13) may then be separated and integrated to yield

$$R_m \delta_c + \hat{R}_c \frac{\delta_c^2}{2} = \frac{\phi_s \Delta p}{(\phi_c - \phi_s)\eta_0} t. \quad (32\text{--}17)$$

In performing the required integration, R_m is assumed to be constant (no significant membrane fouling or compaction over time), as are ϕ_c and $\hat{R}_c$ (no significant changes in cake compression over time). This quadratic equation may be solved for the cake thickness to yield

$$\delta_c(t) = \frac{R_m}{\hat{R}_c}\left[\left(1 + \frac{2\hat{R}_c \phi_s \Delta p t}{(\phi_c - \phi_s)\eta_0 R_m^2}\right)^{1/2} - 1\right].$$

$$(32\text{--}18)$$

By combining this with Eq. (32–11), the flux expression is

$$J(t) = J_0\left(1 + \frac{2\hat{R}_c \phi_s \Delta p t}{(\phi_c - \phi_s)\eta_0 R_m^2}\right)^{-1/2}, \quad (32\text{--}19)$$

where the initial flux J_0 is given by

$$J_0 = \frac{\Delta p}{\eta_0 R_m}. \quad (32\text{--}20)$$

The total permeate volume produced after a time t may be determined by integrating Eq. (32–19) or by combining Eqs. (32–11), (32–13), and (32–18):

$$V(t) = \frac{A(\phi_c - \phi_s)}{\phi_s} \delta_c(t)$$

$$= \frac{A(\phi_c - \phi_s)R_m}{\phi_s \hat{R}_c}$$

$$\cdot \left[\left(1 + \frac{2\hat{R}_c \phi_s \Delta p t}{(\phi_c - \phi_s)\eta_0 R_m^2}\right)^{1/2} - 1 \right]. \quad (32\text{–}21)$$

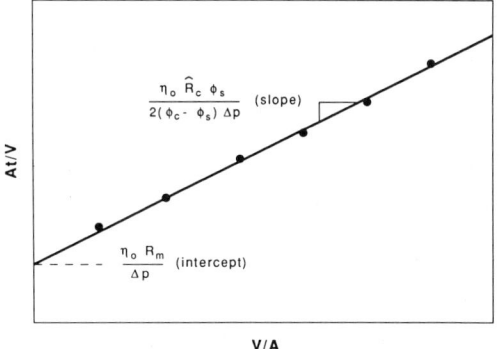

FIGURE 32–2. Examples of graph used to find the membrane resistance and the specific cake resistance.

As seen from Eq. (32–19), the permeate flux starts at its initial value for a clean membrane and then decreases linearly with time for short times due to cake buildup. As the flux declines, the rate of cake buildup also declines. For long times, the flux is inversely proportional to the square root of time, and the cake thickness and total permeate volume are directly proportional to the square root of time. More specifically, when $2\hat{R}_c \phi_s \Delta p t \gg (\phi_c - \phi_s)\eta_0 R_m^2$, the *cake resistance dominates* and Eqs. (32–18), (32–19), and (32–21) reduce to, respectively,

$$\delta_c(t) = \left(\frac{2\phi_s \Delta p t}{(\phi_c - \phi_s)\eta_0 \hat{R}_c}\right)^{1/2}, \quad (32\text{–}22)$$

$$J(t) = \left(\frac{(\phi_c - \phi_s)\Delta p}{2\hat{R}_c \phi_s \eta_0 t}\right)^{1/2}, \quad (32\text{–}23)$$

$$V(t) = A\left(\frac{2(\phi_c - \phi_s)\Delta p t}{\phi_s \eta_0 \hat{R}_c}\right)^{1/2}. \quad (32\text{–}24)$$

In contrast, when $2\hat{R}_c \phi_s \Delta p t \ll (\phi_c - \phi_s)/\eta_0 R_m^2$, the *membrane resistance dominates*. Thus, initially with a clean membrane, the permeate volume, cake thickness, and permeate flux are given by Eqs. (32–14), (32–15), and (32–20), respectively, for short times.

The theory developed above may be used to predict the time course of filtration when the cake and membrane properties are known from previous measurements or from correlations such as Eqs. (32–4) and (32–5). Since this is often not the case, a simple method for determining these properties is presented as follows. In a deadend batch filtration experiment at constant pressure drop, the total permeate volume per unit membrane area, $V(t)/A$, is measured as a function of time. By substituting Eq. (32–21) in Eq. (32–17), and then rearranging the result, the following expression is achieved:

$$\left(\frac{A}{V}\right)t = \frac{\eta_0 \hat{R}_c \phi_s}{2(\phi_c - \phi_s)\Delta p}\left(\frac{V}{A}\right) + \frac{\eta_0 R_m}{\Delta p}. \quad (32\text{–}25)$$

As depicted in Figure 32–2, a linear plot of At/V versus V/A then yields an intercept of $\eta_0 R_m/\Delta p$ and a slope of $\eta_0 \hat{R}_c \phi_s/[2(\phi_c - \phi_s)\Delta p]$. In this manner, the slope is used to determine the specific cake resistance, and the intercept is used to determine the membrane resistance.

The procedure described above may be repeated at different pressure drops to determine if significant cake compression or membrane compaction occurs with increasing pressure drop. If cake compression is suspected, then a common and simple approach is to correlate the specific cake resistance according to Eqs. (32–9) and (32–10):

$$\hat{R}_c = \rho_s \phi_c R'_c = \rho_s \phi_c \alpha_0 (\Delta p)^s. \quad (32\text{–}26)$$

Although the specific cake resistance is a function of pressure drop for a cake that obeys Eq. (32–26), it will be constant during the course of a constant pressure batch filtration. As such, all of the equations developed in this section still apply; the only change that is required is for $\hat{R}_c$ to be replaced by $\rho_s \phi_c \alpha_0 (\Delta p)^s$, as indicated by Eq. (32–26).

STEADY OPERATION OF CONTINUOUS ROTARY DRUM FILTERS

Rotary vacuum drum filters are widely used for difficult filtrations involving the removal of microbial cells and other particles that form compressible cakes with large resistances. A filter aid, such as diatomaceous earth, is usually added to reduce the cake resistance and to help remove the particles being filtered.

A schematic of rotary vacuum filtration is shown in Figure 32–3. A cylindrical drum, around which a filter cloth is placed, is made to rotate at constant angular velocity through a reservoir containing the suspension to be filtered. A vacuum is applied to the interior space of the drum. As the drum rotates, four steps occur. The first step is cake formation, during which the drum surface is in contact with the suspension reservoir. When a given location on the drum surface first enters the reservoir, it has either no cake or only a thin cake known as *precoat*. As the drum surface rotates through the suspension reservoir, the vacuum sucks permeate perpendicular through the drum surface. This leads to buildup of a cake layer with distance from the reservoir entrance. The second step is cake washing, during which particle-free liquid is sprayed onto the cake to remove the solution remaining in the void spaces. This step is usually employed when there is a valuable dissolved product in the solution; a common example is the recovery of antibiotics from fermentation broths containing filamentous fungi. The third step is cake drying, during which the applied vacuum is used to remove most of the remaining liquid from the cake. The final step is cake discharge, which uses a knife blade or other device to remove the cake from the drum surface.

The analysis of the cake formation step of continuous rotary vacuum filtration is analogous to the analysis of deadend batch filtration, despite the fact that the former is a steady-state process, whereas the latter is a transient process. The analogy is accomplished by replacing

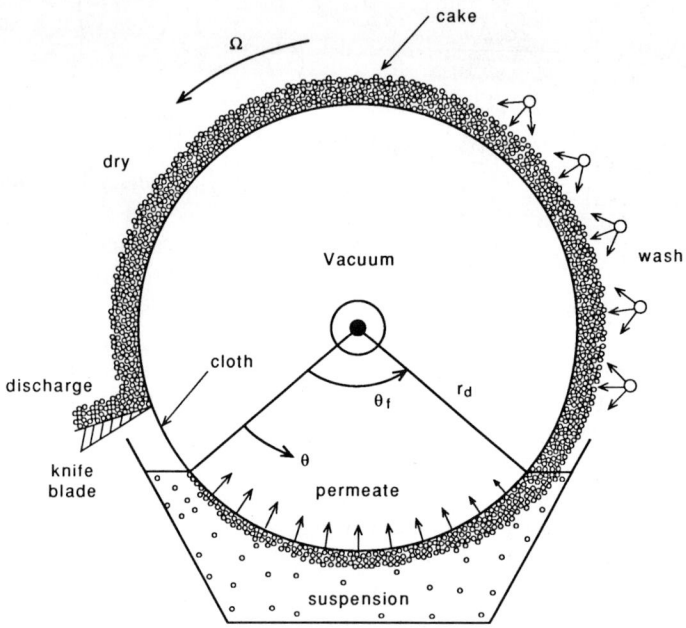

FIGURE 32–3. Schematic of a rotary vacuum drum filter.

the time variable t for deadend batch filtration with the time passed since a given location on the drum surface entered the suspension reservoir in rotary vacuum filtration. This time since entrance is equal to θ/Ω, where θ is the angular distance traveled by the drum since entering the suspension and Ω is the angular velocity of the drum. Using this analogy, the cake thickness and the filtration flux at the location θ are given by transforming Eqs. (32–18) and (32–19), respectively:

$$\delta_c(\theta) = \frac{R_m}{\hat{R}_c}\left[\left(1 + \frac{2\hat{R}_c\phi_s\Delta p\theta}{(\phi_c - \phi_s)\eta_0 R_m^2\Omega}\right)^{1/2} - 1\right],$$

(32–27)

$$J(\theta) = \frac{\Delta p}{\eta_0 R_m}\left(1 + \frac{2\hat{R}_c\phi_s\Delta p\theta}{(\phi_c - \phi_s)\eta_0 R_m^2\Omega}\right)^{-1/2},$$

(32–28)

where R_m is the resistance of the filter cloth plus any precoat.

The volumetric rate of permeate formation is determined by integrating the flux expression over the entire portion of the drum area in contact with the reservoir:

$$Q = Wr_d \int_0^{\theta_f} J(\theta)\,d\theta$$

$$= \frac{Wr_d(\phi_c - \phi_s)R_m\Omega}{\hat{R}_c\phi_s}$$

$$\cdot \left[\left(1 + \frac{2\hat{R}_c\phi_s\Delta p\theta_f}{(\phi_c - \phi_s)\eta_0 R_m^2\Omega}\right)^{1/2} - 1\right],$$

(32–29)

where r_d is the drum radius, W is the drum width, and θ_f is the angle subtended by the portion of the drum's surface that is in contact with the suspension reservoir. This equation is useful for design calculations because it relates the filter capacity to the filter dimensions, the material properties, and the operating conditions.

In practice, rotary vacuum filters are usually designed so that the cake resistance dominates. This requires that $2\hat{R}_c\phi_s\Delta p\theta_f \gg (\phi_c - \phi_s)\eta_0 R_m^2\Omega$, in which case Eq. (32–29) reduces to

$$Q = Wr_d\theta_f\left(\frac{2(\phi_c - \phi_s)\Delta p^{1-s}\Omega}{\rho_s\phi_s\phi_c\alpha_0\eta_0\theta_f}\right)^{1/2},$$

(32–30)

where it is noted that $Wr_d\theta_f$ is the filter area in contact with suspension and that θ_f/Ω is the time per cycle during which a given portion of the drum's surface is in contact with suspension. Also, Eq. (32–26) has been used for the specific cake resistance, since the cakes that are subject to rotary vacuum filtration are often compressible. The quantity $\rho_s\phi_s$ is the mass concentration (mass solids/volume suspension) of the particles in the suspension to be filtered, and the quantity $(\phi_c - \phi_s)/\phi_c$ is close to unity for dilute suspensions.

Rotary vacuum filters employ deadend filtration in the sense that the primary flow of suspension is in the direction perpendicular to the filter surface. The small amount of tangential shear stress that arises from the drum rotation is insufficient to affect the cake buildup, as described in Chapter 33.

PARTICLE CAPTURE IN GASES

We now turn our attention to depth mechanisms of microfiltration. Models of depth filtration of particles from gases are reviewed first, and this is followed by a review of the subject for liquids. The recent text by Tien (1989) presents a good overview of deadend microfiltration of both hydrosols and aerosols by depth filtration.

Capture Mechanisms

In gases, particles are brought into contact with a filter medium by six different mechanisms. Five of these are depth filtration mechanisms (gravitational settling, electrostatic deposition, impaction, interception, and diffusion) and are shown in Figure 32–4. This figure depicts a cross section of a portion of the filter medium, the fluid streamlines around the medium, and

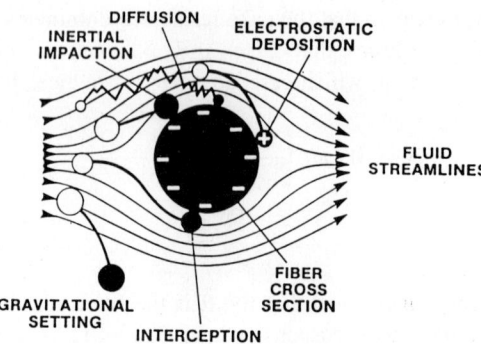

FIGURE 32-4. Particle capture mechanisms (reprinted from Grant et al. 1989 with permission).

the path of particles removed by each of the mechanisms described. Under most conditions, the particles are removed once they contact the medium.

Gravitational settling is simply the removal of particles from the flow stream due to their large masses and low velocities through the filter medium. *Electrostatic deposition* results from attractive forces between charged particles and an oppositely charged filter matrix. *Impaction* occurs when particles depart from the fluid streamlines due to their inertia as the flow is diverted around the filter structure. Impaction is favored by high velocities, relatively high particle densities, and large particle sizes. Particles are removed by *interception* if the location of the gas streamline and the particle size result in contact between the filter matrix and the particle. This occurs when the minimum distance between the gas streamline and the filter matrix is less than or equal to the particle radius. *Diffusive capture* results from the random motion of very small particles. This random or Brownian motion is caused by collisions of the particles with gas molecules and results in a particle following a zigzag path as it passes through the filter medium. The zigzag path increases the probability of particle capture. The last mechanism, *sieving*, occurs when particles are physically too large to pass through the openings in the filter medium.

The effectiveness of a membrane for removal of particles depends on the membrane structure. For example, membranes made by phase inversion casting are more efficient at removing particles than filters made by the track-etch process (Liu, Pui, and Rubow 1983).

The theory of particle capture by phase inversion membranes is well developed. Rubow (1981) and Rubow and Liu (1986) have shown that the two most important mechanisms for the removal of submicrometer particles from gases are interception and diffusion. The other mechanisms usually play negligible roles. Their theory models the filter membrane as a series of parallel fibers with dimensions determined from measured void volume and fiber diameters.

Definitions of Particle Capture Efficiency

The removal effectiveness of filters is often expressed as either fractional penetration (P) or removal efficiency ($1 - P$), where

$$P = \phi_{out}/\phi_{in}, \qquad (32\text{-}31)$$

and ϕ_{in} and ϕ_{out} are the particle volume fractions at the filter inlet and outlet, respectively.

Another expression of filter removal effectiveness is the filter log reduction value (LRV), which describes the number of orders of magnitude by which the feed concentration is reduced:

$$\text{LRV} = \log_{10}(\phi_{in}/\phi_{out}) = \log_{10}(1/P). \qquad (32\text{-}32)$$

A fourth measurement of filter efficiency is the single collector efficiency (η_s). This is simply the removal efficiency of a single layer or unit thickness of collection medium.

Theory for Single Collector Efficiency

The single fiber efficiency can be estimated by adding the efficiencies of each of the operative mechanisms. Simple addition of the mechanisms results in an overestimation of filter efficiency because the same particle cannot be removed by more than one mechanism. However, because the efficiencies for each of the individual mechanisms for a given particle size are generally very different, the error is slight.

The two particle mechanisms included in the Rubow (1981) theory are diffusion and interception. Hence,

$$\eta_s = \eta_d + \eta_i, \quad (32\text{-}33)$$

where η_d is the single fiber efficiency due to diffusion and η_i is the single fiber efficiency due to interception. These single fiber efficiencies are given by

$$\eta_d = 2.86 \left(\frac{1-\epsilon}{K}\right)^{1/3} \text{Pe}^{-2/3}$$
$$\cdot \left[1 + 0.389 \left(\frac{(1-\epsilon)\text{Pe}^{1/3}}{K}\right)\right], \quad (32\text{-}34)$$

$$\eta_i = \frac{1-\epsilon}{K} \frac{R_i^2}{(1+R_i)} \left(1 + \frac{2}{R_i}\right), \quad (32\text{-}35)$$

where
- Pe = Peclet number = vd_f/D
- d_f = diameter of the filter fibers
- v = velocity in the filter medium = $v_s/(1-\epsilon)$
- v_s = superficial face velocity
- D = particle diffusion coefficient
- ϵ = filter void fraction
- R_i = interception parameter = d_p/d_f
- d_p = diameter of the particles
- K = Kuwabara hydrodynamic factor
  = $-0.75 - 0.5 \ln\epsilon + \epsilon - 0.25\epsilon^2 + C_s(-0.5 - \ln\epsilon + 0.5\epsilon^2)$
- C_s = constant describing fraction of molecules reflected diffusely from the surface
  = 1.14 for air.

Relationship between Single and Overall Efficiency

According to Rubow (1981), the overall fractional penetration of particles through a membrane filter can be described using the properties of the filter and the single collector efficiency. Since Rubow modeled the filter as an array of cylindrical fibers, the appropriate relationship is

$$P = \exp\left(-\frac{4\epsilon l \eta_s}{\pi d_f \sigma}\right), \quad (32\text{-}36)$$

where l is the filter thickness and σ is the inhomogeneity factor (an empirically determined factor correlating theoretical and measured pressure drop).

Results and Discussion for Airborne Particle Capture

Typical theoretical performance of a membrane aerosol filter based on this theory is shown in Figure 32–5. Curves describing capture by interception, diffusion, and the combination of interception and diffusion are shown. When particle size increases, the probability of capture by interception increases, while the probability of capture by diffusion decreases. Hence,

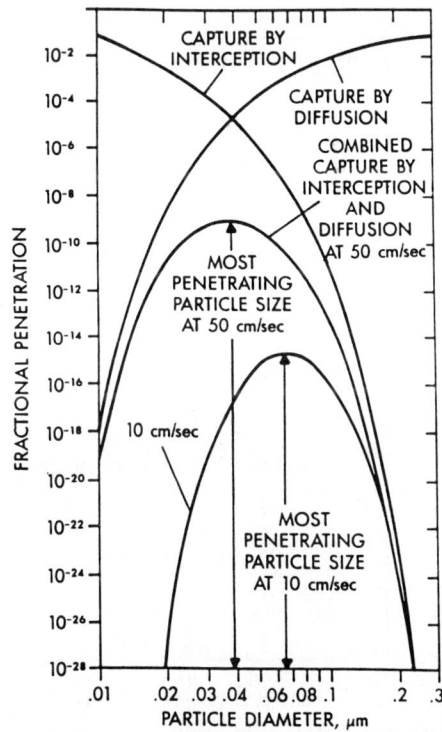

FIGURE 32–5. Theoretical representation of the most penetrating particle size.

capture by the combined mechanisms results in a particle size that exhibits maximum penetration. This is referred to as the most penetrating particle size (MPPS).

Particle retention is a function of the fluid velocity through the filter. Increasing the velocity increases particle penetration and decreases the MPPS. Both effects result because capture by interception is independent of velocity while diffusive capture decreases as velocity is increased. Capture by diffusion decreases because the residence time within the filter is shortened, thereby decreasing the probability of particle/filter contact.

The existence of a MPPS has been shown for various types of porous phase inversion membranes, including those made of cellulose blends (Rubow 1981) and ultrahigh efficiency polyvinylidenefluoride (PVDF) membrane gas filters (Rubow, Liu, and Grant 1987; Liu, Rubow, and Pui 1985). A good review of the performance of many filter types is found in a paper by Liu, Pui, and Rubow (1983). This paper describes the retention characteristics of 76 different types of filtration media.

Accomazzo, Rubow, and Liu (1984) investigated the removal of dioctylphthalate (DOP) particles by membrane filters with liquid-rated pore sizes of 0.2, 0.45, and 0.8 μm. They were only able to measure penetration through the 0.8-μm filter. The fractional penetration of the other filters was $<10^{-9}$. The results they obtained with the 0.8-μm filter are reproduced in Figure 32–6. This figure indicates that this filter is also very efficient at removing particles. Even at a velocity of 247 cm/s, the maximum fractional penetration was only 7×10^{-4}. The figure also shows that the MPPS and the magnitude of penetration are a function of the velocity through the membrane.

The applicability of the Rubow theory for filters with small pore sizes (down to 0.2 μm) was verified using a specially prepared PVDF membrane filter with a thickness of 23 μm, one-fifth that of standard production media (Figure 32–7). Because filter thickness affects the degree of penetration but not the MPPS (Rubow 1981), production media can be expected to have the same MPPS but significantly less penetration. The fractional penetration for

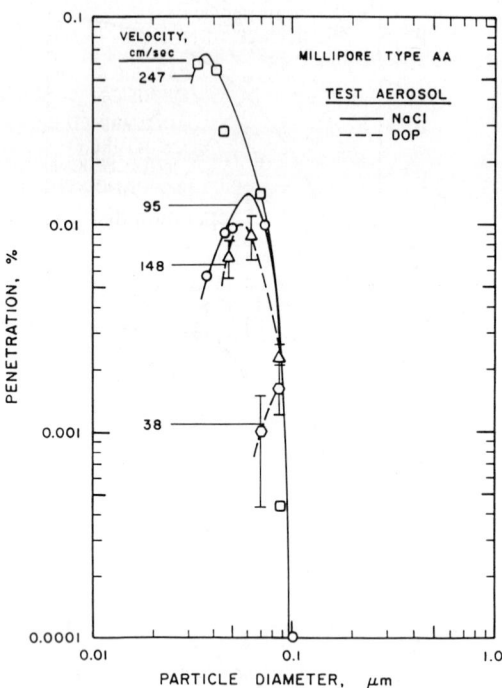

FIGURE 32–6. Gasborne particle penetration through a 0.81-μm liquid-rated filter (reprinted from Accomazzo, Rubow, and Liu 1984 with permission).

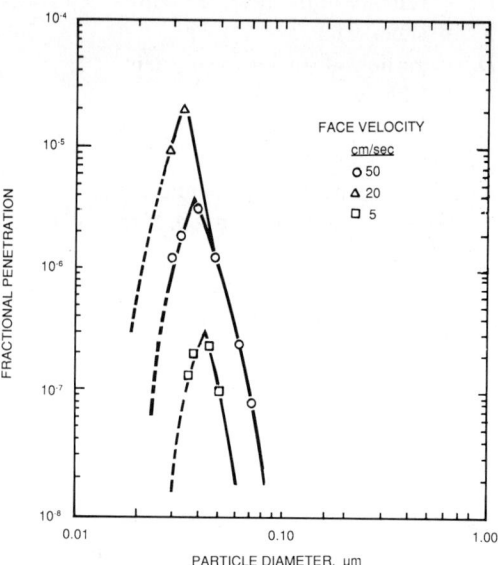

FIGURE 32–7. Gasborne particle penetration through an ultrathin PVDF membrane medium (reprinted from Grant et al. 1989 with permission).

the thicker production medium is estimated to be 6×10^{-32} at the MPPS of 0.045 μm for a typical face velocity of 5 cm/s (Rubow, Liu, and Grant 1987). Hence, small pore size filters formed by phase inversion remove essentially all particles of all sizes from gases in most applications.

PARTICLE CAPTURE IN LIQUIDS

The analysis presented above indicates that 0.2-μm liquid-rated membrane filters of typical thickness are extremely efficient at removing particles from gases; virtually no particles of any size pass through the filter. However, in liquids the situation is very different. The physical chemistry of the particle/fluid/membrane interface determines which of the particle capture mechanisms described above are operative. Under favorable chemical conditions, membranes have particle removal efficiencies in liquids similar to those in gases. However, under unfavorable conditions, particle capture is limited to sieving and membranes become orders of magnitude less efficient for particles smaller than the pore size (Grant et al. 1989).

DLVO Theory for Particle Attraction and Repulsion

The physical chemistry that determines which particle capture mechanisms are operative can be described using the DLVO theory of Derjaguin-Landau-Verwey-Overbeek (Verwey and Overbeek 1948). This theory describes the combined effects of van der Waals intermolecular forces and electrical double layer forces on solid surfaces immersed in a liquid medium.

van der Waals forces between two molecules in a vacuum have both attractive and repulsive components. The attractive component is a result of molecular dipole interactions, which can be divided into orientation, induction, and dispersion (or London) effects. The repulsion component (also called Born repulsion) arises from the interaction of the electronic atmospheres and nuclei of the atoms in molecules that are very close together. The attractive force decreases with the sixth power of separation distance while the repulsive force decreases with the twelfth power. Thus, the attractive force acts over a much longer distance than the repulsive force. The combined attractive and repulsive components of the van der Waals forces between two molecules are attractive, except at very close approaches.

Electrical double layer forces occur at solid/liquid interfaces because ions do not distribute themselves equally between phases. This unequal distribution causes one side of an interface to acquire a net charge and the other side to acquire a net charge of opposite sign. This leads to a potential across the interface known as the *electrical double layer*. Electrical double layer forces between two surfaces immersed in a liquid are attractive when the surfaces carry opposite charges and repulsive when the surfaces carry like charges. Solid materials with similar chemical compositions adsorb ions with the same charge so that the electrical double layer force is usually repulsive.

Under the majority of circumstances when solids are immersed in liquids, van der Waals forces are attractive and the double layer forces are repulsive. However, both the van der Waals forces and the double layer forces can be either attractive or repulsive (Visser 1981). The net effect of these often opposing forces is an energy potential relationship such as the one shown in Figure 32–8. Note that the Born repulsion that occurs at very short separation distances is not included in the figure.

Figure 32–8 shows a deep primary mini-

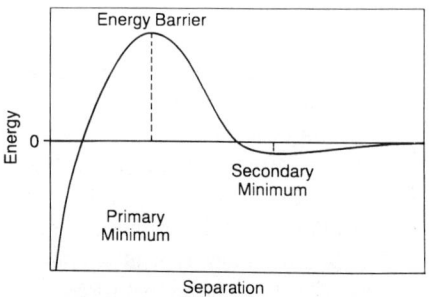

FIGURE 32–8. Schematic representation of interaction energy versus separation for combined van der Waals attraction and double layer repulsion (reprinted from Grant et al. 1989 with permission).

mum, an energy barrier, and a shallow secondary minimum. The surfaces are attracted to one another at separation distances that are in both the primary and secondary minima and repulsed at other separations. The attraction is very strong in the primary minimum and weak in the secondary minimum. Note that in order for the surfaces to reach the primary minimum where they are tightly held together, the energy barrier between the two minima must be overcome.

The magnitude of the energy barrier is very important in determining whether the surfaces approach each other closely enough to reach the primary minimum. The magnitude of the barrier is affected by many variables including the charge density on the two surfaces, the intensity of the intermolecular attraction force between the two surfaces, and the ionic strength of the liquid. Higher charges, lower van der Waals attraction, and lower ionic strength all increase the magnitude of the energy barrier.

The magnitude and sign of the combined intermolecular and double layer forces act like a switch that turns various collection mechanisms on and off. When the switch is off (forces are repulsive or unfavorable), particle capture is limited to sieving. However, when the switch is on (forces are attractive or favorable), particles are captured by interception, diffusion, gravitational settling, and sometimes electrostatic attraction, as well as by sieving.

Results and Discussion for Liquidborne Particle Capture

The theory of liquidborne particle capture by phase inversion-formed membranes under favorable chemical conditions can be estimated using Rubow's model (1981) describing airborne particle capture by interception and diffusion. Slight modifications of the model are required to eliminate corrections for fluid slip (Grant 1988). The model (Figure 32–9) predicts that the MPPS for a 0.2-μm liquid-rated filter will be 0.065 μm under typical operating conditions with favorable attractive forces between the particles and membrane material. The predicted fractional penetration of particles of that size is on the order of 10^{-30}. These predictions

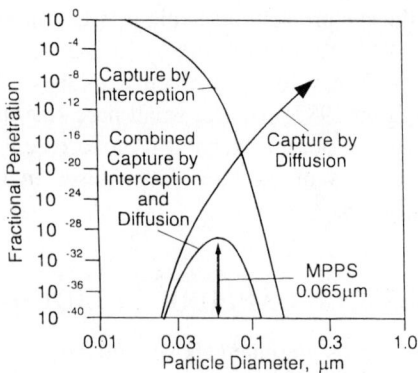

FIGURE 32–9. Predicted liquidborne particle capture by interception and diffusion based on the Rubow (1981) model (reprinted from Grant et al. 1989 with permission).

indicate that when favorable chemical conditions exist, virtually all particles will be removed, regardless of size.

Very high removal efficiencies for particles smaller than the rated pore size have been obtained using microporous filters by modifying the membrane surfaces to yield a charge opposite that of the particles. These filters have been shown to remove essentially all particles (Carazzone et al. 1985).

As indicated earlier, membrane filters are extremely efficient for particle removal under favorable chemical conditions. However, under unfavorable chemical conditions, capture of small particles is significantly reduced because particles are only captured by sieving. Larger particles are still removed extremely efficiently, with particles larger than the pore size retained quantitatively.

The retention of smaller particles is a function of filter loading. Grant (1988) showed that as filter loading increased, the retention decreased. The rate of decrease was shown to depend on a number of factors including particle size, filter pore size, filter thickness, particle size dispersity, and the degree of filter loading.

The retention of monodispersed particles by heavily loaded filters was examined by Simonetti, Schroeder, and Meltzer (1986), who measured the retention of monodisperse polystyrene latex (PSL) spheres in the presence of a surfactant. These conditions are extremely chemically unfavorable for two reasons. First, the PSL particles have a strong negative charge (Fitzpatrick and Spielman 1973); hence, strong double layer repulsion occurs when the membrane medium also has a negative charge. Since the membrane materials they tested had a negative charge when immersed in water, repulsion was prevalent in their experiments. In addition, the surfactant adsorbed onto both the membrane and the particle surfaces, resulting in steric hindrance, which prevented the particles from approaching the filter medium to the point where capture could occur.

Figure 32-10 shows the particle penetration results published by Simonetti and Schroeder (1986) for a 0.2-μm filter compared to predicted performance under favorable chemical conditions. As indicated, sieving is far less effective at removing particles than the other mechanisms, especially for heavily loaded filters.

Grant (1988) measured the retention of polystyrene latex (PSL) monodisperse by clean and lightly loaded 0.45-μm filters. The HVLP-series membranes used were made of polyvinylidenefluoride, modified so that they were hydrophilic. Grant found particle removal efficiency to be a strong function of particle diameter, filter thickness, and filter loading. The removal efficiency (in terms of the log reduction-value, LRV) increased with increasing particle size (Figure 32-11) and filter thickness (Figure 32-12). It decreased as the filter was loaded with small particles (Figure 32-11). For larger particles, retention initially decreased with increased loading and subsequently increased (Figure 32-13). The presence of larger particles had no effect on the initial retention of smaller particles but decreased the rate of particle breakthrough (Figure 32-14). In these figures, the horizontal axis refers to the volume of particles that has entered the filter medium per unit area ($\phi_{in}v_s t$), divided by the pore volume of the filter medium per unit area ($l\epsilon$). Also, the particle concentrations in the feed were varied from 1.0×10^7 to 1.4×10^9 particles/L, whereas the superficial velocity or flux was $v_s = 3.7$ cm/min for the experiments shown. Retention efficiency was found to be independent of filtration velocity and concentration over the limited ranges tested in the study.

Grant's experimental procedure was modified to evaluate the initial particle retention of membrane filtration cartridges (Millipore 1990; Grant and Zahka 1990). The filtration devices were challenged with a mixture of monodispersed latex beads from 0.109 to 0.726 μm. The concentration of the beads was about 10^5 particles/mL. This corresponds to a particle concentration that is 100 to 1000 times the levels typically found in electronics chemicals. The retention for a number of particle sizes was measured with a PMS HSLIS/μLPS-16 optical particle counter. Figures 32-15 and 32-16 show particle retention (log reduction value) versus particle size for devices using polytetrafluoroethylene (PTFE) and polyvinylidenefluoride (PVDF) membranes, respectively. They show that particle retention is related to particle size and the pore size rating of the membrane as determined by bubble point measurements.

Grant (1988) modified a model developed by Soo (1983) and described by Soo and Radke (1986a, 1986b, 1984) to predict the retention of small monodisperse particles by 0.45-μm membrane filters as a function of particle loading. The model states that the particle concentration of fluid exiting the filter medium can be described by three parameters: the initial reduced

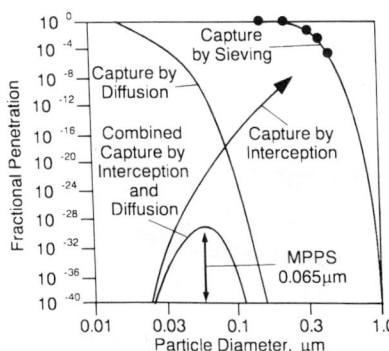

FIGURE 32-10. Comparison of liquidborne particle capture by sieving at high filter loadings to predicted capture by diffusion and interception. The solid circles represent experimental data (reprinted from Grant et al. 1989 with permission).

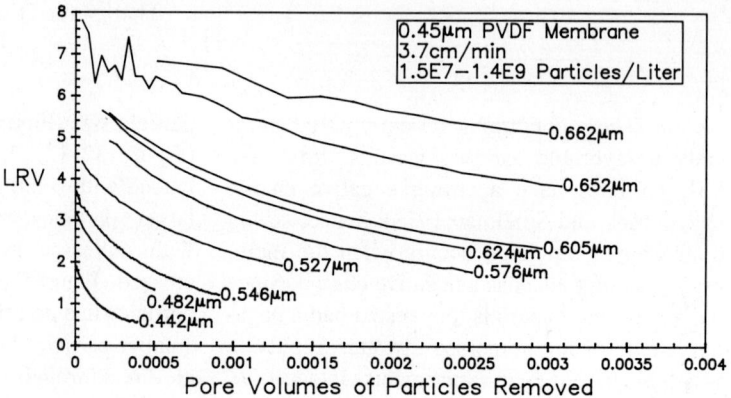

FIGURE 32–11. Filter removal efficiency as a function of particle size and filter loading.

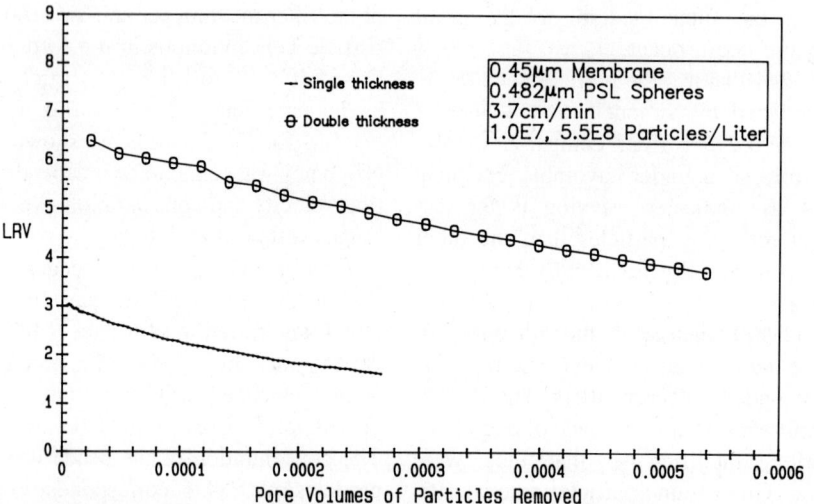

FIGURE 32–12. The effect of filter thickness on particle removal efficiency.

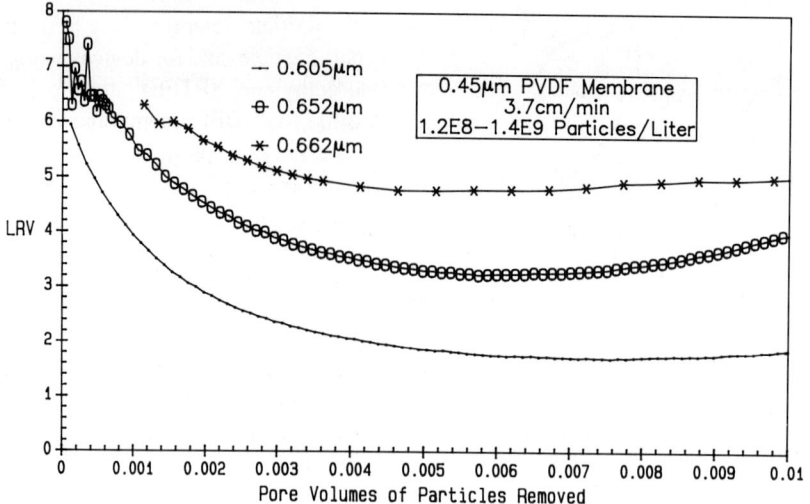

FIGURE 32–13. Retention of larger particles at high filter loadings.

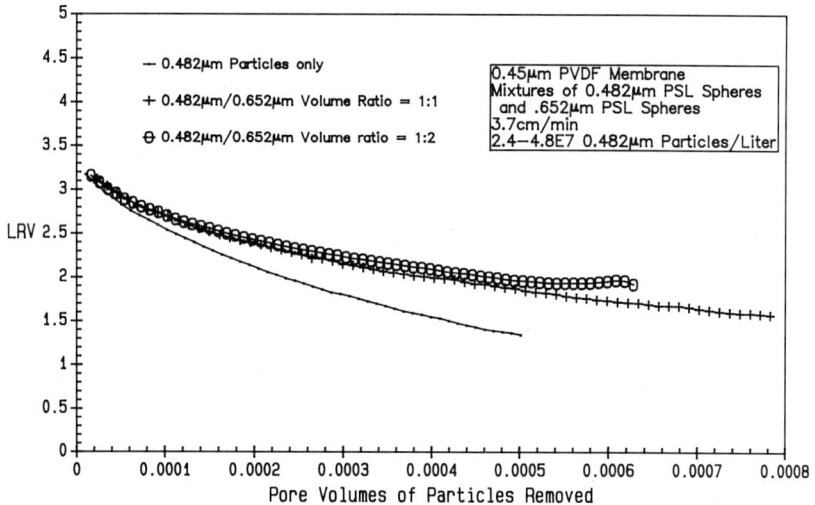

FIGURE 32–14. Effect of the presence of 0.652-μm particles on 0.482-μm particle retention.

dimensionless filter coefficient (Λ), the flow redistribution parameter (α), and the flow redistribution decrement parameter (ξ).

The model predicts that the fractional penetration of particles through the filter is given by

$$P = \frac{\{\exp[\alpha\Lambda\phi_{in}(\tau - 1)] - 1\}\exp(\alpha\Lambda\phi_{in}\tau)}{\{1 - \exp\Lambda - \exp[\alpha\Lambda\phi_{in}(\tau - 1)]\}[1 - \exp(\alpha\Lambda\phi_{in}\tau)]},$$

(32–37)

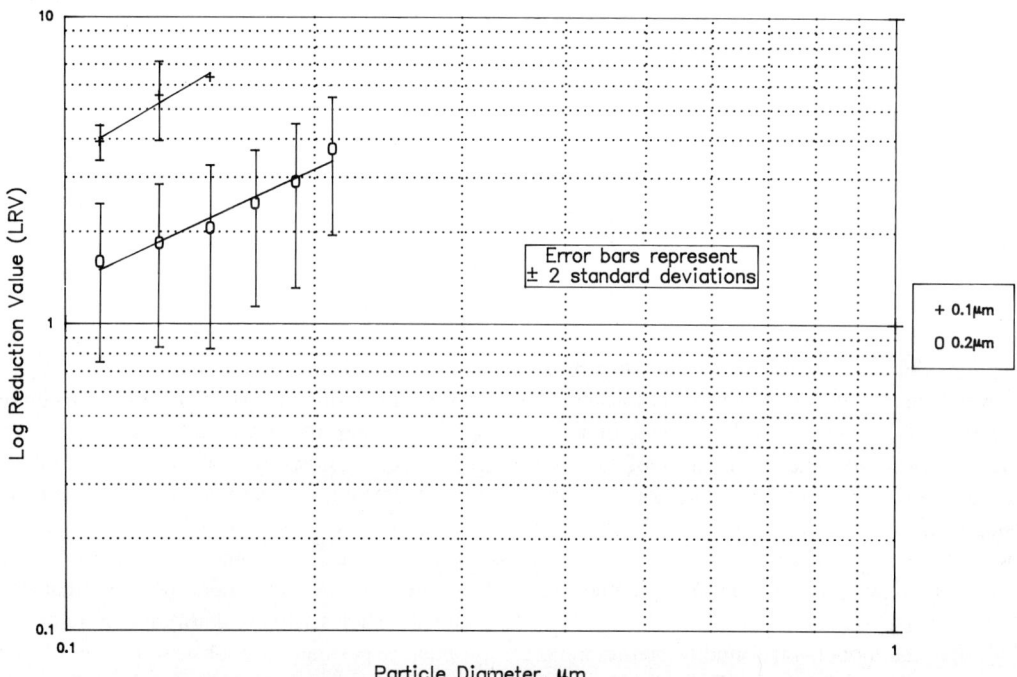

FIGURE 32–15. Initial particle retention of PTFE membrane filters (reprinted from Grant and Zahka 1990 with permission).

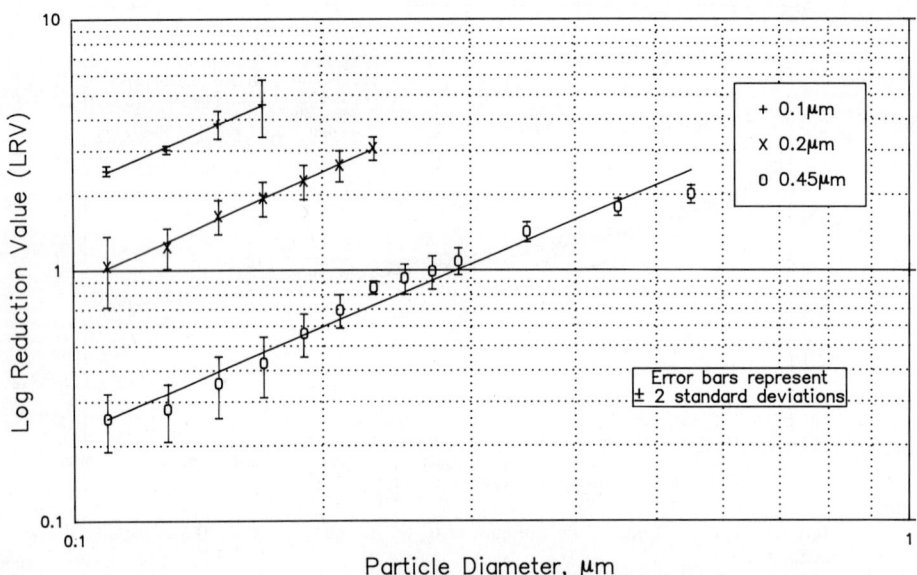

FIGURE 32–16. Initial particle retention of PVDF membrane filters (reprinted from Grant and Zahka 1990 with permission).

where

$\tau = v_s t/l\epsilon$, the number of pore volumes of fluid passed through the filter medium
v_s = fluid superficial velocity
t = time
l = filter thickness.

The flow redistribution parameter in the equation shown above is a function of the initial flow redistribution parameter (α_i) and the flow redistribution decrement parameter as defined by

$$\alpha = \alpha_i - \xi\tau(\phi_{in} - \phi_{out}). \qquad (32\text{--}38)$$

The model was shown to predict the breakthrough of monodisperse particles through the filter medium very well, as shown in Figure 32–17. However, the predictions of the modified model were only accurate for the initial stages of breakthrough. The model parameters were found to be a strong function of particle size. The model was not applied to other pore sizes.

Also, the model was found to predict a more rapid breakthrough of smaller particles and a less rapid breakthrough of larger particles than was experimentally observed (Figure 32–18). The differences between the predicted and the experimental results probably arise because the model parameters are functions of both the size and the size distribution of the particles being captured, while the parameters were determined using monodisperse particles.

This model obviously has limitations. In applications where the removal of monodisperse particles of known size is required, the model can be used to predict filter performance accurately. In cases for which the particles are polydisperse, the data show that breakthrough of smaller particles will occur more slowly and larger particles more rapidly than predicted. However, there are usually more small particles present in industrial suspensions than large particles. Also, the larger particles are removed much more efficiently than the smaller particles. Therefore, the particles in the suspension exiting the filter medium will consist mainly of the smaller particles in the suspension entering the filter. Since the model predicts that the smaller particles will break through more rapidly than experimentally observed, use of the model to predict the breakthrough of the total number of particles in a polydisperse suspen-

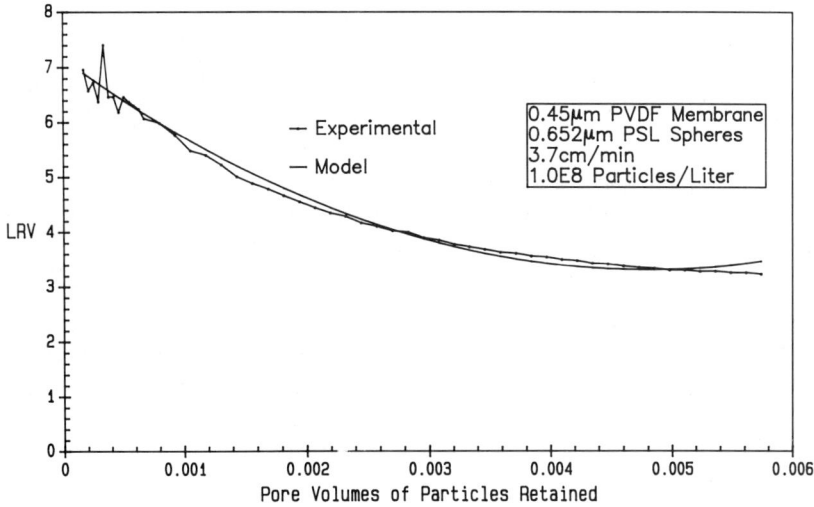

FIGURE 32–17. Comparison of measured and calculated filter performance using the modified sieving model (reprinted from Grant and Zahka 1990 with permission).

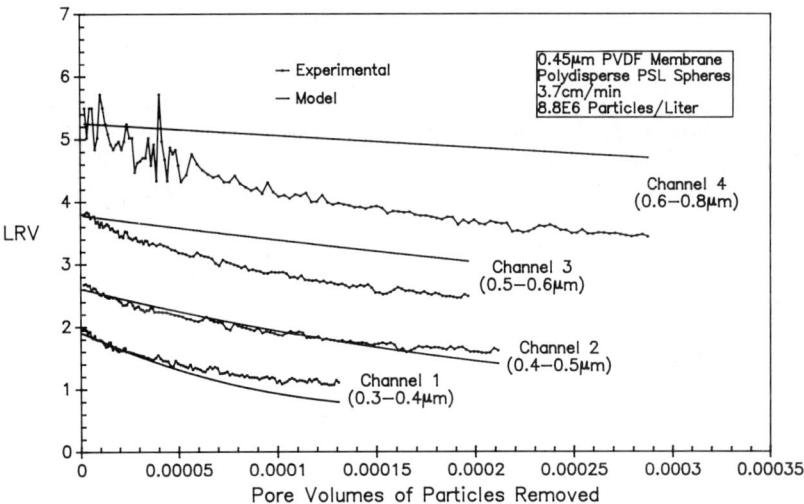

FIGURE 32–18. Comparison of measured and predicted filter performance during a polydisperse challenge experiment (reprinted from Grant and Zahka 1990 with permission).

sion represents "worst case" performance of the filter medium.

NOTATION

General Notation

See the General Notation section at the beginning of this handbook.

Special Notation

A membrane area, L^2
d_p particle diameter, L
J volumetric flux, L/t
J_0 initial flux, L/t
K hydrodynamic constant, dimensionless
P fractional penetration, dimensionless
Pe   Peclet number, vd_f/D, dimensionless

Q axial flow rate, L^3/t
r particle radius, L
R_i interception parameter, dimensionless
V total permeate volume, L^3

Greek Letters

α flow distribution parameter, dimensionless
α_i initial flow distribution parameter, dimensionless
α_0 constant in cake compressibility equation, L/Mp^s
ϵ filter void fraction, dimensionless
η viscosity, M/Lt
η_s single collector efficiency, dimensionless
θ angular distance, dimensionless
ξ flow redistribution decrement parameter, dimensionless
σ inhomogeneity factor, dimensionless
τ number of pore volumes of fluid passed through filter, dimensionless
ϕ solids volume fraction, dimensionless

REFERENCES

Accomazzo, M. A., K. L. Rubow, and B. Y. H. Liu. 1984. Ultrahigh efficiency membrane filters for semiconductor process gases. *Solid State Technol.* 27(3):141–146.

Belter, P. A., E. L. Cussler, and W.-S. Hu. 1988. *Bioseparations—Downstream Processing for Biotechnology.* New York: John Wiley & Sons.

Carazzone, M., D. Arecco, M. Fava, and P. Sancin. 1985. A new type of positively charged filter. *J. Parenteral Sci. Technol.* 39(2):69–74.

Carman, P. C. 1938. Fundamental principles of industrial filtration. *Trans. Inst. Chem. Eng.* 16:168–187.

Doshi, M. R., and D. R. Trettin. 1981. Ultrafiltration of colloidal suspensions and macromolecular solutions in an unstirred batch cell. *Ind. Eng. Chem. Fundam.* 20:221–229.

Fitzpatrick, J. A., and L. A. Spielman. 1973. Filtration of aqueous latex suspensions through beds of glass beads. *J. Colloid Interface Sci.* 43(2):350–369.

Grace, H. P. 1953. Resistance and compressibility of filter cakes. *Chem. Eng. Prog.* 49:303–318.

Grant, D. C. 1988. Sieving capture of particles by microporous membrane filtration media, M.S. thesis, Univ. of Minnesota, Mechanical Engineering Department, Minneapolis, MN.

Grant, D. C., B. Y. H. Liu, W. G. Fischer, and R. A. Bowling. 1989. Particle capture mechanisms in gases and liquids: an analysis of operative mechanisms. *J. Environ. Sci.* 42(4):43–51.

Grant, D. C., and J. G. Zahka. 1990. Sieving capture of particles by membrane filters from clean liquids. *Swiss Contamination Control* 3(4a):160–164.

Gutman, R. G. 1987. *Membrane Filtration: The Technology of Pressure-Driven Crossflow Processes.* Bristol, U.K.: Adam Hilger.

Liu, B. Y. H., D. Y. H. Pui, and K. L. Rubow. 1983. Characteristics of air sampling filter media. In *Aerosols in the Mining and Industrial Work Environments, Instrumentation,* ed. V. A. Marple and B. Y. H. Liu, Vol. 3, Ann Arbor, MI: Ann Arbor Science.

Liu, B. Y. H., K. L. Rubow, and D. Y. H. Pui. 1985. Performance of HEPA and ULPA filters. In *Proc. Annual Meeting of the Institute for Environmental Sciences,* pp. 25–28.

Millipore Electronics Division. 1990. Process chemical filter performance as characterized by a continuous challenge latex bead test. Technical Brief TB042. Millipore Corporation.

Porter, M. C. 1977. What, when, and why of membranes—MF, UF and RO. In *What the Filter Man Needs to Know About Filtration, AIChE Symp. Ser. No. 171,* ed. W. Shoemaker, pp. 83–103. New York: American Institute of Chemical Engineers.

Rubow, K. L. 1981. Submicron aerosol filtration characteristics of membrane filters, Ph.D. thesis, Univ. of Minnesota, Mechanical Engineering Department, Minneapolis, MN.

Rubow, K. L., and B. Y. H. Liu, 1986. Characterization of membrane filters for particle collection. In *Fluid Filtration: Gas,* ed. R. R. Raber. Vol. IU, ASTM STP 975, pp. 74–94. Philadelphia, PA.: American Society for Testing and Materials.

Rubow, K. L., B. Y. H. Liu, and D. C. Grant. 1987. Characteristics of ultra-high efficiency membrane filters. In *Proc. 33rd Annual Meeting of the Institute of Environmental Sciences,* pp. 383–387.

Simonetti, J. A., and H. G. Schroeder, 1986. Particle retention of submicrometer membranes. In *Fluid Filtration: Liquids,* ed. P. R. Johnston and H. G. Schroeder. Vol. II, ASTM STP 975, pp. 37–50. Philadelphia, PA: American Society for Testing and Materials.

Simonetti, J. A., H. G. Schroeder, and T. H. Meltzer. 1986. A review of latex sphere retention work: its application to membrane pore size rating. *Ultrapure Water* 3(4):46–51.

Soo, H. 1983. Flow of dilute, stable emulsions in porous media, Ph.D. thesis, Univ. of California, Department of Chemical Engineering, Berkeley, California.

Soo, H., and C. J. Radke. 1984. The flow mechanism of dilute, stable emulsions in porous media. *Ind. Eng. Chem. Fundam.* 23:342–347.

Soo, H., and C. J. Radke. 1986a. A filtration model for the flow of dilute, stable emulsions in porous media—I. theory. *Chem. Eng. Sci.* 41(2):263–272.

Soo, H., and C. J. Radke. 1986b. A filtration model for the flow of dilute, stable emulsions in porous media—II. parameter evaluation and estimation. *Chem. Eng. Sci.* 41(2):263–272.

Tien, C. 1989. *Granular Filtration of Aerosols and Hydrosols*. Boston: Butterworths.

Verwey, E. J. W., and J. T. G. Overbeek. 1948. *Theory of the Stability of Lyophobic Colloids*. Amsterdam: Elsevier Scientific Publishing Co.

Visser, J. 1981. The concept of negative Hamaker coefficients: 1. history and present status. *Adv. Colloid Interface Sci.* 15:157–169.

Ward, A. S. 1987. Liquid filtration theory. In *Filtration Principles and Practices*, ed. M. J. Matteson and C. Orr, 2nd ed., pp. 132–161. New York: Marcel Dekker.

33

Theory for Crossflow Microfiltration

Robert H. Davis
University of Colorado

OPERATIONAL CONFIGURATIONS
  FOR CROSSFLOW MICROFILTERS
  Analysis of Total Recycle Configuration
  Analysis of Dialysis (Diafiltration) Configuration
  Analysis of Batch Recycle Configuration
  Analysis of the Feed-and-Bleed Configuration
STEADY OPERATION OF
  CROSSFLOW MICROFILTERS
  Concentration-Polarization Model
  Similarity Solution
  Integral Model
  Inertial Lift Theory
TRANSIENT OPERATION OF
  CROSSFLOW MICROFILTERS
SUMMARY OF FLUX MODELS FOR
  CROSSFLOW MICROFILTERS
COMPARISON WITH
  EXPERIMENTS
CONCLUSIONS
ACKNOWLEDGMENTS
NOTATION
REFERENCES

In this chapter, recent theories describing crossflow microfiltration behavior are presented. First, the use of macroscopic balances to describe the overall behavior of various microfiltration module configurations is briefly reviewed. A major portion of this chapter is then devoted to recent models that predict the steady-state and transient permeate flux for crossflow microfiltration. These models are summarized in a brief section that describes the predicted dependence of the permeate flux on the material properties of the suspensions and the operating conditions of the filter. The focus is on the use of microporous membranes, which accomplish the desired separation using the sieving mechanism of surface filtration. The assumption is made that the membrane completely rejects the particles reaching its surface. The chapter concludes with a review of crossflow filtration experiments and their comparison with theory.

OPERATIONAL CONFIGURATIONS FOR CROSSFLOW MICROFILTERS

Crossflow filters can be used and studied in a variety of operating configurations. Four common configurations are shown in Figure 33–1. In total recycle, no net concentration of particles occurs because both the permeate and retentate are returned to the feed reservoir. This configuration is usually used in research to

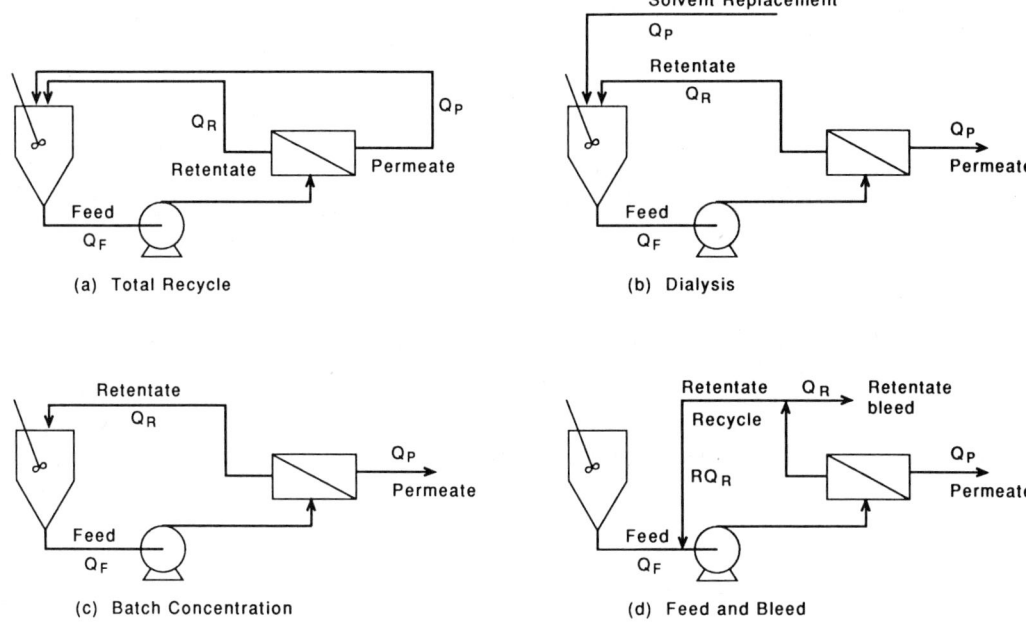

FIGURE 33–1. Schematics of common operational configurations for crossflow microfiltration.

measure the permeate flux at different operating conditions. During *dialysis,* the retentate is recycled to the feed reservoir but the permeate is replaced by an equal amount of fresh solvent. This confuguration is also called *diafiltration* and typically used to wash the particles or to remove an acid or a dissolved solute species. In batch concentration, the retentate is returned to the feed reservoir and the permeate is removed without replacement. As a result, the concentration of particles increases with time. In contrast, feed-and-bleed configurations are designed for steady-state operation. The permeate is removed from the system, as is a small portion of the retentate referred to as the *bleed.* Most of the retentate is recycled, however, to maintain a high tangential velocity in the filter module. In the following sections, these four configurations are analyzed using macroscopic mass balances.

Analysis of Total Recyle Configuration

The analysis of the total recycle configuration is straightforward. A macroscopic mass balance on total suspension (liquid plus solids) about the filter module yields

$$Q_f = Q_r + Q_p, \qquad (33\text{-}1)$$

where Q is the volumetric flow rate, and the subscripts f, r, and p refer to the feed, retentate, and permeate streams, respectively. The use of volumetric flow rates instead of mass flow rates is justified as long as the densities of the liquid and solid materials remain constant. The permeate flow rate is equal to the membrane area multiplied by the area-averaged permeate flux:

$$Q_p = \langle J \rangle A. \qquad (33\text{-}2)$$

A macroscopic mass balance on the particles entering and exiting the filter yields

$$\phi_f Q_f = \phi_r Q_r + \phi_p Q_p, \qquad (33\text{-}3)$$

where ϕ is the volume fraction of the solids being filtered. In deriving Eq. (33-3), we assumed that the accumulation of particles inside the filter due to transient cake buildup is negligible. Also, both sides of Eq. (33-3) may be multiplied by the particle density ρ_s to yield an expression in terms of the mass concentration, $\rho_s \phi$.

Most microporous membranes are chosen with pore sizes sufficiently small so that virtually all of the suspended particles are rejected. In this case, $\phi_p = 0$, and Eqs. (33–1), (33–2), and (33–3) may be combined to give

$$\frac{\phi_r}{\phi_f} = \frac{Q_f}{Q_r} = \frac{Q_f}{Q_f - \langle J \rangle A}. \quad (33\text{-}4)$$

The quantity ϕ_r/ϕ_f is called the *concentration factor,* and Eq. (33–4) provides a relationship between this quantity and the length-averaged permeate flux $\langle J \rangle$.

Analysis of Dialysis (Diafiltration) Configuration

The macroscopic mass balances for the total recycle configuration also apply to the dialysis configuration, provided that the replacement solvent is added at the same rate that permeate is removed. This is easily accomplished by using a level controller on the feed reservoir. An additional mass balance is needed for the solute that is being dialyzed out of the solution. Denoting the solute concentration in the solution as c_i, the unsteady macroscopic mass balance for this species about the entire system (reservoir and filter module) is

$$V_s \frac{dc_i}{dt} = -Q_p c_i, \quad (33\text{-}5)$$

where V_s is the total volume of suspension. The right-hand side of Eq. (33–5) represents the rate of removal of solute species i with the permeate stream. (We assume here that the microporous membrane does not reject any of the solute molecules.) The left-hand side represents the rate of decrease in the concentration of species i in the system due to the replacement of solute-containing permeate by solute-free solvent. This equation together with Eq. (33–2) may be integrated to yield the solute concentration as a function of time:

$$c_i(t) = c_i(0) \exp\left(-\frac{A}{V_s} \int_0^t \langle J \rangle \, dt\right), \quad (33\text{-}6)$$

where $c_i(0)$ is the initial concentration. This result takes the form of a simple exponential decay for the solute concentration if the permeate flux is constant at its steady-state value for most of the duration of the dialysis.

Analysis of Batch Recycle Configuration

The batch recycle configuration is used to concentrate a suspension in a transient, batchwise manner. To accomplish this, the retentate is returned to the feed reservoir, whereas the permeate is removed from the system without replacement. The unsteady macroscopic mass balance on total material in the system is

$$\frac{dV_s}{dt} = -Q_p. \quad (33\text{-}7)$$

This may be integrated with the aid of Eq. (33–2) to yield

$$V_s(t) = V_s(0) - A \int_0^t \langle J \rangle \, dt, \quad (33\text{-}8)$$

where $V_s(0)$ is the initial volume of suspension to be concentrated. The average permeate flux $\langle J \rangle$ usually decreases over time because the increase in the retained solids concentration with time leads to thicker cake layers on the membrane surfaces.

The increase in the retained solids concentration in the feed reservoir may be determined by a simple mass balance:

$$\phi_f(t) = \phi_f(0) V_s(0)/V_s(t), \quad (33\text{-}9)$$

where $\phi_f(0)$ is the initial solids volume fraction. In deriving Eq. (33–9), the assumption is that the reservoir is sufficiently large such that the accumulation of solids in the filter cake may be neglected. Together with Eq. (33–8), this equation may be used to predict the time required to achieve a desired solids volume fraction.

Analysis of the Feed-and-Bleed Configuration

The feed-and-bleed configuration is also used to concentrate a suspension, but in a steady-state, continuous manner. This is accomplished by recycling the majority of the retentate stream while bleeding off a small portion of this stream as product. The macroscopic mass balances on total material and on solids only about the filter unit including the recycle loop are the same as those for the total recycle configuration. Thus, the concentration factor is given by Eq. (33–4).

In most applications, the desired concentration factor and feed flow rate are specified, and it is the required permeate flux and/or membrane area that is to be determined. Rearranging Eq. (33–4) yields

$$\langle J \rangle A = Q_f(1 - \phi_f/\phi_r), \quad (33\text{--}10)$$

where ϕ_r is the solids volume fraction in the retentate product stream, and Q_f and ϕ_f are, respectively, the flow rate and solids volume fraction of the feed stream *before* it is joined by the recycle stream. The area-averaged permeate flux $\langle J \rangle$ is a function of the flow rate and the solids volume fraction of the feed stream entering the filter module *after* it has been joined by the recycle stream. Thus, mass balances on total material and on solids only about the junction of the feed and recycle streams are needed, yielding

$$Q_{in} = Q_f + RQ_r, \quad (33\text{--}11)$$

$$\phi_{in} = (\phi_f Q_f + \phi_r RQ_r)/Q_{in}, \quad (33\text{--}12)$$

where Q_{in} is the flow rate of the stream entering the filter module, ϕ_{in} is the solids volume fraction in this stream, and R is the recycle ratio.

In typical applications, large recycle ratios of R on the order of 10^2 are used, indicating that only a small fraction of the fluid entering the filter is removed during a single pass. These are necessary because large circulation velocities through the filter modules are required to reduce cake layer buildup and to maintain high permeate fluxes. As a result, the solids concentration of the stream entering the filter is only slightly less than the solids volume fraction in the retentate stream exiting the filter, $\phi_{in} \approx \phi_r$. Unfortunately, a high solids concentration in the stream entering the filter leads to a low permeate flux. Thus, when the use of a feed-and-bleed system is desired to produce a concentrated slurry, it is often more efficient to use multiple filters in series. Each filter then is required to achieve only a portion of the desired concentration factor. The feed-and-bleed filters that are nearer the beginning of the cascade have lower product concentrations and, hence, higher permeate fluxes.

STEADY OPERATION OF CROSSFLOW MICROFILTERS

The macroscopic mass balances that describe the operation of various crossflow microfiltration configurations require knowledge of the area-averaged permeate flux $\langle J \rangle$. In this section, recent theories for predicting this quantity at steady or quasi-steady state are reviewed. By quasi-steady state, we refer to the situation in which the short-term flux decline due to cake buildup has already occurred, but where there is still a long-term flux decline that is occurring due to phenomena such as membrane fouling or gradual increases in the particle concentration in the feed stream. In the next section, transient models for the decline in permeate flux due to cake buildup and other phenomena are described.

The use of crossflow microfiltration to concentrate colloidal, cellular, and fine particles has become common only within the past two decades, with early research reviewed by Blatt et al. (1970), Henry (1972), and Porter (1972a, 1986). At first, it was thought that the analogy with ultrafiltration of macromolecules would allow the concentration-polarization model to predict the steady-state microfiltration flux. However, predicted fluxes using the Brownian diffusivity given by the Stokes-Einstein relationship were found to be one or two orders of magnitude less than those observed in practice.

This finding (obvious, in hindsight!) follows from the fact that the Brownian diffusivities of micron-sized particles are much lower than the molecular diffusivity of macromolecules (yielding lower *predicted* fluxes), whereas the membrane and cake permeabilities for microfiltration are higher than the corresponding permeabilities for ultrafiltration (yielding higher *observed* fluxes). Green and Belfort (1980) refer to this discrepancy as the "flux paradox for colloidal suspensions."

One proposed resolution of the flux paradox is that the back-diffusion of particles away from the membrane is supplemented by a lateral migration of particles due to inertial lift—the "tubular-pinch effect" (Madsen 1977; Green and Belfort 1980; Altena and Belfort 1984; Belfort 1989). If the conditions are such that the inertial lift velocity is sufficient to offset the opposing permeate velocity, then the particles are not expected to be deposited on the membrane (Otis et al. 1986). Unfortunately, the inertial life velocity is often much less than the permeation velocity in typical crossflow microfiltration systems. When this is true, a concentrated layer of deposited particles forms on the membrane surface. If this cake layer built up indefinitely, it would plug the tube or channel. Instead, Blatt et al. (1970), Kraus (1974), and Schneider and Klein (1982) have hypothesized that the cake layer accumulates only until the hydrodynamic shear exerted by the flow of suspension causes the cake to flow tangentially along the membrane surface at a rate that balances the deposition of particles.

A simple convective-flow mathematical model that describes the simultaneous deposition of particles onto the cake layer and the sweeping of this layer along the membrane surface has been presented by Leonard and Vassilieff (1984). These authors were able to obtain analytical solutions for the cake layer thickness by assuming that the velocity profile in the vicinity of the cake layer was linear, the permeate flux was constant, and the cake layer could be treated either as an immobile solid or as a Newtonian fluid with the same effective viscosity as that of the bulk suspension. Davis and Birdsell (1987) developed a steady stratified flow model that relaxes some of these restrictions, and they performed experiments that yielded cake layer thickness profiles that are in good agreement with the model predictions.

In an alternative approach, Zydney and Colton (1986) proposed that the concentration-polarization model could be applied to microfiltration provided that the Brownian diffusivity was replaced by the shear-induced hydrodynamic diffusivity first measured by Eckstein, Bailey, and Shapiro (1977). Shear-induced hydrodynamic diffusion of particles occurs because particles undergo displacements from the streamlines in a shear flow as they interact with and tumble over other particles. The shear-induced hydrodynamic diffusivity is proportional to the square of the particle size multiplied by the shear rate. Unlike Brownian diffusion, shear-induced diffusion increases with particle size. The Brownian diffusivity of a micron-sized particle is only on the order of 10^{-9} cm^2/s in magnitude, whereas its shear-induced hydrodynamic diffusivity in typical crossflow microfiltration applications is approximately four orders of magnitude greater (Davis and Leighton 1987; Romero and Davis 1988). Zydney and Colton (1986; 1987) showed that the shear-induced diffusion model provides good agreement with blood plasmapheresis experiments. The model assumes a steady state and negligible membrane resistance. These assumptions are reasonable for plasmapheresis because a thin cake layer of deformed red blood cells with a very high resistance quickly forms on the membrane surface.

For small particles that are not highly deformable and form cake layers that have low hydraulic resistances, the approach described above is inadequate. Instead, the author's research team has developed an integral theory that includes both the cake resistance and the membrane resistance, and uses concentration-dependent shear-induced diffusivities and shear viscosities appropriate for particulate suspensions (Davis and Leighton 1987; Romero and Davis 1988, 1990). In the limit of cake-dominated resistance, Davis and Sherwood (1990) have also performed a similarity solution to predict the steady-state permeate flux. Experiments by Romero and Davis (1991) with latex suspensions are in good agreement with both the steady-state and transient model predictions for cake buildup and flux decline.

In the following sections, we first examine the concentration-polarization model employing shear-induced hydrodynamic diffusion. The similarity solution, which relaxes the assumptions of constant viscosity and diffusivity, is then described. This is followed by a summary of the integral solution, which includes both the cake and membrane resistances. For comparison, the predictions of the inertial lift theory are also described.

Concentration-Polarization Model

When the colloidal particles being filtered are rejected by a microporous membrane used in crossflow microfiltration, they build up as a cake layer on the membrane surface. This cake layer accumulates and reduces the permeate flux until steady state is reached. At steady state, the convection of particles toward the cake layer by the permeate flow is balanced by diffusion and inertial lift of particles away from the cake layer, and by convection of particles toward the filter exit due to the tangential flow of suspension. A concentration-polarization boundary layer forms adjacent to the cake layer in which there is a rapid decrease in particle concentration from that at the edge of the cake layer to that in the bulk suspension.

In the traditional concentration-polarization model, convection of particles toward the cake layer is assumed to be balanced by particle diffusion away from the cake layer. This is assumed to occur in a thin boundary layer of thickness δ. This is described mathematically by

$$-D \frac{\partial \phi}{\partial y} = J\phi, \qquad (33\text{-}13)$$

where D is the diffusion coefficient and y is the coordinate normal to the cake surface. If the diffusion coefficient is assumed constant, then this expression may be integrated subject to the boundary conditions $\phi = \phi_w$ at $y = 0$ (the cake surface) and $\phi = \phi_s$ at $y = \delta$, yielding

$$J = \frac{D}{\delta} \ln\left(\frac{\phi_w}{\phi_s}\right). \qquad (33\text{-}14)$$

As described by Porter (1972b), the mass transfer coefficient D/δ is evaluated using theoretical or empirical correlations developed for convective heat or mass transfer in tubes or channels having nonporous walls. For laminar flow, the length-averaged mass transfer coefficient is determined from the Lévêque solution, resulting in

$$\langle J \rangle = 0.807 \left(\frac{\dot{\gamma}_0 D^2}{L}\right)^{1/3} \ln\left(\frac{\phi_w}{\phi_s}\right), \qquad (33\text{-}15)$$

where $\dot{\gamma}_0$ is the shear rate in the boundary layer and L is the length of the filter tube or channel. Note that this result implies that the permeate flux at steady state is independent of the imposed pressure drop. This is because the cake layer thickness, and hence the permeate flux, is controlled by the concentration-polarization boundary layer. If the pressure drop were to be increased, then the cake layer would thicken until convection and diffusion in the boundary layer were again in balance.

The Brownian diffusivity of a particle of radius r is given by the Stokes-Einstein relationship.

$$D_b = \frac{kT}{6\pi \eta_0 r}, \qquad (33\text{-}16)$$

where $k = 1.38 \times 10^{-16} \text{g·cm}^2/\text{s}^2 \cdot ^\circ\text{K}$ is the Boltzmann constant and T is the absolute temperature. However, using this relationship in Eqs. (33-14) and (33-15) yields predicted fluxes that are much lower than those observed in practice. Instead, Zydney and Colton (1986) suggested that this should be replaced by the shear-induced diffusivity first measured by Eckstein, Bailey, and Shapiro (1977). By measuring the lateral displacement of a tagged particle in a sheared suspension in a Couette device, Eckstein, Bailey, and Shapiro (1977) observed that the shear-induced diffusivity is proportional to the shear rate and increases with increasing concentration. Although their data had considerable scatter, they estimated that

$$D_s \approx 0.1 \phi r^2 \dot{\gamma} \quad \text{for } 0 < \phi < 0.2, \quad (33\text{--}17)$$

$$\approx 0.025 r^2 \dot{\gamma} \quad \text{for } 0.2 < \phi < 0.5, \quad (33\text{--}18)$$

where $\dot{\gamma}$ is the local shear rate.

Zydney and Colton (1986) used $D = 0.03 r^2 \dot{\gamma}_0$, so that Eq. (33–15) becomes

$$\langle J \rangle = 0.078 \left(\frac{r^4}{L} \right)^{1/3} \dot{\gamma}_0 \ln \left(\frac{\phi_w}{\phi_s} \right). \quad (33\text{--}19)$$

According to this relationship, the length-averaged permeate flux increases in proportion to the applied shear rate, which is much more rapid than the one-third power dependence predicted by Eqs. (33–15) and (33–16) combined. This is a direct result of the shear-induced diffusivity being proportional to the shear rate.

Several approximations were made in the derivation of Eq. (33–19). One is that the shear-induced diffusivity is treated as a constant. Leighton and Acrivos (1987a) have shown that the experiments of Eckstein, Bailey, and Shapiro (1977) were limited by the presence of the walls of their Couette device, thereby giving diffusivities that are too low at high particle concentrations, and that the correct shear-induced diffusivity for rigid particles increases montonically with concentration. A second approximation is that the Lévêque solution is based on a linear velocity profile described by simple shear flow in the boundary layer. This is only true if the suspension viscosity is constant. However, the effective viscosity of a suspension depends strongly on the particle volume function (Leighton and Acrivos 1987b) and so is expected to vary across the concentration-polarization boundary layer. Fortunately, the two approximations described above have compensating effects on the predicted filtration flux (Romero and Davis 1988). A third approximation is that the Lévêque solution is strictly valid only when the permeate flux becomes small. As a result, it only applies when the bulk suspension is concentrated so that $1 - \phi_s/\phi_w << 1$ (Trettin and Doshi 1980).

Zydney and Colton (1987) also derived a more detailed model specifically for membrane plasmapheresis of blood suspensions. They relaxed the approximation of a linear velocity profile and small permeate flux implied by the Lévêque solution. The correlation of Eckstein, Bailey, and Shapiro (1977) was still used for the shear-induced diffusivity. Although this correlation underpredicts the shear-induced diffusivity for rigid particles, it may be a reasonable approximation for the highly deformable red blood cells. In the next section, we consider a similarity solution that relaxes all three of the approximations described above and presents results appropriate for suspensions of rigid particles.

Similarity Solution

Davis and Sherwood (1990) have performed an exact similarity solution for the convective-diffusion equation governing the steady-state concentration-polarization boundary layer in crossflow microfiltration of fine particles, under conditions in which a thin stagnant layer of particles deposited on the microporous membrane surface provides the controlling resistance to filtration. The analysis assumes that the flow of suspension is steady state, laminar, and fully developed, and that it may be treated as an incompressible Newtonian fluid with a shear viscosity and a shear-induced particle diffusivity that depend on the local particle concentration. The continuity equations (microscopic mass balances) for bulk suspension and particles in the sheared concentration-polarization boundary layer are, respectively,

$$\frac{\partial v_x}{\partial x} + \frac{\partial v_y}{\partial y} = 0, \quad (33\text{--}20)$$

$$\frac{\partial (v_x \phi)}{\partial x} + \frac{\partial (v_y \phi)}{\partial y} - \frac{\partial}{\partial y} \left(D \frac{\partial \phi}{\partial y} \right) = 0,$$

$$(33\text{--}21)$$

where x is the distance from the filter entrance, y is the distance from the edge of the fouling

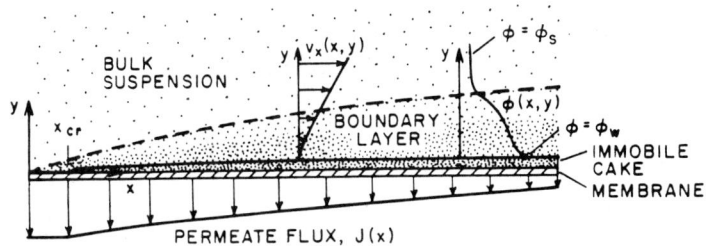

FIGURE 33-2. Schematic of the concentration-polarization boundary layer (with the y scale greatly expanded) in crossflow microfiltration.

layer on the membrane surface (see Figure 33-2), v_x and v_y are the axial and transverse velocities, respectively, of the bulk flow, ϕ is the solids volume fraction, and $D(\phi)$ is the particle diffusivity. As is usual in boundary layer or quasi-parallel flows, the axial diffusion term is relatively small and has been omitted in Eq. (33–21). Also, it is assumed that the particles are sufficiently small such that the gravity sedimentation and inertial lift velocities of the particles are small relative to the permeate velocity.

Since the boundary layers in crossflow filtration are typically thin relative to the channel half-height or diameter, the Cartesian coordinates used are appropriate for both flat-channel and tubular geometries. Moreover, the shear stress in the boundary layer exerted by the tangential flow of bulk suspension may be set equal to the wall shear stress, τ_{w0}, so that

$$\eta \frac{\partial v_x}{\partial y} = \tau_{w0}, \quad (33\text{–}22)$$

where $\eta(\phi)$ is the effective shear viscosity of the suspension. For parabolic laminar flow in narrow tubes and channels, the wall shear stress is given by, respectively,

$$\tau_{w0} = 4\eta(\phi_s)Q/(\pi r_t^3) \quad (33\text{–}23)$$

$$= 3\eta(\phi_s)Q/(2WH_0^2), \quad (33\text{–}24)$$

where Q is the axial flow rate through the tube or channel, r_t is the tube radius, H_0 is the channel half-height, and W is the channel width. It is assumed that the boundary layer is sufficiently thin relative to the tube radius or channel half-height so that the shear-stress in the boundary layer is constant. It is also assumed that the total permeate flow rate is small compared to the axial flow rate so that Q is approximately constant along the length of the tube or channel.

The appropriate boundary conditions are

$$\phi = \phi_s \qquad \text{for } y \gg \delta, \quad (33\text{–}25)$$

$$v_x = 0, \quad v_y\phi = D\frac{\partial \phi}{\partial y} \quad \text{at } y = 0, \quad (33\text{–}26)$$

$$\phi = \phi_w, \quad -v_y = J \qquad \text{at } y = 0. \quad (33\text{–}27)$$

Outside of the boundary layer, defined approximately by $y = \delta(x)$, the particle volume fraction is equal to its value in the bulk suspension, ϕ_s, as given by Eq. (33–25). Equation (33–26) represents the no-slip condition at the edge of the immobile cake on the membrane and the requirement of no flux of particles through the membrane (complete rejection). Equation (33–27) states that the particle volume fraction at the edge of the immobile cake on the membrane is ϕ_w, and that the flux of fluid through the membrane is J, a positive quantity. However, these two quantities cannot be given independently without overspecifying the system of governing equations. Typically, either a constant wall flux or a constant wall concentration is assumed in filtration problems. In the present problem, a thin stagnant fouling layer of particles is assumed to have formed on the membrane, and so the volume fraction of particles in equilibrium with this layer must be constant and equal to the maximum random packing volume fraction for a sheared suspension, $\phi_{\max}$, assuming that the particles do not irreversibly stick on the membrane or stagnant layer when deposited. As noted by Romero and Davis (1988), this maximum packing is equal to $\phi_{max} \approx 0.6$ for monodispersed rigid spheres, but its exact value is not important in the analysis because the suspension becomes extremely viscous as ϕ approaches ϕ_{max}. Thus, the wall flux, $J(x)$, must be determined as part of the solution.

Davis and Sherwood (1990) have extended the similarity analysis of ultrafiltration by Brown, Tulin, and Van Dyke (1971) and Shen and Probstein (1977) and defined the following nondimensional similarity variables:

$$\bar{y} = \frac{y}{(3\eta_0 D_0 x/\tau_{w0})^{1/3}}, \quad (33\text{--}28)$$

$$\bar{J} = \frac{J}{[\tau_{w0} D_0^2/(3\eta_0 x)]^{1/3}}, \quad (33\text{--}29)$$

where η_0 is the pure fluid viscosity and $D_0 \equiv r^2 \tau_{w0}/\eta_0$ is a characteristic diffusivity for shear-induced hydrodynamic diffusion.

Using these definitions, Davis and Sherwood (1990) show that the transformed versions of Eqs. (33–20) through (33–27) may be combined to yield

$$\left(\bar{J} + 2\bar{y}\int_0^{\bar{y}} \frac{d\bar{y}'}{\bar{\eta}[\phi(\bar{y}')]} - 2\int_0^{\bar{y}} \frac{\bar{y}' \, d\bar{y}'}{\bar{\eta}[\phi(\bar{y}')]}\right) \frac{d\phi}{d\bar{y}}$$
$$+ \frac{d}{d\bar{y}}\left(\bar{D}[\phi(\bar{y}')] \frac{d\phi}{d\bar{y}}\right) = 0, \quad (33\text{--}30)$$

where $\bar{\eta}(\phi) \equiv \eta(\phi)/\eta_0$ is the relative viscosity and $\bar{D}(\phi) \equiv D(\phi)/D_0$ is the dimensionless diffusivity.

To solve Eq. (33–30), functional forms for $\bar{\eta}(\phi)$ and $\bar{D}(\phi)$ must be specified. For constant properties ($\bar{\eta} = \bar{D} = 1$), the solution as $\bar{y} \to \infty$ yields

$$\frac{\phi_s}{\phi_w} = 1 - \bar{J}\int_0^\infty \exp\left(-\frac{\bar{y}^3}{3} - \bar{J}\bar{y}\right) d\bar{y}. \quad (33\text{--}31)$$

Trettin and Doshi (1980) present graphical results for $\bar{J}$ as a function of ϕ_s/ϕ_w from a numerical solution of Eq. (33–31), as well as the following asymptotic expansions for the solution in the limits of concentrated and dilute suspensions, respectively:

$$\bar{J} \sim 0.776\left(1 - \frac{\phi_s}{\phi_w}\right) \sim 0.776 \ln\left(\frac{\phi_w}{\phi_s}\right),$$
$$1 - \phi_s/\phi_w \ll 1 \quad (33\text{--}32)$$

$$\sim \left(\frac{2\phi_w}{\phi_s}\right)^{1/3}, \quad \phi_s/\phi_w \ll 1. \quad (33\text{--}33)$$

They noted that the classical concentration-polarization model, which gives the same result as Eq. (33–32), is valid for concentrated suspensions but greatly underpredicts the steady-state flux for dilute suspensions.

For the present application of microfiltration, the viscosity and diffusivity relationships used by Davis and Leighton (1987) and Romero and Davis (1988) are

$$\bar{\eta}(\phi) = \left(\frac{0.58 - 0.13\phi}{0.58 - \phi}\right)^2, \quad (33\text{--}34)$$

$$\hat{D}(\phi) \equiv \bar{\eta}(\phi)\bar{D}(\phi) = 0.33\phi^2(1 + 0.5e^{8.8\phi}).$$
$$(33\text{--}35)$$

Using these functions, Eq. (33–30) was solved numerically for the dimensionless permeate flux $\bar{J}$.

Figure 33–3 shows the numerical results for this permeate flux as a function of the particle volume fraction in the bulk suspension. As anticipated, the permeate flux decreases as the particle concentration in the bulk feed suspension is increased, with this decrease being quite dramatic for highly concentrated suspensions. For comparison, the approximate result of Romero and Davis (1988) in the limit when the cake resistance determines the filtration flux is also shown in Figure 33–3. This result is described in more detail in the next section, and it is seen to be in very good agreement with the exact similarity solution for all particle volume fractions in the feed suspension. Moreover, the two solutions are identical for dilute suspensions:

$$\bar{J}(\phi_s) = 0.0581\phi_s^{-1/3}, \quad \phi_s \ll 1. \quad (33\text{--}36)$$

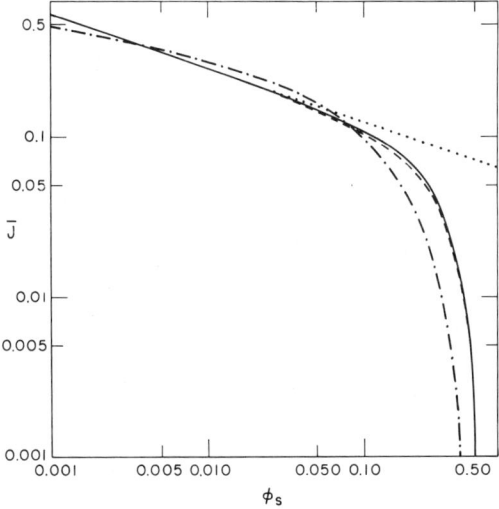

FIGURE 33–3. The dimensionless steady-state permeate flux from various shear-induced diffusion models as a function of the particle volume fraction in the bulk suspension for cake-dominated resistance. The solid line is the similarity solution of Davis and Sherwood (1990), the dashed line is the integral solution of Romero and Davis (1988), and the dashed-dotted line is the concentration-polarization model of Zydney and Colton (1986). The dotted line is the limit given by Eq. (33–36) for dilute suspensions.

As seen in Figure 33–3, Eq. (33–36) is a valid approximation for $\phi_s < 0.10$.

The similarity solution applies under conditions where a thin stagnant layer of particles has been deposited on the membrane surface and provides the controlling resistance to filtration. Near the filter entrance, the resistance of the membrane prevents the permeate flux from reaching the value given by the similarity solution (infinite at $x = 0$), and so no immobile cake forms and the similarity solution does not hold in this region. However, if the filter is long compared to the entrance length, x_{cr}, where no immobile cake is able to form (Romero and Davis 1988), then the similarity solution is valid over most of the filter length so that the length-averaged permeate flux is given by

$$\langle J \rangle \equiv \frac{1}{L}\int_0^L J(x)\,dx = \frac{3}{2}\dot\gamma_0\left(\frac{r^4}{3L}\right)^{1/3}\bar\eta(\phi_s)\bar J(\phi_s),$$

(33–37)

where $\dot\gamma_0 \equiv \tau_{w0}/\eta(\phi_s)$ is the shear rate in the bulk suspension just outside the boundary layer. For dilute suspensions, this takes on a particularly simple form

$$\langle J \rangle = 0.0604\dot\gamma_0\left(\frac{r^4}{\phi_s L}\right)^{1/3}, \quad (33\text{–}38)$$

where the value of the leading constant is 0.0604 for the particular concentration-dependent shear-viscosity and shear-induced-diffusivity relationships given by Eqs. (33–34) and (33–35), respectively. If other relationships are used for suspensions having different properties, then the form of Eq. (33–38) still holds, but the value of the leading constant will be different. Equations (33–37) and (33–38) are of similar form to Eq. (33–19), the prediction made by Zydney and Colton (1986) on the basis of the classical Lévêque solution, except for the functional dependence of the filtration rate on concentration. For comparison, the dimensionless version of Eq. (33–19) is also shown in Figure 33–3. It agrees with the exact similarity solution to within 15% for $0.002 < \phi_s < 0.10$ but becomes sharply lower outside of this range.

Integral Model

Both the concentration-polarization model and the similarity solution are restricted to situations in which a thin cake layer, which provides the dominant resistance to filtration, forms over most of the membrane area. For many crossflow microfiltration applications, this is not the case. For example, if the tangential shear flow along the membrane surface is sufficiently strong to convect the particles toward the filter exit as fast as they are convected by the permeate flow toward the membrane surface, then a stagnant cake layer will not form. On the other hand, if the tangential shear flow is relatively weak, and rigid particles are used so that the cake resistance is not dominant, then a thick cake layer will build up on the membrane surface.

Romero and Davis (1988) have developed an integral model for steady-state crossflow microfiltration that incorporates both the cake resistance and the membrane resistance and also predicts whether or not a stagnant cake layer will form and what its thickness profile will be. The key feature of their model is the integration of Eq. (33–21) across the concentration-polarization boundary layer, and application of the associated boundary conditions to yield

$$\frac{\partial}{\partial x} \int_0^\delta v_x(\phi - \phi_s) dy = J\phi_s. \quad (33\text{–}39)$$

The right-hand side represents the flux of particles carried into the boundary layer by the transverse flow, whereas the left-hand side represents the convection of particles along the boundary layer by the axial flow. Diffusion does not explicitly enter this expression because the concentration gradient and the diffusive flux are zero at the top edge of the boundary layer. However, it does appear implicitly because the velocity profile $v_x(y)$ and the concentration profile $\phi(y)$ are affected by diffusion.

Equation (33–39) may be integrated along the filter to yield

$$Q' \equiv \int_0^{\delta(x)} v_x(\phi - \phi_s)\, dy = \int_0^x J\phi_s\, dx, \quad (33\text{–}40)$$

where the quantity Q' is defined by Davis and Leighton (1987) as the "excess particle flux." This equation indicates that, for steady state, the rate at which particles are convected downstream in the flowing boundary layer at a given distance x from the filter entrance must equal the rate at which particles are carried into the boundary layer by the permeate flow everywhere upstream of the location x. The term $(\phi - \phi_s)$ appears rather than ϕ in the left-hand side of Eqs. (33–39) and (33–40) to account for the presence at the bulk volume fraction ϕ_s of particles, which are present in the boundary layer independent of the permeate flow. Davis and Leighton (1987) also defined a dimensionless excess particle flux $\hat{Q}$, which will be useful later in our analysis:

$$\hat{Q} \equiv \frac{\eta_0^3 J^2 Q'}{\tau_w^3 r^4}. \quad (33\text{–}41)$$

Equation (33–40) requires expressions for the particle concentration profile $\phi(y)$ and the axial velocity profile $v_x(y)$ in the concentration-polarization boundary layer. These are found in an approximate manner by neglecting the axial convection term in the microscopic particle mass balance and by assuming that the transverse velocity is constant across the boundary layer. By comparing the resulting approximate profiles with the exact profiles given by the similarity solution for cake-limited filtration, Davis and Sherwood (1990) showed that this approximation is exact for dilute suspensions and is accurate to within a few percent for concentration suspensions. With axial convection neglected, Eq. (33–21) may be integrated to yield

$$D(\phi)\frac{\partial \phi}{\partial y} + J\phi = 0, \quad (33\text{–}42)$$

where the boundary conditions specified by Eqs. (33–26) and (33–27) have been applied.

Davis and Leighton (1987) describe how Eqs. (33–42) and (33–22), respectively, may be integrated across the boundary layer to find the required profiles of $\phi(y)$ and $v_x(y)$. When these profiles are substituted into Eq. (33–40), and the variable of integration is transformed from y to ϕ using Eq. (33–42), a double integral for the dimensionless excess particle flux results:

$$\hat{Q}(\phi_s, \phi_w) = \int_{\phi_s}^{\phi_w} \left(\int_\phi^{\phi_w} \frac{\hat{D}(\phi')\, d\phi'}{\phi'\, \bar{\eta}^2(\phi')} \right)$$
$$\cdot \frac{(\phi - \phi_s)\hat{D}(\phi)}{\phi\bar{\eta}(\phi)}\, d\phi, \quad (33\text{–}43)$$

where $\hat{D}(\phi) \equiv D(\phi)/(r^2|\partial v_x/\partial y|)$ is a dimensionless particle diffusivity. This double integral has a maximum value of $\hat{Q}(\phi_s, \phi_w) \equiv \hat{Q}_{cr}(\phi_s)$ when $\phi_w = \phi_{max}$, indicating that a stagnant layer has formed on the membrane surface. From this, Romero and Davis (1988) have presented a comprehensive picture of steady-state crossflow microfiltration, which is summarized as follows.

There is a region near the filter entrance, $0 \leq x < x_{cr}$, where the shear flow prevents a stagnant cake layer from forming. In this region, the permeate flux is equal to that controlled by the membrane resistance alone, $J = J_0$, as given by Eq. (32–20). Since J is constant, Eqs. (33–40) and (33–41) may be combined to yield

$$\hat{Q} = \frac{\eta_0^3 J_0^3 \phi_s x}{\tau_{w0}^3 r^4}; \quad J = J_0, \quad 0 \leq x < x_{cr}, \tag{33-44}$$

where τ_{w0} is the wall shear stress given by Eqs. (33–23) and (33–24) when there is no constriction of the tube or channel due to cake buildup. This result is then used together with Eq. (33–43) to find the volume fraction of particles at the membrane surface, $\phi_w(x)$. When ϕ_w reaches ϕ_{max}, a stagnant layer forms on the membrane surface. This occurs for all $x \geq x_{cr}$, where the critical length is found by rearranging Eq. (33–44):

$$x_{cr} = \frac{\tau_{w0}^3 r^4 \hat{Q}_{cr}(\phi_s)}{\eta_0^3 J_0^3 \phi_s}. \tag{33-45}$$

For $x \geq x_{cr}$, the presence of the stagnant cake layer reduces the permeate flux and constricts the channel. The reduction in permeate flux due to the cake layer buildup is described by Darcy's law. By combining Eqs. (32–11) and (32–20), Darcy's law may be expressed as

$$J = \frac{J_0}{(1 + \hat{R}_c \delta_c / R_m)}. \tag{33-46}$$

A small correction for curvature of the cake layer in cylindrical tubes is neglected here. The constriction of the effective channel half-height or tube radius due to cake buildup also leads to an increase in the shear stress exerted by the axial flow of suspension. Provided that the axial flow rate is constant, then the modified wall shear stress is

$$\tau_w = \tau_{w0} r_t^3 / (r_t - \delta_c)^3 \tag{33-47}$$

$$= \tau_{w0} H_0^2 / (H_0 - \delta_c)^2, \tag{33-48}$$

for a tube and a two-dimensional channel, respectively.

The essence of the integral solution described by Romero and Davis (1988) is that the *dimensionless* excess particle flux $\hat{Q}$ has a constant value of $\hat{Q}_{cr}(\phi_s)$ for $x \geq x_{cr}$. Since the *dimensional* excess particle flux Q' increases with x according to Eq. (33–40), then Eq. (33–41) indicates that J must decrease with x and/or τ_w must increase with x due to cake buildup. The analysis assumes that the permeate flow rate is small compared to the flow rate of suspension entering the tube or channel, and that the axial pressure drop is sufficiently small that a constant, average value for the transverse pressure drop across the cake and membrane may be used. Combining Eqs. (33–40), (33–41), (33–45), (33–46), and (33–48) yields

$$\frac{\eta_0^3 J_0^3 \phi_s (1 - \delta_c/H_0)^6}{\tau_{w0}^3 r^4 (1 + \hat{R}_c \delta_c / R_m)^2} \int_0^x \frac{dx}{1 + \hat{R}_c \delta_c / R_m} = \hat{Q}_{cr}(\phi_s), \tag{33-49}$$

which describes the cake thickness profile in a two-dimensional, rectangular channel. The corresponding result for a tubular filter is

$$\frac{\eta_0^3 J_0^3 \phi_s (1 - \delta_c/r_t)^9}{\tau_{w0}^3 r^4 (1 + \hat{R}_c \delta_c / R_m)^2} \int_0^x \frac{dx}{1 + \hat{R}_c \delta_c / R_m} = \hat{Q}_{cr}(\phi_s). \tag{33-50}$$

These equations may be solved numerically for the cake thickness profile $\delta_c(x)$. The steady-state flux profile $J(x)$ may then be determined from Eq. (33–46). The solution applies for $x \geq x_{cr}$, whereas $J = J_0$ and $\delta_c = 0$ for $x < x_{cr}$. It proves convenient to first make these equations dimensionless by defining $\hat{J} = J/J_0$, $\hat{x} = x/x_{cr}$, and $\hat{\delta} = \delta_c/H_0$ (channel) or $\hat{\delta} = \delta_c/r_t$ (tube). Equations (33–46), (33–49), and (33–50) then become

$$\hat{J} = \frac{1}{1 + \beta \hat{\delta}}, \tag{33-51}$$

$$\frac{(1 - \hat{\delta})^n}{(1 + \beta \hat{\delta})^2} \left[1 + \int_1^{\hat{x}} \frac{d\hat{x}}{1 + \beta \hat{\delta}} \right] = 1, \tag{33-52}$$

where $n = 6$ for a rectangular channel, $n = 9$ for a tube, and β is a dimensionless relative

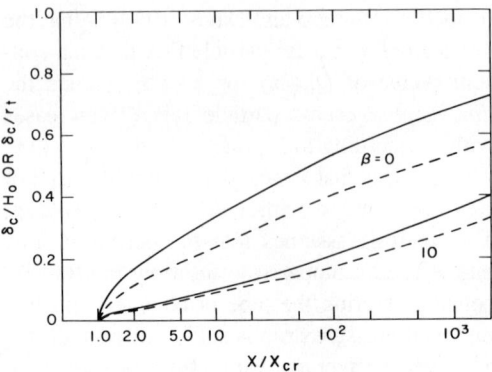

FIGURE 33-4. The dimensionless steady-state stagnant cake thickness for a two-dimensional channel (solid lines) and for a tube (dashed lines). The results are based on the shear-induced diffusion model of Romero and Davis (1988).

cake resistance defined by $\beta \equiv \hat{R}_c H_0/R_m$ (channel) or $\beta \equiv \hat{R}_c r_t/R_m$ (tube).

Numerical solutions to Eqs. (33-51) and (33-52) for various β are presented in Figures 33-4 and 33-5. As β is increased, the flux is decreased due to the increased resistance of the cake layer. The cake thickness also decreases with increasing β because the reduction in flux implies that less channel constriction is required to achieve a steady-state balance of the deposition of particles into the boundary layer and the convection of these particles toward the filter exit. The cake buildup is less for a tube than for

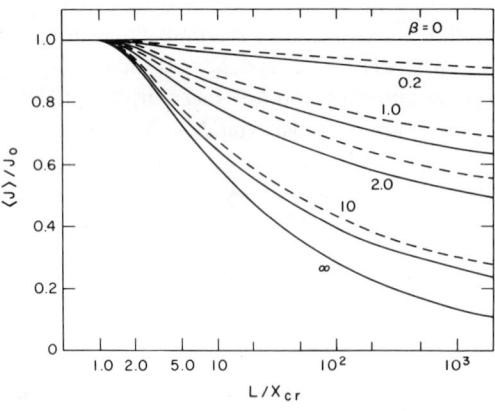

FIGURE 33-5. The dimensionless length-averaged permeate flux at steady state for a two-dimensional channel (solid lines) and for a tube (dashed lines). The results are based on the shear-induced diffusion model of Romero and Davis (1988).

a two-dimensional rectangular channel because the same cake thickness leads to a greater reduction in cross-sectional area (and, hence, increase in shear stress) for a tube than for a channel. As a result, the steady-state flux is greater for a tube than for a rectangular channel. In the limiting cases of $\beta \ll 1$ and $\beta \gg 1$, the flux is independent of geometry. The minimum flux for a finite value of β is $J = J_0/(1 + \beta)$, which corresponds to the tube or channel becoming nearly filled with suspension. This occurs only in the limit $x/x_{cr} \to \infty$.

Analytical solutions to Eqs. (33-51) and (33-52) may be obtained in the limiting cases of membrane-dominated resistance ($\beta \ll 1$) and cake-dominated resistance ($\beta \gg 1$). In particular, for membrane-dominated resistance,

$$\hat{J} = 1, \quad \beta \ll 1, \qquad (33\text{-}53)$$

$$\hat{\delta} = 1 - \hat{x}^{-1/n}, \quad \beta \ll 1, \hat{x} \geq 1, \qquad (33\text{-}54)$$

indicating that the flux remains at its initial value of $J = J_0$, whereas the cake layer thickness builds up to fill an appreciable portion of the tube or channel for $x \gg x_{cr}$. In contrast, for cake-dominated resistance,

$$\hat{J} = \left(\frac{3\hat{x}}{2} - \frac{1}{2}\right)^{-1/3}, \quad \beta \gg 1, \hat{x} \geq 1, \qquad (33\text{-}55)$$

$$\hat{\delta} = \beta^{-1}\left[\left(\frac{3\hat{x}}{2} - \frac{1}{2}\right)^{1/3} - 1\right], \quad \beta \gg 1, \hat{x} \geq 1, \qquad (33\text{-}56)$$

indicating that the flux is inversely proportional to the one-third power of the distance from the channel entrance for $x \gg x_{cr}$, and that the cake layer remains thin due to its high resistance.

The solution for the permeate flux $J(x)$ may be integrated along the length L of a filter in order to find the length-averaged permeate flux. It is this length-averaged flux that is shown in Figure 33-5. For the dual limit of $\beta \gg 1$ and $L/x_{cr} \gg 1$, which implies that the cake resistance is dominant over most of the filter length, this yields $\langle \hat{J} \rangle = (3/2)^{2/3}(x_{cr}/L)^{1/3}$ or, in dimensional form,

$$\langle J \rangle = 1.31 \dot{\gamma}_0 \bar{\eta}(\phi_s) \left(\frac{r^4 \hat{Q}_{cr}(\phi_s)}{\phi_s L} \right)^{1/3} \quad (33\text{--}57)$$

Of particular importance is the fact that Eq. (33–57) predicts that the permeate flux is proportional to the shear rate and independent of the pressure drop for cake-dominated resistance. This is in contrast to membrane-dominated resistance, for which the permeate flux is proportional to the transmembrane pressure drop and independent of the shear rate.

The entire development presented thus far in this section is independent of the specific choices made for the concentration dependence of the dimensionless viscosity $\bar{\eta}(\phi)$ and the dimensionless diffusivity $\hat{D}(\phi)$. However, these functions are needed to perform the integrations specified in Eq. (33–43) for the dimensionless excess particle flux $\hat{Q}$. For the special case of suspensions of rigid spheres, Eqs. (33–34) and (33–35) are used to find that $\hat{Q}_{cr}(\phi_s \to 0) = 9.79 \times 10^{-5}$. Equation (33–57) then becomes identical to Eq. (33–38) in the limit of dilute feed suspensions ($\phi_s \ll 1$). For nondilute suspensions, Eq. (33–57) gives a length-averaged permeate flux that is virtually identical to that predicted by the similarity solution described in the previous section, as shown in Figure 33–3 together with Eq. (33–37). Figure 33–6 is a plot of $\hat{Q}_{cr}$ versus ϕ_s; the result may be approximated by $\hat{Q}_{cr} = 9.79 \times 10^{-5}(1 - 4.38\phi_s)$ for $\phi_s \leq 0.10$.

Inertial Lift Theory

The inertial lift theory is the primary alternative to the shear-induced model of crossflow microfiltration described in the three previous sections. Indeed, inertial lift was the first mechanism to be widely accepted as an explanation for the flux paradox of crossflow microfiltration. It has been reviewed extensively by Belfort, Weigand, and Mahar (1985), Belfort (1989), and Belfort and Nagata (1985). Thus, only a brief summary is included here.

When an isolated, neutrally buoyant particle is present in a duct under laminar flow conditions, it will migrate across the fluid streamlines due to complex interactions with the flow field and the duct walls. The migration occurs as a direct result of inertia and is absent under creeping flow conditions. Cox and Brenner (1968) developed the basic theory for the inertial lift force on a particle in a laminar flow duct having nonporous walls. Explicit expressions for the lateral force and velocity of the particle were derived by Ho and Leal (1974) and Vasseur and Cox (1976) for a two-dimensional channel, and by Ishii and Hasimoto (1980) for a tube. Altena and Belfort (1984) and Weigand, Altena, and Belfort (1985) extended these results to porous channels and tubes, respectively, and showed that the lateral velocity of the particle can be found by vectorial addition of the inertial lift velocity and the fluid permeate velocity under conditions of small permeate velocities.

The general results for the inertial migration velocity are of the form

$$v_{L,0} = \frac{\rho_0 r^3 \dot{\gamma}_0^2 f(\hat{y})}{16 \eta_0}, \quad (33\text{--}58)$$

where ρ_0 is the fluid density, $\dot{\gamma}_0$ is the shear rate at the wall of the tube or channel, and $\hat{y}$ is the distance from the wall made dimensionless by the tube diameter or channel height. In the region near the wall, $f(\hat{y})$ is positive, indicating that the inertial lift velocity is away from the

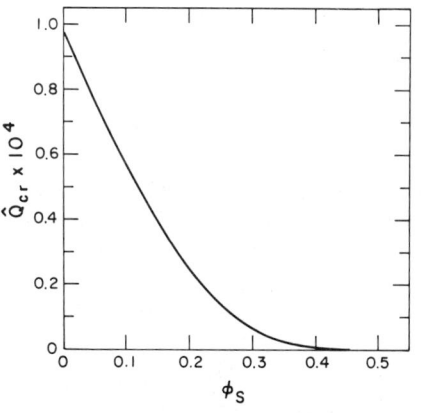

FIGURE 33–6. The maximum dimensionless excess particle flux as a function of the particle volume fraction in the bulk suspension, using Eqs. (33–34) and (33–35) for the shear viscosity and shear-induced diffusivity, respectively.

wall. The maximum value occurs at the wall and is $f(\hat{y}) \approx 1.6$ for a two-dimensional channel (Vasseur and Cox 1976) or $f(\hat{y}) \approx 1.3$ for a tube (Ishii and Hasimoto 1980).

The above results are restricted to small channel (or tube) Reynolds numbers, $Re_c \equiv 2\rho_0 \langle v \rangle H_0/\eta_0 \ll 1$, where $\langle v \rangle$ is the bulk-average axial velocity. In contrast, most crossflow microfilters are operated with large channel Reynolds numbers, typically on the order of 10^2 or 10^3. Schonberg and Hinch (1989) have shown that the function $f(\hat{y})$ decreases as the channel Reynolds number is increased. When $Re_c \gg 1$, the maximum lift velocity for a spherical particle near the wall of a two-dimensional channel is (Drew, Schonberg, and Belfort 1991)

$$v_{L,0} = 0.577 \frac{\rho_0 r^3 \dot{\gamma}_0^2}{16 \eta_0}. \qquad (33\text{-}59)$$

When applied to crossflow microfiltration, the basic premise of the inertial lift theory is that particles are carried to the membrane walls only if the permeate flux exceeds the maximum inertial lift velocity. If this condition is met for a clean membrane, then a stagnant cake layer will form due to particle deposition. As the cake layer builds up, it reduces the permeate flux according to Eq. (33-46). Also, it constricts the tube or channel, thereby leading to an increase in the shear rate and inertial lift velocity, as described by Green and Belfort (1980). The cake layer builds up until the permeate flux and inertial lift velocity are equal. At steady state for fixed axial flow rate, Eq. (33-59) is replaced by

$$J = v_L = \frac{v_{L,0}}{(1-\hat{\delta})^m}, \qquad (33\text{-}60)$$

where $m = 4$ for a two-dimensional channel, $m = 6$ for a tube, and $v_{L,0}$ is the lift velocity for a clean tube or channel. The value of 0.577 for the leading constant in Eq. (33-59) is only strictly valid for a two-dimensional channel, although this value is not expected to differ greatly for a tube. Equations (33-46) and (33-60) are then solved simultaneously for $\hat{\delta}$ and J.

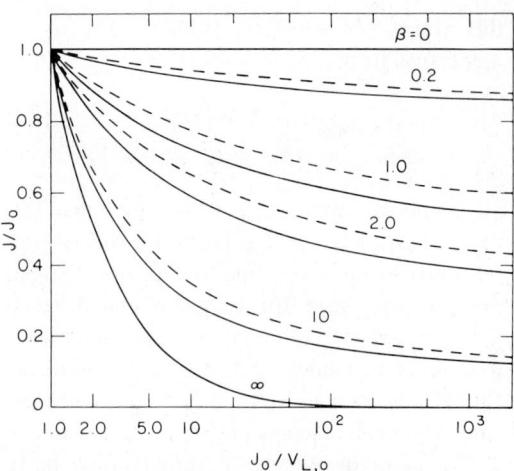

FIGURE 33-7. The dimensionless permeate flux at steady state for a two-dimensional channel (solid lines) and for a tube (dashed lines). The results are based on the inertial lift theory of Green and Belfort (1980) adapted to large channel or tube Reynolds numbers.

The numerical solution to Eqs. (33-46) and (33-60) is given in Figure 33-7 as a plot of J/J_0 versus $J_0/v_{L,0}$ for various values of the relative cake resistance β. As with the shear-induced diffusion model, the steady-state flux decreases with increasing β. In contrast to the shear-induced diffusion model, however, the steady-state flux given by the inertial lift theory is independent of axial position and particle concentration. In the limit of membrane-dominated resistance, the solution yields

$$J = J_0 \; ; \quad \hat{\delta} = 1 - \left(\frac{v_{L,0}}{J_0}\right)^{1/m}, \quad \beta \ll 1,$$
$$(33\text{-}61)$$

indicating that the flux remains at its initial value, independent of the inertial lift. Also, the cake layer fills an appreciable portion of the channel, since $v_{L,0} \ll J_0$ in typical applications. On the other hand, when the cake resistance dominates,

$$J = v_{L,0}; \quad \hat{\delta} = \beta^{-1}\left(\frac{J_0}{v_{L,0}} - 1\right), \quad \beta \gg 1,$$
$$(33\text{-}62)$$

indicating that the flux is equal to the inertial life velocity, independent of the filter geometry, and that the cake layer remains thin.

TRANSIENT OPERATION OF CROSSFLOW MICROFILTERS

The previous sections described several models for the steady or quasi-steady permeate flux in crossflow microfiltration. In practice, flux decline from the initial value to the steady value is observed over time. Typically, there is a short-term flux decline, occurring over time scales of minutes or hours, due to cake buildup. There is also a long-term flux decline, occurring over time scales of hours or days, due to membrane fouling, membrane compaction, and cake compaction. Both short-term and long-term flux decline may be described by Darcy's law, with the resistances being time dependent:

$$J = \frac{\Delta p}{\eta_0[R_m(t) + R_c(t)]}. \quad (33\text{-}63)$$

Currently little is known of how the membrane resistance $R_m(t)$ increases with time due to fouling. Petzny and Quinn (1969) have shown that the membrane permeability (the inverse of the specific resistance) decreases as a result of adsorbed macromolecules decreasing the effective membrane pore radius. Chandavarkar and Cooney (1989) have further investigated the dynamics of flux decline due to protein/membrane interactions. Errede (1984) proposed that the rate of decrease of the membrane permeability is proportional to the flux of filtrate through the membrane, from which it may be inferred that the membrane permeability decreases exponentially with time. This predicts a steady-state flux of zero, which is not usually observed in practice. Instead a phenomenological expression for the membrane resistance, which is based on the experimental data reported by Fane (1986), is proposed:

$$R_m(t) = R_{mi} + (R_{mf} - R_{mi})[1 - \exp(-t/\tau_m)], \quad (33\text{-}64)$$

where R_{mi} is the initial membrane resistance, R_{mf} is the final membrane resistance, and τ_m is a membrane fouling time constant characteristic of the concentration of the fouling agents and their interaction with the membrane material.

The three parameters in Eq. (33-64) may be determined by fitting experimental data. Other expressions may also be proposed and evaluated against experimental data.

The cake resistance increases as the cake thickness increases, as described previously:

$$R_c(t) = \hat{R}_c(t)\delta_c(t). \quad (33\text{-}65)$$

However, when cake compaction over time occurs, then the specific cake resistance $\hat{R}_c(t)$ will not be constant but will instead increase with time. This increase is difficult to predict and is best measured via a laboratory-scale batch filtration test in which a fixed mass of particles is placed on a nonfouling membrane and then pure solvent is filtered through the cake under constant pressure drop over a long time period. Both the cake height and the permeate flux are then measured as functions of time and imposed pressure drop.

Both the membrane and cake resistances contribute to the flux decline during the transient portion of crossflow microfiltration. As such, the integral method provides the best approach to predicting the transient flux decline. As shown by Romero and Davis (1990), Eq. (33-39) may be extended to the time-dependent case by including terms representing the accumulation of particles in the stagnant cake and the flowing boundary layer:

$$\frac{\partial}{\partial t}\left((\phi_c - \phi_s)\delta_c + \int_{\delta_c}^{\delta_c + \delta}(\phi - \phi_s)\,dy\right)$$
$$+ \frac{\partial}{\partial x}\left(\int_{\delta_c}^{\delta_c + \delta} v_x(\phi - \phi_s)\,dy\right) = J\phi_s,$$

$$(33\text{-}66)$$

where $\delta_c(x,t)$ is the thickness of the growing cake layer, and $\delta(x,t)$ is the flowing concentration-polarization boundary layer thickness. For convenience, the transverse coordinate y is now measured from the membrane surface rather than the cake surface.

Romero and Davis (1990) have shown that Eqs. (33-63) through (33-66) may be solved for the cake thickness $\delta_c(x,t)$ and the permeate flux $J(x,t)$ using the method of characteristics.

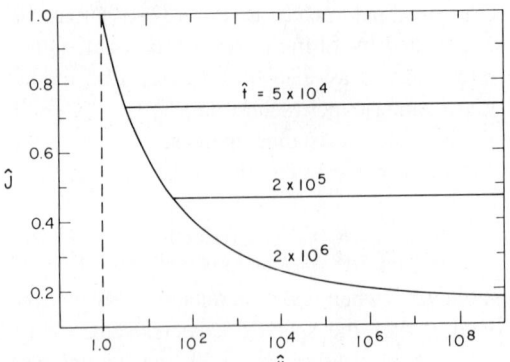

FIGURE 33–8. Transient flux decline predicted by the shear-induced diffusion model of Romero and Davis (1990) for $\phi_s = 0.001$, Pe = 10^3, and $\beta = 5.0$ in a two-dimensional channel.

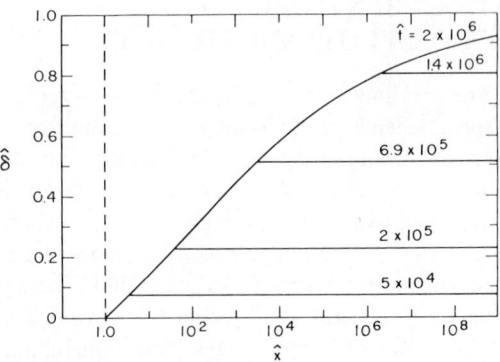

FIGURE 33–9. Transient cake-layer buildup predicted by the shear-induced diffusion model of Romero and Davis (1990) for $\phi_s = 0.001$, Pe = 10^3, and $\beta = 5.0$ in a two-dimensional channel.

They found that the flowing boundary layer develops in a very short time relative to the time scale for stagnant cake layer development. Numerical computations were performed for the special case of negligible membrane fouling and compaction (constant R_m) and negligible cake compaction (constant $\hat{R}_c$). Typical results are shown in Figures 33–8 and 33–9 for a two-dimensional channel with $\phi_s = 0.001$, $\beta = 5.0$, and Pe = 10^3, where Pe $\equiv J_0 H_0/D_0$ is the Peclet number, and $D_0 \equiv \tau_{w0} r^2/\eta_0$ is the characteristic shear-induced diffusivity. In these figures, $\hat{x} \equiv x/x_{cr}$, $\hat{t} \equiv J_0^2 t/D_0$, $\hat{\delta} \equiv \delta_c/H_0$, and $\hat{J} \equiv J/J_0$, where J_0 is the initial flux in the absence of a cake layer. From these figures, it is evident that there are two distinct portions in the filter: a developed region near the filter entrance where the stagnant layer thickness and the permeate flux have reached their steady-state values, and a developing region near the filter exit where the stagnant layer thickness and permeate flux are increasing and decreasing, respectively, with time but are independent of position.

In the developing portion of the filter, the cake growth and associated flux decline are nearly insensitive to the shear flow. As a result, their values are given to a very close approximation by Eqs. (32–18) and (32–19) for transient batch filtration. Moreover, as long as the cake layer is still developing over most of the filter length, then the length-averaged flux will be nearly the same as that given by Eq. (32–19), which may be rearranged to yield

$$\langle J \rangle = \frac{J_0}{(1 + 2t/\tau_c)^{1/2}}, \quad (33\text{–}67)$$

where $\tau_c \equiv R_m(\phi_c - \phi_s)/(J_0 \hat{R}_c \phi_s)$ is the time constant for flux decline due to cake buildup.

Figure 33–10 shows $\langle J \rangle$ versus t for conditions that are typical of crossflow microfiltration for a ceramic tube filter ($r = 0.32$ μm, $r_t = 0.13$ cm, $L = 37$ cm, $\phi_s = 0.0015$, $\beta = 2.2$, $J_0 = 0.0034$ cm/s, $\dot{\gamma}_0 = 2400$ s^{-1}). The solid line is the complete numerical solution for Eq. (33–66), the dotted line is the batch flux decline given by Eq. (33–67), and the dashed line is the steady-state solution given by Figure 33–5. The batch expression falls increasingly below the complete numerical solution with time because of the increasing region over which the cake layer becomes developed. However, the difference is not large, and Figure 33–10 suggests a simple but accurate method for approximating the transient flux decline due to cake buildup. Namely, Eq. (33–67) is used to determine the permeate flux until the time at which the steady-state value is reached. After this time, the steady-state solution is used.

SUMMARY OF FLUX MODELS FOR CROSSFLOW MICROFILTERS

Although crossflow microfiltration is analogous to crossflow ultrafiltration in many ways, its

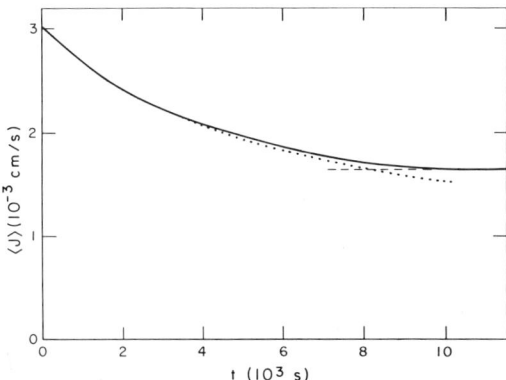

FIGURE 33-10. The predicted length-averaged permeate flux versus time for a tube with $r = 0.32$ μm, $r_t = 0.13$ cm, $L = 37$ cm, $\phi_s = 0.0015$, $\beta = 2.2$, $J_0 = 0.0034$ cm/s, and $\dot{\gamma}_0 = 2400$ s^{-1}. The solid line represents the shear-induced diffusion model of Romero and Davis (1990), the dotted line is the approximate result based on transient deadend filtration, and the dashed line is from the steady-state shear-induced diffusion model of Romero and Davis (1988).

analysis is more complex. Multiple mechanisms are available for transverse motion of particles away from the membrane, and both the membrane resistance and the cake resistance may be important in typical applications. As a result, no single formula may be used to predict the permeate flux under all crossflow microfiltration conditions. This chapter has reviewed several models that apply for different ranges of the material and operating conditions. In this section, some guidelines are presented for choosing the appropriate model for a given application. For each model, the dependence of the permeate flux on the material properties and operating conditions is discussed.

To start with, an assessment must be made of the relative contributions of Brownian diffusion, shear-induced diffusion, and inertial lift to the transverse motion of particles away from the membrane. By comparing Eqs. (33–16) and (33–18), a rough estimate of the ratio of the shear-induced diffusivity to the Brownian diffusivity is $D_s/D_b \approx \eta_0 \dot{\gamma}_0 r^3/kT$. Under typical conditions of $\eta_0 = 10^{-2}$ g/cm·s, $\dot{\gamma}_0 = 10^3$ s^{-1}, $r = 10^{-4}$ cm, and $kT = 4 \times 10^{-14}$ g·cm^2/s^2, this ratio is 250. Thus, shear-induced diffusion is large compared to Brownian diffusion for most

crossflow microfiltration applications, except for those with very small particle sizes and/or low shear rates. By comparing Eqs. (33–38) and (33–59), a rough estimate of the ratio of shear-induced back-diffusion and inertial lift for cake-dominated resistance is $\langle J_s \rangle / \langle J_L \rangle \approx 1.7 \eta_0 / (\rho_0 \dot{\gamma}_0 r^{5/3} L^{1/3} \phi_s^{1/3})$, where the subscripts s and L refer to the shear-induced diffusion model and the inertial lift models, respectively. For the typical conditions of $\eta_0 = 10^{-2}$ g/cm·s, $\rho_0 = 1$ g/cm^3, $\dot{\gamma}_0 = 10^3$ s^{-1}, $r = 10^{-4}$ cm, $L = 30$ cm, and $\phi_s = 0.1$, this ratio is $\langle J_s \rangle / \langle J_L \rangle \approx 60$. Thus, shear-induced diffusion is also large compared to inertial lift for most crossflow microfiltration applications, except for those with very high shear rates and/or large particle sizes. When the cake resistance is not dominant, then cake buildup constricts the tube or channel and increases the effective shear rate. This increases the steady-state flux predicted by both the shear-induced diffusion and inertial lift models, although the effect on the inertial lift model is greater due to its dependence on the square of the shear rate.

An assessment must also be made of the relative contributions of the membrane resistance and the cake resistance to controlling the permeate flux specified by Darcy's law. The cake resistance dominates over most of the filter length only when $L/x_{cr} \gg 1$ and $\beta \gg 1$. For typical conditions of $r = 10^{-4}$ cm, $\eta_0 = 10^{-2}$ g/cm·s, $\tau_{w0}/\eta_0 = 10^3$ s^{-1}, $\Delta p = 7 \times 10^5$ g/cm·s^2 (10 psi), $\phi_s = 0.1$, $R_m = 10^9$ cm^{-1}, $H_0 = 0.1$ cm, and $L = 30$ cm, Eqs. (32–20) and (33–45) are used to show that $L/x_{cr} \approx 2 \times 10^8$, indicating that a stagnant layer forms on virtually the entire membrane surface. Using the Carman-Kozeny equation for the specific cake resistance [Eq. (32–5) with $K = 5$, $\epsilon_c = 0.4$, and $S_c = 3/r$] yields $\hat{R}_c = 250/r^2$, or $\hat{R}_c = 2.5 \times 10^{10}$ cm^{-2} for $r = 10^{-4}$ cm. Thus, $\beta \equiv \hat{R}_c H_0/R_m = 2.5$ for the conditions quoted above. This result indicates that both the cake resistance and the membrane resistance are important in typical crossflow microfiltration applications. Of course, the Carman-Kozeny equation for the cake resistance only applies for rigid particles. For biological and other deformable particles often encountered in crossflow microfiltration applications, the cake resistance may be much higher and provide the controlling resistance.

When the membrane provides the dominant resistance ($L/x_{cr} < 1$ and/or $\beta \ll 1$), then the permeate flux is given by Eq. (32–20),

$$\langle J \rangle = J_0 \equiv \frac{\Delta p}{\eta_0 R_m}, \quad (33\text{–}68)$$

regardless of the mechanism of particle motion away from the membrane. In this limit, the flux is proportional to the pressure drop and independent of shear rate, particle size, particle concentration, and position. It is also independent of time, unless significant membrane fouling occurs. The flux increases with increasing temperature because of the temperature dependence of the suspending fluid viscosity.

When the cake provides the dominant resistance ($L/x_{cr} \gg 1$ and $\beta \gg 1$), then the flux decline is very rapid, and a steady-state flux controlled by a thin cake layer is quickly reached. For this case, the mechanisms of particle motion away from the cake layer must be considered. For shear-induced diffusion as the dominant mechanism, the steady-state flux is predicted by the solution to Eq. (33–37), as shown in Figure 33–3 for the particular shear viscosity and shear-induced diffusivity relationships for suspensions of rigid spheres. For dilute suspensions ($\phi_s < 0.1$), this solution is given by Eq. (33–38):

$$\langle J \rangle = 0.0604 \, \dot{\gamma}_0 \left(\frac{r^4}{\phi_s L} \right)^{1/3}. \quad (33\text{–}69)$$

In this limit, the flux is proportional to the shear rate and independent of the pressure drop and fluid viscosity. It also increases in proportion to the four-thirds power of the particle size (a direct result of the shear-induced diffusivity being proportional to the square of the particle size) and decreases in proportion to the one-third power of the bulk particle concentration and to the one-third power of the filter length.

For cake-dominated resistance with very small particles and/or low shear rates, for which Brownian diffusion is the dominant mechanism of transverse particle motion, then the similarity solution of Trettin and Doshi (1980) is expected to be valid. For dilute suspensions, Eqs. (33–16), (33–33), and (33–37) may be combined to yield

$$\langle J \rangle = 0.185 \left(\frac{\dot{\gamma}_0 k^2 T^2}{\eta_0^2 r^2 L} \right)^{1/3} \left(\frac{\phi_w}{\phi_s} \right)^{1/3}, \quad \frac{\phi_s}{\phi_w} \ll 1.$$

$$(33\text{–}70)$$

The corresponding result for concentrated suspensions is

$$\langle J \rangle = 0.114 \left(\frac{\dot{\gamma}_0 k^2 T^2}{\eta_0^2 r^2 L} \right)^{1/3} \ln \left(\frac{\phi_s}{\phi_w} \right), \quad 1 - \frac{\phi_s}{\phi_w} \ll 1.$$

$$(33\text{–}71)$$

These steady-state expressions show that the flux increases in proportion to only the one-third power of the shear rate. It also increases rapidly with increasing temperature and decreasing particle size because the Brownian diffusivity is directly proportional to the temperature and inversely proportional to the fluid viscosity and particle size. As in the shear-induced diffusion model, the permeate flux decreases with increasing filter length and particle concentration.

For cake-dominated resistance with very large particles and/or high shear rates, for which inertial lift is the dominant mechanisms of particle motion away from the cake, then the permeate flux is determined by combining Eqs. (33–59) and (33–62):

$$\langle J \rangle = 0.0361 \rho_0 r^3 \dot{\gamma}_0^2 / \eta_0. \quad (33\text{–}72)$$

This expression predicts that the steady-state permeate flux is independent of pressure drop, particle concentration, and filter length. It also predicts a stronger dependence on the shear rate and particle size than do the other models.

When both the membrane resistance and the cake resistance are important in determining the permeate flux, as is typical for microfiltration of rigid or nearly rigid particles, then Figure 33–5 (shear-induced diffusion) or Figure 33–7 (inertial lift) should be used to predict the steady-state flux. There will be a complex dependence of this flux on the shear rate, pressure drop, etc. Moreover, the rate of flux decline may be relatively slow as the cake layer builds up toward its steady-state thickness. The transient flux is given approximately by Eq. (33–67), the batch filtration result. The time constant for flux decline due to cake buildup is

$$\tau_c \equiv \frac{R_m(\phi_c - \phi_s)}{J_0 \hat{R}_c \phi_s} = \frac{\eta_0 R_m^2(\phi_c - \phi_s)}{\Delta p \hat{R}_c \phi_s}.$$

(33–73)

This time constant decreases with increasing temperature, pressure drop, and particle concentration, but it is independent of the shear rate.

COMPARISON WITH EXPERIMENTS

Early experiments that measured the steady-state flux for crossflow microfiltration of microbial and colloidal suspensions have been summarized by Porter (1972a, 1972b), Henry (1972), and Blatt et al. (1970). Additional flux measurements have been reported by Zahka and Leahy (1985), Baker et al. (1985), and Rautenbach and Schock (1988), among others. The primary conclusion of the comparison of experiments with existing theories during the early 1970s was that the concentration-polarization model using the Stokes-Einstein Brownian diffusivity underpredicts the microfiltration flux data by one or more orders of magnitude. Moreover, when the steady-state flux data were plotted versus shear rate on a log-log scale, the slope was found to vary between 0.49 and 1.33 (Henry 1972). These values are higher than the value of one-third predicted by the concentration-polarization model with Brownian diffusion and indicate that multiple mechanisms for particle back-diffusion or migration away from the membrane come into play. Also, when the steady-state flux data were plotted against the imposed transverse pressure drop, it was found that the flux increases linearly with pressure drop at low values and then becomes independent of pressure drop at high values (Porter 1977). The pressure-independent value of the flux was found to increase with increasing shear rate and decreasing particle concentration in the feed. These findings indicate that the membrane resistance is dominant for low pressure drops, whereas the cake resistance with boundary layer control is dominant for high pressure drops.

It is difficult to make quantitative comparisons of many of the flux measurements reported in the literature with recent theories of crossflow microfiltration. This is due, in part, to the multiple mechanisms that play roles in crossflow microfiltration. Also, key material properties, such as the membrane and specific cake resistances, are sometimes not reported. In perhaps the most complete comparison of experiments with theory at the time, Zydney and Colton (1986) examined length-averaged flux data from 12 different studies of crossflow microfiltration of blood, bacterial, latex, clay, platelet, and electroprimer suspensions. They assumed that the data were for the steady state and for cake-dominated resistance, although it was not possible to verify these assumptions in some cases. It was shown that the concentration-polarization model using shear-induced diffusion provides a good fit of the crossflow microfiltration flux data over the full range of experimental conditions. As true of earlier findings, Zydney and Colton (1986) showed that the concentration-polarization model incorporating a Brownian diffusion coefficient grossly underpredicts the flux and gives a weaker dependence on shear rate than observed.

Otis et al. (1986) have performed careful single-particle trajectory experiments that show very good agreement with the predictions of inertial lift theory, under conditions of small channel Reynolds numbers. Comparisons of crossflow microfiltration flux data with inertial lift theory have not been as successful. When not corrected for channel constriction due to cake buildup, the inertial lift theory underpredicts flux data by one or more orders of magnitude, except at very high shear rates or large particle sizes (Madsen 1977, Green and Belfort 1980). Moreover, the predictions of inertial lift theory that the steady-state flux is independent of particle concentration and increases with the square of the shear rate, when the resistance is dominated by a thin cake layer, are not borne out by experiments (Zydney and Colton 1986).

When the inertial lift theory is corrected for channel constriction due to cake buildup, then Green and Belfort (1980) and Zydney and Colton (1986) have shown that the predicted flux is in order of magnitude agreement with data from latex and blood suspensions, respectively. Nevertheless, the inertial lift model was found to give a dependence on shear rate, particle concentration, and transverse pressure drop different from observation (Zydney and Colton 1986). Inertial lift can play an important role in crossflow microfiltration, but only under conditions of high shear rates and/or large particle sizes. It may also be important for adhesive cakes that do not allow particles to freely diffuse away from their surfaces. It should not be used to interpret flux data unless it has first been established, using the procedures described in the previous section, that it is dominant over the Brownian and shear-induced diffusion mechanisms for a given application. Also, since nearly all crossflow microfiltration experiments are carried out with channel Reynolds numbers much greater than unity, Eq. (33–59) should be used. This expression predicts inertial lift velocities that are a factor of 2 to 3 smaller than do the often-used expressions that are based on very small channel Reynolds numbers.

During the past decade, several Ph.D. theses have been devoted to comparing crossflow microfiltration flux measurements with theoretical predictions based on shear-induced hydrodynamic diffusion as the primary mechanism of transverse particle motion away from the filter cake and membrane. Zydney (1985) performed crossflow plasmapheresis experiments with whole blood suspensions. He found that a very rapid flux decline occurred over a time scale of the order of 1 minute, during which a very thin layer of red blood cells formed on the membrane surface. The steady-state flux was controlled by the high resistance of this compressed layer. Ofsthun (1989) performed crossflow microfiltration experiments with yeast suspensions and latex particle suspensions. She found that the flux decline occurred over a time scale of the order of 1 hour. This longer time scale occurs because the higher permeability of latex and yeast cakes, compared to that of a layer of red blood cells, requires that thicker cake layers build up before the steady state is reached. A freeze substitution technique was used by Ofsthun (1989), as described by Ofsthun and Colton (1987), to observe directly that this is true. Ofsthun (1989) also presents an excellent review of crossflow microfiltration applications in biotechnology. Romero (1989) performed transient and steady-state experiments with latex particle suspensions under a variety of conditions. These experiments are summarized in the following paragraphs. Although some scatter and anomalies exist in the data, all three theses report good agreement between measured permeate fluxes and model predictions based on the mechanism of shear-induced hydrodynamic diffusion.

Romero (1989) performed two types of crossflow microfiltration experiments, as described by Romero and Davis (1991). The first used rectangular channels with glass sidewalls so that cake layer buildup could be directly observed. For ease of visualization, the experiments were carried out with relatively large particles. This resulted in thick cake layers with low permeabilities. The steady-state and transient cake thickness profiles were found to be in good agreement with the predictions of Romero and Davis (1988, 1990). In particular, the rate of cake buildup increased with increasing particle concentration and transmembrane pressure drop. The steady-state cake thickness increased with increasing transmembrane pressure drop, particle concentration, and distance from the filter exit, and decreased with increasing shear rate and particle size. The steady-state data from all experiments are summarized in Figure 33–11. The solid line is the theoretical prediction given by Eq. (33–54), with a modification made to account for the fact that the filter used had only one porous wall rather than two. Although there is considerable scatter in the data, the measured cake thicknesses follow closely the predictions of the shear-induced hydrodynamic diffusion theory.

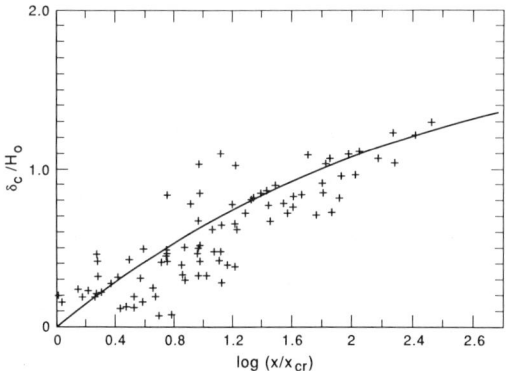

FIGURE 33–11. Dimensionless cake layer thickness versus dimensionless axial distance from the entrance of a two-dimensional channel. The solid line is the theoretical prediction of the shear-induced diffusion model of Romero and Davis (1988) for negligible cake resistance, and the symbols represent the data of Romero and Davis (1991).

The second type of experiment performed by Romero (1989) used 37-cm-long ceramic tubes with 0.13-cm inside diameters. The pore size ratings were 0.2 and 0.45 μm, and the monodispersed polyvinyltoluene particles used had diameters of 0.45, 0.64, 0.95, and 1.37 μm. Key parameter ranges investigated are typical of those of crossflow microfiltration: 700 s^{-1} ≤ $\dot{\gamma}_0$ ≤ 2800 s^{-1}, 5 psi ≤ Δp ≤ 12 psi, 600 ≤ Re$_c$ ≤ 2300, and 2 ≤ β ≤ 43. The values of the relative resistance β indicate that both cake and membrane resistances were important in most of the experiment. Transient and steady-state length-averaged permeate fluxes were measured by collecting the permeate on a microbalance interfaced with a microcomputer. Typical results are shown in Figure 33–12. The upper set of curves corresponds to larger particles ($r = 0.70$ μm) than those for the lower set of curves ($r = 0.32$ μm). The shear rate for the upper set was also nearly twice that of the lower set; the other conditions for the two runs were the same. A more rapid flux decline occurred for the smaller particles due to their large cake resistance. Also, the larger particles had a higher steady-state permeate flux due to their greater shear-induced diffusivity. The theoretical predictions (dashed lines) from the shear-induced diffusion model of Romero and Davis (1990) are in good agreement with the flux decline

measurements, except that the steady-state fluxes are lower than predicted. Membrane fouling was suspected of causing gradual flux decline after steady state was predicted to be reached.

The observed steady-state permeate fluxes for all tube experiments are compared in Figure 33–13 with the values predicted by the shear-induced diffusion model of Romero and Davis (1990). The diamonds refer to experiments with a newer membrane, whereas the plus symbols refer to experiments with an older membrane. The newer membrane resistance was approximately a factor of 10 lower than that of the older membrane. Although again some scatter occurs in the data, the agreement with predictions is very good. On average, the theory predicts steady-state permeate fluxes that are only 11% higher than the measured fluxes. In contrast, the concentration-polarization model with Brownian diffusivity, Eqs. (33–15) and (33–16), underpredicts the steady-state fluxes by two orders of magnitude. The concentration-polarization model with a constant and variable shear-induced diffusivity, Eqs. (33–19) and (33–68), respectively, underpredicts the steady-state fluxes by one order of magnitude. This is because these expressions apply only in the limit of cake-dominated resistance. The inertial

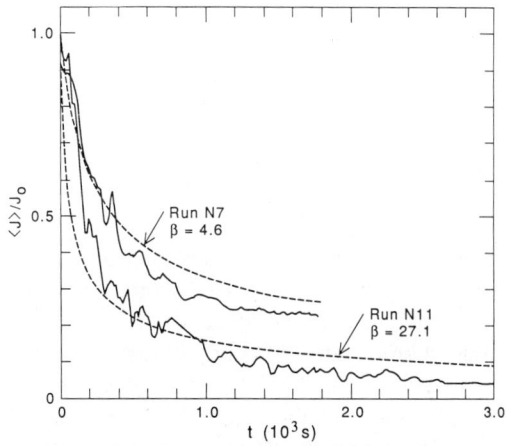

FIGURE 33–12. Permeate flux decline in a tubular filter for two suspensions having different-sized particles. The dashed line is from the shear-induced diffusion model of Romero and Davis (1990), and the solid lines are the measurements by Romero and Daviss (1991).

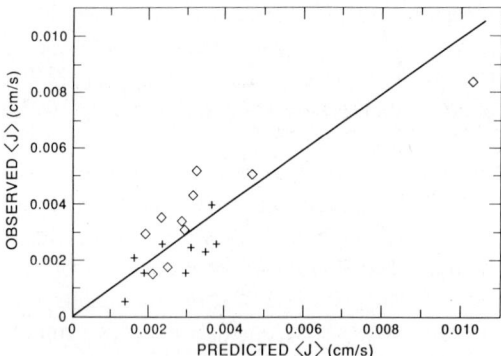

FIGURE 33-13. Observed steady-state permeate flux for a tubular filter versus the steady-state permeate flux predicted by the shear-induced diffusion model of Romero and Davis (1988). The solid line is the theoretical prediction, and the symbols represent the data of Romero and Davis (1991).

lift theory in the limit of cake-dominated resistance, Eq. (33-72), underpredicts the steady-state fluxes by three orders of magnitude. When corrected for the constriction of the channel due to cake buildup, the inertial lift velocities were still found by Romero and Davis (1991) to be a factor of 10 lower, on average, than the observed steady-state fluxes shown in Figure 33-13.

CONCLUSIONS

Membrane microfiltration refers to surface filtration of particles in the approximate size range of 0.02 to 10 μm. The analysis of microfiltration is complex, and the interpretation of flux data is complicated. Both the membrane and cake layer resistances are important in typical applications, except for those with highly deformable particles that form very thin cake layers having high resistances, and there are multiple mechanisms of particle transport away from the membrane during crossflow microfiltration. Under a broad range of conditions, the recently elucidated phenomenon of shear-induced particle diffusion is the dominant mechanism. However, for large particles and shear rates, inertial lift may provide an important contribution, whereas in the opposite limit of small particles and shear rates, Brownian diffusion must also be considered.

This chapter has presented a comprehensive review of recent theories for crossflow membrane microfiltration. Considerable experimental and theoretical research remains, however, to complete our understanding of crossflow microfiltration. For example, issues of direct membrane fouling by the attachment of particles and precipitates to the membrane pores and surface have not been adequately addressed. This author is pursuing a research program in which phenomenological expressions such as Eq. (33-64) are used to describe the increase in the membrane resistance with time due to fouling. These expressions are being incorporated into transient models of crossflow microfiltration, and then flux decline data are used to evaluate the parameters in the phenomenological expressions. Fundamental studies of membrane fouling mechanisms are also needed, as are studies of cake compression and particle adhesion to cake surfaces.

Crossflow microfilters are occasionally operated in the turbulent regime, whereas all of the models described in this chapter are restricted to laminar flow. As described by Porter (1977), Henry (1972), and others, use of a turbulent mass-transfer correlation is straightforward for the concentration-polarization film theory model with Brownian diffusion and cake-limited resistance. However, this approach is not expected to be appropriate for the shear-induced diffusion or inertial lift mechanisms since the transverse particle motion is directly dependent on the local flow fields. Careful experiments are needed for turbulent flow so that new models may be proposed and tested.

Flux decline due to cake layer buildup in crossflow microfiltration may be reduced in practice by periodic backflushing. Several studies have demonstrated the utility of blackflushing in maintaining high average flux rates. However, the frequency and duration of backflushing are chosen empirically. Models of cake removal during backflushing are needed so that the process may be optimized in a rational and predictive manner.

Several improvements are suggested for the models described in this chapter. For situations in which the cake resistance is not dominant, the effects of the axial pressure drop on the length-averaged permeate flux should be considered. These are expected to be particularly important when channel constriction due to cake buildup is significant. Also, there are intermediate parameter ranges for which multiple mechanisms of particle motion away from the membrane and cake occur simultaneously. Brownian and shear-induced diffusion may be considered simultaneously by adding the diffusion coefficients, although recent simulations have shown that the diffusion coefficients are not strictly additive (Brady and Bossis 1988). Inertial lift may be incorporated into the shear-induced diffusion model of crossflow microfiltration by replacing the flux J on the right-hand sides of Eqs. (33-39) and (33-66) with $J - v_L$, where v_L is the inertial lift velocity given by Eqs. (33-59) and (33-60). This approach is also expected to be approximate because it does not account for the effects of particle interactions on the inertial lift velocity in the concentration-polarization boundary layer, nor does it account for the fact that inertial lift will change the concentration profile and, hence, the diffusive flux in the boundary layer. Finally, the models described in this chapter are based on idealized suspensions of equisized spheres, which do not irreversibly stick to the membrane or cake surfaces but rather are free to diffuse or lift away. Further experiments and models are needed to study Brownian diffusion, shear-induced diffusion, and inertial lift in real suspensions of nonspherical, deformable particles having both narrow and broad size distributions.

ACKNOWLEDGMENTS

The author thanks Kay Shams for assistance with the literature review, and Cecily Romero and Sanjeev Redkar for assistance with numerical computations.

NOTATION

General Notation

See the General Notation section at the beginning of this handbook.

Special Notation

A — membrane area, L^2
D_0 — characteristic diffusivity, $r^2 \tau_{w0}/\eta_0$, L^2/t
f — inertial lift function, dimensionless
J — volumetric flux, L/t
J_0 — initial flux, L/t
k — Boltzmann's constant, 1.38×10^{-16} g·cm^2/s^2·°K
L — membrane length, L
m — exponent, $m = 4$ for rectangular channel, $m = 6$ for tube, dimensionless
n — exponent, $n = 6$ for rectangular channel, $n = 9$ for tube, dimensionless
Pe — Peclet number, $J_0 H_0/D_0$, dimensionless
Q — axial flow rate, L^3/t
r — particle radius, L
R — recycle ratio, dimensionless
Re$_c$ — Reynolds number, $\rho_0 \langle v \rangle r_t / \eta_0$, dimensionless

Greek Letters

β — relative cake resistance, dimensionless
δ — polarization boundary layer thickness, L
η — viscosity, M/Lt
ϕ — solids volume fraction, dimensionless
ϕ' — dummy variable of integration, dimensionless
$\phi_{\max}$ — maximum random packing fraction of solids in cake, dimensionless

Diacritical Marks

¯ — dimensionless value
^ — dimensionless value

REFERENCES

Altena, F. W., and G. Belfort. 1984. Lateral migration of spherical particles in porous flow channels: application to membrane filtration. *Chem. Eng. Sci.* 39(3):343–355.

Baker, R. J., A. G. Fane, C. J. D. Fell, and B. H. Yoo. 1985. Factors affecting flux in crossflow filtration. *Desalination* 53:81–93.

Belfort, G. 1989. Fluid mechanics in membrane filtration: recent developments. *J. Membr. Sci.* 40:123–147.

Belfort, G., and N. Nagata. 1985. Fluid mechanics and cross-flow filtration: some thoughts. *Desalination* 53:57–79.

Belfort, G., R. J. Weigand, and J. T. Mahar. 1985. Particulate membrane fouling and recent developments in fluid mechanics of dilute suspensions. In *ACS Symp. Ser. No. 281: Reverse Osmosis and Ultrafiltration,* ed. S. Sourirajan and T. Matsuura, pp. 383–401. Washington, DC: American Chemical Society.

Blatt, W. F., A. Dravid, A. S. Michaels, and L. Nelson. 1970. Solute polarization and cake formation in membrane ultrafiltration: causes, consequences, and control techniques. In *Membrane Science and Technology,* ed. J. E. Flinn, pp. 47–97. New York: Plenum Press.

Brady, J. F., and G. Bossis. 1988. Stokesian dynamics. *Ann. Rev. Fluid Mech.* 20:111–158.

Brown, C. E., M. P. Tulin, and P. Van Dyke. 1971. On the gelling of high molecular weight impermeable solutes during ultrafiltration. *Chem. Eng. Prog. Symp. Ser.* 67(114):174–180.

Chandavarkar, A., and C. L. Cooney. 1989. Dynamics of flux decline during microfiltration caused by protein-membrane interactions. Paper read at 198th National Meeting of the American Chemical Society, 10–15 September 1989, Miami Beach, FL.

Cox, R. G., and H. Brenner. 1968. The lateral migration of solid particles in Poiseuille flow—I theory. *Chem. Eng. Sci.* 23:147–173.

Davis, R. H., and S. A. Birdsell. 1987. Hydrodynamic model and experiments for crossflow microfiltration. *Chem. Eng. Commun.* 49:217–234.

Davis, R. H., and D. T. Leighton. 1987. Shear-induced transport of a particle layer along a porous wall. *Chem. Eng. Sci.* 42:275–281.

Davis, R. H., and J. D. Sherwood. 1990. A similarity solution for steady-state crossflow microfiltration. *Chem. Eng. Sci.* 45:3203–3209.

Drew, D. A., J. A. Schonberg, and G. Belfort. 1991. Lateral inertial migration of a small sphere in fast laminar flow through a membrane duct. *Chem. Eng. Sci.* 46:3219–3224.

Eckstein, E. C., P. G. Bailey, and A. H. Shapiro. 1977. Self-diffusion of particles in shear flow of a suspension. *J. Fluid Mech.* 79:191–208.

Errede, L. A. 1984. Effect of organic anion adsorption on water permeability of microporous membranes. *J. Colloids. Interface Sci.* 100:414–422.

Fane, A. G. 1986. Ultrafiltration: factors influencing flux and rejection. In *Progress in Filtration and Separation,* ed. R. J. Wakeman, Vol. IV, pp. 101–79. New York: Elsevier Scientific Publishing Co.

Green, G., and G. Belfort. 1980. Fouling of ultrafiltration membranes: lateral migration and the particle trajectory model. *Desalination* 35:129–147.

Henry, J. D. 1972. Cross flow filtration. In *Recent Developments in Separation Science* ed. N. N. Li, Vol. 2, pp. 205–225. Boca Raton, FL: CRC Press.

Ho, B. P., and L. G. Leal. 1974. Inertial migration of rigid spheres in two-dimensional unidirectional flows. *J. Fluid Mech.* 65:365–400.

Ishii, K., and H. Hasimoto. 1980. Lateral migration of a spherical particle in flows in a circular tube. *J. Phys. Soc. Japan* 48:2144–2153.

Kraus, K. A. 1974. Cross-flow filtration and axial filtration. In *Proceedings of the 29th Industrial Waste Conference,* 7–9 May 1974, Purdue University, West Lafayette, IN, p. 1059.

Leighton, D. T., and A. Acrivos. 1987a. Measurement of the shear induced coefficient of self-diffusion in concentrated suspensions of spheres. *J. Fluid Mech.* 177:109–131.

Leighton, D. T., and A. Acrivos. 1987b. The shear-induced migration of particles in concentrated suspension. *J. Fluid Mech.* 181:415–439.

Leonard, E. F., and C. S. Vassilieff. 1984. The deposition of rejected matter in membrane separation processes. *Chem. Eng. Commun.* 30:209–217.

Madsen, R. E. 1977. *Hyperfiltration and Ultrafiltration in Plate-and-Frame Systems.* New York: Elsevier Scientific Publishing Co.

Ofsthun, N. J. 1989. *Crossflow Membrane Filtration of Cell Suspensions,* Ph.D. thesis, Massachusetts Institute of Technology, Chemical Engineering Department, Cambridge, MA.

Ofsthun, N. J., and C. K. Colton. 1987. Visual evidence of concentration polarization in crossflow membrane plasmapheresis. *Am. Soc. Artif. Intern. Organs J.* 10:510–517.

Otis, J. R., F. W. Altena, J. T. Mahar, and G. Belfort. 1986. Measurement of single spherical particle trajectories with lateral migration in a slit with a porous wall under laminar flow conditions. *Expl. Fluids* 4:1–10.

Petzny, W. J., and J. A. Quinn. 1969. Calibrated membranes with coated pore walls. *Science* 166:751–753.

Porter, M. C. 1972a. Concentration polarization with membrane ultrafiltration. *Ind. Eng. Chem. Prod. Res. Dev.* 11:233–248.

Porter, M. C. 1972b. Ultrafiltration of colloidal suspensions. In *Recent Advances in Separation Techniques, AIChE Symp. Ser. No. 120* pp. 21–30. New York: American Institute of Chemical Engineers.

Porter, M. C. 1977. What, when, and why of membranes-MF, UF and RO. In *What the Filter Man Needs to Know About Filtration, AIChE Symp. Ser. No. 171,* ed. W. Shoemaker, pp. 83–103. New York: American Institute of Chemical Engineers.

Porter, M. C. 1986. Microfiltration. In *Synthetic Membranes: Science, Engineering and Applications,* ed. P. M. Bungay, H. K. Lonsdale, and M. N. de Pinho, pp. 225–246. Dordrecht: D. Reidel Publishing Co.

Rautenbach, R., and G. Schock. 1988. Ultrafiltration of macromolecular solutions and cross-flow microfiltration of colloidal suspensions. A contribution to permeate flux calculations. *J. Membr. Sci.* 36:231–242.

Romero, C. A. 1989. Modeling and laboratory observations of shear induced hydrodynamic diffusion in crossflow microfiltration, Ph.D. thesis, Univ. of Colorado, Chemical Engineering Department, Boulder, CO.

Romero, C. A., And R. H. Davis. 1988. Global model of crossflow microfiltration based on hydrodynamic particle diffusion. *J. Membr. Sci.* 39:157–185.

Romero, C. A., and R. H. Davis. 1990. Transient model of crossflow microfiltration. *Chem. Eng. Sci.* 45:13–25.

Romero, C. A., and R. H. Davis. 1991. Experimental verification of the shear-induced hydrodynamic diffusion model of crossflow microfiltration. *J. Membr. Sci.* 62:249–273.

Schneider, K., and W. Klein. 1982. The concentration of suspensions by means of crossflow-microfiltration. *Desalination* 41:263–275.

Schonberg, J. A., and E. J. Hinch. 1989. Inertial migration of a sphere in Poiseuille flow. *J. Fluid Mech.* 203:517–524.

Shen, J. J. S., and R. F. Probstein. 1977. On the prediction of limiting flux in laminar ultrafiltration of macromolecular solutions. *Ind. Eng. Chem. Fundam.* 16:459–465.

Trettin, D. R., and M. R. Doshi, 1980. Limiting flux in ultrafiltration of macromolecular solutions. *Chem. Eng. Commun.* 4:507–522.

Vasseur, P., and R. G. Cox. 1976. Lateral migration of spherical particles in two-dimensional shear flow. *J. Fluid Mech.* 78(2):385–413.

Weigand, R. J., F. W. Altena, and G. Belfort. 1985. Lateral migration of spherical particles in laminar porous tube flows: application to membrane filtration. *Phys. Chem. Hydrodyn.* 6:393–413.

Zahka, J., and T. J. Leahy. 1985. Practical aspects of tangential flow filtration in cell separations. *ACS Symp. Ser. No. 172–Purification of Fermentation Products,* pp. 51–69. Washington, DC: American Chemical Society.

Zydney, A. L. 1985. Crossflow plasmapheresis: an analysis of flux and hemolysis, Ph.D. thesis, Massachusetts Institute of Technology, Chemical Engineering Department, Cambridge, MA.

Zydney, A. L., and C. K. Colton. 1986. A concentration polarization model for the filtrate flux in crossflow microfiltration of particulate suspensions. *Chem. Eng. Commun.* 47:1–21.

Zydney, A. L., and C. K. Colton. 1987. Fundamental studies and design analysis for crossflow membrane plasmapheresis. In *Artificial Organs: Proceedings of International Symposium on Artificial Organs, Biomedical Engineering and Transplantation,* ed. J. D. Andrade et al., pp. 343–358. New York: VCH Publishers.

34

Deadend Microfiltration: Applications, Design, and Cost

Vinay Goel, Mauro A. Accomazzo*, Anthony J. DiLeo, Peter Meier, Aldo Pitt, and Malcolm Pluskal
Millipore Corporation

Robert Kaiser
ARGOS Associates, Inc.

INTRODUCTION
   Background
   Membrane Types
   Membrane Modules
   Membrane Performance Characteristics
   Microfiltration Systems
STERILIZATION IN THE PHARMACEUTICAL INDUSTRY
   Pharmaceutical Filtration Needs
   Types of Devices
   Filter Sizing
   System Design
   Case Studies
BEVERAGE APPLICATIONS OF MICROFILTRATION MEMBRANES
   Introduction
   Applications
   System Design

PURIFICATION OF PROCESS FLUIDS IN SEMICONDUCTOR MANUFACTURE
   Introduction
   Process Gas Filtration
   Process Liquids Filtration
ANALYTICAL APPLICATIONS OF MICROFILTRATION MEMBRANES
   Historical Perspective
   Microbiological Assay
   Monitoring of Particulate Contamination
   Blotting of Macromolecules to Microporous Membrane Substrates
   Cell Growth Studies on Microporous Membranes
ACKNOWLEDGMENTS
NOTATION
REFERENCES

INTRODUCTION

Background

The previous three chapters covered a general description of microfiltration and the theories of deadend and crossflow microfiltration. This chapter focuses on the practical aspects of microfiltration with special emphasis on deadend microfiltration using commercial membranes, important applications, design criteria, and cost estimates.

By far the largest use of microfiltration membranes is the quantitative separation of suspended matter in the 0.1- to 10-μm size range from liquids and gases. Although the separations are carried out predominantly on the basis of size, there are some membranes and some applications in which adsorption also

---
*Present affiliation: Ionpure Technologies Corporation.

plays an important role. In general, the applications fall into one of the following major categories:

1. Purification
2. Clarification
3. Sterilization
4. Concentration
5. Analysis.

(Another, somewhat different, type of application involves the use of hydrophobic microporous membranes in waterproof breathable fabrics. This application is considered outside the scope of this handbook.)

Before discussing specific applications in more detail, it should prove useful to familiarize the reader with the various types of commercial membranes, their principal manufacturing methods, characteristics and characterization methods, types of membrane modules, and the commercial entities engaged in the development, manufacture, and marketing of microfiltration systems.

Membrane Types

Membranes can be classified according to one or more of the following characteristics:

1. *Material of construction:* organic, polymer, ceramic (including glass and porcelain), or metal
2. *Structure:* homogeneous, asymmetric, or composite
3. *Method of manufacture:* phase inversion, sintering, stretching, or track etching
4. *Geometry:* flat sheet, hollow fiber, or tubular
5. *Unsupported or reinforced*
6. *Hydrophobic or hydrophilic*
7. *Surface charge:* negative, neutral, or positive.

Figure 34–1 shows scanning electron micrographs of a number of different types of membranes. (Other media such as paper, fiberglass, nonwovens, fabric screens, fibrous string wound, and pressure-formed fibers or particles are also used for filtration, but the particles are usually coarser and thus are not described in detail because they are considered outside the scope of this handbook.)

As the long list implies, a very large number of membranes can be made, and indeed are available from an ever-increasing number of membrane suppliers. This provides the user with many options that should enable him/her to choose the best one for a specific application. This choice, however, can also lead to a fair amount of confusion arising from conflicting performance claims by competing suppliers. A fairly comprehensive list of microporous membrane companies and their commercial membranes is presented in Table 34–1.

Membrane Modules

The very early polymeric microporous membranes were produced in flat-sheet form and used in metal or plastic holders with elastomeric seals. The holder is reused but the membrane is replaced after each use. In recent years, driven by convenience and economics, the trend has shifted toward fabricated devices and elements. These devices can range in size from less than 1 cm^2 of active membrane area to pleated cartridges housing approximately 30 ft^2 ($\sim$3 m^2) of membrane area. The design and construction of these modules varies depending on whether they are intended for deadend or crossflow mode of operation, as well as specific performance requirements and manufacturers' preferences.

The separation of suspended matter from a fluid by filtration can be accomplished by one of two basic modes: deadend or crossflow, as indicated in Figure 34–2. In deadend, or through-flow filtration, pressure drives the entire fluid stream through the porous filter medium. The particles are trapped by the medium and accumulate with time. When the pores or surface becomes plugged with retained material, flow stops and the medium must be renewed or replaced.

Some representative flat sheet-devices intended for deadend operation are illustrated in Figure 34–3. As can be seen, the large variety of module shapes and connectors is a result of the emphasis on ease of use. The very small filter devices typically consist of a small disk or a strip of membrane sealed to a plastic support with luer-type connectors that fit easily into the end of syringes for small volume filtrations.

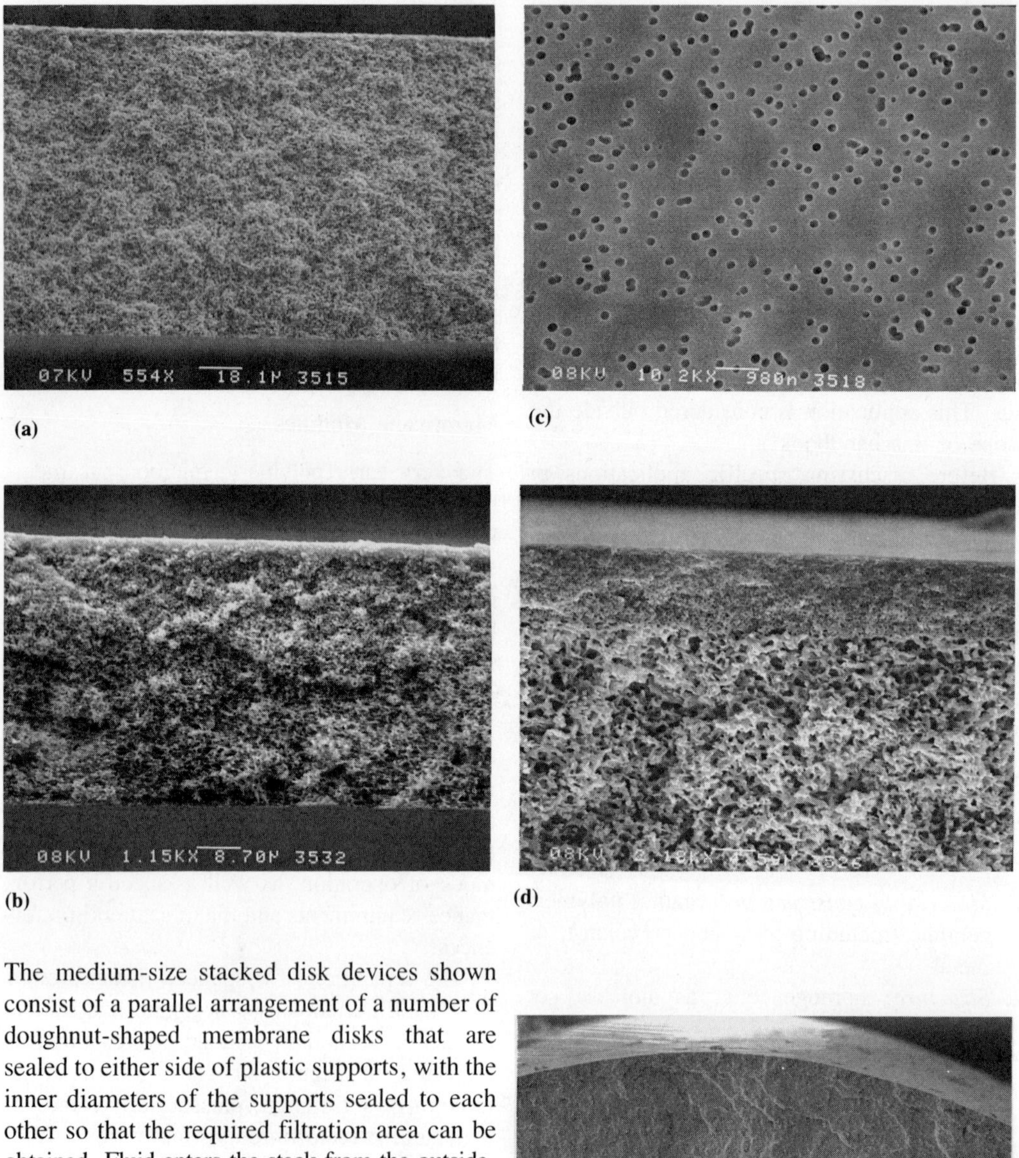

The medium-size stacked disk devices shown consist of a parallel arrangement of a number of doughnut-shaped membrane disks that are sealed to either side of plastic supports, with the inner diameters of the supports sealed to each other so that the required filtration area can be obtained. Fluid enters the stack from the outside, and filtrate is collected in the inner diameter.

The pleated cartridge design also shown in Figure 34–3 enables a larger membrane area to be packed in a given cylindrical shape. Typically, the membrane is sandwiched between two nonwoven fabric supports and pleated. A cylinder of the pleated sandwich is formed by sealing the two ends of the membrane. A plastic core is inserted inside the cylinder and the whole assembly is inserted into a sleeve. With appropriate connections, a pleated cartridge can be used either as a single element or as a multiple element by stacking elements on ends.

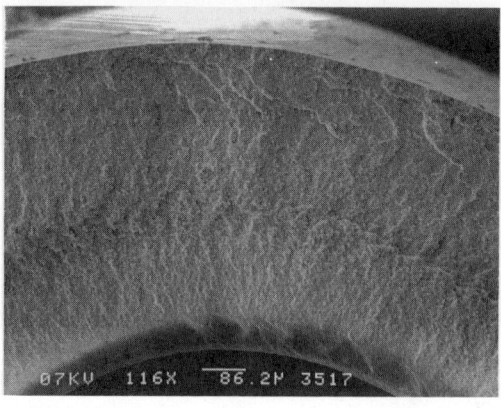

FIGURE 34–1. Scanning electron micrographs of a variety of membranes: (a) symmetric, (b) asymmetric, (c) track etched, (d) composite, and (e) hollow fiber (reprinted from Millipore Corporation with permission).

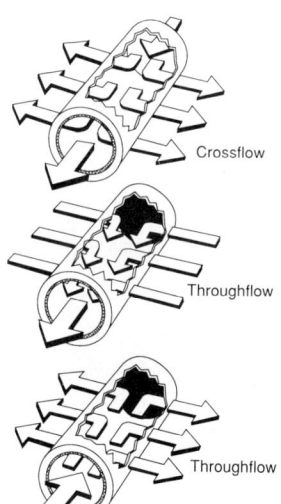

FIGURE 34-2. Filters can operate in the deadend or crossflow mode (reprinted from Michaels 1989 with permission).

Membrane Performance Characteristics

Important membrane performance characteristics for filtration applications are:
1. Retention efficiency
2. Permeability
3. Dirt handling capacity or throughput
4. Chemical resistance
5. Surface energy or wetting behavior
6. Temperature limits
7. Mechanical strength
8. Cleanliness
9. Adsorption characteristics

It would be very helpful to the user if there were standard measures and methods for these characteristics. Unfortunately, reality is quite complex. Although attempts have been made by various organizations to develop standards, the results are somewhat fragmented and incomplete. Nevertheless, a partial list of standard methods developed by the American Society for Testing and Materials (ASTM) is presented in Table 34-2.

Retention efficiency is by far the most important performance characteristic of a microporous membrane. It is governed by the pore size and/or the pore size distribution of a membrane. The most common technique used to measure the pore size of a membrane is the so-called "bubble point." Bubble point measurement involves filling the pores of a membrane with a wetting liquid and then applying gas pressure on one side of the membrane. The lowest pressure at which gas bubbles are seen on the other side of the membrane is referred to as the bubble point. The relationship between pore size and bubble point is based on application of the Young-Laplace equation to liquid rise in capillaries. A commonly used form of this equation is

$$p_{BP} = \frac{4\gamma \cos\theta}{d}, \quad (34\text{-}1)$$

where

d = pore diameter
p_{BP} = the bubble point pressure
θ = contact angle between liquid and membrane surface
γ = surface tension of liquid.

The smaller the pore diameter, the greater the bubble point pressure. This equation can be used in the above form for membranes with straight-through capillaries such as the track-etched type, but needs to be modified for use with other membranes that look more like sponge-like polymeric matrices. For these membranes, a tortuosity correction factor K_t is often incorporated into the above equation:

$$p_{BP} = \frac{K_t 4\gamma \cos\theta}{d}. \quad (34\text{-}2)$$

In practice, this tortuosity factor is less than 1 and needs to be determined experimentally for each family of membranes. The procedure for determining bubble point is described in ASTM Standard Method F316. Note that if the bubble point is defined as the lowest pressure at which gas bubbles are first seen on the low-pressure side, then measurement is for the size of the largest pore in the membrane, not the mean pore size. In critical applications such as sterilization, which requires absolute retention of microbial species, the largest pore size is precisely what needs to be determined. In noncritical applications, average pore size and pore size distribution are more desirable characteristics needing quantitation. The two common

TABLE 34-1. Commercial Microporous Membranes.

Company	Membrane Material	Membrane Type	Remarks
A/G Technology	Polysulfone	Hollow fibers	Recent startup
AKZO (ENKA Division)	Polypropylene	Hollow fibers and flat sheet	Thermal phase inversion membranes
	Nylon 6		
ALCAN	Ceramic	Flat-sheet devices	Anodized aluminum membranes
ALCOA	Ceramic	Tubesheets	Purchased Ceraver
AMICON (W. R. Grace)	Polysulfone	Hollow fibers	Microfiltration membranes are recent
ASAHI	Polypropylene	Hollow fibers	
	PTFE	Hollow fibers	
HOECHST CELANESE	Polypropylene	Flat sheet	
CERAMEM	Ceramic	Monolith	Recent startup
CUNO (Commercial Intertech)	Nylon 66	Flat sheet	
	PTFE	Flat sheet	
DDS	Polysulfone	Flat sheet	
	PVDF	Flat sheet	
DOMINICK HUNTER	CA/CN	Flat sheet	
	Nylon 66	Flat sheet	
	PTFE	Flat sheet	
FILTERITE (MEMTEC)	Polysulfone	Flat sheet	Used to be part of Brunswick Technetics
	PTFE		
FUJI	Polysulfone	Flat sheet	
GELMAN SCIENCES	CTA	Flat sheet	
	Cellulose	Flat sheet	
	CA/CN	Flat sheet	
	PVC	Flat sheet	
	Polysulfone	Flat sheet	
	PVDF	Flat sheet	
	PTFE	Flat sheet	
	Nylon 66	Flat sheet	
	Polypropylene	Flat sheet	
	Acrylic Copolymer	Flat sheet	Bought membranes
W. L. GORE	PTFE	Flat sheet and Tubular	Pioneered stretched PTFE membranes
KINETEK SYSTEMS	Polysulfone	Hollow fibers	Startup involved in small-scale bioseparations

Company	Material	Form	Notes
KURARAY	PVA	Hollow fibers	
MICROGON	CA/CN	Hollow fibers	Recent startup
MITSUBISHI RAYON	Polypropylene	Hollow fibers	
	Polyethylene	Hollow fibers	
MILLIPORE	CA/CN	Flat sheet	Leading supplier of microfiltration products
	PVDF	Flat sheet	
	Polycarbonate	Flat sheet	
	PTFE	Flat sheet	
	Ceramic	Tubesheet	
MFS	CN	Flat sheet	
	CA	Flat sheet	
	CA/CN	Flat sheet	
	Cellulose	Flat sheet	
	PTFE	Flat sheet	
MSI	CA/CN	Flat sheet	Recent startup
	Nylon 66	Flat sheet	
NITTO DENKO	PTFE	Flat sheet	
NUCLEPORE (Costar)	Polycarbonate	Flat sheet	Pioneered track-etched membranes
	Polyester	Flat sheet	
PALL	Nylon 66	Flat sheet	
	PTFE	Flat sheet	
	PVDF	Flat sheet	
PORETICS	Polycarbonate	Flat sheet	Track-etched membranes
	Polyester	Flat sheet	
SARTORIUS	CA	Flat sheet	
	CN	Flat sheet	
	Cellulose	Flat sheet	
	PTFE	Flat sheet	
SCHLEICHER-SCHUELL	CA	Flat sheet	
	CN	Flat sheet	
	CA/CN	Flat sheet	
	Cellulose	Flat sheet	
	PTFE	Flat sheet	
WHATMAN	CN	Flat sheet	
	CA/CN	Flat sheet	
	PTFE	Flat sheet	
X-FLOW B.V.	Polysulfone	Hollow fibers	Startup
	Polyetherimide	Hollow fibers	

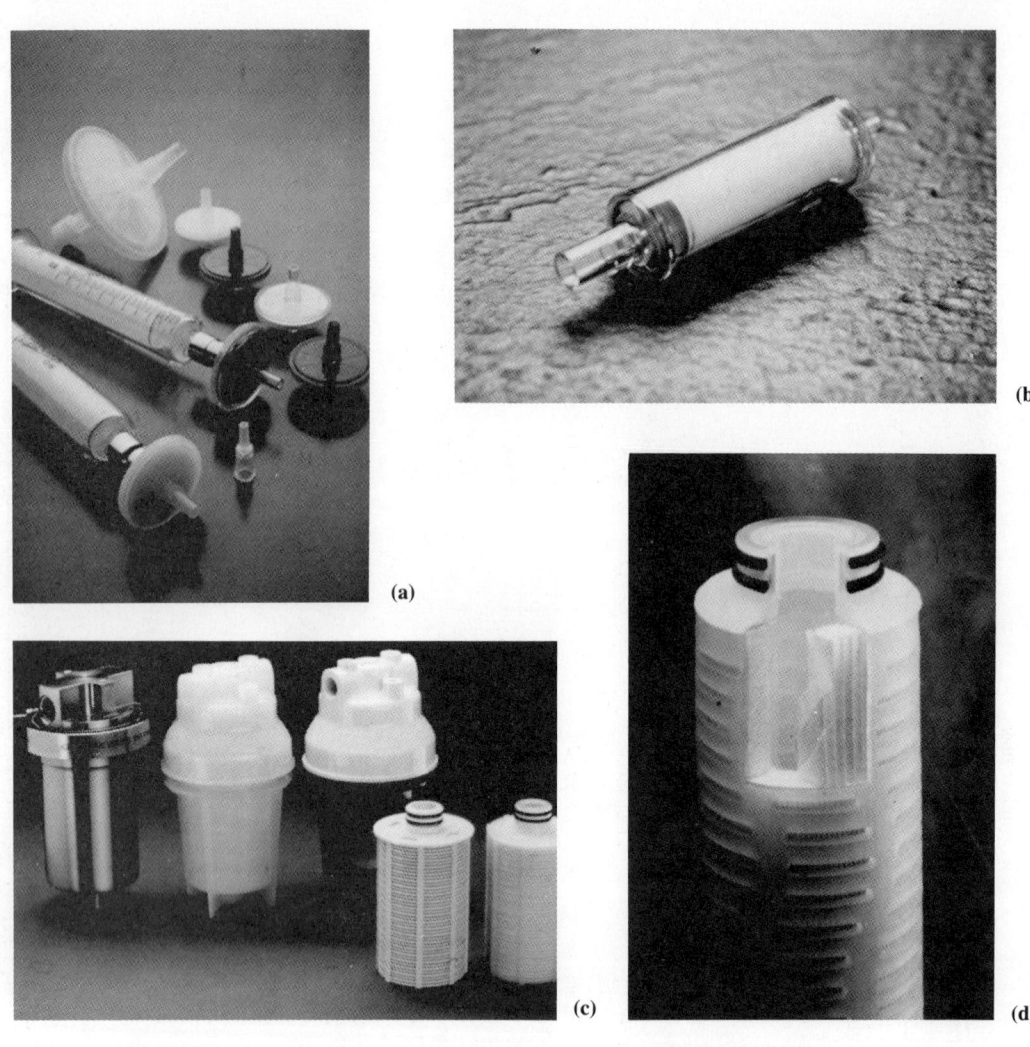

FIGURE 34-3. Flat-sheet microfiltration devices: (a) disposable syringe filters, (b) IV filter, (c) stacked disk cartridges, and (d) pleated cartridge (reprinted from Millipore Corporation with permission).

methods used to determine these are mercury intrusion (Honold and Skau 1954) and air flow porosimetry. Both these methods rely on the phenomenon of capillarity. The mercury intrusion method involves intrusion of mercury into dry membrane pores by the application of pressure. The air flow method involves measuring air flow as a function of pressure through a dry membrane compared to a wet membrane. The Young-Laplace equation is then used to relate the bubble point distribution, obtained from the test, to pore size distribution. Due to the need for very large pressures in the mercury intrusion method (the surface tension of mercury is 480 dynes/cm), the air flow method is finding more use. Commercial instruments based on this method are now available from at least one manufacturer, Coulter Electronics Inc., Hialeah, FL.

Permeability of membranes is a function of pore size, void volume or porosity, thickness, and asymmetry [permeability = filtration rate/(area × pressure difference)]. Membrane permeability has a significant effect on design of membrane modules and systems, and on process economics. For homogeneous or nonasym-

TABLE 34–2. ASTM Standard Methods Relevant to the Characterization of Microfiltration Membranes.

D3861-84	Standard Test Method for Quantity of Water-Extractable Matter in Membrane Filters
D3862-80	Standard Test Method for Retention Characteristics of $0.2 \mu m$ Membrane Filters Used in Routine Filtration Procedures for the Evaluation of Microbiological Water Quality
D3863-87	Standard Test Method for Retention Characteristics of 0.40 to $0.45 \mu m$ Membrane Filters Used in Routine Filtration Procedures for the Evaluation of Microbiological Water Quality
D4196-82	Standard Test Method for Confirming the Sterility of Membrane Filters
D4197-82	Standard Test Method for Percent Porosity of Membrane Filters
D4198-82	Standard Methods for Evaluating Absorbent Pads Used with Membrane Filters for Bacteriological Analysis and Growth
D4200-82	Standard Test Method for Evaluating Inhibitory Effects of Ink Grids on Membrane Filters
E1294-89	Standard Test Method for Pore Size Characteristics of Membrane Filters Using Automated Liquid Porosimeter
F316-86	Standard Test Methods for Pore Size Characteristics of Membrane Filters by Bubble Point and Mean Flow Pore Test
F317-72	Standard Test Method for Liquid Flow Rate of Membrane Filters

metric membranes, permeability decreases with decreasing pore size, decreasing void volume, and increasing thickness. Asymmetric or skinned membranes behave differently because the retentive layer is quite thin and has a smaller pore size then the rest of the membrane. These membranes, in general, have higher permeabilities than homogeneous membranes of the same pore size. However, the consequences of damage to the retentive surface are far more severe than in the case of homogeneous membranes.

Dirt holding capacity, or *throughput,* is the volume of liquid that can be passed through a filter before the filtration rate drops to an economically unacceptable value. The drop in filtration rate occurs because the particles retained by the filter occlude it. Because the rate of plugging and the plugging mode are a function of the characteristics and concentration of the particles, and also the filter structure, quantitative determination of the dirt holding capacity is application-specific and, therefore, has to be obtained by experimentation.

Chemical resistance is an inherent property of the material or materials from which the membrane is made. It should be one of the first properties probed when selecting a membrane for a particular application. A word of caution is warranted here. Determination of suitability of a material in a particular chemical environment should not be based solely on properties listed in a polymer handbook or polymer manufacturers' manuals. Membranes have 100 to 1000 times more exposed surface areas than solid objects of comparable size. This feature can have a significant impact on their performance characteristics. It would be advisable to consult membrane manufacturers or to test membranes under actual conditions.

The same words of caution applies to *high-* or *low temperature* conditions of use. The glass transition temperature and the heat deflection temperature of the polymer provide starting guidelines, but the final determination should be based on actual tests. This is especially true if filters are to be used for long periods of time at elevated temperatures. In these situations, polymer creep becomes an important consideration.

Surface energy is also an inherent property of the membrane material, although it can be changed by surface chemical modifications to suit a given end use.

The *mechanical strength* of a membrane depends to some extent on the inherent strength of the membrane material, but to a larger extent on the integrity of the structure. Strength is an important design parameter for manufacturers in the design of the support structures, but is rarely an area of concern for the user as long as the user operates the system within recommended limits. As noted in the previous paragraph, one area of concern is the effect of extended high-temperature service on creep of polymeric membranes.

Cleanliness of membrane filters has taken on

increased importance in recent years as users, particularly in the semiconductor industry, have sought higher and higher levels of cleanliness in their fluid streams. In some cases, filters that shed particles or have extractable contaminants can make fluid streams dirtier than they originally were. Although no standard tests or standard measures exist for cleanliness, important categories of contaminants include particles, ions, and organics. On-line instruments are now available to measure all three down to extremely low levels.

The *adsorption* characteristics of membranes are important because removal or alteration of the desirable constituents in a fluid stream is almost as bad as the introduction of contaminants. In practice, binding of proteins by polymeric membranes is one example in which adsorption has played an important undesirable role in biological and biochemical filtrations. In other cases, such as blotting, it is an important desirable characteristic.

Microfiltration Systems

Components

Microfilters are not stand-alone components; they are always used as an element, usually the key element, of a microfiltration system. A typical microfiltration system is illustrated in Figure 34–4. Other necessary elements or components of such a system include the *filter housing,* which provides the means for

1. Containing the fluid to be filtered
2. Mechanically supporting the filter
3. Sealing the filter so that the challenge solution does not bypass the filter and mix with the filtrate, thus rendering the filtration step inoperative.

Additional components are a *source of transmembrane pressure differential,* which is needed to provide the driving force for passage of the fluid through the filter; *fluid reservoirs* and *recipients* to store the filter feed and the filtrate; and *piping* to convey the feed from storage to the other components of the system and, finally, to convey the filtrate to storage. While microfiltration systems can be operated without them, other components that are commonly used (Figure 34–4) include *prefilters* to extend the life of a filter, and *instrumentation,* such as pressure flow sensors and readouts, to monitor and control the operation.

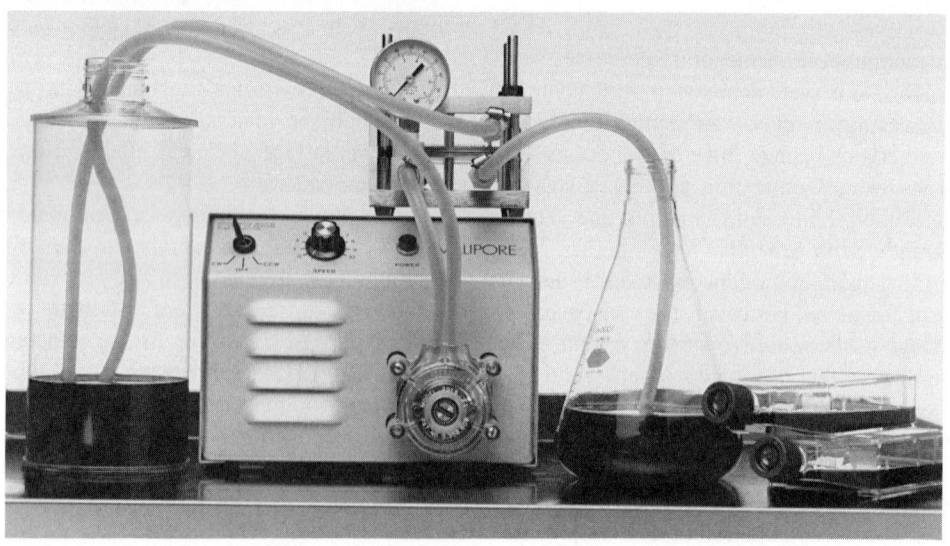

FIGURE 34–4. Representative microfiltration system (reprinted from Millipore Corporation with permission).

General Design Considerations and Modes of Operation

Given the variety of situations where microfiltration is or could be used, no one unique design would satisfy all user requirements. The needs and constraints that apply to microanalytical particle counts performed on microfilters in a research laboratory are obviously different from those that are pertinent to the clarification of fermentation broths in the manufacture of antibiotics. The fluids handled, the particulates recovered, the scale of operations, the economic justifications, and the operational environments are totally different. Because of the plurality of significant factors that needs to be considered, the design and mode of operation of a microfiltration system are application-specific.

In general, the user's operational goal is the most important factor in establishing system design. Critical questions that need to be addressed are:

1. What needs to be accomplished? Does a liquid need clarifying or do suspended solids need to be recovered?
2. What are the competing technologies that could achieve the desired goals?
3. What is the perceived added value of performing the separation?
4. Is absolute separation of suspended matter larger than a certain size required?
5. What value is placed on product quality and process reliability?
6. What value is placed on product yield?
7. What are the financial constraints on installing (capital investment) and operating the system?
8. What, if any, legal, regulatory, or environmental constraints need to be considered?

A number of the important process parameters that have to be considered in the development of an optimum design are given in Table 34–3. The user must determine that the filter chosen for the applications meets the membrane performance characteristics listed previously. The characteristics of the other components of the system must also be established, and their compatability with the fluids being processed and with the microfiltration membrane confirmed. Special consideration has to be given to balancing the filter area required with the capacity and pressure head of the pump.

TABLE 34–3. Process Parameters of Importance to the Design of a Microfiltration System.

1. Fluid to be Filtered:
   State-liquid or gas
   Composition of fluid to be filtered
   Quantity of fluid to be filtered
   Viscosity
   Temperature
   Shear sensitivity of fluid to be filtered
2. Suspended Material to be Retained:
   Composition
   State-liquid or solid
   Initial concentration
   State of dispersion
   Dispersion stability
   Surface properties
   Mechanical properties
3. System Considerations:
   Specifications of processes and materials
   Constancy/variability of feed
   Need for sterile conditions
   Time available for processing
   Batch or continuous operations

The application of microfiltration systems to different commercial systems is discussed in the following sections of this chapter to illustrate how these general design considerations are applied to specific applications of current industrial interest. The topics to be discussed are:

1. Sterilization of pharmaceutical solutions
2. Clarification of beverages
3. Purification of fluids in the semiconductor manufacturing industry
4. Selected analytical applications.

STERILIZATION IN THE PHARMACEUTICAL INDUSTRY

Pharmaceutical Filtration Needs

Need for Sterile Filtration

Microporous membrane have found wide application in the pharmaceutical industry in the preparation and processing of parenteral (injectable) drugs, and of their constituent components, both liquids and gases, in order to achieve three primary goals:

1. Final sterilization, i.e., the total removal of bacteria from heat labile pharmaceutical products
2. The reduction of bacterial burden for the explicit purpose of maintaining low pyrogen levels in parenteral formulations that will be autoclaved for final sterilization
3. The removal of particles, both organic and inorganic, from parenteral solutions, the feedstock fluids that are used to prepare and process these solutions, and of any aerosols generated during processing.

Drug sterility is a critical requirement in the pharmaceutical industry; consequently, manufacturers and the Food and Drug Administration (FDA) focus attention in this area. Drug products that are labeled as being sterile are processed by one of two methods: terminal sterilization and aseptic processing. Terminal sterilization is a process whereby a drug product is subjected to a final, "terminal" sterilization process after having been formulated, prefiltered, filled, and sealed into containers. Although the sterilization is accomplished through the use of steam or an autoclave, membrane filters are used in the upstream part of the process to reduce bioburden and for pyrogen management.

Aseptic processing, on the other hand, involves the filling of presterilized containers with presterilized drugs. These filled containers are closed with presterilized closures. Filter sterilization, an integral component of aseptic processing, is the method of choice for rendering the drug product free of microbiological contaminants prior to filling. This type of processing is the only practical method of producing and processing drugs that cannot withstand the rigors of terminal sterilization.

Applications
Applications for membrane filtration in the pharmaceutical industry can be categorized into the following five areas:

1. General services, including WFI (water for injection), gases and steam, and vents
2. Synthetic parenterals including large-volume parenterals (LVPs), both sterile and nonsterile, and small-volume parenterals (SVPs)
3. Ophthalmic solutions
4. Fermentation products, including antibiotics, vaccines, and bioengineered proteins
5. Serum and plasma processing.

The unique features and important requirements for each of these areas are listed in Table 34–4. The manner in which they affect choice of membrane material, pore size, device configuration and membrane area is dealt with in more detail in the remainder of this section.

Design Considerations
Membrane Pore Size
The specification of the filtration system for applications in the pharmaceutical environment generally requires consideration of the nature of the fluid to be processed, the ultimate quality required in the filtered product, and the operating constraints under which the fluids are to be processed. Consequently, the detailed specification of a filtration system is largely application-dependent. Ultimately, each filtration operation must be assessed in its actual setting, and in those critical applications that require validation, to satisfy regulatory requirements. The general description of pharmaceutical applications in terms of these factors is given in Table 34–4. Three factors, however, generally govern the specification of a filtration system:

1. The removal efficiency of the final filter
2. Final filter service life: the volume of fluid that can be processed prior to plugging
3. Throughput rate.

The removal efficiency of the final filter for the intended application is the most important design consideration in selecting the proper filter pore size to remove the particulate contaminants of concern. These filters are specified on the basis of their pore-size rating, materials of construction, and scale of use to endow the treated fluid with the ultimate degree of purification sought for it (Meltzer 1987). General selection criteria based on the performance objective of the final filter are listed in Table 34–5.

Mechanism of Removal

Experimental studies to date have supported the sieve retention model of bacterial retention for sterilizing grade 0.22-μm rated filters. Larger pore size, nonsterilizing grade filters also retain bacteria, but do so by combined sieve and adsorption mechanisms. Both hydrophilic and hydrophobic microporous membranes are used in sterilizing applications. Hydrophilic membranes are applied to aqueous sterilizations because of their wettability, while hydrophobic membranes are used to sterilize gas and solvent streams because they resist water wetting and have broader chemical compatibility. The bacterial retention performance of 0.22-μm rated sterilizing grade filters is independent of the filtration pressure and the number of organisms used to challenge the filter. There is a relationship between physical integrity test measurements of a filter, such as the bubble point, and the bacterial retention, and which is dependent on the type of filter. These subjects are essential to the utility of filtration in the pharmaceutical industry and are discussed in more detail below.

Sterilizing grade filters are totally retentive, even at operating pressures exceeding those recommended by their manufacturers for efficient operation. Testing with 0.22- and 0.45-μm rated cartridge filters at 60 psi has demonstrated that the sterilizing grade 0.22-μm filters completely retain *Pseudomonas diminuta*, and that 0.45-μm rated filter elements retain at least 9 logs of *P. diminuta* (Aicholtz, Wilkins, and Gabler 1987) at these same test pressures.

Nonsterilizing grade filters retain organisms with a combination of sieve and adsorption mechanisms. Consistent with these mechanisms, retention with nonsterilizing grade filters decreases with increasing concentration of test microorganisms, and with increasing applied filtration pressure.

Qualification of Membrane Bacterial Retention

The validation of a filter's ability to sterilize is critical to the pharmaceutical industry. This validation needs to be accomplished in a nondestructive manner by performing a physical measurement of integrity prior to and after the use of the filters. The value of an integrity test is based solely on empirical correlations between integrity test values and corresponding organism retention values for both membrane and the fabricated filter elements. Sterilizing filters must perform to total removal levels of 10^{10} to 10^{11} total microorganisms since even one microorganism emerging from a sterilizing filter can render the entire filtrate nonsterile.

Membrane filters retain or remove bacteria from fluid streams primarily on the basis of size exclusion. *Pseudomonas diminuta*, 0.3 μm in diameter with an aspect ratio of 1.5:1, is widely recognized as the definitive bacterium upon which filter integrity is standardized. Since the mechanism of bacterial removal is size exclusion, the objective of the integrity test becomes the determination of the size of the largest hole in the filter element. Bubble point determinations and diffusion tests are used for this purpose. The theory and practice of these tests were recently reviewed (Emory 1989a, 1989b; Meltzer 1989b, 1989c). The bubble point and diffusion tests are related in that they are both based, at least in part, on the capillary pressure required to displace a wetting liquid from the porous structure of the membrane with a gas.

For membranes with effective surface areas of less than 2 ft^2 (e.g., a 293-mm disk), a bubble point test is recommended. The bubble point is measured by recording the pressure at which gas bubbles flow from the downstream side of the membrane or filter. [See Eq. (34–2) for the definition of bubble point pressure.] Water and water-alcohol mixtures are usually used for hydrophilic and hydrophobic membranes, respectively, and the gas is usually air or nitrogen. This pressure is expressed as the maximum difference between the gas and liquid

TABLE 34-4. Representative Pharmaceutical Applications of Microfiltration.

Application	Definition	Process Fluid	Product Quality	Operating Constraints
WFI	Water for injection	Water	Low pyrogen Low particles	High discharge flow rates Recirculation loops: frequent changeout
Vents	Storage tank "breathe"	Gas Water vapor	Sterile product tanks Aerosol retentive filters	Pressure equilibrium, post steaming: Post steam during filling and emptying, environmental fluctuations High gas flow rate
LVPs	Parenterals: 100- to 1000-mL quantities	Large volume dextrose: particles variable Amino acids Emulsions: high viscosity low viscosity	Low particle load Low pyrogen load Low bacteria count (<1/mL)	Time ≤8–16 h High throughput: continuous operation No downtime low-value added products Operations done in sterile area Variable pressure feed pump Filling machine pressure derived from centrifugal pump
SVPs	Parenterals: <100 mL	Small volume Some: quarantined sugar particles	Low bacteria Low pyrogen Some: sterile	Time ≤8 h High throughput: batch operation Only final sterilization and filling performed in sterile area Batch operation
Fermentor	Vent	Offgas	Sterile	High air flow rates Micro-organism containment Warm temperatures
	Air/oxygen	Gas	Sterile Particle free	High flow rates Presterilized lines

Category	Type	Composition	Requirements	Characteristics
	Nutrient feed	Aqueous Particles Serum	Sterile Particle free	Continuous feed up to weeks
	Liquid product	Nutrient broth Multicomponent Cell debris Emulsions Antifoam agents	Sterile Particle free Low pyrogen	Time: rapid processing
Antibiotics	Viral/bacterial antigens	High solids Aqueous in fermentor, crystallized in solvents	Sterile Low pyrogen Low particles	Crystal particles Aggressive solvents High pyrogen adsorption Manufacture in nonsterile area Fast to minimize pyrogens
Recombinant proteins	Therapeutic proteins	Aqueous Highly purified	Sterile Low pyrogen Low particles High product activity recovery	Low protein binding; low losses High throughput Nonsterile area
Serum	Animal/human serum	Aqueous Multicomponent Colloids High particles Proteinaceous gels Bacteria Cell debris Mycoplasma Highly variable	Sterile In cell culture applications: high cell propagation Low pyrogens Low particles No mycoplasma (FCS)[a]	Low pressure Cool: 10 to 15°C Constant volume flow operation Low protein binding Low losses

[a]FCS = fetal calf serum.

TABLE 34–5. Selection Criteria for Pharmaceutical Filters.

Application	Performance Objective	Rated Pore Size
Clarification and visible particle removal	Visual/optical clarity	5-μm depth or surface filter
Large organism/particle removal	Yeast and mold removal	0.65-μm final filter
Bacteria reduction	Pyrogen management	0.45-μm final filter
Bacteria retention	Sterilization	0.22-μm final filter
Small organism removal	Mycoplasma removal	0.1-μm final filter

TABLE 34–6. Relationship of LRV Versus Water Bubble Point for Hydrophilic Durapore®.

Water Bubble Point (psi)	LRV[a]	Water Bubble Point (psi)	LRV[a]
52.0	>10[b]	39.5	>10.5
52.0	>10	38.5	>9.2
52.0	>10	38.0	>9.4
52.0	>10	37.5	>10.5
52.0	>10	37.5	>9.4
50.0	>10	37.5	>9.4
50.0	>10	36.5	>9.2
50.0	>10	36.0	>9.4
50.0	>10	35.5	>9.4
50.0	>10	35.5	>9.2
45.0	>10	35.0	>9.4
45.0	>10	34.5	>9.2
45.0	>10	34.0	10.0
45.0	>10	33.0	5.3
45.0	>10	31.0	3.8
45.0	>9.5	31.0	3.7
45.0	>9.5	31.0	3.4
45.0	>9.5	29.0	3.3
44.0	>9.2	29.0	4.2
44.0	>9.5	28.0	3.9
44.0	>9.5	28.0	2.1
43.5	>10.5	28.0	2.1
42.5	>10.5	24.0	1.9
40.5	>10.5	24.0	1.9
40.0	>10.0	20.0	0.5
40.0	>9.4	20.0	0.4
40.0	>9.4		

[a] $LRV = \log_{10} \dfrac{\text{organisms in challenge}}{\text{organisms in effluent}}$

[b] "Greater than" signs indicate sterile effluent.

TABLE 34-7. Relationship of LRV Versus Methanol Bubble Point for Hydrophobic Durapore®.

Methanol Bubble Point (psi)	LRV[a]	Methanol Bubble Point (psi)	LRV[a]
19	>10.54[b]	15	>10.45
19	>10.84	15	>10.46
19	>10.91	15	>10.52
19	>10.86	15	>10.46
18	>10.83	14	10.18
18	>10.85	14	9.79
18	>10.79	14	9.69
18	>10.65	14	9.62
17	>10.84	12	8.90
17	>10.00	12	8.93
16	>10.72	12	8.75
16	>10.52	12	8.87
16	>10.52	12	8.97
16	>10.52	12	8.30
16	>10.34	12	8.61
16	>10.46	12	8.34
16	>10.46	12	8.23
16	>10.40	12	8.18
16	>10.40	12	8.04
16	>10.49	12	8.50
15	>10.40	12	8.08

[a] $LRV = \log_{10} \frac{\text{organisms in challenge}}{\text{organisms in effluent}}$.

[b] "Greater than" signs indicate sterile effluent.

pressures ($p_G - p_L$) that can be resisted by the liquid's suface tension, which holds the liquid in the pore. For the case of membrane filters, with their nonideal pores (noncircular pores with broken perimeters), the capillary pressure is called the bubble point pressure (p_{BP}), and usually includes a tortuosity factor or "shape" correction factor (K_t) to account for the real, nonideal structure (Emory 1989a, 1989b) in the bubble point equation. (For sterilizing grade membranes, K_t is in the neighborhood of 0.2 to 0.3, depending on the membrane in question and the value assumed for θ.)

An inverse relationship exists between the bubble point pressure and the pore size. An oversized pore is identified by a bubble point below the manufacturer's minimum specification for the integral membrane. The proportionality between the bubble point and the surface tension of the test fluid is of critical concern and, as a result, can make the measurement of integrity very difficult in product fluids where surface properties may vary considerably.

The relationship between filter element bubble point and P. diminuta retention for both hydrophilic and hydrophobic Durapore® membranes are illustrated in Tables 34-6 and 34-7. Hydrophilic filters are totally retentive (log reduction values greater than 9) if they exhibit a water bubble point higher than 34.5 psi. Typically, sterilizing grade 0.22-μm Durapore® has a water bubble point of 52 psi with a minimum specification of greater than 45 psi. Therefore, a 10-psi safety margin is inherent in the Durapore® membrane specifications. Similarly, the bubble point in methanol has to be above 15 psi for hydrophobic Durapore® to be retentive.

There is a 3-psi (methanol) margin of safety for membranes that are specified as having a minimum methanol bubble point of 18 psi.

For larger area filters, the diffusion test is the preferred integrity test. In this test, the flow rate of a gas is measured as it diffuses through the water in a wetted integral filter or convectively flows through defects in a nonintegral or defective filter. If a diffusion-test pressure is selected appropriately by the filter manufacturer (typically near 80% of the normal bubble point pressure), then defective filters can be identified as those that have measured gas flow rates that are higher than the acceptable flow rates specified by the manufacturer. The diffusion rate increases with both increasing pressure and filter area. Also important is that the diffusion rate can vary significantly for different systems, depending on the effects of specific gases, liquids, and secondary solutes on the diffusivity and solubility of the these gases.

The equation for diffusion through a filter is derived from the ideal-gas law, Fick's first law of diffusion, and Henry's law for the pressure-dependent solubility of gases in liquids (Emory 1989a, 1989b):

$$Q = \left(\frac{\rho_l}{M_l}\right)(DH)\left(\frac{p_t - p_a}{p_a}\right)\frac{RTA}{l}\epsilon, \quad (34\text{-}3)$$

where

Q = volumetric gas flow rate at standard (atmospheric) pressure
ρ_l = density of the liquid
D = diffusivity of the test gas through the liquid
p_t = upstream test pressure (absolute)
p_a = atmospheric pressure
R = ideal-gas-law constant
T = temperature (absolute)
A = total membrane area
ϵ = membrane porosity needed
M_l = molecular weight of the liquid
H = Henry's law constant for the test gas and liquid
l = diffusion path length (membrane thickness)

Large-area fabricated devices are made using only membranes that have been qualified via the bubble point criterion. The retention characteristics of these fabricated devices are then 100% tested with a diffusion test. The retentivity of these units, and of the diffusion test methodology, are confirmed on a lot release basis by destructively testing, under Health Industries Manufacturers Association (HIMA 1982) test conditions with a *P. diminuta* challenge, a statistically significant number of units. The standard of acceptance is zero organisms in the effluent. The correlation between air diffusion test values—at several test pressures and after up to 30 steamings—and bacterial retention for Durapore® TP cartridges is shown in Table 34–8. At 30 psi the normal range of air diffusion values is between 4.9 and 7.9 cm^3/min and bacterial integrity is maintained up to diffusion rates of 11 cm^3/min.

Grow-Through

Processing time limitations also need to be addressed. Sterility must be ensured even under conditions of protracted filter processing, including assurance against bacterial grow-through. The evaluation of bacterial grow-through of a Durapore® 0.22-μm filter under continuous operation is summarized in Table 34-9. *P. diminuta* did not penetrate 0.22-μm Durapore® cartridges under worst case processing conditions for periods of up to 96 h. Nonsterilizing grade 0.45-μm filters did show *P. diminuta* penetration under these processing conditions after just 8 hours of continuous operation. Even if no filter grow-through occurs, in protracted use applications, pyrogen formation should be of utmost concern.

Membrane Materials

Membrane material selections are made by considering several factors:

1. *Chemical compatibility*
2. *Wetting characteristics:* hydrophilic for aqueous streams and hydrophobic for gases and organic solvents
3. *Thermal resistance:* able to withstand steam or autoclaving temperatures
4. *Inertness:* noninteractive with parenteral or its preservatives

TABLE 34-8. Representative Integrity Test Results Obtained with Durapore® Membranes.

Integrity Test (Diffusion) Results for Lot 1, Durapore® CVGL 0.22-μm TP Cartridge Filters Challenged with P. diminuta at 2.1 × 10^8 Per cm^2 After 1× Steaming

Cartridge Number	@ 30 psig (cm^3/min)	@ 35 psig (cm^3/min)	@ 40 psig (cm^3/min)	Number of Organisms in Filtrate after 1× Steam	Post Test Diffusion (Air) @ 30 psig (cm^3/min)
2354	6.5	8.2	9.5	0	6.6
2339	6.3	8.3	9.4	0	6.7
2371	6.4	8.1	9.4	0	6.6
2394	6.3	8.0	9.3	0	6.4
2404	6.4	8.0	9.0	0	6.7
2325	6.3	8.0	9.2	0	6.8
2388	6.5	7.9	9.4	0	6.6
2362	5.6	6.6	7.2	0	7.6
2408	4.9	6.4	7.5	0	7.2
2348	6.0	7.3	8.4	0	7.5
2373	5.0	6.4	7.5	0	7.0
2352	5.2	6.5	7.5	0	7.3
2398	4.9	6.5	7.7	0	7.0
2334	5.4	6.9	7.2	0	6.8
2401	5.6	6.8	7.6	0	6.8
2364	5.1	6.5	7.5	0	6.0
2393	5.6	6.9	8.0	0	5.8
2343	5.2	6.5	7.6	0	6.1

Integrity Test (Diffusion) Results for Lots 2 and 3, Durapore® CVGL 0.22-μm TP Cartridge Filters Challenged with P. diminuta at 2.1 × 10^8 Per cm^2 After 1× and 30× Steaming

Cartridge Number	Number of Organisms in Filtrate after 1× Steam	Number of Organisms in Filtrate After 30× Steam	Diffusion (Air) @ 30 psig (cm^3/min) 30× Steam
3740	0	0	5.6
3711	0	0	5.9
3723	0	0	5.5
3741	0	0	5.4
3737	0	0	5.4
3728	0	0	5.7
3736	0	0	5.7
3745	0	0	5.6
3735	0	0	5.9
3715	0	0	5.6
6064	0	0	5.4
6076	0	0	5.4
6035	0	0	5.2
6023	0	0	5.9
6019	0	0	11.9
6020	0	0	6.4
6022	0	0	7.2
6074	0	TNTC[a]	Failed
6048	0	0	7.0
6062	0	0	6.8

[a]TNTC = too numerous to count.

TABLE 34-9. Extent of Bacterial Grow-Through in Stabilizing Grade Filters Due to a Continuous Challenge.

Membrane Type	Bubble Point (psi)	Challenge Flow Rate (mL/min)	Challenge Duration (h)	Average Total Challenge (CFU)	Downstream Bacterial Counts (CFU)
GVPP	24.5 ±0.77	2.0	8	3.3×10^8	No growth
GVPP	25.0 ±0.47	3.0	48	5.3×10^8	No growth
		1.0	48	6.2×10^8	No growth
GVPP	24.5 ±0.77	1.0	96	5.5×10^8	No growth
GVPP	25.0 ±0.47	1.0	96	2.1×10^8	No growth
Controls					
HAWG	33	2.0	8	3.5×10^8	>200 CFU
		3.0	48	5.3×10^8	TNTC
		6.0	48	6.2×10^8	TNTC
		2.0	96	5.5×10^8	TNTC
		8.0	96	2.1×10^8	TNTC

Notes: Results are average of seven membrane filters. CFU = colony forming units. GVPP = Durapore® 0.22-μm membrane. HAWG = cellulose acetate 0.45-μm membrane. TNTC = too numerous to count.

5. *Shedding:* no particle generation during operation especially intermittent flow conditions
6. *Toxicity:* must pass USP Class VI toxicity testing
7. *Extractables:* demonstrate low extractables in the solvent, preflushing may be desirable.

Filter Adsorption and Reduction in Biological Activity

Adsorption of product materials to the filter medium is undesirable not only in terms of product loss and the resultant increase in hydraulic resistance, but also in terms of product degradation. Published studies on the adsorption of plasma proteins to polymer surfaces have demonstrated that the protein adsorption process involves a dynamic equilibrium between native proteins in bulk solution and denatured proteins adsorbed to the surface. In equilibrium with these entities is a conformationally altered state that is reversibly held on or near the surface (Beissinger and Leonard 1981). This is of great importance to the pharmaceutical industry since conformationally altered parenteral proteins may present unwanted clinical reactions.

Protein adsorption to microporous membranes can vary greatly and is a strong function of the surface properties of the membrane polymer, as indicated in Table 34-10 (Pitt 1987). The amount of protein adsorbed on a membrane, which can be significant due to its vast internal surface area, is directly related to protein concentration in a Langmuir-type adsorption isotherm. Hydrophobic and charged surfaces such as polysulfone and nylon have higher protein adsorption capacities than hydrophilic surfaces such as polyhydroxypropylacrylate grafted PVDF filters.

TABLE 34-10. Range of Relative Binding Capacities for Microporous Membranes.

Membrane	Binding Capacity ($\mu g/cm^2$)
Hydrophobic Durapore®	300 to 400
Nitrocellulose	80 to 100
PVC	10 to 100
PTFE	5 to 10
Cellulose acetate	1 to 5
Hydrophilic Durapore®	0.1 to 1

Note: The data presented were recorded after ambient exposure to ^{125}I labeled IgG, 200 μg/mL in Tris buffer for 30 min at pH 7.2, 25°C followed by washing. Binding of protein to microporous membranes is dependent on salt concentration, pH, and saturation level of other solutes.

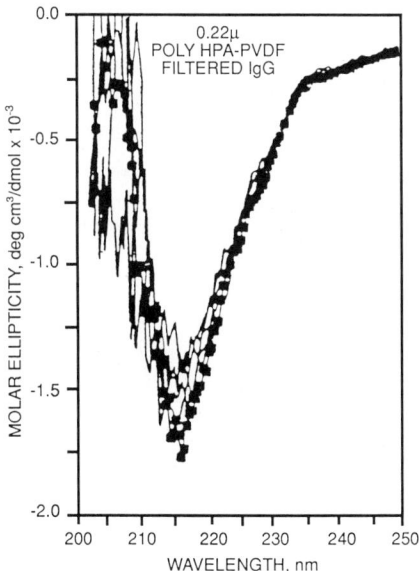
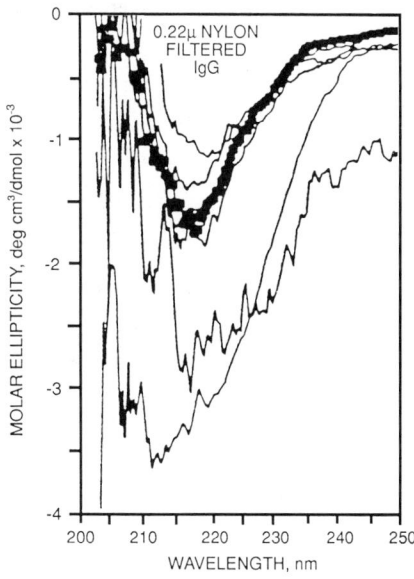

FIGURE 34-5. Replica circular dichroism spectra for human IgG controls and samples filtered through 0.22-μm poly-HPA-PVDF and nylon membranes (reprinted from Truskey et al. 1989 with permission).

Studies performed with microporous filters and three proteins have also demonstrated that changes in circular dichroism spectra (a measure of protein conformation) can be detected in some protein solutions after filtration through some microporous filters, as indicated in Figure 34–5. These spectral changes are protein and polymer material specific. In all cases, spectral changes did not seem to correlate to either pore size or applied pressure, but only to polymer material. No spectral changes were measured with any of the proteins filtered through polyhydroxypropylacrylate grafted PVDF filters. These changes seem to be most closely related to the membrane's ability to bind protein, as indicated in Figure 34–6 (Truskey et al. 1989).

Particulate Contamination and Filter Shedding

Even though the health significance of particulate matter introduced into parenterals is at best controversial, injectable drugs should not contain particulates that could be introduced into the blood stream. Major sources of particulates include people, clothing, containers, closures, processing equipment such as pumps and val-

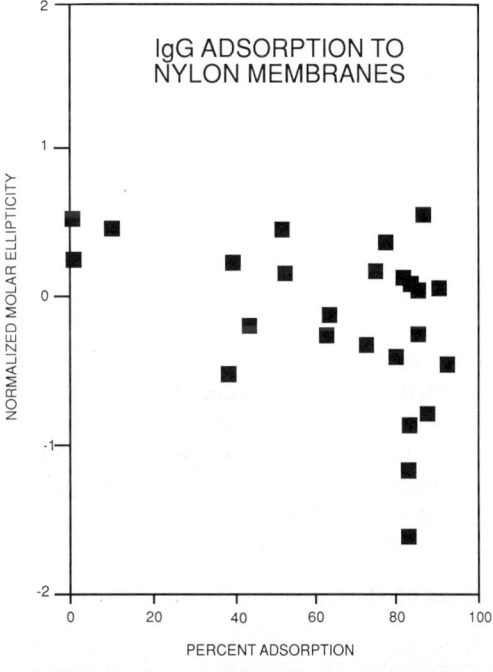

FIGURE 34-6. IgG adsorption to nylon membranes (reprinted from Truskey et al. 1989 with permission).

TABLE 34-11. Comparison of Alternative Filter Configurations.

Configuration	Advantages	Disadvantages
Flat sheet	Simple, easy to use	O-ring compression seal
	Small volume processing	Inconvenient scale-up
		Poor fluid distribution
		Heavy housing
		Difficult to clean and sterilize
Cartridges	Prefabricated	Housing required
	Large surface areas	Multiple materials of construction
	Low operating pressures	
	Easy scale-up	Center core holdup volume
	Reliable redundant O-ring/ groove seal	
	Good fluid distribution	
	Rapid installation & replacement	
	Thorough sterilization	
	Audit trail	
	Pretested	
	Economical	
	Disposable filter	
Capsules	Fully disposable	
	Integral housing	
	Convenient	
Stacked Disk	Low particle generation	
	Low downstream holdup volume	
	No membrane flexing: no drip	

ves, and the drug components themselves. Cleaning and preparation of containers and closures, aseptic processing, and isolation technology have improved the particulate quality of drugs. Yet the sterilizing filter itself may contribute to particle contamination. Prefilters and some final filters shed particles from their matrices; shedding is related to the polymers used and the mode of construction. Stacked disk cartridges, which provide the membrane with a rigid substrate rather than a woven or nonwoven polymeric support material, are designed to be free from particle generation and are preferred in extended operations, particularly under intermittent flow conditions, where particle generation is a concern.

Types of Devices

Sterilizing filters are used in all available device configurations in the pharmaceutical industry. Cartridge filters have gained the most widespread use in the industry. The advantages and disadvantages of these devices are summarized in Table 34–11.

Fabricated filter elements provide large membrane surface areas in a small-volume device, thus achieving reliable performance while attaining the engineering benefits of low applied differential pressure and large volumetric flows. In addition, these devices are optimized to minimize the downstream fluid holdup volume. These large surface area units are simple to install and maintain, and the associated labor costs are therefore low. Finally, these units can be delivered presterilized and with documentation and product part numbers that facilitate accurate record keeping.

Filter Sizing

Gas Vent Applications

Vent filters are often mounted on containers used for storage of liquids prior to final dispensing. Typical uses for these filters include carboys for small volumes, 5 to 50 L, and large stainless steel tanks used for storing sterile product, WFI (water for injection), or purified water. These filters are intended to protect the contents of the storage vessel from bacterial and particle contamination while allowing the tank to maintain pressure equilibrium during operation. Consequently, these filters are made from hydrophobic materials, such as PTFE, PVDF, or polypropylene, that will not wet but will allow the flow of gas. These filters are pore size rated in accordance with the intended application. As a general rule, the closer the storage vessel is to the final sterile product, the more stringent the vent sterility assurance requirements. Product storage tanks are protected with validated bacterially retentive 0.22-μm rated filters. WFI tanks are protected with either 0.22-μm rated or 0.45-μm rated filters depending on whether complete bacterial retention is desired. Other general service storage tanks are vented with 0.45-μm rated elements.

In many cases, the containers are steam sterilized or autoclaved. Hence, the vent filter is required to not only withstand repeated sterilizations, but also allow the container to "breathe" after the sterilization is complete. Hydrophobic filters must be used, and allowances made for the possible "wetting out" of the filter.

When the container is steamed or autoclaved, its volume contains water vapor, with a specific volume that will depend on temperature and pressure, but is of the order of 22.4 L/mol, that will condense to 18 mL/mol on cooling. Air must enter the vessel rapidly through the vent to make up this volume and avoid an implosion. The accurate design of the vent filter should account for the wall thickness of the vessel, the vessel diameter and height, the rate at which the liquid contents will be removed, and whether the vessel will be steamed or autoclaved.

The importance of sizing based on the air flow requirements accompanying steam condensation is illustrated in Table 34–12. This table presents the results of filter sizing calculations for two types of filters in sterile and nonsterile service. In sterile service, air flow requirements are increased by one order of magnitude due to condensate volume reduction. Consequently, for both filter types listed, a proportionately larger filter is required. The effect of membrane permeability on filter area is also shown in this example. The filter made with the higher permeability Aerex Teflon® membrane is smaller than the one made with a Durapore® PVDF membrane.

Filter vendors can provide detailed design calculations to size tank vents correctly. Filters should be mounted parallel to the ground so that condensate will drain into the housing and will not block the filter. Larger housings are equipped with drain ports for this purpose. To prevent condensation of liquid in the filter pores, the filter can be purged with compressed gas after steaming. Steam jacketed filter housings are also used to keep the filter temperature above ambient temperature, or at that of the tank contents, to avoid condensation.

TABLE 34–12. Sizing of Vent Filters.

Tank	Durapore®		Aerex®	
	Area (ft^2)	Safety Factor	Area (ft^2)	Safety Factor
Nonsterilized	7.4	3.2	7.4	3.6
Sterilized	51.8	2.3	43	2.2

Note: Process conditions:
  Tank diameter = 1.4 m.
  Tank volume = 4.3 m^3.
  Air flow due to liquid pumping = 2.65 SCFM.
  Air flow due to steaming = 26.3 SCFM.

Liquid Filter Systems

The key to sizing a filtration system to achieve the right balance between economy and performance is to have a basic understanding of small particles and how they interact with a filter matrix. Particles in suspension have a number of properties that affect the way they are retained in a filter. The smaller the particle, the more difficult it is to remove from a fluid. Smaller particles require finer filter media and tend to shorten the service life of a filter. Several key characteristics of suspended particles contribute to filtration performance:

1. Particles of natural origin exist in suspension in a distribution of particle sizes that is strongly skewed to the size of the smallest particles in the distribution. The population of large particles is low. Exceptions to this property exist with latexes and bacterial cultures. Chemical precipitates and crystallization products tend to exhibit a tightly defined distribution.
2. Particles can exist as rigid nondeformable entities that are easily filterable or as deformable or gelatinous materials. Deformable materials pass more easily through filters with pores of approximately their size or larger. Nondeformable materials, especially colloids, tend to close off the filtration area by spreading over the filter surface.
3. Particles tend to aggregate or flocculate.

Similarly, microorganisms, although their size distribution is narrower, tend to deform and aggregate. Consequently, filter clogging can occur as retained microorganisms accumulate on the filter surface.

The optimal choice of filters depends on achieving a balance between the retention efficiency, the ability to remove particles above a certain size, and the filtration throughput, the amount of fluid that can be processed before system clogging.

The volume of fluid that can be processed prior to the plugging of the final filter element by retained material dictates filter service life, and it can also control system sizing, especially prefilter sizing. Due to their very nature, surface retentive final filters can accommodate only relatively small quantities of particulate material before becoming blocked or plugged. Clearly, the filter application lifetime requires knowledge of the particle load in the fluid as well as the nature of the fluid itself. All too frequently, this information is not available and must be determined experimentally prior to implementation. Where the particulate load is excessive for the area of the final filter that is being used, prefilters and in some cases depth filters are used in combination with the final filter to remove part of the particle burden and prolong the service life of the final filter. In many applications, prefiltration is the key to a successful filtration process. In fact, prefiltration combinations allow for a significant level of flexibility in the design of filtration systems. The use of these filters was recently reviewed by Meltzer (1989a) and by Levy and Leahy (1990).

In addition, system sizing is also affected and can be designed on the basis of throughput rate. Generally, this criterion is controlled by the pore size of the filter employed, membrane plugging, and the differential transmembrane pressure that is available as a result of pumping constraints. As an operative rule, it is customary to use the most open filter that will provide the particle retention required by the intended operation. The selection of a final filter pore size below that necessary would add assurance that particles of that and smaller diameter would be removed, but at the expense of increased pressure drop across the filter element (which increases as the fourth power of pore radius). Accumulation of these additional particles onto the filter area not only also shortens filter life, but increases the resistance to fluid flow through the membrane as well.

Screen filters retain 100% of particles equal to or larger than their rated pore size quite efficiently. However, there is little or no retention of smaller sized particles.

Depth and surface filters are not absolute; the retention of particles equal to or larger than the nominal pore size approaches but never achieves 100% efficiency. This is due to the presence of the larger pores in the filter media. Yet, due to their depth retention mechanism, these filters are capable of removing substantial quantities of smaller particles as well, primarily due to adsorption. The efficiency of depth and surface filters can be regulated by selecting filters of different nominal pore ratings or different materials.

Typically, a depth or screen filter is used to remove particles that would clog the final filter. As a rule of thumb, depending on stream complexity, the area of the prefilter should be two to ten times that of the final filter.

Three mechanisms have been used to model the volume flow rate reduction that accompanies filter blockage: cake filtration, gradual pore blockage, and complete pore blockage. These postulates all describe volumetric flow rate with the Darcy equation, and they differ in their description of the resistance of the clogging material, as discussed in more detail in Chapter 33. By conducting experiments at either constant pressure or constant flow rate, relationships can be derived that empirically determine a clogging layer resistance. More importantly, under conditions of constant pressure, as is typically the case during flow decay experiments, the gradual pore blockage and the cake clogging models both predict straight-line relationships for the ratio of time to cumulative volume, one with time, the other with cumulative volume. Both can be used to conveniently size a filter system. Typically, experiments are performed in which the final filter is specified first, and then, depending on the results obtained, different prefilters in ascending pore size order are added to achieve the target throughput and service life. For screening experiments conducted with a small membrane area A_{fd}, and in which volume of filtrate V is measured as a function of time t, these relationships are given below.

Cake Filtration Theory Model at Constant Pressure

As discussed in Chapter 32, the time-volume relationship for constant cake and membrane properties and no membrane fouling is given by Eq. (32–25) and

$$t/V = b_3 V + b_4, \quad (34\text{--}4)$$

where b_3 is an empirical constant proportional to the specific cake resistance and b_4 is an empirical constant proportional to the membrane resistance [see Eq. (32–25)].

System Sizing Based on the Cake Filtration Model Using constant pressure data, conventional flow decay calculations are used as described below.

Continuous Operation Sizing Continuous operation would justify oversizing to eliminate frequent filter changes. Sizing for either maximum throughput or flow rate is based on the measured throughput at 80% flow decay, the cumulative volume that is processed until the throughput has diminished by 80% of the initial value. These values are expressed as follows: System area sizing for maximum throughput:

$$A = \frac{1.5 V_b A_{fd}}{V_{80\%}}, \quad (34\text{--}5)$$

where

$V_{80\%}$ = throughput at 80% flow decay
V_b = volume of the batch to be processed
A_{fd} = membrane area used in the flow decay experiment.

System area sizing for flow rate

$$A = \frac{1.5 A_{fd} p_t Q_{min}}{p_{max} Q_{80\%}}, \quad (34\text{--}6)$$

where

p_t = test pressure
Q_{min} = minimum acceptable flow rate
$Q_{80\%}$ = flow rate at 80% decay
p_{max} = maximum system differential pressure.

Batch Operation Sizing The filters used in a batch process should be sized close to the projected needs; the filter should be 80% plugged at the end of the filtration. Therefore, the ratio of filter area to that used in the flow decay studies is proportional to the ratio of the batch volume to the volume processed to reach 80% flow decay.

In many cases, the membrane is significantly fouled during filtration and Eq. (32–25) does not adequately model system performance. In this case, two other models have been proposed on which system sizing can be performed.

Gradual Pore Plugging Model at Constant Pressure

In this case, it is assumed that the effective diameter of the pores is reduced by particles of radius r deposited on the walls. This gives a relationship for the effective volume of the pores, V, as

$$V = \pi l N_p (r_0 - 2r)^2, \qquad (34-7)$$

where

N_p = total number of pores
r = radius of particle
r_0 = pore radius at $t = 0$.

Incorporating this into the Hagen-Poiseuille equation and integrating it leads to the following relationship:

$$\frac{t}{V} = b_1 t + b_2, \qquad (34-8)$$

where

$$b_1 = \frac{1}{\pi l} \left(\frac{\pi}{8 \mu l N_p} \right)^{1/2} \left(\frac{\Delta p}{(dV/dt)_0} \right)^{1/2},$$

and $b_2 = 1/(dV/dt)_0$ and $(dV/dt)_0$ is the volumetric flow rate at $t = 0$.

System Sizing Based on the Gradual Pore Plugging Model The maximum batch volume that can be processed prior to plugging, V_{max}, is given by the following relation:

$$V_{max} = 1/b_1. \qquad (34-9)$$

Based on this relationship, an operative V/A ratio is empirically defined on the basis of the experimental value of V_{max}/A_{fd} as

$$V/A = 0.68(V_{max}/A_{fd}). \qquad (34-10)$$

For continuous operation, the filter area needed to obtain optimal throughput is given by the following relation:

$$A_{ot} = V_p/(V/A), \qquad (34-11)$$

where V_p is the volume to be processed continuously.

The area required to obtain optimal flow rate A_{of} is equal to the area required to achieve 10% of the initial flow rate:

$$A_{of} = 0.1 b_2 Q A_{fd}. \qquad (34-12)$$

For batch operation, the area required for optimal throughput is given by

$$A = V_b/(V/A), \qquad (34-13)$$

where V_b is the volume of batch to be processed.

The operating time t_{op} required to filter a batch of volume V_b is given by

$$t_{op} = V_b(b_1 t + b_2). \qquad (34-14)$$

Complete Pore Plugging Model at Constant Pressure

This model is derived simply by assuming that the number of open pores is equal to the number of pores at initial time minus the number of pores that are plugged at any point in time. By incorporating this assumption into the Hagen-Poiseuille equation the following relationship is derived:

$$V = b_5[1 - \exp(-b_6 t)], \qquad (34-15)$$

where

$$b_5 = \frac{\text{number of pores at } t = 0}{\text{number of plugged pores per unit volume of filtrate } (K)}$$

and

$$b_6 = \pi r_0^4 \, \Delta p K / 8 \eta l.$$

This model is rarely used to calculate system sizing.

When flow decay studies cannot be performed, other rules of thumb are used, including:

1. *Sizing on the basis of pressure:* The volumetric flow rate that results in a pressure drop of 1 to 3 psi across the filter train is established on a pilot unit. Scale-up is performed by extrapolating linearly from this flow rate.
2. *Sizing on the basis of flux:* For a prefilter, a flux of 0.25 gal/min/ft^2 is used. For a final filter, a flux of 0.50 gal/min/ft^2 is used.

In practice, sterile filtration systems have been routinely sized to operate at differential pressures below 20 psi.

System Design

Sanitary Design

In critical sterile applications such as parenteral production, pumps, filter housings, lines, valves, and tubing are required to be capable of maintaining sterility. Therefore, devices used in the applications are of sanitary design. This will require the use of triclover instead of threaded fittings. Sanitary design also requires that the components be fabricated from ASI-300 stainless steel. All rigid pipes have welded joints and a smooth #4 finish. Horizontal lines must be self-draining and have pitched drain ports. There must be no stagnant areas. The fluid velocity through the pipes and fittings in the system must be at least 5 ft/s. These requirements are intended to minimize the possibility of microbiological holdup and to facilitate steam sterilization of all wetted surfaces in the system.

Sanitary filter housings must be pressure rated, be equipped with an air vent at their highest point, and have a liquid drain at their lowest point. Two types of designs dominate the industry, straight-through and T-type design. The latter is preferred because it is easier to drain and vent when horizontally mounted. Integrity testing with these housings can be more reliable since liquid can be more easily drained. Even if the liquid is retained, it is not in contact with the filter. Dual O-rings are used to seal cartridges in place.

Finally, the design must be integrity testable *in situ* without downstream contamination.

Validation of Filters and Processes

Validation guides, written by filter manufacturers, commonly contain information on materials of construction, USP XXII extractables (United States Pharmacopeia 1990), USP XXII Class VI plastics, maximum operating pressures, integrity test/bacterial retention correlations, hydraulic resistance (e.g., back pressure and pulsation), thermal stability, chemical compatibility, LAL pyrogen testing, USP XXII particle testing (for prefilters), and flow rate performance.

Final filters should be validated to remove microorganisms under worst case processing conditions and be demonstrated to remove bacteria under simulated processing conditions in their drug products. Other validation data including in-drug product bubble point, filterability, compatibility, and filter sizing based on product throughput data can be provided.

The filtration process must also be validated. The validation of aseptic pharmaceutical processes has been extensively reviewed (Carleton and Agalloco 1986). Technical and compliance issues were reviewed by Goldsmith and Grundelman (1985) and by Lee (1989). These include microbial challenge testing, filtration parameters, compatibility, assembly and sterilization of the filter, integrity test methods, and specifications which were discussed previously.

Customer In-Process Validation

Each filter manufacturer makes recommendations on the appropriate integrity test and test conditions for evaluating their filters. Filter users can obtain quantitative results that, when compared to the manufacturer's integrity test specifications, determine the integrity of the filter. Furthermore, to eliminate human subjectivity inherent in manual integrity testing, many manufacturers now market sensitive, automated equipment to perform both the bubble point and diffusion tests.

In aqueous stream filtration, many filter users find it convenient to conduct post-use tests with the filter wet with the processed liquid rather than with water. The obvious issues in product testing are the differences in the bubble point and diffusion rates between the product liquid and water. Equally important, however, but often overlooked, are the normal batch-to-batch variations in product chemistry that can affect surface tension, solubility, and diffusion.

One rational approach to setting specifications for the bubble point and diffusion rate in product fluids is to obtain experimentally the product-to-water bubble point and diffusion ratios with integral flat-stock membrane and filters, respectively, and then multiply the ratios by the specifications for water. This (and similar methods) will result in legitimate specifications when the product chemistry (i.e., solute concentration) does not change significantly from batch to batch. When trying to evaluate the significance of concentration variations, keep in mind, for example, that surface active components are defined as those that make relatively large changes in the surface tension of water and solutions when present in only trace quantities. Extremely small variations in the concentration of such solutes can drastically change the bubble point and lead to misinterpretation of tests results if not acknowledged. Similar changes in the diffusion rate can also be expected due to changes in the gas solubility and/or diffusivity, as well as to significant reductions in the bubble point if such shifts are not accounted for.

The bacterial retentivity of Durapore® cartridges has been confirmed in aqueous injectables ranging in surface tension from 51.6 to 71.9 dynes/cm and nonaqueous injectables, both with and without a variety of additives, fetal bovine serum protein solutions, and opthalmics. In all, 100% of the tests demonstrated complete retention of *P. diminuta* at $10^7/cm^2$.

End-User Validation Soaking
Typically, when a membrane filter is put in place with a new parenteral, the filter unit is soaked in the drug and the solution monitored for a period of several months for activity, safety, and efficacy.

Operational Considerations
Cleaning
Filtration systems used in aseptic processing must be cleaned to ensure that no foreign substances become part of the drug product. Cleaning takes two major forms: clean-out-of-place (COP) and clean-in-place (CIP).

In COP, equipment is disassembled, cleaned with detergents either manually or with a mechanical cleaner, and rinsed several times. CIP uses an automated distribution and pumping system to circulate several fluids, generally detergents and water, through fully assembled filtration systems (Adams and Agarwal 1989). All internal parts of the system are usually cleaned including tanks, lines, pumps, and filter housings. Filters may or may not be in place.

CIP is generally not practiced in sterile filtration for systems that use reusable filters installed in housings. The nature of the cleaning agents chosen will depend on the compatibility of these agents with system parts. If prefilters and sterilizing filters are cleaned in place and exposed to these agents, compatibility with filter materials must be demonstrated. The cleaning procedure must be validated, and documentation provided to demonstrate that it achieves a defined level of cleanliness. This is verified by performing specific chemical and microbiological tests.

Sterilization of Filters
Filters used in aseptic processing must be rendered sterile. A number of methods are used for this purpose.

1. *Steam autoclaving:* Mounted filters are exposed to steam for a minimum of 15 min at 121°C under 15 psi (1 bar) within a validated autoclave and aseptically attached to the remainder of the presterilized processing system.
2. *Steam-in-place (SIP):* This is the preferred method for *in situ* sterilization of filters because of the lower risk of contamination. Filters are exposed to flowing steam within the processing system, which is sterilized concurrently. One caveat is to guard against filter blockage as water condenses in the membrane pores. This can be overcome by raising the upstream system pressure to the bubble point of the filter medium and forc-

ing the water out. In addition, care must be taken to maintain always a slightly higher pressure upstream of unsupported media to prevent rupturing of the filter. Validation of SIP processes has been extensively described and reviewed (Berman, Meyers, and Chrai 1986; Myers and Chrai 1981, 1982; Kovary, Agalloco, and Gordon 1983).

3. *Gas sterilization:* Used for disposable filters, such as stacked disk products or capsules. Ethylene oxide (ETO), chlorofluorocarbon (CFC), and moisture are used at high temperature (50°C). This method requires that the product be quarantined for a considerable period of time to allow for adequate contamination of the gas sterilant.

4. *Irradiation:* This technique is displacing gas sterilization techniques, but requires that the filter element be compatible with gamma radiation.

Economics

Representative sterile filtration applications are listed in Table 34–13. In most of these applications, the filtration systems are sized to maximize throughput or flow rate to minimize product degradation and minimize costs. One example where this is not true is with antibiotics where the flow rate is restrained to avoid aeration of the product. The broad range of applications listed illustrates that sterile filtration is used across the industry for a variety of fluid

TABLE 34–13. Representative Examples and Costs of Sterile Filtration.

Application	Volume (L)	Viscosity (cp)	Flow Rate (L/min)	Membrane Area (ft^2)	Additional Requirements	Operating Cost ($/L)
Liquids						
Antibiotic:						
Extraction alcohol	500	1	25	10	SIP, 10 uses	0.02–0.04
Glycol antibiotic	5000	6	14	90	SIP, 130°C	0.20–0.70
LVP:						
Radiodiagnostic injectable	2500	4	10	15		0.10–0.25
D5W (5% dextrose solution)	3600	50	50	22	0.45 μm	0.01–0.02
SVP:						
Aqueous	4000	1	2	7	SIP[a], 130°C	0.02–0.07
Virus vaccine	500	1	10[b]	15	5°C, 1-h limit, No adsorption losses	0.40–1.00
Tissue culture media	100	1	5	7		0.10–0.25
Plasma AHF (Antihemophylic factor)	100	2	2	7	1-h process limit, No adsorption losses	1.00–2.50
Fetal calf serum	400	1.5	5	14	0–15°C, 6-h limit Prefilters	2.00–3.00
Gases						
Product tank vent			50[c]	7	Steam and air	
Fermentor Air Feed[d]			2000[e]	266	1-yr life, SIP, p_{dif} (1 psi)	3.00[f]

[a]SIP = steam in place.
[b]Average value.
[c]scfm.
[d]Assumes two fermentations per week.
[e]CFM.
[f]Cost per fermentation.

conditions. In large-volume continuous operation, the cost of sterile filtration is tens of cents per liter fluid processed for a single use application.

Serum or plasma products are exceptions and cost more to filter. These fluids must be operated at constant volumetric flow rate due to membrane plugging, and they require prefiltration. These applications cost $2 to $3 per liter processed.

Gases are sterile filtered in applications such as fermentor air feeds. Sterile filters are used to protect and vent tanks in which sterility of the contents must be assured. In these applications hydrophobic sterilizing grade filters, which have been validated as bacterial retentive in liquids, are used. The large membrane surface areas used cost about $3 per fermentation, based on a 1-yr service life.

Case Studies

Case studies of SVP production and antibiotic production are presented in the following to illustrate the uses of filtration in the production environment.

Small-Volume Parenterals (SVPs)

SVPs include a wide variety of parenteral solutions that are administered in single doses of less than 100 mL in volume. SVP solutions vary considerably in composition and complexity. Generally, "current good manufacturing practices" (CGMPs) intended for the preparation of LVPs have been applied to SVPs. Batch sizes of SVPs are generally smaller than those for LVPs, of the order of 5 to 500 L, and generally, therefore, pose fewer filtration problems with regard to processing the batch within the 8-hour CGMP guidelines.

A representative SVP manufacturing process is depicted in Figure 34–7. The compounded SVP is pumped, typically with peristaltic pumps, through the filtration system to a holding tank. The SVP can be quarantined in this tank during the quality assessment period. Finally, the contents of the holding tank are transferred to the filling machine. There are also sterile vent filters on the storage tank and on the filling machine. In some cases, the compounding tank is pressurized to reduce the potential for contamination. In these cases, the tank is vented with a 0.22-μm hydrophobic vent filter, and the gas is passed through a 0.45-μm filter.

Two filtration strategies are commonly practiced in SVP production. In the first, the SVP is terminally sterilized by autoclaving. In these applications, filtration practices tend to follow those applied to LVP production, and 0.45-μm final filtration is used. In other applications, SVP is sterilized with a 0.22-μm filter element.

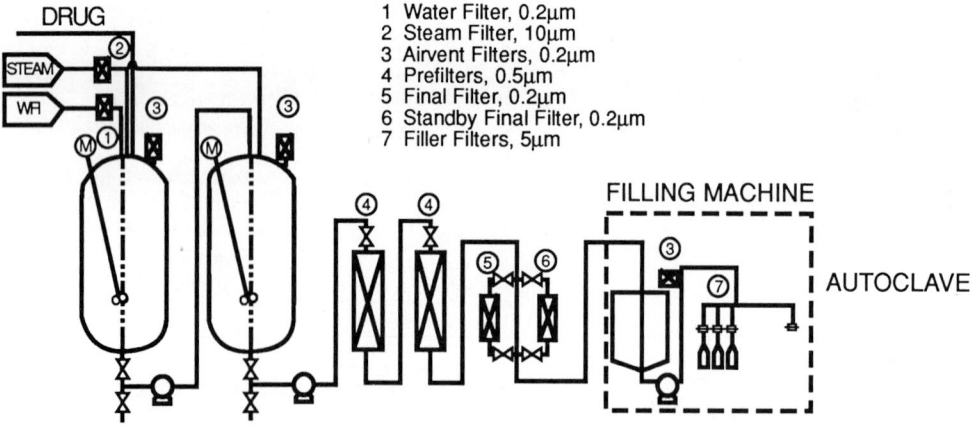

FIGURE 34–7. Flow diagram of current pharmaceutical and biological process system: 12% amino acid injection (reprinted from Millipore Corporation with permission).

Typically, redundant 0.22-μm elements are serially employed. This results in two benefits:

1. The compounding and upstream operations need not be conducted in a sterile Class 100,000 area.
2. Manufacturing costs are lower because terminal sterilization via autoclaving is not needed.

In these applications, the first of the two filters is not presterilized. The SVP is passed through a sterile barrier through piping to the second presterilized filter. In this arrangement, the first of the final filters serves to protect the second final filter in the sterile area, thus increasing its service life. The advantage of this arrangement is that the upstream final filter can be easily changed after blockage without breaching the sterility of the SVP line, or compromising the sterility of the final filter in the sterile area (Meltzer 1987).

Filtration is typically performed at 3 to 6 L/min at pressures of 15 to 20 psi. The filling machine is equipped with scavenging filters with a 0.5- to 0.8-μm rated pore size.

The total filtration needs for the depicted process are two sterilizing grade 0.22-μm rated filters, two prefilters, three vent filters, and one particle filter on the filling machine.

Antibiotics

Antibiotics, vitamins, interferons, vaccines (both viral and bacterial antigens), antitumor drugs, and alkaloids are all produced by fermentation processes that can be generally grouped into this category of applications.

Antibiotics and therapeutic peptides are generally purified by crystallization in aggressive solvents. Sterile filtration of this final product must be performed in these solvents. The filter material must be compatible with these solvents. Generally, hydrophobic filters made of PVDF, PTFE, or polypropylene are employed.

The fermented antibiotic undergoes several processing operations prior to bulk preparation. These can include harvesting from the cells, extraction and other purification steps, and chemical modification. Few of these operations are performed in an aseptic manner; therefore, they must be performed quickly to minimize the risk of contamination.

Bulk sterile preparation of the antibiotic batch precedes the antibiotics crystallization process. The filtration system is designed to remove microorganisms that may have been introduced in the previous steps, and pyrogens and particles that would provide undesirable nucleation sites that would affect product purity or crystal size during the crystallization process.

A schematic of the bulk processing operation is shown in Figure 34–8. This entire operation is performed in a sterile environment. The protection of the fermentor is of paramount importance. Oxygen and nutrients added are sterile filtered to prevent microorganisms from entering the growth environment. The fermentor vents must also be of sterilizing grade quality to prevent microorganism introduction. Finally, certain pathogenic, and some genetically engineered, organisms require that the entire fermentation process be contained in a closed system. Therefore, fermenter offgases need to be filtered to prevent the discharge of harmful organisms.

A typical filtration system includes filtration through:

1. Activated carbon or other material to remove traces of color
2. Diatomaceous earth to remove large particles
3. Prefiltration through 3- to 10-μm rated filter elements
4. Two 0.22-μm sterilizing filters in series before entering the sterile crystallization vessels.

Bulk antibiotic filtration systems are designed to operate at moderate throughputs and at low temperatures. This is done to minimize product degradation by oxidation that could occur as a result of excessive aeration during filtration, or from downtime if the filter elements were to plug.

Sterile solvents are used to crystallize the

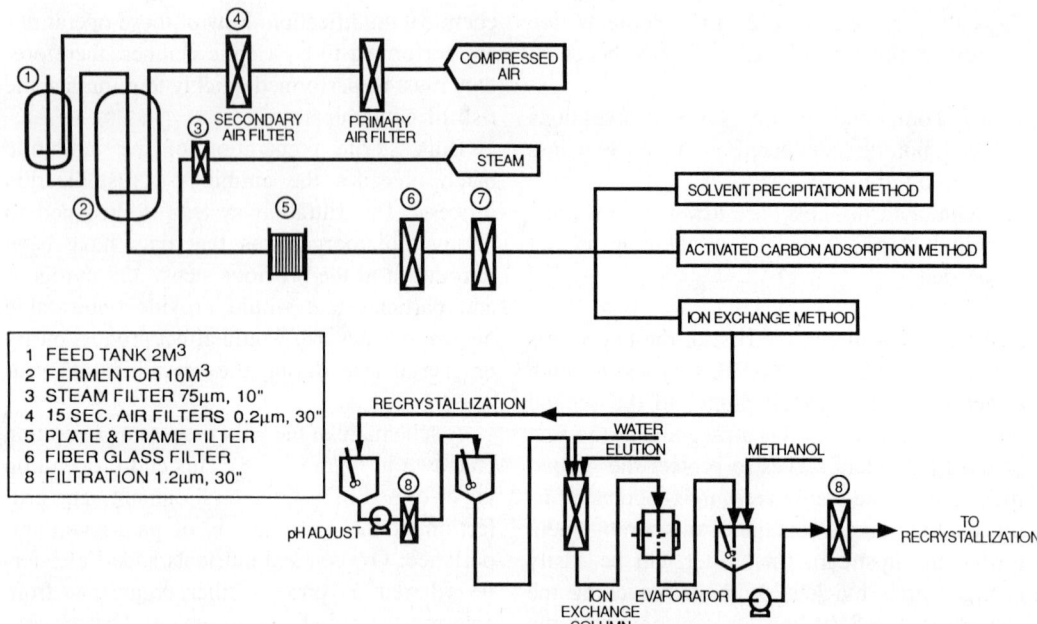

FIGURE 34–8. Flow diagram of current pharmaceutical and biological process system: typical antibiotic bulk production (reprinted from Millipore Corporation with permission).

product. The resulting suspension of antibiotic microcrystals (which are typically smaller than 0.2 μm) can then be filtered through a positively charged depth filter to depyrogenate it, and finally through a 0.22-μm sterilizing filter. The filters used in these operations must be compatible with and have low extractables in the aggressive solvents such as acetone, methylene chloride, or 70% alcohols, which are used for crystallization. The suspended precipitate is then dried and pulverized.

In some cases, water is used to crystallize the product. These solutions are passed at high throughputs (up to 100 L/min) through a 0.22-μm filter under pressures of from 20 to 30 psi. This optimization is intended to minimize aeration and oxidative degradation of the antibiotic (Meltzer 1987).

When aqueous suspensions are to be prepared, the solid antibiotic is added to sterile water for injection (WFI) that is generally filtered through a 0.22-μm filter at flow rates of 25 to 40 L/min. The suspension is taken to the fill line such as that depicted in Figure 34–9. Because the antibiotic solution is a suspension, no filter is used on the filling machine (Meltzer 1987).

Total filtration requirements during the bulk preparation phase are six filter systems, selected for chemical compatibility and large particle removal capacity. During the preparation and filling stage, three filters are used for liquid handling; two prefilters, one sterilizing grade filter, and four vent filters are required. In addition, seven sterilizing grade filters are used on vents and gas lines.

BEVERAGE APPLICATIONS OF MICROFILTRATION MEMBRANES

Introduction

Microfiltration membranes are used increasingly in the beverage industry for clarification and biological stabilization. This industry traditionally used filtration with adsorptive depth media (such as asbestos and cellulose) and thermal pasteurization. Although depth fil-

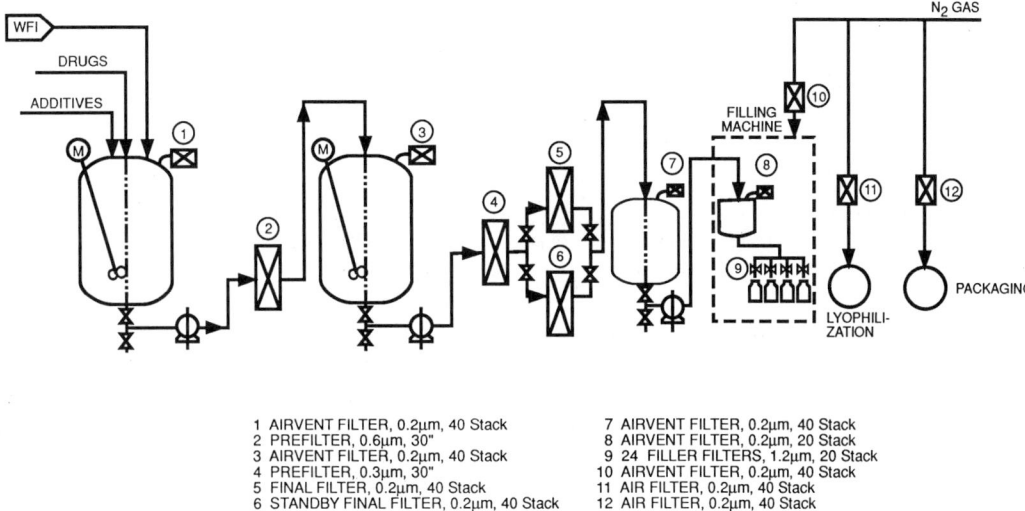

FIGURE 34–9. Flow diagram of current pharmaceutical and biological process system: penicillin (reprinted from Millipore Corporation with permission).

tration is still the most economical method for initial clarification, use of membrane filtration as the final step is growing because it offers validatable absolute microbial removal without the use of chemicals or heat. The organoleptic properties are thus preserved in a completely natural state until the product is opened—weeks, months, or years after filling.

Beverage applications in which membrane filtration is commonly used are beer, wine, wine coolers, bottled water, fruit-flavored beverages, and seltzers. Of these, beer and wine account for more than 90% of the membrane usage.

Applications

Beer Stabilization

The first use of polymeric membranes for beer stabilization was reported in the early 1960s (Haffenreffer 1962; Anonymous 1963; Bush 1964). In 1963, four U.S. brewing companies—Haffenreffer (Boston), Peter Hand (Chicago), Pittsburgh Brewing, and Duquesne (Pittsburgh)—were selling beer that was "cold filtered" with microporous membranes. The number of companies continued to grow through the 1960s and reports began to appear that analyzed factors affecting filter choice, efficiency, plugging, and economics (Brenner and Iffland 1966; Jesukawicz 1967; Dwyer 1968; Haffenreffer 1968; Haffenreffer et al. 1968). The "fad," however, ended in the United States around 1970, and it did not reappear until almost 16 years later when the Miller Brewing Company began to promote their "cold-filtered" beer quite heavily. Today, a growing number of companies in the United States, Japan, Europe, and South Africa report the use of membrane filters for stabilization in preference to pasteurization (Reid et al. 1990).

Wine Stabilization

Although accompanied by less fanfare than beer filtration, wine stabilization with membrane microfiltration is now accepted in every part of the world as the premier method of achieving microbiological stability.

It is odd that beer rather than wine was the first beverage product of consequence to be filtered with membranes. Perhaps this is because vintners already knew in the 1960s that hot bottling or pasteurization damaged the organoleptic properties and occasionally the appearance of fine wines. Wine has a higher alcohol concentration than beer (10 to 13% versus 3 to 6%), and for that reason it is sometimes less susceptible to microbial contamination.

TABLE 34-14. Beverage Spoilage Organisms.

Type	Example	Beverage	Typical Size	Results of Growth
Yeast	*Saccharomyces bayanus*	wine	3 μm	off flavors
	Saccharomyces diastaticus	beer	3 μm	off flavors
	Candida mycoderma	wine	3 μm	acetaldehyde
Bacteria	*Acetobacter oxydans*	beer and wine	0.5 × 1 μm pairs	acetic acid fumethanol
	Lactobacillus pastorinus	wine	0.5 × 1 μm suds	mousy aromas
	Lactobacillus breuis	beer	0.8 μm	off tastes/flavors
	Pediococcus damnosa	beer	0.6 μm	off tastes/flavors
	Zymomonas anaerobia	beer	—	Acetaldehyde and H_2S
	Pectinatus/ cerevisiiphilus	beer	0.7 × 2-μm rods	Acetic and propionic acids and H_2S
	Megasphaera/species	beer	0.8 μm	butyric, caproic, acetic acids
	Coliform bacteria	water	0.5 μm	indicators of infection

Nevertheless, since the nineteenth century, wine bottlers have used adsorptive depth filters rather than pasteurization to extend the shelf life of fine wines, particularly whites and those with residual sugar.

When the beer market for membrane stabilization in the United States collapsed, it was a natural extension of the technology to address wine stabilization. The results have been spectacular. Nearly all bottlers who desire shelf-life insurance against microbial contamination now filter their products through 0.45- or 0.65-μm microporous membrane filters.

The choice of pore size is quite critical, and it is made based on a balanced trade-off between assurance of microbial retention and economics. Common microorganisms found in beverages are listed in Table 34-14. A 0.45-μm membrane will remove all these organisms quantitatively, but in some processes, it can be expensive because it plugs more rapidly than a 0.65-μm or larger pore size membrane. In addition, some wine producers believe that small pore size membranes remove micron-sized colloidal substances, which are organoleptically important in red wines. These producers, hence, prefer to use 0.65-μm membranes.

Two genera (*Pectinatus* and *Megasphaera*) have only recently been identified (Jansen 1990) as important beer spoilage bacteria (see Table 34-14). Spoilage molds are a problem only upstream of the bottling operation where they can contaminate fruit and processing equipment. Because molds need oxygen, and because CO_2 and alcohol are good mold inhibitors, they will not spoil bottled beer or wine. Mold spores can be both small (<0.5 μm) and resistant to heat. Bottlers of mineral water and fruit-flavored beverages will occasionally choose a 0.2-μm pore sized membrane to guard against these spores and small coliform bacteria that survive in partially nutrient-deficient liquids (water).

System Design

Design Considerations

The design of filtration systems for beverage applications is governed by the need for microbial control, hence sanitary design. Economics are also important because the beverage business is highly competitive.

The need for sanitary design dictates that all housings and plumbing be amenable to easy cleaning and sanitation. There should not be any dead or stagnant areas. Pumps should not be placed between the final filters and filling stations, and the length of downstream piping should be minimized. The materials of choice for filter housings and downstream plumbing are 304 or 316 grade stainless steel. Sampling valves are located at strategic locations to monitor microbiological counts during filling.

Prefilter and final filter housings should be located adjacent to each other so that the effluent from the prefilter flows immediately into the final filter. The reason for this requirement is that submicron filtration removes many "protective" colloids and particulates. New colloidal aggregates can form soon after filtration as a result of molecular rearrangements. Sometimes, these new colloids form within minutes. The best examples are the highly branched beta-glucan molecules which can form loose but gummy molecular aggregates with proteins. A tight "prefiltration" will break up some of the aggregates, but these can reform again within minutes and plug the final filters. Thus, if final membrane filtration follows tight prefiltration and interim storage, the final membrane may become plugged prematurely by new colloids or precipitates formed during storage as a consequence of prefilter clarification.

This same phenomenon is responsible for the occasional clouds, hazes, or precipitates that may appear in the bottle even though the beverage has perfect microbiological stability and was of high optical clarity when bottled. Examples of these precipitates are proteins, protein-tannin complexes (chill hazes), and tartarates. If a haze or precipitate forms after submicron filtration, additional upstream fining or filtration, enzyme treatment, or cold stabilization may be needed prior to final membrane filtration.

Finally, when a filtration step is added to a filling operation, a new pump is often required because the head requirement for the filler pump is greatly increased. The additional head is required to accommodate the pressure drop across two filters in series, which can attain 50 psi per filter at plugging. A typical filtration system schematic is shown in Figure 34–10.

The other important driving force in the design is overall economics, which influences

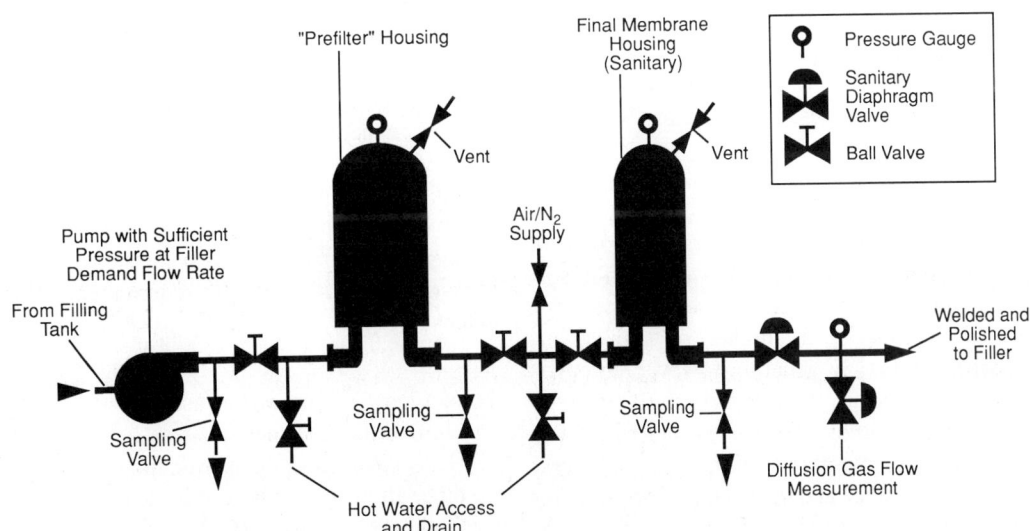

FIGURE 34–10. Elements in a cartridge filtration system for aseptic filling (reprinted from Millipore Corporation with permission).

choice of pretreatment and filter sizing. Membrane filters have relatively low dirt holding capacity and are more expensive than pads or prefilters. Although their absolute microbial retention properties justify their usage, it is economically prudent to prolong membrane life by proper pretreatment.

Even though a beverage may have sparkling clarity, it can plug a tight membrane quickly if it contains colloidal carbohydrates. Examples are late-harvest wines and beers with a high glucan concentration. Pretreatment is accomplished by upstream fining operations, and by pad and/or diatomaceous earth filtration. Sometimes, a special set of cartridge filters (commonly known as *prefilters*) is used just before the final membrane.

Prefilters are often placed immediately upstream of the final membrane to remove any remaining colloids. These prefilters have nominal or approximate pore size ratings that are usually much closer to the absolute rating of the final filter than the nominal ratings of diatomaceous earth and inexpensive depth filters, i.e., some pads, string-wound, or molded cartridges.

Prefilters are carefully selected and tested by the supplier to improve the overall economy of the process. Suppliers of membrane filters will usually recommend methods to test the "filterability" or the expected membrane throughput of the beverage. If such tests show the presence of colloids that may prematurely plug final membranes, a prefilter or a very tight pad would be recommended.

Filter Sizing

The sizes of the prefilter and final filter depend on the plugging characteristics of the filters in the liquids being filtered. These characteristics are influenced by a number of application-specific factors. In general, however, throughputs are greater at lower superficial velocities, and a good rule of thumb is to stay below 0.5 gal/min/ft^2 of membrane area for final filters, and between 0.17 and 0.25 gal/min/ft^2 of membrane area for prefilters. The prefiltration area, therefore, is two to three times greater than that required for final filtration.

Costs of Beverage Filtration Systems
Capital (Fixed) Costs

Installed costs for stainless steel housings for membrane areas run from about $60/ft^2 at 60 ft^2 (about three 30-in. cartridges) to $23/ft^2 at 2000 ft^2 (about ninety 30-in. cartridges).

Installed capital costs may be estimated using a 0.7 power rule:

$$\frac{\text{cost (desired)}}{\text{cost (known)}} = \left(\frac{\text{desired ft}^2}{\text{known ft}^2}\right)^{0.7}. \quad (34\text{--}16)$$

As an example, for flow rates of up to 30 gal/min, a 30-in. sanitary six-round cartridge prefilter housing would be selected followed by a three-round 30-in. sanitary final filter housing. These would provide approximately 120 ft^2 of prefiltration area and 60 ft^2 of final filtration area. Based on sanitary housings, the installed capital costs would be about $10,200 ($6300 + $3900).

Small users (boutique wineries and microbreweries) pay more for hardware on a per square foot basis. The cost of the smallest 10-in. stainless steel sanitary cartridge housing, suitable for flows of up to about 3 gal/min, runs about $120/ft^2.

Installation costs usually comprise 10 to 15% of the total capital cost, since foundations are rarely needed. Installation typically involves minimal piping (between housings and drain connections) and fittings, except for very large systems where overhead mechanical lifts, foundations, and more extensive plumbing are sometimes required.

Auxiliary equipment is required for aseptic filling. This equipment, which can significantly increase the initial capital requirements, includes

1. A filter for the water used to flush and sanitize the housings
2. Heat sanitizable housings for gases and other fluids, at or near the filter
3. Optionally, a system to spray sanitizing fluids on equipment surfaces in contact with the process stream.

Operating Costs

The only significant contributor to operating costs is the replacement of plugged membranes. All other costs are small in comparison. The combined costs of labor to sanitize and integrity test, energy, maintenance, storage chemicals, and flushing water rarely contribute as much as $0.01/gal. By comparison, membrane replacement costs are over $0.02/gal for the best microporous membrane filtration operations, with prefiltration and final filtration throughputs of 500 gal/ft^2 and of over 1000 gal/ft^2, respectively.

In many instances, the annual membrane replacement costs can be significantly higher than the installed costs of the initial hardware. Under these circumstances, it is cost-effective to increase the filter area, thereby decreasing the rate of flow per unit filter area and increasing throughput or filter life.

The sensitivity of costs to filter throughput is outlined in the examples presented in Table 34–15. The assumption is that the throughput of both prefilter and final filter can be increased by 30% if the filtration velocity is reduced by a factor of 2. As indicated in this table, it is cost-effective to install additional filter area so as to prolong filter life. Total annual costs for case 2 are approximately $6000, or 18%, less than for case 1. Unit filtration costs are reduced by approximately $0.005/gal. For larger wine or beer bottlers that process over 10 million gallons per year, these savings can exceed $50,000 per year.

PURIFICATION OF PROCESS FLUIDS IN SEMICONDUCTOR MANUFACTURE

Introduction

Since the development of the semiconductor and later the integrated circuit (IC) in the 1960s,

TABLE 34–15. Effect of Filtration Throughput on the Economics of Beverage Filtration.

	Case 1	Case 2
Process Conditions		
Flow rate, gal/min	20	20
On-stream time, h/yr	1,000	1,000
Total annual flow, kgal	1,200	1,200
Prefilter		
Cartridge length, in.	30	30
Number of cartridges	6	12
Prefiltration area, ft^2	120	240
Assumed throughput, gal/ft^2	500	650
Filter consumption, ft^2/yr	2,400	1,846
Final Filter		
Cartridge length, in.	30	30
Number of cartridges	3	6
Filtration area, ft^2	60	120
Assumed throughput, gal/ft^2	1,000	1,300
Filter consumption, ft^2/yr	1,200	923
Economics		
Installed equipment cost	$10,200	$18,000
Annualized capital related costs (@ 30% of installed equipment cost)	$3,060	$5,400
Annual Membrane Replacement Costs		
Prefilters @ $8/ft^2	$19,200	$14,769
Final filters @ $14/ft^2	$16,800	$12,923
Subtotal	$36,000	$27,692
Total Annualized Costs	$39,060	$33,092
Total Annualized Costs/gal	$0.0326	$0.0276

membrane microfiltration has been used to remove particles from the fluids that are used to manufacture these devices. The first ICs contained circuit dimensions in the 5- to 10-μm range. As a rule of thumb, the industry has determined that the presence of foreign particles that are between one-fifth and one-tenth the size of the critical circuit dimensions will disrupt these circuits. To achieve economical production yields, particles in this size range must be removed from process fluids.

The first membrane filters used were rated at 0.5 μm and were similar in construction to those developed for the pharmaceutical industry. However, since the dimensions of current memory devices (>1 Mbit dynamic random access memory) are below 1 μm, microfilters in the 0.05- to 0.1-μm range are now utilized. Circuit dimensions continue to shrink (Table 34–16), and the industry has begun to use ultrafiltration for deionized (DI) water in order to remove even smaller particles.

Both gaseous and liquid process fluids are used by the electronics industry, and the appropriate filters can be broadly classified into gas filters and liquid filters. Gas filters are used to remove particles from the so-called "bulk" gases (nitrogen, argon, hydrogen, and oxygen) and specialty gases such as silanes, arsine, phosphine, ammonia, etc. (Accomazzo and Grant 1986, 1989). These gas filters are also classified by their location in the manufacturing process. Larger filters are required during manufacture of the gases and also during their bulk storage and distribution at the semiconductor plant. These types of filters are called *central gas filters*. Smaller filters are also used at the point of use (POU) to remove any of the particles that may have been generated in the gas distribution system, consisting of piping, valves, regulators, and instrumentation. These filters are called *POU filters*.

Liquid filters can be divided into three categories:

1. Chemical filters other than photoresist filters
2. Photoresist filters
3. DI water filters.

The chemical filters are used to remove particles from acids, bases, solvents, and photoresists. The DI water filters are used to remove particles and bacteria from DI water. As with gas filters, two general categories of filters are utilized: so-called central filters and POU filters.

In the following, we discuss the various types of membrane filter devices by application in terms of their important performance and cost characteristics.

Process Gas Filtration

Filter Performance

Membrane filters can very effectively remove particles from gas streams. However, the performance of a membrane filter for a specific application is also dependent on the other materials of construction that comprise the filter device. The ideal filter will remove all of the particles above its pore size rating and not add any contaminants (particles or volatiles) to the filtered gas.

TABLE 34–16. Device Technology Trends. (Reprinted from Larrabee 1990 with permission.)

	1987	1989	1991	1994	1997	2000
Dram integration	1M	4M	16M	64M	256M	1024M
Minimum feature size (μm)	1.0	0.8	0.6	0.35	0.25	0.15
Critical particle size (μm)	0.2	0.15	0.10	0.08	0.05	0.03
Defects/cm^2	0.8	0.5	0.4	0.3	0.08	0.05
Mask levels	16	18	20	25	28	32

FIGURE 34-11. Central gas filtration system for a semiconductor manufacturing facility (reprinted from Millipore Corporation/Texas Instruments Corporation with permission).

Central Gas Filters

Typical flow rates for bulk gases range from 300 to 30,000 standard L/min and therefore the surface area of central gas filters is large. The electronics industry utilizes pleated cartridge filters placed in stainless steel housings for this application (see Figure 34-11). The membranes for these applications can be made of polysulfone, nylon, polyvinylidenefluoride (PVDF), or polytetrafluoroethylene (PTFE). Typical particle removal ratings of these filters are 0.05 μm and below. Other materials of construction used in the filters are polyester or polypropylene for the membrane supports, and polypropylene for the molded cores, sleeves, and endcaps. These filters may need to be replaced occasionally due to plugging from particles generated by the gas purifiers. When these filters are changed, extensive flushing of the filters and housing is required to remove atmospheric contaminants (particles, oxygen, carbon dioxide, and moisture). Use of redundant filtration is common at this location. This approach protects the downstream filters from plugging and hence these are rarely changed.

Point-of-Use Filters

Current semiconductor gas distribution systems are fabricated from electropolished 316L stainless steel tubing that is welded together under inert gas conditions. POU filters and semiconductor processing equipment are connected with special leak-resistant fittings. Figure 34-12 shows several types of POU filters capable of flow rates up to 300 standard L/min. POU filters need to exhibit low particle shedding.

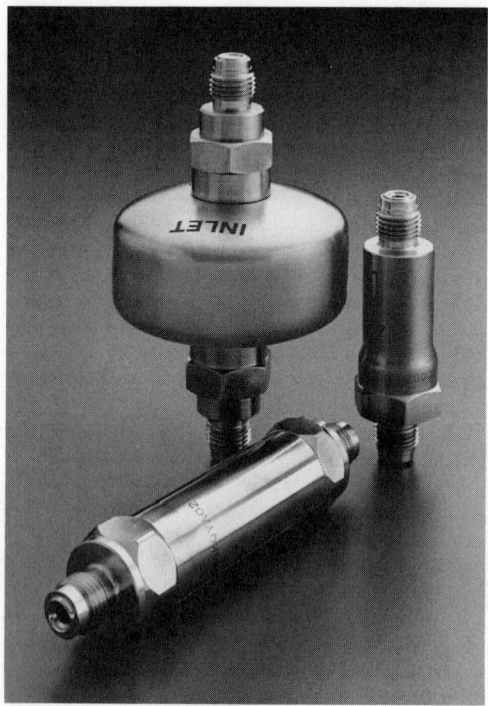

FIGURE 34–12. POU gas filters for semiconductor manufacturing: mini POU filters (reprinted from Millipore Corporation with permission).

Concern about adsorbed contaminants has led one manufacturer to offer Mini-POU™ filters that are prepackaged in a dry nitrogen atmosphere, and which are also shown in Figure 34–12.

Gas Filtration Costs

The cost of gas filters is dependent on the application. Central pleated cartridge filters typically cost between $100 and $150 per 10-in. element. Based on a filtration area of 7.5 ft^2 per element, this corresponds to a filter cost of approximately $13 to $20/ft^2. Housing costs range from $300 to $10,000 depending on their size, materials of construction, and surface finish requirements. The first filters after the gas purifiers are typically changed on an annual basis or when the pressure drop has increased by a factor of 2 or 3.

POU filters, which have surface areas of about 0.01 to 0.5 ft^2, typically cost between $100 and $300 (approximately $600 to >$1000/ft^2). These filters are rarely replaced and remain in place for the life of the line until the plant is refurbished (5 to 10 yr). In general, the cost of gas filtration is small compared to the value of the products being produced by the electronics plant. An index is the cost of downtime, which can exceed $100,000 per hour for a modern semiconductor plant.

Process Liquids Filtration

Chemical Filters
Performance

Chemicals are used by the semiconductor industry to clean the wafer surfaces, etch circuits, and remove unwanted process materials, such as photoresist (a thin layer of photosensitive polymer). These chemicals are filtered by the chemical manufacturers to semiconductor industry (Semiconductor Equipment Manufacturing Institute or SEMI) specifications, filtered after bulk storage at the semiconductor factory, and again at the POU. Depending on the application, different types of membrane microfilters are utilized.

Because of their chemical resistance, two membrane polymers, PVDF and PTFE in order of increasing resistance, are almost exclusively used for chemical filtration. These membranes are available in pore sizes ranging from 0.05 to 5 μm. Pleated cartridges are typically used for chemical manufacturing and bulk storage applications. For most of these applications, the other components of the cartridge (supports, core, sleeve, and endcaps) are made of polypropylene. However, strong oxidizing acids and some solvents (i.e., xylenes) may require that all the components of a pleated cartridge be made from PTFE.

The chemicals are transferred to the POU (i.e., the semiconductor processing equipment) either in bottles or through PVDF or Teflon® tubing. The chemicals are again filtered at the POU, using either pleated or stacked disk membrane cartridges. Some users, especially in Japan, prefer to use disposable filters that include the housing as shown in Figure 34–13. The first fully disposable chemical cartridge utilized an ethylene-chlorotrifluoroethylene

errors in unstable and high-viscosity chemicals, respectively. The removal efficiency also appears to be a function of the fluid flow rate.

The best way to maintain low particle counts in a chemical solution is to recirculate the solution continuously from a reservoir through the filter (Grant and Schmidt 1989). This approach is also used in recirculation etch baths since particles may be formed during the etching process.

Costs

The cost of chemical filtration is significantly higher than that of gas filtration. First, the filters are more expensive due to their materials of construction. The cost of a 10-in. pleated cartridge used for bulk filtration applications ranges from about $200 (about $25/ft^2) for one with a PTFE membrane and polypropylene construction to $800 (about $100/ft^2) for an all-PTFE cartridge. Such a cartridge is usually replaced after filtering between 1,000 to 10,000 L of chemical. Disposable POU filters cost between $400 and $2,500 (about $400/ft^2), and typically are replaced only once or twice a year.

Photoresist Filters
Performance

Photoresists are radiation-sensitive polymers that are applied to the surfaces of wafers in order to transfer the desired circuit features with appropriate processing steps. Photoresists are typically applied to the wafer by dispensing a puddle of polymer solution to the center of the wafer; the wafer is then spun at high speed (about 4000 rpm) to remove excess material. The film thickness can range from 0.3 to 3.0 μm. The viscosity of photoresist solutions typically ranges from 10 to 100 cp, but can exceed 1000 cp in the case of some high molecular weight polymer formulations. The high viscosity of photoresist polymer solutions, as well as their shear sensitivity, makes them more difficult to filter than other liquids used in the semiconductor industry. However, filtration of these solutions by their manufacturers and at the POU is necessary to obtain high manufacturing yields.

All of the suppliers of photoresist filter their

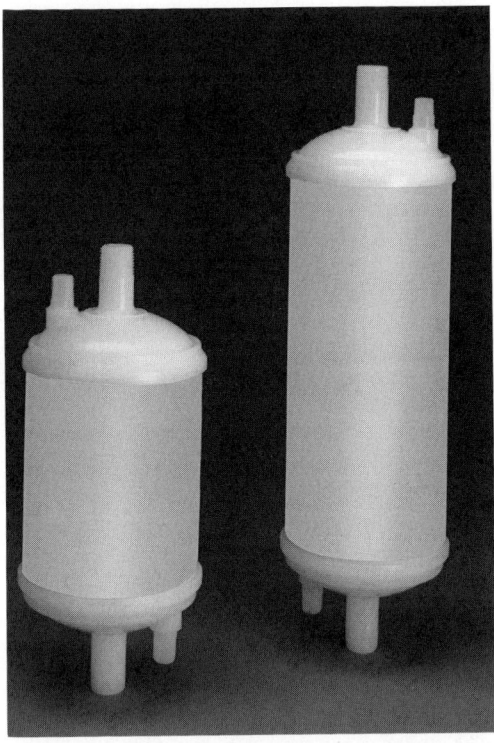

FIGURE 34–13. Disposable filters with housing (reprinted from Millipore Corporation with permission).

copolymer (ECTFE) housing, with ECTFE stacked disks and a PTFE filter (0.2 μm). Today, all-Teflon® disposable filters with pore size ratings down to 0.05 μm are available.

The critical performance characteristics of these filters are similar to gas filters. These filters should remove all of the particles below their pore size ratings, not shed particles, nor add extractables to the filtrate. The early performance studies on particle shedding (Peacock et al. 1985) were conducted in DI water. These have been complemented by recent filter performance studies conducted in electronic chemicals (Schmidt et al. 1987; Grant 1989). The removal efficiency of particles >0.3 μm by a stacked disk 0.2-μm filter was measured to be >99.99% for both 49% HF and 37% HCl. However, the measured particle removal efficiency was only 99.8% in 30% hydrogen peroxide and 98% in concentrated sulfuric acid. These apparently lower efficiencies are believed to be the result of particle measurement

products with 0.1- to 0.2-μm rated membrane filters. They typically use pleated cartridges (7.5 ft^2 per 10-in. element) in multiple cartridge housings (stainless steel) with either nylon or PTFE membrane filters. The other cartridge materials of construction (supports, core, sleeve, and endcaps) are usually polypropylene.

The semiconductor manufacturers also filter (0.1 to 0.2 μm) the photoresist to remove foreign particles generated by the shipping containers, and gel particles that result from polymer instability (Blazka and Hegde 1989). These gel particles are difficult to filter because at high differential pressures they can be extruded through the porous filter matrix.

POU filtration of photoresist is crucial to high-yield processing. As previously mentioned, the ideal filter for this application should remove particles greater than its pore size rating and not add contaminants (particles or extractables). Because most photoresist solutions contain good polymer solvents, filters of all fluoropolymer construction are preferred. POU filters also should be constructed in such a way that no air entrapment occurs as the liquid flows through the filter. This is necessary to ensure both bubble-free photoresist and the prevention of dripping after completion of each dispense cycle.

One manufacturer offers an all-Teflon® photoresist pump (Cross and Lucas 1987) with an integral stacked disk filter placed within the pump housing (Figure 34–14). This all-Teflon® pump and filter design ensures delivery of contaminant-free photoresist to the wafer surface. Depending on their output, pumps of this design utilize stacked disk filters that range from 200 to 2000 cm^2 in surface area.

Costs

The costs associated with filtering photoresist polymer solutions are significantly higher than gas filtration, and somewhat higher than chemical filtration. The higher costs are due to materials of construction and the need for increased filter area due to viscosity and gel particle considerations.

The cost of pleated cartridges (per 10-in. element = 7.5 ft^2) ranges from $100 to $200

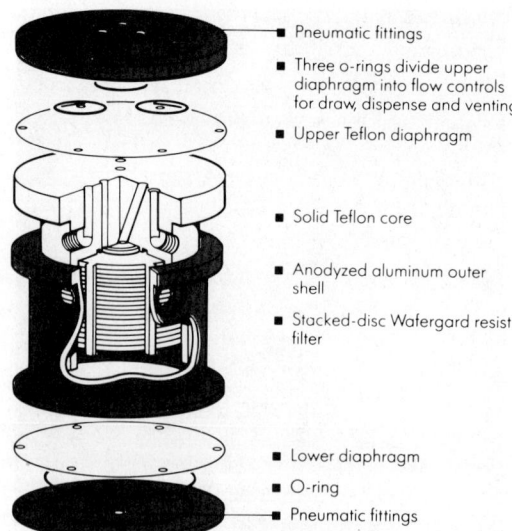

FIGURE 34–14. Photoresist pump and filter assembly (reprinted from Millipore Corporation with permission).

($15 to $30/ft^2) for ones that use polypropylene support structures to $800 ($100/ft^2) for an all-Teflon® pleated cartridge. A photoresist manufacturer can typically filter 100 to 300 L of solution with one 10-in. pleated cartridge element. Membrane filtration costs are estimated to be in the range of $1 to $10/L of photoresist solutions. While seemingly high, these costs are significantly less than the value of these solutions, which can exceed $100/L.

All-Teflon® stacked disk cartridges are available with surface areas ranging from 4000 to 8000 cm^2, and at a cost of from $600 to $1100. These membrane filters are typically used by the semiconductor manufacturers in their central storage and distribution systems. Their throughputs are typically in the 500- to 1500-L/cartridge range since the photoresist has already been filtered by the supplier.

POU photoresist filters range in cost from $50 to $250. These filters are usually associated with the photoresist dispensing system. Their operating life ranges from one to six months. Typical throughputs range from 50 to 250 L, resulting in POU filtration costs in the range of $0.10/L.

Deionized Water Filters

Performance

Because a wafer may be rinsed with DI water many times during the manufacturing of semiconductor devices, it may come into contact with from 100 to 1000 L of DI water during the manufacturing process. Because of this repetitive exposure, DI water quality is critical to the successful operation of a semiconductor manufacturing operation. Current SEMI (Semiconductor Equipment Manufacturing Institute) guidelines for pure water for semiconductor processing are listed in Table 34–17. The attainable column represents the best achievable quality attainable with current state-of-the-art components. The acceptable column represents levels of water quality most often found in fabrication plants with acceptably high yields. The quality of water during a "yield bust" provided the limits for the critical column. The alert condition is representative of conditions in which a growing level of malfunction exists but the consequences had not yet become evident in the processing area (Balazs and Poirier 1984).

Most wafer fabricators obtain their water from local utilities that are concerned with delivering potable water from natural sources. While fit for human consumption, as-received water is not suitable for the production of microelectronic devices, and extensive water purification systems, such as the one outlined in Figure 34–15, are an integral part of semiconductor fabrication facilities. There are usually three levels of water treatment. Primary and secondary treatment water operations, which are beyond the scope of this chapter, have been recently reviewed by Faylor and Gorski (1988).

Microfiltration equipment of present interest is an essential component of the DI water recirculation system, and is represented as tertiary treatment in Figure 34–15. The purpose of the tertiary water system is to remove all contaminants from the recirculating water. Traces of ions are stripped by the mixed bed deionizers. Bacteria are controlled in part by UV sterilizers, as well as by periodic treatment with

TABLE 34–17. SEMI Deionized Water Purity Guideline.

Test Parameter	Attainable	Acceptable	Alert	Critical
Residue, ppm	0.1	0.3	0.3	0.5
TOC, ppm	0.020	0.050	0.100	0.400
Particulates, counts/L	500	1000	2500	5000
Bacteria, counts/100 mL	0	6	10	50
Dissolved silica (SiO_2), ppb	3	5	10	40
Resistivity, megohm-cm	13.3	17.9	17.5	17
Cations, ppb				
Aluminum (Al)	0.2	2.0	5.0	[a]
Ammonium (NH_4)	0.3	0.3	0.5	[a]
Chromium (Cr)	0.02	0.1	0.5	[a]
Copper (Cu)	0.002	0.1	0.5	[a]
Iron (Fe)	0.02	0.1	0.2	[a]
Manganese (Mn)	0.05	0.5	1.0	[a]
Potassium (K)	0.1	0.3	1.0	4.0
Sodium (Na)	0.05	0.2	1.0	5.0
Zinc (Zn)	0.02	0.1	0.5	[a]
Anions, parts per billion				
Bromide (Br)	0.1	0.1	0.3	[a]
Chloride (Cl)	0.05	0.2	0.8	[a]
Nitrate (NO_2)	0.05	0.1	0.3	[a]
Nitrate (NO_3)	0.1	0.1	0.5	[a]
Phosphate (PO_4)	0.2	0.2	0.3	[a]
Sulfate (SO_4)	0.05	0.3	1.0	[a]

[a] Values not assignable at this time.

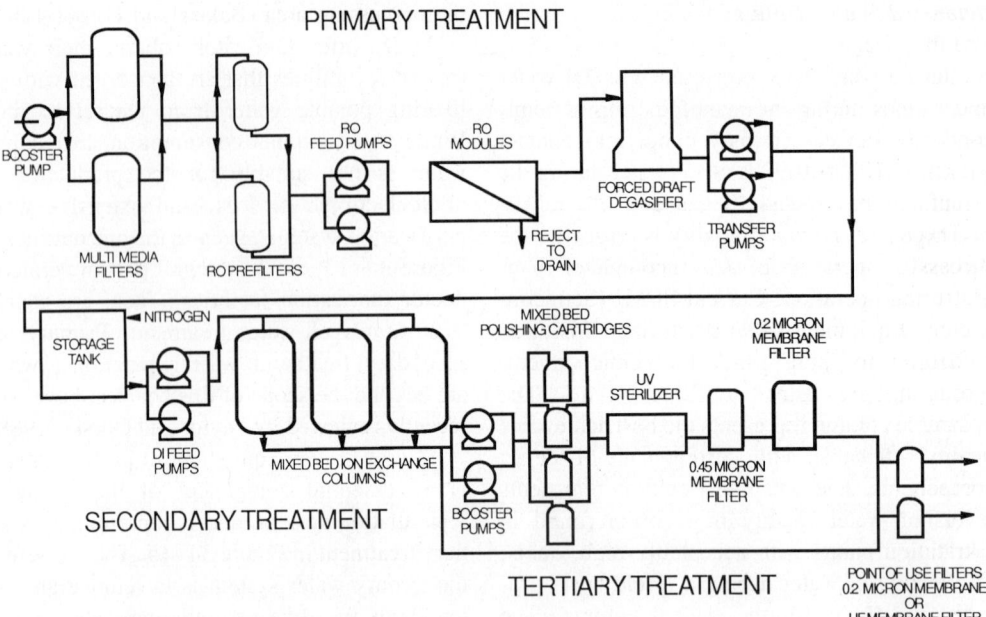

FIGURE 34–15. Typical treatment process to produce high purity water for electronics manufacturing (reprinted from Accommazzo, Ganzi, and Kaiser 1988 with permission).

sanitizing agents such as hydrogen peroxide (1 to 5%) or ozone. The nominal requirements placed on DI water filtration equipment are that it totally remove any suspended particles that may be present in the incoming water stream without itself introducing any suspended or dissolved contaminants into the filtered product water stream.

Particulate contamination that is removed by microfiltration in the recirculation loop arises from several sources. Primary and secondary treatment do not completely remove all traces of the organic colloids or silica present in the system feed. Ion-exchange resins in the deionizers break down with time into smaller particles that need to be removed. Filters are also a means of removing bacteria, which can grow even in such an inhospitable environment, and the pyrogens formed by bacteria decomposition in the sanitizers. The piping, pumps, and instrumentation used to control flow circulation all generate and add particles that need to be removed.

Membrane microfilters are used to control particulate contamination of the water being recirculated in the DI loop and of the water withdrawn at an individual POU. DI water is recirculated at a high rate. Depending on the size of the facility and the type of device produced, the flow rate in the DI water loop ranges from more than 100 gal/min to more than 1000 gal/min. DI water is consumed by many individual machines (i.e., POU) that draw intermittently from the recirculation loop. The DI water requirements of any POU are relatively low, of the order of 10 gal/min or less.

For the recirculation loop, the semiconductor industry has relied on pleated membrane cartridge filters with a rated pore size of 0.2 μm or less, even though some of the newer facilities are using ultrafiltration instead of microfiltration for particulate control. The characteristics of some commercial microporous pleated membrane cartridges recommended for the filtration of electronic DI water are listed in Table 34–18. In addition to particle retention, other important performance characteristics include the ability to test filter integrity *in situ,* and chemical stability in the DI water loop environment. DI water is considered an aggressive solvent because of the requirements for high purity. The filter cartridge must not contain any materials that are not absolutely stable in water, both under normal operations and under sanitizing conditions (i.e., at elevated temperatures in the presence of oxidizing agents). Extractables are essentially not tolerated.

The effect of flow rate on the pressure drop across some membrane microfilters that are currently used in DI water loops is presented in Figure 34–16. Filter area is typically based on an operating level of 2 psi per gal/min of throughput. Thus, because of the high volumetric flow, the required filtration area ranges from several hundred to several thousand square feet. This is usually achieved by using parallel banks of long (i.e., 30-in.) filtration cartridges in stainless steel housings.

Whereas DI water loop filters are used in a high continuous flow environment, POU filters are used in low-flow, intermittent service. In addition to meeting all the previously discussed performance requirements of DI water loop filters, POU filters must not shed particles when subjected to pressure pulsations and flow variations. As already indicated in the section on chemical filtration, particle shedding is a function of cartridge structure. Some cartridge shedding data obtained by Grant, Peacock, and Accomazzo (1987) under conditions of cyclic pulsed flow are presented in Figure 34–17. When pulsed, all the pleated cartridges shed significantly more particles than did the stacked disk cartridge tested. The superior performance of the stacked disk PVDF cartridge under transient flow conditions may be interpreted as being due to the greater mechanical support offered to the membrane by the plate-and-frame assembly, which results in less membrane flexure than probably occurs with a pleated cartridge under the stresses caused by the transient flow conditions. The characteristics of a commercially available stacked disk cartridge are also listed on the next spread.

Cost

The costs of DI loop microfilter cartridges range from about $100 to $400 each depending on their materials of construction. The cost per unit area of filtration is in the range of $10 to $20/ft^2. The cost of stainless steel housings ranges from $500 to $1500 per cartridge (about $50 to $100/ft^2), depending on cartridge size and finishing details. These cartridges are rarely changed and have an expected life of 1 to 5 years.

The cost and life expectancy of POU filters in DI water service are similar to those used in chemical service, as already discussed.

ANALYTICAL APPLICATIONS OF MICROFILTRATION MEMBRANES

Historical Perspective

While the previous sections of this chapter emphasized process applications of microfiltration membranes, it should be noted that analytical applications provided the initial impetus for their commercialization. The whole microfiltration industry descends from the development of an improved method of microbiological examination of water in Germany during World War II. In this initial application, bacteria were removed, cultured, and analyzed sequentially on a microfiltration membrane. Among the disruptions brought about by this conflict was the destruction of supplies of potable drinking water, and there was a great impetus to being able to identify rapidly and accurately safe sources of drinking water as a

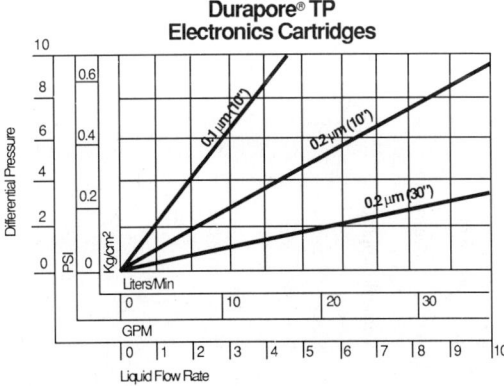

FIGURE 34–16. Typical flow rate of water through microporous filters (reprinted from Millipore Corporation with permission).

TABLE 34-18. Characteristics of Commercial Microporous Membrane Filter Cartridges Recommended for Filtration of Deionized Water in the Electronics Industry.

Description									
Manufacturer	Brunswick	Gelman Acroflow	Millipore Durapore	Millipore Durapore	Millipore Wafer-gard-40	Nuclepore Polycarbonate	Pall N-66 Polsidyne	Pall M-66 Ultipore	Sartorius Sartobran
Filter Name	BTSM	SUPER E	CVDI	CVVI	WGVL	QR	MIZE	MIE	II
Rated pore size, μm	0.1	0.2	0.2	0.1	0.1	0.1	0.1	0.1	0.1
Cartridge configuration	Pleated	Pleated	Pleated	Pleated	Stacked disk	Pleated	Pleated	Pleated	Pleated
Height, in.	10	9.8	10	10	4.0	10	10	10	10
Diameter, in.	2.7	2.6	2.9	2.9	2.5	2.8	9	7.5	2.8
Filtration area, ft^2	5.8	5.2	7.4	7.4	2.2	18			6.4
Materials of contruction									
Membrane	PS	ACR+/PES	MPVDF	MPVDF	MPVDF	PC	HYL/PET	HYL/PET	CA
Vertical supports	PES	PES/PP	PET	PET		PP	PES	PES	PP
Core	PS	ACE+	PP	PP		PP	PP	PP	PP
Sleeves	PS		PP	PP		PP			PP
End caps	PS	PS	PP	PP	PS	PP	PES	PES	PP
Support disks					PS				
O-rings	S	EP	V	V	V	BM	S	S	S
Potting adhesive	PU	PU	TP	TP	TP		TP	TP	

Performance properties									
Integrity testable	Yes	Yes	Yes	Yes	Yes	Yes	Yes	Yes	Yes
Bubble point test value, psi	80	80	40	70	60		80	80	74
Maximum differential pressure									
psig	20	88	50	50	60	100	50	50	20
@ Temperature, °C			23	23	20	25			
Maximum operating temperature, °C									
Continuous exposure	126	88	80	80	100	79	50	50	
Intermittent				145			125	125	
Sanitizable	Yes	Yes	Yes	Yes	Yes	Yes	Yes	Yes	Yes
Water flow, ΔP (psi)/gal	0.7	0.9	0.9	2.0	2.3	1.25	1.5	0.6	1.25
Permeability, gpm/psi ΔP-ft^2	0.259	0.226	0.150	0.268	0.200		0.074	0.240	0.125
Extractables, mg/ft^2	2.5						4		

Key:
- ACE+ = acetal copolymer
- ACR+ = acrylic copolymer
- BM = BUMA-M
- CA = cellulose acetate
- EP = ethylene-propylene copolymer
- MPVDF = modified polyvinylidene fluoride
- MYL = nylon
- PBT = polybutylterephthalate
- PC = polycarbonate
- PES = polyester
- PET = polyethyleneterephthalate
- PFA = Polyfluoroalkoxy
- PP = polypropylene
- PS = polysulfone
- PTFE = polytetrafluoroethylene
- PU = polyurethane
- S = silicone
- TP = thermoplastic
- V = viton

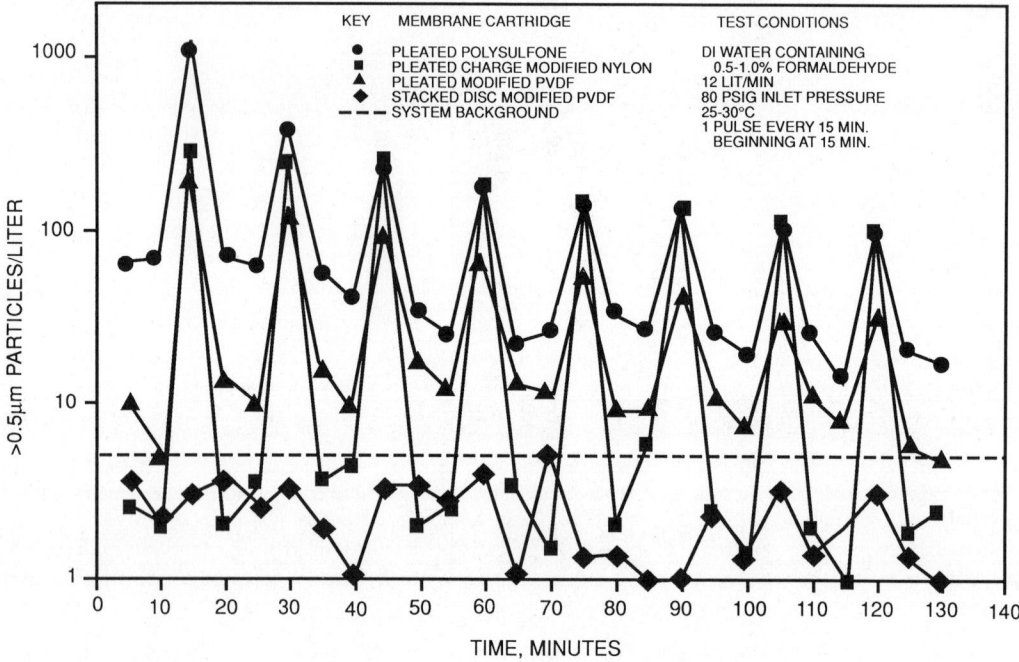

FIGURE 34–17. Comparison of cartridge shedding in DI water during initial pulsing cycle (reprinted from Grant, Peacock, and Accommazzo 1987 with permission).

matter of maintaining some minimum level of public health, which this microfiltration-based method provided.

The general utility of microfiltration membranes as a means of removing particulate materials from a fluid stream, which could also serve as a platform for quantitative assay, was noted by analysts who rapidly adapted them to many other analytical problems. Microfiltration membranes are now used to assess particulate contamination in a wide variety of liquids and gases. Many of the process applications of microfiltration discussed earlier were originally based on prior analytical use of microfiltration membranes.

The argument can be made that the development of new analytical applications represents the leading edge of microfiltration technology. In these typically small-scale applications, economics and utility are driven primarily by achieving superior improved technical performance. Interesting examples of novel applications of microfiltration membranes are to be found in the developing area of molecular and cell biology.

Microbiological Assay

Microporous membranes have played an important role in the detection of microorganisms in potable water sources, pharmaceutical process streams and products, and in food and beverages. The general technique involves filtering a sample of known volume through a microfiltration membrane that traps the suspended microorganisms, culturing the trapped bacteria on the membrane, and then counting the resulting colonies with a low-magnification (10 to 20×) stereomicroscope.

Water Microbiology

Epidemics such as typhoid fever, dysentery, and cholera are caused by pathogenic bacteria transmitted via polluted drinking water. These organisms are difficult to culture *in vitro*. Waterborne intestinal parasites and viruses are an even greater challenge to laboratory analysis.

Coliform organisms, while relatively harmless themselves, are almost always present in water containing enteric pathogens. Because they are relatively easy to isolate and because they survive longer than the disease-producing organisms, coliforms are a useful indicator of the possible presence of enteric pathogenic bacteria and viruses. In most cases, water that is free of total coliforms is considered free of disease-producing bacteria.

The term *coliform* refers to a group of gram-negative nonspore-forming, rod-shaped bacteria that ferment lactose at 35°C in 24 to 48 h. These are widely distributed in nature and many are native to the gut of warm-blooded animals and man.

A more definitive test for recent fecal pollution than the total coliform test is fecal coliform analysis. *Fecal coliforms* are distinguished from the total coliforms in that they ferment lactose at higher temperatures than the 35°C that is optimal for coliforms. The best temperature to select coliforms specifically of fecal origin has been found to be 44.5°C.

For years the classical test for the examination of coliform density in water supplies and water disposal systems has been the tube fermentation or MPN (most probable number) system (Peterson 1974). This test takes advantage of the ability of the coliform group of organisms to ferment lactose at 35°C in the appropriate growth medium, producing CO_2 gas in a 48-h period. When it was introduced, this method was accepted as a means of assessing the biological quality of water because

1. The coliform group was known to be always present in water when enteric pathogens are present.
2. Coliforms satisfy the major criteria of the ideal indicator, allowing for relatively simple and convenient testing procedures for their identification.

The major limitation of this method is that the completed procedure requires four to five days before pertinent data can be obtained. This is a major drawback when the quality of drinking water has to be assessed under emergency conditions, when time is of the essence. This shortcoming provided the impetus for the development of the membrane filter technique.

The membrane filter technique employs a 0.45-μm pore size plastic membrane filter as a collection device. By the capillary action of its pores, the membrane filter draws a selective medium for coliform growth onto its surface. When the coliform organisms ferment lactose, one of the intermediate materials produced is an aldehyde. This aldehyde formation, indicative of coliform growth, is identified by the appearance of a metallic sheen on the surface of the coliform colonies. This sheen is caused by an aldehyde complex formed from the interaction with basic fuchsin and sodium sulphite contained in the growth medium. At an incubation temperature of 35°C, about 18 to 24 hours are required to complete the procedure.

Since its introduction as a tentative method for coliform enumeration in the 10th edition of *Standard Methods* (American Public Health Association 1955), the membrane filter method has gained wide acceptance not only for total coliform, but also for fecal coliform, total bacteria, and a variety of other bacterial tests. The unique advantage of the membrane filter method over other test methods is its ability to concentrate and localize bacteria from large samples. Hence, the method increases the sensitivity of quantitative bacteriology into the range of well below one organism per milliliter. As already mentioned, another advantage of the method is that results are obtained in approximately 24 hours, compared to 48 to 96 hours for the MPN system. The membrane method also has the practical advantage of being easier and less costly to run than the MPN method.

The major limitations of the membrane filter method are that factors such as elevated turbidity, injured coliforms, high numbers of noncoliform bacteria, and membrane filter type may severely influence the sensitivity of the procedure (LeChevalier, Cameron, and McFeters 1983).

The earliest techniques for bacteriological analysis with membrane filters involved direct microscopic examination of bacteria trapped on the membrane surface. Here the optimum structure required pores smaller than the organisms being trapped for examination so that they would lie in a single microscopic plane to facilitate their detection with a high-power microscope. The above requirements evolved naturally to the practice of retaining organisms on the membrane surface for various culture techniques. As was originally noted by Sladek et al. (1975), the pore size of the membranes chosen on the basis of high-power microscopic examination was too small to provide an optimal structure for colony growth.

Membranes for quantitative bacteriology are characterized as follows. They have pores small enough to retain bacteria but open enough not to interfere with growth of the organism, and they also must not cause aberrant colony morphology.

Surface pore morphology was found to have a critical effect on organism recovery (Sladek et al. 1975; Green, Clausen, and Litsky 1975; Lin 1977). The crucial step in the development of a colony from a single bacterium is the onset of cellular division, and it is not unreasonable to expect that this delicate process is influenced by the extent and nature of the contact of the organism with the solid, and of the nutrient film surrounding it. Further, nutrient supply by diffusion of the medium and removal of subsequent metabolic waste products are functions of membrane structure and pore morphology.

Pore morphology plays an important role in trapping the organisms and maintaining them, at or below the surface of the membrane, surrounded by nutrient medium. Capillary action provides transport of nutrient solution to the surface of the membrane. The transport rate must be such that evaporation is minimized so as to prevent concentration of nutrients and desiccation. The best bacterial recoveries are obtained on membranes that have funnel-shaped pores: wide at the surface with a finer pore mesh below to maintain capillary flow of the medium.

Growth and colony formation are also affected by a number of other factors, including surfactant content, sterilization procedures, and adsorption of heavy metals to the membrane matrix. Significant surfactant concentrations must be avoided, although small amounts of nonionic detergents do not usually interfere with bacterial growth. Residual amounts of ethylene oxide, which can be toxic to microorganisms, are closely controlled by membrane manufacturers to guard against sterilant toxicity. Membranes that do not bind metals are preferred because of cytotoxic or other effects.

In general, gridded (to ease colony counting) membranes made from mixed esters of cellulose with average pore sizes of either 0.45 or 0.7 μm are used in this application. The smaller pored membrane is commonly used for total coliform analyses, and the larger one for fecal coliform analyses. The latter, due to its 2.4-μm surface, funnel-like opening, and 0.7-μm restrictive interior pore, allows for improved recovery of stressed fecal coliform organisms, especially those found in chlorinated effluents.

Over the years, the various membrane suppliers have also developed specialized device configurations to facilitate these microbiological analyses. Traditionally, the samples were vacuum filtered through 47-mm membranes that were held in nondisposable diameter glass or stainless filter holders, such as the system shown in Figure 34–18. If many analyses need to be performed, it becomes cost-effective to use presterilized, self-contained disposable holders that combine a 100-mL funnel with a 0.45-μm membrane filter, such as the Milliflex-100™ system shown in Figure 34–19. The advantages of this system include:

1. No time-consuming preparation
2. Simplicity of use
3. Faster filtration
4. Consistent recoveries
5. Simplified colony counting
6. Choice of media
7. Compliance with worldwide standards.

Alternatively, integrated devices, such as the Sterifil™ aseptic system shown in Figure 34–20,

FIGURE 34-18. Standard laboratory glass vacuum filtration system (reprinted from Millipore Corporation with permission).

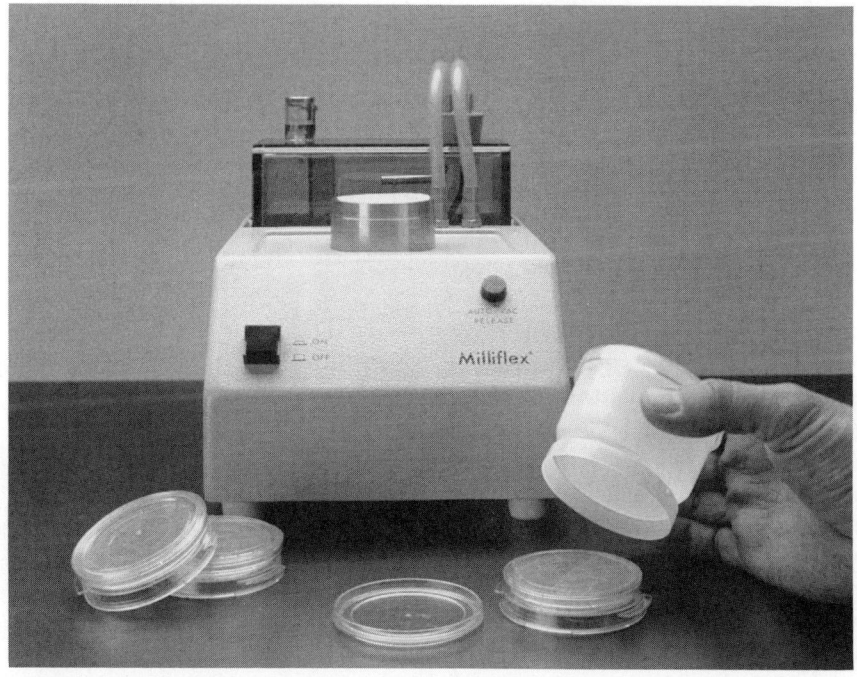

FIGURE 34-19. The Milliflex-100™ system for the microbiological examination of aqueous fluids (reprinted from Millipore Corporation with permission).

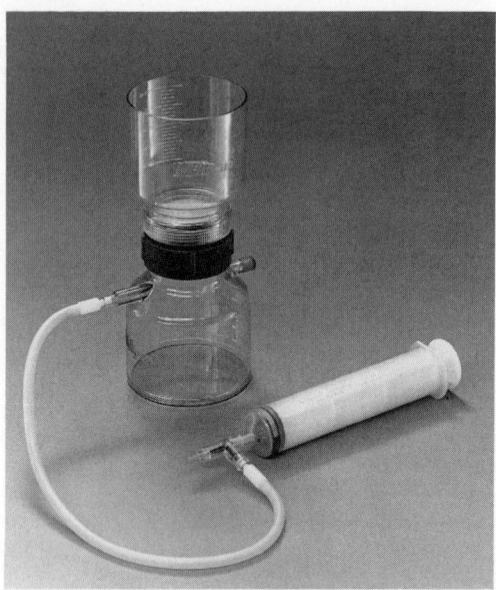

FIGURE 34-20. Sterifil™ aseptic system, 47 mm, with hand vacuum pump assembly (reprinted from Millipore Corporation with permission).

have been developed to facilitate the filtration of occasional samples.

To facilitate field sampling, integrated samplers exist that can integrate a 0.45-μm membrane filter, the growth medium, the incubating dish, and sample filtration in one device. Such a sampler is illustrated in Figure 34-21. When it is immersed in the water sample, the absorbent pad rapidly (in less than 30 seconds) draws 1 mL of fluid through the microporous filter and into the pad. When fluid passes through the filter, all organisms above the rated pore size are firmly affixed to the filter surface. As the fluid is drawn into the pad, it immediately hydrates the nutrient medium, which then diffuses up through the pores up to the surface of the filter. Nutrients are provided to the trapped organisms, allowing development during incubation into visible colonies that can be readily examined and counted. With these samplers, there is no need to filter samples or set up or sterilize equipment. Operators do not have to be trained in aseptic techniques. Because of their small size, samplers can easily be transported to the sampling site and back to the laboratory for subsequent incubation and examination. The major limitation of these samplers is their small sampling volume, which limits the applicability of the test to samples with relatively high bacterial populations.

Other Microbiological Assays

The general membrane filter method of identifying and assaying coliform bacteria in water streams, which was described earlier, has found general microbiological use. The membrane filter method is now used to analyze for the presence and number of numerous different organisms in a wide variety of liquids, such as foods and beverages, pharmaceutical products, oils and oil-based products, and process streams such as deionized water in the semiconductor manufacturing industry.

The principal difference between the various tests is in the choice of the culturing nutrient, medium, and conditions, which are specific to a particular microorganism. Such a discussion is beyond the scope of this handbook (Millipore 1967).

Most microbiological analyses employ mixed cellulose ester membrane filters with an average pore size of 0.45 or 0.7 μm. These

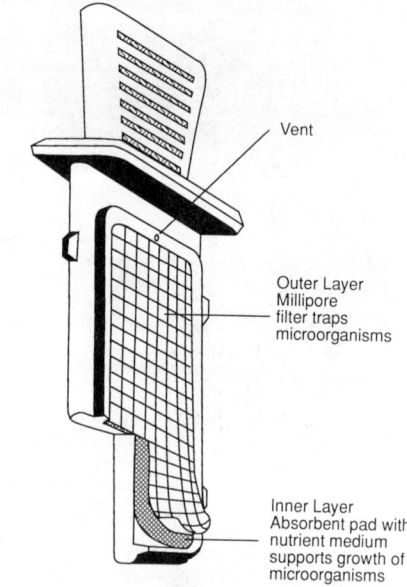

FIGURE 34-21. The Millipore field sampler for determining microbiological contamination (reprinted from Millipore Corporation with permission).

generally have a grid-marked surface. Depending on the organism, a white or black membrane may be used. For example, it is easier to identify colonies against a black rather than a white background (Millipore 1981).

Monitoring of Particulate Contamination

There is an increasing need for the monitoring of the presence and concentration of suspended particles, both in liquids and gases.

Prolonged workplace and environmental exposure to airborne contaminants has been linked to a number of occupational disabilities and diseases. In recognition of these health hazards, national and international regulatory agencies have established threshold limits and standard methods for a large number of airborne contaminants such as, for example, asbestos.

The increased precision and lower tolerances for electronic and mechanical components used in aircraft and aerospace systems, fossil fuel and atomic power plants, and in a growing number of other fields also demand that particulate contamination be carefully controlled. A host of military and commercial specifications, as well as voluntary standards, now exist that address the multifaceted aspects of this issue.

Microfiltration membranes have proven to be important tools in the separation, collection, and analysis of particulate matter. Minute quantities of solid separated from gases and liquids and concentrated on the filter surface may be weighed, counted, optically inspected, and chemically analyzed. Table 34–19 provides an overview of some of the many analytical procedures performed on microfiltration membranes. The particular method used is often dictated by specifications and/or standards such as those issued by U.S. government agencies such as the Environmental Protection Agency (EPA), National Institute of Occupational Safety and Health (NIOSH), Occupational Safety and Health Administration (OSHA), and the Department of Defense (MIL-Specs), professional societies such as ASTM, Institute for Environmental Sciences (IES), and Society of Automotive Engineers (SAE). A partial list of ASTM standard methods that specify the use of membrane filtration as part of the procedure is given in Table 34–20.

In those instances for which no established method exists, and a new analytical application involving the use of a microfiltration membrane needs to be developed, the following considerations should be kept in mind.

Pore Size

The pore size of the membrane should be small enough to retain all of the particulates of interest to the analysis. Usually, a finer pored filter will be required to retain particles from a liquid phase rather than a gas phase. Of course, the smaller the pore size of the filter, the lower its permeability, which is of importance in establishing the conditions of sample collection (sample size, sampling time, transmembrane pressure drop, and filtration area).

The pore size should also be compatible with the proposed method of examination. For example, if the particles are to be examined microscopically, the pore size should be small enough so that the membrane acts as a surface filter, and pore features do not interfere with optical examination of the particles.

Materials of Construction

The membrane filter has to be chemically and physically compatible not only with the particle sampling environment, but also with the analytical environment. For example, if the particles on a membrane are to be examined at high magnification by scanning electron microscopy (SEM), the membrane should be stable in the electron beam of the SEM for the duration of time needed to obtain a photomicrograph. In that regard, polycarbonate membranes are more stable than those based on cellulose esters.

Needless to say, the membrane should be clean and free of particulates that would interfere with the measurements. If the membranes are to be used as substrates for chemical analysis, they should not contain any chemicals that would interfere with the proposed method of particle assay.

In some cases, it may be desirable to dissolve or physically modify the membrane filter as part of the analysis process. This establishes

TABLE 34-19. **Analytical Applications of Microporous Membranes.**

Method	Visible Light Microscopy	Electron Microscopy	Microchemical Analysis	Ring Oven Analysis	Infrared Absorption	Visible Light Absorption	Ultraviolet Absorption
Time of analysis (Long = 2 h) (Short = 0.5 h)	Short	Long	Short	Long to Short	Short	Short	Short
Specificity—To one substance in a mixture	Excellent	Excellent	Fair to Good	Excellent	Fair to Good	Poor to Fair	Poor to Fair
Qualitative analysis — What is measured?	Visible appearance, size, shape, color, texture	Size, shape, texture	Specific chemical reaction	Specific chemical reaction	Wave lengths of absorbed energy	Wave lengths of absorbed energy	Wave lengths of absorbed energy
How good is it for this use?	Fair to Good	Good	Good	Good	Good	Poor	Poor
Quantitative analysis — What is measured?	Size and quantities of particles	Size and quantities of particles	Change in physical property: color, etc.	—	Amount of radiation transmitted	Amount of light transmitted	Amount of radiation transmitted
How good is it for this use?	Excellent	Excellent	Poor	—	Good in medium concentration range	Good in lower, poor in higher ranges	Excellent in trace ranges
Precision/reproducibility of results	As good as sampling accuracy	As good as sampling accuracy	Good	Good	Good	Good	Excellent to poor depending on range
Instrument price (High > $20,000) (Low < $10,000)	Low	Medium to High	Low	Low	Low to Medium	Low to Medium	Low to Medium

Flame Photometry	Emission Spectroscopy	X-Ray Fluorescence	X-Ray Diffraction	Radioactivity Monitoring	Autoradiography	Activation Analysis	Gravimetric Analysis
Short after calibration	Short after calibration	Short after calibration	Short	Long to Moderate	Long	Short after activation	Moderate
Good	Excellent	Excellent	Excellent	Poor	Excellent for individual particles	Excellent	—
Wave length of emitted light	Wave length of emitted light	Wave length of emitted x-radiation	X-ray diffraction angles	Radioactivity emission	Specific emission from particles	Energy and decay rate of induced activity	—
Excellent	Excellent	Excellent	Excellent	Poor	Excellent for determining individual particle activity	Excellent	—
Amount of emitted light	Amount of emitted light	Amount of emitted x-radiation	Amount of radiation diffracted	Amount of radioactive emission	Amount of radioactivity of single particles	Amount of induced radioactivity	Weight
Fair in low ranges	Good in trace ranges, fair in higher ranges	Good in midrange concentrations	Fair	Good	Poor	Excellent in trace ranges	Excellent
Good in proper ranges	Good in proper ranges	Good in proper ranges	Poor for quantitative use	Good	Good	Excellent in trace ranges	Excellent
Low to Medium	High	Medium	Medium	Low	Low	Medium to High	Low

TABLE 34-20. **A Partial List of ASTM Standards That Specifiy Microporous Membrane Filtration in the Assessment of Particulate Contamination.**

	Method Number	Date	Title
Gases			
	D 3267-88	1988	Standard Test Method for Separation and Collection of Particulate and Water Soluble Gaseous Fluorides in the Atmosphere (Filter and Impinger Method)
	D 3268-89	1989	Standard Test Method for Separation and Collection of Particulate and Water Soluble Gaseous Fluorides in the Atmosphere (Sodium Bicarbonate-Coated Glass Tube and Particulate Filter Method)
	D 4240-83	1989	Standard Test Method for Airborne Asbestos Concentration in Workplace Atmosphere
	D 4765-88	1988	Standard Test Method for Fluorides in Workplace Atmosphere
	F 25-68	1988	Standard Test Method for Sizing and Counting Airborne Particulate Contamination in Clean Rooms and Other Dust Controlled Areas Designated for Electronic and Similar Applications
	F 318-78	1989	Standard Practice for Sampling Airborne Particulate Contamination in Clean Rooms for Handling Aerospace Fluids
Liquids			
	D 2276-88	1988	Standard Test Method for Particulate Contaminant in Aviation Fuel
	D 3830-79	1984	Standard Practice for Filter Membrane Color Ratings of Aviation Turbine Fuels
	F 311-78	1983	Standard Practice for Processing Aerospace Liquid Samples for Particulate Contamination Analysis Using Membrane Filters
	F 312-69	1980	Standard Methods for Microscopial Sizing and Counting Particles from Aerospace Fluids on Membrane Filters
	F 313-78	1983	Standard Test Method for Insoluble Contamination for Hydraulic Fluids by Gravimetric Analysis
Other			
	F 24-65	1983	Standard Method for Measuring and Counting Particulate Contamination on Surfaces
	F 51-68	1984	Standard Test Method for Sizing and Counting Particulate Contaminants in and on Clean Room Garments

different constraints. For example, particles cannot be examined on a nontransparent membrane by transmitted light microscopy. Cellulosic filters, which are normally white, are suitable substrates for microscopic work because they can be rendered microscopically clear with immersion in a suitable oil that has an index of refraction of 1.51.

Accurate particulate analysis depends as much on proper sampling as on particle collection. The user has to ensure that the stream being sampled is a representative one, and that there is satisfactory contact between the stream being sampled and the collection membrane. In particular, it is important that adherence of particles to the walls of the collection device be minimized and that the particles be distributed evenly over the collection membrane. For example, the filter membrane in the aerosol monitor shown in Figure 34-22 is sealed between the monitor sections with a thin cellulose pad under the filter to distribute the sample flow evenly over the filter surface. This monitor meets the specifications of the NIOSH method for airborne asbestos fiber monitoring.

Blotting of Macromolecules to Microporous Membrane Substrates

Microporous membrane substrates have been widely used in the study of complex macro-

number of analytical procedures and to confer considerable advantages in a number of important applications, such as solid-phase protein immunoassays or hybridization analysis with nucleic acids.

Transfer or blotting of macromolecules can be achieved by various methods such as direct application, dot blotting, passive diffusion, solvent flow, and electrophoretic blotting. The most popular of these, the electrophoretic transfer of proteins to membranes, is often referred to as *Western blotting* (Towbin, Staehelin, and Gordon 1979). In nucleic acid applications, solvent flow transfers are more common and are referred to as *Southern* for DNA (Southern 1975) or *Northern* for RNA (Thomas 1980).

Blotting Methods

Electrophoretic Blotting
This most widely used methodology employs the use of an electric field to elute macromolecules from the separation medium and on to the solid phase. This process is fast, efficient, and retains the original analytical separation such as would be achieved by gel electrophoresis. The two current configurations of electroblotting apparatus are:

1. Tanks of electrolyte with platinum wire electrode arrays (Gershoni, Davis, and Palade 1985)
2. Flat-plate type, the so called *semidry* transfer devices (Kyhse-Anderson 1984).

FIGURE 34–22. Aerosol analysis monitors, 37 mm in diameter, with and without gridded membranes (reprinted from Millipore Corporation with permission.)

Both types of devices offer unique advantages, but the latter is rapidly gaining popularity.

Solvent-Flow Blotting
In this application a mass flow of solvent is used to elute macromolecules from the separation medium on to the solid phase. In the case of nucleic acids, this is usually an overnight process and is driven by the capillary "wicking" of absorbent filter paper. More recently solvent flow transfers have been facilitated by application of vacuum suction. Such transfers can be achieved in 1 hour (Peferoen, Huybrechts, and DeLoof 1982).

molecules in biochemistry for over a decade. In most cases, the macromolecule is immobilized by adsorption or by the formation of a covalent linkage of biomolecules or cells to reactive groups on the solid phase. The general term *blotting* refers to a process by which macromolecules are brought into contact with the surface of the solid phase to promote the above interaction. As a consequence, this application exploits the physical and chemical characteristics of the microporous membrane. This blotting process has been found to facilitate a

Solid-Phase Immobilization Matrices

In blotting applications, the physical and chemical characteristics of the solid phase are very important. Two types of membranes are used for this application: synthetic polymeric films and fibrous papers. The latter were first used in the form of ion-exchange papers, such as DEAE cellulose (McLellan and Ramshaw 1981), phosphocellulose (Reiss, Sprengel, and Will 1984), or as activated papers such as diazobenzyloxymethyl (DBM)- and diazophenylthioether (DPT)-cellulose papers (Alwine, Kemp, and Parker 1979).

These have been largely replaced by polymeric film membrane substrates. *Nitrocellulose* was first used for protein and nucleic acid blotting. Pure nitrocellulose is preferred to mixed ester formulations, which offer better handling but lower binding capacity. The mechanism of protein binding to nitrocellulose is not clearly understood. It is thought to involve a combination of ionic interactions with the net negative charge on the surface conferred by the strong dipolar character of the nitrate group. In addition, the binding of proteins may also involve hydrophobic interactions, as inferred by the ability of nonionic detergents to bring about protein desorption from this surface (Batteiger, Newhall, and Jones 1982). A membrane with a standard pore size of 0.45 μm has been used for many applications. However, low molecular weight (<20,000 daltons) proteins can fail to be retained. To promote better retention of these molecules, membranes with smaller pore sizes can be used, or methanol can be included in the transfer buffer. The latter is thought to promote hydrophobic interactions with the membrane surface. It will also facilitate removal of SDS (sodium dodecyl sulfate) from proteins that could interfere with adsorption to the surface. Retention of proteins on the surface can be promoted by drying, treatment with base, glutaraldehyde, cyanogen bromide, or divinyl sulfone. In the case of nucleic acids, a similar additional "fixation" step is often used to enhance further the retention of nucleic acid by the surface. In most cases, this step involves baking under vacuum at 80°C for 2 hours. This presumably dehydrates the membrane and promotes hydrophobic interactions with the surface. In addition, the adsorbed nucleic acid may also undergo denaturation of the secondary structure, which, on cooling, could renature around the fibers of the microporous membrane, further entrapping the molecule on the surface.

UV cross-linking has also been applied under carefully controlled conditions. Nitrocellulose is widely accepted but can be brittle and difficult to handle. It is not durable enough for multiple cycles of hybridization as is needed in the case of nucleic acids.

More recently, *cationic nylon membranes* have been introduced for blotting applications. Various nylon substrates are commercially available and fall into two distinct classes: nylon 66 films and charge-modified nylon 66. The latter are made by incorporating tertiary and quaternary amines into preformed nylon 66, either during casting or as a post-treatment coating. These charge-modified substrates offer higher ion-exchange capacity, better binding, and improved retention characteristics than unmodified nylon 66 films. With such positively charged nylon membranes, the attachment of proteins or nucleic acids occurs mainly through electrostatic attraction. This interaction is very stable under most of the conditions the adsorbed molecule would be exposed to during detection after blotting, i.e., immunostaining and hybridization. However, due to the high charge density, visualization of blotted macromolecules with common (anionic) dyes is not possible. The suppression of nonspecific adsorption to unoccupied binding sites during immunostaining is also much more difficult than with nitrocellulose membranes. Very few immunostaining applications have been developed on these charged nylon membranes for this reason. However, more recently many nonstaining detection strategies have been developed so some of these disadvantages of nylon are being overcome. They are used extensively in nucleic acid applications where radioactive detection methods are employed. In the nucleic acid application, post-blotting fixation by UV cross-linking is frequently carried out. This process appears to work better on nylon 66 films than on charge-modified surfaces. It has been suggested (Church and Gilbert 1980) that UV irradiation induces the formation of thymidine free radicals, which react with primary amine groups on the nylon 66 surface. Charge-modified nylon does not contain as many such groups accessible on the surface. Clearly, in this case, the proper surface chemical composition is crucial for successful formation of covalent links to the surface. This post-blotting fixation of nucleic acid leads to retention of nucleic acid through many cycles of hybridization analysis.

In addition to the above solid phases, several other modified polymeric microporous membranes have been described for blotting applications. The use of a *hydrophobic polyvinylidenefluoride (PVDF)* based substrate (Immobilon™-P) has been reported (Pluskal et al. 1986), and it has found rapid acceptance as an alternative to nitrocellulose in many protein blotting applications. Its high binding capacity and physical robustness, together with its chemical stability, are major attributes of this new substrate. Protein molecules appear to be well retained on its surface by a combination of hydrophobic and electrostatic interactions with the strong dipolar character of the molecules. Several unique applications have been demonstrated for this substrate, including:

1. Peptide microsequencing (Matsudaira 1987), which exploits the excellent chemical stability of the substrates during gas- or liquid-phase sequencing reactions
2. A family of protein microsequencing applications developed at Milligen/Biosearch to exploit the chemical reactivity of the PVDF surface to enhance retention of protein for covalent pulsed liquid microsequencing.

Another interesting application of the PVDF surface is the technique called *cell blotting* (Kendall and Hymer 1988) in which cells are grown on the hydrophobic surface. These cells secrete a peptide product, which diffuses from the media into the microporous membrane substrate. The product then is "wicked" away from the cells by the capillarity of the membrane structure. The amount of protein can be quantified by measuring the distance the product "front" moves in the membrane. Clearly, this application exploits the unique hydrophobic and microporous character of the PVDF substrate.

PVDF membranes have also been charge modified to introduce surface cationic groups. These Immobilon®-N (Millipore Corporation) membranes offer a hybrid hydrophobic and ionic surface. They have received early acceptance as an alternative to charge-modified nylon membranes and have been considered to offer some advantages (Mann, Venkatraj, and Auerbach 1989). It appears that the unique ionic/hydrophobic character of these membranes works well in nucleic blotting applications, and these show excellent retention and blotting performance compared to charge-modified nylon membranes.

Several other immobilization matrices have been proposed, but have received limited acceptance. Activated glass-fiber filters have been used to recover blotted peptides for microsequencing (Aebersold, Teplow, and Hood 1986). Silica-impregnated polyvinyl chloride microporous membranes have been marketed for blotting applications.

New Developments in Solid Phases for Blotting Applications

In the field of macromolecular blotting, the physical and chemical characteristics of the solid-phase substrate clearly play an important role. As this application becomes applied in new areas, the solid phase will continue to evolve. The availability of microporous membranes in a range of pore sizes may offer surface area and retention advantages in blotting of low molecular weight proteins. The process of blotting is still an art at the present stage of membrane development. Some of the present limitations might be circumvented in the future by use of composite membrane structures.

Cell Growth Studies on Microporous Membranes

This is a relatively new and exciting application involving the passive use of microporous mem-

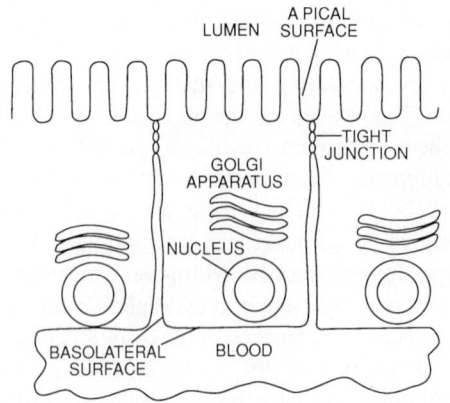

FIGURE 34–23. An epithelial cell schematic diagram depicting the apical membrane surface (reprinted from Pitt and Gabriels 1986 with permission.)

branes as substrates to which cells can attach and grow. Both physiologically and anatomically, microporous membranes are an ideal artificial surface for growing mammalian cells. This is particularly true for epithelial cells, which constitute more than 60% of all the recognized cell types in the human body. Epithelial cell layers typically line all body cavities and serve to separate two very different body compartments, which are all critical in sustaining life. *In vivo,* these cells display a distinctly polarized anatomy physiology, and generally exhibit some form of vectorial transport from one side to the other (Taub 1985). This is shown schematically in Figure 34–23 (Pitt and Gabriels 1986). The apical or luminal surface is separated by "tight junctions," which serve as an impermeable barrier to material on the opposite or basolateral side of the cellular sheet. *In vivo,* cells are surrounded by tissue fluids and extracellular matrix components that allow the cells to achieve and maintain their characteristic morphological and physiological functions.

Until recently, cells grown *in vitro* have almost always been grown on impermeable "tissue culture" (plasma-discharge) treated polystyrene. The solid, impermeable nature of the plastic forces the cells to attach, grow, obtain nutrients, discharge metabolic wastes, and be exposed to fluid on only one side. Fortunately,

most established tissue culture lines have been "immortalized" and have adapted well to this standard *in vitro* tissue culture technology. One frequently used and well-characterized cell line is the Madin Darby Canine Kidney (MDCK), which has retained the ability to express many of its original polarized cell functions (Handler 1983).

Growing these cells on a microporous membrane allows the cells to express differentiated properties such as measurable electrical resistance (Gumbiner and Simons 1986), polarized receptor localization (Mostov and Dietcher 1986; Fuller and Simons 1986), distinct columnar anatomical changes (Byers et al. 1986), and specific ion transport (Perkins and Handler 1981). Many of these newly expressed differentiation markers are not observed when these MDCK cells are grown on the traditional nonporous plastic. Growing cells *in vitro* on a porous polymeric membrane substrate allows the cells to achieve and express many markers typically only observed when the cells are grown *in vivo* (Byers et al. 1986; Pitt and Gabriel 1986).

Although this application of membranes seems relatively novel, it was first reported in 1953 by Grobstein (1953). Only after the commercialization of flat stack membranes, about 20 to 25 years later, did the number of literature references start to increase. The first commercial introduction of presterilized cell culture inserts occurred in 1985 or 32 years later [Millipore Corporation's Millicell®-HA cell culture inserts, which contain a surfactant-free, mixed cellulose ester (acetate and nitrate), 0.45-μm microporous membrane)]. This product represents a major convenience to cell biologists who are now able to use the technology to their benefit without first having to make and sterilize their own tissue culture inserts. The added convenience and continued recognition by the scientific community of the numerous advantages of growing cells on a microporous membrane is increasing, as evidenced by the rapidly increasing number of scientific papers on this topic.

The idealized properties of a membrane for

cell culture are given below. No one membrane is perfect for every application; therefore, a variety of polymeric membranes are currently used for cell culture. These include:

1. Biocompatibility; no extractables; inertness; nontoxic
2. High porosity
3. Wide range of pore sizes (0.2 to 30 μm)
4. Superior cell adhesion characteristics
5. No protein binding
6. No autofluorescence
7. Gamma sterilizable
8. Reproducible (on a micron scale)
9. Extremely thin for rapid diffusion
10. Microscopically transparent
11. Thin sections for both light and electron microscopy
12. Mechanically strong and stable
13. Easy to handle
14. Inexpensive to use
15. Mixed cellulose ester (MF) membranes
16. Hydrophilized PTFE membranes
17. Track-etched polycarbonate membranes.

The main advantages of MF or mixed cellulose ester membranes are a strong historical precedent, ready-to-use features, and biocompatibility. Their main disadvantages are that they are opaque and tend to bind proteins.

Hydrophilized PTFE microporous membranes (Biopore® Millicell™-CM, Millipore Corporation) have been developed that become microscopically transparent when wet. A key benefit of this membrane is that it allows growing cells to be observed microscopically in a nondestructive fashion on a porous substrate that allows the cells to differentiate (Pitt et al. 1987). In addition, the Biopore® membrane has negligible protein binding and background fluorescence, making it an ideal substrate for these analytical techniques.

The third polymeric substrate frequently used for cell culture is track-etched polycarbonate membranes (Nuclepore membranes, Nuclepore Corporation; Millicell™-PC and Millicell™-PCF, Millipore Corporation). The main advantage of these membranes is that they are extremely thin (about 10 μm), which results in rapid diffusion and equilibration times. Their main disadvantages are their low porosity and microscopic opacity.

Although no single, ideal membrane exists, the ability to form membranes with different characteristics from alternative polymeric materials offers the user a wide selection of substrates for various applications, as outlined in Table 34–21. The ultimate goals of these applications of microporous membranes are multifaceted, but all involve the development of *in vitro* cellular models that mimic *in vivo* states. Such improved models will help to advance the scientific progress made in a diverse number of areas from novel cancer therapies and gene insertion techniques to accepted *in vitro* toxicology assays. Many of the current and future directions of microporous membrane cell culture (generally improved in vitro models) are given below.

1. Toxicology
    Toxicity testing
    Animal testing alternatives
    Artificial tissue or organ equivalents
2. Cancer
    Metastatic inhibition
    Chemotherapy susceptibility testing
    Gene alterations
3. Hormone receptors
4. Pharmacology
5. Drug discovery and mechanisms
6. Cellular transport mechanisms
7. Inflammation
    Chemotaxis
8. Cellular aging
9. Immunology
10. Neurology
11. Cell/cell interactions
12. Mechanisms of virus infection and propagation (e.g., HIV, influenza, etc.).

ACKNOWLEDGMENTS

The authors acknowledge the assistance given by Millipore Corporation in allowing its staff to spend the time that was required to prepare this

TABLE 34-21. Millicell-PC and Millicell-PCF Selection Chart.

Millicell Unit	Pore Size	Materials	Applications	
Millicell-PC insert: nontissue culture treated	3.0 μm	Polycarbonate membrane in a polystyrene plastic holder	Coculture Suspension culture Cell motility Cell invasion Transport studies	Tumor cell metastasis Chemotaxis Cell/cell interactions Plant cells
Millicell-PC insert: nontissue culture treated	0.4 μm	See above	Transport studies Plant cells Suspension culture	Cell/cell interactions Coculture
Millicell-PCF insert: tissue culture treated	3.0 μm	See above	Enhanced differentiation Electrophysiology Virus infection In vitro toxicology Permeability Endothelial cell penetration	Polarized functions Transport studies Cell invasion Tumor cell metastasis Plant cells Chemotaxis
Millicell-PCF insert: tissue culture treated	0.4 μm	See above	Enhanced differentiation Electrophysiology Polarized functions In vitro toxicology	Transport studies Virus infection Permeability Plant cells
Millicell-HA	0.45 μm	Triton-free mixed esters of cellulose nitrate and acetate membrane	Enhanced differentiation Electrophysiology Polarized functions In vitro toxicology	Transport studies Virus infection Permeability Plant cells
Millicell-CM	0.4 μm	Biopore™ membrane (hydrophilized PTFE)	Enhanced differentiation Electrophysiology Polarized functions In vitro toxicology Requires ECM coating Microscopically transparent	Transport studies Virus infection Permeability Plant cells

chapter. Special thanks are due to Ms. Katherine Joy for her invaluable assistance in the preparation of this manuscript.

NOTATION

General Notation

See the General Notation section at the beginning of this handbook.

Special Notation

A_{fd}	area used in flow decay tests, L^2
A_{of}	area for optimal flow rate, L^2
A_{ot}	area for optimal throughput, L^2
b_1	empirical constant, Eq. (34–8), L^{-3}
b_2	empirical constant, Eq. (34–8), t/L^3
b_3	empirical constant, Eq. (34–4), t/L^6
b_4	empirical constant, Eq. (34–4), t/L^3
b_5	empirical constant, Eq. (34–15), L^3
b_6	empirical constant, Eq. (34–15), t^{-1}
d	pore diameter, L
D	diffusivity of gas through a liquid, L^2/t
K_t	tortuosity factor, dimensionless
l	membrane thickness, L
M_l	molecular weight of liquid, M/mol
N_p	number of pores
p_a	atmospheric pressure, M/Lt^2 or p
p_{dif}	differential pressure, M/Lt^2 or p

p_{max} maximum system differential pressure, M/Lt^2 or p
p_{BP} bubble point pressure, M/Lt^2 or p
p_t upstream test pressure (absolute), M/Lt^2 or p
P permeability = filtration rate/(area · pressure difference), L^2t/M or L/tp
Q volumetric gas flow at standard atmospheric pressure, L^3/t
Q_{min} minimum acceptable flow rate, L^3/t
$Q_{80\%}$ flow rate at 80% decay, L^3/t
r radius of particle, L
r_0 pore radius at $t = 0$, L
t_{op} operation time for batch filtration, t
V_b volume of batch, L^3
V_p volume to be processed continuously, L^3
$V_{80\%}$ throughput at 80% flow decay, L^3

Greek Letters

γ surface tension of liquid, M/t^2 or E/L^2
ϵ membrane porosity, dimensionless
η viscosity, M/Lt
θ contact angle between liquid and membrane surface
ρ_l density of liquid, M/L^3

REFERENCES

Accomazzo, M. A., G. Ganzi, and R. Kaiser. 1988. Deionized (DI) water filtration technology. In *Handbook of Contamination Control in Microelectronics*, ed. D. L. Tolliver, pp. 210–256. New York: Noyes Publications.

Accomazzo, M. A., and D. L. Grant. 1986. Mechanisms and devices for filtration of critical process gases. In *Fluid Filtration: Gas,* ed. R. R. Raber, Vol. I. ASTM STP 975, pp. 402–420. Philadelphia, PA: American Society for Testing and Materials.

Accomazzo, M. A., and D. L. Grant. 1989. Particle retention and downstream cleanliness of point-of-use filters for semiconductor process gases. In *Particles in Gases and Liquids,* ed. K. L. Mittal, Vol. 1, pp. 223–233. New York: Plenum Press.

Adams, D. G., and D. Agarwal. 1989. Clean-in-place system design. *Biopharm.* 2(6):48–57.

Aebersold, R., D. Teplow, and L. Hood. 1986. Electroblotting onto activated glass: high efficiency preparation of proteins from analytical SDS-polyacrylamide gels for direct sequence analysis. *J. Biol. Chem.* 261:4, 229–224, 238.

Aicholtz, P., R. Wilkins, and R. Gabler. 1987. Sterile filtration under conditions of high pressure and bacterial challenge levels. *J. Parenteral Sci. Technol.* 41(4):117–120.

Alwine, J. C., D. J. Kemp, and B. A. Parker. 1979. Detection of specific RNAs or specific fragments of DNA by fractionation in gels and transfer to diazobenzyl-oxymethyl paper. *Meth. Enzymol.* 68:220–242.

American Public Health Association. 1955. *Standard Methods for the Examination of Water and Waste Water,* 10th ed. Washington, D.C.

Anonymous. 1963. Microfiltration reduces beer spoilage. *Chem. Eng. News* 41(December):48.

Balazs, M. K., and S. J. Poirier. 1984. Pure water specifications: clearing up the confusion. In Proc. 3rd. Semiconductor Pure Water Conference, 12–13 January 1984 in San Jose, CA. pp. 153–162.

Batteiger, B., W. J. V. Newhall, and R. B. Jones. 1982. The use of Tween 20 as blocking agent in the immunological detection of proteins transferred to nitrocellulose membranes. *J. Immunol. Meth.* 55:297–307.

Beissinger, R., and E. Leonard, 1981. Sorption kinetics of binary protein solutions: general approach to multicomponent systems. *J. Colloid Interface Sci.* 85(2):521–533.

Berman, D., T. Meyers, and S. Chrai. 1986. Factors involved in cycle development of a steam-in-place system. *J. Parenteral Sci. Technol.* 40(4):119–121.

Blazka, S., and R. Hegde. 1989. Effect of microfiltration on photoresist quality and integrity. *Microelec. Manufac. Testing* 12(7):33–35.

Brenner, M. W., and H. Iffland. 1966. Economics of microbiological stabilization of beer. *MBAA Technical Quarterly* 3(3):193.

Bush, J. W. 1964. Beer stability and controlled filtration. Brewers Digest, February.

Byers, S. W., M. A. Hadley, M. Dym, and D. Djakiew. 1986. Growth and characterization of polarized monolayers of epidimyal cells in transfilter metanephric culture. *Am. J. Pathol.* 116:289–296.

Carleton, F. J., and J. P. Agalloco. Eds. 1986. *Validation of Aseptic Pharmaceutical Processes.* New York: Marcel Dekker.

Church, G. M., and W. Gilbert. 1980. Genomic sequencing. *Proc. Natl. Acad. Sci. USA* 81:1991–1995.

Cross, R. S., and A. J. Lucas. 1987. Advances in point-of-use photochemical dispense and filtration. *Microelec. Manufac. Testing* 10(1):5–7.

Dwyer, J. L. 1968. The technology of absolute microfiltration. *MBAA Technical Quarterly* 5(4):243.

Emory, S. 1989a. Principles of integrity testing hydrophilic microporous membranes. *Pharm. Technol.* 13(9):68–77.

Emory, S. 1989b. Principles of integrity testing hydrophilic microporous membranes, part II. *Pharm. Technol.* 13(10):36–46.

Faylor, T. L., and J. J. Gorski. 1988. Ultra high purity water—new frontiers. In *Handbook of Contamination Control in Microelectronics,* ed. D. L. Tolliver, pp. 185–209. New York: Noyes Publications.

Fuller, S. D., and K. Simons. 1986. Transferrin receptor polarity and recycling accuracy in "tight" and "leaky" strains of madin-darby canine kidney cells. *J. Cell Biol.* 103:1767–1779.

Gershoni, J. M., F. E. Davis, and G. E. Palade. 1985. Protein blotting in uniform or gradient electric fields. *Anal. Biochem.* 144:32–40.

Goldsmith, S. H., and G. P. Grundelman. 1985. Validation of pharmaceutical filtration products. *Pharm. Manufac.* 2(11):31–37.

Grant, D. C. 1989. Improved methodology for measurement of particle concentrations in semiconductor process chemicals. In *Particles in Gases and Liquids,* ed. K. L. Mittal, Vol. 1, pp. 121–134. New York: Plenum Press.

Grant, D. C., S. L. Peacock, and M. A. Accomazzo. 1987. Shedding characteristics of filters in liquids. *Microcontam.* 5(7):30–36. (Canon Communications, Inc.)

Grant, D. C., and W. R. Schmidt. 1989. Particle performance of a central chemical delivery system. Paper read at 7th Annual Microelectronics Technical Symposium, Bedford, MA, Millipore Corporation.

Green, B. L., E. Clausen, and W. Litsky. 1975. Comparison of the new Millipore HC with conventional membrane filters for the enumeration of fecal coliform bacteria. *Appl. Microbiol.* 30(4):697–699.

Grobstein, C. 1953. Morphogenic interaction between embryonic mouse tissues separated by a membrane filter. *Nature* 172:860–872.

Gumbiner, B., and K. Simons. 1986. A functional assay for proteins involved in establishing an epithelial occluding barrier: identification of a uvomorulin-like polypeptide. *J. Cell Biol.* 102:457–468.

Haffenreffer, A. H. 1962. A practical approach to sterile filtration. In *Technical Proc. Master Brewers Association of America.*

Haffenreffer, A. H. 1968. A review of current methods for stabilizing packaged malt beverages. *Contam. Control* 7(11):14.

Haffenreffer, A. H., J. W. Dwyer, C. Fifield, and J. Hall. 1968. A study of the colloidal material affecting the filtration of beer. *MBAA Technical Quarterly* 5(3):171.

Handler, J. S. 1983. Use of cultured epithelia to study transport and its regulation. *J. Exp. Biol.* 106:55–69.

Health Industry Manufacturers Association. 1982. Microbiological evaluation of filters for sterilizing liquids. HIMA Document 3, Vol. 4, April.

Honold, E., and E. L. Skau. 1954. Application of mercury intrusion method for determination of pore-size distribution of membrane filters. *Science* 120:805–806.

James, J., P. D. Blanden, V. Krygier, and G. Howard, Jr. 1988. Bacterial and endotoxin retention by sterilizing grade filters during long term use in pharmaceutical high purity water systems. Scientific and Technical Report PUF03 (February). Glen Cove, NY: Pall Corporation.

Jansen, G. 1990. Assistant Quality Assurance Manager, Pabst Brewing. Personal communication.

Jesukawicz, J. J. 1967. Processing conditions affecting the efficiency of Millipore filter systems. *MBAA Technical Quarterly* 4(4):257.

Kendall, M., and W. C. Hymer. 1988. Cell blotting: a new approach to quantify hormone secretion from individual rat pituitary cells. *Endocrinology* 121:2260–2262.

Kovary, S. J., J. P. Agalloco, and B. M. Gordon. 1983. Validation of the steam-in-place sterilization of disc filter housings and membranes. *J. Parenteral Sci. Technol.* 37(2):55–64.

Kyhse-Anderson, J. 1984. Electroblotting of multiple gels: a simple apparatus without buffer tank for rapid transfers of proteins from polyacrylamide to nitrocellulose. *J. Biochem. Biophys. Meth.* 10:203–209.

Larrabee, G. 1990. Understanding the challenge of future device technologies. Paper read at 8th Annual Microelectronics Technical Symposium, San Jose, CA, Millipore Corporation, Bedford, MA.

LeChevalier, M. W., S. C. Cameron, and G. A. McFeters. 1983. Comparison of verification procedures for the membrane filter total coliform technique. *Appl. Environ. Microbiol.* 45(3):1126–1128.

Lee, J. Y. 1989. Auditing an aseptic filtration process. *Pharm. Technol.* 13(2):66–72.

Levy, R. V., and T. J. Leahy. 1990. Sterilization filtration. In *Disinfection Sterilization and Pres-*

ervation, ed. S. S. Block (in press). Philadelphia, PA: Lea and Febiger.

Lin, S. D. 1977. Comparison of membranes for fecal coliform recovery in chlorinated effluents. *J. WPCF* 68:2255–2264.

Mann, V., V. S. Venkatraj, and A. D. Auerbach. 1989. Two hour DNA hybridizations using a new transfer membrane. *Nucleic Acid Res.* 17:5, 410.

Matsudaira, P. 1987. Sequence from picomole quantities of protein electroblotted onto polyvinylidene difluoride membranes. *J. Biol. Chem.* 262:10,035–10,038.

McLellan, T., and J. A. M. Ramshaw. 1981. Serial electrophoretic transfers: a technique for the identification of numerous enzymes from single polyacrylamide gels. *Biochem. Genet.* 19:648–654.

Meltzer, T. H. 1987. *Filtration in the Pharmaceutical Industry*. New York: Marcel Dekker.

Meltzer, T. H. 1989a. Filtration: the practice of prefiltration and its related considerations. *Ultrapure Water* 6(3):24–34.

Meltzer, T. H. 1989b. Filtration: a critical review of filter integrity testing: part I. the bubble point method, assessing filter compatibility, initial and final testing. *Ultrapure Water* 6(4):40–51.

Meltzer, T. H. 1989c. Filtration: a critical review of filter integrity testing: part II. the diffusive air flow and pressure-hold methods, assessing filter compatibility, initial and final testing. *Ultrapure Water* 6(5):44–56.

Meyers, T., and S. Chrai. 1981. Design considerations for development of steam-in-place sterilization processes. *J. Parenteral Sci. Technol.* 35(1):8–12.

Meyers, T., and S. Chrai. 1982. Steam-in-place sterilization of cartridge filters in-line with a receiving tank. *J. Parenteral Sci. Technol.* 36(3):108–112.

Michaels, S. L. 1989. Crossflow microfilters: the ins and outs. *Chem. Eng.* 96(1):84–91.

Millipore Corporation. 1967. Techniques for microbiological analysis. ADM 40. Bedford, MA.

Millipore Corporation. 1981. Detection of bacteria, yeast and mold in liquid food products using the Millipore black membrane filter. Technical Service Brief TS064. Bedford, MA.

Mostov, K., and D. L. Deitcher. 1986. Polymeric immunoglobulin receptor expressed in MDCK cell transcytoses IGA. *Cell* 46:613–621.

Peacock, S. L., M. A. Accomazzo, D. L. Grant, and P. M. Meier. 1985. A comparison of particle shedding from different chemical filtration products. In Proc. 1st Annual Microcontamination Conference 20–22 November 1985 in San Jose, CA.

Peferoen, M., R. Huybrechts, and A. DeLoof. 1982. Vacuum blotting: a new simple and efficient transfer of proteins from sodium dodecyl sulfate-polyacrylamide gels to nitrocellulose. *FEBS Lett.* 145:369–372.

Perkins, F. M., and J. S. Handler. 1981. Transport properties of toad kidney epithelia in culture. *Am. J. Physiol.* 241:C154–C159.

Peterson, J. 1974. Comparison of MF technique and MPN technique for the estimation of coliforms in water. *Public Health Lab.* 32(6):182–193.

Pitt, A. 1987. The nonspecific binding of polymeric microporous membranes. *J. Parenteral Sci. Technol.* 41(3):110–113.

Pitt, A., and J. Gabriels. 1986. Epithelial cell culture on microporous membranes. *Am. Biotechnol. Lab.* (September/October) 4(5):38. (International Scientific Communications, Inc.)

Pitt, A., J. Gabriels, F. Badmington, J. McDowell, L. Gonzales, and M. E. Waugh. 1987. Cell culture on a microscopically transparent microporous membrane. *Biotechniques* 5(2):162–171.

Pluskal, M. G., M. B. Przekop, M. R. Kavonian, C. Vecoli, and D. A. Hicks. 1986. Immobilon™ PVDF transfer membrane: a new membrane substrate for Western blotting of proteins. *Biotechniques* 4:272–283.

Reid, G., A. Hwang, R. H. Meisel, and E. R. Allcoch. 1990. The sterile filtration and packaging of beer into PET containers. *ASBC J.* 48(3):85.

Reiss, B., R. Sprengel, and H. Will. 1984. A new sensitive method for qualitative and quantitative assay of neomycin phosphotransferase in crude cell extracts. *Gene* 30:211–218.

Schmidt, W. R., C. Becker, J. Mehta, R. Novak, D. C. Grant, K. A. Foster, and C. P. Myhaver. 1987. The effect of point-of-use chemical filtration in a centrifugal spray system on wafer particle counts. Paper read at 5th Annual Microelectronics Technical Symposium, Zurich, Switzerland, Millipore Corporation, Bedford, MA.

Sladek, K. J., R. V. Suslavich, B. I. Sohn, and F. W. Dawson. 1975. Optimum membrane structures for growth of fecal coliform organisms. *Appl. Microbiol.* 30:685.

Southern, E. 1975. Detection of specific sequences among DNA fragments separated by gel electrophoresis. *J. Mol. Biol.* 98:503–517.

Taub, M. 1985. Ed. *Tissue Culture of Epithelial Cells*. New York: Plenum Press.

Thomas, P. S. 1980. Hybridization of denatured RNA and small DNA fragments transferred to nitrocellulose. *Proc. Natl. Acad. Sci. USA* 77:5201.

Towbin, H., T. Staehelin, and J. Gordon. 1979. Electrophoretic transfer of proteins from polyacrylamide gels to nitrocellulose sheets. *Proc. Natl. Acad. Sci. USA* 76:4350–4354.

Truskey, G. A., R. Gabler, A. DiLeo, and T. Manter. 1989. The effect of membrane filtration upon protein conformation. *J. Parenteral Sci. Technol.* 41(6):180–193.

United States Pharmacopeia. 1990. United States Pharmacopeia XXII and National Formulary NFXVII. Rockville, MD: United States Pharmacopeial Convention, Inc.

35
Crossflow Microfiltration: Applications, Design, and Cost

Leon Mir, Stephen L. Michaels, and Vinay Goel
Millipore Corporation

Robert Kaiser
ARGOS Associates, Inc.

INTRODUCTION
    Process Definition
    Crossflow Filtration Modules
    General System Design Considerations
    Economics of Crossflow Filtration
    Application Window
ANTIBIOTICS CLARIFICATION
    Process Description
    Comparison of Solids Removal Processes
    Process Design
    System Design
    Economic Analysis
MICROFILTRATION OF MAMMALIAN CELLS
    Process Description
    Comparison of Cell Removal Processes
    Process Design
    System Design
    General Design Standards
    Economic Analysis
ACKNOWLEDGMENTS
NOTATION
REFERENCES

INTRODUCTION

Process Definition

Compared to deadend filtration, in crossflow filtration (also called tangential flow or inertial filtration) pressure drives only part of the feed through the medium; the remaining feed flows tangentially along the surface of the medium, continuously sweeping particles from the medium's surface back into the feed. Generally, crossflow filters are operated as surface filters and have pores that are smaller than the particles to be removed.

The principles of crossflow filtration (see Chapter 33) have been well established in the related technologies of reverse osmosis and ultrafiltration, which are concerned with the removal of soluble compounds from solutions. By using a microfiltration membrane as the separation medium, particles in the 0.1- to 10-μm range can be removed by crossflow filtration.

Crossflow filtration is inherently a more complex operation than deadend filtration. In the latter case, one is concerned with only two process streams of essentially equal flow rate: the feed and the filtrate. With crossflow filtration, in addition to the feed and the permeate (i.e., the filtrate), there is also the retentate, the nonfiltered effluent laden with suspended material. The relative flow rates of permeate and retentate are established by controlling the back pressure on the retentate to establish the pressure drop across the filtration membrane. To maintain a high liquid velocity parallel to the filtration surface, and thus prevent retained components from accumulating on the surface, the flow rate of process feed is significantly greater than that of the permeate. As a consequence, the retentate must usually be recirculated. The material balances for these types of systems with recirculation of retentate are treated in Eqs. (33–1) through (33–12). A schematic diagram of a typical crossflow filtration system is presented in Figure 35–1. In addition to the filtration module(s), the additional basic elements of a crossflow filtration system include a feed tank, a recirculation pump, associated piping, and control/monitors of pressure, temperature, and flow rates.

Crossflow Filtration Modules

Crossflow filtration modules are available in a wide range of materials and geometries, as indicated in Table 35–1. Polymeric microfilters can be configured in a variety of geometries (Figure 35–2). The most common configurations of crossflow modules are the plate-and-frame, spiral, hollow-fiber, and tubular types.

Flat-sheet membrane filters normally are assembled in plate-and-frame devices similar to filter presses, as indicated in Figure 35–3. The space between sheets ranges from 0.25 to 2.50 mm. Pleated sheet configurations are enclosed by cylindrical cartridges, in a manner similar to through-flow cartridge filters. The spacing between the sheets is again 0.25 to 2.50 mm. In spiral-wound units, feed enters the end of a spiral module and is contained in a 0.25- to 0.50-mm-wide space between the membranes. Concentrate exits at the opposite end of the spiral-wound cylinder, and permeate leaves the unit via a collection tube at its center axis.

Hollow fine-fiber units employ a multitude of fibers with an internal diameter of from 0.25 to 1.00 mm. The fibers are arranged in a shell-and-tube geometry. Feed enters the tubes at one end of the unit, is concentrated, and exits at the other end. Permeate is removed from the shell side. Tubular filters are placed in shell-and-tube configurations similar to hollow-fiber devices, but the tube diameters are larger, generally 2.5 to 25 mm.

Polymeric crossflow filtration modules are suitable for many bioengineering applications. However, if high-temperature capability and/or resistance to corrosive chemicals are required, modules made from inorganic materials, especially ceramics, are better suited because of their superior environmental resistance. While ceramic microfilters are available in flat-sheet or tubular configurations, multichannel monoliths, such as the Ceraflo™ module shown in Figure 35–4, are best suited for crossflow filtra-

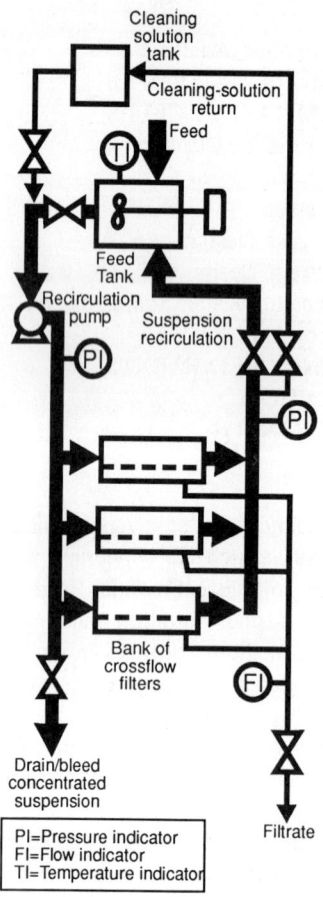

FIGURE 35–1. Typical crossflow operation including recirculation loop (reprinted from Michaels 1989 with permission).

TABLE 35-1. Partial List of Commercial Crossflow Microfilter Media (reprinted from Michaels 1989 with permission).

Material	Geometries					
	Pleated sheet	Tubular	Spiral wound	Tubular MC*	Hollow fiber	Flat sheet
1. *Polymers*						
Cellulosics	■				■	■
Polysulfone		■	■		■	■
Polyvinylidenefluoride		■	■			
Acrylic		■			■	
Polytetrafluoroethylene						■
Polybenzimadazole			■			
Polypropylene					■	■
Nylon	■				■	■
2. *Ceramics*						
Alumina				■		
Zirconia/alumina				■		
Zirconia/sintered metal		■				
Zirconia/carbon		■				
Silica		■				
Silicon carbide				■		
3 *Sintered metal*						
Type 316 stainless steel		■				
Other alloys		■				

*MC = multichannel monolithic elements.

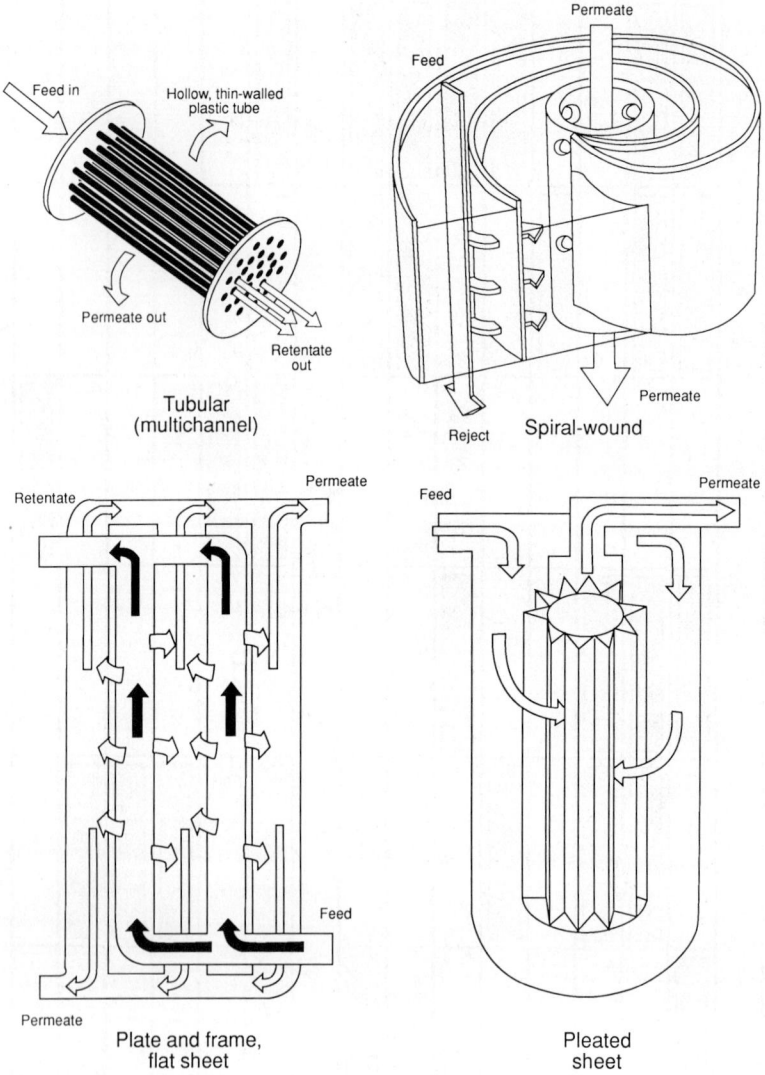

FIGURE 35–2. Microfilters can be configured in a variety of geometries (reprinted from Michaels 1989 with permission).

tion because of their open channel configuration. Each Ceraflo™ ceramic filter element contains nineteen 3-mm-i.d. coaxial flow channels (lumens) through which feed is recirculated. Under pressure, filtrate is pushed through the microporous walls of each element and fills the shell of the housing surrounding the element from which it can be removed. The concentrated suspension or solution is removed from the opposite end of each element. Ceraflo™ filter elements are made from alpha-alumina, with surface pore sizes of 0.2, 0.45, and 1.0 μm. Depending on the composition of the seals in the system, they can operate at pressures up to 18 bars and temperatures up to 200°C.

Other than particle retention, the key characteristic of a filter module is its permeability, the filtration rate per unit filtration area and unit pressure differential. The permeability depends on the construction of the module. For a liquid of given viscosity, permeability is in-

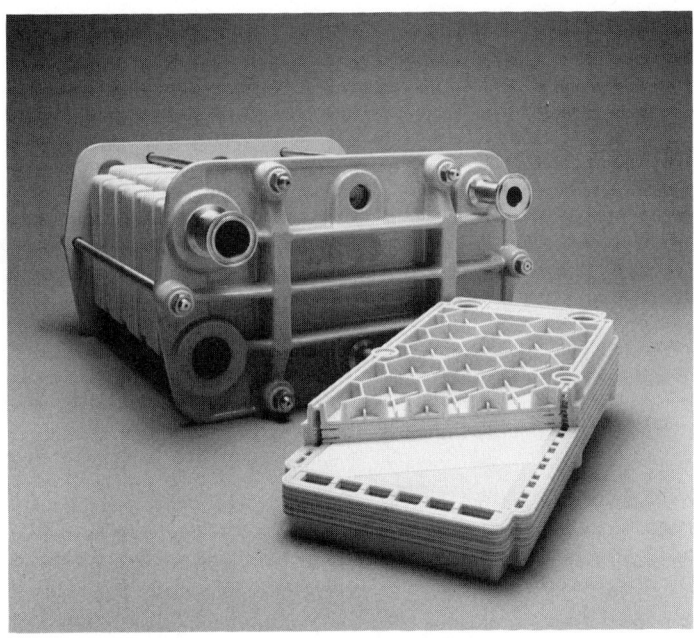

FIGURE 35–3. The Prostak™ system; an example of a plate-and-frame crossflow filtration assembly (reprinted from Millipore Corporation with permission).

fluenced by the pore size, pore size distribution, porosity, and depth of the medium. The water permeabilities of typical commercial filter media are given in Table 35–2. Since pore size is usually set by process requirements, improvements in permeability are obtained by altering filter porosity and average pore length. Asymmetric filters, which are made of a thin surface layer of a rated pore size that is supported by a thick layer that has a much larger pore size, may have a greater permeability than symmetric media of comparable pore size because the pore length (thickness of the membrane layer) is shorter than that of symmetric filters. However, the porosity of this surface layer is usually not as high as that of a symmetric filter.

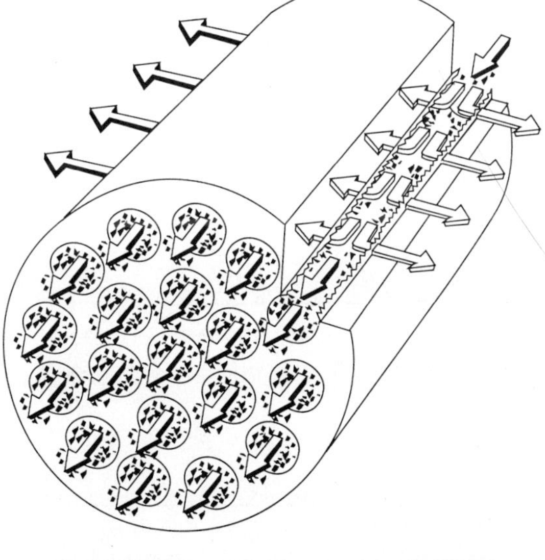

The most porous media are the symmetric polymers, such as the polyolefins, fluoropolymers, and nylons, which may have porosity as high as 85%. The porosities of ceramic media, which are asymmetric, range up to a maximum of 30% to 40% in the thin membrane layer. The porosities of sintered metals, which are symmetric in structure, are lower, of the order of 25% to 30%. The porosities of the thin membrane layers of asymmetric polymer filters are still lower, usually less than 10%.

General System Design Considerations

While crossflow microfiltration systems can be used to accomplish a variety of different separations tasks, these systems are principally tools

FIGURE 35–4. Ceraflo™ ceramic filter element (reprinted from Millipore Corporation with permission).

TABLE 35–2. Permeabilities of Representative Commercially Available Crossflow Filtration Membranes.

Supplier	Commercial Designation	Material	Mean Pore Size (μm)	Membrane Porosity (%)	Permeability ($m^3/m^2/bar/h$)
Millipore	Durapore™	PVDF	0.22	65	7.0
Millipore	Durapore™	PVDF	0.45	65	17.1
Millipore	Ceraflo™	Alumina	0.2	35	4.0
Millipore	Ceraflo™	Alumina	0.45	35	5.5
Alcoa	Membralox™	Alumina	0.2	35	3.0

for separating suspended solids from a liquid. Crossflow microfiltration is a complex operation that depends on the interplay of many variables. Filtration procedures are affected by a host of application-specific parameters, including but not limited to feed composition and temperature, physical properties such as viscosity and surface tension, the volume of the feed to be treated, particle size distribution of the feed, compressibility of the solids, and the final desired concentration of solids in the retentate. Design parameters include the characteristics of the filter modules, filtration area, filtration pressure (transmembrane pressure), the linear velocity of the feed across the membrane, and the retentate recirculation rate, among others.

The most critical and expensive components of a crossflow microfiltration system are the microfiltration modules, which accomplish the separation, and the pump, which provides fluid circulation and the driving force for filtration. Optimal permeate *flux,* the filtration rate obtained per unit filter area, is obtained by balancing circulation and filtration costs. The higher the flux, the smaller the required filter area. However, for a given system, a higher flux is obtained by increasing the transmembrane pressure and fluid circulation rate, which, in turn, requires a larger pump and driver.

Under constant pumping conditions, flux can be viewed as a useful measure of the productivity of a crossflow microfiltration system. As indicated by Shiloach, Martin, and Moes (1988), the batch-concentration operation of a crossflow filtration system can be characterized by a three-step decay in flux with time: an initial polarization phase, a period of essentially constant flux, and a period of decreasing flux, as indicated in Figure 35–5. The modeling of the constant flux phase is treated exhaustively in Chapter 33. The initial polarization phase is also treated in Chapter 33. Note, however, that the models presented are not of sufficient accuracy to be used for *ab initio* design.

The initial polarization phase is inherent in the technology and cannot be avoided. During the initial stages of separation, particles concentrate at the surface while liquid is filtered through the membrane. These accumulated particles form an additional resistance to flow, which reduces the effective permeability of the filter to a value lower than its initial value.

The applied shear rate to the slurry in the flow channel of the microfilter controls the layer of retained solids; the higher the rate,

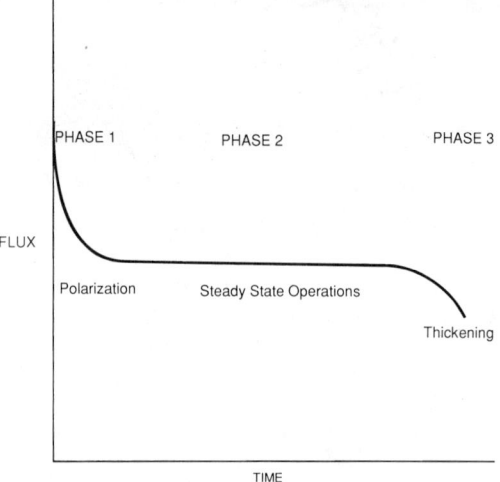

FIGURE 35–5. Flux regimes as a function of time in crossflow filtration.

the thinner the layer. This shear rate is a function of the geometry of the flow channel. The best geometries are those that produce relatively high shear rates at low feed flow rates (flux is kept high while circulation rate is kept low). Narrow channels generate higher velocities and greater shear rates. Thus, for streams with suspended solids that do not plug thin channels, thin (narrow)-channel flat-plate, spiral-wound, or hollow-fiber microfilters are preferred. When plugging presents a potential problem, wider channel flat-plate or larger diameter tubular microfilters are needed.

Because of their greater rigidity, ceramic microfilters can accommodate fluxes two to three times those of symmetric polymer and symmetric sintered-metal media, and fluxes five to ten times those of asymmetric polymerics.

The decay in flux in the final stage is also attributable to a noticeable increase in the viscosity of the fluid being filtered as a result of the increasing concentration of suspended particles. See Eqs. (33–32) and (33–33) for the theoretical variation of flux with concentration of suspended solids. This limits the concentration of particles that can be produced by crossflow filtration.

Flux and the maximum suspended solids concentration that can be attained are also influenced by the mechanical properties of the retained solids. Obviously, nonplugging, uniformly sized solids will result in less flux reduction than would gelatinous, nonuniform particles. Also, the maximum suspended solids content that can be achieved is less if the particles are gelatinous, of the order of 100 g/L, compared to rigid particle suspensions where concentrations of the order of 700 g/L can be attained.

Adsorption of solids, or of components dissolved in the liquid, on the membrane surface or in the pores will result in a reduction in filter permeability. Preliminary experimental tests are usually performed to determine if unwanted adsorption occurs. If observed, adsorption is usually controlled by altering the composition of the feed stream, pretreating the medium, or selecting a nonadsorbing filter medium.

TABLE 35–3. Types of Solutions Used in the Cleaning of Crossflow Filtration Membranes (reprinted from Michaels 1989 with permission).

Type of Solution	What It Removes
Acidic	Inorganic scale
Alkaline	Proteins and other biological deposits
Oxidizing	Oxidizable organic deposits
Detergent	Insoluble particulates, colloids, and emulsions
Organic solvent	Organic-chemical deposits that are insoluble in aqueous solutions

Since the viscosity of a liquid decreases with temperature, increasing the filtration temperature can result in a significant increase in flux. Temperature can also influence membrane fouling by changing the adsorptive properties of the foulants present in a liquid stream.

All microfilters periodically need **cleaning** because of the formation of deposits on the filter surface and the inevitable plugging of pores by depth filtration of trace amounts of particles smaller than the design pore size. Typical cleaning problems and solutions are summarized in Table 35–3. Unless the filter can be cleaned by one of these or another solution, it will become useless after some time.

Generally, the greater the corrosion resistance of the microfilter, the easier it is to find suitable cleaning solutions for it. Particular advantages of ceramic membrane modules over polymeric modules in bioprocessing are their resistance to corrosive cleaning and sterilization chemicals, and their ability to be steam sterilized in line and autoclaved repeatedly.

Another advantage of ceramic modules over those made of other materials is a significantly longer service life. With time, the cleaning agents used will eventually degrade polymeric microfilters, so that they have to be replaced periodically. For most polymer filters, typical lifetimes vary from about 1 year or less for hydrophilic materials, 1 to 2 years for hydrophobic ones, and up to 4 years for fluoropolymers. Ceramic filters, because they are stron-

TABLE 35-4. Costs of Crossflow Microfiltration Systems (reprinted from Michaels 1989 with permission).

Material	Cost of Medium ($/ft^2)	Cost of Installed System ($/ft^2)	Relative Area Required	
			Low-fouling	High-fouling
Polymers	6–90	160–650	8–10	2–4
Ceramics	80–300	230–700	1–2	1–2
Metals	60–120	220–900	3–6	1–3

ger, and thus less subject to flow and creep than plastics, and because they can also be cleaned more aggressively, can remain in service for up to 10 years.

Economics of Crossflow Filtration

Both the *installed costs* and the *life-cycle costs* of a crossflow filtration system will depend on the type of filtration medium used.

The installed costs of a crossflow filtration system will depend primarily on the type of medium used and on its surface area. Typical unit costs ($/ft^2) are presented for various crossflow microfiltration systems in Table 35–4. As can be noted from this table, the cost of the filter media represents only a small fraction (15% to 25%) of the total installed cost of a system.

The installed cost of a specific system is obtained by multiplying the cost per unit area by the required filtration area. The required area is a function of the volume of liquid to be filtered, the time available for filtration, and the design value of the membrane flux. Because of differences in filter characteristics, significantly different operating conditions may be recommended to optimize flux for different filter systems. Because of differences in achievable flux, the required surface area for a given application will be a function of the specific type of filter medium used. In general, higher fluxes can be achieved with ceramic filters than with filters made of alternative materials. Relative areas for low-fouling and high-fouling liquids are also presented in Table 35–4.

These areal comparisons take into account the fact that the cost of the filter medium represents only a relatively small fraction of the total installed cost of a system, and that careful consideration needs to be given to those factors that influence the cost of nonfilter hardware, such as pumps. The lowest cost system design is usually not based on minimum membrane costs per unit area, but on obtaining the best balance of system components that results in the lowest capital cost per unit volume of liquid to be treated.

Life-cycle operating costs, which may be a truer economic measure than installed costs, take into account direct operating costs as well as capital costs annualized over the projected life of the system. These direct costs include the costs of product losses (i.e., yield), pumping the solution, cleaning and replacing membrane modules (which can vary significantly with module type), and maintaining the system. The optimal economic choice thus also depends on the duration of the intended project and will be very different if a short-term campaign is planned compared to a long-lived bread and butter production application.

Application Window

The separation of suspended matter from process liquids occurs frequently in bioengineering processes. As indicated in Figure 35–6, this problem can be addressed by a variety of physical separation methods other than crossflow filtration. While the optimal separation method for a particular situation depends primarily on the characteristic size of the suspended matter that is to be separated, as implied

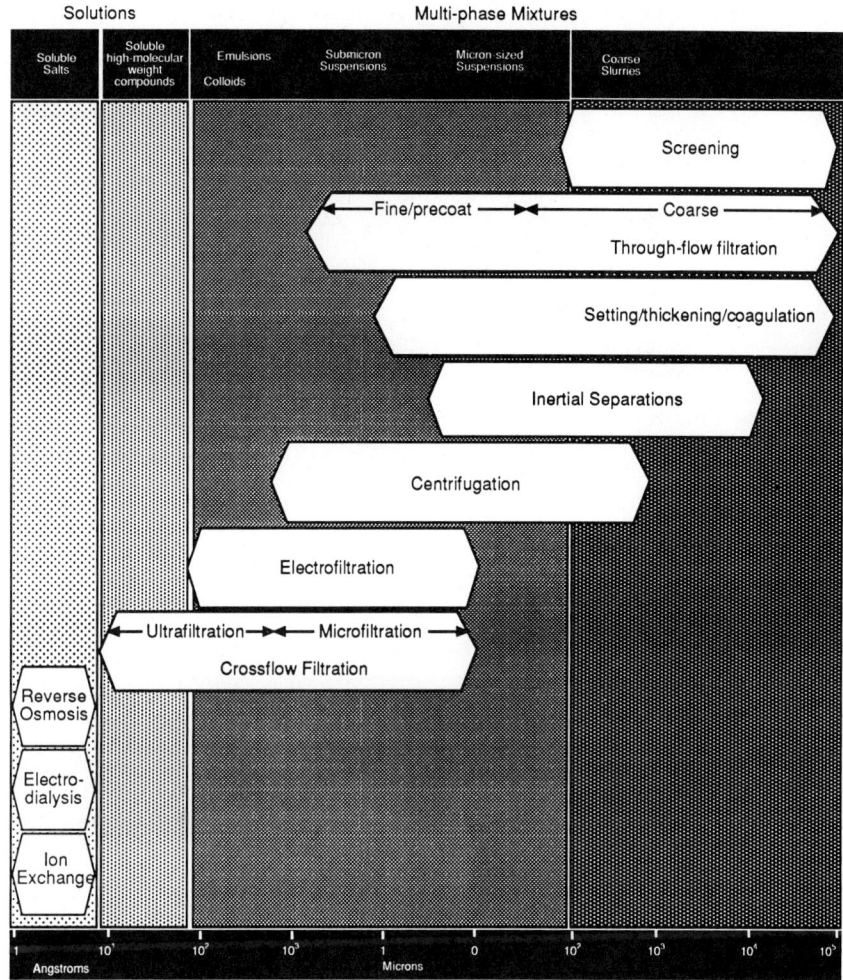

FIGURE 35–6. Crossflow filtration occupies a key niche in the separations spectrum (reprinted from Michaels 1989 with permission).

by Figure 35–6, other factors, such as the physical characteristics and concentration of the suspended matter, the degree of separation required, achievable product yield, and the need for maintaining sterile conditions, also need to be taken into consideration.

As indicated in Figure 35–6, a number of separation processes are capable of removing particles in the same size range (0.1 to 10 μm) as crossflow filtration. The major competing processes of industrial interest (Miller et al. 1984; Belter, Cussler, and Hu 1988) are centrifugation (Brunner and Hemfort 1988), precoat filtration with diatomaceous earth, and deadend filtration. The relative merits of these competing technologies are summarized in Table 35–5.

The major factors that need to be considered when selecting a separation process are:

1. The objective of the separation
2. Characteristics of the process fluid
3. System economics.

The separation objective is the most important factor in process selection. Table 35–6 lists some common separation objectives as well as the ability of the competing separation methods of interest to achieve these objectives.

TABLE 35–5. Relative Merits of Competing Technologies of Current Interest.

Technology	Advantages	Disadvantages
Centrifugation	Can handle high solids No disposables Closed system Proven technology Economical Many vendors.	High maintenance due to high-speed moving parts Capacity sensitive to particle size and density Incomplete separation.
Diatomaceous earth filtration	Can handle high solids Tailorable by DE grade Not sensitive to particle density Proven technology Economical Many vendors.	Open to the environment DE is a lung carcinogen Capacity is sensitive to feed properties Cannot handle small particles Plugging is sensitive to variations in feed properties Purchase and disposal of DE can result in high operating costs.
Deadend filtration	Very low capital cost Wide range of filter products available for the removal of a wide range of particles Well suited to dilute (low solids content) solutions Closed system.	Particle removal capacity and operating costs are both sensitive to solids concentration in feed.
Crossflow filtration	With suitably designed process and equipment: Insensitive to size, density, and concentration of suspended particles in feed High product yields Clear filtrate with wide range of filter media Closed system No disposables Limited maintenance Low operating costs.	High capital cost Cannot produce a "cake" Limited to solids contents of up to 50 vol% High yields may require dilution of feed (diafiltration) Requires periodic cleaning of filter medium.

TABLE 35–6. Capability of Crossflow Filtration and Competing Technologies to Meet Representative Separation Objectives.

	Separation Technology Performance			
Objective	Crossflow Filtration	Deadend Filtration	Diatomaceous Earth Filtration	Centrifugation
High yield of dissolved product	Superior	Superior	Acceptable	Acceptable
Solids purification	Superior	Acceptable	Poor or not feasible	Acceptable
Recovery of highly concentrated solids	Poor or not feasible	Acceptable	Poor or not feasible	Superior
Solids classification	Poor or not feasible	Acceptable	Poor or not feasible	Superior
Maintain sterile conditions	Superior	Superior	Poor or not feasible	Varies
Continuous culture/cell recycle	Superior	Poor or not feasible	Poor or not feasible	Acceptable
Minimize wastes	Superior	Acceptable	Poor or not feasible	Acceptable

TABLE 35–7. Representative Biotechnical Applications of Crossflow Microfiltration.

Liquid-Phase Clarification:
  Clarification of antibiotic fermentation broths
  Clarification of bacterial lysates
  Clarification of protein solutions
  Clarification of yeast lysates
Solid-Phase Recovery:
  Continuous culture fermentation and cell recycle
  Harvesting and washing of bacterial cells
  Harvesting and washing of mammalian cells
  Mycelia concentration
  Yeast concentration

A partial list of the growing number of biotechnical process applications of crossflow filtration is presented in Table 35–7. In addition, crossflow filtration has found extensive clinical use in applications, such as plasmapheresis, which have been recently reviewed in depth by Colton (1987). To illustrate further the versatility of crossflow filtration, two very different applications—the preparation of antibiotics and the harvesting of mammalian cells—are examined in more detail to illustrate the variety of factors that influences the choice and design of a crossflow filtration system. The preparation of antibiotics represents the introduction of a new technology into a mature and established large-scale (i.e., treatment of more than 100,000 L) process where crossflow filtration is being applied primarily to improve process yields and to lower costs. The harvesting of mammalian cells is becoming increasingly important in the production of therapeutic proteins. It is a small-scale operation that produces an extremely sensitive and valuable product, and a premium is placed on the high recoveries attainable with crossflow filtration.

ANTIBIOTICS CLARIFICATION

Process Description

Antibiotics are secondary metabolites produced by such organisms as *Penicillium chrysogenum* (Penicillin G or V) and *Cephalosporium acremonium* (Cephalosporin C). The recovery and purification of the antibiotics from the fermentation broths is a complex process (Vandamme 1984) involving several separation steps and is outlined in general terms in Figure 35–7. The objective of the filtration step is the separation of the microbial biomass and other particulates from the dissolved antibiotics. In some processes the filtration step is avoided by extraction of the antibiotic from the whole broth; these processes are not treated here. The removal of the primary biomass is clearly needed if the subsequent primary isolation steps are to function properly. However, residual colloids and dissolved polymeric contaminants in the filtrates, such as proteins and starches, can also interfere with subsequent isolation and purification steps by forming emulsions and interfacial sludges that reduce the efficiency of chromatographic and extractive processes, and thus reduce yields. The removal of these micro-

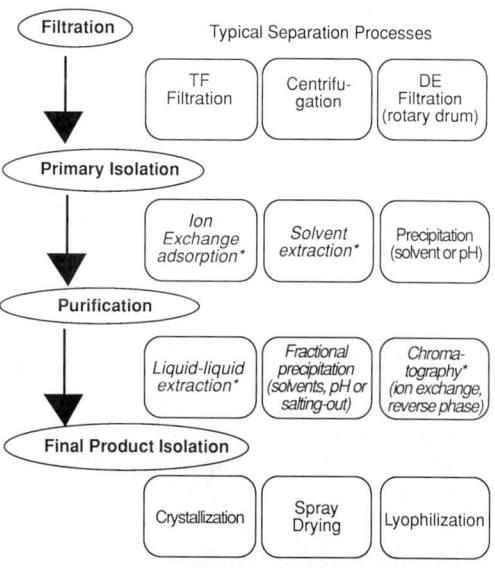

FIGURE 35–7. Downstream processing of fermentation products (reprinted from Millipore Corporation with permission).

contaminants influences the choice of filters and centrifuges.

The quantity and composition of the fermentor broths can vary widely. The range of typical values is:

Volume of broth: 5 to 300 m^3
Solids concentration: 40 to 100 kg/m^3
Antibiotic concentration: 1 to 30 kg/m^3.

Comparison of Solids Removal Processes

The most common methods of removing the microbial biomass are centrifugation and rotary drum vacuum filtration with added diatomaceous earth body feed. A general comparison of these traditional processes with crossflow microfiltration was given in the introductory section of this chapter. For a specific process, a choice of one of these options has to be made on the basis of the technical and economic factors that pertain to the specific process. In broad terms, however, the choice involves weighing the major relative advantages and disadvantages of each of these operations.

Centrifugation
Advantages:

1. Established process with predictable performance.
2. No disposal of spent diatomaceous earth (DE).

Disadvantage:

1. Supernatant can be heavily contaminated with colloids and small particles.

Diatomaceous Earth Filters
Advantage:

1. Low capital cost.

Disadvantages:

1. The weight of DE added is typically 30 kg/m^3 of broth and the weight of filter cake can be up to 200 kg/m^3 of broth.
2. The disposal of waste DE is becoming increasingly more difficult and expensive.
3. The yield loss associated with the filtration step is typically between 10% and 20%.

Crossflow Microfiltration
Advantages:

1. Complete removal of suspended solids.
2. Complete removal of polymeric contaminants if ultrafilters are used.
3. Closed system resulting in yields of 95% to 98%.

Disadvantage:

1. Higher capital cost.

Process Design

General Process Description
Other sections of this chapter describe the variety of crossflow process configurations available, ranging from simple batch to elaborate stages in series systems with permeate recycle. This section will be confined to simple batch processes, which illustrate the key design considerations. Large filtration systems may, however, be based on continuous stages in series. Figure 35–8 shows the essential elements of a two-stage series/parallel batch crossflow filtration (CFF) system.

The recovery of antibiotics involves three distinct phases: concentration of the microbial mass, diafiltration of this concentrate at constant volume, and final concentration. An example of a material balance resulting from this three-part process is shown in Table 35–8. Note

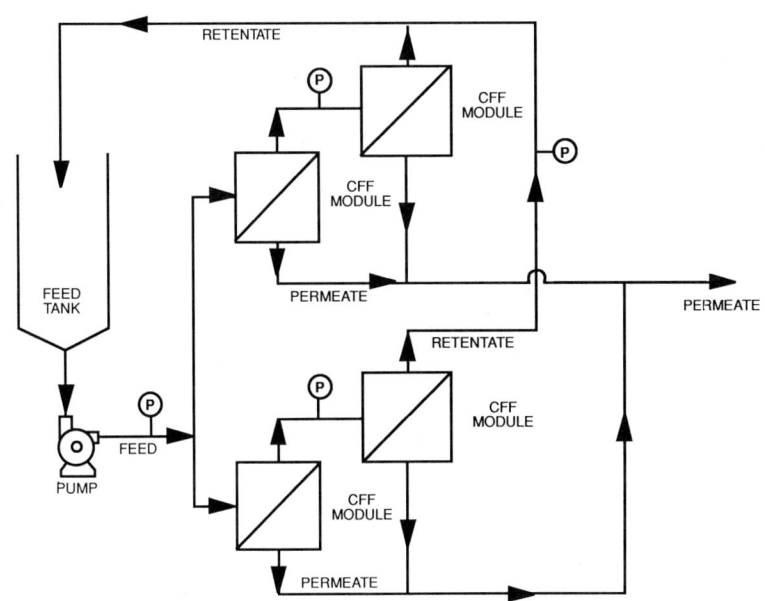

FIGURE 35–8. Schematic diagram of a two-stage series/parallel batch crossflow filtration system.

the high yield obtained and that the diafiltration phase results in about a twofold dilution of the antibiotic.

The material balances governing this process are given as:

Overall Recovery of Antibiotic in Filtrate Y_p:

$$Y_p = 1 - X_t^{(R-1)} \exp[X_d(R - 1)]. \quad (35\text{--}1)$$

Concentration of Antibiotic in Filtrate c_p:

$$c_p = (Y_p X_t)/[(X_t - 1) + X_d], \quad (35\text{--}2)$$

where

X_t = the ratio of starting volume to final volume of concentrate
X_d = the ratio of diafiltration volume to volume of concentrate during the constant volume of diafiltration step
R = the rejection coefficient of the antibiotic.

TABLE 35–8. Materials Balance: Crossflow Filtration Process for Clarification of Antibiotic Broth.

Step	Starting Volume (L)		Final Volume (L)		Mass of Antibiotic in Liquid (gram)		
	Total	Liquid	Total	Liquid	Start	Final	Permeate
Concentration (3×)	30,000	28,800	10,000	8,800	28,800	12,660	16,140
Diafiltration (3×)	10,000	8,800	10,000	8,800	12,660	1,550	11,110
Concentration (5/3×)	10,000	8,800	6,000	4,800	1,550	930	620
						Recovery =	27,870
							(96.8%)

Notes:

c_p = average permeate concentration = $\dfrac{27{,}870}{30{,}000 + 25{,}200}$ = 0.50 g/L.

c_{initial} = 1 g/L.
V_{initial} = 30,000 L.
R = 0.3.
$x_{\text{solids initial}}$ = 0.04 (devoid of antibiotic).

The rejection coefficient is defined in terms of X_t, c_p, and c_f, the feed concentration, as follows:

$$R = \frac{c_p}{X_t c_f}. \qquad (35\text{-}3)$$

Process Parameters

Crossflow microfiltration processes are designed on the basis of pilot trials. These are needed to establish the following key parameters of a crossflow filtration process.

Flux needs to be established as a function of the circulation velocity through the modules, the concentration of suspended solids, and the transmembrane pressure drop.

Rejection of the antibiotic needs to be established as a function of the pore size of the membrane, circulation velocity, and transmembrane pressure drop.

No reliable methods are available for predicting these key parameters because of the complexity of the fermentation broths. A typical pilot plant for doing this work is shown in Figure 35–9.

The most common crossflow filtration modules used are hollow fibers (acrylic and polysulfone), tubes of from 0.5 to 2.5 mm in diameter, and ceramic filters of from 2 to 7 mm in diameter. Plate-and-frame modules with a channel height of from 1 to 3 mm are also used. Under virtually all circumstances, the flow through these elements is in the laminar to slightly turbulent regime: starting out in turbulent conditions at the beginning of the process, and ending up in laminar flow at the end because of high solids concentrations.

The general guidelines for ceramic (Ceraflo) filtration modules follow:

Flux The empirical equation giving flux as a function of concentration of the microbial mass may be expressed as follows. (See Chapter 33 for a detailed discussion of the theoretical basis of this relation, as well as its limitations at low suspended solids concentrations.)

$$J = J_0 \ln(c_w/c_s), \qquad (35\text{-}4)$$

where

J = flux in liters per square meter of filter per hour (L/m²h)
J_0 = an empirical constant, ranging from 50 to 150 L/m²h
c_w = an empirical constant, ranging from 300 to 500 kg/m³
c_s = the concentration of microbial mass in kg/m³.

Rejection The rejection of antibiotic is usually low, ranging from 0.05 to 0.20 for filters rated at 0.2 μm, the most commonly used filters.

Optimization of Process Design

Two major design choices need to be optimized in terms of antibiotics recovery and minimization of capital and operating costs.

Diafiltration Point It is well known that for a flux equation of the type shown above, the minimum membrane area results when diafiltration is carried out at a microbial solids concentration equal to c_w/e, or about 130 kg/m³. About 95% recovery of antibiotic is obtained when the value of X_d is between 2 and 3.

Circulation Rate Later sections of this chapter show that the capital cost of crossflow microfiltration systems depends strongly on the

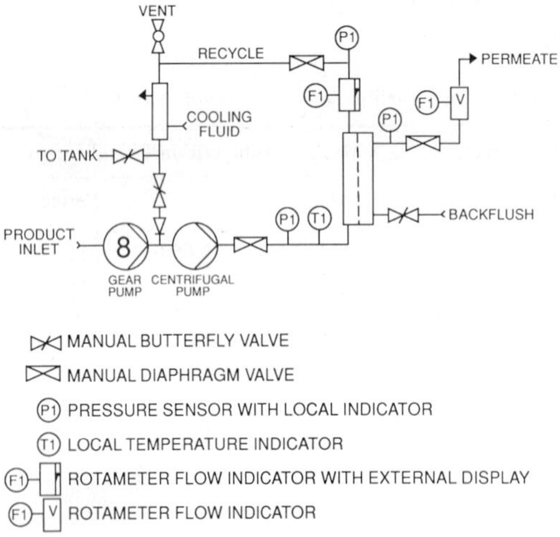

FIGURE 35–9. Crossflow filtration pilot plant system (filter area = 0.14 m²) (reprinted from Millipore Corporation with permission).

total circulation rate through the modules. This implies that for a given total area of filter the maximum number of modules in series should be used. This maximum number is constrained by the maximum inlet pressure that the microfiltration modules can tolerate and by the tendency of the antibiotic rejection coefficient R to increase as the transmembrane pressure increases. The overall choice of the optimuum velocity is complex in that modules of different lumen diameters have different pressure drops and different dependencies of flux on circulation velocity. For ceramic (e.g., Ceraflo™) modules, circulation velocities of from 2.5 to 4.0 m/s are most satisfactory. Typical transmembrane pressure drops are 3 to 5 bars.

System Maintenance

At the conclusion of the final concentration step of the operating cycle, the concentrated microbial mass is pumped out of the feed tank and displaced from the piping and filtration modules with water. The filtration modules are then cleaned of accumulated foulant with an appropriate clean-in-place (CIP) agent. For polymeric filters, dilute solutions of sodium hydroxide and sodium hypochlorite or compounded surfactants are used. A balance has to be struck between cleaning effectiveness and damage to the filters. Ceramic modules are typically cleaned with the same CIP solutions that are used in the rest of the fermentation facility, hot (60 to 80°C) dilute (2% to 5%) sodium hydroxide and sodium hypochlorite. If desired, the ceramic filters can be steamed, although this is seldom necessary.

System Design

This section describes the major components used in crossflow microfiltration systems for antibiotics and the basis for choosing among them.

Filtration Modules

The major choice to be made is between polymeric and ceramic modules. Polymeric modules are less expensive than ceramic modules, $500 to $1000/m^2 versus $1200 to $2000/m^2. Polymeric modules need to be replaced about once a year while ceramic modules seldom need to be replaced. This represents a choice between higher investment costs versus lower operating costs. Table 35–9 shows a comparison of specifications of typical ceramic modules from two suppliers.

TABLE 35–9. Element Characteristics of Two Commercially Available Ceramic Microfilters.

Ceramic Membrane Supplier	Ceraflo™ Millipore	Membralox™ Alcoa
Substrate	Alpha alumina	Alpha alumina
Membrane	Alpha alumina	Alpha alumina
Pore sizes, μm[a]	0.2, 0.45, 1.0	0.2, 0.45, 0.8
		1.2, 1.5, 3, 5
Element geometry		
Cross section	Circular	Hexagonal
Length, cm	88	85
Diameter, cm	1.6	2.8
Number of lumens	19	19, 37
Lumen diameter, mm	2.0, 2.7, 4.0, 6.0	3.0, 4.0, 6.0
Available surface area, m^2	0.14	0.20, 0.30
Pressure limit, bar	21[b]	>100[c]
Water permeability at 20°C		
Pore size = 0.2 μm, m^3/h/m^2/bar	4.0	3.0

[a]Smaller pore ultrafiltration ceramic membranes are also available from both suppliers.
[b]Operating pressure.
[c]Burst pressure.

Pumps

Typical circulation requirements for systems based on Ceraflow™ modules are between 4 and 7 m³/h/m² of filter area, which results in total circulation rates of between 100 and 400 m³/h at a head of 3 to 6 bars. Single-stage stainless steel centrifugal pumps are appropriate for this service.

Hydraulic Components

These components, which do not need to meet "sanitary" standards, are designed to satisfy common industrial standards for stainless steel. If flow meters are to be used on the concentrate, they must be of a nonplugging design, such as magnetic or sonic types.

Heat Exchange

Since antibiotics can be thermally labile, the temperature of the broth needs to be controlled. Most of the energy input from the circulation pump is removed by the permeate or can be compensated for by using chilled diafiltration water. A small auxiliary heat exchanger or a jacketed feed tank is adequate.

Process Configuration

To reduce the energy consumption of the process, a feed-and-bleed circulation loop can be used instead of the simpler and cheaper standard batch configuration. This can reduce energy requirements by 30% to 50% at the expense of a somewhat lower average flux. The two configurations are compared in Figure 35–10.

Crossflow filtration systems are usually supplied in skid-mounted form. These skids provide filtration modules ranging from <100 to 200 m² of filter area with the associated circulation pumps and hydraulic controls. Minimal expense is incurred in connecting the skid to the feed tank and permeate collection tank.

Economic Analysis

Capital Costs

A reliable model for capital costs is given by

$$C_t = C_c + AC_m + C_h Q_r N_p, \tag{35-5}$$

where

- C_t = the total cost of a system
- C_c = a constant factor related to engineering and automation costs
- A = the total filter area, m²
- C_m = the filter and housing cost per unit of membrane area
- Q_r = the circulation rate per filter pass, m³/h
- N_p = the number of filter passes in the system
- C_h = the cost of circulation components per m³/h.

The largest of the three factors on the right-hand side of Eq. (35–5) is the one related to circulation requirements.

The above equation can be put into a more illuminating form:

$$C_t = C_c + \left(\frac{Q_p}{\bar{J}}\right)\left(C_m + \frac{C_h Q_r}{N_s A_s}\right), \tag{35-6}$$

Simple Batch System

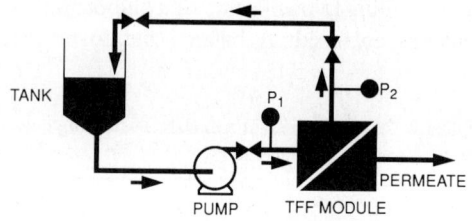

Single State Feed and Bleed Continuous System

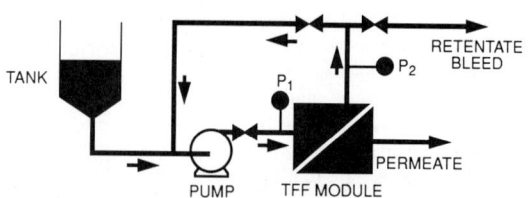

FIGURE 35–10. Comparison of a simple batch and a continuous feed and bleed crossflow filtration system (reprinted from Millipore Corporation with permission).

where

N_s = the number of modules in each pass
A_s = the filter area per module, m²
Q_p = the required permeation rate, L/m² h
$\overline{J}$ = the average flux, L/m² h.

This equation shows that for a given total membrane area, many modules in series are desirable. High flux is an obvious need, but it is obtained at the expense of high circulation rates. This balancing of flux and circulation rate is the central problem in optimizing the cost of the system.

Typical values of some of the key parameters for well-designed ceramic crossflow filtration systems are

A_s = 3.3 m²
Q_r = between 30 and 45 m³/h
N_s = from one to four modules in series
C_m = from \$1500 to \$2500 per m²
C_h = from \$800 to \$1400 per m³/h.

Operating Costs

Operating costs of crossflow microfiltration systems are dominated by the allocation of capital cost-related items and the cost of filter replacement. The operating costs at three different fluxes for the small system described in Table 35–10 (which is rated at about 2 m³/h) are presented in Table 35–11. For these cases, the operating costs are dominated by depreciation and interest charges. Had polymeric filters been specified, the cost of filter replacement would have been significantly larger. In general, the cost of the process will be between \$5 and \$10/m³ of permeate produced.

MICROFILTRATION OF MAMMALIAN CELLS

Process Description

Mammalian cells have been used for several decades for the production of vaccines (van Wezel, van Herwaarden, and van de Heuvel-de Rijk 1979; Plotkin 1980; van Wezel et al. 1984; Plotkin and Mortimer 1988). Since the invention of hybridomas (Kohler and Milstein 1975) and the development of methods for inclusion of foreign genetic material into mammalian cells, mammalian cells have become the most important means of producing therapeutic proteins (Mizrahi 1988; Morrison et al. 1988).

The separation of mammalian cells from their liquid environment by centrifugation or filtration is done for two distinctly different reasons.

The liquid medium in which the cells grow or produce may need to be exchanged in order to supply cells with nutrients or to change the cells' environment as they are induced to transition from the growth phase to the production phase (Arathoon et al. 1987). This process is called *perfusion*.

At the conclusion of a mammalian cell fermentation, the expressed proteins are present in a dilute (30 to 100 mg/L) and impure form in the growth medium. The concentration and purification of these proteins by ultrafiltration and chromatography requires the removal of the mammalian cells and cell fragments. This process is called *clarification*.

The perfusion process may be carried out in a batch mode or in a continuous mode depend-

TABLE 35–10. Technical Characteristics of Ceramic Crossflow Filtration System Used in Economic Analysis.

Module Characteristics	
Module type	Millipore-33
Lumen diameter, m	0.0027
Lumens/module	19
Module length, m	0.84
Area per module, m²	3.08
Module life, yr	5
System Characteristics	
Modules in series	2
Area per pass, m²	6.16
Flow per pass, m³/s	0.01
Superficial velocity, m/s	4.0
Permeate flow, m³/h	1.85
Retentate flow, m³/h	34.46
Pressure drop/module, N/m²	7.72×10^4
Pressure drop/pass, N/m²	1.54×10^5
Power/pass (at 60% efficiency), kW	2.465

TABLE 35-11. Economic Analysis of Ceramic Crossflow Filtration Systems as a Function of Flux.

System Parameters			
Flux, L/m^2h	50	100	150
Passes in system	6	3	2
Operating hours per year	6,000	6,000	6,000
Annual permeate flow, m^3	11,088	11,088	11,088

System Capital Investment			
Hardware	$317,301	$158,650	$105,767
Hardware plus membrane modules	$357,070	$178,535	$119,023

Operating Cost Summary

Operating Cost Component	Units	Unit Cost	Units/yr	Annual Cost	% Total Cost	Units/yr	Annual Cost	% Total Cost	Units/yr	Annual Cost	% Total Cost
Direct Costs											
Electricity	kWh	$0.05	88,740	$4,437	3.7%	44,370	$2,219	3.0%	29,580	$1,479	2.5%
Membrane	$/m^2	$538.00	7	$3,977	3.3%	4	$1,988	2.7%	2	$1,326	2.3%
Cleaning chemicals	kg	$2.50	1,500	$3,750	3.1%	1,500	$3,750	5.1%	1,500	$3,750	6.4%
Labor	man-hrs	$30.00	800	$24,000	20.0%	800	$24,000	32.4%	800	$24,000	41.0%
Subtotal				$36,164	30.1%		$31,956	43.2%		$30,555	52.2%
Capital Related Costs											
Depreciation	$/$system	0.14		$44,422	36.9%		$22,211	30.0%		$14,807	25.3%
Interest	$/$system	0.10		$31,730	26.4%		$15,865	21.4%		$10,577	18.1%
Maintenance	$/$system	0.025		$7,933	6.6%		$3,966	5.4%		$2,664	4.5%
Subtotal				$84,085	69.9%		$42,042	56.8%		$28,028	47.8%
Total Operating Costs				$120,249	100.0%		$73,999	100.0%		$58,583	100.0%
Permeate Operating Cost	$/m^3			$10.84			$6.67			$5.28	

ing on whether the supply of nutrients or the complete exchange of the growth medium is required. The clarification process is most commonly used with suspension cultures and is usually a batch process. The scale of operation varies over a wide range with batch sizes ranging from 20 L for suspension cultures of potent therapeutics in the preclinical phase to 10,000 L in the production of some animal vaccine viruses.

Comparison of Cell Removal Processes

The criteria used in selecting methods for separating cells from the medium depend on whether the goal is perfusion or clarification.

Perfusion requires that the cells be recovered with minimal loss of viability since they will continue to be used. The separator must be sterile to preclude the contamination of the fermentation broths.

Clarification does not place such a high demand on cell viability but disruption of the cells is to be avoided to prevent the intracellular proteins from contaminating the extracellular medium and rendering the subsequent separation steps more difficult.

Governmental regulations on containment of recombinant organisms are an important consideration in choosing means of separation. These regulations vary with locality. In general, it is important to prevent the organisms from escaping the confines of the process equipment into the plant environment. It is also important to achieve complete removal of cells from the clarified medium in order to eliminate the necessity, in subsequent steps, of having to process the medium under conditions mandated for materials containing recombinant organisms.

In light of these constraints, filtration processes have significant advantages over the traditional centrifugation processes as follows:

1. Filtration processes are inherently closed and therefore present a much lower risk of contaminating the environment.
2. Filtration processes achieve a virtually complete removal of cells because the available filter media, ranging in nominal pore size from 0.2 to 1.0 μm, are much smaller than the mammalian cells, which range in size from 4 to 8 μm.
3. For very small batches, small disposable filters are more convenient to use than batch centrifuges (Brunner and Hemfort 1988; Miller et al. 1984).
4. Filtration processes can be operated under conditions of very low cell disruption whereas, inevitably, there are zones of high shear in centrifuges that result in cell disruption.

Process Design

General Considerations for Clarification Processes

Although a wide range of process configurations is available for carrying out crossflow filtrations, for mammalian cell clarification the simple batch approach is universally used (see Chapter 33). A diafiltration step may be added for increasing the recovery of the extracellular proteins. (See Chapter 33 where the term *dialysis* is used for diafiltration.) Figure 35–11 shows the degree of recovery achievable without a diafiltration step. For properly operated mammalian cell clarification, the rejection coefficient of the extracellular proteins to be recovered is close to zero. If the degree of product recovery as calculated from this chart is not adequate, then a diafiltration step may be added and its increase on recovery estimated

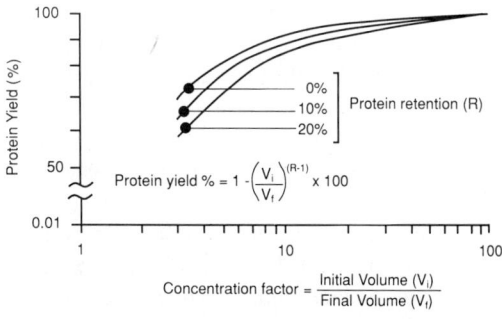

FIGURE 35–11. Effect of protein retention on protein yield for clarification applications (reprinted from Millipore Corporation with permission).

TABLE 35–12. Recommended Evaluation Protocol for Crossflow Filtration Apparatus.

Objective

To establish optimal flux, recovery, and operating regimes for harvesting mammalian cell lines with a Prostak™ system (Figure 35–12).

Procedures
1. Set water flux regimes:
   a. Introduce isotonic buffer into feed tank.
   b. Divert retentate and permeate to feed tank.
   c. Start recirculation pump at 2.0 Lpm/channel.
   d. Start permeate pump. Set flux to 25 L/m²h. Record inlet, outlet and permeate pressures.
   e. Repeat step (d) at fluxes of 75, 100, 200, and 300 L/m²h.
2. Set Cell Culture Regimes:
   a. *Optimize flux:* Recycle isotonic buffer throughout the system. Introduce culture to process vessel, volume-to-surface area ratio should be >1.0 L/ft². Divert permeate and retentate to process vessel. Recirculate fluid at a given flow rate. Record inlet, outlet, and permeate pressures. Slowly ramp permeate pump to a given flux level. Record inlet, outlet, and permeate pressures for 30 min. Monitor stability of transmembrane pressure (TMP). A rapid increase in TMP indicates deposition of cells (i.e., membrane fouling). If TMP is stable, ramp flux to the next level. Repeat the last three steps and monitor TMP. If TMP is not stable, decrease flux to prior level. Recommended flux rates are 20, 50, 70, 90, and 120 L/m²h.
   b. *Optimize wall shear rate:* Wall shear rate will affect TMP stability to a certain degree. The above procedures can be repeated at various recirculation rates in the range of about 1 to 2 L/min/stack.
   c. *Optimize cell concentration:* TMP stability is also affected by cell concentration; safe operating flux will decrease with increasing cell concentration. Repeat the above procedures at 2, 5, and 10 times the original cell density. Concentration can be performed with a pilot unit. The system should be cleaned prior to flux and shear excursions for experimental accuracy.

from the equations in the previous section on antibiotics clarification.

Process Parameters

The sizing of mammalian cell crossflow filtration equipment is very simple because the flux does not depend to any major degree on the fraction of the feed that has permeated during the course of the process. This insignificant variation of flux with cell concentration is due to the very low volumetric concentration of cells [see Eq. (33–33)]. The allowable flux is determined by the need to avoid retention of a cell layer on the filter surface. These cells are subjected to a pressure drop that is almost equal to the total transmembrane pressure drop, which is large enough to disrupt most cells. This maximum flux is determined by testing the microfiltration apparatus with the cells of interest, as shown in Table 35–12.

As a general guide, microfiltration of mammalian cells should be carried out at shear rates of 500 to 2000 s^{-1} and fluxes of 30 to 100 L/m² h.

System Maintenance

At the conclusion of the concentration or diafiltration step, the system is drained and flushed with water. Proper disposal of this waste material should follow local regulations. The modules are washed, typically with an alkaline sodium hypochlorite solution at 40 to 50°C. After the washing solution is drained and flushed, the filtration equipment should be stored in a sanitizing solution of 50 to 100 ppm NaOCl at pH 6 to 7. In perfusion applications the system needs to be steamed prior to use.

System Design

This section describes the major components of crossflow microfiltration systems and criteria for making choices.

Filtration Modules

In light of the discussion in the preceding sections, the most satisfactory modules for this application should have the following characteristics:

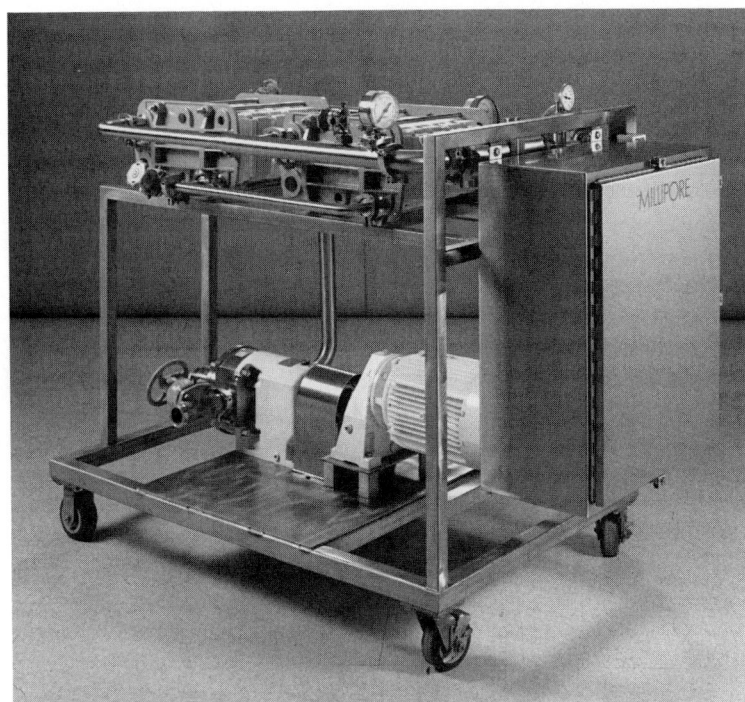

FIGURE 35–12. Millipore Prostak™ PKS-2 TFF system (reprinted from Millipore Corporation with permission).

1. The flow channel should be between 0.5 and 1.0 mm to be able to achieve the required shear rates without excessive circulation requirement. Circulation pumps can be quite expensive for this service (vide infra).
2. The flow path should be reasonably short so that all parts of the module can operate below the critical, cell damaging flux. This path length is typically between 30 and 100 cm. Shorter path lengths result in excessive pumping requirements. A cutaway diagram and dimensional details of a Prostak™ microfiltration module are shown in Figure 35–13, which illustrates some of these principles of design.
3. The module should be steamable for applications requiring sterile operation, and its materials of construction need to meet regulatory standards.

Pumps

To avoid excessive damage to the cells, peristaltic pumps are used on small-volume batches and rotary lobe pumps for large batches. The latter are operated in a deaerated mode, at from 200 to 400 rpm.

Flux Control

Since it is so important to control the flux to a level below that resulting in cell disruption, a metering pump is usually used on the permeate as illustrated in Figure 35–14. The Prostak™ module shown in this figure incorporates two pumps, one for circulation of the process stream and the second for control of filtrate pressure and flow. This two-pump crossflow system allows for gentler processing of delicate cells with a lower rate of wall shear and transmembrane pressure, and with greater product yield.

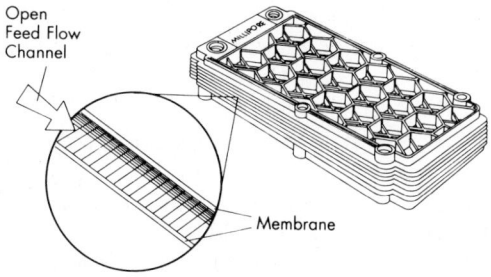

FIGURE 35–13. Prostak™ module with open flow channel (reprinted from Millipore Corporation with permission).

FIGURE 35–14. Two-pump Prostak™ crossflow filtration system (reprinted from Millipore Corporation with permission).

General Design Standards

Operating a crossflow filtration system in a sanitary manner is an important consideration in the processing of mammalian cells. Sanitary system design is not an exact science. Rather, it is a compilation of generally accepted practices, which, when used correctly, result in a sanitary system. Table 35–13 lists materials that are generally regarded as acceptable for use in sanitary crossflow filtration designs. An example of a sanitary crossflow filtration system that utilizes materials in the manner outlined in Table 35–13 is shown in Figure 35–12.

Economic Analysis

The operating costs of mammalian cell crossflow filtration systems are dominated by the costs of membrane replacement and capital-related charges. When steaming is not used, the expected life of crossflow filtration modules can be expected to range from 200 to 500 runs. When steaming is used, the expected life is between 50 and 150 runs. Module cost is between $400 and $800/m².

The following equation gives the expected filter replacement costs:

$$C_p = C_f/(t_r n Q), \qquad (35\text{–}7)$$

TABLE 35–13. Acceptable Materials of Construction for Sanitary Systems.

Item	Construction	Comments
Piping	Stainless steel 316 or 316L grade	180 grit inside diameter finish or better
		Electropolish optional, satin outside diameter finish
		No pits or cavities in welds, welds 100% penetrated
Fittings	Stainless steel	Ladish® type
Elastomers	EPDM, silicone	Ladish® type for fittings
Valves	Stainless steel	Sanitary design (butterfly, diaphragm preferred)
Pumps	Stainless steel	Sanitary design (i.e., centrifugal, rotary lobe)
Instrumentation	—	Sanitary design

where

C_p = the filter replacement cost per liter of permeate
C_f = the filter cost per square meter (this value differs from C_m previously defined in that C_m includes the housing cost per square meter)
t_r = the average run duration in hours
n = the module life expectancy in runs
Q = the permeation rate, L/m² h.

For an average run duration of 6 hours, an average flux of 65 L/m² h, and an average module cost of $600/m², the following cost ranges are obtained:

1. For systems with steaming, between $0.03 and $0.01/L.
2. For systems without steaming, between $0.003 and $0.008/L.

The capital costs of crossflow filtration systems for this application are extremely dependent on such specific factors as the finish specified and the degree of instrumentation. No general correlations are available but a range of costs for systems using between 5 and 30 m² of filter area is $4000 to $8000/m².

For fluxes ranging from 30 to 100 L/m² h, assuming a 5-year depreciation schedule and a 10% interest charge, the capital-related operating costs are outlined based on 1000 hours/year of operation:

Capital Investment ($/m²)	4000	8000
Assumed Flux (L/m²h)	Capital-Related Costs ($/L)	
30	0.13	0.23
100	0.04	0.08

ACKNOWLEDGMENT

The authors hereby acknowledge the help and support of Millipore Corporation, which provided them with the time and resources necessary to prepare this chapter.

NOTATION

General Notation

See the General Notation section at the beginning of this handbook.

Special Notation

A total filter area, L^2 or m²
A_s filter area per module, L^2 or m²
c_f concentration of recoverable species in feed, M/L^3 or kg/m³
c_p concentration of recoverable species in product, M/L^3 or kg/m³
c_s concentration of microbial mass, M/L^3 or kg/m³
c_w empirical constant ranging from 300 to 500 kg/m³, M/L^3 or kg/m³
C_c constant factor related to engineering and automation costs
C_f filter cost per square meter (this value differs from C_m defined below in that C_m includes the housing cost per square meter)
C_h cost of circulation components per m³/h
C_m filter and housing cost per unit of membrane area
C_p filter replacement cost per liter of permeate
C_t total cost of a system
J flux, L/t or L/m²h
$\bar{J}$ average flux, L/t or L/m²h
J_0 empirical constant ranging from 50 to 150 L/m²h, L/t or L/m²h
n module life expectancy in runs, dimensionless
N_p number of filter passes in the system
N_s number of modules in each pass
Q permeation rate, L/t or L/m²h
Q_p required permeation rate, L/t or L/m²h
Q_r circulatation rate per filter pass, L^3/t or m³/h
R rejection coefficient of recoverable product, dimensionless
t_r average run duration in hours, t or h
X_d ratio of diafiltration volume to volume of concentrate during the constant volume diafiltration step, dimensionless

X_t ratio of starting volume to final volume of concentrate, dimensionless

Y_p overall recovery of product in filtrate, dimensionless

REFERENCES

Arathoon, W. R., S. E. Builder, A. S. Lubiniecki, and R. D. Vanreis. 1987. Process for producing biologically active plasminogen activator. European Patent 0,248,675.

Belter, P. A., E. L. Cussler, and W.-S. Hu. 1988. *Bioseparations—Downstream Processing for Biotechnology.* New York: John Wiley & Sons.

Brunner, K. H., and H. Hemfort. 1988. Centrifugal separation in biotechnical processes. In *Downstream Processes: Equipment and Techniques—Advances in Biotechnical Processes,* ed. A. Mizrahi, Vol. 8, pp. 3–50. New York: Allan R. Liss.

Colton, C. K. 1987. Analysis of membrane processes for blood purification. *Blood Purif.* 5:202–251.

Kohler, G., and C. Milstein. 1975. Continuous cultures of fused cells secreting antibody of predefined specificity. *Nature* 256:495–497.

Michaels, S. L. 1989. Crossflow microfilters: the ins and outs. *Chem. Eng.* 96(1):84–91.

Miller, S. A., C. M. Ambler, R. C. Bennett, D. A. Dahlstrom, J. D. Darji, R. C. Emmett, J. B. Gray, C. F. Gurnham, L. J. Jacobs, R. P. Klepper, A. W. Michalson, J. Y. Oldshue, C. E. Silverblatt, J. C. Smith, and D. B. Todd. 1984. Liquid-solid systems. In *Perry's Chemical Engineers' Handbook,* ed. R. H. Perry and D. W. Green, 6th ed., Sec. 19. New York: McGraw Hill Book Co.

Mizrahi, A. 1988. *Monoclonal Antibodies: Production and Application—Advances in Biotechnological Processes,* Vol. 11, New York: John Wiley & Sons.

Morrison, S. L., S. Canfield, S. Porter, L. K. Tan, M. H. Tao, and L. A. Wims. 1988. Production and characterization of genetically engineered antibody molecules. *Clin. Chem.* 34(9):1668–1675.

Plotkin, S. A. 1980. Rabies vaccine prepared in human cell cultures: progress and perspective. *Rev. Infectious Diseases* 2(3):433–447.

Plotkin, S. A., and E. A. Mortimer. 1988. *Vaccines.* New York: W. B. Saunders Company.

Shiloach, J., N. Martin, and H. Moes. 1988. Tangential flow filtration. In *Downstream Processes: Equipment and Techniques—Advances in Biotechnical Processes,* ed. A. Mizrahi, Vol. 8, pp. 102–125. New York: Allan R. Liss.

Vandamme, E. J. 1984. *Biotechnology of Industrial Antibiotics.* New York: Marcel Dekker.

van Wezel, A. L., J. A. M. van Herwaarden, and E. W. van de Heuvel-de Rijk. 1979. Large scale concentration and purification of virus suspensions from microcarrier culture for the preparation of inactive virus vaccines. *Develop. Biol. Standard* 42:65–69.

van Wezel, A. L., G. van Steenis, P. van der Marel, and A. D. Osterhaus. 1984. Inactivated polio virus vaccine: current production methods and new developments. *Rev. Infectious Diseases* 6(S2):S335–S340.

IX

Emulsion Liquid Membranes

36. Definitions	597
37. Theory	611
38. Design Considerations	656
39. Applications	701
40. Capital and Operating Costs	718

IX

Emulsion Liquid Membranes

36

Definitions

W. S. Winston Ho
Exxon Research and Engineering Company

Norman N. Li
Allied Signal, Inc.

INTRODUCTION
GENERAL DESCRIPTION OF EMULSION LIQUID MEMBRANES
FACILITATED MECHANISMS AND DRIVING FORCES
Type 1 Facilitation
Type 2 Facilitation
REFERENCES

INTRODUCTION

The emulsion liquid membrane process is unique and different from the membrane processes discussed above. The membrane is a liquid phase involving an emulsion configuration. Emulsion liquid membranes, also called surfactant liquid membranes or liquid surfactant membranes, are essentially double emulsions, i.e., water/oil/water (W/O/W) systems or oil/water/oil (O/W/O) systems. For the W/O/W systems, the oil phase separating the two aqueous phases is the liquid membrane. For the O/W/O systems, the liquid membrane is the water phase that is between the two oil phases.

Since their discovery by Li (1968) just over two decades ago, emulsion liquid membranes have demonstrated considerable potential as effective tools for a wide variety of separations (Li and Frankenfeld 1988; Maugh 1976; Halwachs, Flaschel, and Schugerl 1980; Halwachs and Schugerl 1980; Marr and Kopp 1982; Stroeve and Varanasi 1982; Bargeman and Smolders 1986; Noble and Way 1987; Noble, Way, and Bunge 1988; Marr, Bart, and Draxler 1990; Ho 1990). These separations include:

1. The removal of zinc from wastewater in the viscose fiber industry (Draxler, Fürst, and Marr 1988; Draxler, Marr, and Prötsch 1988; Prötsch and Marr 1983; Draxler and Marr 1986; Lorbach and Marr 1987; Ruppert, Draxler, and Marr 1988)
2. The removal of phenol from wastewater (Zhang, Liu, and Lu 1987; Zhang et al. 1988a, 1988b; Li and Shrier 1972; Cahn and Li 1974; Matulevicius and Li 1975; Halwachs, Flaschel, and Schugerl 1980; Marr and Kopp 1980; Ho et al. 1982; Terry, Li, and Ho 1982; Chang and Li 1983; Teramoto et al. 1983a; Boyadzhiev, Bezenshek, and Lazarova 1984; Ho and Li 1984; Miao, Li, and Zhang 1985; Kataoka, Nishiki, and Kimura 1989; Kataoka et al. 1990)
3. The recovery of nickel from electroplating

solutions (Draxler and Marr 1986; Marr, Lackner, and Bart 1989)
4. The removal of heavy metals, such as zinc, cadmium, copper, and lead, from wastewater in metallurgical plants (Marr, Bart, and Draxler 1990)
5. The removal of heavy metal, such as zinc, cadmium, copper, lead, and mercury, in incineration plants (Marr, Bart, and Draxler 1990)
6. The removal of heavy metals, such as zinc, lead, chromium, cadmium, copper, and mercury, from wastewater (Schiffer et al. 1974b; Kitagawa et al. 1977; Frankenfeld and Li 1977, 1979; Boyadzhiev and Kyuchoukov 1980; Weiss, Griegoriev, and Mühl 1982; Fuller and Li 1984)
7. The removal of ammonia from wastewater (Li and Shrier 1972; Schiffer et al. 1974a; Maugh 1976; Frankenfeld and Li 1977; Cahn, Li, and Minday 1978; Halwachs and Schugerl 1978; Downs and Li 1981; Marr and Kopp 1982; Schlosser and Kossaczky 1980; Frankenfeld and Li 1987; Marr and Koncar, 1990; Lee and Chan 1990)
8. Hydrometallurgical extraction applications including the recovery of uranium from wet process phosphoric acid (WPPA) (Hayworth et al. 1983; Bock et al. 1982; Hayworth 1981) and from mine leachate and aqueous solutions (Hirato et al. 1991), the enrichment of copper from mine leachate and aqueous solutions (Cahn et al. 1981; Frankenfeld, Cahn, and Li 1981; Martin and Davies 1977, 1980; Lee, Evans, and Cussler 1978; Kondo et al. 1979; Cussler and Evans 1980; Vokel, Halwachs, and Schugerl 1980; Strzelbicki and Charewicz 1980; Miyake, Takenoshita, and Teramoto 1983; Teramoto et al. 1983b, 1983c; Lorbach, Bart, and Marr 1986; Kataoka et al. 1987, 1989; Bart et al. 1988; Goto, Kondo, and Nakashio 1989; Kataoka et al. 1990; Matsumoto et al. 1990; Nilsen, Jong, and Stubbs 1991), the extraction of chromium from aqueous solutions (Hochhauser and Cussler 1975; Li and Chang 1982; Qi et al. 1982; Vohra, Kaur, and Sharma 1989), the extraction of cobalt from aqueous solutions (Strzelbicki 1978; Strzelbicki and Charewicz 1980; Gu et al. 1982; Wasan, Gu, and Li 1984; Gu, Wasan, and Li 1985, 1986, 1988; Abou-Nemeh and Van Peteghem 1989; Strzelbicki and Schlosser 1989; Kumamaru et al. 1990; Okamoto et al. 1990), the recovery of rare earth metals (Teramoto et al. 1986; Goto et al. 1989; Zhang and Wang 1989; Zhang and Xiao 1989, 1990), and the recovery and removal of other metals from aqueous solutions (Li 1972, 1978a, 1978c; Cussler and Evans 1980; Osseo-Asare and Kenney 1980; Strzelbicki and Charewicz 1980; Lamb et al. 1980, 1981; Biehl et al. 1982; Boyadzhiev and Bezenshek 1982; Izatt et al. 1982, 1983a, 1983b; Christensen et al. 1983; Osseo-Asare, Lin, and Chaiko 1983; Perez de Ortiz 1986; Kataoka et al. 1987, 1989, 1990; Degener 1988; Zheng et al. 1988; Abou-Nemeh and Van Peteghem 1989; Hirato et al. 1990; Yan, Shi, and Su 1990; Wodzki, Wyszynska, and Narebska 1990; Goto et al. 1991; Hayashita et al. 1991; Mikulaj and Vasekova 1991; Shiau 1991)
9. The removal of alkali metal cations, such as Na^+, K^+, Li^+, and Cs^+, from wastewater (Reusch and Cussler 1973; Schiffer et al. 1974b; Cussler and Evans 1980; Bartsch et al. 1987)
10. The removal of organic and inorganic acids, such as acetic and propionic acids, cresols, and hydrocyanic acid, from wastewater (Halwachs, Flaschel, and Schugerl 1980; Terry, Li, and Ho 1982; Yan, Huang, and Shi 1987; Qian, Ma, and Shi 1989)
11. The removal of anions, such as chloride, sulfate, phosphate, and chromate, from wastewater (Schiffer et al. 1974b; Frankenfeld and Li 1977; Cussler and Evans 1980)
12. Biochemical processing applications including the separation of amino acids, antibiotics, and phospholipids from fermentation broth and the recovery of organic acids, such as lactic, acrylic, and acetic acids, from fermentation broth (Thien, Hatton, and Wang 1986; Ho and Cowan 1987; Thien and Hatton 1988)
13. Biomedical applications including blood

oxygenation (Li and Asher 1973; Wallace, Asher, and Li 1973; Li 1976; Frankenfeld, Asher, and Li 1978; Li 1980; Frankenfeld and Li 1982), the preparation of artificial red blood cells (Davis and Asher 1983; Davis, Asher, and Wallace 1987; Zheng, Beissinger, and Wasan, 1991), the extraction of cholesterol from blood (Yagodin et al. 1983), the removal of toxins from blood (Halwachs, Vokel, and Schugerl 1980), the treatment of chronic uremia (Asher et al. 1975, 1976, 1977; Frankenfeld, Asher, and Li 1978; Asher et al. 1979, 1980; Asher and Vogler 1980; Frankenfeld and Li 1982), the treatment of drug overdose (Frankenfeld, Fuller, and Rhodes 1976; Rhodes, Frankenfeld, and Fuller 1976; Chiang et al. 1978), the slow release of drugs (Brodin and Frank 1978; Asher, Li, and Shrier 1980; Yang and Rhodes 1980), and other pharmaceutical applications (Chilamkurti and Rhodes 1980; Davis 1981)

14. The fractionation of hydrocarbons (Li 1968, 1971a, 1971b, 1971c, 1971d, 1977; Shah and Owens 1972; Cahn and Li 1976a, 1976b; Stelmaszek and Borkowska 1977; Casamatta, Chavarie, and Angelino 1978; Kikic, Alessi, and Orlandini-Visalberghi 1978; Alessi, Kikic, and Orlandini-Visalberghi 1980; Halwachs, Flaschel, and Schugerl 1980; Halwachs and Schugerl 1980; Goswami and Rawat 1984a, 1984b; Krishna, Goswami, and Sharma 1987; Hung, Chen, and Lee 1989; Gupta, Goswami, and Rawat 1990; Garti and Kovacs 1991; Ulbrich, Marr, and Draxler 1991).

The first two applications in the preceding list have been commercialized. The commercialization of the first application was for the removal of zinc from wastewater at Lenzing AG, Austria, in 1986 (Draxler, Fürst, and Marr 1988; Draxler, Marr, and Prötsch 1988). The second application for the removal of phenol from wastewater was at Nanchung Plastic Factory, Guangzhou, People's Republic of China, around 1986 (Zhang, Liu, and Lu 1987; Zhang et al. 1988a, 1988b). These two commercial applications and other potential separation/purification applications are described in more detail in Chapters 39 and 40.

In addition to these two commercial applications, another use has been as a well control fluid since 1985 (Exxon 1985). This well control fluid is a pumpable water-in-oil emulsion containing clay particles in the oil phase. The emulsion is thickened to viscous, high-strength paste by high shear at drill bit nozzles. The high shear ruptures the oil film separating the water droplets and clay particles, and this allows direct contact between the water and clay particles, resulting in tremendous swelling of clay particles and the formation of the high-strength paste. The well control fluid is for preventing well blowout and sealing loss zones in oil and gas wells. Besides the well control fluid, other oil field applications have been proposed for liquid membrane emulsions. The potential applications include the injection of viscous emulsions into oil or gas wells to achieve hydraulic fracturing, the injection of acid-containing emulsions into wells to dissolve formation rock (e.g., carbonates) for fracturing, and the use of multiple liquid membrane emulsions for simultaneous injection of reactive components, which must be kept separate until they reach the formation where fracturing is needed (Li 1978b; Salathiel et al. 1980, 1982; Dawson, Li, and O'Brien 1983; Li and Frankenfeld 1988). These applications in the oil field as an encapsulation means are different from those described above as separation and purification tools.

Emulsion liquid membranes (ELM) also have potential utility as membrane reactors incorporating simultaneous separation and reaction processes (Ollis, Thompson, and Wolynic 1972; Li 1973; Wolynic and Ollis 1974; Cussler and Evans 1980; Scheper, Halwachs, and Schugerl 1983; Scheper et al. 1987). This utility includes the use of ELM for controlling chemical reactions, such as the heterogeneous catalytic oxidation of ethylene to acetaldehyde over the catalyst system of $PdCl_2$ and $CuCl_2$ (Ollis, Thompson, and Wolynic 1972; Wolynic and Ollis 1974), and for carrying out catalytic reactions using enzymes (Li 1973, 1975; May and Li 1972, 1974, 1977; Mohan and Li 1974, 1975).

The scope of the following discussion is limited to the use of ELM for separation and purification.

GENERAL DESCRIPTION OF EMULSION LIQUID MEMBRANES

Emulsion liquid membranes are usually prepared by first forming an emulsion between two immiscible phases, and then dispersing the emulsion in a third (continuous) phase by agitation for extraction. The membrane phase is the liquid phase that separates the encapsulated, internal droplets in the emulsion from the external, continuous phase, as shown schematically in Figure 36–1. In general, the internal, encapsulated phase and the external, continuous phase are miscible. However, the membrane phase must not be miscible with either of these two phases in order to be stable. Therefore, the emulsion is of the W/O type if the external, continuous phase is water, and it is of the O/W type if the external, continuous phase is oil. To maintain the integrity of the emulsion during the extraction process, the membrane phase generally contains some surfactant(s) and additive(s) as stabilizing agents, and it also contains a base material that is a solvent for all the other ingredients.

Typically, the encapsulated, internal droplets in the emulsion are 1 to 3 μm in diameter to provide a good emulsion stability for ELM extraction. When the emulsion is dispersed by agitation in the external, continuous phase during the extraction process, many small globules of the emulsion are formed. The size of the globules depends on the characteristics and concentration of the surfactant(s) in the emulsion, the viscosity of the emulsion, and the intensity and mode of mixing (Ohtake et al. 1987; Rautenbach and Machhammer 1988; Ho 1986). Usually, the globule size is controlled in the range of 100 to 2000 μm in diameter. Thus, a very large number of emulsion globules can be formed easily to produce a very large mass transfer area adjacent to the external, continuous phase. Each emulsion globule contains many 1- to 3-μm internal droplets. Thus, the internal mass transfer surface area, typically 10^6 m^2/m^3, is even much larger than the external mass transfer surface area. Therefore, a rapid mass transfer in the ELM process can occur from either the external, continuous phase to the internal, encapsulated phase or vice versa.

An example of an ELM system is shown schematically in Figure 36–2 for the removal of phenol from wastewater. The wastewater feed containing phenol is the external, continuous phase. The encapsulated droplets containing NaOH in water make up the internal phase. The liquid phase between the external and internal phases is the membrane phase containing an oil as the solvent in which surfactant(s) and additive(s) are dissolved. The internal reagent NaOH reacts with phenol and converts it to sodium phenolate, which is insoluble in the membrane phase and trapped in the internal phase.

Separation of mixtures can be achievable by selective diffusion of one component through the membrane phase into the receiving phase of lower equivalent concentration. Surfactant(s) and additive(s) included in the membrane phase can control the selectivity and permeability of the membrane. An individual component can be trapped and concentrated in the internal phase for later disposal or recovery. Once separation is achieved, the emulsion and external, continuous phases are separated, usually by settling

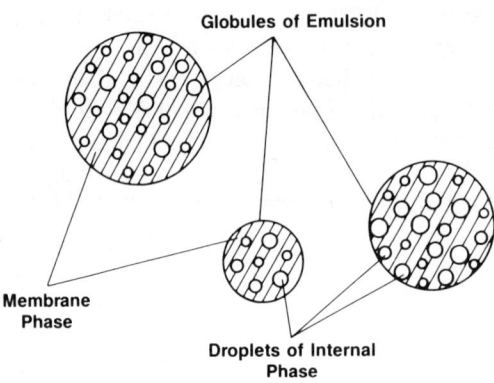

FIGURE 36–1. Schematic of an emulsion liquid membrane system.

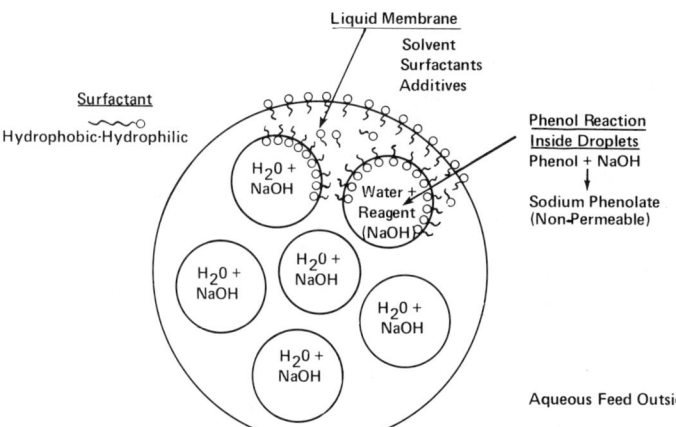

FIGURE 36–2. Schematic of an emulsion liquid membrane system for phenol removal.

as in conventional solvent extraction. The extracted component can be recovered from the "loaded" internal phase of the emulsion by breaking the emulsion, usually by the use of an electrostatic coalescer (Hsu, Li, and Hucal 1983a, 1983b; Hsu and Li 1985; Feng, Wang, and Zhang 1988; Marr, Bart, and Draxler 1990; Kataoka and Nishiki 1990). From breaking the emulsion, the membrane phase recovered can then be recycled to the emulsification step for the preparation of the emulsion with a regenerated or fresh internal reagent phase.

Figure 36–3 shows a schematic of a continuous ELM process. As discussed above, this process includes four steps: (1) emulsification, (2) dispersion of the emulsion in contact with the external, continuous phase for extraction, (3) settling to separate the emulsion from the external phase, which is the raffinate if the internal phase becomes the extract, and (4) breaking the emulsion to recover the internal phase as the extract and the membrane phase for recycle.

FACILITATED MECHANISMS AND DRIVING FORCES

The effectiveness of emulsion liquid membranes is a result of two facilitated mechanisms: Type 1 and Type 2 facilitations (Matulevicius and Li 1975; Li 1978c, 1981). Both Type 1 and Type 2 facilitations can maximize both the extraction rate, i.e., the flux through the membrane phase, and the capacity of the receiving phase (the internal phase in the case with an external feed phase) for the diffusing species.

Type 1 Facilitation

In this type of facilitation, the reaction in the receiving phase (the internal phase if the external phase is a feed) maintains a solute concentration of effectively zero. This is the minimization of the diffusing species in the receiving phase. The reaction of the diffusing species with a chemical reagent in the receiving phase forms a product incapable of diffusing

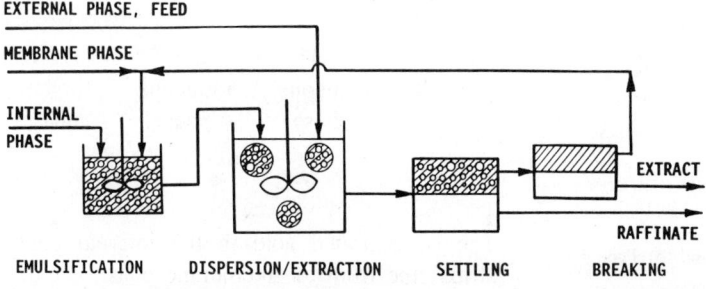

FIGURE 36–3. Schematic of a continuous emulsion liquid membrane process.

back through the membrane. This type of facilitation can be illustrated by the extraction of phenol from wastewater (Li and Shrier 1972; Cahn and Li 1974; Matulevicius and Li 1975; Halwachs, Flaschel, and Schugerl 1980; Marr and Kopp 1980; Terry, Li, and Ho 1982; Chang and Li 1983; Teramoto et al. 1983a; Zhang, Liu, and Lu 1987; Zhang et al. 1988a, 1988b; Kataoka, Nishiki, and Kimura 1989; Ho et al. 1982; Ho and Li 1984; also see Chapter 39) as shown in Figure 36–4(a), which is a simplified model of the ELM globule shown in Figure 36-2. In Figure 36-4(a), the phenol in the external aqueous phase dissolves in a membrane oil phase. Then, phenol diffuses across the membrane phase into the NaOH-containing internal phase, where it reacts with NaOH to form sodium phenolate. Since the ionic sodium phenolate is not soluble in the membrane oil phase, it is trapped in the internal phase. The reaction maintains the phenol concentration at effectively zero in the internal phase, giving a high driving force and thus a high extraction rate. The driving force is the phenol concentration difference between the external feed phase and the internal reagent phase. As discussed earlier, the driving force is enhanced via the reaction with NaOH. This reaction also increases the extraction capacity via incorporating a sufficient amount of NaOH in the internal phase for this reaction.

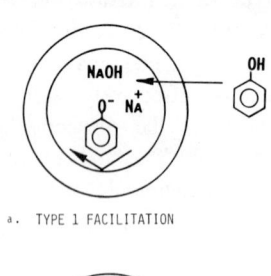

a. TYPE 1 FACILITATION

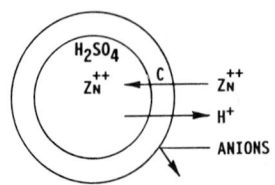

b. TYPE 2 FACILITATION (CARRIER FACILITATED TRANSPORT)

FIGURE 36–4. Schematic of two facilitated mechanisms: (a) Type 1 facilitation and (b) Type 2 facilitation.

Type 2 Facilitation

This facilitation is also called carrier facilitated transport. The diffusing species is carried across the membrane phase by incorporating a "carrier" compound (complexing agent or extractant) in the membrane phase. In this type of facilitation, the carrier compound carries the diffusing species, and reactions take place both at the external interface between the external and membrane phases and the internal interface between the membrane and internal phases. This facilitation can be illustrated by the removal of zinc from wastewater (Prötsch and Marr 1983; Draxler and Marr 1986; Lorbach and Marr 1987; Draxler, Fürst, and Marr 1988; Draxler, Marr, and Prötsch 1988; Ruppert, Draxler, and Marr 1988; also see Chapter 39) as shown in Figure 36–4(b).

In Figure 36–4(b), a zinc ion (Zn^{++}) in the external aqueous phase reacts at the external interface with the carrier compound (extractant), HR, in the membrane phase to form a complex, ZnR_2, as follows:

$$Zn^{++} + 2HR \rightarrow ZnR_2 + 2H^+$$
$$\text{(aqueous} \quad \text{(organic} \quad \text{(organic} \quad \text{(aqueous}$$
$$\text{phase)} \quad \text{phase)} \quad \text{phase)} \quad \text{phase)}$$

$$(36\text{--}1)$$

The carrier, HR, is the protonated form of a liquid ion-exchange extractant, and it is denoted as C in Figure 36–4(b). The reaction forms the zinc complex in the membrane phase and releases protons to the external aqueous phase. Then the zinc complex diffuses across the membrane phase to the concentrated H_2SO_4-containing internal phase, where the stripping reaction takes place at the internal interface as follows:

$$ZnR_2 + 2H^+ \rightarrow Zn^{++} + 2HR$$
$$\text{(organic} \quad \text{(aqueous} \quad \text{(aqueous} \quad \text{(organic}$$
$$\text{phase)} \quad \text{phase)} \quad \text{phase)} \quad \text{phase)}$$

$$(36\text{--}2)$$

The concentrated acid in the internal phase strips zinc from the membrane phase into the internal phase to become the zinc ion, and it donates protons to the extractant in the membrane phase. That is, the protons are exchanged for the zinc ions.

The concentrated acid drives the stripping reaction as shown in Eq. (36–2) to the right and maintains a low concentration of the zinc complex, ZnR_2, at the interface adjacent to the internal phase, giving a high driving force in terms of the zinc complex concentration difference between the external and internal interfaces and thus a high extraction rate. The concentration profiles for this facilitation are shown schematically in Figure 36–5, where A^{++} is the metal ion, i.e., Zn^{++} for this example. The high extraction rate is achieved via the driving force of continuous transport of protons from the internal phase to the external phase. The concentrated acid allows the zinc ion to be effectively concentrated in the internal phase, resulting in high extraction capacity. The zinc ion concentration can be more than 70 times that in the external feed phase (Draxler, Fürst, and Marr 1988; Draxler, Marr, and Prötsch 1988; also see Chapter 39). This is the case for which the driving force of proton transport "pumps" the transport of metal ion against its own concentration difference between the receiving and feed phases (Cussler 1971; Cussler, Evans, and Matesich 1971; Schiffer et al. 1974a, 1974b) as shown in Figure 36–5.

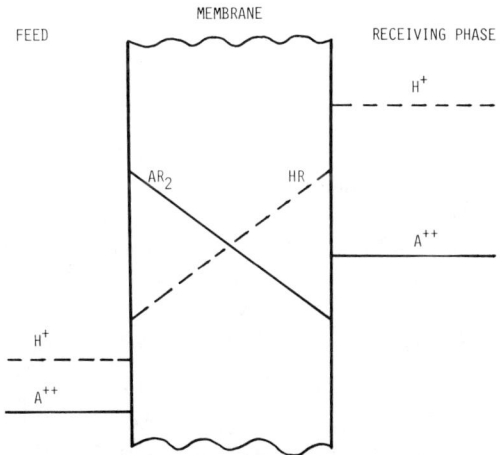

FIGURE 36–5. Schematic of concentration profiles for carrier facilitated transport (Type 2 facilitation).

Also shown in Figure 36–4(b), anions in the external feed phase cannot complex with the extractant and are thus rejected for counter-transport, in which the flux of the solute is in the opposite direction from the flux of the proton. For counter-transport, the extractants for metals form anion ligands; thus, these extractants do not complex with anions. In the other cases, co-transport of the solute with the proton or its counter-ion occurs. The co-transport cases include the extraction of dichromate with an amine complexing agent and the extraction of alkali metal ions with crown ethers, in which the fluxes of dichromate and alkali metal ions are in the same direction as those of the proton and anionic counter-ions, respectively.

For both types of the facilitated mechanisms, simultaneous extraction and stripping take place in one single step rather than two separate steps as required by solvent extraction. This is one of important advantages of ELM extraction versus solvent extraction. In addition, because of the simultaneous extraction and stripping, the ELM process drives both the extraction and stripping reaction equilibria as shown in Eqs. (36–1) and (36–2) to the right by the removal of the complex (ZnR_2, AR_2) and the extractant (HR) as formed, which is shown in Figure 36–5. The ELM feature of simultaneous extraction and stripping removes the equilibrium limitation inherent in solvent extraction. Therefore, the complete removal of the solute from a feed can be achievable with single-step ELM extraction. For example, chromium concentration in an aqueous feed is reduced from 100 to nearly 0 ppm in about 4 minutes via an ELM system (Hochhauser and Cussler 1975; Cussler and Evans 1980; Ho and Li 1984; Li and Frankenfeld 1988). Furthermore, another advantage resulting from the nonequilibrium feature of an ELM process is the significant reduction of the extractant inventory required for the ELM extraction versus solvent extraction. The reduction by a factor of more than 10 is possible for the extraction of copper from dilute aqueous solutions (Martin and Davies 1977; Li and Frankenfeld 1988).

REFERENCES

Abou-Nemeh, I., and A. P. Van Peteghem. 1989. Extraction of cobalt and manganese from an industrial effluent by liquid emulsion membranes. In *Proc. 2nd International Conference on Separations Science and Technology,* ed. M. H. I. Baird and S. Vijayan, Vol. 2, pp. 416–423. Ottawa, Ontario: Canadian Society of Chemical Engineering.

Alessi, P., I. Kikic, and M. Orlandini-Visalberghi. 1980. Liquid membrane permeation for the separation of C_8 hydrocarbons. *Chem. Eng. J.* 19:22.

Asher, W. J., K. C. Bovee, J. W. Frankenfeld, R. W. Hamilton, L. W. Henderson, P. C. Holtzapple, and N. N. Li. 1975. Liquid membrane system directed toward chronic uremia. *Kidney Int.* 7:S-409.

Asher, W. J., K. C. Bovee, J. W. Frankenfeld, R. W. Hamilton, J. W. Henderson, P. G. Holtzapple, and N. N. Li. 1976. Liquid membrane system directed toward chronic uremia. *Kidney Int.* 10:S-254.

Asher, W. J., K. C. Bovee, T. C. Vogler, R. W. Hamilton, and P. G. Holtzapple. 1979. Liquid membrane capsules for removal of urea from the blood. *Clin. Nephrol.* 11:92.

Asher, W. J., K. C. Bovee, C. T. Vogler, R. W. Hamilton, and P. G. Holtzapple. 1980. Secretion modified release of urease from liquid membrane capsules (LMC). *Trans. Am. Soc. Artif. Intern. Organs* 26:120.

Asher, W. J., N. N. Li, and A. L. Shrier. 1980. Detoxifying-medical emulsions. U.S. Patent 4,183,918.

Asher, W. J., and T. C. Vogler. 1980. Detoxification by means of the controlled, in vivo secretion triggered rupture of liquid membrane capsules. U.S. Patent 4,183,960.

Asher, W. J., T. C. Vogler, K. C. Bovee, P. G. Holtzapple, and R. W. Hamilton. 1977. Liquid membrane capsules for chronic uremia. *Trans. Am. Soc. Artif. Intern. Organs* 23:73.

Bargeman, D., and C. A. Smolders. 1986. Liquid membranes. In *Synthetic Membranes: Science, Engineering and Applications,* ed. P. M. Bungay, H. K. Lonsdale, and M. N. de Pinho, NATO Advanced Science Institute Series C, Vol. 181, pp. 567–579. Dordrecht, Holland: D. Reidel Publishing Co.

Bart, H. J., R. Wachter, R. J. Marr, and H. J. Müller. 1988. Mass transfer in a permeation column. *J. Membr. Sci.* 36:413–423.

Bartsch, R. A., W. A. Charewicz, S. I. Kang, and W. Walkowiak. 1987. Proton-coupled transport of alkali metal cations across liquid membranes by ionizable crown ethers. In *Liquid Membranes: Theory and Applications,* ed. R. D. Noble and J. D. Way, ACS Symp. Ser. No. 347, pp. 86–97. Washington, DC: American Chemical Society.

Biehl, M. P., R. M. Izatt, J. D. Lamb, and J. J. Christensen. 1982. Use of a macrocyclic crown ether in an emulsion (liquid surfactant) membrane to effect rapid separation of Pb^{2+} from cation mixtures. *Sep. Sci. Technol.* 17(2):289–294.

Bock, J., R. R. Klein, P. L. Valint, and W. S. Ho. 1982. Liquid membrane extraction of uranium from wet process phosphoric acid. In *Sulfuric/Phosphoric Acid Plant Operations,* pp. 175–183. New York: American Institute of Chemical Engineers.

Boyadzhiev, L., and E. Bezenshek. 1982. Zink-Rückgewinnung aus wässrigen Lösungen mittels Extraktion mit einer carrier-phase. *Chem.-Ing.-Tech.* 54:506–508.

Boyadzhiev, L., E. Bezenshek, and Z. Lazarova. 1984. Removal of phenol from waste water by double emulsion membranes and creeping film pertraction. *J. Membr. Sci.* 21:137–144.

Boyadzhiev, L., and G. Kyuchoukov. 1980. Further development of carrier-mediated extraction. *J. Membr. Sci.* 6:107–112.

Brodin, A. F., and S. G. Frank. 1978. Drug release from O/W/O multiple emulsion systems. *Acta Pharm. Suec.* 15(1):111.

Cahn, R. P., J. W. Frankenfeld, N. N. Li, D. Naden, and K. N. Subramanian. 1981. Extraction of metal ions by liquid membrane. In *Recent Developments in Separation Science,* ed. N. N. Li, Vol. 6, p. 51. Boca Raton, FL: CRC Press.

Cahn, R. P., and N. N. Li. 1974. Separation of phenol from waste water by the liquid membrane technique. *Sep. Sci.* 9(6):505–519.

Cahn, R. P., and N. N. Li. 1976a. Hydrocarbon separation by liquid membranes. In *Membrane Separation Processes,* ed. P. Meares, pp. 327–349. Amsterdam: Elsevier Scientific Publishing Co.

Cahn, R. P., and N. N. Li. 1976b. Separations of organic compounds by liquid membrane processes. *J. Membr. Sci.* 1(2):129.

Cahn, R. P., N. N. Li, and R. M. Minday. 1978. Removal of ammonium sulfide from industrial sour water by a liquid membrane process. *Environmental Sci. Technol.* 12(9):1051.

Casamatta, G., C. Chavarie, and H. Angelino. 1978. Hydrocarbon separation through a liquid water membrane: modeling of permeation in an emulsion drop. *AIChE J.* 24(6):945–949.

Chang, Y. C., and S. P. Li. 1983. A study of emulsified liquid membrane treatment of phenolic waste water. *Desalination* 47:351–361.

Chiang, C.-C., G. C. Fuller, J. W. Frankenfeld, and C. T. Rhodes. 1978. Potential of liquid membranes for drug overdose treatment: in vitro studies. *J. Pharm. Sci.* 67(1):63–66.

Chilamkurti, R. N., and C. T. Rhodes. 1980. Transport across liquid membranes: effect of molecular structure. *J. Appl. Biochem.* 2:17–24.

Christensen, J. J., S. P. Christensen, M. P. Biehl, S. A. Lowe, J. D. Lamb, and R. M. Izatt. 1983. Effect of receiving phase anion on macrocycle-mediated cation transport rates and selectivities in water-toluene-water emulsion membranes. *Sep. Sci. Technol.* 18(4):363–373.

Cussler, E. L. 1971. Membranes which pump. *AIChE J.* 17(6):1300–1303.

Cussler, E. L., and D. F. Evans. 1980. Liquid membranes for separations and reactions. *J. Membr. Sci.* 6:113–121.

Cussler, E. L., D. F. Evans, and M. A. Matesich. 1971. Theoretical and experimental basis for a specific countertransport system in membranes. *Science* 172:377–379.

Davis, S. S. 1981. Liquid membranes and multiple emulsions. *Chem. Ind. (London)* 19:683–687.

Davis, T. A., and W. J. Asher. 1983. Process for preparing artificial red cells. U.S. Patent 4,376,059.

Davis, T. A., W. J. Asher, and H. W. Wallace. 1987. Artificial red cells with polyhemoglobin membranes prepared by liquid membrane techniques. *Trans. Am. Soc. Artif. Intern. Organs* 28:404–407.

Dawson, C. R., N. N. Li, and D. E. O'Brien. 1983. Method of using a well treating fluid. U.S. Patent 4,397,354.

Degener, W. 1988. Liquid membrane extraction—a method for processing of wastewaters. *Metall. (Berlin)* 42(4):391–396.

Downs, H. H., and N. N. Li. 1981. Extraction of ammonia from waste water by the liquid membrane process. *J. Sep. Process Technol.* 2(4):19–24.

Draxler, J., W. Fürst, and R. J. Marr. 1988. Separation of metal species by emulsion liquid membranes. *J. Membr. Sci.* 38:281–293.

Draxler, J., and R. J. Marr. 1986. Emulsion liquid membranes: part I. phenomenon and industrial application. *Chem. Eng. Process.* 20:319–329.

Draxler, J., R. J. Marr, and M. Prötsch. 1988. Commercial-scale extraction of zinc by emulsion liquid membranes. In *Separation Technology*, ed. N. N. Li and H. Strathmann, pp. 204–214. New York: United Engineering Trustees.

Exxon Production Research Company and Exxon Research and Engineering Company. 1985. How Exxon broke new ground in complex fluids to solve a serious drilling problem. In *Chem. Eng. News* (December 9):28–29, and *The Lamp* (an Exxon publication), Summer 1985.

Feng, Z.-L., X.-D. Wang, and X.-J. Zhang. 1988. A new high voltage electrostatic coalescer EC-1 applied to liquid membrane separation. *Water Treatment* 3:320–328.

Frankenfeld, J. W., W. J. Asher, and N. N. Li. 1978. Biochemical and biomedical separation using liquid membranes. In *Recent Developments in Separation Science*, ed. N. N. Li, Vol. 4, pp. 39–50. Boca Raton, FL: CRC Press.

Frankenfeld, J. W., R. P. Cahn, and N. N. Li. 1981. Extraction of copper by liquid membranes. *Sep. Sci. Technol.* 16(4):385.

Frankenfeld, J. W., G. C. Fuller, and C. T. Rhodes. 1976. Potential use of liquid membranes for emergency treatment of drug overdose. *Drug Develop. Commun.* 2:405–419.

Frankenfeld, J. W., and N. N. Li. 1977. Waste water treatment by liquid ion exchange in liquid membrane systems. In *Recent Developments in Separation Science*, ed. N. N. Li, Vol. 3, pp. 285–292. Boca Raton, FL: CRC Press.

Frankenfeld, J. W., and N. N. Li. 1979. Liquid membrane systems. In *Ion Exchange for Pollution Control*, ed. C. Calmon and H. Gold, Vol. 2, pp. 163–172. Boca Raton, FL: CRC Press.

Frankenfeld, J. W., and N. N. Li. 1982. Biochemical and biomedical separations using liquid membranes. In *Treatise on Analytical Chemistry*, Part I, Vol. 5, *Theory and Practice*, pp. 251–280. New York: Wiley-Interscience.

Frankenfeld, J. W., and N. N. Li. 1987. Recent advances in liquid membrane technology. In *Handbook of Separation Process Technology*, ed. R. W. Rousseau, p. 840. New York: John Wiley & Sons.

Fuller, E. J., and N. N. Li. 1984. Extraction of chromium and zinc from cooling tower blowdown by liquid membranes. *J. Membr. Sci.* 18:251–271.

Garti, N., and A. Kovacs. 1991. Facilitated emulsion liquid membrane separation of complex hydrocarbon mixtures. *J. Membr. Sci.* 56:239–246.

Goswami, A. N., and B. S. Rawat. 1984a. Studies on permeation of hydrocarbons through liquid surfactant membranes. *J. Membr. Sci.* 24:145.

Goswami, A. N., and B. S. Rawat. 1984b. Permeation of benzene through liquid surfactant membranes. *J. Chem. Tech. Biotech.* 34A:174.

Goto, M., K. Kondo, and F. Nakashio. 1989. Extraction kinetics of copper with liquid surfactant membranes containing LIX65N and nonionic surfactant. *J. Chem. Eng. Japan* 22(1):71–78.

Goto, M., H. Yamamoto, K. Kondo, and F.

Nakashio. 1991. Effect of new surfactants on zinc extraction with liquid surfactant membranes. *J. Membr. Sci.* 57:161–174.

Goto, M., N. Yoshii, K. Kondo, and F. Nakashio. 1989. Separation of lanthanum and neodymium by liquid surfactant membranes. In *Proc. Symp. Solvent Extr.*, pp. 113–118. Fukuoka, Japan: Kyushu University.

Gu, Z. M., R. M. Kurzeja, D. T. Wasan, and N. N. Li. 1982. Separation of metal ions by liquid membranes ligand accelerated transport. Paper read at AIChE Annual Meeting, 14–19 November 1982, Los Angeles, CA.

Gu, Z. M., D. T. Wasan, and N. N. Li. 1985. Interfacial mass transfer in ligand accelerated metal extraction by liquid surfactant membranes. *Sep. Sci. Technol.* 20(7&8):599–612.

Gu, Z. M., D. T. Wasan, and N. N. Li. 1986. Ligand-accelerated liquid membrane extraction of metal ions. *J. Membr. Sci.* 26:129–142.

Gu, Z. M., D. T. Wasan, and N. N. Li. 1988. Liquid surfactant membranes for metal extractions. *Surfactant Sci. Ser.* 28:127–168.

Gupta, T. C. S. M., A. N. Goswami, and B. S. Rawat. 1990. Mass transfer studies in liquid membrane hydrocarbon separations. *J. Membr. Sci.* 54:119–130.

Halwachs, W., E. Flaschel, and K. Schugerl. 1980. Liquid membrane transport—a highly selective separation process for organic solutes. *J. Membr. Sci.* 6:33.

Halwachs, W., and K. Schugerl. 1978. The liquid-membrane technique—a promising extraction process. *Chem.-Ing.-Tech.* 50:767.

Halwachs, W., and K. Schugerl. 1980. The liquid membrane technique—a promising extraction process. *Int. Chem. Eng.* 20(4):519–528.

Halwachs, W., W. Vokel, and K. Schugerl. 1980. Removal of toxins from plasma and blood with liquid surfactant membranes. In *Proc. Int. Solv. Extr. Conf., ISEC'80,* 6–12 September 1980, Liege, Belgium, Paper 80–88.

Hayashita, T., R. A. Bartsch, T. Kurosawa, and M. Igawa. 1991. Selective concentration of lead (II) chloride complex with liquid anion-exchange membranes. *Analy. Chem.* 63(10):1023–1027.

Hayworth, H. C. 1981. A case of technology transfer. *Chemtech* (June):342.

Hayworth, H. C., W. S. Ho, W. A. Burns, Jr., and N. N. Li. 1983. Extraction of uranium from wet process phosphoric acid by liquid membranes. *Sep. Sci. Technol.* 18(6):493–521.

Hirato, T., I. Kishigami, Y. Awakura, and H. Majima. 1991. Concentration of uranyl sulfate solution by an emulsion-type liquid membrane process. *Hydrometallurgy* 26(1):19–33.

Hirato, T., K. Koyama, Y. Awakura, and H. Majima. 1990. Concentration of Mo(VI) from aqueous sulfuric acid solution by an emulsion type liquid membrane process. *Mater. Trans. JIM* 31(3):213–218.

Ho, W. S. 1986. The size of liquid membrane emulsion globules in an agitated continuous phase: a dynamic measuring technique. Paper read at the 191st ACS National Meeting and First International Conference on Separations Science and Technology, 16–18 April 1986, New York, NY.

Ho, W. S. 1990. Emulsion liquid membranes: a review. In *Proc. 1990 International Congress on Membranes and Membrane Processes,* 20–24 August 1990, Chicago, IL, Vol. I, pp. 692–694.

Ho, C. S., and R. M. Cowan. 1987. Separating lactic acid from fermentation broth with liquid surfactant membranes. Paper read at the 194th ACS National Meeting, 30 August–4 September 1987, New Orleans, LA.

Ho, W. S., T. A. Hatton, E. N. Lightfoot, and N. N. Li. 1982. Batch extraction with liquid surfactant membranes: a diffusion controlled model. *AIChE J.* 28(4):662–670.

Ho, W. S., and N. N. Li. 1984. Modeling of liquid membrane extraction processes. In *Hydrometallurgical Process Fundamentals,* ed. R. G. Bautista, pp. 555–597. New York: Plenum Press.

Hochhauser, A. M., and E. L. Cussler. 1975. Concentrating chromium with liquid surfactant membranes. *AIChE Symp. Ser.* 71:136–142.

Hsu, E. C., and N. N. Li. 1985. Membrane recovery in liquid membrane separation processes. *Sep. Sci. Technol.* 20(2&3):115–130.

Hsu, E. C., N. N. Li, and T. Hucal. 1983a. Electrodes for electrical coalescence of liquid emulsions. U.S. Patent 4,415,426.

Hsu, E. C., N. N. Li, and T. Hucal. 1983b. Electrical coalescence of liquid emulsion. U.S. Patent 4,419,200.

Hung, T.-M., C.-S. Chen, and C.-J. Lee. 1989. Effect of process conditions on the separations of toluene and p-xylene with liquid surfactant membrane. *J. Chin. Inst. Chem. Eng.* 20(6):319–325.

Izatt, R. M., M. P. Biehl, J. D. Lamb, and J. J. Christensen. 1982. Rapid separation of Tl^+ and Pb^{2+} from various binary cation mixtures using dicyclohexano-18-crown-6 incorporated in emulsion membranes. *Sep. Sci. Technol.* 17:1351–1360.

Izatt, R. M., D. V. Dearden, P. R. Brown, J. S. Bradshaw, J. D. Lamb, and J. J. Christensen. 1983a. Cation fluxes from binary Ag^+-Mn^{2+} mixtures in a H_2O-$CHCl_3$-H_2O liquid membrane system containing a series of macrocyclic ligand carriers. *J. Am. Chem. Soc.* 105:1785.

Izatt, R. M., J. D. Lamb, J. L. Oscarson, and J. J. Christensen. 1983b. Metal separations using emulsion liquid membranes. Paper read at 3rd Symp. Sep. Sci. Technol. Energy Appl., 28 June–1 July 1983, Gatlinburg, TN.

Kataoka, T., and T. Nishiki. 1990. Development of a continuous electric coalescer of W/O emulsions in liquid surfactant membrane process. *Sep. Sci. Technol.* 25(1&2):171–185.

Kataoka, T., T. Nishiki, and S. Kimura. 1989. Phenol permeation through liquid surfactant membrane—permeation model and effective diffusivity. *J. Membr. Sci.* 41:197–209.

Kataoka, T., T. Nishiki, S. Kimura, and Y. Tomioka. 1989. Batch permeation of metal ions using liquid surfactant membranes. *J. Membr. Sci.* 46:67–80.

Kataoka, T., T. Nishiki, S. Kimura, and Y. Tomioka. 1990. A model for mass transfer through liquid surfactant membranes. *Water Treatment* 5:136–149.

Kataoka, T., T. Nishiki, M. Yamauchi, and Y. Zhong. 1987. A simulation for liquid surfactant membrane permeation in a continuous countercurrent column. *J. Chem. Eng. Japan* 20(4):410–415.

Kikic, I., P. Alessi, and M. Orlandini-Visalberghi. 1978. Liquid membrane permeation for the separation of xylenes. *Inst. Chem. Eng. Symp. Ser.* 54:153–164.

Kitagawa, T., Y. Nishikawa, J. W. Frankenfeld, and N. N. Li. 1977. Waste water treatment by liquid membrane process. *Environmental Sci. Technol.* 11(6):602–605.

Kondo, K., K. Kita, I. Koida, J. Irie, and F. Nakashio. 1979. Extraction of copper with liquid surfactant membranes containing benzoylacetone. *J. Chem. Eng. Japan* 12(3):203–209.

Krishna, R., A. N. Goswami, and A. Sharma. 1987. Effect of emulsion breakage on selectivity in the separation of hydrocarbon mixtures using aqueous surfactant membranes. *J. Membr. Sci.* 34:141.

Kumamaru, T., Y. Okamoto, M. Yamamoto, Y. Obata, and K. Onizuka. 1990. High enrichment method for the determination of ultratrace levels of cobalt by liquid-liquid extraction using water/oil/water emulsions as liquid surfactant membranes. *Anal. Chim Acta.* 232(2):389–391.

Lamb, J. D., J. J. Christensen, J. L. Oscarson, B. L. Nielsen, B. W. Asay, and R. M. Izatt. 1980. The relationship between complex stability constants and rates of cation transport through liquid membranes by macrocyclic carriers. *J. Am. Chem. Soc.* 102:6820.

Lamb, J. D., R. M. Izatt, D. G. Garrick, J. S. Bradshaw, and J. J. Christensen. 1981. The influence of microcyclic ligand structure on carrier-facilitated cation transport rates and selectivities through liquid membranes. *J. Membr. Sci.* 9:83–107.

Lee, C. J., and C. C. Chan. 1990. Extraction of ammonia from a dilute aqueous solution by emulsion liquid membranes: 1. experimental studies in batch system. *Ind. Eng. Chem. Res.* 29:96–100.

Lee, K. H., D. F. Evans, and E. L. Cussler. 1978. Selective copper recovery with two types of liquid membranes. *AIChE J.* 24:860.

Li, N. N. 1968. Separating hydrocarbons with liquid membranes. U.S. Patent 3,410,794.

Li, N. N. 1971a. Removal of organic compounds by liquid membranes. U.S. Patent 3,617,546.

Li, N. N. 1971b. Permeation through liquid surfactant membranes. *AIChE.J.* 17(2):495.

Li, N. N. 1971c. Separation of hydrocarbons by liquid membrane permeation (I). *Ind. Eng. Chem. Process Des. Dev.* 10:215.

Li, N. N., 1971d. Separation of hydrocarbons by liquid membrane permeation (II). In *Membrane Processes in Industry and Biomedicine,* ed. M. Bier, p. 175. New York: Plenum Press.

Li, N. N. 1972. Removal of inorganic species by liquid membranes. U.S. Patent 3,647,488.

Li, N. N. 1973. Process for the reaction and separation of components utilizing a liquid surfactant membrane and an enzyme catalyst. U.S. Patent 3,740,315.

Li, N. N. 1975. Immobilized enzymes and methods for preparation thereof. U.S. Patent 3,897,308.

Li, N. N. 1976. Blood oxygenation process. U.S. Patent 3,942,527.

Li, N. N. 1977. Separating hydrocarbon mixtures by emulsification. U.S. Patent 4,056,462.

Li, N. N. 1978a. Metal extraction by combined solvent and liquid membrane extraction. U.S. Patent 4,086,163.

Li, N. N. 1978b. Liquid membrane encapsulated reactive products. U.S. Patent 4,098,736.

Li, N. N. 1978c. Facilitated transport through liquid membranes. *J. Membr. Sci.* 3:265.

Li, N. N. 1980. Detoxifying medicinal emulsions. U.S. Patent 4,183,918.

Li, N. N. 1981. Encapsulation and separation by liquid surfactant membranes. *Chem. Eng.* (Rugby, England) 370:325–327.

Li, N. N., and W. J. Asher. 1973. Blood oxygenation by liquid membrane permeation. In *ACS Advances in Chemistry Series, Chemistry Engineering in Medicine,* ed. R. F. Gould, p. 1. New York: American Chemical Society.

Li, S. P., and Y. C. Chang. 1982. Extraction of chromium ions from aqueous solution. In *Proc.*

Joint CIESC and AIChE Meeting, 19–22 September 1982, Beijing, China, Chemical Industry Press, Vol. II, pp. 571–582.

Li, N. N., and J. W. Frankenfeld. 1988. Liquid membranes. In *Encyclopedia of Chemical Processing and Design,* ed. J. C. McKetta, Vol. 28, pp. 273–303. New York: Marcel Dekker.

Li, N. N., and A. L. Shrier. 1972. Liquid membrane water treating. In *Recent Developments in Separation Science,* ed. N. N. Li, Vol. 1, p. 163. Boca Raton, FL.: CRC Press.

Lorbach, D., H. J. Bart, and R. J. Marr. 1986. Mass transfer in liquid membrane permeation. *Ger. Chem. Eng.* 9:321–327.

Lorbach, D., and R. J. Marr. 1987. Emulsion liquid membranes: part II. modeling mass transfer of zinc with bis(2-ethylhexyl)dithiophosphoric acid. *Chem. Eng. Process.* 21:83–93.

Marr, R. J., H. J. Bart, and J. Draxler. 1990. Liquid membrane permeation. *Chem. Eng. Process.* 27:59–64.

Marr, R. J., and M. Koncar. 1990. Rückgewinnung von Ammoniak aus Industrieabwasser. *Chem.-Ing.-Tech.* 62(3):175–182.

Marr, R. J., and A. Kopp. 1980. Flüssigmembran-Technik-Übersicht über Phänomene, Transportmechanismen und Modellbildunden. *Chem.-Ing.-Tech.* 52:399.

Marr, R. J., and A. Kopp. 1982. Liquid membrane technology—a survey of phenomena, mechanisms and models. *Int. Chem. Eng.* 22(1):44–60.

Marr, R. J., H. Lackner, and H. J. Bart. 1989. Verfahren zur Abtrennung von Nickel aus Nickellonen enthaltenden verdünnten wässrigen Lösungen. European Patent Application 89112656.7.

Martin, T. P., and G. A. Davies. 1977. The extraction of copper from dilute aqueous solutions using a liquid membrane process. *Hydrometallurgy* 2:315–334.

Martin, T. P., and G. A. Davies. 1980. Extraction of copper from aqueous solutions using a liquid membrane process: a model to simulate the process. In *Proc. Int. Solv. Extr. Conf., ISEC'80,* 6–12 September 1980, Liege, Belgium, Paper 80-230.

Matsumoto, M., K. Ema, K. Kondo, and F. Nakashio. 1990. Copper extraction with liquid surfactant membrane in Mixco extractor. *J. Chem. Eng. Japan* 23(4):402–407.

Matulevicius, E. S., and N. N. Li. 1975. Facilitated transport through liquid membranes. *Sep. Purif. Methods* 4(1):73–96.

Maugh, T. H. 1976. Liquid membrane—new techniques for separation, purification. *Science* 193:134–150.

May, S. W., and N. N. Li. 1972. The immobilization of urease using liquid surfactant membranes. *Biochem. Biophys. Res. Commun.* 47(5):1179.

May, S. W., and N. N. Li. 1974. Encapsulation of enzymes in liquid membrane emulsions. In *Enzyme Engineering,* ed. E. K. Pye and L. B. Wingaard, Vol. 1, p. 77. New York: Plenum Press.

May, S. W., and N. N. Li. 1977. Liquid membrane encapsulated enzymes. In *Biomedical Applications of Immobilized Enzymes and Proteins,* ed. T. M. S. Chang, Vol. 1, p. 171. New York: Plenum Press.

Miao, F.-D., X.-P. Li, and Y.-Q. Zhang. 1985. The mathematical modeling of the removal of phenol with emulsified liquid membranes. *Desalination* 56:355–366.

Mikulaj, V., and L. Vasekova. 1991. Emulsion membrane extraction of strontium and calcium using 18-crown-6, picric acid and halogenated hydrocarbon membrane. *J. Radioanal. Nucl. Chem.* 150(2):281–285.

Miyake, Y., Y. Takenoshita, and M. Teramoto. 1983. Extraction rates of copper with SME 529: mechanism and effects of surfactants. In *Proc. Int. Solv. Extr. Conf., ISEC '83,* 26 August–2 September 1983, Denver, CO, pp. 301–302.

Mohan, R. R., and N. N. Li. 1974. Enzymes encapsulated by liquid membranes. *Biotech. Bioeng.* 16:513.

Mohan, R. R., and N. N. Li. 1975. Nitrite reduction by liquid membrane encapsulated whole cells. *Biotech. Bioeng.* 17(8):1137–1156.

Nilsen, D. N., B. W. Jong, and A. M. Stubbs. 1991. Copper extraction from aqueous solutions with liquid emulsion membranes: a preliminary laboratory study. *Bur. Mines Rep. Invest.* RI 9375.

Noble, R. D., and J. D. Way. 1987. Applications of liquid membrane technology. In *Liquid Membranes: Theory and Applications,* ed. R. D. Noble and J. D. Way, ACS Symp. Ser. No. 347, pp. 110–122. Washington, DC: American Chemical Society.

Noble, R. D., J. D. Way, and A. L. Bunge. 1988. Liquid membranes. *Solvent Extr. Ion Exch.* 10:63–103.

Ohtake, T., T. Hano, K. Takagi, and F. Nakashio. 1987. Effects of viscosity on drop diameter of W/O emulsion dispersed in a stirred tank. *J. Chem. Eng. Japan* 20(5):443–447.

Okamoto, Y., T. Takahashi, K. Isobe, and T. Kumamaru. 1990. Graphite furnace atomic absorption spectrometric determination of ultratrace levels of cobalt after solvent extraction using W/O/W emulsions as liquid surfactant membranes. *Analy. Sci.* 6(3):401–405.

Ollis, D. F., J. B. Thompson, and E. T. Wolynic. 1972. Liquid membrane reactor concept and preliminary experiments in acetaldehyde synthesis. *AIChE J.* 18:457–458.

Osseo-Asare, K., and M. E. Kenney. 1980. Sulfonic acids: catalysts for the liquid-liquid extraction of metals. *Sep. Sci. Technol.* 15(4):999.

Osseo-Asare, K., K. L. Lin, and D. J. Chaiko. 1983. Mass transfer in catalytic systems: micelles, liquid membranes and single drops. In *Proc. Int. Solv. Extr. Conf., ISEC '83*, 26 August–2 September 1983, Denver, CO, pp. 313–314.

Perez de Ortiz, R. S. 1986. The surfactant liquid membrane applications to metal extraction and pollution control. *NATO Adv. Sci. Inst. Series E* 107:575–584.

Prötsch, M., and R. J. Marr. 1983. Development of a continuous process for metal ion recovery of liquid membrane permeation. In *Proc. Int. Solv. Extr. Conf., ISEC '83*, 26 August–2 September 1983, Denver, CO, pp. 66–67.

Qi, Q. J., M. X. Xu, B. L. Tang, Z. M. Gu, L. Y. Zhu, and F. J. Cui. 1982. Removal of Cr(VI) from waste water by liquid surfactant membranes. Paper read at Symposium of Liquid Membrane Separations, 24–28 September 1982, Dalian, China.

Qian, X.-L., X.-S. Ma, and Y.-J. Shi. 1989. Removal of cyanide from wastewater with liquid membranes. *Water Treatment* 4:99–111.

Rautenbach, R., and O. Machhammer. 1988. Modeling of liquid membrane separation processes. *J. Membr. Sci.* 36:425–444.

Reusch, C. F., and E. I. Cussler. 1973. Selective membrane transport. *AIChE J.* 19:736.

Rhodes, C. T., J. W. Frankenfeld, and G. C. Fuller. 1976. Use of liquid membrane technology in the oral treatment of drug overdose. Paper read at Symposium on Separation and Encapsulation by Liquid Membranes, American Chemical Society Centennial Meeting, 6 April 1976, New York, NY.

Ruppert, M., J. Draxler, and R. J. Marr. 1988. Liquid-membrane-permeation and its experiences in pilot-plant and industrial scale. *Sep. Sci. Technol.* 23 (12&13):1659–1666.

Salathiel, W. M., T. W. Muecki, C. E. Cooke, Jr., and N. N. Li. 1980. Well treatment with emulsion dispersions. U.S. Patent 4,233,265.

Salathiel, W. M., T. W. Muecki, C. E. Cooke, Jr., and N. N. Li. 1982. Well treatment with emulsion dispersions. U.S. Patent 4,359,391.

Scheper, T., W. Halwachs, and K. Schugerl. 1983. Preparation of L-amino acids by means of continuous enzyme-catalyzed D,L-amino acid ester hydrolysis inside liquid surfactant membranes. In *Proc. Int. Solv. Extr. Conf., ISEC '83*, 26 August–2 September 1983, Denver, CO, pp. 389–390.

Scheper, T., Z. Likidis, K. Makryaleas, Ch. Nowattny, and K. Schugerl. 1987. Three different examples of enzymatic bioconversion in liquid membrane reactors. *Enzyme Microb. Technol.* 9:625–631.

Schiffer, D. K., E. M. Choy, D. F. Evans, and E. L. Cussler. 1974a. More membrane pumps. *AIChE Symp. Ser.* 70:150–156.

Schiffer, D. K., A. M. Hochhauser, D. F. Evans, and E. L. Cussler. 1974b. Concentrating solutes with membranes containing carriers. *Nature* 250:484–486.

Schlosser, S., and E. Kossaczky. 1980. Comparison of pertraction through liquid membranes and double liquid-liquid extraction. *J. Membr. Sci.* 6:83–105.

Shah, N. D., and T. C. Owens, 1972. Separation of benzene and hexane with liquid membrane technique. *Ind. Eng. Chem. Prod. Res. Dev.* 11:58.

Shiau, C. Y. 1991. Analysis of transport rate of zinc extraction through liquid surfactant membrane. *Sep. Sci. Technol.* 26(12):1519–1530.

Stelmaszek, J., and B. Borkowska. 1977. Separation of mixtures using liquid membranes. *Int. J. Chem. Eng.* 17:566.

Stroeve, P., and P. P. Varanasi. 1982. Transport processes in liquid membranes: double emulsion separation systems. *Sep. Purif. Methods* 11 (1):29–69.

Strzelbicki, J. 1978. Separation of cobalt by liquid surfactant membranes. *Sep. Sci. Technol.* 13:141.

Strzelbicki, J., and W. Charewicz. 1980. The liquid surfactant membrane separation of copper, cobalt and nickel from multicomponent aqueous solutions. *Hydrometallurgy* 5:243.

Strzelbicki, J., and S. Schlosser. 1989. Influence of surface-active substances on pertraction of cobalt (II) cations through bulk and emulsion liquid membranes. *Hydrometallurgy* 23:67–75.

Teramoto, M., T. Sakai, K. Yanagawa, and Y. Miyake. 1983c. Modeling of the permeation of copper through liquid surfactant membranes by continuous operations. *Sep. Sci. Technol.* 18 (11):985–997.

Teramoto, M., T. Sakai, K. Yanagawa, M. Ohsuga, and Y. Miyake. 1983b. Modeling of the permeation of copper through liquid surfactant membranes. *Sep. Sci. Technol.* 18(8):735–764.

Teramoto, M., T. Sakuramoto, T. Koyama, H. Matsuyama, and Y. Miyake. 1986. Extraction of lanthanoids by liquid surfactant membranes. *Sep. Sci. Technol.* 21(3):229–250.

Teramoto, M., H. Takihana, M. Shibutani, T. Yuasa, and N. Hara. 1983a. Extraction of phenol and cresol by liquid surfactant membrane. *Sep. Sci. Technol.* 18(5):397–419.

Terry, R. E., N. N. Li, and W. S. Ho. 1982. Extraction of phenolic compounds and organic acids by liquid membranes. *J. Membr. Sci.* 10:305–323.

Thien, M. P., and T. A. Hatton, 1988. Liquid emulsion membranes and their applications in biochemical processing. *Sep. Sci. Technol.* 23 (8&9):819–853.

Thien, M. P., T. A. Hatton, and D. I. C. Wang. 1986. Liquid emulsion membranes and their applications in biochemical separations. *ACS Symp. Ser.* 314:67–77.

Ulbrich, M., R. J. Marr, and J. Draxler. 1991. Selective separation of organic solutes by aqueous liquid surfactant membranes. *J. Membr. Sci.* 59:189–203.

Vohra, D. K., S. Kaur, and A. Sharma. 1989. Extraction of Cr(VI) from acid (sulfate) aqueous medium using liquid surfactant membrane emulsions. *Indian J. Technol.* 27:574–576.

Vokel, W., W. Halwachs, and K. Schugerl. 1980. Copper extraction by means of a liquid surfactant membrane process. *J. Membr. Sci.* 6:19–31.

Wallace, H. W, W. J. Asher, and N. N. Li. 1973. Liquid-liquid oxygenation: a new approach. *Trans. Am. Soc. Artif. Intern. Organs* 21:80.

Wasan, D. T., Z. M. Gu, and N. N. Li. 1984. Separation of metal ions by ligand-accelerated transfer through liquid surfactant membranes. *Faraday Discuss. Chem. Soc.* 77:67–74.

Weiss, S., V. Griegoriev, and P. Mühl. 1982. The liquid membrane process for the separation of mercury from waste water. *J. Membr. Sci.* 12:119–129.

Wodzki, R., A. Wyszynska, and A. Narebska. 1990. Two-component emulsion liquid membranes with macromolecular carriers of divalent ions. *Sep. Sci. Technol.* 25(11 & 12):1175–1187.

Wolynic, E. T., and D. F. Ollis. 1974. The catalytic liquid membrane reactor making acetaldehyde. *Chemtech* (February):111.

Yagodin, G., Y. Lopukhin, E. Yurtov, T. Guseva, and V. Sergiemko. 1983. Extraction of cholesterol from blood using liquid membranes. In *Proc. Int. Solv. Extr. Conf., ISEC '83*, 26 August–2 September 1983, Denver, CO, pp. 385–386.

Yan, N.-X., S.-A. Huang, and Y.-J. Shi. 1987. Removal of acetic acid from wastewater with liquid surfactant membranes: an external boundary layer and membrane diffusion controlled model. *Sep. Sci. Technol.* 22(2&3):801–818.

Yan, N.-X., Y.-J. Shi, and Y.-F. Su. 1990. A study of gold extraction by liquid membranes. *Water Treatment* 5(2):190–201.

Yang, T. T., and C. T. Rhodes. 1980. Transport across liquid membranes: effect of formulation variables. *J. Appl. Biochem.* 2:7–16.

Zhang, R.-H., and D.-X. Wang. 1989. Extraction of mixed rare earth from aqueous solution with emulsion liquid membrane. *Water Treatment* 4:165–176.

Zhang, R.-H., and L. Xiao. 1989. Design of a liquid membrane system for extracting rare earths. *Water Treatment* 4:473–481.

Zhang, R.-H., and L. Xiao. 1990. Design of a liquid membrane system for extracting rare earths. *J. Membr. Sci.* 5(3):249–258.

Zhang, X.-J., Q.-J. Fan, X.-T. Zhang, and Z.-F. Liu. 1988a. New surfactant LMS-2 used for industrial application in liquid membrane separation. In *Separation Technology,* ed. N. N. Li and H. Strathmann, pp. 215–226. New York: United Engineering Trustees; *Water Treatment* 3:233–240.

Zhang, X.-J., J.-H. Liu, Q.-J. Fan, Q.-T. Lian, X.-T. Zhang, and T.-S. Lu. 1988b. Industrial application of liquid membrane separation for phenolic wastewater treatment. In *Separation Technology,* ed. N. N. Li and H. Strathmann, pp. 190–203. New York: United Engineering Trustees.

Zhang, X.-J., J.-H. Liu, and T.-S. Lu. 1987. Industrial application of liquid membrane separation for phenolic wastewater treatment. *Water Treatment* 2:127–135.

Zheng, S., R. L. Beissinger, and D. T. Wasan. 1991. The stabilization of hemoglobin multiple emulsion for use as a red blood cell substitute. *J. Colloid Interface Sci.* 144(1):72–85.

Zheng, X.-C., L.-X. Li, J.-J. Guo, and F.-S. Long. 1988. Extraction of vanadium from waste water with emulsion liquid membrane. In *Proc. 1st International Conference on Hydrometallurgy,* 1–15 October 1988, Beijing, China, pp. 508–583.

37

Theory

W. S. Winston Ho
Exxon Research and Engineering Company

Norman N. Li
Allied Signal, Inc.

DIFFUSION-TYPE MASS TRANSFER
  MODELS FOR TYPE 1
  FACILITATION
    Spherical Shell Approach
    Emulsion Globule Approach
CARRIER-FACILITATED
  TRANSPORT MODELS FOR TYPE
  2 FACILITATION
    Spherical Shell Approach
    Emulsion Globule Approach
APPENDIX A
APPENDIX B
NOTATION
REFERENCES

This chapter describes the theory behind batch extraction with emulsion liquid membranes (ELMs). The extension of the theory to continuous ELM processes is given in Chapter 38 on design considerations. The theory of batch extraction may be classified into two categories: (1) diffusion-type mass transfer models for Type 1 facilitation and (2) carrier facilitated transport models for Type 2 facilitation (Lorbach and Marr 1987; Ho 1990). The definitions of Type 1 and Type 2 facilitations (Matulevicius and Li 1975; Li 1978, 1981) are given in Chapter 36.

DIFFUSION-TYPE TRANSFER MODELS FOR TYPE 1 FACILITATION

In Type 1 facilitation, reaction in the receiving phase of an ELM system (the internal phase if the external phase is a feed) maintains a solute concentration of effectively zero. This is the minimization of the diffusing species in the receiving phase, resulting in the maximization of the driving force for the diffusion of the solute in the membrane phase from the feed phase to the receiving phase. The diffusion process may be described by two methods: the spherical shell approach and the emulsion globule approach (Hatton, Lightfoot, and Li 1982; Ho 1990).

Spherical Shell Approach

This approach assumes that the mass transfer resistance is diffusion in the spherical "shell" of the membrane phase of constant thickness between the external and internal phases. The shell is shown schematically in Figure 36–4(a) for the example of phenol extraction with the internal phase containing NaOH for the conversion of phenol to sodium phenolate, maintaining the concentration of phenol at effectively zero. Several mass transfer models use this approach (Hatton, Lightfoot, and Li 1982; Marr and Kopp 1982; Gladek, Stelmaszek, and Szust 1982; Chan and Lee 1984, 1987; Noble, Way, and Bunge 1988). Some models consider only the extraction of the solute from the feed phase into the receiving phase (Cahn and Li 1974, 1976a, 1976b; Matulevicius and Li 1975; Kremesec 1981; Kremesec and Slattery 1982). However, the other models have taken into account both this extraction and the leakage of the components (extracted solute and chemical reagent) from the receiving phase to the feed phase (Boyadzhiev, Sapundzhiev, and Bezenshek 1977; Ho and Li 1984).

Models with Overall Mass Transfer Coefficient for Extraction

The models assume negligible leakage of the components from the receiving phase to the feed phase. One of the models developed by Cahn and Li (1974) assumes the mass transfer rate to be directly proportional to the average solute concentration difference between the continuous feed phase and the internal reagent phase and the proportionality, the overall mass transfer coefficient, to be constant. However, contrary to the assumption, the thickness of the shell is not constant, and the effective overall mass transfer coefficient that they obtained varied with time. They also used this model for the analysis of pure permeation processes in the separation of hydrocarbons, wherein the membrane phase separating two hydrocarbon liquid phases was an aqueous medium (Cahn and Li 1976a, 1976b).

The model of Matulevicius and Li (1975) suggests that the extraction of phenol in an ELM globule can be represented mathematically by the diffusion in a single drop. For the spherical shell model, they formulated and solved the unsteady-state equations. In the model, phenol diffuses from the surface of the globule to some fixed, interior position where it is removed by the reaction with the internal reagent. In assuming that an ELM globule can be represented by a single drop, an effective membrane thickness needs to be assigned to account for the mass transfer resistance in the globule. Gladek, Stelmaszek, and Szust (1981, 1982) took a similar modeling approach and applied the model to the separation of benzene from hexane (1981) and the extraction of phenol (1982). The other model using an integral mass balance technique accounts for the effect of the circulation of internal phase droplets on overall mass transfer (Kremesec 1981; Kremesec and Slattery 1982).

Models with Overall Mass Transfer Coefficients for Extraction and Leakage

Like those discussed above, most of the models for ELM extraction available in the literature have not taken the breakage of the internal phase into account. Boyadzhiev, Sapundzhiev, and Bezenshek (1977) were the first to consider the breakage in the modeling of ELM extraction. However, Ho and Li (1984) developed a more general mathematical model with overall mass transfer coefficients for extraction and leakage. Later, Krishna, Goswami, and Sharma (1987) investigated the effect of the breakage on selectivity in the separation of hydrocarbon mixtures. In the model of Ho and Li (1984), two extraction cases are considered: Case I has a chemical reagent in the external feed phase to convert the leaked species to an extractable form for re-extraction into the ELM globules; Case II does not have the reagent in the external phase. Most ELM extraction systems can be classified into these two cases. This model takes care of the leakage mechanisms of solute, from the internal phase to the external phase, not only by breakage but also by diffusion in the membrane phase.

Mass Transfer Mechanisms

In the model of Ho and Li (1984), solute can transfer from the internal phase of an ELM system to the external phase by two mechanisms: diffusional transport and breakage. But, solute can transfer from the external phase to the internal phase only by diffusional transport. These mechanisms are shown schematically in Figure 37–1 in which the globule of the

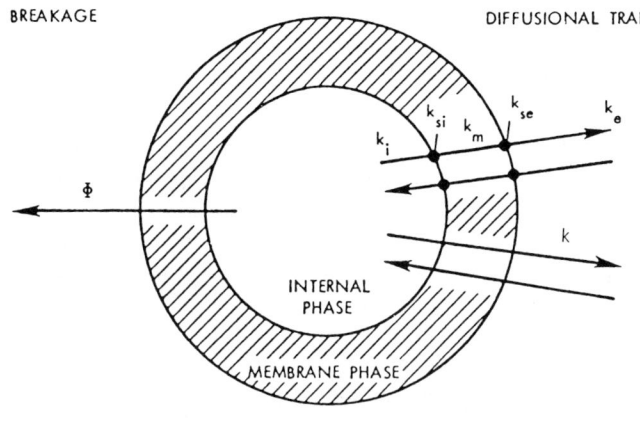

FIGURE 37-1. Schematic of simplified mass transfer mechanisms (reprinted from Ho and Li 1984 with permission).

emulsion is represented by a hollow sphere with the membrane phase separating the internal phase from the external phase.

Diffusional Transport

If the concentration driving force of solute is higher for the internal phase than for the external phase, the direction of the diffusional transport will be from the internal phase to the external phase. This diffusional transport includes five steps:

1. Mass transfer from the internal phase to the interface adjacent to the internal phase
2. Interfacial mass transfer across this interface
3. Diffusion in the membrane phase from this interface to the second interface, adjacent to the external phase
4. Interfacial mass transfer across the second interface
5. Mass transfer from the second interface to the bulk of the external phase.

The mass transfer rate j_k can be expressed in the following with the five individual mass transfer coefficients k_i, k_{si}, k_m, k_{se}, and k_e for these five steps, respectively:

$$j_k = k_i A_i (c_i - c_{im})$$
$$= k_{si} A_i \left(c_{im} - \frac{c_{mi}}{K_i} \right)$$
$$= k_m A_e (c_{mi} - c_{me})$$
$$= k_{se} A_e \left(\frac{c_{me}}{K_e} - c_{em} \right)$$
$$= k_e A_e (c_{em} - c_e). \quad (37\text{--}1)$$

Rearrangement of Eq. (37–1) leads to

$$j_k = \frac{A_i A_e k_i k_{si} k_m k_{se} k_e K_i [c_i - (K_e/K_i) c_e]}{[A_i k_i k_{si}(k_{se} k_e + k_m k_e K_e + k_m k_{se} K_e) + A_e k_m k_{se} k_e K_i (k_{si} + k_i)]}. \quad (37\text{--}2)$$

As described in Chapter 36, the encapsulated droplets of the internal phase are considerably smaller in size (1 to 3 μm in diameter) than the emulsion globules, which usually have diameters in the range of 0.1 to 2 mm (100 to 2000 μm). Also, each emulsion globule contains many encapsulated droplets. Consequently, the internal phase mass transfer area A_i is much larger than the external phase mass transfer area A_e. Thus, Eq. (37–2) reduces to:

$$j_k = \left[\left(\frac{k_m k_{se} k_e K_i}{k_{se} k_e + k_m k_e K_e + k_m k_{se} K_e}\right) A_e\right]\left(c_i - \frac{K_e}{K_i} c_e\right). \tag{37-3}$$

Hence, the mass transfer rate for the diffusion transport can be represented as

$$j_k = k' A_v (V_m + V_i)\left(c_i - \frac{K_e}{K_i} c_e\right)$$

$$= k(V_m + V_i)\left(c_i - \frac{K_e}{K_i} c_e\right), \tag{37-4}$$

where k' is the overall mass transfer coefficient based on the mass transfer area of the emulsion globules in the external phase, A_v the external phase mass transfer area per unit volume of the emulsion, V_m the membrane phase volume, V_i the total volume of the internal phase, and k the overall mass transfer coefficient based on the volume of the emulsion $(V_m + V_i)$. This assumes that the mass transfer area is proportional to the volume of the emulsion under a given mixing condition for contact with the external phase.

The assumption that the mass transfer area is proportional to the emulsion volume under a given mixing condition is justified with the following reasoning. Under a given mixing condition, the Sauter mean diameter did not change significantly as a result of varying the emulsion holdup of a given ELM system (Ho 1986). The Sauter mean diameter was found sufficient to characterize the emulsion globule size of an ELM system (Teramoto et al. 1983a, 1983b). Thus, for a given ELM system, the mass transfer areas A_{e1} and A_{e2} for two different holdup volumes of the emulsion, $(V_m + V_i)_1$ and $(V_m + V_i)_2$, respectively, can be expressed in terms of the Sauter mean diameter d_{32} as follows:

$$A_{e1} = \frac{\pi}{4} d_{32}^2 N_1, \tag{37-5}$$

$$A_{e2} = \frac{\pi}{4} d_{32}^2 N_2, \tag{37-6}$$

where N_1 and N_2 are the numbers of emulsion globules for the two emulsion volumes $(V_m + V_i)_1$ and $(V_m + V_i)_2$, respectively. These volumes can also be expressed in terms of d_{32}:

$$(V_m + V_i)_1 = \frac{\pi}{6} d_{32}^3 N_1, \tag{37-7}$$

$$(V_m + V_i)_2 = \frac{\pi}{6} d_{32}^3 N_2. \tag{37-8}$$

Equations (37-5) through (37-8) give the following relationship:

$$\frac{A_{e1}}{A_{e2}} = \frac{(V_m + V_i)_1}{(V_m + V_i)_2}. \tag{37-9}$$

As shown in Eq. (37-9), the mass transfer area is proportional to the volume of the emulsion.

If the concentration driving force is higher for the external phase than for the internal phase, the five individual mass transfer steps described above are reversed.

Breakage

The breakage of the internal phase in terms of the change of the internal phase volume (or mass) with time can be assumed to be proportional to the internal phase volume (or mass) under a given mixing condition (Boyadzhiev, Sapundzhiev, and Bezenshek 1977; Ho and Li 1984; Krishna, Goswami, and Sharma 1987).

$$-\frac{dV_i}{dt} = \Phi V_i, \tag{37-10}$$

where Φ is the breakage coefficient and t is the extraction time. The mass transfer rate due to the breakage, j_Φ, is

$$j_\Phi = \Phi V_i c_i. \tag{37-11}$$

Integration of Eq. (37-10) gives the internal phase volume as a function of time:

$$V_i = V_{i0} e^{-\Phi t}, \tag{37-12}$$

where V_{i0} is the initial volume of the internal phase. The breakage decreases the internal phase volume but increases the external phase volume, V_e, which is given in the following equation:

$$V_e = V_0 - V_i = V_0 - V_{i0}e^{-\Phi t}, \quad (37\text{--}13)$$

where $V_0 = V_{e0} + V_{i0}$, V_0 is the sum of the external and internal phase volumes, and V_{e0} the initial external phase volume.

Case I Model

In Case I of the model of Ho and Li (1984), a chemical reagent in the external phase converts the leaked species into an extractable form. This model is shown schematically in Figure 37–2. Solute A in the external phase is extracted, via the diffusion mechanism associated with the mass transfer coefficient k_A, into the internal phase where the solute reacts with the reagent in the internal phase to become solute B. Solute B can transfer from the internal phase, via the diffusion mechanism associated with the mass transfer coefficient k_B and the breakage mechanism associated with the coefficient Φ, to the external phase where the external reagent converts B into the extractable solute A. In this case, A exists only in the external phase, whereas B exists only in the internal phase. For example, phenol, the solute A, can be extracted into the internal phase where it reacts with the internal reagent NaOH to become sodium phenolate, the solute B. The phenolate can leak from the internal phase, via breakage (the ionic species B cannot diffuse through an oil-type membrane, i.e., $k_B = 0$), into the external phase where H_2SO_4, the external reagent, converts phenolate into phenol, which can then be re-extracted.

The governing differential equation for the Case I model in a batch extraction operation is as follows:

$$\frac{d(V_e c_{eA})}{dt} = k_{A1}(V_m + V_i)\left(c_{iA} - \frac{K_{eA}}{K_{iA}}c_{eA}\right)$$

$$+ k_B(V_m + V_i)\left(c_{iB} - \frac{K_{eB}}{K_{iA}}c_{eB}\right)$$

$$+ \Phi V_i c_{iB}. \quad (37\text{--}14)$$

Since the concentration of A in the internal phase, c_{iA}, and the concentration of B in the external phase, c_{eB}, are zero, this equation reduces to Eq. (37–15):

$$\frac{d(V_e c_{eA})}{dt} = -k_A^\circ c_{eA} + (k_B^\circ + \Phi V_i)c_{iB},$$

$$(37\text{--}15)$$

where

$$k_A^\circ = k_A(V_m + V_i), \quad (37\text{--}16)$$

$$k_B^\circ = k_B(V_m + V_i), \quad (37\text{--}17)$$

$$k_A = k_{A1}\frac{K_{eA}}{K_{iA}}, \quad (37\text{--}18)$$

where k_A and k_B are the overall mass transfer coefficients for solutes A and B, respectively, and c_{eA} and c_{iB} are the concentrations for A in the external phase and for B in the internal phase, respectively. The k_A° and k_B° terms are assumed to be constant, i.e., the mass transfer area or the emulsion volume ($V_m + V_i$) does not change significantly even though breakage of the internal phase to the internal phase occurs. For typical ELM systems with a reasonably good emulsion stability, the breakage is not appreciable, and the volume of the membrane phase (V_m) is much larger than the volume of the internal phase (V_i), e.g., $V_m \geq 2V_i$. Thus, the emulsion volume does not change significantly.

The initial conditions for Eq. (37–15) are

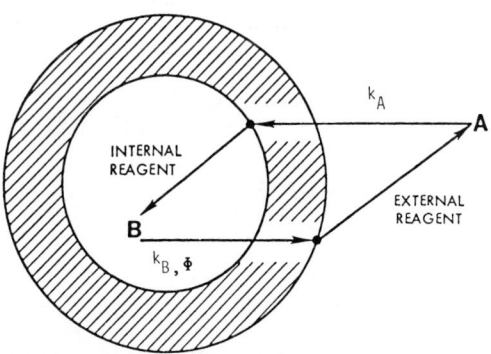

FIGURE 37–2. Schematic of Case I model (reprinted from Ho and Li 1984 with permission).

$c_{eA} = c_{eA0}$ and $c_{iB} = c_{iB0}$. With the assumption that solute accumulation in the membrane phase is negligible compared to the solute amounts in the external and internal phases, c_{iB} is given in the following equation from the conservation of mass:

$$c_{iB} = \frac{1}{V_i}(V_{e0}c_{eA0} + V_{i0}c_{iB0}) - \frac{V_e}{V_i}c_{eA}. \tag{37-19}$$

Equation (37-15) can be solved for the two cases of small breakage or large breakage.

Small Breakage

For small breakage, which is needed for the ELM systems of practical interest, i.e., $\Phi = 0$ or $\Phi \leq 1.4 \times 10^{-5}$ s^{-1}, the change of the internal phase volume is less than 5% as determined from Eq. (37-12) for a 1-hour extraction time. Thus, V_i and V_e can be assumed constant and approximated by their initial volumes, respectively. Substituting Eq. (37-19) into Eq. (37-15) and integrating the resultant equation with the initial condition $c_{eA} = c_{eA0}$ give the solution of c_{eA}:

$$c_{eA} = \left(\frac{(V_e c_{eA0} + V_i c_{iB0})}{V_e}\frac{(k_{B\Phi}/V_i)}{(k_A/V_e + k_{B\Phi}/V_i)}\right)$$

$$\cdot\left\{1 + \frac{1}{k_{B\Phi}/V_i}\left(\frac{k_A c_{eA0} - k_{B\Phi} c_{iB0}}{V_e c_{eA0} + V_i c_{iB0}}\right)\exp[-(k_A/V_e + k_{B\Phi}/V_i)(V_m + V_i)t]\right\}, \tag{37-20}$$

where

$$k_{B\Phi} = k_B + \Phi\left(\frac{V_i}{V_m + V_i}\right), \tag{37-21}$$

with $k_{B\Phi}$ as the overall mass transfer coefficient for leakage. Equations (37-19) and (37-20) can then give the solution for c_{iB}.

Large Breakage

For large breakage, substituting Eqs. (37-12), (37-13), and (37-19) into Eq. (37-15) results in the solution for c_{eA}:

$$c_{eA} = \frac{V_{e0}c_{eA0} + V_{i0}c_{iB0}}{V_0}$$

$$+ \frac{V_{i0}(c_{eA0} - c_{iB0})}{V_0}\left(\frac{V_0 - V_{i0}}{V_0 e^{\Phi t} - V_{i0}}\right)^{\{1+[k_A^\circ/\Phi V_0]\}}\exp[(-k_B^\circ/\Phi V_{i0})(e^{\Phi t} - 1)]$$

$$- \frac{\Psi_1(k_A^\circ - k_B^\circ)(V_{e0}c_{eA0} + V_{i0}c_{iB0})}{\Phi V_0(V_0 - V_{i0})}\left(\frac{V_0 - V_{i0}}{V_0 e^{\Phi t} - V_{i0}}\right)^{[1+(k_A^\circ/\Phi V_0)]}\exp[(-k_B^\circ/\Phi V_{i0})e^{\Phi t}] \tag{37-22}$$

where

$$\Psi_1 = \int_1^{e^{\Phi t}} \left(\frac{V_0 X - V_{i0}}{V_0 - V_{i0}}\right)^{k_A^\circ/(\Phi V_0)} \exp[(k_B^\circ/\Phi V_{i0})X]dX. \tag{37-23}$$

A special case in the Case I model is that the solute B cannot diffuse through the membrane phase, i.e., $k_B = 0$. Ionic species, such as phenolate and ammonium (NH_4^+) ions, generally cannot diffuse through oil-type membranes unless the membranes contain carriers that can complex with the ionic species. For $k_B = 0$, the solutions for c_{eA} (Ho and Li 1984) are given in Appendix A.

Chemical Reagent Consumption

The total consumption amount of the chemical reagent (such as H_2SO_4 for the conversion of phenolate into phenol) in the external phase W_e for the conversion of the leaked species B into the extractable solute A and for the reaction with the leaked internal reagent can be calculated from

$$W_e = \int_0^t [(k_B^\circ + \Phi V_i)c_{iB} + \Phi V_i c_{ir}]\, dt, \qquad (37\text{-}24)$$

where c_{ir} is the concentration of the internal reagent. The differential equation for the material balance of the internal reagent is

$$\frac{d(V_i c_{ir})}{dt} = -k_A^\circ c_{eA} - \Phi V_i c_{ir}. \qquad (37\text{-}25)$$

The initial value for c_{ir} is c_{ir0}. Substitution of Eq. (37–20) into Eq. (37–25) and integration of the resultant equation give the solution of c_{ir} for small Φ values:

$$c_{ir} = c_{ir0} e^{-\Phi t} - \left(\frac{V_e c_{eA0} + V_e c_{iB0}}{V_e} \cdot \frac{k_B \Phi / V_i}{k_A V_e + k_B \Phi / V_i}\right)$$

$$\cdot \left[\frac{k_A(V_m + V_i)}{\Phi V_i}(1 - e^{-\Phi t}) + \left(\frac{k_A c_{eA0} - k_B \Phi c_{iB0}}{(k_A V_i + k_B \Phi V_e)}\right)\right]$$

$$\cdot \left(\frac{k_A(V_m + V_i)}{(k_A/V_e + k_B \Phi/V_i)(V_m + V_i) - \Phi}\right) \{\exp[-(k_A/V_e + k_B \Phi/V_i)(V_m + V_i)t] - \exp(-\Phi t)\}.$$

$$(37\text{-}26)$$

Substituting Eqs. (37–19), (37–20), and (37–26) into Eq. (37–24) and integrating the resultant equation give the total consumption amount of the external reagent for small Φ values:

$$W_e = V_i c_{ir0}(1 - e^{-\Phi t}) + \left(\frac{k_A(V_m + V_i)(1 - e^{-\Phi t})}{V_e(k_A/V_e + k_B \Phi/V_i)}\right)$$

$$\cdot \left(\frac{k_B \Phi(V_e c_{eA0} + V_i c_{iB0})}{\Phi V_i} - \frac{(k_A c_{eA0} - k_B \Phi c_{iB0})}{(k_A/V_e + k_B \Phi/V_i)(V_m + V_i) - \Phi}\right)$$

$$+ \left(\frac{(k_A c_{eA0} - k_B \Phi c_{iB0})}{(k_A/V_e + k_B \Phi/V_i)^2}\right)\left(\frac{k_A/V_e}{(k_A/V_e + k_B \Phi/V_i)(V_m + V_i) - \Phi} - \frac{k_B \Phi}{V_i}\right)$$

$$\cdot \{1 - \exp[-(k_A/V_e + k_B \Phi/V_i)(V_m + V_i)t]\}.$$

$$(37\text{-}27)$$

The total consumption amount of the chemical reagent (such as NaOH for the conversion of phenol into phenolate) in the internal phase W_i for the conversion of solute A into solute B and for the loss due to breakage can be computed from the following equation:

$$W_i = \int_0^t (k_A^\circ c_{eA} + \Phi V_i c_{ir}) \, dt. \tag{37-28}$$

Substitution of Eqs. (37–20) and (37–26) into Eq. (37–28) and integration of the resultant equation lead to the result of W_i for small Φ values:

$$W_i = V_i c_{ir0}(1 - e^{-\Phi t}) + \left[\frac{k_A(V_m + V_i)(1 - e^{-\Phi t})}{V_e(k_A/V_e + k_B\Phi/V_i)}\right]$$

$$\cdot \left(\frac{k_B\Phi(V_e c_{eA0} + V_i c_{iB0})}{\Phi V_i} - \frac{(k_A c_{eA0} - k_B\Phi c_{iB0})}{(k_A/V_e + k_B\Phi/V_i)(V_m + V_i) - \Phi}\right)$$

$$+ \left\{\frac{k_A(V_m + V_i)(k_A c_{eA0} - k_B\Phi c_{iB0})}{V_e(k_A/V + k_B\Phi/V_i)[(k_A/V_e + k_B\Phi/V_i)(V_m + V_i) - \Phi]}\right\}$$

$$\cdot \{1 - \exp[-(k_A/V_e + k_B\Phi/V_i)(V_m + V_i)t]\}. \tag{37-29}$$

Case II Model

In Case II of the model of Ho and Li (1984), there is no chemical reagent in the external phase for the conversion of the leaked species into an extractable form. This model is shown schematically in Figure 37-3. Solute A in the external phase is extracted, via the diffusion mechanism associated with the mass transfer coefficient k_A, into the internal phase where the solute reacts with the internal reagent to become solute B. Solute B can transfer from the internal phase, via both the diffusion mechanism associated with the mass transfer coefficient k_B and the breakage mechanism associated with the coefficient Φ, to the external phase where B remains as the same species. In this case, both solutes A and B can exist in the external phase, whereas B can also exist in the internal phase, but not A. The internal reagent leaked out to the external phase can convert A into B in the external phase. Solute A can exist in the external phase until it is exhausted owing to both extraction and internal reagent leakage. For example, phenol, the solute A, can be extracted into the internal phase where it reacts with the internal reagent NaOH to become sodium phenolate, the solute B. The phenolate can leak out from the internal phase, via breakage (the ionic species B cannot diffuse through an oil-type membrane, i.e., $k_B = 0$), into the external phase where it remains as phenolate. NaOH leaked out from the internal phase to the external phase can convert phenol into phenolate in the external phase. Phenol can exist in the external phase until it is exhausted owing to both extraction and NaOH leakage.

The governing equations for the Case II model in a batch operation are as follows:

$$\frac{d(V_e c_{eA})}{dt} = -k_A^\circ c_{eA} - \Phi V_i c_{ir}, \qquad 0 \le t \le t_1, \tag{37-30}$$

$$\frac{d(V_e c_{eB})}{dt} = k_B^\circ \left(c_{iB} - \frac{K_{eB}}{K_{iB}} c_{eB}\right) + \Phi V_i c_{iB} + \Phi V_i c_{ir}, \qquad 0 \le t \le t_1, \tag{37-31}$$

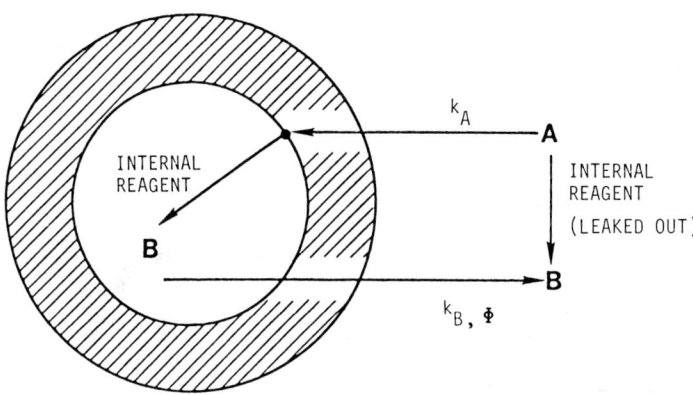

FIGURE 37-3. Schematic of Case II model (reprinted from Ho and Li 1984 with permission).

$$\frac{d(V_i c_{ir})}{dt} = -k_A^\circ c_{eA} - \Phi V_i c_{ir}, \qquad 0 \le t \le t_1, \qquad (37\text{-}32)$$

$$c_{iB} = \frac{1}{V_i}(V_{eo}c_{eA0} + V_{eo}c_{eB0} + V_{i0}c_{iB0}) - \frac{V_e}{V_i}(c_{eA} + c_{eB}), \qquad 0 \le t \le t_1, \qquad (37\text{-}33)$$

$$c_{eA} = 0, \qquad t \ge t_1, \qquad (37\text{-}34)$$

$$\frac{d(V_e c_{eB})}{dt} = k_B^\circ \left(c_{iB} - \frac{K_{eB}}{K_{iB}} c_{eB}\right) + \Phi V_i c_{iB}, \qquad t \ge t_1, \qquad (37\text{-}35)$$

$$\frac{d(V_i c_{ir})}{dt} = -\Phi V_i c_{ir}, \qquad t \ge t_1, \qquad (37\text{-}36)$$

$$c_{iB} = \frac{1}{V_i}(V_{eo}c_{eA0} + V_{eo}c_{eB0} + V_{iB0}c_{iB0}) - \frac{V_e}{V_i} c_{eB}, \qquad t \ge t_1, \qquad (37\text{-}37)$$

where t_1 is the time at which solute A in the external phase is just exhausted as expressed in Eq. (37–34), K_{eB} is the distribution coefficient for the solute B between the membrane and external phases at equilibrium, and K_{iB} is the distribution coefficient for B between the membrane and internal phases at equilibrium. Substituting Eq. (37–30) into (37–32) and integrating the resultant equation lead to the following relationship:

$$V_i c_{ir} = V_e c_{eA} + (V_{i0} c_{ir0} - V_{eo} c_{eA0}), \qquad 0 \le t \le t_1. \qquad (37\text{-}38)$$

Equations (37–30) through (37–38) can be solved for the two cases of small breakage or large breakage.

Small Breakage
For small breakage, i.e., $\Phi = 0$ or $\Phi \le 1.4 \times 10^{-5}$ s^{-1}, as mentioned earlier, V_i and V_e can be assumed constant and approximated by their initial volumes. The solutions for c_{eA}, c_{eB}, and c_{ir} for $0 \le t \le t_1$ are (Ho and Li 1984)

$$c_{eA} = \frac{k_A^\circ c_{eA0} + \Phi V_i c_{ir0}}{k_A^\circ + \Phi V_e} \exp[-(k_A^\circ/V_e + \Phi)t] - \frac{\Phi(V_i c_{ir0} - V_e c_{eA0})}{k_A^\circ + \Phi V_e}, \qquad 0 \le t \le t_1, \qquad (37\text{-}39)$$

$$c_{eB} = (c_{eB0} - \Psi_2 - \Psi_3) \exp\{-[k_B^\circ K_{eB}/(V_e K_{iB}) + (k_B^\circ/V_i) + \Phi]t\}$$

$$+ \Psi_2 \exp[-(k_A^\circ/V_e + \Phi)t] + \Psi_3, \qquad 0 \leq t \leq t_1, \qquad (37\text{--}40)$$

where

$$\Psi_2 = \frac{k_B^\circ V_e (k_A^\circ c_{eA0} + V_i c_{ir0})}{(k_A^\circ + \Phi V_e)[k_A^\circ V_i - k_B^\circ V_e - k_B^\circ V_i(K_{eB}/K_{iB})]}, \qquad (37\text{--}41)$$

$$\Psi_3 = \left(\frac{1}{(k_B^\circ V_i K_{eB}/K_{iB}) + k_B^\circ V_e + \Phi V_i V_e}\right)$$

$$\cdot \left(\frac{\Phi(V_i c_{ir0} - V_e c_{eA0})(k_A^\circ V_i + k_B^\circ V_e + \Phi V_i V_e)}{k_A^\circ + \Phi V_e} + (k_B^\circ + \Phi V_i)(V_e c_{eA0} + V_e c_{eB0} + V_i c_{iB0})\right), \qquad (37\text{--}42)$$

$$c_{ir} = \frac{V_e(k_A^\circ c_{eA0} + V_i c_{ir0})}{V_i(k_A^\circ + \Phi V_e)} \exp[-(k_A^\circ/V_e + \Phi)t] + \frac{k_A^\circ(V_i c_{ir0} - V_e c_{eA0})}{V_i(k_A^\circ + \Phi V_e)}, \qquad 0 \leq t \leq t_1. \qquad (37\text{--}43)$$

The time t_1 is obtained from Eq. (37–39) by letting $c_{eA} = 0$:

$$t_1 = \frac{V_e}{k_A^\circ + \Phi V_e} \ln\left(\frac{k_A^\circ c_{eA0} + \Phi V_i c_{ir0}}{\Phi(V_i c_{ir0} - V_e c_{eA0})}\right). \qquad (37\text{--}44)$$

For $t \geq t_1$, i.e., $c_{eA} = 0$, the solutions for c_{eB} and c_{ir} from Eqs. (37–35), (37–36), and (37–37) are (Ho and Li 1984)

$$c_{eB} = \left(c_{eB}|_{t_1} - \frac{(k_B^\circ + \Phi V_i)(V_e c_{eA0} + V_e c_{eB0} + V_i c_{iB0})}{(k_B^\circ V_i K_{eB}/K_{iB}) + k_B^\circ V_e + \Phi V_e V_i}\right)$$

$$\cdot \exp\{-[k_B^\circ K_{eB}/(V_e K_{iB}) + (k_B^\circ/V_i) + \Phi](t - t_1)\}$$

$$+ \frac{(k_B^\circ + \Phi V_i)(V_e c_{eA0} + V_e c_{eB0} + V_i c_{iB0})}{(k_B^\circ V_i K_{eB}/K_{iB}) + k_B^\circ V_e + \Phi V_e V_i}, \qquad t \geq t_1, \qquad (37\text{--}45)$$

$$c_{ir} = c_{ir}|_{t_1}, \qquad t \geq t_1, \qquad (37\text{--}46)$$

where $c_{eB}|_{t_1}$ is the value of c_{eB} at t_1, which can be obatined from Eq. (37–40) by letting $t = t_1$, and $c_{ir}|_{t_1}$ is the concentration of the internal reagent at the time t_1. This concentration can be obtained from Eq. (37–43) by letting $t = t_1$. Equation (37–46) shows the fact that c_{ir} will not change for $t \geq t_1$ since there is no solute A to be extracted to react with the internal reagent.

Large Breakage
As mentioned above, for large breakage, V_i and V_e can be expressed by Eqs. (37–12) and (37–13), respectively. By the use of these two equations, the solutions for c_{eA}, c_{eB}, and c_{ir} from Eqs. (37–30) through (37–33) for $0 \leq t \leq t_1$ are (Ho and Li 1984):

$$c_{eA} = \frac{(k_A^\circ c_{eA0} + \Phi V_{i0} c_{ir0})}{k_A^\circ + \Phi V_0} \left(\frac{V_0 - V_{i0}}{V_0 e^{\Phi t} - V_{i0}}\right)^{\{1+[k_A^\circ/(\Phi V_0)]\}} - \frac{\Phi(V_{i0} c_{ir0} - V_{e0} c_{eA0})}{k_A^\circ + \Phi V_0}, \qquad 0 \leq t \leq t_1,$$

$$(37\text{--}47)$$

$$c_{eB} = \Psi_4 + \left\{(c_{eB0} - \Psi_4)V_{e0}^{\{1+[k_B^\circ K_{eB}/(\Phi V_0 K_{iB})]\}} \exp[k_B^\circ/(\Phi V_{i0})] + \Psi_4\Psi_6\left(1 - \frac{K_{eB}}{K_{iB}}\right)\frac{k_B^\circ}{\Phi}\right.$$

$$\left. - \Psi_5\left(\frac{k_B^\circ V_0}{\Phi V_{i0}}\right)\frac{(k_A^\circ c_{eA0} + \Phi V_{i0}c_{ir0})}{k_A^\circ + \Phi V_0} + \Psi_6(k_A^\circ - k_B^\circ)\frac{(V_{i0}c_{ir0} - V_{e0}c_{eA0})}{k_A^\circ + \Phi V_0}\right\}$$

$$\cdot \{(V_0 e^{\Phi t} - V_{i0})^{-\{1+[k_B^\circ K_{eB}/(\Phi V_0 K_{iB})]\}} \exp[-(k_B^\circ/\Phi V_{i0})\exp(\Phi t)]\}, \quad 0 \le t \le t_1,$$

(37–48)

where

$$\Psi_4 = \frac{V_{e0}c_{eA0} + V_{e0}c_{eB0} + V_{i0}c_{iB0}}{V_0} + \frac{\Phi(V_{i0}c_{ir0} - V_{e0}c_{eA0})}{k_A^\circ + \Phi V_0}, \quad (37\text{–}49)$$

$$\Psi_5 = \int_1^{e^{\Phi t}} \left(\frac{V_0 - V_{i0}}{V_0 X - V_{i0}}\right)^{k_A^\circ/V_0} (V_0 X - V_{i0})^{k_B^\circ K_{eB}/(\Phi V_0 K_{iB})} \exp\{[k_B^\circ/(\Phi V_{i0})]X\} \, dX, \quad (37\text{–}50)$$

$$\Psi_6 = \int_1^{e^{\Phi t}} (V_0 X - V_{i0})^{k_B^\circ K_{eB}/(\Phi V_0 K_{iB})} \exp\{[k_B^\circ/(\Phi V_{i0})]X\} \, dX, \quad (37\text{–}51)$$

$$c_{ir} = \frac{V_{e0}(k_A^\circ c_{eA0} + \Phi V_{i0}c_{ir0})}{V_{i0}(k_A^\circ + \Phi V_0)}\left(\frac{V_0 - V_{i0}}{V_0 e^{\Phi t} - V_{i0}}\right)^{k_A^\circ/(\Phi V_0)}$$

$$+ \frac{V_{i0}c_{ir0} - V_{e0}c_{eA0}}{(k_A^\circ + \Phi V_0)}\left(\frac{k_A^\circ}{V_{i0}}e^{\Phi t} + \Phi\right), \quad 0 \le t \le t_1. \quad (37\text{–}52)$$

The time t_1 is obtained from Eq. (37–47) by letting $c_{eA} = 0$:

$$t_1 = \frac{1}{\Phi}\ln\left[\frac{V_{i0}}{V_0} + \left(1 - \frac{V_{i0}}{V_0}\right)\left(\frac{k_A^\circ c_{eA0} + \Phi V_{i0}c_{ir0}}{\Phi V_{i0}c_{ir0} - \Phi V_{e0}c_{eA0}}\right)^{\Phi V_0/(k_A^\circ + \Phi V_0)}\right]. \quad (37\text{–}53)$$

For $t \ge t_1$, i.e., $c_{eA} = 0$, the solution for c_{eB} from Eq. (37–35) is (Ho and Li 1984)

$$c_{eB} = \frac{V_{e0}c_{eA0} + V_{e0}c_{eB0} + V_{i0}c_{iB0}}{V_0}$$

$$+ \left(c_{eB}|_{t_1} - \frac{(V_{e0}c_{eA0} + V_{e0}c_{eB0} + V_{i0}c_{iB0})}{V_0}\right)\left[\left(\frac{V_0 e^{\Phi t_1} - V_{i0}}{V_0 e^{\Phi t} - V_{i0}}\right)^{\{1+[k_B^\circ K_{eB}/(\Phi V_0 K_{iB})]\}}\right]$$

$$\cdot \exp\{-[k_B^\circ/(\Phi V_{i0})](e^{\Phi t} - e^{\Phi t_1})\} + \left[\Psi_7\left(1 - \frac{K_{eB}}{K_{iB}}\right)\frac{k_B^\circ(V_{e0}c_{eA0} + V_{e0}c_{eB0} + V_{i0}c_{iB0})}{\Phi V_0}\right]$$

$$\cdot \{(V_0 e^{\Phi t} - V_{i0})^{-\{1+[k_B^\circ K_{eB}/(\Phi V_0 K_{iB})]\}}\exp[-(k_B^\circ/(\Phi V_{i0}))e^{\Phi t}]\}, \quad t \ge t_1,$$

(37–54)

where

$$\Psi_7 = \int_{e^{\Phi t_1}}^{e^{\Phi t}} (V_0 X - V_{i0})^{k_B^\circ K_{eB}/(\Phi V_0 K_{iB})} \exp\{[k_B^\circ/(\Phi V_{i0})]X\} \, dX. \qquad (37\text{-}55)$$

The solution for c_{ir} is $c_{ir}|_{t_1}$, i.e., Eq. (37-46). This value can be determined from Eq. (37-52) by the use of t_1, given by Eq. (37-53).

Similar to the Case I model, a special case in the Case II model is that the solute B cannot diffuse through the membrane phase, i.e., $k_B = 0$. Note that the Case I and Case II models are the same for $k_B = 0$ and $\Phi = 0$. This means that no external external reagent is needed. For ELM systems of practical interest with Type 1 facilitation, $k_B = 0$ and Φ should be negligible, leading to the Case II model which is essentially identical to the Case I model.

The other special case in the Case II model is that there are no chemical conversions of A into B both in the internal and external phases. Solute B is the only species both in the internal and external phases. That is, solute A concentration is zero. Because of no chemical conversion of A into B in the internal phase, the internal reagent for this conversion is not needed. However, an internal reagent such as an acid is still necessary for the ion exchange between protons and metal ions for the extraction of a metal, such as zinc or copper. Mathematically, this special case is given by $k_A = 0$, $c_{eA} = 0$, and $c_{ir} = 0$ (no internal reagent for the conversion of A into B). ELM systems for extraction of zinc (Zn^{2+}), copper (Cu^{2+}), and hydrocarbons are examples of this case. For hydrocarbon extractions, the ratio of K_{eB} to K_{iB} is normally close to unity. For these special cases, the solutions for c_{eB} (Ho and Li 1984) are given in Appendix A.

Chemical Reagent Consumption
The total consumption amount of the chemical reagent in the internal phase W_i for the conversion of solute A into solute B and for the loss due to breakage can be calculated from Eq. (37-28). For small Φ values, the solution of this equation for $0 \leq t \leq t_1$ is (Ho and Li 1984)

$$W_i = \frac{k_A^\circ c_{eA0} + \Phi V_i c_{ir0}}{k_A^\circ + \Phi V_e} \{1 - \exp[-(k_A^\circ/V_e + \Phi)t]\}, \qquad 0 \leq t \leq t_1, \qquad (37\text{-}56)$$

where t_1 is given by Eq. (37-44).

For $t \geq t_1$, the consumption of the internal reagent is due to breakage only since $c_{eA} = 0$. The total consumption amount from $t = 0$ to $t \geq t_1$ is (Ho and Li 1984)

$$W_i = \frac{V_e(k_A^\circ c_{eA0} + \Phi V_i c_{ir0})}{k_A^\circ + \Phi V_e} + \frac{\Phi k_A^\circ (V_i c_{ir0} - V_e c_{eA0})}{k_A^\circ + \Phi V_e}(t - t_1)$$

$$+ \frac{V_e(k_A^\circ c_{eA0} + \Phi V_i c_{ir0})}{k_A^\circ + \Phi V_e} \exp[-(k_A^\circ/V_e + \Phi)t_1][\Phi(t_1 - t) - 1], \quad t \geq t_1. \qquad (37\text{-}57)$$

Validity of Model
The validity of the model has been checked by fitting this model to experimental data (Ho and Li 1984). The Case I model fits the experimental data of Hochhauser (1974) and Hochhauser and Cussler (1975) quite well for batch extraction of chromium from an external aqueous phase for a pH adjusted with H_2SO_4 to 1.6. Under the pH condition, the chromium in the external feed phase presumably existed as dichromate, i.e., $HCr_2O_7^-$, which is solute A in this model. This solute was extracted with an ELM system containing a chromium-complexing agent (carrier), tridodecylamine (TDDA), in the oil phase. The dichromate-TDDA complex was then stripped and the di-

chromate converted to CrO_4^{2-} by the internal reagent NaOH in the internal phase. In this case, CrO_4^{2-} is solute B in the Case I model. From the fitting of the data by the use of Eq. (37–20), the adjustable model parameters obtained are $k_A = 3.84 \times 10^{-2}$ s^{-1} and $k_{B\Phi} = 3.8 \times 10^{-7}$ s^{-1} (indicating a small leakage rate).

The Case I model has also been shown to fit the data quite well for the extraction of uranium from wet process phosphoric acid (Hayworth et al. 1983). The overall mass transfer coefficient k_A as a function of mixing rate and temperature has been obtained and is given in Chapter 38.

The Case II model fits the experimental data of Li and Shrier (1972) quite well for batch extraction of phenol from an external phase of wastewater (Ho and Li 1984). The emulsion used in the extraction contained an oil-type membrane phase and an aqueous internal phase containing NaOH to convert phenol to sodium phenolate. The phenolate leaked from the internal phase, via breakage, to the external phase and remained as the same species since there was no external reagent for the conversion of the phenolate into phenol. In this case, phenol is solute A and phenolate is solute B for the Case II model, in which $k_B = 0$ since phenolate cannot diffuse through the oil-type membrane phase. Li and Shrier's data (1972) indicate a strong surfactant (SPAN 80, sorbitan monooleate manufactured by ICI Americas) concentration effect on extraction results. The fit of Eqs. (37–47), (37–48) [or Eq. (37–114) in Appendix A], and (37–54) [or Eq. (37–115) in Appendix A] to the extraction data gives the adjustable model parameters: $k_A = 2.4 \times 10^{-3}$ s^{-1} and $\Phi = 0.30 \times 10^{-3}$ s^{-1} for a surfactant concentration of 0.1% in the oil phase, $k_A = 50 \times 10^{-3}$ s^{-1} and $\Phi = 0.085 \times 10^{-3}$ s^{-1} for a surfactant concentration of 0.5%, and $k_A = 3 \times 10^{-3}$ s^{-1} and $\Phi = 0.007 \times 10^{-3}$ s^{-1} for a 2% surfactant concentration. Clearly, the breakage coefficient (Φ) decreased significantly with increasing surfactant concentration. The relatively low overall mass transfer coefficient for extraction (k_A) for the lowest surfactant concentration was presumably due to a low mass transfer area associated with poor emulsion stability. For the two higher surfactant concentrations, the mass transfer coefficient appeared to decrease with increasing surfactant concentration. This was presumably due to the diffusivity reduction associated with a viscosity increase for the highest surfactant concentration.

The Case II model has also been applied successfully to the data of Terry, Li, and Ho (1982) for the extraction of phenol. They have shown that the overall mass transfer coefficient k_A can be estimated by fitting Eq. (37–39) with the assumption of $\Phi = 0$ to the theoretical extraction curve predicted from the pseudo-steady-state solution (zero-order perturbation solution) of the advancing front model of Ho et al. (1982), which is described later.

Although this model can fit experimental data reasonably well, it can only provide a first-order estimate of an extraction rate via the use of the model parameters obtained under the same extraction conditions, i.e., the same emulsion, external phase, stirrer tip speed, mixer geometry, and temperature. This is due to the fact that this model does not take the diffusion of solute in emulsion globules into account. This can result in variation of the mass transfer coefficient with time, particularly for a long contact time, as mentioned before. In addition, this model cannot offer a means to probe the actual transport process during the extraction. To overcome these shortcomings, a more rigorous and detailed model based on first principles has been developed, which is called the *advancing front model* (Ho et al. 1982). As mentioned earlier, this rigorous model is described later.

Estimation of Overall Mass Transfer Coefficients for Extraction and Leakage

In the absence of experimental extraction data, the overall mass transfer coefficient k (k_{A1} or k_B) defined in Eq. (37–4) may be estimated from

$$k = \frac{D_m}{l} A_v, \qquad (37\text{–}58)$$

where D_m is the diffusion coefficient of the solute (A or B) in the membrane phase and l the effective membrane thickness between the external and internal phases. The diffusion coefficient can be estimated from the Wilke-Chang correlation (Wilke and Chang 1955) and other correlations (Reid, Prausnitz, and Sherwood 1977; Cussler 1985). The Wilke-Chang correlation written here in terms of SI units for diffusivity D is

$$D = 1.17 \times 10^{-16} \frac{T(\psi_s \tilde{M}_s)^{0.5}}{\eta_s \tilde{V}_A^{0.6}} \quad (m^2/s),$$

(37–59)

where T is the absolute temperature, ψ_s the solvent association factor, $\tilde{M}_s$ the solvent molecular weight, η_s the solvent viscosity, and $\tilde{V}_A$ the molar volume of the solute at normal boiling point.

A rough estimate of the effective membrane thickness may be obtained from the following equation, which gives the interdroplet distance in the liquid membrane emulsion (Kataoka, Nishiki, and Kimura 1989; Gupta, Goswami, and Rawat 1990):

$$l = \left(\frac{4\pi}{3}\right)^{1/3} r_\mu (\phi^{-1/3} - 1), \quad (37\text{–}60)$$

where r_μ is the Sauter mean radius of the internal phase droplets in the emulsion, and ϕ the volume fraction of the internal phase in the emulsion, i.e., $\phi = V_i/(V_m + V_i)$. Kataoka, Nishiki, and Kimura (1989) have obtained the interdroplet distance, Eq. (37–60), based on the Russell (1954) model that visualizes a dispersed system (e.g., an emulsion) as an assembly of cubes of the internal dispersed phase within cubic elements of the continuous phase (e.g., the membrane phase in an emulsion). They have used the interdroplet distance as the thickness of the peripheral liquid membrane layer between the external and internal phases.

The external phase mass transfer area per unit volume of the emulsion A_v can be expressed in the following equation by definition:

$$A_v = \frac{\frac{1}{4}\pi d_{32}^2 N}{\frac{1}{6}\pi d_{32}^3 N} = \frac{3}{2d_{32}}, \quad (36\text{–}61)$$

where N is the number of emulsion globules in the ELM system, and d_{32} the Sauter mean diameter of the emulsion globules. The Sauter mean diameter has to be determined experimentally or estimated from correlation equations, which are given in Chapter 38.

As mentioned above and demonstrated by Terry, Li, and Ho (1982), the overall mass transfer coefficient k_A for the Case II model with no diffusional transport of solute B can be estimated by fitting Eq. (37–39) with the assumption of $\Phi = 0$ to the theoretical extraction curve predicted from the pseudo-steady-state solution (zero-order perturbation solution) of the advancing front model (Ho et al. 1982). Similarly, the overall mass transfer coefficient k_A for the Case I model with no diffusional transport of solute B can also be estimated from the fitting by the use of Eq. (37–20) or (37–110) (in Appendix A) with the assumption of $\Phi = 0$.

In addition to k_B, the overall mass transfer coefficient for leakage $k_{B\Phi}$, defined in Eq. (37–21), also needs the breakage coefficient Φ. Theoretical prediction of the breakage coefficient is difficult. However, as mentioned earlier, the breakage of the internal phase to the external phase is not appreciable for typical ELM systems with a reasonably good emulsion stability. Thus, the breakage coefficient can be assumed to be zero for a first-order estimate of an ELM extraction rate.

Classification of ELM Extraction Cases

In addition to the examples discussed above, some classification examples of emulsion liquid membrane extraction cases are given in Table 37–1 (Ho and Li 1984). In Case I, phenol, cresols (*o*-, *m*-, and *p*-cresols), acetic acid, propionic acid (Terry, Li, and Ho 1982), acrylic acid, lactic acid, benzoic acid, or hydrogen sulfide as solute A in the external phase can be extracted into the internal phase, where solute A reacts with the internal reagent, a base, and becomes ionic species, ϕO^- (where ϕ represents the benzene ring), $CH_3 \phi O^-$, $CH_3 COO^-$, $CH_3 CH_2 COO^-$, $CH_2 = CHCOO^-$, $CH_3 CH(OH)COO^-$, ϕCOO^-, or HS^-, respectively, as solute B. The ionic species that leaks from the internal phase, via breakage, to the external phase can be converted to the original solute A by the external reagent, an acid. Solute

A can then be re-extracted. On the other hand, a base, such as ammonia or amine, as solute A in the external phase can be extracted into the internal phase, where it reacts with the internal reagent, an acid, and becomes an ionic species, such as NH_4^+ or RNH_3^+, as solute B. The ionic species that leaks from the internal phase, via breakage, to the external phase can be converted back to the original solute A by the external reagent, a base. Solute B, an ionic species, cannot diffuse through an oil-type membrane phase and thus k_B can be considered zero unless B can complex with a carrier in the membrane phase.

As mentioned earlier, Case I also includes uranium extraction. In the uranium extraction (Bock et al. 1982; Hayworth et al. 1983), the solute A, uranyl ion (UO_2^{2+}), from the external feed phase of wet process phosphoric acid (WPPA) forms a complex with the complexing agents, di(2-ethylhexyl)phosphoric acid (D2EHPA) and trioctylphosphine oxide (TOPO) (Hurst, Crouse, and Brown 1972), in the oil-type membrane phase. The complex is transported across the membrane to the internal phase containing the reductant, Fe^{2+}, in phosphoric acid, where the uranium is stripped and converted to U^{4+}, solute B. Any U^{4+} that leaks from the internal phase to the external phase is oxidized and converted back to the extractable form, UO_2^{2+}, by an oxidant, such as hydrogen peroxide, introduced into the WPPA feed during a pretreatment step. Since D2EHPA-TOPO does not effectively complex the U^{4+} ion (solute B), this ion is trapped and concentrated in the internal phase, and the overall mass transfer coefficient for U^{4+}, k_B, is much smaller than that for UO_2^{2+}, k_A.

In Case II, phenol (Zhang, Liu, and Lu 1987; Zhang et al. 1988a, 1988b; Li and Shrier 1972; Cahn and Li 1974; Matulevicius and Li 1975; Halwachs, Flaschel, and Schugerl 1980; Marr and Kopp 1980; Ho et al. 1982; Terry, Li, and Ho 1982; Chang and Li 1983; Teramoto et al. 1983a; Boyadzhiev, Bezenshek, and Lazarova 1984; Ho and Li 1984; Miao, Li, and Zhang 1985; Kataoka, Nishiki, and Kimura 1989; Kataoka et al. 1990), cresols (Terry, Li, and Ho 1982; Teramoto et al. 1983a), acetic acid (Terry, Li, and Ho 1982; Thien, Hatton, and Wang 1986; Yan, Huang, and Shi 1987; Thien and Hatton 1988), propionic acid (Terry, Li, and Ho 1982), acrylic acid (Thien, Hatton, and Wang 1986; Thien and Hatton 1988), lactic acid (Thien, Hatton, and Wang 1986; Ho and Cowan 1987; Thien and Hatton 1988), benzoic acid (Marr and Kopp 1980, 1982), hydrogen sulfide, ammonia (Li and Shrier 1972; Schiffer et al. 1974a; Asher et al. 1975, 1976, 1977, 1979; Maugh 1976; Frankenfeld and Li 1977, 1987; Cahn, Li, and Minday 1978; Halwachs and Schugerl 1978; Downs and Li 1981; Marr and Kopp 1982; Schlosser and Kossaczky 1980; Marr and Koncar 1990; Lee and Chan 1990), or amine as solute A can be extracted, in a manner similar to Case I, into the internal phase where solute A is converted into its ionic species as solute B, which cannot diffuse through the membrane phase, i.e., $k_B = 0$. But the solute B that leaks into the external phase remains as the same species because no external reagent exists for the conversion of solute B into A.

Also belonging to Case II is the general class of heavy metal removal and hydrometallurgical extractions by the use of acids as the reagents in the internal phase, such as zinc (Draxler, Fürst, and Marr 1988; Draxler, Marr and Prötsch 1988; Schiffer et al. 1974b; Boyadzhiev and Kyuchoukov 1980; Boyadzhiev and Bezenshek 1982; Prötsch and Marr 1983; Fuller and Li 1984; Draxler and Marr 1986; Lorbach and Marr 1987; Kataoka et al. 1987, 1990; Ruppert, Draxler, and Marr 1988; Marr, Bart, and Draxler 1990; Goto et al. 1991; Shiau 1991), nickel (Draxler and Marr 1986; Marr, Lackner, and Bart 1989; Cussler and Evans 1980; Strzelbicki and Charewicz 1980; Wodzki, Wyszynska, and Narebska 1990), copper (Cahn et al. 1981; Frankenfeld, Cahn, and Li 1981; Schiffer et al. 1974b; Martin and Davies 1977, 1980; Kitagawa et al., 1977; Lee, Evans, and Cussler 1978; Kondo et al. 1979; Boyadzhiev and Kyuchoukov 1980; Cussler and Evans 1980; Vokel, Halwachs, and Schugerl 1980; Strzelbicki and Charewicz 1980; Miyake, Takenoshita, and Teramoto 1983; Teramoto et al. 1983b, 1983c; Lorbach, Bart, and Marr 1986; Kataoka et al. 1987, 1989, 1990; Bart et al. 1988; Li and

TABLE 37–1. Examples in Classification of Emulsion Liquid Membrane Extraction Cases.

		External Phase		Internal Phase		
Case	Extraction	External Reagent	Solute A	Solute B	Internal Reagent	Solute B	Mass Transfer Coefficients
I	Phenol	Acid	ϕOH	None	Base	ϕO$^-$	$k_B = 0$
	Cresols	Acid	CH$_3\phi$OH	None	Base	CH$_3\phi$O$^-$	$k_B = 0$
	Acetic acid	Acid	CH$_3$COOH	None	Base	CH$_3$COO$^-$	$K_B = 0$
	Propionic acid	Acid	CH$_3$CH$_2$COOH	None	Base	CH$_3$CH$_2$COO$^-$	$k_B = 0$
	Acrylic acid	Acid	CH$_2$=CHCOOH	None	Base	CH$_2$=CHCOO$^-$	$k_B = 0$
	Lactic acid	Acid	CH$_3$CH(OH)COOH	None	Base	CH$_3$CH(OH)COO$^-$	$k_B = 0$
	Benzoic acid	Acid	ϕCOOH	None	Base	ϕCOO$^-$	$k_B = 0$
	Hydrogen sulfide	Acid	H$_2$S	None	Base	HS$^-$	$k_B = 0$
	Ammonia	Base	NH$_3$	None	Acid	NH$_4^+$	$k_B = 0$
	Amines	Base	RNH$_2$	None	Acid	RNH$_3^+$	$k_B = 0$
	Chromium	Acid	HCr$_2$O$_7^-$	None	Base	CrO$_4^{2+}$	$k_B = 0$
	Uranium	Oxidant in Acid	UO$_2^{2+}$	None	Reductant in Acid	U^{4+}	$k_B \ll k_A$
	Phenol	None	ϕOH	ϕO$^-$	Base	ϕO$^-$	$k_B = 0$
	Cresols	None	CH$_3\phi$OH	CH$_3\phi$O$^-$	Base	CH$_3\phi$O$^-$	$k_B = 0$
	Propionic acid	None	CH$_3$CH$_2$COOH	CH$_3$CH$_2$COO$^-$	Base	CH$_3$CH$_2$COO$^-$	$k_B = 0$

	Acrylic acid	None	CH_2=CHCOOH	CH_2=CHCOO$^-$	Base	$k_B = 0$
	Lactic acid	None	$CH_3CH(OH)COOH$	$CH_3CH(OH)COO^-$	Base	$k_B = 0$
	Benzoic acid	None	$\phi COOH$	ϕCOO^-	Base	$k_B = 0$
	Hydrogen sulfide	None	H_2S	HS^-	Base	$k_B = 0$
	Ammonia	None	NH_3	NH_4^+	Acid	$k_B = 0$
	Amines	None	RNH_2	RNH_3^+	Acid	$k_B = 0$
	Zinc	None	None	Zn^{2+}	Acid	$k_A = 0$
	Nickel	None	None	Ni^{2+}	Acid	$k_A = 0$
	Copper	None	None	Cu^{2+}	Acid	$k_A = 0$
	Cadmium	None	None	Cd^{2+}	Acid	$k_A = 0$
II	Lead	None	None	Pb^{2+}	Acid	$k_A = 0$
	Mercury	None	None	Hg^{2+}	Acid	$k_A = 0$
	Chromium	None	None	$Cr_2O_7^{2-}$	Acid	$k_A = 0$
	Cobalt	None	None	Co^{2+}	Acid	$k_A = 0$
	Manganese	None	None	Mn^{2+}	Acid	$k_A = 0$
	Molybdenum	None	None	Mo^{4+}	Acid	$k_A = 0$
	Thallium	None	None	Tl^+	Acid	$k_A = 0$
	Gold	None	None	Au^+	Acid	$k_A = 0$
	Silver	None	None	Ag^+	Acid	$k_A = 0$
	Rare earth	None	None	M^{3+}	Acid	$k_A = 0$
	Uranium	None	None	UO_2^{2+}	Base	$k_A = 0$
	Hydrocarbons	None	None	Hydrocarbons	None	$k_A = 0$

Frankenfeld 1988; Goto, Kondo, and Nakashio 1989; Marr, Bart, and Draxler 1990; Matsumoto et al. 1990; Nielsen, Jong, and Stubbs 1991), cadmium (Kitagawa et al. 1977; Boyadzhiev and Kyuchoukov 1980; Marr, Bart, and Draxler 1990), lead (Schiffer et al. 1974b; Boyadzhiev and Kyuchoukov 1980; Biehl et al. 1982; Izatt et al. 1982; Marr, Bart, and Draxler 1990; Hayashita et al. 1991), mercury (Schiffer et al. 1974b; Kitagawa et al. 1977; Weiss, Griegoriev, and Mühl 1982; Marr, Bart, and Draxler 1990), chromium (Kitagawa et al. 1977; Frankenfeld and Li 1977, 1979; Fuller and Li 1984), cobalt (Strzelbicki 1978; Strzelbicki and Charewicz 1980; Wasan, Gu, and Li 1984; Gu, Wasan, and Li 1985, 1986, 1988; Abou-Nemeh and Van Peteghem 1989; Strzelbicki and Schlosser 1989; Kumamaru et al. 1990; Okamoto et al. 1990), manganese (Izatt et al. 1983; Abou-Nemeh and Van Peteghem 1989), molybdenum (Hirato et al. 1990), thallium (Izatt et al. 1982), gold (Yan, Shi, and Su 1990), silver (Izatt et al. 1983), and rare earth metals including La^{3+}, Nd^{3+}, Sm^{3+}, Eu^{3+}, Gd^{3+}, Dy^{3+}, and Yb^{3+} (Teramoto et al. 1986; Goto et al. 1989; Zhang and Wang 1989; Zhang and Xiao 1989, 1990). Another hydrometallurgical extraction class belonging to Case II, for example, the recovery of uranium from mine leachate and aqueous solutions (Hirato et al. 1991), utilizes a base (such as sodium carbonate) as the internal phase reagent.

Case II also includes the separation of hydrocarbons (Li 1968, 1971a, 1971b, 1971c, 1971d, 1977; Shah and Owens 1972; Cahn and Li 1976a, 1976b; Stelmaszek and Borkowska 1977; Casamatta, Charvarie, and Angelino 1978; Kikic, Alessi, and Orlandini-Visalberghi 1978; Alessi, Kikic, and Orlandi-Visalberghi 1980; Halwachs, Flaschel, and Schugerl 1980; Halwachs and Schugerl 1980; Goswami and Rawat 1984a, 1984b; Krishna, Goswami, and Sharma 1987; Hung, Chen, and Lee 1989; Gupta, Goswami, and Rawat 1990; Garti and Kovacs 1991; Ulbrich, Marr, and Draxler 1991). These hydrocarbon and hydrometallurgical extraction examples belong to the special case of the Case II model in which there are no chemical conversions of solute A into B both in the internal and external phases. Solute B is the only species in these phases, i.e., $k_A = 0$, $c_{eA} = 0$, and $c_{ir} = 0$ (no internal reagent for the conversion of solute A into B). The difference between hydrocarbon and hydrometallurgical extractions is that the ratio of distribution coefficients, K_{eB}/K_{iB}, is generally close to unity for hydrocarbon extractions but not for hydrometallurgical extractions (Ho and Li 1984).

Note that the extraction of chromium ($Cr_2O_7^{2-}$) by the use of a base (e.g., NaOH) as the internal phase and no external reagent also belongs to Case II (Schiffer et al. 1974b; Kitagawa et al. 1977; Frankenfeld and Li, 1977, 1979). In this case, solute A in the external phase is $Cr_2O_7^{2-}$; solute B existing in the internal phase from the conversion of solute A by the base and possibly in the external phase via breakage is CrO_4^{2-}, and k_B is very small, nearly zero.

Effects of Leakage and External Reagent on ELM Extraction

Breakage decreases the recovery of solute such as phenol in Case II of emulsion liquid membrane extraction, in which there is no external reagent for the conversion of the species, phenolate, leaked into the extractable form, phenol, in the external phase. However, an external reagent in Case I of ELM extraction can play a significant role in the solute recovery process. For the extraction of solute from the external phase into the internal phase, solute recovery for Case II is defined by

$$\text{solute recovery} = 1 - \frac{V_e(c_{eA} + c_{eB})}{V_{e0}(c_{eA0} + c_{eB0})}.$$

(37–62)

In Case I, i.e., $c_{eB} = 0$, this equation becomes

$$\text{solute recovery} = 1 - \frac{V_e c_{eA}}{V_{e0}\, c_{eA0}}. \quad (37\text{–}63)$$

The effects of leakage and an external reagent on the recovery of solute in ELM extraction are shown in Figure 37–4 (Ho and Li 1984). The curves for the extraction with an

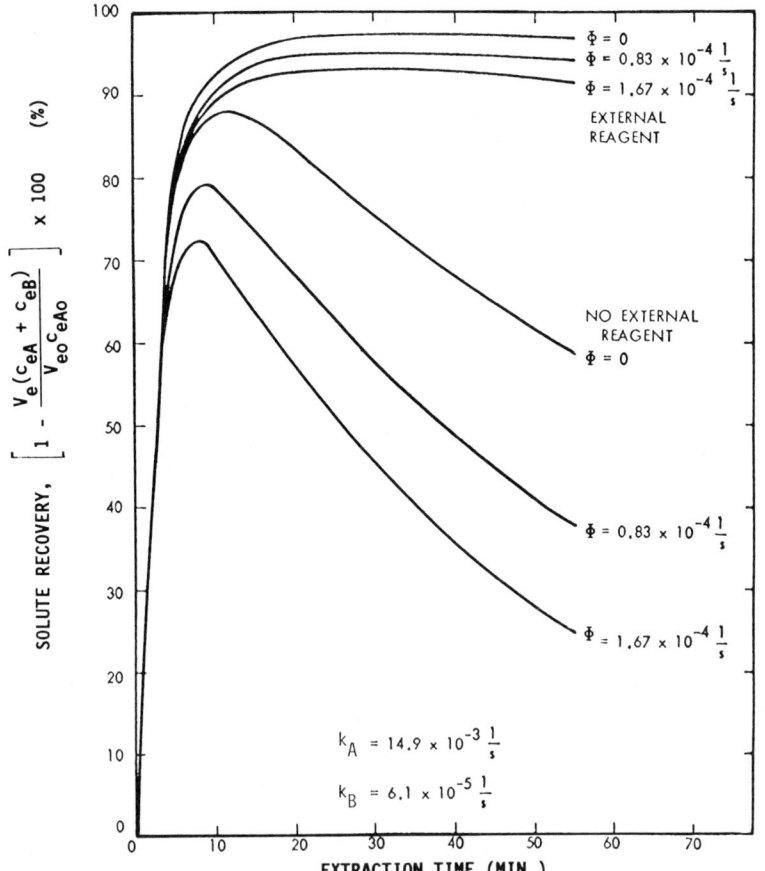

FIGURE 37–4. Effects of leakage and external reagent on emulsion liquid membrane extraction (reprinted from Ho and Li 1984 with permission).

external reagent are obtained from Eqs. (37–20) and (37–22) of the Case I model, whereas the curves for the extraction with no external reagent are calculated results from Eqs. (37–39), (37–40), (37–45), (37–47), (37–48), and (37–54) of the Case II model. In the calculation, the extraction conditions used are $k_A = 14.9 \times 10^{-3}$ s^{-1}, $k_B = 6.1 \times 10^{-5}$ s^{-1}, $V_{eO}/(V_m + V_{iO}) = 3$, $V_{iO}/(V_m + V_{iO}) = 0.333$, $c_{eAO} = 1.68 \times 10^{-3}$ N, $c_{eBO} = 0$, $c_{iBO} = 0$, $c_{irO} = 3.6 \times 10^{-2}$ N, and $K_{eB}/K_{iB} = 1$. This figure shows that the effect of breakage on solute recovery is not very significant for the extraction with an external reagent. However, for the extraction without an external reagent, solute recovery is poor for a long extraction time, and the breakage influences the extraction significantly. Even for cases without breakage, the leakage due to diffusional transport (k_B) can decrease the solute recovery greatly for the extraction without an external reagent but not for the extraction with an external reagent. Thus, an external reagent can affect the ELM extraction greatly by enhancing solute recovery and minimizing the effect of leakage including breakage.

Chemical Reagent Consumption

Although an external reagent can minimize leakage effects and increase solute recovery for ELM extraction, the cost of the external reagent should be considered. The amount of total consumption of the external reagent can be calculated from Eqs. (37–24) and (37–27) (Ho and Li 1984). The extraction conditions used in the calculation are the same as those just mentioned for Figure 37–4. As shown in Figure 37–5, the external reagent consumption increases with increasing extraction time and breakage.

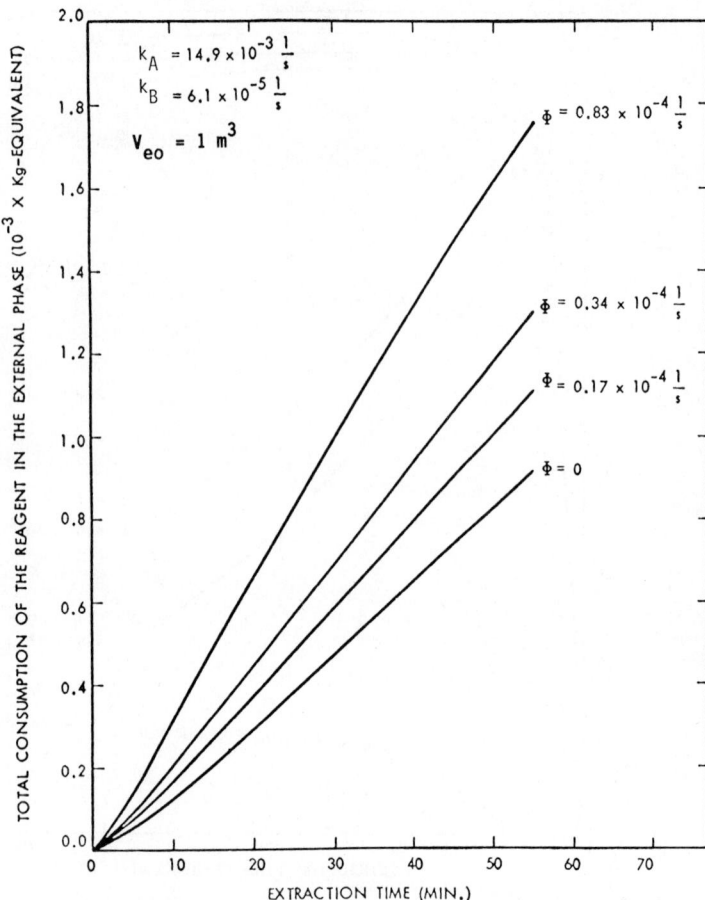

FIGURE 37–5. Consumption of the external reagent for the conversion of leaked species into extractable solute (reprinted from Ho and Li 1984 with permission).

Emulsion Globule Approach

As described earlier, shortcomings are associated with the spherical shell models for the extraction of solute by emulsion liquid membranes. These models do not take the diffusion of solute in emulsion globules into account. This can result in variation of the mass transfer coefficient with time, particularly for a long contact time (Cahn and Li 1974; Hatton, Lightfoot, and Li 1982). These models do not account for the effect of the rate at which the internal reagent is consumed either. Kopp, Marr, and Moser (1978) recognized this problem; they proposed that the diffusion process be described in terms of a boundary at which the reaction occurs and which moves toward the center of the emulsion globule as the internal reagent is consumed. However, their use of the solution for the equivalent planar problem to represent the transport in a spherical geometry limits the range of applicability of their work. Also, their neglect of changes in the external phase concentration further limits the applicability range. In addition, the spherical shell models cannot offer a means to probe the actual transport process during the extraction. To overcome these shortcomings, Ho et al. (1982) have developed a more rigorous and accurate model based on first principles, as explained in the next section.

Advancing Front Model

The advancing front model (AFM) of Ho et al. (1982) assumes the following:

1. The ELM system is a monodisperse collection of spherical emulsion globules in an

external, continuous feed phase. Although it is almost certain that the dispersion of the emulsion in the continuous phase exhibits a nonuniform globule size distribution, this model assumes that the system can be described adequately as being monodispersed with globules of some suitably defined average size, e.g., the Sauter mean diameter. In addition to the good agreement between this model and experimental data, the use of the Sauter mean diameter was further supported later. Teramoto et al. (1983a, 1983b) showed that the Sauter mean diameter was sufficient to characterize globule size, and it was not necessary to use the size distribution.

2. The ELM system is also a noncoalescing collection of the spherical globules. This ignores the possibility of coalescence and redispersion of emulsion globules since these processes can be expected to be largely inhibited as a result of the low interfacial tension produced by the surfactant in the membrane phase (Tavlarides et al. 1970).

3. There is no internal circulation in the globules. This assumption is justified in view of the strong presence of the surfactant in the membrane phase (Rumscheidt and Mason 1961; Levich 1962). This ensures that the internal reagent in the encapsulated droplets is immobilized.

4. The solute taken from the external phase via dissolution in the membrane phase diffuses through the globule to a reaction front, where it is removed by reaction with the internal reagent. This reaction is assumed to be instantaneous and irreversible. As shown in Figure 37–6, the reaction front separates the inner region containing no solute from the outer region in which the internal reagent has been consumed and contains no reagent. This reaction front advances toward the globule center as the internal reagent is consumed.

5. For the solute, there is local equilibrium between the internal and membrane phases in the reacted region (whose internal reagent has been consumed) of the globule. The solute concentration within the globule can be described in terms of the average local concentration. That is, the composite nature of the emulsion can be disregarded, and the emulsion can be treated as though it were a continuum. This assumption is reasonable in view of the fact that the encapsulated internal droplets are considerably smaller in size (1 to 3 μm in diameter) than the emulsion globules (100 to 2000 μm). Consequently, the diffusion time constants for the droplets are considerably less than those for the globules.

6. At the globule surface, the solute concentrations in the membrane phase and the emulsion are in equilibrium with that in the external phase. This is the typical assumption for diffusion-controlled mass transfer.

7. External phase mass transfer resistance is negligible. This assumption is usually justified for a well-agitated system with a high extraction rate.

8. Leakage of the internal reagent and the reaction product in the internal phase into the external phase is negligible. In general, this assumption is justified for typical ELM systems with a reasonably good emulsion stability (with $\leq$3-μm internal phase droplet diameters) and a reaction product that is incapable of dissolution and diffusion in the membrane phase (e.g., sodium phenolate does not have diffusional leakage through an oil-type membrane phase).

Mathematical Description

The equations describing the solute concentrations in the emulsion globules and the external phase are (Ho et al. 1982) as follows:

Globules:

$$\frac{\partial c}{\partial t} = \frac{D'_{\text{eff}}}{r^2} \frac{\partial}{\partial r}\left(r^2 \frac{\partial c}{\partial r}\right), \quad r_f(t) \leq r \leq r_o, \tag{37-64}$$

$$t = 0, \quad c = 0, \quad r < r_o, \tag{37-65}$$

$$r = r_o, \quad c = K'_e c_e, \quad t > 0, \tag{37-66}$$

$$r = r_f(t), \quad c = 0, \quad t > 0. \tag{37-67}$$

FIGURE 37–6. Schematic of the advancing front model (reprinted from Ho et al. 1982 with permission).

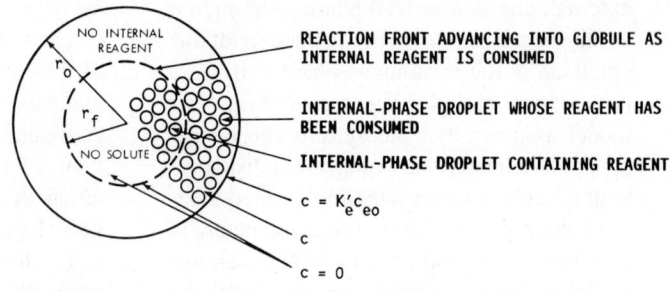

External phase:

$$-V \frac{dc_e}{dt} = N(4\pi r_o^2) D'_{eff} \left.\frac{\partial c}{\partial r}\right|_{r=r_o}$$

$$= \frac{3}{r_o}(V_m + V_i) D'_{eff} \left.\frac{\partial c}{\partial r}\right|_{r=r_o}, \quad (37\text{–}68)$$

$$t = 0, \quad c_e = c_{e0}. \quad (37\text{–}69)$$

The material balance over the reaction front gives

$$-\left(\frac{V_i}{V_m + V_i}\right) c_{ir0} \frac{dr_f}{dt} = D'_{eff} \left.\frac{\partial c}{\partial r}\right|_{r=r_f(t)}, \quad (37\text{–}70)$$

$$t = 0, \quad r_f = r_o, \quad (37\text{–}71)$$

where

$c(r)$	= solute concentration in the reacted region of the emulsion globule, averaged over the membrane and internal phases
D'_{eff}	= solute effective diffusivity, based on the average concentration c in the emulsion mixture
K'_e	= distribution coefficient of the solute between the reacted emulsion mixture and the external phase at equilibrium
r_f	= position of the advancing reaction front
r_o	= emulsion globule radius
V_e, V_m, V_i	= the external phase, membrane phase, and internal phase volumes, respectively
N	= the total number of emulsion globules dispersed in the external phase.

These equations are inherently nonlinear, and they cannot be solved analytically. Ho et al. (1982) have solved them by the use of a perturbation method. The zero-order perturbation solutions are (Ho et al. 1982)

$$\hat{\phi} = \frac{E}{3}(\chi^3 - \beta^3)(1 - \delta), \quad (37\text{–}72)$$

$$h = \frac{E}{3}(\chi^3 - \beta^3), \quad (37\text{–}73)$$

$$\tau = \frac{1}{E}\left[\left(1 + \frac{1}{2\beta}\right)\ln\left(\frac{\chi^3 - \beta^3}{1 - \beta^3}\right) - \frac{3}{2\beta}\ln\left(\frac{\chi - \beta}{1 - \beta}\right)\right]$$

$$- \frac{\sqrt{3}}{E\beta}\left[\tan^{-1}\left(\frac{2\chi + \beta}{\sqrt{3}\beta}\right) - \tan^{-1}\left(\frac{2 + \beta}{\sqrt{3}\beta}\right)\right], \quad (37\text{–}74)$$

where

$$\beta = \left(1 - \frac{3}{E}\right)^{1/3}, \quad (37\text{–}75)$$

$$\hat{\phi} = \zeta g, \quad (37\text{–}76)$$

$$\delta = \frac{1 - \zeta}{1 - \chi}, \quad (37\text{–}77)$$

$$\zeta = \frac{r}{r_o}; \quad \chi = \frac{r_f}{r_o}; \quad \tau = \frac{\epsilon D'_{eff}}{r_o^2};$$

$$g = \frac{c}{K'_e c_{e0}}; \quad h = \frac{c_e}{c_{e0}};$$

$$\epsilon = \frac{K'_e c_{e0}}{[V_i/(V_m + V_i)]\, c_{ir0}}; \quad E = 3\left(\frac{c_{ir0} V_i}{c_{e0} V_e}\right),$$

(37–78)

and ϵ is the ratio of the solute concentration in the emulsion mixture (unreacted) in equilibrium with the initial external phase concentration to the internal reagent concentration averaged over the emulsion mixture. That is, ϵ is a measure of the globule capacity for unreacted solute relative to the reaction capacity provided by the internal reagent. In typical ELM systems, the solute in the external feed phase is normally present in small amounts, whereas the internal reagent concentration is generally high. Thus, ϵ is considerably less than unity, and it serves as the ideal perturbation parameter. The $E/3$ term represents the ratio of the equivalents of the internal reagent to the equivalents of the solute introduced to the external feed phase. Thus, $E/3$ is a measure of the capacity of the ELM system for performing the desired separation.

From a numerical standpoint, the above solution for τ breaks down for $E = 3$ ($\beta = 0$), at which point a number of terms become indeterminate. These indeterminate terms are readily resolved using L'Hospital's rule. The zero-order perturbation solutions for $E = 3$ are

$$\hat{\phi} = \chi^3(1 - \delta),$$

(37–79)

$$h = \chi^3,$$

(37–80)

$$\tau = \ln \chi + \left(\frac{1}{\chi} - 1\right).$$

(37–81)

In typical ELM systems of practical interest, for which the values of ϵ are somewhat less than unity, the zero-order perturbation solutions (pseudo-steady-state solutions) alone are sufficient, and there is no need to use the higher-order solutions (Ho et al. 1982). For example, the deviations between the zero-order and higher-order solutions for $E = 2.0$ and $\epsilon = 0.2$ are probably well within the bounds of experimental errors, and thus there is no advantage to be gained by going beyond the zero-order solutions. However, for higher values of ϵ, appreciable errors can occur if higher-order terms are neglected. The higher-order perturbation solutions along with dimensionless equations for this model (Ho et al. 1982) are given in Appendix B. The range of ϵ values for which the zero-order solutions are adequate appears to increase with increasing E, which corresponds to a decreasing penetration of the reaction front into the emulsion globule.

Estimation of Physical Properties

Two physical properties—the equilibrium distribution coefficient for the solute between the reacted emulsion mixture and the external feed phase (K'_e) and the effective diffusivity of the solute in the emulsion mixture—are necessary if we are to use the AFM for the prediction of the extraction of the solute. Below are the estimation methods of these physical properties.

The average concentration c of the diffusing solute in the reacted region of the emulsion mixture ($r > r_f$) is (Ho et al. 1982)

$$c = \frac{V_m c_m + V_i c_i}{V_m + V_i} = \left(\frac{V_m + V_i/K_i}{V_m + V_i}\right) c_m,$$

(37–82)

where c_m and c_i are the solute concentrations in the membrane phase and the reacted internal phase, respectively, and K_i is the distribution coefficient for the solute between the membrane and reacted internal phases at equilibrium. As described above, the model assumes the equilibrium. Also, this model assumes equilibria among the membrane phase, emulsion mixture, and external phase at the globule surface.

$$c_m = K_e c_e \quad \text{at } r = r_o,$$

(37–83)

$$c = K'_e c_e \quad \text{at } r = r_o,$$

(37–84)

where K_e is the equilibrium distribution coefficient for the solute between the membrane and external phases. As mentioned before, K'_e is the equilibrium distribution coefficient for the solute between the reacted region of the emulsion mixture and the external phase. Equations (37–82), (37–83), and (37–84) give

$$K'_e = \frac{K_e V_m + V_i K_e/K_i}{V_m + V_i}. \quad (37-85)$$

For a typical ELM system, K_e/K_i is about unity. Thus, K'_e may be approximated by

$$K'_e = \frac{K_e V_m + V_i}{V_m + V_i}. \quad (37-86)$$

The effective diffusivity D_{eff} of the solute in the emulsion mixture, based on a concentration driving force defined in terms of the membrane phase concentration c_m, can be estimated from the Jefferson-Witzell-Sibbett equation (Jefferson, Witzell, and Sibbett 1958; Crank 1975; Ho et al. 1982). This equation, in our notation, reads

$$D_{\text{eff}} = D_m \left(\frac{4(1 + 2p)^2 - \pi}{4(1 + 2p)^2} \right)$$
$$+ \frac{\pi}{4(1 + 2p)^2} \left(\frac{(1 + 2p) D_A D_m}{D_m + 2p D_A} \right), \quad (37-87)$$

where

$$D_A = \frac{2(D_i/K_i) D_m}{(D_i/K_i) - D_m}$$
$$\cdot \left[\frac{D_i/K_i}{(D_i/K_i) - D_m} \ln \left(\frac{(D_i/K_i)}{D_m} \right) - 1 \right], \quad (37-88)$$

$$p = 0.403 \left(\frac{V_i}{V_m + V_i} \right)^{-1/3} - 0.5, \quad (37-89)$$

D_i is the solute diffusivity in the reacted internal phase, and as mentioned earlier, D_m is the solute diffusivity in the membrane phase.

The effective diffusivity D'_{eff} based on the average concentration c in the emulsion mixture can be related to D_{eff} through (Ho et al. 1982)

$$D'_{\text{eff}} \frac{dc}{dr} = D_{\text{eff}} \frac{dc_m}{dr}. \quad (37-90)$$

Substituting Eq. (37–82) into Eq. (37–90) and using Eq. (37–85) lead to

$$D'_{\text{eff}} = \left(\frac{K_e}{K'_e} \right) D_{\text{eff}}. \quad (37-91)$$

The diffusivities of the solute in the membrane and internal phases, D_m and D_i, can be estimated from the Wilke-Chang correlation. This correlation, in our notation, is Eq. (37–59).

For the phenol extraction data (at 23°C) of Ho et al. (1982), the distribution coefficient for phenol between the emulsion and the external phase, K'_e, estimated from Eq. (37–85) [or Eq. (37–86)] with the experimentally determined value of 0.52 for K_e (the distribution coefficient for phenol between the membrane and external phases) and the assumption of K_i (the distribution coefficient for phenol between the membrane and reacted internal phases) equal to K_e, was 0.694. The phenol diffusivities estimated from the Wilke-Chang correlation were $D_m = 0.65 \times 10^{-10}$ m²/s and $D_i = 9.98 \times 10^{-10}$ m²/s. The effective diffusivity of phenol in the emulsion, D'_{eff}, estimated from the Jefferson-Witzell-Sibbett equation was 1.23×10^{-10} m²/s.

With the distribution coefficient K'_e and effective diffusivity D'_{eff} plus the experimentally determined Sauter mean globule radii by a photographic technique, 0.50 mm for a mixing rate of 400 rpm and 0.30 mm for 600 rpm, Figure 37–7 shows the comparison between the predictions from the advancing front model (with $E = 3.32$ and $\epsilon = 0.0418$ for the experimental conditions) and the experimental data for phenol extraction (Ho et al. 1982). As shown in this figure, this model agrees quite well with the data. This model is the first to provide reasonably accurate predictions without the use of adjustable parameters. The only information required, in addition to equilibrium distribution coefficients and diffusivities, is the Sauter mean diameter of the suspended emulsion globules.

As shown in Figure 37–7, the differences between the theoretical and experimental results may be due to (1) the error in the estimation of the effective diffusivity, (2) some membrane rupturing, or (3) secondary effects neglected

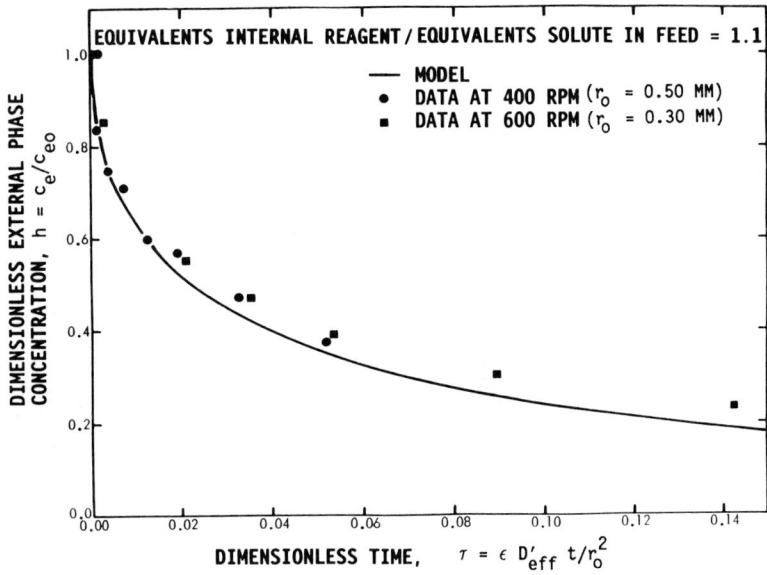

FIGURE 37-7. Advancing front model agrees quite well with data for phenol removal (reprinted from Ho et al. 1982 with permission).

because of the assumptions of the model. However, the differences are not very significant. To take care of the secondary effects, some modifications of the model have been proposed, which are described in the following section.

Modifications to the Advancing Front Model

The advancing front model of Ho et al. (1982) has been modified to account for the effects on the extraction rate of solute neglected because of its assumptions. The effects may be caused by the following factors: (1) external phase mass transfer resistance, (2) reaction reversibility in the internal phase, (3) mass transfer resistance in the peripheral thin membrane layer containing no internal phase droplets (an extra layer to the emulsion, which is treated as though it were a continuum), and (4) leakage of the internal phase into the external phase (Ho 1990). These effects are secondary for typical emulsion liquid membrane systems, for which the assumptions of the advancing front model that neglect these factors are reasonably justified as discussed above.

Fales and Stroeve (1984) and Stroeve and Varanasi (1984) extended the advancing front model to include external phase mass transfer resistance. They obtained zero- and first-order perturbation solutions, and they showed that their results reduce to those of the advancing front model when the external mass transfer resistance becomes negligible. Later, Yan, Huang, and Shi (1987) also considered the external mass transfer resistance in the model for extraction of acetic acid from wastewater. The external mass transfer resistance becomes significant if the Biot number is less than 20. The Biot number, Bi, is

$$\mathrm{Bi} = \frac{k_e r_o}{K'_e D'_{\mathrm{eff}}}, \qquad (37\text{-}92)$$

where k_e is the external phase mass transfer coefficient. This coefficient can be estimated from several available correlations (Calderbank and Moo-Young 1961; Rowe, Claxton, and Lewis 1965; Levins and Glastonbury 1972; Boon-long, Laquerie, and Couderc 1978; Skelland and Lee 1981). In other words, if the Biot number is greater than 20, the differences between their results and those of the advancing front model are within 5% in terms of the degree of extraction or within 10% in terms of external phase concentration. Nevertheless, for typical ELM systems, good mixing in the external phase is used to decrease globule size for high mass transfer surface area. Therefore, the Biot number is usually greater than 20, and the external mass transfer resistance is generally negligible. If the external mass transfer resistance is significant, the polydispersity of emulsion globule size may be considered, as indicated in Eq. (37–92) that the Biot number depends on the globule size. However, as mentioned before, Teramoto et al. (1983a, 1983b) showed that the Sauter mean diameter was sufficient to characterize the globule size, and it was not necessary to use the size distribution.

In addition to external phase mass transfer resistance, Teramoto et al. (1981, 1983a) considered reaction reversibility in the internal phase. Bunge and Noble (1984) also considered the reaction reversibility. Their models have incorporated the reaction equilibrium between the solute and internal reagent throughout the emulsion globule. As pointed out by Noble, Way, and Bunge (1988), one complicating feature of the reaction reversibility is that the apparent diffusion of the solute through the emulsion globule depends on the amount of the solute reacted, and thus it varies with local solute concentration. Teramoto et al. were forced to determine the enhanced diffusivity experimentally. They also investigated the effect of globule size distribution on extraction rate in this work. As mentioned above, they found that the Sauter mean diameter was sufficient to characterize globule size, and it was not necessary to use the size distribution. Bunge and Noble proposed a way to estimate the enhancement effect. In addition, all the models with the reaction reversibility have equations so complicated that even quasi-analytical solutions cannot be developed (Ho 1990). For the model of Bunge and Noble (1984), Baird, Bunge, and Noble (1987) provided experimental data on the amine extraction with hydrochloric acid as the internal phase reagent.

Kim, Chol, and Ihm (1983) and Teramoto and Matsuyama (1986) included an additional mass transfer resistance in the peripheral thin membrane layer of the emulsion next to the external phase. This additional resistance is necessary for compensating their technique of volume-averaging the membrane and internal phase diffusivities, which underestimates the diffusion resistance in the emulsion globule (Noble, Way, and Bunge 1988). Earlier, Casamatta, Chavarie, and Angelino (1978) also considered additional mass transfer resistance in the peripheral thin membrane layer for hydrocarbon separation. Recently, Kataoka, Nishiki, and Kimura (1989) and Kataoka et al. (1990) estimated the thickness of the peripheral layer from internal phase droplet size and volume fraction as shown in Eq. (37–60). They also considered the reaction reversibility in the internal phase and the mass transfer resistance in the external phase. Miao, Li, and Zhang (1985) also considered the external phase resistance and the additional peripheral layer resistance. However, the resulting systems of equations in the models of Kataoka, Nishika, and Kimura (1989), Kataoka et al. (1990) and Miao, Li, and Zhang (1985) were too complicated to develop even quasi-analytical solutions.

Chan and Lee (1987) and Borwankar et al. (1988) took into account the leakage of the internal phase into the external phase in addition to external phase mass transfer resistance. However, their resulting systems of equations were also too complicated to develop even quasi-analytical solutions.

CARRIER FACILITATED TRANSPORT MODELS FOR TYPE 2 FACILITATION

In Type 2 facilitation (carrier facilitated transport), a "carrier" compound (complexing agent or extractant) incorporated in the membrane phase carries the solute across the membrane phase. Reactions involving the carrier and solute take place both at the external interface between the external and membrane phases and the internal interface between the membrane and internal phases. An example of this facilitation is the recovery/removal of metal ions in hydrometallurgical extraction and from wastewater. In the case of an aqueous feed solution containing metal ions as the external phase, the reactions at the external and internal interfaces are ion exchange between metal ions and protons, and a high extraction rate is achieved via the driving force of continuous transport of protons from the internal phase to the external phase. A concentrated acid in the internal phase allows the metal ion to be concentrated effectively in this phase, resulting in a high extraction capacity. The driving force of proton transport "pumps" the transport of the metal ion against its own concentration gradient between the internal and external phases (Cussler 1971; Cussler, Evans, and Matesich 1971; Schiffer et al. 1974a, 1974b). The transport process may be described by two methods: the spherical shell approach and the emulsion globule approach (Lorbach and Marr 1987; Ho 1990).

Spherical Shell Approach

This approach assumes a constant thickness of the shell in the same way as the spherical shell approach discussed above for Type 1 facilitation. The shell is shown schematically in Figure 36–4(b) for the example of zinc removal from wastewater by an ELM system containing a carrier (denoted as C in this figure) in the membrane phase and a concentrated H_2SO_4 solution in the internal phase. Similar to Type 1 facilitation, this approach for Type 2 facilitation includes models with an overall mass transfer coefficient for extraction and models with overall mass transfer coefficients for extraction *and* leakage.

Models with Overall Mass Transfer Coefficient for Extraction

The model assumes negligible leakage of the components from the receiving phase to the feed phase. Cussler (1971), Cussler, Evans, and Matesich (1971), and Caracciolo, Evans, and Cussler (1975) discussed the carrier facilitated transport systems of metal ions with the assumption of steady-state flux. Hochhauser and Cussler (1975) used the same assumption to analyze the concentration of chromium in the internal phase via extraction from the external feed phase by means of co-transport, in which the chromium flux is in the same direction as the proton flux. They also used the typical facilitated transport assumption of the same diffusivity for the free carrier and chromium-carrier complex, resulting in the constant sum of free and complexed carrier concentrations in the membrane phase. Kondo et al. (1979), Volkel, Halwachs, and Schugerl (1980), and Martin and Davies (1980) used a similar modeling approach for the analysis of copper extraction. Lee, Evans, and Cussler (1978) employed a simple model of constant shell thickness for the selective separation of copper from nickel. Noble (1983) gave analytical expressions for shape factors to correct the facilitation factor of planar geometry for cylindrical or spherical geometry (Noble, Way, and Bunge 1988). Stroeve, Varanasi, and Hoofd (1984) modeled facilitated transport in spherical shell membranes and obtained analytical solutions in good agreement with earlier numerical solutions.

Models with Overall Mass Transfer Coefficients for Extraction and Leakage

Similar to Type 1 facilitation, Type 2 facilitation has few models that take the breakage of the internal phase into account. As mentioned before, Boyadzhiev, Sapundzhiev, and Bezenshek (1977) were the first to consider the breakage. Frankenfeld, Cahn, and Li (1981) assumed a constant leakage rate (independent of internal phase volume) to analyze the extraction of copper.

The general model of Ho and Li (1984) with overall mass transfer coefficients for extraction and leakage is also applicable to Type 2 facilitation, as evidenced from the use of this model for the extraction of chromium and uranium discussed above. As described earlier, this model considers two cases: Case I has a chemical reagent in the external feed phase to convert the leaked species to the extractable form for re-extraction; Case II does not have the reagent in the external phase. In the Case I model, Eq. (37–20) for the external phase concentration of solute A, Eq. (37–19) for the internal phase concentration of solute B, Eq. (37–27) for external reagent consumption, and Eq. (37–29) for internal reagent consumption are applicable for small breakage values ($\Phi \leq 1.4 \times 10^{-5}$ s^{-1}), whereas Eq. (37–22) for the external phase concentration of solute A, Eq. (37–19) for the internal phase concentration of solute B, Eq. (37–24) for external reagent consumption, and Eq. (37–28) for internal reagent consumption are suitable for large breakage values. In the Case II model, Eqs. (37–39) and (37–34) for the external phase concentration of Solute A, Eqs. (37–40) and (37–45) for the external phase concentration of solute B, Eqs. (37–33) and (37–37) for the internal phase concentration of solute B, and Eq. (37–56) for internal reagent consumption are applicable for small breakage values, whereas Eqs. (37–47) and (37–34) for the external phase concentration of solute A, Eqs. (37–48) and (37–54) for the external phase concentration of solute B, Eqs. (37–33) and (37–37) for the internal phase concentration of solute B, and Eq. (37–28) for internal reagent consumption are suitable for large breakage values. [However, as mentioned before, hydrometallurgical (e.g., Zn^{2+} and Cu^{2+}) and hydrocarbon extractions belong to a special case of the Case II model, in which solute B is the only species in both the internal and external phases. In this special case, Eqs. (37–116) and (37–117) in Appendix A for the external phase concentration of solute B are applicable for small and large breakage values, respectively.]

Emulsion Globule Approach

The models described above with the spherical shell approach for Type 2 facilitation have the same shortcomings as those for Type 1 facilitation. Models to overcome the shortcomings have to account for the diffusion of carrier and metal-carrier complex in emulsion globules and reversible reactions at external and internal interfaces.

Models with Additional Resistance in Peripheral Thin Membrane Layer
Teramoto et al. (1983b) and Kataoka et al. (1989; 1990) included external phase mass transfer resistance and additional mass transfer resistance in the peripheral thin membrane layer of the emulsion globule in their models. The Teramoto et al. model also considered leakage.

These models have complicated equations and many parameters. Teramoto et al. evaluated their model parameters experimentally, which was quite tedious. They also studied the effect of globule size distribution on copper extraction rates. In a finding similar to their earlier one (Teramoto et al. 1983a), they showed that the Sauter mean diameter was sufficient to characterize globule size, and it was not necessary to use the size distribution. Later, Teramoto et al. (1986) applied the model to the extraction of lanthanoids. Recently, Kataoka et al. (1989, 1990) took the same approach as in their modeling of Type 1 facilitation discussed above to estimate the thickness of the peripheral thin membrane layer [Eq. (37–60)]. They estimated all their model parameters except reaction rate constants, emulsion globule diameter, and internal phase droplet diameter, which were determined experimentally. However, the resulting systems of equations for these models were too complicated to develop even quasi-analytical solutions.

Lorbach and Marr Model
Lorbach, Bart, and Marr (1986) developed a simpler carrier facilitated transport model with fewer equations and parameters than those of Teramoto et al. (1983b) and Kataoka et al. (1989, 1990). The simpler model eliminates the additional mass transfer resistance in the peripheral membrane layer of emulsion globules. This is justified since the peripheral layer is an integral part of the emulsion globule as in the treatment of the advancing front model of Ho et al. (1982). In addition, as mentioned above, the additional mass transfer resistance is only necessary as compensation for the technique of volume-averaging the membrane and internal phase diffusivities, which underestimates the diffusional resistance in the emulsion globule (Noble, Way, and Bunge 1988). If this technique is not used, this additional resistance is not needed. Lorbach, Bart, and Marr assumed the proton concentration in the external feed phase to remain constant since the external phase volume is generally larger than the volumes of the internal and membrane phases, and the pH change in the external phase is usually not appreciable. They also assumed that

any transport resistance in the internal phase is negligible since the diffusion time constants for the internal phase droplets are much less than those for the emulsion globules as in the treatment of the advancing front model. They applied the model to the extraction of copper.

Lorbach and Marr (1987) further simplified the system of equations in the model of Lorbach, Bart, and Marr (1986) by the use of the typical facilitated transport assumption of constant sum of free and complexed carrier concentrations, i.e., the same diffusivity for the free carrier and the metal-carrier complex. Thus, the model of Lorbach and Marr (1987) represents a simplified, but practical, model for Type 2 facilitation. The following summarizes the assumptions of the Lorbach and Marr model:

1. The emulsion liquid membrane system is a monodisperse collection of spherical emulsion globules in an external, continuous feed phase.
2. The ELM system is also a noncoalescing collection of the spherical globules.
3. There is no internal circulation in the globules.
4. For diffusion in the emulsion globule, the composite nature of the emulsion can be disregarded, and the emulsion can be treated as though it were a continuum. The diffusion can be described by the use of effective diffusivity. There is no transport resistance in the internal phase.
5. Breakage of the internal phase into the external phase is negligible.
6. The sum of free and complexed carrier concentrations in the membrane phase is constant, i.e., the free carrier and the metal-carrier complex have the same diffusivity.
7. The concentration of protons in the external phase is constant.
8. The amount of the carrier dissolved in the external and internal phases is negligible.

The first five assumptions are the same as in the treatment of the advancing front model.

As mentioned above, there are reversible reactions at the external interface between the external aqueous feed phase and the organic membrane phase and at the internal interface between the membrane phase and the internal aqueous phase. The extraction reaction of the metal ion, A^{z+}, with the carrier, HR, at the external interface releases protons as

$$A^{z+} + z\,HR \rightleftarrows AR_z + z\,H^+$$
(aqueous phase) (organic phase) (organic phase) (aqueous phase),

(37-93)

where z is the valence of the metal ion, and AR_z the metal-carrier complex. The stripping reaction of the metal-carrier complex at the internal interface regenerates the carrier and releases the metal ion as follows:

$$AR_z + z\,H^+ \rightleftarrows A^{z+} + z\,HR$$
(organic phase) (aqueous phase) (aqueous phase) (organic phase).

(37-94)

Mathematically, the equations describing the solute concentrations in the emulsion globules, external phase, and internal phase for the Lorbach and Marr model are as follows:

Globules:

$$\left(\frac{V_m}{V_m + V_i}\right)\frac{\partial c_m}{\partial t} = D_{\text{eff}}\frac{1}{r^2}\frac{\partial}{\partial r}\left(r^2\frac{\partial c_m}{\partial r}\right) - \left(\frac{A_i}{V_m + V_i}\right)R_r,$$

(37-95)

$$t = 0, \quad c_m = 0, \qquad (37\text{-}96)$$

$$r = r_o, \quad D_{\text{eff}}\left.\frac{\partial c_m}{\partial r}\right|_{r=r_o} = R_f, \qquad (37\text{-}97)$$

$$r = r_o, \quad \frac{\partial c_m}{\partial r} = 0, \qquad (37\text{-}98)$$

$$c_C = c_{C0} - c_m. \qquad (37\text{-}99)$$

External phase:

$$-V_e \frac{dc_e}{dt} = N(4\pi r_o^2)k_e(c_e - c_{em})$$

$$= \frac{3}{r_o}(V_m + V_i)k_e(c_e - c_{em}) \quad (37\text{--}100)$$

$$= \frac{3}{r_o}(V_m + V_i)R_f, \quad (37\text{--}101)$$

$$t = 0, \quad c_e = c_{e0} \quad (37\text{--}102)$$

Internal phase:

$$V_i \frac{\partial c_i}{\partial t} = A_i R_r, \quad (37\text{--}103)$$

$$t = 0, \quad c_i = c_{i0}, \quad (37\text{--}104)$$

$$c_i + c_{iH} = c_{i0} + c_{iH0}, \quad (37\text{--}105)$$

where c_m is the concentration of the metal-carrier complex in the membrane phase; D_{eff} the effective diffusivity of the complex, based on the membrane phase concentration c_m in the emulsion mixture; R_r the reaction rate for stripping per area of the internal interface; R_f the reaction rate for extraction per area of the external interface; c_C the concentration of the free carrier in the membrane phase; c_{C0} the initial concentration of the free carrier in the membrane phase; c_{iH} the concentration of the hydrogen ion (proton) in the internal phase; c_{iH0} the initial concentration of the hydrogen ion in the internal phase, and the rest of the symbols have the same definitions described above. Note that all the concentrations are in the units of N, g-equivalent/L (kg-equivalent/m^3). Thus, the valence of the metal ion, z, does not have to appear in Eqs. (37–99) and (37–105). Equation (37–99) shows a constant sum of the free and complexed carrier concentrations in the membrane phase, which results from assuming the diffusivity of the free carrier to be the same as that of the metal-carrier complex. Equation (37–105) is the material balance for the hydrogen and metal ions in the internal phase.

The above system of differential equations possesses no analytical solutions. It has to be solved numerically. For solving this system of equations, Lorbach and Marr (1987) used a finite difference technique, in which they discretized the spatial derivative of Eq. (37–95) and applied an implicit ordinary-differential-equation routine to solve the resulting system of equations.

To solve this system of equations, the parameters for interfacial areas (r_o and A_i), reaction kinetics (R_f and R_r), and mass transfer (k_e and D_{eff}) have to be known or assigned. Lorbach and Marr (1987) determined r_o and A_i experimentally via the Sauter mean globule radius and internal phase droplet radius (r_μ), respectively. Their values for zinc removal were $r_o = 1.3$ to 1.45×10^{-3} m and $r_\mu = 1.5 \times 10^{-6}$ m.

They expressed the reaction kinetics for zinc removal as

$$R_f = k_f \left(\frac{c_{em} c_C}{c_{eH}} - \frac{c_{eH} c_m}{K_{eq} c_C} \right), \quad (37\text{--}106)$$

$$R_r = k_r \left(c_{iH} c_m - K_{req} \frac{c_i c_C^2}{c_{iH}} \right), \quad (37\text{--}107)$$

where

$$K_{eq} = \frac{c_m c_{eH}^2}{c_{em} c_C^2}, \quad (37\text{--}108)$$

$$K_{req} = \frac{c_m c_{iH}^2}{c_i c_C^2}, \quad (37\text{--}109)$$

and k_f and k_r are the forward and reverse reaction rate constants for the reaction shown in Eq. (37–93), c_{eH} is the hydrogen ion concentration in the external phase, K_{eq} the equilibrium constant for the extraction reaction shown in Eq. (37–93) with $z = 2$ (e.g., Zn^{2+} and Cu^{2+}), and K_{req} the equilibrium constant for the stripping or re-extraction reaction. The K_{req} term is defined in the same way as K_{eq}, as shown in Eqs. (37–108) and (37–109), except the internal phase concentrations are used for K_{req}, whereas the external phase concentrations are used for K_{eq}. They determined the reaction kinetics parameters k_f, k_r, K_{eq}, and K_{req} experimentally. Their values for zinc removal were $k_f = 2.5 \times 10^{-5}$ m/s, $k_r = 1.1 \times 10^{-6}$ m^4/(kg-equivalent · s), $K_{eq} = 106$, and $K_{req} = 10$. As noted by them, the large difference between the extraction and stripping equilibrium constants was the result of using concentrations instead of activities.

Similarly, reaction kinetics for the recovery of copper were expressed by Lorbach, Bart, and Marr (1986) as Eqs. (37–106) and (37–107). Their values for the reaction kinetics parameters were $k_f = 1.72 \times 10^{-6}$ m/s, $k_r = 2.0 \times 10^{-5}$ m^4/(kg-equivalent · s), and $K_{eq} = K_{req} = 17$.

For mass transfer parameters, Lorbach and Marr (1987) determined the external phase mass transfer coefficient k_e for zinc removal experimentally. Their k_e value was 3.0×10^{-5} m/s. However, they used the effective diffusivity D_{eff} as an adjustable parameter for fitting their model to the extraction data for zinc removal. The D_{eff} obtained was 1.7×10^{-10} m^2/s. Nevertheless, the fit was quite good.

Areas for Improvement

As mentioned above, the difference between the extraction and stripping equilibrium constants in the modeling work of Lorbach and Marr (1987) for zinc removal was very large. In principle, the forward and reverse reaction rate constants should be sufficient to define the reaction rates and equilibrium constants. Thus, accurate reaction rate constants are crucial to the success of the modeling. Rather than being used as an adjustable parameter as in the modeling of Lorbach and Marr (1987), effective diffusivity may be estimated reasonably accurately as described in the advancing front model of Ho et al. (1982). The external phase mass transfer coefficient may also be estimated from available correlations mentioned above, or it may be eliminated since external phase mass transfer resistance is generally negligible for typical ELM systems, as in the treatment of the advancing front model. These appear to be the areas of improvement that can be made to the modeling.

Appendix A

Special Cases in the Ho and Li Model with Overall Mass Transfer Coefficients for Extraction and Leakage

CASE I MODEL: NO DIFFUSIONAL TRANSPORT OF SOLUTE B

A special case in the Case I model of Ho and Li (1984) is that solute B cannot diffuse through the membrane phase, i.e., $k_B = 0$. For $k_B = 0$, Eq. (37–20) in the case of small Φ values reduces to

$$c_{eA} = \frac{V_e c_{eA0} + V_i c_{iB0}}{k_A^\circ + \Phi V_e} \left\{ \Phi + \left(\frac{k_A^\circ c_{eA0} - \Phi V_i c_{iB0}}{V_e c_{eA0} + V_i c_{iB0}} \right) \exp\left[-\left(\frac{k_A^\circ}{V_e} + \Phi \right) t \right] \right\}. \quad (37\text{–}110)$$

In the case of large Φ values, Eq. (37–22) reduces to

$$c_{eA} = \frac{\Phi(V_{e0} c_{eA0} + V_{i0} c_{iB0})}{k_A^\circ + \Phi V_0}$$

$$+ \left(\frac{(k_A^\circ + \Phi V_{i0}) c_{eA0} - \Phi V_{i0} c_{iB0}}{k_A^\circ + \Phi V_0} \right) \left(\frac{V_0 - V_{i0}}{V_0 e^{\Phi t} - V_{i0}} \right)^{\{1 + [k_A^\circ/(\Phi V_0)]\}} \quad (37\text{–}111)$$

CASE II MODEL: NO DIFFUSIONAL TRANSPORT OF SOLUTE B

Similar to the Case I model, a special case in the Case II model of Ho and Li (1984) is that the solute B cannot diffuse through the membrane phase, i.e., $k_B = 0$. For $k_B = 0$, Eqs. (37–40) and (37–45) in the case of small Φ reduce to Eqs. (37–112) and (37–113), respectively.

$$c_{eB} = \frac{1}{V_e}(V_e c_{eB0} + V_i c_{iB0} + V_i c_{ir0}) - \frac{V_i}{V_e}(c_{iB0} + c_{ir0})e^{-\Phi t}, \quad 0 \leq t \leq t_1, \quad (37\text{–}112)$$

$$c_{eB} = \frac{1}{V_e}(V_e c_{eA0} + V_e c_{eB0} + V_i c_{iB0}) - \frac{V_i}{V_e}(c_{iB0} + c_{ir0})e^{-\Phi t}$$

$$+ \frac{1}{V_e}(V_i c_{ir0} - V_e c_{eA0})\exp[-\Phi(t - t_1)], \quad t \geq t_1, \quad (37\text{–}113)$$

where t_1 is given by Eq. (37–44).

In the case of large Φ values, Eqs. (37–48) and (37–54) reduce to Eqs. (37–114) and (37–115), respectively:

$$c_{eB} = \frac{1}{(V_0 - V_{i0}e^{-\Phi t})}[(V_{e0}c_{eB0} + V_{i0}c_{iB0} + V_{i0}c_{ir0}) - V_{i0}(c_{iB0} + c_{ir0})e^{-\Phi t}], \quad 0 \le t \le t_1, \quad (37\text{--}114)$$

$$= \frac{V_{e0}c_{eA0} + V_{e0}c_{eB0} + V_{i0}c_{iB0}}{V_0 - V_{i0}e^{-\Phi t}} + \frac{(V_{i0}c_{ir0} - V_{e0}c_{eA0})e^{\Phi t_1} - V_{i0}(c_{iB0} + c_{ir0})}{V_0 e^{\Phi t} - V_{i0}}, \quad t \ge t_1, \quad (37\text{--}115)$$

where t_1 is given by Eq. (37–53).

CASE II MODEL: NO CHEMICAL CONVERSIONS IN INTERNAL AND EXTERNAL PHASES

The other special case in the Case II model of Ho and Li (1984) is that there are no chemical conversions of solute A into B in both the internal and external phases. Solute B is the only species in both the internal and external phases. As mentioned earlier, mathematically, this special case is given by $k_A = 0$, $c_{eA} = 0$, and $c_{ir} = 0$. In this case, the solution of c_{eB} for small Φ values is obtained from Eq. (37–45) by letting $t_1 = 0$:

$$c_{eB} = \left(c_{eB0} - \frac{(k_B^\circ + \Phi V_i)(V_e c_{eB0} + V_i c_{iB0})}{(k_B^\circ V_i K_{eB}/K_{iB}) + k_B^\circ V_e + \Phi V_e V_i}\right)\exp\left[-\left(\frac{k_B^\circ K_{eB}}{V_e K_{iB}} + \frac{k_B^\circ}{V_i} + \Phi\right)t\right]$$

$$+ \frac{(k_B^\circ + \Phi V_i)(V_e c_{eB0} + V_i c_{iB0})}{(k_B^\circ V_i K_{eB}/K_{iB}) + k_B^\circ V_e + \Phi V_e V_i}. \quad (37\text{--}116)$$

Similarly, the solution of c_{eB} for large Φ values is obtained from Eq. (37–54) by letting $t_1 = 0$:

$$c_{eB} = \frac{V_{e0}c_{eB0} + V_{i0}c_{iB0}}{V_0} + \left[c_{eB0} - \left(\frac{V_{e0}c_{eB0} + V_{i0}c_{iB0}}{V_0}\right)\right]$$

$$\cdot \left[\left(\frac{V_0 - V_{i0}}{V_0 e^{\Phi t} - V_{i0}}\right)^{\{1+[k_B^\circ K_{eB}/(\Phi V_0 K_{iB})]\}} \exp\left(-\frac{k_B^\circ}{\Phi V_{i0}}(e^{\Phi t} - 1)\right)\right]$$

$$+ \left[\Psi_6\left(1 - \frac{K_{eB}}{K_{iB}}\right)\frac{k_B^\circ(V_{e0}c_{eB0} + V_{i0}c_{iB0})}{\Phi V_0}\right]$$

$$\cdot \left[(V_0 e^{\Phi t} - V_{i0})^{-\{1+[k_B^\circ K_{eB}/(\Phi V_0 K_{iB})]\}} \exp\left(-\frac{k_B^\circ}{\Phi V_{i0}}e^{\Phi t}\right)\right], \quad (37\text{--}117)$$

where Ψ_6 is expressed by Eq. (37–51).

For hydrocarbon extractions, the ratio of K_{eB} to K_{iB} normally equals unity, which reduces Eq. (37–117) to

$$c_{eB} = \frac{V_{e0}c_{eB0} + V_{i0}c_{iB0}}{V_0}$$

$$+ \left[\left(c_{eB0} - \frac{V_{e0}c_{eB0} + V_{i0}c_{iB0}}{V_0}\right)\left(\frac{V_0 - V_{i0}}{V_0 e^{\Phi t} - V_{i0}}\right)^{\{1+[k_B^\circ/(\Phi V_0)]\}} \cdot \exp\left(-\frac{k_B^\circ}{\Phi V_{i0}}(e^{\Phi t} - 1)\right)\right].$$

$$(37\text{--}118)$$

Appendix B
Dimensionless Equations and Higher Order Perturbation Solutions for the Advancing Front Model

The dimensionless equations for the advancing front model of Ho et al. (1982) corresponding to Eqs. (37–64) through (37–71) with the dimensionless terms defined by Eq. (37–78) are as follows:

Globules:

$$\epsilon \frac{\partial g}{\partial \tau} = \frac{1}{\zeta^2} \frac{\partial}{\partial \zeta} \left(\zeta^2 \frac{\partial g}{\partial \zeta} \right), \quad \chi \leq \zeta \leq 1, \quad (37\text{–}119)$$

$$\tau = 0, \quad g = 0, \quad (37\text{–}120)$$

$$\zeta = 1, \quad g = h, \quad (37\text{–}121)$$

$$\zeta = \chi, \quad g = 0. \quad (37\text{–}122)$$

External phase:

$$\frac{dh}{d\tau} = -E \left. \frac{\partial g}{\partial \zeta} \right|_{\zeta=1}, \quad (37\text{–}123)$$

$$\tau = 0, \quad h = 1. \quad (37\text{–}124)$$

Reaction front:

$$\frac{d\chi}{d\tau} = -\left. \frac{\partial g}{\partial \zeta} \right|_{\zeta=\chi(\tau)}, \quad (37\text{–}125)$$

$$\tau = 0, \quad \chi = 1.$$

As mentioned earlier, these equations are inherently nonlinear, and they cannot be solved analytically. Ho et al. (1982) have solved them by the use of the perturbation method with the standard transformation shown in Eq. (37–76) and the Landau transformation given in Eq. (37–77). The perturbation solutions are expressed in terms of the perturbation parameter ϵ as

$$\hat{\phi} = \hat{\phi}_0 + \epsilon\hat{\phi}_1 + \epsilon^2\hat{\phi}_2 + \ldots, \quad (37\text{--}126)$$

$$h = h_0 + \epsilon h_1 + \epsilon^2 h_2 + \ldots, \quad (37\text{--}127)$$

$$\tau = \tau_0 + \epsilon\tau_1 + \epsilon^2\tau_2 + \ldots. \quad (37\text{--}128)$$

As mentioned before, ϵ is considerably less than unity for typical ELM systems; therefore, it serves as the ideal perturbation parameter.

As described before, the zero-order perturbation solutions are Eqs. (37–72), (37–73), and (37–74) [Eqs. (37–79), (37–80), and (37–81) for $E = 3$]. The higher-order perturbation solutions are as follows:

First-order solutions:

$$\hat{\phi}_1 = \frac{E^2}{54\chi}[a_0(1-\delta) + a_1(1-\delta)$$

$$+ a_2(1-\delta^2) + a_3(1-\delta^3)], \quad (37\text{--}129)$$

$$h_1 = \frac{E^2}{54\chi}a_1$$

$$= \frac{E^2}{18}(\chi^3 - \beta^3)(\chi + 2)(\chi - 1), \quad (37\text{--}130)$$

$$\tau_1 = \frac{1}{6}\left[\chi^2 - 8\chi + 7 + \left(\beta + 4 + \frac{1}{\beta}\right)\ln\left(\frac{\chi^3 - \beta^3}{1 - \beta^3}\right) - 3\left(\beta + \frac{1}{\beta}\right)\ln\left(\frac{\chi - \beta}{1 - \beta}\right)\right]$$

$$+ \frac{1}{\sqrt{3}}\left(\beta - \frac{1}{\beta}\right)\left[\tan^{-1}\left(\frac{2\chi + \beta}{\sqrt{3}\,\beta}\right) - \tan^{-1}\left(\frac{2 + \beta}{\sqrt{3}\,\beta}\right)\right], \quad (37\text{--}131)$$

where

$$a_0 = 7\chi^6 - 6\chi^5 - 8\beta^3\chi^3 + 6\beta^3\chi^2 + \beta^6, \quad (37\text{--}132)$$

$$a_1 = 3\chi^6 + 3\chi^5 - 6\chi^4 - 3\beta^3\chi^3 - 3\beta^3\chi^2 + 6\beta^3\chi, \quad (37\text{--}133)$$

$$a_2 = -9\chi^6 + 9\chi^5 + 9\beta^3\chi^3 - 9\beta^3\chi^2, \quad (37\text{--}134)$$

$$a_3 = 2\chi^6 - 3\chi^5 - \beta^3\chi^3 + 3\beta^3\chi^2 - \beta^6. \quad (37\text{--}135)$$

Second-order solutions:

$$\hat{\phi}_2 = \frac{E^3}{9720\chi^3}[b_0(1-\delta) + b_1(1-\delta) + b_2(1-\delta^2) + b_3(1-\delta^3) + b_4(1-\delta^4) + b_5(1-\delta^5)],$$

$$(37\text{--}136)$$

$$h_2 = \frac{E^3}{9720\chi^3}b_1$$

$$= \frac{E^3}{3240\chi}(\chi^3 - \beta^3)(\chi - 1)[\chi^4 + 80\chi^3 + 24\chi^2 + (8\beta^3 - 120)\chi + 7\beta^3], \quad (37\text{--}137)$$

$$\tau_2 = \frac{E}{540}\left[-7\chi^4 + 40\chi^3 - 100\chi^2 + (144 + 4\beta^3)\chi - (77 + 8\beta^3)\right.$$

$$+ 4\beta^3 \frac{1}{\chi} + 6(2\beta + 1)(\beta + 2)(\beta - 1)\ln\left(\frac{\chi^3 - \beta^3}{1 - \beta^3}\right)$$

$$\left. - 54\beta(\beta - 1)\ln\left(\frac{\chi - \beta}{1 - \beta}\right)\right] - \frac{E}{5\sqrt{3}}\beta(\beta + 1)\left[\tan^{-1}\left(\frac{2\chi + \beta}{\sqrt{3}\,\beta}\right) - \tan^{-1}\left(\frac{2 + \beta}{\sqrt{3}\,\beta}\right)\right],$$

(37–138)

where

$$b_0 = -(b_2 + b_3 + b_4 + b_5)$$
$$= (\chi^3 - \beta^3)[33\chi^7 + 764\chi^6 - 1536\chi^5 + (720 + 87\beta^3)\chi^4$$
$$- 253\beta^3\chi^3 + 204\beta^3\chi^2 - 12\beta^6\chi - 7\beta^6], \quad (37\text{--}139)$$

$$b_1 = (\chi^3 - \beta^3)[3\chi^7 + 237\chi^6 - 168\chi^5 + (24\beta^3 - 432)\chi^4 + (360 - 3\beta^3)\chi^3 - 21\beta^3\chi^2],$$

(37–140)

$$b_2 = (\chi^3 - \beta^3)[-270\chi^7 - 630\chi^6 + 1980\chi^5 - 1080\chi^4 + 90\beta^3\chi^3 - 90\beta^3\chi^2], \quad (37\text{--}141)$$

$$b_3 = (\chi^3 - \beta^3)[360\chi^7 - 410\chi^6 - 300\chi^5 + (360 - 90\beta^3)\chi^4 + 220\beta^3\chi^3 - 150\beta^3\chi^2 + 10\beta^6],$$

(37–142)

$$b_4 = (\chi^3 - \beta^3)[-135\chi^7 + 315\chi^6 - 180\chi^5 - 45\beta^3\chi^3 + 45\beta^3\chi^2], \quad (37\text{--}143)$$

$$b_5 = (\chi^3 - \beta^3)[12\chi^7 - 39\chi^6 + 36\chi^5$$
$$+ 3\beta^3\chi^4 - 12\beta^3\chi^3 - 9\beta^3\chi^2 + 12\beta^6\chi - 3\beta^6]. \quad (37\text{--}144)$$

From a numerical standpoint, the above solutions break down for $E = 3$ ($\beta = 0$) where a number of terms become indeterminate. These indeterminate terms are readily resolved using L'Hospital's rule. The corresponding results for $E = 3$ are as follows:

First-order solutions:

$$\hat{\phi}_1 = \frac{1}{6}\chi^3[\chi(7\chi - 6)(1 - \delta) + 3(\chi + 2)(\chi - 1)(1 - \delta)$$
$$+ 9\chi(1 - \chi)(1 - \delta^2) + \chi(2\chi - 3)(1 - \delta^3)], \quad (37\text{--}145)$$

$$h_1 = \frac{1}{2}\chi^3(\chi + 2)(\chi - 1), \quad (37\text{--}146)$$

$$\tau_1 = \frac{1}{6}[\chi^2 - 8\chi + 7] + 2\ln\chi + \left(\frac{1}{\chi} - 1\right). \quad (37\text{--}147)$$

Second-order solutions:

$$\hat{\phi}_2 = \frac{1}{360} \chi^3 [b'_0(1-\delta) + b'_1(1-\delta) + b'_2(1-\delta^2) + b'_3(1-\delta^3)$$

$$+ b'_4(1-\delta^4) + b'_5(1-\delta^5)], \qquad (37\text{-}148)$$

$$h_2 = \frac{1}{120} \chi^3 (\chi - 1)(\chi^3 + 80\chi^2 + 24\chi - 120), \qquad (37\text{-}149)$$

$$\tau_2 = \frac{1}{180} [-7\chi^4 + 40\chi^3 - 100\chi^2 + 144\chi - 77 - 36 \ln\chi], \qquad (37\text{-}150)$$

where

$$b'_0 = 33\chi^4 + 764\chi^3 - 1536\chi^2 + 720\chi, \qquad (37\text{-}151)$$

$$b'_1 = 3\chi^4 + 237\chi^3 - 168\chi^2 - 432\chi + 360, \qquad (37\text{-}152)$$

$$b'_2 = -270\chi^4 - 630\chi^3 + 1980\chi^2 - 1080\chi, \qquad (37\text{-}153)$$

$$b'_3 = 360\chi^4 - 410\chi^3 - 300\chi^2 + 360\chi, \qquad (37\text{-}154)$$

$$b'_4 = -135\chi^4 + 315\chi^3 - 180\chi^2, \qquad (37\text{-}155)$$

$$b'_5 = 12\chi^4 - 39\chi^3 + 36\chi^2. \qquad (37\text{-}156)$$

The perturbation solutions expressed in Eqs. (37-126), (37-127), and (37-128) are not convergent for large values of ϵ. Although large ϵ values are unlikely, the Shanks transformations extend the range of applicability of the perturbation solutions (Ho et al. 1982):

$$h^* = \frac{h_0 h_1 - \epsilon(h_0 h_2 - h_1^2)}{h_1 - \epsilon h_2}, \qquad (37\text{-}157)$$

$$\tau^* = \frac{\tau_0 \tau_1 - \epsilon(\tau_0 \tau_2 - \tau_1^2)}{\tau_1 - \epsilon \tau_2}. \qquad (37\text{-}158)$$

NOTATION

General Notation

See the General Notation section at the beginning of this handbook.

Special Notation

A   solute existing only in the external phase

A_e total mass transfer area of the emulsion globules in the external phase, L^2 (m^2)

A_i total mass transfer area of the internal phase, L^2 (m^2)

A_v external phase mass transfer area per unit volume of the emulsion, L^{-1} (m^{-1})

B   only solute in the internal phase

B_i Biot number ($k_e r_o / K'_e D'_{\text{eff}}$), dimensionless

c	solute concentration in the reacted region of emulsion globules, N	D_{eff}	effective diffusivity of the solute or metal-carrier complex, based on the membrane phase concentration c_m, in the emulsion mixture, L^2/t (m²/s)
c_C	concentration of the free carrier in the membrane phase, N		
c_{C0}	initial concentration of the free carrier in the membrane phase, N	D'_{eff}	effective solute diffusivity, based on the average concentration c, in the emulsion mixture, L^2/t (m²/s)
c_e	solute concentration in the external phase, N	D_m	diffusivity of the solute or metal-carrier complex in the membrane phase, L^2/t (m²/s)
c_{e0}	initial solute concentration in the external phase, N		
c_{eA}	solute A concentration in the external phase, N	E	$3[V_i c_{ir0}/(V_e c_{e0})]$, $E/3$ = ratio of internal reagent equivalents to solute equivalents in the feed, dimensionless
c_{eA0}	initial solute A concentration in the external phase, N		
c_{eB}	solute B concentration in the external phase, N	g	$c/K'_e c_{e0}$, normalized solute concentration in emulsion globules, dimensionless
c_{eB0}	initial solute B concentration in the external phase, N	h	c_e/c_{e0}, normalized solute concentration in the external phase, dimensionless
c_{em}	solute concentration, in the external phase, at the interface between the external and membrane phases, N		
		h_0, h_1, h_2	zero-, first-, and second-order terms in perturbation expansion for h
c_i	solute concentration in the internal phase, N	h^*	Shanks nonlinear transformation of perturbation solution for h
c_{iB}	solute B concentration in the internal phase, N	k	overall mass transfer coefficient, t^{-1} (s⁻¹)
c_{iB0}	initial solute B concentration in the internal phase, N	k_A	overall mass transfer coefficient for solute A, t^{-1} (s⁻¹)
c_{im}	solute concentration, in the internal phase, at the interface between the internal and membrane phases, N	k_A°	$k_A(V_m + V_i)$, L^3/t (m³/s)
		k_{A1}	$k_A K_{iA}/K_{eA}$, t^{-1} (s⁻¹)
		k_B	overall mass transfer coefficient for solute B, t^{-1} (s⁻¹)
c_{iH}	hydrogen ion concentration in the internal phase, N		
c_{iH0}	initial hydrogen ion concentration in the internal phase, N	k_B°	$k_B(V_m + V_i)$, L^3/t (m³/s)
		$k_{B\Phi}$	$k_B + \Phi[V_i/(V_i + V_m)]$, overall leakage coefficient for solute B, t^{-1} (s⁻¹)
c_{ir}	concentration of the reagent in the internal phase, N	k_e	individual mass transfer coefficient in the external phase, L/t (m/s)
c_{ir0}	initial concentration of the reagent in the internal phase, N	k_f	forward reaction rate constant, L/t (m/s) or various units
c_m	solute concentration or metal-carrier complex concentration in the membrane phase, N	k_i	individual mass transfer coefficient in the internal phase, L/t (m/s)
		k_m	individual mass transfer coefficient in the membrane phase, L/t (m/s)
c_{me}	solute concentration, in the membrane phase, at the interface between the membrane and external phases, N		
		k_r	reverse reaction rate constant L^4/mol t [m⁴/(kg-equivalent · s)] or various units
c_{mi}	solute concentration, in the membrane phase, at the interface between the membrane and internal phases, N		
		k_{se}	interfacial mass transfer coefficient at the interface adjacent to the external phase, L/t (m/s)
d_{32}	Sauter mean diameter of emulsion globules, L (m)		
		k_{si}	interfacial mass transfer coefficient at

	the interface adjacent to the internal phase, L/t (m/s)
K_e	distribution coefficient for the solute between the membrane and external phases at equilibrium, dimensionless
K'_e	distribution coefficient for the solute between the reacted emulsion mixture and the external phase at equilibrium, dimensionless
K_{eA}	distribution coefficient for solute A between the membrane and external phases at equilibrium, dimensionless
K_{eB}	distribution coefficient for solute B between the membrane and external phases at equilibrium, dimensionless
K_i	distribution coefficient of the solute between the membrane and internal phases at equilibrium, dimensionless
K_{iB}	distribution coefficient for solute B between the membrane and internal phases at equilibrium, dimensionless
l	effective membrane thickness between external and internal phases, L (m)
$\tilde{M}_s$	solvent molecular weight, M/mol (kg/kg-mol)
N	number of emulsion globules dispersed in the external phase
r_f	reaction front position, L (m)
r_o	emulsion globule radius, L (m)
r_μ	Sauter mean radius of internal phase droplets in the emulsion, L (m)
$\tilde{V}_A$	molar volume of the solute at normal boiling point, L^3/mol (m^3/kg-mol)
V_{e0}	initial volume of the external phase, L^3 (m^3)
V_{i0}	initial, total volume of the internal phase, L^3 (m^3)
V_0	$V_{e0} + V_{i0} = V_e + V_i$, L^3 (m^3)
z	valence, dimensionless

Greek Letters

β	$[1 - (3/E)]^{1/3}$, dimensionless
δ	$(1 - \zeta)/(1 - \chi)$, dimensionless
ϵ	$K'_e(V_m + V_i)c_{e0}/(V_i c_{ir0})$, perturbation parameter, dimensionless
ζ	r/r_0, normalized radial coordinate, dimensionless
η_s	solvent viscosity, M/Lt (Ns/m^2)
τ	$\epsilon D'_{\text{eff}}/r_o^2$, dimensionless time
τ_0, τ_1, τ_2	zero-, first-, and second-order terms in perturbation expansion for τ
τ^*	Shanks nonlinear transformation of perturbation solution for τ
ϕ	volume fraction of the internal phase in the emulsion, i.e., $\phi = V_i/(V_m + V_i)$, dimensionless
$\hat{\phi}$	ζg
$\hat{\phi}_0, \hat{\phi}_1, \hat{\phi}_2$	zero-, first-, and second-order terms in perturbation expansion of $\hat{\phi}$
Φ	breakage coefficient, t^{-1} (s^{-1})
χ	r_f/r_o, normalized reaction front position, dimensionless

REFERENCES

Abou-Nemeh, I., and A. P. Van Peteghem. 1989. Extraction of cobalt and manganese from an industrial effluent by liquid emulsion membranes. In *Proceedings of the 2nd International Conference on Separations Science and Technology*, ed. M. H. I. Baird and S. Vijayan, Vol. 2, pp. 416–423. Ottawa, Ontario: Canadian Society of Chemical Engineering.

Alessi, P., I. Kikic, and M. Orlandini-Visalberghi. 1980. Liquid membrane permeation for the separation of C_8 hydrocarbons. *Chem. Eng. J.* 19:22.

Asher, W. J., K. C. Bovee, J. W. Frankenfeld, R. W. Hamilton, L. W. Henderson, P. C. Holtzapple, and N. N. Li. 1975. Liquid membrane system directed toward chronic uremia. *Kidney Int.* 7:S-409.

Asher, W. J., K. C. Bovee, J. W. Frankenfeld, R. W. Hamilton, J. W. Henderson, P. G. Holtzapple, and N. N. Li. 1976. Liquid membrane system directed toward chronic uremia. *Kidney Int.* 10:S-254.

Asher, W. J., K. C. Bovee, T. C. Vogler, R. W. Hamilton, and P. G. Holtzapple. 1979. Liquid membrane capsules for removal of urea from the blood. *Clin. Nephrol.* 11:92.

Asher, W. J., T. C. Vogler, K. C. Bovee, P. G. Holtzapple, and R. W. Hamilton. 1977. Liquid membrane capsules for chronic uremia. *Trans. Am. Soc. Artif. Intern. Organs* 23:73.

Baird, R. S., A. L. Bunge, and R. D. Noble. 1987. Batch extraction of amines using emulsion liquid membranes: importance of reaction reversibility. *AIChE J.* 33(1):43–53.

Bart, H. J., A. Bauer, and R. Marr. 1987. Calculation of reactive extraction in countercurrent columns. *Chem. Eng. Technol.* 10:291–296.

Bart, H. J., R. Wachter, R. J. Marr, and H. J. Müller. 1988. Mass transfer in a permeation column. *J. Membr. Sci.* 36:413–423.

Biehl, M. P., R. M. Izatt, J. D. Lamb, and J. J. Christensen. 1982. Use of a macrocyclic crown ether in an emulsion (liquid surfactant) membrane to effect rapid separation of Pb^{2+} from cation mixtures. *Sep. Sci. Technol.* 17(2):289–294.

Bock, J., R. R. Klein, P. L. Valint, and W. S. Ho. 1982. Liquid membrane extraction of uranium from wet process phosphoric acid. In *Sulfuric/Phosphoric Acid Plant Operations*, pp. 175–183. New York: American Institute of Chemical Engineers.

Boon-long, S., C. Laquerie, and J. P. Couderc. 1978. Mass transfer from suspended solids to liquid in agitated vessels. *Chem. Eng. Sci.* 33:813.

Borwankar, R. P., C. C. Chan, D. T. Wasan, R. M. Kurzeja, Z. M. Gu, and N. N. Li. 1988. Analysis of the effect of internal phase leakage on liquid membrane separations. *AIChE J.* 34(5):753–762.

Boyadzhiev, L., and E. Bezenshek. 1982. Zink-Rückgewinnung aus wässrigen Lösungen mittels Extraktion mit einer carrier-phase. *Chem.-Ing.-Tech.* 54:506–508.

Boyadzhiev, L., E. Bezenshek, and Z. Lazarova. 1984. Removal of phenol from waste water by double emulsion membranes and creeping film pertraction. *J. Membr. Sci.* 21:137–144.

Boyadzhiev, L., and G. Kyuchoukov. 1980. Further development of carrier-mediated extraction. *J. Membr. Sci.* 6:107–112.

Boyadzhiev, L., T. Sapundzhiev, and E. Bezenshek. 1977. Modeling of carrier-medicated extraction. *Sep. Sci.* 12:541.

Bunge, A. L., and R. D. Noble. 1984. A diffusion model for reversible consumption in emulsion liquid membranes. *J. Membr. Sci.* 21:55.

Cahn, R. P., and N. N. Li. 1974. Separation of phenol from waste water by the liquid membrane technique. *Sep. Sci.* 9(6):505–519.

Cahn, R. P., and N. N. Li. 1976a. Hydrocarbon separation by liquid membranes. In *Membrane Separation Processes*, ed. P. Meares, pp. 327–349. Amsterdam: Elsevier Scientific Publishing Co.

Cahn, R. P., and N. N. Li. 1976b. Separations of organic compounds by liquid membrane processes. *J. Membr. Sci.*, 1(2):129.

Cahn, R. P., J. W. Frankenfeld, N. N. Li, D. Naden, and K. N. Subramanian. 1981. Extraction of metal ions by liquid membrane. In *Recent Developments in Separation Science*, ed. N. N. Li, Vol. 6, p. 51. Boca Raton, FL: CRC Press.

Cahn, R. P., N. N. Li, and R. M. Minday. 1978. Removal of ammonium sulfide from industrial sour water by a liquid membrane process. *Environmental Sci. Technol.* 12(9):1051.

Calderbank, P. H., and M. B. Moo-Young. 1961. The continuous phase heat and mass transfer properties of dispersion. *Chem. Eng. Sci.* 16:39.

Caracciolo, F., D. F. Evans, and E. L. Cussler. 1975. Membranes with common ion pumping. *AIChE J.* 21(1):160–167.

Casamatta, G., C. Chavarie, and H. Angelino. 1978. Hydrocarbon separation through a liquid water membrane: modeling of permeation in an emulsion drop. *AIChE J.* 24(6):945–949.

Chan, C. C., and C. J. Lee. 1984. Mechanistic models of mass transfer across a liquid membrane. *J. Membr. Sci.* 20:1–24.

Chan, C. C., and C. J. Lee. 1987. A mass transfer model for the extraction of weak acids/bases in emulsion liquid-membrane systems. *Chem. Eng. Sci.* 42(1):83–95.

Chang, Y. C., and S. P. Li. 1983. A study of emulsified liquid membrane treatment of phenolic waste water. *Desalination* 47:351–361.

Crank, J. 1975. *The Mathematics of Diffusion*, 2nd ed., Chap. 12. Oxford: Clarendon Press.

Cussler, E. L. 1971. Membranes which pump. *AIChE J.* 17(6):1300–1303.

Cussler, E. L. 1985. *Diffusion: Mass Transfer in Fluid Systems*, Chap. 5. Cambridge: Cambridge University Press.

Cussler, E. L., and D. F. Evans. 1980. Liquid membranes for separation and reactions. *J. Membr. Sci.* 6:113–121.

Cussler, E. L., D. F. Evans, and M. A. Matesich. 1971. Theoretical and experimental basis for a specific countertransport system in membranes. *Science* 172:377–379.

Downs, H. H., and N. N. Li. 1981. Extraction of ammonia from waste water by the liquid membrane process. *J. Sep. Process Technol.* 2(4):19–24.

Draxler, J., W. Fürst, and R. J. Marr. 1988. Separation of metal species by emulsion liquid membranes. *J. Membr. Sci.* 38:281–293.

Draxler, J., and R. J. Marr. 1986. Emulsion liquid membranes: part I. phenomenon and industrial application. *Chem. Eng. Process.* 20:319–329.

Draxler, J., R. J. Marr, and M. Prötsch. 1988. Commercial-scale extraction of zinc by emulsion liquid membranes. In *Separation Technology*, ed.

N. N. Li and H. Strathmann, pp. 204–214. New York: United Engineering Trustees.

Fales, J. L., and P. Stroeve. 1984. A perturbation solution for batch extraction with double emulsions: role of continuous-phase mass transfer resistance. *J. Membr. Sci.* 21:35.

Folkner, C. A., and R. D. Noble. 1983. Transient response of facilitated transport membranes. *J. Membr. Sci.* 12:289–301.

Frankenfeld, J. W., R. P. Cahn, and N. N. Li. 1981. Extraction of copper by liquid membranes. *Sep. Sci. Technol.* 16(4):385.

Frankenfeld, J. W., and N. N. Li. 1977. Waste water treatment by liquid ion exchange in liquid membrane systems. In *Recent Developments in Separation Science,* ed. N. N. Li, Vol. 3, pp. 285–292. Boca Raton, FL: CRC Press.

Frankenfeld, J. W., and N. N. Li. 1979. Liquid membrane systems. In *Ion Exchange for Pollution Control,* ed. C. Calmon and H. Gold, Vol. 2, pp. 163–172. Boca Raton, FL: CRC Press.

Frankenfeld, J. W., and N. N. Li. 1987. Recent advances in liquid membrane technology. In *Handbook of Separation Process Technology,* ed. R. W. Rousseau, p. 840. New York: John Wiley & Sons.

Fuller, E. J., and N. N. Li. 1984. Extraction of chromium and zinc from cooling tower blowdown by liquid membranes. *J. Membr. Sci.* 18:251–271.

Garti, N., and A. Kovacs. 1991. Facilitated emulsion liquid membrane separation of complex hydrocarbon mixtures. *J. Membr. Sci.* 56:239–246.

Gladek, L., J. Stelmaszek, and J. Szust. 1981. Modeling of the mass transport through a liquid membrane. In *Recent Developments in Separation Science,* ed. N. N. Li, Vol. 6, pp. 29–49. Boca Raton, FL: CRC Press.

Gladek, L., J. Stelmaszek, and J. Szust. 1982. Modeling of mass transport with a very fast reaction through liquid membranes. *J. Membr. Sci.* 12:153–167.

Goswami, A. N., and B. S. Rawat. 1984a. Studies on permeation of hydrocarbons through liquid surfactant membranes. *J. Membr. Sci.* 24:145.

Goswami, A. N., and B. S. Rawat. 1984b. Permeation of benzene through liquid surfactant membranes. *J. Chem. Tech. Biotech.* 34A:174.

Goto, M., K. Kondo, and F. Nakashio. 1989. Extraction kinetics of copper with liquid surfactant membrane containing LIX 65N and nonionic surfactant. *J. Chem. Eng. Japan* 22(1):71–78.

Goto, M., H. Yamamoto, K. Kondo, and F. Nakashio. 1991. Effect of new surfactants on zinc extraction with liquid surfactant membranes. *J. Membr. Sci.* 57:161–174.

Goto, M., N. Yoshii, K. Kondo, and F. Nakashio. 1989. Separation of lanthanum and neodymium by liquid surfactant membranes. *Proc. Symp. Solvent Extr.,* pp. 113–118. Fukuoka, Japan: Kyushu University.

Gu, Z. M., R. M. Kurzeja, D. T. Wasan, and N. N. Li. 1982. Separation of metal ions by liquid membranes ligand accelerated transport. Paper read at AIChE Annual Meeting, 14–19 November 1982, Los Angeles, CA.

Gu, Z. M., D. T. Wasan, and N. N. Li. 1985. Interfacial mass transfer in ligand accelerated metal extraction by liquid surfactant membranes. *Sep. Sci. Technol.* 20(7&8):599–612.

Gu, Z. M., D. T. Wasan, and N. N. Li. 1986. Ligand-accelerated liquid membrane extraction of metal ions. *J. Membr. Sci.* 26:129–142.

Gu, Z. M., D. T. Wasan, and N. N. Li. 1988. Liquid surfactant membranes for metal extractions. *Surfactant Sci. Ser.* 28:127–168.

Gupta, T. C. S. M., A. N. Goswami, and B. S. Rawat. 1990. Mass transfer studies in liquid membrane hydrocarbon separations. *J. Membr. Sci.* 54:119–130.

Halwachs, W., E. Flaschel, and K. Schugerl. 1980. Liquid membrane transport—a highly selective separation process for organic solutes. *J. Membr. Sci.* 6:33.

Halwachs, W., and K. Schugerl. 1978. The liquid-membrane technique—a promising extraction process. *Chem.-Ing.-Tech.* 50:767.

Halwachs, W., and K. Schugerl. 1980. The liquid membrane technique—a promising extraction process. *Int. Chem. Eng.* 20(4):519–528.

Hatton, T. A., E. N. Lightfoot, and N. N. Li. 1982. The mathematical modeling of liquid surfactant membranes. In *Proceedings of Joint CIESC and AIChE Meeting,* 19–22 September 1982, Beijing, China, Chemical Industry Press, Vol. II, pp. 558–570.

Hayashita, T., R. A. Bartsch, T. Kurosawa, and M. Igawa. 1991. Selective concentration of lead (II) chloride complex with liquid anion-exchange membranes. *Analy. Chem.* 63(10):1023–1027.

Hayworth, H. C., W. S. Ho, W. A. Burns, Jr., and N. N. Li. 1983. Extraction of uranium from wet process phosphoric acid by liquid membranes. *Sep. Sci. Technol.* 18(6):493–521.

Hirato, T., I. Kishigami, Y. Awakura, and H. Majima. 1991. Concentration of uranyl sulfate solution by an emulsion-type liquid membrane process. *Hydrometallurgy* 26(1):19–33.

Hirato, T., K. Koyama, Y. Awakura, and H. Majima. 1990. Concentration of Mo(VI) from aqueous sulfuric acid solution by an emulsion type liquid membrane process. *Mater. Trans. JIM* 31(3):213–218.

Ho, W. S. 1986. The size of liquid membrane emulsion globules in an agitated continous phase: a dynamic measuring technique. Paper read at the 191st ACS National Meeting and First International Conference on Separations Science and Technology, 16–18 April 1986, New York.

Ho, W. S. 1990. Emulsion liquid membranes: a review. In *Proceedings of the 1990 International Congress on Membranes and Membrane Processes*, 20–24 August 1990, Chicago, IL, Vol. I, pp. 692–694.

Ho, C. S., and R. M. Cowan. 1987. Separating lactic acid from fermentation broth with liquid surfactant membranes. Paper read at the 194th ACS National Meeting, 30 August–4 September 1987, New Orleans, LA.

Ho, W. S., T. A. Hatton, E. N. Lightfoot, and N. N. Li. 1982. Batch extraction with liquid surfactant membranes: a diffusion controlled model. *AIChE J.* 28(4):662–670.

Ho, W. S., and N. N. Li. 1984. Modeling of liquid membrane extraction processes. In *Hydrometallurgical Process Fundamentals*, ed. R. G. Bautista, pp. 555–597. New York: Plenum Press.

Hochhauser, A. M. 1974. Concentrating chromium with liquid surfactant membranes. Ph.D. thesis, Carnegie-Mellon University, Pittsburgh, Pa.

Hochhauser, A. M., and E. L. Cussler. 1975. Concentrating chromium with liquid surfactant membranes. *AIChE Symp. Ser.* 71:136–142.

Hung, T.-M., C.-S. Chen, and C.-J. Lee. 1989. Effect of process conditions on the separations of toluene and p-xylene with liquid surfactant membrane. *J. Chin. Inst. Chem. Eng.* 20(6):319–325.

Hurst, F. J., D. S. Crouse, and K. B. Brown. 1972. Recovery of uranium from wet process phosphoric acid. *Ind. Eng. Chem. Process Des. Dev.* 11(1):122–128.

Izatt, R. M., M. P. Biehl, J. D. Lamb, and J. J. Christensen. 1982. Rapid separation of Tl^+ and Pb^{2+} from various binary cation mixtures using dicyclohexano-18-crown-6 incorporated in emulsion membranes. *Sep. Sci. Technol.* 17:1351–1360.

Izatt, R. M., D. V. Dearden, P. R. Brown, J. S. Bradshaw, J. D. Lamb, and J. J. Christensen. 1983. Cation fluxes from binary Ag^+-Mn^{2+} mixtures in a H_2O-$CHCl_3$-H_2O liquid membrane system containing a series of macrocyclic ligand carriers. *J. Am. Chem. Soc.* 105:1785.

Jefferson, T. B., O. W. Witzell, and W. L. Sibbett. 1958. Thermal conductivity of graphite-silicon oil and graphite-water separations. *Ind. Eng. Chem.* 50:1589.

Kataoka, T., T. Nishiki, and S. Kimura. 1989. Phenol permeation through liquid surfactant membrane—permeation model and effective diffusivity. *J. Membr. Sci.* 41:197–209.

Kataoka, T., T. Nishiki, S. Kimura, and Y. Tomioka. 1989. Batch permeation of metal ions using liquid surfactant membranes. *J. Membr. Sci.* 46:67–80.

Kataoka, T., T. Nishiki, S. Kimura, and Y. Tomioka. 1990. A model for mass transfer through liquid surfactant membranes. *Water Treatment* 5:136–149.

Kataoka, T., T. Nishiki, M. Yamauchi, and Y. Zhong. 1987. A simulation for liquid surfactant membrane permeation in a continuous countercurrent column. *J. Chem. Eng. Japan* 20(4):410–415.

Kikic, I., P. Alessi, and M. Orlandini-Visalberghi. 1978. Liquid membrane permeation for the separation of xylenes. *Inst. Chem. Eng. Symp. Ser.* 54:153–164.

Kim, K. S., S. J. Chol, and S. K. Ihm. 1983. Simulation of phenol removal from wastewater by liquid membrane emulsion. *Ind. Eng. Chem. Fundam.* 22:167.

Kitagawa, T., Y. Nishikawa, J. W. Frankenfeld, and N. N. Li. 1977. Waste water treatment by liquid membrane process. *Environmental Sci. Technol.* 11(6):602–605.

Kondo, K., K. Kita, I. Koida, J. Irie, and F. Nakashio. 1979. Extraction of copper with liquid surfactant membranes containing benzoylacetone. *J. Chem. Eng. Japan* 12(3):203–209.

Kopp, A. G., R. J. Marr, and F. E. Moser. 1978. A new concept for mass transfer in liquid surfactant membranes without carriers and with carriers that pump. *Inst. Chem. Eng. Symp. Ser.* 54:279.

Kremesec, V. J. 1981. Modeling of dispersed emulsions systems. *Sep. Purif. Methods* 10(2):319.

Kremesec, V. T., and J. C. Slattery. 1982. Analysis of batch, dispersed-emulsion separation systems. *AIChE J.* 28:492.

Krishna, R., A. N. Goswami, and A. Sharma. 1987. Effect of emulsion breakage on selectivity in the spearation of hydrocarbon mixtures using aqueous surfactant membranes. *J. Membr. Sci.* 34:141.

Kumamaru, T., Y. Okamoto, M. Yamamoto, Y. Obata, and K. Onizuka. 1990. High enrichment method for the determination of ultratrace levels of cobalt by liquid-liquid extraction using water/oil/water emulsions as liquid surfactant membranes. *Anal. Chim. Acta* 232(2):389–391.

Lee, C. J., and C. C. Chan. 1990. Extraction of ammonia from a dilute aqueous solution by emulsion liquid membranes: 1. experimental studies in batch system. *Ind. Eng. Chem. Res.* 29:96–100.

Lee, K. H., D. F. Evans, and E. L. Cussler. 1978. Selective copper recovery with two types of liquid membranes. *AIChE J.* 24:860.

Levich, V. G. 1962. *Physicochemical Hydrodynamics.* Englewood Cliffs, NJ: Prentice-Hall, Inc.

Levins, D. M., and J. R. Glastonbury. 1972. Particle-liquid hydrodynamics and mass transfer in a stirred vessel: part II. mass transfer. *Trans. Inst. Chem. Eng.* 50:132.

Li, N. N. 1968. Separating hydrocarbons with liquid membranes. U.S. Patent 3,410,794.

Li, N. N. 1971a. Removal of organic compounds by liquid membranes. U.S. Patent 3,617,546.

Li, N. N. 1971b. Permeation through liquid surfactant membranes. *AIChE J.* 17(2):495.

Li, N. N. 1971c. Separation of hydrocarbons by liquid membrane permeation (I). *Ind. Eng. Chem. Process Des. Dev.* 10:215.

Li, N. N. 1971d. Separation of hydrocarbons by liquid membrane permeation (II). In *Membrane Process in Industry and Biomedicine,* ed. M. Bier, p. 175. New York: Plenum Press.

Li, N. N. 1977. Separating hydrocarbon mixtures by emulsification. U. S. Patent 4,056,462.

Li, N. N. 1978. Facilitated transport through liquid membranes. *J. Membr. Sci.* 3:265.

Li, N. N. 1981. Encapsulation and separation by liquid surfactant membranes. *Chem. Eng. (Rugby, England)* 370:325–327.

Li, N. N., and J. W. Frankenfeld. 1988. Liquid membranes. In *Encyclopedia of Chemical Processing and Design,* ed. J. C. McKetta, Vol. 28, pp. 273–303. New York: Marcel Dekker.

Li, N. N., and A. L. Shrier. 1972. Liquid membrane water treating. In *Recent Developments in Separation Science,* ed. N. N. Li, Vol. 1, p. 163. Boca Raton, FL: CRC Press.

Lorbach, D., H. J. Bart, and R. J. Marr. 1986. Mass transfer in liquid membrane permeation. *Ger. Chem. Eng.* 9:321–327.

Lorbach, D., and R. J. Marr. 1987. Emulsion liquid membranes: part II. modeling mass transfer of zinc with bis(2-ethylhexyl)dithiophosphoric acid. *Chem. Eng. Process.* 21:83–93.

Marr, R. J., H. J. Bart, and J. Draxler. 1990. Liquid membrane permeation. *Chem. Eng. Process.* 27:59–64.

Marr, R. J., and M. Koncar. 1990. Rückgewinnung von Ammoniak aus Industrieabwasser. *Chem.-Ing.-Tech.* 62(3):175–182.

Marr, R. J., and A. Kopp. 1980. Flüssigmembran-Technik-Übersicht über Phänomene, Transportmechanismen und Modellbildungen. *Chem.-Ing.-Tech.* 52:399.

Marr, R. J., and A. Kopp. 1982. Liquid membrane technology—a survey of phenomena, mechanisms and models. *Int. Chem. Eng.* 22(1):44–60.

Marr, R. J., H. Lackner, and H. J. Bart. 1989. Verfahren zur Abtrennung von Nickel aus Nickellonen enthaltenden verdünnten wässrigen Lösungen. European Patent Application 89112656.7.

Martin, T. P., and G. A. Davies. 1977. The extraction of copper from dilute aqueous solutions using a liquid membrane process. *Hydrometallurgy* 2:315–334.

Martin, T. P., and G. A. Davies. 1980. Extraction of copper from aqueous solutions using a liquid membrane process: a model to simulate the process. In *Proc. Int. Solv. Extr. Conf., ISEC '80,* 6–12 September 1980, Liege, Belgium, Paper 80–230.

Matsumoto, M., K. Ema, K. Kondo, and F. Nakashio. 1990. Copper extraction with liquid surfactant membrane in Mixco extractor. *J. Chem. Eng. Japan* 23(4):402–407.

Matulevicius, E. S., and N. N. Li. 1975. Facilitated transport through liquid membranes. *Sep. Purif. Methods* 4(1):73–96.

Maugh, T. H. 1976. Liquid membrane—new techniques for separation, purification. *Science* 193:134–150.

Miao, F.-D., X.-P. Li, and Y.-Q. Zhang. 1985. The mathematical modeling of the removal of phenol with emulsified liquid membranes. *Desalination* 56:355–366.

Miyake, Y., Y. Takenoshita, and M. Teramoto. 1983. Extraction rates of copper with SME 529: mechanism and effects of surfactants. In *Proc. Int. Solv. Extr. Conf., ISEC '83,* 26 August–2 September 1983, Denver, CO, p. 301.

Nielsen, D. N., B. W. Jong, and A. M. Stubbs.

1991. Copper extraction from aqueous solutions with liquid emulsion membranes: a preliminary laboratory study. *Bur. Mines Rep. Invest.* RI 9375.

Noble, R. D. 1983. Shape factors in facilitated transport through membranes. *Ind. Eng. Chem. Fundam.* 22(1):139–144.

Noble, R. D., and J. D. Way. 1987. Applications of liquid membrane technology. In *Liquid Membranes: Theory and Applications,* ed. R. D. Noble and J. D. Way, ACS Symposium Series No. 347, pp. 110–122. Washington, DC: American Chemical Society.

Noble, R. D., J. D. Way, and A. L. Bunge. 1988. Liquid membranes. *Solvent Extr. Ion Exch.* 10:63–103.

Okamoto, Y., T. Takahashi, K. Isobe, and T. Kumamaru. 1990. Graphite furnace atomic absorption spectrometric determination of ultratrace levels of cobalt after solvent extraction using W/O/W emulsions as liquid surfactant membranes. *Analy. Sci.* 6(3):401–405.

Prötsch, M., and R. J. Marr. 1983. Development of a continuous process for metal ion recovery of liquid membrane permeation. In *Proc. Int. Solv. Extr. Conf., ISEC '83,* 26 August–2 September 1983, Denver, CO, pp. 66–67.

Reid, R. C., J. M. Prausnitz, and T. K. Sherwood. 1977. *The Properties of Gases and Liquids,* 3rd ed., Chap. 11. New York: McGraw-Hill Book Co.

Rowe, P. N., K. T. Claxton, and J. B. Lewis. 1965. Heat and mass transfer from a single sphere in an extensive flowing fluid. *Trans. Inst. Chem. Eng.* 43:T14–T31.

Rumscheidt, F. D., and S. G. Mason. 1961. Particle motions in sheared suspensions: XI. internal circulation in fluid droplets. *J. Colloid Sci.* 16:210.

Ruppert, M., J. Draxler, and R. J. Marr. 1988. Liquid-membrane-permeation and its experiences in pilot-plant and industrial scale. *Sep. Sci. Technol.* 23(12&13):1659–1666.

Russel, H. W. 1954. Principles of heat flow in porous insulators. *J. Am. Ceram. Soc.* 18:1–5.

Schiffer, D. K., E. M. Choy, D. F. Evans, and E. L. Cussler. 1974a. More membrane pumps. *AIChE Symp. Ser.* 70:150–156.

Schiffer, D. K., A. M. Hochhauser, D. F. Evans, and E. L. Cussler. 1974b. Concentrating solutes with membranes containing carriers. *Nature* 250:484–486.

Schlosser, S., and E. Kossaczky. 1980. Comparison of pertraction through liquid membranes and double liquid-liquid extraction. *J. Membr. Sci.* 6:83–105.

Shah, N. D., and T. C. Owens. 1972. Separation of benzene and hexane with liquid membrane technique. *Ind. Eng. Chem. Prod. Res. Dev.* 11:58.

Shiau, C. Y. 1991. Analysis of transport rate of zinc extraction through liquid surfactant membrane. *Sep. Sci. Technol.* 26(12):1519–1530.

Skelland, A. H. P., and J. M. Lee. 1981. Drop size and continuous phase mass transfer in agitated vessels. *AIChE. J.* 27:99.

Stelmaszek, J., and B. Borkowska. 1977. Separation of mixtures using liquid membranes. *Int. J. Chem. Eng.* 17:566.

Stroeve, P., and P. P. Varanasi. 1982. Transport processes in liquid membranes: double emulsion separation systems. *Sep. Purif. Methods.* 11(1):29–69.

Stroeve, P., and P. P. Varanasi. 1984. Extraction with double emulsions in a batch reactor: effect of continuous-phase resistance. *AIChE J.* 30:1007.

Stroeve, P., P. P. Varanasi, and L. J. C. Hoofd. 1984. Facilitated transport in spherical shells: application of the combined Damköhler number technique. *J. Membr. Sci.* 19:155–172.

Strzelbicki, J. 1978. Separation of cobalt by liquid surfactant membranes. *Sep. Sci. Technol.* 13:141.

Strzelbicki, J., and W. Charewicz. 1980. The liquid surfactant membrane separation of copper, cobalt and nickel from multicomponent aqueous solutions. *Hydrometallurgy* 5:243.

Strzelbicki, J., and S. Schlosser. 1989. Influence of surface-active substances on pertraction of cobalt (II) cations through bulk and emulsion liquid membranes. *Hydrometallurgy* 23:67–75.

Tavlarides, L. L., C. A. Coulaloglou, M. A. Zeitlin, G. E. Klinzing, and B. Gal-Or. 1970. Bubble and drop phenomena. *Ind. Eng. Chem.* 62:6.

Teramoto, M., and H. Matsuyama. 1986. Effect of facilitated diffusion in internal aqueous droplets on effective diffusivity and extraction rate of phenol in emulsion liquid membranes. *J. Chem. Eng. Japan* 19(5):469–472.

Teramoto, M., T. Sakai, K. Yanagawa, and Y. Miyake. 1983c. Modeling of the permeation of copper through liquid surfactant membranes by continuous operations. *Sep. Sci. Technol.* 18(11):985–997.

Teramoto, M., T. Sakai, K. Yanagawa, M. Ohsuga, and Y. Miyake. 1983b. Modeling of the permeation of copper through liquid surfactant membranes. *Sep. Sci. Technol.* 18(8):735–764.

Teramoto, M., T. Sakuramoto, T. Koyama, H. Matsuyama, and Y. Miyake. 1986. Extraction of lanthanoids by liquid surfactant membranes. *Sep. Sci. Technol.* 21(3):229–250.

Teramoto, M., H. Takihana, M. Shibutani, T. Yuasa, and N. Hara. 1983a. Extraction of phenol and cresol by liquid surfactant membrane. *Sep. Sci. Technol.* 18(5):397–419.

Teramoto, M., H. Takihana, M. Shibutani, T. Yuasa, Y. Miyake, and H. Teranishi. 1981. Extraction of amine by W/O/W emulsion system. *J. Chem. Eng. Japan* 14:122.

Terry, R. E., N. N. Li, and W. S. Ho. 1982. Extraction of phenolic compounds and organic acids by liquid membranes. *J. Membr. Sci.* 10:305–323.

Thien, M. P., and T. A. Hatton. 1988. Liquid emulsion membranes and their applications in biochemical processing. *Sep. Sci. Technol.* 23(8&9):819–853.

Thien, M. P., T. A. Hatton, and D. I. C. Wang. 1986. Liquid emulsion membranes and their applications in biochemical separations. *ACS Symp. Ser.* 314:67–77.

Ulbrich, M., R. J. Marr, and J. Draxler. 1991. Selective separation of organic solutes by aqueous liquid surfactant membranes. *J. Membr. Sci.* 59:189–203.

Vokel, W., W. Halwachs, and K. Schugerl. 1980. Copper extraction by means of a liquid surfactant membrane process. *J. Membr. Sci.* 6:19–31.

Wasan, D. T., Z. M. Gu, and N. N. Li. 1984. Separation of metal ions by ligand-accelerated transfer through liquid surfactant membranes. *Faraday Discuss. Chem. Soc.* 77:67–74.

Weiss, S., V. Griegoriev, and P. Mühl. 1982. The liquid membrane process for the separation of mercury from waste water. *J. Membr. Sci.* 12:119–129.

Wilke, C. R., and P. Chang. 1955. Correlation of diffusion coefficients in dilute solutions. *AIChE J.* 1(2):264–270.

Wodzki, R., A. Wyszynska, and A. Narebska. 1990. Two-component emulsion liquid membranes with macromolecular carriers of divalent ions. *Sep. Sci. Technol.* 25(11&12):1175–1187.

Yan, N.-X, S.-A. Huang, and Y.-J. Shi. 1987. Removal of acetic acid from wastewater with liquid surfactant membranes: an external boundary layer and membrane diffusion controlled model. *Sep. Sci. Technol.* 22(2&3):801–818.

Yan, N.-X., Y.-J. Shi, and Y.-F. Su. 1990. A study of gold extraction by liquid membranes. *Water Treatment* 5(2):190–201.

Zhang, X.-J., Q.-J. Fan, X.-T. Zhang, and Z.-F. Liu. 1988a. New surfactant LMS-2 used for industrial application in liquid membrane separation. In *Separation Technology,* ed. N. N. Li and H. Strathmann, pp. 215–226. New York: United Engineering Trustees; also see *Water Treatment* 3:233–240.

Zhang, X.-J., J.-H. Liu, Q.-J. Fan, Q.-T. Lian, X.-T. Zhang, and T.-S. Lu. 1988b. Industrial application of liquid membrane separation for phenolic wastewater treatment. In *Separation Technology,* ed. N. N. Li and H. Strathmann, pp. 190–203. New York: United Engineering Trustees.

Zhang, X.-J., J.-H. Liu, and T.-S. Lu. 1987. Industrial application of liquid membrane separation for phenolic wastewater treatment. *Water Treatment* 2:127–135.

Zhang, R.-H., and D.-X. Wang. 1989. Extraction of mixed rare earth from aqueous solution with emulsion liquid membrane. *Water Treatment* 4:165–176.

Zhang, R.-H., and L. Xiao. 1989. Design of a liquid membrane system for extracting rare earths. *Water Treatment* 4:473–481.

Zhang, R.-H., and L. Xiao. 1990. Design of a liquid membrane system for extracting rare earths. *J. Membr. Sci.* 51(3):249–258.

38

Design Considerations

Zhongmao Gu
China Institute of Atomic Energy

W. S. Winston Ho
Exxon Research and Engineering Company

Norman N. Li
Allied Signal, Inc.

EMULSIFICATION
   Membrane Formulation
   Emulsion Preparation
   Emulsion Liquid Membrane Systems
DISPERSION/EXTRACTION/
   SETTLING
   Modes of Dispersion Operation
   Emulsion Globule Size
   Extraction
   Leakage
   Swelling
DEMULSIFICATION
   Effect of Applied Voltage
   Effects of Frequency and Wave
     Form
   Other Effects on Demulsification
NOTATION
REFERENCES

As discussed in Chapter 36, an emulsion liquid membrane (ELM) process includes four steps: (1) emulsification, (2) dispersion and extraction, (3) settling, and (4) demulsification (breaking of the emulsion). These four steps are shown schematically in Figure 36–3. This chapter presents the design considerations for these steps in the ELM process.

EMULSIFICATION

Membrane Formulation

Emulsion liquid membranes can be "tailor-made" to meet the requirements of different separations, and thus they can offer versatile processes capable of separating a wide range of liquid mixtures (Li and Frankenfeld 1988; Halwachs, Flaschel, and Schugerl 1980; Noble, Way, and Bunge 1988; Marr, Bart, and Draxler 1990; Ho 1990; see also Chapters 36 and 39). On the other hand, this versatility sometimes brings about difficulties in the design of membrane formulation. Choosing the appropriate membrane formulation for a particular separation task is often quite complex. For Type 1 facilitation, the membrane phase consists only of a diluent and a surfactant to stabilize the primary emulsion. No extractant is needed for this type of facilitation because the solute transport across the membrane is accomplished through its physical solubility and then diffusion in this membrane. However, for Type 2 facilitation, an extractant and its associated stripping agent must be incorporated into the membrane and internal phases, respectively, in order to achieve a coupled extraction/stripping process.

Extractants/Stripping Agents

Screening of the extractant/stripping agent system requires considerable chemical insight. The experience accumulated in solvent extraction can often be used as a guide. When choosing extractants, keep in mind that the selected extractant and its complex must be soluble in the membrane phase, but insoluble in the external and internal phases. Precipitates are also not allowed to form either within the membrane or at the interfaces. Otherwise, the membrane process will fail (Cussler and Evans 1974).

Generally, the selection of the extractant/stripping agent system is based on the thermodynamic and kinetic considerations. Thermodynamically, the selected extractant should favor the distribution of the solute from the external phase to the membrane phase. On the contrary, a stripping agent must be thus selected to partition the solute from the membrane phase to the internal phase. Take Cd^{2+} extraction as an example; in the effluent from a Cd^{2+} plating bath, $Cd(CN)_4^{2-}$ is readily extracted from the external aqueous feed phase to the membrane phase by methyltrioctylammonium chloride (Aliquat 336). However, stripping is quite difficult because of the high stability of the $Cd(CN)_4^{2-}$ complex under basic conditions. The usual stripping agents are ineffective even at high concentrations. Ethylenediaminetetraacetic acid (EDTA), which has a stronger complexing ability with $Cd(CN)_4^{2-}$ than Aliquat, can be used to strip $Cd(CN)_4^{2-}$ from the membrane phase to the internal phase (Kitagawa et al. 1977). This example also tells us that the stability of the solute-extractant complex should be moderate so as to ease the stripping process.

A thermodynamic condition exists under which the liquid membrane process can be operated while solvent extraction cannot. A solvent extraction process needs a high distribution ratio for extraction so as to increase the extraction ability. The nonequilibrium feature of emulsion liquid membranes allows the selected extractant to have a lower distribution ratio for extraction than solvent extraction. For instance, an extractant, which has a low distribution ratio for Cr(VI) extraction and thus is unsuitable for a solvent extraction process, was successfully used to concentrate Cr(VI) in an ELM system (Yan et al. 1990a).

Kinetically, the selected extractant and stripping agent should usually exhibit fast reactions for both extraction and stripping. But it is interesting to note that because of its much higher interfacial area for stripping than that for extraction, an emulsion liquid membrane process is capable of coping with the situation in which the extractant has relatively fast extraction kinetics but the stripping agent has extremely low stripping kinetics. Draxler and Marr (1986; see also Chapter 39) have found that di(2-ethylhexyl)dithiophosphoric acid (DTPA) and di(2,2,4-trimethylphenyl)dithiophosphinic acid (Cyanex 301 produced by American Cyanamid) are very strong extractants for metal ions but nearly unknown in solvent extraction due to the extremely slow stripping kinetics of the extractants. However, they have identified these extractants to concentrate Zn^{2+} and Ni^{2+} effectively in ELM systems. With the advantage of the extremely high interfacial area for stripping, the ELM process can be enhanced by accelerating an originally slow extraction reaction (Boyadzhiev and Kyuchoukov 1980). For example, Co^{2+} extraction by di(2-ethylhexyl)phosphoric acid (D2EHPA) is known to be slow. By introducing certain anionic ligands, such as carboxylates, to the continuous aqueous phase containing the cations to be extracted, the extraction reaction rate is increased and thus the overall liquid membrane mass transfer is enhanced (Gu, Wasan, and Li 1985).

According to functional groups, the extractants are generally divided into three classes: (1) acidic extractants, (2) basic extractants, and (3) neutral extractants (Gu, Wasan, and Li 1988).

Acidic Extractants and Their Associated Stripping Agents

To extract a cation from an aqueous solution, it must be combined with an anionic species to form an uncharged species. Acidic extractants are most effective for extracting cations by exchanging their protons for the cations for liquid membrane cation extraction. Commonly used acidic extractants can be classified into three groups:

1. Chelating extractants, which include (a) hydroxyoximes such as LIX 63, LIX 64N, LIX 65N, LIX 70 (Henkel); SME 529 (Shell); P 17, P 50, and PT 5050 (Acorga); (b) β-hydroxyquinolines (oximes) such as Kelex 100 (Ashland Chemical); and (c) β-diketones such as acetylacetone (AA) and benzoylacetone (BA).
2. Alkylphosphorous compounds, which include (a) organophosphoric acids such as di(2-ethylhexyl)phosphoric acid (D2EHPA), and dibutylphosphoric acid (DBP); (b) organophosphonic acids such as mono(2-ethylhexyl) ester of 2-ethylhexylphosphonic acid (PC 88A, Daihachi Chemical Industry); (c) organophosphinic acids such as di-(2,4,4-trimethylpentyl)phosphinic acid (Cyanex 272, American Cyanamid); (d) thiophosphoric acids such as di(2-ethylhexyl)dithiophosphoric acid (DTPA, Hoechst); and (e) thiophosphinic acids such as di-(2,2,4-trimethylphenyl)dithiophosphinic acid (Cyanex 301, American Cyanamid).
3. Ionizable crown ethers, such as crown ether carboxylic acid and crown ether phosphonic acid monoalkyl ester (Bartsch et al. 1987), and naturally derived antibiotic macrocycles such as monesin (Cussler and Evans 1974).

In general, the coordination complex of the chelating extractant with positively charged metal ions is very specific. Therefore, the chelating extractants can be used to achieve selective separation of metals. Although alkylphosphorous compounds, which are often called *liquid cation exchangers,* are less selective, they are less expensive and their metal complexes are more soluble in organic solvents than metal chelates (Sekine and Hasegawa 1977). Therefore, they are also widely used in hydrometallurgical extraction processes.

In recent years, a series of new extractants has been synthesized to improve extraction processes. For example, D2EHPA is a commonly used organophosphoric acid for Co/Ni separation in a sulfate medium. However, the separation factor, $\alpha_{Co/Ni}$, is very low (2.2). Chuei, Yu, and Zhang (1983) used a new acidic phosphonic extractant [P 5709, mono(2-ethylhexyl)phosphonate] to replace D2EHPA. They obtained the $\alpha_{Co/Ni}$ of 435.4, which was more than 100 times higher than that of the D2EHPA system. Rikelton, Flett, and West (1983) found that dialkylphosphinic acids, such as Cyanex 272, exhibited an even better Co/Ni separation factor.

With substitution of an alkyl-oxo group by an alkyl group, the acidities of the alkylphosphorous compounds show the order of phosphoric > phosphonic > phosphinic, and thus their extraction abilities to cations give the same order. But their stripping tendency for cations is in the opposite order: phosphoric < phosphonic < phosphinic. This means that the relatively low extraction ability of phosphonic or phosphinic extractants can be partially compensated by their high stripping tendency. In their study of rare earth extraction, Zhang and Xiao (1989, 1990) found that phosphonic acid [P 507, mono(2-ethylhexyl) 2-ethylhexylphosphonate] had a lower extraction but higher stripping ability than D2EHPA. As a result, the net liquid membrane extraction efficiency of both extractants was similar.

The acidity of the external aqueous phase sometimes can seriously affect the choice of a suitable extractant. For instance, the zinc effluent arising from spin baths in the viscose production has an acidity up to a pH of 0.5. Under such a feed condition, D2EHPA, Cyanex 272, and PC 88A are unable to extract zinc efficiently. However, Draxler and Marr (1986) found that DTPA could extract zinc effectively even at a pH of about 0.5, and they solved the problem satisfactorily. They have described the zinc extraction in detail in Chapters 39 and 40.

Ionizable crown ethers bear ionizable carboxylic or phosphonic groups on the macrocycles (Izatt et al. 1986; Bartsch et al. 1987; Tang and Wai 1989). These macrocyclic compounds with ionizable protons are capable of facilitating cation transport by a cation-exchange mechanism without the need for an anion to accompany the cation-macrocycle complex across the liquid membrane or for an auxiliary complexing agent in the internal phase. This type of extractant seems to have potential applications.

The extraction and stripping reactions for solute transport across the membrane by an acidic extractant can be represented as follows (consider divalent cations to be transported):

Extraction:

$$M^{2+} + 2\,HR \rightarrow MR_2 + 2\,H^+,$$
aqueous phase, organic phase, organic phase, aqueous phase

(38-1)

Stripping:

$$2\,H^+ + MR_2 \rightarrow 2\,HR + M^{2+},$$
aqueous phase, organic phase, organic phase, aqueous phase

(38-2)

where HR is the protonated form of an acidic extractant, M^{2+} is the metal ion, and H^+ the hydrogen ion. From these equations, it is clear that the proton is the source of the driving force, which "pumps" the metal ion against its own concentration gradient between the internal and external phases (Cussler 1971; Cussler, Evans, and Matesich 1971; Schiffer et al. 1974a, 1974b). Therefore, acids are generally used in the internal phase for stripping the solute from the membrane phase.

In selecting suitable stripping acids, the effect of the acids on the liquid membrane extraction efficiency should be considered. In their study of rare earth extraction by D2EHPA in an ELM system with a polyamine as a surfactant, Zhang and Xiao (1989, 1990) found that the maximum extraction efficiency of rare earths was in the order of HCl > HNO_3 > H_2SO_4 at the same acid concentration (6N) in the internal phase. They observed that in the case of H_2SO_4 stripping, after the transfer of rare earths through the membrane phase into the internal phase, the rather stable complex anions $RE(SO_4)_2^-$ and $RE(SO_4)_3^{3-}$ were formed and extracted by the amine surfactant from the internal phase back to the external phase. This partially canceled the amount of rare earths transferred by D2EHPA from the external phase to the internal phase, thus lowering the overall efficiency of liquid membrane extraction. However, in the case of HCl stripping, no stable complex $RECl_n^{3-n}$ (n = 1–6) was formed, and thus the stripped rare earth ions would stay in the internal phase instead of being extracted by the amine surfactant. But when HNO_3 was used as a stripping acid, it partially damaged the polyamine surfactant, causing an unstable membrane and thus the poor extraction efficiency.

Basic Extractants and Their Associated Stripping Agents

Basic extractants, i.e., the high molecular weight primary amines such as Primene JMT, secondary amines such as Amberlite LA-2, tertiary amines such as trioctylamine (TOA), tri-*n*-octylamine (TNOA), and Alamine 336, and the quarternary alkylammonium salts such as Aliquat 336, are used for the extraction of anionic or neutral metal complexes. Similar to the alkylphosphoric acids, these extractants can be regarded as *liquid anion exchangers*.

In aqueous solutions many metal ions can form a variety of anionic complexes with sulfate, halide, cyanate, thiocyanate, and a number of other anionic ligands. Examples of anionic metal complexes that commonly exist in the solutions in hydrometallurgical and electroplating processes are $Cd(CN)_4^{2-}$, $Cr_2O_7^{2-}$, $AuCl_4^-$, $UO_2(SO_4)_2^{2-}$, WO_4^{2-}, MoO_4^{2-}, $V_3O_9^{3-}$, ReO_4^-, etc. The ease of formation of anionic complexes of different metal ions with the anion ligand in the aqueous solution varies greatly. For this reason, the selective extraction is obtained by the use of suitable amines. For ELM metal extractions, primary and secondary amines show poor emulsion stability and are usually not used (Draxler and Marr 1986). Thus, the commercial tertiary amine Alamine 336 is widely used as extractant for acidic brines, and the commercial quaternary ammonium salt Aliquat 336 is employed for the solutions with high pH.

The transport of metal anions through liquid membrane containing tertiary amines follows two possible mechanisms depending on the stripping agents employed:

1. *Co-transport:* If a basic solution is used as the stripping agent in the internal aqueous phase, the permeation of metal anions proceeds according to the co-transport mechanism, in which the transport of metal anions across the membrane is coupled to the transport of hydrogen ions (H^+) in the same direction with the following extraction and stripping reactions:

Extraction:

$$R_3N \text{ (organic phase)} + H^+ + A^- \text{ (aqueous phase)} \rightarrow R_3NHA \text{ (organic phase)}, \quad (38\text{--}3)$$

Stripping:

$$R_3NHA \text{ (organic phase)} + Na^+ + OH^- \text{ (aqueous phase)} \rightarrow R_3N \text{ (organic phase)} + Na^+ + A^- \text{ (aqueous phase)} + H_2O, \quad (38\text{--}4)$$

where R_3N is the tertiary amine and A^- is the metal anion. From Eq. (38–4), we can see that stripping with a base regenerates the free amine in the membrane, which then re-extracts additional metal ions from the external aqueous phase.

2. *Counter-transport:* If an acid serves as the stripping agent in the internal aqueous phase, the permeation of metal anions obeys the counter-transport mechanism, in which metal anions are transported in the direction opposite to the coupled anions with the following extraction and stripping reactions:

Extraction:

$$R_3NHX \text{ (organic phase)} + H^+ + A^- \text{ (aqueous phase)} \rightarrow R_3NHA \text{ (organic phase)} + H^+ + X^- \text{ (aqueous phase)}, \quad (38\text{--}5)$$

Stripping:

$$R_3NHA \text{ (organic phase)} + H^+ + X^- \text{ (aqueous phase)} \rightarrow R_3NHX \text{ (organic phase)} + H^+ + A^- \text{ (aqueous phase)}, \quad (38\text{--}6)$$

where R_3NHX is the amine salt, A^- is the metal anion, and X^- the coupled anion. According to Eq. (38–6), acid stripping regenerates the amine salt for re-extracting metal anions from the external aqueous phase.

Good examples showing the above transport mechanisms are the dichromate extractions by (1) tertiary amines with the co-transport mechanism and (2) quaternary ammonium salts with the counter-transport mechanism in ELM systems (Kitagawa et al. 1977; Frankenfeld and Li 1979):

1. Extraction by amine neutralization with the co-transport mechanism:

$$2\ R_3N + 2\ H^+ + Cr_2O_7^{2-} \rightarrow (R_3NH)_2Cr_2O_7, \qquad (38\text{-}7)$$
$$\text{organic} \qquad \text{aqueous} \qquad \text{organic}$$
$$\text{phase} \qquad \text{phase} \qquad \text{phase}$$

Basic stripping:

$$(R_3NH)_2Cr_2O_7 + 4(Na^+ + OH^-) \rightarrow 2R_3N + 2(2Na^+ + CrO_4^{2-}) + 3\ H_2O. \qquad (38\text{-}8)$$
$$\text{organic} \qquad \text{aqueous} \qquad \text{organic} \qquad \text{aqueous}$$
$$\text{phase} \qquad \text{phase} \qquad \text{phase} \qquad \text{phase}$$

2. Extraction by salt formation with the counter-transport mechanism:

$$(R_3NH)_2X + 2\ H^+ + Cr_2O_7^{2-} \rightarrow (R_3NH)_2Cr_2O_7 + 2\ H^+ + X^{2-}, \qquad (38\text{-}9)$$
$$\text{organic} \qquad \text{aqueous} \qquad \text{organic} \qquad \text{aqueous}$$
$$\text{phase} \qquad \text{phase} \qquad \text{phase} \qquad \text{phase}$$

Acid stripping:

$$(R_3NH)_2Cr_2O_7 + 2\ H^+ + X^{2-} \rightarrow (R_3NH)_2X + 2\ H^+ + Cr_2O_7^{2-}, \qquad (38\text{-}10)$$
$$\text{organic} \qquad \text{aqueous} \qquad \text{organic} \qquad \text{aqueous}$$
$$\text{phase} \qquad \text{phase} \qquad \text{phase} \qquad \text{phase}$$

where

$$X^{2-} = SO_4^{2-} \text{ or } 2\ Cl^-.$$

Kitagawa et al. (1977) believe that both mechanisms can be used successfully in ELMs. But Hirato et al. (1990) have found that the co-transport type extraction can give a far better result than that of counter-transport for the extraction of U(VI) and Mo(VI) by the tertiary amine, tri-*n*-octylamine (TNOA), in the ELM systems.

The stripping agents are usually the same as in solvent extraction with some restrictions. Ammoniacal solutions with a pH of greater than 11 cannot be used because of the quick transport of ammonia across the membrane. Instead of ammonia, NaOH and Na_2CO_3 can be used (Draxler and Marr 1986). But for a membrane stabilized by SPAN 80 (sorbitan monooleate), Na_2CO_3 rather than NaOH is preferred as a stripping agent (Hirato et al. 1990).

Usually, the concentration of amine extractants in the membrane phase should not be very high. High concentrations of amine extractants in the membrane phase have been observed to lead to high osmotic swelling and high rates of membrane breakdown (Draxler and Marr 1986; Hirato et al. 1990).

A major drawback for the extraction of metal anions is the poor selectivity of amine extractants to different anions. For instance, during the co-transport permeation of Mo(VI) anions through the membrane phase from a sulfuric acid medium, there exists competitive extraction between the Mo(VI) anions and sulfate anions, and every sulfate anion transported by the amine extractant from the external aqueous phase to the internal aqueous phase will be coupled with two hydrogen ions. This will consume excess hydroxide ions in the internal phase (Hirato et al. 1990). When the metal anions in the external aqueous phase become very low (e.g., 3 to 5 ppm), the competitive extraction of sulfate anions to metal anions will be even more significant (Draxler and Marr 1986).

Neutral Extractants and Their Associated Stripping Agents

Neutral extractants often extract uncharged metal complexes or cations together with the coupled anions in order to maintain the electrical neutrality. In this type of extraction, the metal species is coordinated with two different types of ligands, i.e., a water-soluble anion and an organic-soluble electron-donating functional group. Most of the neutral extractants that have been investigated in the liquid membrane studies are organo-phosphoryl compounds such as tri-n-butyl phosphate (TBP), tri-n-butylphosphine oxide (TBPO), and tri-n-octylphosphine oxide (TOPO). These compounds are used extensively as extractants for the separations of actinides and lanthanides. They are especially useful for the recovery of uranium and plutonium in the spent fuel reprocessing of nuclear plants.

In recent years, studies using macrocyclic compounds as metal carriers in liquid membranes have become a quite active area (Izatt et al. 1983; Bartsch, Charewicz, and Kang 1984; Izatt et al. 1987a, 1987b; Tang and Wai 1988). Macrocyclic ligands include crown ethers and their derivatives. These compounds are cyclic or polycyclic organic molecules that contain hetero atoms capable of forming electron-rich interior cavities. They possess the ability to complex ions or molecules in the electron-rich cavity via ion-dipole or dipole-dipole interactions. In many cases, they have the remarkable property of selectively complexing particular ions (Christensen et al. 1981; Lamb et al. 1983). When these macrocyclic molecules are incorporated as carriers into liquid membranes, the flux of different species can differ enormously. Reusch and Cussler (1973) found that the flux of potassium ions through a liquid membrane containing dibenzo-18-crown-6 ether was 4000 times higher than the flux of lithium ions.

Izatt, Christensen, and their coworkers studied extensively the incorporation of a number of macrocyclic carriers in liquid membrane systems to facilitate the transport of several alkali, alkaline earth, and transition metal cations (Christensen et al. 1983; Izatt et al. 1986; Izatt, Bruening, and Christensen 1987). They have undertaken systematic investigations of the influences on carrier facilitated transport rates and selectivities with respect to macrocyclic ligand structure (Lamb et al. 1981), ratio of cation and macrocycle cavity diameters (Izatt et al. 1984), and equilibrium constant for macrocycle/cation interaction (Lamb et al. 1980). Recently, their studies have extended to the transport of the metal anionic complexes (Izatt, Clark, and Christensen 1987). One of the most attractive results in their studies is the remarkably high selectivity for Pb^{2+} over the other cations with didecyl-1, 10-diaza-18-crown-6 and dicyclohexano-18-crown-6 (DC18C6) in the membrane phase even if the concentration ratio of Pb^{2+} to Ca^{2+} in the feed mixture is as low as 1:100 (Izatt et al. 1987b).

Since separations among the alkali, alkaline earth, and trivalent lanthanide and actinide cations have traditionally been difficult to achieve, membranes containing macrocyclic ligand carriers offer an inviting alternative for selective separations of these cations. At present, the initial costs of the macrocyclic compounds are still very high, and in some cases the aqueous distribution of these ligands prevents them from being suitable for large-scale processes. However, compared to solvent extraction, the inventory of these expensive compounds in the membrane is much lower. This would make these extractants quite competitive in a liquid membrane process for separation of precious metals. One of the important tasks for chemists is to develop economical methods of synthesizing these compounds with the necessary phase distribution and metal coordinating properties.

As discussed earlier, in the liquid membrane systems containing acidic or basic extractants as solute (metal) carriers, hydrogen ions or hydroxide ions are used to generate the driving force for solute permeation across the membrane. For liquid membranes with neutral carriers, however, some other driving forces have to be used instead of pH gradient. The general means of accomplishing this is to incorporate strong metal complexing agents in the internal aqueous phase. The stability constant of the metal complex for stripping must be sufficiently higher than that of the metal complex for extraction so as to give the high driving force for metal transport.

The transport mechanism of solute across the liquid membrane incorporated with a neutral extractant can be expressed with the following extraction and stripping reactions:

Extraction:

$$M^+ + A^- + L \rightarrow MLA, \quad (38\text{--}11)$$
aqueous   organic   organic
phase   phase   phase

where M^+ is the metal ion, A^- is the co-anion, and L the neutral extractant.

Stripping:

$$MLA + M_s^+ + R^- \rightarrow L + MR + M_s^+ + A^-, \quad (38\text{--}12)$$
organic   aqueous   organic   aqueous
phase   phase   phase   phase

where R^- is the anion complexing agent and M_s^+ the co-cation.

The neutral extractant serves as a carrier to shuttle cations across the membrane phase, while the anion complexing agent in the internal aqueous phase provides a sink for cations by complexation. This complexation lowers the concentration of the free (unbound) cations in the internal phase and maintains the concentration-gradient driving force across the membrane phase, which is necessary for the transport. Under favorable conditions, the free cation concentration in the internal phase during transport will be essentially zero, and a concentration gradient will be maintained until the external phase is depleted of the cations (Izatt et al. 1984).

An example of metal transport governed by this mechanism is the extraction of Pb^{2+} through ELMs (Christensen et al. 1983; Izatt et al. 1984, 1986, 1987a; Izatt, Clark, and Christensen 1987; Izatt, Lamb, and Bruening 1988). When dicyclohexano-18-crown-6 (DC18C6) was incorporated into the membrane phase as a carrier and $P_2O_7^{4-}$ was used as a stripping agent in the internal phase, the measured stability constants were $10^{4.27}$ (Izatt et al. 1987a) and $10^{11.24}$ (Christensen et al. 1983) for the Pb^{2+}-DC18C6 and Pb^{2+}-$P_2O_7^{4-}$ complexes, respectively. Obviously, such a high difference between these two complexes created an enormous driving force to transport Pb^{2+} from the external phase through the membrane phase to the internal phase successfully.

The use of hydrophilic macrocycles as stripping agents for alkali cations is of interest because of their strong interactions with these cations. For the K^+ transport through an ELM with DC18C6 as the extractant in the membrane phase, the K^+ transport efficiency with a depronated carboxylic acid crown ether, $(CO_2H)_4$-18-crown-6, in the internal phase as a stripping agent was much higher than that obtained with $P_2O_7^{4-}$ as the stripping agent (Izatt, Lamb, and Bruening 1988). However, the macrocycle incorporated into the internal phase must be extremely hydrophilic, or it will quickly equilibrate (e.g., 18-crown-6) through the organic membrane phase between the internal and external phases, resulting in the lowering of the membrane transport efficiency.

Note that some other mechanisms can be employed to generate or enhance the driving force for solute permeation across the membrane. These mechanisms involve reversible electrochemical reduction-oxidation reactions (Grimaldi and Lehn 1979; Kaifer et al. 1983), photochemical reactions (Shinkai et al. 1982), temperature changes (Pannell et al. 1982), etc.

Surfactants

The surfactant is the key component for forming a stable emulsion. In emulsion liquid membranes, the overwhelming majority involves W/O/W double emulsion systems. Water soluble surfactants used to form O/W emulsions are seldom employed, so they are not discussed here.

In the early laboratory studies of ELMs attention was given mainly to the stability of the emulsion. But as membranes have become commercialized, their industrial applications have required more of the surfactants than their major contribution to membrane stability. An ideal surfactant should possess the following properties:

1. It carries virtually no water during operation so as to alleviate osmotic swelling.
2. It does not react with the extractant in the membrane phase; if any, the reaction should promote the extraction process rather than catalyze the decomposition of the extractant.
3. It has a low interfacial resistance to mass transfer.
4. It does not inhibit demulsification.
5. It is soluble in the membrane phase but insoluble in the external and internal phases.
6. It is stable against acids, bases, and bacteria.

In addition, the selected surfactant should be cheap and nontoxic for economic and environmental considerations.

Unfortunately, few attempts up to now have been made to search or synthesize specialized surfactants for ELMs (Goto et al. 1987; Nakashio et al. 1988; Zhang et al. 1988). The most popular surfactants used thus far in different laboratories are SPAN 80, ECA 4360, and their derivatives.

SPAN 80 (sorbitan monooleate) is a nonionic surfactant with a molecular weight of 428. Its structure is:

$$CH_3(CH_2)_7-CH=CH-(CH_2)_7-\overset{O}{\underset{\|}{C}}-O-CH_2-\underset{\underset{OH}{|}\underset{OH}{|}}{\overset{\overset{O}{\diagup\diagdown}}{}}-OH$$

The commercial product of SPAN 80 contains some impurities, such as dioleate and trioleate (Goto et al. 1987; Nakashio et al. 1988). SPAN 80 is a fairly good emulsifier, so it is widely used in ELM studies. During extraction operations, the membranes incorporated with SPAN 80 show less resistance to mass transfer than those with other surfactants (Draxler and Marr 1986; Strzelbicki and Schlosser 1989; Lee and Chan 1990).

SPAN 80, however, suffers from some drawbacks. It is a good carrier for water molecules and therefore favors the osmotic swelling of emulsion (Martin and Davies 1977; Colinart et al. 1984; Draxler and Marr 1986; Nakashio et al. 1988; Hirato et al. 1990; Abou-Nemeh and Van Peteghem 1990). Another disadvantage of SPAN 80 as an emulsifer lies in its poor chemical stability, especially when the NaOH is incorporated into the internal phase (Zhang et al. 1988; Hirato et al. 1990). Abou-Nemeh and Van Peteghem (1990) proposed that the instability of SPAN 80 was caused by its decomposition due to hydrolysis and by macroemulsion formation due to the presence of H_2O in SPAN 80 and other membrane components. In their study to concentrate Mo(VI) by emulsion liquid membranes containing SPAN 80, Hirato et al. (1990) found that the emulsion breakdown was serious when the solution containing NaOH was used as the internal phase. They improved the membrane stability using Na_2CO_3 solution as the internal phase. In addition, SPAN 80 also suffers from its high solubility in an aqueous phase, which causes a high reagent loss (Draxler and Marr 1986; Zhang and Xiao 1989, 1990).

For liquid membrane extraction of copper with Acorga PT 5050, Draxler, Fürst, and Marr (1988) screened a number of surfactants. They observed that although most SPAN reagents (SPAN 20 is sorbitan monolaurate, SPAN 40 sorbitan monopalmitate, SPAN 60 sorbitan monostearate, SPAN 65 sorbitan tristearate, SPAN 80 sorbitan monooleate, and SPAN 85 sorbitan trioleate) were very good emulsifying agents, all had one or more deleterious effects on the copper extraction, as shown in Table 38–1. In this table, PX 100 is a polyamine surfactant and similar to ECA 4360, which is described in the following paragraphs. TWEEN

TABLE 38-1. Effects of Surfactants on Copper Permeation (reprinted from Draxler, Fürst, and Marr 1988 with permission).

Surfactant	Mass Transfer Resistance	Osmosis	Emulsion Stability	Decomposition of Extractant
PX100	++	+	++	--
SPAN 20	+	-	+	-
SPAN 40	-	--	++	--
SPAN 60	--	+	++	--
SPAN 80	+	-	++	-
SPAN 65	--	++	++	++
SPAN 85			--	+
TWEEN 65	-	--	+	-
TWEEN 85	-	--	+	-
PA 18	++	++	++	++

Legend: ++ = very suitable for copper extraction with Acorga PT 5050.
+ = suitable.
- = not suitable.
-- = absolutely not suitable.

65 is polyoxyethylene sorbitan tristearate, TWEEN 85 is polyoxyethylene sorbitan trioleate, and PA 18 is a copolymer of maleic anhydride and 1-octadecene. Marr and his coworkers (Draxler and Marr 1986; Ruppert, Draxler, and Marr 1988) also found during pilot plant experiments that SPAN 80 could be destroyed by bacteria.

ECA 4360 is a nonionic polyamine with a molecular weight of about 1800 (Strzelbicki and Schlosser 1989). It has a structure as shown below (Nakashio et al. 1988):

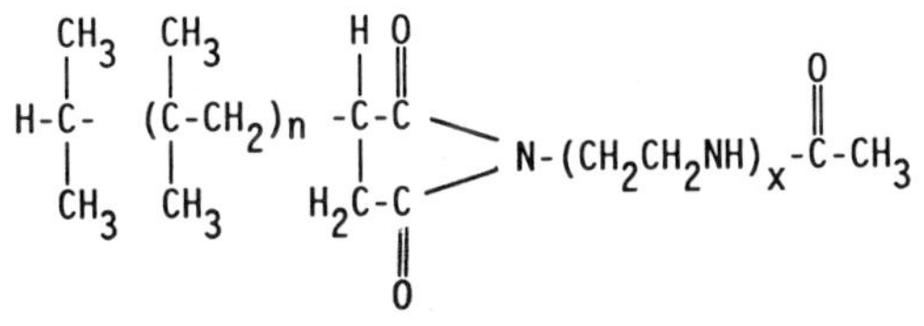

$$\begin{array}{c} CH_3 \quad CH_3 \quad H \quad O \\ | \quad | \quad | \quad \| \\ H-C-(C-CH_2)_n-C-C \\ | \quad | \quad | \\ CH_3 \quad CH_3 \quad H_2C-C \\ \| \\ O \end{array} \rangle N-(CH_2CH_2NH)_x - \overset{O}{\overset{\|}{C}} - CH_3$$

where n = 10 to 60 and x = 3 to 10.

In ECA 4360 and its derivatives, polar groups are placed onto a polymeric molecular chain so that the resultant molecule is both a surfactant due to the polar groups and a membrane-strengthening additive due to the polymetric backbone (Li 1981b). ECA 4360 can form very stable emulsions compared to SPAN 80 (Goto et al. 1987; Qian, Ma, and Shi 1989), and it causes less osmotic swelling of the emulsions than SPAN 80 (Nakashio et al. 1988). It is sparingly soluble in water (Draxler, Fürst, and Marr 1988; Zhang and Xiao 1989, 1990), and it shows high chemical stability, especially against bases (Zhang and Xiao 1989, 1990). In addition, it can be used at temperatures up to 120°C (Li 1981b). The above properties of ECA 4360 have made clear that polyamines are better emulsifiers for ELMs than sorbitol esters (e.g., SPAN 80).

One of ECA 4360's shortcomings is its interactions with organic or inorganic acids (Draxler and Marr 1986). Zhang et al. (1988) found that when nitric acid was encapsulated as the internal phase, the membrane with ECA 4360 showed poor stability. Another drawback of ECA 4360 is its high resistance to interfacial mass transfer (Wasan, Gu, and Li 1984; Strzelbicki and Schlosser 1989). This may be attributed to the formation of a densely packed, rigid film of polyamine molecules at the membrane interfaces. This interfacial film can be regarded as a barrier through which the solute must pass. For the extraction of Co^{2+} by D2EHPA, Gu, Wasan, and Li (1985) showed that the interfacial resistance due to the ECA 4360 surfactant was as high as 2500 s/cm. Strzelbicki and Schlosser (1989) found that the addition of alkylphenol sulfides to the membrane containing ECA 4360 lowered the mass transfer resistance significantly. Attempting to explain this phenomenon, they speculated the formation of a liquid absorption film of alkylphenol sulfides to expand the polyamine film.

One phenomenon that should not be neglected is that in some cases ECA 4360 can enhance the decomposition of the extractant in the membrane phase. For the Cu^{2+} extraction by Acorga PT 5050, Draxler and Marr (1986) found that ECA 4360 acted as an accelerator for the decomposition of the hydroxyoxime extractant.

In recent years, some investigators have noted that the combination of surfactants may be a remedy to overcome the shortcoming of the sole use of a sorbitol ester or polyamine surfactant. For example, membranes with ECA 4360 alone exhibit low osmotic swelling but high resistance to mass transfer. In contrast, membranes with SPAN 80 show high swelling but low mass transfer resistance. By adding 2% SPAN 80 to the membrane with ECA 4360, the mass transfer resistance was substantially lowered while the swelling was still kept to a tolerable level (Draxler and Marr 1986). In addition, Zhang and Wang (1989) improved membrane stability and extraction efficiency by the use of a mixture of SPAN 80 and polyamine (Shang An 205).

Note that some surfactants offer synergistic effects with extractants. Miyake, Takenoshita, and Teramoto (1983) found that for copper extraction by means of the extractant SME 529, the extraction rate was accelerated by anionic surfactants. Zhang and Xiao (1989, 1990) also observed that a polyamine surfactant (N 205) in the membrane phase could enhance the extraction of rare earths existing in the form of anion complexes with NO_3^-, Cl^-, and CH_3COO^- ions.

The industrial applications of ELMs call for further search and synthesis of new surfactants as emulsifiers for the preparation of the membranes. Nakashio and his coworkers (Goto et al. 1987, 1989; Nakashio et al. 1988; Nakashio, Goto, and Kondo 1990) have synthesized two new types of surfactants, both of which have two long alkyl chains. One is a series of derivatives of nonionic glutamic acid dialkyl esters, and the other is a series of derivatives of cationic quaternary ammonium salts. Their structures are as follows:

1. Glutamic acid dialkyl esters abbreviated as 2RGE:

$$\begin{array}{c} O \\ \| \\ RO-C- \end{array} \begin{array}{c} H\ H \\ |\ | \\ C-N-C \\ | \\ RO-C-(CH_2)_2 \\ \| \\ O \end{array} \begin{array}{c} O \\ \| \\ -(CHOH)_4-CH_2-OH \end{array}$$

$R = C_8H_{17}CH=CHC_8H_{16} (C_{18}\Delta^9)$, or

$C_9H_{19}CH=CHC_7H_{14} (C_{18}\Delta^{10})$

2. Quaternary ammonium salts abbreviated as 2R'QA:

$$\begin{array}{c} R' \diagdown \quad \diagup CH_3 \\ N^+\ X^- \\ R' \diagup \quad \diagdown CH_3 \end{array}$$

$R' = C_{12}-C_{18}$, $X^- = Br^-$ or Cl^-

and the other quaternary ammonium salts abbreviated as 2RGEC$_2$QA:

$$\begin{array}{c} O \\ \| \\ RO-C- \end{array} \begin{array}{c} H\ H \\ |\ | \\ C-N-C-CH_2 \\ | \\ RO-C-(CH_2)_2 \\ \| \\ O \end{array} \diagdown \begin{array}{c} N^+\ X^- \\ CH_3\ \ CH_3 \end{array} \diagup \begin{array}{c} CH_2CH_2-OH \end{array}$$

Experiments on the Cu^{2+} extraction by the extractant of LIX 65N (2-hydroxy-5-nonylbenzophenone oxime) showed that glutamic acid dialkyl ester was a better surfactant than SPAN 80 and ECA 4360 for ELMs with respect to membrane stability, swelling, extraction rate, and demulsification.

Zhang et al. (1988) have also synthesized a new surfactant called LMS-2, which is a copolymeric anion surfactant containing a C_4 alkene group. It is an oil-type material with a brown color, and its average molecular weight is 5000 with a viscosity of 8000 to 10,000 cp (8 to 10 Ns/m^2) (at 25°C) and a specific gravity of 0.83 to 0.86. LMS-2 can be used to form very stable emulsion with minimum swelling. It is suitable for encapsulating both an acid and a base. It is also compatible with acidic, neutral, and basic extractants in the membrane phase. In addition, the emulsion with LMS-2 can easily be demulsified by electrostatic coalescence.

On the basis of LMS-2, Liu, Zhang, and Fan (1990) have synthesized an anionic surfactant (sulfonic type) called EM-301. It is a dark brown viscous oil with an average molecular weight of 10,000. They reported that it offered lower leakage and swelling rates for ELMs than LMS-2.

In an attempt to correlate membrane swelling behavior with surfactant properties, Ding et al. (1990) estimated the area occupied by each surfactant molecule and calculated the dipole moment for each monolayer for several surfactants, such as LMS-2, EM-301, SPAN 80, and polyamines. They found that the surfactants that offered small membrane swelling (such as LMS-2 and EM-301) exhibited high values of both the occupied area and the dipole moment. These results may provide inviting information for screening and synthesizing surfactants for improving emulsion liquid membranes.

Diluents

The diluent is the main membrane component in which both extractant and surfactant are dissolved. Although the diluent is normally regarded as an "inert" component, it does affect the membrane properties, such as distribution coefficient and diffusion coefficient, and can have significant impact on the effectiveness of the membrane system. From the viewpoint of industrial application, an ideal diluent should:

1. Have low solubility in the internal and external aqueous phases so as to minimize the solvent loss.
2. Be compatible with extractant and surfactant without the formation of new phases.
3. Have a moderate viscosity. (In solvent extraction, an as-low-as-possible viscosity is desired for the diluent for fast mass transfer. For emulsion liquid membranes, however, a much lowered diluent viscosity would reduce the membrane strength, resulting in membrane instability.)
4. Have a sufficient density difference from the aqueous phase for the fast settling operation.
5. Be both cheap and readily available from a number of alternative sources.
6. Have low toxicity and a high flash point for safety reasons.

Based on the above considerations, aliphatic diluents are generally preferred to aromatic diluents because the aliphatic diluents usually can meet most of the above-mentioned requirements.

The commonly used solvents are LOPS (Low Odor Paraffin Solvent, Exxon), S100N (Solvent 100 Neutral, Exxon), Shellsol T (Shell), etc. The typical composition of LOPS is 52% paraffins, 45% naphthenes, and 3% aromatics with an average molecular weight of about 180. Its physical properties are: specific gravity = 0.796 at 60°F (15.6°C), kinematic viscosity = 2.6 cs (2.6×10^{-6} m^2/s) at 15.6°C and 1.7 cs at 38°C, and flash point = 152°F (66.7°C) (Terry, Li, and Ho 1982; Hayworth et al. 1983). S100N is a dewaxed middle distillate and an isoparaffinic solvent having an average molecular weight of 386.5, a cloud point of 33.9°C, a pour point of 32.2°C, a specific gravity of 0.85 at 25°C, and a kinematic viscosity of 22.6 cs at 38°C (Ho et al. 1982; Terry, Li, and Ho 1982).

Emulsion Preparation

To prepare a stable emulsion, the mean diameter of the dispersed internal droplets should be as small as 1 to 3 μm, which requires a high input of energy density to the water-oil system for emulsification. In laboratory studies, emulsions are usually made by high-speed agitators such as commercially available emulsifiers Tekmar Homogenizer (Itoh et al. 1990) and Waring Blender (O'Brien and Senske 1989) with stirring rates up to 20,000 rpm. Ultrasonic emulsifiers are also used in some laboratories. Shere and Cheung (1988a, 1988b) found that more stable emulsions could be prepared by the use of an ultrasonic emulsifier.

For large-scale preparation of emulsions, commercial devices such as colloidal mills are often used. For example, in the removal of phenol from wastewater, the emulsion was made by mixing the caustic solution with the oil membrane solution in a mechanical preagitator for 1 minute at 1500 rpm to get a rough emulsion, then transferring it into a colloidal mill (Type JTM-50) for 3 minutes to make a very stable emulsion (Zhang, Liu, and Lu 1987). Phenol removal is described further in Chapter 39.

In pilot- and industrial-scale operations, the use of dynamic homogenizers suffered from the corrosion problems caused by aggressive stripping acids. The use of corrosion-resistant material would increase the cost. To solve this problem, a static homogenizer was developed (Marr, Lackner, and Hütter 1988; Marr, Bart, and Draxler 1990).

The schematic of the static homogenizer is shown in Figure 38-1 (Marr, Bart, and Draxler 1990). Pre-emulsification is performed at an orifice, denoted as 3, and the final emulsification is done by turbulent eddies caused by fine blades in the main chamber, 5. A pilot device can produce emulsion up to 600 L/h with a system pressure of about 4 bars (nozzle diameter = 2 mm). A scale-up to 2000 L/h is achievable with an increased number of blades in the main chamber. With this static homogenizer, corrosion problems are minimized since there are no moving parts inside the device. This static homogenizer has been used for the preparation of the emulsion employed for the removal of zinc from wastewater in the viscose industry, which is described in Chapter 39.

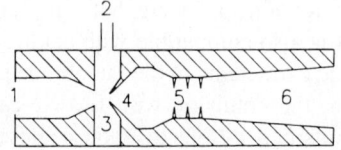

FIGURE 38–1. Schematic of a static homogenizer. 1. feed phase, 2. second phase, 3. pre-emulsification, 4. homogenization nozzle, 5. homogenization blades, 6. diffuser (reprinted from Marr, Bart, and Draxler 1990 with permission).

Emulsion Liquid Membrane Systems

The typical ELM systems are summarized from the literature in Tables 38–2 and 38–3 for Type

TABLE 38–2. Emulsion Liquid Membrane Systems for Type 1 Facilitation.

		Membrane Phase				
Extraction	External Phase	Surfactant	Diluent	Internal Phase	Efficiency	Reference
Acetic acid	CH_3COOH (1–5 g/L)	ECA 4360 (4%)	LOPS	NaOH	>90%	1
Alkaloid	Berberine or ephedrine pH 10	SPAN 80 (2%)	Kerosene	HCl (0.2 N)	~100%	2
Ammonia	NH_3 (4.0 g/L) pH 10.5	SPAN 80 (2%)	S100N	H_2SO_4 (20%)		3
	NH_3 (1.0 g/L) pH ≥12	SPAN 80 (4%)	Paraffin	H_2SO_4 (20%)	99.5%	4
Cholesterol	Blood	SPAN 80	Paraffin	Saponin	80–85%	5
Cyanide	HCN (0.1 g/L) pH 6–7	Polyamine E 644 (1%)	Kerosene or paraffin	NaOH	99%	6
Nitrate	$NaNO_3$ (0.22 g/L)	SPAN 80 (2%)	S100N	H_2SO_4 (50%) $FeSO_4$ (20%) H_2O (30%)	84%	7
Phenol	øOH (0.2–1.0 g/L)	SPAN 80 (2%)	S100N	NaOH (0.1–20%)	98%	7,8
	øOH (1.67 g/L)	LMS-2 (3–5%)	Kerosene + paraffin (6.7%)	NaOH	99.98%	9

References: (1) Terry, Li, and Ho 1982; (2) Tang, Ma, and Liu 1990; (3) Downs and Li 1981; (4) Lee and Chan 1990; (5) Yagodin et al. 1983; (6) Qian, Ma, and Shi 1989; (7) Li, Cahn, and Shrier 1973; (8) Cahn and Li 1974; (9) Zhang et al. 1988.

TABLE 38-3. Emulsion Liquid Membrane Systems for Type 2 Facilitation.

Extraction	External Phase	Membrane Phase			Internal Phase	Efficiency	Reference
		Extractant	Surfactant	Diluent			
Barium	Ba^{2+} (Mg^{2+}, Ca^{2+}, Sr^{2+}) (0.01 N each), pH 8.5	Carboxylic crown ether (0.01 M)	SPAN 80 (5%)	Mineral oil/Toluene	0.2 M HCl	—	1
Cadmium	$Cd(CN)_4^{2-}$	Aliquat 336	SPAN 80 (0.1%) Polyamine (3%)	S100N	EDTA, pH 4–6	99%	2,3
Cerium	Ce^{3+}	TOPO (0.1 M)	SPAN 80/20 (3%)	Cyclohexane	Sodium citrate, pH 8	98.5%	4
Cesium	Cs^+ (0.001 M)	8,8'-dibromo-bis(1,2-dicarbolly) Co(III) (Br_2DCC) (0.002 M)	SPAN 80/85 (3.3%, 2:1)	Nitrobenzene	Sodium citrate (0.05 M)	96%	4
Chromium	$Cr_2O_7^{2-}$ (0.3 g/L)	Alamine 336	SPAN 80 (0.1%) Polyamine (3%)	Isoparaffin	NaOH (10–20%)	99.7%	3
	$Cr_2O_7^{2-}$ (0.06 g/L)	TOA (1%)	LMS-2 (4%)	Kerosene	NaOH (5%)	99.8%	5
	$Cr_2O_7^{2-}$ 0.5N H_2SO_4	TBP (20%)	SPAN 80 (4–5%)	n-Hexane	NaOH (0.1N)	>99%	6
Cobalt	$Co(NO_3)_2$, pH 3.1 0.5M KNO_3	D2EHPA (6.3%)	SPAN 80 (2%)	Cyclohexane	2 M HNO_3	~90%	7
	$CoSO_4$ (1.0 g/L) 0.1 M NaAc, pH 5	D2EHPA	ECA 4360	LOPS	H_2SO_4 (50–200 g/L)	~95%	8
	Co^{2+}, pH 7.95	LIX 64N 0.36%)	SPAN 80	Toluene	EDTA (0.005 M) pH 7.9	99%	4
Cobalt/ manganese	Co^{2+} (0.82–0.86 g/L)/ Mn^{2+} (1.38–1.44 g/L)	TBP	SPAN 80	—	—	Co: 64–95% Mn: 98.6%	9

TABLE 38-3. Emulsion Liquid Membrane Systems for Type 2 Facilitation (continued).

Extraction	External Phase	Extractant	Membrane Phase Surfactant	Diluent	Internal Phase	Efficiency	Reference
Copper	Cu^{2+} (Mg^{2+}, Al^{3+}, Fe^{2+}, Fe^{3+})	LIX 64N (2.5%)	Polyamine (2%)	S100N	H_2SO_4 (20%)	99%	10,11
	Cu^{2+} (8 g/L)/ Zn^{2+} (100 g/L)	Acorga PT 5050 (5%)	PA 18 (2%)	Shellsol T	H_2SO_4 (250 g/L)	—	12,13
	$CuSO_4$ (0.3 g/L)	LIX 64N or SME 529	SPAN 80	Shellsol T	H_2SO_4 (250 g/L)	—	14
Europium	Eu^{3+} (1.3×10^{-3} M) pH 2–3	D2EHPA (0.5%)	SPAN 80 (2%)	Kerosene/ Polybutadiene (10%)	4N HNO_3	—	15
Gold	$AuCl_4^-$ (Pt,Pd,Ag,Cu,Pb,Fe), pH 2.54	N 503	Polyamine E 644 (4%)	Kerosene/Paraffin	Na_2SO_3	>97%	16
Iron	$FeCl_4^-$ (5 mM)	$CH_3(C_8H_{17})_3NCl$ (5 mM)	SPAN 80 (2%)	Toluene	1 N HCl	—	17
Lanthanum/ neodymium	La^{3+} (8.0×10^{-4} M)/ Nd^{3+} (3.5×10^{-4} M) pH 1.5–2.5	PC 88A	SPAN 80, PX100, $2C_{18}\Delta^9GE$, or $2C_{18}\Delta^9GEC_2QA$	n-Heptane	0.5 M H_2SO_4	$\alpha_{Nd/La} =$ 4.1–7.0	18
Lead	Pb^{2+} (Na^+, K^+, Rb^+, Cs^+, Ag^+, Tl^+, Mg^{2+}, Ca^{2+}, Sr^{2+}, Ba^{2+}, Zn^{2+})	DC18C6	SPAN 80 (3%)	Toluene	$Li_4P_2O_7$ (0.05 M)	—	19
Mercury	Hg^{2+} (2.5–190 ppm)	1,1-Dibutyl-3-benzoyl-thiourea	Rofetan OM (Fatty ester) (3.5%)	Decane	0.2 N HCl 7.6 g/L Thiourea	96%	2
	$HgCl_4^{2-}$ (1.1 g/L)	Alamine 336	SPAN 80 (0.1%) Polyamine (3%)	S100N	NaOH	>99%	3,20
Molybdenum	$Mo_7O_{24}^{6-}$ (1.06 g/L) H_2SO_4 (16 g/L)	TNOA (0.02M)	SPAN 80 (5%)	Kerosene	Na_2CO_3 (2M)	99.5%	21
Nickel	Ni^{2+} (2.2 g/L)	DTPA (5%)	ECA 4360 (3%) SPAN 80 (0.2%)	Shellsol T	H_2SO_4 (250 g/L)	—	12

Species	Feed	Extractant	Surfactant	Diluent	Stripping	Yield	Ref.
Phosphate	PO_4^{3-} (0.27–0.57%)	Amines	SPAN 80 (1–2%) Polyamine	Isoparaffin	$CaCl_2 + NH_4OH$ or $Ca(OH)_2$	91–98%	22
Rare earths (single and mixture)	$La^{3+}, Eu^{3+}, Lu^{3+}$ (1×10^{-4} M each)	Carboxylic crown ether	SPAN 80 (5%)	Toluene (50%) Mineral oil (45%)	HNO_3 pH 2	$\alpha_{Lu/La} = 1.25$	23
Rare earth mixture	RE^{3+} (1.0 g/L) NaCl (7%), pH 6	D2EHPA (6%)	Polyamine/SPAN 80 (3:1)	Kerosene	6N HCl	99%	24
Silver	Ag^+ ($Na^+, K^+, Rb^+, Cs^+, Tl^+, Mg^{2+}, Sr^{2+}, Ba^{2+}, Pb^{2+}, Zn^{2+}$)	DC18C6	SPAN 80 (3%)	Toluene	$Li_2S_2O_3$ (0.05 M)	—	19
Sodium	Na^+ (K^+, Rb^+, Li^+) (0.01 M each)	Carboxylic crown ether (0.01 M)	SPAN 80 (5%)	Mineral oil/Toluene	0.2 M HCl	—	1
Strontium	Sr^{2+} (10^{-4} M) 0.01 N HCl	Br_2DCC (2×10^{-3} M)	SPAN 80/85 (3.3%, 2:1)	Nitrobenzene	Sodium citrate (0.05 M)	96%	4
Technetium	TcO_4^- 0.1 N HNO_3	Aliquat 336 (0.5%)	SPAN 80 (3%)	Cyclohexane	$NaClO_4$ (1 M)	92%	4
Uranium	UO_2^{2+} (~0.16 g/L) ~6 M H_3PO_4	D2EHPA/TOPO	Polyamine	LOPS	Reductant[Fe(II)]	>90%	25
	UO_2^{2+} 0.34 M H_2SO_4	TNOA	SPAN 80	Kerosene	Na_2CO_3	95%	26
Vanadium	VO_3^- (16–604 ppm) H_2SO_4	TNOA (2%)	Succimide derivative (4%)	Kerosene	Na_2CO_3 (2.5–5%)	98–99.3%	27
Zinc	Zn^{2+} (0.5 g/L)	DTPA (2–4%)	ECA 4360 (2%)	Shellsol T	H_2SO_4 (250 g/L)	99.5%	12
	Zn^{2+} (6 g/L)/Ca^{2+} (0.014 g/L)/Pb^{2+} (0.004 g/L)	DTPA (5%)	PX 100 (2%)	Shellsol T	H_2SO_4 (250 g/L)	—	13

References: (1) Bartsch, Charewicz, and Kang 1984; (2) Weiss, Grigoriev, and Mühl 1982; (3) Kitagawa et al. 1977; (4) Macásek et al. 1984; (5) Zhang et al. 1988; (6) Vohra, Kaur, and Sharma 1989; (7) Strzelbicki 1978; (8) Gu, Wasan, and Li 1986; (9) Abou-Nemeh and Van Peteghem 1989; (10) Frankenfeld, Cahn, and Li 1981; (11) Cahn et al. 1981; (12) Draxler and Mar 1986; (13) Draxler, Fürst, and Marr 1988; (14) Martin and Davies 1977; (15) Yu, Jiang, and Zhu 1986; (16) Yan, Shi, and Su 1990; (17) Ohki et al. 1983; (18) Goto et al. 1989; (19) Christensen et al. 1983; (20) Frankenfeld and Li 1977; (21) Hirato et al. 1990; (22) Li, Cahn, and Shrier 1973; (23) Tang and Waii 1989; (24) Zhang and Wang 1989; (25) Hayworth et al. 1983; (26) Hirato et al. 1991; (27) Zheng et al. 1988.

1 and Type 2 facilitations, respectively. It is hoped that this information may provide readers with a guide to their liquid membrane studies.

DISPERSION/EXTRACTION/SETTLING

The separation operation for an emulsion liquid membrane process includes dispersion, by which the coupled extraction/stripping is achieved, and settling, which realizes the phase separation between the loaded emulsion and the aqueous raffinate because of their density difference. Before dispersing the emulsion into the feed, the pretreatment of the feed is required as in most membrane separation processes. The pretreatment is typically done by the use of 1–10-μm filters, and it sometimes includes flocculation and sedimentation steps before filtration. During the dispersion operation, the emulsion is dispersed by agitation in the external, continuous phase, and many small globules of emulsion are formed. Normally, the size of emulsion globules is controlled in the range of 1×10^{-4} to 2×10^{-3} m in diameter. Each emulsion globule contains many tiny encapsulated droplets with a typical size of 1 to 3 μm in diameter. Such a large number of emulsion globules together with the numerous pre-encapsulated droplets provides large interfacial areas for both extraction and stripping. After the separation is completed, phase separation of the loaded emulsion from the external raffinate takes place in the settler. The settling is similar to that for conventional solvent extraction and is not discussed further. The modes of dispersion operation, emulsion globule size, extraction, leakage, and swelling are discussed in the following sections.

Modes of Dispersion Operation

The dispersion operation can be carried out either in a batch or continuous mode. Batch operation is normally performed in laboratory work for screening suitable membrane materials, elucidating mass transfer mechanisms, and studying the factors that influence the membrane stability, swelling, and mass transfer rate. The study of hydrodynamic conditions in a batch process also offers useful information for achieving effective separation. On the basis of batch performance, the optimum membrane formulation and operational conditions can be determined, and the basic data necessary for the scale-up of the separation process can be obtained.

Continuous operation is normally used in pilot and industrial processes, and it is carried out with either countercurrent or cocurrent flow. Countercurrent flow is preferred for the contact between the feed and the emulsion phase because it can make full use of the internal phase reagent and achieve high extraction efficiency.

As with solvent extraction processes, the extraction devices for emulsion liquid membranes may either be mixer-settlers or column extractors. Mixer-settlers offer considerable flexibility in terms of only slow changes in performance with phase flow ratios and rates, thus allowing adaptation to changing feed compositions. Their drawbacks are the large equipment size and the potential loss of reagents. Ma and Shi (1987) have observed that the repeated dispersion and coalescence of emulsion globules can lead to serious swelling caused by the entrainment of water from the external phase to the internal phase. They believe that the application of multistage mixer-settlers can increase the swelling.

Mixer-settlers have been used in ELM processes by some investigators (Kitagawa et al. 1977; Cahn et al. 1981; Hayworth et al. 1983). Downs and Li (1981) built a two-stage, countercurrent, continuous mixer-settler unit for ammonia removal from municipal wastewater as shown in Figure 38–2. While the feed and the liquid membrane emulsion flowed cocurrently through each mixer, they flowed countercurrently between mixing stages.

Column contactors are compact, efficient extraction devices that are more suitable for large-scale operations than mixer-settlers (Li and Zhang 1982; Prötsch and Marr 1983). For the removal of phenol from wastewater, Zhang, Liu, and Lu (1987) constructed extraction equipment with two-stage rotary disk columns

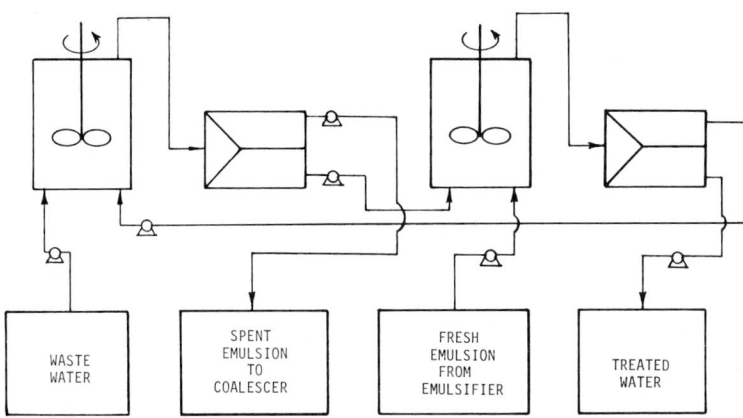

FIGURE 38–2.
Schematic of a two-stage, counter current, continuous mixer-settler unit (reprinted from Downs and Li 1981 with permission).

(22 cm in inner diameter and 3.5 m in height) with a treating capacity of 0.5 ton/h. As shown in Figure 38–3, the pretreated acidic wastewater flows from the top of the first column extractor and contacts the globules of emulsion countercurrently in the two column extractors. These column extractors have been suitable for phenol removal, which is described in more detail in Chapter 39.

Emulsion Globule Size

For a given volume of an emulsion dispersed in an external phase, the size of emulsion globules

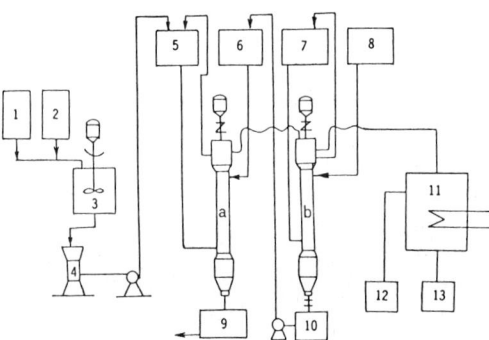

FIGURE 38–3. Schematic of two-stage countercurrent columns for phenol removal. 1. NaOH tank, 2. oil phase tank, 3. preagitator, 4. colloid mill, 5 and 7. high-level tanks for emulsion, 6 and 8. high-level tanks for wastewater, 9. oil-water separator, 10. intermediate tank, 11. demulsifier, 12. stock tank for recovered oil phase, 13. stock tank for recovered internal phase (reprinted from Zhang, Liu, and Lu 1987 with permission).

determines the mass transfer surface area between the emulsion and the external phase. The smaller the globule size, the larger the number of globules and the larger the surface area. Generally, the Sauter mean diameter (or radius) is used to characterize the globule size. As mentioned in Chapter 37, Teramoto et al. (1983a, 1983b) showed that the Sauter mean diameter was sufficient to characterize globule size, and it was not necessary to use the size distribution.

As noted in Chapter 36, the size of emulsion globules depends on the viscosity of the emulsion, the characteristics and concentration of the surfactant(s) in the emulsion, and the intensity and mode of mixing in the dispersion operation (Ohtake et al. 1987; Rautenbach and Machhammer 1988; Ho 1986). Ho (1986) developed a dynamic technique that utilizes the Fraunhofer diffraction principle for the measurement of the size of emulsion globules in an external phase in a standard mixer (equipped with a marine propeller located at the tank center with a distance of 0.5 tank diameter from the bottom of the tank and four baffles each with a width of 0.1 tank diameter, and with a tank-to-impeller diameter ratio of 2). He found that the controlling viscosity is the viscosity of the emulsion, not the membrane phase viscosity, and that the Sauter mean globule diameter correlates well with impeller (stirrer) tip speed. The globule size increases with increasing emulsion viscosity for a given mixing condition, i.e., for a given impeller tip speed for this

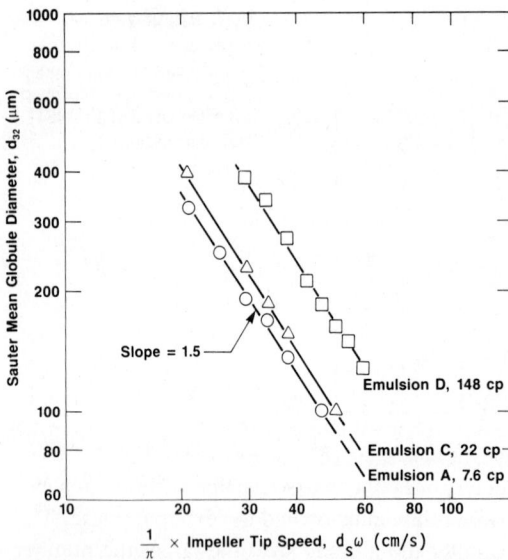

FIGURE 38–4. Emulsion globule size increases with increasing emulsion viscosity (adapted from Ho 1986).

mixer, as shown in Figure 38–4. This figure shows globule size results for three emulsions (designated emulsions A, C, and D) with different emulsion viscosities: 7.6, 22, and 148 cp (1 cp = 10^{-3} Ns/m^2), dispersed in an external phase of 6 M H_3PO_4 (simulating wet process phosphoric acid for uranium extraction). The emulsions also contained 6 M H_3PO_4 as the internal phase, and they had a ratio of 2 between the membrane and internal phases. He also found that the effects of the holdup of emulsion and the number of impellers on globule size are insignificant. This suggests insignificant coalescence of emulsion globules.

Ohtake et al. (1987) investigated the effects of the modes of mixing on the entrainment of the external phase of water into emulsion globules and the size of the globules. They used two modes of mixing: (1) the emulsion was settled on the water layer before stirring and (2) the emulsion was added dropwise into the center of the external phase during stirring. The first mode of mixing gave entrainment as high as 5 to 30%, and it produced much more entrainment than the second mode. For the first mode of mixing, they gave the following correlations for the Sauter mean globule diameter d_{32}, dependent on the viscosity of the emulsion η, in our notation:

$$d_{32}/d_s = 0.12 \text{We}^{-0.5},$$

$$\text{for } \eta < 0.16 \text{ Ns/m}^2, \quad (38\text{–}13)$$

$$d_{32}/d_s = 0.50\eta^{0.8}\text{We}^{-0.5},$$

$$\text{for } \eta > 0.16 \text{ Ns/m}^2, \quad (38\text{–}14)$$

where

$$\text{We} = \omega^2 d_s^3 \rho_e / \gamma. \quad (38\text{–}15)$$

In these equations, We is the Weber number, ω the stirring rate, d_s the stirrer or impeller diameter, ρ_e the density of the external phase, and γ the interfacial tension between the membrane and external phases. They also gave the Sauter mean globule diameter results for the second mode of mixing, which also showed the diameter proportional to the -0.5 power of the Weber number and dependent on the viscosity of the emulsion.

Rautenbach and Machhammer (1988) gave the following globule size correlation for a stirred tank equipped with a six-blade turbine and four baffles (in our notation):

$$\frac{d_{32}}{d_s} = 0.11 \left(\frac{\eta \rho_e}{\eta_e \rho}\right)^{0.32} \left(\frac{V_m + V_i}{V_e + V_m + V_i}\right)^{0.1} \text{We}^{-0.7},$$

$$(38\text{–}16)$$

where η_e is the viscosity of the external phase, ρ_e is the density of the external phase, ρ is the density of the emulsion, and V_e, V_m, and V_i are the volumes for the external, membrane, and internal phases, respectively. They noted that the emulsion viscosity η can be expressed as

$$\frac{\eta}{\rho} = \left(\frac{\eta_m}{\rho_m}\right) \exp\left\{5.32 \left[\left(\frac{V_i}{V_m + V_i}\right) - 0.1\right]\right\},$$

$$(38\text{–}17)$$

where η_m and ρ_m are the viscosity and density of the membrane phase, respectively. They also established the globule size correlation for a countercurrent stirred column extractor equipped with six-blade turbines (seven evenly distributed turbines in a 1.8-m column) and four baffles. Their correlation for stirring conditions at Reynolds numbers above 6000 is

$$\frac{d_{32}}{d_s} = 0.024 \left(\frac{\omega d_s^2 \rho_e}{\eta_e}\right)^{0.6} \text{We}^{-1.2}. \quad (38-18)$$

Extraction

Some of the models for batch extraction, described in Chapter 37, have been extended to continuous process operations. These models may be used for the design of ELM processes. Similar to the treatment of the theory in Chapter 37, the models for continuous process operations may be classified into two categories: (1) diffusion-type mass transfer models for Type 1 facilitation and (2) carrier facilitated transport models for Type 2 facilitation (Ho 1990). The definitions of Type 1 and Type 2 facilitations (Matulevicius and Li 1975; Li 1978, 1981a) are given in Chapter 36.

Diffusion-Type Mass Transfer Models for Type 1 Facilitation

As described before, in Type 1 facilitation, reaction in the receiving phase of an ELM system (the internal phase if the external phase is a feed) maintains the solute concentration at effectively zero, resulting in the maximization of the driving force for the diffusion of the solute in the membrane phase from the feed phase to the receiving phase. Similar to the batch extraction theory discussed above, the diffusion process in continuous extraction operations may be described by two methods: the spherical shell approach and the emulsion globule approach.

Spherical Shell Approach
As discussed in the batch extraction theory, the spherical shell approach assumes that the mass transfer resistance is diffusion in the spherical "shell" [shown schematically in Figure 36–4(a) for the example of phenol extraction] of the membrane phase of constant thickness between the external and internal phases. Based on this approach, the general model with overall mass transfer coefficients for extraction and leakage, described in Chapter 37, can be extended to continuous operations (Ho and Li 1984). Ho and Li (1984) have extended the Case I of this model with small breakage values ($\Phi = 0$ or $\Phi \leq 1.4 \times 10^{-5}$ s^{-1}) to two types of continuous operations, namely, multistage mixer-settlers and mechanically agitated columns. The Case I model has an external reagent to convert the solute B leaked from the internal phase back to the extractable solute A in the external phase. As pointed out in Chapter 37, the Case II model that does not have the external reagent for the conversion is the same as the Case I model for the case in which $\Phi = 0$ and the solute B cannot diffuse through the membrane phase, i.e., its mass transfer coefficient k_B is zero (ionic species, such as phenolate and ammonium ions, cannot diffuse through oil-type membranes). For ELM systems of practical interest with Type 1 facilitation, $k_B = 0$ and Φ should be negligible, leading to a Case II model that is essentially identical to the Case I model. Thus, for ELM systems of practical interest, the following equations developed by Ho and Li (1984) for the Case I model with small breakage values for the two types of continuous operations can also be applied to the Case II model. In each type of continuous operation, both cocurrent and countercurrent modes are considered.

Extraction with Multistage Mixer-Settlers: Cocurrent Mode. Figure 38–5(a) shows the schematic for the cocurrent mode of the continuous extraction operation with n-stage mixer-settlers with the mixer volume V for each stage. The volumetric flow rates for the external, internal, and membrane phases are Q_e, Q_i, and Q_m, respectively. The flow rates are assumed constant in the case under consideration. The material balance for the first stage can be expressed in the following two equations (Ho and Li 1984):

$$Q_e(c_{eA,0} - c_{eA,1}) = Q_i(c_{iB,1} - c_{iB,0}) \quad (38-19)$$

$$Q_e(c_{eA,0} - c_{eA,1}) = k_A V \left(\frac{Q_m + Q_i}{Q_e + Q_m + Q_i}\right) c_{eA,1} - k_B \Phi V \left(\frac{Q_m + Q_i}{Q_e + Q_m + Q_i}\right) c_{iB,1}, - \quad (38-20)$$

where c_{eA} and c_{iB} are the concentrations for solute A in the external phase and solute B in the internal phase, respectively. The subscripts 0 and 1 refer to the inlet and outlet for the stage 1 mixer-settler, respectively. Rearrangement of Eq. (38–20) leads to

$$c_{eA,0} = \frac{V}{Q_e}\left(\frac{Q_m + Q_i}{Q_e + Q_m + Q_i}\right)(k_A c_{eA,1} - k_{B\Phi} c_{iB,1}) + c_{eA,1}. \qquad (38\text{–}21)$$

Substitution of Eq. (38–20) into Eq. (38–19) gives

$$c_{iB,0} = -\frac{V}{Q_i}\left(\frac{Q_m + Q_i}{Q_e + Q_m + Q_i}\right)(k_A c_{eA,1} - k_{B\Phi} c_{iB,1}) + c_{iB,1}. \qquad (38\text{–}22)$$

Multiplication of Eq. (38–21) by k_A and of Eq. (38–22) by $k_{B\Phi}$ and subtraction of one resultant equation from the other give

$$\frac{k_A c_{eA,0} - k_{B\Phi} c_{iB,0}}{k_A c_{eA,1} - k_{B\Phi} c_{iB,1}} = V\left(\frac{Q_m + Q_i}{Q_e + Q_m + Q_i}\right)\left(\frac{k_A}{Q_e} + \frac{k_{B\Phi}}{Q_i}\right) + 1. \qquad (38\text{–}23)$$

The same type of equation can be derived from the other stages. Multiplication of these equations together results in

$$\frac{k_A c_{eA,0} - k_{B\Phi} c_{iB,0}}{k_A c_{eA,n} - k_{B\Phi} c_{iB,n}} = \left[V\left(\frac{Q_m + Q_i}{Q_e + Q_m + Q_i}\right)\left(\frac{k_A}{Q_e} + \frac{k_{B\Phi}}{Q_i}\right) + 1\right]^n. \qquad (38\text{–}24)$$

This equation can be used to calculate the outlet concentrations $c_{eA,n}$ and $c_{iB,n}$ or the degree of solute recovery for a given extractor volume V for each stage with the aid of the overall material balance given by

$$Q_e(c_{eA,0} - c_{eA,n}) = Q_i(c_{iB,n} - c_{iB,0}). \qquad (38\text{–}25)$$

Rearrangement of Eq. (38–24) allows the calculation of the total extractor volume of the n-stage mixers, V_t, from given inlet and outlet concentrations or a given solute recovery.

$$V_t = nV = \frac{n\left[\left(\dfrac{k_A c_{eA,0} - k_{B\Phi} c_{iB,0}}{k_A c_{eA,n} - k_{B\Phi} c_{iB,n}}\right)^{1/n} - 1\right]}{\left(\dfrac{Q_m + Q_i}{Q_e + Q_m + Q_i}\right)\left(\dfrac{k_A}{Q_e} + \dfrac{k_{B\Phi}}{Q_i}\right)}.$$

$$(38\text{–}26)$$

This equation allows the calculation of the total residence time or contact time for extraction t as follows:

$$t = \frac{V_t}{Q_e + Q_m + Q_i}. \qquad (38\text{–}27)$$

Extraction with Multistage Mixer-Settlers: Countercurrent Mode. Figure 38–5(b) shows the schematic for the countercurrent mode of the continuous extraction operation with n-stage mixer-settlers with the mixer volume V for each stage. Following the similar algebraic procedure described for the cocurrent mode, Ho and Li (1984) have obtained the following equation:

$$\frac{k_A c_{eA,0} - k_{B\Phi} c_{iB,1}}{k_A c_{eA,n} - k_{B\Phi} c_{iB,0}} = \left[\frac{V\left(\dfrac{Q_m + Q_i}{Q_e + Q_m + Q_i}\right)\dfrac{k_A}{Q_e} + 1}{V\left(\dfrac{Q_m + Q_i}{Q_e + Q_m + Q_i}\right)\dfrac{k_{B\Phi}}{Q_i} + 1}\right]^n \qquad (38\text{–}28)$$

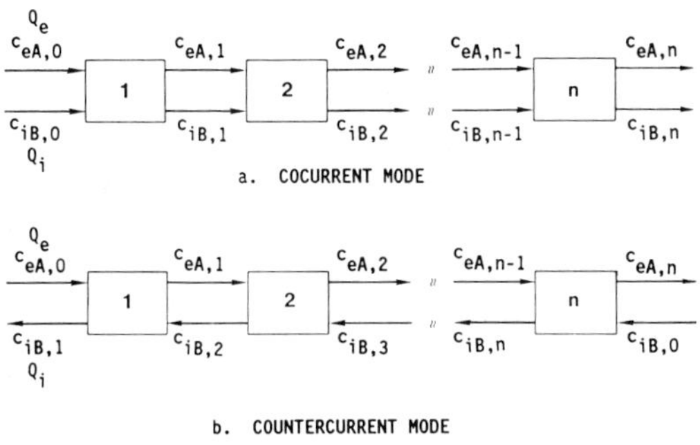

FIGURE 38–5. Schematic of continuous extraction operations with n-stage mixer-settlers (reprinted from Ho and Li 1984 with permission).

This equation allows the calculation of the outlet concentrations $c_{eA,n}$ and $c_{iB,1}$ or the degree of solute recovery for a given extractor volume V for each stage with the aid of the overall material balance:

$$Q_e(c_{eA,0} - c_{eA,n}) = Q_i(c_{iB,1} - c_{iB,0}). \tag{38-29}$$

Rearrangement of Eq. (38–28) allows the calculation of the total extractor volume of the n-stage mixers from given inlet and outlet concentrations or a given solute recovery:

$$V_t = \frac{n\left[\left(\dfrac{k_A c_{eA,0} - k_{B\Phi} c_{iB,1}}{k_A c_{eA,n} - k_{B\Phi} c_{iB,0}}\right)^{1/n} - 1\right]}{\left(\dfrac{Q_m + Q_i}{Q_e + Q_m + Q_i}\right)\left[\dfrac{k_A}{Q_e} - \dfrac{k_{B\Phi}}{Q_i}\left(\dfrac{k_A c_{eA,0} - k_{B\Phi} c_{iB,1}}{k_A c_{eA,n} - k_{B\Phi} c_{iB,0}}\right)^{1/n}\right]}. \tag{38-30}$$

Equations (38–27) and (38–30) can be used to calculate the total residence time or contact time for extraction.

Extraction with a Mechanically Agitated Column: Cocurrent Mode. Figure 38–6(a) shows the schematic for the cocurrent mode of the continuous extraction operation with a mechanically agitated column of the total extraction volume V_t. Ho and Li (1984) have given the following equation:

$$\frac{k_A c_{eA,0} - k_{B\Phi} c_{iB,0}}{k_A c_{eA,1} - k_{B\Phi} c_{iB,1}} = \exp\left[V_t\left(\frac{Q_m + Q_i}{Q_e + Q_m + Q_i}\right)\left(\frac{k_A}{Q_e} + \frac{k_{B\Phi}}{Q_i}\right)\right], \tag{38-31}$$

where $c_{eA,1}$ and $c_{iB,1}$ are the outlet concentrations for the external and internal phases, respectively. This equation allows the estimation of these outlet concentrations or the solute recovery for a given total extractor volume V_t with the aid of the overall material balance:

$$Q_e(c_{eA,0} - c_{eA,1}) = Q_i(c_{iB,1} - c_{iB,0}). \tag{38-32}$$

Rearrangement of Eq. (38–31) gives the expression for the estimation of the total extractor volume from given inlet and outlet concentrations or a given solute recovery.

$$V_t = \frac{1}{k_A/Q_e + k_{B\Phi}/Q_i}\left(\frac{Q_e + Q_m + Q_i}{Q_m + Q_i}\right)\ln\left(\frac{k_A c_{eA,0} - k_{B\Phi} c_{iB,0}}{k_A c_{eA,1} - k_{B\Phi} c_{iB,1}}\right). \tag{38-33}$$

Equations (38–27) and (38–33) can be used to estimate the total residence time or contact time for extraction.

Extraction with a Mechanically Agitated Column: Countercurrent Mode. Figure 38–6(b) shows the schematic for the countercurrent mode of the continuous extraction operation with a mechanically agitated column. Ho and Li (1984) have given the following equation:

$$\frac{k_A c_{eA,0} - k_{B\Phi} c_{iB,1}}{k_A c_{eA,1} - K_{B\Phi} c_{iB,0}} = \exp\left[V_t\left(\frac{Q_m + Q_i}{Q_e + Q_m + Q_i}\right)\left(\frac{k_A}{Q_e} + \frac{k_{B\Phi}}{Q_i}\right)\right]. \quad (38\text{–}34)$$

This equation allows the estimation of the outlet concentrations $c_{eA,1}$ and $c_{iB,1}$ or the solute recovery for a given total extractor volume V_t with the aid of Eq. (38–32) corresponding to the overall material balance for the entire column.

Rearrangement of Eq. (38–34) leads to the expression for the estimation of the total extractor volume from given inlet and outlet concentrations or a given solute recovery:

$$V_t = \frac{1}{k_A/Q_e + k_{B\Phi}/Q_i}\left(\frac{Q_e + Q_m + Q_i}{Q_m + Q_i}\right)\ln\left(\frac{k_A c_{eA,0} - k_{B\Phi} c_{iB,1}}{k_A c_{eA,1} - k_{B\Phi} c_{iB,0}}\right). \quad (38\text{–}35)$$

Equations (38–27) and (38–35) can be employed to estimate the total residence time or contact time for extraction.

Overall mass transfer coefficients k_A and $k_{B\Phi}$ are generally determined experimentally, e.g., from batch experiments, under the same extraction conditions (the same emulsion, external phase, stirrer tip speed, mixer geometry, and temperature). In the absence of experimental data, they may also be estimated from Eqs. (37–58) through (37–61) and the pseudo-steady-state solution of the advancing front model (Ho et al. 1982), as described in Chapter 37. As noted in that chapter, the model can only provide a first-order estimate of the extraction rate and, thus, the extractor size and extraction contact time.

By the use of the model, Ho and Li (1984) have shown that the countercurrent mode of continuous operation can be more effective than the cocurrent mode for both extractions with multistage mixer-settlers and mechanically agitated columns. They have also shown that in some cases the countercurrent column operation can be more effective than the countercurrent operation with two-stage mixer-settlers.

Gladek, Stelmaszek, and Szust (1981, 1982) extended the spherical shell approach similar to the model of Matulevicius and Li (1975) with unsteady-state concentration equations to cocurrent and countercurrent column operations. They applied the model to the separation of benzene from hexane (1981) and the extraction of phenol (1982). Wankat (1980) and Wankat and Noble (1984) developed analysis procedures for both multistage mixer-settler and continuous column operations with three-phase systems. They showed that design methods analogous to those for two-phase systems can be used provided that transport at one interface is limiting.

Emulsion Globule Approach

As mentioned above, the spherical shell approach can only provide a first-order estimate of the extraction rate, extractor size, and extrac-

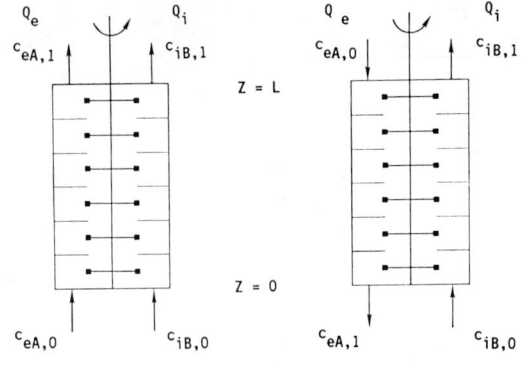

FIGURE 38–6. Schematic of continuous extraction operations with mechanically agitated columns (reprinted from Ho and Li 1984 with permission).

tion contact time. In addition, there are several shortcomings of this approach as discussed in Chapter 37. Thus, Ho et al. (1982) developed the advancing front model to overcome the shortcomings for batch extraction. Later, Hatton and Wardius (1984) and Wardius and Hatton (1985) extended this model with the pseudo-steady-state solution (zero-order perturbation solution) to continuous cocurrent and countercurrent operations with multistage mixer-settlers, and Rautenbach and Machhammer (1988) extended it to a continuous countercurrent column operation.

Extraction with Multistage Mixer-Settlers. As discussed in Chapter 37, the pseudo-steady-state solution (zero-order perturbation solution) of the advancing front model is adequate for typical ELM systems of practical interest (Ho et al. 1982). Based on the pseudo-steady-state solution, Hatton and Wardius (1984) developed equations for the design of cocurrent and countercurrent multistage mixer-settler operations. For the cocurrent mixer-settler operation shown in Figure 38–5(a), their design equations for a typical stage i, in our notation, are

$$\Gamma_{i-1} = \frac{Q_i c_{ir,i-1}}{Q_e c_{e,i-1}} = \frac{\theta_i}{\lambda + \theta_i f_i}, \quad (38\text{--}36)$$

$$\Gamma_i = \frac{1}{\lambda}(1 - f_i)\theta_i, \quad (38\text{--}37)$$

where

$$\theta_i = \left(\frac{c_{ir,i-1}}{c_{e,i}}\right)\left(\frac{r_o^2 V_i (Q_m + Q_i)}{K'_e D'_{\text{eff}}(V_m + V_i)^2}\right), \quad (38\text{--}38)$$

$$\lambda = \frac{r_o^2 Q_e}{K'_e D'_{\text{eff}}(V_m + V_i)}, \quad (38\text{--}39)$$

$$f_i = \frac{Q_e}{Q_i}\left(\frac{c_{e,i-1} - c_i}{c_{ir,i-1}}\right), \quad (38\text{--}40)$$

and c_{ir} is the concentration of internal reagent in the internal phase, c_e the solute concentration in the external phase, r_o the emulsion globule radius, K'_e the distribution coefficient for the solute between the reacted emulsion mixture and the external phase at equilibrium, and D'_{eff} the effective solute diffusivity, based on the average concentration c, in the emulsion mixture.

The parameter θ_i is from the advancing front model defined in Eq. (38–38), and it is the reciprocal of the dimensionless time related to the reaction capacity provided by the internal reagent versus the globule capacity for unreacted solute; λ is the second parameter of the advancing front model defined in Eq. (38–39), and it is the reciprocal of the dimensionless time related to the treat ratio, the ratio of the emulsion volumetric flow rate to the external phase volumetric flow rate; and f_i is, by definition, the fractional utilization of the internal reagent for stage i. The θ_i and f_i terms are related in a unique fashion for a given external emulsion recycle ratio ξ [the volumetric flow rate of the recycled emulsion stream is $\xi(Q_m + Q_i)$] and a given mode of the recycle. For the recycle with the premixed emulsion entering stage i, the relation is (Hatton and Wardius 1984)

$$\frac{f_i}{1 + (1 - f_i)\xi} = 3\int_0^1 \chi^2 \exp\left(-[1 + (1 - f_i)\xi] \cdot \frac{\theta_i}{6}(1 - 3\chi^2 + 2\chi^3)\right) d\chi. \quad (38\text{--}41)$$

For the recycle with the nonpremixed emulsion entering stage i, the relation is (Hatton and Wardius 1984)

$$\frac{f_i}{1 + (1 - f_i)\xi}$$

$$= \frac{3}{(1 + \xi)} \cdot \left[\int_0^1 \chi^2 \exp\left(-(1 + \xi)\frac{\theta_i}{6}(1 - 3\chi^2 + 2\chi^3)\right) d\chi\right.$$

$$\left. + \xi \int_0^1 \chi^2 \exp\left(-(1 + \xi)(1 - f_i)\frac{\theta_i}{6}(1 - 3\chi^2 + 2\chi^3)\right) d\chi\right]. \quad (38\text{--}42)$$

In Eqs. (38–41) and (38–42), ξ is zero for two cases: (1) There is no external emulsion recycle and (2) the holdup ratio of the emulsion volume to the external phase volume can be enhanced independently of inlet flow rates as encountered in the internal recycle mixer proposed by Hatton et al. (1983), which circumvents the need for the introduction of an external recycle stream.

From Eq. (38–36), a set of curves can be prepared giving Γ_{i-1} as a function of f_i (or θ_i) with λ as a parameter. From these curves, the f_i values can be determined for given values of Γ_{i-1} and λ. The value of f_i can now be used to determine Γ_i from Eq. (38–37). Thus, Γ_i can be used with Eq. (38–36) to determine f_{i+1}. The entire procedure described above for stage i can now be repeated for stage $i+1$.

This procedure starts with the feed condition

$$\Gamma_0 = \Gamma^\circ = \frac{Q_i c_{ir,0}}{Q_e c_{e,0}}, \quad (38\text{--}43)$$

where Γ° is the inlet ratio of the internal reagent equivalents to the feed solute equivalents. It is possible to move from stage to stage for one of the two cases: (1) a specified number of stages is reached or (2) until the desired effluent concentration levels corresponding to

$$\frac{\Gamma_n}{\Gamma^\circ} = \frac{1 - f_T}{1 - (\Gamma^\circ f_T)} \quad (38\text{--}44)$$

are obtained. For the first case, the overall fraction utilization of the internal reagent f_T is calculated by the use of the material balance:

$$f_T = \frac{1 - (\Gamma_n/\Gamma^\circ)}{1 - \Gamma_n}. \quad (38\text{--}45)$$

For the second case, the number of stages required to obtain the desired separation is obtained directly.

As pointed out by Hatton and Wardius (1984), the above procedure can be carried out graphically, and it is also readily implemented on a computer. It is analogous to the well-known McCabe-Thiele method for staged equilibrium processes (King 1980), for which Eq. (38–36) is the effective operating line and Eq. (38–37) serves as the equilibrium curve. If f_i is constant (f_0), independent of stage i, the number of stages n can be calculated from the following equation (Hatton and Wardius 1984):

$$n = \frac{\ln(1 - f_T)}{\ln(1 - f_0)}. \quad (38\text{--}46)$$

For the countercurrent mixer-settler operation shown in Figure 38–5(b), the design equations of Hatton and Wardius (1984) for a typical stage i, in our notation, are

$$\Gamma_{i-1} = \frac{Q_i c_{ir,i}}{Q_e c_{e,i-1}} = \left(\frac{\theta_i}{\lambda + \theta_i f_i}\right)(1 - f_i), \quad (38\text{--}47)$$

$$\Gamma_i = \frac{1}{\lambda}\theta_i, \quad (38\text{--}48)$$

where

$$\theta_i = \left(\frac{c_{ir,i+1}}{c_{e,i}}\right)\left(\frac{r_o^2 V_i(Q_m + Q_i)}{K_e^! D'_{eff}(V_m + V_i)^2}\right). \quad (38\text{--}49)$$

As described by Hatton and Wardius (1984), Eqs. (38–47) and (38–48) are the operating and equilibrium curves, respectively. To begin the stage-to-stage design procedure similar to that described above, we must know Γ_0, which can be calculated from

$$\Gamma_0 = \frac{Q_i c_{ir,1}}{Q_e c_{e,0}} = (1 - f_T)\Gamma^\circ. \quad (38\text{--}50)$$

However, f_T in this equation is unknown at this point, and an initial guess must be made in order to obtain Γ_0. The accuracy of this initial guess can be checked after the design procedure has been carried out for the desired number of stages, and, if correct, it should agree with the value obtained from the material balance:

$$f_T = \frac{\Gamma_n}{\Gamma^\circ} - \frac{\Gamma^\circ}{\Gamma_n}. \quad (38\text{--}51)$$

In general, several iterations on f_T are required to ensure convergence.

Also based on the pseudo-steady-state solution of the advancing front model (Ho et al. 1982), Wardius and Hatton (1985) developed equations for the analysis of cocurrent, cascaded mixers without interstage settling. They found that the absence of this interstage settling can lead to significant reductions in extraction efficiency. This is due to the fact that the diffusional resistances built up within the emulsion globules in any given mixer are cumulative in going from one stage to the next unless they are removed by interstage settling.

Recently, Reed, Bunge, and Noble (1987) incorporated reaction reversibility in the internal phase into the advancing front model for a single-stage mixer operation (a continuous stirred-tank reactor). They developed a simple way to extend the pseudo-steady-state solution of the model to this operation.

Extraction with a Column Extractor. Based on the pseudo-steady-state advancing front model, Rautenbach and Machhammer (1988) developed the equations describing solute concentrations in the emulsion globules and the external feed phase for the modeling of the countercurrent extraction operation in a mechanically agitated column, shown schematically in Figure 38–6(b) (where c_{eA} and c_{iB} are c_e and c_i of c_T in this case, respectively). Their equations, in our notation, are as follows:

Globules:

$$(Q_m + Q_i)\frac{dc_T}{dz} - A_s \left(\frac{V_m + V_i}{V_e + V_m + V_i}\right) D_L \frac{d^2 c_T}{dz^2}$$

$$= \frac{3A_s}{r_o}\left(\frac{V_m + V_i}{V_e + V_m + V_i}\right) J, \quad (38\text{--}52)$$

$$J = \left(\frac{K'_e D'_{\text{eff}}}{r_o}\right)\left(\frac{r_f}{r_o - r_f[1 - (K'_e D'_{\text{eff}}/k_e r_o)]}\right) c_e, \quad (38\text{--}53)$$

$$r_f = r_o \left[1 - \left(\frac{V_m + V_i}{V_i}\right) \frac{c_T}{c_{ir,0}}\right]^{1/3}, \quad (38\text{--}54)$$

$$z = 0, \quad c_T = 0, \quad (38\text{--}55)$$

$$z = L, \quad c_T = \left(\frac{Q_e}{Q_m + Q_i}\right)(c_{e,0} - c_{e,1}). \quad (38\text{--}56)$$

External phase:

$$Q_e \frac{dc_e}{dz} + A_s \left(\frac{V_e}{V_e + V_m + V_i}\right) D_{Le} \frac{d^2 c_e}{dz^2}$$

$$= \frac{3A_s}{r_o}\left(\frac{V_m + V_i}{V_e + V_m + V_i}\right) J, \quad (38\text{--}57)$$

$$z = 0, \quad c_e = c_{e,1}, \quad (38\text{--}58)$$

$$z = L,$$

$$c_e = c_{e,0} - \left(\frac{A_s}{Q_e}\right)\left(\frac{V_e}{V_e + V_m + V_i}\right) D_{Le} \frac{dc_e}{dz}\bigg|_{z=L}. \quad (38\text{--}59)$$

where c_T is the concentration of reacted and unreacted solute species averaged over the emulsion mixture, z the vertical coordinate with $z = 0$ at the bottom of the mechanically agitated column, A_s the cross section of the column, D_L the axial dispersion coefficient of the emulsion, D_{Le} the axial dispersion coefficient of the external phase, J the flux, r_f the reaction front position, k_e the external phase mass transfer coefficient, $c_{e,1}$ the outlet solute concentration in the external phase, and the rest of the symbols have the same definitions described earlier.

They established the correlations for the Sauter mean globule radius r_o [Eq. (38–18)], the axial dispersion coefficients of the emulsion D_L and the external phase D_{Le}, and the emulsion holdup $(V_m + V_i)/(V_e + V_m + V_i)$ from the countercurrent operation in a 1.8-m column extractor equipped with seven evenly distributed six-blade turbines and four baffles. They estimated the external phase mass transfer coefficient k_e from the correlation of Bibaud and Treybal (1966). As pointed out in Chapter 37, the external phase mass transfer resistance can be negligible for typical ELM systems of practical interest. Thus, the term with k_e in Eq. (38–53) can be dropped out.

With the parameters from these correlations, they solved the above equations numerically. The model results agreed reasonably well with the extraction data of ammonia from the countercurrent column extractor. Although axial backmixing is included in this model, as noted by them, the negative effect of the backmixing due to increasing the stirring speed is overcompensated by the positive effects: (1) an increase in mass transfer surface area, (2) an increase of the volume fraction of the emulsion (holdup), and consequently (3) an increase in the mean residence time of the emulsion. This model may be used for designing/sizing column extractors for the removal of phenol and ammonia described in Chapter 39.

Carrier Facilitated Transport Models for Type 2 Facilitation

As described in Chapter 36, in Type 2 facilitation, a "carrier" compound (complexing agent or extractant) incorporated in the membrane phase carries the solute across the membrane phase, i.e., the solute-carrier complex diffuses in this phase. Similar to the batch extraction theory discussed in Chapter 37, the diffusion process in continuous extraction operations may be described by two methods: the spherical shell approach and the emulsion globule approach.

Spherical Shell Approach
As discussed earlier, this approach assumes that the mass transfer resistance is diffusion in the spherical "shell" of the membrane phase of constant thickness between the external and internal phases. Similar to the batch extraction theory discussed in Chapter 37, the general model of Ho and Li (1984) with overall mass transfer coefficients for extraction and leakage may be applicable not only for the Type 1 facilitation described earlier but also for Type 2 facilitation in continuous extraction operations.

Extraction with Multistage Mixer-Settlers. In the cocurrent mode of continuous extraction with mixer-settlers for typical ELM systems of practical interest, Eqs. (38–24) and (38–25) from the model of Ho and Li (1984) may be used to calculate the outlet concentrations $c_{eA,n}$ and $c_{iB,n}$ for stage n or the solute recovery for a given extractor volume of the n-stage mixers, and Eqs. (38–26) and (38–27) allow the estimation of the total extractor volume of the n-stage mixers and the total residence time (contact time for extraction), respectively, from given inlet and outlet concentrations or a given solute recovery.

For countercurrent multistage mixer-settler operations with typical ELM systems of practical interest, Eqs. (38–28) and (38–29) allow the estimation of the outlet concentrations $c_{eA,n}$ and $c_{iB,1}$ for stage n or the solute recovery for a given extractor volume, and Eqs. (38–30) and (38–27) may be used to calculate the total extractor volume of the n-stage mixers and the total residence time, respectively, from given inlet and outlet concentrations or a given solute recovery. [For the special case of the Case II model in which solute B is the only species both in the internal and external phases, such as hydrometallurgical (e.g., Zn^{2+} and Cu^{2+}) and hydrocarbon extractions, Eqs. (38–24) through (38–30) may be used by substituting $k_B K_{eB}/K_{iB}$ for k_A, $c_{eB,0}$ for $c_{eA,0}$, and $c_{eB,n}$ for $c_{eA,n}$.]

Bock et al. (1982) and Hayworth et al. (1983) applied this model successfully to the extraction of uranium from wet process phosphoric acid in pilot plant operations using single-stage and countercurrent two-stage mixer-settler systems. With this model, they investigated the effects of mixing rate and temperature on the overall mass transfer coefficient for extraction. Figure 38–7 shows the effect of mixing rate on the overall mass transfer coefficient for extraction k_A (Hayworth et al. 1983).

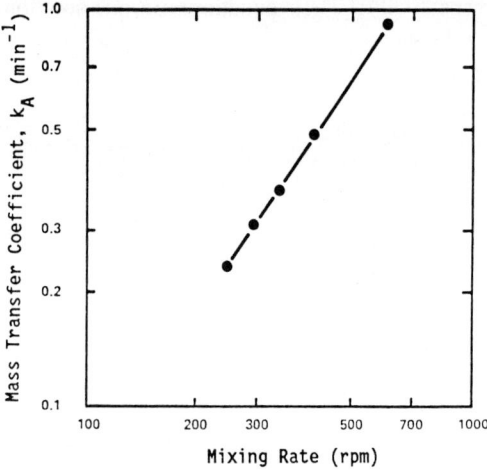

FIGURE 38–7. Effect of mixing rate on mass transfer coefficient for uranium extraction (reprinted from Hayworth et al. 1983 with permission).

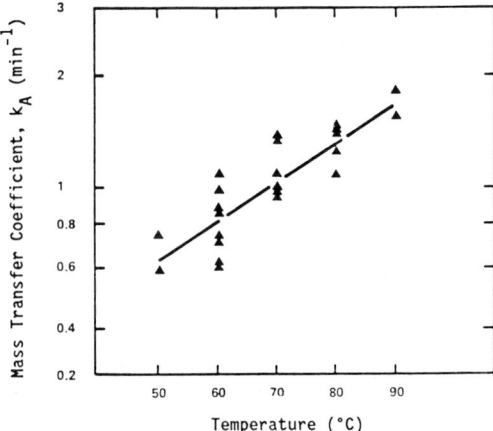

FIGURE 38-8. Effect of temperature on mass transfer coefficient for uranium extraction (reprinted from Hayworth et al. 1983 with permission).

As shown in this figure, this effect is very dramatic. When the mixing rate was increased 2.4-fold, i.e., from 250 to 600 rpm, a factor of about 4 increase in the mass transfer coefficient was obtained. Figure 38-8 presents the effect of temperature on the mass transfer coefficient (Hayworth et al. 1983). This effect is also very pronounced. When the temperature was increased from 50 to 90°C, the mass transfer coefficient was approximately tripled.

Although this model can be applied successfully, as pointed out earlier and in Chapter 37, it can only provide a first-order estimate of the extraction rate and, thus, the extractor size and residence time. A model using the emulsion globule approach, to be discussed in the next section, can determine the extraction rate more accurately.

Way and Noble (1982) modeled the extraction of copper in a continuous stirred tank reactor (one-stage mixer). They included residence time and globule size distributions in their analysis.

Extraction with Column Extractors. In the cocurrent mode of continuous extraction with a mechanically agitated column for ELM systems of practical interest, Eqs. (38–31) and (38–32) from the model of Ho and Li (1984) may be used to estimate the outlet concentrations $c_{eA,1}$ and $c_{iB,1}$ or the solute recovery for a given total extractor volume, and Eqs. (38–33) and (38–27) allow the estimation of the total extractor volume and the total residence time, respectively, from given inlet and outlet concentrations or a given solute recovery. [For the special case of the Case II model mentioned earlier, Eqs. (38–31) through (38–35) and (38–27) may be used by substituting $k_B K_{eB}/K_{iB}$ for k_A, $c_{eB,0}$ for $c_{eA,0}$, and $c_{eB,1}$ for $c_{eA,1}$.]

For countercurrent, mechanically agitated column operations with typical ELM systems of practical interest, Eqs. (38–34) and (38–32) allow the estimation of the outlet concentrations $c_{eA,1}$ and $c_{iB,1}$ or the solute recovery for a given total extractor volume, and Eqs. (38–35) and (38–27) may be used to estimate the total extractor volume and the total residence time, respectively, from given inlet and outlet concentrations or a given solute recovery.

The model of Ho and Li (1984) described for extraction with a mechanically agitated column has not accounted for the effect of axial backmixing on extraction column performance. However, Kataoka et al. (1987), Bart et al. (1988), and Matsumoto et al. (1990) considered axial backmixing in stirred column extractors, and Kataoka et al. (1987) also investigated it in a spray column.

Kataoka et al. (1987) developed a model that takes axial backmixing into account and assumes a pseudo-homogeneous emulsion system with simplified first-order extraction reaction kinetics for continuous extraction in a column extractor. They assumed that the mass transfer process involves three steps: (1) diffusion in the film of the external feed phase with the mass transfer coefficient k_e, (2) extraction reaction at the globule surface with the forward reaction rate k_f, and (3) diffusion in the peripheral membrane layer with the mass transfer coefficient k_m (D_m/l where D_m is the diffusivity of the metal-carrier complex in the membrane phase and l the thickness of the peripheral layer). They further assumed that the extraction reaction kinetics is first-order dependent on the metal solute concentration in the external phase, and that the metal-carrier complex concentration in the membrane phase can be approximated to be zero since the complex formed by the extraction reaction may be stripped into the internal phase droplets present near the surface of the emulsion globule.

For this case, the extraction flux at the globule surface, J, is

$$J = kc_e, \quad (38\text{--}60)$$

where

$$\frac{1}{k} = \frac{1}{k_e} + \frac{1}{k_f} + \frac{1}{K_e k_m}, \quad (38\text{--}61)$$

k is the overall mass transfer coefficient for extraction, and K_e is the distribution coefficient for the metal solute between the membrane and external phases at equilibrium, and it is also the equilibrium constant for the extraction reaction under consideration, K_{eq}. The case under consideration is that the metal solute concentration in the external feed phase is low. In this case, the governing equation for the external phase concentration c_e is Eq. (38–57), containing the axial backmixing term characterized with the axial dispersion coefficient D_{Le} for the external phase, with the boundary conditions described by Eqs. (38–58) and (38–59). The system of the equations, Eqs. (38–57) through (38–60), has an analytical solution (Kataoka et al. 1987) as follows:

$$\frac{c_e}{c_{e,0}} = \frac{2(1 + \Psi)\exp[-\text{Pe}(1 - \Psi)Z/2] - 2(1 - \Psi)\exp[-\text{Pe}(1 + \Psi)Z/2]}{(1 + \Psi)^2 \exp[-\text{Pe}(1 - \Psi)/2] - (1 - \Psi)^2 \exp[-\text{Pe}(1 + \Psi)/2]}, \quad (38\text{--}62)$$

where

$$\text{Pe} = \frac{v_e L}{D_{Le}} = \left(\frac{Q_e}{A_s}\right)\left(\frac{V_e + V_m + V_i}{V_e}\right)\left(\frac{L}{D_{Le}}\right), \quad (38\text{--}63)$$

$$\Psi = \left[1 + \left(\frac{12 A_s}{r_o Q_e}\right)\left(\frac{V_m + V_i}{V_e + V_m + V_i}\right)\left(\frac{kL}{\text{Pe}}\right)\right]^{1/2}, \quad (38\text{--}64)$$

$$Z = z/L, \quad (38\text{--}65)$$

and Pe is the Peclet number, v_e the velocity of the external phase, Z the dimensionless vertical distance from the bottom of the column, and the rest of the symbols were described earlier.

Kataoka et al. (1987) gave the correlations for the axial dispersion coefficient for the external phase D_{Le}, the emulsion holdup $(V_m + V_i)/(V_e + V_m + V_i)$, and external phase mass transfer coefficient k_e for spray and Mixco column extractors. They determined the Sauter mean globule radius r_o experimentally. As described above, Rautenbach and Machhammer (1988) established the correlations for the Sauter mean globule radius, the axial dispersion coefficients for the external phase and the emulsion, and the emulsion holdup for a countercurrent column extractor equipped with six-blade turbines. For this column extractor, they estimated the external phase mass transfer coefficient from the correlation of Bibaud and Treybal (1966). As pointed out above, the external phase mass transfer resistance can be negligible for typical ELM systems of practical interest.

Bart et al. (1988) developed a similar model, in which they assumed that the mass transfer process is reaction controlled, and that the overall extraction reaction kinetics is first-order dependent on the metal solute concentration in the external feed phase for the forward reaction and zero order with respect to the metal-carrier concentration in the membrane phase for the reverse (backward) reaction. They applied the model to copper extraction and found insignificant influence of axial backmixing on column performance. They also noted that axial backmixing effects are usually superimposed on the chemical reaction kinetics and lie within the error boundaries of the reaction kinetics model used (Bart, Bauer, and Marr 1987). The models of Bart et al. (1988) and Kataoka et al. (1987) [Eq. (38–62)] may be used for designing/sizing column extractors for the removal of metals, such as zinc and nickel, as described in Chapter 39.

Recently, Matsumoto et al. (1990) also developed a similar model taking backmixing into account for a multicell Mixco column extractor except for additional mass transfer resistance in the peripheral membrane layer was not included. For the simulation of copper extraction in a 15-cell Mixco extractor, they used independently, experimentally determined parameters for the model except they estimated the external phase mass transfer coefficient from the correlation of Bibaud and Treybal (1966), which was also employed by Rautenbach and Machhammer (1988) as mentioned. Without an adjustable parameter, the model agreed approximately with the experimental results of the copper extraction.

Emulsion Globule Approach

As noted above, the emulsion globule approach can model the extraction rate of ELMs more accurately than the spherical shell approach. Based on the emulsion globule approach, Teramoto et al. (1983c) extended their batch extraction model of carrier facilitated transport for Type 2 facilitation (1983b) to the continuous extraction operation with a single-stage mixer-settler unit. Their system of equations describing the concentrations in the emulsion globules, external phase, and internal phase are the same as that for their batch extraction model except the concentration of the solute (metal) in the external phase for the continuous operation is maintained at a steady-state value determined by its mass balance equation, which contains the terms expressing its extraction rate. Thus, their model has complicated equations and many parameters, which require nearly as tedious efforts as those of their batch extraction model.

The Lorbach and Marr model (1987) described in Chapter 37 [Eqs. (37-95) through (37-105)], which has fewer equations and parameters than the Teramoto et al. models (1983b, 1983c), may be used for the single-stage mixer-settler operation provided that the external phase solute concentration is set to the steady-state value ($c_{e,1}$). As described in Chapter 37, this model still requires numerical solution for its system of equations.

As pointed out by Teramoto et al. (1983c), the residence time distribution of emulsion globules in a continuously stirred tank reactor, i.e., a mixer, must be considered. They used the following residence time distribution function, $f(t)$:

$$f(t) = \frac{\exp(-t/\tau^\circ)}{\tau^\circ}, \qquad (38\text{-}66)$$

where τ° is the mean residence time of the globules, and $\exp(-t/\tau^\circ)$ the fraction of the globules having residence times greater than t. This distribution function is the same as that used by Hatton and Wardius (1984) for Eqs. (38-41) and (38-42). With this function, the extraction flux of the solute is

$$J = \int_0^\infty J(t) \left(\frac{\exp(-t/\tau^\circ)}{\tau^\circ} \right) dt, \qquad (38\text{-}67)$$

where

$$J(t) = D_{\text{eff}} \left. \frac{\partial c_m}{\partial r} \right|_{r=r_o} \qquad (38\text{-}68)$$

and D_{eff} is the effective diffusivity of the metal-carrier complex, based on the membrane phase concentration c_m, in the emulsion mixture. The material balance equation of the metal solute in the external phase with its volumetric flow rate Q_e is

$$Q_e(c_{e,0} - c_{e,1}) = \frac{3}{r_o}(V_m + V_i)J. \qquad (38\text{-}69)$$

Equations (38-67), (38-68), and (38-69) are part of the system of equations for modeling the extraction.

It appears that there are no available models that use the emulsion globule approach of Type 2 facilitation for continuous extraction operations with multistage mixer-settlers and column extractors. This appears to be an area for improvement in the modeling work. In addition, the areas for improvement described in Chapter 37 for modeling of Type 2 facilitation in batch extraction should also be applicable to that in continuous extraction operations. These areas include the determination and use of accurate reaction rate constants and the estimation of effective diffusivity.

Leakage

The degree of membrane leakage or breakdown is an important parameter in assessing the stability of the liquid membranes. Membrane leakage in emulsion liquid membrane systems includes the rupture of the emulsion, leading to the short circuiting of the reagent and extracted solute in the internal phase to the external phase, and the leakage of the internal reagent and extracted solute through the membrane phase to the external phase. As a result, the leakage causes a decrease of the driving force for mass transfer and an increase of the raffinate concentration, thus lowering the extraction efficiency. The main factors governing membrane stability include the membrane formulation, the method of emulsion preparation, and the condition under which the emulsions are contacted with the feed solution. The leakage rate data reported thus far from different laboratories are quite different, ranging from 0.2 to 10% (Zhang et al. 1988).

Effect of Membrane Formulation
As far as membrane formulation is concerned, the properties of the surfactant, diluent, and internal phase and its volume fraction have significant effects on the membrane leakage. The surfactant plays the dominant role for stabilizing the membrane, and the properties of the surfactant affect membrane stability considerably. Membranes with polyamine surfactants (such as ECA 4360), which have one long alkyl chain, have been reported to be more stable than those with sorbitol oleates (such as SPAN 80) (Draxler and Marr 1986; Qian, Ma, and Shi 1989; Zhang and Xiao 1989, 1990). Recently, studies have shown that the incorporation of the surfactant with two long alkyl chains (such as glutamic acid dialkyl esters) to the membrane phase gives even better membrane stability (Goto et al. 1987, 1991; Nakashio et al. 1988).

A reasonably high surfactant concentration is favorable to membrane stability (Takahashi, Ohtsubo, and Takeuchi 1981; Terry, Li, and Ho 1982; Fujinawa et al. 1985; Ma and Shi 1987; Nakashio et al. 1988; Hirato, Suyama, and Majima 1989; O'Brien and Senske 1989). A critical surfactant concentration exists at which the number of surfactant molecules oriented at the membrane interface is sufficient to make the membrane stable. Above this critical concentration, a further increase in surfactant concentration appears unnecessary for membrane stability (Goto et al. 1987; Ma and Shi 1987; Qian, Ma, and Shi 1989; Hirato et al. 1990). On the contrary, preferential adsorption of surfactant molecules at the interface caused by excessively high surfactant concentration can hinder the mass transfer process (Mikucki and Osseo-Asare 1986a, 1986b). Therefore, the suitable surfactant concentration commonly used in practice lies in the 2 to 5% range (Braun and Frag 1978; Frankenfeld, Cahn, and Li 1981; Terry, Li, and Ho 1982; Nakashio et al. 1988).

Strictly speaking, the concept of the surfactant concentration per unit interfacial area should be used to describe its contribution to membrane stability. For the same surfactant content in the membrane phase, smaller internal droplets and emulsion globules result in a lower interfacial concentration of the surfactant. Kinugasa, Watanabe, and Takeuchi (1990) found that for the membrane with the SPAN 80 surfactant, a stable liquid membrane was formed by about 1.6 layers of the surfactant molecules adsorbed at the oil/water interfaces.

Membrane viscosity, determined mainly by the diluent, can also affect membrane stability. The use of a diluent with low viscosity formulates a thin but weak membrane, which offers a high initial extraction rate but shows serious leakage over longer contact times. The membrane with a high viscosity diluent is strong. Although this kind of membrane offers a lower extraction rate, it is more stable and thus has better overall extraction efficiency (Frankenfeld, Cahn, and Li 1981). Therefore, oils with relatively high viscosity are preferred as the membrane diluent for maintaining the necessary membrane stability (Hochhauser and Cussler 1975; Terry, Li, and Ho 1982; Shere and Cheung 1988b; Imai and Furusaki 1990; Yan, Shi, and Su 1990).

Sometimes, the internal reagent also affects membrane stability. For example, when NaOH is used as the internal reagent, the membrane with the SPAN 80 surfactant will be less stable due to the decomposition of SPAN 80 by NaOH (Vohra, Kaur, and Sharma 1989; Abou-Nemeh and Van Peteghem 1990).

Increasing the volume fraction of the internal phase will decrease the membrane thickness and thus increase the leakage rate (Shere and Cheung 1988a, 1988b). However, the low volume fraction of the internal phase will slow down the mass transfer process (Qian, Ma, and Shi 1989). The suitable value of the volume fraction of the internal phase should be about 0.3 to 0.5 (Hirato et al. 1990).

Effect of Emulsion Preparation

In general, an emulsion prepared with a high-energy density input (such as by an ultrasonic method) will have very small droplets. This will enhance membrane stability if the surfactant concentration is high enough. Meanwhile, the small droplet size gives a very large interfacial area for mass transfer (Martin and Davies 1977; Frankenfeld, Cahn, and Li 1981; Hanna and Larson 1985; Shere and Cheung 1988a; Hirato, Suyama, and Majima 1989).

Keep in mind that an ultrastable emulsion should be avoided because of possible difficulties later during the demulsification step. Usually, a membrane leakage rate of about 0.1% is allowable for a practical process (Draxler and Marr 1986).

Effect of Mixing Conditions in the Extraction Vessel

In the extraction vessel, the emulsion is contacted generally with a large excess of the external, continuous phase. Increasing the stirring speed will increase the shear energy, and it will also increase the probability of membrane breakdown (Martin and Davies 1977; Halwachs and Schugerl 1980; Stroeve and Varanasi 1982; Ohtake et al. 1987). In general, a relatively high stirring speed is desired in order to create a large interfacial area for mass transfer.

Swelling

Emulsion swelling that increases the internal phase volume is a troublesome problem associated with the use of emulsion liquid membranes. It is a process by which water is transferred from the external, continuous phase into the internal droplet phase. The water transfer will (1) dilute the solute that has been concentrated in the internal droplets, thus preventing a highly concentrated solute solution from being obtained; (2) reduce the driving force for solute extraction (Draxler and Marr 1986; Yan, Huang, and Shi 1987; Thien and Hatton 1988; Mukkolath, Gadekar, and Tiwari 1990); (3) make the membrane thinner, thereby leading to a less stable emulsion (Matsumoto et al. 1980; Magdassi and Garti 1984; Ma and Shi 1987); and (4) change the rheological properties of the emulsion to cause difficulties in emulsion transport and phase separation (Martin and Davies 1977; Draxler and Marr 1986). Draxler, Fürst, and Marr (1988) have estimated for copper extraction that the economic advantage of ELMs over solvent extraction is lost when the swelling rate is higher than 30 to 40%.

The two types of emulsion swelling that can occur are osmotic swelling and swelling attributed to the entrainment of the external aqueous phase. Osmotic swelling is driven by differences in the osmotic pressure between the external and internal phases. There are two possible mechanisms for osmotic swelling in ELMs (Thien and Hatton 1988; Itoh et al. 1990). The first mechanism suggests the formation of complexes between water and surfactant molecules (hydration of surfactant molecules) (Colinart et al. 1984). A schematic of this mechanism [Figure 38–9(a)] indicates that the hydrophilic portion of the surfactant hydrates at the interface between the oil membrane phase and the external, aqueous phase, which has high water activity. The hydrated surfactant then diffuses across the membrane to the interface between the oil membrane phase and the internal droplet phase, which has low water activity, and it is finally dehydrated (Noguchi, Nakazawa, and Yoshii 1980; Colinart et al. 1984; Thien and Hatton 1988). Besides surfactants, extractants can also serve as the water carrier (Draxler annd Marr 1986; Yan, Huang, and Shi 1987; Thien and Hatton 1988).

The other osmotic swelling mechanism proposes the transport of water by reversed micelles [Figure 38–9(b)]. The reversed micelles are surfactant aggregates that solubilize water in nonpolar solvents. The hydrophilic heads of

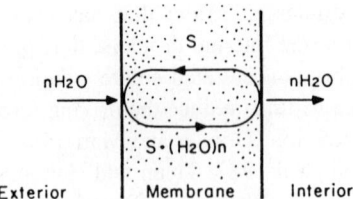

a. Mechanism via Hydrated Surfactant Molecules

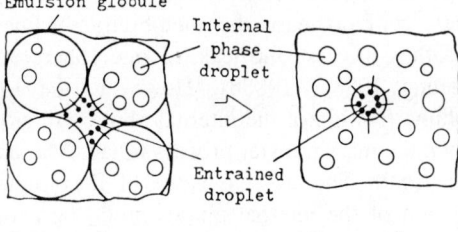

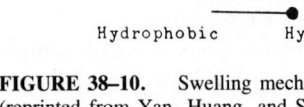

b. Mechanism via Reversed Micelles

FIGURE 38–9. Mechanisms for osmotic swelling in emulsion liquid membranes (reprinted from Thien and Hatton 1988 with permission).

FIGURE 38–10. Swelling mechanism via entrainment (reprinted from Yan, Huang, and Shi 1987 with permission).

the surfactant molecules, along with the water, form the interior of the aggregate, while the hydrophobic tails of the surfactant molecules point toward the nonpolar solvent and form the exterior of the micellar aggregate. Similar to the first mechanism, the micelle forms at the interface next to the external aqueous phase with high water activity and is dehydrated at the interface next to the internal droplet phase with low water activity (Chilamkurti and Rhodes 1980; Matsumoto et al. 1980). These micelles, apart from transporting water, can also accommodate small solutes (Wiencek and Qutubuddin 1986).

The swelling mechanism via the entrainment of the external phase into the emulsion is shown schematically in Figure 38–10 (Yan, Huang, and Shi 1987). This mechanism accounts for the swelling due to the repeated coalescence and redispersion of emulsion globules during the dispersion operation. The evidence supporting this mechanism is the rapid increase of swelling with the increasing stirring speed (Yan, Huang, and Shi 1987; Ding et al. 1988;

Ding and Xie 1991). The reported membrane swelling data from different sources are rather scattered due to the different membrane systems and operation conditions used: 10% reported by Cahn et al. (1981), 20 to 83% by O'Brien and Senske (1989), Thien, Hatton, and Wang (1986), and Terry, Li, and Ho (1982), and 50 to 100% by Martin and Davies (1977). According to Ding et al. (1988) and Ding and Xie (1991), the swelling due to the osmotic effect was below 30%, while that due to entrainment could be up to 500%. As mentioned above, swelling due to entrainment can depend on the mode of mixing in the dispersion operation (Ohtake et al. 1987).

Numerous studies in ELMs have shown that practically all surfactants and extractants carry water to some extent; thus, swelling is unavoidable in most ELM systems. However, several methods are available to suppress the swelling:

1. Screen suitable surfactants to minimize surfactant hydration and micelle formation (Thien and Hatton 1988). For example, the newly synthesized surfactants, glutamic acid dialkyl esters (Goto et al. 1987, 1989; Nakashio et al. 1988; Nakashio, Goto, and Kondo 1990), LMS-2 (Zhang et al. 1988), and EM-301 (Liu, Zhang, and Fan 1990), can be superior to polyamines and sorbitol esters in minimizing membrane swelling.
2. Increase membrane viscosity so as to increase membrane strength and resistance

against water permeation (Terry, Li, and Ho 1982; Colinart et al. 1984).
3. Decrease the volume fraction of the internal aqueous phase to increase the membrane thickness and thus the resistance for water permeation (Ma and Shi 1987).
4. Add nontransportable salts to the external phase in an effort to match the water activity in the external and internal phases (Lorbach, Bart, and Marr 1986).
5. Add a reagent to the membrane phase to have preferential micellization of the surfactant with the added reagent rather than with water, e.g., adding cyclohexanone to the oil membrane phase containing SPAN 80 at about unity volume ratio of cyclohexanone to SPAN 80 (Mukkolath, Gadekar, and Tiwari 1990).

DEMULSIFICATION

The breaking of the loaded emulsion is one of the key steps in the liquid membrane extraction process. After liquid membrane extraction, the membrane phase must be recycled repeatedly, and the enriched internal phase is usually recovered. Therefore, demulsification of the loaded emulsion is unavoidable for the use of this separation process, although a few exceptions use this technology without breaking the emulsion in some special cases (Dines 1982).

Two principal approaches for the demulsification of the loaded emulsion are chemical and physical treatments. Chemical treatment involves the addition of a demulsifier to the emulsion. This method seems to be very effective. However, the added demulsifier will change the properties of the membrane phase and thus prohibit its reuse. In addition, the recovery of the demulsifier by distillation is rather expensive. Therefore, chemical treatment is usually not suitable for breaking liquid membrane emulsions, although few examples of chemical demulsification have been reported for certain liquid membrane systems (Zhang, Huang, and Chen 1988).

Physical treatment methods include heating, centrifugation, ultrasonics, solvent dissolution, high shear, and the use of high-voltage electrostatic fields. Liquid membrane emulsions (both O/W and W/O types) can be effectively broken by the use of a specially formulated solvent mixture and high shear (Borwankar et al. 1988). The specially formulated solvent mixture breaks the emulsion without damaging the surfactant. The solvents used are low-boiling compounds, and they can easily be recovered later by evaporation. The method of demulsification by high shear includes the use of centrifugation as a first step, followed simply by pumping the half-broken emulsion through a high shear device.

In their study to break O/W emulsions, Kato and Kawasaki (1988) found that the emulsions could be easily broken as they were with a coexisting oil and exposed to high shear. They reported that a demulsification yield of 95% was reached in 0.063 second using an agitation speed of 20,000 rpm.

Demulsification with electrostatic fields appears to be the most efficient and economic way for breaking W/O emulsions in liquid membrane processes (Martin and Davies 1977; Hsu, Li, and Hucal 1983a, 1983b; Hsu and Li 1985; Draxler, Fürst, and Marr 1988; Feng, Wang, and Zhang 1988; Yan and Wang 1988; Draxler and Marr 1990; Kataoka and Nishiki 1990).

Electrostatic coalescence is a technique widely used to separate dispersed aqueous droplets from nonconducting oils. The petroleum industry has used it to separate the brine emulsified in crude oil, and the chemical industry has employed it to resolve W/O emulsions generated during extraction (such as removal of naphthenic acids from gas oils). Since this type of demulsification is strictly a physical coalescence process, the technique is most suitable for breaking emulsion liquid membranes to recover the oil membrane phase for reuse.

Note that the conventional electrostatic coalescers for crude oil dewatering cannot be directly adopted for breaking W/O emulsions for liquid membrane processes, because ELM systems are quite different from crude oil. First, liquid membrane emulsions contain much more water (as high as 50% versus roughly 5%), and the recovered oil may contain more water (about 1%) than pipeline-quality dehydrated crude oil (about 0.3%). Second, since liquid membrane emulsions are made intentionally, they contain high concentrations of potent emulsifying surfactants. Finally, the properties of the oil phase for liquid membrane emulsions must be preserved for re-emulsification. The above-mentioned differences call for the development of new electrostatic coalescers designed especially to break liquid membrane emulsions.

The mechanisms of electrostatic demulsification are not fully understood. The general understanding is that the electrical field can polarize and elongate water droplets. The neighboring water droplets, after acquiring induced charge from the electrical field, will attract each other and coalesce to a big drop. The mutual attractive force F between water droplets in the electrical field can be expressed as follows (Marr, Bart, and Draxler 1990):

$$F = \frac{1}{4} \pi \epsilon_0 \frac{\epsilon_d - \epsilon_c^2 d^6 E^2}{\epsilon_d + 2\epsilon_c L^4}, \qquad (38\text{--}70)$$

where E is the electrical field strength, d the diameter of water droplets, L the distance between two droplets, and ϵ_d and ϵ_c are the dielectric constants of the dispersed aqueous phase and the continuous oil phase, respectively. This equation indicates that the high electric field instead of the conduction current is the main cause of the droplet coalescence. Several types of dc and ac electrostatic coalescers with different configurations have already been developed.

Hsu and Li (1985) and Hsu, Li, and Hucal (1983a, 1983b) described a continuous coalescer equipped with a horizontal glass-insulated electrode as shown in Figure 38–11 (where IR stands for internal reagent). In this unit, the glass-insulated electrode was made by bending a 7-mm-o.d., 5-mm-i.d. Pyrex glass tube to form a horizontal grid with two vertical stems. The external surface of the glass grid was coated with a very thin layer of FEP (fluorinated ethylene-co-propylene) to reduce the water wettability of the surface. The horizontal portion of the grid was filled with an aqueous electrolyte to serve as a conductor for transmitting high voltage. The vertical stems of the grid were filled with transformer oil. The glass-insulated electrode was placed inside a cylindrical plexiglass coalescer body with the two vertical stems protruding through two holes at the top of the coalescer cylinder. A flexible wire, inserted through the oil to the electrolyte, connected the electrolyte conductor of the elec-

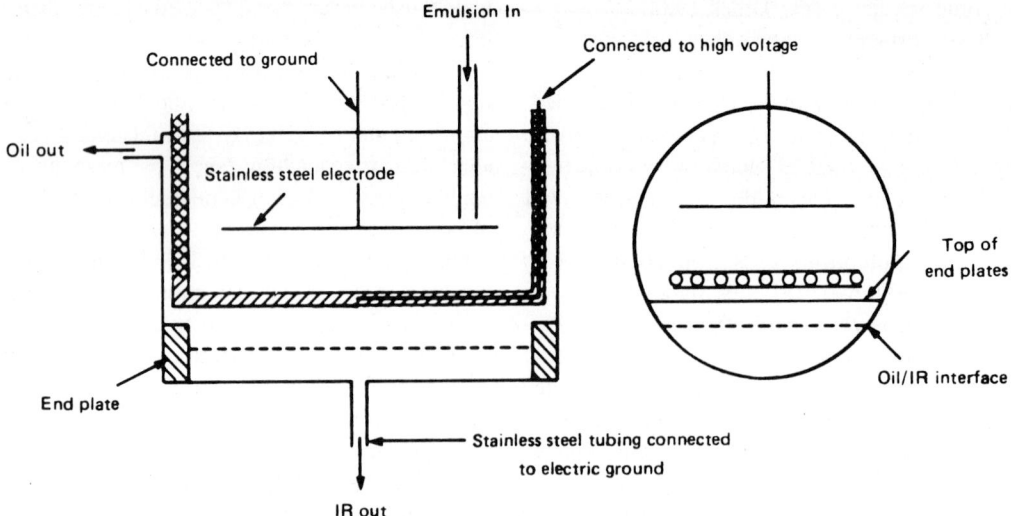

FIGURE 38–11. Schematic of a continuous insulated electrode coalescer (reprinted from Hsu and Li 1985 with permission).

trode to the high-voltage terminal of a transformer. A metal electrode was placed above the glass electrode. This metal electrode and the coalesced aqueous phase at the bottom of the coalescer were grounded.

Draxler and Marr (1986) developed an electrostatic coalescer with two planar insulated electrodes placed in parallel. The energy consumption in this apparatus lay between 0.5 and 5kWh/m^3 emulsion. The use of an insulated electrode prevents sparking under high voltage, lowers energy consumption, precludes any electrolysis, and protects the electrode material from chemical corrosion by an acid or a base. They have used this coalescer in demulsification for the removal of zinc from wastewater in the viscose fiber industry, as described in Chapters 39 and 40.

Recently, Marr and coworkers (Draxler and Marr 1990; Marr, Bart, and Draxler 1990) constructed a circular coalescer with the aim of producing an inhomogeneous field between the electrodes and a good hydrodynamic flow in the coalescer. It consisted of a central electrode and a double glass shell filled with electrolyte as the second electrode. Scale-up can easily be accomplished by increasing the number of such tubes. The concentric manner of the electrode arrangement (nonuniform electric field) was expected to increase the demulsification rate.

Feng, Wang, and Zhang (1988) built a continuous coalescer with a capacity for 32 to 55 L/h emulsion and specially designed a high-voltage power supply for the coalescer. The coalescer was 20 cm wide, 30 cm long, and 60 cm high; inside, two insulated electrodes (7 cm apart) were installed horizontally. A stirrer was also mounted inside the coalescer to agitate the emulsion during operation. The power supply rated 300 VA offered the maximum voltage and current of 20 kV and 50 mA, respectively. Frequency ranged from 0.6 to 2.5 kHz. Yan et al. (1990a) also built a continuous coalescer with a capacity of 33 L/h emulsion and a pulsed high-voltage supply, which could generate a pulse wave with a duration of 0.2 to 0.5 ms, a peak voltage of 15 kV, and a frequency ranging from 200 to 300 Hz.

Several factors that influence the demulsification rate are discussed in the following sections.

Effect of Applied Voltage

According to the mechanism of electrostatic coalescence, raising the applied field strength will increase the mutual attractive force between the water droplets in the electrical field and thus accelerate the coalescence process. This has been verified by several investigators. Kataoka and Nishiki (1990) have reported the effect of applied voltage on demulsification in batch and continuous operations. Both operations showed enhanced demulsification with increasing voltage. Their data also indicated that the emulsion with the polyamine, ECA 4360, was more difficult to break in comparison to SPAN 80.

Note that excessive high voltages should not be applied because the high voltage is liable to damage the insulating layer and thus cause sparking. Furthermore, according to the basic principle of electrostatics, a critical field strength exists above which a single water drop can disintegrate into more smaller droplets (Gu 1990).

Effects of Frequency and Wave Form

Feng, Wang, and Zhang (1988) observed that under 5 kV for 10 minutes, the coalescence efficiency increased from 50 to 100% when the frequency increased from 0.6 to 5 kHz. Draxler, Fürst, and Marr (1988) also found the voltage could be reduced when the frequency was enhanced for the same degree of demulsification (50%).

Bailes and Larkai (1982) and Yan et al. (1990b, 1990c) found that under the constant voltage, a critical (optimum) frequency existed below which the demulsification rate was enhanced with increasing frequency, but above which the rate decreased with increasing frequency. Yan et al. (1990b, 1990c) suggested that under the high-voltage electric field, the equivalent circuit of the coalescer containing an emulsion to be broken could be simplified to a resistor-capacitor series model, in which the resistance came from the chains of water droplets in the W/O emulsion and the insulation layer of the electrode could be regarded as a capacitor. In the low-frequency region, the time needed for completely charging and discharging the capacitor was sufficiently long. Each pulse would cause one step of collision of water droplets (the response of water droplets to pulses has been proven to be very fast during the demulsification process). Therefore, raising the frequency would increase the collision number per unit time, thus enhancing the coalescence rate. As the frequency reached its critical value, the further increase of frequency would cause the incomplete charging of the capacitor and decrease the force acting on water droplets, resulting in a reduction of the demulsification rate.

Draxler, Fürst, and Marr (1988) and Yan et al. (1987) investigated the influence of the wave shape of alternating current, and they found that the square wave form was much better than a triangle or sinus form. This might be explained by noting that under the same peak voltage, the square wave form offers a longer time to act on the water droplets to be coalesced.

Other Effects on Demulsification

Feng, Wang, and Zhang (1988) pointed out that during the demulsification process, stirring the emulsion could accelerate the coalescence rate. Using a 1-L coalescer, they found that the demulsification rate increased five to six times as the stirring speed increased from 0 to 200 rpm.

Hsu and Li (1985) and Yan et al. (1987) suggested that coalescence at an elevated temperature appeared to be a desirable option for the oils having high viscosities. Since it is relatively easy to reduce the viscosity by heating the emulsion, the reduced viscosity of the oil would aid the settling rate.

Concerning the effect of insulation material on demulsification, Hsu and Li (1985) found that the insulation material having a dielectric constant above 4 offered better demulsification performance. This might be attributed to the fact that by use of the high value of the dielectric constant of the insulation layer, a portion of the applied voltage consumed by the insulation layer can be reduced and this in turn can increase the effective voltage across the emulsion, thus enhancing the demulsification process.

One troublesome problem during the demulsification operation is the formation of a stable and viscous sponge-like emulsion with a high water content, which suspends between the oil and aqueous layers (Hsu and Li 1985; Feng, Wang, and Zhang 1988). To minimize this sponge emulsion, an electrostatic field with a sufficiently high field strength acting perpendicularly to the oil/water interface has to be maintained. For this reason, Hsu and Li (1985) suggested a horizontal insulated electrode configuration, which could generate an intense vertical electrical field. Also, the associated electrohydrodynamic turbulence at the oil/water interface could suppress the accumulation of the sponge emulsion.

NOTATION

General Notation

See the General Notation section at the beginning of this handbook.

Special Notation

A	solute existing only in the external phase
A_s	cross-section area of a column extractor, L^2 (m^2)
B	only solute in the internal phase
c_e	solute concentration in the external phase, N
$c_{e,0}$	inlet solute concentration in the external phase, N
$c_{e,1}$	outlet solute concentration in the external phase from stage 1 mixer-settler or from a column extractor, N
$c_{e,i}$	outlet solute concentration in the external phase from stage i mixer-settler, N
c_{eA}	solute A concentration in the external phase, N
$c_{eA,0}$	inlet solute A concentration in the external phase, N

$c_{eA,1}$	outlet solute A concentration in the external phase from stage 1 mixer-settler or from a column extractor, N
$c_{eA,n}$	outlet solute A concentration in the external phase from a stage n mixer-settler, N
c_{eB}	solute B concentration in the external phase, N
$c_{eB,0}$	inlet solute B concentration in the external phase, N
$c_{eB,1}$	outlet solute B concentration in the external phase from stage 1 mixer-settler or from a column extractor, N
$c_{eB,n}$	outlet solute B concentration in the external phase from stage n mixer-settler, N
c_{iB}	solute B concentration in the internal phase, N
$c_{iB,0}$	inlet solute B concentration in the internal phase, N
$c_{iB,1}$	outlet solute B concentration in the internal phase from stage 1 mixer-settler or from a column extractor, N
$c_{iB,n}$	outlet solute B concentration in the internal phase from stage n mixer-settler, N
c_{ir}	concentration of the reagent in the internal phase, N
$c_{ir,0}$	inlet concentration of the reagent in the internal phase, N
$c_{ir,1}$	outlet concentration of the internal phase reagent from stage 1 mixer-settler or from a column extractor, N
$c_{ir,i}$	outlet concentration of internal phase reagent from stage i mixer-settler, N
c_m	solute concentration or metal-carrier complex concentration in the membrane phase, N
c_T	concentration of reacted and unreacted solute species averaged over the emulsion mixture, N
d_{32}	Sauter mean diameter of emulsion globules, L (m)
D_{eff}	effective diffusivity of the solute or metal-carrier complex, based on the membrane phase concentration c_m, in the emulsion mixture, L^2/t (m²/s)
D'_{eff}	effective solute diffusivity, based on the average concentration c in the emulsion mixture, L^2/t (m²/s)
D_L	axial dispersion coefficient of the emulsion, L^2/t (m²/s)
D_{Le}	axial dispersion coefficient of the external phase, L^2/t (m²/s)
D_m	diffusivity of the solute or metal-carrier complex in the membrane phase, L^2/t (m²/s)
f_i	fractional utilization of the internal reagent for stage i, dimensionless
f_T	total fractional utilization of the internal reagent, dimensionless
f_0	constant fractional utilization of the internal reagent each stage, dimensionless
J	flux, kg-equivalent/$L^2 t$ (kg-equivalent/m²s)
k	overall mass transfer coefficient, t^{-1} (s^{-1})
k_A	overall mass transfer coefficient for solute A, t^{-1} (s^{-1})
k_B	overall mass transfer coefficient for solute B, t^{-1} (s^{-1})
$k_{B\Phi}$	$k_B + \Phi[V_i/(V_i + V_m)]$, overall leakage coefficient for solute B, t^{-1} (s^{-1})
k_e	individual mass transfer coefficient in the external phase, L/t (m/s)
k_f	forward reaction rate constant, L/t (m/s) or various units
K_e	distribution coefficient for the solute between the membrane and external phases at equilibrium, dimensionless
K'_e	distribution coefficient for the solute between the reacted emulsion mixture and the external phase at equilibrium, dimensionless
K_{eB}	distribution coefficient for solute B between the membrane and external phases at equilibrium, dimensionless
K_{iB}	distribution coefficient for solute B between the membrane and internal phases at equilibrium, dimensionless
l	effective membrane thickness between external and internal phases, L (m)
L	height of a column extractor, L (m)
n	number of mixer-settler stages
Pe	Peclet number, $v_e L/D_{Le}$, dimensionless
r_f	reaction front position, L (m)
r_o	emulsion globule radius, L (m)
V	extractor volume each stage in n-stage mixer-settlers, L^3 (m³)
V_t	total extractor volume, L^3 (m³)

z	z coordinate (vertical distance from the bottom of a mechanically agitated column), L (m)
Z	z/L, dimensionless

Greek Letters

Γ°	inlet ratio of the internal reagent equivalents to the feed solute equivalents, dimensionless
Γ_i	ratio of the internal reagent equivalents to the external phase solute equivalents for stage i, $\Gamma_i = Q_i c_{ir,i}/Q_e c_{e,i}$ for cocurrent, $\Gamma_i = Q_i c_{ir,i+1}/Q_e c_{e,i}$ countercurrent, dimensionless
η	emulsion viscosity, M/Lt (Ns/m^2)
η_s	solvent viscosity, M/Lt (Ns/m^2)
θ	parameter of the advancing front model, dimensionless
λ	parameter of the advancing front model, dimensionless
ξ	external emulsion recycle ratio, dimensionless
ρ	emulsion density, M/L^3 (kg/m^3)
τ°	mean residence time of emulsion globules in a mixer, t (s)
Φ	breakage coefficient, t^{-1} (s^{-1})
χ	r_f/r_o, normalized reaction front position, dimensionless
ω	stirring rate, t^{-1} (s^{-1})

REFERENCES

Abou-Nemeh, I., and A. P. Van Peteghem. 1989. Extraction of cobalt and manganese from an industrial effluent by liquid emulsion membranes. In *Proceedings of the Second International Conference on Separations Science and Technology,* ed. M. H. I. Baird and S. Vijayan, Vol. 2, pp. 416–423. Ottawa, Ontario: Canadian Society of Chemical Engineering.

Abou-Nemeh, I., and A. P. Van Peteghem. 1990. Some aspects of emulsion instability on using sorbitan monoleate (SPAN 80) as a surfactant in liquid emulsion membranes. *Chem.-Ing.-Tech.* 62(5):420–421.

Bailes, P. J., and S. K. L. Larkai. 1982. Liquid phase separation in pulsed DC fields. *Trans. I. Chem. E.* 60:115–121.

Bart, H. J., A. Bauer, and R. Marr. 1987. Calculation of reactive extraction in countercurrent columns. *Chem. Eng. Technol.* 10:291–296.

Bart, H. J., R. Wachter, R. J. Marr, and H. J. Müller. 1988. Mass transfer in a permeation column. *J. Membr. Sci.* 36:413–423.

Bartsch, R. A., W. A. Charewicz, and S. I. Kang. 1984. Separation of metals by liquid surfactant membranes containing crown ether carboxylic acids. *J. Membr. Sci.* 17:97–107.

Bartsch, R. A., W. A. Charewicz, S. I. Kang, and W. Walkowiak. 1987. Proton-coupled transport of alkali metal cations across liquid membranes by ionizable crown ethers. In *Liquid Membranes: Theory and Applications,* ed. R. D. Noble and J. D. Way, ACS Symp. Ser. No. 347, pp. 86–97. Washington, DC: American Chemical Society.

Bibaud, R. E., and R. E. Treybal. 1966. Axial mixing and extraction in a mechanically agitated liquid extraction tower. *AIChE J.* 12(3):472–477.

Bock, J., R. R. Klein, P. L. Valint, and W. S. Ho. 1982. Liquid membrane extraction of uranium from wet process phosphoric acid. In *Sulfuric/Phosphoric Acid Plant Operations,* pp. 175–183. New York: American Institute of Chemical Engineers. Paper read at AIChE National Meeting, 8–12 November 1981, New Orleans, LA.

Borwankar, R. P., C. C. Chan, D. T. Wasan, R. M. Kurzeja, Z. M. Gu, and N. N. Li. 1988. Analysis of the effect of internal phase leakage on liquid membrane separations. *AIChE J.* 34(5):753–762.

Boyadzhiev, L., and G. Kyuchoukov. 1980. Further development of carrier-mediated extraction. *J. Membr. Sci.* 6:107–112.

Braun, T., and A. B. Frag. 1978. Liquid sandwich-transfer extraction: optimization and analytical use. *Anal. Chim. Acta.* 100:619.

Cahn, R. P., J. W. Frankenfeld, N. N. Li, D. Naden, and K. N. Subramanian. 1981. Extraction of metal ions by liquid membrane. In *Recent Developments in Separation Science,* ed. N. N. Li, Vol. 6, p. 51. Boca Raton, FL: CRC Press.

Cahn, R. P., and N. N. Li. 1974. Separation of phenol from waste water by the liquid membrane technique. *Sep. Sci.* 9(6):505–519.

Chilamkurti, R. N., and C. T. Rhodes. 1980. Transport across liquid membranes: effect of molecular structure. *J. Appl. Biochem.* 2:17–24.

Christensen, J. J., S. P. Christensen, M. P. Biehl, S. A. Lowe, J. D. Lamb, and R. M. Izatt. 1983. Effect of receiving phase anion on macrocycle-mediated cation transport rates and selectivities in water-toluene-water emulsion membranes. *Sep. Sci. Technol.* 18(4):363–373.

Christensen, J. J., J. D. Lamb, P. R. Brown, J. L. Oscarson, and R. M. Izatt. 1981. Liquid membrane separations of metal cations using macrocyclic carriers. *Sep. Sci. Technol.* 16:1193.

Chuei, B.-Y., J.-F. Yu, and L.-P. Zhang. 1983. Some new liquid extraction systems for separation of cobalt from nickel and mathematical modeling of these extraction processes. In *Proc. Int. Solv. Extr. Conf., ISEC'83*, 26 August–2 September 1983, Denver, CO, pp. 193–194.

Colinart, P., S. Delepine, G. Trouve, and H. Renon. 1984. Water transfer in emulsion liquid membrane process. *J. Membr. Sci.* 20:167–187.

Cussler, E. L. 1971. Membranes which pump. *AIChE J.* 17(6):1300–1303.

Cussler, E. L., and D. F. Evans. 1974. How to design liquid membrane separations. *Sep. Purif. Methods* 3(2):399–421.

Cussler, E. L., D. F. Evans, and M. A. Matesich. 1971. Theoretical and experimental basis for a specific countertransport system in membranes. *Science* 172:377–379.

Dines, M. B. 1982. Regeneration of liquid membrane without breaking emulsion. U.S. Patent 4,337,225.

Ding, X.-C., and F.-Q. Xie. 1991. Study of the swelling phenomena of liquid surfactant membranes. *J. Membr. Sci.* 59:183–188.

Ding, X.-C., F.-Q. Xie, C. Ding, and W.-J. Qin. 1990. A study on swelling properties of emulsion liquid membrane. *Membr. Sci. Technol. (Chinese)* 10(2):21–25.

Ding, X.-C., F.-Q. Xie, X.-Y. Xie, and J.-Z. He. 1988. Study of the liquid membrane swelling. *Proc. First Sino-Jap. Symp. Liq. Membr.* 24–27 September 1988, Hangzhou, China, p. 10.

Downs, H. H., and N. N. Li. 1981. Extraction of ammonia from municipal waste water by the liquid membrane process. *J. Sep. Process Technol.* 2(4):19–24.

Draxler, J., W. Fürst, and R. J. Marr. 1988. Separation of metal species by emulsion liquid membranes. *J. Membr. Sci.* 38:281–293.

Draxler, J., and R. J. Marr. 1986. Emulsion liquid membranes, part I: phenomenon and industrial application. *Chem. Eng. Process.* 20:319–329.

Draxler, J., and R. J. Marr. 1990. Design criteria for electrostatic de-emulsifiers. *Chem.-Ing.-Tech.* 62(7):525–530.

Feng, Z.-L., X.-D. Wang, and X.-J. Zhang. 1988. A new high voltage electrostatic coalescer EC-1 applied to liquid membrane separation. *Water Treatment* 3:320–328.

Frankenfeld, J. W., R. P. Cahn, and N. N. Li. 1981. Extraction of copper by liquid membranes. *Sep. Sci. Technol.* 16(4):385.

Frankenfeld, J. W., and N. N. Li. 1977. Waste water treatment by liquid ion exchange in liquid membrane systems. In *Recent Developments in Separation Science*, ed. N. N. Li, Vol. 3, pp. 285–292. Boca Raton, FL: CRC Press.

Frankenfeld, J. W., and N. N. Li. 1979. Liquid membrane systems. In *Ion Exchange for Pollution Control*, ed. C. Calmon and H. Gold, Vol. 2, pp. 163–172. Boca Raton, FL: CRC Press.

Fujinawa, K., T. Morishita, M. Hozawa, and H. Ino. 1985. A study of operation conditions in the extraction of copper by liquid surfactant membranes. *Kagaku Kogaku Ronbunshu.* 11:293.

Gladek, L., J. Stelmaszek, and J. Szust. 1981. Modeling of the mass transport through a liquid membrane. In *Recent Developments in Separation Science*, ed. N. N. Li, Vol. 6, pp. 29–49. Boca Raton, FL: CRC Press.

Gladek, L., J. Stelmaszek, and J. Szust. 1982. Modeling of the mass transport with a very fast reaction through liquid membranes. *J. Membr. Sci.* 12:153–167.

Goto, M., M. Matsumoto, K. Kondo, and F. Nakashio. 1987. Development of new surfactant for liquid surfactant membrane process. *J. Chem. Eng. Japan* 20(2):157–164.

Goto, M., H. Yamamoto, K. Kondo, and F. Nakashio. 1991. Effect of new surfactants on zinc extraction with liquid surfactant membranes. *J. Membr. Sci.* 57:161–174.

Goto, M., N. Yoshii, K. Kondo, and F. Nakashio. 1989. Separation of lanthanum and neodymium by liquid surfactant membranes. *Proc. Symp. Solvent Extr.*, pp. 113–118. Fukuoka, Japan: Kyushu University.

Grimaldi, J. J., and J. M. Lehn. 1979. Multicarrier transport: coupled transport of electrons and metal cations mediated by an electron carrier and a selective cation carrier. *J. Am. Chem. Soc.* 101:1333–1334.

Gu, Z. M. 1990. A new liquid membrane technology—electrostatic pseudo liquid membrane. *J. Membr. Sci.* 52(1):77–88.

Gu, Z. M., D. T. Wasan, and N. N. Li. 1985. Interfacial mass transfer in ligand accelerated metal extraction by liquid surfactant membranes. *Sep. Sci. Technol.* 20(7&8):599–612.

Gu, Z. M., D. T. Wasan, and N. N. Li. 1986. Ligand-accelerated liquid membrane extraction of metal ions. *J. Membr. Sci.* 26:129–142.

Gu, Z. M., D. T. Wasan, and N. N. Li. 1988. Liquid surfactant membranes for metal extractions. *Surfactant Sci. Ser.* 28:127–168.

Halwachs, W., E. Flaschel, and K. Schugerl. 1980. Liquid membrane transport—a highly selective separation process for organic solutes. *J. Membr. Sci.* 6:33.

Halwachs, W., and K. Schugerl. 1980. The liquid membrane technique—a promising extraction process. *Int. Chem. Eng.* 20(4):519–528.

Hanna, G. J., and K. M. Larson. 1985. Influence of

preparation parameters on internal droplet size distribution of emulsion liquid membranes. *Ind. Eng. Chem. Prod. Res. Dev.* 24:269–274.

Hatton, T. A., E. N. Lightfoot, R. P. Cahn, and N. N. Li. 1983. An internal recycle mixer for solvent extraction: mass transfer characterization with liquid surfactant membranes. *Ind. Eng. Chem. Fundam.* 22(12):27–35.

Hatton, T. A., and D. S. Wardius. 1984. Analysis of staged liquid surfactant membrane operations. *AIChE J.* 30(6):934–944.

Hayworth, H. C., W. S. Ho, W. A. Burns, Jr., and N. N. Li. 1983. Extraction of uranium from wet process phosphoric acid by liquid membranes. *Sep. Sci. Technol.* 18(6):493–521.

Hirato, T., I. Kishigami, Y. Awakura, and H. Majima. 1991. Concentration of uranyl sulfate solution by an emulsion-type liquid membrane process. *Hydrometallurgy* 26(1):19–33.

Hirato, T., K. Koyama, Y. Awakura, and H. Majima. 1990. Concentration of Mo(VI) from aqueous sulfuric acid solution by an emulsion type liquid membrane process. *Mater. Trans. JIM* 31(3):213–218.

Hirato, T., T. Suyama, and H. Majima. 1989. Stability of liquid (kerosene) membrane containing tri-n-octylamine and SPAN 80 in aqueous sulfuric acid solutions. *Nippon Kinzoku Gakkaishi.* 53(10):1041–1046.

Ho, W. S. 1986. The size of liquid membrane emulsion globules in an agitated continuous phase: a dynamic measuring technique. Paper read at the 191st ACS National Meeting and 1st International Conference on Separations Science and Technology, 16–18 April 1986, New York.

Ho, W. S. 1990. Emulsion liquid membranes: a review. In *Proceedings of the 1990 International Congress on Membranes and Membrane Processes,* 20–24 August 1990, Chicago, IL, Vol. I, pp. 692–694.

Ho, W. S., T. A. Hatton, E. N. Lightfoot, and N. N. Li. 1982. Batch extraction with liquid surfactant membranes: a diffusion controlled model. *AIChE J.* 28(4):662–670.

Ho, W. S., and N. N. Li. 1984. Modeling of liquid membrane extraction processes. In *Hydrometallurgical Process Fundamentals,* ed. R. G. Bautista, pp. 555–597. New York: Plenum Press.

Hochhauser, A. M., and E. L. Cussler. 1975. Concentrating chromium with liquid surfactant membranes. *AIChE Symp. Ser.* 71:136–142.

Hsu, E. C., and N. N. Li. 1985. Membrane recovery in liquid membrane separation processes. *Sep. Sci. Technol.* 20(2&3):115–130.

Hsu, E. C., N. N. Li, and T. Hucal. 1983a. Electrodes for electrical coalescence of liquid emulsions. U.S. Patent 4,415,426.

Hsu, E. C., N. N. Li, and T. Hucal. 1983b. Electrical coalescence of liquid emulsion. U.S. Patent 4,419,200.

Imai, M., and S. Furusaki. 1990. Mean diameter and leakage of inner aqueous phase of W/O/W [water-in-oil] emulsion with highly viscous organic solvents. *Water Treatment* 5(2):179–189.

Itoh, H., M. P. Thien, T. A. Hatton, and D. I. C. Wang. 1990. Water transport mechanism in liquid emulsion membrane process for the separation of amino acids. *J. Membr. Sci.* 51:309–322.

Izatt, R. M., R. L. Bruening, and J. J. Christensen. 1987. Use of coanion type and concentration in macrocycle-facilitated metal cation separations with emulsion liquid membranes. In *Liquid Membranes: Theory and Applications,* ed. R. D. Noble and J. D. Way, ACS Symp. Ser. No. 347, pp. 98–108. Washington, DC: American Chemical Society.

Izatt, R. M., R. L. Bruening, G. A. Clark, J. D. Lamb, and J. J. Christensen. 1987a. Effect of macrocycle type on Pb^{2+} transport through an emulsion liquid membrane. *Sep. Sci. Technol.* 22(2&3):661–675.

Izatt, R. M., R. L. Bruening, W. Geng, M. H. Cho, and J. J. Christensen. 1987b. Separation of bivalent cadmium, mercury, and zinc in a neutral macrocycle-mediated emulsion liquid membrane system. *Analy. Chem.* 59:2405.

Izatt, R. M., G. A. Clark, J. B. Bradshaw, J. D. Lamb, and J. J. Christensen. 1986. Macrocycle-facilitated transport of ions in liquid membrane systems. *Sep. Purif. Methods* 15(1):21–72.

Izatt, R. M., G. A. Clark, and J. J. Christensen. 1987. Transport of $AgBr_2^-$, $PdBr_4^{2-}$, and $AuBr_4^-$ in an emulsion membrane system using K^+-dichlohexano-18-crown-6 as carrier. *Sep. Sci. Technol.* 22(2&3):691–699.

Izatt, R. M., D. V. Dearden, D. W. McBride, Jr., J. L. Oscarson, J. D. Lamb, and J. J. Christensen. 1983. Metal separations using emulsion liquid membranes. *Sep. Sci. Technol.* 18:1113.

Izatt, R. M., D. V. Dearden, E. R. Witt, D. W. McBride, Jr., and J. J. Christensen. 1984. Cation selectivity in a toluene emulsion membrane system. *Solvent Extr. Ion Exch.* 2(3):459–477.

Izatt, R. M., J. D. Lamb, and R. L. Bruening. 1988. Comparison of bulk, emulsion, thin sheet supported, and hollow fiber supported liquid membranes in macrocycle-mediated cation separations. *Sep. Sci. Technol.* 23(12&13):1645–1658.

Kaifer, A., L. Echegoyen, D. A. Gustowski, D. M. Goli, and G. W. Gokel. 1983. Enhanced sodium cation binding by electrochemically reduced nitrobenzene-substituted lariat ethers. *J. Am. Chem. Soc.* 105:7168–7169.

Kataoka, T., and T. Nishiki. 1990. Development of a continuous electric coalescer of W/O emulsions in liquid surfactant membrane process. *Sep. Sci. Technol.* 25(1&2):171–185.

Kataoka, T., T. Nishiki, M. Yamauchi, and Y. Zhong. 1987. A simulation for liquid surfactant membrane permeation in a continuous countercurrent column. *J. Chem. Eng. Japan* 20(4):410–415.

Kato, S., and J. Kawasaki. 1988. Mechanical demulsification of O/W emulsions. *Proc. First Sino-Jap. Symp. Liq. Membr.*, 24–27 September 1988, Hangzhou, China, pp. 86–88.

King, C. J. 1980. *Separation Processes,* 2nd ed. New York: McGraw-Hill Book Co.

Kinugasa, T., K. Watanabe, and H. Takeuchi. 1990. Stability of (W/O) emulsion drops and water permeation through its liquid membrane in (W/O)/W dispersion. In *Proceedings of the 1990 International Congress on Membranes and Membrane Processes,* 20–24 August 1990, Chicago, IL, Vol. 1, pp. 706–708.

Kitagawa, T., Y. Nishikawa, J. W. Frankenfeld, and N. N. Li. 1977. Waste water treatment by liquid membrane process. *Environmental Sci. Technol.* 11(6):602–605.

Lamb, J. D., P. R. Brown, J. J. Christensen, J. S. Bradshaw, D. G. Garrick, ad R. M. Izatt. 1983. Cation transport at 25°C from binary Na^+-M^{n+}, Cs^+-M^{n+}, and Sr^{2+}-M^{n+} nitrate mixtures in a H_2O-$CHCl_3$-H_2O liquid membrane system containing a series of macrocyclic carriers. *J. Membr. Sci.* 13:89.

Lamb, J. D., J. J. Christensen, J. L. Oscarson, B. L. Nielsen, B. W. Asay, and R. M. Izatt. 1980. The relationship between complex stability constants and rates of cation transport through liquid membranes by macrocyclic carriers. *J. Am. Chem. Soc.* 102:6820.

Lamb, J. D., R. M. Izatt, D. G. Garrick, J. S. Bradshaw, and J. J. Christensen. 1981. The influence of macrocyclic ligand structure on carrier-facilitated cation transport rates and selectivities through liquid membranes. *J. Membr. Sci.* 9:83–107.

Lee, C. J., and C. C. Chan. 1990. Extraction of ammonia from a dilute aqueous solution by emulsion liquid membranes: 1. experimental studies in batch system. *Ind. Eng. Chem. Res.* 29:96–100.

Li, N. N. 1978. Facilitated transport through liquid membranes. *J. Membr. Sci.* 3:265.

Li, N. N. 1981a. Encapsulation and separation by liquid surfactant membranes. *Chem. Eng. (Rugby, England)* 370:325–327.

Li, N. N. 1981b. Novel liquid membrane formulations. U.S. Patent 4,259,189.

Li, N. N., R. P. Cahn, and A. L. Shrier. 1973. Liquid membrane process for the separation of aqueous mixtures. U.S. Patent 3,779,907.

Li, S. P., and Y. C. Chang. 1982. Extraction of chromium ions from aqueous solution. In *Proceedings of Joint CIESC and AIChE Meeting,* 19–22 September 1982, Beijing, China, Chemical Industry Press, Vol. II, pp. 571–582.

Li, N. N., and J. W. Frankenfeld. 1988. Liquid membranes. In *Encyclopedia of Chemical Processing and Design,* ed. J. C. McKetta, Vol. 28, pp. 273–303. New York: Marcel Dekker.

Liu, Z.-F., X.-T. Zhang, and Q. J. Fan. 1990. Liquid membrane properties and applications of surfactant EM-301. *Membr. Sci. Technol. (Chinese)* 10(3):38–43.

Lorbach, D., H. J. Bart, and R. J. Marr. 1986. Stoffübergang in der flüssigmembran-permeation. *Chem.-Ing.-Tech.* 58(2):156.

Lorbach, D., and R. J. Marr. 1987. Emulsion liquid membranes: part II. Modeling mass transfer of zinc with bis(2-ethylhexyl)dithiophosphoric acid. *Chem. Eng. Process.* 21:83–93.

Ma, X.-S., and Y.-J. Shi. 1987. Study of operating condition affecting mass transfer rate in liquid surfactant membrane process. *Sep. Sci. Technol.* 22(2&3):819–829.

Macásek, F., P. Rajec, R. Kopunec, and V. Mikulaj. 1984. Membrane extraction in preconcentration of some uranium fission products. *Solvent Extr. Ion. Exch.* 2(2):227–252.

Magdassi, S., and N. Garti. 1984. Release of electrolytes in multiple emulsions: coalescence and breakdown or diffusion through oil phase? *Colloids Surf.* 12:367–373.

Marr, R. J., H. J. Bart, and J. Draxler. 1990. Liquid membrane permeation. *Chem. Eng. Process.* 27:59–64.

Marr, R. J., H. Lackner, and K. Hütter. 1988. Apparatus for mixing media capable to flow. European Patent Application 88810472.6.

Martin, T. P., and G. A. Davies. 1977. The extraction of copper from dilute aqueous solutions using a liquid membrane process. *Hydrometallurgy* 2:315–334.

Matsumoto, M., K. Ema, K. Kondo, and F. Nakashio. 1990. Copper extraction with liquid surfactant membrane in Mixco extractor. *J. Chem. Eng. Japan* 23(4):402–407.

Matsumoto, S., T. Inoue, M. Kohda, and K. Ikura. 1980. Water permeability of oil layers in W/O/W emulsions under osmotic pressure gradients. *J. Colloid Interface Sci.* 77(2):555–563.

Matulevicius, E. S., and N. N. Li. 1975. Facilitated transport through liquid membranes. *Sep. Purif. Methods* 4(1):73–96.

Mikucki, B. A., and K. Osseo-Asare. 1986a. The liquid surfactant membrane process: effect of the emulsion type on copper extraction by LIX64N-LIX63 mixtures. *Hydrometallurgy* 16:209.

Mikucki, B. A., and K. Osseo-Asare. 1986b. Effects of the emulsifier in copper extraction by LIX65N-SPAN 80 liquid surfactant membranes. *Solvent Extr. Ion Exch.* 4:503.

Miyake, Y., Y. Takenoshita, and M. Teramoto. 1983. Extraction rates of copper with SME 529: mechanism and effects of surfactants. In *Proc. Solv. Extr. Conf., ISEC'83*, 26 August–2 September 1983, Denver CO, pp. 301–302.

Mukkolath, A. V., P. T. Gadekar, and K. K. Tiwari. 1990. A new method for the reduction of swelling in liquid emulsion membrane systems. *Chem. Ind.* 6:192–193.

Nakashio, F., M. Goto, and K. Kondo. 1990. Role of surfactant in liquid surfactant membrane process. *Water Treatment* 5:157–169.

Nakashio, F., M. Goto, M. Matsumoto, J. Irie, and K. Kondo. 1988. Role of surfactants in the behavior of emulsion liquid membranes: development of new surfactants. *J. Membr. Sci.* 38:249–260.

Noble, R. D., J. D. Way, and A. L. Bunge. 1988. Liquid membranes. *Solvent Extr. Ion Exch.* 10:63–103.

Noguchi, S., F. Nakazawa, and K. Yoshii. 1980. Pulsed NMR study on the behavior of water in oil-in-water emulsions containing nonionic surfactants. *Chem. Soc. Japan* 7:1073–1076.

O'Brien, D. J., and G. E. Senske. 1989. Separation and recovery of low molecular weight organic acids by emulsion liquid membranes. *Sep. Sci. Technol.* 24(9&10):617–628.

Ohki, A., H. Hinoshita, M. Takagi, and K. Ueno. 1983. Transport of iron and cobalt complex ions through liquid membrane mediated by methyltrioctyl-ammonium ion with the aid of redox reaction. *Sep. Sci. Technol.* 18(11):969–983.

Ohtake, T., T. Hano, K. Takagi, and F. Nakashio. 1987. Effects of viscosity on drop diameter of W/O emulsion dispersed in a stirred tank. *J. Chem. Eng. Japan* 20(5):443–447.

Pannell, K. H., B. J. Rodriguez, S. Chiocca, L. P. Jones, and J. Molinar. 1982. Dibenzo-crown facilitated transport across a $CHCl_3$ liquid membrane. *J. Membr. Sci.* 11:169–175.

Prötsch, M., and R. J. Marr. 1983. Development of a continuous process for metal ion recovery of liquid membrane permeation. In *Proc. Int. Solv. Extr. Conf., ISEC '83*, 26 August–2 September 1983, Denver, CO, pp. 66–67.

Qian, X.-L., X.-S. Ma, and Y.-J. Shi. 1989. Removal of cyanide from wastewater with liquid membranes. *Water Treatment* 4:99–111.

Rautenbach, R., and O. Machhammer. 1988. Modeling of liquid membrane separation processes. *J. Membr. Sci.* 36:425–444.

Reed, D. L., A. L. Bunge, and R. D. Noble. 1987. Influence of reaction reversibility on continuous flow extraction by emulsion liquid membranes. In *Liquid Membranes: Theory and Applications*, ed. R. D. Noble and J. D. Way, ACS Symp. Ser. No. 347, pp. 62–83. Washington, DC: American Chemical Society.

Reusch, C. F., and E. L. Cussler. 1973. Selective membrane transport. *AIChE J.* 19:736.

Rickelton, W. A., D. S. Flett, and D. W. West. 1983. Cobalt-nickel separation by solvent extraction with dialkylphosphinic acid. In *Proc. Int. Solv. Extr. Conf., ISEC'83*, 26 August–2 September 1983, Denver, CO, pp. 195–196.

Ruppert, M., J. Draxler, and R. J. Marr. 1988. Liquid-membrane-permeation and its experiences in pilot-plant and industrial scale. *Sep. Sci. Technol.* 23(12&13):1659–1666.

Schiffer, D. K., E. M. Choy, D. F. Evans, and E. L. Cussler. 1974a. More membrane pumps. *AIChE Symp. Ser.* 70:150–156.

Schiffer, D. K., A. M. Hochhauser, D. F. Evans, and E. L. Cussler. 1974b. Concentrating solutes with membranes containing carriers. *Nature* 250:484–486.

Sekine, T., and Y. Hasegawa. 1977. *Solvent Extraction Chemistry: Fundamentals and Applications*. New York: Marcel Dekker.

Shere, A. J., and H. M. Cheung. 1988a. Effect of preparation parameters on leakage in liquid surfactant membrane systems. *Sep. Sci. Technol.* 23(6&7):687–701.

Shere, A. J., and H. M. Cheung. 1988b. Modeling of leakage in liquid surfactant membrane systems. *Chem. Eng. Comm.* 68:143–164.

Shinkai, S., T. Ogawa, Y. Kusano, O. Manabe, K. Kikukawa, T. Goto, and T. Matsuda. 1982.

Photoresponsive crown ethers: 4. influence of alkali metal cations on photoisomerization and thermal isomerization of azobis(benzocrown ether)s. *J. Am. Chem. Soc.* 104:1960–1967.

Stroeve, P., and P. P. Varanasi. 1982. Transport processes in liquid membranes: double emulsion separation systems. *Sep. Purif. Methods* 11(1):29–69.

Strzelbicki, J. 1978. Separation of cobalt by liquid surfactant membranes. *Sep. Sci. Technol.* 13:141.

Strzelbicki, J., and S. Schlosser. 1989. Influence of surface-active substances on pertraction of cobalt (II) cations through bulk and emulsion liquid membranes. *Hydrometallurgy* 23:67–75.

Takahashi, K., F. Ohtsubo, and H. Takeuchi. 1981. A study of the stability of W/O/W-type emulsion using a tracer technique. *J. Chem. Eng. Japan* 14:416.

Tang, H., Z.-L. Ma, and L.-J. Liu. 1990. Alkaloid extraction from plants with liquid membrane. *Water Treatment* 5(2):214–221.

Tang, J., and C. M. Wai. 1988. Transport of trivalent lanthanides in a H_2O-$CHCl_3$-H_2O liquid membrane system containing a crown ether carboxylic acid. *J. Membr. Sci.* 35:339–345.

Tang, J., and C. M. Wai. 1989. Transport of trivalent lanthanides through a surfactant membrane containing an ionizable macrocyclic polyether. *J. Membr. Sci.* 46:349–356.

Teramoto, M., T. Sakai, K. Yanagawa, and Y. Miyake. 1983c. Modeling of the permeation of copper through liquid surfactant membranes by continuous operations. *Sep. Sci. Technol.* 18(11):985–997.

Teramoto, M., T. Sakai, K. Yanagawa, M. Ohsuga, and Y. Miyake. 1983b. Modeling of the permeation of copper through liquid surfactant membranes. *Sep. Sci. Technol.* 18(8):735–764.

Teramoto, M., H. Takihana, M. Shibutani, T. Yuasa, and N. Hara. 1983a. Extraction of phenol and cresol by liquid surfactant membrane. *Sep. Sci. Technol.* 18(5):397–419.

Terry, R. E., N. N. Li, and W. S. Ho. 1982. Extraction of phenolic compounds and organic acids by liquid membranes. *J. Membr. Sci.* 10:305–323.

Thien, M. P., and T. A. Hatton. 1988. Liquid emulsion membranes and their applications in biochemical processing. *Sep. Sci. Technol.* 23(8&9):819–853.

Thien, M. P., T. A. Hatton, and D. I. C. Wang. 1986. Liquid emulsion membranes and their applications in biochemical separations. *ACS Symp. Ser.* 314:67–77.

Vohra, D. K., S. Kaur, and A. Sharma. 1989. Extraction of Cr(VI) from acid (sulfate) aqueous medium using liquid surfactant membrane emulsions. *Indian J. Technol.* 27:574–576.

Wankat, P. C. 1980. Calculations for separations with three phases: 1. staged systems. *Ind. Eng. Chem. Fundam.* 19(4):358–363.

Wankat, P. C., and R. D. Noble. 1984. Calculations for separations with three phases: 2. continuous contact systems. *Ind. Eng. Chem. Fundam.* 23(2):137–143.

Wardius, D. S., and T. A. Hatton. 1985. A model for liquid membrane extraction with instantaneous reaction in cascaded mixers. *Chem. Eng. Comm.* 37:159–171.

Wasan, D. T., Z. M. Gu, and N. N. Li. 1984. Separation of metal ions by ligand-accelerated transfer through liquid surfactant membranes. *Faraday Discuss. Chem. Soc.* 77:67–74.

Way, J. D., and R. D. Noble. 1982. A macroscopic model of a continuous emulsion liquid membrane extraction system. In *Residence Time Distribution Theory in Chemical Engineering,* ed. A. Petho and R. D. Noble, pp. 247–254. Weinheim, Germany: Verlag-Chemie Publishing Co.

Weiss, S., V. Griegoriev, and P. Mühl. 1982. The liquid membrane process for the separation of mercury from waste water. *J. Membr. Sci.* 12:119–129.

Wiencek, J. W., and S. Qutubuddin. 1986. Separation of organics using microemulsions. Paper read at AIChE National Meeting, April 1986, New Orleans, LA.

Yagodin, G., Y. Lopukhin, E. Yurtov, T. Guseva, and V. Sergiemko. 1983. Extraction of cholesterol from blood using liquid membranes. In *Proc. Int. Solv. Extr. Conf., ISEC '83,* 26 August–2 September 1983, Denver, CO, pp. 385–386.

Yan, N.-X., S.-A. Huang, and Y.-J. Shi. 1987. Removal of acetic acid from wastewater with liquid surfactant membranes: an external boundary layer and membrane diffusion controlled model. *Sep. Sci. Technol.* 22(2&3):801–818.

Yan, N.-X., Y.-J. Shi, and Y.-F. Su. 1990. A study of gold extraction by liquid membranes. *Water Treatment* 5(2):190–201.

Yan, Z., S.-Y. Li, Y. Chu, and B.-J. Liu. 1990a. Treatment of the wastewater containing high concentration of Cr(VI) using liquid membrane. In *Proceedings of the 1990 International Congress on Membranes and Membrane Processes,* 20–24 August 1990, Chicago, IL, Vol. I, pp. 718–720.

Yan, Z., S.-Y. Li, L.-C. Wang, and F.-Y. Yang. 1990b. Demulsification of liquid membranes in electric fields. *Water Treatment* 5:1–6.

Yan, Z., S.-Y. Li, Y.-C. Yu, and X.-L. Zheng. 1987. An investigation into the breaking-down of water-in-oil type emulsions by means of pulsed voltage. *Desalination* 62:323–328.

Yan, Z., S.-Y. Li, W.-H. Zhang, L.-C. Wang, and P.-Y. Liu. 1990c. An investigation into the breakdown of W/O type emulsions by pulsed voltage. *Water Treatment* 5:127–135.

Yan, Z., and L.-C. Wang. 1988. Frequency effect of demulsification with insulated electrode. In *Separation Technology,* ed. N. N. Li and H. Strathmann, pp. 254–260. New York: United Engineering Trustees.

Yu, J.-H., C.-Y. Jiang, and Y.-J. Zhu. 1986. Separation of europium with liquid surfactant membranes. In *Recent Developments in Separation Science,* ed. N. N. Li, and J. M. Calo, Vol. 9, pp. 197–208. Boca Raton, FL: CRC Press.

Zhang, R.-H., and D.-X. Wang. 1989. Extraction of mixed rare earth from aqueous solution with emulsion liquid membrane. *Water Treatment* 4:165–176.

Zhang, R.-H., and L. Xiao. 1989. Design of a liquid membrane system for extracting rare earths. *Water Treatment* 4:473–481.

Zhang, R.-H., and L. Xiao. 1990. Design of a liquid membrane system for extracting rare earths. *J. Membr. Sci.* 51(3):249–258.

Zhang, X.-J., Q.-J. Fan, X.-T. Zhang, and Z.-F. Liu. 1988. New surfactant LMS-2 used for industrial application in liquid membrane separation. In *Separation Technology,* ed. N. N. Li and H. Strathmann, pp. 215–226. New York: United Engineering Trustees; see also *Water Treatment* 3:233–240.

Zhang, X.-J., P.-Y. Huang, and X.-X. Chen. 1988. Chemical demulsification in liquid membrane separation. *Proc. First Sino-Jap. Symp. Liq. Membr.,* 24–27 September 1988, Hangzhou, China, pp. 83–85.

Zhang, X.-J., J.-H. Liu, and T.-S. Lu. 1987. Industrial application of liquid membrane separation for phenolic wastewater treatment. *Water Treatment* 2:127–135.

Zheng, X.-C., L.-X. Li, J.-J. Guo, and F.-S. Long. 1988. Extraction of vanadium from waste water with emulsion liquid membrane. In *Proceedings of the First International Conference on Hydrometallurgy,* 12–15 October 1988, Beijing, China, pp. 508–583.

39
Applications

Rolf J. Marr
Josef Draxler
Technical University of Graz

REMOVAL OF ZINC FROM WASTEWATER IN THE VISCOSE FIBER INDUSTRY
   Introduction
   Application of Emulsion Liquid Membranes
RECOVERY OF NICKEL FROM ELECTROPLATING SOLUTIONS
OTHER APPLICATIONS
   Removal of Phenols
   Removal of Ammonia
   Removal of Heavy Metals in Metallurgical Plants
   Removal of Heavy Metals in Incineration Plants
   Biochemical Applications
REFERENCES

REMOVAL OF ZINC FROM WASTEWATER IN THE VISCOSE FIBER INDUSTRY

Introduction

In the viscose rayon industry, zinc ions are used in the spinbaths to improve the spinning process and the properties of the fibers. In the subsequent rinsing steps, zinc is lost with the wastewater. In addition to the economic disadvantage, the loss of zinc causes pollution problems. Therefore, most viscose companies in the industry have installed plants for zinc removal. The mere removal of zinc from the wastewater, e.g., by precipitation, merely represents a shift from a wastewater problem to a landfill problem. Therefore, some new processes have been tested and installed recently that enable the recovery of zinc. For the reuse of zinc in the viscose industry, a sufficient selectivity of zinc to calcium is needed to avoid any precipitation of gypsum in evaporation steps.

To date, the following processes have been used and tested for the removal of zinc in the viscose industry:

1. *Precipitation as $Zn(OH)_2$*: This process is widely used in the viscose industry. In most cases the wastewater is neutralized by lime in one stage. The precipitated zinc sludge is filtered, dried, and disposed of in a landfill. This precipitated zinc is not suitable for reuse in the spinbaths, mainly because of the high calcium content, but also because of the flocculation agents used in the process.

A better selectivity can be obtained when precipitation is done in two stages (with lime to pH 5 to 6 and afterward with NaOH to pH 9.5

to 10). But because of the flocculation agents, reuse remains limited.

2. *Precipitation as ZnS*: Precipitation as ZnS results in better zinc removal but is much more expensive when compared to precipitation as $Zn(OH)_2$.

3. *Precipitation as $ZnCO_3$*: This process (pellet reactor) was developed by Dutch engineers (van Dijk, de Moel, and Schöller 1985) for the electroplating industry, and it was tested by AKZO for application in the viscose industry. Although the zinc removal was sufficient, reuse in the spinbaths was not possible. If the precipitated pellets can be treated in the zinc mining industry, this process might become important.

4. *Ion-Exchange Resins*: Some companies in the viscose industry use strongly acidic cation-exchangers for the removal of zinc, which suffers from an insufficient selectivity versus calcium. Although zinc-specific ion-exchangers (e.g., Lewatit TP 207) are now available, some problems remain, such as:
   a. dependence of loading capacity on the concentrations of zinc, sodium, and sulfuric acid,
   b. contamination of the ion-exchanger by organic modifiers needed for the spinning process, and
   c. significant quantity of NaOH needed for regeneration.

5. *Solvent Extraction (SX)*: The use of liquid ion-exchangers in solvent extraction causes problems similar to that of ion-exchange resins, particularly the zinc/calcium selectivity and the influence of the modifiers. The AKZO plant at Arnhem is reported to have solved this problems after several optimizations. The main disadvantage is the high cost due to the high solubility of the aromatic diluent used.

6. *Emulsion Liquid Membranes (ELM)*: The description of ELM is given in this chapter and Chapter 40.

7. *Other Possibilities*: Other possible processes tested for the removal of zinc include:
   a. reverse osmosis: insufficient separation, not enough selectivity,
   b. evaporation: inadequate selectivity and expensive, and
   c. biological treatment: based on a biological reduction of SO_4^{2-} to S^{2-} and a subsequent precipitation to ZnS, but recovery and reuse of zinc in the spinbaths are not possible.

Application of Emulsion Liquid Membranes

Introduction

To date, there have been four industrial applications of ELM on various scales:

1. Lenzing AG, Austria, 75 m^3/h (75,000 L/h) (Draxler, Marr, and Prötsch 1988)
2. Glanzstoff Austria AG, 0.7 m^3/h (700 L/h)
3. CFK Schwarza, Germany, 0.2 m^3/h (200 L/h)
4. AKZO/Ede, The Netherlands, 0.2 m^3/h (200 L/h).

These have proven the feasibility of this process for the viscose industry. But prior to installing full-scale ELM plants (larger than those described above) using this process, some complications have to be solved, as briefly outlined in the following paragraphs.

The total wastewater stream in the viscose industry is usually very large, and it has a low zinc concentration. Treatment of the total wastewater stream with ELM (or SX and ion-exchange resins, respectively) is therefore highly ineffective. It is much more effective to treat only some highly concentrated side streams, which will be reduced in the zinc concentration in the total wastewater stream by about 80%. However, this is not enough and requires an additional precipitation step prior to or simultaneous to the subsequent biological treatment. This means that an ELM plant must work on a pure economic basis. This will be difficult in view of the relatively low zinc concentration even in the highly concentrated wastewater streams (about 500 mg/L) and the relatively low price of zinc, especially when companies already have cheap landfills available for the disposal of the zinc sludge. The more expensive the landfills become, the more interesting the recovery of zinc by ELM.

Another reason for the reluctance to install full-scale ELM plants in the viscose industry is that zinc is one among many contaminants in the viscose wastewaters, and the companies are more and more obliged by local authorities to create processes for the complete treatment of all wastewaters. In the scheme of complete treatment, a zinc recovery plant is only one small part, and will be materialized only when it fits into the overall concept. In view of these situations, full-scale ELM plants can be expected around 1993.

Extractants

As in solvent extraction, liquid ion-exchangers are used in ELM to exchange the zinc ions for protons. But these extractants are not the same. ELM can allow the use of the extractants which are not common in solvent extraction and offer a very high selectivity for zinc over calcium. These extractants are the sulfur-containing derivatives of di(2-ethylhexyl)phosphoric acid (D2EHPA), used in SX for the extraction of Zn^{2+}. In particular, four such extractants are as follows:

1. di(2-ethylhexyl)thiophosphoric acid [MTPA (monothiophosphoric acid), e.g., HOE F 3787 produced by Hoechst AG]:

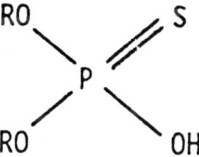

2. di(2-ethylhexyl)dithiophosphoric acid [DTPA (dithiophosphoric acid), e.g., HOE F 3541 by Hoechst AG]:

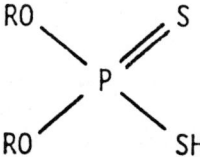

3. di(2,2,4-trimethylpentyl)thiophosphinic acid (e.g., Cyanex 302 by American Cyanamid):

4. di(2,2,4-trimethylpentyl)dithiophosphinic acid (e.g., Cyanex 301 by American Cyanamid):

Compared to D2EHPA, these extractants have two major advantages: extraordinarily better selectivity versus calcium and higher acidity. According to the equilibrium isotherms (Figure 39–1), the zinc/calcium selectivity should be sufficient with D2EHPA, but in practice there were many problems. Furthermore, using D2EHPA demands a strict pH control; at too high a pH value calcium will be extracted,

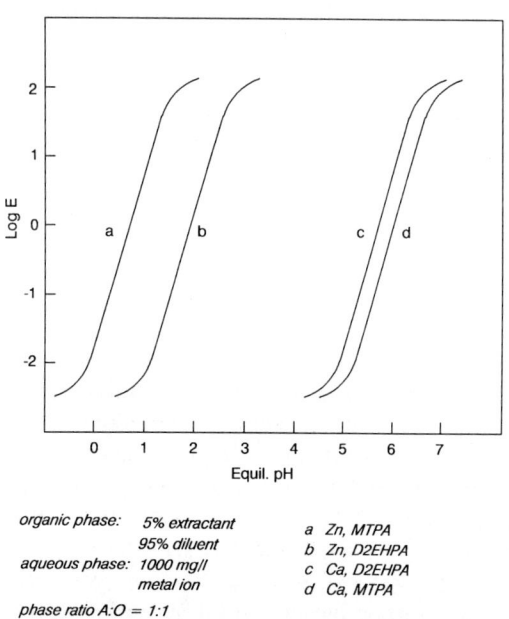

FIGURE 39–1. Extraction isotherms for zinc and calcium (where A and O stand for the aqueous and organic phases, respectively).

whereas at too low a pH value the zinc extraction will deteriorate.

Due to the higher acidity of the sulfur-containing extractants, the wastewater can be treated at a lower pH value, where no calcium is extracted at all. The wastewater streams in the viscose industry, which are to be treated for zinc removal, usually contain 1 to 8 g/L of sulfuric acid. In this range no pre-equilibration of the extractants and no pH control are necessary when using the sulfur-containing extractants. Generally, often in solvent extraction, the higher the acid concentration in the wastewater, the lower the zinc extraction. However, this is not very pronounced in the case of the extraction with ELM as shown in Table 39–1.

Usually DTPA is preferred to MTPA due to its slightly stronger acidity and cheaper price. MTPA is only used when other metals (namely, Fe, Co, Pb) are present in the wastewater. Furthermore, DTPA is an industrial product manufactured in large quantities (as additive for motor oils and as a flotation reagent). Therefore, it is much cheaper than the phosphinic acids, e.g., Cyanex 301 and 302. The major disadvantage of DTPA is its insufficient chemical stability. In contact with acid or water, it is subjected to hydrolysis:

$$\underset{RO}{\overset{RO}{>}}P\underset{SH}{\overset{S}{<}} + 4\,H_2O \rightleftarrows 2\,ROH$$

$$+\ H_3PO_4\ +\ 2\,H_2S$$

MTPA is an intermediate in this overall reaction. The mean half-life was determined at 30°C to be about 2 months, and it is reduced drastically at higher temperatures. Addition of surfactants results in prolongation of this lifetime. Under conditions often prevalent in viscose wastewaters, the half-life is about 6 months. This is long enough and does not justify the use of the expensive phosphinic acids, especially when considering the inevitable losses due to solubility and the necessary makeup with new extractants. However, with Cyanex 301 absolutely no decomposition could be found under conditions characteristic for the viscose wastewaters.

TABLE 39–1. Influence of H_2SO_4 in Wastewater on ELM Extraction.

H_2SO_4 in Wastewater	Zn Concentration in Raffinate in mg/L			
	DTPA		MTPA	
	2 min	5 min	2 min	5 min
1 g/L	0.2	0.2	0.4	0.3
2 g/L	0.3	0.2	0.8	0.5
3 g/L	0.5	0.5	1.8	1.0
4 g/L	1.1	0.7	2.7	1.2
5 g/L	1.1	0.8	2.6	1.3
6 g/L	1.4	1.1	2.6	1.4
7 g/L	1.9	1.4	3.9	2.0
8 g/L	2.6	1.7	8.1	2.7

Operating conditions: (1) 500 ml wastewater with 150 mg/L Zn. (2) 100 ml organic phase containing 5% extractant (DTPA, or MTPA), 3% surfactant (ECA 11522), and 92% diluent (Shellsol T). (3) 20 ml stripping phase (250 g/L H_2SO_4).

The remaining question then is why these extractants are especially suitable for ELM and are not used in SX. The answer lies in the very slow stripping kinetics. In conventional extraction equipment the residence time for sufficient stripping would be uneconomically high. With ELM the very large inner interfacial area (10^6 m²/m³) provides fast stripping kinetics. For the extraction of zinc with the sulfur-containing extractants, SX would be possible, albeit expensive; but for other metals, such as Ni, Cd, Pb, and Hg, only ELM can be used.

Surfactants and Diluents

Although a myriad of surfactants have been tested for ELM, only a few are suitable. For the viscose industry long-chain polyamines (such as ECA 11522) are suitable surfactants. There are no special requirements on the choice of the diluent. Aliphatic diluents are preferred to aromatic ones because of the lower losses. A suitable diluent is, for instance, Shellsol T (Shell) with an aromatic content of less than 0.5%. For safety reasons diluents with a flash point above 100°C should be preferred. Escaid 120 (EXXON) would be one example (flash point = 101°C, compared to 54°C for Shellsol T).

Influence of Spinbath Additives

The viscose wastewaters contain different amounts of spinbath additives, mainly polyethyleneglycol (PEG) and epoxidized fatty amines. These are surface active agents and they therefore have a decisive influence on the feasibility of a SX process and an ELM process. The surface tension between the wastewater and the organic phase is reduced drastically; that is why mixers with low shearing forces have to be used in SX. The influence of the additives on ELM is manifold. They might be partly coextracted (especially the fatty amines), resulting in a higher sulfuric acid consumption. Then, they might act as an additional mass transfer resistance due to their aggregation at the interface. Finally and most importantly, they can strengthen or weaken the activity of the surfactants, resulting in the increase or decrease of emulsion stability.

Although similar additives are used in the viscose industry, these additives differ somewhat. One carbon atom more or less in the chain length of the fatty amines can change the influence on ELM significantly. Therefore, the ELM process can be applied in the viscose industry only after extensive pilot tests in which all parameters have to be optimized and adapted to the additives present in the wastewater.

Usually the additives act as an additional mass transfer resistance, thereby decelerating zinc transport. Compared to ELM processes without any additives in the wastewater, zinc transport in the viscose industry is slower by a factor of about 1.5.

While zinc transport is slowed by all additives, the effects of the additives on emulsion stability are different. Up to the maximum concentration of 2 g/L investigated, PEG was not found to have an influence on emulsion stability. Above a concentration of about 100 mg/L, fatty amines show a significant influence, but the tendency cannot be assessed. Some kinds of fatty amines destabilize the emulsion, causing losses of sulfuric acid. In this case, higher surfactant concentrations have to be used and lower shear forces have to be applied when dispersing the emulsion in the wastewater.

Some kinds of fatty amines may also stabilize the emulsion, causing problems in the subsequent emulsion breaking stage. In this case use of lower surfactant concentrations is not advisable; it is more convenient to optimize an electrostatic coalescer for splitting the emulsion, i.e., voltage, frequency, insulation material, distance between electrodes, and residence time have to be optimized.

Plant Conception

Figure 39–2 shows a schematic of a one-stage ELM plant for the viscose industry. The main parts are the countercurrent column extraction equipment, the oil separator to remove any entrained organics, the emulsion preparation device, and the emulsion splitter. The size of the extraction column may be estimated by the use of one of the models of Bart et al. (1988), Kataoka et al. (1987) [Eq. (38–62)] and Ho and Li (1984) [Eq. (38–35)] described in Chapter 38. The emulsion is prepared by the use of a static homogenizer equipped with a special nozzle, shown schematically in Figure 38–1, through which both the organic phase and the stripping phase are pumped at a pressure of about 1 MPa (Marr, Lackner, and Hütter 1988; Marr, Bart, and Draxler 1990). The emulsion is split electrostatically between parallel plated electrodes that are insulated with enamel. The AC is used at a voltage of 2000 V and a frequency of 10 kHz. The electrostatic coalescer is described in Chapter 38.

Results

Figures 39–3 to 39–6 show some typical results of pilot plant tests at AKZO/Netherlands (performed during October and November 1989). Figures 39–3 and 39–4 show concentration profiles for zinc (two columns were used) for two different wastewater streams. The first, wastewater A, had an initial zinc concentration of 500 mg/L on average, 6 g/L H_2SO_4, and 150 mg/L additives. Zinc was removed in the first column to about 40 mg/L and the additives to about 120 mg/L. By means of ion-exchange, the sulfuric acid concentration was enhanced to nearly 7 g/L. At that time, the second column did not work very well yet, and the zinc concentration could only be decreased to 15 mg/L.

The second wastewater, wastewater B, in-

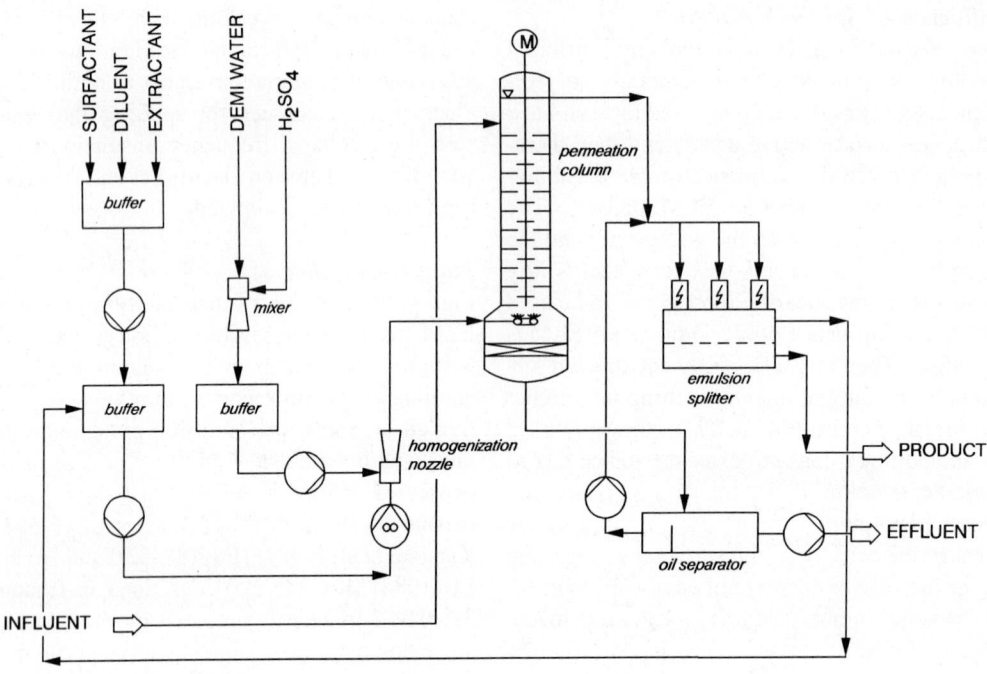

FIGURE 39–2. Schematic of a one-stage ELM plant for zinc removal.

itially contained about one-third of these components. In the first column, zinc could be removed to 15 mg/L. At this time, the second column was already optimized, and the zinc concentration could be further reduced to 0.3 mg/L.

The stripping phase for the second column consisted of 250 g/L H_2SO_4. The initial phase ratio of the wastewater to the stripping phase varied between 300 and 400.

The stripping phase for the first column was prepared from the stripping phase leaving the second stage by upgrading with 96% H_2SO_4 and water to a concentration of 250 g/L. The phase ratio of the wastewater to the stripping phase varied between 150 and 200. With these

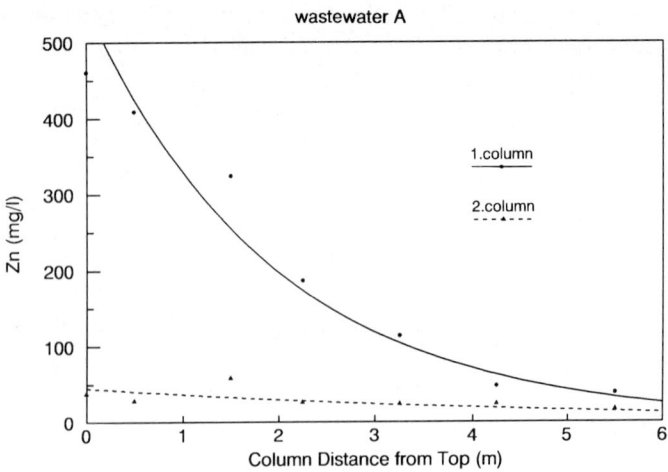

FIGURE 39–3. Extraction column profiles for zinc (wasterwater A).

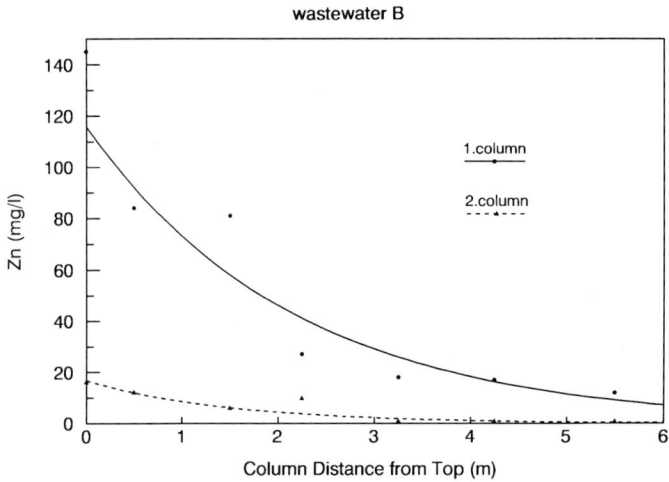

FIGURE 39–4. Extraction column profiles for zinc (wastewater B).

phase ratios a zinc concentration in the stripping phase of 25 to 50 g/L could be achieved. At higher phase ratios higher concentrations would be possible, but in this case the zinc removal in the wastewater would be deteriorated.

According to the phase ratios used, higher zinc concentrations in the stripping phase could be expected. But the stripping phase is diluted due to water transport from the wastewater to the stripping phase (osmosis). Water transport, which is inherent in all membrane processes, is affected by the kinds of extractants and surfactants, as discussed in Chapter 38. This explains why not all surfactants that stabilize the emulsion are suitable for ELM processes.

Besides zinc, the stripping phase also contains up to 3 g/L additives. These will be partly separated when the stripping phase is concentrated further by evaporation prior to reuse in the spinbaths.

Figures 39–5 and 39–6 show all analyses for calcium and magnesium, which are indepen-

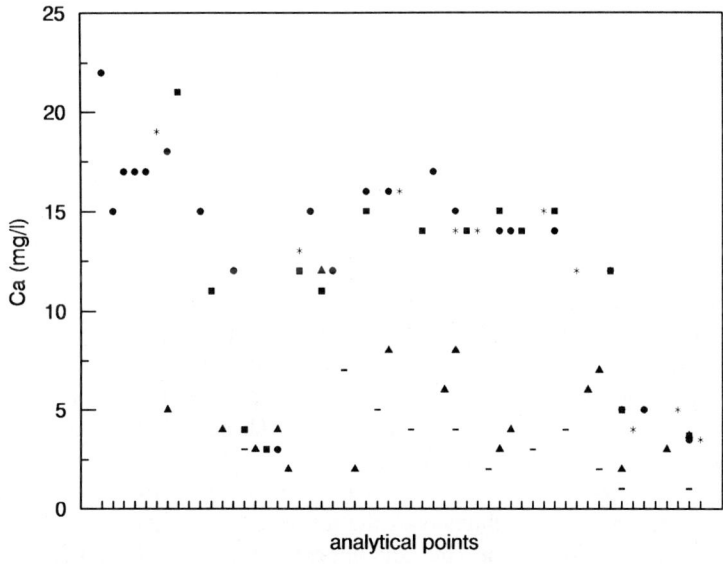

* feed
• raffinate 1
■ raffinate 2
▲ stripping phase 1
− stripping phase 2

FIGURE 39–5. Results for calcium in the ELM extraction of zinc.

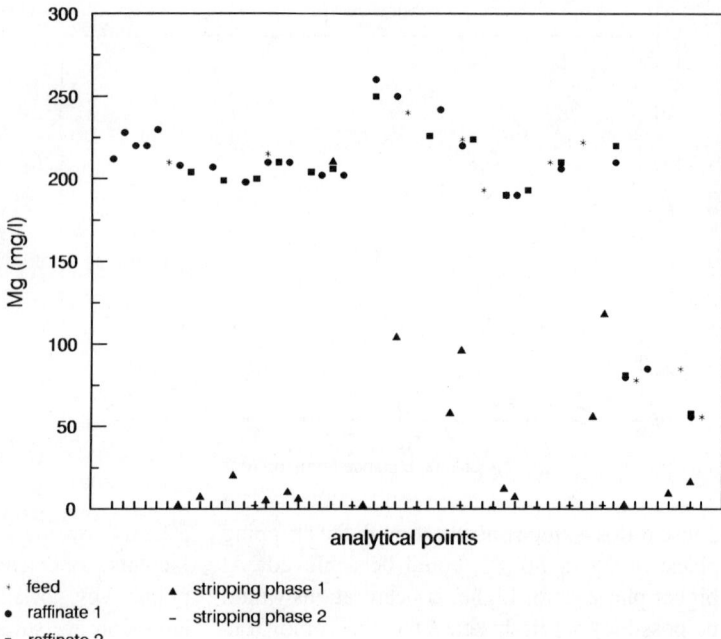

FIGURE 39–6. Results for magnesium in the ELM extraction of zinc.

· feed
● raffinate 1
■ raffinate 2
▲ stripping phase 1
- stripping phase 2

dent of the wastewaters investigated. The concentrations in the raffinates for both columns are about the same as those in the feed phase. In addition, the concentrations in the stripping phase are much lower than those in the feed phase. This is another clear indication that almost no calcium and magnesium are extracted.

RECOVERY OF NICKEL FROM ELECTROPLATING SOLUTIONS

The separation of metal ions from wastewaters in the electroplating industry can be achieved in several ways. Most common is the precipitation of the metals as hydroxides and as sulfides, respectively. Precipitation as sulfides is used when the metals in the wastewater form stable complexes, e.g., NH_3- or EDTA-complexes.

The metal sludges are dried and usually sent to a landfill. That is why recovery processes have become more important in recent years. Reverse osmosis and electrodialysis processes are already widely used; they allow a reuse of the metal concentrates in the plating baths. The disadvantages of these processes are extensive pretreatment and insufficient metal separation in the permeate. If possible, the permeate may be reused in the rinsing baths, but it may not be drained without further treatment.

Solvent extraction with D2EHPA has also been suggested for the extraction of nickel. But if the metal separation by reverse osmosis and electrodialysis is insufficient, it is even more insufficient with SX. SX must work on a pure economic basis, which is determined by the value of the metals recovered. The raffinate still contains metal ions and must be further treated prior to drainage.

For the complete extraction of nickel from wastewaters in the electroplating industry, more powerful extractants than D2EHPA are necessary that allow sufficient extraction even when nickel is complexed. These extractants are the same as those used in the viscose industry, namely, organic dithiophosphoric and dithiophosphinic acids. The extraction equilibria of zinc and nickel are also very similar, as shown in Figure 39–7.

The important difference is that the stripping

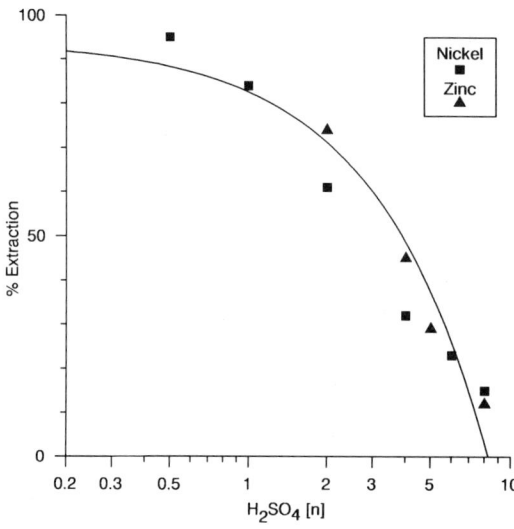

aqueous phase: 1 g/l Zn, Ni
organic phase: 5 wt% DTPA
95 wt% Shellsol T
phase ratio: 1:1

FIGURE 39–7. Extraction equilibria for zinc and nickel.

kinetics is still much slower for nickel than for zinc. Using a mixer-settler device, many hours of residence time would be necessary to partly strip nickel. This led to the development of an ELM process (analogous to Figure 39–2) in which the necessary residence time for the stripping of nickel is enormously reduced due to the large interfacial area. But despite this effect, a residence time of more than 0.5 hour is necessary to reduce nickel sufficiently (<1 mg/L) in the wastewater.

For two reasons this concept was not satisfactory. First, because of the residence time, the column would be too high for the small plants in the electroplating industry, and second, at this long residence time osmosis would be so high that only a dilute stripping phase could be achieved (not more than 20 g/L in the stripping phase at an initial concentration of 400 mg/L in the wastewater). In addition, the physical properties of the emulsion were greatly changed by this large water transport. Therefore, the operation of the plant became difficult.

To overcome the problem of osmosis, a concept already suggested by Cahn and Li (1978) was adapted (Figure 39–8) by Marr et al.

(1988). The wastewater is not in contact with the emulsion until the metals are *entirely* removed; instead, the process of ELM is stopped when the metals are removed to about 90%. For this, a much shorter residence time is necessary, and osmosis is not dramatic. Then the emulsion is split and the wastewater that still contains about 10% of the initial metal concentration is contacted in an extraction step (where no osmosis can occur) with the split, stripped organic phase. At this stage the remaining 10% metals are extracted. The organic phase is not stripped now in a separate stage, but it is directly used for the preparation of the emulsion for the ELM stage. Of course, during the emulsification step the metals extracted in the extraction step will be stripped.

This process was tested extensively, and its feasibility was proven for different kinds of nickel-containing wastewaters. Contrary to the zinc separation in the viscose industry, the organic additives did not interfere with the emulsion, although they are usually present in considerably higher concentrations compared to the viscose industry. Again, part of the organic additives was transported through the membrane phase and concentrated in the stripping phase. Depending on the type of additives, this might affect the direct reuse of the stripping phase in the plating baths. In most cases, the concentrate will be reusable without further treatment. But in cases in which the concentrate will not be reusable, another utilization (e.g., in the metal refining industry) ought to be no problem.

Although this process worked very well, it could be improved further (Marr, Lackner, and Bart 1989). The latest concept is shown in Figure 39–9. It contains the important elements of the ELM process, but it is not a membrane process anymore. The advantages of the ELM process (e.g., large interfacial area) could be preserved, whereas the disadvantages of membrane processes (e.g., osmosis) could be avoided.

The wastewater is contacted countercurrently in a two-stage mixer-settler unit with the organic phase containing 5% of a dithiophosphoric acid in an aliphatic diluent and a small

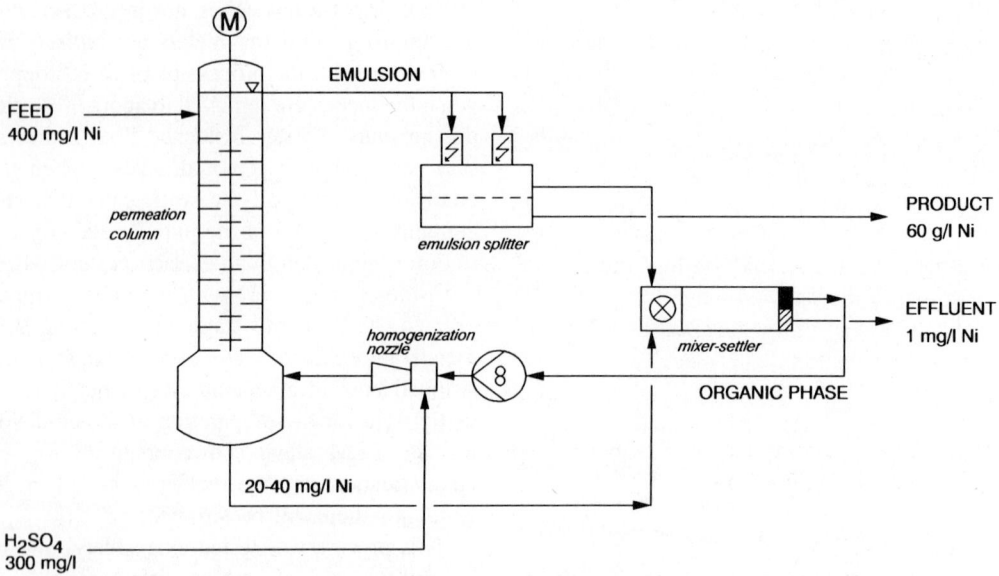

FIGURE 39–8. Combined SX-ELM process for Ni recovery.

amount of surfactant. The surfactant is necessary for the following stripping stage; however, its concentration must remain so low that there is no disadvantageous effect on the extraction step. Due to the strongly acidic extractant DTPA, nickel can be removed in the wastewater down to less than 1 mg/L, independent of the initial concentration (at least up to about 6 g/L).

Stripping of nickel from the organic phase is done in a homogenizer to obtain a large interfacial area. But despite very fine droplets and a specific interfacial area of about 10^6 m^2/m^3, the residence time must be about 15 min to obtain sufficient stripping. Therefore, a small amount of surfactant is necessary to maintain such small droplets for the required time. Separation of these droplets after the stripping step by gravity settling would take too long. Therefore, an electrostatic splitting device is necessary.

The organic additives have negligible effects

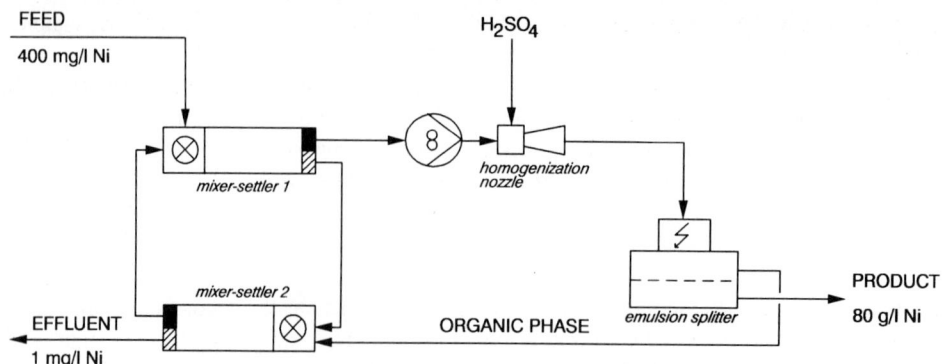

FIGURE 39–9. Improved SX-ELM process for Ni recovery.

on the behavior of the extraction and stripping stages. There is no crud formation, and the settling of the dispersed droplets is only slightly retarded. Acculumation of these additives in the organic phase does not occur.

However, the organic additives have a deleterious effect on the stability of the extractant DTPA. Hydrolysis of DTPA is accelerated by the organic additives, and the half-life at 30°C is less than 2 months. Although DTPA is very cheap, the use of dithiophosphinic acids (e.g., Cyanex 301) has to be considered. Preliminary investigations have shown that the extraction efficiency is about the same; monothiophosphoric and monothiophosphinic acids, however, are not suitable for this process.

OTHER APPLICATIONS

During the 1970s and early 1980s, many possible applications were investigated and proposed for ELM. The investigations centered on the liquid membrane extraction of metals, namely, Cu, Zn, Pb, Cd, Ni, Co, W, V, Cr, Hg, and U. In pilot plant runs for copper (Frankenfeld, Cahn, and Li 1981) and uranium (Hayworth et al. 1983), the technical feasibility of ELM and its economic superiority to SX could be proven. But despite this, no large plant was built, probably because of the decline of metal prices at that time. Then there was simply no need for new copper and uranium plants. Most activities were therefore terminated. A short time later, however, ELM technology was revived when its importance for wastewater treatment was realized.

Removal of Phenols

Besides metal ions, other inorganic and organic substances can be removed from wastewaters by ELM. One example is the separation of phenol, which is based on simple diffusion through the membrane phase without reaction with an extractant. The mechanism is shown schematically in Figure 39–10 [see also Figures 36–4(a) and 37–6].

The pH value of the wastewater must be below 9 in order to keep the phenol in the undissociated form. In this form it is soluble in

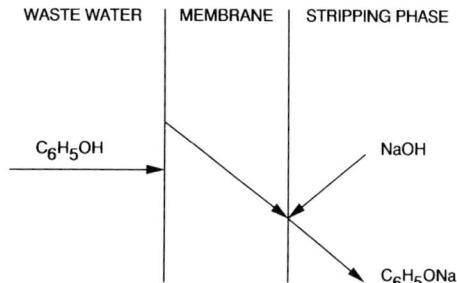

FIGURE 39–10. Mechanism of phenol removal.

the organic membrane phase, which consists only of a surfactant and a diluent. In the stripping phase its reaction with NaOH occurs to form sodium phenolate. The dissociated form of phenol, phenolate, is not soluble in the membrane phase anymore, and it is concentrated in the stripping phase.

Zhang, Liu, and Lu (1987) and Zhang et al. (1988b) have reported on an industrial plant at the Nanchung Plastic Factory in Guangzhou, People's Republic of China, to remove 1000 mg/L phenol in two-stage rotary disk column extractors down to less than 0.5 mg/L. The column extractors are shown (Figure 38.3) and described in Chapter 38. These extractors may be designed/sized by the use of the model of Rautenbach and Machhammer (1988) [Eqs. (38–52) through (38–59)] or the model of Ho and Li (1984) [Eq. (38–35)] described in Chapter 38. The flow rate of the wastewater is 0.25 m^3/h (250 L/h).

The emulsion in this plant consists of a 5 wt.% NaOH solution emulsified in an organic phase consisting of 3.5 wt.% surfactant (LMS-2) and 6.7 wt.% liquid paraffin in kerosene, at a phase ratio of 1:1. The emulsion has been successfully prepared by use of a mechanical preagitator and a colloid mill (Zhang, Liu, and Lu 1987), as described in Chapter 38. It is reported that no osmosis occurs, which is attributed to the new surfactant LMS-2 (Zhang et al. 1988a).

In the stripping phase phenol is concentrated to 55 g/L. Since there is no direct reuse for sodium phenolate in the plant, this stripping phase is incinerated. The economy is reported to be excellent and superior to all other phenol removal processes.

Removal of Ammonia

The mechanism of removal by ELM for ammonia is exactly the same as that for phenols with the only difference being that ammonia is an alkaline substance. Therefore, the pH of the wastewater must be above 10 in order to convert ammonium ions to free ammonia, which is soluble in the membrane phase. Naturally, the stripping phase is now acidic (e.g., sulfuric acid or phosphoric acid), and ammonia is trapped there as ammonium sulfate (or phosphate).

Many publications are available on this subject (e.g., Downs and Li 1981; Schlosser and Kossaczky 1986; Marr and Koncar 1990) in which this process is described in detail together with economic comparisons to other processes. Although the ELM process is clearly advantageous versus other processes, such as air- or steam-stripping, one important disadvantage remains. Contrary to air- or steam-stripping, the product phase can never be an NH_4OH solution; it can only be a solution of an ammonium salt. Should a use arise for the ammonium salt solution (e.g., in fertilizer plants), ELM is a very suitable process for the removal of ammonia from wastewaters.

Removal of Heavy Metals in Metallurgical Plants

Wastewaters in metallurgical plants usually contain a number of different metals that are to be removed all together. As long as there is a use for the stripping phase, ELM is a very effective technique for this task. In metallurgical plants this is usually the case since the stripping phase can be used for leaching of the metal ores.

Table 39–2 shows the results of a two-stage pilot plant of 0.2 m^3/h (200 L/h), which was operated for 6 months in the zinc-producing industry. For these experiments MTPA was used instead of DTPA. These two extractants have about the same extraction efficiency for all metals [except Ni and to a minor extent Fe(III), Co, and Mn]. The main difference lies again in the stripping behavior. Under the conditions used, Cd, Cu, and Pb would not be stripped from DTPA by diluted sulfuric acid.

As shown in Table 39–2, the important metals, Zn, Cd, Cu, and Pb, were removed sufficiently and concentrated in the stripping phase. Pb was also completely stripped by H_2SO_4, but it precipitated as $PbSO_4$ in a finely dispersed, crystalline form. Fe(III) was partly removed, but all other metals were not ex-

TABLE 39–2. Results of an ELM Pilot Plant in the Zinc-Producing Industry.

Metal	Feed (mg/L)	Raffinate 1 (mg/L)	Raffinate 2 (mg/L)	Stripping Phase (mg/L)
Zinc	230	5	0.2	22,000
Cadmium	2.7	0.1	0.02	270
Copper	1.1	0.1	0.007	90
Lead	0.5	0.02	0.01	6
Iron (II)	90	90	90	20
Iron (III)	20	15	12	700
Cobalt	0.4	0.4	0.38	0.1
Nickel	0.3	0.3	0.3	0.1
Manganese	25	24	23	20
Magnesium	37	37	37	4
Calcium	220	220	220	44
Sodium	60	60	60	6

Operating conditions: (1) 0.2 m^3/h (200 L/h) wastewater, pH 3.4. (2) 0.01 m^3/h (10 L/h) organic phase containing 5% extractant (MTPA), 3% surfactant (ECA 11522), and 92% diluent (Shellsol T). (3) 0.001 m^3/h (1L/h) stripping phase (250 g/L H_2SO_4).

tracted at all. The fact that Ca, Mg, and Na are not extracted is advantageous, but it might be necessary for Co, Ni, and Mn to be removed. Two possibilities exist to achieve this:

1. The pH in the second column is increased and maintained above 4. At this pH, Co, Ni, and Mn are extracted, but Ca and Mg are also partly extracted.
2. The pH is only 3 to 3.2 and DTPA is used in the second step. This situation is more convenient because Ca and Mg are not extracted at this pH. Under this situation, Cd, Cu, and Pb are not stripped, but their initial concentrations in the second stage (the raffinate of first stage) are so low that accumulation of the unstripped Cd, Cu and Pb complexes in the organic phase could only occur over a long period of time. Actually, there will be no accumulation at all because the slight loss of the extractant due to degradation and solubility has to be replaced by the fresh extractant, and this is more than the blockage of the extractant by nonstrippable Cd, Cu, and Pb complexes. However, this would not be the case if DTPA were to be used in the first stage.

Removal of Heavy Metals in Incineration Plants

By leaching the solid residues in incineration plants with the acidic waters of the gas purification plant, metal-containing wastewaters will result. The solid residues can then be disposed as nonhazardous wastes, but the metal-containing wastewaters will have to be treated. This is another promising application for ELM since the same metals are present as in metallurgical plants, in addition to mercury. What makes it more difficult is that there is no direct reuse for the concentrated stripping phase. It has to be sold to the metallurgical industry or treated otherwise.

To facilitate the sale of the stripping phase, selective separation of the metal ions was tried (Draxler, Bart, and Marr 1988). This leads to the concept shown in Figure 39–11. As shown in this figure, in the first stage mercury is removed selectively by an ion-exchange resin. By stripping with HCl, a pure mercury solution can be obtained. Then the wastewater flows to the ELM plant, where all other heavy metals are extracted simultaneously by an emulsion consisting of H_2SO_4 (200 g/L) as the stripping phase emulsified in an organic membrane phase

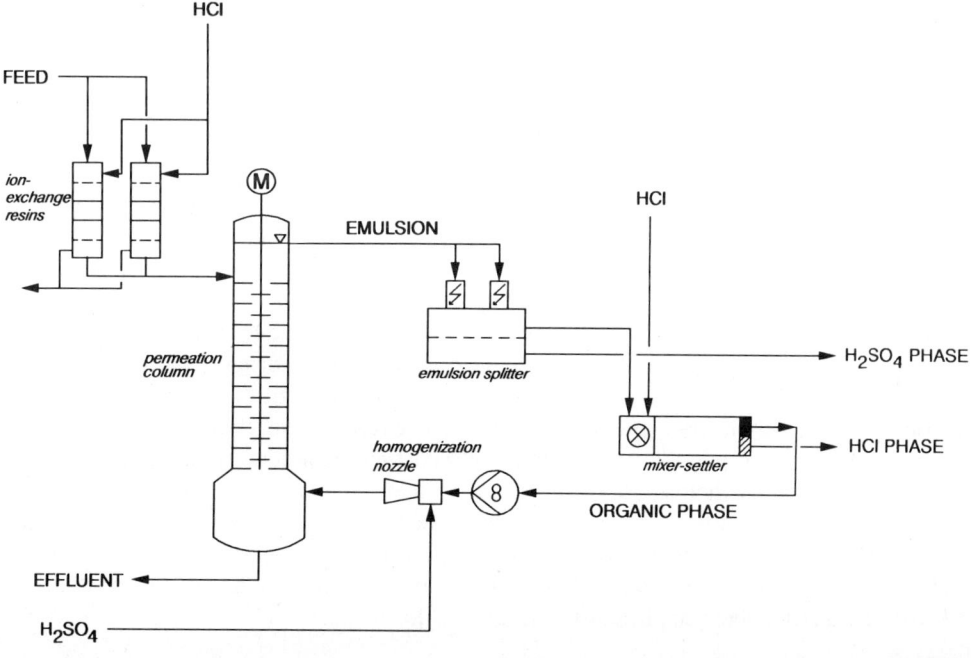

FIGURE 39–11. ELM plant with selective stripping.

TABLE 39-3. Results of an ELM Pilot Plant for Waste Incineration.

Metal	Feed (mg/L)	Raffinate (mg/L)	H_2SO_4 Phase (mg/L)	HCl Phase (mg/L)
Zinc	600	90	30,000	200
Cadmium	27	<0.1	320	1800
Copper	14	<0.1	23	1550
Lead	34	<0.1	6	14
Aluminium	171	169	34	
Iron	18	17	45	
Calcium	2230	2200	80	160

containing MTPA as the extractant. When MTPA is only partly loaded, zinc and lead will be stripped in preference to cadmium and copper by diluted H_2SO_4 (cadium and copper will only be stripped when MTPA is already loaded to more than 50%). Therefore, a stripping phase containing mainly zinc and only traces of the other metals is obtained after splitting the emulsion. This stripping phase can be sold to the metallurgical industry.

The split organic phase still remains loaded with cadmium and copper. In the pilot tests in the hydrometallurgical industry, these metals accumulated to a certain level and were then stripped. This accumulation is avoided here by stripping these metals continuously by hydrochloric acid in a mixer-settler stage. There will probably be no use for this hydrochloric stripping phase; therefore, it might be necessary to neutralize the stripping acid and precipitate the metals. The overall economics nevertheless look very promising because this metal hydroxide sludge, which has to be disposed as hazardous waste, is very small in volume. Table 39-3 shows some results of pilot plant studies in a waste incineration plant.

Biochemical Applications

In addition to the above-mentioned applications, a wide range of possible applications exist for ELM in the biochemical and biomedical field. Although no pilot plant experiments have been reported so far, a number of interesting publications have shown the feasibility of the ELM technique for these applications. Among these are the work of Thien, Hatton, and Wang (1986), Plucinski et al. (1986), Yagodin, Yurtov, and Golubkov (1986), and Thien and Hatton (1988) for the separation of amino acids; Terry, Li, and Ho (1982) for the recovery of acetic acid and propionic acid; Boey, del Cerro, and Pyle (1987) for the extraction of citric acid; Ho and Cowan (1987) and Reisinger, Marr, and Preitschopf (1990) for the recovery of lactic acid; Hano et al. (1990) and Scheper et al. (1986) for the separation of penicillin; Davis (1981) for drug delivery systems; and Yagodin et al. (1983) for the removal of cholesterol from blood. The immobilization of live cells and enzymes in the stripping phase was investigated by Mohan and Li (1975) and Makryaleas et al. (1985).

A major part of the costs of fermentation products is the recovery of these products from fermentation broths. Typically, the recovery comprises many steps such as precipitation, filtration, bleaching, purification, and evaporative crystallization. The ELM technique could therefore offer an attractive alternative to the conventional recovery of fermentation products. ELM can be applied as an *in situ* technique during the fermentation step or afterward as a downstream process.

The use of ELM *in situ* with a fermentation or enzymatic reaction is very attractive since continuous product removal could relieve the kinetic inhibition that is often encountered in such systems. But it seems that the disadvantages outweigh the advantages of an *in situ* operation. Although some authors such as Mohan and Li (1975) and Boey, del Cerro, and

Pyle (1987) found a good biocompatibility of the ingredients of ELM with various microorganisms, the toxicity of the ingredients to microorganisms remains a severe problem. Wang (1988) examined potential extractants for the *in situ* extraction of lactic acid. He found that Alamine 336 (or trioctylamine) was extremely toxic to the growth of *Lactobacillus delbrueckii;* 100% growth inhibition resulted at 30% Alamine dissolved in oleyl alcohol.

There are also some other drawbacks of an *in situ* operation. The pH required for the liquid membrane extraction is rarely the same as the one required for the fermentation. Coextraction of nutrient media might also occur as well as the loading and subsequent blocking of the extractant by denaturated proteins.

Because of these restrictions, a downstream process should be preferred to an *in situ* process. From the publications cited at the beginning of this section, one can summarize two major problems that have to be faced when using ELM as a downstream process: the first is the concentration of the solute in the feed phase and the second the required selectivity.

Few authors used real fermentation broths for their experiments and even fewer used solute concentrations of commercial interest. As could be seen in the applications for metal separations, ELM is particularly suitable for low-concentration solutions, and many authors (e.g., Terry, Li, and Ho 1982 and Plucinski et al. 1986) therefore used low-concentration (<1 g/L) synthetic solutions for their investigations. Real fermentation broths, however, usually contain much higher concentrations (e.g., 3 to 13% for acidic acid and up to 70% for lactic acid). The feasibility of the ELM technique has never been investigated for these high concentrations.

The most interesting application of ELM seems to be the separation of low-concentration bioproducts and by-products. To compete with other processes for this type of application, the selectivity of ELM is of decisive importance. The extractants used for bioseparations are the same as those for metal separations. These extractants have been developed specifically for metal separations, thus they are not selective enough for bioseparations. A main task for the future therefore is the development of new selective extractants for bioproducts. Another way to enhance the selectivity is the use of solute-specific driving forces. By the enzymatic conversion of the solute to be separated to the form that cannot diffuse through the membrane phase of the ELM in the stripping phase, only this solute can be concentrated in this phase.

For applications in biotechnology, ELM is still in its infancy. Pilot plant tests with realistic source phases are urgently needed to obtain better comparisons versus the well-established methods of bioseparations. What can be deduced from the published work is that ELM is not a universal process for bioseparations and that the systems where ELM can be applied advantageously have to be chosen with care. New applications for ELM will have to be examined as well as new methods. Such new methods could be the use of solid membranes in the fermentor to separate the fermentation and ELM steps. In this case the solid membrane must be permeable for the product and impermeable for the microorganisms, by-products, and liquid membrane components. Another improvement could be the transport of solutes through the organic membrane phase by reversed micelles as shown in Figure 39–12 (Thien and Hatton 1988). The use of the re-

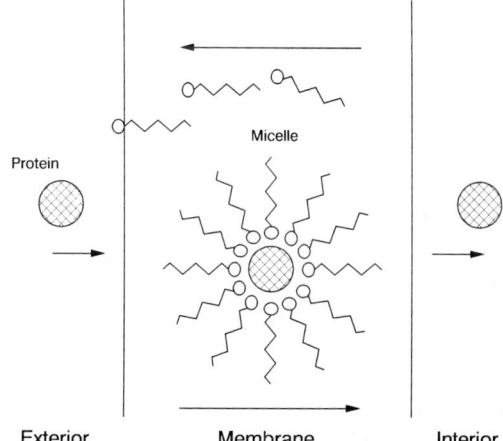

FIGURE 39–12. Combined reversed micelle and ELM system for protein extraction (reprinted from Thien and Hatton 1988 with permission).

versed micelles is suggested for the separation and purification of polyionic solutes such as proteins and polypeptides (Rahaman and Hatton 1988). Without the micelles, ELM is not suitable for extracting proteins because the proteins would be denatured when transported through the organic membrane phase.

REFERENCES

Bart, H. J., R. Wachter, R. J. Marr, and H. J. Müller. 1988. Mass transfer in a permeation column. *J. Membr. Sci.* 36:413–423.

Boey, S. C., M. C. G. del Cerro, and D. L. Pyle. 1987. Extraction of citric acid by liquid membrane extraction. *Chem. Eng. Res. Des. (I. Chem. E.)* 65:218–223.

Cahn, R. P., and N. N. Li. 1978. Metal extraction by combined solvent and LM extraction. U.S. Patent 4,086,163.

Davis, S. S. 1981. Liquid membranes and multiple emulsions. *Chem. Ind.* 10:683–687.

Downs, H. H., and N. N. Li. 1981. Extraction of ammonia from municipal wastewater by the liquid membrane process. *J. Sep. Process Technol.* 2(4):19–24.

Draxler, J., H. J. Bart, and R. J. Marr. 1988. Separation of zinc, cadmium and lead by a combined extraction/liquid membrane process. In *Proceedings of the First International Conference on Hydrometallurgy*, ed. Y. Zheng and J-Z. Xu, pp. 560–563. Beijing: International Academic Publishers.

Draxler, J., R. J. Marr, and M. Prötsch. 1988. Commercial-scale extraction of zinc by emulsion liquid membranes. In *Separation Technology*, ed. N. N. Li and H. Strathmann, pp. 204–214. New York: United Engineering Trustees.

Frankenfeld, J. W., R. P. Cahn, and N. N. Li. 1981. Extraction of copper by liquid membranes. *Sep. Sci. Technol.* 16(4):385–402.

Hano, T., T. Ohtake, M. Matsumoto, F. Hori, and F. Nakashio. 1990. Separation and concentration of fermented products by liquid surfactant membrane process. In *Proceedings of Asia-Pacific Biochemical Engineering Conference '90*, 22–25 April, 1990, Kyungju, Korea, pp. 428–431.

Hayworth, H. C., W. S. Ho, W. A. Burns, and N. N. Li. 1983. Extraction of uranium from wet process phosphoric acid by liquid membranes. *Sep. Sci. Technol.* 18(6):493–521.

Ho, C. S., and R. M. Cowan. 1987. Separating lactic acid from fermentation media with liquid surfactant membranes. Paper read at the 194th National ACS Meeting, 30 August–4 September 1987, New Orleans, LA.

Ho, W. S., and N. N. Li. 1984. Modeling of liquid membrane extraction processes. In *Hydrometallurgical Process Fundamentals*, ed. R. G. Bautista, pp. 555–597. New York: Plenum Press.

Kataoka, T., T. Nishiki, M. Yamauchi, and Y. Zhong. 1987. A simulation for liquid surfactant membrane permeation in a continuous countercurrent column. *J. Chem. Eng. Japan* 20(4):410–415.

Makryaleas, K., T. Scheper, K. Schugerl, and M. G. Kula. 1985. Enzymkatalysierte Darstellung von 1-aminosäure mit kontinuierlicher Coenzymregenerierung mittels flüssig-Membran-Emulsionen. *Chem.-Ing.-Tech.* 57(4):362–363.

Marr, R. J., H. J. Bart, and J. Draxler, 1990. Liquid membrane permeation. *Chem. Eng. Process.* 27: 59–64.

Marr, R. J., and M. Koncar. 1990. Rückgewinnung von Ammoniak aus Industrieabwasser. *Chem.-Ing.-Tech.* 62(3):175–182.

Marr, R. J., H. Lackner, and H. J. Bart. 1989. Verfahren zur Abtrennung von Nickel aus Nickelionen enthaltenden verdünnten wäßrigen Lösungen. European Patent Application 89112656.7.

Marr, R. J., H. Lackner, H. J. Bart, and A. Nickl. 1988. Verfahren zur Abtrennung von Metallionen aus wäßrigen Lösungen. European Patent 88/00637.

Marr, R. J., H. Lackner, and K. Hütter. 1988. Apparatus for mixing media capable to flow. European Patent Application 88810472.6.

Mohan, R. R., and N. N. Li. 1975. Nitrate and nitrite reduction by liquid membrane encapsulated whole cells. *Biotech. Bioeng.* 17:1137–1156.

Plucinski, P., P. Kafarski, B. Lejczak, and M. Cichocki. 1986. The permeation of phosphonodipeptides through liquid membranes. In *Proceedings ISEC'86*, Munich, Vol. III, pp. 669–676. Frankfurt: DECHEMA.

Rahaman, R. S., and T. A. Hatton. 1988. The separation of polyionic solutes using reversed micelles. In *Separation Technology*, ed. N. N. Li and H. Strathmann, pp. 168–189. New York: United Engineering Trustees.

Rautenbach, R., and O. Machhammer. 1988. Modeling of liquid membrane separation processes. *J. Membr. Sci.* 36:425–444.

Reisinger, H. H., R. J. Marr, and W. Preitschopf. 1990. Liquid membrane permeation in the fields of biotechnology. In *Proceedings of the 1990 International Congress on Membranes and Membrane Processes,* 20–24 August, 1990, Chicago, IL, Vol. I, pp. 715–717.

Scheper, T., K. Makryaleas, Z. Likidis, C. Nowottny, E. R. Meyer, and K. Schugerl. 1986. Liquid surfactant membrane emulsions—a new technique for enzyme engineering. In *Proceedings ISEC'86,* Munich, Vol. III, pp. 695–702. Frankfurt: DECHEMA.

Schlosser, S., and E. Kossaczky. 1986. Pertraction of ammonia from wastewaters. In *Proceedings ISEC'86,* Munich, Vol. III, pp. 935–942. Frankfurt: DECHEMA.

Terry, R. E., N. N. Li, and W. S. Ho. 1982. Extraction of phenolic compounds and organic acids by liquid membranes. *J. Membr. Sci.* 10(2&3):305–323.

Thien, M. P., and T. A. Hatton. 1988. Liquid emulsion membranes and their applications in biochemical processing. *Sep. Sci. Technol.* 23(8&9):819–853.

Thien, M. P., T. A. Hatton, and D. I. C. Wang. 1986. Liquid emulsion membranes and their applications in biochemical separations. *ACS Symp. Ser.* 314:67–77.

van Dijk, J. C., P. J. de Moel, and M. Schöller. 1985. Recovery of heavy metals by means of a pellet reactor. DHV Consulting Engineers, 3800 AB Amersfoort, The Netherlands.

Wang, D. I. C. 1988. Integrated bioprocesses. In *Proceedings Biotechnica,* 20–22 September 1988, Hannover, Germany, pp. 37–40.

Yagodin, G., Y. Lopukhin, E. Yurtov, T. Guseva, and V. Sergienko. 1983. Extraction of cholesterol from blood using liquid membranes. In *Proc. Int. Solv. Extr. Conf., SEC'83,* 26 August–2 September 1983, Denver, CO, pp. 385–386.

Yagodin, G. A., E. V. Yurtov, and A. S. Golubkov. 1986. Liquid membrane extraction of aminoacids. In *Proceedings ISEC'86,* Munich, Vol. III, pp. 677–683. Frankfurt: DECHEMA.

Zhang, X.-J, Q.-J. Fan, X.-T. Zhang, and Z.-F. Liu. 1988a. New surfactant LMS-2 used for industrial application in liquid membrane separation. In *Separation Technology,* ed. N. N. Li and H. Strathmann, pp. 215–226. New York: United Engineering Trustees; see also *Water Treatment* 3:233–240.

Zhang, X.-J., J.-H. Liu, Q.-J. Fan, Q.-T. Lian, X.-T. Zhang, and T.-S. Lu. 1988b. Industrial application of liquid membrane separation for phenolic wastewater treatment. In *Separation Technology,* ed. N. N. Li and H. Strathmann, pp. 190–203. New York: United Engineering Trustees.

Zhang, X.-J., J.-H. Liu, and T.-S. Lu. 1987. Industrial application of liquid membrane separation for phenolic wastewater treatment. *Water Treatment* 2:127–135.

40

Capital and Operating Costs

Rolf J. Marr
Josef Draxler
Technical University of Graz

CAPITAL AND OPERATING COSTS
   Basis of the Calculation
   Capital Costs
   Operating Costs
   Return on Investment

EFFECTS OF OPERATING VARIABLES ON CAPITAL AND OPERATING COSTS
NOTATION
REFERENCES

CAPITAL AND OPERATING COSTS

Contrary to most other membrane processes, emulsion liquid membrane (ELM) technology is not based on modules but is tailor-made for each specific problem. Therefore, it does not make sense to give prices, for example, in dollars per square meter of membrane area. Even figures in dollars per cubic meter of treated wastewater cannot be given in a general way because it depends on a number of parameters. To supply an estimate of the costs of an ELM plant, a complete calculation of a plant for the viscose industry according to Figure 39–2 is presented in this chapter (prepared in cooperation with VOEST-Alpine Industrieanlagenbau GmbH). Some effects of flow rate and metal concentration on capital and operating costs are given in the next section.

Other calculations are given, for instance, by Frankenfeld, Cahn, and Li (1981) for a copper plant and by Hayworth et al. (1983) for a uranium plant.

Basis of the Calculation

By combining several different wastewater streams in a viscose plant, the following mixed stream will be obtained as influent for the ELM plant:

Flow rate	75 m^3/h
Zn concentration	300 mg/L
H$_2$SO$_4$ concentration	4–5 g/L
Mg concentration	150 mg/L
Ca concentration	10 mg/L
Fe concentration	2 mg/L
Suspended solids	20 mg/L
Spinbath additives	50 mg/L

Removal of 95% of the zinc and none of the calcium and magnesium is required. Therefore, the effluent (raffinate) will contain the following concentrations:

Zn	15 mg/L
H$_2$SO$_4$	5–6 g/L

Mg	150 mg/L
Ca	10 mg/L
Fe	1 mg/L
Spinbath additives	35 mg/L
Organic membrane phase	3 mg/L

At an initial stripping acid concentration of 250 g/L, the following concentrations will be obtained in the product phase:

Zn	40–50 g/L
H_2SO_4	150 g/L
Mg, Ca, Fe	Traces
Spinbath additives	2 g/L

Capital Costs

All of the following data are based on Spring 1990 quotations. The conversion to US$ is done by assuming these exchange rates: (1) 1 US$ = 1.8 DM (German marks) and (2) 1 US$ = 13 ATS (Austrian shillings). The capital costs are as follows in US$:

Extraction column including internals, fiberglass-reinforced material, plus all storage tanks	230,000
Filters for feed stream and oil separator	30,000
Splitting apparatus (enamel-insulated electrodes) plus electronic device	150,000
Pumps, homogenizer, and static mixers	77,000
Process measuring and control equipment including measuring station	230,000
Electrotechnical devices	100,000
Pipes	210,000
Buildings (assumption: everything new, no existing buildings available)	420,000
Steel structure	130,000
Insulation and paintings	53,000
Erection and assembly	380,000
Supervision for buildings and erection	130,000
Startup	70,000
Engineering	600,000
Licenses	80,000
TOTAL	2,890,000

Operating Costs

For the assessment of the industrial economics, the following costs and credits are accounted for:

1. Direct operating costs of the plant: included are costs for the further treatment of the product phase (evaporation); not included are costs for the neutralization of the raffinate, which possibly has to be accounted for in other plants.
2. Staff and other (analysis) costs.
3. Credits for the recycled zinc.
4. Credits due to opportunity costs for other treatments to reduce the zinc concentration.

These costs and credits are discussed in the following.

Direct Operating Costs

Sulfuric Acid
Sulfuric acid is needed for ion exchange with the zinc ions. Theoretically, for every 1 kg of zinc, 1.5 kg of sulfuric acid is needed. In practice, more is needed because other substances are also extracted, particularly spinbath additives, which also consume sulfuric acid, and because of a small part of emulsion breakup (the release of stripping acid to the wastewater due to membrane instability). As could be seen in various pilot tests, 1.8 kg of sulfuric acid per 1 kg of zinc is a realistic value. This yields 75 m^3/h × 0.285 kg/m^3 × 8760 h/yr × 1.8 = 337,000 kg/yr sulfuric acid, i.e., 337,000 kg/yr × 0.082 US$/kg = 27,600 US$/yr.

The stripping phase contains more sulfuric acid (0.375 m^3/h × 250 kg/m^3 × 8760 h/yr = 820,000 kg/yr) than is needed for the ion exchange, but this additional amount (820,000 − 337,000 = 483,000 kg/yr) is not accounted for in the operating costs because it is recovered in the product phase and it is reused in the spinbaths.

Water
Water is needed only for dilution and cooling of sulfuric acid preparation. It is neglected in the cost estimation.

Organic Membrane Phase
The organic membrane phase is lost due to entrainment and solubility and due to the degradation of the extractant. Furthermore, for the cost estimation, we assume that the entire organic phase is replaced every other year.

Entrainment of the organic phase to the wastewater is neglected because efficient oil separators are able to prevent any entrainment. Several independent analyses [our own analysis, Dutch authorities for the pilot tests (internal report 1989), and Poppe 1987] have shown that the total physical loss of the organic phase is about 2 to 3 mg/L (1 mg/L = 1 g/m^3). The assumption that this is due to solubility and not entrainment therefore seems to be justified. Thus, the maximum physical loss is

$$3 \text{ g/m}^3 \times 75 \text{ m}^3/\text{h} \times 8760 \text{ h/yr} =$$
$$2000 \text{ kg/yr organic phase.}$$

According to the initial composition at 0.95$/kg, the maximum physical loss is

$$2000 \text{ kg/yr} \times 0.95 \text{ US\$/kg} = 1900 \text{ US\$/yr.}$$

The total input of the organic phase is

diluent	9,300 kg (0.83 US$/kg)
extractant	500 kg (2.7 US$/kg)
surfactant	200 kg (2.3 US$/kg)
TOTAL	10,000 kg (0.95 US$/kg)

For the replacement every 2 years, half of this has to be calculated for the annual consumption:

$$5000 \text{ kg/yr} \times 0.95 \text{ US\$/kg} = 4750 \text{ US\$/yr.}$$

For the degradation of the extractant, the half-life of the extractant at 30°C is about 6 months. This means that an additional 500 kg/yr has to be replaced:

$$500 \text{ kg/yr} \times 2.7 \text{ US\$/kg} = 1350 \text{ US\$/yr.}$$

Electric Energy
The total electric power of this plant is about 40 kWh. The energy consumption for 1 year is therefore 350,000 kWh:

$$350,000 \text{ kWh/yr} \times 0.054 \text{ US\$/kWh} =$$
$$18,900 \text{ US\$/yr.}$$

Further Treatment
The product phase containing 40 to 50 g/L of zinc and about 150 g/L of sulfuric acid is further concentrated to saturation by evaporation. The costs of 0.085 US$/kg zinc are based on a steam price of 15.4 US$/1000 kg:

$$0.085 \text{ US\$/kg Zn} \times 21.375 \text{ kg/h} \times$$
$$8760 \text{ h/yr} = 15,900 \text{ US\$/yr.}$$

(The 21.375 kg/h Zn results from the assumption of 95% separation: 300 mg/L × 0.95 × 75 m^3/h = 21.375 kg/h.) The total direct operating costs are therefore 70,000 US$/yr.

Staff and Analysis Costs
Staff Costs
Due to the high automation of the plant and the integration of the measuring station into an existing measuring station, the staff costs are only one-quarter person per shift, that is 2190 h/yr:

$$2190 \text{ h/yr} \times 20 \text{ US\$/h} = 43,800 \text{ US\$/yr.}$$

Analysis
Only a few analyses are needed for the operation of the plant: how many analyses will be made depends very much on the requirements of the local environmental protection authorities. For this plant, we calculate that 33 US$/day will be necessary for analyses, that is,

$$33 \text{ US\$/day} \times 365 \text{ days/yr} = 12,000 \text{ US\$/yr.}$$

Credits for the Recycled Zinc
The following amount of zinc,

$$0.285 \text{ kg Zn/m}^3 \times 75 \text{ m}^3/\text{h} \times 8760 \text{ h/yr} =$$
$$187,200 \text{ kg Zn/yr}$$

will be recovered and reused in the spinbaths. At a zinc price of 1.61 US$/kg Zn, this yields

187,200 kg Zn/yr × 1.61 US$/kg Zn = 301,400 US$/yr.

Credits Due to Opportunity Costs

All over the world viscose plants are no longer allowed to drain wastewater containing zinc without any treatment. The costs for other treatments have to be compared with the ELM process. In particular, the costs for the deposition of a precipitated zinc sludge in a landfill have to be accounted for in the calculations for an ELM plant. Assuming the costs for the deposition are 0.69 US$/kg Zn, the credits for the ELM process are

187,200 kg Zn/yr × 0.69 US$/kg Zn = 129,000 US$/yr.

Return on Investment

The return on investment in US$ is as follows:

Direct operating costs	− 70,400
Staff costs	− 43,800
Analysis costs	− 12,000
Credits for the recovered zinc	+ 301,400
Credits due to opportunity costs	+ 129,000
TOTAL SAVINGS	+ 304,200
Insurance (0.1% of capital costs)	− 2,890
Maintenance (2% of capital costs)	− 57,800
Interests (5% of capital costs)	− 144,500
NET SAVINGS	+ 99,000

The payback time is, therefore, 2,890,000/99,000 = 29 years.

Comments

An ELM plant is an environmental protection plant for many applications; therefore, calculation of a payback time is not convenient. The costs for such a plant ought to be integrated into a production plant and accounted for in the calculation of the payback time of the production plant. The calculation described above serves only for comparison versus other processes, and the absolute figure of the payback time is of no importance. But contrary to many other environmental protection processes, at least the payback time for the ELM process is positive.

If an ELM process were considered as a production plant, the payback time has to be short. From the previous calculations, we can assume that the payback time is much shorter when the metal concentration is higher (see the next section) or the value of the metal is higher, as shown by Frankenfeld, Cahn, and Li (1981) and Hayworth et al. (1983).

EFFECTS OF OPERATING VARIABLES ON CAPITAL AND OPERATING COSTS

It is evident that a number of parameters influence capital and operating costs. In this section the effects of flow rate and zinc concentration on savings and on payback time are investigated. These effects are based on the calculations of the viscose plant in the previous section.

The feed flow rate Q is varied from 20 to 200 m^3/h and the zinc concentration c in the feed from 20 to 1000 mg/L (g/m^3). The following scale-up considerations are for the calculation of different plant sizes and zinc concentrations:

1. *Capital costs:* For a scale-up factor of 0.55:

capital costs for plant with the flow rate Q = $(Q/75)^{0.55}$ × capital costs for the plant with the flow rate of 75 m^3/h.

2. *Operating costs:*
   a. sulfuric acid: directly proportional to flow rate and zinc concentration
   b. organic membrane phase:
      • solubility: independent of zinc concentration, directly proportional to flow rate
      • input organic phase: independent of zinc concentration, proportional to $(Q/75)^{0.9}$

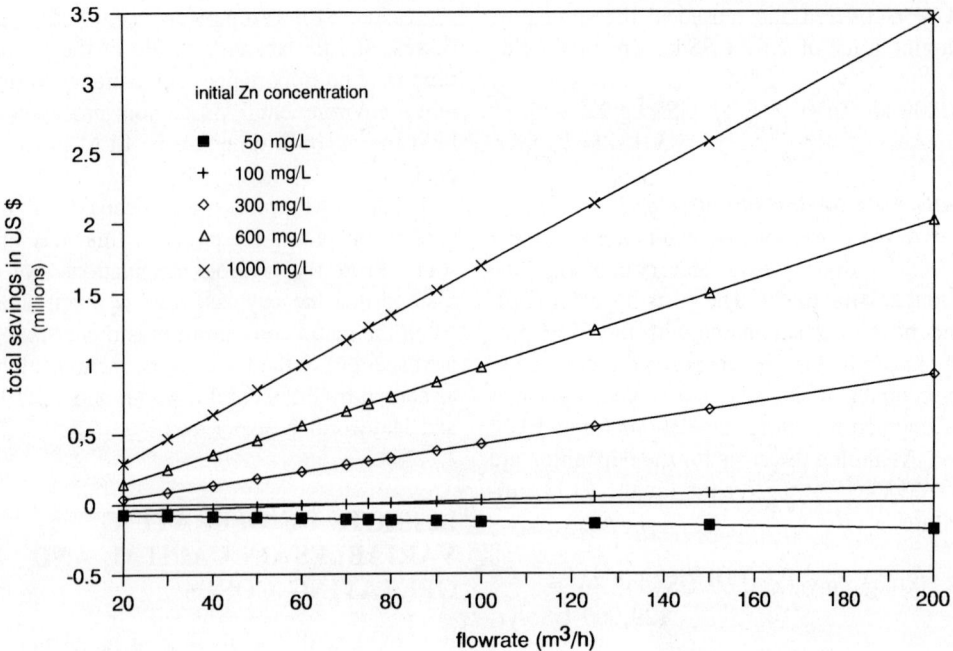

FIGURE 40–1. Dependence of total savings on flow rate and zinc concentration.

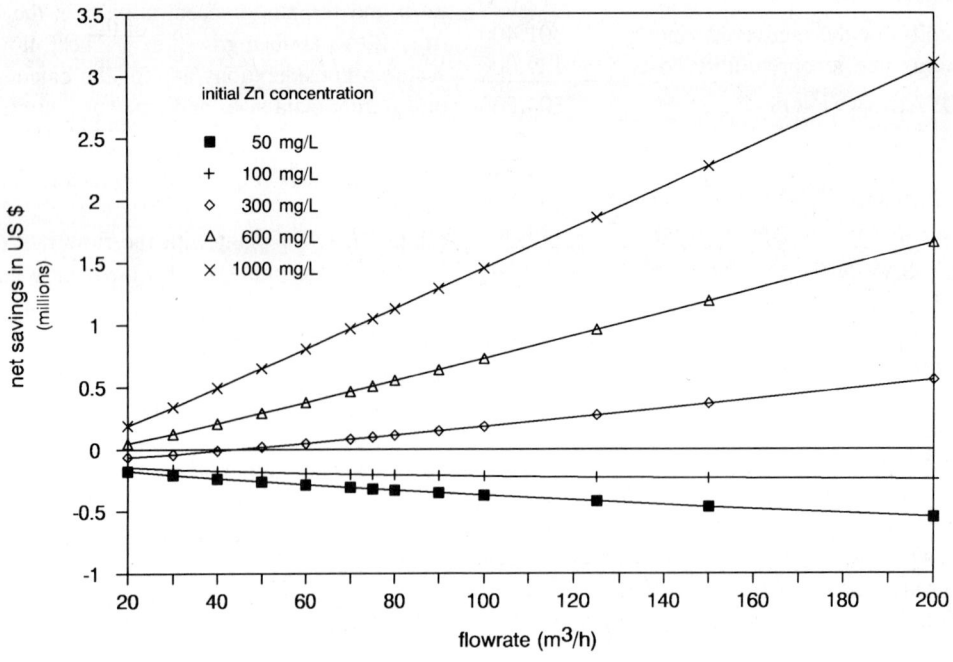

FIGURE 40–2. Dependence of net savings on flow rate and zinc concentration.

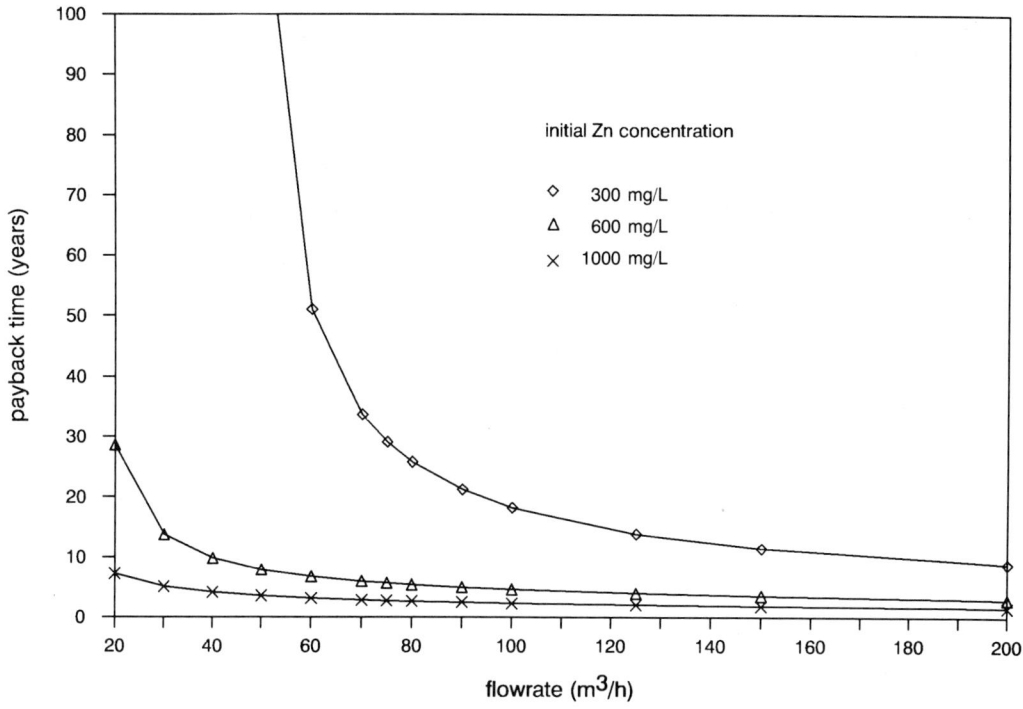

FIGURE 40–3. Dependence of payback time on flow rate and zinc concentration.

- degradation of extractant: independent of zinc concentration, proportional to $(Q/75)^{0.9}$
c. electric energy: independent of zinc concentration, proportional to $(Q/75)^{0.8}$
d. evaporation: directly proportional to flow rate, indirectly proportional to zinc concentration
e. staff and analysis costs: independent of flow rate and zinc concentration.
3. *Credits for the recycled zinc:* Directly proportional to flow rate and zinc concentration.
4. *Credits due to opportunity costs:* Directly proportional to flow rate and zinc concentration.

Figure 40–1 shows the dependence of total savings on flow rate and zinc concentration. As can be expected, the higher the flow rate and zinc concentration, the higher the total savings.

Figure 40–2 illustrates the dependence of net savings on flow rate and zinc concentration. In this figure, the lines for the net savings show smaller slopes and smaller differences between them than those in Figure 40–1. This is due to the fact that the dependence of the capital costs on the flow rate is accounted for in Figure 40–2.

Finally, Figure 40–3 shows the dependence of payback time on flow rate and zinc concentration. As calculated in the previous section, the payback time for a plant with 75 m³/h and 300 mg/L Zn is about 29 years. At a smaller plant size, the payback time will increase sharply, whereas the payback time for larger plants will decrease only gradually. Doubling the zinc concentration in the feed would decrease the payback period from 29 to 7 years, whereas the payback time for the zinc concentrations of less than 100 mg/L is always more than 100 years, if there is any at all.

NOTATION

General Notation

See the General Notation section at the beginning of this handbook.

Special Notation

Q feed volumetric flow rate, L^3/t (m³/h)

REFERENCES

Frankenfeld, J. W., R. P. Cahn, and N. N. Li. 1981. Extraction of copper by liquid membranes. *Sep. Sci. Technol.* 16(4):385–402.

Hayworth, H. C., W. S. Ho, W. A. Burns, and N. N. Li. 1983. Extraction of uranium from wet process phosphoric acid by liquid membranes. *Sep. Sci. Technol.* 18(6):493–521.

Poppe, W. 1987. Aufbereitung saurer Grubenwässer durch Flüssig Membranextraktion. Paper read at Metallurgisches Seminar der GDMB, 19–21 November, 1987, Bad Ems, Germany.

X

New Membrane Processes under Development

41. Membrane-Based Solvent Extraction	727
42. Hollow-Fiber Contained Liquid Membrane	764
43. Membrane Reactors	809
44. Facilitated Transport	833
45. Electrostatic Pseudo-Liquid-Membrane	867
46. Other New Membrane Processes	885

41

Membrane-Based Solvent Extraction

Ravi Prasad*
Kamalesh K. Sirkar**
Stevens Institute of Technology

GENERAL DESCRIPTION OF THE PROCESS
THEORY
    Mass Transfer Relations at a Point
    Mass Transfer in a Solvent Extraction Device
    Mass Transfer with Chemical Reaction
    Breakthrough Pressure
    Asymmetric Membranes
    Mass Transfer in an Extraction Module and a Back Extraction Module
MEMBRANE DEVICES AND TRANSPORT CORRELATIONS
    MHF Devices Commercially Available
    Transport Correlations in MHF Devices

DESIGN OF MODULES AND MODULAR CASCADES
    Selection of Membrane/Module and Operating Mode
    Design of a Hollow-Fiber Module
    Design of a Modular Cascade
POTENTIAL APPLICATIONS
    Metal Extraction
    Organic Pollutant Extraction
    Aromatics Extraction
    Extraction of Pharmaceuticals
    Fermentation Product Extraction
    Extractive Bioreactors
ACKNOWLEDGMENTS
NOTATION
REFERENCES

GENERAL DESCRIPTION OF THE PROCESS

Solvent extraction is a common industrially used equilibrium-based separation process. In such a process, a solute (or solutes) in a solution, aqueous or organic, is extracted into an immiscible solvent, organic or aqueous, by dispersing one of the immiscible phases as drops in the other phase. This creates a large interfacial area and increases the extraction rate considerably. After the extraction is over, the phases are separated and the dispersed phase coalesced. There are two general categories of equipment for solvent extraction. A mixer-settler arrangement provides a single equilibrium stage; a number of them connected together provide multistage extraction. Continuous countercur-

---

*Present affiliation: Hoechst Celanese Corporation
**Present affiliation: New Jersey Institute of Technology

rent contacting equipment whether in the form of columns or centrifugal devices can generate the equivalent of many stages in one device (Treybal 1963).

Regardless of the dispersion-based contacting devices used, conventional equipment has many disadvantages: the need for dispersion and coalescence, problems of emulsification, flooding and loading limits in continuous countercurrent devices, the need for a density difference between the phases, and the high initial, operating, and maintenance costs of centrifugal devices, etc. In addition, scale-up is always difficult. Membrane-based solvent extraction devices, recently developed, appear to eliminate all such problems in addition to providing very high volumetric mass transfer rates. The basic concept behind membrane-based solvent extraction is discussed here.

Consider a *microporous hydrophobic membrane* with an organic phase on one side (Figure 41–1). The organic phase will, in all likelihood, spontaneously wet the membrane and come out through the membrane pores to the other side of the membrane. The appearance of the organic phase on this other side of the membrane can be prevented if an immiscible aqueous phase is maintained on this side at a pressure equal to or higher than that of the organic phase (Kiani, Bhave, and Sirkar 1984). The aqueous-organic interface will be immobilized at each pore mouth on the aqueous side of the membrane. Unless the excess aqueous phase pressure exceeds a critical value Δp_{cr}, called the *breakthrough pressure*, the organic phase in the pores will not be displaced by the aqueous phase.

Thus, over a given range of excess aqueous phase pressure for a given extraction system and a given hydrophobic microporous membrane, the two phases contact each other at the pore mouths; neither phase can go to the other side of the membrane. The membrane serves to immobilize the phase interfaces under conditions of appropriate difference between the pressures of the two phases. Solute or solutes transfer through the phase interfaces from one phase to another and then to the bulk of the latter phase. Extraction is achieved without dispersing one phase as drops in another phase.

Coalescence is unnecessary. No mechanical moving parts are present in the extractor. The flow rates of both aqueous and organic phases can be varied over wide ranges without flooding or loading as long as the correct relative phase pressure conditions are maintained. Systems having a tendency to emulsify spontaneously can be handled without emulsification. If the membrane is in hollow fiber form, the extractor can have a very high membrane surface area per unit equipment volume leading to very high extraction rates/volume (Kiani, Bhave, and Sirkar 1984; Frank and Sirkar 1985; D'Elia, Dahuron, and Cussler 1986; Sirkar 1988, 1991).

Such dispersion-free membrane-based solvent extraction can also be implemented using microporous hydrophilic membranes whose pores are filled with an aqueous phase preferentially wetting the hydrophilic membrane. As shown in Figure 41–2, the appearance of the aqueous solution on the organic side of the hydrophilic membrane and its dispersion as drops in the organic solvent can be prevented by maintaining the organic phase pressure higher than the aqueous phase pressure. The aqueous-

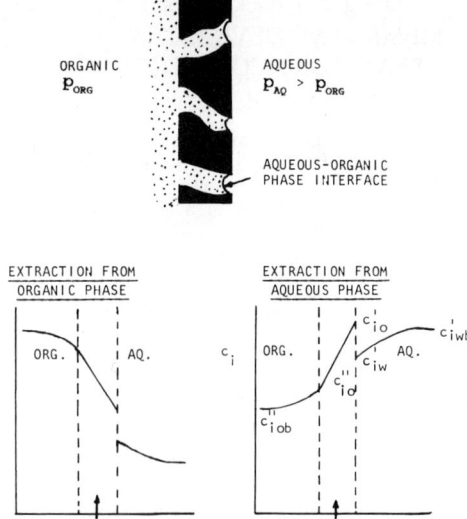

FIGURE 41–1. Solute concentration profiles for aqueous-organic interface immobilized in the pores of a hydrophobic membrane.

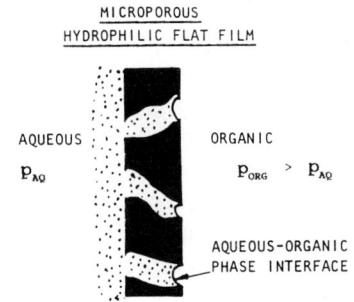

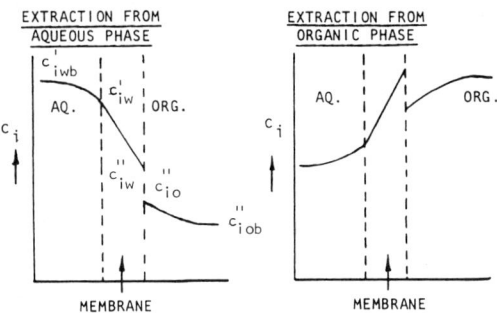

FIGURE 41–2. Solute concentration profiles for aqueous-organic interface immobilized in the pores of a hydrophilic membrane.

critical pressure needed to displace the other phase in the pores of the contiguous section of the membrane (Prasad and Sirkar 1987a, 1987b; Sirkar 1990).

Nondispersive solvent extraction has been experimentally demonstrated with two immiscible phase systems other than aqueous-organic: nonpolar organic-polar organic and two-phase aqueous extraction systems. Toluene was preferentially extracted from a 50-50% v/v toluene-n-heptane hydrocarbon mixture into n-methyl pyrrolidone (NMP), a polar organic solvent using a hydrophobic microporous membrane containing the preferentially wetting hydrocarbon mixture in the pores (Prasad and Sirkar 1987b). The NMP phase not in the pores was maintained at a higher pressure. A two-phase aqueous extraction system containing polyethylene glycol and potassium phosphate was successfully adopted for protein extraction using microporous hydrophobic membranes having the polyethylene glycol phase in the membrane pores and at a lower pressure (Dahuron and Cussler 1988). Recent studies by

organic interface is immobilized at each pore mouth on the organic side of the membrane. Nondispersive liquid extraction can be carried out easily as long as the excess organic phase pressure does not exceed a critical value Δp_{cr} (Prasad and Sirkar 1987a, 1987b; Wald, Lopez, and Matson 1989).

The aqueous-organic interface can also be immobilized in a microporous hydrophobic-hydrophilic membrane, one part of which is hydrophobic, the other part being hydrophilic (Figure 41–3). The hydrophobic section of the composite membrane is wetted by the organic phase and the hydrophilic section by the aqueous phase, the aqueous-organic phase interface is at the hydrophobic-hydrophilic interface. One can operate dispersion-free with such a membrane and effect solvent extraction with either the organic phase on the hydrophobic membrane side having a higher pressure or the aqueous phase on the hydrophilic membrane side having a higher pressure. However, the excess phase pressure should not exceed the

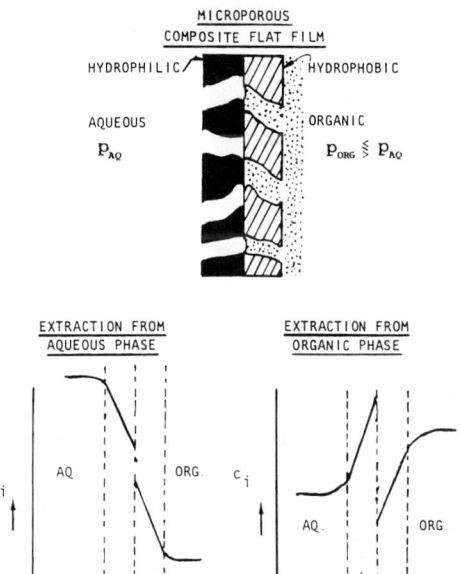

FIGURE 41–3. Solute concentration profiles for aqueous-organic interface immobilized in a composite hydrophobic-hydrophilic membrane.

Prasad et al. (1990) have demonstrated a broader set of choices for liquid-in-pore and membrane wetting characteristics.

Using microporous membranes, hydrophobic or hydrophilic, nondispersive solvent extraction has already been applied to a large variety of systems. These include metal extraction, organic acid extraction, alcohol extraction, protein extraction, priority pollutant extraction, pharmaceutical products extraction, aromatics extraction, etc. Although hydrophilic membranes have been used in a few cases, hydrophobic membranes have been used more often.

In nondispersive membrane-based solvent extraction, maintaining the phase not in the pores at a pressure above that of the phase in the pores without exceeding Δp_{cr} is crucial. Kim (1984) reported the performance of hydrophobic microporous hollow-fiber-based extraction of copper using LIX reagents. The direction of the difference in the pressures of the two flowing phases was reversed at the two ends of this hollow-fiber module. Consequently, dispersion could not be prevented and Kim had to use an aqueous-organic separator in addition to a hollow-fiber extractor. Nondispersive solvent extraction has also been performed using solvent-swollen polymeric membranes (Lee, Ho, and Liu 1976; Ho, Lee, and Liu 1976). Here a solvent swelling a polymeric dense membrane is on one side with the other immiscible phase on the other side. There is no direct phase-to-phase contact; the pressure conditions were not identified for these essentially nonporous membranes. Obviously, the rate of extraction in such a process will be reduced by the slow diffusion of solute through the solvent-swollen gel membrane.

Membrane-based solvent extraction can be carried out in modules with microporous membranes in flat or hollow-fiber form. In the case of hollow fibers, the module may be of the simple shell-and-tube form without baffles or have a crossflow pattern. Flat membranes may be used in spiral-wound form, tangential flow filtration devices, or rotating cylindrical devices. A simplified theory of mass transfer in solvent extraction with microporous membranes is considered now for the purpose of developing module design procedures later. In the development of such a theory, keep in mind that solvent extraction is an equilibrium-based separation process and membrane-based solvent extraction as considered here continues to be an equilibrium-based process regardless of the complications introduced by a microporous membrane. Although the separation that can be ultimately achieved is therefore limited by the conditions of equilibrium between the extract and the raffinate, the membrane-based extractor design is going to be based on the rates of extraction as in any conventional extractor design.

THEORY

Microporous membranes may be symmetric, asymmetric, or composite. They may have a uniform pore size or a distribution of pore sizes. They may be thick, thin, or ultrathin, with or without charges on external and internal surfaces. The theory for membrane-based solvent extraction has been developed primarily for thin uncharged symmetric membranes with no variation in porosity and pore sizes across the membrane thickness since almost all membranes studied and used in extraction devices are of this type. The theory, although far from being exact, is simple and useful.

Mass Transfer Relations at a Point

In a conventional extractor, two immiscible phases flow past each other. In continuous contact devices, this flow is often globally countercurrent. In an agitated mixer prior to a settler, a variety of flow patterns exists. The local rate of extractive mass transfer has to be determined first before overall flow patterns, staging, continuous contact, etc., are taken into account. The same approach is valid also for a membrane-based extractor; therefore, the local mass transfer rate across a microporous membrane is considered now. We start with hydrophobic membranes and then consider hydrophilic and composite membranes and other aspects of mass transfer. The ground rules for selecting hydrophobic over hydrophilic or vice versa are provided in a later section on design of modules.

Hydrophobic Membranes

Figure 41-1 illustrates the concentration profile of a solute i being extracted through a *hydrophobic flat* membrane from an aqueous phase into an organic phase. Assume:

1. Steady-state mass transfer through the phase interfaces at equilibrium with no immiscible displacement taking place through the pores.
2. Aqueous-organic interface at membrane-water interface.
3. Uniform pore size and wetting characteristics throughout the membrane.
4. Negligible influence of aqueous-organic interface curvature on solute distribution coefficient, rate of solute transfer, and interfacial area.
5. No chemical reaction.
6. Solute transport in the aqueous phase and the organic phase outside the membrane may be described by simple film-type mass transfer coefficients without bulk flow correction.
7. The diffusion of solute i through the organic-filled pore may be described by means of a membrane transfer coefficient k_{imo} based on the total area of the membrane (including porous and nonporous sections).
8. No solute transport through nonporous sections of the membrane.
9. The aqueous and organic film resistances may be combined with the membrane pore diffusion resistance as a one-dimensional series of diffusion resistances (Keller and Stein 1967).
10. Solute distribution coefficient m_i is constant over the relevant concentration range and the two liquids are essentially insoluble in each other.

For microporous membranes with porosity greater than 5%, Keller and Stein (1967) and Malone and Anderson (1977) have shown that two dimensional effects are negligible; this justifies assumption 9. Then we can describe the solute flux from the aqueous phase bulk to the organic phase bulk in terms of individual mass transfer coefficients by

$$N_i = k_{iw}(c'_{iwb} - c'_{iw}) = k_{imo}(c'_{io} - c''_{io})$$
$$= k_{io}(c''_{io} - c''_{iob}). \quad (41\text{-}1)$$

Using an organic-phase-based overall mass transfer coefficient K_o for the solute,

$$N_i = K_o(c^*_{io} - c''_{iob}) \quad (41\text{-}2)$$

where c^*_{io} is a hypothetical organic phase concentration in equilibrium with the bulk aqueous phase concentration c'_{iwb}. Further, these two concentrations are related by a solute distribution coefficient m_i. According to assumptions 1 and 10, then

$$m_i = \frac{c^*_{io}}{c'_{iwb}} = \frac{c'_{io}}{c'_{iw}} = \frac{c''_{iob}}{c^*_{iw}}, \quad (41\text{-}3)$$

where c^*_{iw} is a hypothetical aqueous phase concentration in equilibrium with the bulk organic phase concentration c''_{iob}. Using Eqs. (41-1), (41-2), and (41-3), a relation between K_o and the individual mass transfer coefficients is obtained:

$$\frac{1}{K_o} = \frac{m_i}{k_{iw}} + \frac{1}{k_{imo}} + \frac{1}{k_{io}}. \quad (41\text{-}4)$$

The validity of such a relation has been demonstrated by Kiani, Bhave, and Sirkar (1984) and Prasad et al. (1986). For an aqueous-phase-based overall transfer coefficient K_w for the solute, the corresponding relations are

$$N_i = K_w(c'_{iwb} - c^*_{iw}) \quad (41\text{-}5)$$

and

$$\frac{1}{K_w} = \frac{1}{k_{iw}} + \frac{1}{m_i k_{imo}} + \frac{1}{m_i k_{io}}. \quad (41\text{-}6)$$

If the extraction of the solute is from organic phase to aqueous phase, Eqs. (41-4) and (41-

6) between an overall mass transfer coefficient and the individual coefficients remain unchanged. Only the signs in the flux expressions change since the transport direction has changed. If the extraction system is polar organic–nonpolar organic with the nonpolar organic present in the membrane pores, then the relations for aqueous-organic systems are valid with the polar organic phase used instead of the aqueous phase (Prasad and Sirkar 1987b). Similarly, for a biphasic aqueous system, relations for aqueous-organic systems can be used with the salt-containing aqueous phase used instead of the aqueous phase (Dahuron and Cussler 1988).

The previous developments are valid at any location of the extractor for a flat hydrophobic microporous membrane. When the microporous membrane is in the form of a *hollow fiber* (Figure 41–4), the interfacial areas on the two sides of the hollow fiber are different. Further, the lumen side may have the aqueous phase or the organic phase. The overall mass transfer coefficient may be defined based on the surface area calculated using either the inside diameter (ID) or the outside diameter (OD) of the hollow fiber. For calculating an overall mass transfer coefficient, the interfacial area should be based on the diameter where the aqueous-organic phase interface is located (Frank and Sirkar 1985; Prasad and Sirkar 1987b, 1988). Consider, for example, the aqueous phase in hydrophobic fiber lumen (tube side) and organic phase on the shell side. At any location, the rate of solute extraction per unit fiber length is given by

$$N_i \pi d_{ti} = k_{iwt} \pi d_{ti}(c'_{iwb} - c'_{iw})$$
$$= k_{imo} \pi d_{tlm}(c'_{io} - c''_{io})$$
$$= k_{ios} \pi d_{to}(c''_{io} - c''_{iob})$$
$$= K_o \pi d_{ti}(c^*_{io} - c''_{iob})$$
$$= K_w \pi d_{ti}(c'_{iwb} - c^*_{iw}). \quad (41\text{--}7)$$

Using standard procedures, we obtain

$$\frac{1}{K_o d_{ti}} = \frac{1}{d_{to} k_{ios}} + \frac{1}{d_{tlm} k_{imo}} + \frac{m_i}{d_{ti} k_{iwt}} \quad (41\text{--}8)$$

and

$$\frac{1}{K_w d_{ti}} = \frac{1}{m_i d_{to} k_{ios}} + \frac{1}{m_i d_{tlm} k_{imo}} + \frac{1}{d_{ti} k_{iwt}}, \quad (41\text{--}9)$$

with the aqueous-organic interface located on the fiber ID. The utility of such relations has been demonstrated by Prasad and Sirkar (1987b, 1988). Table 41–1 provides different relations for different conditions in hydrophobic hollow fibers. D'Elia, Dahuron, and Cussler (1986) provide a few relations without diameter corrections.

A major difference between membrane-based extraction and conventional extraction is that the interfacial area in membrane-based processes is known (e.g., $\pi d_{ti} N$/unit length in a module with N fibers). In conventional processes, generally the product $K_o a_s$ or $K_w a_s$ (where a_s is the interfacial area per unit volume)

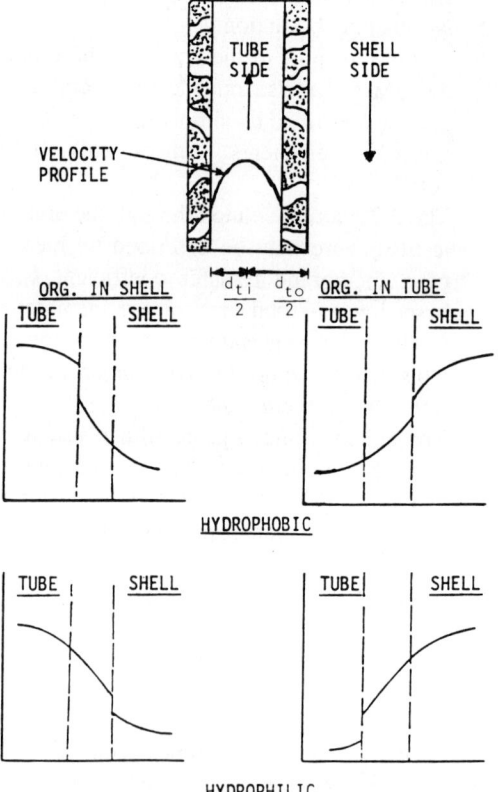

FIGURE 41–4. Solute concentration profiles for solvent extraction using a microporous hollow fiber (reprinted from Prasad and Sirkar 1988 with permission).

is known. This makes scaleup in conventional processes quite difficult since a_s is very hard to determine.

Hydrophilic and Composite Membranes

In a microporous *hydrophilic flat* membrane with the aqueous phase in the pores (Figure 41–2), the steady-state rate of extraction from an aqueous phase into the organic solvent can be modeled by means of assumptions 1 to 10 used earlier for hydrophobic membranes. Assumptions 2 and 7 have to be changed to account for aqueous phase in the pores. The flux of solute i is given by

$$N_i = k_{iw}(c'_{iwb} - c'_{iw}) = k_{imw}(c'_{iw} - c''_{iw})$$
$$= k_{io}(c''_{io} - c''_{iob})$$
$$= K_o(c^*_{io} - c''_{iob}) = K_w(c'_{iwb} - c^*_{iw}) \quad (41\text{--}10)$$

Using Eq. (41–3), the relation between the overall mass transfer coefficients K_o and K_w and the individual transfer coefficients are obtained:

$$\frac{1}{K_o} = \frac{1}{k_{io}} + \frac{m_i}{k_{imw}} + \frac{m_i}{k_{iw}}, \quad (41\text{--}11)$$

$$\frac{1}{K_w} = \frac{1}{k_{iw}} + \frac{1}{k_{imw}} + \frac{1}{m_i k_{io}}. \quad (41\text{--}12)$$

For *hydrophilic hollow fibers* with aqueous phase in the pores and in the fiber lumen (Figure 41–4), the corresponding relations are

$$\frac{1}{K_o d_{to}} = \frac{1}{k_{ios} d_{to}} + \frac{m_i}{k_{imw} d_{tlm}} + \frac{m_i}{k_{iwt} d_{ti}}, \quad (41\text{--}13)$$

$$\frac{1}{K_w d_{to}} = \frac{1}{k_{iwt} d_{ti}} + \frac{1}{k_{imw} d_{tlm}} + \frac{1}{m_i k_{ios} d_{to}}. \quad (41\text{--}14)$$

Here the aqueous-organic phase interface is located on the OD of the fiber. Such relations for

TABLE 41-1. Overall Mass Transfer Coefficients and Individual Coefficients for Different Conditions.

Membrane Type and Form	$\dfrac{1}{K_o} =$	$\dfrac{1}{K_w} =$
Hydrophobic flat	$\dfrac{1}{k_{io}} + \dfrac{1}{k_{imo}} + \dfrac{m_i}{k_{iw}}$	$\dfrac{1}{m_i k_{io}} + \dfrac{1}{m_i k_{imo}} + \dfrac{1}{k_{iw}}$
Hydrophilic flat	$\dfrac{1}{k_{io}} + \dfrac{m_i}{k_{imw}} + \dfrac{m_i}{k_{iw}}$	$\dfrac{1}{m_i k_{io}} + \dfrac{1}{k_{imw}} + \dfrac{1}{k_{iw}}$
Composite flat	$\dfrac{1}{k_{io}} + \dfrac{1}{k_{imo}} + \dfrac{m_i}{k_{imw}} + \dfrac{m_i}{k_{iw}}$	$\dfrac{1}{m_i k_{io}} + \dfrac{1}{m_i k_{imo}} + \dfrac{1}{k_{imw}} + \dfrac{1}{k_{iw}}$
Hydrophobic MHF Aqueous in tube	$\left(\dfrac{d_{ti}}{d_{to} k_{ios}} + \dfrac{d_{ti}}{d_{tlm} k_{imo}} + \dfrac{m_i d_{ti}}{d_{ti} k_{iwt}}\right)$	$\left(\dfrac{d_{ti}}{m_i d_{to} k_{ios}} + \dfrac{d_{ti}}{m_i d_{tlm} k_{imo}} + \dfrac{d_{ti}}{d_{ti} k_{iwt}}\right)$
Organic in tube	$\left(\dfrac{d_{to}}{d_{ti} k_{iot}} + \dfrac{d_{to}}{d_{tlm} k_{imo}} + \dfrac{m_i d_{to}}{d_{to} k_{iws}}\right)$	$\left(\dfrac{d_{to}}{m_i d_{ti} k_{ios}} + \dfrac{d_{to}}{m_i d_{tlm} k_{imo}} + \dfrac{d_{to}}{d_{to} k_{iws}}\right)$
Hydrophilic MHF Aqueous in tube	$\left(\dfrac{d_{to}}{d_{to} k_{ios}} + \dfrac{m_i d_{to}}{d_{tlm} k_{imw}} + \dfrac{m_i d_{to}}{d_{ti} k_{iwt}}\right)$	$\left(\dfrac{d_{to}}{m_i d_{to} k_{ios}} + \dfrac{d_{ti}}{d_{tlm} k_{imw}} + \dfrac{d_{to}}{d_{ti} k_{iwt}}\right)$
Organic in tube	$\left(\dfrac{d_{ti}}{d_{ti} k_{iot}} + \dfrac{m_i d_{ti}}{d_{tlm} k_{imw}} + \dfrac{m_i d_{ti}}{d_{to} k_{iws}}\right)$	$\left(\dfrac{d_{ti}}{m_i d_{ti} k_{iot}} + \dfrac{d_{ti}}{d_t d_{lm} k_{imw}} + \dfrac{d_{ti}}{d_{to} k_{iws}}\right)$

the different configurations are provided in Table 41–1.

For a composite flat membrane with one part hydrophobic and the other part hydrophilic, and having the organic phase in hydrophobic pores and the aqueous phase in hydrophilic pores, the relations between the overall mass transfer coefficients K_o and K_w and the individual transfer coefficients are (Prasad and Sirkar 1987a; 1987b)

$$\frac{1}{K_o} = \frac{1}{k_{io}} + \frac{1}{k_{imo}} + \frac{m_i}{k_{imw}} + \frac{m_i}{k_{iw}}, \quad (41–15)$$

$$\frac{1}{K_w} = \frac{1}{k_{iw}} + \frac{1}{k_{imw}} + \frac{1}{m_i k_{io}} + \frac{1}{m_i k_{imo}}. \quad (41–16)$$

The corresponding relations for a composite microporous hollow-fiber membrane can also be developed easily.

Membrane Transfer Coefficient

The mass transfer rate in solvent extraction of a solute i has been characterized so far using the concept of a one-dimensional series of diffusion resistances, which included the membrane resistance. The diffusional solute transfer rate through an uncharged membrane having a solvent-filled pore or an aqueous-solution-filled pore was conveniently illustrated using solute mass transfer coefficient k_{imo} or k_{imw}. To estimate k_{imo}, for example, unhindered diffusion of solute i through organic solvent-filled symmetric microporous membrane is assumed with an effective diffusivity D_{ieff} given by $(D_{io}\epsilon_m/\tau_m)$. Here D_{io} is the free diffusion coefficient of solute i in organic solvent, ϵ_m is the membrane porosity, and τ_m is the membrane tortuosity. Then,

$$k_{imo} = \frac{D_{io}\epsilon_m}{\tau_m l} = \frac{D_{ieff}}{l}, \quad (41–17a)$$

where l is the membrane thickness. The reduction in the area for diffusion by the impermeable sections of the microporous membrane is accounted for by ϵ_m, whereas the increase in diffusion path length over membrane thickness in the tortuous membrane pores is compensated for by τ_m. Similarly, for an aqueous solution-filled symmetric hydrophilic membrane

$$k_{imw} = \frac{D_{iw}\epsilon_m}{\tau_m l}, \quad (41–17b)$$

where D_{iw} is the free diffusion coefficient of solute i in the aqueous solution.

The validity of such expressions for flat membranes has been demonstrated by Kiani, Bhave, and Sirkar (1984), Prasad et al. (1986), and Prasad and Sirkar (1987a). When the membrane thickness l is replaced by $(d_{to} - d_{ti})/2$, the above expressions can be used for hollow fibers (Prasad and Sirkar 1987b, 1988).

Equations (41–17a) and (41–17b) are useful when the following assumption hold:

1. There is unhindered diffusion of solute.
2. The membrane is symmetric and completely wetted by the designated phase.
3. No two-dimensional effects occur.

Further elaborations on these assumptions will provide a better perspective.

For unhindered diffusion of solute, the solute dimensions should be about two orders of magnitude smaller than the pore dimensions (Beck and Schultz 1970). Prasad and Sirkar (1987a, 1988) have observed that the membrane resistance was increased drastically for larger solutes (e.g., succinic acid) in hydrophilic Cuprophan® membranes, which have comparatively much smaller pore sizes compared to other hydrophilic or hydrophobic membranes studied. The Cuprophan® membranes used were designed for an artificial kidney and therefore have pore diameters in the 30- to 50-Å range.

When the microporous membrane is asymmetric, the asymmetry across the membrane thickness may be in terms of pore sizes or porosity; usually both are present simultaneously. The applicability of Eqs. (41–17a) and (41–17b) to asymmetric membranes has not been investigated. If there is no hindered diffusion in the membrane section with the smallest pore sizes, then free diffusion will exist throughout the membrane. Expressions (41–17a) and (41–17b) will be locally valid at any membrane section. To obtain a valid result for the whole membrane, diffusional flux has to be integrated across the membrane thickness with both ϵ_m and τ_m being functions of membrane thickness. Thus, interpretation of membrane resistance in solvent extraction with microporous asymmetric membranes needs more study.

A key assumption in the previous theoretical developments is that the hydrophobic membrane pore is completely wetted by the organic phase; similarly the hydrophilic membrane pore is completely filled with the aqueous phase. If the excess phase pressure is around the breakthrough pressure Δp_{cr}, part of the membrane may have the organic phase while the rest has the aqueous phase. Since the location of the interface of the aqueous-organic phases in the pore becomes uncertain, the membrane transfer coefficient expressions are unknown (Prasad et al. 1986). Thus, if the operation is near or at the breakthrough pressure, the solute transfer rate may not be predictable. However, the direction of change in mass transfer rate can be predicted under these conditions since the changed configuration resembles a composite membrane whose mass transfer relations are indicated in Eqs. (41–15) and (41–16).

In membrane solvent extraction, the solute passes from a bulk solution through its boundary layer, the liquid in the membrane pore, and then to another boundary layer and its bulk solution. Since the membrane surface is heterogeneous, consisting of pores and impermeable sections, the flux lines are distorted near the pore entrance and exit. This distortion can lead to additional solute transport resistance, which can be determined from a two-dimensional analysis of the transport at the pore-boundary layer region (Keller and Stein 1967). Such an analysis for membrane-based solvent extraction is not available. The value of m_i and the relative role of the two boundary layer resistances are expected to be critical. However for $m_i \sim O(1)$, the two-dimensional effects are likely to be small with membranes of high porosity. Most membranes used so far have $\epsilon_m \geq 0.37$.

Effect of Phase Pressure Difference on Mass Transfer

A pressure difference exists between the two immiscible phases as the solute is extracted nondispersively from one phase to the other. Prasad et al. (1986) have shown that the $\bar{V}_i \Delta p$ component of the chemical potential difference $\Delta \mu_i$ for solute i is negligible compared to the $RT \Delta \ln c_i$ component. Thus, the magnitude of the Δp employed ought not to influence the rate of mass transfer. This has been verified by Kiani, Bhave, and Sirkar (1984) and Prasad et al. (1986) for flat hydrophobic membranes and aqueous/organic systems. Prasad and Sirkar (1987a, 1987b) have also verified the Δp independence of the mass transfer coefficient in polar organic-nonpolar organic-hydrophobic membrane systems as well as in aqueous-organic-hydrophilic membrane systems. These studies were carried out with flat membranes adequately supported against the applied Δp. Figure 41–5 provides an illustration of such pressure independence for a variety of systems. Figure 41–6 provides additional data for one system.

The Δp independence of K values disappears, however, when Δp is close to Δp_{cr} or exceeds Δp_{cr}. With the onset of immiscible displacement of the liquid-in-the-pores by the liquid phase outside, the membrane resistance will change considerably due to m_i (see the next section). Additionally, in hollow-fiber systems, certain levels of Δp and beyond will start deforming fiber walls as well as pore dimensions. This is especially true when the phase in the fiber lumen has a higher pressure.

Fiber deformation in tension creates higher membrane pore sizes, reducing the value of Δp_{cr}. Fiber deformation in compression may create smaller membrane pore sizes, thus increasing the value of Δp_{cr}. At any given value of Δp, such deformations are increased by an increase in d_{ti} and ϵ_m and a reduction in $(d_{to} - d_{ti})$. Basu, Prasad, and Sirkar (1990) have observed such behavior with large-pore, large-bore, high-porosity nylon fibers.

Role of Solute Distribution Coefficient

In conventional solvent extraction, there are only two film mass transfer coefficients: the extract phase coefficient and the raffinate phase coefficient. For aqueous-organic systems with, say, solute being extracted from the aqueous phase (raffinate) to the organic phase (extract),

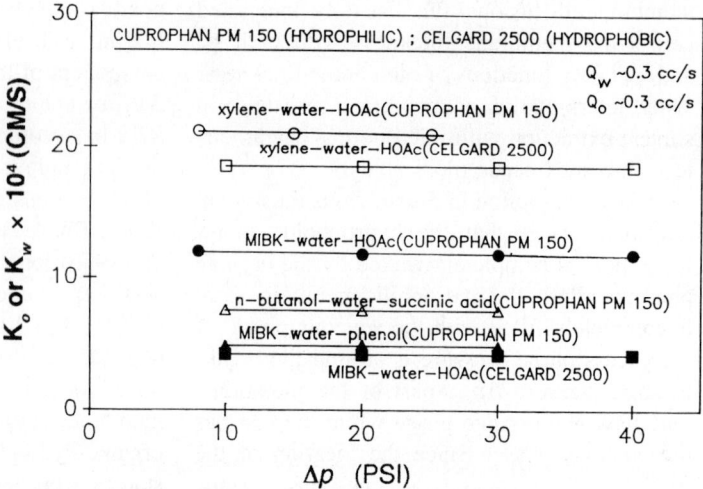

FIGURE 41–5. Effect of excess phase pressure on the overall solute mass transfer coefficient for different membranes and various extraction systems.

there will be two coefficients, the aqueous phase coefficient and the organic phase coefficient (Treybal 1963). Membrane-based solvent extraction is no different. Consider Eqs. (41–4) and (41–11) between the overall mass transfer coefficient and the individual mass transfer coefficients for a flat membrane:

Hydrophobic membrane; organic-in-pore
$$\frac{1}{K_o} = \underbrace{\frac{m_i}{k_{iw}}}_{\text{aqueous}} + \underbrace{\frac{1}{k_{imo}} + \frac{1}{k_{io}}}_{\text{organic}},$$

Hydrophilic membrane; aqueous-in-pore
$$\frac{1}{K_o} = \underbrace{\frac{m_i}{k_{iw}} + \frac{m_i}{k_{imw}}}_{\text{aqueous}} + \underbrace{\frac{1}{k_{io}}}_{\text{organic}}.$$

Here also are two film coefficients or resistances, an aqueous film resistance and an organic film resistance. The only difference is that an extra membrane resistance appears in that phase resistance which occupies the membrane pores. Thus, if k_{iw} and k_{io} values are identical between a conventional solvent extractor and a membrane-based extractor, the value of K_o or K_w in a membrane-based extractor will be lower due to the extra membrane resistance (Kiani, Bhave, and Sirkar 1984).

Such a conclusion is valid when the solute distribution coefficient m_i is $O(1)$. However, when $m_i \gg 1$ or $m_i \ll 1$, the following simplified results hold for different flat membranes provided k_{iw} and k_{io} are of the same order of magnitude (for hollow fibers, area effects need to be included):

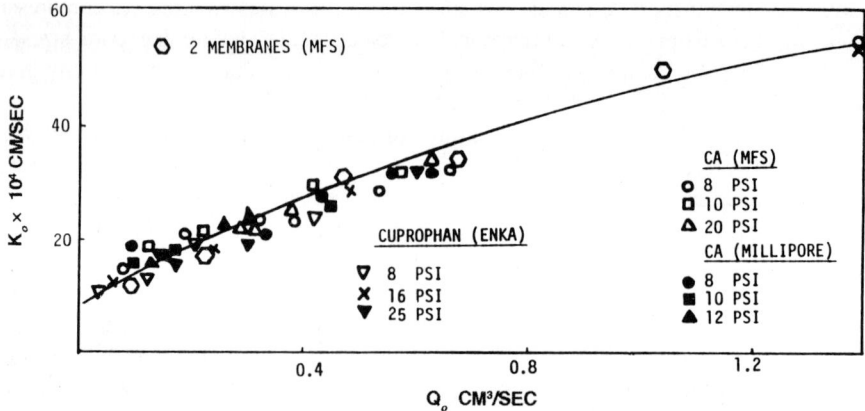

FIGURE 41–6. Effect of organic flow rate and excess organic phase pressure on the overall solute mass transfer coefficient for various hydrophilic membranes for xylene-water-acetic acid system (reprinted from Prasad and Sirkar 1987a with permission).

$$
\begin{array}{c}
 m_i \gg 1 m_i \ll 1 \\[4pt]
\textit{Hydrophobic membrane; organic-in-pore} \left\{
\begin{array}{ll}
\dfrac{1}{K_o} \cong \dfrac{m_i}{k_{iw}}, & \dfrac{1}{K_o} \cong \dfrac{1}{k_{imo}} + \dfrac{1}{k_{io}}, \\[10pt]
\dfrac{1}{K_w} \cong \dfrac{1}{k_{iw}}, & \dfrac{1}{K_w} \cong \dfrac{1}{m_i k_{imo}} + \dfrac{1}{m_i k_{io}}; \\[6pt]
\underbrace{}_{\text{aqueous}} & \underbrace{}_{\text{organic}}
\end{array}
\right. \\[30pt]
\textit{Hydrophilic membrane; aqueous-in-pore} \left\{
\begin{array}{ll}
\dfrac{1}{K_o} \cong \dfrac{m_i}{k_{imw}} + \dfrac{m_i}{k_{iw}}, & \dfrac{1}{K_o} \cong \dfrac{1}{k_{io}}, \\[10pt]
\dfrac{1}{K_w} \cong \dfrac{1}{k_{iw}} + \dfrac{1}{k_{imw}}, & \dfrac{1}{K_w} \cong \dfrac{1}{m_i k_{io}}, \\[6pt]
\underbrace{}_{\text{aqueous}} & \underbrace{}_{\text{organic}}
\end{array}
\right.
\end{array}
$$

(41–18)

The condition $m_i \gg 1$ implies that the solute strongly prefers the organic phase. From conventional mass transfer theory, the aqueous phase resistance will then be controlling; thus, $K_w \cong k_{iw}$ exactly as shown above (for hydrophobic membrane). For systems where $m_i \ll 1$, i.e., the solute strongly prefers the aqueous phase, the organic phase resistance will become the controlling one. The only difference in membrane-based solvent extraction is that if the membrane pore has a particular liquid phase, the membrane resistance will show up if that phase resistance is controlling. For an $m_i \gg 1$ system with hydrophobic membranes, then, membrane resistance is unimportant just as for an $m_i \ll 1$ system with a hydrophilic membrane, membrane resistance plays no role (Prasad and Sirkar 1985a, 1985b). Thus, a large number of systems exist in which the membrane may not affect the mass transfer rate at all. Figure 41–6 illustrates this for hydrophilic flat membranes. However, if there is hindered diffusion of the solute and the membrane thickness is large, the membrane resistance may become important even under these conditions. On the other hand, for an $m_i \gg 1$ system with a hydrophilic membrane or an $m_i \ll 1$ system with a hydrophobic membrane, membrane resistance will be quite important even in the absence of hindered diffusion.

What kind of membrane is better from a mass transfer point of view, hydrophobic or hydrophilic, for given values of m_i, k_{iw}, and k_{io} in a membrane-based solvent extraction system? A comparison of Eqs. (41–4) and (41–11) suggests that for systems with $m_i \ll 1$, a hydrophilic membrane is preferred, whereas for $m_i \gg 1$ a hydrophobic membrane is much better (Prasad and Sirkar 1985a, 1985b; D'Elia, Dahuron, and Cussler 1986). Because $m_i \gg 1$ is desirable in a solvent extraction system with solute extracted from aqueous to organic, mass transfer advantages suggest a hydrophobic membrane. However, for extracting a solute from an organic to an aqueous phase $m_i \ll 1$ is desirable; a hydrophilic membrane would be advantageous here from a mass transfer point of view. The usefulness of a hydrophobic membrane against a hydrophilic membrane for different ranges of values of the distribution coefficient m_i has been demonstrated by Prasad and Sirkar (1988) for hollow fibers.

We may conclude from the above that a hydrophobic membrane should be used when the solute strongly prefers the organic phase and a hydrophilic membrane is better when the solute prefers the aqueous phase. Recent investigations by Prasad et al. (1990) suggest a broader scenario and a larger set of choices. They have indicated that a hydrophilic membrane may be operated with the organic phase in the pores since organics can wet hydrophilic membranes also. In such a case, the mass transfer relations for organic-in-pore hydrophobic membranes will be valid for the hydrophilic membrane also. Obviously, the aqueous phase side had to be maintained at a pressure higher than that of the organic side. They have, however, indicated that the window of excess aqueous phase pressure was significantly lower in this case. For nonpolar organic–polar organic systems, Prasad et al. (1990) have also operated a hydrophobic membrane with the polar organic in the membrane pores and the nonpolar organic phase at a higher pressure since both phases can wet the hydrophobic membrane. Again, the excess nonpolar organic phase pressure window was significantly smaller. These and a number of other factors discussed later in the design section have to be taken into account in selecting the membrane type and the pore-liquid membrane-wetting-characteristic combination.

Effect of Interfacial Tension on Mass Transfer

In conventional dispersion-based solvent extraction devices, a high value of interfacial tension γ_{wo} (e.g., extraction of acetone in a toluene-water system with $\gamma_{wo} \sim 23.1$ dyne/cm) will produce large drops. A system with a very low value of interfacial tension (e.g., extraction of succinic acid in an *n*-butanol-water system with $\gamma_{wo} \sim 1.5$ dyne/cm) will produce correspondingly small drop sizes. A common observation is that extraction efficiency in dispersion-based devices is therefore strongly influenced by γ_{wo} (Seibert and Fair 1988). The lower interfacial tension system invariably yields a higher volumetric mass transfer coefficient. In membrane-based solvent extraction without dispersion, the interfacial tension γ_{wo} does not in general influence any of the mass transfer coefficients. It merely influences the value of Δp_{cr}.

Mass Transfer in a Solvent Extraction Device

For large-scale membrane-based extraction, the only equipment type currently available is a hollow-fiber module. In Figure 41–7(a), only a block diagram of a hollow-fiber module is shown for calculation purposes; a commercially available Liqui-Cel® module is schematically shown in Figure 41–7(b). The latter are built essentially like a shell-and-tube heat exchanger with one shellside pass and one tubeside pass without any baffles on the shell side. The steady-state total rate of transfer of a solute being extracted from a tubeside or shellside feed to a shellside or tubeside extract may be determined in basically two different ways keeping in mind that it is a continuous-contact extractor:

1. Assume or calculate an overall mass transfer coefficient for the extractor using correlations, and then using an overall logarithmic mean or arithmetic mean concentration driving force across the extractor, calculate the total rate of solute extraction.
2. Solve numerically two differential equations for concentration profiles on the tube side and the shell side coupled through the boundary conditions of diffusive transport through the liquid-filled membrane pores.

Since the results from approach 2 are not yet available (Yun, Prasad, and Sirkar (1991)), approach 1 has so far been the basis for determining the mass transfer rate in the hollow-fiber extractor. This approach may be adopted for tubeside or shellside feed, for aqueous solution on the tube side or shell side in aqueous-organic systems, and for a hydrophobic fiber with organic-in-pores or a hydrophilic fiber with aqueous-in-pores. Approach 1 is now illustrated for a particular configuration with hydrophobic hollow fibers, namely, aqueous feed in fiber lumen, organic solvent entering the shell side in countercurrent flow, equilibrium

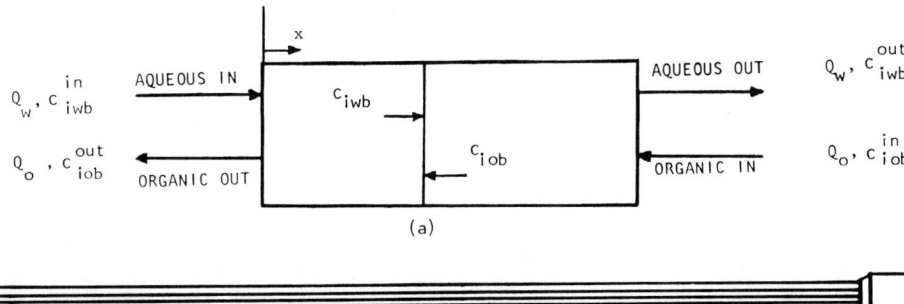

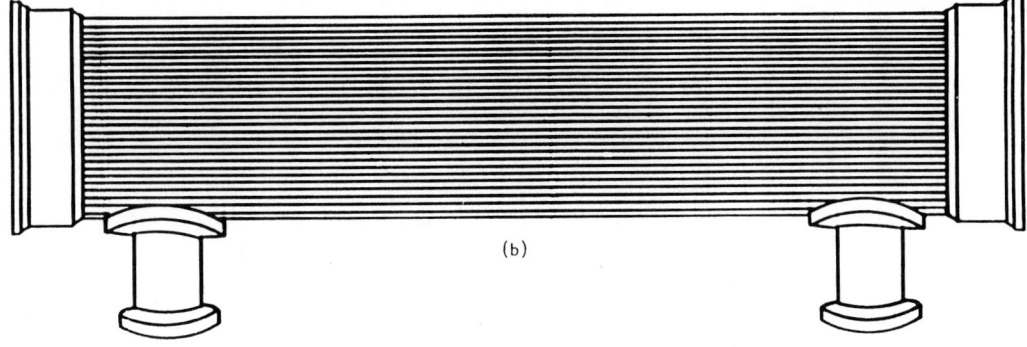

FIGURE 41–7. Schematics of (a) a hollow-fiber module stage for once-through countercurrent operation and (b) a commercial hollow-fiber module. (reprinted from Hoechst Celanese Separations Products Division with permission).

solute distribution over the relevant solute concentration range described by a constant m_i, and a module-averaged overall mass transfer coefficient $\overline{K}_w$ (Prasad, Frank, and Sirkar 1988). The module has N hollow fibers of internal diameter d_{ti} and length L [Figure 41–7(a)]:

Overall solute balance:

$$Q_o(c_{iob}^{out} - c_{iob}^{in}) = Q_w(c_{iwb}^{in} - c_{iwb}^{out}). \qquad (41\text{--}19)$$

Differential solute balance at any axial location x:

$$v_{wt}\left(\frac{\pi}{4}d_{ti}^2 N\right)\frac{dc_{iwb}}{dx} = -K_w(\pi d_{ti}N)\left(c_{iwb} - \frac{c_{iob}}{m_i}\right) \qquad (41\text{--}20)$$

for $x = 0$, $c_{iwb} = c_{iwb}^{in}$; $x = L$, $c_{iwb} = c_{iwb}^{out}$. $\qquad (41\text{--}21)$

Note that $Q_w = v_{wt}(\pi d_{ti}^2 N/4)$. Rearranging (41–20) and integrating along the length,

$$\int_0^L K_w\,dx = \overline{K}_w L = \frac{Q_w}{\pi d_{ti}N}\int_{c_{iwb}^{out}}^{c_{iwb}^{in}}\frac{dc_{iwb}}{c_{iwb} - c_{iob}/m_i}. \qquad (41\text{--}22)$$

Now a solute balance from the aqueous feed inlet to any section may be utilized to obtain

$$Q_w(c_{iwb}^{in} - c_{iwb}) = Q_o(c_{iob}^{out} - c_{iob}). \qquad (41\text{--}23)$$

Use the expression for c_{iob} from here in the integral in Eq. (41–22) and integrate to get

$$L = \frac{Q_w}{\overline{K}_w(\pi d_{ti}N)}\frac{1}{\{1 - [Q_w/(m_iQ_o)]\}}\ln\left[\frac{(c_{iwb}^{in} - (c_{iob}^{out}/m_i))}{(c_{iwb}^{out} - (c_{iob}^{in}/m_i))}\right]. \qquad (41\text{--}24)$$

For a given c_{iwb}^{in}, c_{iob}^{in}, $\bar{K}_w$, Q_w, Q_o, m_i, L, N, and d_{ti}, this equation may be turned inside out using (41–19) for c_{iob}^{out} to obtain

$$c_{iwb}^{out} = \frac{c_{iwb}^{in}\left(1 - \dfrac{Q_w}{Q_o m_i}\right) - \dfrac{c_{iob}^{in}}{m_i}\left[1 - \exp\left(\dfrac{L\{1 - [Q_w/(m_i Q_o)]\}}{LTU}\right)\right]}{\exp\left(\dfrac{L\{1 - [Q_w/(Q_o m_i)]\}}{LTU}\right) - \dfrac{Q_w}{Q_o m_i}},$$

(41–25a)

where the quantity LTU (length of transfer unit) is defined by

$$LTU = \frac{Q_w}{\bar{K}_w \pi d_{ti} N}.$$

(41–25b)

Thus, the total rate of solute extraction in the hollow-fiber module may be determined from Eq. (41–19) knowing c_{iwb}^{out} from, Eq. (41–25a).

An inverse problem of frequent interest is to know the length L of a hollow-fiber module needed to reduce c_{iwb}^{in} to a specified low value of c_{iwb}^{out}. For such a case, the result of Eq. (41–24) is needed. Since all terms on the right side are known, L is easily determined. On the basis of chemical engineering practice, Eq. (41–24) may also be written as (Treybal 1963)

$$L = LTU \times NTU, \qquad (41\text{–}26)$$

where the number of transfer units (NTU) is defined by

$$NTU = \frac{1}{\{1 - [Q_w/(m_i Q_o)]\}} \ln\left(\frac{c_{iwb}^{in} - (c_{iob}^{out}/m_i)}{c_{iwb}^{out} - (c_{iob}^{in}/m_i)}\right).$$

(41–27)

The quantity $\pi d_{ti} N$ is the total interfacial area of the hollow fibers in the module per unit module length. The interfacial area is represented as interfacial area per unit module volume a_{si} based on fiber ID or as a_{so} based on fiber OD. On this basis, LTU may be expressed as

$$LTU = \frac{Q_w/A}{\bar{K}_w \pi d_{ti} N/A} = \frac{v_{sw}}{\bar{K}_w a_{si}}, \qquad (41\text{–}28)$$

where

v_{sw} = superficial velocity of aqueous phase, Q_w/A, (41–29)

A = empty cross-sectional area of module, and a_{si} = interfacial area per unit module volume based on fiber ID.

Conventionally, HTU (height of transfer unit) is the term used to describe what has been identified here as LTU. Hollow-fiber modules need not be operated vertically; in fact, horizontal orientation is equally useful (Prasad and Sirkar 1990). Hence, the name LTU is appropriate. The expressions for LTU and NTU for a number of different configurations are provided in Table 41–2.

The method suggested above is based on a major assumption, namely, that a module-averaged overall transfer coefficient $\bar{K}_w$ is available. Alternately, if K_w is available, it can be integrated with distance x along the flow direction to provide a $\bar{K}_w$. The expression for K_w at any location in the module in this particular case is Eq. (41–9). Thus, K_w as a function of x will be available if k_{ios} and k_{iwt} are each available as a function of x with a known value of membrane coefficient k_{imo}.

Exact analytical expressions for k_{iwt} as a function of tube length are available under laminar flow conditions for either constant wall con-

TABLE 41-2. Expressions for LTU and NIU for Different Systems and Configurations.

Membrane	m_i		LTU	NTU
Hydrophobic	$\ll 1$	Aqueous in tube	$v_{so}/\bar{K}_o a_{si}$	$\ln\left(\dfrac{c_{iob}^{in} - c_{iwb}^{out} m_i}{c_{iob}^{out} - c_{iwb}^{in} m_i}\right) \bigg/ \left(1 - \dfrac{m_i Q_o}{Q_w}\right)$
		Organic in Tube	$v_{so}/\bar{K}_o a_{so}$	
	1	Aqueous in tube	$v_{so}/\bar{K}_o a_{si}$	' '
		Organic in Tube	$v_{so}/\bar{K}_o a_{so}$	' '
	$\gg 1$	Aqueous in tube	$v_{sw}/\bar{K}_w a_{si}$	$\ln\left[\dfrac{c_{iwb}^{in} - (c_{iob}^{out}/m_i)}{c_{iwb}^{out} - (c_{iob}^{in}/m_i)}\right] \bigg/ \left(1 - \dfrac{Q_w}{m_i Q_o}\right)$
		Organic in tube	$v_{sw}/\bar{K}_w a_{so}$	
Hydrophilic	$\ll 1$	Aqueous in tube	$v_{so}/\bar{K}_o a_{so}$	$\ln\left(\dfrac{c_{iob}^{in} - c_{iwb}^{out} m_i}{c_{iob}^{out} - c_{iw}^{in} m_i}\right) \bigg/ \left(1 - \dfrac{m_i Q_o}{Q_w}\right)$
		Organic in tube	$v_{so}/\bar{K}_o a_{si}$	
	1	Aqueous in tube	$v_{so}/\bar{K}_o a_{so}$	' '
		Organic in tube	$v_{so}/\bar{K}_o a_{si}$	' '
	$\gg 1$	Aqueous in tube	$v_{sw}/\bar{K}_w a_{so}$	$\ln\left[\dfrac{c_{iwb}^{in} - (c_{iob}^{out}/m_i)}{c_{iwb}^{out} - (c_{iob}^{in}/m_i)}\right] \bigg/ \left(1 - \dfrac{Q_w}{m_i Q_o}\right)$
		Organic in tube	$v_{sw}/\bar{K}_w a_{si}$	

centration or constant wall flux (Skelland 1974). However, neither of them represents the actual condition in a hollow-fiber extractor. There are models for flow and mass transfer on the shellside of the hollow-fiber device. But they are not exact and are based on particular boundary conditions at the wall not valid in an actual extraction module. In view of the unsatisfactory situation and the absence of an exact numerical solution for the problem, the following strategy may be adopted.

Consider the *limiting condition* of $m_i \gg 1$. From Eq. (41–9), then $K_w \cong k_{iwt}$. The existing exact analytical Graetz solution for laminar parabolic flow and developing mass transfer for uniform wall concentration may be used (Skelland 1974). Two forms of this solution are given below, the first one for the local Sherwood number and the second one for the average Sherwood number based on an arithmetic-mean driving force:

$$\frac{(k_{iwt})_x d_{ti}}{D_{iw}} = (Sh)_x = \frac{\sum_{j=1}^{j=\infty} \dfrac{B_j}{2}\left(\dfrac{d\phi_j}{dr_+}\right)_{r_+=1} \exp\left(\dfrac{-\beta_j^2(x/r_{ti})}{Re_t Sc_i}\right)}{\sum_{j=1}^{j=\infty} \dfrac{B_j}{-\beta_j^2}\left(\dfrac{d\phi_j}{dr_+}\right)_{r_+=1} \exp\left(\dfrac{-\beta_j^2(x/r_{ti})}{Re_t Sc_i}\right)},$$

(41–30a)

$$\frac{(k_{iwt})_a d_{ti}}{D_{iw}} = (Sh)_a = \frac{1}{2}\left(\frac{d_t}{L}\right)Re_t Sc_i \frac{1 - \sum_{j=1}^{j=\infty} \dfrac{-4B_j}{\beta_j^2}\left(\dfrac{d\phi_j}{dr_+}\right)_{r_+=1} \exp\left(\dfrac{-\beta_j^2(x/r_t)}{Re_t Sc_i}\right)}{1 + \sum_{j=1}^{j=\infty} \dfrac{-4B_j}{-\beta_j^2}\left(\dfrac{d\phi_j}{dr_+}\right)_{r_+=1} \exp\left(\dfrac{-\beta_j^2(x/r_t)}{Re_t Sc_i}\right)},$$

(41–30b)

where

$$\beta_j = 4(j-1) + 8/3; \quad j = 1, 2, 3 \ldots,$$

$$B_j = (-1)^{j-1} \times 2.84606 \beta_j^{-2/3}, \quad (41\text{-}30c)$$

$$\frac{-B_j}{2} \left(\frac{d\phi_j}{dr_+}\right)_{r_+=1} = 1.01276 \beta_j^{-1/3},$$

$$\text{Re}_t = \frac{d_{ti} v_w}{\nu_w},$$

$$\text{Sc}_i = \frac{\nu_w}{D_{iw}}.$$

The average mass transfer coefficient $(k_{iwt})_a$ (based on an arithmetic average driving force) can be obtained from the extraction data as

$$(k_{iwt})_a \pi d_{ti} NL \frac{(\Delta c_{i1} + \Delta c_{i2})}{2} = Q_w (c_{iwb}^{in} - c_{iwb}^{out}). \quad (41\text{-}30d)$$

Here the concentration driving forces at the two ends of the extractor in countercurrent flow are

$$\Delta c_{i1} = \left(c_{iwb}^{in} - \frac{c_{iob}^{out}}{m_i}\right), \quad \Delta c_{i2} = \left(c_{iwb}^{out} - \frac{c_{iob}^{in}}{m_i}\right), \quad (41\text{-}30e)$$

Further

$$\Delta c_{lm} = \frac{\Delta c_{i1} - \Delta c_{i2}}{\ln(\Delta c_{i1}/\Delta c_{i2})} \quad (41\text{-}30f)$$

is the logarithmic mean concentration driving force. We can then determine whether the experimentally observed mass transfer coefficient for small extents of solute extraction (wall concentration changes a little) in a module under conditions of $m_i \gg 1$ can be described by Eq. (41–30b). If so, the solution (41–30a) in length-averaged form or solution of Eq. (41–30b) may be used as $\overline{K}_w$ in Eq. (41–24) for $m_i \gg 1$ cases.

A similar procedure can be adopted for organic flow in the fiber lumen with an $m_i \ll 1$ system and a hydrophilic fiber. From Table 41–1, it is seen that $K_o \cong k_{iot}$. Thus, the utility of the Graetz solution in this case can be verified and an approximate solution for $\overline{K}_o$ obtained for use in $L = \text{LTU} \times \text{NTU}$ for this case. For determining the shellside mass transfer coefficients, a similar strategy may be adopted. The procedures are given in detail in Prasad and Sirkar (1988).

For an average mass transfer coefficient $(k_{iwt})_{lm}$ based on Δc_{lm}, the following equation is valid:

$$(k_{iwt})_{lm} \pi d_{ti} NL \Delta c_{lm} = Q_W (c_{iwb}^{in} - c_{iwb}^{out}) \quad (41\text{-}30g)$$

An expression of $(k_{iwt})_{lm}$ similar to Eq. (41–30b) is available in Skelland (1974).

The strategy of using limiting conditions, e.g., $m_i \gg 1$ or $m_i \ll 1$, may not always lead to the overall coefficient being equal to k_{iw} or k_{io}. For example, in a hydrophilic hollow-fiber module having aqueous flow on the tube side, locally

$$\frac{1}{K_w d_{to}} = \frac{1}{k_{iwt} d_{ti}} + \frac{1}{k_{imw} d_{lm}} + \frac{1}{m_i k_{ios} d_{to}}, \quad (41\text{-}31)$$

which for $m_i \gg 1$ reduces to

$$\frac{1}{K_w d_{to}} \cong \frac{1}{k_{iwt} d_{ti}} + \frac{1}{k_{imw} d_{lm}}. \quad (41\text{-}32)$$

Thus,

$$\int_0^L K_w \, dx = \int_0^L \frac{dx}{\left(\frac{1}{k_{iwt}}\right)\frac{d_{to}}{d_{ti}} + \left(\frac{1}{k_{imw}}\right)\frac{d_{to}}{d_{lm}}} = \overline{K}_w L \quad (41\text{-}33)$$

so that the average value of k_{iwt} along the tube will not provide the average value $\overline{K}_w$. A separate integration would be needed with k_{iwt} as a function of x in the above integral with known k_{imw}. Alternately,

$$\int_0^L \frac{dx}{K_w} = L\left(\overline{\frac{1}{K_w}}\right) = \int_0^L \frac{dx}{k_{iwt}\,(d_{ti}/d_{to})} + \frac{d_{to}}{d_{lm}}\frac{L}{k_{imw}},$$

(41–34)

where note that, in general,

$$\left(\overline{\frac{1}{K_w}}\right) \neq \frac{1}{\overline{K_w}}.$$

(41–35)

The procedure indicated by Eq. (41–33) has been followed by Frank (1986); he had incorporated the Newman extension (Newman 1969) of the Lévêque solution for k_{iot} for solvent extraction with a single hydrophobic hollow fiber in a glass tube. The solvent flowed in fiber lumen and the aqueous solution flowed cocurrently in the glass tube. He was able to describe the observed data quite well for an $m_i \ll 1$ system.

Although Eq. (41–30b) appears to describe the mass transfer better over a wider range of conditions (Prasad and Sirkar 1988), a simpler expression of $\overline{K_w}$ from the Lévêque solution (valid only for Gr > 400) may be adopted to obtain an analytical expression for LTU in tubeside controlled extraction. From the Lévêque solution, for a tube of length L and ID d_{ti},

$$\frac{\overline{k_{iwt}} d_{ti}}{D_{iw}} = 1.6151 \left[\left(\frac{d_{ti}}{L}\right) \mathrm{Re}_t\, \mathrm{Sc}_i\right]^{0.33},$$ (41–36)

where

$$\int_0^L k_{iwt}\, dx = \overline{k_{iwt}} L.$$ (41–37)

If $\overline{K_w} \cong \overline{k_{iwt}}$, then by substituting such an expression in Eq. (41–28), an analytical expression for LTU may be obtained. In such an expression,

$$\text{LTU} \propto d_{ti}^{4/3} v_w^{2/3}.$$ (41–38)

Such LTU dependences on d_{ti} and v_w have been experimentally verified (Prasad and Sirkar 1990). As the fiber ID and the tubeside velocity decrease, LTU decreases. Such predictive abilities are not available in dispersion-based extractors. Further LTU values as low as 3 cm were achieved by Prasad and Sirkar (1990). The lowest HTU values reported in conventional dispersion-based devices are an order of magnitude higher. Correspondingly, Wald, Lopez, and Matson (1989) have reported that a membrane-based hollow-fiber extractor had about four times more volumetric throughout (150 L/min·m^3) than that of Karr column in the extraction of penicillin G at a solvent flow rate of 50L/h.

Equation (41–36) indicates that, other conditions remaining constant, a longer hollow-fiber module will lead to a lower value for the tubeside average mass transfer coefficient. If the tubeside average coefficient is important in a given problem, a higher transfer coefficient will be obtained by joining two modules in series instead of a long module. Prasad and Sirkar (1990) have observed that the overall mass transfer coefficient in two modules-in-series was the same as that of any one of the modules since the concentration boundary layer begins again in the second module after the first module.

Mass Transfer with Chemical Reaction

Chemical reaction is often employed in liquid-liquid extraction processes to enhance the extraction rates. The reaction can be instantaneous, fast, or slow and can be reversible or irreversible. The enhanced extraction rate can come about due to enhancement of the mass transfer coefficient and an increased concentration driving force. In membrane-based extraction, this can lead to a variety of conditions: membranes can be hydrophobic or hydrophilic; the reaction front (if any) can be inside the membrane or in the liquid film outside the membrane; liquid-in-pore and membrane wetting characteristics can vary. The effect of chemical reaction is illustrated below for an instantaneous reaction whereby the back extrac-

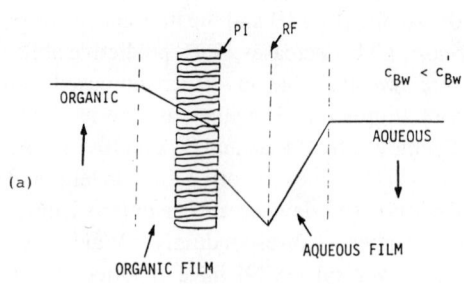
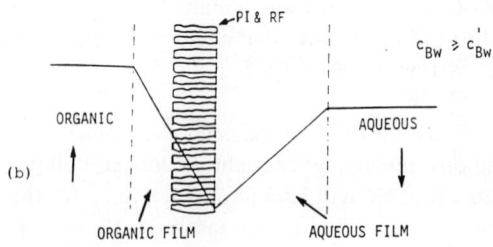

FIGURE 41-8. Solute concentration profiles for a mass transfer with instantaneous chemical reaction using a hydrophobic membrane for phenol extraction from MIBK with a NaOH solution: (a) stoichiometrically deficient caustic solution and (b) stoichiometrically sufficient caustic solution (reprinted from Basu, Prasad, and Sirkar 1990 with permission).

tion rate of phenol from methyl isobutyl ketone (MIBK) into water was significantly enhanced due to the presence of NaOH in water (Basu, Prasad, and Sirkar 1990). To this end, the Hatta approach of a two-film model with simultaneous instantaneous chemical reaction (Astarita 1967) is utilized.

Example: Phenol-Caustic Reaction

Consider a hydrophobic microporous flat membrane wetted by the organic phase containing phenol and an aqueous caustic solution on the other side (Figure 41-8). If phenol is species A and caustic is species B in the reaction

$$C_6H_5OH + NaOH \rightarrow C_6H_5ONa + H_2O \quad (41\text{-}39)$$

then, depending on the level of caustic concentration c_{Bw}, the instantaneous reaction front (RF) will either be in the aqueous boundary layer or at the phase interface (PI). This is determined by the value of the critical caustic concentration c_{Bw}^t:

$$c_{Bw}^t = \left(\frac{D_{Aw}}{D_{Bw}}\right) \frac{k_{Ao}}{k_{Aw}} c_{Ao}. \quad (41\text{-}40)$$

For $c_{Bw} < c_{Bw}^t$, the reaction front will not coincide with the phase interface. Therefore, phenol diffuses not only in the organic phase but also in the aqueous phase up to RF. For $c_{Bw} \geq c_{Bw}^t$, the reaction front is at the phase interface. Phenol diffuses only in the organic phase. This regime is identified as the excess caustic regime.

The organic-phase-based overall mass transfer coefficient K_o in the back extraction of phenol from MIBK into water would be described simply by relation (41-4):

$$\frac{1}{K_o} = \underbrace{\frac{1}{k_{io}} + \frac{1}{k_{imo}}}_{\text{organic resistance}} + \underbrace{\frac{m_i}{k_{iw}}}_{\text{aqueous resistance}}.$$

For $c_{Bw} \geq c_{Bw}^t$, it is obvious that m_i/k_{iw} is negligibly small so that

$$\frac{1}{K_o} \cong \frac{1}{k_{io}} + \frac{1}{k_{imo}}. \quad (41\text{-}41)$$

Remember that $m_i \gg 1$ for phenol between MIBK and water. Thus, whereas in nonreactive cases from Eq. (41-18)

$$K_o \cong \frac{k_{iw}}{m_i} \quad (41\text{-}42)$$

due to the instantaneous reaction under conditions of $c_{Bw} \geq c_{Bw}^t$, K_o has been increased to a much larger K_o:

$$K_o \cong \frac{1}{(1/k_{io}) + (1/k_{imo})}. \quad (41\text{-}43)$$

Further, the phenol flux is given by

$$N_A = K_o c_{Ao} \quad (41\text{-}44)$$

TABLE 41–3. Some Mass Transfer Relations in Extraction with An Instantaneous Reaction* in Aqueous Film.

	$c_{Bw} \geq c_{Bw}'$ **	$c_{Bw} < c_{Bw}'$ †
Hydrophobic film Organic-in-pore	$\dfrac{1}{K_o} \cong \dfrac{1}{k_{io}} + \dfrac{1}{k_{imo}}$	$\dfrac{1}{K_o} = \dfrac{1}{k_{io}} + \dfrac{1}{k_{imo}} + \dfrac{m_i}{k_{iw}}$
Hydrophilic film Aqueous-in-pore	$\dfrac{1}{K_o} \cong \dfrac{1}{k_{io}}$	$\dfrac{1}{K_o} = \dfrac{1}{k_{io}} + \dfrac{m_i}{k_{mw}} + \dfrac{m_i}{k_{iw}}$
Composite film Aqueous-in-Hydrophilic Organic-in-Hydrophobic	$\dfrac{1}{K_o} \cong \dfrac{1}{k_{io}} + \dfrac{1}{k_{imo}}$	$\dfrac{1}{K_o} = \dfrac{1}{k_{io}} + \dfrac{1}{k_{imo}} + \dfrac{m_i}{k_{imw}} + \dfrac{m_i}{k_{iw}}$
Phenol flux	$N_i = K_o\, c_{io}$	$N_i = K_o \left(c_{io} + \dfrac{D_{Bw}}{D_{Aw}} c_{Bw} \right)$

*Reaction is aA + bB → pP + qQ
For example, A is phenol and B is NaOH.
**Reaction front at phase interface.
†Reaction front somewhere in the aqueous film.

since $c_{Aw} = 0$. Table 41–3 identifies the effective relations between overall mass transfer coefficient K_o and the individual transfer coefficients for phenol back extraction into a caustic solution for hydrophobic, hydrophilic, and composite membranes. Observe that the hydrophilic membrane for the excess caustic regime will have the highest mass transfer coefficient, $K_o \cong k_{io}$. These relations are valid for any other instantaneous reaction in the aqueous phase as the solute is extracted into it.

Another class of reactions of great importance in solvent extraction of metals and organic acids is the interfacial reaction. When the solute to be extracted is sparingly soluble or insoluble in the extract phase, chelating or ion-pair-forming extractants are used to increase drastically the amount of solute extracted. The extractants added usually have limited aqueous phase solubility. Reactions take place in the interfacial region with or without an interfacial adsorption step. Matsumoto et al. (1987) have studied copper extraction with a chelating agent in a hydrophobic microporous hollow-fiber device made from PTFE hollow fibers. Basu (1989) has studied citric acid extraction with hydrophobic microporous hollow fibers into an organic phase containing tri-*n*-octylamine. Whatever the nature of the chemical reaction during extraction (instantaneous, fast, interfacial, etc.), differential equations have to be developed along the hollow-fiber extractor length and solved to estimate the rate of extraction. The two references mentioned above may be consulted for such equations and their solutions.

Breakthrough Pressure

To ensure nondispersive solvent extraction with a microporous hydrophobic or hydrophilic membrane, the pressure of the phase not in the membrane pores must not exceed the pressure of the phase in the membrane pores by a critical amount Δp_{cr}. This maximum allowable value of differential pressure is identified as the *breakthrough pressure*. Measured values of Δp_{cr} for four different hydrophilic membranes of different pore sizes obtained in the extraction of acetic acid from water into xylene or methyl isobutyl ketone are available in Prasad and Sirkar (1987a). Additional information on Δp_{cr} is available for different systems and membranes in Prasad et al. (1986), Prasad, Frank, and Sirkar (1988), and Prasad et al. (1990). A few Δp_{cr} values are given in Table 41–4. Kim and Harriott (1987) have experimentally determined the value of Δp_{cr} for a number of aqueous-

TABLE 41-4. Breakthrough Pressure for Some Membrane Extraction Systems.

Membrane	Average Pore Size	System	Liquid in Pore	$c'_{iwb} \times 10^3$ (gmol/cm^3)	Breakthrough Pressure (Δp_{cr})
Celgard® 2400 (polypropylene) (Hoechst Celanese)	0.02 μm	n-heptane-NMP-toluene	NMP	9.4	137 kPa[a]
		Do	n-Heptane	9.4	>241 kPa[a]
		n-butanol-water-succinic acid	n-Butanol	0.044	212 kPa[b]
Cuprophan® 150 PM (regenerated cellulose) (ENKA)	~40Å	MIBK-water-acetic acid	Water	2.33	414 kPa[c]
		n-butanol-water-succinic acid	Water	0.044	>414 kPa[b]
		Xylene-water-acetic acid	Xylene	0.042	68.9 kPa[a]
		Do	Water	0.042	>482 kPa[a]
Regenerated cellulose (MFS)	0.45 μm	MIBK-water-acetic acid	Water	2.33	83 kPa[c]
Cellulose acetate (MFS)	0.2 μm	Xylene-water-acetic acid	Water	0.042	215 kPa[c]
Cellulose acetate (Millipore) EG Series	0.2 μm	Xylene-water-acetic acid	Water	0.042	110 kPa[c]
Nylon (ENKA)	0.2 μm	Xylene-water-acetic acid	Water	0.042	172 kPa[a]
		Do	Xylene	0.042	124 kPa[a]
Gore-Tex® 2	0.2 μm	Xylene-water-acetic acid	Xylene	0.042	158 kPa[d]

[a] Prasad et al. (1990).
[b] Prasad, Frank, and Sirkar (1988).
[c] Prasad and Sirkar (1987a).
[d] Prasad et al. (1986).

organic systems with a large-pore hydrophobic polytetrafluoroethylene flat membrane.

If the microporous membrane could be modeled as a collection of parallel cylindrical pores of radius r_p, then the breakthrough pressure is related to other relevant variables by the Young-Laplace equation

$$\Delta p_{cr} = \frac{2\gamma_{wo} \cos\theta_c}{r_p}, \quad (41\text{-}45)$$

where γ_{wo} is the interfacial tension between the aqueous-organic system and θ_c is the contact angle measured from the pore wall to the tangent of the liquid-liquid interface drawn from the three-phase contact point into the pore liquid. The liquid-liquid interface is concave from inside the pore liquid. Models for membranes with noncylindrical pores have been studied by Kim and Harriott (1987) and Δp_{cr} correlated with $\gamma_{wo} \cos\theta_{\text{eff}}$ where θ_{eff} is an effective contact angle. However, it is obvious from the expression for Δp_{cr} given above that it increases as r_p is reduced and γ_{wo} is increased for a given θ_c.

A number of typical aqueous-organic two immiscible phase systems with significant mutual solubility exist for which the value of γ_{wo} is quite small. For example, a water-n-butanol system has a γ_{wo} of 1.5 dyne/cm with the solute succinic acid present at 0.8% level. [For a list and description of test systems recommended for mass transfer studies in solvent extraction, see Misek (1978).] The presence of surfactants will lower the interfacial tension drastically. Some solutes to be extracted behave as surfactants. For example, the pharmaceutical compound MK-819, mevinolinic acid, was suspected of reducing the value of γ_{wo} in an isopropyl acetate-water system to 0.3 dyne/cm from a value of 10.3 dyne/cm. The value of the breakthrough pressure would naturally be quite low. Membranes with small pore sizes are preferable in such cases so that a comfortable value of Δp_{cr} is available. Alternately, selected compounds e.g., some alcohols such as isopropanol, may be added to the system to increase the interfacial tension and the magnitude of Δp_{cr} (Prasad and Sirkar 1989).

There is an interplay between reducing pore size to increase Δp_{cr} in low interfacial tension systems and the possibility of hindered diffusion for large molecules, e.g., proteins. Protein extraction is carried out in systems (biphasic aqueous or reverse micelle) with low interfacial tension. If the pore size is reduced, Δp_{cr} can be increased but the protein diffusion rate through the pores will be decreased drastically.

Asymmetric Membranes

Almost all microporous membranes investigated for solvent extraction are essentially symmetric in nature. Sirkar, Prasad, and Khare (1990) have presented solvent extraction data from microporous hydrophilic alumina tubes whose pore diameter on the inside surface was either 40 or 100 Å, while that on the outside surface was as high as 100,000 Å. Since Δp_{cr} is inversely proportional to the pore radius r_p, they were able to immobilize the aqueous-organic interface at various locations inside the pore depending on the value of Δp between the phases. In a symmetric membrane having the same r_p all across the membrane thickness, once the Δp_{cr} [Eq. (41–45)] is exceeded, the phase having higher pressure will break through and get dispersed in the phase having lower pressure as drops. In an asymmetric membrane with r_p varying across membrane thickness, if the higher pressure phase is on the side of the membrane with larger r_p, then as Δp exceeds Δp_{cr} for a particular pore size r_{p_1}, the interface shifts to the location where the applied Δp equals the Δp_{cr} for that location with lower $r_p = r_{p_2}$. Thus, such a membrane can be used for a range of Δp_{cr} values corresponding to the largest and the smallest membrane pore sizes.

A number of advantages may accrue from such control of interface location in the pore of a hydrophilic asymmetric membrane. If the organic membrane phase resistance is high (e.g., for $m_i \ll 1$ systems), the interface may be shifted to reduce the thickness of the organic-filled section of the membrane. Similarly, for $m_i \gg 1$ systems, the interface may be shifted to reduce the thickness of the aqueous-filled section of the membrane. Khare (1991) has demonstrated how the mass transfer coefficient may be increased or decreased by shifting the location of the aqueous-organic interface along the thickness of an asymmetric microporous ceramic membrane tube.

In systems having a low value of γ_{wo}, a practical value of Δp_{cr} requires a low value of r_p. For large molecular weight solutes, this implies considerable hindered diffusion in a symmetric membrane. An asymmetric membrane with an appropriate pore size variation across the membrane thickness can ensure a reasonable Δp_{cr} for low γ_{wo} systems and yet drastically reduce the hindered diffusion of large molecular weight solutes. For bioseparations of larger molecules, extraction using asymmetric membranes may thus be particularly advantageous.

Mass Transfer in an Extraction Module and a Back Extraction Module

In many solvent-extraction-based separation processes, the solute extracted into a solvent may have to be recovered by back extraction into a liquid miscible with the original feed and immiscible with the extracting solvent. Many such examples are available in Hanson, Lo, and Baird (1983). Two membrane-based solvent extraction devices may also be utilized for the same goal: one device for solvent extraction of solute and the other for back extraction of solute. Figure 41-9 illustrates such an arrangement.

The question of interest here is the follow-

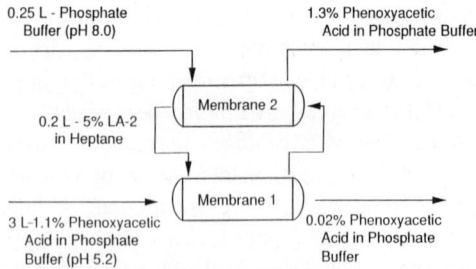

FIGURE 41–9. Schematic of a pH swing extraction/purification of phenoxyacetic acid (adapted from Wald, Lopez, and Matson (1989) with permission).

ing: Other conditions remaining constant, what fiber and liquid-in-pore should be used for this two-module extraction and back extraction? Suppose that, in the extraction module, conditions are such that $m_i \gg 1$ favoring extraction into the organic solvent from an aqueous feed. Further, let the conditions in the back extraction module be such that $m_i \ll 1$ favoring solute back extraction into an aqueous solution from the solvent phase. The answers are easily obtained from basic mass transfer characteristics of the microporous membranes in solvent extraction pointed out earlier.

For example, for an $m_i \gg 1$ system, an organic-in-hydrophobic membrane has a higher K value than an aqueous-in-hydrophilic membrane. Therefore, for extraction, a hydrophobic fiber module with organic-in-pore should be used since $m_i \gg 1$. Similarly, for back extraction with $m_i \ll 1$, an aqueous-in-pore hydrophilic membrane has a higher K value than an organic-in-pore hydrophobic membrane, suggesting the use of a hydrophilic membrane for back extraction. The efficiency of this configuration of hydrophobic-for-extraction and hydrophilic-for-back extraction in a two-module system over other configurations of hydrophilic-hydrophilic and hydrophilic-hydrophobic with an aqueous feed, has been demonstrated by Wald, Lopez, and Matson (1989) for phenoxyacetic acid extraction. The additional configuration of hydrophobic-hydrophobic was studied by Alexander and Callahan (1987) for gold extraction and back extraction. Kim (1984) had also studied copper extraction and back extraction; however, his operation was not dispersion-free.

Obviously, if both modules use hydrophilic fibers with the first one for $m_i \gg 1$ having organic-in-pores and the second one for $m_i \ll 1$ having aqueous-in-pores, then such a combination could also provide the highest mass transfer rates based on the studies by Prasad et al. (1990). What combination of liquid-in-pore and fiber should be optimal for extraction and back extraction is not yet resolved.

MEMBRANE DEVICES AND TRANSPORT CORRELATIONS

A number of types of membrane devices can be used for carrying out solvent extraction: hollow-fiber, spiral-wound, plate-and-frame, and rotating annular devices. It appears that only hollow-fiber devices are currently available on a commercial basis. A brief description of microporous hollow-fiber (MHF) devices is provided first. The transport correlations for mass transfer and pressure drop are then presented. At the end of this section, any information available on other types of modules is given.

MHF Devices Commercially Available

MHF devices for solvent extraction are available from two sources. The Liqui-Cel® modules of Hoechst Celanese (Separations Products Division, Charlotte, North Carolina) are simple shell-and-tube type devices without any shell-side baffles. The shell casing can be of the following plastics: nylon, polycarbonate, polysulfone, and polypropylene. The potting compound used to make the tubesheet is either epoxy or polyurethane. Celgard® microporous hydrophobic polypropylene hollow fibers are used. Table 41–5 provides details of some available Liqui-Cel® modules, e.g., fiber dimensions, fiber number, effective module length, interfacial area, etc. Figure 41–10 is a photograph of a commercially available module. Modules with epoxy potting and polypropylene casings can withstand most chemicals except chlorinated solvents, ketones, and DMF.

The second source of MHF hollow-fiber devices is Sepracor Inc., Marlborough, Massa-

TABLE 41–5. Details of Liqui-Cel® Hollow-Fiber Modules.

Fiber Diameter (μm)		Number of Fibers	Shell Dimensions (cm)		Area Per Unit Volume (cm^{-1})
ID	CD		Diameter	Length	
405	464	900	1.9	15.8	46.3
244	298	7500	4.7	24.1	40.4
405	464	3200	4.7	24.1	26.8
244	298	7500	4.7	54.6	40.4
405	464	3200	4.7	54.6	26.8

Note: Area per unit volume based on fiber OD.
Source: Hoechst Celanese Separations Products Div., Charlotte, NC

chusetts. The technology trademark is MSX. Shell-and-tube modules available are of three types: Model LP containing hydrophilic microporous membrane for the pH range of 4 to 10; Model LG containing hydrophilic gel membrane that can operate in excess of 80°C; and Model BP contains a hydrophobic microporous membrane and tolerates pH extremes and temperatures in excess of 80°C. MSX Sepracor modules are not compatible with ethyl acetate, all chlorinated solvents, DMSO, DMF, dimethylacetamide, 1.0 N solutions of HCL, H$_2$SO$_4$, NaOH, or KOH. Only short time compatibility ($\leq$60 min) is observed with acetone, methyl ethyl ketone, and 0.1 N solutions of HCl, H$_2$SO$_4$, NaOH, or KOH per Sepracor brochures. The hydrophilic hollow fibers are presumably of polyacrylonitrile.

Two types of microporous hydrophilic hollow fibers are available from AKZO (Asheville, NC): Cuprophan® hollow fibers and nylon hollow fibers. Hollow-fiber devices of these fibers have been prepared and studied by Prasad and Sirkar (1988) and Basu, Prasad, and Sirkar (1990). Commercially available single-ceramic (alumina; ALCOA, Warrendale, PA) membrane tubes with an asymmetric structure and microporous glass tubes (ASAHI, New York) have also been used for nondispersive extraction (Prasad et al. 1990).

Transport Correlations in MHF Devices

Available correlations, their sources and experimental conditions are identified below for mass transfer coefficients and pressure drops.

Tubeside Mass Transfer Correlations

Only a limited amount of work has been done to develop correlations for mass transfer coefficients in hollow-fiber devices having solvent extraction. These are summarized below first for the tube side and then for the shell side.

Although the flow regime in the tube side of conventional shell-and-tube heat exchangers can vary from laminar to turbulent, the tubeside flow regime in MHF devices is most likely to be laminar due to the very small lumen diameter.

FIGURE 41–10. Photograph of a commercially available hollow-fiber solvent extraction module (reprinted from Hoechst Celanese Separations Products Division with permission).

Otherwise, the pressure drop along the fiber would be exceedingly high. In any such correlation developed and indicated below, the effects of any maldistribution of flow on the tube side resulting from particular designs of inlet and outlet headers are hidden (Park and Chang 1986). Similarly, the effects of any fiber inlet crimping or fiber inlet plugging at the tube sheet are unknown.

For extraction of solutes from an aqueous solution to a solvent under conditions of no more than a quarter of the solute being extracted, Prasad and Sirkar (1988) have found that the following correlation [see Eq. 41-30b)] described the tubeside coefficient for species i quite well for hollow fibers of ID d_{ti} and length L:

$$\text{Sh}_i = \frac{k_{it}d_{ti}}{D_{it}} = 0.5 \left(\frac{d_{ti}}{L}\right) \text{Re}_t \text{Sc}_i \frac{1-\zeta}{1+\zeta}, \quad (41\text{-}46)$$

where

$$\text{Re}_t = \frac{d_{ti}v_t}{\nu_t}, \quad (41\text{-}47)$$

$$\text{Sc}_i = \frac{\nu_t}{D_{it}}, \quad (41\text{-}48)$$

$$\zeta = \sum_{n=1}^{\infty} -4 \left(\frac{B_n}{\beta_n^2}\right)\left(\frac{d\phi}{dr_+}\right)_{r_+=1} \exp\left(\frac{-2\beta_n^2 L}{d_{ti}\text{Re}_t\text{Sc}_i}\right), \quad (41\text{-}49)$$

$$\beta_n = 4(n-1) + 2.666, \quad n = 1,2,3 \ldots, \quad (41\text{-}50a)$$

$$B_n = -(1)^{n-1} \times 2.84606 \beta_n^{-2/3}, \quad (41\text{-}50b)$$

$$-B_n\left(\frac{d\phi}{dr_+}\right)_{r_+=1} = 2(1.01276 \beta_n^{-0.33}). \quad (41\text{-}51)$$

These correlations were developed with both hydrophobic and hydrophilic hollow fibers for $300 < \text{Sc}_i < 1000$ and $0 < \text{Re}_t < 60$. Prasad and Sirkar (1990) have found this correlation to be valid also for simultaneous extraction of 4-methylthiazole (MT) and 4-cyanothiazole (CNT) from aqueous phase with solute recoveries as high as 99%. Both systems have $m_i \gg 1$. Yun, Prasad, and Sirkar (1989) have also arrived at the same conclusion for priority organic pollutants having $m_i \gg 1$. Note that an arithmetic average driving force [see, for example, Eqs. (41-30d) and (41-30e)] should be used instead of a logarithmic average driving force for high solute recoveries when using such a correlation. For low solute recoveries, the two driving forces yield similar results.

Dahuron and Cussler (1988) have found that the tubeside mass transfer coefficients for the extraction of small molecules and proteins into a solvent can be correlated by

$$\text{Sh}_i = \frac{k_{it}d_{ti}}{D_{it}} = 1.5 \left(\frac{d_{ti}v_t}{LD_{it}}\right)^{1/3} \quad (41\text{-}52)$$

for hydrophobic hollow fibers. The ranges of Re_t and Sc_i are not available; however, the Sh_i was in the range of 8 to 40.

A similar correlation developed by Yang and Cussler (1986) for hydrophobic hollow fibers from absorption of gases into water flowing in the fiber lumen is as follows:

$$\text{Sh}_i = \frac{k_{it}d_{ti}}{D_{it}} = 1.64 \left[\left(\frac{d_{ti}}{L}\right)\text{Re}_t\text{Sc}_i\right]^{0.33}. \quad (41\text{-}53)$$

This was developed for a Sc_i of 464 and a modified Graetz number $(v_t d_{ti}^2/D_{it}L)$ varying between 3 and 500 (the Graetz number for a tube is $(\pi/4)v_t d_{ti}^2/D_{it}L)$. This correlation is simply the length-averaged form of the Lévêque solution [Eq. (41-36)] if the constant is changed to 1.615.

Shellside Mass Transfer Correlations

Two types of gross flow patterns have been used on the shell side of a hollow-fiber device: parallel flow and crossflow. Using solvent extraction studies, correlations for parallel flow have been developed. Crossflow correlations have been formulated from gas absorption studies. Shellside flow in existing hollow-fiber devices is likely to be significantly influenced by bypassing, backmixing, and channeling. Thus, any correlation developed probably reflects the specific system and conditions used; extrapolation to other conditions may be undertaken only with caution.

Prasad and Sirkar (1988) have obtained the following correlation for parallel flow on the shell side in solvent extraction with simple shell-and-tube devices having hydrophobic fibers:

$$\mathrm{Sh}_i = \frac{k_{is} d_e}{D_{is}} = 5.85(1 - \phi)\left(\frac{d_e}{L}\right)\mathrm{Re}_s^{0.66}\,\mathrm{Sc}_s^{0.33}, \quad (41\text{-}54)$$

where

$$\mathrm{Re}_s = \frac{d_e v_s}{\nu_s}, \quad (41\text{-}55)$$

$$\mathrm{Sc}_s = \frac{\nu_s}{D_{is}}, \quad (41\text{-}56)$$

ϕ is the packing fraction of hollow fibers in shell and d_e is the hydraulic diameter of the shell side, 4 × cross-sectional area/wetted perimeter. For hydrophilic fibers, the constant 5.85 is to be changed to 6.1.

From solvent extraction studies in hydrophobic fiber modules, Dahuron and Cussler (1988) have suggested the following shellside correlation in parallel flow:

$$\mathrm{Sh}_i = \frac{k_{is} d_e}{D_{is}} = 8.8\left(\frac{d_e}{L}\,\mathrm{Re}_s\right)(\mathrm{Sc}_s)^{0.33}. \quad (41\text{-}57)$$

On the basis of gas absorption studies, Yang and Cussler (1986) recommend the following correlations for the shell-side in parallel flow:

$$\mathrm{Sh}_i = 1.25\left(\frac{d_e}{L}\,\mathrm{Re}_s\right)^{0.93}(\mathrm{Sc}_s)^{0.3} \quad \text{for } \phi \le 0.26 \quad (41\text{-}58)$$

and

$$\mathrm{Sh}_i = 0.80, \quad \text{for } \phi \ge 0.40. \quad (41\text{-}59)$$

When the liquid (water) is in crossflow, Yang and Cussler (1986) have found that

$$\mathrm{Sh}_i = 0.9(\mathrm{Re}_s)^{0.4}(\mathrm{Sc}_s)^{0.33} \quad (41\text{-}60)$$

is valid for low values of ϕ (≤ 0.07), whereas

$$\mathrm{Sh}_i = 1.38(\mathrm{Re}_s)^{0.34}(\mathrm{Sc}_s)^{0.33} \quad (41\text{-}61)$$

for higher packing fractions. The data were obtained at $\mathrm{Sc}_s = 476$. Kang et al. (1988) have compared the two shellside correlations [Eqs. (41-54) and (41-58)] for a low value of ϕ (~0.03) for absorption of O_2 in water. They have observed that correlation (41-54) described their data much better. Basu, Prasad, and Sirkar (1990) have found some difficulty in describing their data on shellside mass transfer coefficients using the parallel flow correlations of Eqs. (41-54) and (41-58). This suggests further studies to resolve the problem.

Pressure Drop Correlations

Laminar flow correlations for pressure drop in tubes can be used to estimate the lumen side pressure drop in hollow-fiber bundles. The Hagen-Pouseuille law for pressure drop in an incompressible Newtonian fluid is particularly useful:

$$\Delta p = \frac{32\eta_t L v_t}{d_{ti}^2}. \quad (41\text{-}62)$$

Very little work has been done to develop general pressure drop correlations for the shell-side liquid flow in hollow-fiber extraction devices. The pressure drop on the shell side is a complex function of the flow velocity, fiber diameter, fiber packing fraction, flow pattern (parallel or cross), and module fabrication. Instead of developing generic equations of pressure drop in the shell side of a module, it is suggested that module manufacturers should be contacted directly.

DESIGN OF MODULES AND MODULAR CASCADES

Two types of problems result if a feed containing a solute or solutes is to be subjected to solvent extraction for recovery or elimination of the solutes. In a design problem, we need to know the type, number, and dimensions of the modules needed to recover the solute or purify the feed to a specified extent. In a rating problem, we calculate the solvent extraction capabilities of a given module or a given set of modules for a given feed stream. Any such design or rating calculation should be preceded by some preliminary considerations regarding the selection of membrane/module and the operating mode. These preliminary considerations may need some experimental input about the system *vis-à-vis* the module/membrane to develop a design that would be successful in practice.

Selection of Membrane/Module and Operating Mode

A typical set of questions in membrane-based solvent extraction design/application for a process problem are as follows:

1. Which membrane, hydrophobic or hydrophilic, is to be selected?
2. Should the extract flow in the fiber lumen or the shell side of the hollow-fiber module?
3. Should the extraction be carried out in an unsteady batch recirculation mode or in continuous contact extractor mode?

That, other conditions remaining constant, a module with a higher interfacial area/module volume should be selected is obvious.

Selection of Membrane/Module

Since the breakthrough pressure is much larger with organic-in-hydrophobic membrane (or aqueous-in-hydrophilic membrane), such a membrane wetting mode is to be chosen unless extraction conditions suggest otherwise. Hydrophobic membranes with organic-in-pore have the following advantages:

1. Higher mass transfer coefficient for the $m_i > 1$ system.
2. Generally higher pH and chemical stability.
3. Reduced fouling with whole cells.
4. Most available membranes do not have very small pores (<100 Å). Thus, for larger molecules, hindered diffusion may be avoided.
5. Easier sterilizability.

Hydrophilic membranes are particularly advantageous when the following conditions exist:

1. System has $m_i < 1$, thus providing a higher mass transfer coefficient under comparable boundary layer mass transfer conditions and membrane transfer coefficient.
2. System has lysed cells, proteins, etc. They are likely to foul the hydrophilic membrane surface to a lesser extent. There are wide variations, however, in fouling characteristics among different hydrophilic membranes.
3. Since most available hydrophilic membranes for solvent extraction have small pore sizes, very large size molecules can be preferentially excluded during solvent extraction, due to hindered diffusion.

In a recent study on extraction of a pharmaceutical compound from an isopropyl acetate solution into water at high pH where $m_i < 1$, a hydrophobic membrane was successfully utilized since the hydrophilic membrane pores were plugged by an unstable precipitate in the aqueous phase (Prasad and Sirkar 1989). In this case the hydrophilic membrane with aqueous-in-pore was a natural *initial* choice since $m_i < 1$. If a hydrophilic membrane has to be used, an alternative approach would be to employ the organic phase in pore to avoid precipitation in pore. This suggests that, in spite of general recommendations, specific choices have to be made based on particular system properties.

Consistent with the process objectives, the pore size is to be chosen after the type of membrane, hydrophobic or hydrophilic, is selected. Unless the solute mass transfer rate in extraction is reduced by hindered diffusion, it is preferable to select as small a pore size as possible to increase the value of Δp_{cr}. Pore size selection may also be influenced by the need to exclude whole cells, lysed cell fragments, and bacteria from the membrane pores in applications of solvent extraction from fermentation broth. Cleaning properties and sterilization capabilities are additional items needing attention. Ceramic modules, if available, are inherently advantageous in this respect because of their high-temperature capabilities.

The fouling characteristics of fermentation broths, their solids content and rheology may dictate the choice of module type among hollow-fiber module, tangential flow filtration type device, rotary device, etc., if unclarified broth is to be used for extraction. Broths containing filamentous fungi are probably better handled in extractors with tangential flow type devices. However, very little is known about the performance of such slurries in solvent extraction. Fouling experiences of ultrafiltration (UF) and microfiltration (MF) devices are not automatically transferable to membrane-based solvent extraction devices since solvent flux in UF and MF is totally absent in solvent extraction.

Module Operation Mode

After the selection of membrane/module has been made, the module operation mode has to be selected. Two aspects need to be addressed:

1. Tubeside feed or shellside feed.
2. Batch recirculation or continuous contact mode; continuous contact will usually have countercurrent flow but cocurrent mode is practiced, especially in systems with chemical reaction.

In the hollow-fiber modules currently available for solvent extraction, a distinct possibility exists for channeling, bypassing, and backmixing on the shell side (Prasad and Sirkar 1990). A high degree of solute extraction from the feed stream (introduced to the shell side) into the extract stream, if desired, cannot be achieved in a hollow-fiber module with significant shellside bypassing, etc. Therefore, it is suggested that the solute-containing feed stream should flow in fiber lumen and the extract flow on the shell side unless a number of modules are being used in series. Such choices have to be tempered with considerations of solids or particulate content and particle size in each stream. Streams with particles having dimensions at least one to two orders of magnitude smaller than the fiber lumen diameter may flow in the fiber lumen. In the absence of experimental data, filtration prior to extraction is recommended. A well-packed shell side may be more prone to particle deposition in dead spots. No such problem has been encountered in fermentation broths with whole cells on the shell side of relatively open hydrophobic hollow-fiber devices (Frank and Sirkar 1985, 1986; Kang 1989).

An additional consideration is the nature of bonding between the hollow-fiber outside surface and the tubesheet. If the quality of bonding is poor, then a high value of excess phase pressure on the shell side may lead to leakage through the tubesheet. The phase with the excess pressure should then preferably flow in the fiber lumen. Alternately, the value of Δp between the phases may be kept at a low level. The unbonding effect is observed at lower Δp-s as the fiber diameter is increased.

Once a selection is made between tubeside feed and shellside feed, attention should shift to batch recirculation versus the continuous-contact operating mode. That membrane-based solvent extraction is an equilibrium-based separation process will be the primary consideration here.

In the single-pass continuous-contact countercurrent extractor of Figure 41-7, the exiting extract stream composition c_{iob}^{out} can at the most be in equilibrium with the incoming feed concentration c_{iwb}^{in} regardless of the length of the continuous-contact device. On the other hand, since the solvent stream often comes in with c_{iob}^{in} essentially equal to zero, the value of c_{iwb}^{out} can be reduced to very low levels if feed flow rates are sufficiently low. As Eq. (41–25a) indicates, c_{iwb}^{out} will be exponentially reduced with the length subject to the thermodynamic limit of equilibrium with c_{iob}^{in}.

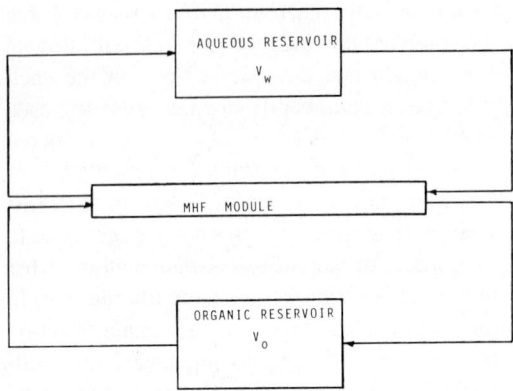

FIGURE 41-11. Schematic of the batch recirculation mode of operation using a hollow-fiber module.

In a batch recirculation arrangement (Figure 41-11) a volume V_w of aqueous solution of composition c_{iw}^{in} is contacted with an organic solvent of volume V_o of composition c_{io}^{in} through the membrane extraction device in countercurrent flow. Only a part of the aqueous solution is contacting a part of the organic solvent in the device in continuous fashion. However, the contacted liquids are returned continuously to each liquid reservoir, which also continuously feeds the device. The composition in each reservoir changes with time slowly but asymptotically to final equilibrium values c_{iw}^f and c_{io}^f related by

$$\frac{c_{io}^f}{c_{iw}^f} = m_i. \qquad (41\text{-}63)$$

If the feed aqueous solution is to be purified from c_{iw}^{in} to a value less than $c_{iw}^f = (c_{io}^f/m_i)$, the batch recirculation mode cannot achieve it,

while continuous countercurrent contact can. The total fractional solute extraction in batch recirculation is

$$\frac{V_o c_{io}^f}{V_w c_{iw}^{in}} = 1 - \frac{1}{[1 + (m_i V_o/V_w)]}, \qquad (41\text{-}64)$$

where the assumption is that $c_{io}^{in} = 0$. Thus, the limit of fractional solute extraction is determined by m_i, V_o, and V_w (unlike that in continuous-contact countercurrent extractor for $c_{iob}^{in} = 0$). Obviously, the continuous countercurrent contact mode is far more efficient than the batch recirculation mode, which provides at the most one equilibrium stage.

Expressions for the time-dependent concentration of solute in aqueous and organic reservoirs are available in Dahuron and Cussler (1988) and Prasad, Frank, and Sirkar (1988).

Design of a Hollow-Fiber Module

Relations describing the rate of transfer of a solute in a hollow-fiber extractor were provided in an earlier section. If the length of a hollow-fiber module needed to reduce a given c_{iwb}^{in} to a given c_{iwb}^{out} for given conditions is desired, Eq. (41-26) can be used.

In this section, the actual design of a hollow-fiber extractor is illustrated. The model system for design is the countercurrent extraction of solute 4-cyanothiazole (CNT) from an aqueous solution using benzene as an extractant. The system has been studied by Prasad and Sirkar (1990), which should be consulted for further details. The design conditions for the hollow-fiber extractor are as follows:

Test system	= benzene-water-CNT
Direction of solute transfer	= water to benzene
Aqueous flow rate, Q_w	= 30 gal/min (US gallons)
Organic flow rate, Q_o	= 15 gal/min
Aqueous solute concentration, c_{iwb}^{in}	= 13.0 mg/L
Aqueous concentration out, c_{iwb}^{out}	= 0.4 mg/L
Organic concentration in, c_{iob}^{in}	= 0.0
Distribution coefficient, m_i	= 6.5
Solute diffusion coefficient in organic phase, D_{io}	= 3.2×10^{-5} cm²/s
Solute diffusion coefficient in aqueous phase, D_{iw}	= 1.2×10^{-5} cm²/s

Viscosity of aqueous phase, η_w = 1.0 cp (centipoise)
Viscosity of organic phase, η_o = 0.6 cp
Density of aqueous phase, ρ_w = 0.998 gm/cm^3
Density of organic phase, ρ_o = 0.8 gm/cm^3

Using Eq. (41–27), we can calculate NTU, the total number of transfer units required for this separation. This works out to 5.22.

We pointed out already that due to backmixing on the shell side, the feed stream should preferably be allowed to flow on the tube side. In this case, the aqueous phase should flow on the tube side while benzene will flow on the shell side in a countercurrent fashion.

For this particular illustration, assume operation with a hydrophobic Celgard® X-20 type of hollow fiber whose details are as follows:

Fiber inside diameter, d_{ti} = 240 μm
Fiber wall thickness = 25 μm
Porosity of the hollow fiber, ϵ_m = 0.4
Tortuosity of the hollow fiber, τ_m = 2.4

The calculations are started by assuming a flow rate of aqueous phase per fiber. Assume this number to be 3.6×10^{-4} cm^3/s. This means that for the total aqueous flow rate of 30 gal/min, 5.5×10^6 fibers are needed. If a packing fraction of 0.45 of the hollow fibers in a shell is assumed, this would lead to a shell diameter of 101.38 cm (~3.3 ft). The contact area of the two phases per unit equipment volume can be estimated to be 51.4 cm^{-1}.

The overall mass transfer coefficient based on the aqueous phase can now be estimated using Eq. (41–8). The film transfer coefficient on the tube side, k_{iwt}, is estimated from Eq. (41–46) for k_{it}. The shell and membrane transfer coefficients are estimated from Eqs. (41–54) and (41–17a), respectively. These expressions, except for Eq. (41–17a) are functions of the module length. Next, substitute these functions of the length in Eq. (41–24) to solve for the length L. This works out to a value of 63 cm (2.1 ft) as the length of the extractor. The hollow-fiber extractor thus designed for the above duty will be *3.3 ft in diameter* and *2.1 ft long* containing 5.5×10^6 fibers of Celgard® X-20 type with a 240-μm inside diameter.

A conventional packed column extractor for the same duty would be an order of magnitude taller. This is due to the high contacting efficiency in a hollow-fiber device, which in turn leads to a very low value of the total length required for the separation. However, a single hollow-fiber module for the extraction of CNT from water using benzene would not be a very practical solution at this time since the available hollow-fiber modules have much smaller diameters. This forces us to consider alternative design strategies using available hollow-fiber extractors or to develop alternative hollow-fiber modules.

Design of a Modular Cascade

Prasad and Sirkar (1989) have proposed a modular assembly for hollow-fiber extractors using standard modules commercially available. Their proposed scheme is shown in Figure 41–12. The modular cascade consists of a number of parallel limbs (PM), each containing the same number of modules connected in series (SM). The computation procedure consists of assuming a fixed number of modules in series (SM) and then iterating on the total number of parallel limbs needed for the desired separation. The total number of modules needed for the specified solvent extraction duty will be SM × PM. This is equivalent to a certain amount of membrane area.

In general, in any such design calculation, the number of parallel limbs, PM, will decrease as the number of modules in series, SM, increases. This does not imply that the value of SM × PM remains constant. Prasad and Sirkar (1990) have shown that the increase in SM can lead to a very large reduction in PM, thereby reducing the total number of modules required for extraction. However, this sharp reduction in

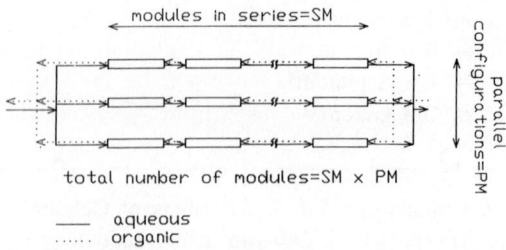

FIGURE 41-12. Schematic of a series-parallel connection of hollow-fiber modules for large-scale operation. (adapted from Prasad and Sirkar 1990 with permission).

the total module requirement leads to increased flow on the tube side and shell side of each module in the assembly and, hence, increased pressure drop in each stream in a module. The trade-off in the modular design, therefore, is the increased mass transfer coefficient and decreased total membrane area of the hollow-fiber modules due to the increase in flow at the expense of an increased pressure drop in each module.

A modular assembly such as that shown in Figure 41-12 has been used to design the system of benzene-water-CNT considered earlier. The hollow-fiber module used for this design calculation was as follows:

Length of the module = 54.6 cm
Shell diameter = 4.7 cm
Total number of fibers = 7500
Fiber ID = 244 μm
Fiber OD = 294 μm
Membrane porosity = 0.4
Membrane tortuosity = 2.6

The design results for the modular assembly are presented in Figure 41-13. As the number of modules in series increases, the total number of modules decreases rapidly in the beginning and then levels off; however, the pressure drop on the tube side keeps increasing with the increase in the number of modules in series. Thus, as the capital cost (related to the module number or membrane surface area) decreases with increased flow/module, the operating cost increases due to an increase in pressure drop. Obviously, there is room for an optimum value of SM × PM that minimizes total cost unless the pressure drop encountered under optimum conditions is unacceptable from the point of view of an acceptable Δp_{cr}.

POTENTIAL APPLICATIONS

Nondispersive membrane-based solvent extraction can replace conventional dispersion-based solvent extraction in almost all applications. A number of applications of this technique have already been made. These recent applications of microporous membrane solvent extraction are summarized below in six distinct categories: metal extraction, organic pollutant extraction, aromatics extraction, pharmaceuticals extraction, fermentation product extraction, and extractive bioreactors.

Metal Extraction

Gold present as $HAuCl_4 \cdot 3H_2O$ was efficiently extracted from an aqueous 0.43 M HCl solution containing $PdCl_2$ and $CuCl_2 \cdot 2H_2O$ using di-

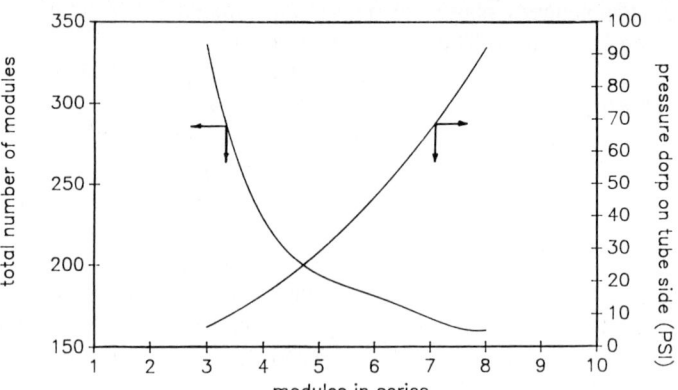

FIGURE 41-13. The total number of modules required for extraction of CNT from an aqueous stream and total pressure drop over the whole modular assembly.

ethylene glycol dibutyl ether (DGDE) as solvent in a microporous hydrophobic hollow-fiber (Celgard® X-20, 400-μm inside diameter) module with a nylon casing (Alexander and Callahan 1987). The stripping of gold from DGDE was also achieved by the same authors using an aqueous 2% KCN solution buffered with 0.01 M $Na_2B_4O_7$ at pH 9 in the same hollow-fiber device. Kim (1984) has presented information on extraction of metals such as Cu and Fe using LIX64N. The extractant was regenerated by acid solutions. He had studied modules having microporous hydrophobic and hydrophilic hollow fibers of different kinds. Since he did not apply the correct pressure difference, he had to use an aqueous-organic separator in addition to an extractor and a back extractor. The same suggestion was made by De Haan, Bartels, and de Graauw (1989) who studied extraction of Cu into LIX 84 and decane in one polypropylene hollow-fiber device and stripping into aqueous H_2SO_4 in another similar device. Temmink, van Maanen, and Sprang (1989) had studied separation of copper and zinc using 10% DEHPA in kerosene in hollow-fiber modules of Romicon HF-53-20-PM-10. Both extraction (pH, 3.3) and back extraction (pH, 0.3) were studied.

Matsumoto et al. (1987) studied the kinetics of copper extraction by the chelating agent Kelex 100 in toluene and copper stripping by sulfuric acid using a single microporous hollow fiber of polytetrafluoroethylene. Yoshizuka, Kondo, and Nakashio (1986) had earlier used the same fiber configuration for studies on copper extraction using N-8-quinolyl-p-dodecylbenzenesulfonamide in toluene. Sato et al. (1989) have studied the extraction rate of molybdenum using 2-ethylhexyl phosphoric acid mono-2-ethylhexyl ester dissolved in n-heptane in a single hollow-fiber configuration used in the previous two studies.

Organic Pollutant Extraction

Cooney and Jin (1985) had extracted phenol from an aqueous solution into a stagnant solvent on the shell side of a hydrophilic hollow-fiber module (C-DAK 135 sce; CORDIS-DOW, Miami, FL). Since there was no solvent flow, phenol broke through after some time in the aqueous solution in the tube side because it operated like a chromatographic column. The module had cellulose acetate blend fibers and a silicone tubesheet in a polycarbonate shell, which was attacked by solvents MIBK and n-butyl acetate. The authors did not recommend a procedure for dispersion-free flow of solvent on the shell side.

Phenol was continuously extracted from an aqueous solution into MIBK efficiently using hydrophobic Celgard® hollow fibers as well as hydrophilic regenerated cellulose Cuprophan® hollow fibers in small hollow-fiber modules by Prasad and Sirkar (1988). Back extraction of phenol from MIBK into an aqueous caustic solution has been experimentally investigated and modeled by Basu, Prasad, and Sirkar (1990) in modules of hydrophobic as well as hydrophilic fibers. Recently, Yun, Prasad, and Sirkar (1989) have presented experimental results and design calculations for simultaneous extraction of phenol, 2-chlorophenol, toluene, nitrobenzene, and acrylonitrile from waste aqueous streams into a number of organic solvents, e.g., MIBK, hexane, and isopropyl acetate. They had used a hydrophobic Liqui-Cel® polypropylene hollow-fiber module containing 240-μm ID fibers. Temmink, van Maanen, and Akkerhuis (1989) have studied removal of benzene, trichloroethane, or tetrachloromethane in water using an extractant of kerosene, silicone oil, or both in an Amicon H1P 10-20 hollow-fiber module.

Aromatics Extraction

Prasad and Sirkar (1987b) have studied the extraction of toluene from a 50–50% v/v toluene-n-heptane mixture into the solvent NMP using microporous hydrophobic flat Celgard® 2400 film with the nonpolar organic phase in the pores of the membrane. In a later study, Prasad et al. (1990) have carried out the same extraction using the same membrane with the polar organic NMP in the pores instead of the nonpolar organic toluene-n-heptane mixture. This is an example of a case where both members of the immiscible two-phase system are organic and either of them may be present in the pores of a hydrophobic membrane. No studies of aromatics extraction with hollow-fiber modules are available in the literature.

Extraction of Pharmaceuticals

Using a number of different microporous hydrophobic hollow-fiber modules of the Liqui-Cel® variety, Prasad and Sirkar (1990) have studied the simultaneous extraction of 4-methylthiazole (MT) and 4-cyanothiazole (CNT) from an aqueous ammoniacal solution into either benzene or toluene. These two compounds, intermediates for an animal health product, could be extracted to the extent of 99%+ by using only two 15-cm-long (1.9-cm-diameter) modules in series in the once-through mode. *Further, in a 64-day study with an actual plant feed stream containing MT and CNT, extraction behavior remained unchanged with solvent benzene.* The hollow fibers were later tested under electron microscope to determine any change. No change was observed suggesting *highly reliable and stable operation of this membrane-based solvent extraction technique* (Figure 41–14).

Prasad and Sirkar (1989) have also demonstrated the utility of the technique in an extraction-purification scheme used to purify mevinolinic acid using a pH swing solvent extraction scheme. The solute of interest is present in an isopropyl acetate (IPAc) extract from a lysed fermentation broth. This IPAc extract containing a lot of suspended cell debris is contacted with an alkaline solution (pH ≅ 9) into which the product is extracted. Later, the product is back extracted into a pure IPAc stream at a much lower pH. This reduction in aqueous pH is achieved by simultaneous extraction of acetic acid from the organic extract phase into the aqueous feed (or raffinate) phase. Wald, Lopez, and Matson (1989) have studied the extraction of penicillin V in a phosphate buffer (pH = 5.0) using a hydrophilic module and then its back extraction in another hydrophilic module using a phosphate buffer at pH = 8.0. They have also studied a similar scheme for extraction of phenoxyacetic acid.

Fermentation Product Extraction

Although the last two applications involved compounds produced by fungal fermentation, they were not included here since they were pharmaceutically important products. Among other fermentation products, extraction of organic acids has been investigated in a number of studies. Kiani, Bhave, and Sirkar (1984) and Prasad et al. (1986) have studied the extraction of acetic acid into xylene and MIBK using flat hydrophobic microporous membranes. D'Elia, Dahuron, and Cussler (1986) and Dahuron and Cussler (1988) have used hydrophobic Celgard® hollow-fiber modules to study extraction of acetic acid. The extraction behaviors of acetic acid and succinic acid have been experimentally obtained by Prasad and Sirkar (1987a; 1987b; 1988) with a variety of flat membranes and hollow fibers. Basu and Sirkar (1991) have studied the extraction of citric acid by trioctylamine in a number of organic solvents in a hollow-fiber module.

Membrane solvent extraction has also been applied to extract proteins that can be obtained from genetic engineering. Dekker et al. (1987) have studied the extraction of α-amylase using a reversed micellar phase of iso-octane and surfactants through a MHF polypropylene module

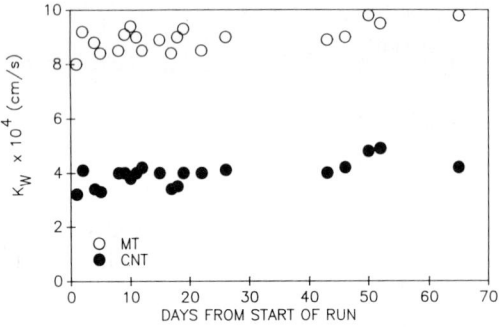

FIGURE 41–14. Overall mass transfer coefficients for MT and CNT as a function of time for long-term stability studies with a commercial hollow-fiber solvent extraction module (reprinted from Prasad and Sirkar 1990 with permission).

(ENKA, pore size of 0.2 μm). They have observed that the system was stable over a very small pressure range ($\Delta P \sim 8$ to 20 cm H_2O); it was suggested that scale-up of the process may be difficult. Dahuron and Cussler (1988) have extracted proteins using two types of systems: aqueous biphasic systems formed using polyethyleneglycol and potassium phosphate for the extraction of proteins such as cytochrome-C, myoglobin, α-chymotrypsin, catalase, and urease, and reverse micelle systems with aerosol OT (AOT) in octane with phosphate buffer for cytochrome-C and α-chymotrypsin.

Extractive Bioreactors

A novel use of nondispersive membrane solvent extraction has appeared in fermentation processes. The microporous membrane has been incorporated directly inside the fermentor to achieve *in situ* nondispersive extraction of fermentation products inhibitory to the fermentation process. Frank and Sirkar (1985, 1986) incorporated microporous hydrophobic hollow fibers along the length of a tubular fermentor producing ethanol on the shell-side by yeast fermentation. When required, solvent dibutyl phthalate was passed through the fiber lumen to extract ethanol from the broth. A theoretical analysis of the above system has been made by Frank (1986) and Fournier (1988). Naser and Fournier (1988) have provided an economic analysis for alcohol production via such an extractive bioreactor. Kang (1989) and Kang, Shukla, and Sirkar (1990) have demonstrated considerable improvement in alcohol productivity in such a fermentor using oleyl alcohol as an extractant and a 300 gm/L glucose solution as feed.

Extractive ethanol fermentation using microporous hydrophobic flat membranes in a multimembrane fermentor has been studied by Cho and Shuler (1986) using tributyl phosphate (TBP) as an extractant. Employing the hollow-fiber-based extractive bioreactor configuration of Frank and Sirkar (1985), Shukla, Kang, and Sirkar (1989) have demonstrated considerable productivity improvement in acetone-butanol-ethanol (ABE) fermentation by using 2-ethyl-1-hexanol extractant in the fiber lumen.

Another novel use of membrane solvent extraction involves enzymatic conversion followed by solvent extraction through a membrane, which itself has immobilized enzymes. For example, olive oil was hydrolyzed by lipase enzyme immobilized on hydrophobic hollow fibers; the product fatty acid was extracted into the oil phase on one side of the membrane while glycerol was extracted into the aqueous phase on the other side of the membrane (Hoq, Yamane, and Shimizu 1985). Pronk et al. (1986) as well as van der Padt, van Dorp, and van't Riet (1986) have also studied enzymatic oil hydrolysis in a similar reactor. Prasad, Frank, and Sirkar (1988) have provided a wider perspective on the use of a membrane solvent extraction technique in bioprocessing.

ACKNOWLEDGMENTS

We acknowledge the support provided to us during the writing by HSMRC, a NSF/IUCC at NJIT, Newark, NJ, and by the Center for Membranes and Separation Technologies at Stevens Institure of Technology.

NOTATION

General Notation

See the General Notation section at the beginning of this handbook.

Special Notation

a_{si}	aqueous-organic interfacial area per unit volume based on hollow-fiber internal diameter, L^{-1}
a_{so}	same as a_{si} except based on fiber outer diameter, L^{-1}
A	empty cross-sectional area of module, L^{-2}
c'_{Bw}	critical caustic concentration in

	aqueous solution, Eq. (41–40), mol/L^3
c_{io}, c_{iw}	organic and aqueous phase concentrations of species i, respectively, mol/L^3
c_{io}^*, c_{iw}^*	defined by Eq. (41–3), mol/L^3
d_e	hydraulic diameter of shell side, Eq. (41–55), L
d_{ti}	hollow-fiber inside diameter, L
d_{tlm}	logarithmic mean diameter of hollow fiber, L
d_{to}	hollow-fiber outside diameter, L
D_{Aw}, D_{Bw}	diffusion coefficients of species A and B in water, L^2/t
D_i	diffusion coefficient of species i, L^2/t
k_{im}	membrane mass transfer coefficient for species i, L/t
k_{io}	local mass transfer coefficient of species i in organic phase, L/t
k_{iw}	local mass transfer coefficient of species i in aqueous phase, L/t
k_{ios}, k_{iws}	values of k_{io} and k_{iw} on the shell side, L/t
k_{iot}, k_{iwt}	values of k_{io} and k_{iw} on the tube side, L/t
$\bar{k}_{iw}$	length averaged value of local aqueous phase mass transfer coefficient k_{iw}, L/t
K_o	overall mass transfer coefficient based on organic phase, L/t
K_w	overall mass transfer coefficient based on aqueous phase, L/t
$\bar{K}_w$	length averaged value of K_w, L/t
L	length of hollow-fiber extractor, L
m_i	distribution coefficient for species i, Eq. (41–3), dimensionless
N	number of hollow fibers, dimensionless
N_i	molar flux of species i in a fixed reference frame, mol/$L^2 t$
Q_w	volumetric aqueous solution flow rate, L^3/t
Q_o	volumetric organic solvent flow rate, L^3/t
r_p	pore radius of membrane, L
R	universal gas constant, ML^2/t^2T mol
Re_t	tubeside Reynolds number, dimensionless
Re_s	shellside Reynolds number, dimensionless
Sc_i	Schmidt number for species i, dimensionless
Sh	Sherwood number, dimensionless
T	temperature, T
v_s	superficial velocity of liquid, L/t
v_{so}	value of v_s for organic phase based on empty shell cross-sectional area, L/t
v_{sw}	superficial velocity of aqueous phase based on empty shell cross-sectional area, L/t
v_t	velocity through fiber bore, L/t
v_w	water velocity through fiber bore, L/t
$\bar{V}_i$	partial molar volume of species i, L^3/mol
V_o	organic solvent volume, L^3
V_w	aqueous solution volume, L^3
x	distance from aqueous stream inlet in extractor, L

Greek Letters

γ_{wo}	interfacial tension, aqueous-organic system, M/t^2 or E/L^2
ϵ_m	membrane porosity, dimensionless
η	liquid dynamic viscosity, M/Lt
θ_c	contact angle, Eq. (41–45), dimensionless
ν	liquid kinematic viscosity, L^2/t
ϕ	packing fraction of fibers in a shell, dimensionless
τ_m	tortuosity of pores in a membrane, dimensionless

Subscripts

a	arithmetic average-based quantity
A, B	species A, species B
b	bulk
eff	effective value
i	species i; inside surface area of fiber
lm	logarithmic mean
m	membrane
o	organic, outside surface
s	shell side
t	tube side
w	aqueous

Superscripts

in   at inlet location
f   final value
out   at outlet location
*   equilibrium value

REFERENCES

Alexander, P. R., and R. W. Callahan. 1987. Liquid-liquid extraction and stripping of gold with microporous hollow fibers. *J. Membr. Sci.* 35:57–71.

Astarita, Gianni. 1967. *Mass Transfer With Chemical Reaction.* Amsterdam: Elsevier Publishing Co.

Basu, R. 1989. Mass transfer enhancement in phase barrier membrane separators for reactive systems. Ph.D. diss., Stevens Institute of Technology, Hoboken, NJ.

Basu, R., R. Prasad, and K. K. Sirkar. 1990. Nondispersive membrane solvent back extraction of phenol. *AIChE J.* 36:450–460.

Basu, R., and K. K. Sirkar. 1991. Hollow fiber contained liquid membrane separation of citric acid. *AIChE J.* 37:383–393.

Beck, R. E., and J. S. Schultz. 1970. Hindered diffusion in microporous membranes with known pore geometry. *Science* 170:1302–1305.

Cho, T., and M. L. Schuler. 1986. Multimembrane bioreactor for extractive fermentation. *Biotech. Prog.* 2(1):53–60.

Cooney, D. O., and C. L. Jin. 1985. Solvent extraction of phenol from aqueous solution in a hollow fiber device. *Chem. Eng. Comm.* 37:173–191.

Dahuron, L., and E. L. Cussler. 1988. Protein extractions with hollow fibers. *AIChE J.* 34(1):130–136.

D'Elia, N. A., L. Dahuron, and E. L. Cussler. 1986. Liquid-liquid extractions with microporous hollow fibers. *J. Membr. Sci.* 29:309–319.

De Haan, A. B., P. V. Bartels, and J. de Graauw. 1989. Extraction of metal ions from waste water. Modelling of the mass transfer in a supported-liquid-membrane process. *J. Membr. Sci.* 45:281–297.

Dekker, M., K. Van't Riet, J. M. G. M. Wijnans, J. W. A. Baltussen, B. H. Bijsterbosch, and C. Laane. 1987. Membrane based liquid/liquid extraction of enzymes using reversed micelles. Paper read at the 1st International Conference on Membranes and Membrane Processes, 8 June 1987, Tokyo, Japan.

Fournier, R. L. 1988. Mathematical model of microporous hollow-fiber membrane extractive fermentor. *Biotech. Bioeng.* 31:235–239.

Frank, G. T. 1986. Membrane solvent extraction with hydrophobic microporous hollow fiber and extractive bioreactor development for fuel ethanol production. Ph.D. diss., Stevens Institute of Technology, Hoboken, NJ.

Frank, G. T., and K. K. Sirkar. 1985. Alcohol production by yeast fermentation and membrane extraction. *Biotech. Bioeng. Symp. Ser.* 15:621–631.

Frank, G. T., and K. K. Sirkar. 1986. An integrated bioreactor-separator: In-situ recovery of fermentation products by a novel dispersion-free solvent extraction technique. *Biotech. Bioeng. Symp. Ser.* 17:303–316.

Hanson, C., T. C. Lo, and M. H. I. Baird. 1983. *Solvent Extraction Handbook.* New York: John Wiley & Sons.

Ho, W. S., L. T. C. Lee, and K. J. Liu. 1976. Membrane hydrometallurgical extraction process. U.S. Patent 3,957,504.

Hoq, M. M., T. Yamane, and S. Shimizu. 1985. Continuous hydrolysis of olive oil by lipase in a microporous hydrophobic membrane bioreactor. *JAOCS* 62(6):1016–1021.

Kang, W. 1989. Hollow fiber membrane-based extractive bioreactors and a whole cell immobilization technique. Ph.D. diss., Stevens Institute of Technology, Hoboken, NJ.

Kang, W., R. Shukla, G. T. Frank, and K. K. Sirkar. 1988. Evaluation of O_2 and CO_2 transfer coefficients in a locally integrated tubular hollow fiber bioreactor. *Appl. Biochem. Biotech.* 18:35–51.

Kang, W., R. Shukla, and K. K. Sirkar. 1990. Ethanol production in a microporous hollow-fiber based extractive fermentor with immobilized yeast. *Biotech. Bioeng.* 34:826–833.

Keller, K. H., and T. R. Stein. 1967. A two-dimensional analysis of porous membrane transport. *Math. Biosci.* 1:421–437.

Khare, S. 1991. Nondispersive membrane solvent extraction using asymmetric ceramic membranes. M. Engg. thesis, Stevens Institute of Technology, Department of Chemistry and Chemical Engineering, Hoboken. NJ.

Kiani, A., R. R. Bhave, and K. K. Sirkar. 1984. Solvent extraction with immobilized interfaces in a microporous hydrophobic membrane. *J. Membr. Sci.* 20:125–145.

Kim. B. M. 1984. Membrane-based extraction for

selective removal and recovery of metals. *J. Membr. Sci.* 21:5–19.

Kim, B. S. and P. Harriott. 1987. Critical entry pressure for liquids in hydrophobic membranes. *J. Colloid Interface Sci.* 115(1):1–8.

Lee, L. T. C., W. S. Ho, and K. J. Liu. 1976. Membrane solvent extraction. U.S. Patent 3,956,112.

Malone, D. M., and J. L. Anderson. 1977. Diffusional boundary-layer resistance for membranes with low porosity. *AIChE J.* 23(2):177–184.

Matsumoto, M., H. Shimauchi, K. Kondo, and F. Nakashio. 1987. Kinetics of copper extraction with Kelex 100 using a hollow fiber membrane extractor. *Solvent Extr. Ion Exch.* 5(2):301–323.

Misek, T. 1978. *Recommended Systems For Liquid Extraction Studies.* Rugby, U.K.: Institute of Chemical Engineers.

Naser, S. F., and R. L. Fournier. 1988. A numerical evaluation of a hollow fiber extractive fermentor process for the production of ethanol. *Biotech. Bioeng.* 32:628–638.

Newman, J. 1969. Extension of Leveque solution. *Trans. Am. Soc. Mech. Eng.* Series C, *J. Heat Trans.* 91:177–178.

Park, J. K., and H. N. Chang. 1986. Flow distribution in the lumen side of a hollow-fiber module. *AIChE J.* 32:1937–1947.

Prasad, R., R. R. Bhave, A. K. Kiani, and K. K. Sirkar. 1986. Further studies on solvent extraction with immobilized interfaces in a microporous hydrophobic membrane. *J. Membr. Sci.* 26:79–97.

Prasad, R., G. T. Frank, and K. K. Sirkar. 1988. Nondispersive solvent extraction using microporous membranes. *AIChE Symp. Ser.* 84 (261):42–53.

Prasad, R., S. Khare, A. Sengupta, and K. K. Sirkar. 1990. Novel liquid-in-pore configurations in membrane solvent extraction. *AIChE J.* 36:1592–1596.

Prasad, R., and K. K. Sirkar. 1985a. Microporous membrane solvent extraction. Paper read at the 4th Symposium on Separation Science and Technology for Energy Applications, 20–24 October, 1985, Knoxville, TN.

Prasad, R., and K. K. Sirkar. 1985b. Solvent extraction with microporous hydrophilic and composite membranes. Paper read at the AIChE Annual Meeting, 10–15 November, 1985, Chicago, IL.

Prasad, R., and K. K. Sirkar. 1987a. Solvent extraction with microporous hydrophilic and composite membranes. *AIChE J.* 33(7):1057–1066.

Prasad, R., and K. K. Sirkar. 1987b. Microporous membrane solvent extraction. *Sep. Sci. Technol.* 22(2,3):619–640.

Prasad, R., and K. K. Sirkar. 1988. Dispersion-free solvent extraction with microporous hollow fiber modules. *AIChE J.* 34(2):177–188.

Prasad, R., and K. K. Sirkar. 1989. Hollow fiber solvent extraction of pharmaceutical products: a case study. *J. Membr. Sci.* 47:235–259.

Prasad, R., and K. K. Sirkar. 1990. Hollow fiber solvent extraction: performances and design. *J. Membr. Sci.* 50:153–175.

Pronk, W., A. W. Knol, E. Le Clercq, C. van Helden, and K. van't Riet. 1986. Kinetics and stability of enzymatic fat hydrolysis in a membrane bioreactor. Personal communication.

Sato, Y., Y. Akiyoshi, K. Kondo, and F. Nakashio. 1989. Extraction kinetics of copper with 2-ethylhexyl phosphonic acid mono-2-ethylhexyl ester. *J. Chem. Eng. Japan* 22(2):182–189.

Seibert, A. F., and J. R. Fair. 1988. Hydrodynamics and mass transfer in spray and packed liquid-liquid extraction columns. *Ind. Eng. Chem. Res.* 27:470–481.

Shukla, R., W. Kang, and K. K. Sirkar. 1989. Acetone-butanol-ethanol (ABE) production in a novel hollow fiber fermentor-extractor. *Biotech, Bioeng.* 34:1158–1166.

Sirkar, K. K. 1988. Immobilized interface solute transfer apparatus. U.S. Patent 4,789,468.

Sirkar, K. K. 1990. Asymmetrically-wettable porous membrane process. U.S. Patent 4,921,612.

Sirkar, K. K. 1991. Immobilized interface solute transfer processes. U.S. Patent 4,997,569.

Sirkar, K. K., R. Prasad, and S. K. Khare. 1990. Ceramic membranes for novel separation processes. Paper read at AIChE Annual Meeting. 13 November 1990, Chicago, IL.

Skelland, A. H. P. 1974. *Diffusional Mass Transfer.* New York: John Wiley & Sons.

Temmink, H. M. G., H. C. H. J. van Maanen, and J. J. Akkerhuis. 1989. Pertraction of (chlorinated) Hydrocarbons. Personal communication.

Temmink, H. M. G., H. C. H. J. van Maanen, and W. J. A. M. Sprang. 1989. Pertraction of heavy metal ions. Personal communication.

Treybal, Robert. 1963. *Liquid Extraction.* New York: McGraw-Hill Book Co.

Van der Padt, A., L. J. van Dorp, and K. van't Riet. 1986. Membrane reactor for immobilized enzymatic ester synthesis of mono-di-and triacylglycerols. Personal communication.

Wald, S. A., J. L. Lopez, and S. L. Matson. 1989. Membrane-mediated antibiotic extraction using liquid ion exchangers. Paper read at the 3rd Annual Meeting of the North American Membrane Society, 19 May 1989, Austin, TX.

Yang, M. C., and E. L. Cussler. 1986. Designing hollow-fiber contactors.. *AIChE J*. 32(11):1910–1916.

Yoshizuka, K., K. Kondo, and F. Nakashio. 1986. Effect of interfacial reaction on rates of extraction and stripping in membrane extractor using a hollow fiber. *J. Chem. Eng. Japan* 19(4):312–318.

Yun, C. H., R. Prasad, and K. K. Sirkar. 1989. Solvent extraction of priority organic pollutants using hollow fiber membranes. Paper read at AIChE National Meeting, 21–23 August 1989, Philadelphia, PA.

Yun, C. H., R. Prasad, and K. K. Sirkar. 1991. A two-dimensional model for solvent extraction in a microporous hollow fiber module. Paper read at the Fourth National Meeting of the North American Membrane Society, 31 May 1991, San Diego, CA.

42

Hollow-Fiber Contained Liquid Membrane

Sudipto Majumdar*
Kamalesh K. Sirkar*
Stevens Institute of Technology

Amitava Sengupta
SRI International

GENERAL DESCRIPTION OF THE
  PROCESS
  Liquid Membranes
  Various SLM Structures and Their
    Limitations
  Basic Concept of HFCLM for Gas
    Separation
  Basic Concept of HFCLM for Liquid
    Separation
  Pervaporation Separation by HFCLM
THEORY
  Effective Membrane Thickness in
    HFCLM Permeation
  HFCLM Permeator Model for Gas
    Separation without Reaction
  Liquid Separation in a HFCLM Permeator
  Purification Limits in Liquid Membrane Permeators
  HFCLM Permeator versus Two Separate Hollow-Fiber Contactors
  Liquid Separation
  Gas Separation
  Comparison of HFCLM Permeator
    with Conventional Polymeric Membrane Permeator for Gas Separation
  Pressure Conditions in HFCLM Permeators
HFCLM PERMEATOR DEVICE AND
  DESIGN CONSIDERATIONS
  Gas Separation
  Liquid Separation
  Module Availability
APPLICATIONS
  Liquid Separations
  Gas Separations
ACKNOWLEDGMENTS
NOTATION
REFERENCES

GENERAL DESCRIPTION OF THE PROCESS

Liquid Membranes

In gas separations by permeation through a nonporous polymeric membrane, gas molecules undergo dissolution in the membrane at the feed gas/membrane interface. The dissolved species diffuse through the membrane and are desorbed at the other membrane surface, the permeate gas/membrane interface. For permanent gases at temperatures greater than the critical temperature of the gases, the dissolution/desorption behavior of a gas has been found to obey Hen-

---

*Present affiliation: New Jersey Institute of Technology

ry's law if the membrane is made of a rubbery material; the gas species dissolve in the membrane or desorb from it as if the membrane were a liquid. It is then obvious that a thin liquid layer ought to be able to function as a selective membrane provided it can withstand the pressure difference between the feed gas and the permeate gas and preserve itself. Not only a pure liquid but any solution could then be used as a permselective membrane.

Ward and Robb (1967) soaked a highly porous cellulose acetate film with a 6.4 N cesium bicarbonate solution and developed an *immobilized liquid membrane* (ILM). The liquid was held in the membrane pores by capillary forces [Figure 42–1(a); for details of configuration, see Table 42–1]. A highly selective permeation of CO_2 from an O_2-CO_2 feed mixture was achieved with a high CO_2 permeability through the liquid in the pores. A significant part of the enhanced CO_2 permeability was due to its facilitated transport in the bicarbonate solution (Chapter 44). A number of studies on facilitated transport of gases such as CO_2, O_2, H_2S, NO, SO_2, and CO through ILMs are available in the literature. [See reviews by Schultz, Goddard, and Suchdeo (1974), Smith, Lander, and Quinn (1977), Sengupta and Sirkar (1986a), Noble and Way (1987).] Such ILMs are often called SLMs (supported liquid membranes).

Separation of a liquid solution by a liquid membrane has attracted even greater attention. The much earlier studies by Harber and Beutner (Kuo and Gregor 1983; Sollner 1984) used a thin film of oil between two aqueous phases. Two distinctly different techniques were adopted in later studies: the emulsion liquid membrane (ELM) and the supported liquid membrane (SLM).

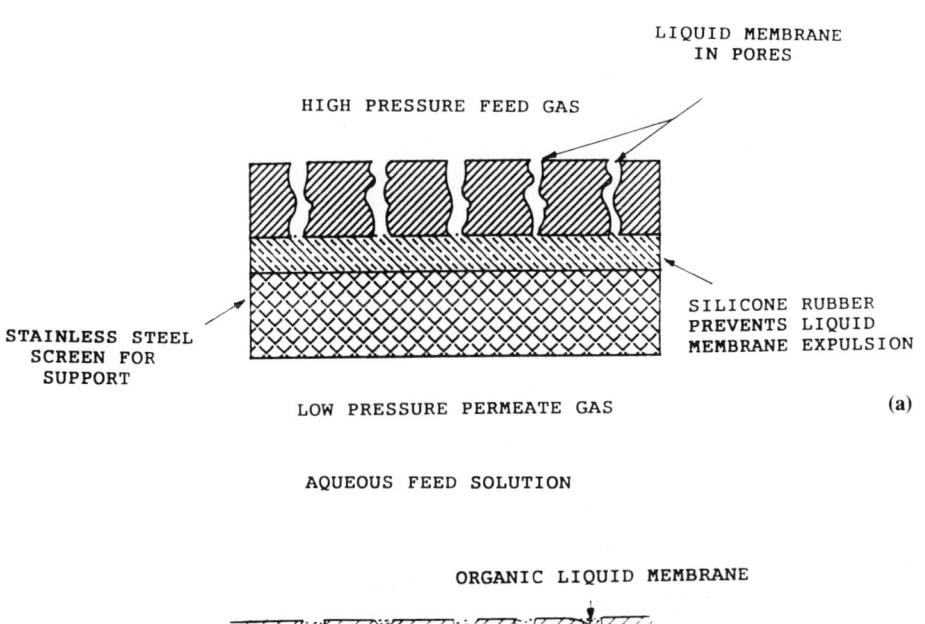

FIGURE 42–1. (a) Immobilized liquid membrane in a microporous film for gas separation. (b) Supported organic liquid membrane in a microporous film for separation from an aqueous solution.

In the SLM technique [Figure 42–1(b)], an organic liquid is immobilized in the pores of a microporous membrane interposed between two aqueous solutions. At the interface between the feed solution and the membrane, the solute is extracted into the membrane liquid; it then diffuses by itself and/or in a complexed form to the other side of the membrane where the strip aqueous solution flows. At the strip solution/ membrane interface, the solute is back extracted into the strip solution. Often the solute obtained is in a highly concentrated form in the strip solution. The earliest studies were conducted by Bloch (1970) and Cussler (1971). Extensive reviews are available (Marr and Kopp 1982; Way et al. 1982; Danesi 1984–85; Noble, Koval, and Pellegrino 1989). A review of ELM techniques is available in Chapters 36 through 40.

Because these SLMs have reversibly reacting chemical extractants or absorbents, they can be extraordinarily selective for particular solutes. Further, their capacity to remove solutes from feed gas or feed solution literally improves as the solute concentration goes down because solute permeability increases rapidly with decreasing solute concentration due to facilitated solute transport. They can also reduce the solute concentration in feed to very low levels. Reduction of metal concentration in a dilute feed by two or more orders of magnitude has been demonstrated (Danesi 1984–85). Using counter-transport or co-transport, the metal or solute species could also be concentrated in the strip solution. SLMs thus have many advantages: a high separation factor in each stage; low capital, operating, and energy costs; very low inventory of extractants (or membrane liquid); no extractant loss due to poor coalescence as in solvent extraction, and fewer moving parts resulting in lower maintenance costs. Compact and modular hollow-fiber devices can be used with exceptionally high mass transfer area per unit equipment volume. Despite such obvious advantages, SLMs have not been adopted for larger scale industrial processes. A major reason is their lack of long-term stability.

Various SLM Structures and Their Limitations

The detailed structures of a number of different SLM configurations are briefly considered here in the context of their stability problems and other shortcomings. Table 42–1 identifies a number of studies and the SLM structures used therein for gas separation. The SLM structure reported with each study is representative of the particular SLM variety and may have been used in other studies not identified.

Consider now the limitations of such techniques. Unless the gas streams were properly humidified, none of the techniques can *prevent volatilization* of the solvent in the pores of the membrane. Technique 6 merely reduces the rate of volatilization via vapor pressure reduction in small pores. Among the techniques, only 2 and 5 can prevent *flooding* of the membrane from condensation of water vapor sure to take place in permeators with a significant amount of gas permeation. If the membrane liquid constituents are *poisoned*, say, by some unwanted components in feed or sweep gas streams, the techniques of Table 42–1 offer no provisions to renew or replace the liquid membrane *in situ* without shutting down the process (Matson, Lopez, and Quinn 1983; Sengupta and Sirkar 1986a). Most of the techniques do not use hollow fibers to pack a high membrane surface area in the permeator; when hollow fibers are used, the fluxes are quite low due to dense skin (technique 4), a dense, thick ion-exchange membrane (technique 7), and low porosity and high tortuosity of the hollow-fiber support wall (technique 5).

The variety of SLM structures for *liquid separation* is somewhat limited. Table 42–2 lists typical configurations of SLM membranes. In general, SLMs have been found to be unstable. Danesi (1984–85) has identified the following possible causes for SLM instability:

1. Loss of extractant by solubility in mobile feed and strip solutions
2. Progressive wetting of the support pores by surface-active carrier molecules

TABLE 42–1. Different SLM (ILM) Structures Used in Gas Separation.

Serial No.	Source	SLM Structure
1.	Ward and Robb (1967)	Porous cellulose acetate film impregnated with alkali bicarbonate solution. For positive ΔP across film, it was backed by a nonporous highly permeable 0.5-mil silicone rubber membrane [Figure 42–1(a)].
2.	Otto and Quinn (1971)	Aqueous solution in the holes of a stainless steel plate covered and sealed on both sides by 20- to 25-μm-thick nonporous silicone copolymer films, each of the latter being supported by a plexiglass plate with holes matching those in the stainless steel plate.
3.	Kimura, Matson, and Ward (1979); LeBlanc et al. (1980)	Porous membranes, e.g., Cuprophan®, ion-exchange membranes, etc., containing aqueous solution are backed by a nonwetting porous polymeric membrane of fluorocarbon (Gore-Tex®) supported on a stainless steel screen. Maximum ΔP achieved was around 300 psi.
4.	Hughes, Mahoney, and Steigelmann (1986)	Commercial anisotropic cellulose ester RO hollow fiber with dense skin and porous substructure. Water in the membrane exchanged with aqueous $AgNO_3$ solution. Limited pressure differential allowed across hollow-fiber wall.
5.	Bhave and Sirkar (1986, 1987)	Aqueous solution incorporated in the pores of a hydrophobic microporous film or hollow fiber (Celgard®) by an exchange process. Maximum ΔP used for operation was 550 cm Hg with hollow fibers. No support was used unlike that in sources 1, 2, and 3.
6.	Deetz (1987)	Cellulose acetate/nitrate membranes having small pore size (0.025-μm radius) containing aqueous solutions of LiBr. Liquid volatilization and flooding reduced via Kelvin effect in small pores. Porous asymmetric cellulose acetate membrane with the skin region only containing LiBr solution acting as membrane.
7.	Way et al. (1987)	Complexing agent in aqueous solution held in an ion-exchange membrane by electrostatic forces.

3. Pressure differential across the membrane exceeding capillary forces holding the liquid.

Danesi, Reichley-Yinger, and Rickert (1987) have further demonstrated that a large osmotic pressure gradient across the SLM due to an ionic concentration variation between the feed and the strip tends to make SLMs unstable and leads to considerable flow of water across the membrane. Further, low SLM stability was shown to be highly likely in a system with low interfacial tension and high aqueous-organic mutual solubility. Note that the most efficient carrier molecules incorporated in solvents for selective transport often have some surface-active properties. Further, as Babcock et al. (1980a) discovered, thickening or gelation of the liquid membrane in membrane pores did not eliminate instability.

Therefore, the exploration of new liquid membrane structures that are stable and can

TABLE 42–2. Different SLM Structures Used in Liquid Separation in Liquid-Liquid Systems.

Serial no.	Source	SLM structure
1.	Bloch (1970)	A solvent membrane of essentially tributyl phosphate (TBP) formed in a polyvinylchloride matrix by casting a solution of PVC, TBP, and cyclohexane and supported further on a paper support.
2.	Cussler (1971)	Filter paper soaked in an organic solution. Alternately the organic soaked filter paper held between two sheets of cellophane (normally used for dialysis) for transport from a feed aqueous to a strip aqueous solution.
3.	Lee, Evans, and Cussler (1978)	Porous 25-μm-thick hydrophobic polypropylene Celgard® 2500 film wetted by organic liquid kept between two aqueous solutions.
4.	Babcock et al. (1980a)	Microporous hollow-fiber pores wetted by solvent and carrier thickened by polymeric additives such as polyisoprene or polyvinylchloride; hollow fiber kept between two aqueous solutions.
5.	Danesi and Rickert (1986)	Vertical single hollow-fiber module having membrane solvent reservoir at top of the module with the solvent soaking continuously into the porous support by capillary action. SLM kept between two aqueous solutions as in 4.
6.	Nakano, Takahashi, and Takeuchi (1987)	Single hydrophobic Gore-Tex® hollow fiber with organic liquid in the pores held vertically with a pool of membrane liquid in the bottom of fiber. This liquid soaks into the pore and moves up the pore-continuum or lumen side surface by capillary action and buoyancy.
7.	Armstrong and Jin (1987)	Cellulose filter paper containing aqueous solution in the pores acting as membrane between two organic solutions.

retain the inherent SLM advantages is worthwhile. The recently developed hollow-fiber contained liquid membrane (HFCLM) techniques for gas separation (Majumdar, Guha, and Sirkar 1988) and liquid separation (Sengupta, Basu, and Sirkar 1988) provide practical solutions in that direction.

Basic Concept of HFCLM for Gas Separation

Consider an aqueous solution in contact with a microporous hydrophobic hollow fiber. If this solution does not wet the fiber pores, it cannot enter the membrane pores unless its pressure exceeds a critical value identified as the *bubble point* or *entry pressure*. Below this critical value, the pores are filled with gas. If a gas mixture flows on one side of the fiber while the aqueous solution flows on the other side, then the gas/liquid interface is immobilized at each gas-filled pore mouth on the liquid side of the fiber. The gas pressure must be lower than the liquid pressure to prevent bubbling and maintain a stable gas/liquid interface (Callahan 1988). Such is the basis of blood oxygenation by microporous hydrophobic hollow fibers (Tsuji et al. 1981); through the immobilized interfaces, both gas absorption and stripping can occur (see Chapter 46 for further details).

Now consider two (instead of one) microporous hydrophobic hollow fibers (Figure 42–2). The feed gas mixture flows through the lumen of one hollow fiber. On the outside, an

Hollow-Fiber Contained Liquid Membrane

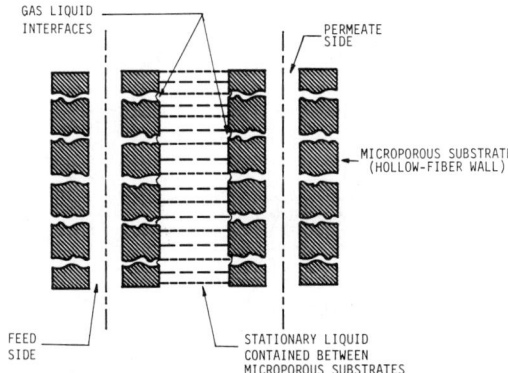

FIGURE 42–2. Aqueous liquid membrane between two hydrophobic microporous hollow fibers with immobilized gas/liquid interfaces.

aqueous nonwetting solution is present at a pressure higher than the feed gas pressure to immobilize the gas/liquid interfaces. Species from the feed gas are absorbed in the stationary aqueous solution on the outside. The second hollow fiber is placed close to the first one with a sweep gas flowing through the fiber lumen at a pressure less than that of the aqueous solution on the outside of the fiber. Any dissolved species in the aqueous solution will now be stripped into the sweep gas in the second fiber lumen. Thus, the aqueous solution acts as a liquid membrane. At the first gas/liquid interface in the first fiber, species are selectively absorbed; then the dissolved species diffuse to the second gas/liquid interface at the outside diameter of the second fiber where they are desorbed into the sweep gas flowing in the second fiber lumen.

Such a process is achieved by packing thousands of microporous hydrophobic hollow fibers in a permeator shell filled later with the aqueous solution that acts as the membrane (Figure 42–3). However, the fibers are present in two distinct sets; the feed set and the sweep set with the ends of each set being separated. Moreover, the fibers are distributed in the bundle such that a feed-gas-carrying fiber is very likely to be immediately adjacent to a sweep-gas-carrying fiber. Since the fibers contain the aqueous liquid membrane on the shell side, the latter is identified as the contained liquid membrane (CLM), in contrast to a supported liquid membrane; the separator is described as the hollow-fiber contained liquid membrane (HFCLM) permeator.

The aqueous liquid membrane, maintained in the shell side at a pressure higher than both gas pressures, is connected to an external membrane liquid reservoir which is maintained at a higher pressure (Figure 42–4). As a result, any membrane liquid lost by evaporation or otherwise into dry gas streams in the permeator is automatically and continuously replaced by fresh membrane liquid from the reservoir. When the membrane liquid is pure, such automatic renewal keeps the membrane liquid unchanged on the shell side as was demonstrated for pure water by Majumdar, Guha, and Sirkar (1986). When a solution is used as a membrane such that different components have different evaporation rates, different procedures are adopted (see the following section) to keep the liquid membrane unchanged.

The HFCLM permeator described above for separating the feed gas mixture using a sweep gas stream has the following advantages:

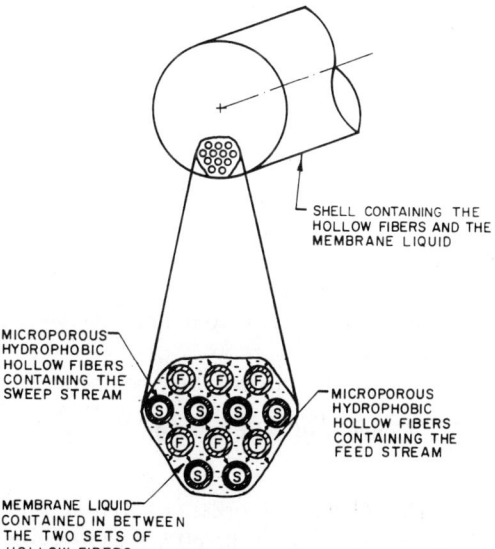

FIGURE 42–3. Configuration of the hollow-fiber contained liquid membrane in a permeator shell (reprinted from Majumdar, Guha, and Sirkar 1988 with permission).

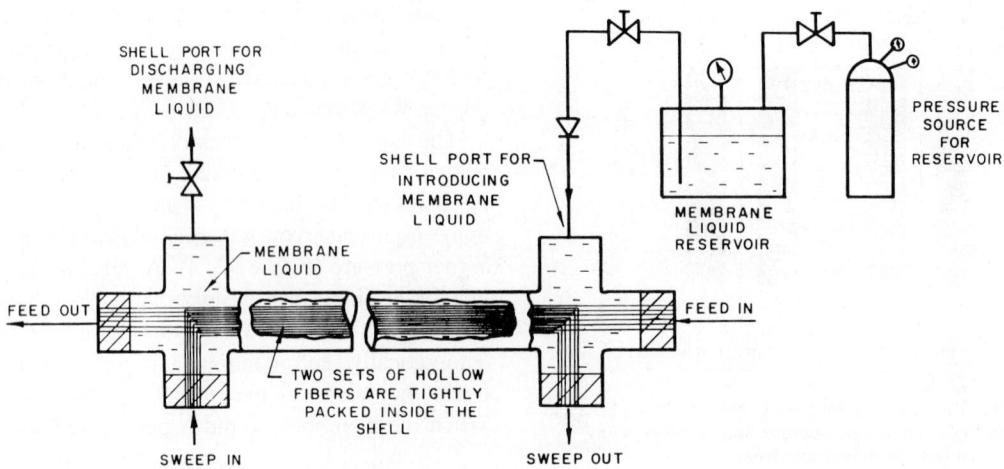

FIGURE 42–4. HFCLM permeator with membrane liquid reservoir (reprinted from Majumdar, Guha, and Sirkar 1988 with permission).

1. Fiber wall defects lead only to a loss of membrane liquid through the fiber lumen. Unlike SLM systems and polymeric membrane permeators, feed does not leak into the permeate gas.
2. Gas humidification is unnecessary for a pure liquid.
3. The membrane is stable. Membrane liquid replenishment is automatic and easy.
4. The hollow-fiber porosity and the pore tortuosity do not influence the gas flux in general.
5. Hydrophobic fibers prevent membrane flooding from moisture condensation in the fiber pore. Condensed moisture is mostly swept away by the gas stream.
6. The use of hollow fine fibers leads to a high membrane surface area per unit permeator volume and low effective liquid membrane thickness.
7. The HFCLM permeator may be operated in a variety of modes much more easily than a SLM permeator.

Majumdar, Guha, and Sirkar (1988) have studied the separation of a 40–60 v/v CO_2-N_2 gas mixture at 45 psig using helium as the sweep gas and pure water as a membrane liquid. They have used a 5-ft-long permeator containing 300 feed fibers and 300 sweep fibers in a permeator shell 1.27 cm. in diameter. These hydrophobic fibers had a 150-μm o.d. and a 100-μm i.d. An effective liquid membrane thickness of 111 μm was observed. The active membrane surface area per unit permeator volume was about 44.6 cm^{-1}. With dry feed gas, the membrane was absolutely stable and the purification capabilities were excellent. The purified gas stream contained only 0 to 3% CO_2 when the feed flow rate was varied between 100 and 300 std cm^3/min. The CO_2 permeation rate (P_{CO_2}/l_e) through pure water was found to be 6.3×10^{-9} mol/$m^2 \cdot s \cdot Pa$ and the separation factor for the CO_2-N_2 system through pure water was about 37.7. The advantages claimed in the previous paragraph were clearly demonstrated in their studies.

Modes of Operation of a HFCLM Permeator

A HFCLM permeator can be operated with other modes of permeate side operation beside sweep gas. Guha (1989) has demonstrated successful HFCLM operation in the following three modes: conventional polymeric membrane mode, vacuum mode, and sweep liquid mode. Any such mode must provide a partial pressure driving force if the species is to permeate through the liquid membrane to the sweep side.

In the *conventional polymeric membrane*

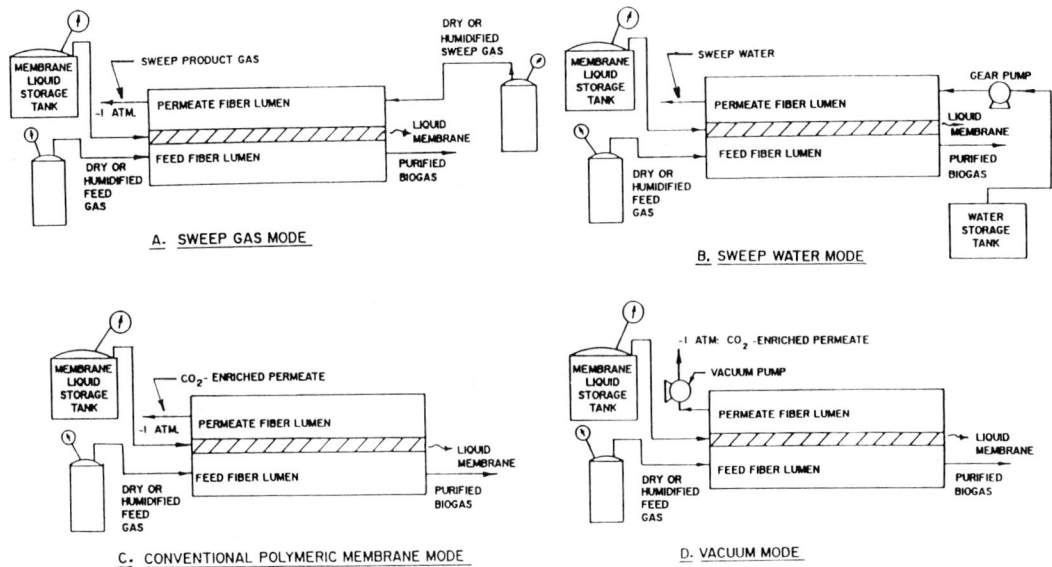

FIGURE 42–5. Four separation modes for purification of biogas using the HFCLM permeation technique (reprinted from Guha, Majumdar, and Sirkar 1991a with permission).

mode, no sweep gas is introduced in the sweep fiber lumen. Permeation of species through the liquid membrane generates a permeate gas stream in the sweep fiber lumen; this permeate stream exits through the fiber outlet under its own axial pressure gradient. [See Figure 42–5 for a schematic of all four modes (Guha, Majumdar, and Sirkar 1991a).] Thus, the liquid membrane permeator acts as if it were a conventional polymeric membrane permeator.

Alternatively, a *vacuum* could be pulled through the sweep fiber lumen and provide enough partial pressure driving force. Obviously, the partial pressure difference in the vacuum mode would be higher than that in the conventional polymeric membrane mode for given feed conditions. A third mode of operation was illustrated by Guha (1989). An aqueous liquid flowing through the sweep fiber lumen absorbed the permeating gases and provided the partial pressure driving force for permeation. The aqueous liquid did not wet the fiber pores and contaminate the liquid membrane (first used for flat films by Matson, Herrick, and Ward 1977). Such a *sweep liquid mode* may also be operated with an organic sweep liquid that wets the fiber pores but is immiscible with the aqueous liquid membrane. A variety of such configurations as well as many gas separation systems that can be studied in a HFCLM device are identified in a patent (Sirkar 1988a).

When a pure membrane liquid is being used, dry feed gas can be used in any one of the four modes (sweep gas, polymeric membrane, vacuum, and sweep liquid); yet the liquid membrane has been found to be completely stable over long periods of time (Guha 1989) since any loss is automatically replaced from the external membrane liquid reservoir. When an aqueous solution is used as a membrane, additional steps are often needed to ensure long-term membrane stability even though automatic replacement of membrane liquid from the external reservoir occurs. These steps are described in Table 42–3, which is based on extended gas separation experiments conducted by Guha (1989) and Majumdar et al. (1990a) using liquid membranes such as water, aqueous diethanolamine, and aqueous K_2CO_3 solutions.

Table 42–3 identifies two basic procedures for ensuring the stability of aqueous liquid membranes. These generally correspond to two extreme conditions of membrane evaporation. When dry gas streams are used or a vacuum is

TABLE 42-3. Recommended Procedures for HFCLM Stability with Aqueous Solutions as Membranes for Gas Separation.

Operational Mode	Gas Humidification	Membrane Liquid
Sweep gas mode	Humidify both gas streams.	Withdraw small fractions of membrane liquid via second shellside port in permeator as needed, e.g., once every few days.
	Dry gas streams.	The membrane liquid should be withdrawn very slowly via second shellside port and components: e.g., water, reactive volatile carrier should be replenished and put back into membrane liquid reservoir.
Sweep liquid mode	Dry feed gas. Water is sweep liquid.	Follow procedure for sweep gas mode with humidified gas streams.
	Dry feed gas. Aqueous solution is sweep liquid.	Follow procedure for sweep gas mode with dry gas streams.
	Dry feed gas. Nonaqueous solution is sweep liquid.	Follow procedure for sweep gas mode with dry gas streams.
Polymeric membrane mode	Humidified feed gas stream.	Follow procedure for sweep gas mode with humidified gas streams
	Dry feed gas.	Follow procedure for sweep gas mode with dry gas streams.
Vacuum mode	Humidified or dry feed gas.	Follow procedure for sweep gas mode with dry gas streams.

pulled through the sweep fibers, the membrane composition is likely to change rapidly. It would then be desirable to withdraw the concentrated membrane solution through the second shellside port (Figure 42-4) at a rate consistent with the rate of change of the membrane solution, add water and other volatile solution components to it (if needed), and recycle to the membrane liquid reservoir. In one case, namely, the sweep liquid mode with water as the sweep, it was found, however, that even with a dry feed gas, the membrane composition change was very small since the sweep water was supplying the moisture needed.

In any of the modes (except vacuum), if the gas streams are humidified, the rate of change of membrane composition, if any, is quite small unless extremely volatile carrier components are used in solution. A nonzero rate of change will be due to either evaporative loss of carrier components in solution, deterioration or poisoning of carrier species, or any loss of moisture due to inadequate gas humidification or process fluctuations. Liquid membrane stability may be ensured in such cases by occasional withdrawal (once every few days) of a fraction of the membrane liquid via the second shell side port (Majumdar et al. 1990a).

Additional HFCLM Configurations for Gas Separations

The most common HFCLM configuration studied is based on hydrophobic microporous hollow fibers (MHF) and an aqueous membrane liquid that does not wet the hydrophobic fibers. Other possible HFCLM configurations for gas separation are:

1. *Hydrophobic MHF:* nonwetting organic membrane liquid
2. *Hydrophobic MHF:* wetting organic membrane liquid

3. *Hydrophilic MHF:* organic or aqueous membrane liquid.

In the case of a nonwetting organic membrane liquid and hydrophobic MHFs, the permeator is operated in a manner identical to aqueous solutions. Majumdar et al. (1990a) operated a HFCLM permeator to separate a 40–60 v/v CO_2-N_2 gas mixture using hydrophobic fibers and *n*-methylpyrollidinone (NMP) containing 30% water as a membrane liquid for a significant length of time. If the organic membrane liquid wets the hydrophobic fibers, the pressure conditions need to be reversed. The feed and sweep gas pressures must be higher than the liquid membrane pressure. Since the support pores are now wetted, the membrane resistance is significantly increased. An identical condition exists when hydrophilic MHFs are used since both organic and aqueous membrane liquids will wet the pores of a hydrophilic MHF.

The contained liquid membrane in the shell side of a HFCLM permeator operated in any of the above configurations is generally stationary except for any movement caused by membrane liquid replenishment. Unlike SLM, a HFCLM configuration offers the additional possibility of operating a module with the liquid membrane in uniform motion; the liquid exiting from one end of the permeator may be recirculated to the other end of the shell side of the permeator. The movement of the liquid membrane may be beneficial in those situations where most of the mass transfer resistance lies in the membrane liquid. Such a liquid membrane may be identified as a mobile liquid membrane (MLM) or a flowing liquid membrane (FLM). The behavior of such a hollow-fiber contained mobile liquid membrane has not been studied.

The liquid membrane can be contained between two flat hydrophobic or hydrophilic membranes (instead of hollow fibers) with appropriate pressure control (Sirkar 1988a). Gas separation with the liquid membrane flowing in between two such microporous membranes packed in a spiral-wound configuration has been studied by Teramoto et al. (1989a). Liquid membrane convection increases the membrane transport coefficient in gas separation; the membrane surface packing density is, however, considerably lower in a spiral-wound system. Pump and pumping costs are add-ons.

Basic Concept of HFCLM for Liquid Separation

Suppose an aqueous solution is contacted with one surface of a microporous *hydrophilic* membrane. The solution will spontaneously wet the membrane, fill the pores, and appear on the other side of the membrane. If, however, there is an organic solvent or solution on the other side of the membrane at a higher pressure, the aqueous/organic interface will be immobilized at the pore mouths on the organic side of the membrane (Prasad and Sirkar 1987; see also Chapter 41); the aqueous phase will not be dispersed in the organic phase and vice versa. Through the aqueous/organic phase interface at the pore mouth, solute extraction or back extraction can be achieved.

Consider Figure 42–6 where one-half of a *hydrophilic* microporous hollow fiber is shown next to one-half of another hydrophilic microporous hollow fiber. Let feed aqueous solution flow through the lumen of one fiber and a strip aqueous solution through the lumen of the other. The space between the two fibers on the outside is occupied by the organic solvent. If the organic solvent pressure is higher than the pressure of both feed and strip solutions, an immobilized aqueous/organic interface will exist on the outside surface of each hollow fiber. At the first interface, solute from the feed solution in the feed fiber will be partitioned into the organic solvent outside. The solute will then diffuse through the organic solvent to the second interface on the outside of the strip fiber where it will be partitioned into the strip aqueous solution. The organic solvent on the outside of the fibers now acts as an organic liquid membrane between two aqueous solutions.

The HFCLM permeator for separation from an *aqueous feed* to an *aqueous strip* consists of a cylindrical shell with a dense population of

FIGURE 42–6. Immobilized phase interfaces in a CLM for *hydrophilic* hollow-fiber substrates, aqueous feed and strip solutions, and organic liquid membrane (reprinted from Sengupta, Basu, and Sirkar 1988 with permission).

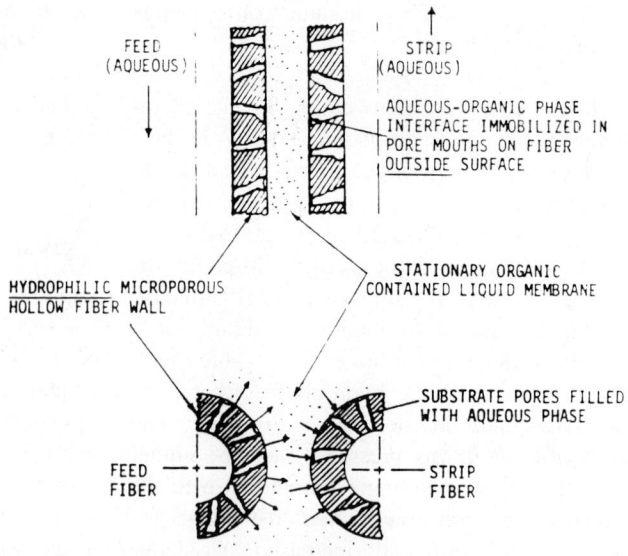

two sets of hydrophilic MHFs separated at the ends. The fiber configurations and assembly are similar to those used for gas separation (Figure 42–4). The shell side is filled with the organic membrane liquid connected to an external membrane liquid reservoir at a pressure higher than the pressure of both feed and strip solutions. If any membrane liquid is lost due to fiber defect or by dissolution into the feed or strip solution, it is automatically replenished from the external membrane liquid reservoir. Feed and the strip aqueous solutions flow in a countercurrent or cocurrent fashion through the lumen of two sets of hollow fibers and liquid membrane separation is achieved (Sengupta, Basu, and Sirkar 1988; Sengupta et al. 1988; Basu and Sirkar 1989). Note that aqueous feed and strip solutions occupy the pores in the respective hollow-fiber wall and the aqueous-organic interface is on the outside diameter of each fiber.

The basic HFCLM permeator for separation from an aqueous feed to an aqueous strip solution can also be made out of *hydrophobic* microporous hollow fibers provided the required conditions for aqueous/organic interface immobilization in hydrophobic membrane pore mouths are implemented; i.e., the aqueous phase pressure must be equal to or greater than that of the organic phase (Kiani, Bhave, and Sirkar 1984). In a HFCLM permeator containing hydrophobic fibers or substrates, the organic membrane liquid phase on the shell side wets the fiber pores. The aqueous/organic interfaces are located on the inside diameter of each fiber as the aqueous feed and strip solutions flow through the lumen of the respective fiber set (Figure 42–7).

The overall HFCLM permeator setup remains unchanged for *hydrophobic* hollow fibers with the exception that the membrane liquid reservoir is maintained at a pressure lower than those of both feed and strip solutions. If there is a surge in feed and/or strip pressure, they may leak into the membrane phase. However, the shellside liquid can be removed from the other port on the shell side through a valve without interrupting the separation (Sengupta, Basu, and Sirkar 1988).

The performances of HFCLM permeators with hydrophobic or hydrophilic hollow fibers have been studied by Sengupta, Basu, and Sirkar (1988) and Sengupta et al. (1988) for various pure membrane liquids. They have in-

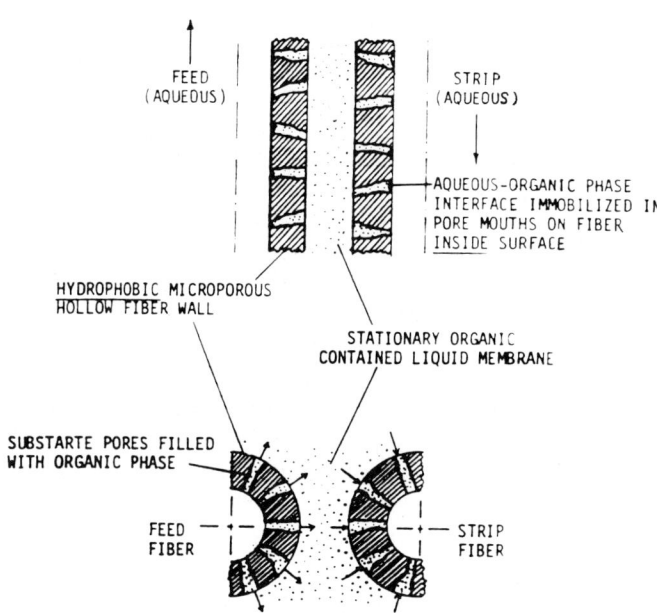

FIGURE 42–7.
Immobilized phase interfaces in a CLM for *hydrophobic* hollow-fiber substrates, aqueous feed and strip solutions, and organic liquid membrane (reprinted from Sengupta, Basu, and Sirkar 1988 with permission).

vestigated the separation of phenol or acetic acid from a feed aqueous solution to a strip aqueous solution via liquid membranes of either methyl isobutyl ketone (MIBK) or decanol or xylene. In one set of experiments on phenol removal, a 2% caustic solution was used as the strip. Otherwise, pure water was used. The performance was always stable. Of particular note is the result with MIBK as the membrane. The small volume of MIBK present as membrane in the permeator shell side would have disappeared by dissolution in the feed and strip solutions in a few hours without automatic and continuous addition from the membrane liquid reservoir. In fact, this run was continued for 44 hours without any problem. While the dissolution of the membrane liquid in the feed and strip liquids may or may not be acceptable in an actual process situation, the example of MIBK as a membrane liquid demonstrates the versatility and robustness of the HFCLM technique. A regular SLM would be completely unworkable with MIBK as a membrane liquid since the membrane would disappear within minutes of starting.

The volumetric transfer rate of the solute species in such devices is quite high due to high membrane surface packing density. Sengupta, Basu, and Sirkar (1988) have calculated the $K_w a$ value for phenol removal with a caustic strip in their device and found it (0.58 min^{-1}) to be about two-thirds of the rate achieved in ELM systems (Cahn and Li 1974). The $K_w a$ value can be increased further using higher surface packing densities.

Basu and Sirkar (1991) have studied the facilitated transport separation of citric acid from an aqueous solution through a HFCLM of xylene containing tri-n-octylamine (TOA) as the carrier and a 0.5 M caustic strip. At 25°C, they had obtained a citric acid flux of around 27 μg/cm^2 · min for 40% TOA in xylene in a HFCLM permeator removing around 97% of the citric acid present in the 10% citric acid-containing feed solution. This flux level is comparable to the average flux levels of 15 to 29 μg/cm^2 · min obtained by Friesen et al. (1991) using a SLM of 38 vol% trilaurylamine and 15 vol% dodecanol in Shell Sol 71 in a 25-μm-thick Celgard 2400 microporous polypropylene film with an 18-cm^2 area at a high temperature of 60°C (which was found to be the optimum temperature). Moreover, the flux levels in Friesen et al. (1991) decreased with time. Basu and Sirkar (1991) have also demonstrated a completely stable performance over a month-long period in the 48-cm-long permeator. The only additional step they had taken was to withdraw a very small volume of membrane liquid every day (2 to 3 cm^3/day) from the second shellside port. If this step is not taken, the multicomponent membrane liquid becomes concentrated in the less soluble amine; a steady state is reached with a flux value somewhat lower (about 50%) than the initial flux value as was observed in a two-month-long run by Basu and Sirkar (1991). Since Basu and Sirkar (1991) have shown that this resulted from increasing the concentration of TOA, recirculation of the membrane liquid may be achieved by adding xylene outside the permeator.

HFCLM separation can be easily implemented with an *organic feed solution* and an *organic strip solution*. The liquid membrane is of necessity an aqueous or polar organic liquid immiscible with feed and strip phases (Cahn and Li 1976). Either hydrophobic or hydrophilic fiber may be used. For a hydrophilic fiber and an aqueous or polar organic liquid membrane, the liquid membrane is in the membrane pores. The organic feed and strip pressures should be higher than the CLM pressure. If the fibers are hydrophobic, the liquid membrane phase is excluded from the fiber pores; further, the CLM pressure should be higher than those of feed and strip streams to immobilize the phase interfaces (Sengupta et al. 1988). These basic HFCLM structures and the required pressure conditions are summarized in Table 42–4.

A variety of HFCLM structures and separation systems are identified for aqueous or organic feeds in Sirkar (1988b, 1991). He has also indicated that the liquid membrane can be contained between two hydrophobic or hydrophilic flat membranes by appropriate pressure control between the adjacent phases. Separation of a liquid solution using a liquid membrane flowing between two microporous flat membranes packed in the module in a spiral-wound configuration has been studied by Teramoto et al. (1989b) for chromium recovery. The velocity of the flowing liquid membrane was not found to influence chromium recovery or transport. Compared to the HFCLM configuration, a spiral-wound arrangement generally packs a much lower membrane surface area/equipment volume. Such a flowing liquid membrane has some similarities to the creeping film perstractor (Boyadzhiev, Bezenshek, and Lazarova 1984).

TABLE 42–4. HFCLM Structures and Pressure Conditions for Different Porous Substrates, Feeds, and Membranes for Liquid Separations.

Hollow Fiber Substrate Nature	Feed and Strip Solutions	Membrane Liquid	Pore Liquid	Pressure Conditions[†]	
				Feed Fiber	Strip Fiber
Hydrophobic	Aqueous	Nonpolar organic	Nonpolar organic	$p_F > p_M$	$p_S > p_M$
	Nonpolar organic	Aqueous and/or polar organic	Nonpolar organic	$p_M > p_F$	$p_M > p_S$
Hydrophilic	Aqueous	Nonpolar organic	Aqueous	$p_M > p_F$	$p_M > p_S$
	Nonpolar organic	Aqueous and/or polar organic	Aqueous and/or polar organic	$p_F > p_M$	$p_S > p_M$

[†] p_F-feed pressure; p_M-membrane pressure; p_S-strip pressure

Pervaporation Separation by HFCLM

In the membrane separation technique of pervaporation (Chapters 7 through 10), the feed is a liquid solution and the permeate is gaseous. The permeate side either has vacuum or a sweep gas to provide the necessary driving force. Although both are equivalent, current pervaporation practices favor application of vacuum on the permeate side. Since microporous hollow fibers are being used in HFCLM techniques, application of vacuum is possible only if the liquid membrane does not wet the hollow-fiber pores; otherwise, the applied vacuum would draw the liquid continuously into the fiber bore. Aqueous solutions and many polar organics do not wet *hydrophobic* MHFs and therefore may be used as contained liquid membranes for separating organic liquid mixtures. The organic feed solution will wet the feed fiber pores; however, the pressure of the aqueous or polar organic contained liquid membrane phase will be higher thus immobilizing the phase interface at pore mouths on the fiber outside diameter. On the permeate side, the liquid membrane/gas interface will be immobilized at the outside surface of the permeate or strip fiber with gas in the fiber pores. A shortcoming of such a technique is the contamination of feed and permeate phases by the liquid membrane constituents via partitioning/evaporation when compared with polymer membrane-based pervaporation for certain types of purification applications.

Vacuum mode pervaporation in the HFCLM configuration using *hydrophilic* fibers has also been carried out (Sirkar 1986) for separating alcohol from a dilute aqueous solution using decanol as the liquid membrane. Properly dried Cuprophan® fibers were used for both sets. On the strip side, this allowed vacuum without breakthrough of the membrane liquid.

THEORY

A hollow-fiber contained liquid membrane permeator is a membrane device to separate the constituents of a feed fluid mixture. Given the properties of the membrane *vis-à-vis* the feed mixture components, the dimensions of the separator and the conditions of operation, it should be possible to predict the extent of separation achieved in the HFCLM device. There are, however, no general solutions to this problem for any combination of feed mixture, liquid membrane, and operating condition. Approximate solutions are available; some are quite broad in scope. This section will provide an analysis of mass transport and separation in HFCLM permeators first for a feed gas mixture and then for a feed liquid solution. Such analyses will be preceded by identifying some distinguishing characteristics of HFCLM devices in the following section.

Effective Membrane Thickness in HFCLM Permeation

Unlike conventional polymeric membranes or supported liquid membranes, the liquid membrane in a HFCLM device at any axial location is not a geometrically well-defined entity; thus, the membrane thickness is *a priori*, unknown. The relative positions between a feed fiber and the surrounding strip fibers (or vice versa) are not unique. Radial and circumferential variations can occur in such configurations. Further, since concentrations are changing in the z-direction also due to z-direction (axial) motion of feed and strip streams, the most general governing equation for permeation in the membrane can be written in the absence of any chemical reaction and convection as

$$\frac{\partial^2 c}{\partial x^2} + \frac{\partial^2 c}{\partial y^2} + \frac{\partial^2 c}{\partial z^2} = 0. \quad (42\text{--}1\text{a})$$

In general, the length scales for transport in the x- and y-directions are much smaller than that in the z-direction; the term $\partial^2 c/\partial z^2$ is therefore neglected unless the permeator length is very small:

$$\frac{\partial^2 c}{\partial x^2} + \frac{\partial^2 c}{\partial y^2} = 0. \quad (42\text{--}1\text{b})$$

Solution of such an equation requires well-defined boundaries. A unit cell (Figure 42–8) of

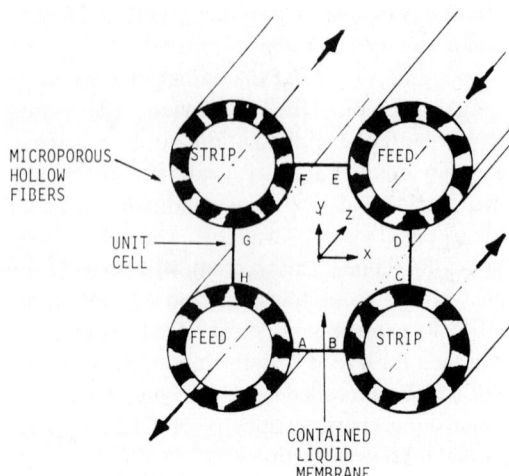

FIGURE 42–8. A unit cell of two feed and two strip fibers in a particular configuration.

For gaseous feed and gaseous permeate, the solute distribution equilibria at the feed fiber outside surface and strip fiber outside surface are, respectively,

Surfaces HA and ED: $c = Hp'$, (42–2c)

Surfaces BC and GF: $c = Hp''$. (42–2d)

Here H is Henry's constant for the species being transferred. Majumdar (1986) and Majumdar et al. (1989) have solved Eq. (42–1b) subject to boundary conditions of the type of Eqs. (42–2a) through (42–2d) for four possible configurations (Figure 42–9). Assuming each con-

two feed and two sweep (or strip) fibers in a particular configuration in a square array provides well-defined boundaries. Four such possible configurations of feed-sweep fibers in a square array are shown in Figure 42–9. This figure also illustrates how the hollow-fiber bundle in the permeator can be made out of a repeating pattern of these four configurations. Configurations of either four feed fibers or four sweep fibers are neglected. Further, all four possible configurations are assumed equally probable.

It is now possible to define an *effective membrane thickness* (EMT) in a HFCLM as the thickness l_e of a hypothetical liquid film that exhibits the same mass transfer resistance as the contained liquid membrane for the same amount of mass transfer area in both cases. *For gas separation,* such a value can be estimated by first solving the *two-dimensional diffusion equation* (42–1b) for a nonreactive liquid membrane subject to the following boundary conditions (Figure 42–8):

Planes AB and EF: $\partial c/\partial y = 0$ for all x, (42–2a)

Planes CD and GH: $\partial c/\partial x = 0$ for all y. (42–2b)

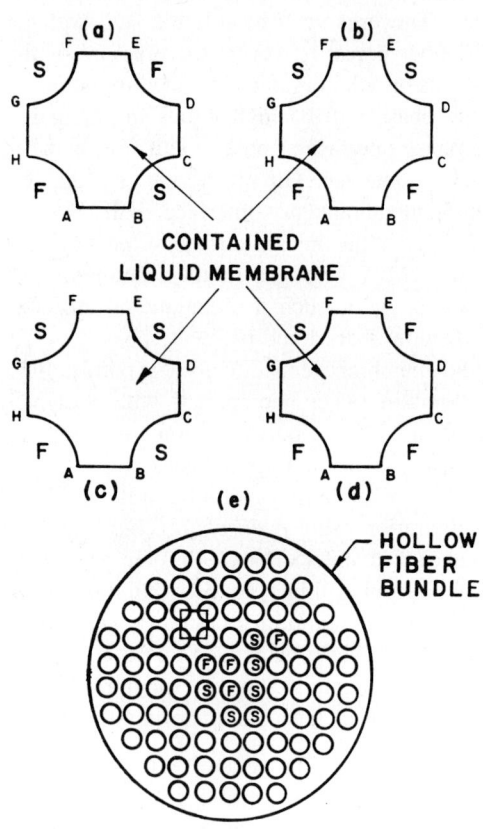

F: FEED FIBER; S: SWEEP FIBER

FIGURE 42–9. Four possible useful configurations of fibers in a unit square cell (reprinted from Majumdar et al. 1989 with permission).

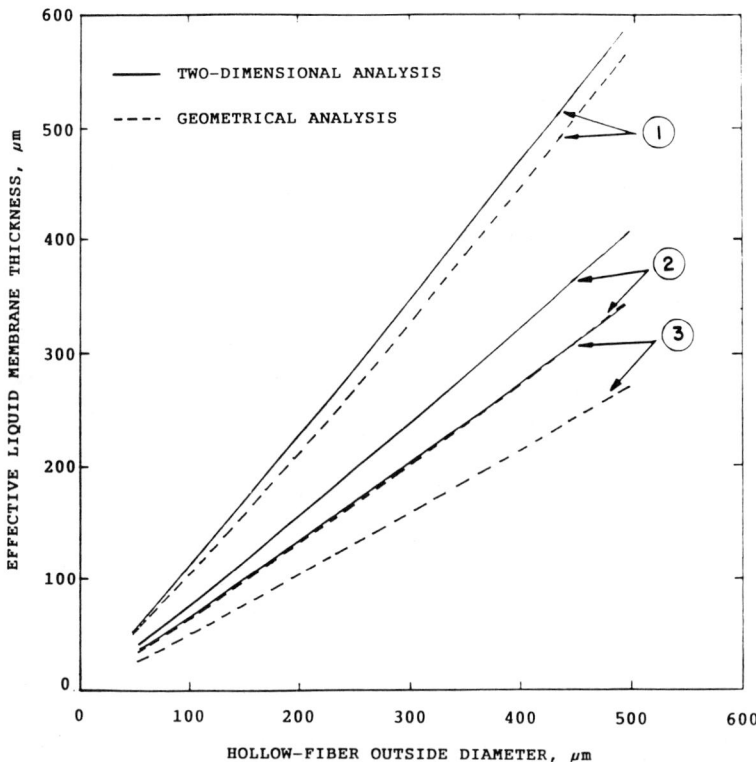

FIGURE 42–10. Effective liquid membrane thickness as a function of fiber outside diameter and packing density (%) for square cell arrangement; (1) 20%; (2) 33%; (3) 40% (reprinted from Majumdar et al. 1989 with permission).

figuration to be equally probable, they have determined the value of the EMT as a function of the fiber outside diameter and the fiber packing density. As shown in Figure 42–10, this liquid membrane thickness was found to be essentially linearly proportional to the fiber outside diameter. However, as the fiber packing density increased, the EMT was reduced significantly. They calculated EMT to be 111.5 μm for 150-μm-o.d. fibers and a packing density of 0.3. Their experimental measurements of pure gas and gas mixture permeation through water as liquid membrane in a number of HFCLM permeators built of 150-μm-o.d. fibers could be explained satisfactorily using HFCLM permeator models if an EMT value of 111.5 μm was assumed.

An alternative procedure for determining the EMT in a fiber bundle is a simple geometrical method based on calculating the equivalent diameter d_e of a circle containing the membrane liquid around any feed fiber of outside diameter d_{to} in a unit cell such as that shown in Figure 42–11 (Majumdar, Guha, and Sirkar 1988). This figure identifies three possible configurations of the unit cell. The EMT, l_e, is defined as

$$l_e = \tfrac{1}{2}(d_e - d_{to}). \qquad (42\text{--}3\text{a})$$

For a square cell, the value of pitch t_p (Figure 42–11) is given by

$$t_p = \frac{d_b - d_{to}}{4(N_T' + N_T'')} \cdot \{\pi + \sqrt{[\pi^2 + 4\pi(N_T' + N_T'')]}\} \qquad (42\text{--}3\text{b})$$

and is related to d_e by

$$d_e = [8(t_p^2/\pi) - d_{to}^2]^{1/2}. \qquad (42\text{--}3\text{c})$$

Knowing the diameter of fiber bundle d_b, fiber outside diameter (d_{to}), and the total number of feed (N_T') and sweep (N_T'') fibers, l_e may be calculated. Figure 42–10 provides the relation between l_e and d_{to} for a few cases. Although

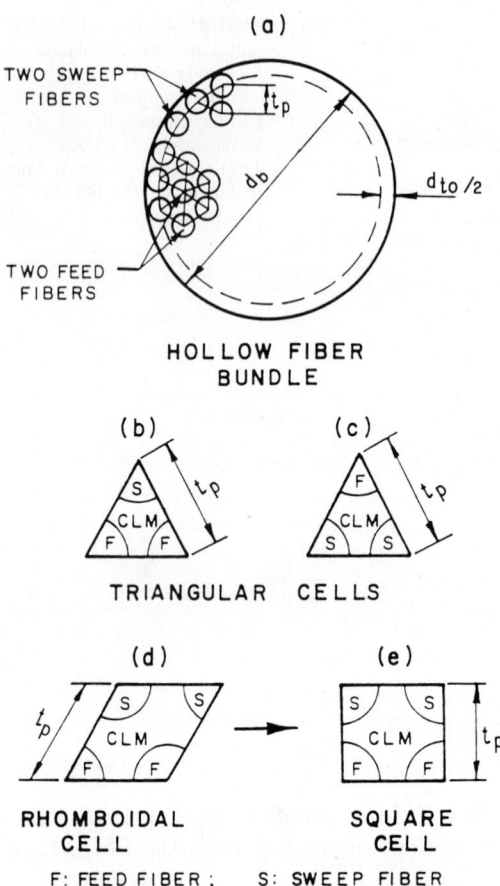

FIGURE 42–11. Unit cell configurations for geometrical method of calculating effective liquid membrane thickness (reprinted from Majumdar, Guha, and Sirkar 1988 with permission).

this calculation method is simple, the theoretical basis for it is unsound. Further, it cannot accommodate fibers of different diameters; the two-dimensional analysis can account for fibers of different diameters (Majumdar et al. 1990a). The two-dimensional analysis, based on first principles, may be modified to incorporate reactive diffusion systems for facilitated transport where the local variation of diffusion path length can lead to a spectrum of facilitation factors for given feed-sweep conditions.

HFCLM Permeator Model for Gas Separation without Reaction

Majumdar, Guha, and Sirkar (1988) have modeled the separation of a binary feed gas mixture flowing countercurrent to a sweep gas stream in a HFCLM permeator having simple liquid membrane permeation through an effective liquid membrane thickness. Both gas streams were subjected to a pressure drop in the axial flow direction. The above analysis was extended to the separation of a gas mixture involving n components by Majumdar et al. (1989). The following assumptions were used in their analysis:

1. The gas species permeates from one side of the liquid membrane to the other purely by a solution-diffusion mechanism; no reaction occurs within the membrane.
2. The permeability coefficients P_i of the gas components are the same as those of pure gases and are independent of pressure.
3. The effective liquid membrane thickness l_e exists along the length of the permeator as well as at all radial locations inside the fiber bundle.
4. No mass transfer resistance occurs in the bulk gas phase and in the stagnant gas layers inside the pores.
5. Diffusion is negligible along the mean gas flow path compared to the bulk gas flow.
6. A plug flow model can be used for both gas streams.
7. The Hagen-Poiseuille law governs pressure drop inside the fiber.
8. The viscosity of the gas mixture depends only on composition.
9. On both ends of a module, two sets of fibers are separated leading to a very large l_e and negligible separation. Such end effects in the permeator and deformation of fibers under external pressure are neglected.
10. The permeator contains an equal number of feed and sweep fibers of identical size.
11. The effect of a curved gas/liquid interface at the pore mouth is neglected.

Figure 42–12 provides the schematic of a HFCLM permeator for gas separation with a sweep gas stream and shows only one feed and one sweep fiber. A general formulation of equations for n components in terms of dimensionless variables are

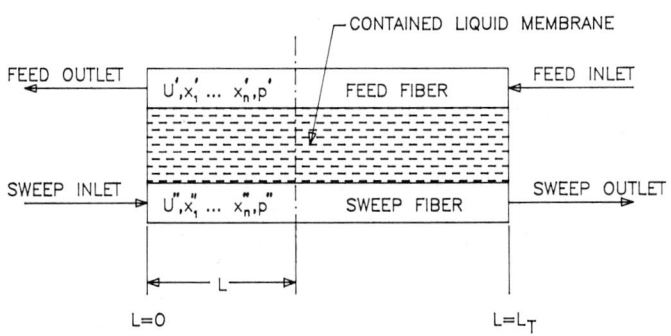

FIGURE 42–12. Schematic of HFCLM permeator with a sweep gas stream for calculating multicomponent gas separation (reprinted from Majumdar et al. 1989 with permission).

$$\frac{dU'_d}{dA_d} = \sum_{j=1}^{n} P^*_j(p'_d x'_j - p''_d x''_j), \quad (42\text{--}4)$$

$$\frac{dp'_d}{dA_d} = \beta \eta'_d \frac{U'_d}{p'_d}, \quad (42\text{--}5)$$

$$\frac{dx'_i}{dA_d} = [P^*_i(p'_d x'_i - p''_d x''_i) - x'_i \sum_{j=1}^{n} P^*_j(p'_d x'_i - p''_d x''_i)]/U'_d \quad \text{for } i = 1, n, \quad (42\text{--}6)$$

$$\frac{dU''_d}{dA_d} = \sum_{j=1}^{n} P^*_j(p'_d x'_j - p''_d x''_j), \quad (42\text{--}7)$$

$$\frac{dp''_d}{dA_d} = -\beta \eta''_d \frac{U''_d}{p''_d}, \quad (42\text{--}8)$$

$$\frac{dx''_i}{dA_d} = [P^*_i(p'_d x'_i - p''_d x''_i) - x''_i \sum_{j=1}^{n} P^*_j(p'_d x'_i - p''_d x''_i)]/U''_d \quad \text{for } i = 1, n, \quad (42\text{--}9)$$

where

$$A_d = \pi d_{to} \left(\frac{P_r}{l_e}\right)\left(\frac{p_r}{U_r}\right) L, \quad (42\text{--}10a)$$

$$p'_d = \frac{p'}{p_r}; \quad p''_d = \frac{p''}{p_r}, \quad (42\text{--}10b)$$

$$U'_d = \frac{U'}{U_r}; \quad U''_d = \frac{U''}{U_r}, \quad (42\text{--}10c)$$

$$\eta'_d = \frac{\eta'}{\eta_r}; \quad \eta''_d = \frac{\eta''}{\eta_r}, \quad (42\text{--}10d)$$

$$P^*_i = \frac{P_i}{P_r} \quad \text{for } i = 1, n, \quad (42\text{--}10e)$$

$$\beta = \frac{128 RT U^2_r \eta_r}{\pi^2 d^4_{ti} d_{to}(P_r/l_e) p^3_r}. \quad (42\text{--}10f)$$

Here U' and U'' are feed gas flow rate/fiber and strip gas flow rate/fiber at any location. The choices of reference parameters of pressure p_r, flow rate U_r, permeability P_r, and viscosity η_r in the above equations are arbitrary. Note that Eqs. (42–4) to (42–9) are derived for countercurrent flow pattern with the dimensionless membrane area A_d increasing in the direction of the sweep gas flow and the sweep inlet end is considered to be the starting point for calculation.

The permeability P_i of any species in the liquid membrane may be estimated from the product of the diffusivity D_{im} and solubility H_{im} of species i in the membrane liquid:

$$P_i = D_{im} H_{im}. \quad (42\text{--}11)$$

The gas mixture viscosity in the feed or strip side can be determined from the equation (Wilke 1950):

$$\eta = \sum_{i=1}^{n} \frac{y_i \, \eta_i}{\sum_{j=1}^{n} \lambda_{ij} y_j}, \qquad (42\text{-}12\text{a})$$

where

$$\lambda_{ij} = \frac{[1 + (\eta_i/\eta_j)^{1/2} (\tilde{M}_j/\tilde{M}_i)^{1/4}]^2}{[8(1 + \tilde{M}_i/\tilde{M}_j)]^{1/2}}. \qquad (42\text{-}12\text{b})$$

For n components, there are a total of $2n + 4$ nonlinear coupled ordinary differential equations. Using $\Sigma x_i = 1$ and $\Sigma y_i = 1$, the total number can be reduced to $2n + 2$. Normally, feed gas flow rate (U_d^l) and composition (x_i^ls) are known at the feed inlet end ($A_d = A_T$) and the strip gas flow rate (U_d^{ll}) and composition (x_i^{ll}s) are known at the strip inlet end ($A_d = 0$). Depending on the problem, feed and strip stream pressures (p_d^l and p_d^{ll}) would be specified at either of the two ends. It is a two-point boundary value problem, the solution of which can be obtained by applying various numerical integration techniques (Majumdar 1986; Guha 1989). From the gas flow rates and compositions at the feed outlet stream, the separation achieved can be easily estimated.

When no sweep gas is present, as for example in gas mixture separation under the vacuum mode or conventional polymeric membrane mode, all of the above equations remain unchanged. However, the value of the permeate flow rate U_d^{ll} is zero at $A_d = 0$. In addition, the values of x_i^{ll}'s are not explicitly specified at $A_d = 0$. They are determined by solving the following set of nonlinear algebraic equations at that particular point:

$$x_i'' = \frac{P_i^*(p_d^l x_i' - p_d^{ll} x_i'')}{\sum_{j=1}^{n} P_j^*(p_d^l x_j' - p_d^{ll} x_j'')} \quad \text{for } i = 1,n. \qquad (42\text{-}13)$$

Further, the derivatives dx_i''/dA_d are indeterminate at $A_d = 0$ and have to be determined by applying L'Hôspital's rule (Guha 1989).

Guha (1989) has also modeled the sweep liquid mode of operation where gas absorption takes place at the sweep liquid/pore gas interface in addition to gas-liquid desorption at the membrane liquid/pore gas interface.

The governing equations for pure gas permeation in the absence of a strip gas can be readily derived from the previous equations. The equations for countercurrent flow are given by (Majumdar et al. 1989):

$$\frac{dU_d^l}{dA_d} = P^*(p_d^l - p_d''), \qquad (42\text{-}14)$$

$$\frac{dp_d^l}{dA_d} = \frac{\beta \eta_d^l U_d^l}{p_d^l}, \qquad (42\text{-}15)$$

$$\frac{dU_d''}{dA_d} = P^*(p_d^l - p_d''), \qquad (42\text{-}16)$$

$$\frac{dp_d''}{dA_d} = -\frac{\beta \eta_d'' U_d''}{p_d''}. \qquad (42\text{-}17)$$

The solution of such a set of equations was used by Majumdar et al. (1989) to determine the value of l_e that would describe the observed pure component permeation behavior.

Facilitated Transport Systems for Gas Separation

In facilitated transport permeation-separation of a gas species (permeant) from a gas mixture, the liquid membrane contains a nonvolatile carrier that reacts reversibly with the particular species. Any free species in the liquid membrane diffuses across the membrane; simultaneously the carrier-permeant complex formed at the feed/membrane interface diffuses through the membrane and dissociates into the carrier and the permeant at the sweep/membrane interface. Simultaneous diffusion of the free permeant species and the permeant-carrier complex enhances the total mass flux of the permeant. Such systems, although complex, have been analyzed for SLMs by solving the reaction-diffusion equations in one dimension along the film. Reasonable success has been achieved in describing the observed behavior (Chapter 44).

In a HFCLM, the reaction-diffusion equations have to be solved in two dimensions, x

and y, with curved boundaries. For a reversible reaction of permeant A and carrier B producing a nonvolatile complex C

$$A + B \underset{k_r}{\overset{k_f}{\rightleftarrows}} C. \quad (42\text{--}18)$$

The equations to be solved at any location in the permeator across the membrane are

$$D_A\left(\frac{\partial^2 c_A}{\partial x^2} + \frac{\partial^2 c_A}{\partial y^2}\right) - k_f c_A c_B + k_r c_C = 0,$$

(42–19a)

$$D_C\left(\frac{\partial^2 c_C}{\partial x^2} + \frac{\partial^2 c_C}{\partial y^2}\right) + k_f c_A c_B - k_r c_C = 0.$$

(42–19b)

At the feed fiber and sweep fiber boundaries (Figures 42–8 and 42–9), c_A has to be related to the partial pressure of species A in the gas phases. Local solution of the above two equations with such and other boundary conditions for species A [e.g., Eqs. (42–2a) and (42–2b)] and species B and C will yield a value of P_A, the permeability of species A, needed (as P_i) to solve Eqs. (42–4) to (42–9) for the HFCLM permeator. No such solution is currently available. The problem is complex for the following reasons:

1. Equations (42–19a) and (42–19b) are non-linear partial differential equations and the boundary conditions are complex.
2. The partial pressures of species A in the feed and sweep fiber boundaries are unknown in general.
3. Simultaneous solution of these equations is needed along with those of the nonlinear z-directional mass balance equations, Eqs. (42–4) to (42–9).

A number of simplifications, if allowable, are:

1. A large excess of species B, which reduces $k_f c_A c_B$ to $k_f' c_A$.

2. The HFCLM locally can be modeled in one dimension using the effective membrane thickness l_e.
3. Use assumptions such as the equilibrium approximation (Ward 1970), which will provide an analytical solution to the problem if simplification (2) is acceptable. An analytical expression for P_A will then be available in terms of p_A^I and p_A^{II} at any z for incorporation in the z-directional mass balance equations.

This last approach is somewhat similar to that adopted by Roberts, Gottschlich, and Way (1988): except in the HFCLM permeator model illustrated earlier, pressure drops in the two gas streams via Eqs. (42–5) and (42–8) make the numerical solution more realistic but difficult. To this end, Guha (1989) has adopted an equilibrium approximation with excess carrier (Otto and Quinn 1971) for facilitated CO_2 transport through an aqueous K_2CO_3 solution and solved the model for CO_2 permeation and separation from N_2 in a HFCLM permeator.

Liquid Separation in a HFCLM Permeator

Preliminary Considerations

In the separation of a gas mixture by a HFCLM permeator made of hydrophobic fibers and a nonwetting liquid membrane, the only resistance to mass transfer is due to the liquid membrane; the boundary layer resistance in the gas flowing in the fiber lumen and the diffusional resistance in the gas-filled pores of the fibers are almost always negligible. In the separation of a liquid mixture or solution by a HFCLM permeator, however, all three types of resistances need to be considered because diffusional rates in liquids are orders of magnitude slower than those in gases. Further, there is a larger variety of usable HFCLM structures (Table 42–4) in liquid separations since both hydrophobic and hydrophilic fibers can be easily used. In gas separation with aqueous membranes, hydrophilic structures will introduce additional resistance due to the liquid-filled pores, whereas there are no such limitations in general in liquid separations.

Unlike gas separations where the bulk gas flow rate changes along the permeator, the bulk liquid flow rate does not change along the HFCLM permeator. Although bulk gas flow pressure drop is usually significant along the permeator and influences the species driving force in gas separation, flow pressure drop in liquid separations has no effect on separation unless the immobilization of the aqueous/organic phase interface is affected. The ill-defined boundaries of the contained liquid membrane are, however, common to both gas and liquid separations. Consequently, the notion of an effective membrane thickness l_e, illustrated earlier using gas permeation, is equally applicable to liquid separations. In liquid separations, the liquid membrane often occupies the hollow-fiber pores beside the interstitial space on the shell side between fibers. Diffusion in hollow-fiber pores and in the contained liquid membrane of thickness l_e needs to be taken into account in liquid separations.

Quantitative analysis of the role of such resistances in a HFCLM system is facilitated by considering *effective radial concentration profiles* at any axial location in the permeator. For the removal of a solute from an aqueous feed to an aqueous strip solution via an organic liquid membrane, Figure 42–13 illustrates such profiles for hydrophilic and hydrophobic MHFs. Note that the hydrophilic MHFs have aqueous solution in membrane pores and the hydrophobic MHFs have the organic liquid membrane in pores (Sengupta, Basu, and Sirkar 1988). For organic feed and strip solutions, similar profiles are illustrated in Sengupta et al. (1988).

Mass transfer coefficients that describe the observed mass transfer rates in a HFCLM device using the effective radial concentration profiles may be considered *effective mass transfer coefficients*. Such an approach is needed since mass transfer coefficients can vary from fiber to fiber due to liquid flow maldistributions, radial and circumferential variations in the fiber bundle, etc. Additionally, it is useful to work with a mass transfer coefficient for each liquid stream to develop a first-order model of HFCLM separation. Solution of a partial differential equation for each feed fiber lumen and strip fiber lumen coupled via diffusion through pore and membrane liquid can provide these mass transfer coefficients. Such solutions are, however, not yet available.

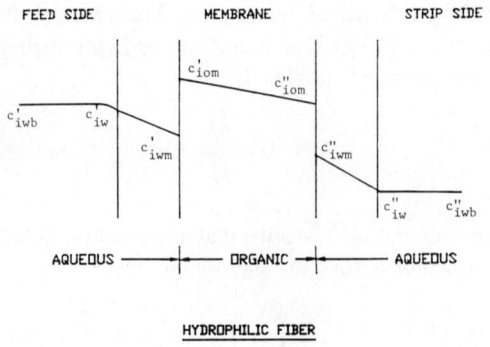

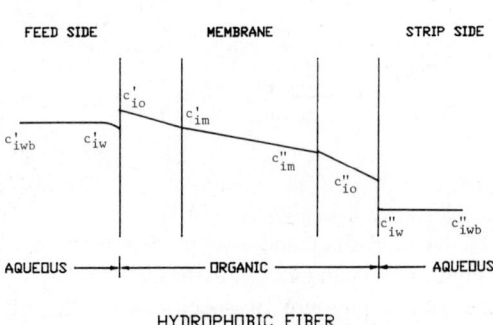

FIGURE 42–13. Effective radial concentration profiles for aqueous/organic/aqueous systems in HFCLM permeators with two different fibers (reprinted from Sengupta, Basu, and Sirkar 1988 with permission).

Mass Transport in Simple Permeation

A first-order analysis of mass transport in a HFCLM permeator for the separation of a solute from a feed solution to a strip solution through the contained liquid membrane is provided below for simple permeation without any chemical reaction. Table 42–4 suggests that there are only two general cases to be analyzed:

1. *Case I:* Contained liquid membrane does not penetrate the pores of the hollow-fiber wall. This includes:

a. hydrophilic fibers with aqueous feed and strip and therefore aqueous phase in the pores (Sengupta, Basu, and Sirkar 1988);
b. hydrophobic fibers with nonpolar organic feed and strip solutions and the same phase in the pores (Sengupta et al. 1988).
2. *Case II:* Contained liquid membrane penetrates the pores of the hollow-fiber wall. This includes:
a. hydrophobic fibers with aqueous feed and strip and therefore nonpolar organic CLM phase present in the pores (Sengupta, Basu, and Sirkar 1988);
b. hydrophilic fibers with nonpolar feed and strip phases and aqueous CLM phase present in the pores (Sengupta et al. 1988). There can be configurations in which the feed fiber has a *Case I* situation but the strip fiber has a *Case II* situation or vice versa. Such cases are not treated here.

Each such case can be analyzed using the *resistances-in-series* approach. The total mass transfer resistance is the sum of five individual resistances:

$$\begin{pmatrix}\text{total mass}\\\text{transfer}\\\text{resistance}\end{pmatrix} = \begin{pmatrix}\text{feedside}\\\text{feed film}\\\text{resistance}\end{pmatrix} + \begin{pmatrix}\text{substrate}\\\text{resistance}\end{pmatrix} + \begin{pmatrix}\text{CLM}\\\text{resistance}\end{pmatrix} + \begin{pmatrix}\text{permeate film}\\\text{resistance}\end{pmatrix} + \begin{pmatrix}\text{permeate-side}\\\text{substrate}\\\text{resistance}\end{pmatrix} \quad (42\text{--}20)$$

Each term in this equation would be associated with one mass transfer coefficient. The total mass transfer resistance will be associated with an overall mass transfer coefficient K. Although such an approach is essentially correct at any cross section in the permeator, we will use it for the whole permeator using permeator-averaged transfer coefficients. This and other assumptions needed to develop quantitative relations representing Eq. (42–20) are given below:

1. Permeator-averaged mass transfer coefficients can be used.
2. Permeation is governed by molecular diffusion and simple solute partitioning. There is no interfacial resistance at the aqueous/organic phase interfaces. There are no chemical reactions anywhere.
3. A single effective membrane thickness l_e, known through independent experiment, is available and is valid for the whole permeator. The length of the permeator is to be determined by eliminating the sections at the two ends where the two sets of fibers are separated.
4. Mass flux is not affected by any two-dimensional effect (Keller and Stein 1967) as the permeant diffuses from the small pores in the substrate into continuous phases, e.g., feed, membrane, and strip, and vice versa.
5. Hollow-fiber substrate mass transfer resistance can be obtained from permeant distribution coefficients m_i, hollow-fiber substrate porosity ϵ_m, tortuosity τ_m, and thickness.
6. The film transfer coefficients inside the feed fibers and strip fibers can be obtained from corresponding mass transfer analyses for flow through nonporous tubes.
7. The membrane liquid is stagnant.
8. The permeator may have an unequal number of feed fibers (N'_T) and strip fibers (N''_T). Further, their diameters may be different ($d'_{ti}, d'_{to}, d''_{ti}, d''_{to}$).

Consider now Case IIa with *hydrophobic fibers*, *aqueous feed-strip*, and *organic CLM membrane in fiber pores*. The total rate of transfer of solute i per unit length of the permeator R_T may be described by

$$R_T = Q'_w \frac{dc'_{iwb}}{dL}$$

$$= K_w(\pi d'_{ti} N'_T c'_{iwb} - \pi d''_{ti} N''_T c''_{iwb}).$$

(42–21)

Here K_w is the overall transfer coefficient based on the two aqueous phase bulk concentrations (Figure 42–13). This form of mass transfer equation is being used to accommodate different values of feed-membrane transfer area and strip-membrane transfer area. When the two areas are equal, the form of Eq. (42–21) becomes familiar with only the concentration difference $(c'_{iwb} - c''_{iwb})$ becoming the driving force.

This solute transfer rate R_T may also be related to the individual mass transfer coefficients by the following relations:

$$R_T = k'_w \pi d'_{ti} N'_T (c'_{iwb} - c'_{iw}) \quad (42\text{–}22a)$$

$$= k'_{so} \pi d'_{tlm} N'_T (c'_{io} - c'_{im}) \quad (42\text{–}22b)$$

$$= k_{mo}(\pi d'_{to} N'_T c'_{im} - \pi d''_{to} N''_T c''_{im}) \quad (42\text{–}22c)$$

$$= k''_{so} \pi d''_{tlm} N''_T (c''_{im} - c''_{io}) \quad (42\text{–}22d)$$

$$= k''_w \pi d''_{ti} N''_T (c''_{iw} - c''_{iwb}). \quad (42\text{–}22e)$$

The solute concentrations in the two phases at the two phase interfaces are related by

$$m'_i = \frac{c'_{io}}{c'_{iw}} \quad \text{and} \quad m''_i = \frac{c''_{io}}{c''_{iw}}. \quad (42\text{–}23)$$

Using the above relations (42–21), (42–22), and (42–23), the following general relation between K_w and the individual coefficients has been developed for a *hydrophobic substrate, aqueous feed-aqueous strip*, and *organic CLM in the fiber pores*:

$$\frac{1}{d'_{ti} K_w} + \frac{\pi}{R_T}\left[\left(1 - \frac{d''_{to} m''_i}{d'_{to} m'_i}\right) N''_T \frac{c''_{im}}{m''_i} - \left(1 - \frac{d''_{ti}}{d'_{ti}}\right) N''_T c''_{iwb}\right]$$

$$= \frac{1}{d'_{ti} k'_w} + \frac{1}{d'_{tlm} m'_i k'_{so}} + \frac{1}{d'_{to} m'_i k_{mo}} + \frac{1}{d''_{tlm} m''_i k''_{so}} + \frac{1}{d''_{ti} k''_w}. \quad (42\text{–}24)$$

Under the conditions

$$d'_{to} = d''_{to} = d_{to}; \quad d'_{ti} = d''_{ti} = d_{ti}; \quad d'_{tlm} = d''_{tlm} = d_{tlm}; \quad m'_i = m''_i = m_i, \quad (42\text{–}25)$$

Eq. (42–24) is simplified to

$$\frac{1}{K_w} = \left(\frac{1}{k'_w} + \frac{1}{k''_w}\right) + \frac{d_{ti}}{d_{tlm}} \frac{1}{m_i}\left(\frac{1}{k'_{so}} + \frac{1}{k''_{so}}\right) + \frac{d_{ti}}{d_{to}} \frac{1}{m_i} \frac{1}{k_{mo}}.$$

| overall resistance | feed + strip film resistances | feed + strip substrate resistances | contained liquid membrane resistance | (42–26) |

This result of Eq. (42–26) is available in Sengupta, Basu, and Sirkar (1988). Correlations to estimate k'_w and k''_w are provided by Eqs. (41–30a) and (41–36) in Chapter 41. The following relations may be used to determine k_{mo}, k'_{so}, and k''_{so}:

$$k_{mo} = D_{io}/l_e, \quad (42\text{–}27a)$$

$$k'_{so} = 2D_{io}\epsilon_m/[\tau'_m (d'_{to} - d'_{ti})], \quad (42\text{–}27b)$$

$$k''_{so} = 2D_{io}\epsilon_m/[\tau''_m (d''_{to} - d''_{ti})]. \quad (42\text{–}27c)$$

A few useful conclusions may be drawn from Eq. (42–26) by using an order-of-magnitude analysis. Suppose the individual transfer coefficients are of similar order. If $m_i \gg 1$, only the feed and strip film resistances will be controlling. If, however, $m_i \ll 1$, then the CLM resistance and the hollow-fiber substrate resistances will control $(1/K_w)$.

Another aspect requiring attention is that the same value of K_w may be obtained by different combinations of k'_w and k''_w (and therefore different combinations of Re'_w and Re''_w). Sengupta, Basu, and Sirkar (1988) therefore studied the variation of $(1/K_w)$ as a function of $[(1/Re'_w) + (1/Re''_w)]$ when analyzing their data.

Case Ia for hydrophilic hollow fibers with aqueous feed-strip, aqueous phases in fiber pores, and organic CLM (Figure 42–13) are now treated briefly; a complete derivation is available in Sengupta, Basu, and Sirkar (1988). For an overall mass transfer coefficient K_w defined by Eq. (42–21), the general relation between K_w and the individual coefficients are

$$\frac{1}{d'_{to}K_w} + \frac{\pi}{R_T}\left[\left(1 - \frac{d''_{to}m''_i}{d'_{to}m'_i}\right)N''_T c''_{iwm} - \left(1 - \frac{d''_{to}}{d'_{to}}\right)N''_T c''_{iwb}\right]$$

$$= \frac{1}{d'_{ti}k'_w} + \frac{1}{d'_{tlm}k'_{sw}} + \frac{1}{d'_{to}m'_i k_{mo}} + \frac{1}{d''_{tlm}k''_{sw}} + \frac{1}{d''_{ti}k''_w}.$$

(42–28)

If the simplifying conditions of Eq. (42–25) hold, then

$$\frac{1}{K_w} = \frac{d_{to}}{d_{ti}}\left(\frac{1}{k'_w} + \frac{1}{k''_w}\right) + \frac{d_{to}}{d_{tlm}}\left(\frac{1}{k'_{sw}} + \frac{1}{k''_{sw}}\right) + \frac{1}{m_i}\frac{1}{k_{mo}}.$$

| overall resistance | feed + strip film resistances | feed + strip substrate resistances | contained liquid membrane resistance |

(42–29)

From an order-of-magnitude analysis, contained liquid membrane resistance is observed to be negligible if $m_i \gg 1$. For $m_i \ll 1$, on the other hand, the CLM resistance may control. Comparing Case Ia with Case IIa for the same feed, same liquid membrane, and similar fiber dimensions, it is obvious that the hydrophobic fiber (Case IIa) will be better than the hydrophilic fiber (Case Ia) for $m_i \gg 1$. On the other hand, for $m_i \ll 1$, hydrophilic fiber (Case Ia) will be superior to hydrophobic fiber (Case IIa) by providing a higher overall solute transfer coefficient.

Although the simplified relations of Eqs. (42–26) and (42–29) are developed at any axial location in the permeator, it is suggested that they *may* be used for the whole permeator since an exact analysis is not yet available. This implies that values of k'_w and k''_w for the whole length of the hollow fibers, if available via suitable correlations or otherwise, may be used in Eqs. (42–26) and (42–29); the resulting K_w is likely to represent the total HFCLM permeator behavior.

Experimental determination of K_w may be made using logarithmic-mean concentration driving force or arithmetic-mean driving force over the whole permeator. Assume d_{ti}, d_{to}, and N_T to be the same for both fiber sets. Overall solute balances for aqueous feed-strip systems in countercurrent flow are

Hydrophobic fibers: $Q'_w(c'_{iw}|_{in} - c'_{iw}|_{out}) = K_w \pi d_{ti} L N_T \Delta c_{lm} = Q''_w(c''_{iw}|_{out} - c''_{iw}|_{in})$ (42–30a)

Hydrophilic fibers: $Q'_w(c'_{iw}|_{in} - c'_{iw}|_{out}) = K_w \pi d_{to} L N_T \Delta c_{lm} = Q''_w(c''_{iw}|_{out} - c''_{iw}|_{in})$ (42–30b)

The logarithmic mean concentration driving force Δc_{lm} is

$$\Delta c_{lm} = \frac{(c'_{iw}|_{in} - c''_{iw}|_{out}) - (c'_{iw}|_{out} - c''_{iw}|_{in})}{\ln[(c'_{iw}|_{in} - c''_{iw}|_{out})/(c'_{iw}|_{out} - c''_{iw}|_{in})]}.$$

(42–31)

For cocurrent flow, the mass balances in Eqs. (42–30a) and (42–30b) are changed to

$$Q'_w(c'_{iw}|_{in} - c'_{iw}|_{out}) = Q''_w(c''_{iw}|_{out} - c''_{iw}|_{in})$$

(42–32)

without changing the $K_w \times$ (area) $\times \Delta c_{lm}$ expression. However, Δc_{lm} is to be calculated from

$$\Delta c_{lm} = \frac{(c'_{iw}|_{in} - c''_{iw}|_{in}) - (c'_{iw}|_{out} - c''_{iw}|_{out})}{\ln[(c'_{iw}|_{in} - c''_{iw}|_{in})/(c'_{iw}|_{out} - c''_{iw}|_{out})]}.$$

(42–33)

Experimental results for removal of phenol from an aqueous feed to an aqueous strip through a MIBK liquid membrane are shown in Figure 42–14 for Cases Ia and IIa along with the predictions from Eqs. (42–29) and (42–26), respectively (Sengupta, Basu, and Sirkar 1988).

Both Case Ia and Case IIa involved an aqueous feed and an aqueous strip solution. Cases Ib and IIb involve a nonpolar organic feed solution and a nonpolar organic strip solution. They may be analyzed exactly as before except aqueous is to be replaced by organic and the distribution coefficient relations of Eq. (42–23) suitably defined. Using the earlier distribution coefficient definition of organic phase concentration over aqueous phase concentrations, the results for Case Ib are briefly presented here (Sengupta et al. 1988). If assumptions (42–25) hold, then the relation equivalent to Eq. (42–29) is

$$\frac{1}{K_o} = \frac{d_{to}}{d_{ti}}\left(\frac{1}{k'_o} + \frac{1}{k''_o}\right)$$

$$+ \frac{d_{to}}{d_{tlm}}\left(\frac{1}{k'_{so}} + \frac{1}{k''_{so}}\right) + \frac{m_i}{k_{mw}}.$$

(42–34)

For nonpolar organic feed and strip, in general, $m'_i \neq m''_i$ since the basic constituents of the two organic streams for bulk separations will be different. A general expression equivalent to Eq. (42–28) for this case is

$$\frac{1}{d'_{to}K_o} + \frac{\pi}{R_T}\left[\left(1 - \frac{d''_{to}m''_i}{d'_{to}m'_i}\right)N''_T c''_{iom} - \left(1 - \frac{d''_{to}}{d'_{to}}\right)N''_T c''_{iob}\right]$$

$$= \frac{1}{d'_{ti}k'_o} + \frac{1}{d'_{tlm}k'_{so}} + \frac{m'_i}{d'_{to}k_{mw}} + \frac{1}{d''_{tlm}k''_{so}} + \frac{1}{d''_{ti}k''_o},$$

(42–35)

where R_T is now defined as

$$R_T = Q'_o \frac{dc'_{iob}}{dL} = K_o(\pi d'_{to}N'_T c'_{iob} - \pi d''_{to}N''_T c''_{iob}).$$

(42–36)

A similar treatment for Case IIb is available in Sengupta et al. (1988) for hydrophilic fibers. Emphasis has been placed here on hydrophobic fibers because among the currently available fibers, hydrophobic ones have much higher chemical stability with respect to both pH and solvent resistance.

Mass Transport with Chemical Reaction
The three broad classes of permeation mechanisms in liquid membranes are simple permeation, facilitated transport, and coupled transport. The last two are based on reversible chemical complexation. Variations are introduced by reversible or irreversible chemical

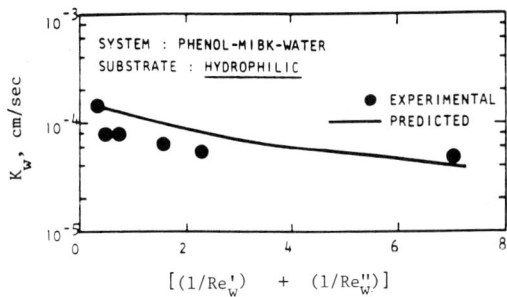

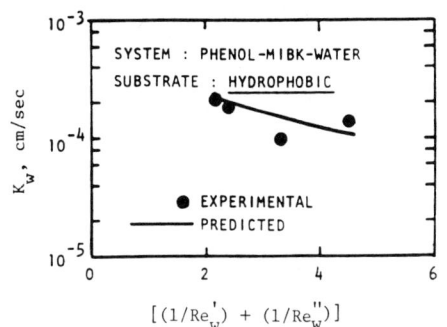

FIGURE 42-14. Experimental values of overall mass transfer coefficients compared with predicted values for phenol-MIBK-water system and two kinds of substrates (reprinted from Sengupta, Basu, and Sirkar 1988 with permission).

reactions in the strip or feed phase, dissociation of one or more species in the feed and/or strip solution, solvent in the membrane acting as a complexing agent, etc. The nature of the chemical complexation may involve ion-pair formation, liquid ion exchange, host-guest complexation, solvation, etc., in addition to other interactions such as hydrogen bonding, hydrophobic interactions, etc. The new technique of HFCLM has been applied to only a few such cases.

The species to be transported across the membrane may or may not exist as free species in the liquid membrane. For aqueous feed-strip solutions and organic liquid membranes, metal ions or ions containing a metal can exist in the membrane only as a complex formed at the aqueous/organic interface by an interfacial reaction. The complex diffuses through the membrane. This is valid whether chelating extractants or basic extractants (long-chain alkylamine) are used. The same is true of strong inorganic mineral acids, e.g., HCl, HNO$_3$; there may be additional complexation here with solvating solvents such as TBP (tributyl phosphate). In the case of carboxylic acids, however, both free carboxylic acid and a complex (say, carboxylic acid-long chain amine) are possible in the membrane. Analysis of such a case, namely, transport of citric acid through an organic liquid membrane containing a long-chain alkylamine is summarized here (Basu and Sirkar 1991). The hollow-fiber substrate chosen is hydrophobic with the pores containing the liquid membrane.

The complexation of citric acid with a long-chain alkyl amine may be represented as an interfacial reaction:

$$C_6H_8O_7(aq) + 3R_3N(org) \rightleftarrows (R_3NH)_3C_6H_5O_7(org), \quad (42\text{-}37)$$

as well as an organic phase complexation reaction

$$\underset{A}{C_6H_8O_7(org)} + \underset{3\ B}{3R_3N(org)} \rightleftarrows \underset{C}{(R_3NH)_3C_6H_5O_7(org)}. \quad (42\text{-}38)$$

Citric acid is present both as free acid and the complex $(R_3NH)_3C_6H_5O_7$ in the organic phase. The reaction of the basic amine with citric acid is considered to be sufficiently fast for reaction (42-38) to be in equilibrium everywhere in the membrane. If the amine R_3N (species B) and the amine-complex (species C) are assumed insoluble in the aqueous phases, then the conventional equilibrium approximation of facilitated transport (Ward 1970; Olander 1960) is valid. Assuming diffusion coefficients for B and C in the membrane to be equal, i.e., $D_{Bo} = D_{Co}$, it follows from the basic reaction-diffusion equations that

$$c_{Bo} + 3c_{Co} = c_{BT} \quad (42\text{-}39)$$

everywhere in the membrane where c_{BT} is the total amine concentration in the membrane. The equilibrium constant $K_{eq,o}$ for the organic phase reaction (42-38),

$$K_{eq,o} = \frac{c_{Co}}{c_{Ao}(c_{Bo})^3}, \quad (42\text{-}40)$$

can be related to an interfacial reaction equilibrium constant $K_{eq,int}$ at the feed interface, for example, by

$$K_{eq,int} = \frac{c'_{Co}}{c'_{Aw}(c'_{Bo})^3} = K_{eq,o} \frac{c'_{Ao}}{c'_{Aw}} = K_{eq,o} m'_A,$$

$$(42\text{-}41)$$

where m'_A is the distribution coefficient of species A at the feed interface. If $K_{eq,int}$ is found to hold for the strip solution also, then

$$K_{eq,int} = \frac{c''_{Co}}{c''_{Aw}(c''_{Bo})^3}. \quad (42\text{-}42)$$

The total flux of penetrant species A in facilitated transport through a liquid membrane of thickness l is due both to free species A and complex C in the membrane:

$$J_{AT} = \frac{D_{Ao}}{l}(c'_{Ao} - c''_{Ao}) + \frac{D_{Co}}{l}(c'_{Co} - c''_{Co}).$$

$$(42\text{-}43)$$

At any permeator location, species A diffuses through the feed solution boundary layer, organic liquid membrane in feed fiber pores, the contained liquid membrane, organic liquid membrane in strip fiber pores, and strip solution boundary layer. Basu and Sirkar (1991) have found that the feed and strip solution boundary layer resistances did not influence the acid transport. Relation (42-26) may now be expressed in terms of an overall transfer coefficient K_o (based on the organic phase) for any species as

$$\frac{1}{K_o} = \frac{d_{ti}}{d_{tlm}}\left(\frac{1}{k'_{so}} + \frac{1}{k''_{so}}\right) + \frac{d_{ti}}{d_{to}}\frac{1}{k_{mo}}.$$

Use Eqs. (42-27a), (42-27b), and (42-27c) for species A and species C in the above expression and obtain

$$\frac{1}{K_{o,A}} = \frac{d_{ti}}{d_{tlm}} \frac{2}{D_{Ao}\epsilon_m/[\tau_m(d_{to} - d_{ti})]}$$

$$+ \frac{d_{ti}}{d_{to}} \frac{l_e}{D_{Ao}} = \frac{l_{eff}}{D_{Ao}}, \quad (42\text{-}44)$$

$$\frac{1}{K_{o,C}} = \frac{d_{ti}}{d_{tlm}} \frac{2}{D_{Co}\epsilon_m/[\tau_m(d_{to} - d_{ti})]}$$

$$+ \frac{d_{ti}}{d_{to}} \frac{l_e}{D_{Co}} = \frac{l_{eff}}{D_{Co}}, \quad (42\text{-}45)$$

The effective value of l, l_{eff}, here includes the effective contained liquid membrane thickness, l_e and the contribution from the pores of the substrate. Expression (42-43) is rewritten in terms of l_{eff}:

$$J_{AT} = K_{o,A}(c'_{Ao} - c''_{Ao}) + K_{o,C}(c'_{Co} - c''_{Co}).$$

$$(42\text{-}46)$$

Since the boundary layer resistances of feed and strip are negligible, using the definition of m_i

$$m'_i = \frac{c'_{Aw}}{c'_{Ao}} = \frac{c''_{Aw}}{c''_{Ao}} = m''_i = m_i, \quad (42\text{-}47)$$

where we have assumed $m'_i = m''_i$,

$$J_{AT} = m_i K_{o,A}(c'_{Aw} - c''_{Aw})$$
$$+ K_{o,C}(c'_{Co} - c''_{Co}). \quad (42\text{-}48)$$

It is obvious that $K_{o,A}$ and $K_{o,C}$ do not vary along the HFCLM permeator length. For *cocurrent flow* of the feed and strip stream in the permeator, differential mass balances on A provide

$$-Q'_w \frac{dc'_{Awb}}{dL} = \pi d_{ti} N_T[m_i K_{o,A}(c'_{Awb} - c''_{Awb})$$

$$+ K_{o,C}(c'_{Co} - c''_{Co})], \quad (42\text{-}49a)$$

$$Q_w'' \frac{dc_{Awb}''}{dL} = \pi d_{ti} N_T [m_i K_{o,A}(c_{Awb}' - c_{Awb}'')$$
$$+ K_{o,C}(c_{Co}' - c_{Co}'')], \quad (42\text{-}49b)$$

where $c_{Awb}' = c_{Aw}'$ and $c_{Awb}'' = c_{Aw}''$ since the feed and strip boundary layer resistances are negligible. Further, from Eqs. (42–39), (42–41), and (42–42), at the feed aqueous/organic interface

$$c_{Bo}' + 3c_{Co}' = c_{BT}, \quad (42\text{-}50a)$$

$$K_{eq,int} = \frac{c_{Co}'}{c_{Awb}'(c_{Bo}')^3}, \quad (42\text{-}50b)$$

and at the strip aqueous/organic interface

$$c_{Bo}'' + 3c_{Co}'' = c_{BT}, \quad (42\text{-}51a)$$

$$K_{eq,int} = \frac{c_{Co}''}{c_{Awb}''(c_{Bo}'')^3}. \quad (42\text{-}51b)$$

At $L = 0$, $c_{Awb}' = c_{Aw,in}'$ and $c_{Awb}'' = 0$.

Solution of this set of equations is facilitated by noting that Eqs. (42–50a) and (42–50b) can be combined to form the cubic equation

$$c_{Bo}'^3 + (1/3K_{eq,int}c_{Awb}')c_{Bo}'$$
$$- c_{BT}/3K_{eq,int}c_{Awb}' = 0, \quad (42\text{-}52)$$

whose solution provides c_{Bo}' at any permeator location in terms of $K_{eq,int}$, c_{Awb}', and c_{BT}. By Eq. (42–50a), c_{Co}' is available. Similarly, c_{Co}'' is obtained. A numerical solution of Eqs. (42–49a) and (42–49b) is then easily achieved, providing an estimate of the separation capability of the HFCLM permeator.

In the case described above, the strip solution was simply water. When a basic aqueous solution is used, e.g., an aqueous solution of NaOH, there will be an instantaneous reaction of caustic with citric acid. When the amount of caustic is sufficient, the reaction interface and the aqueous/organic phase interface will coincide (Basu, Prasad, and Sirkar 1990) and $c_{Aw}'' = 0 = c_{Awb}''$. This condition can be valid throughout the length of the permeator for sufficient amounts of caustic. Simultaneously $c_{Co}'' = 0$. Equations (42–49a) and (42–49b) are simplified; Eqs. (42–51a) and (42–51b) are no longer needed. A numerical solution can be developed as before (Basu and Sirkar 1991) to relate the change in c_{iwb}' with L or membrane area $\pi d_{ti} N_T L$.

Expressions (42–44), (42–45), and (42–27a) to (42–27c) ignore the possibility of any interfacial resistance to the extraction or back extraction of the solute. There is significant evidence to the contrary for carboxylic acids (Basu and Sirkar 1991). Expressions (42–44) and (42–45) need to be corrected; otherwise, the values of $K_{o,A}$ and $K_{o,C}$ would be higher. In the absence of any formula for interfacial resistance, the following strategy may be adopted. Measure experimentally the value of $K_{o,A}$ without any complexation in the membrane by using only the diluent. The experimental value of $K_{o,A}|_{diluent}$ may be used in Eq. (42–44) to define a new value of l_{eff}. The value of $K_{o,A}$ under other conditions (with complexing agent) is obtained by correcting for D_{Ao}:

$$\frac{K_{o,A}|_{diluent}}{K_{o,A}|_{diluent+carrier}} = D_{Ao}|_{diluent}/D_{Ao}|_{diluent+carrier}. \quad (42\text{-}53)$$

Assuming that the solute-carrier complex C encounters a similar interfacial resistance,

$$\frac{K_{o,A}|_{diluent+carrier}}{K_{o,C}|_{diluent+carrier}} = \frac{D_{Ao}|_{diluent+carrier}}{D_{Co}|_{diluent+carrier}}. \quad (42\text{-}54)$$

Thus, $K_{o,A}$ and $K_{o,C}$ values in the carrier containing liquid membrane are known. By using the above procedure, Basu and Sirkar (1991) were able to describe the facilitated flux in the permeator in their citric acid study quite well over an amine concentration range of 0 to 40%.

Most studies on supported liquid membranes utilized very small membrane areas. There is practically no model to predict very large extents of solute removal in a SLM permeator. The model developed by Basu and Sirkar (1991) was able to predict up to 97% removal of citric acid from the feed in a 48-cm-long HFCLM permeator. Unless a few more systems have been studied in HFCLM permeators and the data compared with models, caution should be exercised when using the current models for large-scale design and development of HFCLM permeators.

Purification Limits in Liquid Membrane Permeators

What is the limit of purification achievable in a HFCLM permeator (or, in general, any liquid membrane permeator) for a given feed gas mixture (or feed solution) and a sweep (or permeated) gas stream (or a strip solution)? Such a question should be considered before a HFCLM permeator is employed to achieve the required level of purification. The answer is provided by the theoretical limits and the practical ranges of the forces driving the transport. Since a variety of situations can be dictated by the transport mechanism, we focus briefly on a few important cases.

Gas Separation

The species can be transported by either simple permeation (i.e., solution-diffusion) or by facilitated transport.

Simple Permeation
From the boundary conditions of Eqs. (42–2c) and (42–2d), the concentration driving force in the liquid membrane for any species i is $(H_i p_i' - H_i p_i'')/l_e$. Thus, as long as $p_i' - p_i''$ is positive, separation will continue. The desired level of a species such as H_2S in the feed gas after purification is very low, around 10 ppm. In the conventional polymeric membrane permeation mode, the permeate gas will be enriched in H_2S and will always be at a total pressure higher than atmospheric. The lowest value of p_i'' is then substantial, limiting the lowest possible level of p_i'. If the vacuum mode is used, the lowest possible value of p_i' can be considerably lower since the lowest possible value of p_i'' will be reduced by vacuum. Due to H_2S being enriched in permeate, p_i'' will still be significant.

Identical arguments hold for the sweep mode in countercurrent flow if a sweep gas or vapor is available and usable. *Theoretically,* p_i'' at the sweep gas inlet location can approach zero. Whether a sweep gas/vapor strategy is used to remove H_2S is a different issue. Similarly, in the sweep liquid mode in countercurrent flow with wetted pores, p_i'' will approach zero at the sweep liquid inlet end. Thus, both sweep modes can provide the condition of $p_i'' = 0$ at the sweep inlet end; this will allow p_i' to achieve very low values consistent with an acceptable level of driving force at the exit end of the permeator for a countercurrent flow pattern. In the sweep liquid mode in countercurrent flow with nonwetted pores, however, p_i'' will not approach zero at the sweep liquid inlet end.

Facilitated Transport
Although the carrier-penetrant complex may be the dominant species being transported in the liquid membrane, $(H_i p_i' - H_i p_i'')/l_e$ still continues to be the overall driving force across the liquid membrane. Thus, the conclusions for simple permeation obtained above are also valid for facilitated transport. The overall species permeability is, however, considerably enhanced at low levels of $(H_i p_i' - H_i p_i'')$ compensating the low driving force. This is a major advantage of facilitated transport.

Liquid Separation

A variety of transport mechanisms can be present: simple permeation, facilitated transport, and coupled transport. An additional aspect (not present in gas separation) is a strong chemical reaction at the strip/membrane interface with perhaps an acid or alkali in the strip stream. Such a reaction with an appropriate level of acid or alkali can easily reduce the concentration of the penetrant species to essentially zero at the strip/membrane interface. Thus, as long as the species is present in the feed solution, there will be a driving force across the liquid membrane regardless of the mechanism of transport. Theoretically, the limits of purification then are very high. If the species transport is facilitated or coupled, the concentration level of the complex in the membrane will also be high, leading to a high level of flux even at low levels of feed concentration.

If there is no such reaction at the strip/membrane interface, the driving force across the liquid membrane exists only as long as $m'_i c'_{iw} - m''_i c''_{iw}$ is nonzero for simple permeation and facilitated transport. The need to maintain a reasonable driving force in this case will then set a lower limit to the value of c'_{iw} that can be achieved.

HFCLM Permeator Versus Two Separate Hollow-Fiber Contactors

One set of microporous hollow fibers packed inside a shell as a shell-and-tube device can act as a contactor between two immiscible phases, gas/liquid or liquid/liquid. Chapter 41 covers the technique of solvent extraction using such a device. The subject of gas absorption or stripping in such a device is briefly treated in Chapter 46. As pointed out in Chapter 41, two such single-fiber devices could be used, one for extraction, say, from an aqueous to an organic solvent and the second one for back extraction from the organic solvent to a strip aqueous solution (Figure 41–9). One HFCLM permeator for an aqueous feed and an aqueous strip achieves the same goal. Which approach is more efficient? The same question is equally relevant for a HFCLM permeator for gas separation and two hollow-fiber devices, one for absorption and the other for stripping.

Liquid Separation

The issue has to be analyzed simultaneously with respect to the following aspects: mass transfer resistances and capital and operating costs. Obviously, when two extractors are used instead of one HFCLM device with a stationary liquid membrane, an extra pump is needed to circulate the extractant from the extraction device to the back extraction device and back. The capital and operating costs of pumping are not insignificant. The inventory of the extracting solvent also will be somewhat higher in the "two extraction devices" mode. The number of modules is reduced in the HFCLM mode to half that in the contacting mode. Although HFCLM module size would be larger and module fabrication is somewhat more difficult, reduction in module numbers, possibly by a factor of 2, is a significant cost advantage.

Which configuration has higher mass transfer rate or lower overall resistance can be judged by considering the different component resistances for different extraction mechanisms and conditions. The distribution coefficient m_i for simple nonreactive extraction may be >1 or <1. Reactive extraction may involve an instantaneous, fast or slow interfacial or bulk reaction; there may be complexation in the extractant phase. A limited and brief qualitative analysis is provided below.

Consider the case in which the interfacial reaction, if any, is controlling as is often encountered in metal extraction [e.g., Cu extraction studies by Komasawa, Otake, and Yamashita (1983) and by Nguyen and Callahan (1989)]. Increased convection of either bulk phase will not influence the mass transfer rate. Thus, the overall extraction rate for the metal is governed by chemical compositions and membrane area. A HFCLM device will provide the same transfer rate as a combination of two single hollow-fiber extraction devices. Sometimes, either extraction or back extraction may benefit from larger area, i.e., larger number of fibers. A HFCLM device may have different numbers of fibers for the two functions (Guha 1989). Should these numbers vary greatly, the two-device mode is more efficient.

There are conditions in extraction (of metals, say) using a SLM where the diffusional resistance of the microporous membrane dominates, as was observed by Babcock et al. (1980b) in uranium separation by coupled transport using a tertiary amine in a diluent. Such a resistance will exist with the same magnitude in a single fiber solvent extraction device also. Thus, two extraction devices do not provide any mass transfer advantage when hollow-fiber substrate resistance controls.

Basu (1989) has studied citric acid extraction using a tertiary amine in an organic diluent as a membrane in a HFCLM device. He has also studied the same extraction using a single fiber extractor and then investigated its back extraction later in the same hollow-fiber extractor. He has found that the transfer rate observed in a HFCLM device is also achieved in the two-device system in the particular reactive extraction system employed. This is due to the dominance of the substrate resistance and the interfacial resistance. Both resistances are independent of the mode, CLM or single fiber device, as long as the same hollow fiber is used.

Feed or strip boundary layer resistances often control the mass transfer rate. For systems with $m_i \gg 1$ and aqueous feed/strip, the boundary layer resistances are quite important. If the feed and strip phases flow in the fiber lumen in both modes, CLM or single fiber device, then the two modes have identical resistances. If the feed phase flows in the shell side of a single fiber device, there is considerable scope for bypassing, channeling, etc. (see Chapter 41), leading to lower solute transfer rates. With better shellside design of hollow-fiber devices, such bypassing, etc., may be reduced. Only in the case of the strip phase flowing in the shell side of a single fiber back extraction device is there a possibility that the mass transfer coefficient will be higher than that in the strip fiber lumen of a HFCLM device. However, the strip side often is subjected to a reaction; for example, phenol is stripped with NaOH, which drastically reduces the phase boundary layer resistance (Basu, Prasad, and Sirkar 1990). Thus, there appears to be limited justification for adopting the two-device mode over the HFCLM permeator mode if the liquid membrane mechanism is an efficient one. Only in cases where $m_i \ll 1$ and the contained liquid membrane resistance is very important, will the two-device mode be justified. Further, if the HFCLM module fiber configuration is inefficient leading to a high l_e, the two-device mode will be more efficient due to the high CLM resistance in HFCLM device.

Gas Separation

Feed and sweep boundary layer resistances, hollow-fiber substrate resistances, and interfacial reactions are generally absent in HFCLM devices for gas separation using hydrophobic fibers and aqueous membranes. The contained liquid membrane resistance is controlling. Thus, in the two-device mode, a flowing absorbent or stripping liquid may provide a higher mass transfer coefficient.

Majumdar, Guha, and Sirkar (1988) have obtained the mass transfer coefficient for CO_2 through pure water membrane in their HFCLM device. They have found this value to be five to six times larger than that calculated from the Yang and Cussler (1986) correlation for water flowing outside the hollow fibers in parallel flow. That a HFCLM permeator with a stationary liquid has a mass transfer advantage over a single fiber-based gas absorber with a flowing absorbent is most likely due to considerable channeling in the shellside liquid flow of a well-packed single fiber-based device (Yang and Cussler 1986).

A combination of two simple parallel flow type hollow-fiber devices is therefore unlikely to have radical mass transfer advantages over a HFCLM device for gas separation. To this, should be added the disadvantages of a pump, the added pumping cost, the increase in the number of modules, and the attendant system cost and complexities of the two-device mode. However, if the species partial pressure is very low in feed, a HFCLM process *may* suffer from lack of driving force.

Comparison of HFCLM Permeator with Conventional Polymeric Membrane Permeator for Gas Separation

A comparison of a HFCLM permeator with a conventional polymeric membrane permeator can be made based on the gas permeation rates and separation factors of the liquid membrane systems with those of polymeric membrane systems used commercially. In a small module, Majumdar, Guha, and Sirkar (1986) observed that the CO_2 permeation rate through pure water liquid membrane was 6.3×10^{-9} mol/m^2 · s · Pa and the separation factor for a CO_2-N_2 system was about 38. With an aqueous 30 wt.% K_2CO_3 solution as a liquid membrane, the overall permeation rate of CO_2 in the same module was estimated to be 3.7×10^{-9} mol/m$^2 \cdot$ s $\cdot$ Pa; the overall separation factor for the CO_2-N_2 system was around 180. For an aqueous 20 wt.% DEA solution as a liquid membrane, the specific permeation rates for CO_2 and CH_4 were found to be 0.61×10^{-8} and 0.78×10^{-10} mol/m$^2 \cdot$ s $\cdot$ Pa, respectively, yielding a separation factor of 78 for a CO_2-CH_4 system (Guha, Majumdar, and Sirkar 1991a); the separation factor for a CO_2-N_2 system with the same liquid membrane was about 108.

In a larger module with an aqueous 20 wt.% DEA membrane, the permeation rates for CO_2 varied from 0.77×10^{-9} to 0.71×10^{-8} mol/m$^2 \cdot$ s $\cdot$ Pa. The lowest value was observed for a high feed pressure (411.5 kPa) and high feed CO_2 concentration (41%), whereas the highest value was obtained for a low feed pressure (135.8 kPa) and a low feed CO_2 concentration (4%). The specific permeation rates for N_2 and CH_4 were found to be around 1.8×10^{-11} and 2.2×10^{-11} mol/m$^2 \cdot$ s $\cdot$ Pa, respectively. Thus, depending on the CO_2 composition in the feed gas, the separation factor for CO_2-CH_4 systems would vary from 35 to 323 (Guha, Majumdar, and Sirkar 1991b). Note that the effective liquid membrane thickness in the large module was around 280 μm, suggesting significant room for improvement. On the other hand, the CO_2 permeation rate through commercial polysulfone membrane (used in Monsanto's Prism separator) and cellulose ester membranes are 1.4×10^{-8} and 3.3 to 5.0×10^{-8} mol/m$^2 \cdot$ s $\cdot$ Pa, respectively. The separation factors for a CO_2-CH_4 system are around 17 to 24 and 26 to 30, respectively (Kulkarni et al. 1983). Therefore, the CO_2 permeation rates in HFCLM systems are lower but the CO_2-CH_4 separation factor values are significantly higher than those in commercial polymeric membrane systems.

A higher specific permeation rate can be achieved in a HFCLM device if the effective liquid membrane thickness can be reduced to a smaller value. This may be achieved by utilizing higher fiber packing density and by making modules with smaller diameter hollow fibers (Majumdar et al. 1989).

Pressure Conditions in HFCLM Permeators

The hollow-fiber contained liquid membrane is confined to the shell side or to the shell side and the fiber pores by the right combination of phase pressure differences and fiber pore wetting characteristics. As indicated in Chapter 41, the interface of an immiscible liquid/liquid system will remain immobilized at the membrane pore mouth until Δp_{cr} is reached. Once this limit is exceeded, either the membrane phase will break through into the feed-strip phases or vice versa, depending on which phase is at a higher pressure. Thus, the pressure differential between the membrane phase and the feed phase as well as that between the membrane phase and the strip phase need to be controlled. This in effect implies controlling the feed-strip pressure difference as long as the individual phase pressure differential *vis-à-vis* the membrane phase is in the right direction. But the ability to change the membrane phase pressure is an advantage in HFCLM systems compared to a SLM system. Typical pressure conditions are provided in Sengupta, Basu, and Sirkar (1988) and Basu and Sirkar (1991).

From Eq. (41–45) in Chapter 41, we see that a high value of Δp_{cr} is achieved when the interfacial tension is high and the membrane pore size is low. The same equation holds also for a gas/liquid system except the interfacial tension γ_{wo} is replaced by the surface tension. In general, the surface tension for a gas/liquid system is considerably higher than the interfacial tension for an aqueous/organic (liquid/liquid) system. Thus, the allowable pressure difference between the liquid membrane phase and the gas phase is considerably higher. However, the feed gas pressure is higher than the sweep or permeate side pressure *often* by a considerable amount. With hydrophobic fibers and aqueous membrane liquids, sweep or permeate side conditions become critical to preventing membrane breakthrough into the sweep fibers since the feed pressure can be slightly below the membrane pressure. Very small pores in the microporous fibers are therefore essential to operation at a very high pressure differential.

Concurrently, fiber strength must be sufficient to prevent substantial fiber deformation. Current hydrophobic microporous hollow fibers of polypyropylene (Celgard® fibers, Hoechst Celanese, Charlotte, NC) have pore sizes around 0.03 μm. The X-10 variety has a 100-μm-i.d., 150-μm-o.d. fiber. For this fiber, the breakthrough pressures for air-water systems are around 10 atm. The fiber collapse pressures are considerably higher (around 300 to 400 psi). Bhave and Sirkar (1987) have continuously operated these fibers in SLM mode up to 200 psi with minor deformation. A smaller pore size and a higher fiber strength would be needed for operation at higher pressure differentials.

HFCLM PERMEATOR DEVICE AND DESIGN CONSIDERATIONS

The HFCLM permeation technique is relatively new. Device analysis and design procedures are at best exploratory in nature. No more than a few broad guidelines are identified here first for gas separations and then for liquid separations. Information on the commercial availability of such devices is provided at the end.

Gas Separation

In a HFCLM permeator using the same type and number of hollow fibers for each fiber set, either set may be used for feed or sweep. For a device that has two different dimensions for fibers in the two different sets (Guha 1989), the larger bore fiber is used for the gas stream requiring the lower pressure drop. This requirement is to be balanced against the compression load imposed on any hydrophobic fiber by the higher pressure of the aqueous liquid membrane not wetting the fiber pores. The excess pressure of the aqueous liquid membrane over either gas phase must not exceed the breakthrough pressure for the given gas-liquid-membrane system and membrane pore size of the hydrophobic fiber. Fibers with low pore size, significant strength, and lower porosity are more suited to high pressure differential operations. Hollow-fiber selection is thus linked directly to the operating pressure level, maximum allowable axial gas pressure drop, and the surface tension of the system.

For a fiber of given outside diameter d_{to}, the total number of fibers ($N_T' + N_T''$) packed in a permeator is determined by the inside diameter of the permeator shell, the fiber packing fraction ϕ in the shell, and d_{to}:

$$N_T' + N_T'' = \frac{d_b^2}{d_{to}^2} \phi. \qquad (42\text{--}55)$$

Here d_b, the diameter of the fiber bundle, is taken to be the shell inside diameter. The fiber packing fraction can be varied between 0.3 to 0.6. Lower values will result in higher EMTs, whereas the highest values of ϕ could lead to accidental local inter-fiber contact points without membrane liquid in between. The latter would allow mixing of feed gas and permeate gas, which is to be avoided.

Knowing d_b, d_{to}, N_T', and N_T'', the effective membrane thickness, l_e may be obtained from simple geometrical considerations using Eqs. (42-3a), (42-3b), and (42-3c) for a square cell arrangement. Alternately, the results of two-dimensional analysis plotted in Figure 42–10 may be used to read off l_e. This EMT should be considered as very close to the *optimum* or *minimum value*. In practice, the value of l_e, strongly influenced by the fiber bundle making procedure, will probably be higher. The actual membrane thickness may be determined by the pure gas permeation procedure described in Majumdar et al. (1989). Table 42–5 provides details of some typical HFCLM devices (Guha 1989; Majumdar et al. 1990a). All of them pack a very high membrane surface area.

Once these basic device parameters are fixed, a gas separation design may be initiated for a particular gas separation problem after the permeabilities of each gas species through the liquid membrane are available. For nonreactive permeation, solution of Eqs. (42-4) through (42-9) may now be implemented. The actual calculation will depend on whether "design" or "rating" is the goal with modules of fixed fiber length. Procedures developed for polymeric membrane permeators for gas separation are

TABLE 42–5. Geometrical Characteristics of Some HFCLM Permeator Modules.

Module	Active Length (in.)	Feed Fiber Total No.	Feed Fiber ID/OD (μm)	Permeate Fiber Total Number	Permeate Fiber ID/OD (μm)	Void Space (%)	Area[a] Volume (m^2/m^3)	EMT[b] l_e (μm)
A	62	300	100/150	300	100/150	66.5	4464	111.5
B	62	300	100/150	300	100/150	63.7	4844	111.5
C	24	300	100/150	155	240/290	46.7	4844	254.0
D	24	300	100/150	155	240/290	46.7	4844	—
E	17	300	100/150	300	100/150	63.7	4844	123.1
F	24.5	3504	100/150	3504	100/150	56.5	5793	317.5
G	24.5	3504	100/150	3504	100/150	56.5	5793	279.4

[a]Area per volume indicates the ratio of active membrane surface area to equipment volume
[b]Determined by pure gas permeation measurements (Majumdar et al., 1989). Value of P_{CO_2}/l_e for module A or B with l_e = 111.5 μm and pure water liquid membrane is 6.3 × 10^{-9} mol/m^2 · s · Pa.
Source: From Guha (1989) and Majumdar et al. (1990a).

almost directly usable here for any given set of operating conditions, gas flow rates, and compositions (Sengupta and Sirkar 1986a). The pressure drops in the lumen of both fiber sets are distinctive of HFCLM permeators and provide points of departure from polymeric membrane-based gas permeator design. Computer programs for such are available in two theses (Majumdar 1986; Guha 1989). Fiber bundle making and other procedures for fabricating HFCLM permeators have been illustrated in the same theses and Sirkar (1988a, 1988b).

Liquid Separation

The HFCLM permeator for gas separations can also be used for liquid separations if the module has adequate chemical resistance. Whether it would be optimal can be decided only after a detailed design has been developed. Information on N_T', N_T'', d_b, d_{to}, d_{ti}, and l_e are needed prior to any design effort as in gas separation. Fiber substrate properties such as ϵ_m and τ_m are needed in addition. The design of a permeator for a given problem may then be implemented by solving the appropriate set of equations, e.g., (42–49a), (42–49b), (42–50a), and (42–52).

We illustrate here the calculation of mass transfer coefficient K_w defined by Eq. (42–26) for aqueous feed-aqueous strip, hydrophobic microporous fibers, and an organic nonreactive CLM also present in the fiber pores. The system selected is transport of phenol from feed aqueous solution through an organic liquid membrane of methyl isobutyl ketone (MIBK) to a strip aqueous solution in a 10-cm-long permeator containing 300 feed/300 strip Celgard® X-10 fibers (Sengupta, Basu, and Sirkar 1988). The feed and strip flow rates are, respectively, 1.5155 and 1.1332 cm^3/min. Fiber dimensions and properties are d_{ti} = 100 × 10^{-4} cm; d_{to} = 150 × 10^{-4} cm; $\epsilon_m' = \epsilon_m'' = \epsilon_m = 0.3$; $\tau_m \sim 3.5$. Further l_e = 110 × 10^{-4} cm and m_i = 55.

Feed side: The tubeside velocity in each fiber v_{tw}' is

$$v_{tw}' = \frac{4Q_w'}{\pi d_{ti}'^2 N_T' 60} \frac{cm}{s} = \frac{4 \times 1.5155}{\pi (100 \times 10^{-4})^2 \times 300 \times 60} = 1.07 \frac{cm}{s},$$

$$Re_w' = \frac{v_{tw}' d_{ti}' Q_w'}{\eta_w'} = \frac{1.07 \times 100 \times 10^{-4} \times 1}{0.01} = 1.07,$$

$$Sc_w' = \frac{\eta_w'}{\rho_w' D_{iw}'} = \frac{0.01}{1 \times 1.05 \times 10^{-5}} = 952.38,$$

where the diffusion coefficient for phenol in water is $D'_{iw} = 1.05 \times 10^{-5}$ cm²/s. Now use Eq. (41–46) to calculate k'_w for $L = 10$ cm and obtain $k'_w = 5.36 \times 10^{-4}$ cm/s. Note that Re_i and Sc_i in Eqs. (41–47) and (41–48) are Re'_w and Sc'_w here.

Strip side: The tubeside velocity in each fiber v''_{tw} is

$$v''_{tw} = \frac{4Q''_w}{\pi d''^2_{ti} N''_T 60} \frac{cm}{s} = \frac{4 \times 1.1332}{\pi (100 \times 10^{-4})^2 \times 300 \times 60} = 0.80 \frac{cm}{s},$$

$$Re''_w = \frac{v''_{tw} d''_{ti} \rho''_w}{\eta''_w} = \frac{0.8 \times 100 \times 10^{-4} \times 1}{0.01} = 0.80,$$

$$Sc''_w = 952.38.$$

Again use Eq. (41–46) to calculate k''_w for $L = 10$ cm and obtain $k''_w = 4.01 \times 10^{-4}$ cm/s.

$$\text{Substrate coefficients: } k'_{so} = k''_{so} = k_{so} = \frac{2 D_{io} \epsilon_m}{\tau_m (d_{to} - d_{ti})}.$$

Note that $d''_{to} = d'_{to} = d_{to}$ and $d''_{ti} = d'_{ti} = d_{ti}$.

$$k_{so} = \frac{2 \times 2.65 \times 10^{-5} \times 0.3}{3.5 (150 \times 10^{-4} - 100 \times 10^{-4})} = 9.09 \times 10^{-4} \frac{cm}{s}.$$

Here $D_{io} = 2.65 \times 10^{-5}$ cm²/s is the phenol diffusion coefficient through MIBK present in the substrate pores.

CLM coefficient:
$$k_{mo} = \frac{D_{io}}{l_e} = \frac{2.65 \times 10^{-5}}{110 \times 10^{-4}} = 24.09 \times 10^{-4} \frac{cm}{s},$$

$$d_{tlm} = \frac{d_{to} - d_{ti}}{\ln(d_{to}/d_{ti})} = \frac{150 \times 10^{-4} - 100 \times 10^{-4}}{\ln(150/100)} = 123.32 \times 10^{-4} \text{ cm}.$$

Now by Eq. (42–26),

$$\frac{1}{K_w} = \left(\frac{1}{k'_w} + \frac{1}{k''_w}\right) + \frac{d_{ti}}{d_{tlm}} \frac{1}{m_i} \left(\frac{1}{k'_{so}} + \frac{1}{k''_{so}}\right) + \frac{d_{ti}}{d_{to}} \frac{1}{m_i} \frac{1}{k_{mo}}$$

$$= \left(\frac{10^4}{5.36} + \frac{10^4}{4.01}\right) + \frac{100}{123.32} \times \frac{1}{55} \left(\frac{2 \times 10^4}{9.09}\right) + \frac{100}{150} \times \frac{1}{55} \times \frac{10^4}{24.09} \frac{s}{cm}$$

$$= (1865.6 + 2493.7) + 32.4 + 5.03 = 4396.73 \text{ s/cm},$$

$$K_w = 2.27 \times 10^{-4} \text{ cm/s}.$$

Calculation of concentration changes in such a device for the feed and strip streams can be obtained using Eq. (42–30a).

Module Availability

Large-scale HFCLM modules containing two sets of fibers are difficult to make compared to modules having a single set of fibers. The difficulty lies especially in mixing the two sets of fibers to achieve a low membrane thickness. A semiautomatic spooling arrangement to lay out the fibers for a large HFCLM module containing 7000 fibers has been described by Majumdar et al. (1990a).

HFCLM devices are, however, commercially available from Hoechst Celanese Corporation, SPD, Charlotte, NC, under the trade name Liqui-Cel® contained liquid membrane module. The characteristics of small modules provided in two sizes are given below.

		E-240/1.4	E-400/1.0
Materials of Construction	Fiber	Polypropylene Celgard® X-10	Polypropylene Celgard® X-10
	Casing	Polypropylene, Teflon®	Polypropylene, Teflon®
	Potting	Epoxy	Epoxy
Fiber Characteristics	Wall thickness	30 μm	30 μm
	Diameter	240 μm	400 μm
	Pore size	0.05 μm	0.05 μm
	Porosity	0.30	0.30
Module Dimensions	Effective length	25.4 cm	25.4 cm
	Number of fibers	700 (350/bundle)	300 (150/bundle)
	Module length	34.9 cm	34.9 cm
	Effective surface area/bundle	0.07 m²	0.05 m²
	Module diameter	1.59 cm	1.59 cm
Operating range	Temperature	0–75°C	0–75°C
	Pressure	20 psi	20 psi

The chemical resistances of these modules are similar to those of Liqui-Cel® modules used for solvent extraction (see Chapter 41). Other than the small modules described above, a prototype version containing a total of 15,000 microporous polypropylene hollow fibers is also available from the same manufacturer. The module diameter and effective length are 5.08 and 22.9 cm, respectively. The fiber inside and outside diameters are 240 and 300 μm, respectively. The module has an effective mass transfer area of 1.3 m². A picture of a small and a large HFCLM module is shown in Figure 42–15. The contained liquid membrane thickness of these modules at the time of writing is large. Fiber matting procedures need to be changed to achieve much lower membrane thicknesses (Table 42–5) characteristic of well-mixed bundles (Majumdar et al. 1989). A highly reproducible manufacturing method that provides a desirable distribution of fibers and a low membrane thickness value is essential if the HFCLM technology is to be widely accepted.

APPLICATIONS

HFCLM permeation-separation may be applied to almost any separation problem to which other liquid membrane techniques have been applied. Some applications and potential applications are briefly identified below, first for liquid separations and then for gas separations.

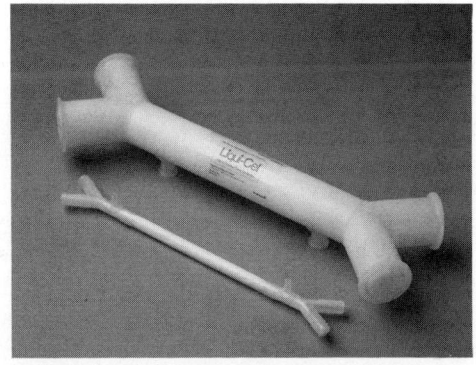

FIGURE 42–15. Photograph showing small and large HFCLM modules that are commercially available (reprinted with permission from Hoechst Celanese Corporation, Charlotte, NC).

Liquid Separations

Applications of HFCLM technique to separation of liquid solutions are considered under the following categories:

1. Wastewater treatment
2. Metal extraction-concentration
3. Fermentation product recovery.

Wastewater Treatment

Phenol is a typical acidic organic pollutant in many waste streams. Removal of phenol from an aqueous waste stream through an organic liquid membrane into an aqueous strip solution containing caustic (Cahn and Li 1974) has been studied in HFCLM devices made out of Celgard® hydrophobic hollow fibers by Sengupta, Basu, and Sirkar (1988). Experimental studies on phenol permeation with or without caustic have also utilized hydrophilic Cuprophan® hollow fibers (Sengupta and Sirkar 1986b; Sengupta et al. 1988).

The solvents used were MIBK and decanol. The values of the distribution coefficient in both solvents are high (35 and 25). According to Eq. (42–26), if all transfer coefficients are of similar order of magnitude, K_w is determined by k'_w and k''_w since m_i is $>>1$. The performance observed by Sengupta, Basu, and Sirkar (1988) was described reasonably by Eq. (42–26) (see Figure 42–14). Since the permeators used were short in length, the extent of phenol removal from the feed was limited. When caustic is present in the strip, the phenol is obtained as sodium phenolate. Using a low strip flow rate and a high caustic concentration, a concentrated sodium phenolate solution is easily obtained. Another aspect to be noted is the low value of the CLM resistance, thus allowing for the possibility of a large EMT.

The HFCLM technique may be conveniently applied to other acidic impurities in the aqueous stream as long as a permeate side reaction allows the species to be concentrated. A similar approach may be adopted for any basic organic pollutant with significant solubility in an organic liquid membrane phase. However, an acid-base reaction is needed in the permeate side, which can be achieved by incorporating an acid in the strip to concentrate the pollutant. Inorganic basic pollutants, e.g., ammonia, may also be treated by such a method. The organic membrane in this case must have an acidic complexing agent forming an ion-pair with NH_4^+ ion to extract ammonia. Alternately, only ammonia will be extracted into the organic membrane if the pH of the solution is high enough to eliminate NH_4^+ ion.

Metal Extraction-Concentration

Nguyen and Callahan (1989) have studied the recovery of copper in preference to zinc from an aqueous ammoniacal solution using LIX 54 (a phenyl alkyl beta-diketone from Henkel Corporation) in a small HFCLM permeator. The device had 180 feed and 180 strip hydrophobic fibers (Celgard® X-10, Hoechst Celanese) of 20 cm effective length; another device used Celgard® X-20 fibers to test the effect of fiber porosity. Copper was selectively recovered over zinc efficiently in such a device. Bubbling of the contained liquid membrane with nitrogen did not influence the rate of copper recovery; this indicated that CLM resistance was not the controlling one. Dilution of LIX 54 with octanol influenced extraction rates substantially and suggested interfacial reaction kinetics to be controlling (Komasawa, Otake, and Yamashita 1983). No long-term stability studies were conducted. An unsteady-state batch recirculation mode was adopted for experiments instead of once-through flow of feed and strip through the respective fiber set. With time, therefore, the copper concentration in the feed reservoir decreased drastically while that in the strip solution of low pH (via strong H_2SO_4 solution) increased correspondingly.

Basu and Sirkar (1989) have studied the recovery of copper from a $CuSO_4$ feed solution (1000 mg/L as Cu, pH ~ 4.25) into a strip solution of very low pH (200 g H_2SO_4/L) through a liquid membrane of n-heptane containing 10% v/v of LIX 84 (anti-2-hydroxy-5-nonyl acetophenone-oxime, Henkel Corporation). The HFCLM permeator had an effective length of 10 cm. It contained 300 feed and 300 strip hydrophobic hollow fibers (100-μm i.d.

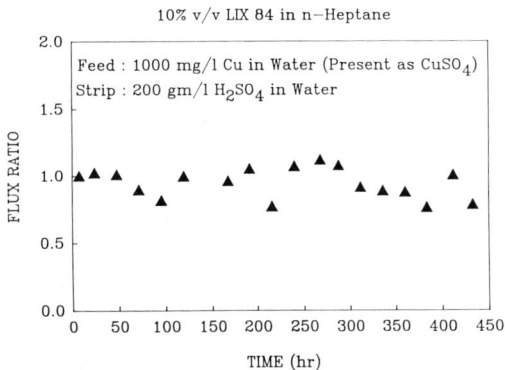

FIGURE 42–16. Extended performance data in copper removal in a HFCLM permeator containing 10% v/v LIX 84 in *n*-heptane as a liquid membrane. The flux ratio is defined as the ratio of copper flux at any time to copper flux at 7.5 h from beginning (reprinted from Basu and Sirkar 1989 with permission).

and 150-μm o.d.) in a 0.61-cm-diameter cylindrical shell. The interfacial area per unit module volume was 32.2 cm^{-1}. The experiment was conducted with once-through flow of feed and strip solution through the respective fibers. The copper flux expressed as a ratio of the copper flux at 7.5 h remained stable over a month-long period. The performance stability data for 450 hours are shown in Figure 42–16. The only step taken was to withdraw approximately 2 cm^3 of membrane liquid every day from the shellside exit port. With pure LIX 84, such a step would be unnecessary. Simple alternative procedures are being developed to eliminate the need for any membrane liquid withdrawal when a solution is used as a membrane and no membrane poisoning is involved. Studies with other metals such as Cr^{6+} and Hg^{2+} are also in progress.

Fermentation Product Recovery

Of the numerous compounds produced by fermentation, recovery of only two carboxylic acids, acetic and citric, from a synthetic broth has been explored using a HFCLM permeator. Basu and Sirkar (1991) have used two permeators of hydrophobic fibers, one 48 cm long and another 10 cm long to study citric acid recovery. The latter permeator was also used for copper recovery. Both permeators were essentially identical except for the membrane area (454 cm^2 for the long one versus 94 cm^2 for the 10-cm permeator).

The feed aqueous solution was either 10 or 3% w/v citric acid solution in water. The strip solution of pure water or 0.5 M aqueous solution of NaOH flowed cocurrently to the feed solution. No glucose was present in the feed solution since an earlier experiment with glucose in citric acid-containing feed indicated no measurable glucose in the strip solution. A variety of liquid membranes were studied; the organic solvents were either MIBK or xylene or *n*-heptane containing 10 vol. % octanol. The complexing agent was tri-*n*-octylamine at various concentrations.

The stability of the HFCLM was studied in the 48-cm-long permeator in two ways using a liquid membrane of 40% v/v tri-*n*-octylamine in xylene, 10% citric acid in feed, and a 0.5 aqueous NaOH strip. In the first method, a small volume of membrane liquid, 2 to 3 cm^3, was withdrawn every day from the shellside exit port. The citric acid flux was found to remain constant over a period of 850 hours (Figure 42–17). In the second method, the shellside exit port was kept closed. Over a two-month period, the citric acid flux first slowly decreased from the initial to a somewhat lower value and then remained steady at that level for a month when the run was stopped. This value was found to correspond to a xylene liquid membrane containing 60% tri-*n*-octylamine.

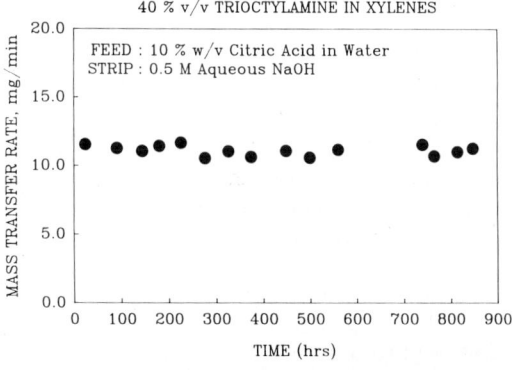

FIGURE 42–17. HFCLM stability studies for citric acid separation: mass transfer rate with time when a small volume of membrane liquid was withdrawn once a day (reprinted from Basu and Sirkar 1991 with permission).

According to Basu and Sirkar (1991), this behavior resulted from a much higher solubility of xylene (compared to the amine) in the feed and strip solutions. The conditions of operation in this permeator were such that about 30% of the citric acid was being recovered in the 48-cm-long permeator. At a lower flow rate, the same study showed a recovery of about 97% of the feed citric acid in the same permeator.

Studies on recovery of acetic acid from a synthetic feed using organic liquid membranes of MIBK or xylene in small HFCLM permeators have been carried out by Sengupta, Basu, and Sirkar (1988) and Sengupta et al. (1988). The permeators utilized in phenol permeation studies were used. No complexing agent was incorporated in the liquid membrane. The fluxes and the transfer coefficients were low due to low values of m_i in the solvents used and the lack of any complexing carrier. As expected from Eqs. (42–26) and (42–29), the CLM resistance was a major component of the overall resistance since $m_i \ll 1$ for xylene or $m_i < 1$ for MIBK.

Separation of chlorophenoxyacetic acid (CPA), a penicillin mimic, has been carried out recently in a Liqui-Cel® HFCLM module by Sorenson and Callahan (1990). The feed solution was at pH = 2.5, the membrane liquid was amyl acetate, and the strip solution pH was 7. This is an example of how the conventional extraction and back extraction steps carried out for antibiotics may be merged into one step by a HFCLM.

Since studies on separation of organic mixtures using HFCLM device were in progress during 1991, no results are reported. A number of other aqueous systems with organic liquid membranes were also under study in several places. These include the use of reverse micelles in the organic phase for protein separation.

Gas Separations

HFCLM permeation-separation may be used to separate a variety of gas mixtures currently separated by polymeric membranes or equilibrium separation processes. To date, only a few systems have been studied. These are N_2-CO_2, CH_4-CO_2, SO_2-CO_2-N_2, NO-N_2-O_2, and SO_2-NO-CO_2-N_2-O_2. The first two systems focus on model mixtures for purification of a landfill gas or biogas. The last three systems arise in flue gas purification.

A very brief account of separation of a 40–60 v/v mixture of CO_2-N_2 gas mixture was provided earlier in this chapter. Details are available in NYSERDA (1987), Majumdar, Guha, and Sirkar (1988), and Guha (1989). The liquid membranes studied were pure water; 7 and 30% aqueous K_2CO_3 solution; 10, 20, and 30% diethanolamine (DEA) in water, and n-methylpyrollidinone (NMP) containing 30% water. Feed gas pressure was usually 45 psig. Sweep fibers had either a helium stream as sweep gas or pure water as a sweep liquid. Alternatively, the permeator was run as a conventional polymeric membrane permeator with or without a vacuum pulled in the lumen of permeate fibers, producing a permeate stream highly enriched in CO_2 (see Figure 42–18 for results for a conventional mode of operation). The feed gas was highly concentrated in N_2 (85 to 98%) at the permeator exit. In the sweep gas mode, a small amount of sweep gas species permeated to the feed side.

The permeators usually had 300 feed and 300 sweep fibers (100-μm i.d., 150-μm o.d., $\epsilon_m = 0.2$) with the permeator length varying between 5 ft to 18 in. The shellside exit port was always kept closed. Stability studies were conducted for as long as 100 hours and up. Operation with water as membrane always showed complete stability. For aqueous solutions as membranes, some of the procedures recommended in Table 42–3 for stable operation were used to obtain stable performance.

Majumdar et al. (1990a) have also studied the separation of the same gas mixture with a sweep helium stream and a 20% aqueous DEA solution as membrane in 24-in.-long permeators having 3504 feed fibers and 3504 sweep fibers. The feed gas flow rate was around 500 cm^3/min whereas the feed gas flow rate in the smaller permeators (300 feed and 300 sweep fibers) of earlier studies was generally around 50 to 100 cm^3/min. The experiment in this study with larger permeators was run continuously for 1

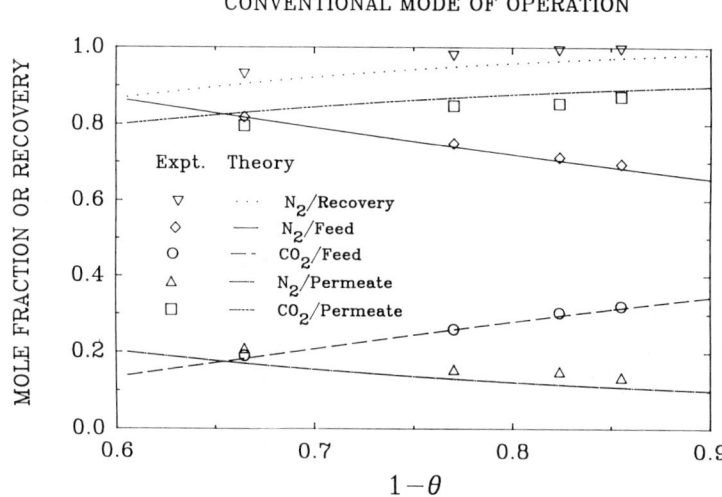

FIGURE 42-18. Separation of CO_2-N_2 gas mixture using conventional mode of operation in module C. The stage cut Θ is defined as the ratio of the permeate flow rate to the feed flow rate. Therefore, the factor $(1-\Theta)$ represents the unpermeated fraction of the feed stream in a single stage (reprinted from Guha, Majumdar, and Sirkar 1991a with permission).

month (Figure 42-19). To maintain performance stability, about 25 cm³ of membrane liquid was withdrawn every week from the shellside exit port. This liquid volume was about one-third of the membrane liquid volume in a permeator. The membrane separation performance was completely stable during the whole month. It is not known whether the membrane liquid quality changed because of deterioration of the amine or due to evaporation of moisture resulting from incomplete humidification of the feed and sweep streams (necessitating the membrane liquid withdrawal).

Although a 40–60 CO_2-N_2 stream can be obtained from a denitrification bioreactor, such a gas mixture served primarily as a model for landfill gas or biogas. Majumdar et al. (1990a) and Guha (1989) have also studied the separa-

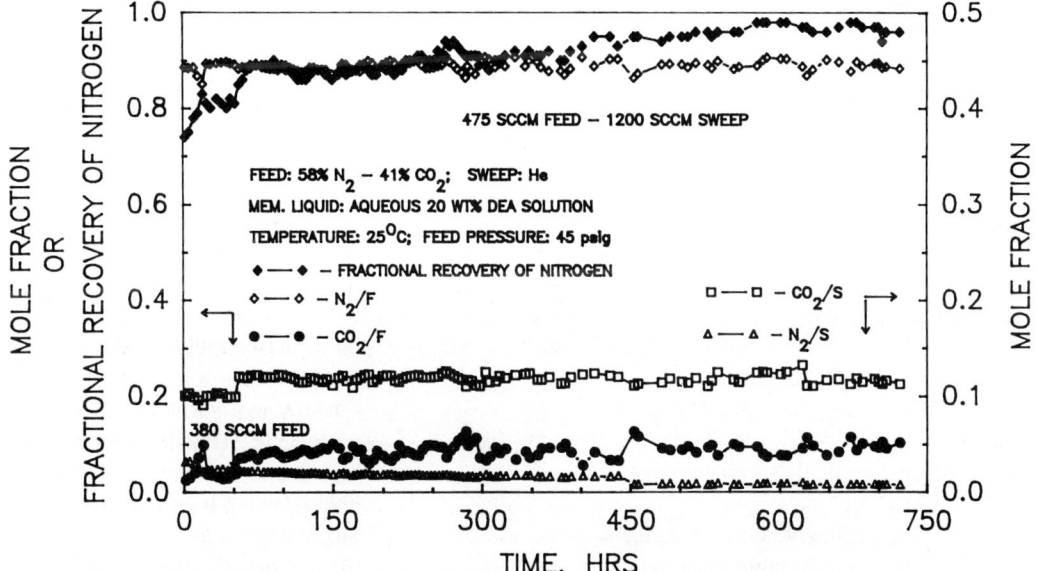

FIGURE 42-19. CLM stability studies: sweep gas mode of operation with aqueous 20 wt% DEA solution as a liquid membrane in modules F and G in series (reprinted from Majumdar et al. 1990a with permission).

tion of a 44–56 v/v CO_2-CH_4 gas mixture at 45 psig (a model mixture for biogas or landfill gas) in large as well as small permeators using nitrogen as a sweep gas. A 10-day-long run in a large permeator with 7000 fibers with 20% aqueous DEA as membrane liquid showed excellent stability. A methane concentration of 93% and a methane recovery of 93% was obtained in the purified feed gas. For the given permeabilities of the gases, calculations show that air may be used as a sweep gas without increasing the O_2 mole fraction in the purified biogas to larger than 0.5%. Water as a sweep liquid always provided even higher methane recovery and higher membrane selectivity between CO_2 and CH_4.

Flue gases from coal combustion power plants contain SO_2 and NO_X in small quantities along with CO_2, N_2, H_2O, and O_2. Sengupta, Raghuraman, and Sirkar (1990) have found that thin liquid membranes of water and aqueous solutions of Na_2SO_3 or $NaHSO_3$ or Fe^{+2}EDTA or Fe^{+3}EDTA have very high permeability for SO_2 as well as being extremely selective for SO_2. The SO_2/CO_2 selectivity is in the range of 50 to 250 and SO_2/N_2 selectivity is in the range of 1500 to 3500 for SO_2 concentrations in the range of 5000 to 500 ppm. The liquid membrane from a 0.01 M aqueous solution of Fe^{+2}EDTA has been found to have a very high selectivity for NO over N_2. Thus SO_2/NO_X, identified as the source of acid rain, can be removed simultaneously using an aqueous Fe^{+2}EDTA or Fe^{+3}EDTA solution in a HFCLM permeator with hydrophobic fibers. Recent studies (Majumdar et al. 1990b) have shown that as much as 80 to 90% of SO_2 and 70% of NO can be removed from a model flue gas having 5000 ppm SO_2, 450 ppm NO, 1.8% O_2, 12% CO_2, with the remainder being N_2 and H_2O in small HFCLM permeators 2 to 3 ft long. The permeate side is operated either under vacuum or with a sweep gas. Under the vacuum mode of operation, the permeate gas is highly enriched in SO_2.

Oxygen enrichment with an aqueous membrane containing a suitable organometallic oxygen carrier, H_2S separation from various gas streams, CO separation from a H_2-N_2 mixture in synthesis gas, and olefin separation from an olefin-paraffin mixture are some other examples of potential applications of the HFCLM technique.

ACKNOWLEDGMENTS

We acknowledge the support provided to us during the writing by DOE contract DE-AC22-87PC79853, NYSERDA 663-RIER-NCRB-84, and EPANHSRC at Newark. We have been greatly influenced by many detailed analyses and experiments carried out recently at Stevens Institute of Technology by A. Guha and R. Basu.

NOTATION

General Notation

See the General Notation section at the beginning of this handbook.

Special Notation

A_d	dimensionless area, defined by Eq. (42–10a)
c_{im}	solute concentration in the bulk membrane phase, mol/L^3
c_{io}	solute concentration in the membrane phase at the aqueous/organic interface, for hydrophobic substrate, mol/L^3
c_{iom}	solute concentration in the membrane phase at the aqueous/organic interface, for hydrophilic substrate, mol/L^3
c_{iw}	solute concentration at the fiber inside wall surface, mol/L^3
c_{iwb}	solute concentration in aqueous bulk phase, mol/L^3
c_{iwm}	solute concentration at the fiber outside wall surface, mol/L^3
d_b	hollow fiber bundle diameter, L
d_e	equivalent diameter of a circle enclosing liquid membrane around any feed fiber, L
d_{ti}, d_{to}, d_{tlm}	inside, outside, and log-mean diameters of microporous hollow fibers, respectively, L

H	Henry's law constant, $t^2\text{mol}/L^2M$
k_{mo}	contained liquid (organic) membrane transfer coefficient, L/t
k_{mw}	contained liquid (aqueous) membrane transfer coefficient, L/t
k_o	boundary layer coefficient in the organic phase, L/t
k_{so}	transfer coefficient for organic-filled hydrophobic substrate, L/t
k_{sw}	transfer coefficient for water-filled hydrophilic substrate, L/t
k_w	boundary layer coefficient in aqueous phase, L/t
K	overall mass transfer coefficient, L/t
l_e	effective membrane thickness, defined by Eq. (42–3a), L
L	membrane length, L
m_i	solute distribution coefficient, ratio of organic phase concentration to aqueous phase concentration, dimensionless
$\tilde{M}_i$	molecular weight of gas species i, M/mol
N_T	number of microporous hollow fibers, dimensionless
p, p_i	total pressure and partial pressure of species i, respectively, M/Lt^2 or p
P, P^*	permeability and dimensionless permeability of a gas species, respectively, mol t/M or mol/Ltp
Q	mobile phase flow rate, L^3/t
R	universal gas constant, ML^2/t^2T mol
R_T	solute transfer rate per unit permeator length, mol/tL
t_p	distance between the centers of two adjacent fibers in a bundle, L
U	molar gas flow rate per fiber, mol/t
v_{tw}	tubeside velocity of aqueous phase, L/t
x_i	mole fraction of gas species i, dimensionless
y_i	mole fraction of i in a multicomponent gas mixture in Eq. (42–12a), dimensionless

Greek Letters

β	dimensionless parameter, defined in Eq. (42–10f)
ϵ_m	substrate porosity, dimensionless
η	gas or liquid viscosity, M/Lt
λ_{ij}	constants arising in gas mixture viscosity estimation, defined in Eq. (42–12b), dimensionless
τ_m	substrate tortuosity, dimensionless

Subscripts

cr	critical value
d	dimensionless quantity
im	pertaining to a solute species in the membrane phase
o	value based on or refers to organic phase
r	reference parameter
w	value based on or refers to aqueous phase

REFERENCES

Armstrong, D. W., and H. L. Jin. 1987. Enrichment of enantiomers and other isomers with aqueous liquid membranes containing cyclodextrin carriers. *Analy. Chem.* 59:2237–41.

Babcock, W. C., R. W. Baker, J. W. Brooke, D. J. Kelly, E. D. Lachapelle, and H. K. Lonsdale. 1980a. Coupled transport membranes for metal recovery—phase II, National Technical Information Service, PB81-179947.

Babcock, W. C., R. W. Baker, E. D. Lachapelle, and K. L. Smith. 1980b. Coupled transport membranes III: the rate-limiting step in uranium transport with a tertiary amine. *J. Membr. Sci.* 7:89–100.

Basu, R. 1989. Mass transfer enhancement in phase barrier membrane separators for reactive systems. Ph.D. diss., Stevens Institute of Technology, Hoboken, NJ.

Basu, R., and K. K. Sirkar. 1989. Selective separations using contained liquid membranes. Paper read at the National Conference on Pollution Prevention for the 1990's: A Chemical Engineering Challenge, 4–5 December 1989, Washington, DC.

Basu, R., and K. K. Sirkar. 1991. Hollow fiber contained liquid membrane separation of citric acid. *AIChE J.* 37:383–393.

Basu, R., R. Prasad and K. K. Sirkar. 1990. Nondispersive membrane solvent back-extraction of phenol. *AIChE J.* 36:450–460.

Bhave, R. R., and K. K. Sirkar. 1986. Gas permeation and separation by aqueous membranes immobilized across the whole thickness or in a thin section of hydrophobic microporous Celgard films. *J. Membr. Sci.* 27:41–61.

Bhave, R. R., and K. K. Sirkar. 1987. Gas permeation and separation with aqueous membranes immobilized in microporous hydrophobic hollow fibers. In *Liquid Membranes Theory and Applications*, ed. Richard D. Noble and J. Douglas Way, pp. 138–151. Washington, DC.: American Chemical Society.

Bloch, R. 1970. Hydrometallurgical separations by solvent membranes. In *Membrane Science and Technology*, ed. J. Flynn, pp. 171–187. New York: Plenum Press.

Boyadzhiev, L., E. Bezenshek, and Z. Lazarova. 1984. Removal of phenol from waste water by double emulsion membranes and creeping film pertraction. *J. Membr. Sci.* 21:137–144.

Cahn, R. P., and N. N. Li. 1974. Separation of phenol from waste water by the liquid membrane technique. *Sep. Sci.* 9:505–519.

Cahn, R. P., and N. N. Li. 1976. Hydrocarbon separation by liquid membrane processes. In *Membrane Separation Processes*, ed. Patrick Meares, pp. 327–349. Amsterdam: Elsevier Scientific Publishing Co.

Callahan, R. W. 1988. Novel uses of microporous membranes: a case study. In *New Membrane Materials and Processes for Separation*, ed. Kamalesh K. Sirkar and Douglas R. Lloyd., pp. 54–63. New York: American Institute of Chemical Engineers.

Cussler, E. L. 1971. Membranes which pump. *AIChE J.* 17:1300–1303.

Danesi, P. R. 1984–85. Separation of metal species by supported liquid membranes. *Sep. Sci. Technol.* 19:857–894.

Danesi, P. R., L. Reichley-Yinger, and P. G. Rickert. 1987. Life-Time of supported liquid membranes: the influence of interfacial properties, chemical composition and water transport on the long term stability of the membranes. *J. Membr. Sci.* 31:117–145.

Danesi, P. R., and P. G. Rickert. 1986. Some observations on the performance of hollow fiber supported liquid membranes for Co-Ni separations. *Solv. Ext. Ion Exch.* 4:149–164.

Deetz, D. W. 1987. Stabilized ultrathin liquid membranes for gas separation. In *Liquid Membranes Theory and Applications*, ed. Richard D. Noble and J. Douglas Way, pp. 152–165. Washington, DC.: American Chemical Society.

Friesen, D. T., W. C. Babcock, D. J. Brose, and A. R. Chambers. 1991. Recovery of citric acid from fermentation beer using supported-liquid membranes. *J. Membr. Sci.* 56:127–141.

Guha, A. 1989. Studies on different gas separation modes with hollow fiber contained liquid membrane. Ph.D. diss., Stevens Institute of Technology, Hoboken, NJ.

Guha, A., S. Majumdar, and K. K. Sirkar. 1991a. Gas separation modes in a HFCLM permeator. *Ind. Eng. Chem. Res.* (in press).

Guha, A., S. Majumdar, and K. K. Sirkar. 1991b. A larger-scale study of gas separation by hollow-fiber-contained liquid membrane permeator. *J. Membr. Sci.* 62:293–307.

Hughes, R. D., J. A. Mahoney, and E. F. Steigelmann. 1986. Olefin separation by facilitated transport membranes. In *Recent Developments in Separation Science*, ed. Norman N. Li and James M. Calo, pp. 174–195. Boca Raton, FL: CRC Press.

Keller, K. H. and T. R. Stein. 1967. A two-dimensional analysis of porous membrane transport. *Math. Biosci.* 1:421–437.

Kiani, A., R. R. Bhave, and K. K. Sirkar. 1984. Solvent extraction with immobilized interfaces in a microporous hydrophobic membrane. *J. Membr. Sci.* 20:125–145.

Kimura, S. G., S. L. Matson, and W. J. Ward III. 1979. Industrial applications of facilitated transport. In *Recent Developments in Separation Science*, ed. Norman N. Li, Vol. V, pp. 11–25. Cleveland, OH: CRC Press.

Komasawa, I., T. Otake, and T. Yamashita. 1983. Mechanism and kinetics of copper permeation through a supported liquid membrane containing a hydroxyoxime as a mobile carrier. *I&EC Fundam.* 22:127–131.

Kulkarni, S. S., E. W. Funk, N. N. Li, and R. L. Riley. 1983. Membrane separation processes for acid gases. *AIChE Symp. Ser. No. 229*, 79:172–178.

Kuo, Y., and H. P. Gregor. 1983. Acetic acid extraction by solvent membrane. *Sep. Sci. Technol.* 18:421–440.

LeBlanc, O. H., W. J. Ward, S. L. Matson, and S. G. Kimura. 1980. Facilitated transport in ion-exchange membranes. *J. Membr. Sci.* 6:339–343.

Lee, K-H, D. F. Evans, and E. L. Cussler. 1978. Selective copper recovery with two types of liquid membranes. *AIChE J.* 24:860–868.

Majumdar, S. 1986. A new liquid membrane technique for gas separation. Ph.D. diss., Stevens Institute of Technology, Hoboken, NJ.

Majumdar, S., A. K. Guha, Y. T. Lee, T. Papadopoulos, S. Khare and K. K. Sirkar. 1990a. Liquid membrane purification of biogas. Final report to New York State Energy Research and Development Authority, 23 January 1990.

Majumdar, S., A. K. Guha, Y. T. Lee, and K. K. Sirkar. 1989. A two-dimensional analysis of membrane thickness in a hollow-fiber-contained liquid membrane permeator. *J Membr. Sci.* 43: 259–276.

Majumdar, S., A. K. Guha, and K. K. Sirkar. 1986. A new liquid membrane technique for gas separation. Paper read at the AIChE Annual Meeting, 3 November 1986, Miami Beach, FL.

Majumdar, S., A. K. Guha, and K. K. Sirkar. 1988. A new liquid membrane technique for gas separation. *AIChE J.* 34:1135–1145.

Majumdar, S., A. Sengupta, J. S. Cha, T. H. Papadopoulos, and K. K. Sirkar. 1990b. Studies on flue gas cleanup by hollow fiber contained liquid membranes. In *Proc. 1990 International Congress on Membranes and Membrane Processes*. Vol. 1, pp 662–663. 20–24 August 1990, Chicago, IL.

Marr, R., and A. Kopp. 1982. Liquid membrane technology–a survey of phenomena, mechanisms and models. *Int. Chem. Eng.* 22:44–60.

Matson, S. L., C. S. Herrick, and W. J. Ward III. 1977. Progress on the selective removal of H_2S from gasified coal using an immobilized liquid membrane. *Ind. Eng. Chem. Process Des. Dev.* 16:370–374.

Matson, S. L., J. Lopez, and J. A. Quinn. 1983. Separation of gases with synthetic membranes. *Chem. Eng. Sci.* 38:503–524.

Nakano, M., K. Takahashi, and H. Takeuchi. 1987. A method for continuous operation of supported liquid membranes. *J. Chem. Eng. Japan* 20:326–328.

Nguyen, K. V., and R. W. Callahan. 1989. Mass transfer with contained liquid membranes. *Polym. Mat. Sci. Eng.* 61, ACS Proceedings, Miami, FL.

Noble, R. D., C. A. Koval and J. J. Pellegrino. 1989. Facilitated transport membrane systems. *Chem. Eng. Prog.* March: 58–70.

Noble, R. D., and J. D. Way. 1987. Liquid membrane technology an overview. In *Liquid Membranes Theory and Applications*, ACS Symp. Ser. No. 347. Washington, DC: American Chemical Society.

NYSERDA. 1987. A new liquid membrane technique to purify landfill gas. Report 87-10 April 1987.

Olander, D. R. 1960. Simultaneous mass transfer and equilibrium chemical reaction. *AIChE J.* 6:233–239.

Otto, N. C., and J. A. Quinn. 1971. The facilitated transport of carbon dioxide through bicarbonate solutions. *Chem. Eng. Sci.* 26:949–961.

Prasad, R., and K. K. Sirkar. 1987. Solvent extraction with microporous hydrophilic and composite membranes. *AIChE J.* 33:1057–1066.

Roberts, D. L., D. E. Gottschlich, and J. D. Way. 1988. A theoretical comparison of the facilitated transport and solution-diffusion membrane modules for gas separation. Paper read at the North American Membrane Society Meeting, 1–3 June 1988, Syracuse, NY.

Schultz, J. S., J. D. Goddard, and S. R. Suchdeo. 1974. Facilitated transport via carrier-mediated diffusion in membranes: Part I. mechanistic aspects, experimental systems and characteristic regimes. *AIChE J.* 20:417–445.

Sengupta, A., R. Basu, and K. K. Sirkar. 1988. Separation of solutes from aqueous solutions by contained liquid membranes. *AIChE J.* 34:1698–1708.

Sengupta, A., R. Basu, R. Prasad, and K. K. Sirkar. 1988. Separation of liquid solutions by contained liquid membranes. *Sep. Sci. Technol.* 23:1735–1751.

Sengupta, A., B. Raghuraman, and K. K. Sirkar. 1990. Liquid membranes for flue gas desulfurization. *J. Membr. Sci.* 51:105–126.

Sengupta, A., and K. K. Sirkar. 1986a. Membrane gas separation. In *Progress in Filtration and Separation*, ed. Richard J. Wakeman, pp. 289–415, Amsterdam: Elsevier Scientific Publishing Co.

Sengupta, A., and K. K. Sirkar. 1986b. Phenol removal from aqueous streams using a new liquid membrane technique. Paper read at the AIChE Annual Meeting, 7 November 1986, Miami Beach, FL.

Sirkar, K. K. 1986. Fuel ethanol production: a novel hollow fiber fermentor-separator with simultaneous ethanol separation and CO_2 removal. Final Contract Report to Governor's Commission NJ H85-990670-3. July 1986.

Sirkar, K. K. 1988a. Selective permeation gas-separation process and apparatus. U.S. Patent 4,750,918, 14 June 1988.

Sirkar, K. K. 1988b. Immobilized interface solute transfer apparatus. U.S. Patent 4,789,468. 6 December 1988.

Sirkar, K. K. 1991. Immobilized interface solute

transfer processes. U.S. Patent 4,997,569. 5 March 1991.

Smith, D. R., R. J. Lander, and J. A. Quinn. 1977. Carrier-mediated transport in synthetic membranes. In *Recent Developments in Separation Science,* ed. Norman N. Li, Vol. 3, pp. 225–241. Cleveland, OH: CRC Press.

Sollner, K. 1984. The basic electrochemistry of liquid membranes. In *Diffusion Processes,* ed. J. N. Sherwood, A. V. Chadwick, W. M. Muir, and F. L. Swinton. Vol. 2, pp. 655–730. New York: Gordon and Breach.

Sorenson, B. V., and R. W. Callahan. 1990. Penicillin separations with contained liquid membranes. In *Proc. 1990 International Congress on Membranes and Membrane Processes,* Vol. 1, pp. 695–697, 20–24 August 1990, Chicago, IL.

Teramoto, M., H. Matsuyama, T. Yamashiro and S. Okamoto. 1989a. Separation of ethylene from ethane by a flowing liquid membrane using silver nitrate as a carrier. *J. Membr. Sci.* 45:115–136.

Teramoto, M., N. Tohno, N. Ohnishi and H. Matsuyama. 1989b. Development of a spiral-type flowing liquid membrane module with high stability and its application to the recovery of chromium and zinc. *Sep. Sci. Technol.* 24:981–999.

Tsuji, T., K. Suma, K. Tanishita, H. Fukazawa, M. Kamo, H. Hasegawa and A. Takahasi. 1981. Development and clinical evaluation of hollow fiber membrane oxygenator. *Trans. Am. Soc. Artif. Intern. Organs* 27:280–284.

Ward, W. J. 1970. Analytical and experimental studies of facilitated transport. *AIChE J.* 16:405–410.

Ward, W. J., and W. L. Robb. 1967. Carbon dioxide-oxygen separation: facilitated transport of carbon dioxide across a liquid film. *Science* 156:1481–1483.

Way, J. D., R. D. Noble, T. M. Flynn, and E. D. Sloan. 1982. Liquid membrane transport: a survey. *J. Membr. Sci.* 12:239–259.

Way, J. D., R. D. Noble, D. L. Reed, and L. A. Jarr. 1987. Facilitated transport of CO_2 in ion exchange membranes. *AIChE J.* 33:480–487.

Wilke, C. R. 1950. A viscosity equation for gas mixtures. *J. Chem. Phys.* 18:517–519.

Yang, M. C., and E. L. Cussler. 1986. Designing hollow-fiber contactors. *AIChE J.* 32:1910–1916.

43

Membrane Reactors

Stephen L. Matson
Arete Technologies, Inc.

John A. Quinn
University of Pennsylvania

INTRODUCTION AND OVERVIEW
MEMBRANE REACTORS: A
  FUNCTIONAL DEFINITION
  Catalysis
  Organization
  Separation
PRODUCT SEPARATION AND
  ENRICHMENT
  Concept
  Experimental Demonstration
SEPARATION OF MULTIPLE
  PRODUCTS
  Concept
  Experimental Demonstration
THERMODYNAMIC
  CONSIDERATIONS
CONCLUSIONS
ACKNOWLEDGMENT
NOTATION
REFERENCES

INTRODUCTION AND OVERVIEW

The term *membrane reactor* first began to appear in the chemical processing literature around 1980. Over the past decade, it has attained a proper niche in the lexicon of membrane technology and the topic is now a regular feature at membrane conferences and symposia as well as being the subject of a growing technical literature. Although there is no commonly accepted definition of a membrane reactor, the term is usually applied to membrane processes/devices whose function is to perform net chemical conversion under conditions in which the unique contacting features of membrane devices are exploited. In particular, the term *membrane reactor* is reserved for those processes wherein the membrane functions as more than simply a *reactive membrane*, i.e., a membrane matrix used for catalyst immobilization. These special features of membrane reactors have been demonstrated with multilayer devices (Matson 1979; Matson and Quinn 1986) and, more recently, with multiphase membrane contactors (Matson 1989a, 1989b; Matson and Lopez 1989; Lopez et al. 1990). These important developments appear to be among the first ones in this emerging new area of reaction engineering and, therefore, a review such as the present one can serve as but a snapshot of the field in 1990, presenting underlying concepts and illustrating typical applications.

Advances in membrane technology emanate from two principal sources: the science of membrane materials and the engineering of membrane contacting devices. The former is usually considered to be the major goal, i.e., synthesize a more permeable, more selective membrane material and a membrane process would be competitive. However, a case can be made that many recent advances in the field have in fact resulted from inventions in engineering devices, e.g., hollow-fiber contactors, asymmetric membranes, ultrathin membranes, etc. It is in this context that membrane reactors have emerged: No new membrane materials are necessarily involved but unique contacting features of membranes have been employed in novel reactor designs.

In this chapter we discuss basic concepts underlying membrane reactors, including a functional definition and a taxonomy of the field. With most development coming within the last decade, it is not surprising that there are few comprehensive published reviews at this time (Matson 1979; Flaschel, Wandrey, and Kula 1983; Cheryan and Mehaia 1986; Matson and Quinn 1986; Belfort 1989; Hsieh 1989; Lopez et al. 1990). However, a considerable patent literature exists (Knazek et al. 1975; Breslau 1981; Michaels, Robertson, and Cohen 1984; Matson and Quinn 1988; Matson 1989a, 1989b), and even a perfunctory search for the term *membrane reactor(s)* in the *Chemical Abstracts* data base for the years 1987 through mid-1990 reveals about 150 published references. Only a few of these most recent papers representative of activity in the field can be cited here. For instance, approximately two-thirds of these (nearly 100 articles) deal with membrane *bioreactors* based on enzymes or whole cells (Lopez 1983; Vasudevan et al. 1987; Berke et al. 1988; Lee and Hong 1988; Molinari, Drioli, and Barbieri 1988; Pronk et al. 1988; Efthymiou and Shuler 1989; Vasic-Racki et al. 1989; Mannheim and Cheryan 1990; Kise and Hayashida 1990; Steckhan et al. 1990); for the most part, these reactors employ synthetic polymeric membranes. A smaller but rapidly growing number of papers (about 20) appeared in this time period describing heterogeneous catalysis as conducted in membrane reactors, the membrane typically being fashioned from ceramic or other inorganic materials (Itoh 1987; Ito et al. 1988; Omata et al. 1989; Uemiya, Matsuda, and Kikuchi 1990). Finally, a few investigators have examined the utility of membrane reactors in homogeneous catalysis (Ollis, Thompson, and Wolynic 1972; Stanley 1986; Stanley and Quinn 1987; Matson and Stanley 1988; Chen and Kao 1990) and various specialty applications such as analytical instrumentation. To illustrate the special attributes of membrane reactors, we focus on a specific example of a multilayer device that demonstrates the unique capabilities of membrane reactors for coupling and combining process operations.

MEMBRANE REACTORS: A FUNCTIONAL DEFINITION

Since the catalytic efficiency of biochemical systems can often be attributed to the action of membrane-bound enzymes, it is not surprising that catalysis in and about both natural and synthetic membranes is an increasingly popular topic of research in the biological and physical sciences and in engineering circles. Many aspects of the structure and function of biological membranes that have been elucidated over the past two decades make it abundantly clear that nature has designed a highly integrated chemical plant that engineers might profitably mimic.

At the same time, chemists have developed an arsenal of techniques for immobilizing catalysts—both enzymatic and inorganic—to polymeric supports. Enzymes can now be covalently bound, cross-linked, entrapped, encapsulated, and adsorbed on or within porous particles and membranes, and industrial applications of this technology are emerging. Complementary techniques have been developed for insolubilizing inorganic homogeneous catalysts with the aid of bifunctional ligands. While porous particles have served as the catalyst support in the majority of experimental studies and in virtually all significant industrial-scale applications of such immobilized catalysts, we consider here the proposition that in certain circumstances immobilizing the catalyst in a membrane structure may be a better approach. Since membranes are generally more expensive than porous particles or pellets, the burden of proof is clearly on those who would advocate the use of membranes as alternative catalyst supports.

The increasingly frequent appearance of the term *membrane reactor* in the chemical engineering literature thus results from a confluence of efforts in the areas of membrane science, immobilized catalyst engineering, and semipermeable membrane technology. Unfortunately, the term has often been used in a *structural* sense to describe a variety of reactor configurations involving either catalytic or permselective films, many of which lead to uninteresting reactor behavior. An objective of this chapter is to advance a *functional* definition of membrane reactors that readily distinguishes between trivial and nontrivial cases on the basis of whether functions or modes of operation unique to membranes are invoked. The term *reactive membranes* is applied where the membrane serves merely as a catalyst support, while the term *membrane reactors* is reserved for cases in which the unique abilities of membranes to organize, compartmentalize, and/or separate are involved.

To assist in identifying promising membrane reactors, the capabilities of membranes are examined here briefly. Emphasis is given to functions unique to membranes since they will form the basis for differentiating membrane reactors both from conventional reactors employing catalyst particles as well as from membranes that serve merely as catalyst supports.

Catalysis

Membranes are generally activated with catalysts either by impregnating porous films with catalyst solutions, by entrapping the catalyst within the membrane, by dispersing or adsorbing a catalytic species throughout the membrane matrix or depositing a normally heterogeneous catalyst on the exterior and/or interior surfaces of a membrane, or by covalently attaching a normally soluble catalyst to the pore wall surfaces of membranes. Many alternative approaches for incorporating catalytic activity are available (Ollis, Thompson, and Wolynic 1972; Zaborsky 1973; Wolynic and Ollis 1974; Updike 1976; Grubbs 1977; Matson 1979; Bernstein et al. 1982; Drioli et al. 1982; Shipman 1985). Most typically in previous studies, such catalytic membranes have functioned simply as catalyst slabs with reactants and products diffusing in or out from both sides. In other cases, a convective flux of reactant solution has been made to occur across a porous membrane by applying a transmembrane pressure difference. However, nontrivial reactor function is rarely realized in either situation.

While it is true that intrinsic catalyst characteristics (e.g., activity, selectivity, and life) are often modified when an enzyme or other synthetic homogeneous catalyst is attached to a support, for the most part these are microscopic chemical phenomena that are unrelated to the geometry of the support. For example, porous catalytic films operated in a forced permeation mode closely resemble exceedingly short packed beds of catalyst particles, whereas membrane catalysts uniformly bathed in a solution of reactants differ trivially from their particulate counterparts. As argued above, membrane reactors must be operationally superior to and not just structurally different from conventional reactors if they are to be useful in practice.

The apparent (cf. intrinsic) kinetics exhibited by immobilized catalysts are, however, sensitive to the geometry of the support because boundary conditions and transport distances associated with heat and mass transfer differ for membrane and particulate catalysts. Effectiveness factors are particularly important in processes that are catalyst cost-intensive or where selectivity is influenced by diffusional resistances, and practical restrictions (such as packed bed pressure drop, fluidized bed carryover, and particle settling velocity) frequently place a lower limit on catalyst particle size. These restrictions are such that membranes can sometimes be made thinner than particles can be made small. Furthermore, by operating in a pressure-driven permeation mode, external and pore diffusion effects may be practically eliminated, and very short catalyst contact times are made possible. The latter might be useful in maximizing selectivity in sequential reactions.

With regard to the former, the use of flow-through enzyme-activated membranes to improve enzyme effectiveness in biocatalytic processes has been amply demonstrated on both the laboratory scale (Kozarek 1975; Gregor and Rauf 1975) and the pilot plant scale. When fluid is pumped across a porous catalytic membrane, axial pore diffusion is eliminated as a potentially rate-controlling step. Flat-sheet immobilized enzyme reactor systems have been piloted for the lactase-catalyzed hydrolysis of whey permeate (Goldberg and Chen 1987) and for amino acid and penicillin biotransformations. These reactor modules consist of multiple layers of microporous PVC-silica membrane potted into a series of reactor plates, with the enzyme covalently coupled to the membrane matrix via standard silica attachment chemistry.

Occasionally, membranes developed primarily for separation purposes also exhibit catalytic activity that may be exploited in membrane reactors. These membranes generally fall into one of two categories: polymeric membranes with pendant functional groups, and metal and metal oxide membranes. Ion-exchange membranes provide examples of the former, with the Nafion® perfluorinated cation-exchange membranes made by Du Pont probably being the most thoroughly studied. Several papers and reviews on the use of ion-exchange membrane catalysis have been published, with emphasis on laboratory-scale synthetic applications (Helfferich 1962; Cares 1977; Aldrich 1986; Olah, Iyer, and Prakash 1986; Sondheimer, Bunce, and Fyfe 1986). Ion-exchange membrane catalysis has also been conducted by incorporating metallic counter-ions (e.g., Pd and Pt for hydrogenations and dehydrogenations) in such membranes (Haag and Whitehurst 1978; Waller 1986).

Other inherently catalytic and permselective membranes are inorganic rather than polymeric in nature. Semipermeable metallic membranes made from palladium and palladium alloys have received considerable attention for their ability to catalyze hydrogenation and dehydrogenation reactions (Gryaznov, Smirnov, and Slin'ko 1973; Gryaznov, Smirnov, and Mischenko 1975; Gryaznov et al. 1979; Gryzanov 1983, 1986; Nagamoto and Inoue 1985; Itoh 1987) as discussed further below. Other workers have investigated a Bi-La-O membrane that is simultaneously catalytic and capable of selectively transporting oxygen via oxide-ion conduction (Di Cosimo, Burrington, and Grasselli 1986a, 1986b). Their "membrane" (actually, a 500-μm-thick disk) catalyzed an oxidative dehydrodimerization reaction between oxygen present on one surface of the disk and reactant/products on the other. By preventing direct contact between organic reactant and oxidant, over-oxidation to CO_2 was reduced, albeit at the expense of reduced productivity.

Organization

Two aspects of organization by membranes are considered here: (1) structural (i.e., intramembrane) organization within the membrane composite itself and (2) compartmentation. The former makes possible the construction of multifunctional membrane sandwiches, while the latter provides an additional degree of freedom for reactant/catalyst contact, for transmembrane contact of two process streams, and for product withdrawal.

Multiple Membrane Elements

Multilayer sandwiches can be readily constructed from several films to which different catalysts have been attached or from combinations of permselective and catalytic membranes. Structures with distributed catalyst activities are potentially useful for multistep reaction sequences, whereas two-layer composites of selective and catalytic membranes are valuable for their ability to control the fluxes of reaction participants as shown subsequently in this chapter.

Membranes permit a much higher order and smaller scale of organization than do particles. As an example, laminates of certain types of polymeric membranes can be fabricated from submicron-thick individual films, and in principle one could imagine immobilizing a different catalyst in each. This is without counterpart with pelleted catalysts, where the smallest scale of organization (of the order of several microns to tens of microns) is much larger.

Good control over the distribution of several catalyst activities is obviously important for directing the course of reaction in a complex network to maximize the selectivity or yield of a desired product. Enzymes associated with biological membranes are thought to be highly organized not only normal to the lipid bilayer (e.g., *vectorial* enzymes) but also in the plane of the bilayer in "clusters" of multienzyme sequences (Welch 1977). Furthermore, catalyst incompatibility—either in preparation or operation—may preclude the homogeneous co-immobilization of two or more catalysts on the same support. In such cases, thin membranes in intimate contact could achieve catalyst isolation at minimum cost in terms of support or fluid phase diffusional resistances.

Compartmentation

The fluids on either side of a membrane sandwich can obviously be segregated by appropriate membrane packaging, thus providing the engineer with an additional degree of freedom in reactor design. While there is only a single interface between a catalyst particle and the solution surrounding it, there exist two membrane/solution interfaces that can be exploited in a number of ways.

For example, a membrane reactor might be used to catalyze reaction between two components present in separate streams that cannot be mixed for some reason—perhaps to avoid a subsequent separation problem or a noncatalytic side reaction. Catalytic hydrocarbon oxidations provide a case in point; they are sometimes conducted in two stages (one vessel for hydrocarbon reaction, a second for catalyst reoxidation) both for safety reasons and to permit the use of air rather than oxygen in the process. Conceivably, the same ends could be achieved by simultaneously conducting reactant conversion and catalyst regeneration in a catalytic liquid membrane, with counterdiffusion of the active and reduced catalytic species between opposite faces of the membrane substituting for the pumping of catalyst solution between separate reaction vessels (Ollis, Thompson, and Wolynic 1972; Wolynic and Ollis 1974).

Gryaznov and coworkers (Gryaznov, Smirnov, and Slin'ko 1973; Gryaznov, Smirnov, and Mischenko 1975; Gryaznov et al. 1979; Gryaznov 1983, 1986) have demonstrated the utility of membrane catalysts in conducting reaction between separate process streams. Their work involved coupling hydrogenation and dehydrogenation reactions that took place at opposite surfaces of a palladium alloy membrane catalyst, which was selectively permeable to the atomic hydrogen formed at one surface and consumed at the other. By conjugating the two reactions, the thermodynamically favorable process assists the thermodynamically less favorable process. In addition, isolation of the two hydrocarbon streams in separate compartments formed by the membrane minimizes the task of product recovery. In related work with similar simultaneously permselective and catalytic metal membranes, Japanese investigators have used a sweep gas to remove permeated hydrogen from an *extractive* membrane reactor (Nagamoto and Inoue 1985; Itoh 1987).

Separation

Perhaps the simplest of reactor designs to be referred to in the literature as a membrane reactor involves the use of catalyst-impermeable selective membranes for catalyst confinement. In these systems, a semipermeable membrane (often an ultrafilter) is used to contain a dissolved or dispersed catalyst in a reaction vessel, while product and unconverted reactant are allowed to leave the reactor. The catalytic reaction typically occurs in solution in a continuous stirred tank reactor (CSTR) coupled to a separate membrane unit that retains and recycles the catalyst. Reactant-containing solution is added continually to the vessel, while product, together with excess solvent, is withdrawn in a feed-and-bleed operation.

In CSTR ultrafiltration (UF) enzyme reactors, the contents of the vessel are circulated continuously through the membrane module. Enzyme-free ultrafiltrate is removed from the membrane module, while fresh substrate for the enzymatic conversion is fed to the reactor. The result is continuous formation and withdrawal of the enzymatically transformed product without loss of biocatalyst. A number of nutritionally important amino acids are now being synthesized or optically resolved on an industrial scale in both Japan and Germany using enzymatic reactions conducted in this type of CSTR/UF membrane reactor (Jandel, Hustedt, and Wandrey 1982). More recently, both single-step and coupled multistep (multienzyme) and cofactor-requiring bioconversions have been conducted on a continuous basis using such membrane reactors. The cofactors (e.g., NAD) are provided with macromolecular tails such as poly(ethylene glycol) to facilitate their retention by ultrafiltration membranes (Wichmann et al. 1981; Katayama, Urabe, and Okada 1983).

A few attempts have been made to use relatively "tight" reverse osmosis (RO) membranes to retain soluble catalytic metal complexes in a reactor while permitting organic products to permeate out (Jones 1969; Gosser, Knoth, and Parshali 1973). However, RO membranes used in this way must have outstanding solvent resistance, and such membrane reactors can operate only at relatively low productivities due to the low fluxes characteristic of RO membranes.

The use of permselective membranes in conjunction with catalytic films makes possible a degree of control over the fluxes of the reaction participants—and, hence, over the course of reaction—that is impossible to achieve with catalyst particles. In the applications described below, membrane reactors consisting of two-layer sandwiches of permselective and catalytic membranes are made to separate product from inert and simultaneously enrich the reaction product, using the excess of the chemical free energy of the reactants over that of the products as the required energy source. Other membrane reactors utilize semipermeable films to manipulate an intermediate species in a thermodynamically unfavorable sequential reaction system (Matson 1979).

In addition to mediating the fluxes of reactants and products to and from a membrane-immobilized catalyst, membranes can also serve as contactors and separators of immiscible process streams. In biphasic aqueous/organic systems, for example, catalytic membranes that are microporous and hydrophilic will be preferentially wetted by the aqueous process stream, with the aqueous/organic interface being on the side of the membrane that is in contact with the organic process stream. In this way, two immiscible liquid phases can be contacted across a membrane without one of the phases having to be dispersed in the other, as is required in most conventional multiphase reaction systems. The membrane functions as an interfacial catalyst, a phase contactor, and a phase separator in this type of multiphase membrane reactor. The concept has applications in both biocatalysis and homogeneous catalysis [e.g., in the enzymatic resolution of sparingly water-soluble substrates (Matson 1989a, 1989b; Matson and Lopez 1989; Lopez et al. 1990) and in phase transfer catalysis (Stanley 1986; Stanley and Quinn 1987; Matson and Stanley 1988)].

In summary, the definition of "membrane reactors" that is adopted here is partly structural but primarily functional. A membrane reactor worthy of the name should accomplish tasks that cannot be carried out in conventional reactors, for only then is it likely that membrane systems will compete successfully with more conventional reactor designs. When membrane reactors are designed to take advantage of some of their unique attributes, however, they promise benefits that are not readily realized with conventional reactor geometries:

- Efficient multiphase contacting
- Elimination of diffusional resistance in rapid reactions
- Integration of separation and purification steps with catalytic reaction
- Integration of heat exchange with catalytic reaction
- Displacement of unfavorable equilibria
- Controlled contact of incompatible coreactants
- Elimination of undesirable side reactions
- Control of reaction pathway in complex networks
- Coupling of sequential or parallel multistep reactions
- Amelioration of catalyst poisoning.

We now discuss in some detail two examples of composite membrane reactors that are drawn from our own experience and illustrate nontrivial membrane reactor operation.

PRODUCT SEPARATION AND ENRICHMENT

Concept

The task of separating and concentrating products from reactor effluent mixtures frequently requires equipment that is more expensive than the reactor itself. Furthermore, an external source of energy is generally needed to accomplish these physical processes of separation and enrichment. In contrast, a membrane reactor can carry out catalytic reaction, product separation, and product concentration in an energy-sufficient operation conducted in a single, passive device (Matson 1979; Matson and Quinn 1986).

Figure 43–1 shows a cross section of one such composite membrane reactor capable of product separation and enrichment. It consists of a two-layer sandwich of permselective and catalytic membranes contacted on one side with a feed mixture of reactant and some inert material from which the product must eventually be separated; the other membrane surface contacts a so-called "sweep" stream, the purpose of which is to carry the product out of the reactor. The permselective film permits reactant to diffuse into the catalytic membrane, but it is impermeable to the reaction product and thus prevents the backdiffusion of product into the feed stream. The retention of the product species on the product side of the membrane sandwich is a first step in its separation from the inert component in the feed.

A second feature of this membrane reactor is its ability to enrich the product—that is, to deliver product at a concentration higher than that of the reactant in the feedstream. If this reaction is essentially irreversible, then product will be formed in the catalytic membrane at a molar rate determined by the local feed and sweep stream concentrations, membrane transport properties, and reaction kinetics. Furthermore, if the selective film is essentially impermeable to the product, then this component will diffuse out of the catalytic membrane into the product stream at the same molar rate. Given these assumptions, the concentration of product in the exiting product stream will vary inversely with the volumetric rate of product stream removal. Hence, high degrees of product enrichment can be realized simply by operating the membrane reactor at high feed-to-product stream flow rate ratios [see Figure 43–2(a)]. Typically, the feed stream flow rate will be adjusted to achieve the desired reactant conversion; the product stream flow rate then controls the degree of product enrichment.

Only one aspect of product/inert separation has been considered thus far—namely, retention of product in the product stream; a second condition for this separation is that most of the inert must be retained in the feed stream. Per-

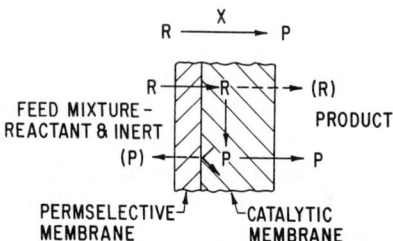

FIGURE 43–1. Product separation and enrichment concept (reprinted from Matson and Quinn 1986 with permission).

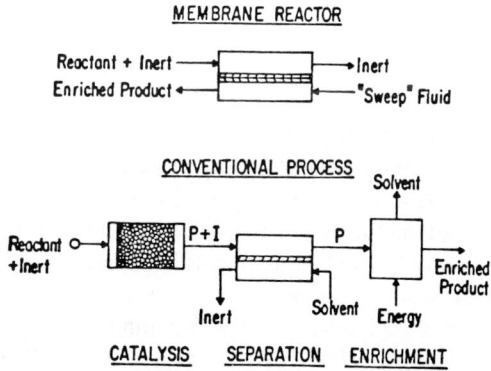

FIGURE 43–2. Comparison of (a) membrane reactor and (b) conventional process (reprinted from Matson and Quinn 1986 with permission).

haps the most obvious means of accomplishing this would be to choose the permselective membrane such that it selectively rejected the inert as well as the product, but this approach would impose an additional constraint on the permselective membrane that would often prove impossible to satisfy. Fortunately, the high feed-to-product stream flow rate ratios required for product enrichment automatically ensure high fractional recoveries of inert in the exiting feed stream. Even in the limiting case where the membrane sandwich is infinitely permeable to the inert (so that the concentrations of inert on both sides of the membrane are the same), the molar ratio of inert recovered in the exiting feed stream to that "lost" with the product stream will be approximately equal to the flow rate ratio (see the following section).

Neither product separation nor enrichment can be accomplished in conventional reactors employing catalyst particles for the fundamental reason that sufficient degrees of freedom are lacking. Continuous separation is impossible because all components exit the reactor via a common stream, and product enrichment is prohibited by the material balance. However, the membrane reactor is a passive device capable of performing these three operations of catalysis, separation, and concentration simultaneously rather than in separate process equipment as indicated in Figure 43–2(b). Moreover, product separation and enrichment are accomplished in the membrane reactor without benefit of the external energy source required for the conventional process.

Experimental Demonstration

Reaction System
The ability of the enzyme α-chymotrypsin to stereospecifically hydrolyze ester derivatives of the L-amino acids while leaving the D-stereoisomers alone is the basis for preparative scale resolution of the aromatic amino acids and more than a hundred related compounds:

D,L-ester + H_2O → L-acid + alcohol + D-ester

One substrate capable of being resolved in this manner by chymotrypsin is N-acetyl-L-tyrosine ethyl ester (ATEE):

$$HO-\langle\bigcirc\rangle-CH_2CHCOOC_2H_5 + H_2O \xrightarrow[pH7.8]{\alpha CHT}$$
$$\qquad\qquad\quad NHCOCH_3$$
D,L - ESTER

$$HO-\langle\bigcirc\rangle-CH_2CHCOO^{\ominus} + H^{\oplus} + C_2H_5OH + D\text{-ESTER}$$
$$\qquad\qquad\quad NHCOCH_3$$
$$\qquad\qquad\underbrace{\qquad\qquad}_{L\text{-ACID}}$$

Recipes for enzymatic resolution based on the esterase function of chymotrypsin have been published for phenylalanine (Clement and Potter 1971) and tryptophan (Huang and Niemann 1951), for precursors of the drug dopa (Matta et al. 1974), and many other compounds (Jones and Beck 1976).

The application of a product separation and enrichment membrane reactor to amino acid resolution is illustrated in Figure 43–3(a). The reactor consists of a two-layer sandwich of an immobilized chymotrypsin film in intimate contact with an immobilized liquid membrane (ILM) that is selectively permeable to electrically neutral species but substantially impermeable to charged species. In operation, the uncharged N-acyl-L-amino acid ester permeates across the selective membrane to the enzyme film, where it undergoes hydrolysis to form the N-acyl-L-amino acid product. In the pH range at which the enzyme operates, the acidic reaction product dissociates and is thus prevented from backdiffusing across the ILM into the feedstream. Thus, the L-acid product leaves the membrane reactor only via the "sweep" stream as desired, and by operating the reactor at high feed-to-product flow rate ratios, most of the inert D-ester will naturally be recovered in the exiting feed stream. Finally, because ester hydrolysis is highly irreversible, the L-acid product can be removed at a higher concentration than that of the L-ester in the racemic feed mixture if the product stream is withdrawn more slowly than the solution of racemic mixture is fed to the device.

Optical resolution results when the reactor of Figure 43–3(b) is operated to give high L-ester conversion (i.e., at a large space time) and high

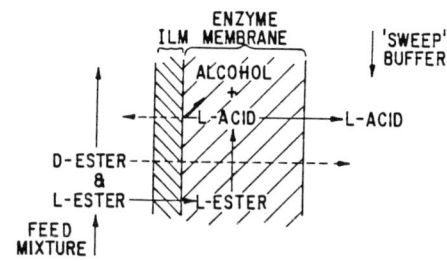

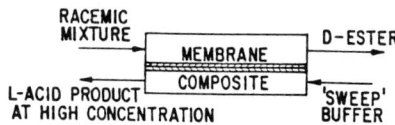

FIGURE 43–3. Product separation and enrichment for amino acid resolution: (a) membrane function and (b) reactor operation (reprinted from Matson and Quinn 1986 with permission).

L-acid enrichment (i.e., at a large flow rate ratio). The former condition results in an exiting feed stream that contains primarily D-ester that has been stripped of most of the L-ester reactant, whereas the latter condition ensures a high ratio of L- to D-material in the product stream.

Catalytic Membrane
Chymotrypsin was covalently bound to Millipore® type MF mixed cellulose ester membrane filters by the cyanogen bromide method. The cyanogen bromide method for covalently binding enzymes to cellulose is a classical enzyme immobilization technique (Axen, Porath, and Ernback 1967). The two-step procedure involves polymer activation followed by enzyme coupling.

A continuous, differential recycle reactor was used for the determination of Michaelis-Menten parameters; the concentration rate profile at pH 7.8 is shown in Figure 43–4. The Michaelis-Menten constant for the immobilized enzyme (0.46 mM) is indistinguishable from that of the dissolved enzyme (0.53 mM), suggesting that chymotrypsin immobilization has little, if any, effect on the strength of L-ATEE binding. Based on published K_I values for the free enzyme of 5.0 mM for D-ATEE and 80 to 110 mM for L-AT, it is likely that membrane reactor performance is influenced little by D-ester or L-acid inhibition.

At substrate concentrations below K_M, the apparent first-order rate constant for the enzyme membrane is

$$k_1 = \frac{V_{\max}}{K_M} = 5.0 \text{ s}^{-1}.$$

The value of K_M is sufficiently large compared to the substrate concentration typically existing in the catalytic membrane during membrane reactor runs that the chymotrypsin film can be considered to operate in the first-order regime. The calculated membrane Thiele modulus of about 20, which corresponds to these kinetics, indicates that the membrane catalytic activity is quite ample, providing a significant buffer against the effects of enzyme deactivation and ensuring that the membrane reactor product stream will remain substantially reactant-free.

The pH/activity profile of the bound enzyme was determined in plug flow integral reactor tests using a low initial substrate concentration (0.1 to 0.2 mM) to ensure first-order kinetics. Bound enzyme activity is at its highest at pH 7.8 ($k_1 = 4.7 \text{ s}^{-1}$) in agreement with the optimum pH of 8.2 determined for this substrate by others (Parks and Plaut 1953) for the enzyme in solution. In subsequent membrane reactor experiments in which the L-acid product was enriched to concentrations as high as 30 mM, the pH of the 0.10 M phosphate buffer sweep stream dropped from 7.8 to 7. While this pH change corresponds to a 30% decrease in catalytic membrane activity, it proved unimportant to membrane reactor performance.

Immobilized Liquid Membrane
Immobilized liquid membranes can be designed to exhibit high permeabilities to the reactant ester while being substantially impermeable to the product acid, the basis for permselectivity being the difference in electrical charge of these

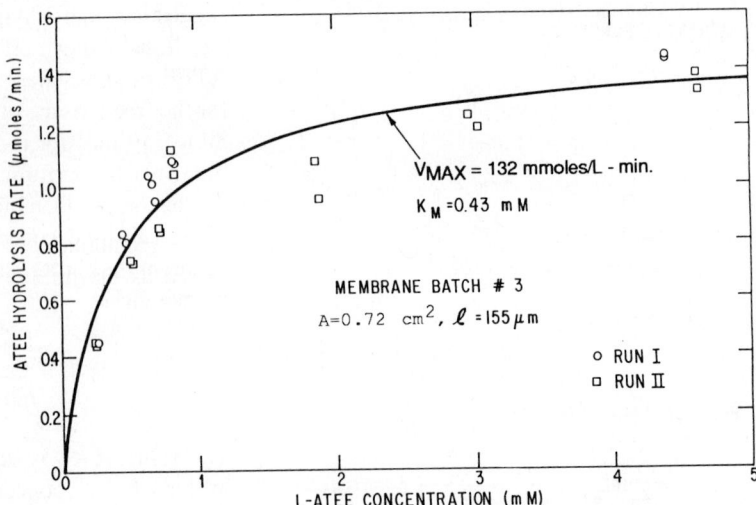

FIGURE 43–4. Kinetics of L-ATEE hydrolysis by membrane-bound chymotrypsin.

species. The immobilized liquid membranes we used consisted of microporous, hydrophobic support membranes impregnated with water-immiscible organic solvents that were confined to the pores of the support film by capillary action. Such membranes combine the desirable permeation properties of the membrane liquid (e.g., high permeant solubility and diffusivity) with the mechanical properties and geometry of the support matrix. Specifically, the ILM solvent required in our work had to dissolve the reactant L-ATEE readily but not the product L-AT. Since organic solvents of low dielectric constant are ineffective in dissolving electrolytes such as the product acid, high reactant solubility is the more demanding of these requirements. Other solvent characteristics pertinent to ILM design include solubility in water, viscosity, and interfacial tension with water.

Batch extractions were conducted with a few of the better ATEE solvents—primarily ketones, esters, and alcohols—in order to determine L-ATEE partition ratios or distribution coefficients. Distribution coefficient measurements with 1-decanol were carried out at a fixed initial concentration of ester in the feed (1.11 mM) by varying the extract-to-raffinate phase ratio (see Figure 43–5). The good solubility of ATEE in decanol, combined with its low volatility and solubility in water (0.004 wt.%), made decanol the solvent of choice for the ILM in this application.

The ATEE:AT separation factor of the ILM can be estimated from solvent and permeant properties using a simple solution-diffusion model of permeation. The critical assumption in this model is that only the ester and the undissociated acid are capable of permeating across the ILM; contributions to the total acid flux due to ions (e.g., acetyl tyrosinate) or to ion pairs are neglected. On this basis, the ATEE:AT separation factor is given as the product of (1) the solubility ratio of the ester and the *un*dissociated acid

$$K_{ATEE}/K_{AT} = 3.2/0.59 = 5.4$$

multiplied by (2) the fraction $(1 - \alpha)$ of the acid present in the undissociated form at the prevailing pH:

$$(1 - \alpha) = \frac{[HA]}{[HA] + [A^-]} = \left(\frac{\gamma_{A^-}}{\gamma_{HA}}\right) 10^{(pK_a - pH)}.$$

Based on the acid's pK_a of 3.4 and the membrane reactor operating pH of 7.5, the predicted ATEE:AT separation factor of the ILM is on the order of 90,000—far in excess of that required for successful reactor operation.

A number of hydrophobic, microporous

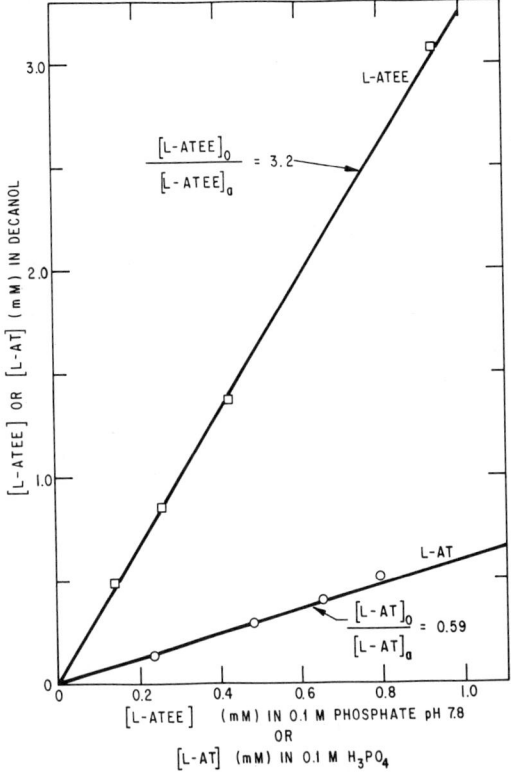

FIGURE 43–5. ATEE and AT partition coefficients.

membranes were evaluated as ILM supports on the basis of their porosity, thickness, tortuosity, bubble point, handleability, and chemical stability. Two types of solvent-resistant membranes were tested in permeation experiments: Celgard® polypropylene films and Gore-Tex® polytetrafluoroethylene membranes. Their physical characteristics are summarized in Table 43–1. ILM permeability varies directly with the porosity and inversely with the thickness of the support, and the Gore-Tex® S10187 and Celgard® 2500 films proved superior in these respects. The Gore-Tex®-supported ILM was found to be more permeable to L-ATEE and was used exclusively in subsequent membrane reactor tests.

The membrane bubble point pressure p_b is defined as the minimum transmembrane pressure difference required to force the ILM solvent from the porous support matrix:

$$-p_b = \frac{4K\gamma \cos\theta}{d_p},$$

where θ is the contact angle, d_p is the effective pore diameter, γ is the interfacial tension, and K is a shape correction factor. Pressure differences can arise in membrane reactor operation due to the resistance afforded by narrow flow passages, and these pressures must not exceed the membrane bubble point. The bubble point of the decanol/Gore-Tex® S10187 ILM was estimated at 5.7 psi.

ILM permeability was measured in stirred, two-compartment Plexiglas® diffusion cells using multiple ILM layers. If P is the permeability of a single ILM to ATEE and k_b is the mass transfer coefficient of the feed or permeate solution boundary layers (assumed identical), then the measured overall permeability P_o is given by

$$\frac{1}{P_o} = \frac{1}{P}(n) + \frac{2}{k_b}.$$

TABLE 43–1. ILM Support Characteristics.

Membrane	Material	Nominal Pore Size (μm)	Porosity (%)	Thickness (μm)
Celgard® 2400	Polypropylene	0.02	38	25
Celgard® 2500	Polypropylene	0.04	45	25
Gore-Tex® S10070	PTFE	0.02–0.03	45–50	70
Gore-Tex® S10187	PTFE	1.0	85–90	42
Gore-Tex® S10187 on Delnet mesh	PTFE	1.0	85–90	95

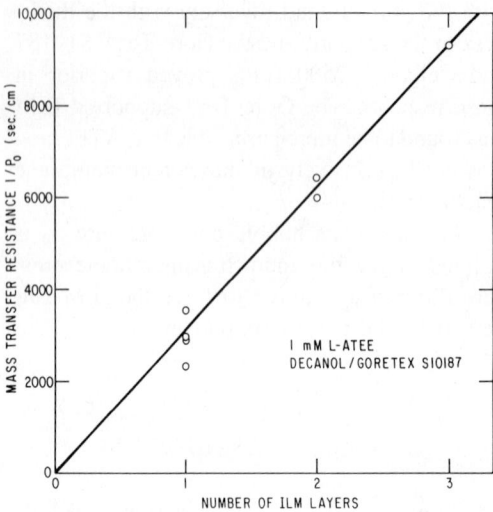

FIGURE 43–6. Permeability of multilayer ILM sandwiches.

The intercept of Figure 43–6 is thus a measure of the solution boundary layer resistance, which is clearly negligible at the conditions of these measurements. The slope yields a single-layer ILM permeability to ATEE of 3.2×10^{-4} cm/s.

An independent determination of the permeability of this ILM is shown in Figure 43–7 for a double thickness of liquid membrane. The single-layer ATEE permeability calculated from the slope of these data is 3.3×10^{-4} cm/s, in excellent agreement with the value from Figure 43–6. These measured permeabilities are consistent with a solution-diffusion model of ILM transport that predicts

$$P = KD\epsilon/l\tau$$

if the Gore-Tex® membrane tortuosity is assumed to be 1.7. The permeabilities of the particular immobilized liquid membranes used to construct composite membrane reactors were not determined experimentally; rather, they were estimated from values measured for other ILMs and corrected using the above equation for variations in support membrane thickness.

L-AT permeability and ILM stability under conditions representative of membrane reactor operation were evaluated in a continuous 360-hour permeation test conducted in a countercurrent flow test cell (see Figure 43–8). Membrane performance was reasonably stable during more than 2 weeks of operation. Small transmembrane pressure differences were applied in the feed-to-sweep direction in order to deduce whether any significant leakage of L-AT was taking place through membrane defects that might have been present. The absence of such holes in the ILM can be inferred from the independence of L-AT flux on pressure difference. Moreover, the average L-AT permeability of 2.1×10^{-8} cm/s is so low as to essentially preclude the existence of significant membrane defects. The calculated ATEE:AT separation factor for the ILM of Figure 43–8 is about 15,000 based on the ATEE permeability of 3.2×10^{-4} cm/s measured in stirred cell tests. This value is lower than but of the same order of magnitude as the value of 90,000 estimated from the simple solution-diffusion model discussed above.

In summary, the properties of the decanol/Gore-Tex® immobilized liquid membrane proved quite satisfactory for membrane reaction purposes in each of three important respects: (1) reactant (ATEE) permeability, (2) reactant:

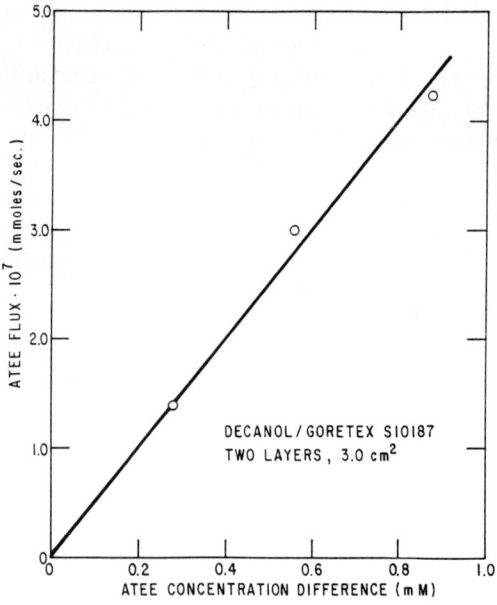

FIGURE 43–7. ATEE flux versus concentration difference.

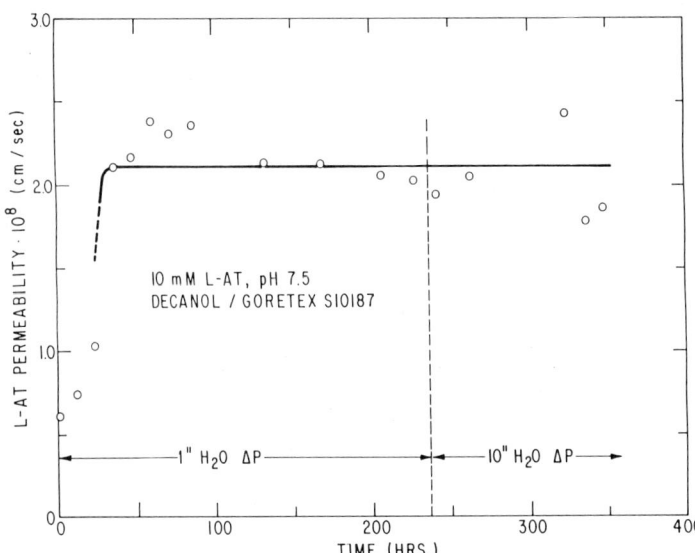

FIGURE 43–8. ILM stability and permeability.

product (ATEE:AT) selectivity, and (3) stability at operating conditions. The ATEE:AT separation factor exceeds by orders of magnitude what is required for high product enrichment, and ILMs subsequently proved stable in membrane reactor tests lasting as long as 6 weeks.

Membrane Reactor Performance

Membrane reactors of from 74 to 145 cm² active area were tested in the countercurrent flow configuration shown in Figure 43–9. Table 43–2 summarizes membrane reactor operating conditions and performance determined for the 1.00 mM L-ATEE feed. Reactant removal X_R and conversion X_C were first determined as a function of the dimensionless membrane reactor space time Φ, which is defined as the intrinsic ILM permeability P to L-ATEE (2.9×10^{-4} cm/s) times the membrane area A divided by the feed flow rate V_F. Subsequently, the feed-to-product flow rate ratio θ (V_F/V_P) was varied at fixed space time and conversion.

The fraction of reactant remaining in the feed stream $(1 - X_R)$ is plotted in Figure 43–10 as a function of membrane reactor space time. The data clustered about a space time of 4 were taken at flow ratios θ ranging from 0.9 to 40.5, demonstrating that the latter parameter had little effect on reactant removal, as is to be expected given the very large Thiele modulus of the catalytic membrane. Practically no L-ATEE survived diffusion across the chymotrypsin film. As a result, reactant conversion and removal were nearly identical, and the fractional loss of reactant to the product stream seldom exceeded about 2%.

The observed rate of reactant removal was

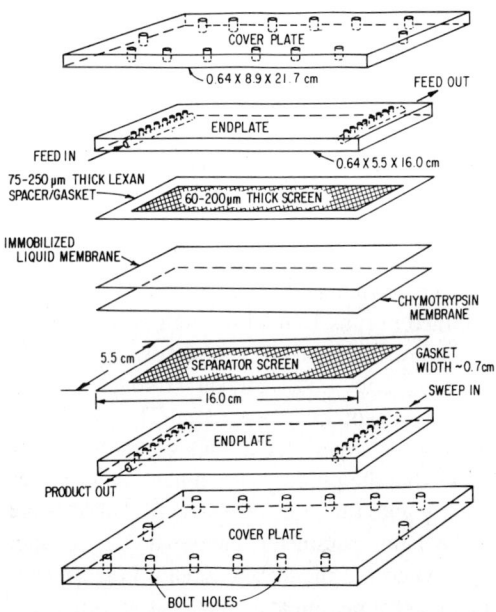

FIGURE 43–9. Flat-sheet membrane reactor test cell.

TABLE 43-2. Membrane Reactor Performance—L-ATEE Feed.

Run	Space Time Φ	Flow Rate Ratio θ	Reactant Removal (%)	Reactant Conversion (%)	Product Enrichment Factor	Product Retention (%)
1	4.24	1.88	86	85	1.49	98
			86	84	1.49	98
2	2.09	1.91	66	66	1.14	98
			68	68	1.11	98
3	6.46	2.47	93	90	1.12	97
			92	88	1.92	98
4	3.84	1.04	86	85	0.72	97
			85	85	0.74	97
5	4.05	3.94	84	82	2.75	97
			84	81	2.77	97
6	4.19	19.6	82	81	13.1	95
			83	81	13.2	95
			82	81	13.2	96
7	3.92	10.3	82	80	7.64	98
			82	80	7.75	98
8	4.10	40.5	80	80	29.9	98
			80	80	30.4	98
			81	81	30.6	98
9	4.51	0.898	86	86	0.77	100
			86	86	0.76	100

lower than predicted by an analytical model (Matson 1979) using ILM and chymotrypsin membrane parameters determined in separate experiments described above. A best fit to the data (line A) was obtained for an apparent overall ATEE mass transfer coefficient of 1.4×10^{-4} cm/s, significantly lower than the intrinsic ILM permeability P of 2.9×10^{-4} cm/s. This implies the existence of an extra-membrane transport resistance comparable to that of the ILM itself, presumably attributable to the feed stream boundary layer and/or channeling.

The effect of feed-to-product flow ratio θ on product enrichment and recovery was determined in runs 1 and 4 through 9. Since these data were obtained over a narrow space time range (3.8 to 4.5), reactant removal X_R was fairly constant at about 83%. The solid line through the experimental data of Figure 43–11 shows that the product enrichment factor varies linearly with flow rate ratio over a 45-fold range in the latter parameter. In run 8, the product acid exited the membrane reactor at a concentration 30 times greater than that at which the reactant ester was fed to the device (see Figure 43–12).

Observed product enrichment factors were in good agreement with the limiting enrichment $X_R \theta$ that corresponds to the ideal case of no reactant loss to the product stream (i.e., $X_R = X_C$) and complete retention of product by the permselective membrane. As shown in Table 43–2, the fractional recovery of L-AT in the product stream was high (>95%) at all membrane reactor operating conditions investigated.

The ability of the membrane reactor to separate optical isomers from a racemic feed mixture of 2.14 mM DL-ATEE in 0.10 M phosphate pH 7.8 was examined in a series of runs wherein the reactor space time Φ was fixed to provide high conversion of L-ATEE (93 to 94%) and the feed-to-product flow rate ratio θ was varied from 10 to 24. The optical purity of the exiting feed stream improves with increasing L-ester conversion (i.e., at large space times), while the optical purity of the product stream increases with the degree of L-acid enrichment and, thus, with the flow rate ratio.

The enantiomeric excess (ee) of the exiting feed stream, a measure of the excess of the D-isomer over the L-isomers (primarily unconverted L-ATEE) is given by

Membrane Reactors 823

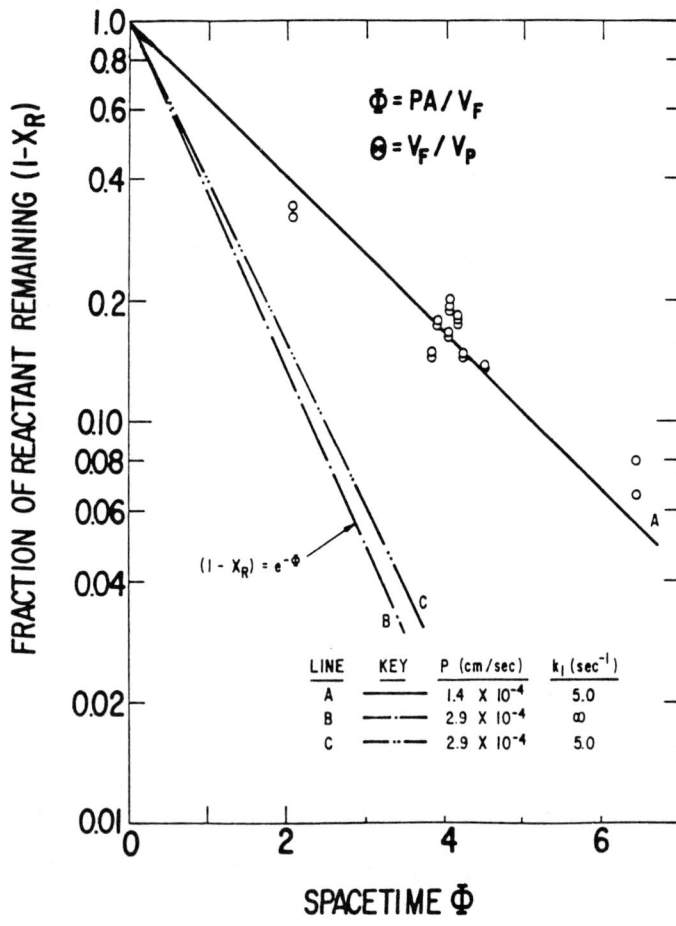

FIGURE 43–10. Effect of membrane reactor space time on degree of reactant removal.

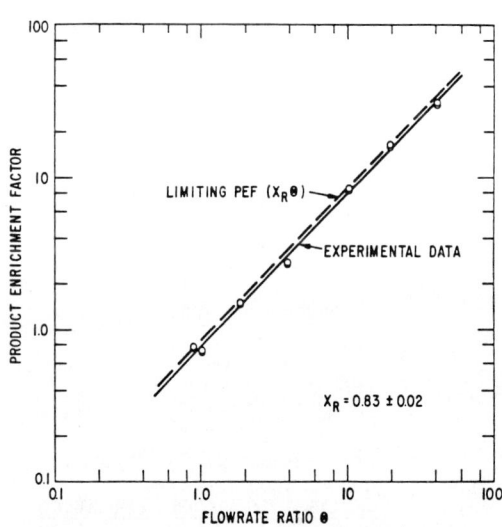

FIGURE 43–11. Effect of feed-to-product flow rate ratio on product enrichment (reprinted from Matson and Quinn 1986 with permission).

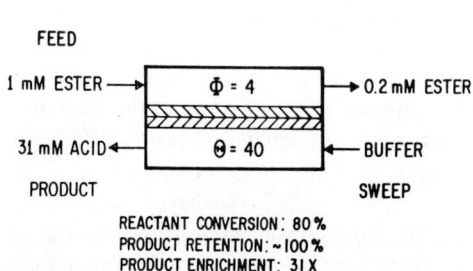

FIGURE 43–12. Product separation and enrichment (reprinted from Matson and Quinn 1986 with permission).

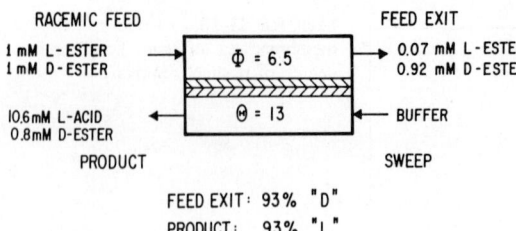

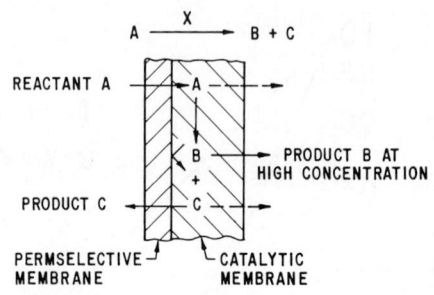

FIGURE 43-13. Amino acid resolution (reprinted from Matson and Quinn 1986 with permission).

FIGURE 43-14. Separation of products by selective enrichment (reprinted from Matson and Quinn 1986 with permission).

$$\text{ee} = \frac{X_D - X_L}{X_D + X_L} = 2X_D - 1,$$

where X_D and X_L are the concentrations of the respective enantiomers. (The enantiomeric purity of the product stream on the opposite side of the membrane sandwich is calculated by reversing X_L and X_D above.) At the reactor operating conditions of Figure 43-13, both exiting streams contained predominantly one or the other isomer at optical purities of about 86%. In other runs conducted at still higher flow rate ratios (Matson 1979), product ee values as high as 95% were achieved.

These results demonstrate the ability of the product separation and enrichment membrane reactor to accomplish optical resolution at high efficiency. Still higher feed and product stream optical purities can be attained by increasing the membrane reactor space time and flow rate ratio, respectively.

SEPARATION OF MULTIPLE PRODUCTS

Concept

Figure 43-14 shows a composite membrane reactor capable of separating two reaction products by selectively enriching one of them. The reactor consists of a two-layer sandwich of catalytic and permselective films, where the latter is chosen to be highly impermeable to the reactant (A) and to one of the reaction products (C) while being highly impermeable to another

reaction product (B). The feed stream flows past the permselective membrane and exits as the raffinate [Figure 43-15(a)]; the opposite face of the membrane sandwich is flushed with a sweep that carries off one of the two reaction products. As shown in Figure 43-15(a), the sweep stream may be supplied by recycling a portion of the raffinate flow.

In operation, reactant A diffuses across the permselective membrane from the feed stream to the catalytic film, where it undergoes reaction to products B and C. The permselective membrane permits product C to permeate across it into the feed stream, but it acts as a barrier to the other product B. As a result, product B is retained in the product stream. If the membrane permeability to C is high, and if the flow rate of the feed stream is large com-

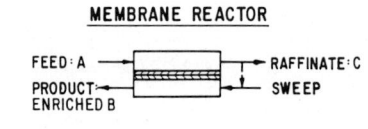

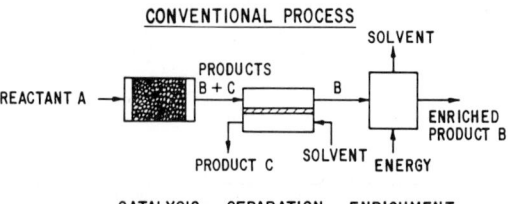

FIGURE 43-15. (a) Membrane reactor and (b) conventional process separating multiple products.

pared to that of the product stream, then the material balance dictates that most of product C will exit the reactor in the raffinate stream, and separation of product B from product C is thereby realized.

Like its predecessor discussed in the previous section, this membrane reactor is also capable of delivering product B at a concentration higher than that of reactant A in the feed stream. Separation (and enrichment) of multiple-reaction products cannot be achieved in conventional reactors because sufficient degrees of freedom are lacking [i.e., both products leave a conventional reactor in a single stream as shown in Figure 43–15(b)]. As discussed above, separation processes invariably require energy, which usually must be supplied by some external source. The membrane reactor avoids this need, however, by coupling the chemical conversion to the separation and concentration processes.

Experimental Demonstration

Demonstration of this multiple products separation concept is convenient using the same reaction system as described above—namely, the chymotrypsin-catalyzed hydrolysis of N-acetyl-L-tyrosine ethyl ester to the corresponding acid and alcohol:

$$HO\text{-}\phi\text{-}CH_2CHCOOC_2H_5 + H_2O \xrightarrow[pH7.8]{\alpha ChT}$$
$$\quad\quad\quad NHCOCH_3$$
$$\quad\quad\quad A$$

$$HO\text{-}\phi\text{-}CH_2CHCOO^{\ominus} + H^{\oplus} + C_2H_5OH$$
$$\quad\quad\quad NHCOCH_3$$
$$\quad\quad\quad B \quad\quad\quad\quad C$$

Figure 43–16 shows an immobilized liquid membrane used to retain the dissociated acid (product B) in the product stream. By withdrawing the product stream slowly, the acid can be concentrated well beyond the initial concentration of reactant ester in the feed. However, since ethanol (product C) can readily permeate across the ILM into the feed stream, most of this product exits in the raffinate stream. In this fashion, alcohol is separated from enriched acid.

The ILM used in our experiments again consisted of decanol, immobilized by capillarity in a microporous PTFE membrane (Gore-Tex® S10187 film). The permeability of the ILM to ethanol (4.0×10^{-5} cm/s) was estimated from ethanol's solubility in decanol, which was determined in batch extraction experiments. Note that ethanol poses no threat to decanol ILM stability as a result of the low ethanol concentrations encountered in our tests (<2 mM) and ethanol's small decanol:water partition coefficient (0.42). The catalytic film was prepared by covalently binding the chymotrypsin enzyme to Millipore® filter membranes using the cyanogen bromide procedure described earlier.

The effect of flow rate ratio on effluent concentrations from a countercurrent products separation reactor operated at high conversion is shown in Figure 43–17. The acid concentration in the product stream increases linearly with flow rate ratio. The ethanol concentrations in the raffinate and product streams also increase with flow rate ratio; however, at high θ values,

• MEMBRANE FUNCTION

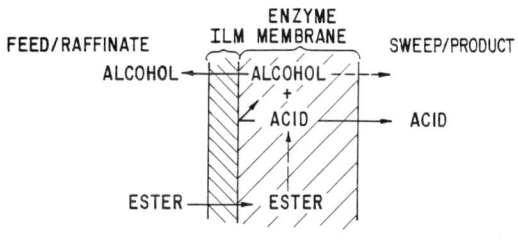

• REACTOR OPERATION

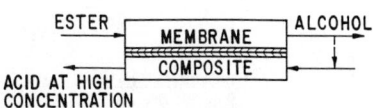

FIGURE 43–16. Separation of products by selective enrichment: ester hydrolysis.

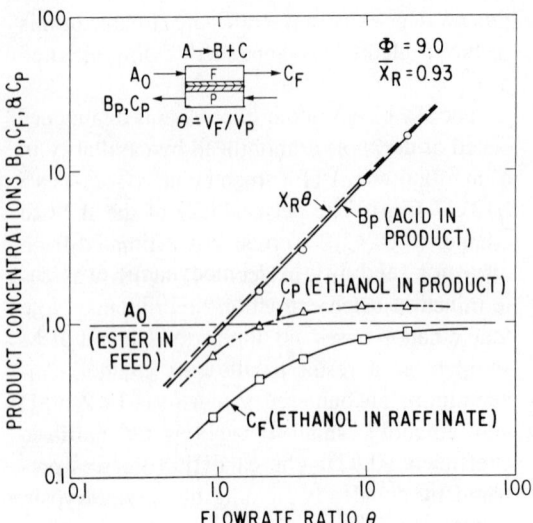

FIGURE 43–17. Effect of feed-to-product flow rate ratio on product exit concentrations.

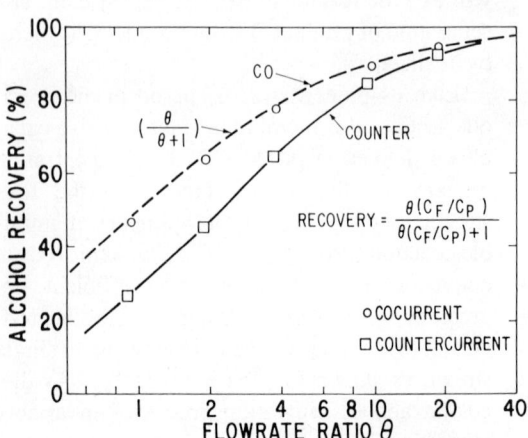

FIGURE 43–19. Comparison of cocurrent versus countercurrent operation: alcohol recovery.

limiting concentrations are approached that are determined by membrane characteristics and the material balance.

Operation at high flow rate ratio also favors high product purities and recoveries as shown in Figure 43–18. Acid recovery in the product stream drops slightly as θ increases due to increased transmembrane leakage of acid, but recovery of this product is still quite acceptable even at the highest flow rate ratio examined.

While countercurrent reactor operation is superior in the case of the product/inert separation and enrichment membrane reactor considered in the previous section, cocurrent operation affords better recovery of the product ethanol in the present case where multiple reaction products are being separated (see Figure 43–19). Figures 43–20 and 43–21 explore why recoveries are better in the cocurrent situation. Simply put, higher recovery can be attributed to the

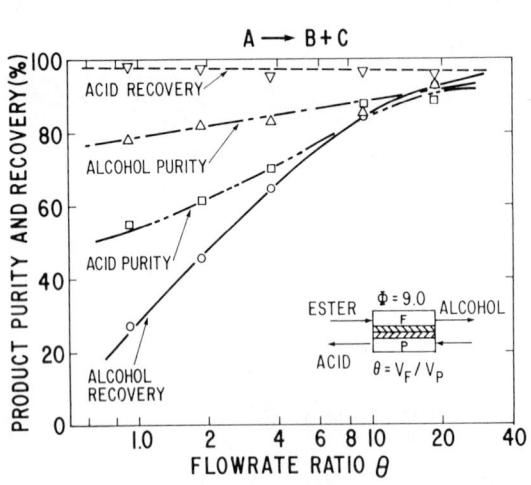

FIGURE 43–18. Product purities and recoveries from a countercurrent membrane reactor.

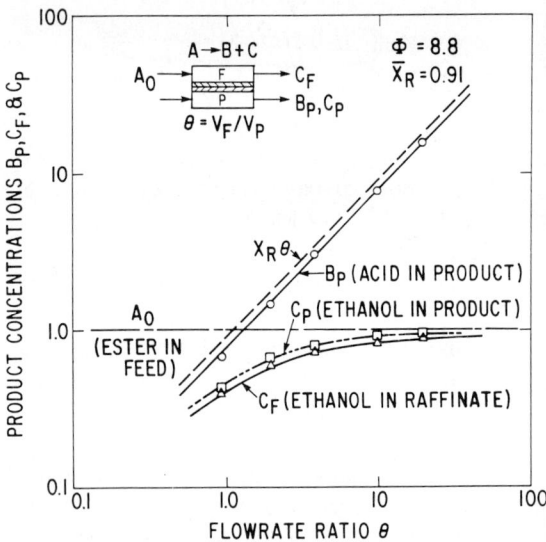

FIGURE 43–20. Effect of feed-to-product flow rate ratio on product concentration (cocurrent operation).

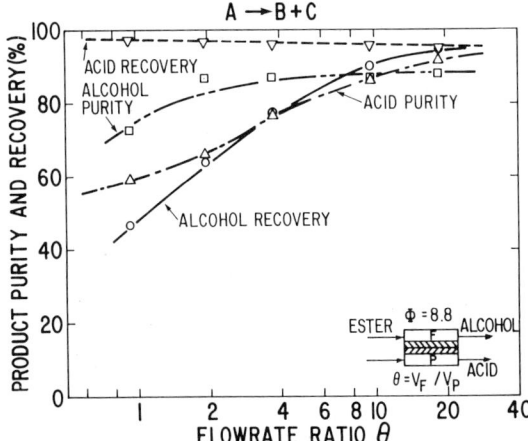

FIGURE 43–21. Product purities and recoveries from a cocurrent membrane reactor.

greater membrane area that is available in cocurrent flow for ethanol to backdiffuse into the feed/raffinate stream from the product side of the membrane sandwich where it is produced. This closer approach to attainment of diffusional equilibrium across the membrane is reflected in the smaller difference between product and raffinate stream ethanol concentrations in the cocurrent case of Figure 43–20 compared to the countercurrent case of Figure 43–17.

THERMODYNAMIC CONSIDERATIONS

Membrane reactors can be compact, passive, and capable of conducting several chemical processing operations at once. However, one of their more intriguing attributes is undoubtedly their ability to "couple" the energetics of the reaction and purification processes in such a way that improved second-law efficiency is realized. Because the membrane reactor conducts a reaction less irreversibly than its conventional counterpart, the membrane reactor produces less entropy and more useful work—namely, work that manifests itself in the processes of product separation and enrichment. In contrast, external work must be performed downstream of a conventional reactor in order to obtain products of the same purity as those that issue directly from a membrane reactor (see Figure 43–22). Interestingly, this higher second-law efficiency stems from the operation of a permselective membrane that itself represents a source of entropy production and irreversibility that is not present in a conventional reactor. The merit of this permselective film is in its reduction of the greater irreversibility associated with the chemical reaction, and overall thermodynamic efficiency is thereby improved (Matson 1979).

Figure 43–22 shows yet another example of a type of reaction/separation "coupling" that can be brought about with the aid of semipermeable films—namely, the extractive membrane reactor. Here, the sense of reaction/separation coupling is just the opposite of that which operates in the composite membrane reactors described earlier. Separation work is *performed on* the extractive reaction system in order to "pull" a thermodynamically reluctant reaction to completion by selective removal of a high-energy reaction product. In the multilayer membrane reactors featured in this chapter, separation work may be viewed as being *performed by* a thermodynamically favorable reaction.

A very basic availability analysis can be performed to indicate the improvement in second-law efficiency that can result from membrane reactor operation. In particular, we have estimated (albeit very roughly!) the decrease in availability of the membrane reactor process corresponding to the experimental conditions shown in Figure 43–23. The supporting free-

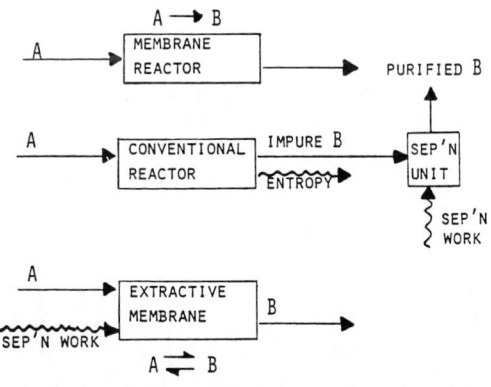

FIGURE 43–22. Thermodynamic considerations in membrane reactors.

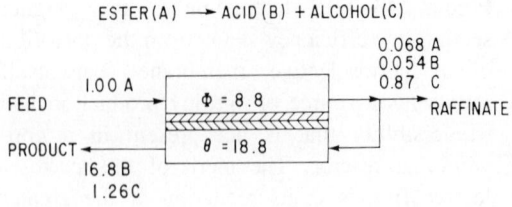

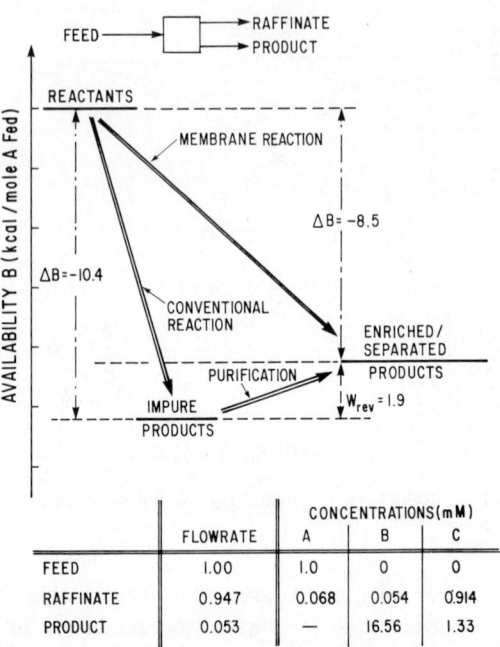

	RECOVERY	PURITY	RELATIVE ENRICHMENT
ACID IN PRODUCT	0.94	0.93	18.0
ALCOHOL IN RAFFINATE	0.92	0.88	0.93

FIGURE 43–23. Products separation by selective enrichment: countercurrent operation with recycle (reprinted from Matson and Quinn 1986 with permission).

	FLOWRATE	CONCENTRATIONS (mM)		
		A	B	C
FEED	1.00	1.0	0	0
RAFFINATE	0.947	0.068	0.054	0.914
PRODUCT	0.053	—	16.56	1.33

FIGURE 43–24. Availability analysis: membrane versus conventional reactor.

energy calculations for the ATEE hydrolysis are summarized in Table 43–3, and ideal solution behavior has been invoked in computing the free energies of mixing. While this latter assumption is undoubtedly poor given the low water solubilities of these amino acid derivatives, the procedure seems justifiable within the context of an illustrative calculation.

As shown in Figure 43–24, the availability decrease associated with membrane reactor operation (–8.5 kcal/mol of reactant fed) compares favorably with the availability loss suffered in a conventional reactor, producing a relatively dilute product mixture (–10.4 kcal/mol). Put another way, reversible purification work in the amount of 2 kcal/mol would be required downstream of a conventional reactor in order to bring the products to membrane reactor purity. This is perhaps a modest savings in terms of reversible work, but it should be kept in mind that real separation and concentration processes are often characterized by relatively poor thermodynamic efficiencies that would amplify the energy savings brought about by the membrane reactor.

TABLE 43–3. Thermodynamics of ATEE Hydrolysis.

$$\text{AcET} + H_2O \underset{(A)}{\overset{1}{\Longleftrightarrow}} \text{AcH} + \text{ETOH} \overset{2}{\Longleftrightarrow} \text{Ac}^- + H^+ + \text{ETOH}$$
$$(B) + (C)$$

Reaction	$\Delta G°$ (kcal/mol)	pK
Hydrolysis (1)	–0.8	0.6
Ionization (2) at pH 7.8	–6.0	4.4 (pH – pK)
Overall (pH 7.8)	–6.8	5.0
$A \Longleftrightarrow B + C$; $K = c_B c_C / c_A = 10^5$		

CONCLUSIONS

Membrane reactors provide a promising technology for a variety of chemical conversion/separation processes. This results from the facts that catalysts immobilized in membranes are fundamentally different from catalysts immobilized in conventional particulate media and that membranes are inherently more versatile and permit control of the fluxes of reaction participants. To date, membrane reactor performance has been demonstrated with reactors in the form of multilayer composite structures and as multiphase, single-membrane contactors.

It is fair to say that this first decade of membrane reactor engineering has but sampled what is proving to be a rich lode of innovative and powerful process technology. In sum, the advent of synthetic membranes into reactor engineering has introduced a new and important dimension into process design. We will undoubtedly see this technology used in an ever wider range of applications in the not-too-distant future.

ACKNOWLEDGMENT

We acknowledge the National Science Foundation (ENG78-13281) for its support of much of the authors' own work described above.

NOTATION

General Notation

See the General Notation section at the beginning of this handbook.

Special Notation

A membrane area, L^2
B availability function, ML^2/t^2 or E/mol
d_p membrane pore diameter, L
ee   enantiomeric excess, dimensionless
k_1 apparent first-order rate constant, $1/t$
k_b boundary layer mass transfer coefficient, L/t
K shape correction factor in bubble point equation, dimensionless, or distribution coefficient, dimensionless
K_a acid dissociation equilibrium constant, mol/L^3
K_I Michaelis constant for inhibitor, mol/L^3
K_M Michaelis constant for substrate, mol/L^3
p_b membrane bubble point pressure, M/Lt^2 or p
P permeability of single ILM, L/t
P_o overall permeability, L/t
V_F feed stream flow rate, L^3/t
V_{max} maximal enzymatic reaction rate, mol/$L^3 t$
V_P product stream flow rate, L^3/t
X_C fractional reactant conversion, dimensionless
X_D mole fraction of D-isomer, dimensionless
X_L mole fraction of L-isomer, dimensionless
X_R fractional removal of reactant from feed stream, dimensionless

Greek Letters

α fraction of acid present in dissociated form, dimensionless
γ species activity coefficient, dimensionless, or interfacial tension, M/t^2
ϵ porosity, dimensionless
θ contact angle, radians, or feed-to-product stream flow rate ratio, dimensionless
τ tortuosity, dimensionless
Φ membrane reactor spacetime $P \times A/V_F$, dimensionless

REFERENCES

Aldrich Chemical Company. 1986. Nafion resins: versatile heterogeneous catalysts. *Aldrichimica Acta* 19(3):76.

Axen, R., J. Porath, and S. Ernback. 1967. Chemical coupling of peptides and proteins to polysaccharides by means of cyanogen halides. *Nature* 214:1302.

Belfort, G. 1989. Membranes and bioreactors: a technical challenge in biotechnology. *Biotech. Bioeng.* 33:1047–1066.

Berke, W., H. J. Schuez, C. Wandrey, M. Morr, G. Denda, and M. R. Kula. 1988. Continuous regeneration of ATP in enzyme membrane reactor for enzymic syntheses. *Biotech. Bioeng.* 32(2):130–139.

Bernstein, P., J. P. Coffey, A. E. Varker, J. T. Arms, W. D. K. Clark, and P. D. Goodell. 1982. Catalyst sheet and preparation: fibrillatable polymer, supporting polymer, pore-former. U.S. Patent 4,332,698.

Breslau, B. R. 1981. Catalytic processes utilizing hollow fiber membranes. U.S. Patent 4,266,026.

Cares, W. R. 1977. Tert-butanol preparation in presence of perfluorosulfonic acid catalyst membrane. U.S. Patent 4,065,512.

Chen, S., and Y. K. Kao. 1990. Direct oxidation of ethylene to acetaldehyde in a hollow fiber membrane reactor. *Chem. Eng. Commun.* 88:31–47.

Cheryan, M., and M. Mehaia. 1986. Membrane bioreactors. In *Membrane Separations in Biotechnology*, ed. W. C. McGregor, pp. 255–295. New York: Marcel Dekker.

Clement, G. E., and R. Potter. 1971. Enzymic resolutions: organic-biochemical laboratory experiment. *J. Chem. Ed.* 48:695.

Di Cosimo, R., J. D. Burrington, and R. K. Grasselli. 1986a. Effecting oxidative dehydrodimerization. U.S. Patent 4,571,443.

Di Cosimo, R., J. D. Burrington, and R. K. Grasselli. 1986b. Oxidative dehydrodimerization of propylene over a bismuth oxide-lanthanum sesquioxide oxide ion-conductive catalyst. *J. Catal.* 102:234.

Drioli, E., G. Iorio, M. Derosa, H. Gambacorta, and B. Nicolaus. 1982. High-temperature immobilized-cell ultrafiltration reactors. *J. Membr. Sci.* 11:365.

Efthymiou, G. S., and M. C. Shuler. 1989. Apparatus and process to eliminate diffusional limitations in a membrane reactor by pressure cycling. U.S. Patent 4,861,483.

Flaschel, E., C. Wandrey, and M. R. Kula. 1983. Ultrafiltration for the separation of biocatalysts. *Adv. Biochem. Eng.* 26:73.

Goldberg, B. S., and R. Y. Chen. 1987. Reactor with immobilized protein on substrate wound as spiral. U.S. Patent 4,689,302.

Gosser, L. W., W. H. Knoth, and G. W. Parshall. 1973. Reverse osmosis in organometallic synthesis. *J. Am. Chem. Soc.* 95:3436.

Gregor, H., and P. W. Rauf. 1975. Enzyme-coupled ultrafiltration membranes. *Biotech. Bioeng.* 17:445.

Grubbs, R. H. 1977. Hybrid-phase catalysts. *Chemtech.* 7:512.

Gryaznov, V. M. 1983. Palladium and its alloys as membranous catalysts. *Kinet. Catal. (USSR)* 82:1151 (translated from *Kinet. Katal.* 1982).

Gryaznov, V. M. 1986. Hydrogen permeable palladium membrane catalysts: an aid to the efficient production of ultra pure chemicals and pharmaceuticals. *Platinum Metals Rev.* 30:68.

Gryaznov, V. M., V. S. Smirnov, and A. P. Mischenko. 1975. Membrane catalysts for carrying out simultaneous processes involving evolution and consumption of hydrogen: hydrogenation catalysts, Pd alloys. U.S. Patent 3,876,555.

Gryaznov, V. M., V. S. Smirnov, and M. G. Slin'ko. 1973. Heterogeneous catalysis with reagent transfer through selectivity permeable catalysts. *Proc. 5th Int. Congr. Catal.* 2:1139. Amsterdam: North Holland Publishing Co.

Gryaznov, V. M., V. S. Smirnov, V. M. Vdovin, B. B. Ermilova, L. D. Gogua, N. A. Pritula, and L. A. Litvinov. 1979. Method of preparing a hydrogen-permeable membrane catalyst on a base of palladium or its alloys for the hydrogenation of unsaturated organic compounds. U.S. Patent 4,132,668.

Haag, W. O., and D. D. Whitehurst. 1978. Insoluble resin-metal compound complex prepared by contacting weak base ion exchange resin with solution of metal-ligand. U.S. Patent 4,111,856.

Helfferich, F. 1962. *Ion Exchange.* New York: McGraw-Hill Book Co.

Hsieh, H. P. 1989. Inorganic membrane reactors—a review. *AIChE Symp. Ser.* 85(268):53–74.

Huang, H. T., and C. Niemann. 1951. The kinetics of the α-chymotrypsin-catalyzed hydrolysis of acetyl- and nicotinyl-L-tryptophanomide in aqueous solutions at 25°C and pH 7.9. *J. Am. Chem. Soc.* 73:1541–1548.

Ito, N., Y. Shindo, K. Haraya, and T. Hakuta. 1988. A membrane reactor using microporous glass for shifting equilibrium of cyclohexane dehydrogenation. *J. Chem. Eng. Japan* 21(4):399–404.

Itoh, N. 1987. A membrane reactor using palladium. *AIChE. J.* 33(9):1576–1578.

Jandel, A.-S., H. Hustedt, and C. Wandrey. 1982. Continuous production of L-alanine from fumarate in a two-stage membrane reactor. *Eur. J. Appe. Microbiol.* 15:59.

Jones, F. N. 1969. Hydrogenation of dicyanobutene with selected rhodium (I) catalysts and a basic promoter. U.S. Patent 3,459,785.

Jones, J. B., and J. F. Beck. 1976. Asymmetric syntheses and resolutions using enzymes. In *Applications of Biochemical Systems in Organic Chemistry: Part I. Techniques of Chemistry X*, ed. J. B. Jones, C. J. Sin, and D. Perlman, pp. 112–401. New York: Wiley-Interscience.

Katayama, N., I. Urabe, and H. Okada. 1983.

Steady-state kinetics of coupled two-enzyme reactor with recycling of poly(ethylene glycol)-bound NAD. *Eur. J. Biochem.* 132:403.

Kise, S., and M. Hayashida. 1990. Two phase system membrane reactor with cofactor recycling. *J. Biotech.* 14(2):221–228.

Knazek, R. A., P. M. Guillino, R. L. Dedrick, and W. R. Kidwell. 1975. Cell culture on semipermeable tubular membranes. U.S. Patent 3,883,393.

Kozarek, R. 1975. The kinetics of reactions catalyzed by enzymes bound to membranes. Ph.D. thesis. Carnegie-Mellon University, Pittsburgh, PA.

Lee, C. K., and J. Hong. 1988. Membrane reactor coupled with electrophoresis for enzymic production of aspartic acid. *Biotech. Bioeng.* 32(5):647–654.

Lopez, J. L. 1983. Carrier-mediated transport in membrane reactors: deacylation of benzylpenicillin. Ph.D. thesis, University of Pennsylvania, Philadelphia, PA.

Lopez, J. L., S. L. Matson, T. J. Stanley, and J. A. Quinn. 1990. Liquid/liquid extractive membrane reactors. In *Extractive Bioconversions, Bioprocess Technologies Series*, ed. B. Mattiasson and O. Holst, pp. 27–66. New York: Marcel Dekker.

Mannheim, A., and M. Cheryan. 1990. Continuous hydrolysis of milk protein in a membrane reactor. *J. Food Sci.* 55(2):381–385.

Matson, S. L. 1979. Membrane reactors. Ph.D. thesis, University of Pennsylvania, Philadelphia, PA.

Matson, S. L. 1989a. Multiphase asymmetric membrane reactor systems. U.S. Patent 4,795,704.

Matson, S. L. 1989b. Multiphase and extractive membrane reactor systems. U.S. Patent 4,800,612.

Matson, S. L., and J. L. Lopez. 1989. Multiphase membrane reactors for enzymatic reaction: diffusional effects on stereoselectivity. In *Frontiers in Bioprocessing*, ed. S. K. Sikdar, P. Todd, and M. Bier, pp. 391–403. Boca Raton, FL: CRC Press.

Matson, S. L., and J. A. Quinn. 1986. Membrane reactors in bioprocessing. *Ann. N.Y. Acad. Sci.* 469:152–165.

Matson, S. L., and J. A. Quinn. 1988. Method and apparatus for conducting catalytic reactions with simultaneous product separation and recovery. U.S. Patent 4,786,597.

Matson, S. L., and T. J. Stanley. 1988. Phase transfer catalysis. U.S. Patent 4,754,089.

Matta, M. S., J. A. Kelley, A. J. Tietz, and M. F. Rohde. 1974. Resolution of some 3-(3,4-dihydroxyphenyl)alanine precursors with alpha-chymotrypsin. *J. Org. Chem.* 39:2291.

Michaels, A. S., C. R. Robertson and S. N. Cohen. 1984. Microbiological methods using hollow fiber membrane reactor. U.S. Patent 4,440,853.

Molinari, R., E. Drioli, and G. Barbieri. 1988. Membrane reactor in fatty acid production. *J. Membr. Sci.* 36:525–534.

Nagamoto, H., and H. Inoue. 1985. A reactor with catalytic membrane permeated by hydrogen. *Chem. Eng. Commun.* 34:315.

Olah, G. A., P. S. Iyer, and G. K. S. Prakash. 1986. Perfluorinated resin sulfonic acid (Nafion-H) catalysis in synthesis. *Synthesis*, No. 7 (July): 513–531.

Ollis, D. F., J. B. Thompson, and E. T. Wolynic. 1972. Catalytic liquid membrane reactor: I. concept and preliminary experiments in acetaldehyde synthesis. *AIChE J.* 18:457.

Omata, K., S. Hashimoto, H. Tominaga, and K. Fujimoto. 1989. Oxidative coupling of methane using a membrane reactor. *Appl. Catal.* 52(1&2):L1–L4.

Parks, R. E., and G. W. E. Plaut. 1953. A manometric assay for chymotrypsin. *J. Biol. Chem.* 203:755–761.

Pizzichini, M., C. Fabiani, A. Adami, and V. Cavazzoni. 1989. Performance of a membrane reactor for cellobiose hydrolysis. *Biotech. Bioeng.* 33(9):955–962.

Pronk, W., P. J. A. M. Kerkhof, C. Van Helden, and K. Van't Riet. 1988. The hydrolysis of triglycerides by immobilized lipase in a hydrophilic membrane reactor. *Biotech. Bioeng.* 32(4):512–518.

Shipman, G. H. 1985. Method of making microporous sheet material and articles made therewith. U.S. Patent 4,539,256.

Sondheimer, S. J., N. J. Bunce, and C. A. Fyfe. 1986. Structure and chemistry of Nafion-H: a fluorinated sulfonic acid polymer. *J. Macromol. Sci. Rev. Macromol. Chem. Phys.* C26(3):351–411.

Stanley, T. J. 1986. Advances in membrane reactors with applications to dilution effects. Ph.D. thesis, University of Pennsylvania, Philadelphia, PA.

Stanley, T. J., and J. A. Quinn. 1987. Phase-transfer catalysis in a membrane reactor. *Chem. Eng. Sci.* 42(10):2313–2324.

Steckhan, E., S. Herrmann, J. Thoemmes, C. Wandrey, and R. Ruppert. 1990. Continuous production of NAD(H) from NAD+ and formate with a molecular weight-enhanced homogeneous catalyst in a membrane reactor. *Angew. Chem.* 102(4):445–447.

Uemiya, S., T. Matsuda, and E. Kikuchi. 1990. Aromatization of propane assisted by palladium membrane reactor. *Chem. Lett.* 8:1335–1338.

Updike, S. J. 1976. Dialysis membrane: catalysis of hydrogen peroxide to hydrogen and oxygen. U.S. Patent 3,996,141.

Vasic-Racki, D., M. Jonas, C. Wandrey, W. Hummel, and M. R. Kula. 1989. Continuous (R)-mandelic acid production in an enzyme membrane reactor. *Appl. Microbio. Biotech.* 31(3):215–222.

Vasudevan, M., T. Matsuura, G. K. Chotani, and W. R. Vieth. 1987. Simultaneous bioreaction and separation by an immobilized yeast membrane reactor. *Sep. Sci. Technol.* 22(7):1651–1657.

Waller, F. J. 1986. Catalysis with metal cation-exchange resins. *Catal. Rev. Sci. Eng.* 28:1.

Welch, G. R. 1977. On the role of organized multienzyme systems in cellular metabolism: a general synthesis. *Prog. Biophys. Molec. Biol.* 32:103–191.

Wichmann, R., C. Wandrey, A. F. Bueckmann, and M. R. Kula. 1981. Continuous enzymic transformation in an enzyme membrane reactor with simultaneous NAD(H) regeneration. *Biotech. Bioeng.* 23:2789.

Wolynic, E. T., and D. F. Ollis. 1974. Catalytic liquid membrane reactor: making acetaldehyde. *Chemtech.* 4:111–117.

Zaborsky, O. 1973. *Immobilized Enzymes.* Boca Raton, FL: CRC Press.

44

Facilitated Transport

J. Douglas Way
Oregon State University

Richard D. Noble
University of Colorado

GENERAL DESCRIPTION
    Introduction
    Terminology
    The Driving Force in Facilitated Transport
THEORY
    Optimization of Reaction Properties
    Competitive Facilitated Transport
    Two-Dimensional Systems and Modules
    Fixed Carrier Membranes
    Transient Effects
    Temperature Effects

POTENTIAL APPLICATIONS
    Separation of Oxygen from Nitrogen
    Acid Gas Removal
    Flue Gas Cleanup
    Separations of Olefins from Saturated Hydrocarbons
    Carbon Monoxide Separation
    Other Applications
    What Research Needs to Be Done?
CONCLUSIONS
ACKNOWLEDGMENTS
NOTATION
REFERENCES

GENERAL DESCRIPTION

Introduction

The commonly accepted mechanism for the transport of a penetrant in nonporous polymer membranes is solution-diffusion (Crank and Park 1968). The penetrant species dissolves in the membrane and diffuses across the membrane due to an imposed concentration gradient. Facilitated transport membranes also involve a reversible complexation reaction in addition to penetrant dissolution and diffusion. The addition of the complexation reaction makes facilitated transport analogous to a chemical absorption process on the feed (high partial pressure) side of the membrane and a stripping process on the product, or permeate, side of the membrane. Facilitated transport membranes, which are similar to emulsion liquid membranes and hollow-fiber contained liquid membranes described in previous chapters, have several general characteristics:

1. They are highly selective.
2. A maximum flux or minimum permeability is reached at high concentration driving forces.
3. Very high permeabilities can be obtained at very low concentration driving forces.

4. They are often unstable in the conventional immobilized liquid membrane configuration.

As shown in Figure 44–1, the complexation reaction in the membrane creates another transport mechanism in addition to solution-diffusion. After the gas molecule of interest dissolves in the membrane, it can diffuse down its concentration gradient or react with the complexation agent or carrier species. Diffusion of the gas molecule across the facilitated transport membrane can take place by two mechanisms: diffusion of the uncomplexed species or diffusion of the carrier-gas complex. The second transport mechanism is not accessible to gases that do not react with the carrier species. Therefore, it is the complexation reaction that makes facilitated transport membranes highly selective.

The total mass transfer rate of the gas that reacts with the carrier is the sum of the flux of the carrier-gas complex and the flux of the uncomplexed penetrant. In the limit of fast reactions, the rate of diffusion controls the rate of transfer, while the reaction rate controls the mass transfer rate when the complexation reaction is slow. In between these two limiting regimes, the contributions of both reaction and diffusion are important.

The reactive carrier mechanism is the reason that the flux of facilitated transport membranes is not always proportional to the concentration driving force across the membrane. At very high driving forces, all of the carrier species are bound to solute molecules and an increase in driving force does not result in an increased flux from the reactive pathway. Under these conditions, known as *carrier saturation,* the carrier is doing all it can. At very low driving force conditions, the flux due to the solution-diffusion pathway is very small, and the majority of the diffusion is due to diffusion of the carrier-gas complex. As the driving force decreases further, the flux of the uncomplexed gas molecules decreases much faster than the carrier transport. Therefore, the flux is again not directly proportional to the driving force when it is small.

The three general configurations for facilitated transport membranes are an immobilized liquid film, a solvent-swollen polymer, and a solid polymer film containing reactive functional groups (a fixed carrier membrane) in addition to the emulsion liquid membranes and hollow-fiber contained liquid membranes described earlier. An immobilized liquid membrane (ILM) is usually prepared by impregnating the pore structure of a very thin, microporous support with a solution of the carrier in a solvent or even pure carrier if it is a liquid. Typical thicknesses range from 5 to 25 μm, which is one to two orders of magnitude thicker than the dense skin layer of an asymmetric polymer membrane. However, liquid phase diffusivities are of the order of 10^{-6} to 10^{-5} cm^2/s, much larger than diffusivities in glassy polymers, which seldom exceed 10^{-8} cm^2/s. Therefore, fluxes for ILMs and polymer membranes are often comparable. The carrier solution is held in place by capillary forces and the solvent concentration of the feed and/or permeate gases must be carefully controlled to avoid drying out the membrane or condensing solvent on the membrane surface (Deetz and Kreevoy 1987). The ILM is generally considered to be the least stable configuration.

A facilitated transport membrane structure that is intermediate between liquid and solid phases can be made by swelling a polymer film in a solvent and introducing the carrier species by diffusion, or by ion exchange in the case of ionomer membranes (LeBlanc et al. 1980; Way et al. 1987). If the solvent used to swell the polymer film is a good physical solvent for the

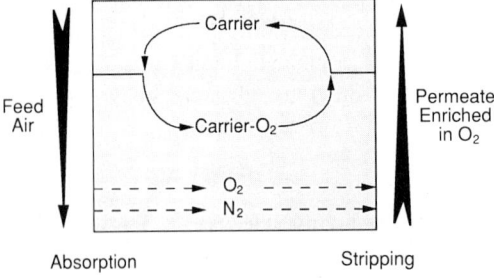

FIGURE 44–1. Schematic diagram of a facilitated transport membrane with a mobile carrier species that reacts reversibly with oxygen.

gas of interest, solvent-swollen polymer films can be used as gas separation membranes without a carrier species present (Matson et al. 1988). Ion-exchange membranes have several unique advantages as supports for facilitated transport membranes. Once the charged carrier species is exchanged into the membrane, the carrier cannot be lost unless it is replaced by another ion. This is quite unlikely in a gas separation application. Second, the carrier loading in an ion-exchange membrane is determined by the ion-exchange site density, not the solubility of the carrier in the solvent. Consequently, the local carrier concentration obtained in heterogeneous materials such as perfluorosulfonic acid ionomers can be very high. Way et al. (1987) reported that local carrier concentrations of up to 8 M were obtained using perfluorosulfonic acid ionomer membranes as supports for ethylenediamine cations. Solvent-swollen polymer membranes are intermediate in stability between ILMs and fixed carrier membranes.

To improve the stability of facilitated transport membranes, complexation agents have been attached to polymer chains and membranes have been made from these reactive polymer materials. This configuration, known as *fixed carrier* or *chained-carrier* membranes, have the potential to be highly selective and stable. Recently, several accounts have appeared in the literature describing efforts to prepare solid facilitated transport membranes containing reactive sites selective for acid gases, oxygen, and nitrogen (Yoshikawa et al. 1986; Nishide et al. 1988, 1989; Sugie 1988). Although controversial, these studies provide some evidence that facilitated diffusion can occur in the solid state. The major topic of controversy is the unknown mechanism of transport in these materials, which is addressed in the Theory section in this chapter.

The stability of facilitated transport membranes is a very important issue that will ultimately determine whether this technology is used for large-scale gas separations. Instability can result from the complexation chemistry, the support configuration, or both. For example, an ILM can dry out, destroying the integrity of the liquid membrane. In the same ILM, the solvent concentration in the feed and permeate gas streams could be perfectly controlled and the liquid membrane itself would be stable, but if the carrier is lost or deactivated, the selectivity is also lost. Stability problems with facilitated transport membranes can be addressed using several methods currently under study. The use of hollow-fiber contained liquid membranes (discussed in Chapter 42) can overcome drying problems with ILMs simply by changing the configuration. As described below, the use of molten salts with very low vapor pressure has also been proposed to overcome physical stability problems with ILMs.

Terminology

The energy of the bond between the solute and carrier molecule for this reversible reaction must fall within a certain range to be effective (King 1980, 1988) (see Figure 44–2). If the binding energy is too weak, very little solute is transported by this reactive pathway; if the binding is too strong, the solute cannot be removed effectively at the downstream side of the membrane. In either extreme, the use of the

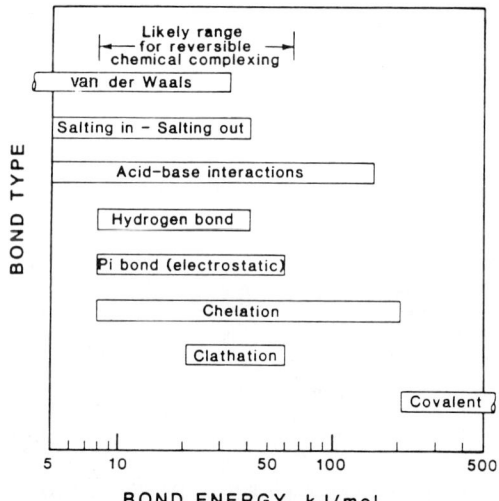

FIGURE 44–2. The range of bond energies for chemical complexation used in separation processes (reprinted from King 1980 with permission).

TABLE 44-1. Important Variables in Facilitated Transport.

k	= mass transfer coefficient based on concentration driving force
D_A	= solute diffusion coefficient
D_{AB}	= diffusion coefficient of solute-carrier complex
k_r	= forward rate coefficient of complexation reaction
c_{A0}	= concentration of gas in the liquid membrane at feed interface
c_T	= total concentration of carrier
l	= membrane thickness

reactive pathway for increased flux and selectivity is inefficient.

The performance of a facilitated transport membrane process is dependent on a number of system properties that determine the rates of mass transport to and through the membrane for the solute of primary interest. The important variables that describe this process are listed in Table 44-1. These physical properties may be combined in a number of dimensionless groupings that allow for a simplified evaluation of the expected performance of particular carrier/solute combinations. These dimensionless variables have physical significance and are presented in Table 44-2. Facilitation factor F is the ratio of the total solute flux with the carrier present to the solute diffusional flux. It can be viewed as a measure of increased selectivity for multicomponent feed mixtures. Also, multiplying the solute diffusion coefficient by F gives an effective "overall" diffusion coefficient. A mobility ratio α can be defined for the reactive versus the diffusive pathway; K is a dimensionless equilibrium constant; and ε is the inverse of a Damköhler number and is a measure of the characteristic reverse reaction time to the characteristic diffusion time. It serves the same function as a Thiele modulus in catalysis. The effect of external mass transfer resistance is incorporated as a Sherwood number, which is a measure of external boundary layer resistance to the diffusional resistance of the membrane.

The physical and chemical properties in these dimensionless variables can be independently measured or estimated. This approach minimizes empiricism and provides a basis for determining some relationships between system properties and performance. The relationship between the physical and chemical properties in a facilitated transport system and system performance has previously been discussed by Koval and Reyes (1987). They provided some examples and indicated areas where improvements could be made.

As noted above, an *overall* diffusion coefficient can be obtained by multiplying the solute diffusion coefficient by facilitation factor F. In this way, standard design equations for membrane systems (Pan and Habgood 1974; Shindo et al. 1985) can be used with this correction. Note that F will vary with the solute feed concentration. So, if the feed concentration is changing along the membrane/feed interface (as in the usual case in hollow-fiber modules, for example), the value of F needs to be adjusted accordingly.

The symbol α is used in this chapter to represent two different quantities, which may be confusing to the reader. When the symbol α is used, it will refer to the mobility ratio described above. When the symbol α_{ij} is used, it will refer to the separation factor or selectivity of a membrane and is defined as the ratio of the permeability of species i divided by the permeability of species j. For example, the symbol α_{O_2/N_2} is the separation factor of oxygen over nitrogen.

TABLE 44-2. Important Dimensionless Numbers in Facilitated Transport.

F	= facilitation factor
	= $\dfrac{\text{solute flux with carrier present}}{\text{solute diffusion flux}}$
α	= mobility ratio, ratio of mobility of carrier to mobility of solute ($D_{AB}c_T/D_A c_{A0}$)
ε	= inverse Damköhler number ($D_{AB}/k_r l^2$), ratio of characteristic times for reverse reaction and diffusion
K	= dimensionless reaction equilibrium constant ($k_f c_{A0}/k_r$)
Sh	= Sherwood number for solute mass transfer (kl/D_A)

The Driving Force in Facilitated Transport

The flux of the free solute that is not complexed with the carrier species is proportional to the concentration gradient of this free solute within the membrane. The total flux is not directly proportional to the concentration gradient due to the existence of two transport mechanisms in the membrane; solution-diffusion and diffusion of the carrier-solute complex. The solute concentration in the membrane can often be related to the gas phase partial pressure using Henry's law or a similar equilibrium relationship. At higher pressures, vapor-liquid equilibrium or gas-polymer absorption data are necessary to determine the concentration gradient in the membrane.

THEORY

The first major review of the mathematical aspects of facilitated transport was done by Goddard, Schultz, and Suchdeo (1974). They focused primarily on the limit of very fast or very slow reactions and discussed techniques for obtaining asymptotic or approximate solutions. Goddard (1977) provided a further survey that focused on systems near the reaction equilibrium limit. Smith, Meldon, and Colton (1973) used a perturbation analysis to obtain solutions for the very fast and very slow reaction limit. Yung and Probstein (1973) used a similarity transform to simplify the differential equations and obtain a numerical solution.

Most modeling and analysis of facilitated transport systems has used the following one-step reaction mechanism:

$$A + B \underset{k_r}{\overset{k_f}{\rightleftarrows}} AB,$$

where

A = solute being transported
B = carrier
AB = solute-carrier complex.

Ward (1970a) provided analytical solutions for the facilitation factor under diffusion-limited (small ε) and reaction-limited (large ε) regimes. In general, operation in the diffusion-limited regime is desirable. In this mode, full utilization of the carrier is obtained. His solution for the diffusion-limited regime is

$$F = 1 + \frac{\alpha K}{1 + K}. \qquad (44-1)$$

To obtain this solution, some assumptions are necessary. The diffusion coefficients of the carrier and solute-carrier complex are assumed equal. This is usually justified when the carrier is normally much larger than the solute. The more difficult problem when the diffusivities of the carrier and the carrier-solute complex are not equal has recently been solved by Basaran, Burban, and Auvil (1989). The downstream (exit) solute concentration is assumed to be zero. This ensures maximization of the reverse reaction at the exit.

Smith and Quinn (1979) extended the range of the analytical solution. They assumed a large excess of carrier. This allowed them to linearize the differential equations describing the transport across the membrane. Their result is

$$F = \frac{1 + \dfrac{\alpha K}{1 + K}}{1 + \dfrac{\alpha K}{1 + K} \dfrac{\tanh \lambda}{\lambda}}, \qquad (44-2)$$

where

$$\lambda = \frac{1}{2}\left(\frac{1 + (\alpha + 1)K}{\varepsilon(1 + K)}\right)^{1/2}. \qquad (44-3)$$

They showed that their solution had the proper behavior in both the diffusion-limited and reaction-limited regimes. Of course, by assuming an excess of carrier, this solution does not display the leveling of performance as the carrier is saturated.

This same result was independently derived by Hoofd and Kreuzer (1979). They applied a combined Damköhler technique to obtain the solution. A large Damköhler number (small ε) solution is assumed to provide a first estimate of the solute concentration. A small Damköhler number (large ε) solution is then solved over the entire membrane using the first estimate of the solute concentration and a correction factor. Solving for the correction factor completes the solution.

Noble, Way, and Powers (1986) extended this model further to incorporate external mass transfer effects adjacent to the membrane boundaries. This external mass transfer effect is described by a Sherwood number (Sh):

$$F = \frac{\left(1 + \dfrac{\alpha K}{1 + K}\right)\left(1 + \dfrac{2}{\text{Sh}}\right)}{\left[1 + \left(\dfrac{\alpha K}{1 + K}\right)\dfrac{\tanh \lambda}{\lambda}\right] + \left(1 + \dfrac{\alpha K}{1 + K}\right)\left(\dfrac{2}{\text{Sh}}\right)}. \quad (44-4)$$

As the external mass transfer resistance decreases, Sh becomes larger. In the limit (Sh $\to \infty$), Eq. (44-4) reduces to Eq. (44-2).

The above results can be used in the following ways. By comparing Eqs. (44-1) and (44-2), the value of $\tanh\lambda/\lambda$ is a measure of the facilitation in the absence of external mass transfer resistance (Sh $\to \infty$). A simple and quick calculation of this one term can provide an estimate of the facilitation effect:

Maximum facilitation: $\quad \dfrac{\tanh \lambda}{\lambda} \to 0,$

Minimum facilitation: $\quad \dfrac{\tanh \lambda}{\lambda} \to 1.$

We can also determine the necessary property modifications to move toward reaction equilibrium by subsequent calculations to move the above quantity toward zero. The property values for the facilitated transport of H_2S in ion-exchange membranes containing monopositive ethylenediamine as a carrier species can be used to illustrate this calculation (Way and Noble 1987):

$D_A = 2.85 \times 10^{-6}$ cm^2/s,

$D_{AB} = 2.52 \times 10^{-8}$ cm^2/s,

$c_{A0} = 8.46 \times 10^{-2}$ M,

$c_T = 8.32$ M,

$K_{eq} = 3.74 \times 10^2$ M^{-1},

$k_f = 1.0 \times 10^{11}$ M^{-1} s^{-1},

$k_r = 2.67 \times 10^8$ s^{-1},

$l = 1.0 \times 10^{-4}$ cm (1 μm),

$K = K_{eq}c_{A0}$
$= (3.74 \times 10^2$ M$^{-1}) (8.46 \times 10^{-2}$ M$)$
$= 31.6,$

$\varepsilon = \dfrac{D_{AB}}{k_r l^2}$

$= \dfrac{2.52 \times 10^{-8} \text{ cm}^2/\text{s}}{(2.67 \times 10^8 \text{ s}^{-1})(1.0 \times 10^{-4} \text{ cm})^2}$

$= 9.4 \times 10^{-9},$

$\alpha = \dfrac{D_{AB}c_T}{D_A c_{A0}}$

$= \dfrac{(2.52 \times 10^{-8} \text{ cm}^2/\text{s})(8.32 \text{ M})}{(2.85 \times 10^{-6} \text{ cm}^2/\text{s})(8.46 \times 10^{-2} \text{ M})}$

$= 0.87,$

$\lambda = \dfrac{1}{2}\left(\dfrac{1 + (\alpha + 1)K}{\varepsilon(1 + K)}\right)^{1/2}$

$= \dfrac{1}{2}\left(\dfrac{1 + (0.87 + 1)(31.6)}{(9.4 \times 10^{-9})(1 + 31.6)}\right)^{1/2}$

$= 7155,$

$\dfrac{\tanh \lambda}{\lambda} = \dfrac{1}{7155}.$

The term $(\tanh \lambda)/\lambda$ approaches zero, which implies reaction equilibrium and maximum facilitation.

In an earlier publication, Noble (1985) defined a kinetic efficiency factor η:

$$\eta = \frac{\text{actual facilitated flux}}{\text{facilitated flux under reaction equilibrium conditions}}.$$

To obtain η, he determined the flux from numerical calculations. As expected, η increased as ε decreased. One observation was that the time to reach steady state increased as η decreased. The explanation is that the diffusion-limited regime ($\eta \approx 1$) should have the minimum time to reach steady state. In the reaction-limited regime ($\eta \ll 1$), the diffusion time is fast compared to the reaction time, and the time required to reach steady state increases. So, η^{-1} can be used as a qualitative measure of the relative time to reach steady state.

Three distinct regions could be identified. For $\varepsilon \ll 1$, η is very close to 1, which corresponds to the reaction equilibrium (diffusion-limited) regime. When η is very small and approximately constant, mass transport is reaction rate limited. In between, there is a mixed regime where η decreases as ε increases.

A rearrangement of Eq. (44-4) can be used to analyze the results of transport experiments and estimate some system properties. Assuming reaction equilibrium $[(\tanh \lambda)/\lambda \to 0]$, Eq. (44-4) can be rearranged as

$$(F - 1)^{-1} = E^{-1} = \left(1 + \frac{2}{\text{Sh}}\right)\alpha^{-1}$$
$$+ \left[\left(\frac{2}{\text{Sh}}\right)\left(1 + \frac{1}{\alpha K}\right) + \frac{1}{\alpha K}\right],$$

(44-5)

where α^{-1} is directly proportional to the solute feed concentration (c_{A0}). A plot of E^{-1} versus c_{A0} can then be constructed. As c_{A0} is reduced, the plot should be linear and Eq. (44-5) is valid. For the straight-line portion of the curve, the slope and intercept can be used to estimate two unknown quantities if all other properties have been independently measured or estimated. Typically, the two unknown quantities are Sh and D_{AB}.

Of course, Eq. (44-4) can be used to estimate F quickly and easily. Properties such as membrane thickness l or carrier concentration c_T can be varied to predict their effect on performance.

Optimization of Reaction Properties

One important consideration in facilitated transport and many other processes is optimization. To address this issue, Kemena, Noble, and Kemp (1983) have numerically solved the governing differential equations for facilitated transport, based on the one-step reaction described at the beginning of this section. They determined the optimal values of K and F for a given ϵ and α. Their results indicated that K had values from 1 to 10 for orders of magnitude change in ε and α. This result has several useful features and allows the determination of an optimal property set. Thus, the set of physical and chemical properties that will result in the maximum facilitation effect can be determined. This information provides the basis for the uses discussed below.

The above results are very useful for screening potential carriers. The narrow range of optimal values for K provides a rapid method of selecting good carriers. A single measurement (equilibrium constant) combined with the solute feed concentration yields K. Therefore, using literature values and/or measurements, a number of candidate carriers can be quickly evaluated. Those carriers for which K falls in or near the optimal range can be used for facilitated transport systems. The other carriers can be eliminated before any transport experiments are started.

An indirect example of this is based on the data given by Smith and Quinn (1980) for CO transport. They report the facilitation factors for 5% CO feed gas mixtures using Cu^+, Co^{2+}, and Fe^{2+} ions as the carrier (see their Table 1). The cuprous ion performed much better than the other two. Based on their data, K for the Cu^+ ion is 0.07. They state that the Cu^+ ion is preferable because it has a forward rate constant that is much larger than the other carriers. This implies that $K \ll 1$ for Co^{2+} and Fe^{2+} and hence these ions would not be good carriers because of very weak binding, which translates to low facilitation.

The range for optimal K can be used as the basis for modifying carriers. If K is too large or too small, this indicates the direction needed to modify the rate constants for the reversible reaction. This can be useful information to the synthetic chemist. Carrier modifications can be chosen to attempt to move toward the optimal range for K.

Koval and Reyes (1987) reported on two modifications to an Fe(TIM) carrier for CO transport. The modifications were undertaken to increase the carrier solubility c_T since the value of K was already in the optimal range. Their results showed an increase in F while maintaining K within the optimal range. Their results also show that much more work is required to fully understand this relationship.

The optimal range of solute feed concentration can be estimated for a given carrier (equilibrium constant). Again, using the data of Smith and Quinn (1980), the value of K_{eq} is 1600 M^{-1}. The optimal feed range (corresponding to $1 < K < 10$) is 0.67 to 6.7 atm. This analysis assumes a single-step reaction, which may not be valid at high CO pressures.

One valuable use of the above factor is in the comparison of actual to optimal performance. This comparison can be used to estimate whether or not large improvements are possible in system performance. Murray and Wyman (1971) give some data on facilitated transport of CO in hemoglobin solutions. Their data correspond to $K = 3900$, $\varepsilon = 0.065$, and $\alpha = 5.34 \times 10^{-3}$. It is apparent that their K value is orders of magnitude too large. This directly supports their conclusion that there was no noticeable CO facilitation under these conditions because the CO is too strongly bound to the carrier. This also explains why CO is toxic to humans. It attaches to the binding sites and cannot be easily removed, which blocks the site of O_2 transport. Hoofd and Kreuzer (1978) show facilitation for very low CO concentrations. This corresponds to a much lower K value and a higher α value. Since this moves K toward the optimal condition, improvement in CO facilitation would be expected.

Recently, Kirkköprü-Dindi and Noble (1989) extended this analysis to multiple-step complexation reactions. Their results indicate that this same range of $1 < K < 10$ holds for each step in multistep reactions in which one complexing agent can bind more than one solute molecule.

Competitive Facilitated Transport

The issue of competitive transport has also been addressed. In competitive transport, two or more solutes can react with the complexing agent. This is demonstrated in Figure 44–3 (Way and Noble 1989). The system is initially transporting H_2S. When CO_2 is introduced into the feed, the CO_2 rapidly displaces the H_2S.

Cussler (1971) used a simplified analysis to describe this effect. He developed an analytical solution for the flux of one solute. His main assumptions were equal diffusion coefficients for each species and reaction equilibrium within the membrane. The solute flux contains three terms: simple diffusion of solute, facilitated transport of solute, and a third term based on the flux of the second solute. He showed that this third term could supply the energy to pump the first solute "uphill" against its concentration gradient. In addition, the magnitude of the third term relative to the first two terms can give a measure of the competitive effect.

Smith, Lander, and Quinn (1977) recognized that both CO_2 and H_2S can reversibly react with an aqueous carbonate solution and showed by means of analytical models that these two reactions could be used as a means to couple the fluxes of CO_2 and H_2S. A gradient in H_2S partial pressure could be used to pump CO_2 against its partial pressure gradient or vice versa.

Niiya and Noble (1985) studied the competitive effect for both transient and steady-state

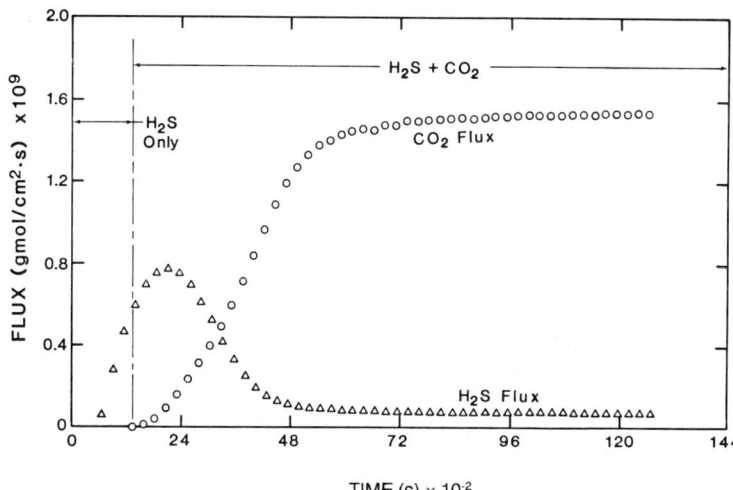

FIGURE 44-3. The effect of CO_2 on the facilitated flux of H_2S through PFSA IEMs containing an ethylenediamine counterion carrier species (reprinted from Way and Noble 1989 with permission).

conditions. They also demonstrated that a solute could be pumped "uphill." In addition, they pointed out that solute 1 would have a competitive advantage over solute 2 if the binding and release rate of solute 1 was greater than that of solute 2. Some comparison with literature data gave very good results.

Way and Noble (1989) published experimental data on the competitive transport of CO_2 and H_2S through ion-exchange membranes with ethylenediamine as the carrier, as shown in Figure 44-3. They used a numerical model based on Niiya and Noble (1985) and found very good agreement between model predictions and experimental results.

Two-Dimensional Systems and Modules

Facilitated transport in two-dimensional systems and modules has been discussed. A hollow-fiber facilitated transport membrane in which the feed flows through the lumen and permeates through the membrane is an example. Membrane contactors and similar devices where the reactive liquid flows on one side of the membrane are considered in Chapter 46.

The simplest analysis was done by Noble (1984). By assuming plug flow through the lumen, he used the following equation to obtain an expression for the fraction of solute removed as a function of the length of the hollow fiber:

$$J_{A_T} = \left(\frac{r_o}{r_i}\right) \frac{SFD_A}{l} (c_{Ai} - c_{Ao}), \quad (44\text{-}6)$$

where S is the shape factor defined by Noble (1983). His analysis showed that the fraction of solute removed showed an exponential asymptote with length. A one-dimensional analysis had a linear relationship.

Kim and Stroeve (1988) recently provided a more detailed analysis for laminar flow through these devices. Their numerical results are presented graphically, which allows easy use and comparison. Their main conclusion was that facilitated transport works best when the membrane mass transfer coefficient is small.

Gottschlich, Roberts, and Way (1988) compared the performance of facilitated transport and commercial polymer membrane modules in a computer simulation study using the separation of CO_2/CH_4 mixtures as an example. Assuming reaction equilibrium and using the facilitated transport membrane properties from Way et al. (1987), the facilitation effect is significant only at CO_2 partial pressures below 10 psia. Below 10 psia CO_2 partial pressure, the membrane area requirement for the facilitated transport module was a factor of 2 less than the polymer membrane module. Above 50 psia CO_2 partial pressure, the membranes function identically.

A more rigorous numerical study of facilitated transport membranes and modules was presented by Basaran, Burban, and Auvil (1989) to investigate the effect of differing carrier and carrier-solute complex diffusivities. Relative to the case when the $D_B = D_{AB}$, the facilitation factor F increases if $D_B < D_{AB}$, and F decreases if $D_B > D_{AB}$. An optimization was also carried out to determine the maximum facilitation factor and dimensionless equilibrium constant extending the results of Kemena, Noble, and Kemp (1983) to cases where $D_B \neq D_{AB}$.

Fixed Carrier Membranes

Recently, interest has arisen in fixed-site carrier membranes. These are solid polymer films into which complexing agents are incorporated. These membranes represent a limit of facilitated transport because there is no liquid phase. Therefore, there is no mobility of the complexing agent since it is anchored in the film.

Cussler, Aris, and Bhown (1989) have proposed a theory for facilitated transport in these membranes based on the chained-carrier or "Tarzan swing" mechanism. Two complexing agents must be close enough to each other to "pass off" a solute molecule from one complexing agent to the next one. As concentration of the complexing agent becomes more dilute, the distance between complexing agents increases. At some concentration (percolation threshold), the distance becomes too large for a pass-off and the solute flux drops precipitously. They derived expressions for the solute flux under diffusion-limited and reaction-limited conditions. They concluded that these equations have the same form as normal facilitated transport. So a key element in the theory is the existence of the percolation threshold.

Noble (1990) recently proposed a theory that does not require the existence of a percolation threshold. Solute molecules can migrate between complexing agent molecules by moving along the polymer chain. The result is an expression analogous to Eq. (44-2). In the limit as (tanh λ)/λ approaches zero, Eq. (44-2) approaches Eq. (44-1). Equation (44-1) is analogous to the dual-mode sorption model, which has been used to describe gas transport in glassy polymers. Using the analysis based on Eq. (44-5), the effective diffusion coefficient D_{AB} (effective mobility of this pathway) can be determined. Literature data for O_2 transport in fixed carrier membranes were used to support the theory.

Transient Effects

A few publications have discussed the transient response of facilitated transport membranes. Spaan (1973) modeled oxygen transport under transient conditions in a stationary film. He used the advancing front hypothesis to measure oxygen movement through the film. Curl and Schultz (1973) used a polynomial approximation for unsteady-state diffusion of oxygen into hemoglobin solutions. Their approximation is more general than previous models. Their model reduces to other cases (i.e., advancing front and linear isotherm cases) when the slope of the lines represented by polynomial approximation are varied for the hemoglobin oxygenation saturation curve. Spaan, Kreuzer, and Hoofd (1980) developed an unsteady-state model for oxygen transport in hemoglobin solutions. Their model assumes chemical equilibrium. Plots of dimensionless oxygenation time versus three dimensionless parameters are obtained.

Folkner and Noble (1983) solved the governing differential equations for planar, cylindrical, and spherical geometry. They produced plots that could be used to predict facilitation as a function of dimensionless time. One interesting observation is the fact that the facilitation factor can drop below 1 at short times. This is due to the fact that the solute-complexing agent molecule diffuses at a slower rate than the solute.

Temperature Effects

Heat transfer can have an effect on facilitated transport. Kemp and Noble (1983) showed that the imposition of a temperature gradient across the membrane can cause a significant increase or decrease in facilitation. In extreme cases, there can even be a reversal in the direction of the facilitated flux.

POTENTIAL APPLICATIONS

Separation of Oxygen from Nitrogen

Air separation is potentially the most important application for membrane gas separations because of its enormous market size. In 1989 nitrogen was the second largest commodity chemical and oxygen was the third largest commodity chemical produced in the United States with a total market size of approximately 2.3 billion dollars (Anonymous 1990). However, as of 1991, only nitrogen production from air was economical using membranes compared to competing separation processes such as cryogenic distillation and pressure swing adsorption (PSA) due to limited selectivity of existing polymeric materials. At that time, probably the best oxygen selective polymer materials, halogen substituted polycarbonates, had α_{O_2/N_2} values of 7.5 (Muruganandam and Paul 1987). The equation below gives the maximum permeate mole fraction, y_{max}, as a function of the membrane separation factor, α_{O_2/N_2}, and the feed gas mole fraction, x_0, which for oxygen in air is approximately 0.21:

$$y_{max} = \frac{\alpha_{O_2/N_2} x_0}{\alpha_{O_2/N_2} x_0 + 1 - x_0}. \quad (44-7)$$

Using Eq. (44-7), the maximum oxygen permeate purity that could be achieved with a substituted polycarbonate membrane is 67%, far below the 95% oxygen purity from cryogenic distillation or PSA. The maximum permeate mole fraction from Eq. (44-7) is plotted as a function of the membrane separation factor in Figure 44-4 for a normal air feed gas. As shown in this figure, the minimum selectivity required to produce 95% O_2 is 70, an order of magnitude larger than the most selective polymer material available in 1991. In a detailed analysis of the economics of oxygen production using membrane technology, Matson et al. (1986) showed that if a membrane with a selectivity of >60 could be developed, a membrane process would be cost competitive with cryogenics for the production of medium purity (>90%) oxygen. Consequently, production of medium and higher purity oxygen using membrane technology is considered by many to be the "holy grail" of membrane applications.

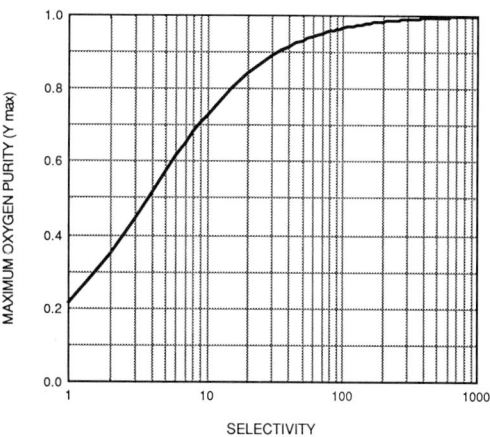

FIGURE 44-4. The minimum membrane selectivity or separation factor required to produce an oxygen-enriched permeate stream using air as the feed gas.

Facilitated transport would appear to be an ideal mechanism to achieve the very high selectivities necessary for the production of even medium-purity oxygen. Indeed, facilitated transport membranes have achieved O_2/N_2 separation factors as high as 30 in the laboratory with O_2 permeabilities greater than or equal to polydimethylsiloxane polymers (Roman and Baker 1985). However, as discussed above, the reason that development has stalled at the laboratory level has been the stability of the complexation chemistry. Since the work of Roman and Baker (1985), many studies of oxygen facilitated transport using either the immobilized liquid membrane or fixed carrier membrane configuration have appeared in the literature. Although most of the published studies describe work with oxygen-selective complexation chemistry, recently Nishide et al. (1989) and Nishide, Kawakami, and Tsuchida (1989) have fabricated nitrogen-selective membranes.

Immobilized Liquid Membranes

Immobilized liquid membranes selective for O_2 typically consist of an anhydrous organic solution of a carrier and various additives (such as Lewis bases) impregnated in a microporous support that is often a polymer microfilter or ultrafilter. For laboratory development studies, ILMs range from 25 to over 100 μm in thickness. A commercial application of an ILM for air separation would likely require a thickness between 1 and 5 μm (Matson and Lonsdale 1987).

A wide variety of synthetic compounds that reversibly bind O_2 can be found in the chemical literature (Niederhoffer, Timmons, and Martell 1984; Busch 1988). The most common organometallic compounds mimic heme in structure. A metal center is complexed within a tetradentate ligand, usually a Schiff-base, porphyrin, or similar chelate. A typical Schiff-base structure, Co(3-MeOSalTmen), is shown in Figure 44–5 (Roberts and Laine 1986). There are generally two O_2 binding sites on the metal center, often above and below the plane of the molecule. For oxygen binding to occur, a Lewis basic ligand must first bind to one of these two sites. This axial base can be an additive to the liquid membrane solution or the solvent itself. In organometallic Schiff-base or porphyrin solutions, various amines have been used such as N-methylimidazole, N,N-dimethylaminopyridine (DMAP), N-methylpyrrolidinone (NMP), or pyridine.

All of these synthetic dioxygen carriers are subject to auto-oxidation by oxygen itself. Although the degradation pathways are the subject of current research, the three general mechanisms include irreversible formation of 2:1 peroxo-bridged dimers by carriers that function reversibly as 1:1 carriers, irreversible ligand oxidation, and central metal atom oxidation (Busch 1988; Martell 1988). A peroxo-dimer is

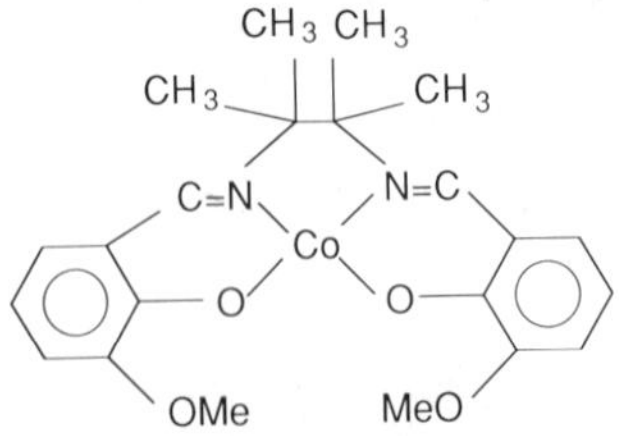

FIGURE 44–5. The chemical structure of a typical cobalt Schiff-base type oxygen carrier, Co(3-MeOSalTmen) (adapted from Roberts and Laine 1986).

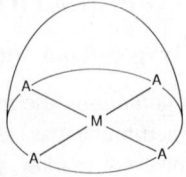

FIGURE 44–6. The basic structure of a lacunar, or caged, organometallic Schiff-base oxygen carrier (reprinted from Busch 1988 with permission).

formed as a "sandwich" of a dioxygen molecule between two carriers that have an axial base bound to one coordination site. These three degradation mechanisms may occur as simultaneous or consecutive reactions. The oxygen carrier design favored by Busch and coworkers has the basic structure shown in Figure 44–6 (Busch 1988; Ramprasad and Busch 1987; Norman, Ramprasad, and Busch 1988). A strap ligand perpendicular to the plane containing the metal center creates a cavity above the metal center. These elaborate strapped (lacunar) Schiff-base complexes reduce or eliminate peroxo-dimer formation but are still subject to other degradation mechanisms.

The first study of facilitated transport of oxygen using hemoglobin as a carrier was carried out by Scholander (1960) using aqueous solutions of hemoglobin immobilized in cellulose acetate supports. In the presence of hemoglobin, the oxygen flux was eight times that of a pure water film. Scholander proposed a bucket-brigade transport mechanism where the O_2 molecule hops from one reactive site to the next. Further experimental (Wittenburg 1966) and theoretical studies (Kreuzer and Hoofd 1970) extended the work of Scholander, and it was recognized that a mobile carrier mechanism and reversible reaction were necessary to describe the O_2 transport data.

Bassett and Schultz (1970) performed the first study of facilitated transport of O_2 using a synthetic carrier, bis(histidinato)-Co(II). The presence of the synthetic carrier at a concentration of 0.05 M increased the total oxygen flux compared to control experiments. A mathematical model was presented for the oxygen flux in terms of the fundamental system parameters found in Table 44–2. Despite carrier degradation problems, this study showed that facilitated transport of oxygen using a synthetic carrier was possible.

A major advance in the development of synthetic membranes for oxygen separation was described by Roman and Baker (1985) and later by Johnson et al. (1987). Immobilized liquid membranes with very high performance were prepared containing relatively high concentrations of synthetic Co Schiff-base O_2 carriers in low volatility organic solvents. In sharp contrast to previous studies, these ILMs were stable for extended time periods measurable in weeks. Roman and Baker (1985) reported an $\alpha_{O_2/N_2} = 30$ and $P_{O_2} = 1500$ Barrers [1 Barrer $= 10^{-10}$ cm^3 (STP) · cm/(cm^2 · s · cm Hg)] at 25°C, over twice the permeability of silicone rubber, for an ILM containing a Co Schiff base, CoSalPr, at 1 M concentration. This ILM produced an initial permeate O_2 purity of 88%. No lifetime data were presented in the patent.

Johnson et al. (1987) presented a comprehensive account of the lifetime and performance of the ILMs initially described by Roman and Baker (1985). Immobilized liquid membranes with the longest lifetime consisted of 0.4 M solutions of the carrier Co(3-MeOSalTmen) shown in Figure 44–5 and 0.6 M DMAP axial base in NMP supported in 125-μm-thick nylon microfilters. Baker, Roman, and Lonsdale (1987) discussed the solvent selection for the facilitated transport studies in a companion study of the O_2 permeability of organic solvents. At 263° K, the typical O_2 permeability was 260 Barrers with an O_2/N_2 separation factor of 26. In a lifetime test of over 120 days, the ILM initially produced greater than 85% O_2, which decreased to 67% O_2 at 100 days. In the 36 days beyond this point, the O_2 purity precipitously decreased from 67 to only 20%. The initial O_2 permeability was 157 Barrers, which decreased to 21.7 Barrers after 126 days. The same carrier solution was immobilized in very thin (2 to 6 μm), 80% porosity, regenerated cellulose membranes supported on Gore-Tex® microfilters. The oxygen permeance for the thin ILMs ranged from 33.3 × 10^5 to 1.03 × 10^6 Barrers/cm, and the O_2 purity was greater than 80%. The lifetime of the thin membranes was not measured beyond 9 days, when the oxygen flux was 85% of the initial value.

The productivity of the thin ILMs can be compared to the most selective polymer membranes for air separations (Muruganandam and Paul 1987) most effectively by computing the pressure normalized flux or permeance, which is defined as the permeability in Barrers divided by the membrane thickness in centimeters. Assuming a skin thickness of 0.2 μm, an asymmetric tetrabromopolycarbonate membrane would have an O_2 permeance of 6.8 × 10^4 Barrers/cm, which is about five times less than the average value for the thin ILM of Johnson and coworkers of 3.68 × 10^5 Barrers/cm. In this case, the higher permeability coefficients associated with ILMs more than compensate for the higher membrane thicknesses compared to asymmetric polymer membranes.

The economics associated with the O_2 facilitated transport membranes of Johnson et al. (1987) was compared to pressure swing adsorption (PSA) and cryogenic distillation for producing high-purity O_2 by Matson and Lonsdale (1987). In a 10 ton/day plant, the liquid membrane process could produce 85% oxygen for $88/ton of equivalent pure oxygen. In calculations for a 100 ton/day plant size, the liquid membrane process could produce O_2 at competitive costs compared to PSA and cryogenic distillation. However, the liquid membrane process would not effectively compete with PSA and cryogenics because the oxygen produced by liquid membranes was only 85% pure compared to 90 to 95% O_2 for PSA and >99% for cryogenic distillation.

Bellobono et al. (1987) prepared facilitated transport membranes for O_2 consisting of the Co Schiff-base CoSalen dissolved in dimethylformamide (DMF) in a support prepared by photografting a microporous layer of an epoxy-diacrylate copolymer onto cellulose. The carrier-oxygen complex was the peroxo-dimer of CoSalen and O_2. Presumably, DMF acted as the axial base in this system. Permeability measurements on these composite liquid membranes showed very high O_2/N_2 separation factors of 50 at 343°K. No lifetime data for the membranes were presented.

Although the majority of the literature dealing with facilitated transport of O_2 has dealt with porphyrin or Schiff-base type carriers, several groups have investigated alternate carrier chemistries. Kawakami et al. (1982) prepared liquid membranes containing Cu(I)-tetraethylenepentamine carrier compounds. The copper salt Cu(SCN) was dissolved in tetraethylenepentamine (tetren) at a molar ratio of approximately 1:12 to form the liquid membrane solution. At 313K and O_2 feed gas partial pressures of less than 2 cm Hg, separation factors greater than 20 were reported for 0.32-cm-thick ILMs. Govind and coworkers (Govind 1987; Kulkarni, Hinegardner, and Govind 1987; Kulkarni and Govind 1988; Kulkarni and Govind 1989) discussed the development of organometallic manganese halides (McAuliffe et al. 1983), $Mn(dppe)_2X_2$, where dppe is 1,2-bis(diethylphosphinoethane), and X is a halide such as Br, Cl, or NCS. The $Mn(depe)_2Br_2$ complex sorbed 26 cm^3 O_2/g complex in solution at 294°K and 1 atm total pressure. This large O_2 capacity compares well to the O_2 capacities of conventional adsorbents such as zeolites and carbon molecular sieves.

Much of the most recent published work on facilitated transport of oxygen was performed in industrial laboratories and has appeared in the patent literature. Pez and Carlin (1986) describe a novel configuration for a facilitated transport membrane for O_2 separation from air that consisted of a molten lithium or sodium nitrate salt immobilized in a porous metal screen. At a temperature of 800°K, they reported an excellent O_2 permeability of 1110 Barrers and an extraordinary O_2/N_2 selectivity of 79. Using a selectivity of 79, Eq. (44–7) predicts a maximum O_2 permeate concentration of 95%. This configuration has the potential to be stable since a molten salt has a very low vapor pressure. Furthermore, the use of a pure reactive phase for the liquid instead of a solution of a carrier in a solvent maximizes the carrier concentration, which can prevent carrier saturation.

A practical configuration for such a system might consist of the molten salt immobilized in the thin, small-pore, top layer of a high surface area ceramic membrane module. However, commercial development of processes using molten lithium and nitrate salts as reactive absorbents for oxygen separation (the Moltox process) has been hampered by the highly corrosive nature of the molten salts. The compatibility of these molten salts with microporous ceramic membrane structures is currently under study.

Norman, Pez, and Roberts (1988) discussed the performance of lacunar or strapped Co Schiff-base O_2 carriers in immobilized liquid membranes at 273°K. Despite the use of lacunar complexes, which effectively eliminate the formation of peroxo-dimer species, in the presence of excess N-methylimidazole (MeIm), the lacunar complexes degraded rapidly. Solutions of the lacunar carrier in dichlorobenzene without additional axial base had longer lifetimes, up to 30 hours, but the O_2/N_2 separation factor was only 5.3. The authors attributed the low separation factor to the low diffusivity of the large carrier molecule and to the low equilibrium constant in the absence of axial base. Addition of two equivalents of N-benzylimidazole axial base to a 0.2 M solution of lacunar carrier increased the O_2/N_2 separation factor to 13 at the cost of increased degradation rate. The separation factor of 13 was observed for an ILM with a thickness of 128 μm. Keeping the liquid membrane composition constant, fluxes were measured as the membrane thickness was decreased to 63 and 25 μm. The separation factor decreased to 11 at a thickness of 63 μm and further decreased to 10 at a thickness of 25 μm. This is strong evidence that the reaction rate was limiting the mass transfer rate, or that the value of ε was approaching unity.

One of the unique aspects to the work disclosed in the Japanese patent literature is the attention given to the development of very thin, fine-pore, polymer top layers on support membranes for the fabrication of very thin ILMs using plasma polymerization. Kagawa, Kawakami, and Okita (1986) described the plasma polymerization of 4-vinylpyridine on a polytetrafluoroethylene filter to produce a 3-μm ILM while Yamada et al. (1987) claimed the formation of a 0.2-μm porous layer by the plasma polymerization of NMP on a porous polyarylene ether hollow fiber. The 0.2-μm top layer was impregnated with a 0.1 M dimethyl sulfoxide (DMSO) solution of N,N'-bis(2-aminobenzal)ethylenediamine and the ILM produced a permeate O_2 concentration of 61% and an O_2 permeance of 1.2×10^5 Barrers/cm.

Kobayashi, Konno, and Matsura (1988a, 1988b, 1988c) disclosed ILMs for O_2 separation consisting of a layer of Co polyamine salt solutions held on top of a polytrimethylvinylsilane supporting membrane. At 303° K and 2 mm Hg permeate pressure, O_2/N_2 separation factors ranged from 8.6 to 14.6 with O_2 permeances of 5.9×10^3 to 5.1×10^4 Barrers/cm.

In a recent patent, Kazuhiro, Yamanochi, and Yamada (1989) disclosed the use of a Co histidine dimer carrier immobilized in a polyvinylidenefluoride filter. Using ethylene glycol as the solvent increased both the O_2 permeance and ideal separation factor to 3.1×10^3 Barrers/cm and 25 compared to 2.4×10^3 Barrers/cm and 4, respectively, for an identical ILM where water was used as the solvent.

Fixed Carrier Solid Membranes

Several similar accounts of the fabrication of solid facilitated transport membranes containing reactive species selective for oxygen appeared in technical journals and the patent literature in 1986 (Sterzel, Sanner, and Neümann 1986; Zupancic and Swedo 1986; Nishide et al. 1986a; Drago and Balkus 1986). These membranes are referred to as *fixed carrier* or *chained-carrier* membranes and are generally assumed to be more stable than an immobilized liquid membrane with a mobile carrier. Although controversial, these studies provide some evidence that facilitated diffusion can occur in the solid state. A major topic of controversy is the unknown mechanism of transport in these materials. As discussed above, several hypotheses for the mass transfer mechanism have been published including dual-mode sorption (Nishide et al. 1986b, 1987a), intersegmental mobility of polymer chains with reactive groups (Cussler, Aris, and Bhown 1989), and diffusion along the polymer chain between fixed sites (Noble 1990).

Sterzel, Sanner, and Neümann (1986) made the remarkable and yet unsubstantiated claim of an $\alpha_{O_2/N_2} > 1000$ with an O_2 permeance between 1×10^4 and 5×10^4 Barrers/cm for a fixed carrier membrane at 30 to 35°C consisting of a copolymer of vinyl monomers such as styrene and 2-vinylpyridine. A cobalt Schiff-base such as CoSalen was covalently attached to the β carbon of the vinyl group of the styrene monomer prior to polymerization. A polypropylene microfiltration membrane was coated with the reactive polymer mixture to fabricate a membrane for permeation tests.

Zupancic and Swedo (1986) prepared a solution of a polymer with Lewis base sites and an organometallic oxygen carrier. This solution was then mixed with a second organic solution of isocyanate-capped polymer and the resulting polymer solution was cast on a porous support. The composite membrane formed has an interpenetrating network structure with the polymer containing the oxygen carrier and axial base sites physically mixed with the isocyanate polymer. A range of performance data was reported for similar membranes and the highest α_{O_2/N_2} reported was 4.22 for a feed gas consisting of air at 75 psia. The corresponding oxygen permeance was 1.1×10^3 Barrers/cm. The unusual nature of the performance data was that the separation factor often increased with increasing feed pressure. This is contrary to the generally observed carrier saturation behavior of facilitated transport membranes.

The largest collection of work on fixed carrier polymer membranes for air separations is from Nishide, Tsuchida, and coworkers who prepared a variety of polyalkylmethacrylate membranes containing Co porphyrins and Co Schiff-bases as oxygen carriers and Mn complexes as nitrogen carriers (Nishide et al. 1986a, 1986b, 1989; Nishide, Kawakami, and Tsuchida 1989). In their earliest work (Nishide et al. 1986a, 1987a), a Co porphyrin and axial base complex (CoPIm) was dissolved in a toluene solution of polybutylmethacrylate (PBMA) under a N_2 atmosphere and the solution was subsequently cast to form a dense membrane. Reversible 1:1 binding of O_2 to the fixed carrier was demonstrated using visible spectroscopy. Pure gas O_2 and N_2 permeabilities were measured at 25°C where the polymer was in the rubbery state. A 20-μm PBMA/CoPIm membrane containing 4.5 wt.% carrier exhibited an O_2 permeability of 23 Barrers and an ideal separation factor of 12 at a very low O_2 partial pressure of 5 mm Hg. The expected carrier saturation behavior was observed and the oxygen permeability and separation factor for a membrane containing 2.5% carrier decreased by approximately 25% as the feed oxygen partial pressure increased to 700 mm Hg. The N_2 permeability was not a function of oxygen partial pressure. This membrane was a physical mixture in which no chemical interaction occurred between the carrier and the polymer chain.

Nishide et al. (1986b) presented another strategy for preparing fixed carrier membranes for oxygen where the axial base was incorporated into the polymer by copolymerizing 2-vinylpyridine with octyl methacrylate. A simple Co Schiff-base, CoSalen, was dissolved in a toluene solution of the copolymer in an O_2-free environment and the solution was cast to form a dense membrane. The structure of a fixed carrier membrane is shown in Figure 44-7. Again, reversible 1:1 binding of oxygen to the carrier in the solid state was demonstrated using visible spectroscopy and the equilibrium constant of the reaction was measured at 30°C in both the liquid and membrane states. The equilibrium constant in the membrane was 16 times larger than in solution. This surprising result was explained by examining the stability of the Schiff-base and axial base complex. In solution a dynamic equilibrium exists between the axial base and the carrier but in the solid state the Schiff-base axial base complex is much more stable and therefore more active. At 30°C, a polyoctylmethacrylate pyridine (POMPy) membrane containing 2.5 wt.% CoSalen had an O_2 permeability of 17 Barrers and an ideal separation factor of 4.25 at 25 mm Hg O_2 feed pressure. The O_2 permeability decreased to 14 Barrers as the feed O_2 pressure increased to

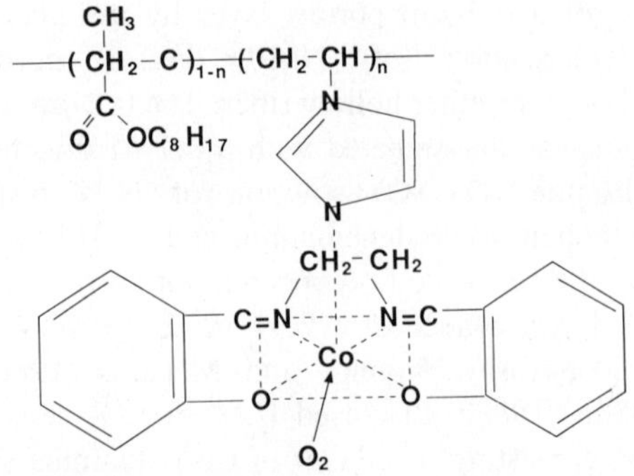

FIGURE 44-7. The chemical structure of a fixed carrier membrane for oxygen transport consisting of the copolymer of octyl methacrylate and vinylimidazole and the cobalt Schiff-base CoSalen (reprinted from Tsuchida et al. 1987 with permission).

200 mm Hg, indicative of typical carrier saturation behavior.

The two preliminary studies described above demonstrated that facilitated transport of oxygen was taking place using permeation experiments and spectroscopy. The problem encountered is that the oxygen carrier species in the solid state saturate very easily at oxygen feed pressures of less than 50 mm Hg, far below the 160 mm Hg O_2 partial pressure in air at 1 atm. Using the theoretical framework presented, one explanation for carrier saturation is that the dimensionless equilibrium constant is too large. Nishide and coworkers studied this problem by systematically decreasing c_{A0}, the O_2 concentration in the membrane. This was accomplished by adding an additional polymer layer on the feed gas side of the membrane using polymers with a range of O_2 solubilities (Nishide et al. 1987b, 1987c) and by fabricating the facilitated transport membrane using polymers with different O_2 solubilities (Ohyanagi et al. 1987, 1988). The conclusion from both studies was that polymers with lower O_2 solubility produced membranes less susceptible to carrier saturation, but at the cost of much lower O_2 permeabilities. For example, a layer of polyhydroxyethylacrylate - co - hydroxyethylmethacrylate (PHAM) was laminated to the feed gas side of a PBMA/CoPIm membrane to increase the ideal separation factor to 12 from 4.7 without the PHAM layer. However, the presence of the PHAM layer decreased the O_2 permeability by a factor of approximately 100.

Ohyanagi et al. (1988) also investigated the effect of changing the metal center in the porphyrin carrier molecule from Co to Fe. The equilibrium constant for the FePIm carrier was 5.0 cm Hg^{-1}, over twice the value for the CoPIm carrier. However, both the forward and reverse reaction rate constants for the FePIm reaction with oxygen were lower than those for the CoPIm carrier. Holding all other experimental parameters constant, the permeation tests showed that the PBMA/CoPIm membrane produced higher degrees of oxygen facilitation at O_2 feed pressures below 400 mm Hg than the PBMA/FePIm membrane. A similar study on the effect of axial base strength on the performance of fixed carrier membranes for oxygen produced some very high separation factors (Tsuchida et al. 1987). Membranes were prepared using CoSalen and copolymers of polyoctylmethacrylate and 1-vinylimidazole (POMIm) and 4-pyridine (POMPy). The equilibrium constant for oxygen binding in the POMIm membrane was larger than in the POMPy membrane at all temperatures tested between 25 and 45°C due to the stronger basicity of the imidazole (Im) group. However, at the conditions of 10 mm Hg O_2 feed pressure and 35°C, a POMPy/CoSalen membrane with 12 wt.% carrier had the best ideal separation factor of 15 with an O_2 permeability of 14.7 Barrers.

Prior parametric studies identified some of the desirable properties of an oxygen carrier for a fixed carrier membrane including a high equilibrium constant and a high reverse rate constant (k_r) (Nishide et al. 1988). Other factors that influence the O_2/N_2 separation factor are the carrier concentration and the O_2 solubility in the membrane. Increasing carrier loading will increase the separation factor, while increasing O_2 solubility in the membrane decreases the separation factor. Nishide et al. (1988) described the synthesis of a new Co porphyrin, CoMP, with both a larger equilibrium constant and larger k_r than CoPIm. The CoMP porphyrin had a very short lifetime in both toluene solution and an alkyl acrylate membrane using methylimidazole as an axial base so CoMP was polymerized with hexyl methacrylate to form a copolymer that reacted with oxygen. The copolymer was much more stable with half-lives at 25°C for 1 day in toluene solution and over 1 month in the membrane state. The half-life of CoPIm in solution at 25°C was 1 week and over 1 month in a PBMA membrane. Surprisingly, the lifetime of the CoMP copolymer was equivalent to the PBMA/CoPIm membrane even though CoMP had an equilibrium constant 660 times larger than CoPIm and a reverse rate constant 5 times larger. This is strong evidence that the fixed carrier configuration is more stable than an immobilized liquid membrane configuration and that reducing the mobility of oxygen carriers in a solid membrane reduces degradation compared to solution. At 30°C and an O_2 feed pressure of 160 mm Hg, the CoPM copolymer membrane containing only 0.17 wt.% carrier had an O_2 permeability of 26 Barrers compared to 11 Barrers for a 2.5 wt.% PBMA/CoPIm membrane at the same temperature. The ideal separation factor for this membrane was not reported.

Sugie (1988) used CoSalen and acrylate copolymers similar to those used by Nishide et al. (1986b) to prepare fixed carrier membranes for oxygen using a novel technique. In an N_2 atmosphere, a porous polymer ultrafilter was dipcoated with a solution of CoSalen. The solvent was allowed to evaporate, leaving a uniform coating of the Co Schiff-base throughout the pore structure. The coated filter was then dipped in a solution of the copolymer polyhexylmethacrylate-co-4-vinylpyridine (PHMPy) to form a continuous layer of a PHMPy/CoSalen complex on the support. Pure gas permeation tests were conducted at 25°C and very high separation factors were observed. For a feed oxygen pressure of 100 mm Hg, the ideal separation factor was a remarkable 26.4. However, the corresponding O_2 permeance was only 3.12 $\times 10^2$ Barrers/cm, approximately two orders of magnitude lower than what would be required for commercial air separations.

Drago and coworkers (Drago and Balkus 1986; Barnes, Drago, and Balkus 1988) prepared polystyrene membranes with covalently bound cobalt Schiff-bases for selective transport of oxygen. A major difference between this work and the work of Nishide and coworkers is the use of a glassy polymer (polystyrene), rather than rubbery polymers (polyalkylmethacrylates). Cobalt and nickel Schiff-bases were covalently attached to polystyrene beads, which were crushed and then added to a solution of polystyrene. The polymer solution was subsequently cast to produce the fixed carrier membrane. The polystyrene/Co Schiff-base (PS/CoSPDT) membrane would react with oxygen while the PS/NiSPDT membrane was inert and served as a blank. Permeabilities were measured at ambient conditions using a constant volume technique. Air at ambient pressure was used as the feed gas and the permeate pressure was measured as a function of time. As expected from theory, the oxygen permeability was observed to be a function of the permeate O_2 partial pressure for the PS/CoSPDT membranes. To determine the extent of the O_2 facilitation by the CoSPDT fixed carrier, the average separation factors were determined for the blank membranes containing the NiSPDT complexes. From the N_2 permeability of the PS/CoSPDT membrane and the separation factor of the blank, a permeate O_2 partial pressure was calculated representing no complexation reaction. This permeate O_2 partial pressure without facilitation was compared with the actual O_2 permeate pressure from the experiments with the PS/CoSPDT membranes. The permeate oxygen partial pressures for the fixed carrier membranes were larger than the values corresponding to no complexation reaction and this difference was attributed to facilitated transport. Unfortunately, both the O_2 permeability and the O_2/N_2 separation factor were reported as an average over time instead of as a function of O_2 partial pressure. Consequently, no comparison can be made with other work in the literature.

More recently, Delaney, Reddy, and Wessling (1990) reported an investigation of fixed carrier membranes selective for oxygen. Using an approach similar to that of Drago and coworkers, they synthesized styrene copolymers with grafted Co(II) Schiff-base complexes. The best permeation results were obtained using a Co(II) complex of a styrene-methacrylamidoethyl - N,N' - disalicylidene-3,4-diaminobenzoate copolymer [PS-IPO-Co(II) Salphen]. Triethylamine axial base was simply added directly to the polymer solution and polymer films were cast in an N_2 atmosphere. The styrene Schiff-base copolymer without Co(II) and axial base was used as one of the controls and had a pure gas P_{O_2} = 2.16 Barrers and an ideal O_2/N_2 separation factor of 5.14. The addition of 3.76 wt.% Co(II) and excess axial base to this copolymer simultaneously increased both the pure gas O_2 permeability to 3.99 Barrers and the ideal separation factor to 12.2 at an O_2 pressure of 283 cm Hg or 54.7 psia. Compared to previous work on O_2 selective fixed carrier membranes, the PS-IPO-Co(II) Salphen · $N(C_2H_5)_3$ membrane did not exhibit the same carrier saturation behavior and maintained a high ideal separation factor even at high O_2 pressures. Although the authors did not measure the O_2 binding constants of the PS-IPO-Co(II) Salphen polymer, their explanation of the improved performance at high O_2 pressures was due to a much lower O_2 binding constant than the membranes of Nishide and coworkers. A series of PS-IPO-Co(II) Salphen · $N(C_2H_5)_3$ membranes was prepared with Co(II) loadings ranging from 0 to 4 wt.%. Oxygen permeation measurements showed no percolation threshold. However, the lifetime of these membranes is quite limited. The ideal separation factor decreased by 51% in just 200 hours of exposure to pure oxygen. The authors hypothesized that carrier oxidation or volatilization of the axial base could be responsible for the rapid performance loss.

Nitrogen is usually considered an inert molecule; however, a large number of transition metal complexes of nitrogen have been synthesized for the purpose of reducing the nitrogen coordinated to the metal center to produce NH_3 (Henderson, Leigh, and Piekett 1983; Yamamoto 1986). Such nitrogen complexes are often unstable in the presence of oxygen. Nishide et al. (1989) and Nishide, Kawakami, and Tsuchida (1989) discussed the synthesis of fixed carrier polymer membranes selective for nitrogen. The membrane was a copolymer of octyl methacrylate and (vinylmethylcyclopentadienyl)tricarbonyl manganese. After casting the polymer solution in an Ar atmosphere, the membrane was exposed to UV radiation to remove one of the CO ligands, creating a binding site for molecular N_2. Reversible 1:1 binding of N_2 to the Mn complex was observed using UV spectroscopy. A membrane containing 6.4 wt.% of $CpMn(CO)_2$ exhibited facilitated transport behavior for N_2. At 45°C, the pure gas N_2 permeability was a function of N_2 feed pressure, decreasing from 10.2 Barrers at a feed pressure of 21 mm Hg to 9.15 Barrers at 740 mm Hg. The pure gas O_2 permeability was a constant value of 21 Barrers at 45°C. Surprisingly, the lowest carrier concentration tested, 6.4 wt.% had the highest N_2 permeability although the higher carrier loadings of 18.5 and 32 wt.% exhibited less carrier saturation at high N_2 feed pressures. The $POM/CpMn(CO)_2$ membrane also reacted with acetylene (Nishide, Kawakami, and Tsuchida 1989), which will be discussed below.

Acid Gas Removal

The facilitated transport of carbon dioxide has been studied from both a biological and an engineering perspective. One system that has been extensively studied is the use of the bicarbonate ion as a carrier. Enns (1967) established the role of the bicarbonate ion in carbon dioxide facilitated transport and demonstrated that the flux could be increased by catalyzing the reaction sequence with the enzyme carbonic anhydrase. Also, anions of weak acids such as selenite, arsenite, tellurite, and hypochlorite have been found to be effective catalysts (Meldon, Smith, and Colton 1977; Kimura, Matson, and Ward 1979). Ward and Robb (1967) used sodium arsenite as a hydration catalyst for removing carbon dioxide from a carbon dioxide-oxygen mixture. They achieved a separation factor of 4100. The low oxygen permeability can be partially attributed to the preferential precipitation of oxidized salts in the 6M $CsHCO_3$ membrane solution. Otto and Quinn (1971) measured the carbon dioxide flux through thin bicarbonate solution films. They found the transport to be reaction rate limited in the absence of catalysts. The facilitation factor increased with increasing layer thickness and alkali metal concentration. They were able to increase the carbon dioxide flux to reaction equilibrium (diffusion-limited) values with the addition of the enzyme carbonic anhydrase. Meldon, Smith, and Colton (1977) have shown that carbon dioxide facilitation is increased by the addition of a weak acid buffer solution. A review of this system has been published by Meldon, Stroeve, and Gregoire (1982). LeBlanc et al. (1980) reported facilitated carbon dioxide transport using ethylenediamine as the carrier in ion-exchange membranes (IEMs). They noted the advantages of ion-exchange membranes over immobilized liquid membranes. They also stated that these advantages could contribute to a longer operating life in practical gas separations under adverse high-temperature and -pressure conditions. Way et al. (1987) also used ethylenediamine as the carrier in a water-swollen perfluorosulfonic acid ion-exchange membrane. They measured separation factors of up to 551 for CO_2 over CH_4 at a CO_2 feed partial pressure of 30 mm Hg.

Matson, Herrick, and Ward (1977) utilized a carbonate solution immobilized in a porous polymer film to study hydrogen sulfide removal for coal gasification applications. Their membrane had a high hydrogen sulfide permeability and greater selectivity than conventional hot carbonate scrubbers. They made permeability measurements at high temperature and pressure (90 to 130°C, total feed pressure 300 psia) and observed a hydrogen sulfide permeability dependence on carbon dioxide partial pressure. This result is reasonable since both hydrogen sulfide and carbon dioxide can compete for the carrier. They also studied membrane lifetime. Their apparatus was operated continuously for periods of up to 1000 hours and no appreciable decrease in membrane permeability was observed. They also noted that carrier deactivation due to the presence of oxygen in coal gas was possible and that fouling due to coal tars and dust would have to be considered in an industrial-scale system.

Kimura, Matson, and Ward (1979) reported progress toward applying facilitated transport membranes to industrial-scale separations. They measured carbon dioxide and hydrogen sulfide permeabilities at industrially significant operating conditions (100 psia CO_2, 90 to 130°C). Under these conditions, the authors determined that carbon dioxide transport is influenced by both diffusion and reaction rates. They also noted that hydrogen sulfide selectivity over carbon dioxide can be improved by introducing gas gaps in a multilayer membrane because of the higher reaction rates of hydrogen sulfide. They found that catalysts increased the carbon dioxide hydration rate by a factor of 2 at low carbon dioxide partial pressure; this effect was observed to increase as carbon dioxide partial pressure decreased. Economic studies based on experimental data indicate that cost savings on the order of 30 to 50% over conventional acid gas scrubbing were possible with immobilized liquid membranes.

Way and Noble (1987) reported facilitated transport of H_2S in perfluorosulfonic acid (PFSA) ion-exchange membranes containing organic diamine cations at ambient temperature and pressure. The IEMs were highly selective for H_2S over CH_4 and separation factors for H_2S/CH_4 up to 1200 were observed. An H_2S permeability of 4122 Barrers was measured at an H_2S feed partial pressure of 18 mm Hg. Similar membranes were used in a study of competitive facilitated transport with feed gas mixtures of CO_2 and H_2S (Way and Noble 1989). Over a wide range of feed gas composition, the transport of CO_2 was favored over the transport of H_2S (see Figure 44-3). Preferential transport of CO_2 was expected since the equilibrium constant of the CO_2-ethylenediamine reaction was 30 times the equilibrium constant of the H_2S reaction with the carrier.

As discussed in earlier chapters, polymeric membranes such as cellulose acetate are now widely used to remove acid gases from mixtures with hydrocarbons. Polysulfone hollow fibers and cellulose acetate spiral-wound modules are also used commercially to recover H_2. Many commercial processes such as steam reforming of hydrocarbons, ammonia production, and the carbon monoxide shift reaction could be operated more efficiently if the acid gases could be selectively removed from process streams. For example, in steam reforming of CH_4 to produce H_2, the reaction products are CO_2 and H_2. A membrane process that could selectively remove CO_2 could be used to shift the equilibrium toward the products and to perform some of the acid gas removal prior to conventional physical and chemical absorption processes in a hybrid configuration. Analogously, in the Haber process to produce ammonia, if NH_3 could be removed selectively from the reaction product mixture leaving the unreacted N_2 and H_2 at high pressure, the equilibrium could be shifted toward the NH_3 product. Glassy polymers often separate gases on the basis of differences in diffusion coefficients of the penetrant species and are usually H_2 selective. Consequently, to remove acid gases selectively from mixtures with H_2 with a membrane process will require new materials that exploit differences in solubility or reactivity of the penetrant species. Candidate materials include rubbery polymers, liquids, solvent swollen or gel polymers, and reactive polymers.

Pellegrino, Nassimbene, and Noble (1988) discussed a technique to modify the structure of facilitated transport PFSA ion-exchange membranes that simultaneously increases both the permeability and the separation factor. PFSA IEMs containing an ethylenediamine carrier swollen in glycerine at temperatures up to 225°C were compared to IEMs without the heat treatment. The hot glycerine treatment increased the water content of the membrane almost by a factor of 4, the facilitated CO_2 flux by a factor of 16, and the diffusive flux (inert counter-ion, no carrier) by a factor of 5 for experiments at ambient temperature and pressure. The observation that the facilitated flux increased much more than the diffusive flux was attributed to an increase in the carrier mobility due to structural changes in the IEM. Such treatments did not drastically change the permselectivities of the gases that did not react with the carrier so that the separation factors also increased due to the treatment. The CO_2/CO separation factor after treatment at 225°C was 126, almost a factor of 3 larger than the CO_2/CO separation factor without treatment of 43.2.

To increase the selectivity of facilitated transport IEMs for CO_2 over H_2, Way and Hapke (1988) swelled PFSA membranes with low volatility aprotic solvents such as propylene carbonate, N-methylpyrrolidinone (NMP), and ethylene glycol. The highest CO_2 permeabilities and CO_2/H_2 separation factors were obtained with NMP swollen films. For a binary mixture feed gas at 165 psia and 25°C, a CO_2/H_2 separation factor of 11.8 was observed with a corresponding CO_2 permeability of 507 Barrers.

Matson et al. (1988) used an analogous approach to prepare composite gel-coated membranes selective for acid gases. An asymmetric cellulose acetate membrane was coated with a 15-μm selective layer consisting of a cross-linked polyurethane gel swollen with N-cyclohexyl-2-pyrrolidone or similar polar, high boiling solvents. Pure gas pressure-normalized fluxes were measured at 115 psia and ambient temperature. The ideal separation factor for CO_2/H_2 was 14.7 and the CO_2 permeance was 1.43×10^5 Barrers/cm. The solvent-swollen membranes were selective for H_2S over CO_2 as well as CH_4 and a very high H_2S permeance of 1.18×10^6 Barrers/cm was observed.

Extending previous work on molten salt liquid membranes for facilitated transport of gases, Laciak et al. (1989) summarized work to develop new materials to prepare CO_2- and NH_3-selective liquid and gel membranes. The CO_2-selective liquid membrane was based on the discovery of the reversible reaction of molten tetramethylammonium fluoride tetrahydrate salt (TMAF) with CO_2 (Quinn, Pez, and Appleby 1988). The melting point of the TMAF salt is between 39 and 42°C and the resulting molten salt has a high CO_2 absorption capacity attributed to a two-step acid-base reaction. A partial pressure of 800 mm Hg CO_2 in equilibrium with TMAF produces a solution that is 2 M in CO_2. A TMAF solution supported on a polytrimethylsilylpropyne (TMSP) membrane exhibited typical facilitated transport behavior in permeation studies with humidified CO_2/H_2 mixtures. The CO_2/H_2 separation factor was 370 at a CO_2 feed partial pressure of 30 mm Hg, and the separation factor decreased to 30 as the CO_2 feed partial pressure increased to 1490 mm Hg. Over the same range of CO_2 feed partial pressure, the CO_2 permeance decreased from 4×10^5 to 0.8×10^5 Barrers/cm.

Ammonia-selective liquid membranes are based on the reversible interaction of NH_3 with ammonium thiocyanate, NH_4SCN. However, an ILM based on this liquid is unstable at NH_3 partial pressures above 760 mm Hg so a polymer analog of NH_4SCN was prepared, polyvinylammonium thiocyanate, abbreviated PVAmSCN (Laciak and Pez 1988; Pez and Laciak 1988). Absorption of NH_3 by PVAmSCN causes the polymer to swell and form a gel-like structure at NH_3 partial pressures over 1500 mm Hg. The high solubility of NH_3 in PVAmSCN was postulated to result from hydrogen bonding between NH_3 and SCN^- and NH_4^+ ions. To measure permeability, a composite membrane was prepared consisting of a thin layer (typically 10 μm) of PVAmSCN between two supporting layers of TMSP. In permeation experiments with a 1:1 mixture of NH_3 and H_2 at feed pressures below 75 psia, the NH_3 permeance increased with increasing NH_3 partial pressure, suggesting a solution diffusion transport mechanism (Cussler 1990). High-pressure multicomponent experiments at ambient temperature were performed with a constant NH_3 partial pressure of 115 psia, with H_2 and N_2 partial pressures in a 1:1 ratio increasing up to a total feed pressure of 835 psia. At a feed pressure of 835 psia, the NH_3 permeance was 3.2×10^5 Barrers/cm while the NH_3/N_2 separation factor was 1390 and the NH_3/H_2 separation factor was 2000. The authors noted that since NH_3 solubility decreases with increasing temperature, the NH_3 permeance will also decrease.

Unlike much of the recent work to fabricate CO_2-selective liquids or gel materials described above, Yoshikawa et al. (1986) prepared a solid, fixed carrier polymer membrane with a basic functional group chosen to interact with

CO_2. A copolymer of 4-vinylpyridine and acrylonitrile was solution cast to form the membrane. The pure gas CO_2 permeability measured at 30°C was a strong function of CO_2 feed gas partial pressure. At 38 mm Hg CO_2, the CO_2 permeability was 0.71 Barrer, which decreased to 0.025 Barrer at a CO_2 partial pressure of 610 mm Hg. Oxygen and nitrogen permeabilities were constant at all feed gas pressures. The CO_2/O_2 separation factor was 8.5 when the CO_2 partial pressure was 38 mm Hg and it decreased presumably due to carrier saturation to 3.0 at 610 mm Hg CO_2. The apparent facilitated transport mechanism was explained in terms of an acid-base interaction between the CO_2 penetrant and the basic pyridine group on the polymer chain. However, an alternate possible mechanism is possible in the presence of water vapor, which could react with the pyridine groups to form hydroxyl ions, which could in turn react with CO_2 to form bicarbonate ion. This mechanism was proposed by Meldon, Kang, and Sung (1985) to explain the CO_2 transport data of Tajar and Miller (1972) for polymer membranes containing basic functional groups.

An application that is often proposed for facilitated transport membranes is the removal of acid gases from natural gas streams. However, wellhead natural gas pressures range from 800 to 1015 psia. If the CO_2 mole fraction of the natural gas stream is appreciable, for example 0.1 or above, then the CO_2 partial pressure is 100 psia or above. The serious problem of carrier saturation must be considered. Facilitated transport may be suitable for removal of H_2S, which is usually present at low concentrations; although if the partial pressure driving force is low the corresponding flux will be low. However, cellulose acetate membranes have been successfully used to reduce H_2S concentrations in natural gas from 50 ppm to the pipeline specification of 4 ppm (Houston 1989).

There are several potential applications for facilitated transport membranes in which moderate to low acid gas partial pressures are encountered. As an example of a gas stream with a moderate acid gas partial pressure, Early, Kilgour, and Medvetz (1983) described a landfill gas with a CO_2 partial pressure of 27 cm Hg. Purification of this gas stream with polymeric membranes would generally require a compression step. With a facilitated transport membrane, compression is not generally required and the gas stream could be treated at atmospheric pressure.

Guha, Majumdar, and Sirkar (1990) prepared ILMs consisting of aqueous solutions of 20% diethanolamine (DEA) immobilized in 25-μm microporous polypropylene supports and reported permeabilities and separation factors for mixed gases with CO_2 partial pressure differences ranging from 12 to 129 cm Hg. The CO_2 permeabilities increased from 974 to 4820 Barrers as the CO_2 partial pressure decreased from 126 to 11.6 cm Hg. Corresponding CO_2/N_2 separation factors increased from 55.8 to 276 over the same CO_2 partial pressure range. For a 25-μm ILM at an intermediate CO_2 partial pressure of 22 cm Hg that is representative of landfill gas, the CO_2 permeance was 9.72×10^5 Barrers/cm and the CO_2/N_2 separation factor was 153. Spillman and Cooley (1987) reported a CO_2 permeance of 8×10^5 Barrers/cm and a CO_2/N_2 separation factor of 31 for a commercial cellulose ester spiral-wound membrane module. The laboratory data for the facilitated transport membrane compare favorably with the commercial membranes and may be competitive in a stable liquid membrane configuration such as a hollow-fiber contained liquid membrane for the removal of CO_2 from moderate partial pressure sources such as landfill gas or biogas.

Another application with high potential for facilitated transport membranes is the simultaneous removal of CO_2 and water from closed breathing environments that will be found in long-duration space missions and permanent orbital and planetary bases. The maximum CO_2 partial pressure allowable in breathing environments is very small, typically 1% or <1 cm Hg. As described above, facilitated transport membranes are known to have extremely high CO_2 permeabilities and separation factors at these conditions. Furthermore, ion-exchange or gel membranes containing CO_2-selective carriers have high water permeabilities and could be used to simultaneously dehydrate breathing atmospheres.

Flue Gas Cleanup

Facilitated transport membranes selective for sulfur and nitrogen oxides, the primary toxic component of flue gas, have been prepared. The conditions of flue gas, atmospheric pressure and very low concentration of SO_2 and NO_x, favor chemically reactive separations such as facilitated transport and chemical absorption. However, the very low partial pressures of SO_2 and NO_x encountered imply low fluxes in a practical application.

Roberts and Friedlander (1980a, 1980b) reported experimental and theoretical studies of the SO_2 flux through aqueous and alkaline salt ILMs. These ionic solutions are complex; however, the primary complexation reaction in pure water is the ionization of SO_2 to HSO_3^-. For liquid membranes containing the ion SO_3^{2-}, water, SO_2, and SO_3^{2-} also react to form HSO_3^-. Very high facilitation factors above 1000 were observed at low SO_2 feed gas mole fractions for alkaline aqueous liquid membranes containing NaOH, $NaHSO_3$, or $Na_2S_2O_5$.

More recently, Sengupta, Raghuraman, and Sirkar (1990) performed screening experiments with ILMs to evaluate several solvent/carrier systems for potential application to flue gas desulfurization using hollow-fiber contained liquid membranes. At 298°K and atmospheric pressure, ILMs were prepared with 25-μm microporous polypropylene supports. The best performance for removal of SO_2 from a synthetic flue gas mixture was observed using water, and 1 N aqueous solutions of $NaHSO_3$ and Na_2SO_3 as membrane liquids. Typical SO_2/CO_2 separation factors ranged from 138 to 190 with corresponding SO_2 permeabilities of 19.6×10^4 to 43.1×10^4 Barrers. The SO_2 concentration in the synthetic flue gas ranged from 340 to 5500 ppm.

Ward and coworkers (Ward 1970a, 1970b; Bdzil et al. 1973) have extensively investigated the facilitated transport of NO in organic solutions of ferrous (Fe^{2+}) ion. Ward (1970a) measured the NO flux through a formamide solution of ferrous ion immobilized between two thin silicone rubber membranes. His mathematical analysis showed that this system was operating somewhere in between the limiting regimes of reaction-limited and diffusion-limited mass transport. In a further investigation, Ward (1970b) observed that NO could be transported across a ferrous chloride film against a nitric oxide partial pressure gradient by applying a voltage difference across the film. Bdzil et al. (1973) presented a mathematical analysis of electrically induced facilitated transport for the NO/Fe^{2+} system.

Separations of Olefins from Saturated Hydrocarbons

There are several reports of facilitated transport of olefins using immobilized $AgNO_3$ solutions (Hughes and Steigelmann 1973; Steigelmann and Hughes 1973; Hughes, Mahoney, and Steigelmann 1981, 1986; Teramoto et al. 1986). The complexation reaction is the well-known reaction of olefins with aqueous Ag^+ ion (Beverwijk et al. 1970). Hughes, Mahoney, and Steigelmann (1981, 1986) presented the results of a bench- and pilot-scale study of ethylene and propylene transport using aqueous silver ion solutions immobilized in asymmetric, porous hollow-fiber reverse osmosis membranes at ambient temperature. In bench-scale tests with the hollow-fiber supports and 2 M $AgNO_3$ solutions, the ethylene permeance was 4.6×10^4 Barrers/cm for an ethylene feed partial pressure of 65 psia. The ethylene/ethane separation factor for a 2 M $AgNO_3$ ILM was 243 at an ethylene partial pressure difference of 95 cm Hg. The ethylene permeance for the hollow-fiber ILMs was an order of magnitude lower than values obtained for thin films of $AgNO_3$ solutions, which led the authors to conclude that the limiting mass transfer resistance was the skin layer of the hollow-fiber supports.

A field test was conducted to recover propylene from the purge gas of a commercial polypropylene facility using modified commercial hollow-fiber reverse osmosis modules with areas ranging from 200 to 400 ft^2. During the longest run of the field test, a permeator processed over 16 tons of feed gas and produced 7 tons of propylene at greater than 97 mol% purity. The present authors calculated an average propylene permeance of 3.5×10^4 Barrers/cm from field test data. Operational problems encountered with the ILM system for propylene recovery included low permeance, temperature rise in the sweep liquid due to the heat of mixing of propylene in hexane, gradual flux decline due to loss of solvent and carrier, and the necessity to remove hydrogen from the feed gas to prevent reduction of the Ag^+ carrier. This work is very significant because it is the first report on the use of facilitated transport membranes for gas separations on a pilot scale.

Teramoto et al. (1986) also studied ethylene transport with immobilized $AgNO_3$ aqueous solutions at ambient temperature. They found selectivity for ethylene over ethane of approximately 1000 when the $AgNO_3$ concentration was 4 M and the ethylene partial pressure was 37 cm Hg. The highest ethylene permeance of 3×10^5 Barrers/cm was observed for a 4 M carrier concentration.

Kawakami et al. (1987) screened polyethyleneglycol (PEG) solutions of Group VIII metal ions at 25°C to ascertain their ability to absorb reversibly light olefins. Only solutions of Rh^{3+} in PEG 300 sorbed ethylene and propylene reversibly. Immobilized liquid membranes were prepared using 0.045 M Rh^{3+}/PEG solutions in glass fiber filters and C_3H_6/N_2 separation factors were 73—1.6 times that of pure water. Addition of 0.5 M KNO_3 further increased the separation factor to 87 due to the salting out effect.

Solvent-swollen ionomer films have been used as supports for ionic complexation agents in order to improve the stability of facilitated transport membranes for gas and liquid phase olefin transport (LeBlanc et al. 1980; Kraus 1986; Koval and Spontarelli 1988; Koval, Spontarelli, and Noble 1989). Use of ionomer materials improves the stability of facilitated transport membranes because the support is nonporous and the carrier cannot be removed from the membrane except by an ion-exchange reaction.

LeBlanc et al. (1980) reported ethylene transport using a Ag^+ counter-ion carrier in a water-saturated sulfonated polydimethylphenyleneoxide cation-exchange membrane at ambient temperature. A pure gas ethylene permeance was reported at atmospheric pressure of 2.1×10^3 Barrers/cm. The ethane permeance at the same conditions was 0.73×10^3 Barrers/cm, corresponding to an ideal separation factor of 288.

Similarly, Kraus (1986) prepared facilitated transport ion-exchange membranes using Nafion® 415 containing an Ag^+ carrier. However, following the ion exchange to introduce the Ag^+ carrier into the IEM, the membrane was dried and soaked in an organic alcohol such as glycerol or octanol. For a Ag/Nafion® IEM soaked in glycerol, the ethylene/ethane separation factor was 10 and the ethylene permeability was 6.9 Barrers at ambient temperature and pressure.

Koval and Spontarelli (1988) reported the liquid phase facilitated transport of hexene and hexadiene between feed and permeate water-saturated decane solutions using water-saturated 25-μm-thick PFSA membranes containing Ag^+ counter-ions. Facilitation factors were computed as the ratio of the penetrant flux with Ag^+ carrier present to the flux with an inert counter-ion such as Na^+. The observed facilitation factors for hexene and hexadiene were 460 and 450, respectively. Koval, Spontarelli, and Noble (1989) recently described the separation of styrene and ethylbenzene mixtures using PFSA membranes containing Ag^+ counter-ions. Ideal separation factors, the ratio of the pure styrene permeability to the pure ethylbenzene permeability, ranged from 10 to 36. The smaller separation factors were observed for higher feed styrene concentrations, which was attributed to carrier saturation. The maximum styrene flux of 1.6×10^{-10} mol/(cm^2 · s) was reported for a feed concentration of 0.1 M styrene.

The separation of olefin and paraffin mixtures using hybrid distillation/membrane processes has been proposed. Membranes could be added to existing cryogenic distillation processes to form more efficient hybrid processes or possibly replace the distillation process in new plants. In a modeling study of the propane/propylene separation using the data of Teramoto et al. (1986), Gottschlich and Roberts (1989) reported that a membrane/distillation hybrid could produce savings of 2 to 4 cents/lb compared to the distillation process alone.

Ho et al. (1988) reported the separation of olefin/paraffin and olefin isomer mixtures by chemical absorption using a chemical system that could easily be used in a liquid or facilitated transport membrane configuration. The chemical solvent consists of cuprous hexafluoroacetylacetonate (diketonate) in an olefinic weakly complexing solvent (WCS) such as α-methyl styrene. The synthesis of these materials is disclosed by Doyle et al. (1984). For example, the olefin to be separated displaces the α-methyl styrene ligand bound to the cuprous diketonate carrier. The WCS is necessary to prevent the disproportionation of the Cu(I) diketonate to the Cu(II) diketonate and copper metal. The weakly complexing solvent was chosen by selecting an olefin that was highly sterically hindered at the double bond such as α-methyl styrene. Separation factors for C_2H_4/C_2H_6 and C_3H_6/C_3H_8 measured at 28°C and 90 cm Hg olefin were 17 and 10, respectively. The separation factors of 1-butene from other butene isomers at 50°C and 81 cm Hg olefin were over 2. The final example of an application for this novel complexing solution was the separation of 1-pentene from other pentenes at 90°C and 60 cm Hg olefin; separation factors over 1.5 were reported.

The POM/CpMn(CO)$_2$ fixed carrier membrane discussed above also reacts with acetylene (Nishide, Kawakami, and Tsuchida 1989). The time lag of the acetylene diffusion is a function of the acetylene partial pressure, which the authors claim is due to reversible interactions with the fixed carrier. However, the acetylene permeability is not a function of acetylene partial pressure although only two data points are reported below 200 mm Hg acetylene partial pressure.

Another recent study of facilitated transport of butenes was reported by Peinemann and Shukla (1989). They prepared composite membranes by casting PFSA and amine-modified polyethylene solutions containing Ag salts on microporous supports. At ambient temperature and pressure, butene permeance values ranged from 8×10^4 to 6×10^5 Barrers/cm with corresponding butene/butane ideal separation factors of 10 to 3000.

Carbon Monoxide Separation

The facilitated transport of carbon monoxide by cuprous chloride in aqueous solution was first studied by Hughes and Steigelmann (1974) and more recently by Smith and Quinn (1980). Smith and Quinn found the cuprous ion to be a very effective carrier for carbon monoxide that increased the flux by a factor of 100 over the purely diffusive case for a modest carrier concentration of 0.2 M. The facilitation factor was measured as a function of carbon monoxide partial pressure, total copper concentration, and membrane thickness. The CO permeability at the same carrier concentration was 1000 Barrers.

Kawakami et al. (1984) prepared a complexing solution for carbon monoxide in order to avoid the stability problems with Cu(I) species, which are often unstable toward oxidation to Cu(II) or by disproportionation to Cu(II) and copper metal in aqueous solution. Cu(SCN) was dissolved in tetraethylenepentamine (tetren) up to a mole ratio of 1:12 Cu(I) to tetren. In this case, tetren was used as a stabilizing ligand for the Cu(I) ion and as a solvent. This complexing solution was also observed to react with O_2 (Kawakami et al. 1982) as previously described. The CO flux was observed to be a strong function of the Cu^+ concentration and the CO feed pressure, supporting a facilitated transport mechanism. At 30°C and a Cu^+ concentration of 0.39 M, the CO permeance was 2.1×10^2 Barrers/cm for a CO feed gas pressure of 22.5 mm Hg. This CO flux corresponds to a facilitation factor of 3.1 and a CO/N_2 ideal separation factor of 3.8.

Recently, Koval et al. (1985) and Reyes (1986) reported CO facilitated transport using an immobilized benzonitrile solution of a ferrous complex derived from the tetraimine macrocyclic ligand 2,3,9,10-tetramethyl-1,3,8,11-tetraazacyclotetradeca-1,3,8,10-tetraene (TIM). They measured the kinetic and diffusional constants and showed selectivity for CO over a variety of other gases. The transport of CO was primarily limited by the solubility of the Fe(II)TIM complex in benzonitrile. The experimental and mathematical procedures they described can be used for any simple complexation reaction.

Okada and coworkers (Okada, Omiyama, and Matsura 1986; Okada et al. 1986, 1987) disclosed a liquid membrane system that also consists of Cu(I) ions in a complexing solvent. For example, Cu(SCN) and N-methylimidazole were dissolved in DMSO and supported on an asymmetric poly(trimethylvinylsilane) membrane to prepare an ILM selective for CO. The CO permeance was 2.4×10^5 Barrers/cm and the ideal separation CO/N_2 separation factor was 191.

Other Applications

Kuo and Gregor (1983) prepared solvent-swollen polymer membranes containing a decalin solution of trioctylphosphine oxide (TOPO) carrier to extract acetic acid from aqueous solution. Facilitated transport of acetic acid was demonstrated although the measured fluxes were very low due to large membrane thicknesses. The selectivity of the TOPO-impregnated membrane was approximately 3 for acetic acid over hydrochloric acid.

Ishikawa et al. (1985) reported the separation of fumaric acid from aqueous mixtures with L-malic acid using ILMs consisting of 1-decanol immobilized in cellulosic or polytetrafluoroethylene microfilters. No carrier species was used and the transport mechanism was solution-diffusion of the undissociated acid across the hydrocarbon liquid membrane. The product phase was maintained at a pH value higher than the pK of the acids in order to use the dissociation reaction to trap the permeated acids in the product phase. This trapping reagent mechanism has been widely used in emulsion liquid membranes described earlier for the recovery of various organic species. In laboratory-scale batch experiments, the product phase was enriched in fumaric acid over L-malic acid by a factor of 3.89.

Citric acid is an important food additive produced commercially by fermentation. Babcock et al. (1986) and Friesen et al. (1991) presented a study of the recovery of citric acid from synthetic and actual fermentation beers using immobilized liquid membranes as a function of temperature, liquid membrane composition, and citric acid feed concentration. A tertiary amine carrier, trilaurylamine, and an organic alcohol modifier were dissolved in an aliphatic hydrocarbon solvent and immobilized in 25-μm microporous polypropylene ultrafilters. Both facilitated transport and coupled transport mechanisms were investigated. In the facilitated transport experiments, a citric acid concentration driving force was used to transport citric acid. In the coupled transport mode, the driving force was the difference in hydrogen ion concentration across the membrane, and monosodium citrate was produced on the product side of the membrane. Transport of the citrate anion is coupled to the opposite flux of the hydrogen ion via an ion-exchange reaction. By maintaining a low pH on the feed side of the membrane and a high pH on the product side, citrate anions can be transported across the membrane from a feed solution of low concentration to a high concentration on the product side. Using a coupled transport mechanism, citrate fluxes of greater than 60 μg/(cm^2 · min) and citric acid to glucose selectivities of over 2000 were observed (see also page 775). Preliminary economic calculations indicated that citric acid could be recovered using coupled transport for approximately one-half of its market value.

This chapter has only concentrated on the separation applications of facilitated transport and immobilized liquid membranes. Other possible applications of selective barriers for gas, vapor, and liquid phase solutes include sensors, controlled release, and membrane reactors, devices that combine separation and reaction functions. Membrane reactors are described in detail in Chapter 43.

Facilitated transport and liquid membranes can be used in at least two ways to prepare gas and vapor sensors. The membrane can be used as a selective barrier between the gas phase and the transducer. Also, a property change in the membrane itself such as resistance, conductivity, or absorbance due to complexation can be measured. Bard and Faulkner (1980) review applications of immobilized liquid and facilitated transport membranes to the design of electrodes for the detection of cations and anions in aqueous solutions. Abdelrahman, Deetz, and Zook (1988) describe the fabrication of liquid membrane-based sensors for organic vapors such as tobacco smoke.

What Research Needs to Be Done?

The great potential of facilitated transport for the development of selective membranes for separations and other applications such as sensors, membrane reactors, and controlled release has been largely unrealized. The primary problem is the stability of the membranes, which is a combination of factors dealing with the stability of the complexation chemistry and the stability of the selective liquid, solid, or gel membrane. Consequently, further research is needed in complexation chemistry, thin-film technology, support materials, and the theory of transport in fixed carrier membranes.

To commercialize chemically based separation processes for gas separations, carrier lifetime must be improved. What is especially needed is an understanding of the fundamental degradation mechanisms and an understanding of the relationship between carrier structure and the binding constant and kinetic parameters. Additionally, the potential application or use of a complex will often drastically affect its lifetime. As described above, an oxygen carrier may have a very short lifetime in solution, but incorporating the same carrier into a solid polymer film dramatically increases its lifetime. Therefore, the development of chemically based gas separation processes is fundamentally a multidisciplinary problem. It is not enough to synthesize complexation agents with acceptable lifetimes. The effects of poisons and impurities that are commonly present in industrial gas streams on carrier performance and lifetime should be studied. Once industrially attractive permeance values and separation factors are achieved at practical operating conditions, lifetime data are needed to justify the substantial development that will be necessary to fabricate modules for industrial application.

Improved thin-film technology would definitely speed industrial application of liquid membrane technology, especially in gas separation applications. Improved supports and immobilization techniques are necessary in order to fabricate ILMs and ion-exchange membranes in thicknesses of a few microns or less to produce economically attractive fluxes. New materials such as microporous ceramic and inorganic membranes may be suitable as supports for new complexation agents such as molten salts to fabricate stable, liquid membranes that may be capable of operation at high temperatures.

Capillary condensation can be used to stabilize a liquid phase in a very small pore support. Because the condensed liquid phase in the pore is in thermodynamic equilibrium, it is inherently stable under condensation conditions. Other physical processes may also be exploited. In addition, very low volatility liquids, such as silicone oils or fluorocarbons, can also be used to improve liquid membrane lifetime.

Related theoretical work could define the set of properties required to optimize the separation and provide guidelines for improving existing complexation reactions. The exact mass transfer mechanism in fixed carrier membranes needs to be identified; the mechanism may be specific to individual carrier chemistries. Once the mass transfer mechanism for fixed carrier membranes has been determined for various stoichiometries, theoretical work needs to be done to define the optimum binding properties for the fixed carrier species.

CONCLUSIONS

The objective of early research on facilitated transport membranes was to prepare simple, well-characterized model systems in order to study transport in biological systems (Cussler 1984). However, because of the high selectivity and large fluxes often observed with facilitated transport membranes, there has been great interest and research activity in the past 20 years to fabricate synthetic, reactive membranes for industrial gas and liquid separations. The importance of industrial air separations is one of the factors driving research to develop facilitated transport membranes selective for oxygen. The reason for the great interest in facilitated transport membranes for air separations is the need for high selectivities. As of 1991, probably the best commercial polymer membrane available had an $\alpha_{O_2/N_2} = 7.5$ (Muruganandam and Paul 1987). However, to produce 95% O_2, a selectivity of over 70 is needed. As described above, very high O_2/N_2 separation factors ranging from 25 to almost 80 have been reported for facilitated transport membranes in the last 5 years.

Other potential industrial gas separation applications for facilitated transport membranes include acid gases (such as CO_2 and H_2S), NH_3, unsaturated (e.g., ethylene) and aromatic hydrocarbons and carbon monoxide. In several cases, a facilitated transport membrane has been shown to perform desirable separations not possible with glassy polymers, the usual membrane material. Examples are the removal of CO_2 and NH_3 from mixtures with H_2, separation of light olefins from mixtures with saturated hydrocarbons, and the removal of CO from mixtures with N_2. It is quite possible that facilitated transport membranes could be used effectively in hybrid configurations with other processes such as distillation, polymer membranes, and physical absorption.

As described in Chapters 36 and 39, some commercial applications of facilitated transport for liquid phase separations have been successful. However, other than several pilot-scale studies, there have been no industrial applications of facilitated transport membranes for gas separations. As discussed above, development of a practical facilitated transport system for gases is a very complex, multidisciplinary problem in chemistry, chemical engineering, and materials science. The demands on the complexation agent are many. It must resist degradation and poisoning and be stable for extended time periods of months and even years. Yet the carrier must also have optimum complexation kinetics and a binding constant to avoid carrier saturation. The selective layer containing the carrier must be mechanically stable and thin enough to produce industrially acceptable permeance values of 10^5 to 10^6 Barrers/cm. Furthermore, the membrane must be manufacturable in high-surface-area flat-sheet or hollow-fiber modules. In spite of these difficult requirements, there have been many very promising research achievements toward the development of practical facilitated transport membranes for gas separations. These include the use of synthetic organometallic carriers, the application of ion-exchange materials to support charged complexation agents, the use of solvent-swollen polymer films as membrane materials, the identification of reactive molten salts for liquid membranes, and the fabrication of polymer membranes containing fixed carriers.

ACKNOWLEDGMENTS

J. D. Way gratefully acknowledges the support for the preparation of this chapter from a Faculty Development Grant from the Dow Chemical Company and the help of his wife, Debra J. Way, who provided expert assistance in the preparation of manuscript and prepared the majority of the figures.

J. D. Way dedicates this chapter to the memory of his friend and colleague from the Department of Chemical Engineering at Oregon State University, Robert V. Mrazek, who died suddenly during its preparation.

NOTATION

General Notation

See the General Notation section at the beginning of this handbook.

Special Notation

E enhancement factor $(F - 1)$, dimensionless

F facilitation factor, ratio of flux with carrier present to flux without carrier, dimensionless

J_i flux of species i, M/L^2t or mol/L^2t

K dimensionless reaction equilibrium constant $(k_f c_{A0}/k_r)$
k mass transfer coefficient based on concentration driving force, L/t
K_{eq} equilibrium constant (k_f/k_r), L^3/mol
k_f forward reaction rate constant, $L^3/\text{mol}\, t$
k_r reverse reaction rate constant, t^{-1}
L membrane length, L
M molarity, mol/L
r radius of a hollow-fiber membrane, L
S shape factor from Noble (1983), correction for flux in a cylindrical geometry, dimensionless
P permeability, Barrers [10^{-10} cm (STP) $\cdot$ cm/(cm$^2 \cdot$ s $\cdot$ cm Hg)]
x feed gas mole fraction
y permeate mole fraction

Greek Letters

α mobility ratio $(D_{AB}c_T/D_A c_{A0})$, dimensionless
ε inverse Damköhler number $(D_{AB}/k_r l^2)$, dimensionless
η kinetic efficiency factor, dimensionless

Subscripts

0 initial value
A solute being transported
AB solute-carrier complex
B carrier species or complexation agent
i solute species i or value in the membrane at the inside radius
o value in the membrane at the outside radius

REFERENCES

Abdelrahman, M., D. W. Deetz, and J. D. Zook. 1988. Membrane-selective vapor sensing. U.S. Patent 4,745,796.

Anonymous 1990. Facts and figures for the chemical industry. *Chem. Eng. News* 68(25):34–83.

Babcock, W. C., D. J. Brose, A. R. Chambers, and D. T. Friesen. 1986. Separation of citric acid from fermentation beer using supported-liquid membranes. Paper read at the AIChE National Meeting, 24–27 August 1986, Boston, MA.

Baker, R. W., I. C. Roman, and H. K. Lonsdale. 1987. Liquid membranes for the production of oxygen-enriched air: I. introduction and passive liquid membranes. *J. Membr. Sci.* 31(1):15–29.

Bard, A. J., and L. R. Faulkner. 1980. *Electrochemical Methods*. New York: John Wiley & Sons.

Barnes, M. J., R. S. Drago, and K. J. Balkus, Jr. 1988. Cobalt (II)-facilitated transport of dioxygen in a polystyrene membrane. *J. Am. Chem. Soc.* 110(20):6780–6785.

Basaran, O. A., P. M. Burban, and S. R. Auvil. 1989. Facilitated transport with unequal carrier and complex diffusivities. *Ind. Eng. Chem. Res.* 28(1):108–119.

Bassett, R. J., and J. S. Schultz. 1970. Nonequilibrium facilitated diffusion of oxygen through membranes of aqueous cobaltodihistidine. *Biochim. Biophys. Acta* 211:194–215.

Bdzil, J., C. C. Carlier, H. L. Frisch, W. J. Ward, and M. W. Breiter. 1973. Analysis of potential difference in electrically induced carrier transport systems. *J. Phys. Chem.* 77(6):846–850.

Bellobono, I. R., F. Moffato, E. Selli, L. Righetto, and R. Tacchi. 1987. Transport of oxygen facilitated by peroxo-bis[N,N'-ethylene bis-(salicylideneiminato) - dimethylformamide - cobalt(III)] embedded in liquid membranes immobilized by photografting onto cellulose. *Gas Sep. Purif.* 1:103–106.

Beverwijk, C. D. M., G. J. M. Van Der Kerk, A. J. Leusink, and J. G. Noltes. 1970. Organosilver chemistry. *Organometal. Chem. Rev. A* 5:215–280.

Busch, D. H. 1988. Synthetic oxygen carriers for dioxygen transport. In *Oxygen Complexes and Oxygen Activation by Transition Metals*, ed. A. E. Martell and D. T. Sawyer. New York: Plenum Press.

Crank, J., and G. S. Park. 1968. *Diffusion in Polymers*. New York: Academic Press.

Curl, R. L., and J. S. Schultz. 1973. Polynomial approximation for unsteady-state diffusion of oxygen into hemoglobin solutions. *Adv. Exp. Med. Biol.* 37B:929.

Cussler, E. L. 1971. Membranes which pump. *AIChE J.* 17(6):1300–1303.

Cussler, E. L. 1984. *Diffusion: Mass Transfer in Fluid Systems*. New York: Cambridge University Press.

Cussler, E. L. 1990. Personal communication.

Cussler, E. L., R. Aris, and A. Bhown. 1989. On the limits of facilitated diffusion. *J. Membr. Sci.* 43:149–164.

Deetz, D. W., and M. M. Kreevoy. 1987. Stabilized liquid films. U.S. Patent 4,710,205.

Delaney, M. S., D. Reddy, and R. A. Wessling. 1990. Oxygen/nitrogen transport in glassy polymers with oxygen-binding pendant groups. *J. Membr. Sci.* 49:15–36.

Doyle, G., R. L. Pruett, D. W. Savage, and W. S. Ho. 1984. Separation of olefin mixtures by Cu(I) complexation. U.S. Patent 4,471,152.

Drago, R. S., and K. J. Balkus. 1986. Cobalt(II)-facilitated transport of dioxygen in a polystyrene membrane. *Inorg. Chem.* 25(6):716–718.

Early, C. L., R. L. Kilgour, and S. S. Medvetz. 1983. Monsanto Prism separators for upgrading landfill gas. Paper read at the 21st Annual International Seminar and Equipment Show of Government Refuse Collection and Disposal Association Meeting, 29 August–1 September 1983, Winnipeg, Manitoba, Canada.

Enns, T. 1967. Facilitation by carbonic anhydrase of carbon dioxide transport. *Science* 155:44–47.

Folkner, C. A., and R. D. Noble. 1983. Transient response of facilitated transport membranes. *J. Membr. Sci.* 12:289–301.

Friesen, D. T., W. C. Babcock, D. J. Brose, and A. R. Chambers. 1991. Recovery of citric acid from fermentation beer using supported-liquid membranes. *J. Membr. Sci.* 56:127–141.

Goddard, J. D. 1977. Further applications of carrier-mediated transport theory: a survey. *Chem. Eng. Sci.* 32:795.

Goddard, J. D., J. S. Schultz, and S. R. Suchdeo. 1974. Facilitated transport via carrier-mediated diffusion in membranes: part II. mathematical aspects and analyses. *AIChE J.* 20(4):625–645.

Gottschlich, D. E., and D. L. Roberts. 1989. Economics and thermodynamics of membrane and hybrid separation processes. Paper read at the 3rd Annual Meeting of the North American Membrane Society, 17–19 May 1989, Austin, TX.

Gottschlich, D. E., D. L. Roberts, and J. D. Way. 1988. A theoretical comparison of facilitated transport and solution-diffusion membrane modules for gas separation. Paper read at the 2nd Annual Meeting of the North American Membrane Society, 1–3 June 1988, Syracuse, NY.

Govind, R. 1987. Adsorption of gases by amine and phosphine complexed manganese (II) compounds. European Patent 220,963.

Guha, A. K., S. Majumdar, and K. K. Sirkar. 1990. Facilitated transport of CO_2 through an immobilized liquid membrane of aqueous diethanolamine. *Ind. Eng. Chem. Res.* 29:2093–2100.

Henderson, R. A., G. J. Leigh, and C. J. Piekett. 1983. The chemistry of nitrogen fixation and models for the reactions of nitrogenase. *Adv. Inorg. Chem. Radiochem.* 27:197–292.

Ho, W. S., G. Doyle, D. W. Savage, and R. L. Pruett. 1988. Olefin separations via complexation with cuprous diketonate. *Ind. Eng. Chem. Res.* 27(2):334–337.

Hoofd, L., and F. Kreuzer. 1979. A new mathematical approach for solving carrier-facilitated steady-state diffusion problems. *J. Math. Biol.* 8:1.

Hoofd, L., and F. Kreuzer. 1978. Calculation of the facilitation of O_2 or CO transport by HB or MB by means of a new method for solving the carrier-diffusion problem. In *Oxygen Transport to Tissue—III*, ed. I. A. Silver, M. Erecinska, and H. I. Bicher, pp. 163–168. New York: Plenum Press.

Houston, C. D. 1989. Gas separation short course. Presentation made at the 3rd Annual Meeting of the North American Membrane Society, 17–19 May 1989, Austin, TX.

Hughes, R. D., J. A. Mahoney, and E. F. Steigelmann. 1981. Olefin separation by facilitated transport membranes. Paper read at the AIChE National Meeting, 5–9 April 1981, Houston, TX.

Hughes, R. D., J. A. Mahoney, and E. F. Steigelmann. 1986. Olefin separation by facilitated transport membranes. In *Recent Developments in Separation Science*, ed. N. N. Li, Vol. 9, pp. 173–199. Boca Raton, FL: CRC Press.

Hughes, R. D., and E. F. Steigelmann. 1973. Process. U.S. Patent 3,758,605.

Hughes, R. D., and E. F. Steigelmann. 1974. Process for separating carbon monoxide. U.S. Patent 3,823,529.

Ishikawa, H., T. Murakami, M. Hata, and H. Hikita. 1985. Separation and enrichment of weak organic acids or bases by immobilized liquid membranes. *Chem. Eng. Commun.* 34:123–136.

Johnson, B. M., R. W. Baker, S. L. Matson, K. L. Smith, I. C. Roman, M. E. Tuttle, and H. K. Lonsdale. 1987. Liquid membranes for the production of oxygen-enriched air: II. facilitated transport membranes. *J. Membr. Sci.* 31(1):31–67.

Kagawa, S., M. Kawakami, and K. Okita. 1986. Liquid membranes. Japanese Patent 61,293,525.

Kawakami, M., H. Iwanaga, M. Iwamoto, and S. Kagawa. 1982. The transport of oxygen facilitated by a Cu(I)-tetraethylenepentamine system. *J. Chem. Soc. Chem. Commun.* No. 24:1396–1398.

Kawakami, M., H. Nagano, M. Iwamoto, and S. Kagawa. 1984. Transport of carbon monoxide facilitated by a Cu(I)-tetraethylenepentamine system. *Chem. Lett.* No. 1:109–112.

Kawakami, M., M. Tateishi, M. Iwamoto, and S. Kagawa. 1987. Selective permeation of ethylene and propylene through Rh^{3+}-polyethylene glycol liquid membranes. *J. Membr. Sci.* 30:105–110.

Kazuhiro, O., S. Yamanochi, and K. Yamada. 1989. Cobalt-histidine complex for gas separation membranes. Japanese Patent 01,242,124.

Kemena, L. L., R. D. Noble, and N. J. Kemp. 1983. Optimal regimes of facilitated transport. *J. Membr. Sci.* 15:259–274.

Kemp, N. J., and R. D. Noble. 1983. Heat transfer effects in facilitated transport liquid membranes. *Sep. Sci. Technol.* 18:1147–1165.

Kim, J.-I., and P. Stroeve. 1988. Mass transfer in separation devices with reactive hollow fibers. *Chem. Eng. Sci.* 43(2):247–257.

Kimura, S. G., S. L. Matson, and W. J. Ward. 1979. Industrial applications of facilitated transport. In *Recent Developments in Separation Science*, ed. N. N. Li, Vol. 5, pp. 11–25. Boca Raton, FL: CRC Press.

King, C. J. 1980. *Separation Processes*. 2nd ed. New York: McGraw-Hill Book Co.

King, C. J. 1988. Separation processes based on reversible chemical complexation. In *Handbook of Separation Processes*, ed. R. W. Rousseau, pp. 760–774. New York: John Wiley & Sons.

Kirkköprü-Dindi, A., and R. D. Noble. 1989. Optimal regimes of facilitated transport for multiple site carriers. *J. Membr. Sci.* 42:13–25.

Kobayashi, Y., I. Konno, and J. Matsura. 1988a. Liquid membrane for selective separation of gases. Japanese Patent 63,011,503.

Kobayashi, Y., I. Konno, and J. Matsura. 1988b. Liquid membrane for selective separation of gases. Japanese Patent 63,011,504.

Kobayashi, Y., I. Konno, and J. Matsura. 1988c. Liquid membrane for selective separation of gases. Japanese Patent 63,017,204.

Koval, C. A., R. D. Noble, J. D. Way, B. Louie, A. Reyes, G. M. Horn, and D. L. Reed. 1985. Selective transport of gaseous CO through liquid membranes using an iron(II) macrocyclic complex. *Inorgan. Chem.* 24(8):1147–1152.

Koval, C. A., and Z. E. Reyes. 1987. Chemical aspects of facilitated transport through liquid membranes. In *Liquid Membranes: Theory and Applications*, ed. R. D. Noble and J. D. Way, ACS Symp. Ser. No. 347, pp. 28–38. Washington, DC: American Chemical Society.

Koval, C. A., and T. Spontarelli. 1988. Condensed phase facilitated transport of olefins through an ion exchange membrane. *J. Am. Chem. Soc.* 110(1):293–295.

Koval, C. A., T. Spontarelli, and R. D. Noble. 1989. Styrene/ethylbenzene separation using facilitated transport through perfluorosulfonate ionomer membranes. *Ind. Eng. Chem. Res.* 28(7):1020–1024.

Kraus, M. A. 1986. Water free hydrocarbon separation membrane and process. U.S. Patent 4,614,524.

Kreuzer, F., and L. Hoofd. 1970. Facilitated diffusion of oxygen in the presence of hemoglobin. *Resp. Phys.* 8:280–302.

Kulkarni, V. B., and R. Govind. 1988. Oxygen sensitive $Mn(depe)_2X_2$ complexes. *Inorg. Chim. Acta* 150:11–12.

Kulkarni, V. B., and R. Govind. 1989. Studies on coordination complexes for separation of oxygen from air. Paper read at the 6th Symposium on Separation Science and Technology for Energy Applications. 22–26 October 1989, Knoxville, TN.

Kulkarni, V. B., J. Hinegardner, and R. Govind. 1987. Separation of oxygen from air using organometallic complexes. Paper read at the 1st Annual Meeting of the North American Membrane Society, 3–5 June 1987, Cincinnati, OH.

Kuo, Y., and H. P. Gregor, 1983. Acetic acid extraction by solvent membrane. *Sep. Sci. Technol.* 18(5):421–440.

Laciak, D. V., and G. P. Pez. 1988. Ammonia separation using ion exchange polymeric membranes and sorbents. U.S. Patent 4,758,250.

Laciak, D.V., R. Quinn, G. P. Pez, J. B. Appleby, and P. S. Puri. 1989. Selective permeation of ammonia and carbon dioxide by novel membranes. Paper read at the 6th Symposium on Separation Science and Technology for Energy Applications, 22–26 October 1989, Knoxville, TN.

LeBlanc, O. H., W. J. Ward, S. L. Matson, and S. G. Kimura. 1980. Facilitated transport in ion exchange membranes. *J. Membr. Sci.* 6:339–343.

Martell, A. E. 1988. Formation and degradation of cobalt dioxygen complexes. In *Oxygen Complexes and Oxygen Activation by Transition Metals*, ed. A. E. Martell and D. T. Sawyer. New York: Plenum Press.

Matson, S. L., C. S. Herrick, and W. J. Ward. 1977. Progress on the selective removal of H_2S from gasified coal using an immobilized liquid membrane. *Ind. Eng. Chem. Process Des. Dev.* 16:370.

Matson, S. L., E. K. L. Lee, D. T. Friesen, and D. J. Kelly. 1988. Acid gas scrubbing by composite solvent-swollen membranes. U.S. Patent 4,737,166.

Matson, S. L., and H. K. Lonsdale. 1987. Liquid membranes for the production of oxygen-enriched air: III. process design and economics. *J. Membr. Sci.* 31(1):69–87.

Matson, S. L., W. J. Ward, S. G. Kimura, and W. R. Browall, 1986. Membrane oxygen enrichment: II. economic assessment. *J. Membr. Sci.* 29:79–96.

McAuliffe, C. A., H. F. Al-Khateeb, D. S. Barrett, J. C. Briggs, A. Challita, A. Hosseiny, M. G. Little, A. G. Mackie, and K. L. Minten. 1983. Coordination chemistry of managanese: part 8. the coordination of small molecules by manganese (II) phosphine complexes: part 1. the reversible coordination of dioxygen by [MnX$_2$(PR$_3$)] (X = Cl, Br, or I; R$_3$ = Bu$_3$ or PhBu$_2$) complexes in solution and an infrared criterion for predicting dioxygen binding activity in the diisothiocyanato(phosphine)manganese (II) complexes. *J. Chem. Soc. Dalton Trans.* (10):2147–2153.

Meldon, J. H., Y. Kang, and N. Sung. 1985. Analysis of transient permeation through a membrane with immobilizing chemical reaction. *Ind. Eng. Chem. Fundam.* 24(1):61–64.

Meldon, J. H., D. A. Smith, and C. K. Colton. 1977. The effect of weak acids upon the transport of carbon dioxide in alkaline solutions. *Chem. Eng. Sci.* 32:939.

Meldon, J. H., P. Stroeve, and C. K. Gregoire. 1982. Facilitated transport of carbon dioxide: a review. *Chem. Eng. Comm.* 16:263–300.

Murray, J. D., and J. Wyman. 1971. Facilitated diffusion: the case of carbon monoxide. *J. Biol. Chem.* 246(19):5903.

Muruganadam, N., and D. R. Paul. 1987. Evaluation of substituted polycarbonates and a blend with polystyrene as gas separation membranes. *J. Membr. Sci.* 34:185–198.

Niederhoffer, E. C., J. H. Timmons, and A. E. Martell. 1984. Thermodynamics of oxygen binding in natural and synthetic dioxygen complexes. *Chem. Rev.* 84(2):137–203.

Niiya, K. Y., and R. D. Noble. 1985. Competitive facilitated transport through liquid membranes. *J. Membr. Sci.* 23:183–198.

Nishide, H., H. Kawakami, Y. Kurimura, and E. Tsuchida. 1989. Reversible coordination and facilitated transport of molecular nitrogen in poly((vinyl-cyclopentadienyl)manganese) membrane. *J. Am. Chem. Soc.* 111(18):7175–7179.

Nishide, H., H. Kawakami, and E. Tsuchida. 1989. Facilitated transport of nitrogen and acetylene in poly(vinylcyclopentadienylmanganese) membrane. *Polym. Mater. Sci. Eng.* 61:492–496.

Nishide, H., M. Ohyanagi, Y. Funada, T. Ikeda, and E. Tsuchida. 1987c. Oxygen transport behavior through the membrane containing a fixed carrier and adhered to a second polymer. *Macromolecules* 20(9):2312–2313.

Nishide, H., M. Ohyanagi, H. Kawakami, and E. Tsuchida. 1986b. Dual-mode transport of molecular oxygen through the membrane of the poly(octyl methacrylate-co-4-vinylpyridine) (N,N'-disalicylideneethylenediamine) cobalt (II) complex. *Bull. Chem. Soc. Japan* 59(10):3213–3216.

Nishide, H., M. Ohyanagi, O. Okada, and E. Tsuchida. 1986a. Highly selective transport of molecular oxygen in a polymer containing a cobalt porphyrin complex as a fixed carrier. *Macromolecules* 19(2):494–496.

Nishide, H., M. Ohyanagi, O. Okada, and E. Tsuchida. 1987a. Dual-mode transport of molecular oxygen in a membrane containing a cobalt porphyrin complex as a fixed carrier. *Macromolecules* 20(2):417–422.

Nishide, H., M. Ohyanagi, O. Okada, and E. Tsuchida. 1987b. Spectroscopic study of oxygen sorption and diffusion in a membrane containing a cobalt porphyrin complex. *Polymer J.* 19(7):839–844.

Nishide, H., M. Ohyanagi, O. Okada, and E. Tsuchida. 1988. Oxygen binding and transport in the membrane of poly[[tetrakis(methacrylamidophenyl)porphinato]cobalt-co-hexyl methacrylate]. *Macromolecules* 21(10):2910–2913.

Noble, R. D. 1983. Shape factors in facilitated transport through membranes. *Ind. Eng. Chem. Fundam.* 22(1):139–144.

Noble, R. D. 1984. Two-dimensional permeate transport with facilitated transport membranes. *Sep. Sci. Technol.* 19(8&9):469–478.

Noble, R. D. 1985. Kinetic efficiency factors for facilitated transport membranes. *Sep. Sci. Technol.* 20(7&8):577–585.

Noble, R. D. 1990. Analysis of facilitated transport in fixed site carrier membranes. *J. Membr. Sci.* 50:207–214.

Noble, R. D., J. D. Way, and L. A. Powers. 1986. Effect of external mass transfer resistance on facilitated transport. *Ind. Eng. Chem. Fundam.* 25(3):450–452.

Norman, J. A. T., G. P. Pez, and D. A. Roberts. 1988. Reversible complexes for the recovery of dioxygen. In *Oxygen Complexes and Oxygen Activation by Transition Metals*, ed. A. E. Martell and D. T. Sawyer, pp. 107–125. New York: Plenum Press.

Norman, J. A. T., D. Ramprasad, and D. H. Busch. 1988. Pillared cobalt complexes for oxygen separation. U.S. Patent 4,735,634.

Ohyanagi, M., H. Nishide, K. Suenaga, T. Nakamura, and E. Tsuchida. 1987. Effect of polymer matrices on oxygen-binding of a cobalt porphyrin complex. *Bull Chem. Soc. Japan* 60(8):3045–3046.

Ohyanagi, M., H. Nishide, E. Suenaga, and E. Tsuchida. 1988. Effect of polymer matrix and metal species on facilitated oxygen transport in metalloporphyrin (oxygen carrier) fixed membranes. *Macromolecules* 21(6):1590–1594.

Okada, M., Y. Kobayashi, T. Omiyama, and J. Matsura. 1986. Selective gas separation. Japanese Patent 61,268,338.

Okada, M., T. Omiyama, Y. Kobayashi, and J. Matsura. 1987. Gas separating materials. Japanese Patent 62,049,928.

Okada, M., T. Omiyama, and J. Matsura. 1986. Gas separation materials. Japanese Patent 61,268,337.

Otto, N. C., and J. A. Quinn. 1971. The facilitated transport of carbon dioxide through bicarbonate solutions. *Chem. Eng. Sci.* 26:949–961.

Pan, C. Y., and H. W. Habgood. 1974. An analysis of the single-stage gaseous permeation process. *Ind. Eng. Chem. Fundam.* 13(4):323–331.

Peinemann, K.-V., and S. K. Shukla. 1989. Olefin separation by fixed carrier membranes. Paper read at the 3rd Annual Meeting of the North American Membrane Society, 17–19 May 1989, Austin, TX.

Pellegrino, J. J., R. Nassimbene, and R. D. Noble. 1988. Facilitated transport of CO_2 through highly swollen ion-exchange membranes: the effect of hot glycerine treatment. *Gas. Sep. Purif.* 2:126–130.

Pez, G. P., and R. T. Carlin. 1986. Method for gas separation. U.S. Patent 4,617,029.

Pez, G. P., and D. V. Laciak. 1988. Ammonia separation using semipermeable membranes. U.S. Patent 4,762,535.

Quinn, R., G. P. Pez, and J. B. Appleby. 1988. Molten salt hydrate membranes for the separation of gases. U.S. Patent 4,780,114.

Ramprasad, D., and D. H. Busch. 1987. Lacunar cobalt complexes for oxygen separation. U.S. Patent 4,680,037.

Reyes, Z. E. 1986. Kinetic and thermodynamic aspects of facilitated transport through liquid membranes. Ph.D. thesis, University of Colorado, Boulder, CO.

Roberts, D. L., and S. K. Friedlander. 1980a. Sulfur dioxide transport through aqueous solutions: part I. theory. *AIChE J.* 26(4):593–602.

Roberts, D. L., and S. K. Friedlander. 1980b. Sulfur dioxide transport through aqueous solutions: part II. experimental results and comparison with theory. *AIChE J.* 26(4):602–610.

Roberts, D. L., and R. M. Laine. 1986. Gas separation process. U.S. Patent 4,605,475.

Roman, I. C., and R. W. Baker. 1985. Method and apparatus for producing oxygen and nitrogen and membrane therefor. U.S. Patent 4,542,010.

Scholander, P. F. 1960. Oxygen transport through hemoglobin solutions. *Science* 131:585.

Sengupta, A., B. Raghuraman, and K. K. Sirkar. 1990. Liquid membranes for flue gas desulfurization. *J. Membr. Sci.* 51:105–126.

Shindo, Y., T. Hakuta, H. Yoshitome, and H. Inoue. 1985. Calculation methods for multicomponent gas separation by permeation. *Sep. Sci. Technol.* 20(5&6):445–459.

Smith, D. R., R. J. Lander, and J. A. Quinn. 1977. Carrier-mediated transport in synthetic membranes. In *Recent Developments in Separation Science*, ed. N. N. Li, Vol. 3, pp. 225–241. Boca Raton, FL: CRC Press.

Smith, D. R., and J. A. Quinn. 1979. The prediction of facilitation factors for reaction augmented membrane transport. *AIChE J.* 21(1):197–200.

Smith, D. R., and J. A. Quinn. 1980. The facilitated transport of carbon monoxide through cuprous chloride solutions. *AIChE J.* 26(1):112.

Smith, K. A., J. H. Meldon, and C. K. Colton. 1973. An analysis of carrier-facilitated transport. *AIChE J.* 19(1):102.

Spaan, J. A. E., 1973. Transfer of oxygen into hemoglobin solution. *Pflügers Arch.* 342:289.

Spaan, J. A. E., F. Kreuzer, and L. Hoofd. 1980. A theoretical analysis of nonsteady-state oxygen transfer in layers of hemoglobin solution. *Pflügers Arch.* 384:231–239.

Spillman, R. W., and T. E. Cooley. 1987. Economic considerations in membrane gas separation process design. Paper read at the AIChE National Meeting, 29 March–2 April 1987, Houston, TX.

Steigelman, E. F., and R. D. Hughes. 1973. Process for separation of unsaturated hydrocarbons. U.S. Patent 3,758,603.

Sterzel, H. J., A. Sanner, and P. Neümann. 1986. Vinyl polymer membranes. U.S. Patent 4,584,359.

Sugie, K. 1988. Highly selective oxygen separating membrane containing Schiff-base chelate bound to polymeric ligand. *Polym. Mater. Sci. Eng.* 59:139–143.

Tajar, J. G., and I. F. Miller. 1972. The permeation of carbon dioxide, oxygen, and nitrogen through weakly basic polymer membranes. *AIChE J.* 18(1):78–83.

Teremoto, M., H. Matsuyama, T. Yamashiro, and Y. Katayama. 1986. Separation of ethylene from ethane by supported liquid membranes containing silver nitrate as a carrier. *J. Chem. Eng. Japan* 19(5):419–424.

Tsuchida, E., H. Nishide, M. Ohyanagi, and H. Kawakami. 1987. Facilitated transport of molecular oxygen in the membranes of polymer-coordinated cobalt Schiff-base complexes. *Macromolecules* 20:1907–1912.

Ward, W. J. 1970a. Analytical and experimental studies of facilitated transport. *AIChE J.* 16(3):405–410.

Ward, W. J. 1970b. Electrically induced carrier transport. *Nature* 227:162.

Ward, W. J., and W. L. Robb. 1967. Carbon dioxide-oxygen separation: facilitated transport of carbon dioxide across a liquid film. *Science* 156:1481–1484.

Way, J. D., and R. L. Hapke. 1988. Facilitated transport membranes for hydrogen production from synthesis gas. Final Report for U.S. Department of Commerce, Contract 50RANB70C093.

Way, J. D., and R. D. Noble. 1987. Hydrogen sulfide facilitated transport in perfluorosulfonic acid membranes. In *Liquid Membranes: Theory and Applications,* ed. R. D. Noble and J. D. Way, ACS Symp. Ser. No. 347, pp. 123–137. Washington, DC: American Chemical Society.

Way, J. D., and R. D. Noble. 1989. Competitive facilitated transport of acid gases in perfluorosulfonic acid membranes. *J. Membr. Sci.* 46:309–324.

Way, J. D., R. D. Noble, D. L. Reed, and G. M. Ginley. 1987. Facilitated transport of CO_2 in ion exchange membranes. *AIChE J.* 33(3):480–487.

Wittenberg, J. B. 1966. The molecular mechanism of hemoglobin-facilitated oxygen diffusion. *J. Biol. Chem.* 241:104.

Yamada, K., K. Okita, S. Asako, and S. Toyooka. 1987. Liquid membranes. Japanese Patent 62,061,619.

Yamamoto, A. 1986. *Organotransition Metal Chemistry.* New York: John Wiley & Sons.

Yoshikawa, M., T. Eyaki, K. Sanui, and N. Ogata. 1986. Synthetic polymer membranes with pyridine moiety for gas separation. *Kobunshi Ronbunshu* 43(11):729–732.

Yung, D., and R. L. Probstein. 1973. Similarity considerations in facilitated transport. *J. Phys. Chem.* 77(18):2201.

Zupancic, J. J., and R. J. Swedo. 1986. Facilitated gas enrichment membrane composites and a process for the preparation thereof. U.S. Patent 4,627,859.

45

Electrostatic Pseudo-Liquid-Membrane

Zhongmao Gu
China Institute of Atomic Energy

INTRODUCTION
THEORY
    Principle of Electrostatic Dispersion
    Mechanism of Mass Transfer
MASS TRANSFER PROCESS
    Typical Experimental Results
    Effect of Electrical Field
    Effect of Extractant Concentration
    Effect of Flow Ratio of Feed to
        Stripping Solution
COMPARISON OF ESPLIM WITH
    ELM AND SLM
    Comparison with ELM
    Comparison with SLM
    Further Development of ESPLIM
SCALE-UP OF AN ESPLIM DEVICE
POTENTIAL APPLICATIONS
    Hydrometallurgy
    Recovery of Metal Values from
        Dilute Effluents
    Removal of Toxic Material from
        Wastewater
ACKNOWLEDGMENTS
NOTATION
REFERENCES

INTRODUCTION

Solvent extraction can be said to have started in the 1940s. With the growing needs of separation engineering (especially in nuclear fuel cycles) and the continuing synthesis of new extractants, solvent extraction technology developed rapidly and has played an increasingly important role in hydrometallurgy.

One of the prominent features in the development of modern science and technology is the cross-fertilization of ideas among different disciplines. Over the past 25 years or so, the combination of solvent extraction with other techniques has generated a number of new separation processes, which include solvent-impregnated resins (Warshawsky 1981), extraction chromatography (Brauwn and Ghershin 1975), electrostatic extraction (Thornton 1968; Scott 1989), supercritical extraction (Humphrey and Fair 1983), membrane solvent extraction (Ho, Lee, and Liu 1976; Lee, Ho, and Liu 1976), reversed-micelle extraction (Goklen and Hatton 1987), gel extraction (Cussler, Stokar, and Varberg 1984; Freitas and Cussler 1987), and liquid membranes (Li 1972; Ward 1972). Among them the liquid membrane technology appears to be an attractive one owing to its high flux, high selectivity, and low consumption of reagents.

Liquid membranes can be divided into two types according to their configuration. One is supported liquid membranes (SLMs) and the other is emulsion liquid membranes (ELMs). SLMs are made with porous solid membranes (e.g., porous hollow fibers) impregnated with membrane materials (extractant and diluent). This process is still under laboratory investigation. An ELM, invented by Li (1968), is essentially a double-emulsion system of either the W/O/W type or O/W/O type. This process has already reached to industrial scale (Draxler and Marr 1986; Zhang, Liu, and Lu 1987; Ruppert, Draxler, and Marr 1988; Zhang et al. 1988a, 1988b; also see Chapters 36, 39, and 40).

One of the important characteristics of the liquid membrane process is the simultaneous extraction and stripping in one single step instead of two separate steps as required by solvent extraction. In this way, the equilibrium limitation inherent in solvent extraction is eliminated. This nonequilibrium feature of liquid membranes results in the reduction of the number of separation stages required, the saving of reagents, and hence a lowering of costs.

Some problems with liquid membranes still remain unsolved. The major problem of SLMs is the poor membrane stability due to the gradual loss of the carrier in the liquid membrane retained in a microporous membrane support by dissolution and spontaneous emulsification during the operation. For ELMs, surfactant is indispensable for maintaining membrane integrity. However, the introduction of surfactant to liquid membrane systems leads to the two additional procedures: emulsification and demulsification, which make the process complicated.

Combining an electrostatic technique with the principle of a liquid membrane, Gu (1988) developed a new type of separation method, which he named *electrostatic pseudo-liquid-membrane* (ESPLIM). This process is a continuous process in which extraction and stripping take place simultaneously in a specially made reaction tank (Figure 45–1). Inside this tank, an oil layer exists within the baffle plates that is rather stagnant and acts as the liquid membrane. This oil layer allows the transport of complex (metal-carrier) from the extraction cell to the stripping cell and carrier (regenerated extractant) from the stripping cell to the extraction cell, while preventing the mixing of the aqueous feed and the stripping solution. In this way, extraction and stripping are coupled inside the reaction tank through this oil layer.

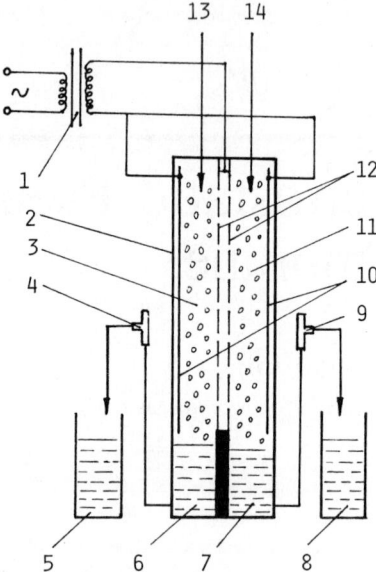

1 HIGH VOLTAGE SUPPLY
2 REACTION TANK
3 EXTRACTION CELL
4 TRITON
5 RAFFINATE TANK
6 EXTRACTION SETTLER
7 STRIPPING SETTLER
8 CONCENTRATE TANK
9 TRITON
10 HIGH VOLTAGE ELECTRODE
11 STRIPPING CELL
12 GROUNDED ELECTRODE (BAFFLE PLATE)
13 FEED
14 STRIPPING SOLUTION

FIGURE 45–1. Schematic of the ESPLIM process (reprinted from Gu 1990a with permission).

Although ESPLIM configuration is totally different from those of SLM and ELM, it is characterized by the nonequilibration mass transfer, which is a basic feature of liquid membranes. Take Co^{2+} extraction by di(2-ethylhexyl)phosphoric acid (D2EHPA) as an example (see Table 45–1) (Gu et al. 1985). Under the steady-state condition, when the concentration of Co^{2+} dropped from 1000 ppm to 10 ppm in the extraction side and increased from 0 to

TABLE 45–1. Typical Experimental Conditions and Results for the Extraction of Cobalt from Aqueous Solutions (reprinted from Gu et al. 1985 with permission).

Process Conditions	
Feed	1000 ppm Co^{2+}, 0.1 M sodium acetate, pH = 5
Stripping solution	1.0 M H_2SO_4
Oil phase	Kerosene containing 10% (v/v) D2EHPA
Operational Conditions	
Voltage	3.0 kV
Current	670 μA
Flow rate of feed	2×10^{-4} m^3/h
Flow rate of stripping solution	1×10^{-5} m^3/h
Mean residence time of aqueous droplets in electrical field	4 s
Results	
Co^{2+} concentration in raffinate	10 ppm
CO^{2+} concentration in stripping solution	19,750 ppm
Concentration of cobalt complex	300 ppm (in stripping cell)
	400 ppm (in extraction cell)
pH in raffinate	4

19,750 ppm in the stripping side in one single step, the concentration of cobalt complex in the continuous oil phase was always as low as 300 to 400 ppm, much lower than its equilibrium concentration. Obviously, such results can never be obtained by solvent extraction in one single step. In view of these aspects, particularly the nonequilibrium mass transfer character, ESPLIM is closely related to liquid membranes, rather than solvent extraction, and, therefore, it retains the main advantages of liquid membranes. Moreover, the unique features of ESPLIM, which are further discussed later in this chapter, overcome some shortcomings of liquid membranes, such as the complicated ELM process and the instability of SLMs. This chapter discusses the theory and basic principle of ESPLIM in addition to its major features. Its potential applications are also outlined.

THEORY

Principle of Electrostatic Dispersion

When a high-voltage electrostatic field is applied across an oil solution with a good dielectric property to which water drops are simultaneously introduced, the water drops will be elongated into ellipsoidal shape by the induced polarization. The ratio of the electrostatic drag force acting on the water drop to the oil/water interfacial tension force can be represented by the electrostatic Weber number We_E (Martin et al. 1983):

$$We_E = 9.0 \times 10^9 \frac{\epsilon_0 \epsilon E^2 d}{\gamma}. \quad (45\text{–}1)$$

As the applied electrical field strength E rises to a certain critical value, We_E reaches 0.409 and the major-to-minor axis ratio of the deformed water drop equals 1.838, which represents the maximum deformation of the water drop. Below the critical field strength, the adjacent water drops attract each other and coalesce into a big one. This is the case of crude oil dewatering, electrostatic demulsification in ELM, and electrostatic phase separation in solvent extraction. When the electrical field strength exceeds the critical value, the water drop can no longer maintain integrity. It disintegrates into numerous droplets in the continuous oil phase. Phase dispersion of ESPLIM is achieved on the basis of this principle.

Mechanism of Mass Transfer

During operation, an ac high-voltage electrostatic field is applied simultaneously across the extraction cell and the stripping cell. Under the electrical field, which is high enough for phase dispersion, the feed solution added to the extraction cell and the stripping solution added to the stripping cell are dispersed into numerous droplets in the continuous organic phase. In the extraction cell, the solute in the aqueous droplets is extracted into the organic phase. The complex formed in the extraction cell, driven by its own concentration gradient, diffuses through the perforated baffle plate into the stripping cell. The extractant is regenerated after the solute is stripped off in the stripping side. The concentration gradient of the extractant causes it to diffuse back to the extraction cell through the perforated baffle plate. With this process, the extraction and stripping are coupled inside the reaction tank, thus continuously transporting solute from the feed to the stripping solution for concentration.

Figure 45–1 shows the schematic of the ESPLIM process. A reaction tank is filled with extractant and diluent. The upper part of the reaction tank is divided into an extraction cell and a stripping cell by a baffle plate, and in its lower part are the extraction and stripping settlers separated by a partition. The extraction cell and its settler are intercommunicated, as are the stripping cell and its settler. Two pairs of electrodes are mounted across the extraction and stripping cells, respectively. The level of the interface between the continuous oil phase and the raffinate (or concentrate) is adjusted by the height of the triton, of which one branch is connected to the atmosphere to avoid the creation of a siphon. Figure 45–2 shows the schematic structure of the electrodes and baffles. The function of the baffles is to provide a passageway for the continuous oil phase and prevent the mixing of feed and stripping solutions, thus coupling extraction and stripping inside the reaction tank. Figure 45–2 shows that the continuous oil can flow freely through the holes of the baffled electrode. However, if the droplets of the aqueous feed or the stripping solution enter the holes, the very weak electric field inside the holes causes them to coalesce. Upon the action of gravity, these coalesced drops will slip away from the holes and back to their respective cells. In our practice, the baffle with a height of 5 mm is adequate to prevent the mixing of feed and stripping solution.

Figure 45–3 depicts the relationship of two chemical equilibria, favorable to solute transport, which are established at the interfaces for extraction and stripping at a certain cross section of the reaction tank. These two chemical

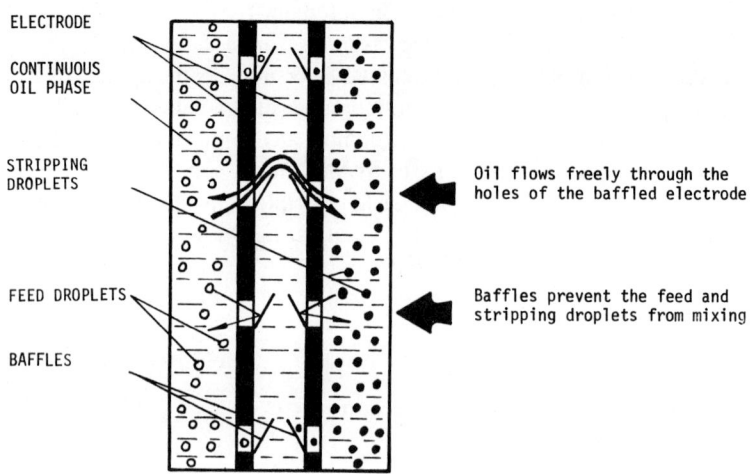

FIGURE 45–2. Schematic of electrode-baffle structure (reprinted from Gu 1990b with permission).

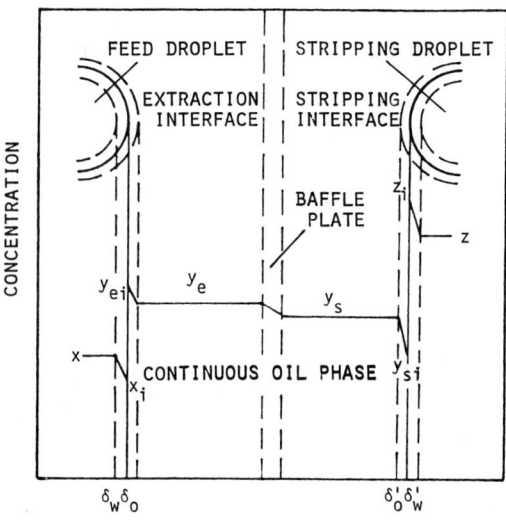

FIGURE 45–3. Relationship of chemical equilibria established at a certain cross section of the reaction tank (reprinted from Gu 1990a with permission).

equilibria result in the chemical potential difference across the baffle plate, which causes the diffusion of both the complex and the extractant. Under steady-state condition, the increments of the total mass transfer rate (dj), chemical reaction rates (dj_e, dj_s), interfacial film diffusion rates (dj_o, dj_w, dj'_o, dj'_w), and baffle plate diffusion rate (dj_m) are equal to one another within a differential height dh at a certain cross section of the reaction tank, i.e.,

$$dj = dj_w = dj_e = dj_o = dj_m = dj'_o = dj_s = dj'_w, \quad (45\text{-}2)$$

where

$$\frac{dj_w}{dh} = 10^{-3} k_w a S (x - x_i), \quad (45\text{-}3)$$

$$\frac{dj_o}{dh} = 10^{-3} k_o a S (y_{ei} - y_e), \quad (45\text{-}4)$$

$$\frac{dj_m}{dh} = 10^{-3} P \left(\frac{\lambda A}{H} \right) (y_e - y_s), \quad (45\text{-}5)$$

$$\frac{dj'_o}{dh} = 10^{-3} k'_o a' S' (y_s - y_{si}), \quad (45\text{-}6)$$

$$\frac{dj'_w}{dh} = 10^{-3} k'_w a' S' (z_i - z), \quad (45\text{-}7)$$

The equations for the reaction rates of extraction and stripping, j_e and j_s, depend on the particular extraction system.

Now let us analyze the vertical mass transfer process inside the reaction tank. The aqueous feed and stripping droplets flow downward to contact the continuous oil phase, which remains still macroscopically. Such kinds of contact can be regarded as countercurrent extraction and stripping from the point of view of relative movement. When these two kinds of aqueous droplets arrive at their own settlers where no (or very weak) electrical field is present, coalescence occurs and the two aqueous phases separate from the continuous oil phase. In this way, the raffinate and the concentrate are collected.

To sum up, the vertical countercurrent extraction and stripping processes, taking place in the extraction and stripping cells, respectively, together with the lateral transport process of the complex and the extractant across the two cells, constitute the whole ESPLIM process.

Figure 45–4 shows schematically the mass

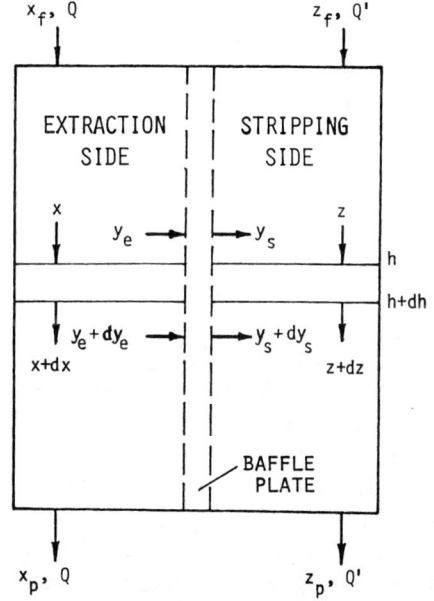

FIGURE 45–4. Mass transfer process inside the reaction tank under the steady-state condition (reprinted from Gu 1990a with permission).

transfer process inside the ESPLIM reaction tank under the steady-state condition. By combining the vertical and lateral mass transfer processes, the following fundamental equations are obtained (Gu 1990a):

$$\text{Extraction side:} \quad -\frac{dx}{dh} = \frac{k_w a S}{Q}(x - x_i), \tag{45-8}$$

$$\text{Stripping side:} \quad \frac{dz}{dh} = \frac{k'_w a' S'}{Q'}(z_i - z), \tag{45-9}$$

Keep in mind that the above equations solely describe the ideal situation in which no backmixing is considered. The permeation constant of solute through the baffle plate P and mass transfer flux J can be expressed as follows (Gu 1990a):

$$P = \frac{Q(x_f - x_p)}{\lambda A(y_e - y_s)_m}, \tag{45-10}$$

$$J = 10^{-3} \frac{Q(x_f - x_p)}{\lambda A}. \tag{45-11}$$

MASS TRANSFER PROCESS

Typical Experimental Results

In the early experimental work aimed at demonstrating the principle of ESPLIM process, Gu et al. (1985) successfully achieved the Co^{2+} extraction by D2EHPA from the dilute aqueous solution by means of a small rectangular setup, which was about 12 cm high, 8 cm wide, and 1.5 cm thick. Table 45-1 gives a set of typical data of their experiment. From Table 45-1, we can see that 99% extraction of Co^{2+} was achieved after the aqueous feed containing 1000 ppm Co^{2+} passed through a 10-cm-high electrostatic field with the residence time of only 4 seconds. At the same time, the Co^{2+} concentration in the stripping solution increased sharply from 0 to nearly 20,000 ppm. The consumption of electrical power was only 2 watts.

Zhou and Gu (1988) have demonstrated the extraction of Eu^{3+} by P 5709 (a phosphonic acid-type extractant synthesized by Beijing Institute of Uranium Ore Processing) using a similar setup. From the feed solution containing 280 ppm Eu^{3+}, the extraction reached 95% while Eu^{3+} was enriched 60 times in the stripping cell. The residence time of aqueous droplets in the electrical field was also about 4 seconds.

The extraction of Y^{3+} by P 5709 (Gu, Zhou, and Wu 1987) has been found to be very efficient. When feed solution containing 100 ppm Y^{3+} was added to the extraction cell, the resulting raffinate contained only 0.05 ppm Y^{3+} (99.95% extraction). Even if increasing the feed concentration to 1000 ppm Y^{3+}, the extraction efficiency of Y^{3+} was still up to 99%. The above experimental data clearly show the high extraction efficiency of the ESPLIM process.

Effect of Electrical Field

According to the principle of electrostatic dispersion, the applied electrical voltage across the cell should be high enough to ensure that the field strength throughout the electrical field is always above its critical value for satisfactory phase dispersion. It is important to point out that the applied voltage required to reach critical field strength varies according to the specific extraction system, the distance between electrodes, and the nature and thickness of the insulation layer.

Using his particular experimental setup for Co^{2+} extraction, Gu (1990a) has obtained the relationship between mean diameter of dispersed aqueous droplets and the applied voltage. As shown in Figure 45-5, the droplet diameter increases with the increasing voltage in the region of low electrical field, and then it reaches its maximum as the applied voltage is raised to 1.3 kV. This implies that the applied field strength, when lower than its critical value, will accelerate the coalescence of aqueous droplets. In the region of 1.3 to 1.9 kV, the droplet diameter, in turn, decreases gradually. This indicates that the field strength

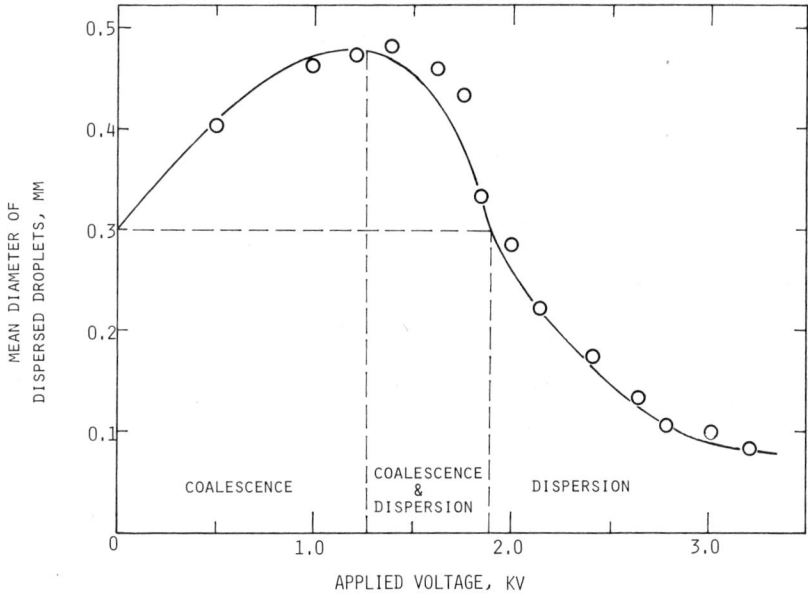

FIGURE 45–5. Mean diameter of dispersed aqueous droplets as a function of applied voltage (reprinted from Gu 1990a with permission).

in certain areas of the electrical field reaches or exceeds the critical value, causing the partial disintegration of aqueous droplets. This is a mixed region called the *coalescence-dispersion region*. When the applied electrical field is equal to 1.9 kV, the droplet diameter reduces to the value at zero voltage. Above 1.9 kV, the aqueous droplets become finely dispersed, which tells us that the field strength exceeds the critical value throughout the electrical field. These observations agree with those discovered by Martin et al. (1983). It is advisable not to apply too high of an electrical voltage because the extra finely dispersed aqueous droplets are liable to be entrained by the continuous oil phase, which will increase leakage and "swelling" of the process and possibly cause emulsification.

Electrodes used for electrostatic dispersion should be coated with an insulation layer to prevent sparking between electrodes (Hsu and Li 1985) and to protect the electrode material from chemical corrosion. The insulating layer can be regarded as a capacitor that consumes part of the applied voltage. Gu (1990b) found that when the electrodes were coated with polytetrafluoroethylene (PTFE) of thickness ~0.1 mm, only a small part of applied voltage was dropped across the oil phase, while most of the electrical energy was dissipated inside the insulating layers (Table 45–2). Reduction of the portion of voltage drop across the insulation

TABLE 45–2. Effect of Electrostatic Field on Co^{2+} Extraction (reprinted from Gu 1990b with permission).

Applied Voltage (kV)	Potential Drop in Oil Phase (kV)	Extraction of Cobalt (%)
1.0	0.25	10
2.0	0.40	58
3.0	0.45	92
3.5	0.50	95
4.0	0.55	(emulsified)

layer would be beneficial. One possible way to do so might be to use an insulating material with higher dielectric constant.

An interesting phenomenon observed by Gu (1990b) was the "bridging" of aqueous drops across the electrodes when a hydrophilic polyester film was used as the insulating layer. When this bridging occurred, the aqueous drops failed to disintegrate into droplets. Thus, the phase dispersion was seriously impaired. To overcome this problem, the polyester film was replaced by a polypropylene or Teflon® film, which is relatively hydrophobic. As a result, the bridging phenomenon disappeared. Therefore, the selected insulating material must be voltage-resistant, oil-proof, and have a hydrophobic surface.

Compared to conventional mechanical agitation, *electrostatic agitation* possesses some unique features. The first advantage of electrostatic agitation is the saving of energy consumption. For the treatment capacities of 0.5 and 50 L/h, the consumed electrical powers for phase dispersion were measured to be 3.1 and 53 W, respectively (Gu 1990a; 1990b). This indicates the negligible energy consumption of electrostatic agitation compared to mechanical stirring. This negligible energy consumption is attributed to the fact that the phase dispersion by mechanical stirring is achieved by vigorous turbulence of the continuous phase, while for electrostatic agitation, the electrostatic force acts directly on the aqueous phase to be dispersed without the necessity of stirring the entire continuous phase.

Another feature of electrostatic agitation is reliable operation due to the absence of mechanical moving parts inside the reaction vessel. This is especially favorable when radioactive effluents are treated. The third advantage of electrostatic agitation is the enhancement of interfacial mass transfer, which is discussed later in this chapter.

Effect of Extractant Concentration

Only a minor effect on Co^{2+} extraction rate was observed when extractant concentration was varied over a fairly wide range (Gu 1990a). When the extractant concentration was lowered from 0.3 to 0.03 mol/L (a tenfold reduction), the Co^{2+} extraction rate only decreased by 10% (Figure 45–6). Practically, satisfactory results were obtained with 0.15 mol/L extractant (less than 5%) in the oil phase, which was much lower than that required for solvent extraction (Gu, Wasan, and Li 1988). Similar results were obtained for Eu^{3+} extraction (Zhou and Gu 1988). The extraction rate of Eu^{3+} remained relatively high (94%) even if the extractant concentration was as low as 6%. These data are comparable to those obtained in ELM experiments (Frankenfeld, Cahn, and Li 1981).

Gu (1990a) observed that changing the extractant concentration did not affect significantly the cobalt complex concentration in the oil

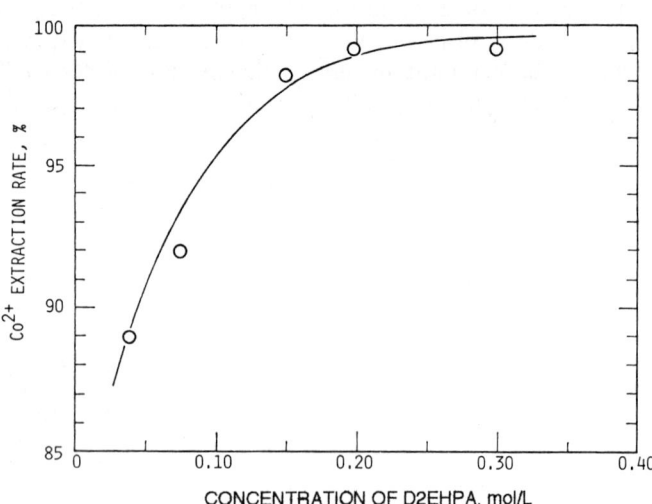

FIGURE 45–6. Effect of extractant concentration on Co^{2+} extraction (reprinted from Gu 1990a with permission).

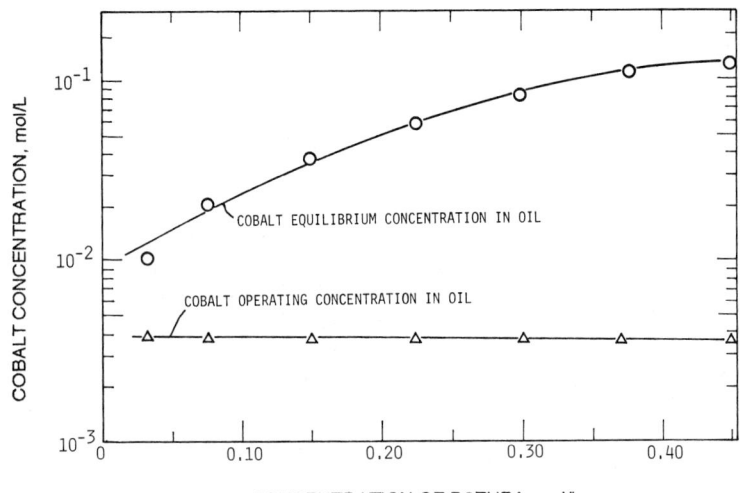

FIGURE 45–7. Comparison of the equilibrium concentration of cobalt to its operating concentration in the oil phase for extraction at the feed inlet under various extractant concentrations (reprinted from Gu 1990a with permission).

phase in the extraction cell, and the cobalt complex concentration was always much lower than its equilibrium concentration (Figure 45–7). This reflects the nonequilibrium feature of the ESPLIM process, in which the organic extractant "shuttles" back and forth as a "mobile carrier" to transport metal ions continuously from the feed to the stripping solution. Under such circumstances, the extractant in the reaction tank will never be fully loaded with metal, resulting in a very high driving force for mass transfer.

Effect of Flow Ratio of Feed to Stripping Solution

The flow ratio of feed to stripping solution (Q/Q') is an important parameter that measures the performance of the extraction system. Usually, a high value of Q/Q' is favorable for increasing the concentration factor of the system. However, an excessively high value of Q/Q' will lead to a rise in the complex concentration in the oil phase, thus raising the raffinate concentration and lowering the extraction efficiency due to the insufficient ability of stripping.

The Co^{2+} extraction data at different Q/Q' ratios are summarized in Table 45–3 (Gu 1990a). Increasing the Q/Q' ratio gives a high concentration factor at the expense of lowering the extraction rate of Co^{2+}. For the feed solution containing 1.69×10^{-2} mol/L Co^{2+} (1000 ppm), the flow ratio of 50 still offers the relatively high extraction rate of Co^{2+} (95.6%), and it gives a Co^{2+} concentration in the con-

TABLE 45–3. Data of Co^{2+} Extraction at Different Flow Rates of Feed to Stripping Solution ($A = 0.0074$ m^2, $\lambda = 14\%$) (reprinted from Gu 1990a with permission).

Q (10^{-6}/m^3)	Q' (10^{-9} m^3/s)	Q/Q'	x_p ($10^{-4} \times$ mol/L)	z_p ($10^{-1} \times$ mol/L)	Extraction (%)	P ($10^{-3} \times$ m/s)	J [$10^{-4} \times$ mol/(m^2·s)]
0.04	6.67	6.00	0.85	1.04	99.5	0.95	6.45
0.04	3.89	10.28	1.02	1.68	99.4	0.95	6.45
0.04	3.05	13.11	1.12	2.20	99.3	0.95	6.45
0.06	3.05	19.67	1.53	3.35	99.1	1.05	9.67
0.08	2.78	28.78	3.57	4.98	97.9	1.94	12.74
0.11	2.78	39.57	4.07	6.62	97.6	2.30	17.46
0.14	2.78	50.36	7.47	8.11	95.6	3.16	21.77
0.19	2.78	68.35	22.92	10.26	86.5	3.20	26.71

centrate approaching 0.811 mol/L (47.75 g/L), which meets the requirement for the electrolytic winning of cobalt. Note that for the extraction of Eu^{3+} by P 5709, an extraction efficiency of 94.3% was obtained for the flow ratio of 61 (Zhou and Gu 1988).

COMPARISON OF ESPLIM WITH ELM AND SLM

Like liquid membranes, the ESPLIM process is characterized by nonequilibrium mass transfer. Therefore, it offers the same advantages as liquid membranes, such as the reduction of required separation stages and the saving of chemical reagents. In addition, it possesses some unique features compared to liquid membranes as discussed below.

Comparison with ELM

The ELM process offers very high mass transfer efficiency because of the very high membrane area per unit volume (Cussler and Evans 1980; Gu, Wasan, and Li 1988). In the ELM system, surfactant is employed to maintain membrane integrity. In other words, the interfacial tension of the system is lowered by chemical means to achieve phase dispersion. Once the surfactant is dissolved in solution, it is difficult to remove from the solution. Therefore, the introduction of surfactant to the ELM system can be regarded as irreversible. This makes the process complicated. As shown in Figure 45–8(a), the whole process consists of four steps, i.e., emulsification, extraction/stripping, settling (not shown in this figure), and demulsification (also see Figure 36–3). Associated with the complexity of the system, the operation cost of ELM increases. During the operation, the handling and transfer of the emulsion will cause an additional loss of the reagents.

In the ESPLIM process, an electrostatic field (i.e., physical means) is applied to change the interfacial tension of the system and hence to achieve phase dispersion or phase separation. As pointed out by Thornton (1968), when an electrostatic field was applied to a water/oil binary system in which the oil phase acted as continuous phase, the surface charge of the water drops gave rise to a mechanical force that was directed outward from the surface, i.e., in

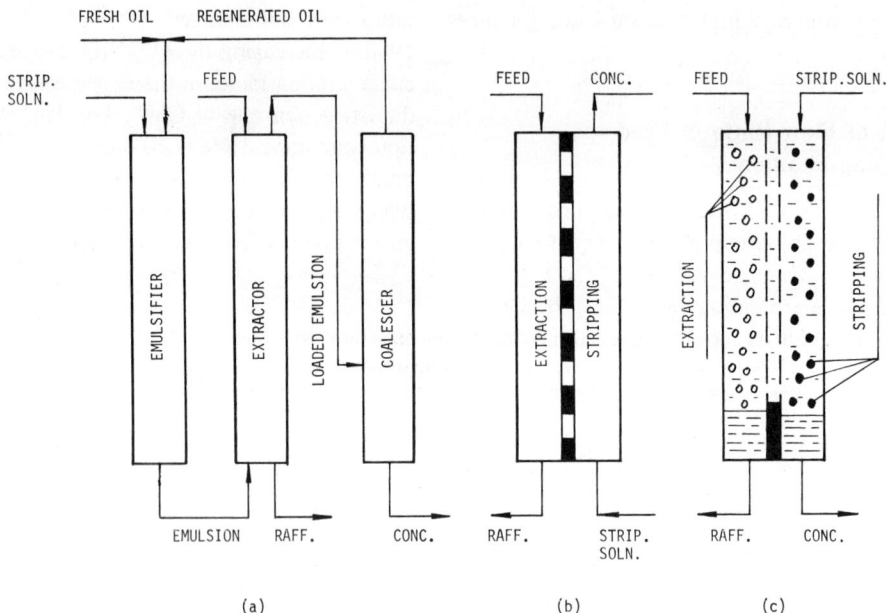

FIGURE 45–8. Schematics of liquid membrane processes: (a) ELM, (b) SLM, and (c) ESPLIM (reprinted from Gu 1990b with permission).

direct opposition to the inward-acting interfacial tension. The water drops therefore suffered a reduction in effective interfacial tension owing to an electrostatic repulsion force at the surface. In brief, the increased electrical field results in the lowering of the interfacial tension of the system.

In an ESPLIM system, an electrical field is applied across the extraction cell and the stripping cell to lower the interfacial tension, thus causing the dispersion of feed and stripping solutions in the organic phase. In the two settlers, no (or very weak) electrical field is applied. Hence, the interfacial tension rises again immediately, and the aqueous droplets coalesce and separate from the organic phase. Such a mode of phase dispersion and phase separation in the ESPLIM system precludes the need for emulsification and demulsification equipment. As a result, the process is remarkably simplified, as shown in Figure 45–8(c).

Because ESPLIM is a simple process, operation is easy and, hence, investment and operation costs are reduced. During operation, only feed and stripping solutions are transported, while the organic solution remains in the reaction tank. Therefore, the loss of organic reagents is minimized.

The degree of membrane leakage is an important parameter in assessing the stability of liquid membranes. Membrane leakage in the ELM system involves the rupture of emulsion or the leakage of the internal phase. This leakage leads to an increase of solute concentration in the raffinate, thus lowering the extraction efficiency. The leakage rate is affected by the membrane formulation (such as properties of surfactant, additive, and diluent), the properties of the internal and/or external phase, the method of emulsification, and the mixing conditions during the extraction process (see Chapter 38). The data on leakage rate reported from different laboratories are different, ranging from 0.2 to 10% (Zhang et al. 1988a).

In the ESPLIM process, leakage refers to the leak of the stripping solution from the stripping cell to the extraction cell during operation. The leakage rate varies as a function of baffle structure, applied voltage, etc. Gu (1990b) reported that the maximum leakage rate under his experimental conditions was 0.06%, which was much lower than any value reported in ELM processes. This result indicates the high extraction efficiency of the ESPLIM process.

In an ELM system, emulsion swelling is a troublesome problem, which dilutes the internal phase and makes it difficult to recycle the emulsion. Emulsion swelling is supposed to be caused mainly by *entrainment*, in which the external solution is partially encapsulated by the membrane phase during the extraction process. Also, the osmotic pressure difference between the internal and external phases contributes to membrane swelling by causing the permeation of water from the external to the internal phase (Yan, Shi, and Su 1988). The degree of emulsion swelling depends strongly on the kinds of surfactants used (Nakashio, Goto, and Kondo, 1988; also see Chapter 38).

In the ESPLIM system, feed droplets could be entrained by the oil solution (continuous phase) from the extraction cell through the baffle channels to the stripping cell. This entrainment is regarded as "swelling" by analogy with that in the ELM process. Here, the swelling rate is defined as the volume ratio of the entrained feed solution per unit time to the flow rate of the stripping solution. Gu (1990b) found in his experiments that by properly adjusting the baffle structure and the operational conditions, the swelling rate in the ESPLIM system could be controlled to ~1%. It is known, however, that the swelling rate due to osmotic effects was below 30%, while that due to entrainment could be up to 500% in ELM systems (Ding et al. 1988).

The rather low swelling rate of the ESPLIM process promises a high potential of this technology in concentrating solutes from dilute solutions. Wu (1988) studied the extraction of Y^{3+} by ESPLIM. In his experiment, the feed solution containing 1000 ppm Y^{3+} was passed through the extraction cell in a once-through mode, while the stripping solution with an initial HCl concentration of 6 N was circulated. In this way, the Y^{3+} concentration in the stripping solution increased gradually during the extraction process. As the Y^{3+} concentration in the stripping solution reached 30 g/L (30,000 ppm), the Y^{3+} concentration in the raffinate was still below 0.3 ppm. This is a good example showing the good ability of the ESPLIM process to concentrate solutes from dilute solutions.

Comparison with SLM

The SLM process [see Figure 45–8(b)] is as simple as the ESPLIM process. However, the mass transfer process of SLM was shown to be much slower than that of ESPLIM. The flux of solutes through an SLM lies in the range of 10^{-10} to 10^{-9} mol/(cm^2·s) [10^{-6} to 10^{-5} mol/(m^2·s)] (Largman and Sifniades 1978; Lee, Evans, and Cussler 1978; Babcock et al. 1980). In their study on the extraction of Co^{2+} and Ni^{2+} by a SLM, Danesi et al. (1984) found that the Co^{2+} flux was 2.8×10^{-9} mol/(cm^2·s) [2.8×10^{-5} mol/(m^2·s)].

For Co^{2+} extraction by ESPLIM, the Co^{2+} flux was reported to be from 6.5×10^{-8} to 2.7×10^{-7} mol/(cm^2·s) [6.5×10^{-4} to 2.7×10^{-3} mol/(m^2·s)] (Gu 1990a). These data indicate that the Co^{2+} flux in the ESPLIM system is one order of magnitude higher than that in an SLM system.

The relatively high resistance to mass transfer for SLM systems stems from the adoption of a porous solid membrane as the membrane support. Solute transport through the SLM is by molecular diffusion. In addition, the stagnant layers adjacent to the membrane also slow down the mass transfer process.

For the ESPLIM process, however, no solid membrane is employed. The organic solution (continuous phase) can flow freely through the baffle channels. Meanwhile, the aqueous droplets have induced surface charge acquired from the electrostatic field. Inside an alternate electrical field, the voltage gradient forces the charged aqueous droplets to oscillate, thus causing the convection of the continuous oil phase as observed in our experiments. This convection constitutes the major form of solute transport through the ESPLIM.

Another factor in interpreting the high flux of the ESPLIM process might be the variation of interfacial charge density between the aqueous droplets and the oil phase caused by the electrical field. This variation produces an interfacial tension gradient, resulting in interfacial turbulence (Marangoni instability), which causes frequent renewal of interface, thus enhancing interfacial mass transfer (Austin, Banczyk, and Savistowski 1971).

One problem that has not yet been satisfactorily solved for the SLM process is membrane instability due to the gradual loss of membrane solution by dissolution and spontaneous emulsification during the operation. Usually, the lifetime of an SLM is one or several weeks. So far, no commercial application of SLM has been reported.

In the ESPLIM process, a baffle plate is used instead of a porous solid membrane. This not only enhances the mass transfer process but also avoids the problems of membrane instability and membrane fouling as encountered in the SLM.

Further Development of ESPLIM

Like all other technologies, ESPLIM has its own limitation. It is only effective when the continuous phase is an organic solution of low polarity and the dispersed phase is aqueous solution. It is thus confined to the extraction and separation of solutes from aqueous solutions. To overcome this problem, some other methods, such as spray and ultrasonic techniques, might be alternatives to the electrical field to achieve phase dispersion (Navratil, Cheng, and Gu 1990; Zhou and Gu 1989).

Zhou and Gu (1989) developed a spray pseudo-liquid-membrane process in which spray dispersion was adopted instead of electrostatic dispersion. Because of the absence of an electrical field in this process, no restriction to the continuous phase is needed, i.e., the continuous phase may be either organic or aqueous solution, and the only requirement is the density difference between the continuous and dispersed phases. As a result, the application area of the pseudo liquid membrane has been greatly widened. Besides, the avoidance of an electrical field further simplifies the process.

Using a spray pseudo-liquid-membrane setup, Zhou and Gu (1989) performed Co^{2+} extraction by D2EHPA. In their experiments, the feed solution contained 1000 ppm Co^{2+} and 0.1 M CH_3COONa with an initial pH of 5, and the stripping solution was 2M H_2SO_4. The organic phase was kerosene containing 15% (by volume) D2EHPA. When the flow rates of the feed and stripping solutions were adjusted to 125 and 28 mL/h, respectively, the extraction rate reached about 98%. Using the same setup, they achieved the extraction of Eu^{3+} from a highly acid solution (3 N HNO_3) by a neutral bidentate organophosphorus extractant, dihexyl diethyl carbamoyl methylene phosphonate (DMDECMP). The extraction rate of Eu^{3+} was 74.6%. This might provide a way to remove and/or recover +3 (e.g., Am^{3+}) as well as +4 and +6 actinides from certain highly acidic nuclear fuel cycle waste liquors. It is believed that removal of long-lived actinides from such radioactive wastes will greatly reduce potential hazards for the long-term storage and/or disposal of nuclear wastes.

SCALE-UP OF AN ESPLIM DEVICE

To build an ESPLIM setup on a commercial basis, the following requirements must be met:

1. The setup is compact so as to obtain a high volumetric throughput.
2. The major components, such as baffle plates and electrodes, are simple in structure and easy to manufacture, mount, and dismount.
3. On the premise of fitting in with the needs of process conditions, the materials constituting the setup must be cheap and easily available so as to lower the investment.

The rectangular multicell reaction tank with the planar electrode-baffle plates is recommended in view of the ease of scale-up. An example of the cell arrangement is shown in Figure 45–9(a). In this arrangement, the basic unit has a two-cell structure, which consists of an extraction cell with its settler and a stripping

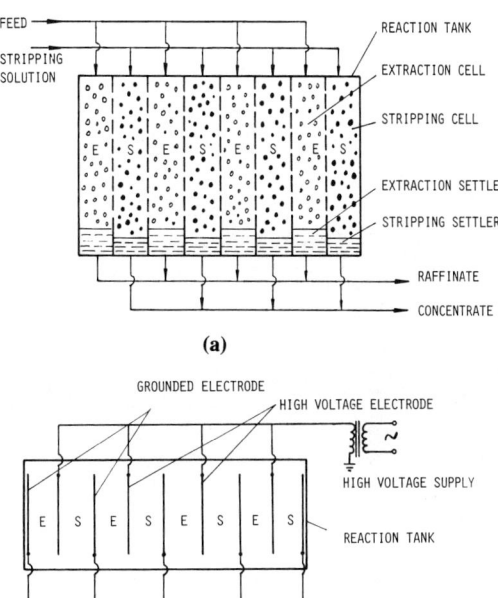

FIGURE 45–9. Schematics of (a) multicell ESPLIM reaction tank and (b) a layout of electrodes.

cell with its settler. The multicell reaction tank is composed of a number of such basic units in parallel. Feed and stripping solutions are uniformly distributed to the top of the individual extraction cells and stripping cells, respectively. The raffinates from individual extraction settlers flow into a main pipe and then discharge to a raffinate receiver. Likewise, the concentrates from individual stripping settlers flow into another main pipe and are collected in a concentrate receiver. Accordingly, the high-voltage electrical power is supplied via a busbar to each high-voltage electrode in parallel, and all the grounded electrodes are parallelly connected to another busbar, which is grounded [Figure 45–9(b)]. A photograph of the entire ESPLIM setup with a capacity of 10 L/h is shown in Figure 45–10.

POTENTIAL APPLICATIONS

The simple process, with easy operation, negligible leakage and swelling, and low consumption of energy, promises ESPLIM as a potentially competitive technology in the fields

FIGURE 45–10. Photo of an ESPLIM setup with a capacity of 10 L/h.

of hydrometallurgy and wastewater treatment. As an emerging technology, its possible application areas are still to be exploited. This section discusses potential applications.

Hydrometallurgy

The South China provinces are rich in ion-sorbed rare earth mineral resources. For such rare earth ores, rare earth ions are leached out by an electrolyte [e.g., $(NH_4)_2SO_4$] solution. The resultant leachate contains ~1 g/L mixed rare earth ions. The current means of concentration is to precipitate rare earth elements from the leachate by oxalic acid. Rare earth oxalate is then filtered and calcinated to obtain mixed rare earth oxides (RE_2O_3). To provide feed solution for a rare earth separation plant, RE_2O_3 needs to be dissolved by acid to form a concentrated rare earth solution with the mixed rare earth concentration up to 100 g/L or more. Because of the complicated process and thus the high cost of the present method, several alternatives are being explored with an attempt to concentrate the mixed rare earth directly from the leachate to a concentration high enough to be the feed for the rare earth separation plant. ELM technology has been reported to extract more than 98% rare earth ions from the leachate and rare earth ions can be concentrated up to 80 g/L. Meanwhile, the operating cost of ELM was reported to be only one-sixth of the oxalate precipitation method (Yu, Wang, and Jiang 1987). However, ELM is still rather complicated due to the use of emulsification and demulsification steps. One troublesome problem of ELM is a relatively high membrane swelling rate, which would limit the further concentration of mixed rare earth ions. As discussed above, both the leakage and swelling rates of ESPLIM are at least one order of magnitude lower than those of ELM. Using an ESPLIM process, Wu (1988) concentrated the mixed rare earth from the leachate. The extraction efficiency of mixed rare earth ions was above 98%, and the rare earth content reached 95.2 g/L in the con-

TABLE 45-4. Typical Data of Ni(II) Recovery from Rinse Water by ESPLIM Under Steady-State Conditions[a] (reprinted from Gu, Zhou, and Jin 1990 with permission).

No.	Applied Voltage (kV)	Current (μA)	Ni(II) Concentration, ppm			Total Efficiency (%)
			in Feed	in Raffinate Step 1	Step 2	
1	3.1	450	408	5.5	0.45	99.89
2	3.1	470	408	6.0	0.50	99.88
3	3.4	490	408	3.1	0.30	99.93
4	3.3	480	408	3.7	0.32	99.92

[a]Flow rate of feed = 1.12×10^{-4} m³/h; flow rate of stripping solution = 3×10^{-6} m³/h; acetate concentration in feed = 0.1 M; initial pH = 4.6.

centrated solution. Navratil, Cheng, and Gu (1990) also demonstrated the utility of ESPLIM for rare earth recovery.

Recovery of Metal Values from Dilute Effluents

A rare earth separation plant produces large-volume effluents containing rare earth ions, such as Y^{3+}, La^{3+}, Eu^{3+}, Yb^{3+}, etc. The rare earth concentration ranges from 100 to 2000 ppm. Gu, Zhou, and Wu (1987) tested an ESPLIM process with actual effluents from a rare earth solvent extraction plant. For the effluent containing 2350 ppm Y^{3+} and 2 M NH_4Cl, a 99.8% recovery of Y^{3+} was obtained after a two-step treatment by the ESPLIM process. The total organic content in the raffinate was measured to be less than 15 ppm, and no entrained oil was observed in the aqueous phases.

Removal of Toxic Material from Wastewater

As a carcinogenic substance, nickel discharged to sewers or public water streams must be strictly limited. Effluents from nickel processing plants must be suitably treated to remove nickel before discharge to the environment. The rinse water from a nickel plating process usually contains less than 1.0 g/L nickel. Obviously, solvent extraction is not a good choice for treating an effluent with such a low concentration.

Gu, Zhou, and Jin (1990) have studied the removal of nickel from the rinse water of a nickel plating process with a bench-scale ESPLIM setup. Typical data are shown in Table 45-4. After two-step treatment by the ESPLIM process, the total removal of nickel reaches 99.9%, and the nickel concentration in rinse water declines to less than 0.5 ppm.

ACKNOWLEDGMENTS

This work was financially supported by the China Natural Science Foundation. The author wishes to thank Dexi Wang and Shuheng Yan for their strong support and helpful suggestions.

NOTATION

General Notation

See the General Notation section at the beginning of this handbook.

Special Notation

a	specific extraction area, m^{-1}
a'	specific stripping area, m^{-1}
A	total area of the perforated baffle plate, m^2

d	diameter of water drop before deformation, m
E	electrical field strength, V/m
h	height of the perforated baffle plate, m
H	total height of the perforated baffle plate, m
j	overall mass transfer rate, mol/s
j_w	aqueous film diffusion rate for extraction, mol/s
j'_w	aqueous film diffusion rate for stripping, mol/s
j_m	solute diffusion rate through the perforated baffle plate, mol/s
J	mass transfer flux, mol/(m^2·s)
k	interfacial mass transfer coefficient for extraction, m/s
k'	interfacial mass transfer coefficient for stripping, m/s
Q	flow rate of feed solution, m^3/s
Q'	flow rate of stripping solution, m^3/s
P	permeation constant of the perforated baffle plate, m/s
S	cross-section area of the extraction cell, m^2
S'	cross-section area of the stripping cell, m^2
We_E	electrostatic Weber number
x	metal concentration in the aqueous phase in the extraction cell, mol/L (kg-mol/m^3)
y	metal complex concentration in the oil phase, mol/L (kg-mol/m^3)
z	metal concentration in the aqueous phase in the stripping cell, mol/L (kg-mol/m^3)
$(y_e - y_s)_m$	mean value of the metal complex concentration difference between the extraction and stripping cells, mol/L (kg-mol/m^3)

Greek Letters

γ	interfacial tension, N/m
δ	thickness of the interfacial film for extraction, m
δ'	thickness of the interfacial film for stripping, m
ϵ	relative dielectric constant of the continuous oil phase, F/m
ϵ_0	vacuum dielectric constant, F/m
λ	porosity of perforated baffle plate, dimensionless

Subscripts

e	extraction
i	interface between two phases
f	inlet
o	oil film
p	outlet
s	stripping
w	aqueous film

REFERENCES

Austin, L. J., L. Banczyk, and H. Savistowski. 1971. Effect of electric field on mass transfer across a plane interface. *Chem. Eng. Sci.* 26:2120.

Babcock, W. C., R. W. Baker, E. D. Lachapelle, and K. L. Smith. 1980. Coupled transport membranes: II. the mechanism of uranium transport with a tertiary amine. *J. Membr. Sci.* 7:89.

Brauwn, T., and G. Ghershin. 1975. *Extraction Chromatography*. Amsterdam: Elsevier Scientific Publishing Co.

Cussler, E. L., and D. F. Evans. 1980. Liquid membranes for separation and reactions. *J. Membr. Sci.* 6:113.

Cussler, E. L., M. R. Stokar, and J. E. Varberg. 1984. Gels as size selective extraction solvent. *AIChE J.* 30(4):578.

Danesi, P. R., L. Reichley-Yinger, C. Cianetti, and P. G. Rickert. 1984. Separation of cobalt and nickel by liquid-liquid extraction and supported liquid membranes with di(2,4,4-trimethylpentyl)phosphinic acid (Cyanex 272). *Solvent Extr. Ion Exch.* 2(6):781.

Ding, X. C., F. G. Xie, X. Y. Xie, and J. Z. He. 1988. Study of the liquid membrane swelling. Paper read at the 1st Sino-Japan Symposium on Liquid Membranes, 24–27 September 1988, Hangzhou, China.

Draxler, J., and R. Marr. 1986. Emulsion liquid membranes: part I. phenomenon and industrial application. *Chem. Eng. Process.* 20:319.

Frankenfeld, J. W., R. P. Cahn, and N. N. Li. 1981. Extraction of copper by liquid membranes. *Sep. Sci. Technol.* 16(4):385–402.

Freitas, R. F. S., and E. L. Cussler. 1987. Temperature sensitive gels as size selective adsorbents. *Sep. Sci. Technol.* 22(2&3):911–919.

Goklen, K. E., and T. A. Hatton. 1987. Liquid-

liquid extraction of low molecular weight proteins by selective solubilization in reversed micelles. *Sep. Sci. Technol.* 22(2&3):831.

Gu, Z. M. 1988. Electrostatic pseudo liquid membrane separation method and its equipment. Chinese Patent 86,101,730.

Gu, Z. M. 1990a. Electrostatic pseudo liquid membrane separation technology. *J. Chem. Ind. Eng.* (China) 5(1):44–55.

Gu, Z. M. 1990b. A new liquid membrane technology—electrostatic pseudo liquid membrane. *J. Membr. Sci.* 52(1):77–88.

Gu, Z. M., D. T. Wasan, and N. N. Li. 1988. Liquid surfactant membranes for metal extractions. In *Surfactants in Chemical Process Engineering*, ed. D. T. Wasan, M. E. Ginna, and D. D. Shah, pp. 127–168. New York: Marcel Dekker.

Gu, Z. M. M. X. Xu, L. Y. Zhu, and L. R. Jin. 1985. A study on electrostatic pseudo liquid membrane separation. *Annual Report of Institute of Atomic Energy*, Beijing, China.

Gu, Z. M., Q. J. Zhou, and Q. F. Wu. 1987. Extraction of Y^{3+} by electrostatic pseudo liquid membrane. Personal communication.

Gu, Z. M., Q. J. Zhou, and L. R. Jin. 1990. Recovery of Ni(II) from rinse water from nickel plating with liquid membranes. *Water Treatment* 5:170–178.

Ho, W. S., L. T. C. Lee, and K. J. Liu. 1976. Membrane hydrometallurgical extraction process. U.S. Patent 3,957,504.

Hsu, E. C., and N. N. Li. 1985. Membrane recovery in liquid membrane separation process. *Sep. Sci. Technol.* 20 (2&3):115–130.

Humphrey, J. H., and J. R. Fair. 1983. Low-energy separations for the process industry. *Sep. Sci. Technol.* 18 (14&15):1765.

Largman, T., and S. Sifniades. 1978. Recovery of copper (II) from aqueous solutions by means of supported liquid membranes. *Hydrometallurgy* 3:153.

Lee, K. H., D. F. Evans, and E. L. Cussler. 1978. Selective copper recovery with two types of liquid membranes. *AIChE J.* 24(5):860.

Lee, L. T. C., W. S. Ho, and K. J. Liu. 1976. Membrane solvent extraction. U.S. Patent 3,956,112.

Li, N. N. 1968. Separating hydrocarbons with liquid membranes. U.S. Patent 3,410,794.

Li, N. N. 1972. Liquid membrane separation process. U.S. Patent 3,696,028.

Martin, J., P. Vignet, C. Fombarlet, and F. Lancelot. 1983. Electrical field contactor for solvent extraction. *Sep. Sci. Technol.* 18(14&15):1455.

Nakashio, F., M. Goto, and K. Kondo. 1988. Role of surfactants in the liquid surfactant membrane process—development of new surfactants. Paper read at the 1st Sino-Japan Symposium on Liquid Membranes, 24–27 September 1988, Hangzhou, China.

Navratil, J. D., H. Y. Cheng, and Z. M. Gu. 1990. Rare earth recovery using pseudo liquid membranes. Paper read at the International Solvent Extraction Conference, 18–21 July 1990, Tokyo, Japan.

Ruppert, M., J. Draxler, and R. Marr. 1988. Liquid-membrane-permeation and its experiences in pilot-plant and industrial scale. *Sep. Sci. Technol.* 23(12&13):1659–1666.

Scott, T. C. 1989. Use of electric fields in solvent extraction: a review and prospectus. *Sep. Purif. Methods* 18(1):65–109.

Thornton, J. D. 1968. The applications of electrical energy to chemical and physical rate processes. *Rev. Pure Appl. Chem.* 18:197–218.

Ward, W. J. 1972. Immobilized liquid membranes. In *Recent Developments in Separation Science*, ed. N. N. Li, pp. 153. Boca Raton, FL: CRC Press.

Warshawsky, A. 1981. Extraction with solvent-impregnated resins. In *Ion Exchange and Solvent Extraction*, ed. J. A. Marinsky and Y. Marcus, pp. 229–310. New York: Marcel Dekker.

Wu, Q. F. 1988. Studies on the extraction of rare earth by electrostatic pseudo liquid membrane. M. S. diss., China Institute of Atomic Energy, Beijing.

Yan, N.-X., Y.-J. Shi, and Y.-F. Su. 1988. Swelling of emulsion in liquid surfactant membrane process. Paper read at the 1st Sino-Japan Symposium on Liquid Membranes, 24–27 September 1988, Hangzhou, China.

Yu, J.-H., S.-Z. Wang, and C.-Y. Jiang. 1987. Extraction of rare earth by liquid surfactant membranes. *Rare Earth (Chinese)* 1:1.

Zhang, X.-J., Q.-J. Fan, X.-T. Zhang, and Z.-F. Liu. 1988a. New surfactant LMS-2 for industrial application in liquid membrane separation. In *Separation Technology*, ed. N. N. Li and H. Strathmann, pp. 215–226. New York: United Engineering Trustees; see also *Water Treatment* 3:233–240.

Zhang, X.-J., J.-H. Liu, Q.-J. Fan, Q.-T. Lian, X.-T. Zhang, and T.-S. Lu. 1988b. Industrial application of liquid membrane separation for phenolic wastewater treatment. In *Separation Technology*, ed, N. N. Li and H. Strathmann, pp. 190–203. New York: United Engineering Trustees.

Zhang, X.-J., J.-G. Liu, and T.-S. Lu. 1987. Industrial application of liquid membrane separation for phenolic wastewater treatment. *Water Treatment* 2:127–135.

Zhou, Q. J., and Z. M. Gu. 1988. Studies on the extraction of Eu^{3+} by means of electrostatic pseudo liquid membrane. *Water Treatment* 3(2):127–135.

Zhou, Q. J., and Z. M. Gu. 1989. Spray pseudo liquid membrane equipment. Chinese Patent 88,213,050.1.

# 46

Other New Membrane Processes

Kamalesh K. Sirkar*
Stevens Institute of Technology

MEMBRANE-BASED GAS
  ABSORPTION AND STRIPPING
    Introduction
    Theoretical Considerations on Mass
      Transfer
    Membrane Devices and Transport
      Correlations
    Aspects of Hollow-Fiber Module Design
    Applications
MEMBRANE DISTILLATION
    Introduction
    Types of Membrane Distillation Processes
    Theoretical Considerations
    Membranes and Devices
    Applications
PERSTRACTION
    Introduction
    Flux and Separation in Perstraction
    Membranes
    Applications
ACKNOWLEDGMENTS
NOTATION
REFERENCES

MEMBRANE-BASED GAS ABSORPTION AND STRIPPING

Introduction

In conventional gas absorption, gas is dispersed in the absorbent liquid as bubbles in a vessel acting as a single stage or in a multistage column with the gas and the liquid in countercurrent flow. The dispersion of the gas increases the gas/liquid contact area and increases the mass transfer rate of the species to be absorbed in the liquid. After contacting is over, the gas bubbles coalesce as they leave the liquid phase except when foaming tendencies appear. Conventional gas absorption is also carried out with liquid dispersed as drops or thin films in the gas flowing countercurrently as in spray towers, packed towers, venturi scrubbers, etc. (Treybal 1980). Current gas stripping processes operate in an identical manner and in the devices used for gas absorption except that species are transferred from the liquid to the gas phase.

Such contacting devices suffer from a number of deficiencies. In tray columns, the individual phase flow rates cannot be varied over wide ranges due to flooding, weeping, priming, dumping, entrainment, etc. Packed columns suffer from flooding and loading at high phase

---
*Present affiliation: New Jersey Institute of Technology

flow rates. Except in systems with high mechanical agitation, the gas/liquid interfacial area per unit equipment volume in conventional contacting devices is not large. Scale-up is demanding. Foaming tendencies and extensive froth formation lead to process inefficiencies and may initiate process interruption. Microporous membrane-based gas absorption/stripping processes to be described can now overcome these shortcomings.

Consider the microporous hydrophobic membrane shown in Figure 46–1(a). An aqueous solution that does not wet the membrane flows on one side of the membrane. A gas mixture flows on the other side of the membrane at a pressure less than that of the aqueous phase. As long as the excess aqueous solution pressure is less than a breakthrough pressure, Δp_{cr}, the aqueous solution will not penetrate the pores. Unless the gas phase pressure is higher than that of the aqueous phase, the gas will not bubble into the aqueous solution. Thus, over the excess aqueous phase pressure range of 0 to Δp_{cr}, the gas/liquid interface is immobilized at the pore mouth of the hydrophobic membrane on the solution side. Through such an interface, one or more gas species may be absorbed into the aqueous solution or one or more gas species may be stripped from the aqueous solution. This nondispersive gas/liquid contacting can be achieved throughout a contacting device.

Nondispersive operation of a membrane-based gas/liquid contactor provides major benefits. The gas flow rate and the liquid flow rate can be varied independently of each other without flooding. loading, weeping, foaming, etc., characteristic of dispersion-based contactors. All of the membrane surface area is available for contacting regardless of how low the flow rates are. The gas/liquid interfacial area is known *a priori* since the membrane surface area is known.

This mode of nondispersive gas absorption or stripping was first introduced in blood oxygenation. Blood flowed on one side of a microporous hydrophobic flat Gore-Tex® membrane of Teflon® (Esato and Eiseman 1975) or hollow-fiber membranes of polypropylene (Tsuji et al. 1981). Oxygen flowed on the other side either as pure oxygen or as air at a pressure slightly lower than the blood pressure. Oxygen was absorbed into the blood as CO_2 from the blood was stripped out to the O_2-containing gas stream. Such a technique is being increasingly investigated for possible application to large-scale gas absorption (Qi and Cussler 1985a) and gas stripping (Semmens, Qin, and Zander 1989).

In the above examples, the microporous membrane functions primarily as a phase barrier preventing the aqueous solution from entering the membrane pores containing the gas. The membrane acts as a phase barrier also for the gas, preventing bubbles from appearing in

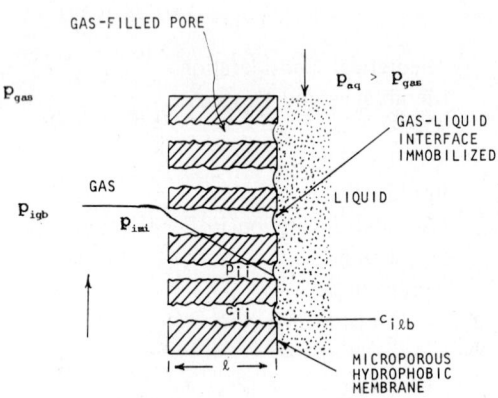
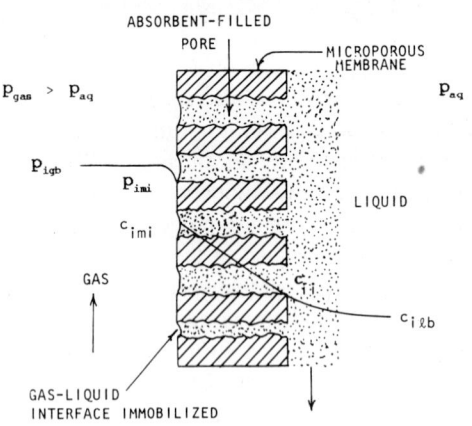

FIGURE 46–1. Solute partial pressure and concentration profiles in gas absorption through a gas/liquid interface immobilized at the pore mouth of a microporous hydrophobic membrane with (a) gas-filled pores and (b) absorbent-filled pores.

the liquid as long as the liquid phase pressure is equal to or higher than the gas pressure and the membrane pores are gas-filled. The conventional membrane function of acting as a species barrier is absent. Equilibrium separation processes of gas absorption and gas stripping take place with one phase interface, the gas/liquid interface, immobilized at the membrane pore mouth in a membrane-based gas/liquid contactor.

Gas/liquid contacting can also be carried out using a microporous *hydrophilic* membrane, which is spontaneously wetted by the aqueous absorbing liquid. The direction of phase pressure differential however is different; the gas phase pressure has to be higher than the liquid phase pressure to immobilize the gas/liquid phase interface at the pore mouths on the gas side of the membrane (Karoor and Sirkar 1991). A *hydrophobic* membrane can be operated in such a mode also. By wetting the hydrophobic membrane via an exchange process (Bhave and Sirkar 1986, 1987) and incorporating an aqueous solution in the membrane pores, Karoor and Sirkar (1991) have studied nondispersive gas absorption with the gas phase at a higher pressure [Figure 46–1(b)].

Membrane gas absorption has been implemented primarily in hollow-fiber-based membrane devices. These devices with highly packed hollow fibers of small dimensions can pack an enormous gas/liquid contact area per unit module volume, as much as 30 to 100 cm^{-1}; this is more than an order of magnitude larger than that possible in packed towers of equivalent size. Thus, the gas/liquid contactor volume may be reduced by at least an order of magnitude if mass transfer coefficient K values are not too small. In this respect, a hollow-fiber absorber or stripper competes with the new HIGEE type of centrifugal contacting device (Keyvani and Gardner 1989). A membrane contactor, however, does not have any moving parts. Further, they are modular in nature and are easier to scale up, in general. An additional advantage lies in having known values of a_{si} in membrane devices; thus $K_g a_{si}$ or $K_l a_{si}$ obtained from data will directly provide K_g or K_l. It is just not possible in dispersion-based devices without an extensive effort to determine the bubble or drop size distribution. Thus, the advantages of a membrane-based gas/liquid contactor parallel those of membrane-based solvent extraction devices studied in Chapter 41 and first pointed out by Kiani, Bhave, and Sirkar (1984).

Using a microporous hydrophobic membrane to carry out either nondispersive gas absorption and/or stripping across the gas/liquid interface on one side of the membrane has been extended to a membrane process in which both sides of the "phase barrier" microporous membrane participate in solute partitioning. Suppose two different nonwetting aqueous solutions are on two sides of a microporous hydrophobic membrane with air-filled pores. If a gas species is dissolved in one of the solutions and is not present in the solution on the other side, then this species desorbs from the first solution into the gas-filled pores, diffuses through the inert gas in the pore, the "*gas membrane,*" and is absorbed in the second solution on the other side of the membrane. Thus, stripping takes place at the first gas/liquid interface and absorption occurs at the second gas/liquid interface (Figure 46–2).

This process was first studied by Imai, Furusaki, and Miyauchi (1982) for the separation of

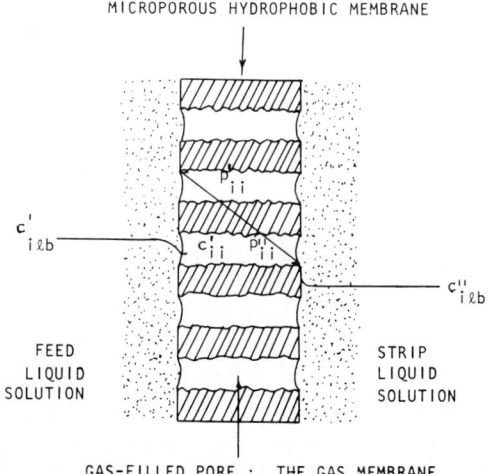

FIGURE 46–2. Solute partial pressure and concentration profiles in transport through a gas membrane in a microporous hydrophobic membrane.

NH$_3$ and I$_2$ from an aqueous solution and their absorption into H$_2$SO$_4$ solution and NaOH solution, respectively. Qi and Cussler (1985b) studied such a process for removal of Br$_2$ from brine into an aqueous solution of NaOH. Kenfield et al. (1988) have studied the removal of HCN vapor from an aqueous solution into a caustic solution. Semmens, Foster, and Cussler (1990) have investigated ammonia removal from water into a H$_2$SO$_4$ solution. Although such a "gas membrane" process has two different phase interfaces, the inert gas in the membrane pore has no distinction characteristic of membranes; any one gas is as good as another one. They are different membranes only to the extent that the diffusion coefficient of the transported species is mildly affected by the nature of the stagnant gas—the gas membrane.

The limited inventory of the inert gas in the gas membrane may be reduced significantly by its dissolution in the external aqueous phases. There are also possibilities of condensation in the pores and pore wetting, which will mix the two aqueous solutions on two sides of the membrane and stop the separation process, although Qi and Cussler (1985c) have claimed that their gas membranes were stable for months. Pore wetting does not stop gas absorption or stripping processes with one gas/liquid interface; thus, they are inherently more stable than gas membrane processes. The subject of gas membranes as such will therefore receive only limited attention here. Gas membrane is, however, an integral part of the membrane distillation process covered in the second part of this chapter.

Theoretical Considerations on Mass Transfer

In the following subsections, simple relations are first developed between an overall mass transfer coefficient and individual coefficients and studied for different membranes and membrane wetting conditions, etc., at any location in a module. Equations describing overall gas absorption or stripping behavior in a hollow-fiber module for nonreactive cases are presented, followed by a brief discussion on the effect of chemical reaction. Subsequent sections cover mass transfer correlations for design of modules and the information on available modules. Examples of different applications are provided at the end.

From a mass transfer point of view, only two conditions of operation are possible in membrane gas absorption/stripping:

1. Gas-filled membrane pore
2. Absorbent-filled membrane pore.

We consider first the more widely studied configuration, the gas-filled membrane pore.

Gas-Filled Membrane Pore

The rate of membrane-based gas absorption or stripping of a species i needs to be determined at two levels: locally at any point in a device and globally for the whole device. For the *local rate* of interphase transport of any species being absorbed from the gas into the liquid (or desorbed from the liquid into the gas) through a microporous *hydrophobic membrane not wetted by the liquid*, consider the concentration profile shown in Figure 46–1(a). Species i diffuses through the gas film, the gas-filled membrane pore, and the liquid film in series. The species i flux expressions for the three regions are

$$N_i = k_{ig}(p_{igb} - p_{imi}) = k_{im}(p_{imi} - p_{ii})$$
$$= k_{il}(c_{ii} - c_{ilb}). \qquad (46\text{–}1)$$

At the gas/liquid interface, concentrations in the two phases are at equilibrium and may be related by Henry's law:

$$c_{ii} = H_i p_{ii}. \qquad (46\text{–}2)$$

(Note this definition of Henry's constant H_i is practiced in membrane processes. In conventional chemical engineering literature, a different definition is used.)

In terms of an overall mass transfer coefficient K_g based on the gas phase, or K_l based on the liquid phase,

$$N_i = K_g(p_{igb} - p_i^*) = K_l(c_i^* - c_{ilb}), \qquad (46\text{–}3)$$

where

$$c_{ilb} = H_i p_i^* \text{ and } c_i^* = H_i p_{igb}. \quad (46\text{-}4)$$

Recognizing that

$$p_{igb} - p_i^* = (p_{igb} - p_{imi}) + (p_{imi} - p_{ii}) + (p_{ii} - p_i^*), \quad (46\text{-}5)$$

one can get

$$\frac{1}{K_g} = \frac{1}{k_{ig}} + \frac{1}{k_{im}} + \frac{1}{k_{il} H_i}. \quad (46\text{-}6)$$

| overall | gas film | membrane | liquid film |
| resistance | resistance | resistance | resistance |

The overall resistance to gas transfer comes from three resistances in series: the gas film resistances, the membrane resistance, and the liquid film resistance. In conventional dispersed-phase absorption or stripping, only resistances due to gas film and liquid film appear. The diffusional resistance through the gas-filled membrane pore is an additional resistance in the membrane-based process.

Relation (46-6) can be simplified if gas species i is sparingly soluble or highly soluble or a rapid chemical reaction occurs. For sparingly soluble gases, H_i is very small. Therefore,

$$\frac{1}{K_g} \cong \frac{1}{H_i k_{il}} \quad (46\text{-}7)$$

if k_{ig}, k_{im}, and k_{il} are of the same order of magnitude.

The overall gas phase-based mass transfer coefficient K_g is then controlled by the liquid film resistance only. Such a behavior is observed for sparingly soluble gases like O_2 absorbing in water or for somewhat more soluble gases like CO_2. For extremely soluble gases, H_i is several orders of magnitude larger and it is conceivable to have a limiting situation where

$$\frac{1}{K_g} \cong \frac{1}{k_{ig}} + \frac{1}{k_{im}}, \quad (46\text{-}8)$$

and the membrane resistance ($1/k_{im}$) is for gas-filled pores. Using small microporous hydrophobic flat films, Kreulen et al. (1990) had observed that in the case of absorption of NH_3 from a NH_3-N_2 mixture into an aqueous sulfuric acid solution, K_g was indeed described by Eq. (46-8) with a significant gas-filled membrane resistance and no liquid phase resistance.

For many gas absorption/stripping situations, it is more useful to develop a relation of the type of Eq. (46-6) in terms of the overall mass transfer coefficient K_l based on the liquid phase. For a hydrophobic membrane with gas-filled pores, such a relation is

$$\frac{1}{K_l} = \frac{H_i}{k_{ig}} + \frac{H_i}{k_{im}} + \frac{1}{k_{il}}. \quad (46\text{-}9a)$$

For sparingly soluble gases,

$$\frac{1}{K_l} \cong \frac{1}{k_{il}}, \quad (46\text{-}9b)$$

indicating liquid film-controlled mass transfer. For highly soluble gases, we have

$$\frac{1}{K_l} \cong H_i \left(\frac{1}{k_{ig}} + \frac{1}{k_{im}} \right) \quad (46\text{-}10)$$

corresponding to Eq. (46-8).

Absorbent-Filled Membrane Pore

Local rates of gas absorption/stripping, carried out with the absorbent liquid in membrane pores, may be described using the relation between an overall mass transfer coefficient and the individual mass transfer coefficients. Such a relation will be independent of whether the membrane is hydrophobic or hydrophilic as long as the pores are wetted. For the concentration profile shown in Figure 46-1(b), the flux expressions for species i are

$$N_i = k_{ig} (p_{igb} - p_{imi}) = k_{im}(c_{imi} - c_{ii})$$
$$= k_{il}(c_{ii} - c_{ilb}) = K_g(p_{igb} - p_i^*)$$
$$= K_l(c_i^* - c_{ilb}). \quad (46\text{-}11)$$

This leads to

$$\frac{1}{K_g} = \frac{1}{k_{ig}} + \frac{1}{k_{im}H_i} + \frac{1}{k_{il}H_i} \quad (46\text{-}12a)$$

and

$$\frac{1}{K_l} = \frac{H_i}{k_{ig}} + \frac{1}{k_{im}} + \frac{1}{k_{il}}. \quad (46\text{-}12b)$$

For sparingly soluble gases, an additional membrane resistance term $[(1/k_{im})H_i]$ exists in addition to the liquid film resistance $[(1/k_{il})H_i]$ in $(1/K_g)$. Kreulen et al. (1990) have found that the membrane resistance was controlling in the absorption of CO_2 and N_2O in water wetting the pores of flat hydrophilic membrane. Karoor and Sirkar (1991) have observed that the gas absorption rates were significantly lower when the hydrophobic hollow-fiber membrane pores were wetted with water by an exchange process for absorption of CO_2 or SO_2 in pure water. For extremely soluble gases undergoing absorption with an instantaneous reaction and a high concentration level of reactive absorbent, there is a possibility, however, that the total resistance will be lower since the membrane resistance may be eliminated:

$$\frac{1}{K_g} \cong \frac{1}{k_{ig}} ; \frac{1}{K_l} \cong \frac{H_i}{k_{ig}}. \quad (46\text{-}12c)$$

Membrane Resistance

If the membrane pore is *wetted* by the *absorbent liquid*, the membrane mass transfer coefficient may be described by the expression

$$k_{im} = \frac{D_{il}\epsilon_m}{\tau_m l}, \quad (46\text{-}13a)$$

where D_{il} is the diffusion coefficient of species i in the absorbent liquid. More details are provided in Chapter 41. If the membrane pore is *gas-filled*, the membrane transfer coefficient expression will depend on the regime of gas diffusion in the membrane pore. If the ratio of the membrane pore radius r_p to the mean free path of the gas λ, r_p/λ, is much smaller than 1, Knudsen flow regime exists. The transfer coefficient k_{im} can then be obtained from Knudsen flux expression as

$$k_{im} = \frac{2r_p}{3}\left(\frac{8RT}{\pi \tilde{M}_i}\right)^{1/2} \frac{\epsilon_m}{\tau_m l}, \quad (46\text{-}13b)$$

where $\tilde{M}_i$ is the molecular weight of the ith gas species. Under near-atmospheric conditions (average pressure around 1.18×10^5 Pa) and membranes with very small pores ($r_p \sim 0.01$ μm), Knudsen flow was found to be dominant (Callahan 1988). For a perspective, the value of λ for N_2 at 50 kPa and 25°C is 0.13 μm (Schofield, Fane, and Fell 1990a).

For larger pore membranes and/or gases at higher pressures, viscous flow exists if r_p/λ is much larger than 1. For ratios of r_p/λ between the two limits, transitional flow regime conditions are operative. For example, Schofield, Fane, and Fell (1990a) have shown that, at subatmospheric pressures, gas flow through microporous membranes of pore radius r_p around 0.1 to 0.45 μm falls in the transition regime between Knudsen and viscous flow. In case of absorption of a species i through a stagnant gas in the membrane pore under conditions of viscous flow, conventional expressions for molecular diffusion may be used for k_{im} with allowances for ϵ_m and τ_m. Such a model describes satisfactorily the membrane transfer coefficients measured by Kreulen et al. (1990) for large-pore nonwetted flat membranes with pore diameters around 0.1 to 0.2 μm.

Silicone-Coated Microporous Membrane

In blood oxygenation with microporous hydrophobic membranes, some studies have indicated greater platelet and red cell damage than those observed with silicone membranes. It was further noted that prolonged use led to partial filling of membrane pores with blood elements and a progressive reduction in gas transport rate (Evren and Piskin 1986). These and other authors have therefore studied the use of hydrophobic hollow fibers with an extremely thin nonporous coating of silicone (usually plasma polymerized) on the blood side. Keller and Shultis (1979) have studied oxygen permeability of ultrathin and microporous hydrophobic membranes with or without such a coating. Such a coated membrane does not require control of the gas/liquid pressure differential for nondispersive processing since it is a membrane process as opposed to a membrane-based phase contact process being considered here.

Obviously, this thin coating will increase the gas transfer resistance. However, it may have two distinct advantages. First, this increase can often be small since silicone is highly permeable to most gases and the coating thickness is very small (~1 to 3 μm). Second, the coating will eliminate the possibility of filling the membrane pores with liquid and thus may provide lower resistance in the longer run for particular situations.

Simultaneous Stripping-Absorption Via Gas Membrane

Consider the solute concentration profiles shown in Figure 46–2 for a gas membrane. The mass flux may be expressed as

$$N_i = k'_{il}(c'_{ilb} - c'_{ii}) = k_{im}(p'_{ii} - p''_{ii})$$
$$= k''_{il}(c''_{ii} - c''_{ilb}), \quad (46\text{--}14)$$

where ' refers to the feed solution side from which species i is stripped, and " refers to the product solution side into which species i is absorbed. Using Eq. (46–2) for Henry's law, we can easily obtain

$$\frac{1}{K_l} = \frac{1}{k'_{il}} + \frac{H_i}{k_{im}} + \frac{1}{k''_{il}}, \quad (46\text{--}15a)$$

where

$$N_i = K_l(c'_{ilb} - c''_{ilb}) \quad (46\text{--}15b)$$

defines an overall coefficient K_l. Usually, an instantaneous chemical reaction takes place on the absorption side at such a rate that

$$\frac{1}{K_l} \cong \frac{1}{k'_{il}} + \frac{H_i}{k_{im}}. \quad (46\text{--}16)$$

This result is often complicated by the presence of chemical reactions on the feedside liquid (Qi and Cussler 1985c).

Mass Transfer in a Microporous Hydrophobic Hollow-Fiber Absorber or Stripper

Nondispersive gas/liquid contacting may be carried out by microporous membranes in a number of membrane module configurations. These include hollow-fiber, plate-and-frame, spiral-wound modules, etc. A number of studies of hollow-fiber modules have been carried out with either a simple shell-and-tube parallel flow configuration (Yang and Cussler 1986; Karoor and Sirkar 1991) or a crossflow configuration (Yang and Cussler 1986). One empirical study of gas stripping has been carried out in a plate-and-frame module (Boswell 1991). None has been reported for spiral-wound modules. An exact numerical solution of the mass transfer rate in a hollow-fiber module with parallel flow has been developed when the liquid flows through the fiber bores (Karoor and Sirkar 1991). A brief analysis of *nonreactive* mass transfer with a dilute gas in a hollow-fiber module using the approach of mass transfer correlations is provided below for *parallel flow*. Crossflow configuration has not been considered here.

The module has N hollow fibers of internal diameter ID (d_{ti}) and length L (Figure 46–3). An aqueous liquid flows at a velocity v_s through the shellside parallel to the fibers *countercurrent* to a gas flowing through the lumen of the fibers at a velocity v_g. Any *bypassing* or *channeling* in the shellside flow is neglected. Assuming that the absorption of gas species i does not substantially change the gas flow rate, an overall mass balance over the module length is

$$Q_g(c_{igb}^{in} - c_{igb}^{out}) = Q_s(c_{ilb}^{out} - c_{ilb}^{in}). \quad (46\text{--}17)$$

A differential mass balance at any axial location x (measured from feed gas inlet location) is

$$v_g\left(\frac{\pi}{4}d_{ti}^2 N\right)\frac{dc_{igb}}{dx} = -K_g(p_{igb} - p_i^*)(\pi d_{to} N)$$

$$(46\text{--}18a)$$

based on the gas phase only. A similar balance based on the liquid phase is

$$v_s \left(\frac{\pi}{4} d_{ts}^2 (1-\phi)\right) \frac{dc_{ilb}}{dx} = -K_l(c_i^* - c_{ilb})(\pi d_{to} N). \quad (46\text{-}18b)$$

Assume now that K_l is not dependent on any fluid phase concentration. Rearranging the above and integrating along the module length yields

$$\frac{Q_s}{\pi d_{to} N} \int_{c_{ilb}^{out}}^{c_{ilb}^{in}} \frac{dc_{ilb}}{c_i^* - c_{ilb}} = -\int_0^L K_l\, dx = -\bar{K}_l L, \quad (46\text{-}19a)$$

where

$$Q_s = v_s(\pi d_{ts}^2/4)(1-\phi) \quad \text{and}$$
$$Q_g = v_g(\pi d_i^2/4)N. \quad (46\text{-}19b)$$

A balance on species i from the feed gas inlet location to location x is

$$Q_g(c_{igb}^{in} - c_{igb}) = Q_s(c_{ilb}^{out} - c_{ilb}). \quad (46\text{-}20a)$$

If ideal gas law and Henry's law [Eq. 46-2] are valid,

$$c_{igb}^{in} = \frac{p_{igb}^{in}}{RT}; \quad c_{igb} = \frac{p_{igb}}{RT} = \frac{c_i^*}{H_i RT}. \quad (46\text{-}20b)$$

These allow us to obtain

$$c_i^* - c_{ilb} = H_i RT c_{igb} - c_{ilb}. \quad (46\text{-}20c)$$

Further from Eq. (46-20a)

$$c_{igb} = c_{igb}^{in} - \frac{Q_s}{Q_g}(c_{ilb}^{out} - c_{ilb}). \quad (46\text{-}21)$$

Substituting this expression for c_{igb} in Eq. (46-20c) and using it in Eq. (46-19a) leads to, after integration and simplification,

$$L = \frac{Q_s}{\bar{K}_l(\pi d_{to} N)} \frac{1}{(1 - (H_i RT Q_s/Q_g))} \times$$

$$\log\left(\frac{H_i RT c_{igb}^{out} - c_{ilb}^{in}}{H_i RT c_{igb}^{in} - c_{ilb}^{out}}\right). \quad (46\text{-}22a)$$

The length L of the module required to achieve a change in gas composition c_{igb}^{in} to c_{ilb}^{out} is thus obtained easily provided $\bar{K}_l$ is known. A more traditional format is

$$L = \text{LTU} \times \text{NTU}, \quad (46\text{-}22b)$$

where the length of transfer unit (LTU) and the number of transfer units (NTU) are defined respectively by

$$\text{LTU} = \frac{Q_s}{\bar{K}_l \pi d_{to} N} = \frac{Q_s/A}{\bar{K}_l(\pi d_{to} N/A)} = \frac{v_{su}}{\bar{K}_l a_{so}}, \quad (46\text{-}22c)$$

$$\text{NTU} = \frac{1}{(1-(H_i RT Q_s/Q_g))} \times$$

$$\log\left(\frac{H_i RT c_{igb}^{out} - c_{ilb}^{in}}{H_i RT c_{igb}^{in} - c_{ilb}^{out}}\right). \quad (46\text{-}22d)$$

Here

v_{su} = superficial velocity of absorbent liquid on shell side, $v_s(1-\phi)$
A = empty cross-sectional area of module
a_{so} = gas liquid interfacial area per unit module volume based on fiber outside diameter.

Since NTU is fixed for a given problem, the LTU of any absorber will decrease as $\bar{K}_l a_{so}$ increases. For given $\bar{K}_l$ and v_s, the higher the value of a_{so}, the lower the LTU and the lower L is. Maximization of $\bar{K}_l a_{so}$ is therefore desirable. Note that the quantity c_{ilb}^{out} in the above relations is obtained from Eq. (46-17).

By turning Eq. (46-22a) inside out in the manner of Eq. (41-25a), an expression for c_{igb}^{out} can be obtained in terms of c_{igb}^{in}, c_{ilb}^{in}, c_{ilb}^{out}, and other relevant quantities.

If the gas absorption is controlled by the liquid film resistance with gas-filled membrane pores, then

$$K_l \cong k_l, \quad (46\text{-}23a)$$

$$\bar{K}_l \cong \bar{k}_l. \quad (46\text{-}23b)$$

FIGURE 46-3.
Schematic for a hollow-fiber module contactor for countercurrent gas absorption or stripping.

If a shellside mass transfer correlation is available, such a correlation may be used in Eq. (46–23b).

When the absorbent liquid flows through the fiber bores countercurrent to the gas flowing on the shell side, the result [Eq. (46–22a)] is unaffected; however, the liquid flow rate Q_s is now changed to Q_t where

$$Q_t = v_t \left(\frac{\pi}{4} d_{ti}^2 N \right), \quad (46\text{–}24a)$$

$$Q_g = v_s \left(\frac{\pi d_{ts}^2}{4} \right)(1 - \phi). \quad (46\text{–}24b)$$

For liquid film-controlled mass transfer, the tubeside transfer resistance will be controlling. An unstated assumption is that there is no bypassing or channeling in the shellside gas flow.

If the contactor functions as a *stripper* (e.g., stripping volatile organic compounds from wastewater), the mass transfer analysis in countercurrent flow is similar; only the direction of transport of gas species i is reversed. Consider liquid flow in fiber lumen and the stripping gas flow on the shell side in a countercurrent fashion. If the total liquid flow rate through the fiber lumen is Q_t $(= v_t \pi d_{ti}^2 N/4)$, then adopting the procedure outlined earlier for a hollow-fiber gas absorber, the following relation is obtained for a stripper of length L:

$$L = \underbrace{\left(\frac{Q_t}{\pi d_{ti} N \bar{K}_l} \right)}_{\text{LTU}} \underbrace{\left[\frac{1}{1 - H_i RT(Q_t/Q_g)} \right] \log \left(\frac{c_{ilb}^{in} - H_i RT c_{igb}^{out}}{c_{ilb}^{out} - H_i RT c_{igb}^{in}} \right)}_{\text{NTU}}.$$

(46–25)

Relations (46–22a) and (46–25) can be used only under the assumption that a suitable module-averaged mass transfer coefficient $\bar{K}_l$ is available. If gas stripping is controlled by the liquid phase coefficient, then $\bar{K}_l = \bar{k}_l$. By appropriate changes, the above equation can be easily adopted to stripping gas flowing in the fiber lumen. Any shellside bypassing or channeling is neglected in this *parallel flow* configuration of a shell-and-tube device.

Effect of Chemical Reaction

Qi and Cussler (1985a) have studied the reactive absorption of gases in hydrophobic microporous hollow fibers for the following systems: NH_3 in H_2SO_4; H_2S in NaOH; SO_2 in NaOH; CO_2 in NaOH; CO_2 in various amines; and H_2S in various amines. The liquid in each case was an aqueous solution of the inorganic acid or base or an organic amine.

For the essentially instantaneous reactions of NH_3 in H_2SO_4, H_2S in NaOH, and SO_2 in NaOH, the overall gas phase-based coefficient K_g was found to be independent of the gas phase flow rate. They have calculated the liquid film coefficient for chemical reaction by using the standard film theory expression for liquid film mass transfer coefficient enhanced by an instantaneous chemical reaction between a gas A and a species B in the liquid phase:

$$k_l^r = k_l \left(1 + \frac{D_B c_{Bb}}{D_A c_{Ai}} \frac{\nu_A}{\nu_B} \right), \quad (46\text{–}26a)$$

$$v_A \, A + v_B \, B \rightarrow \text{products}. \quad (46\text{-}26\text{b})$$

They have stated that the observed mass transfer resistance was much larger than that of the liquid film calculated using Eq. (46-26a). They concluded that the membrane resistance dominated the absorption of all three gases—NH_3, SO_2, and H_2S—in the strong acid and basic solutions. Such a condition could be justified only by postulating that the pores were operating somewhere between the limits of gas-filled pores and liquid-filled pores. How a hydrophobic gas-filled pore with nonwetting aqueous solution on one side at a pressure below Δp_{cr} has part of the pore filled with liquid is not known.

For CO_2 absorption in aqueous NaOH solution, the same authors have found that the observed K_g value was close to the value of $H_i k'_l$ estimated for a fast reaction; the membrane resistance was only a minor component. Results of absorption of CO_2 and H_2S in aqueous solutions of various organic amines led to similar conclusions; namely, that the hydrophobic gas-filled membrane pores provide very little resistance. A general conclusion then is that unless the liquid film coefficients were very large or the membrane suffered from partial pore filling with the absorbent liquid, the membrane resistance was not important.

Cooney and Jackson (1989) have studied the absorption of SO_2 and NH_3 from air and air/CO_2 in a hydrophilic hollow-fiber gas absorber. For SO_2 absorption, dilute basic solutions, e.g., 0.025 M $NaHCO_3$, 0.01 N NaOH, 0.15 M NH_4OH, were used. No formal analysis of the role of reaction in the hydrophilic hollow fiber with pores wetted by the aqueous solution was provided. The solutions were weakly basic to prevent destruction of the cellulose acetate hollow fibers used. Karoor and Sirkar (1991) have experimentally studied and modeled the absorption of SO_2 in water in an absorber made of hydrophobic microporous hollow fibers. The model developed incorporated the extensive ionization of SO_2 in water via

$$SO_2 + H_2O \leftrightarrows H^+ + HSO_3^-. \quad (46\text{-}27)$$

In a study of SO_2 stripping from an aqueous solution into an aqueous solution of NaOH via a *gas membrane,* Qi and Cussler (1985c) recognized that there may be considerable ionization in the feed aqueous solution due to the reaction of Eq. (46-27). Thus, transports of both SO_2 and HSO_3^- have to be considered in the aqueous solution whereas only SO_2 is being transported through the gas membrane. Equation (46-14) then needs to be corrected. Qi and Cussler (1985c) provided the corrected form of Eq. (46-15a) when k''_{il} is very large and $(1/k''_{il})$ is neglected. In studying ammonia removal from water through a gas membrane in a microporous hydrophobic hollow fiber, Semmens, Foster, and Cussler (1990) similarly studied the role of the ionization reaction

$$NH_3 + H_2O \leftrightarrows NH_4^+ + OH^-. \quad (46\text{-}28)$$

They have demonstrated how chemical reaction influences the rate of NH_3 stripping via a nonlinear combination of feed solution resistance and the membrane resistance.

Membrane Devices and Transport Correlations

A few industrial organizations have made hollow-fiber modules commercially available. Membrane devices having other configurations, e.g., spiral wound, are not yet available although Imai, Furusaki, and Miyauchi (1982) had studied the gas membrane process with a spiral-wound module.

Simple shell-and-tube configurations are used in Cel-Life® devices manufactured by Hoechst Celanese Corporation (Charlotte, NC) for gas absorption and desorption (degassing) applications as well as in biomedical gas transfer processes. These devices have a polycarbonate casing, Celgard® polypropylene hollow-fiber membranes, and urethane potting. Available modules span a range of 12,500 to 900 fibers of the X-10 type (effective pore size 0.05 μm) with fibers having 240 or 400 μm inside diameters (30-μm wall thickness), module lengths varying between 61 to 20.3 cm, and module diameters between 6.35 to 2.54 cm. Available membrane surface area/module varies between 0.2 to 3.8 m². The maximum operating pressure range varies between 60 and 20 psi depending on the module. Maximum operating temperature range is 0 to 75°C with a particular type of device autoclavable twice. The chemical resistance of the Cel-Life® modules is generally good with respect to acid or alkalies and poor with respect to most solvents (unlike the generally good solvent resistance of Liqui-Cel® devices manufactured by the same corporation).

A range of modules for degassing of aqueous solutions or absorption of gases is available from AKZO (Enka America, Asheville, NC). There are MD-type modules of polypropylene that are currently called Enka Microdyn® modules. They are based on polypropylene capillary membranes with 0.2-μm pore size, 0.6- to 1.8-mm i.d., and membrane lengths and area varying between 0.5 to 1.0 m and 0.1 to 10 m². The module diameter ranges between 20 to 150 mm. Polypropylene tubular membranes of pore size 0.2 μm and 5.5-mm i.d. are also available with polypropylene potting material; the smaller inside diameter capillaries are potted with polyurethane.

The liquid phase mass transfer coefficient correlations to be used for gas absorption or stripping in a hollow-fiber module are provided in Chapter 41. Some of those correlations (e.g., Yang and Cussler 1986) were in fact developed in gas absorption and gas stripping processes. Correlations for parallel flow and crossflow configurations are also identified there. No correlation of gas phase mass transfer coefficient in hollow-fiber devices has been developed yet. Semmens, Qin, and Zander (1989) have suggested the use of

$$Sh = 0.022 Re^{0.60} Sc^{0.33} \qquad (46\text{--}29a)$$

for predicting the mass transfer coefficient of air on the shell side.

Imai, Furusaki, and Miyauchi (1982) have studied transfer via a gas membrane with a module of hollow-fiber bundles having semicircular baffle plates. The flow on the shell side was essentially crossflow of a liquid phase. The mass transfer coefficient was correlated by

$$Sh = \alpha \, Re^{0.6} Sc^{1/3}, \qquad (46\text{--}29b)$$

where α was found to vary between 0.15 for NH_3 and 0.19 for I_2.

Aspects of Hollow-Fiber Module Design

A comprehensive hollow-fiber module design procedure will not be outlined here. Specifically omitted are considerations of thermal effects, multicomponent absorption/stripping, and gas mixtures that are not dilute in the species being transferred. Before undertaking any module design, it is useful to ensure that the excess pressure of the liquid phase over the gas phase does not exceed the Δp_{cr} value for the membrane material, pore size, and the gas/liquid system. An approximate estimate of Δp_{cr} may be obtained from the relation

$$\Delta p_{cr} = \frac{2\gamma \cos \theta}{r_p}, \qquad (46\text{--}29c)$$

where

γ = surface tension of the absorbent liquid exposed to the gas mixture
θ = contact angle
r_p = pore radius.

If the excess liquid pressure exceeds Δp_{cr}, the pores will no longer be gas-filled and gas phase pressure will have to exceed the liquid pressure for nondispersive operation. Further, liquid-filled pores generally lead to lower mass transfer rates. Additionally, the excess liquid phase pressure, for example, should not deform the hollow fiber. A few guidelines are now provided to initiate a rational module design. The first question to be considered is that of where the feed fluid phase should flow, on the tube side or shell side.

In *gas absorption,* the feed fluid phase is gaseous. If the feed gas flows on the shell side of a hollow-fiber module of conventional design, there may be considerable bypassing and channeling. Ogundiran, LeBlanc, and Varanasi

(1989) have speculated about the possibility of gas on the shell side bypassing and reducing absorption. Achieving a high level of feed gas purification may be difficult in one module for high gas flow rates since existing modules are short. Unless a number of modules are to be used in series, it is suggested that the feed gas flow through the fiber lumen so that channeling is completely eliminated. This would be at the expense of higher operating costs resulting from a higher gas pressure drop (tubeside pressure drop is usually larger than shellside pressure drop). If an upper limit exists for the allowable gas pressure drop (e.g., in flue gas desulfurization), a satisfactory compromise would require larger diameter fiber lumen and/or smaller module length. The extent of any maldistribution of feed gas flow into the tube sheet of the fiber bundle is unknown.

In *gas stripping*, the feed fluid phase is liquid. That there will be some bypassing or channeling if feed liquid flows on the shell side is known from solvent extraction studies (Prasad and Sirkar 1990). Therefore, it is recommended that the fiber lumen should accommodate the flow of the feed liquid phase as the stripping gas passes through the shell side. To reduce the liquid pressure drop and pumping cost, the fiber lumen diameter selected should be larger and the module length should be smaller.

After the selection of tubeside/shellside flow of the feed fluid phase, module design can be undertaken. For a particular problem, the values of the gas flow rate and the liquid flow rate should be given; similarly the inlet and exit compositions of each stream are either specified or calculated (by mass balance). An approach based on LTU-NTU suggests that the NTU for design of a gas absorber [e.g., Eq. (46–22d)] is now known. Similarly, the NTU in gas stripping is also known, e.g., via Eq. (46–25). If a suitable correlation is available for $\overline{K}_l$, the design may be implemented for physical absorption/stripping only.

To illustrate, consider stripping of volatile organic compounds (VOCs) from wastewater by air flowing countercurrently on the shell side of a parallel flow hollow-fiber device. It is known that the liquid film controls the mass transfer. In a hydrophobic hollow-fiber module, with the feed wastewater flowing through the fiber lumen, Semmens, Qin, and Zander (1989) have observed that the Lévêque solution-based correlation, given by Eq. (41–53), describes $\overline{k}_l$ quite well. From relation (46–25), we get

$$\text{NTU} = \frac{1}{1 - H_i RT(Q_t/Q_g)} \log\left(\frac{c_{ilb}^{in} - H_i RT c_{igb}^{out}}{c_{ilb}^{out} - H_i RT c_{igb}^{in}}\right)$$

$$= \frac{\overline{K}_l(\pi d_{ti} N)L}{Q_t} = \frac{4\overline{K}_l(\pi d_{ti} N)L}{v_t \pi d_{ti}^2 N} \cong \frac{4\overline{k}_l L}{v_t d_{ti}}$$

$$= \text{constant} \qquad (46\text{--}30\text{a})$$

since NTU is fixed for the particular design. Introducing the Lévêque solution-based correlation for $\overline{k}_l$ in this relation results in

$$\text{NTU} = 6.56 \left(\frac{LD_{it}}{d_{ti}^2 v_t}\right)^{0.67} = 6.56 \left(\frac{\pi LD_{it} N}{4Q_t}\right)^{0.67}.$$

$$(46\text{--}30\text{b})$$

This design result suggests that, for a given system with specified feed water flow rate through the fiber lumen, the required composition change will be achieved for a fixed value of NL. Fiber length may be decreased simply by increasing the fiber number with the added benefit of reduced pressure drop in the lumen side.

The length of the stripper may be obtained now explicitly as

$$L = \left(\frac{4Q_t}{\pi D_{it} N}\right)\left(\frac{\text{NTU}}{6.56}\right)^{0.67} \qquad (46\text{--}31)$$

for a given system, operating conditions, and total fiber number in the module.

For the design of an absorber scrubbing a dilute gas flowing in the fiber lumen, Eq. (46–22a) can be used in an identical fashion; only a suitable shellside correlation for the liquid flow is to be used for $\overline{k}_l$ assuming that $\overline{K}_l \cong \overline{k}_l$. The absorber design, especially the prediction of length L can be carried out using Eqs. (46–30a) and (46–30b) if the absorbent liquid is flowing in the fiber lumen and only physical absorption is involved.

Design of an absorber or stripper with chemical reaction is not treated here due to lack of any quantitative studies using hollow-fiber modules.

In most large-scale gas absorption or gas stripping problems, a cascade of hollow-fiber modules needs to be used for three reasons. A cascade of modules is more flexible; very large hollow-fiber modules are not made unlike large packed columns. Further, if the tubeside mass transfer coefficient is important, shorter modules will have higher mass transfer coefficients since $\bar{k}_l$ varies approximately as $(L)^{-0.33}$; thus, two modules in series are better than one longer module. A number of modules can be connected together in a number of ways. The simplest cascade consists of a number of parallel limbs, each containing the same number of modules connected in series. Such a cascade is similar to that illustrated for membrane solvent extraction in Figure 41–11. This type of series-parallel cascade has a constant cross-sectional area (Prasad and Sirkar 1989).

Alternative cascade forms will be of the tapered cross-section type wherein the cross-sectional area of the entering section will be much less than that of the exiting section or vice versa. This decrease or increase in module cross section can be sharp or moderate. A decrease in cross-sectional area will lead to an increase in flow velocity, transfer coefficients, the LTU, and the pressure drop but will decrease the total membrane area required. An increase in the cross-sectional area will reverse the above pattern. Cascade design has to be based on optimization of capital and operating costs for a number of such combinations. An additional variable could be the fiber internal diameter. Wicksramsinghe, Semmens, and Cussler (1990) have addressed this issue and developed graphs to determine optimum fiber diameter for different membrane costs.

Applications

Membrane gas absorption and membrane gas stripping may be applied to most systems treated by conventional gas absorption or gas stripping processes. In general, however, such applications have remained restricted to aqueous or organic solutions that do not wet the microporous membranes. This ensures a gas-filled membrane and a low value of membrane resistance. Three general application areas are treated here in the following order: biomedical uses, biochemical applications, and use in chemical and allied industries. The examples where a gas membrane has been used are also included in this section.

Blood oxygenation is the earliest application of gas liquid contacting using microporous membranes. A large number of microporous hollow-fiber oxygenators are in actual use in clinical practice in hospitals. A number of studies have been conducted. The following include some of the earliest studies and later references to aspects of blood oxygenation via an artificial lung: Esato and Eiseman (1975), Tsuji et al. (1981), Evren and Piskin (1986), Leonard (1986), and Callahan (1988). While these biomedical devices are externally present during procedures in hospitals, a new application of such membranes is an intra vascular oxygenator, wherein the hollow-fiber bundle is threaded through a vein in the leg or the neck into the body's main vein to deliver O_2 and remove CO_2.

Another potential application is the artificial gill, which has been studied for small animals (Yang and Cussler 1986, 1989). When the animal exhales into the gill, its breath passes into a small surge volume connected to the lumen of the hollow fibers. The water flowing on the outside of hollow fibers dissolves the CO_2 away from the breath. Oxygen dissolved in the water is transferred to the tube side and is used as the animal inhales the gas from the surge volume.

Matsumoto et al. (1990) have used a gas membrane in a polypropylene hollow-fiber module as an artificial gill connecting fish (carp) habitation tank water flowing on one side of the membrane with the algae *(Chlorella)* cultivation tank water flowing on the other side of the membrane. They were able to transfer O_2 and CO_2 in opposite directions in a stable fashion for 30 days as part of basic tools to study a closed ecological life support system.

Aerobic fermentation processes need O_2 supply to the cells in the fermentation broth and require also removal of CO_2. Frank and Sirkar (1986) had incorporated microporous hydrophobic hollow fibers directly into a tubular yeast-based fermentor producing ethanol; they had supplied O_2 to the broth during cell growth by flowing air in the fiber lumen and simultaneously stripping dissolved CO_2 from the broth. Kang et al. (1988) have evaluated the mass transfer coefficients for O_2 absorption and CO_2 stripping in such a hollow-fiber fermentor configuration. Shukla, Kang, and Sirkar (1989) carried out an anaerobic fermentation in a hollow-fiber-based fermentor device by supplying N_2 through the fiber bore and stripping H_2 and CO_2 from the fermentation broth in acetone-butanol-ethanol fermentation by *C. acetobutylicum*.

Acetic acid from acidic wines produced in some unsuccessful fermentations may be removed through a gas membrane into a caustic solution on the other side of a hydrophobic microporous hollow fiber (Qin et al. 1990). The ammonia level in a mammalian cell-culture media was reduced from 14 to 0.5 mM by circulating the media through the fiber lumen of a separate hollow-fiber gas membrane device where ammonia was selectively eliminated by a H_2SO_4 solution in the shell (Brose and Van Eikeren 1990). A hydrophobic microporous membrane supported in the pores a gas membrane, which allowed the transfer of two volatile substances (acetic acid and ethanol) in opposite directions and enabled an enzyme-catalyzed reaction and a synthetic organic reaction to proceed on two sides of the membrane under widely different conditions (Van Eikeren et al. 1990). Bandini and Gostoli (1990) studied continuous removal of ethanol from a fermentation broth through air trapped in polytetrafluoroethylene (PTFE) membrane pores into an aqueous mixture of ethylene or propylene glycol, which does not wet the membrane.

Applications to chemical and allied industries of gas absorption/stripping and gas membrane processes are identified below. Qi and Cussler (1985a) have studied the absorption of a variety of gases present in an inert gas in acidic or alkaline medium using a hydrophobic polypropylene hollow-fiber device: NH_3 in 1 N H_2SO_4; H_2S in 1 N NaOH; SO_2 in 1 N NaOH; CO_2 in 1 N NaOH; H_2S in solutions of different ethanolamines; and CO_2 in solutions of different ethanolamines. Radovich, Babcock, and Pearson (1986) studied the removal of SO_2 from flue gas in a polysulfone hollow-fiber absorber using an aqueous solution of Na_2SO_3. Cooney and Jackson (1989) reported results for the following systems using hydrophilic fibers: absorption of SO_2 from air and air/CO_2 streams in water, 0.025 M $NaHCO_3$, 0.01 N NaOH, and 0.15 M NH_4OH; and the absorption of NH_3 in water. Ogundiran, LeBlanc, and Varanasi (1989) studied the absorption of SO_2 in microporous hydrophobic hollow-fiber modules using an aqueous sodium sulfite solution (concentration ≥ 0.1 M) and obtained nearly 100% SO_2 removal. Karoor and Sirkar (1991) have obtained nearly complete removal of SO_2 from a pure SO_2 stream as well as from a 0.9% SO_2-containing mixture using water in microporous hydrophobic hollow-fiber modules having wetted or nonwetted pores.

Yang and Cussler (1986) have investigated the stripping of O_2 from water using vacuum or sweep gas in hydrophobic polypropylene hollow-fiber devices having either parallel flow or crossflow. Such stripping is useful for deaeration of boiler feed water and beverage makeup water; this is likely to be an area of immediate commercial application (Reed, Prasad, and Callahan 1990). Semmens, Qin, and Zander (1989) studied and demonstrated air-stripping of volatile organic compounds from water in similar hydrophobic polypropylene hollow-fiber devices with countercurrent flow. A list of *gas membrane* stripping studies from an aqueous solution into an absorbing aqueous solution is given in Table 46–1. Almost all of the examples of gas absorption or gas stripping or gas membranes identified in this table utilized aqueous solutions and hydrophobic membranes. Bonne et al. (1990) have disclosed a membrane dehumidification process using a microporous hydrophobic hollow-fiber module.

TABLE 46–1. List of Studies on Gas Membrane Stripping from an Aqueous Solution into an Absorbing Solution.

System Solute-Reactive Absorbent in Water	Source
NH_3-H_2SO_4	1, 2, 5, 6
H_2S-NaOH	5
SO_2-NaOH	5
Br_2-NaOH	4, 5
I_2-NaOH	2, 5
HCl-NaOH	5
HCN-NaOH	3
CH_3COOH-NaOH	5
Lactic acid-NaOH	5
$(NH_4)_2S$-$AgNO_3$ + HNO_3	5

(1) Brose and Van Eikeren (1990).
(2) Imai, Furusaki, and Miyauchi (1982).
(3) Kenfield et al. (1988).
(4) Qi and Cussler (1985b).
(5) Qi and Cussler (1985c).
(6) Semmens, Foster, and Cussler (1990).

Moisture from air is absorbed in triethylene glycol (TEG) flowing in the fiber lumen with gas-filled pores since TEG does not wet the pores. The absorbing liquid leaving the polypropylene hollow-fiber module is to be heated later and regenerated by air bubbling.

MEMBRANE DISTILLATION

Introduction

Membrane distillation (MD) is a thermally driven evaporation process for separating volatile solvent (or solvents) from solution on one side of a nonwetted microporous membrane. Generally, the evaporated solvent is condensed or removed on the other side of the membrane.

Consider a hydrophobic microporous membrane of the type discussed in Chapters 41 and 42. Suppose a hot aqueous solution is placed on one side of the membrane and a cold aqueous solution is placed on the other side such that neither solution wets the membrane pores [Figure 46–4(a)]. If the water vapor pressure of the hot solution is greater than that of the cold solution, water vapor will diffuse from the hot solution/membrane interface to the cold solution/membrane interface where the water vapor will condense. Water vapor will diffuse in the membrane pores, in general, through air or any other noncondensable gaseous phase trapped in the pores. In the absence of any gaseous phase, the pores will have only water vapor. If the cold solution happens to be only water, this process is achieving evaporation of water from the hot solution and its condensation in the cold liquid in a simple fashion. The two liquids on two sides of the membrane may be at any pressure as long as the membrane pores are not wetted by them. The microporous membrane is functioning here as a liquid phase barrier as water evaporation continues. The membrane distillation process described above is identified as direct contact MD, since the microporous membrane is in contact with liquid on both sides.

This concept was proposed first by Findley (1967). None of the membranes used by Findley was good enough for extended duration MD; Findley had therefore recommended that membranes with strong water repellancy should be used. Recent experiments by Gore (1982), Schneider and Van Gassel (1984), Sarti and Gostoli (1984), Schofield, Fane, and Fell (1987), and Gostoli, Sarti, and Matulli (1984) have successfully demonstrated membrane distillation using microporous hydrophobic membranes of polytetrafluoroethylene (PTFE), polypropylene (PP), and polyvinylidenefluoride (PVDF).

Separation of aqueous mixtures via membrane distillation has many advantages. No entrainment is likely to occur. The device can be horizontal, eliminating the need for a costly structure used to support heavy columns. Low-temperature energy sources may be conveniently used. Heating of the feed solution can take place outside the evaporator; yet the condensation takes place inside and is locally integrated with evaporation. A hydrophobic membrane surface reduces the possibility of scaling. A hollow-fiber configuration may allow highly compact devices as cold liquid for condensation flows through either shell or tube side. Aqueous solution of organic species (e.g., ethanol) can be processed as long as the membrane pores are not wetted. The more volatile organic species

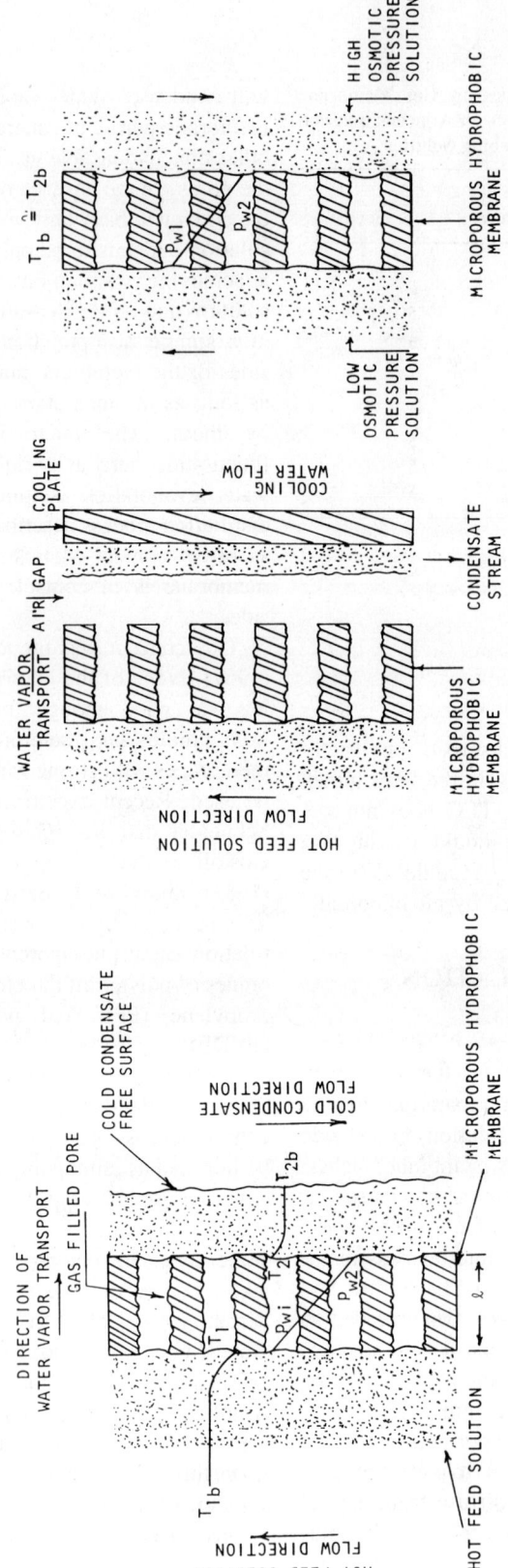

FIGURE 46-4. Schematics of (a) direct contact membrane distillation with a microporous hydrophobic membrane, (b) air gap membrane distillation, and (c) osmotic distillation with a microporous hydrophobic membrane.

(e.g., ethanol) is concentrated in the colder condensate stream. However, the upper limit for polytetrafluoroethylene membranes was 50 wt.% ethanol in permeate (Nakane et al. 1987). Membrane distillation of this type is not suitable for distillation of organic mixtures. On the other hand, with aqueous solutions of nonvolatile solutes, pure water may be obtained as the condensate using very low grade heat.

Types of Membrane Distillation Processes

Membrane distillation may be carried out in a number of different ways (Gostoli and Sarti 1989). These are:

1. *Direct contact MD:* This has been already described.
2. *Air gap MD:* Instead of using the other side of the microporous membrane as a condensing surface, a separate condensation surface is provided. There is now an air gap between the membrane and the condensation surface in addition to the air in the membrane pores [Figure 46–4(b)].
3. *Vacuum MD:* No condensation surface is provided in the device. The vapor is withdrawn by pulling a vacuum on the other side of the membrane. The product side pressure is maintained lower than the saturation pressure of the evaporating stream.
4. *Sweep gas MD:* A sweep gas is used to carry away the vapor from the MD device. In items 3 and 4, the partial pressure difference is provided by vacuum and the sweep gas, respectively.

The membrane distillation process has some similarities with the *osmotic distillation* (OD) process (Lefebvre, Johnson, and Yip 1987). Both use hydrophobic microporous membranes with generally aqueous solutions on the two sides such that the pores are not wetted. Further, water vapor is transferred through the pore in both cases due to its partial pressure gradient from the feed to the other side. Whereas a temperature gradient drives the partial pressure gradient of water in MD in the same direction, in osmotic distillation with both solutions at the same bulk temperature, an osmotic pressure gradient drives the water vapor in the opposite direction [Figure 46–4(c)]. Instead of the condensate stream in MD, a highly concentrated brine solution (for example) with a high osmotic pressure is maintained on the other side at essentially the same temperature. Water vapor is transferred from the feed solution to this high osmotic pressure solution, which has a much lower water vapor pressure.

The gas phase, trapped within the pores of the hydrophobic membrane in direct contact MD, acts as a *gas membrane*. Direct contact MD may then be characterized also as a gas membrane separation process with a temperature difference and a partial pressure difference across the gas membrane.

Theoretical Considerations

The flux of water vapor from the hot aqueous solution to the cold aqueous solution in direct contact MD is driven by the water vapor partial pressure gradient across the air-filled pore [Figure 46–4(a)]. The mechanism of transport is molecular diffusion through a stagnant air film (Sarti and Gostoli 1984). Schofield, Fane, and Fell (1987; 1990a; 1990b) have suggested that a combination of molecular and Knudsen diffusion controls the diffusive flux through membranes having a pore size between 0.1 to 0.45 μm. A simple representation for water vapor flux was used by Schofield, Fane, and Fell (1987):

$$N_w = k_{wm}(p_{w1} - p_{w2}), \quad (46\text{–}32a)$$

where k_{wm} is a membrane mass transfer coefficient. Schofield, Fane, and Fell (1990b) have suggested an expression for k_{wm} based on simple molecular diffusion through a stagnant air film when air is trapped within membrane pores. However, when the membrane is totally deaerated, an expression for k_{wm} was developed by them based on a combination of Knudsen diffusion and Poiseuille flow. They have also utilized a linear addition of both types of resistances for the more general case of low partial air pressure in the pores.

Expression (46–32a) may be written *for dilute solutions with small values of* $(T_1 - T_2)$ as

$$N_w = k_{wm} \left(\frac{dp_w}{dT}\right)_{T_m} (T_1 - T_2). \quad (46\text{–}32b)$$

Here T_m is the mean membrane temperature. The procedure for calculating $(dp_w/dT)_{T_m}$ via the Clausius-Clapeyron equation has been described by Schofield, Fane, and Fell (1987). Note that for stagnant air trapped in pores, k_{wm} will be influenced by the partial pressure of air at locations 1 and 2 along with membrane thickness l, porosity ϵ_m, tortuosity τ_m, water vapor diffusion coefficient, R, and T. For more concentrated solutions, the above flux expression has to be corrected for the effects of solute on boiling point elevation or vapor pressure reduction.

Since the flux of water vapor is related to the temperature difference between the two sides of the membrane, the energy flux between the two liquid solutions also has to be considered. The energy flux q is due to a combination of the conductive heat flux q_c and q_d resulting from the diffusion of water vapor and therefore its enthalpy. Conduction of heat takes place through the gas-filled pore and the nonporous sections of membrane. The expression for the total energy flux is (Schofield, Fane, and Fell 1987; Gostoli, Sarti, and Matulli 1987):

$$q = q_c + q_d = \left(\frac{\epsilon_m k_{gt} + (1 - \epsilon_m)k_{st}}{l}\right)$$
$$\cdot (T_1 - T_2) + N_w[\lambda + C_p(T - T_m)_{\text{avg}}], \quad (46\text{–}33)$$

where k_{gt} and k_{st} are the thermal conductivities, respectively, of the gas in the pore and the membrane material; λ is the latent heat of water vapor; and C_p its specific heat. The sensible heat component of q_d [i.e., $C_p(T - T_m)_{\text{avg}}$] is often neglected.

The temperatures on the two surfaces of the membrane T_1 and T_2 appearing in the expressions of water vapor flux and the heat flux are not known; the bulk temperatures of solutions on the two sides T_{1b} and T_{2b} are more easily used. From definitions of heat transfer coefficients h_1 and h_2 in the two aqueous solutions,

$$q = h_1(T_{1b} - T_1) = h_2(T_2 - T_{2b}). \quad (46\text{–}34a)$$

Using the earlier relations between N_w and $(T_1 - T_2)$ and q and N_w, the following relation is obtained

$$q = h_m(T_1 - T_2), \quad (46\text{–}34b)$$

where

$$h_m = \frac{\epsilon_m k_{gt} + (1 - \epsilon_m)k_{st}}{l}$$
$$+ [\lambda + C_p(T - T_m)_{\text{avg}}]k_{wm}\left(\frac{dp_w}{dT}\right)_{T_m}.$$
$$(46\text{–}34c)$$

An overall heat transfer coefficient U is defined by

$$q = U(T_{1b} - T_{2b}), \quad (46\text{–}35a)$$

where

$$\frac{1}{U} = \frac{1}{h_1} + \frac{1}{h_2} + \frac{1}{h_m}. \quad (46\text{–}35b)$$

The water vapor flux N_w may now be expressed as

$$N_w = k_{wm}\left(\frac{dp_w}{dT}\right)_{T_m} \frac{(T_1 - T_{2b})}{(h_m/h_1) + (h_m/h_2) + 1}.$$
$$(46\text{–}36)$$

When $h_m \gg h_1$ and $h_m \gg h_2$, the water flux N_w is considerably reduced due to a reduction in the value of $T_1 - T_2$ for given $T_{1b} - T_{2b}$. This is sometimes characterized as *temperature polarization*. For an efficient air gap MD process, the film transfer coefficients h_1 and h_2 in the two aqueous solutions should be high and h_m should be comparatively low. Otherwise, the driving force $(T_1 - T_2)$ across the membrane is reduced, lowering the water vapor flux.

A recent study and analysis by Schofield, Fane, and Fell (1990b) showed that deaeration increased the direct contact membrane distillation flux by about 40% although the membrane transfer coefficient increased sevenfold. However, such a flux increase resulted in worse temperature polarization and a fivefold decrease in $(T_1 - T_2)$. Thus, film heat transfer coefficients h_1 and h_2 need to be maximized, suggesting improved design of MD modules.

The expression of Eq. (46–36) for N_w was obtained for dilute solutions and small values of $(T_2 - T_2)$. The corresponding expressions for more concentrated solutions are available in Sarti and Gostoli (1984), Gostoli, Sarti, and Matulli (1987), and Schofield, Fane, and Fell (1987). These expressions identify a minimum value of bulk temperature difference below which the separation process does not work if, for example, pure water is to be taken out of a salt solution.

The analysis of MD with an air gap between the colder side of the microporous membrane and a separate cold condensing surface [Figure 46-4(b)] has been carried out by Kimura, Nakao, and Shimatani (1987). Gostoli and Sarti (1989) have extended the analysis of MD with an air gap to two volatile compounds diffusing through a stagnant film of inert gas using the Stefan-Maxwell formulation of flux expressions.

Vacuum membrane distillation has been studied using a flat microporous membrane (Bandini, Sarti, and Gostoli 1988) and microporous hollow fibers (Sarti, Gostoli, and Cavuoti 1990). No condensation takes place within the membrane module; the transport mechanism from the evaporating surface is one of convection driven by a total pressure difference. As a result, the fluxes are significantly higher than in the other configurations.

Membranes and Devices

Microporous hydrophobic membranes made from polypropylene, polytetrafluoroethylene, and polyvinylidenefluoride polymers have been used to study membrane distillation. The membrane forms investigated include both flat and capillary (hollow fiber). The pore size of most membranes used thus far is in the range of 0.1 to 3 μm, with 0.1 to 0.45 μm being the most common.

A basic requirement to be satisfied by these membranes during operation is that solutions on both sides of the membrane do not spontaneously wet the pores of the hydrophobic membrane. Any such solution pressurized to a certain level will enter the membrane pores. This pressure level is identified as the liquid entry pressure or penetration pressure, etc. For aqueous solutions of organic solutes, this pressure decreases *in general* with increasing solute concentration due to a reduction in surface tension. At a certain organic solute concentration, the liquid entry pressure required is zero and spontaneous membrane wetting occurs. The surface tension at this critical concentration is called the *critical surface tension*. The dependence of liquid entry pressure on the organic solute concentration for a given microporous membrane has been illustrated for an ethanol-water system by Franken et al. (1987) and Gostoli and Sarti (1989). The roles of the pore size and contact angle in estimating the liquid entry pressure have been considered by Franken et al. (1987). See also Chapter 28 on ultrafiltration membranes.

Microporous hollow fibers of polypropylene have been potted in a shell in a shell-and-tube arrangement and utilized for membrane distillation by Drioli, Wu, and Calabro (1987). Polypropylene hollow fibers of pore diameter 0.43 μm, porosity 0.70, and wall thickness 150 μm made by ENKA AG (Wuppertal, Germany) were used in a module of 0.15-m^2 membrane area. Such hollow-fiber devices are appropriate for direct contact MD, vacuum MD, and sweep gas MD; they are not useful for air gap MD. Spiral-wound microporous polytetrafluoroethylene membranes have been used for MD by Gore (1982).

Applications

Membrane distillation application studies have often employed direct contact MD. Drioli, Wu, and Calabro (1987) have studied membrane distillation separation of aqueous solutions of glucose (10 and 30 g/L) as well as aqueous solutions of NaCl (0.05 and 0.50 M). The feed solution temperature entering the hollow-fiber module was varied between 25 to 50°C; the values of $(T_{1b} - T_{2b})$ varied between 2 to 9.5°C. The condensate or permeate obtained was almost pure water. Kimura, Nakao, and Shimatani (1987) have investigated air gap membrane distillation of a variety of aqueous solutions with flat PTFE membranes. The condensate pH and conductivity in the MD of aqueous solutions of nonvolatile solutes H_2SO_4 or NaOH were found to be comparable with those of the condensate from a solution of NaCl. However, when aqueous solutions of volatile solutes HCl or HNO_3 were used in membrane distillation, the condensate concentration of solute increased with feed solute concentration; at concentrations higher than 1 mol/L, they became almost equal. They also studied the separation of aqueous solutions of organic acids, acetic and formic, concentration of orange juice, milk, gelatin solution, and sugar solution.

Concentration of a dilute aqueous ethanol solution by a microporous Gore-Tex® membrane via direct contact MD has been studied by Nakane et al. (1987). Gostoli and Sarti (1989) have numerically and experimentally studied separation of aqueous ethanol by air gap MD using a PTFE flat membrane. The separation factor was found to increase with an increase in the temperature difference between the evaporation and the condensation surface. Useful separation factors are obtained only with appreciable temperature differences. Direct contact MD is not as attractive as air gap MD since the temperature differences across the membrane in the former are quite small unless h_1 and h_2 are quite large. Other conditions remaining constant, separation is highly sensitive to feed composition and decreases as ethanol content increases. The numerical model shows the process to be ethanol-selective at a low ethanol content of feed; at higher ethanol feed concentrations, the process becomes water-selective.

There are very few reports of application studies of *osmotic distillation*. Lefebvre, Sheng, and Johnson (1990) have claimed successful concentration of liquid foods such as milk, fruit juices, and instant coffee and tea. Water flux data have been published by them for concentration of apple, grape, and orange juices.

PERSTRACTION

Introduction

Consider a multicomponent liquid feed flowing on one side of a nonporous membrane (Figure 46–5). Different species in feed establish different sorption equilibriums at the feed/membrane interface. Each dissolved liquid species then diffuses through the membrane material if there is a concentration gradient in the membrane. The dissolved molecules desorb from the membrane at the other side of the membrane into a different flowing liquid phase variously identified as the sweep liquid, purge liquid, extracting liquid, stripping liquid, etc. This liquid provides the concentration gradient for permeation. The interface of this liquid and the membrane is the second interface where solute species undergo partitioning again. The sweep liquid may or may not extract selectively from the penetrating species. Thus, the process can combine selective permeation and extraction and may be called *perstraction*. The word *pertraction* has also been used to describe the same process.

Michaels and Bixler (1968) have identified several requirements in perstraction to be satisfied by the sweep liquid, which is absorbing or extracting the permeating species:

1. This liquid must be incapable of permeation through the nonporous membrane to the feed side.
2. It should have a substantially lower volatility than the permeating components.

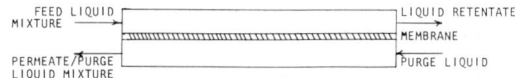

FIGURE 46–5. Schematic of a countercurrent perstractor stage.

3. It must have a substantial solubility for the permeating species. Acharya et al. (1988) have indicated that it must be completely miscible with the permeate.

In the separation of low-boiling hydrocarbon mixtures by perstraction, the sweep liquid is often a relatively high-boiling hydrocarbon oil. Since the permeant species will be minor constituents of the sweep liquid to maintain a permeation driving force, their boiling points should be much lower to reduce energy consumption in distillation used next to strip the permeants and recycle the high-boiling purge liquid. The high-boiling nature of the purge liquid also ensures negligible permeation through the membrane and contamination of the feed liquid. Michaels and Bixler (1968) have further suggested the use of the following as possible purge liquids: concentrated aqueous solution of an active hydrotrope, e.g., sodium toluene sulfonate, or a concentrated aqueous solution of soap or micellizing surfactant. (A hydrotrope is a freely water-soluble organic compound, which, at a concentration sufficient to induce an aqueous solution structure of hydrotrope molecules or aggregates, considerably increases the aqueous solubility of organic substances that are almost insoluble in water under normal conditions.)

Acharya et al. (1988) have studied the perstraction separation of a mixture of benzene and cyclohexane permeating through a polymer alloy membrane into a purge liquid of decalin (a mixture of cis- and trans-decahydronaphthalene). The weight fraction of benzene in the permeate (obtained by separating the product from the purge liquid) is shown in Figure 46–6 plotted against the weight fraction of benzene in the feed liquid. When compared with the separation achieved in a single vapor-liquid equilibrium stage (shown in Figure 46–6 as VLE), the separation by a perstraction membrane in a single cell (which is a *de facto* permeation stage) is remarkable.

However, the objective in the above separation system is to get a very high separation between benzene and cyclohexane. Had it not been for the fact that the benzene-cyclohexane

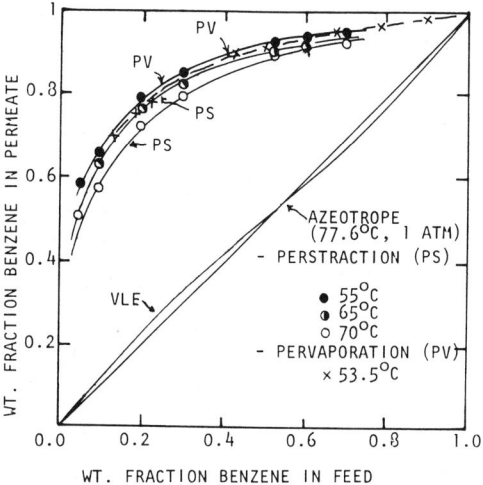

FIGURE 46–6. Permeate composition versus feed composition in perstraction and pervaporation of benzene/cyclohexane mixture through a polymer alloy film (reprinted from Acharya et al. 1988 with permission).

mixture is an azeotrope, one distillation column will have separated the mixture into relatively pure benezene and relatively pure cyclohexane as two liquid product streams. One perstraction device is incapable of achieving such a separation. It is necessary to have a cascade (Figure 46–7) of perstraction stages or cells. The purge liquid product from each stage is fed to a stripper to recycle the purge liquid, stripped of permeants, and then fed back to the stage; the retentate (raffinate) is sent as feed to a lower perstraction stage, whereas the permeate is sent to a higher perstraction stage (Michaels and Bixler 1968).

A perstractive stage can be based on simple permeation through a liquid membrane also. Cahn and Li (1976) have illustrated the separation of a mixture of toluene and *n*-heptane through an aqueous emulsion liquid membrane into a purge stream of kerosene (see Chapter 36 for this and other examples). Toluene was enriched in the permeate over heptane; the separation factor was as high as 21. They have described in detail how a cascade of three aqueous emulsion liquid membrane stages may be connected to produce 99.8% toluene from a 50–50 toluene-*n*-heptane feed using a heavy paraffin, *n*-dodecane, as the purge liquid. The permeate-

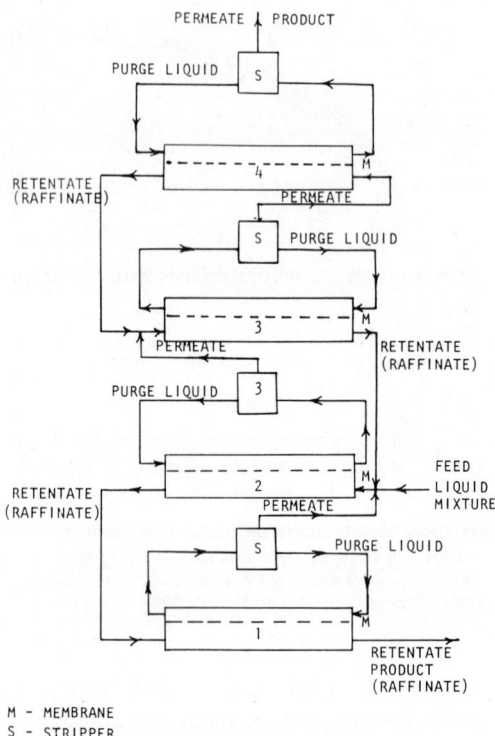

FIGURE 46–7. A four-stage perstraction cascade.

containing purge liquid from each stage was fed to an attached distillation column to recycle the purge liquid to the respective stage; this is an easy separation by distillation since the boiling point of *n*-dodecane is much higher than those of *n*-heptane and toluene.

For some liquid membrane processes, the solute is transferred by mechanisms other than simple permeation. Often constituents of the sweep liquid may actively participate in such a process by providing, say, protons that diffuse into the membrane. Chapter 37 provides a comprehensive theory for these processes. Such processes are not exact analogs of perstraction using polymeric membranes; Schlosser and Kossaczky (1980) have, however, identified them as *pertraction* processes.

The perstraction examples provided earlier had an organic liquid as the feed and an organic liquid purge stream. Perstraction may be implemented also with an aqueous feed and an aqueous purge stream. Boyadzhiev, Be-zenshek, and Lazarova (1984) had investigated removal of phenol from an aqueous feed to an aqueous strip solution via an organic liquid in creeping motion and called it a creeping film *pertractor*. A decade earlier, Cahn and Li (1974) removed phenol from wastewater into a caustic solution via an organic liquid membrane (see Chapters 36 through 39 for details). There is no specific distinguishing characteristic from a membrane separation point of view between these examples and the more traditional perstraction examples of organic liquid separation through a solid or liquid membrane having an organic purge stream (Acharya et al. 1988; Cahn and Li 1976). The major difference lies in the method to be adopted for regenerating the purge stream and the nature of interaction between the permeants and the purge stream.

Schlosser and Kossaczky (1980) have distinguished between the following three types of feed mixtures in perstraction and other such contexts:

1. Two component mixtures, one of which is not soluble in the membrane phase (e.g., phenol-water with an organic liquid membrane in which water is essentially insoluble)
2. A mixture in which all components have varying solubility in the membrane phase (e.g., hydrocarbon mixture as feed)
3. A mixture containing three or more components, at least one of which is not soluble in the membrane phase (e.g., separation of one metal from another in an aqueous solution through an organic membrane containing a carrier complexing with only one metal).

Perstraction is different from pervaporation (Chapters 7 through 10) in which no sweep liquid or purge liquid is used on the permeate side. In pervaporation, the permeate is desorbed from the membrane surface as a vapor into a sweep gas or pulled away as a vacuum is applied to the permeate side. Acharya et al. (1988) had studied the separation of a liquid mixture of benzene and cyclohexane through a polymer alloy membrane in the pervaporation mode as well as the perstraction mode. The selectivity in perstraction was observed to be quite close to that in pervaporation for the benzene-cyclohexane system with the same polymer alloy membrane and same feed. Pervaporation yielded a somewhat higher permeation rate due possibly to mass transport resistance at either the membrane/purge liquid interface or the purge film resistance. However, a supply of thermal energy must exist for volatilization of the permeate species in pervaporation in the membrane stage itself. Beyond the perstraction or pervaporation stage, there are additional energy requirements. Energy is required for the distillation separation of permeate from sweep liquid in perstraction, for compression of permeate vapor in pervaporation in the vacuum mode with condenser, etc. A detailed comparative analysis was not available as of late 1991.

Flux and Separation in Perstraction

The permeation of a penetrant species through a nonporous polymeric membrane in perstraction is usually considered to be based on a solution-diffusion mechanism. If a liquid membrane is being used in the perstraction mode and no chemical reactions occur in the liquid membrane, the solution-diffusion mechanism is equally valid. For one-dimensional diffusion across a membrane of thickness l, the flux of penetrant species i according to Fick's first law is

$$J_i = - D_{im} \frac{dc_{im}}{dz}. \qquad (46\text{-}37)$$

For a polymeric membrane that swells in the feed liquid, D_{im} is the local mutual diffusion coefficient in the polymer-penetrant system and is a function of c_{im}. Using an averaged diffusion coefficient $\bar{D}_{im}$ over the membrane thickness l, Eq. (46-37) may be integrated across z to obtain at steady state

$$J_i = \frac{\bar{D}_{im}}{l} (c'_{im} - c''_{im}), \qquad (46\text{-}38a)$$

where

$$\bar{D}_{im} = \int_{c''_{im}}^{c'_{im}} \frac{D_{im} \, dc_{im}}{c'_{im} - c''_{im}}. \qquad (46\text{-}38b)$$

The membrane concentrations c'_{im} and c''_{im} at the two interfaces are unknown. They can be related to the feed liquid and sweep liquid concentrations at the two interfaces using appropriate partition coefficients K'_i and K''_i:

$$K'_i = \frac{c'_{im}}{c'_{if}} ; \quad K''_i = \frac{c''_{im}}{c''_{ip}}. \qquad (46\text{-}39)$$

If the liquid film resistances on the feed and the permeate side are negligible compared to the membrane resistance, then the liquid concentrations at the two interfaces are equal to the corresponding bulk concentrations

$$c'_{if} = c'_{ifb} ; \quad c''_{ip} = c''_{ipb}. \qquad (46\text{-}40)$$

The steady-state flux relation of Eq. (46-38a) may now be expressed as

$$J_i = \frac{\bar{D}_{im}}{l} (K'_i c'_{ifb} - K''_i c''_{ipb}). \qquad (46\text{-}41)$$

Since the feed liquid and the sweep liquid are invariably different in perstraction, K'_i will not equal K''_i in general. For the particular case in which $K_i = K'_i = K''_i$, a membrane permeability coefficient for species i, P_i, is defined as

$$P_i = \bar{D}_{im} K_i, \qquad (46\text{-}42)$$

$$J_i = \frac{P_i}{l} (c'_{ifb} - c''_{ipb}). \qquad (46\text{-}43)$$

If species i and j are simultaneously permeating through the membrane, then the local separation factor α_{ij} between the two species may be defined as

$$\alpha_{ij} = \frac{J_i}{J_j} \frac{c'_{jfb}}{c'_{ifb}} = \frac{\bar{D}_{im}}{\bar{D}_{jm}} \frac{(K'_i c'_{ifb} - K''_i c''_{ipb}) c'_{jfb}}{(K'_j c'_{jfb} - K''_j c''_{jpb}) c'_{ifb}}.$$

$$(46\text{-}44)$$

Under the limiting conditions of $K'_i = K''_i = K_i$, $K'_j = K''_j = K_j$, $c''_{ipb} \ll c'_{ifb}$, and $c''_{jpb} \ll c'_{jfb}$,

$$\alpha_{ij} = \frac{\bar{D}_{im} K_i}{\bar{D}_{jm} K_j} = \frac{P_i}{P_j} \quad (46\text{-}45)$$

is an expression of the ideal separation factor. Note that definition (46-44) of α_{ij} is based on crossflow due to the ratio (J_i/J_j) being used instead of (c''_{ipb}/c''_{jpb}), which is used in the more traditional definition of α_{ij}.

In perstraction separation of organic mixtures using polymeric membranes, e.g., a polymer alloy of cellulose acetate and poly-(bromophenylene oxide dimethylphosphonate ester), Acharya et al. (1988) noted that the membrane needed to be "conditioned" by exposure to a high benzene-containing feed to develop a high α_{ij} and flux at lower feed concentrations. In general, polymeric membranes get considerably swollen by organic liquids; the extent of swelling depends on the local penetrant concentration and, therefore, local feed concentration. Membrane conditioning may reduce significantly this effect of local feed concentration. Membrane conditioning is likely to influence considerably the values of P_i, P_j, and α_{ij}.

The analysis presented above did not indicate specifically whether the membrane was homogeneous along its thickness l. It is likely that the extent of membrane swelling will vary within the membrane from location to location as a function of the penetrant concentration in the direction of transport. This would be especially true for polymeric membranes and organic feed mixtures.

Such nonhomogeneity in the membrane is to be considered along with an asymmetric structure in which the membrane is a thin skin supported by a microporous backing. This type of asymmetric structure, probably needed in practice to increase permeation rate, will, however, require consideration of diffusion in the microporous backing. The pores in the backing will most likely be filled with the purge liquid. Diffusional rates in liquid streams are not very high; therefore, minimization of the microporous backing thickness will also be required.

Since the perstraction process does not require any ΔP across the membrane as a driving force, a thin microporous backing is acceptable as long as the ΔP across the membrane due to flow pressure drops is tolerable.

Membranes

The number of studies in membrane separation identified as perstraction are very few. Consequently, very little information is available on perstraction membranes. Acharya et al. (1988) used a flat membrane made from a polymer alloy composed of cellulose acetate and a phosphoryl ester derivative of poly(dimethylphenylene oxide). Michaels and Bixler (1968) have illustrated permeation of the following liquid mixtures through flat linear polyethylene membranes: benzene-cyclohexane, o-xylene, m-xylene, and p-xylene. Selectivities as high as 7 were observed for particular feed compositions of benzene-cyclohexane when the membrane was stretched and annealed. Such a selectivity is two to four times lower than those observed by Acharya et al. (1988).

Information on perstraction membrane modules is scarce. Hollow fibers, spiral-wound as well as plate-and-frame structures, may be used. Of these, plate-and-frame structures will be the most expensive but they are likely to have much higher organic solvent resistance than the other modules.

Applications

There are no examples of perstraction in current industrial practice. It is suggested that metal removal by liquid membranes (treated in Chapters 37 and 38) should not be considered perstraction. The following types of organic mixtures appear to be attractive candidates for separation by perstraction (Michaels and Bixler 1968):

1. Azeotropes
2. Similar compounds with close boiling points (difficult to separate by distillation)
3. Isomeric mixtures.

The benzene-cyclohexane system described earlier falls in the second category. Additional examples are toluene-n-heptane, styrene-ethylbenzene, benzene-hexane, etc. Azeotropic systems, e.g., methanol-benzene, isomeric mixtures of xylenes, and normal and isoparaffins are also prospective candidates for perstraction.

ACKNOWLEDGMENTS

The author acknowledges the assistance of the following persons: Kathy Norman for word processing of the manuscript, T. H. Papadopoulos for drawing the figures, and Grace Peck for typing the figure captions.

NOTATION

General Notation

See the General Notation section at the beginning of this handbook.

Special Notation

- a_{si} gas/liquid interfacial area per unit volume based on hollow-fiber internal diameter, L^{-1}
- a_{so} same as a_{si} except based on fiber outer diameter, L^{-1}
- A empty cross-sectional area of module, L^2
- c_i^* defined by Eq. (46–4), mol/L^3
- C_p specific heat at constant pressure, L^2/t^2T or E/MT
- d_{ti} hollow-fiber inside diameter, L
- d_{to} hollow-fiber outside diameter, L
- h_1 heat transfer coefficient in hot liquid, M/t^3T
- h_2 heat transfer coefficient in cold liquid, M/t^3T
- h_m heat transfer coefficient defined by Eq. (46–34b), M/t^3T
- K_g overall mass transfer coefficient based on gas phase, mol t/ML or mol/L^2tp
- k_l local liquid phase mass transfer coefficient, L/t
- $\bar{k}_l$ length averaged value of local liquid phase mass transfer coefficient k_l, L/t
- K_l overall mass transfer coefficient based on liquid phase, L/t
- $\bar{K}_l$ length averaged value of K_l, L/t
- N number of hollow fibers, dimensionless
- N_i molar flux of species i in a fixed reference frame, mol/L^2t
- p_i^* defined by Eq. (46–4), M/Lt^2 or p
- Q_g volumetric gas flow rate, L^3/t
- Q_s volumetric liquid flow rate on shell side, L^3/t
- Q_t volumetric liquid flow rate on tube side, L^3/t
- U overall heat transfer coefficient defined in Eq. (46–35b), M/t^3T
- v_s actual shellside velocity of absorbent liquid, L/t
- v_{su} superficial velocity of liquid based on empty shell cross-sectional area, L/t

Greek Letters

- ϵ_m membrane porosity, dimensionless
- λ mean free path, L, or latent heat of vaporation, ML^2/t^2 mol or E/mol
- ϕ packing fraction of fibers in a shell, dimensionless
- τ_m tortuosity of pores in a membrane, dimensionless

Subscripts

- b bulk
- g gas
- i species i; phase interface location
- l liquid
- m membrane

Superscripts

- in   at inlet location
- out   at outlet location

REFERENCES

Acharya, H. R., S. A. Stern, Z. Z. Liu, and I, Cabasso. 1988. Separation of liquid benzene/cyclohexane mixtures by perstraction and pervaporation. *J. Membr. Sci.* 37:205–32.

Bandini, S., and C. Gostoli. 1990. Continuous removal of ethanol from fermentation broths by gas membrane extraction. In *Proc. International Congress of Membranes,* Vol. 1, pp. 373–375. 20–24 August 1990. Chicago, IL.

Bandini, S., G. C. Sarti, and C. Gostoli. 1988. Membrane distillation through porous hydrophobic membranes. In *Proc. 3rd International Conference on Pervaporation Processes,* pp. 117–

126. Bakish Materials Corporation, Engelwood, NJ.

Bhave, R. R., and K. K. Sirkar. 1986. Gas permeation and separation by aqueous membranes immobilized across the whole thickness or in a thin section of hydrophobic microporous Celgard Films. *J. Membr. Sci.* 27:41–61.

Bhave, R. R., and K. K. Sirkar. 1987. Gas permeation and separation with aqueous membranes immobilized in microporous hydrophobic hollow fibers. In *Liquid Membranes Theory and Applications*, ed. Richard D. Noble and J. Douglas Way, pp. 138–151. Washington, DC: American Chemical Society.

Bonne, U., D. W. Deetz, J. H. Lai, D. J. Odde, and J. D. Zook. 1990. Membrane dehumidification. U.S. Patent 4,915,838, 10 April 1990.

Boswell, S. T. 1991. Membrane air stripping. Ph.D. thesis, Stevens Institute of Technology, Hoboken, NJ.

Boyadzhiev, L., E. Bezenshek, and Z. Lazarova. 1984. Removal of phenol from wastewater by double emulsion membranes and creeping film pertraction. *J. Membr. Sci.* 21:137–144.

Brose, D. J., and P. Van Eikeren. 1990. Membrane-based removal of ammonia from mammalian-cell culture. *Appl. Biochem. Biotech.* 24&25:457–468.

Cahn, R. P., and N. N. Li. 1974. Separation of phenol from wastewater by the liquid membrane technique. *Sep. Sci.* 9:505–519.

Cahn, R. P., and N. N. Li. 1976. Hydrocarbon separation by liquid membrane processes. In *Membrane Separation Processes*, ed. Patrick Meares, pp. 327–349. Amsterdam: Elsevier Scientific Publishing Co.

Callahan, R. W. 1988. Novel uses of microporous membranes: a case study. In *Membrane Materials and Processes for Separation*, ed. Kamalesh K. Sirkar and Douglas R. Lloyd, pp. 54–63. AIChE Symp. Ser. No. 261, Vol. 84. New York: American Institute of Chemical Engineers.

Cooney, D. O., and C. C. Jackson. 1989. Gas absorption in a hollow fiber device. *Chem. Eng. Comm.* 79:153–163.

Drioli, E., Y. Wu, and V. Calabro. 1987. Membrane distillation in the treatment of aqueous solutions. *J. Membr. Sci.* 33:277–284.

Esato, K., and B. Eiseman. 1975. Experimental evaluation of Gore-Tex membrane oxygenator. *J. Thorac. Cardiovascular Surg.* 69(5):690–697.

Evren, V., and E. Piskin. 1986. Membrane oxygenators embodying silicone-coated microporous membranes. In *Progress in Artificial Organs—1985.* ed. Y. Nose', C. Kjellstrand and P. Ivanovich, pp. 566–569. Cleveland, OH: ISAO Press.

Findley, M. E. 1967. Vaporization through porous membranes. *Ind. Eng. Chem. Process Des. Dev.* 6(2):226–230.

Frank, G. T., and K. K. Sirkar. 1986. An integrated bioreactor-separator: in-situ recovery of fermentation products by a novel dispersion-free solvent extraction technique. *Biotech. Bioeng. Symp. Ser.* 17:303–316.

Franken, A. C. M., J. A. M. Nolten, M. H. V. Mulder, D. Bargeman, and C. A. Smolders. 1987. Wetting criteria for the applicability of membrane distillation. *J. Membr. Sci.* 33:315–328.

Gore, D. W. 1982. Gore-Tex membrane distillation. In *Proc. 10th Annual Convention of Water Supply Improvement Association,* 25–29 July 1982, Honolulu, Hawaii.

Gostoli, C., and G. C. Sarti. 1989. Separation of liquid mixtures by membrane distillation. *J. Membr. Sci.* 41:211–224.

Gostoli, C., G. C. Sarti, and S. Matulli. 1987. Low temperature distillation through hydrophobic membranes. *Sep. Sci. Technol.* 22:855–872.

Imai, M., S. Furusaki, and T. Miyauchi. 1982. Separation of volatile materials by gas membranes. *Ind. Eng. Chem. Process Des. Dev.* 21:421–426.

Kang, W., R. Shukla, G. T. Frank, and K. K. Sirkar. 1988. Evaluation of O_2 and CO_2 transfer coefficients in a locally integrated tubular hollow fiber bioreactor. *Appl. Biochem. Biotech.* 18:35–51.

Karoor, S., and K. K. Sirkar. 1991. Microporous hollow fiber gas absorption in reactive and nonreactive systems. Paper presented at the Fourth National Meeting of NAMS, 30 May 1991, San Diego, CA.

Keller, K. H., and K. L. Shultis. 1979. Oxygen permeability in ultrathin and microporous membranes during gas-liquid transfer. *Trans. Am. Soc. Artif. Intern. Organs* 225:469–472.

Kenfield, C. F., R. Qin, M. J. Semmens, and E. L. Cussler. 1988. Cyanide recovery across hollow fiber gas membranes. *Environmental Sci. Technol.* 22:1151–1155.

Keyvani, M., and N. C. Gardner. 1989. Operating characteristics of rotating beds. *Chem. Eng. Prog.* September:48–52.

Kiani, A., R. R. Bhave, and K. K. Sirkar. 1984. Solvent extraction with immobilized interfaces in a microporous hydrophobic membrane. *J. Membr. Sci.* 20:120–145.

Kimura, S., S. I. Nakao, and S. I. Shimatani. 1987.

Transport phenomena in membrane distillation. *J. Membr. Sci.* 33:285–298.

Kreulen, H., G. F. Versteeg, C. A. Smolders, and W. P. M. van Swaaij. 1990. Determination of mass transfer rates in wetted and non-wetted membranes. In *Proc. International Congress of Membranes.* Vol. 1, pp. 52–54. 20–24 August 1990, Chicago, IL. Paper not presented.

Lefebvre, M. S., R. A. Johnson, and V. Yip. 1987. Theoretical and practical aspects of osmotic distillation. Paper 1-0A0806 read at International Congress of Membranes, 8–12 June 1987, Tokyo, Japan.

Lefebvre, M. S., J. Sheng, and R. A. Johnson. 1990. Comparative performances of membrane distillation and osmotic distillation. In *Proc. International Congress of Membranes*, Vol. 1, pp. 68–70. 20–24 August 1990, Chicago, IL.

Leonard, E. F. 1986. The application of artificial organs in biotechnology. In *Membrane Separations in Biotechnology*, ed. W. Courtney McGregor, pp. 319–353. New York: Marcel Dekker.

Matsumoto, H., S. Satoh, N. Shiozi, A. Kakimoto, A. Hamasaki, and Y. Kita. 1990. Development of artificial gill and basic test of small closed environment life support system utilizing it. In *Proc. International Congress of Membranes*, Vol. 1, pp. 55–57. 20–24 August 1990, Chicago, IL.

Michaels, A. S., and H. J. Bixler. 1968. Membrane permeation: theory and practice. In *Progress in Separation and Purification*, ed. E. S. Perry, pp. 143–186. New York: Wiley Interscience.

Nakane, T., H. Yanagishita, M. Tamura, Y. Takahashi, T. Hakuta, and H. Yoshitome. 1987. Concentration of dilute aqueous ethanol solution by membrane distillation. Poster 9-P40 presented at the International Congress of Membranes, 8–12 June 1987, Tokyo, Japan.

Ogundiran, S. O., S. E. LeBlanc, and S. Varanasi. 1989. Membrane contactors for SO_2 removal from flue gases. Paper read at the Pittsburgh Coal Conference, September 1989.

Prasad, R., and K. K. Sirkar. 1989. Hollow fiber solvent extraction of pharmaceutical products: a case study. *J. Membr. Sci.* 47:235–259.

Prasad, R., and K. K. Sirkar. 1990. Hollow fiber solvent extraction: performances and design. *J. Membr. Sci.* 50:153–175.

Qi, Z., and E. L. Cussler. 1985a. Microporous hollow fibers for gas absorption. I. mass transfer in the liquid II. mass transfer across the membrane. *J. Membr. Sci.* 23:321–340.

Qi, Z., and E. L. Cussler. 1985b. Bromine recovery with hollow fiber gas membranes. *J. Membr. Sci.* 24:43–57.

Qi, Z., and E. L. Cussler. 1985c. Hollow fiber gas membranes. *AIChE J.* 31:1548–1553.

Qin, R., A. K. Zander, M. J. Semmens, and E. L. Cussler. 1990. Separating acetic acid from liquids. *J. Membr. Sci.* 50:51–55.

Radovich, J. M., W. C. Babcock and K. R. Pearson. 1986. Removal of SO_2 from flue gas in a hollow fiber membrane contactor. Paper No. 249 read at the American Chemical Society Rocky Mountain Regional Meeting, June 1986, Denver, CO.

Reed, B. W., R. Prasad, and R. W. Callahan. 1990. Oxygenation/deoxygenation of aqueous streams with microporous membranes. Paper presented at Third SPSJ International Polymer Conference, 26–29 November 1990, Nagoya, Japan.

Sarti, G. C., and C. Gostoli. 1984. Use of hydrophobic membranes in thermal separation of liquid mixtures: theory and experiments. Paper presented at Europe-Japan Meeting on Membranes and Membrane Processes, June 1984, Stressa, Italy; in *Membranes and Membrane Processes,* ed. E. Drioli and M. Nakagaki, pp. 349–360. New York: Plenum Press, 1986.

Sarti, G. C., C. Gostoli, and G. Cavuoti. 1990. Vacuum membrane distillation through capillary polypropylene membranes. In *Proc. International Congress of Membranes*, Vol. 1, pp. 65–67. 20–24 August 1990, Chicago, IL.

Schlosser, S., and E. Kossaczky. 1980. Comparison of pertraction through liquid membrane and double liquid-liquid extraction. *J. Membr. Sci.* 6:83–105.

Schneider, K., and T. S. Van Gassel. 1984. Membrane distillation. *Chem. Ing. Tech.* 56(7):514–521.

Schofield, R. W., A. G. Fane, and C. J. D. Fell. 1987. Heat and mass transfer in membrane distillation. *J. Membr. Sci.* 33:299–313.

Schofield, R. W., A. G. Fane, and C. J. D. Fell. 1990a. Gas and vapour transport through microporous membranes. I. Knudsen-Poiseuille transition. *J. Membr. Sci.* 53(1&2):159–171.

Schofield, R. W., A. G. Fane, and C. J. D. Fell. 1990b. Gas and vapour transport through microporous membranes. II. membrane distillation. *J. Membr. Sci.* 53(1&2):173–185.

Semmens, M. J., D. M. Foster, E. L. Cussler. 1990. Ammonia removal from water using microporous hollow fibers. *J. Membr. Sci.* 51(1&2):127–140.

Semmens, M. J., R. Qin, and A. Zander. 1989.

Using a microporous hollow fiber membrane to separate VOCs from water. *Journal AWWA* April:162–167.

Shukla, R., W. K. Kang, and K. K. Sirkar. 1989. Acetone-butanol-ethanol (ABE) production in a novel hollow fiber fermentor-extractor. *Biotech. Bioeng.* 34:1158–1166.

Treybal, R. E. 1980. *Mass Transfer Operations*. 3rd ed. New York: McGraw Hill Book Co.

Tsuji, T., K. Suma, K. Tanishita, H. Fukazawa, M. Kanno, H. Hasegawa, and A. Takahashi. 1981. Development and clinical evaluation of hollow fiber membrane oxygenator. *Trans. Am. Soc. Artif. Intern. Organs* 27:280–284.

Van Eikeren, P., D. J. Brose, D. C. Muchmore, and J. B. West. 1990. Membrane assisted synthesis of chiral drugs and fine chemicals. In *Enzyme Engineering X*, ed. H. Okada and A. Tanaka, Annals of New York Academy of Science, Vol. 613, pp. 796–801. New York: New York Academy of Science.

Wicksramasinghe, S. R., M. J. Semmens, and E. L. Cussler. 1990. The best hollow fiber contactor. In *Proc. International Congress of Membranes*, Vol. 1, pp. 48–49. 20–24 August 1990, Chicago, IL.

Yang, M. C., and E. L. Cussler. 1986. Designing hollow-fiber contactors. *AIChE J.* 32:1910–1916.

Yang, M. C., and E. L. Cussler. 1989. Artificial gills. *J. Membr. Sci.* 42:273–284.

XI
Controlled Release

47
Controlled Release

Kelly L. Smith
Scott M. Herbig
Bend Research, Inc.

GENERAL DESCRIPTION OF
   CONTROLLED-RELEASE
   TECHNOLOGIES
TRANSPORT PRINCIPLES
   Diffusion in Polymers
   Release Kinetics of Controlled-
     Release Systems
APPLICATIONS
   Cattle Insecticide Eartags
   Microporous Beads Containing Pher-
     omones
   Cattle Oral Bolus
   Osmotic Nifedipine Tablet
NOTATION
REFERENCES

GENERAL DESCRIPTION OF CONTROLLED-RELEASE TECHNOLOGIES

Controlled-release technologies are designed to deliver a wide variety of active ingredients (drugs, pesticides, fragrances, etc.) at a specified rate, for a specified period of time, and at a desired location. This might mean steady, small quantities of an active ingredient released over a long period of time, or short bursts of specified quantities released at designated intervals. Regardless of the rate or manner of release, the release kinetics from controlled-release formulations are characteristically dependent on the formulation rather than on the properties of the active ingredient (e.g., solubility or volatility) in the environment of use. In many cases, the release of the active ingredient is controlled by its rate of diffusion through a membrane.

Controlled-release formulations offer (1) greater efficacy because optimal concentrations of active ingredient can be maintained in the environment of use, (2) improved safety from reliable release kinetics and control over the amount of active ingredient available at any one time, and (3) greater convenience because fewer applications or treatments are needed.

Controlled-release formulations that are designed to release an active ingredient at a controlled rate for a desired time period can maintain a constant level of just enough active ingredient to be effective, but no more than necessary at any one time, thus minimizing toxic effects. This is illustrated in Figure 47–1. As illustrated in the figure, several conventional treatments would be required in place of a single dose of a controlled-release formulation to maintain effective levels. Furthermore, the total amount of active ingredient is much higher in

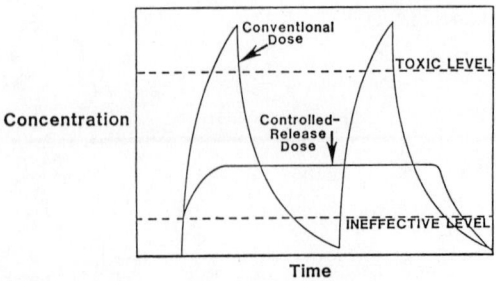

FIGURE 47–1. Concentration profiles for conventional and controlled-release formulations.

conventional treatments, increasing the likelihood of unwanted side effects.

Controlled-release formulations that deliver active ingredients in discrete bursts or pulses at designated times and at a rapid rate offer the same advantages as those that deliver active ingredients at a slow, steady rate—i.e., they deliver the optimum dose at the right time and to the right place.

Of the many controlled-release technologies developed and commercialized since the late 1960s, most rely on membrane technology—i.e., they operate by diffusion of either the active ingredient or water through a rate-controlling membrane or polymer matrix. These controlled-release technologies can be divided into four types: (1) reservoir systems, (2) matrix systems, (3) swelling-controlled systems, and (4) osmotic systems.

Reservoir systems consist of a core containing the active ingredient surrounded by a rate-controlling membrane. The rate-controlling membrane can be either a dense (nonporous) membrane or a microporous membrane. The active ingredient diffuses through the rate-controlling membrane into the environment of use.

In matrix systems, the active ingredient is either dissolved or dispersed in a polymer matrix, from which it is released by diffusion through the polymer matrix. Matrix systems have also been developed that depend on water diffusing into the matrix. The water swells the matrix, which allows the active ingredient to diffuse out of the swollen matrix much more rapidly than it could through an unswollen matrix.

Osmotic systems also rely on the diffusion of water through a rate-controlling membrane to provide controlled release. These systems consist of a core containing the active ingredient surrounded by a rate-controlling membrane. Due to an osmotic gradient, water is imbibed into the core, creating a hydrostatic pressure, which forces a solution containing the active ingredient out of the device.

Controlled-release systems have also been developed that are not primarily dependent on diffusive transport. These systems include erodible polymers, pressure-sensitive devices, magnetic systems, electrically driven pumps, liposomes, and the use of monoclonal antibodies. Many of these systems are used when release is required on demand, when it is triggered by an external factor, or for site-specific delivery. Since our focus in this chapter is on the use of membranes in controlled-release systems, we do not describe any of these "nondiffusion-based" technologies in more detail. However, Table 47–1 provides a list of references for those interested in more thorough descriptions of these technologies.

TRANSPORT PRINCIPLES

Diffusion in Polymers

Transport through polymeric membranes in controlled-release systems occurs by a solution-diffusion process. The active ingredient must first dissolve in the membrane and then diffuse from the surface or area of high active ingredient concentration to the opposite surface or area of low concentration. This diffusive transport phenomenon can be quantified by the equation known as *Fick's first law*:

$$J = -D \frac{dc_m}{dx}, \qquad (47\text{--}1)$$

where J is the active ingredient flux, D is the diffusivity of the active ingredient in the rate-controlling membrane, and dc_m/dx is the concentration gradient of the active ingredient in the membrane.

TABLE 47–1. References Describing Controlled-Release Technologies Not Based on Diffusion Through Membranes.

Type of System	Key References
Bioerodible polymers	Heller 1980
	Pitt and Schindler 1980
	Harris 1984
	Pitt and Schindler 1983
	Heller 1984a
	Heller 1984b
	Leong and Langer 1987
	Baker 1987
	Smith, Schimpf, and Thompson 1990
Pressure-sensitive	Watanabe and Hayashi 1976
Magnetically activated	Hsieh and Langer 1983
	Widder and Senyei 1983
	Kost and Langer 1990
Electrical pumps	Uhlig, Graydon, and Zingg 1983
	Sefton 1984
Liposomes	Knight 1981
	Gregoriadis 1984
	Cullis et al. 1989
	Wright and Huang 1989
Monoclonal antibodies	Vitetta, Fulton, and Uhr 1986
	Hellstrom, Hellstrom, and Goodman 1987
	Wright and Huang 1989

To use Eq. (47–1), the concentration of active ingredient in the membrane must be known, but it is not easily measurable. However, since one can usually measure the concentration of active ingredient on either side of the membrane (either in liquid or vapor phase), another form of *Fick's first law* is much more useful because it includes a partition coefficient between the solution (or vapor) and the rate-controlling membrane. Assuming a single partition coefficient for each side of the membrane, this can be expressed simply by

$$J = -DK \frac{dc}{dx}, \quad (47\text{–}2)$$

where K is the ratio of the concentration in the membrane to the concentration in the solution and dc/dx is the concentration gradient between the solutions on either side of the membrane. In some cases, boundary layers in one or both of the external phases or even in the membrane can lower the driving force for diffusion across the membrane. Depending on the partition coefficient, the concentration of active ingredient in the membrane at the surface can be either higher or lower than the concentration of the active ingredient in the adjacent solution.

In most cases the diffusivity within the membrane can be assumed to be constant, and Eq. (47–2) can be integrated to give the steady-state flux

$$J = DK \frac{\Delta c}{l} = P \frac{\Delta c}{l}, \quad (47\text{–}3)$$

where l is the thickness of the membrane, Δc is the difference in concentration between the source solution on one side and the sink on the other side of the membrane, and P is the permeability of the membrane to the active ingredient (defined as $P = DK$). Thus, the rate of transport or release of active ingredient from a controlled-release system is governed principally by two solute-polymer properties—diffusivity and solubility—as well as by the concentration gradient across the rate-controlling membrane and the geometry of the device. For reservoir systems in which the reservoir contains a pure or saturated active ingredient, all terms in Eq. (47–3) are constant, and the resulting release rate is therefore constant. For matrix systems, the diffusional path length l usually increases with time, resulting in a decreasing release rate with time. These effects are discussed in more detail in later sections of this chapter.

Two possible concentration profiles illustrating steady-state diffusive transport from a solution with a high concentration of active ingredient across a membrane to a solution with a low concentration of active ingredient are shown in Figure 47–2. In the first case, the solubility of active ingredient is lower in the membrane than in the solution, and the partition coefficient is less than 1. In the second case, the solubility is higher in the membrane than in the solution, and the partition coefficient is greater than 1.

The two key properties of membrane systems that influence diffusion and materials selection are *diffusivity* and *solubility*. Diffusivity is a kinetic property and describes the ease with which active-ingredient molecules move within the membrane. Solubility is an equilibrium property and describes the interaction or compatibility between the active ingredient and the membrane material. Thus, understanding the factors that influence these properties is important when designing or developing controlled-release systems.

The key factors that influence diffusivity in polymer systems are the size and shape of the active ingredient molecules, the flexibility of the polymer chains, and the packing density or crystallinity of the polymer chains. Empirical correlations between diffusivity and solute molecular weight are well established for numerous solutes in fluid systems (Cussler 1984; Welty, Wicks, and Wilson 1984). Similar correlations have also been established for diffusion in polymers, as shown in Figure 47–3 (Bixler and Sweeting 1971; Grun 1947; Stannett 1968; Michaels et al. 1975). As the molecular weight of the solute increases, the diffusivity generally decreases, although other factors such as molecular shape also influence the diffusivity (Bixler and Sweeting 1971).

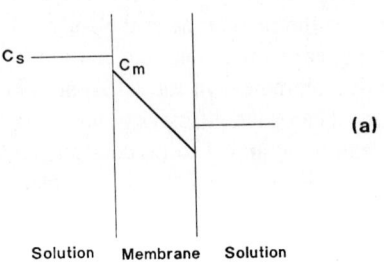

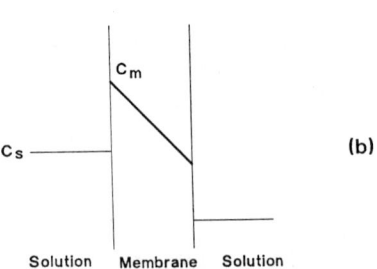

FIGURE 47–2. Concentration profile of active ingredient across membrane for two cases: (a) $K < 1$ and (b) $K > 1$.

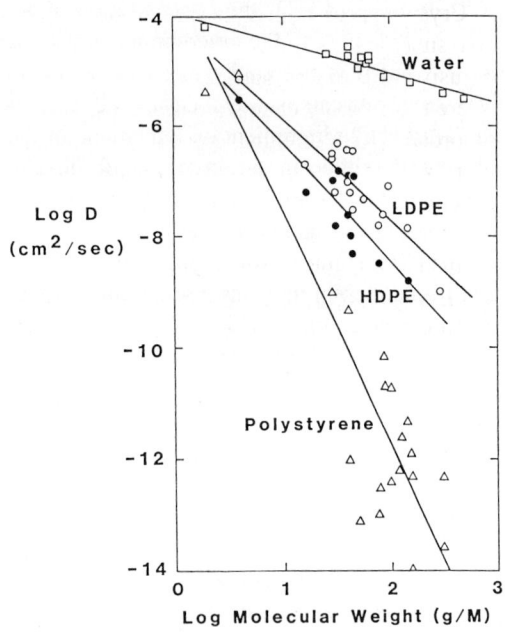

FIGURE 47–3. Dependence of diffusivity on solute molecular weight in water and in three polymers (reprinted from Smith and Lonsdale 1985 with permission).

The flexibility of polymer chains is primarily dependent on temperature, plasticization, and the amount of cross-linking between the polymer chains. Polymers exist in glassy or rubbery states and can go from one state to the other depending on changes in temperature or plasticizer concentration. In the rubbery state (including gels) segments of the polymer backbone can move freely, allowing greater polymer chain flexibility and, therefore, higher diffusivities. This is illustrated in Figure 47-3, where the diffusivities in the two polyethylene polymers, which are rubbery at 25°C, are much greater than that in polystyrene, which is glassy at 25°C (Smith and Lonsdale 1985). In work with polyethyleneterephthalate diffusivities have been shown to increase by more than an order of magnitude when the polymer is in the rubbery state rather than the glassy state (Kesting 1985). Polymer cross-linking also reduces the flexibility of polymer chains. Diffusivities decrease in polymers proportionally as cross-linking density increases and as the length of the cross-linking chains decreases (Barrer and Skirow 1984; Stannett et al. 1962).

Crystalline or semicrystalline polymers contain small crystallites, which act to decrease diffusivity. These crystallites consist of regions where the polymer chains are densely packed in an ordered and oriented configuration. Diffusivity within these crystallites is believed to be extremely low, and essentially all diffusion in semicrystalline polymers occurs in the amorphous regions between crystallites (Michaels and Bixler 1968). Thus, diffusivities are lower in more-crystalline polymers than in less-crystalline polymers.

Solubility of solutes in polymers involves many complex interactions between the solute and polymer such as polarity, hydrogen bonding, and steric effects. Thermodynamically, the solubility is largely determined by the heat of mixing, which has been used to derive solubility parameters of the solute and polymer (Hildebrand and Scott 1964; Hildebrand, Prausnitz, and Scott 1970; Hansen and Beerbower 1971). The solubility parameter is dependent on the extent of hydrogen bonding, polarity, and nonpolar dispersion forces and has been tabulated for many solutes and polymers (Barton 1983; Grulke 1989). Following the general rule in chemistry that "like dissolves like," solutes are most soluble in polymers with similar solubility parameters (i.e., similar chemical functionality). In addition, solute solubility is very low in polymer crystallites; thus, solubility decreases proportionately to increases in polymer crystallinity.

The effects of the polymer characteristics on the release of active ingredients can be summarized with regard to the following factors. Release rates are greater from systems where the rate-controlling membrane is made from polymers that are (1) rubbery rather than glassy, (2) amorphous rather than crystalline, (3) uncross-linked rather than cross-linked, and (4) similar in chemistry to the active ingredient rather than dissimilar.

More detailed explanations of the theory of diffusive transport are readily available in the literature (Crank 1975; Barrer 1968; Cussler 1984; Baker and Lonsdale 1974; Crank and Park 1968; Paul and McSpadden 1976).

Release Kinetics of Controlled-Release Systems

The release of active ingredients from controlled-release systems can follow a wide variety of patterns. However, the release kinetics from most systems that depend on membrane diffusion can be categorized into three types of release profiles: (1) zero-order kinetics, (2) $t^{-1/2}$ kinetics, and (3) first-order kinetics. For systems exhibiting zero-order kinetics, the release rate remains constant until essentially all of the active ingredient has been delivered (the term *zero order* derives from the release kinetics being independent of the quantity of drug remaining, i.e., drug concentration term raised to the zero power). The release rate decreases proportionately to the square root of time in controlled-release systems with $t^{-1/2}$ kinetics. First-order release kinetics occur in systems where the release rate is proportional to the quantity of drug remaining in the system (i.e., proportional to the quantity to the first power). First-order release is common in conventional

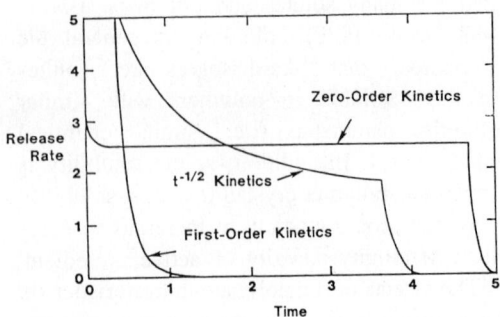

FIGURE 47-4. Comparison of release kinetics from conventional (first-order) and controlled-release formulations.

noncontrolled-release systems as well as some controlled-release systems and is characterized by an exponential decrease in release rate over time. A comparative theoretical release profile for each type of system is shown in Figure 47-4. Since the release rate is expressed as the quantity of active ingredient released over time, a horizontal line on the graph in Figure 47-4 indicates constant release. Note that for both $t^{-1/2}$ and first-order systems most of the active ingredient is released during the first third of the total duration of release. Depending on the specifications for particular controlled-release formulations, either zero-order or $t^{-1/2}$ systems are typically preferred.

Reservoir Systems

Reservoir systems consist of a core containing an active ingredient surrounded by a rate-controlling membrane. The active ingredient is released from these systems by diffusion through the rate-controlling membrane. Coated tablets, beads, and particles, membrane-based pouches (e.g., transdermal drug delivery systems), and microcapsules represent several common types of reservoir systems.

Reservoir systems typically provide a constant rate of release over a substantial portion of the duration of release. The desired rate of release can usually be achieved by properly selecting the rate-controlling membrane. Release rates from reservoir systems are critically dependent on the membrane thickness, area, and permeability. Reservoir systems can contain more active ingredient than other controlled-release systems, which minimizes the cost of formulation materials and allows active ingredients to be released at higher rates and for longer durations. The major disadvantage of reservoir systems is that they are subject to dose dumping should the rate-controlling membrane be ruptured.

Equations describing release kinetics from reservoir systems can be derived easily from Fick's first law. Only factors describing the geometry of the system need to be added to the equation. Release-rate equations for reservoir systems where the rate-controlling membrane is in the form of a flat sheet, a cylinder, or a sphere are

$$\text{Flat sheet:} \quad \frac{dM_t}{dt} = ADK\frac{\Delta c}{l}, \qquad (47\text{-}4)$$

$$\text{Cylinder:} \quad \frac{dM_t}{dt} = 2\pi LDK\frac{\Delta c}{\ln(r_o/r_i)}, \qquad (47\text{-}5)$$

$$\text{Sphere:} \quad \frac{dM_t}{dt} = 4\pi DK\Delta c\frac{r_o r_i}{r_o - r_i}, \qquad (47\text{-}6)$$

where

M_t = mass of agent released up to time t
dM_t/dt = release rate at time t
A = surface area
L = length of the cylinder
l = thickness of the flat sheet
r_o = outside radius
r_i = inside radius (Baker and Lonsdale 1974).

If the diffusivity and concentration gradient across the membrane remain constant, then the release rate will be constant (zero-order release kinetics). To maintain a constant concentration gradient, the activity of the active ingredient within the reservoir must remain constant and transport to and from the membrane surfaces must be rapid with respect to transport through the membrane. (It is typically assumed that the activity of the active ingredient is very low outside the reservoir system since the surrounding environment usually provides an infinite sink.) Thus, reservoir systems are particularly useful when they can be loaded entirely with an active ingredient or when the concentration of active ingredient within the reservoir is at saturation during most of the duration of release.

When the concentration of active ingredient within the core is less than saturation, its concentration continuously decreases as the active ingredient is released. As the concentration of active ingredient in the reservoir decreases, the concentration gradient across the membrane decreases, and first-order release kinetics are observed. First-order release kinetics from reservoir systems without a constant concentration gradient across the membrane can be expressed as

$$\frac{dM_t}{dt} = \frac{c_s ADK}{l} \exp\left(-\frac{ADKt}{Vl}\right), \quad (47\text{-}7)$$

where c_s is the saturation concentration of the active ingredient in the reservoir, t is time, and V is the volume of the reservoir (Baker and Lonsdale 1974). The quantity of active ingredient released from a reservoir system at a constant rate is dependent on the solubility of the active ingredient in the membrane as well as its solubility in the reservoir and on the volume of the reservoir.

The initial release rates from reservoir systems are commonly higher or lower than the steady-state release rate, depending on the storage history of the device. Systems used shortly after being made will exhibit an initially lower release rate until the steady-state concentration profile of active ingredient in the membrane is established. This is known as the *time lag*. Likewise, systems stored long enough for the membrane to become completely saturated with the active ingredient exhibit a higher release rate until the steady-state concentration profile within the membrane is established. This is known as the *burst effect*. Both the time lag and burst effect are dependent on the diffusivity of the active ingredient in the membrane and on the membrane thickness, as shown in the following equations describing the zero-order portion of the release curve for flat-sheet reservoir systems:

Time lag: $\quad M_t = \dfrac{Dc_m}{l}\left(t - \dfrac{l^2}{6D}\right), \quad (47\text{-}8)$

Burst effect: $\quad M_t = \dfrac{Dc_m}{l}\left(t + \dfrac{l^2}{3D}\right), \quad (47\text{-}9)$

where M_t is the quantity of active ingredient released and c_m is the concentration of active ingredient in the membrane at the surface in contact with the reservoir (Baker and Lonsdale 1974). Expressions have also been developed for cylinders and spheres (Crank 1975). The time lag θ is defined as $l^2/6D$ and can be easily determined from a release profile from a freshly made device. The intercept of the steady-state portion of a release profile on the t-axis will be $l^2/6D$ for the time lag and $-l^2/3D$ (or twice the time lag) for the burst effect, as shown in Figure 47–5. Thus, diffusivities can be calculated for active ingredients in membranes by measuring the quantity of active ingredient released over time and then determining the time lag and calculating the diffusivity.

Burst effects are more common in practical applications since products are usually stored for a sufficiently long period of time prior to use to allow the membrane to become saturated with active ingredient. The release of the active ingredient during the initial burst is approximately first order since the release kinetics are dependent on the decreasing concentration of active ingredient within the membrane.

Steady-state release is typically achieved within a period of time equal to about three time lags for both time-lag and burst-effect cases (Baker and Lonsdale 1974). Thus, it is important to know the diffusivity and thickness of the membrane or to measure the time lag to predict approximately when steady-state release rates will be achieved. For systems having high diffusivities ($>10^{-8}$ cm²/s) and thin membranes (<50 μm) steady-state release rates will be achieved within a few minutes, whereas for systems having low diffusivities ($<10^{-9}$ cm²/s) and thick membranes (>100 μm) steady-state release rates might not be achieved for several hours or days.

FIGURE 47–5. Amount released versus time for a reservoir device that has been stored for a long time (burst effect) and for a device that has been freshly made (time lag) (reprinted from Baker and Lonsdale 1974 with permission).

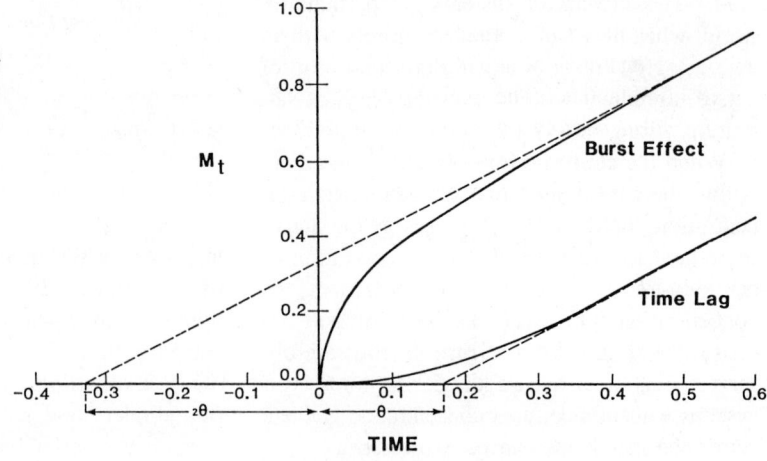

Matrix Systems

Matrix systems consist of active ingredients dissolved or dispersed throughout a polymer. Release of the active ingredient occurs by diffusion of the active ingredient through the polymeric matrix. The release kinetics from these types of systems are dependent on the quantity of active ingredient in the matrix, whether the active ingredient is dissolved or dispersed, the diffusivity of the active ingredient in the polymer, and the geometry of the system. Typically, release rates from matrix systems decrease with time since the path length for diffusion increases with time as the active ingredient near the surface is released first and that in the interior must diffuse farther to be released. The release kinetics from matrix systems are usually not substantially altered by defects in the polymer matrix, contrary to reservoir systems, where defects in the rate-controlling membrane can change the release kinetics dramatically.

Three types of matrix systems are commonly encountered: (1) matrices wherein all of the active ingredient is dissolved in the polymer, (2) matrices wherein relatively small quantities of active ingredient are dispersed in the polymer and are released by diffusion through the polymer, and (3) matrices wherein large quantities of active ingredient are dispersed in the polymer and are released primarily by diffusion through pores that are formed as the active ingredient is released. These three types of matrix systems are shown schematically in Figure 47–6 as films or flat sheets; however, matrix systems can be easily formed in many different geometries. Liquid active ingredients are typically dissolved in polymer matrices, whereas solid active ingredients are typically dispersed in polymer matrices. Expressions describing the release kinetics from matrix systems have been presented by several authors (Higuchi 1961; Baker and Lonsdale 1974; Paul and McSpadden 1976; Cardinal 1984; Baker 1987).

The release kinetics from matrices containing dissolved active ingredient fall into two

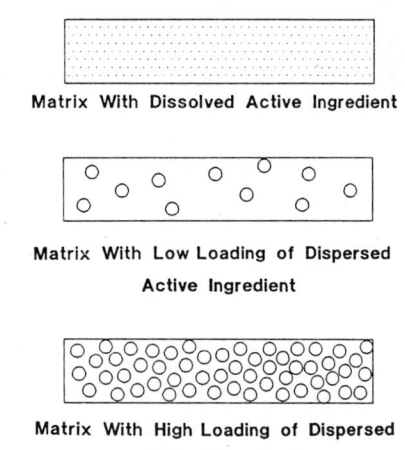

FIGURE 47–6. Illustration of three types of matrix devices.

regimes. Delivery of the first 60% of active ingredient follows $t^{-1/2}$ kinetics, whereas delivery of the last 60% of active ingredient follows first-order kinetics. (The overlapping portion is described adequately by either regime.) Equations describing the release kinetics from films or flat sheets for both regimes are

for $\dfrac{M_t}{M_\infty} < 0.6$,

$$\frac{dM_t}{dt} = 2M_\infty \left(\frac{D}{\pi l^2 t}\right)^{1/2}, \quad (47\text{–}10)$$

for $\dfrac{M_t}{M_\infty} > 0.4$,

$$\frac{dM_t}{dt} = \frac{8DM_\infty}{l^2} \exp\left(-\frac{\pi^2 Dt}{l^2}\right), \quad (47\text{–}11)$$

where M_∞ is the total active ingredient initially loaded into the matrix (Baker and Lonsdale 1974). Similar equations describing release kinetics from cylinders and spheres have also been derived and are presented in Table 47–2 (Baker and Lonsdale 1974; Baker 1987). In general, the initial release rates are faster from spheres and cylinders and decrease faster than do release rates from flat sheets.

Matrices containing active ingredient (less than about 20 vol%) dispersed throughout the polymer exhibit $t^{-1/2}$ kinetics for essentially the entire duration of release, assuming most of the active ingredient is dispersed rather than dissolved. As long as dispersed active ingredient remains in the matrix (i.e., the concentration of active ingredient is greater than the saturation concentration), the release kinetics for flat sheets can be described by

$$\frac{dM_t}{dt} = \frac{A}{2} \left(\frac{Dc_{ms}(2c_\infty - c_{ms})}{t}\right)^{1/2}, \quad (47\text{–}12)$$

where c_{ms} is the saturation concentration of the active ingredient in the polymer, and c_∞ is the concentration of active ingredient (both dis-

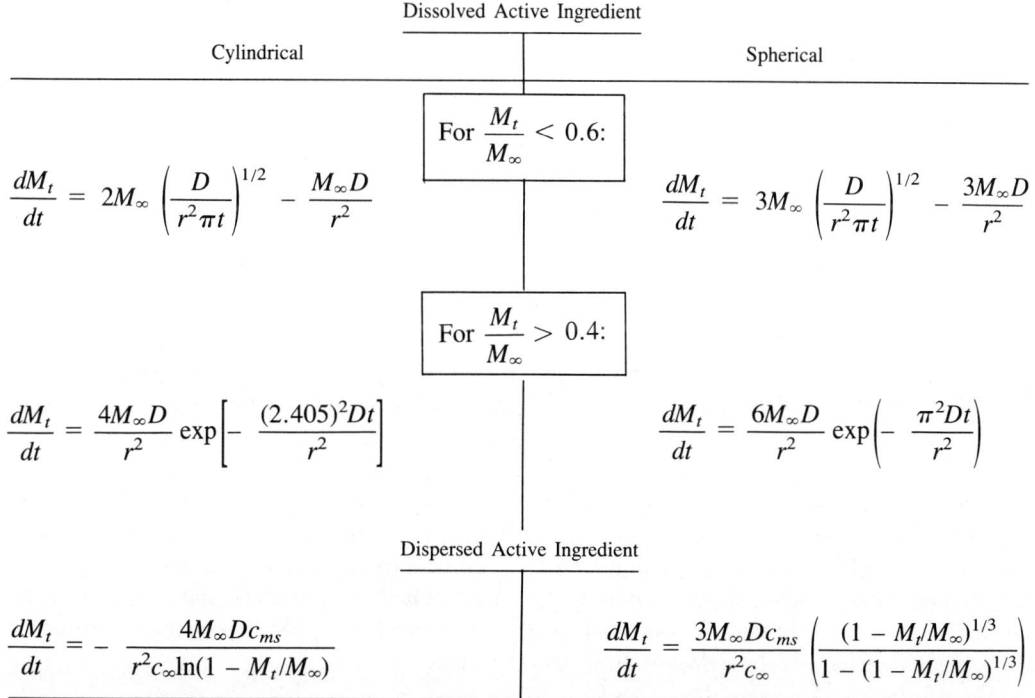

TABLE 47–2. Release-Rate Expressions for Monolithic Matrix Systems: Cylindrical and Spherical Geometries.

Cylindrical	Spherical
Dissolved Active Ingredient	
For $\dfrac{M_t}{M_\infty} < 0.6$:	
$\dfrac{dM_t}{dt} = 2M_\infty \left(\dfrac{D}{r^2 \pi t}\right)^{1/2} - \dfrac{M_\infty D}{r^2}$	$\dfrac{dM_t}{dt} = 3M_\infty \left(\dfrac{D}{r^2 \pi t}\right)^{1/2} - \dfrac{3M_\infty D}{r^2}$
For $\dfrac{M_t}{M_\infty} > 0.4$:	
$\dfrac{dM_t}{dt} = \dfrac{4M_\infty D}{r^2} \exp\left[-\dfrac{(2.405)^2 Dt}{r^2}\right]$	$\dfrac{dM_t}{dt} = \dfrac{6M_\infty D}{r^2} \exp\left(-\dfrac{\pi^2 Dt}{r^2}\right)$
Dispersed Active Ingredient	
$\dfrac{dM_t}{dt} = -\dfrac{4M_\infty Dc_{ms}}{r^2 c_\infty \ln(1 - M_t/M_\infty)}$	$\dfrac{dM_t}{dt} = \dfrac{3M_\infty Dc_{ms}}{r^2 c_\infty}\left(\dfrac{(1 - M_t/M_\infty)^{1/3}}{1 - (1 - M_t/M_\infty)^{1/3}}\right)$

persed and dissolved) initially loaded into the polymer matrix (Baker and Lonsdale 1974; Paul and McSpadden 1976). Once all of the active ingredient contained in the polymer matrix is dissolved, the release kinetics are described by Eqs. (47–10) and (47–11). Active ingredients dispersed in polymer matrices usually have very low solubilities in the polymer, in which case the release of the last portion of active ingredient by first-order kinetics [Eq. (47–11)] is insignificant compared with the total active ingredient released by $t^{-1/2}$ kinetics. Expressions have also been developed for cylindrical and spherical geometries as shown in Table 47–2 (Baker 1987).

Matrices containing large quantities of dispersed active ingredient (greater than about 20 vol%) also follow $t^{-1/2}$ release kinetics; however, the rate of release is higher than that predicted by Eq. (47–12). This is due to a higher effective diffusivity within the matrix caused by fluid-filled holes created by the dissolution and release of active ingredient particles. The higher diffusivity can be approximated by the relationship

$$D = \frac{D_f \epsilon}{\tau}, \quad (47\text{–}13)$$

where D_f is the diffusivity of the active ingredient in the fluid that fills the pores, ϵ is the porosity, and τ is the tortuosity of the matrix (Baker 1987). Substituting this relationship into Eq. (47–12) and assuming that c_∞ is much greater than c_{ms} gives the following equation that approximates the release kinetics from matrices containing high concentrations of active ingredient:

$$\frac{dM_t}{dt} = \frac{A}{2}\left(\frac{2D_f \epsilon c_{ms} c_\infty}{\tau t}\right)^{1/2}. \quad (47\text{–}14)$$

This type of release kinetics would also be observed from matrices that contain a large volume of soluble excipient in addition to the active ingredient. Addition of soluble excipients or *porosigens* to matrix systems allows the active ingredient to be released from the matrix at a higher rate than would be possible with low concentrations of active ingredient alone.

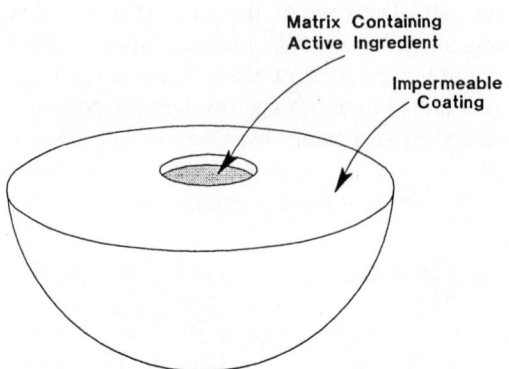

FIGURE 47–7. Illustration of matrix device with special geometry that yields near-zero-order release.

The release kinetics from matrix systems can approach zero-order kinetics if the matrix is produced in special geometries. In particular, matrices that provide larger cross-sectional area as the diffusion path length increases can provide mostly constant release rates (Cobby, Mayersohn, and Walker 1974; Hsieh, Rhine, and Langer 1983; Boettner et al. 1988). An example of such a geometry is shown in Figure 47–7, where the device is covered with an impermeable coating except for a hole in the center of the hemisphere through which active ingredient is released. Thus, the increasing diffusional path length is counteracted by the increasing effective surface area, resulting in nearly constant release rates. Another way in which matrix systems can provide approximately zero-order release is by nonuniform loading of active ingredient, in which the loading is higher in the interior of the device, where the path length is greater. In this case, the diffusional driving force increases with time, counteracting the increasing path length.

Thus, although matrix systems do not inherently provide zero-order release kinetics, they do provide prolonged and controlled release of active ingredients, and in special cases can be made to yield approximate zero-order release.

Osmotic Systems

Osmotic controlled-release systems are based on osmosis—the diffusion of water across a semipermeable membrane from a solution with a low osmotic pressure (high water activity) to a solution with a high osmotic pressure (low water activity). Osmotic systems are usually reservoir-type systems with the distinction that the release of active ingredient is dependent on the diffusion of water through the rate-controlling membrane into the system rather than diffusion of the active ingredient out through the rate-controlling membrane. In these systems, a core with a high osmotic pressure is separated from a solution with a low osmotic pressure by a semipermeable rate-controlling membrane that is much more permeable to water than to the active ingredient or to any osmotic agent that may be present. Due to the osmotic imbibition of water into the core, a hydrostatic pressure is generated within the core that forces or "pumps" a solution or slurry containing the active ingredient out through delivery ports in the rate-controlling membrane. Many different osmotic system designs have been developed, such as tablets, beads, or multiple-compartment devices ranging in size from small particles, capsules, or tablets to devices several inches long or several inches in diameter (Rose and Nelson 1955; Theeuwes 1975; Theeuwes and Yum 1976; Eckenhoff 1981; Zentner, Rork, and Himmelstein 1985; Baker, Smith, and Brooke 1988; Swanson et al. 1987).

Osmotic systems characteristically provide a constant rate of release as long as the osmotic gradient across the rate-controlling membrane is constant and as long as the concentration of active ingredient that remains in the reservoir is constant. These systems can be used to deliver water-soluble or water-insoluble active ingredients; however, more complicated formulations are required to deliver active ingredients that have either very low water solubilities or very high water solubilities. A significant advantage associated with osmotic systems is that the release rate is essentially independent of the pH and of mixing or stirring of the surrounding fluid. In addition, the active ingredient can be completely delivered within hours, days, or weeks depending on the water permeability of the rate-controlling membrane.

Osmotic systems can have single-compartment cores or multiple-compartment cores. Although they are more complicated, multiple-compartment systems provide more formulation flexibility and, consequently, more ability to achieve the desired release kinetics.

In osmotic systems with a single-compartment core, the active ingredient and any soluble fillers in the formulation provide the osmotic driving force and are pumped out of the device, as shown schematically in Figure 47–8. The release kinetics from this type of system can be expressed as

$$\frac{dM_t}{dt} = \frac{AP_w \Delta \pi c_{sa}}{l}, \quad (47\text{--}15)$$

where P_w is the permeability of the rate-controlling membrane to water, $\Delta \pi$ $(= \pi_{core} - \pi_{surr})$ is the difference in osmotic pressure between the core (π_{core}) and the surrounding fluid (π_{surr}), and c_{sa} is the solubility of the active ingredient within the core (Theeuwes et al. 1983). Typically, the permeability, area, and thickness of the rate-controlling membrane remain constant. Thus, as long as the active ingredients and soluble fillers remain saturated within the core, the release kinetics will be zero order. A time lag is usually observed for osmotic systems due to the time required for water to

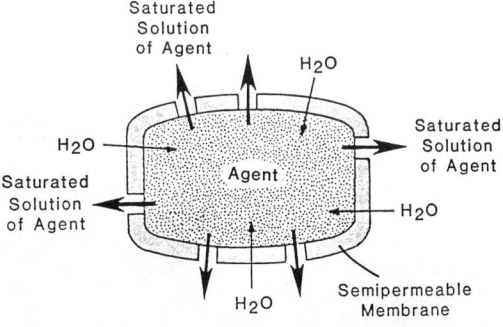

FIGURE 47–8. Illustration of water imbibition into and drug release from an osmotic device.

diffuse across the membrane and produce a hydrostatic pressure inside the reservoir.

The release rate from osmotic systems decreases with time, similar to first-order release kinetics, once all of the active ingredient or soluble filler has been dissolved, causing both $\Delta\pi$ and c_{sa} to decrease with time (Theeuwes 1975). The quantity of active ingredient released at a constant rate (zero-order kinetics) is dependent on the solubility of the active ingredient, the solubility of the soluble filler, and their densities. Assuming that excess soluble filler is loaded in the core (i.e., that some soluble filler remains as a solid once all of the active ingredient has dissolved), one can determine the fraction of total active ingredient delivered at a constant rate from the equation

$$X = 1 - \frac{c_{sa}}{\rho_a} - \frac{c_{sf}}{\rho_f}, \quad (47\text{--}16)$$

where X is the fraction of active ingredient released at a constant rate, c_{sa} is the solubility of the active ingredient within the core, c_{sf} is the solubility of the soluble filler within the core, and ρ_a and ρ_f are the densities of the active ingredient and soluble filler, respectively (Smith et al. 1985). Thus, the more soluble the active ingredient and/or the soluble filler, the less active ingredient will be delivered at a constant rate.

Osmotic systems have also been developed that consist of separate compartments for the osmotic agent and the active ingredient (Theeuwes 1981; Cortese and Theeuwes 1982; Wong et al. 1986). This type of device is illustrated in Figure 47–9. The compartment containing the osmotic agent expands as water is osmotically imbibed, forcing the active ingredient, which can be either a solution or suspension in the adjacent compartment, out of the delivery port. This allows the zero-order delivery of active ingredients that have very low or very high solubilities since the concentration of active ingredient pumped from the device is completely independent of the osmotic gradient across the rate-controlling membrane and is not diluted by imbibed water.

Osmotic systems made with multiple compartments follow the same release kinetics as those described by Eq. (47–15); however, the osmotic-pressure driving force in this case is entirely dependent on the osmotic agent in the expandable compartment, and the concentration of active ingredient in the adjacent compartment remains constant at whatever level it was loaded into the system. Thus, the release kinetics from these osmotic systems can be controlled by adjusting several independent factors: surface area, thickness, membrane permeability, osmotic agent, and the concentration of active ingredient.

Swelling-Controlled Systems

Swelling-controlled systems usually consist of an active ingredient dispersed throughout a swellable polymer, similar to the design of matrix systems. In the unswollen or dry state, the

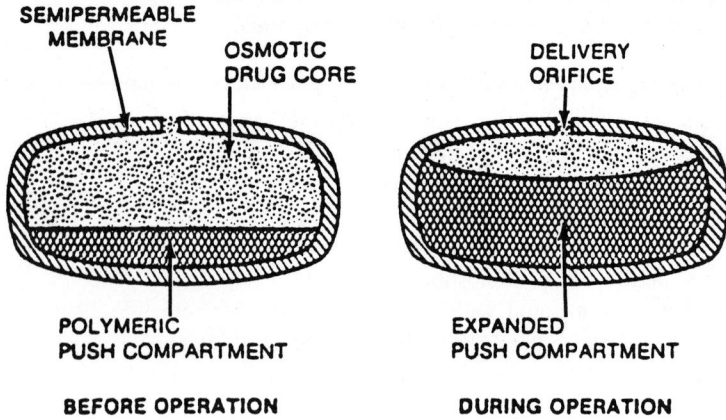

FIGURE 47–9. A two-compartment osmotic device showing an expanding compartment that pushes a drug suspension out at a constant rate (reprinted from Swanson et al. 1987 with permission).

diffusivity of active ingredient is several orders of magnitude lower than its diffusivity in the swollen polymers; consequently, none of the active ingredient is released unless the polymer matrix is swollen (Hopfenberg, Apicella, and Saleeby 1981; Korsemeyer and Peppas 1981; Korsmeyer 1990). Thus, diffusion of the active ingredient out of swelling-controlled systems—and consequently release of the active ingredient—is dependent mostly on the rate and degree of swelling.

Diffusion in systems with polymer swelling due to imbibition of solvent is not well defined by Fick's law since swelling depends on both the diffusion of solvent into the polymer and physical relaxation of the polymer chains. For systems in which the diffusion of the solvent into the polymer matrix is the rate-controlling step, $t^{-1/2}$ kinetics are exhibited for most geometries, as predicted by Fickian diffusion. When the rate of relaxation of the polymer chains is the rate-controlling step in the sorption process, zero-order release kinetics have been observed (Alfrey, Gurnee, and Lloyd 1966; Hopfenberg, Apicella, and Saleeby 1981; Korsmeyer and Peppas 1984). This is called *Case II transport*. Often, polymer swelling and, consequently, the release of active ingredients is dependent on a combination of Fickian diffusion and Case II transport. Thus, release kinetics from swelling-controlled systems vary from $t^{-1/2}$ kinetics to zero-order kinetics (Peppas 1984).

Due to the complexity of the release kinetics from swelling-controlled systems, expressions derived to describe the release kinetics are quite complex and of limited use (Good 1976; Siegel 1990; Korsmeyer 1990). As shown in the following equation, constants describing the particular characteristics of a specific polymer active ingredient solvent system must be utilized (Peppas 1984):

$$\frac{dM_t}{dt} = Ac_\infty nkt^{(n-1)}. \qquad (47\text{--}17)$$

In Eq. (47–17), k and n are constants describing particular swelling-controlled systems. If n is 0.5., then release is governed by Fickian diffusion; if n is 1.0, then release is due to Case II transport. Both k and n must be determined empirically for individual systems; thus, Eq. (47–17) has limited value for predicting release kinetics from new swelling-controlled systems.

A wide range of patterns of release is possible from swelling-controlled systems, many approaching or providing zero-order release kinetics. However, the release profile must be determined experimentally for each new system.

APPLICATIONS

In this section, selected examples of commercial controlled-release devices that demonstrate principles presented in the earlier sections of this chapter are discussed: insecticide-releasing cattle eartags as an example of one type of membrane-coated reservoir, microporous beads containing insect pheromones for agricultural applications as an example of a novel type of membrane-coated reservoir, an oral bolus for cattle as an example of a monolithic matrix, and a cardiovascular drug as an example of an osmotic tablet. Many other illustrative examples are available in the literature (see, for example, Kydonieus 1980; Smith 1985; Langer and Wise 1984).

Cattle Insecticide Eartags

Insecticide-releasing eartags for control of horn flies and face flies on cattle have been available for several years (Miller 1984; Quisenberry and Strohbehn 1984; Tarry 1985; Miller, Oehler, and Kunz 1983). The principal benefit of such eartags is greater weight gain of the cattle due to less annoyance caused by the flies (Quisenberry and Strohbehn 1984). Both eartags with a matrix structure and eartags with a reservoir structure have been commercialized; a comparison of the release kinetics of each is informative.

Most insecticide-releasing eartags utilize a matrix system. It is essentially a monolithic structure, with an insecticide—typically the synthetic pyrethroid, permethrin, usually in combination with a synergist, piperonyl butoxide (PBO)—dissolved and/or dispersed in a

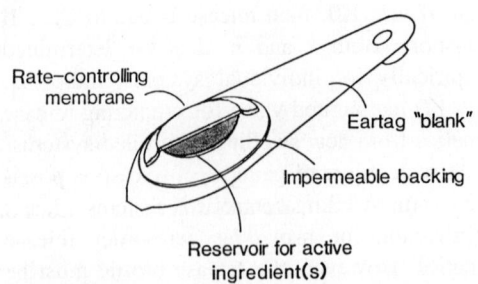

FIGURE 47–10. An insecticide-releasing cattle eartag based on a membrane reservoir design.

polymeric matrix. As explained in the section on transport principles, the release rate from such a device is initially rapid and decreases according to the square root of time. This decreasing release rate results in a limited useful lifetime because the release rate soon falls below the minimum effective level. In addition, ineffective release rates may have also contributed to a buildup of insect resistance to synthetic pyrethroids, which has limited the usefulness of these eartags in recent years.

A membrane-based eartag, commercialized by Fearing Manufacturing (St. Paul, MN) was developed to obtain more constant release rates of the active ingredients (permethrin and PBO) (Speckman 1986). This device consists of a microporous reservoir containing the active ingredients surrounded by a rate-controlling membrane through which the active ingredients must diffuse, all attached to a conventional inert eartag, as illustrated in Figure 47–10.

Since both permethrin and PBO are present in the reservoir and they are miscible, it is not clear what release kinetics to expect. The component that is released more rapidly should decrease in concentration, decreasing its release rate; similarly, the component released more slowly should increase in concentration, thus increasing its release rate. It was found that PBO was released more rapidly than was permethrin (at equal concentrations), and its release rate was lower when permethrin was present, as would be expected by dilution due to the permethrin. It was unexpectedly found that permethrin diffused more rapidly through the membrane in the presence of PBO than when it was present in pure form; this effect was thought to be due to plasticization of the membrane by PBO (Herbig and Smith 1988).

The release kinetics of this membrane-based tag (containing a 5:1 ratio of permethrin to

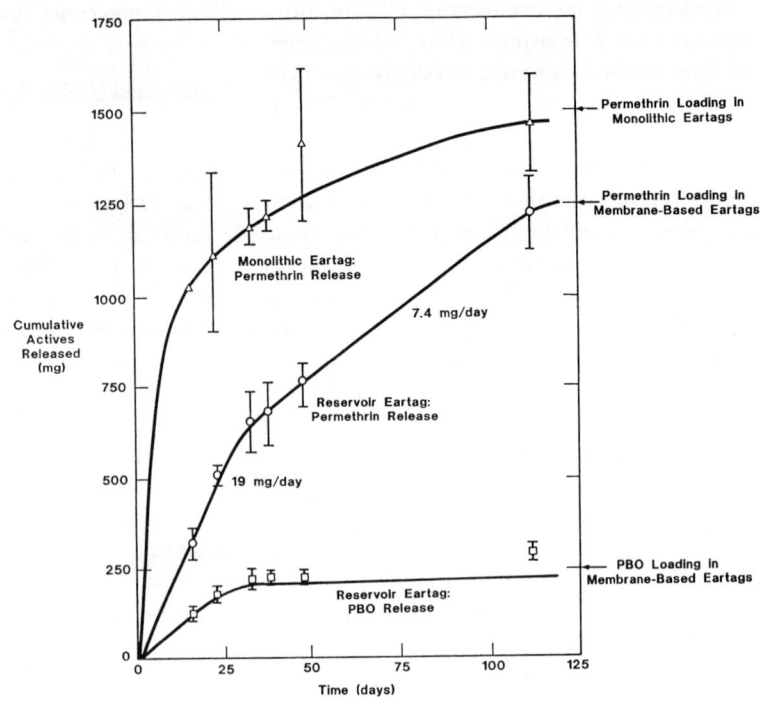

FIGURE 47–11. Insecticide release from eartags of either monolithic or reservoir designs (reprinted from Herbig and Smith 1988 with permission).

PBO) are compared with the release kinetics of a monolithic tag in Figure 47–11. The initially high release rate, decreasing with time, of permethrin from the latter was as expected: the permethrin closest to the surface was released first with a very short diffusion path length; the path length increased with time, slowing the release rate of permethrin from the center of the tag. As can be seen in Figure 47–11, all of the PBO was released from the membrane tag in the first 30 to 40 days. This resulted in two distinct regions of steady-state release of permethrin: a high, constant release rate while PBO was present, and a lower, constant release rate after PBO was depleted. Presumably, the plasticizing effect of the PBO on the membrane resulted in these two regimes (Herbig and Smith 1988).

Microporous Beads Containing Pheromones

Pheromones are chemicals emitted by insects for communication. They have been used by humans for insect control in two ways: by placing them in traps to attract insects to the traps, and by broadcasting them in a field to mask natural pheromone present, thus disrupting communication and preventing mating. They are volatile, expensive liquids, and a controlled-release mechanism is necessary to obtain useful durations of release and economical insect control.

Although microcapsules have been investigated for controlled release of pheromones, it has been difficult to obtain sufficiently long durations of release, primarily due to their small size and thin membranes. For this reason, most controlled-release pheromone formulations have used larger particle sizes. One such formulation is based on asymmetric microporous beads, about 1 mm in diameter. These beads typically have an overall porosity of 70 to 90%, and an asymmetric, or "skinned," structure—i.e., with large pores in the center of each bead and progressively smaller pores toward the surface, as shown in Figure 47–12.

This skinned structure acts as a membrane-coated reservoir since the primary resistance to diffusion is through the relatively dense skin at the surface. The graduated nature of the pores maintains contact between the liquid and the surface membrane, resulting in a constant driving force, and, consequently, a constant release rate.

A commercial product based on these asymmetric beads has been developed by Consep Membranes, Inc. (Bend, OR) for control of the pink bollworm in cotton (Herbig, Smith, and Banfield 1989). Cotton-field temperatures are usually extremely high, and controlled release is essential to achieve effective durations

FIGURE 47–12. Scanning electron photomicrograph of an asymmetric microporous bead.

1 MM

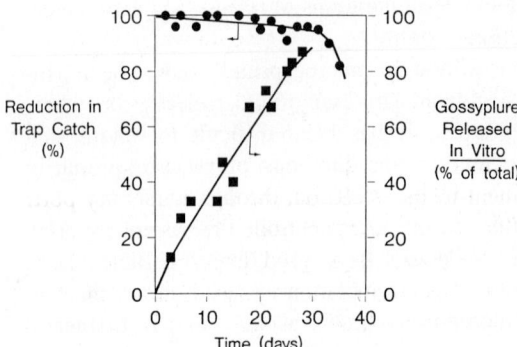

FIGURE 47–13. Laboratory release of the pink bollworm pheromone, gossyplure, and corresponding efficacy in field tests.

of more than a few days. By selecting appropriate polymers and controlling the pore structure, constant release rates of gossyplure for as long as 30 days have been achieved, as shown in Figure 47–13. In this case, about 80% of the pheromone was released at a constant rate, and the laboratory duration of release correlated well with the efficacy in the field. Assuming similar permeabilities, microcapsules with a mean diameter of 50 μm would have exhibited a duration of release of only a couple of days.

Cattle Oral Bolus

Boluses are large drug-releasing devices administered orally to cattle or other ruminants. Because of their size, shape, density, or other characteristics, boluses remain lodged in a portion of the digestive tract, thus serving as a stationary platform from which to release drugs to the animal over a long period of time. A drug-releasing polymeric matrix offers ease of manufacture and tolerance of punctures or other rough handling that might occur during administration or use. However, most matrix devices do not yield constant release rates.

A novel approach to achieving near-constant release rates from matrix devices was developed by Pfizer (Groton, CT) in the form of a perforated laminate (Boettner et al. 1988). In this case, the drug morantel tartrate was incorporated at high loadings in a polymer core matrix. The matrix, in flat-sheet form, was then coated on both sides with a membrane that was essentially impermeable to the drug. Then the entire trilaminate was perforated at regular intervals, as diagrammed in Figure 47–14. The drug is released only from the edges of the laminate and from the holes through the coatings. As the drug is released from the holes, the depleted matrix near the hole becomes porous, offering little resistance to diffusion, and the effective area of release increases around each hole. Although the decreasing rate of release from the uncoated edges of the device counteracts this effect (the rate of release from edges decreases both because the diffusion length increases as release proceeds and because the area of release from the edges decreases as release proceeds), the net result is a near-constant release rate.

By carefully controlling the number and size of holes as well as determining the solubility, diffusivity, and porosity of drug in the matrix, drug release from this device can be modeled. Excellent agreement of theory with ex-

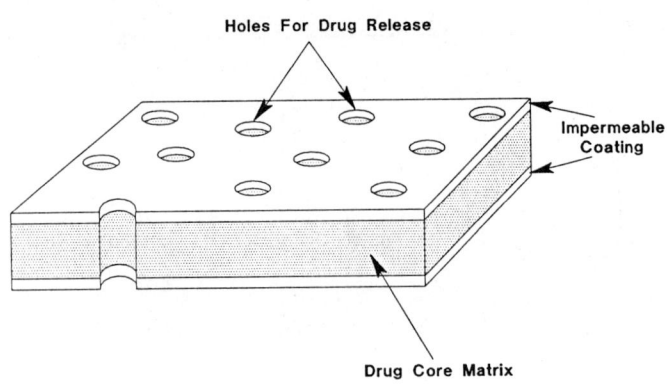

FIGURE 47–14. Illustration of structure of Paratect® bolus for delivery of morantel tartrate to cattle.

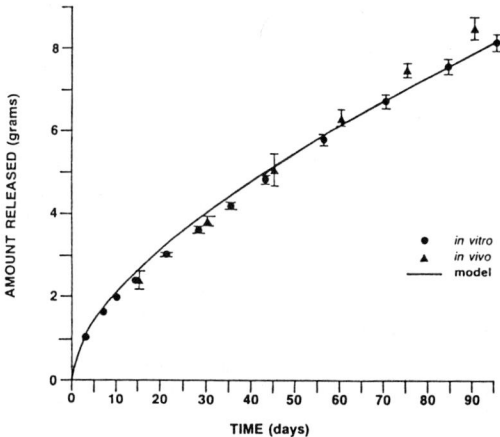

FIGURE 47–15. Release kinetics of morantel tartrate from Paratect® bolus compared with model (reprinted from Boettner et al. 1988 with permission).

perimental release rates was obtained, as shown in Figure 47–15 (Boettner et al. 1988).

Osmotic Nifedipine Tablet

As explained earlier in the section on transport principles, osmotic delivery of compounds with low water solubility is most easily achieved with a multi-compartment osmotic device. In this system, water is imbibed into a compartment with a high osmotic pressure; expansion of this compartment causes a solution or suspension of drug in an adjacent compartment to be forced out through a delivery port. Such a device has been developed by Alza Corporation (Palo Alto, CA) to deliver nifedipine, a calcium-channel blocker that has a very low water solubility (about 10 ppm) (Swanson et al. 1987). In this device, the nifedipine is mixed with polyethyleneglycol in one compartment and an osmotic attractant is in the other compartment; both compartments are coated with a cellulosic membrane. Water diffuses into both compartments, creating a drug suspension in one compartment and an expanding aqueous solution in the other compartment. The osmotic-pressure driving force does not remain constant, however, due to the imbibition of water into the expanding compartment. Zero-order drug-release rates were obtained by controlling the water imbibition rate such that the rate of drug dissolution in the drug compartment was

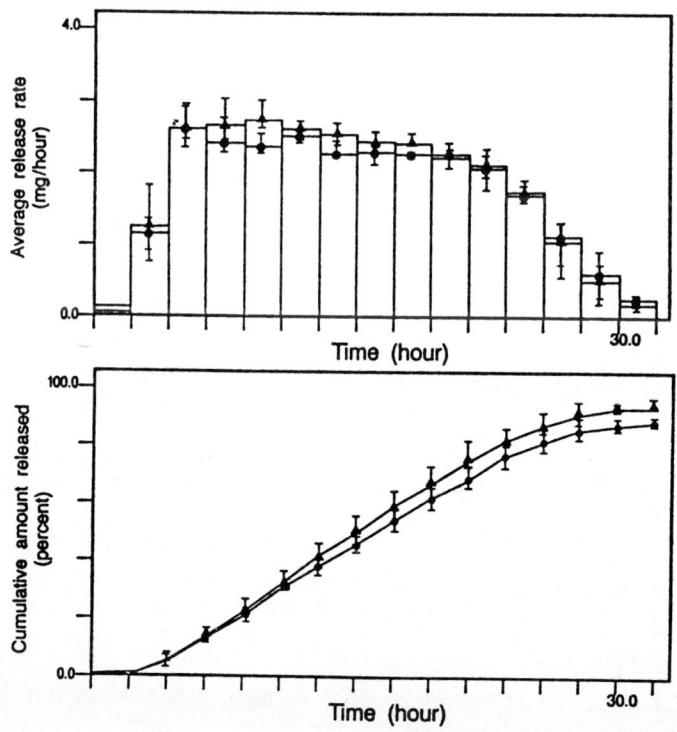

FIGURE 47–16. Release kinetics of nifedipine from two-compartment osmotic tablet (reprinted from Swanson et al. 1987 with permission).

the rate-controlling step. This worked very well indeed, as shown in Figure 47–16. Excellent *in vitro/in vivo* correlation was found as well (Swanson et al. 1987).

NOTATION

General Notation

See the General Notation section at the beginning of this handbook.

Special Notation

c_m	concentration of active ingredient in the membrane, M/L^3
c_{ms}	saturation concentration of active ingredient in the membrane, M/L^3
c_s	saturation concentration in solution, M/L^3
c_{sa}	saturation concentration of active ingredient in the reservoir of an osmotic device, M/L^3
c_{sf}	saturation concentration of soluble filler in the reservoir of an osmotic device, M/L^3
c_∞	total concentration of active ingredient loaded into a polymer matrix, M/L^3
D_f	diffusivity of active ingredient in the fluid contained within a pore of a porous matrix
dc/dx	gradient of concentration of active ingredient, M/L^4
dM_t/dt	release rate of active ingredient from a device, M/t
K	ratio of solubility of active ingredient in the membrane to its solubility in aqueous solution (partition coefficient), dimensionless
k	constant, characteristic of swelling systems, dimensionless
l	membrane thickness or diffusional path length, L
M_t	cumulative mass of active ingredient released to time t, M
M_∞	total mass of active ingredient loaded into matrix, M
n	constant, characteristic of swelling systems, dimensionless
P	permeability of membrane to active ingredient, defined as DK, L^2/t
P_w	permeability of membrane to water, L^2/pt

Greek Letters

ϵ	porosity of porous matrix, dimensionless
π_{core}	osmotic pressure of the solution in the core of an osmotic device, p
π_{surr}	osmotic pressure of the solution surrounding an osmotic device, p
$\Delta\pi$	difference in osmotic pressure between the core of an osmotic device and the solution surrounding it, p
ρ_a	density of active ingredient, M/L^3
ρ_f	density of soluble filler, M/L^3
θ	diffusion time lag, t
τ	tortuosity of porous matrix, dimensionless

REFERENCES

Alfrey, T., E. F. Gurnee, and W. G. Lloyd. 1966. Diffusion in glassy polymers. *J. Polym. Sci.* C(12):249.

Baker, R. W. 1987. *Controlled Release of Biologically Active Agents*, pp. 84–131. New York: John Wiley & Sons.

Baker, R. W., and H. K. Lonsdale. 1974. Controlled release: mechanisms and rates. In *Controlled Release of Biologically Active Agents*, ed. A. C. Tanquary, and R. E. Lacey, pp. 15–72. New York: Plenum Press.

Baker, R. W., K. L. Smith, and J. W. Brooke. 1988. Delivery system. U.S. Patent 4,769,027.

Barrer, R. M. 1968. In *Diffusion In Polymers,* ed. J. Crank and G. S. Park, p. 165. New York: Academic Press.

Barrer, R. M., and G. S. Skirrow. 1984. Transport and equilibrium phenomena in gas-elastomer systems. I. Kinetic phenomena. *J. Polym. Sci.* 3:549.

Barton, A.F.M. 1983. *CRC Handbook of Solubility Parameters and Other Cohesion Parameters.* Boca Raton, FL: CRC Press.

Bixler, H. J., and O. J. Sweeting. 1971. Barrier properties of polymer films. In *The Science and Technology of Polymer Films,* ed. O. J. Sweeting, Vol. 2, pp. 1–130. New York: Wiley Interscience.

Boettner, W. A., A. J. Aguiar, J. R. Cardinal, A. C.

Curtiss, G. R. Ranade, J. A. Richards, and W. F. Sokol. 1988. The morantel sustained release trilaminate: a device for the controlled ruminal delivery of morantel to cattle. *J. Contr. Rel.* 8(1):23–30.

Cardinal, J. R. 1984. Matrix systems. In *Medical Applications of Controlled Release,* ed. R. S. Langer and D. L. Wise, pp. 41–67. Boca Raton, FL: CRC Press.

Cobby, J., M. Mayersohn, and G. C. Walker. 1974. Influence of shape factors on kinetics of drug release from matrix tablets: I theoretical and II experimental. *J. Pharm. Sci.* 63(5):725.

Cortese, R., and F. Theeuwes. 1982. Osmotic device with hydrogel driving member. U.S. Patent 4,327,725.

Crank, J. 1975. *The Mathematics of Diffusion,* 2nd ed. London: Oxford University Press.

Crank, J., and G. S. Park. 1968. In *Diffusion In Polymers,* ed. J. Crank and G. S. Park, p. 1. New York: Academic Press.

Cullis, P. R., L. D. Mayer, M. B. Bally, T. D. Madden, and M. J. Hope. 1989. Generating and loading of liposmal systems for drug-delivery applications. *Adv. Drug. Deliv. Rev.* 3(3):267–282.

Cussler, E. L. 1984. *Diffusion: Mass Transfer in Fluid Systems.* New York: Cambridge University Press.

Eckenhoff, J. B. 1981. Osmotically driven pumps for rate-controlled delivery of solutions and viscous suspensions. *AIChE Symp. Ser.* 77 (206):1–9.

Good, W. R. 1976. Diffusion of water-soluble drugs from initially dry hydrogels. In *Polymeric Delivery Systems,* ed. R. Kostelnik, pp. 139–156. New York: Gordon and Breach.

Gregoriadis, G. ed. 1984. *Liposome Technology,* Vol. I, II, and III. Boca Raton, FL: CRC Press.

Grulke, E. A. 1989. Solubility parameter values. In *Polymer Handbook,* 3rd ed., ed. J. Brandrup and E. H. Immergut, pp. VII 519–559. New York: John Wiley and Sons.

Grun, F. 1947. Measurements of diffusion in rubber. *Experientia* 3:490–492.

Hansen, C. M., and A. Beerbower. 1971. Solubility parameters. In *Encyclopedia of Chemical Technology,* 2nd ed. pp. 889–910. New York: John Wiley & Sons.

Harris, F. W. 1984. Controlled release from polymers containing pendent bioactive substituents. In *Medical Applications of Controlled Release,* ed. R. S. Langer and D. L. Wise, Vol. 1, pp. 103–128. Boca Raton, FL: CRC Press.

Heller, J. 1980. Controlled release of biologically active compounds from bioerodible polymers. *Biomater.* 1:51–57.

Heller, J. 1984a. Biodegradable polymers in controlled drug delivery. *CRC Crit. Rev.* 1:39–90.

Heller, J. 1984b. Bioerodible systems. In *Medical Applications of Controlled Release,* ed. R. S. Langer, and D. L. Wise, Vol. 1, pp. 69–101. Boca Raton, FL: CRC Press, Inc.

Hellstrom, K. E., I. Hellstrom, and G. E. Goodman. 1987. Antibodies for drug delivery. In *Controlled Drug Delivery,* ed. J. R. Robinson and V.H.L. Lee, pp. 623–653. New York: Marcel Dekker.

Herbig, S. M., and K. L. Smith. 1988. A membrane-based cattle insecticide eartag. *J. Contr. Rel.* 8:63–72.

Herbig, S. M., K. L. Smith, and M. G. Banfield. 1989. Porous beads for controlled release of agrichemicals. In *Proc. 16th International Symposium in Controlled Release of Bioactive Materials,* ed. R. Pearlman and J. A. Miller, pp. 261–262. Lincolnshire, IL: Controlled Release Society.

Higuchi, T. 1961. Rate of release of medicaments from ointment bases containing drugs in suspension. *J. Pharm. Sci.* 50:874–875.

Hildebrand, J. H., J. M. Prausnitz, and R. L. Scott. 1970. *Regular and Related Solutions.* New York: Van Nostrand-Reinhold.

Hildebrand, J. H., and R. L. Scott. 1964. *The Solubility of Nonelectrolytes,* 3rd ed. New York: Dover.

Hopfenberg, H. B., A. Apicella, and D. E. Saleeby. 1981. Factors affecting water sorption in and solute release from glassy ethylene-vinyl alcohol copolymers. *J. Membrane Sci.* 8:273–282.

Hsieh, D.S.T., and R. Langer. 1983. Zero-order drug delivery systems with magnetic control. In *Controlled Release Delivery Systems,* ed. T. J. Roseman and S. Z. Mansdorf, pp. 131–132. New York: Marcel Dekker.

Hsieh, D.S.T., W. D. Rhine, and R. Langer. 1983. Zero-order controlled-release polymer matrices for micro- and macromolecules. *J. Pharm. Sci.* 73(1):17–22.

Kesting, R. E. 1985. *Synthetic Polymeric Membranes,* pp. 27–30, 113–120. New York: John Wiley & Sons.

Knight, C. G. (ed.). 1981. *Liposomes: From Physical Structure to Therapeutic Applications.* Amsterdam: Elsevier/North Holland Biomedical Press.

Korsmeyer, R. W. 1990. Diffusion controlled systems. II. Hydrogels. In *Polymers for Controlled Drug Delivery*, ed. P. J. Tarcha, pp. 15–37. Boca Raton, FL: CRC Press.

Korsmeyer, R. W., and N. A. Peppas. 1981. Effect of the morphology of hydrophilic polymeric matrices on the diffusion and release of water soluble drugs. *J. Membr. Sci.* 9:211–227.

Korsmeyer, R. W. and N. A. Peppas. 1984. Solute and penetrant diffusion in swellable polymers. III. Drug release from glassy poly (HEMA-co-NVP) copolymers. *J. Contr. Rel.* 1:89–98.

Kost, J., and R. Langer. 1990. Magnetically and ultrasonically modulated drug delivery systems. In *Pulsed and Self-Regulated Drug Delivery*, ed. J. Kost, pp. 3–15. Boca Raton, FL: CRC Press, Inc.

Kydonieus, A. F. 1980. *Controlled Release Technologies: Methods, Theory and Applications, Vols. I and II*. Boca Raton, FL: CRC Press.

Langer, R. S., and D. L. Wise. 1984. *Medical Applications of Controlled Release*. Boca Raton, FL: CRC Press.

Leong, K. W., and R. Langer. 1987. Polymeric controlled drug delivery. *Adv. Drug Deliv. Rev.* 1:199–233.

Michaels, A. A., and H. J. Bixler. 1968. Membrane permeation: theory and practice. In *Prog. Sep. Purif*, ed. E. S. Perry, Vol. I, pp. 143–186. New York: Wiley Interscience.

Michaels, A. S., P. S. L. Wong, R. Prather, and R. M. Gale. 1975. A thermodynamic method of predicting the transport of steroids in polymer matrices. *AIChE J.* 21:1073–1080.

Miller, B. 1984. Fly tags paint a fascinating picture. *Success. Farming* March:B1–B13.

Miller, J. A., D. D. Oehler, and S. E. Kunz. 1983. Release of pyrethroids from insecticidal ear tags. *J. Econ. Entom.* 76:1335–1340.

Paul, D. R., and S. K. McSpadden. 1976. Diffusional release of a solute from a polymer matrix. *J. Membr. Sci.* 1:33–48.

Peppas, N. A. 1984. Release of bioactive agents from swellable polymers: theory and experiments. In *Recent Advances in Drug Delivery Systems*, ed. J. M. Anderson and S. W. Kim, pp. 279–289. New York: Plenum Press.

Pitt, C. G., and A. Schindler. 1980. The design of controlled drug delivery systems based on biodegradable polymers. In *Progress in Contraceptive Delivery Systems*, ed. E.S.E. Hafez and W. A. van Os, pp. 17–46. Lancaster, UK: M.P.T. Press.

Pitt, C. G., and A. Schindler. 1983. Biodegradation of polymers. In *Controlled Drug Delivery*, ed. S. D. Bruck, pp. 53–80. Boca Raton, FL: CRC Press.

Quisenberry, S. S., and D. R. Strohbehn. 1984. Hornfly control on beef cows with permethrin-impregnated ear tags and effect on subsequent calf weight gains. *J. Econ. Entom.* 77:422–424.

Rose, S., and J. F. Nelson. 1955. A continuous long-term injector. *Austral. J. Exper. Biol.* 33:415.

Sefton, M. V. 1984. Implantable pumps. In *Medical Applications of Controlled Release*, ed. R. S. Langer and D. L. Wise, Vol. 1, pp. 129–158. Boca Raton, FL: CRC Press.

Siegel, R. A. 1990. pH-sensitive gels: swelling equilibria, kinetics, and applications for drug delivery. In *Pulsed and Self-Regulated Drug Delivery*, ed. J. Kost, pp. 129–157. Boca Raton, FL: CRC Press.

Smith, K. L. 1985. Membrane systems: practical applications. In *Methods—Enzymology*, ed. K. J. Widder and R. Green, pp. 504–520. Orlando, FL: Academic Press.

Smith, K. L., and H. K. Lonsdale. 1985. Membrane systems: theoretical aspects. In *Methods in Enzymology*, ed. K. J. Widder and R. Green, pp. 495–504. Orlando, FL: Academic Press.

Smith, K. L., M. E. Schimpf, and K. E. Thompson. 1990. Bioerodible polymers for delivery of macromolecules. *Adv. Drug Deliv. Rev.* 4:343–357.

Smith, K. L., A. C. Shah, K. S. Albert, G. J. Szpunar, and E. A. Pratt. 1985. Controlled drug release with a novel oral osmotic delivery system. In *Proc. 12th International Symposium on Controlled Release of Bioactive Materials*, ed. N. A. Peppas and R. J. Haluska, pp. 231–232. Lincolnshire, IL: Controlled Release Society.

Speckman, C. A. 1986. Pest control in animals. U.S. Patent 4,562,794.

Stannett, V. 1968. Simple gases. In *Diffusion in Polymers*, ed. J. Crank and G. S. Park, pp. 41–73. New York: Academic Press.

Stannett, V., M. Szwarc, R. L. Bhargava, J. A. Meyer, A. W. Myers, and C. E. Rogers. 1962. Simple gases. In *Permeability of Plastic Films and Coated Paper to Gases and Vapors*, Monograph Series No. 23. New York: Technical Association of the Pulp and Paper Industry.

Swanson, D. R., B. L. Burday, P. S. L. Wong, and F. Theeuwes. 1987. Nifedipine gastrointestinal therapeutic system. *Am. J. Med.* 83:3–10.

Tarry, D. W. 1985. Cattle fly control using controlled-release insecticides. *Vet. Parasit.* 18:229–234.

Theeuwes, F. 1975. Elementary osmotic pump. *J. Pharm. Sci.* 64(12):1987–1991.

Theeuwes, F. 1981. Novel drug delivery systems. In *Drug Absorption,* ed. L. F. Prescott and W. S. Nimmo, p. 157. Balgowalah, Australia: ADIS Press.

Theeuwes, F., D. Swanson, P. Wong, P. Bonsen, V. Place, K. Heimlich, and K. C. Kwan. 1983. Elementary osmotic pump for indomethacin. *J. Pharm. Sci.* 72(3):253–258.

Theeuwes, F. and S. I. Yum. 1976. Principles of the design and operation of generic osmotic pumps for the delivery of semisolid or liquid drug formulations. *Ann. Biomed. Eng.* 4:343–353.

Uhlig, E. L. P., W. F. Graydon, and W. Zingg. 1983. The electro-osmotic actuation of implantable insulin micropumps. *J. Biomed. Mater. Res.* 17:931.

Vitetta, E. S., R. J. Fulton, and J. W. Uhr. 1986. Immunotoxins: the development of new strategies for treating B cell tumors. In *Site-Specific Drug Delivery,* ed. E. Tomlinson and S. S. Davis, pp. 69–80. New York: John Wiley & Sons.

Watanabe, A., and T. Hayashi. 1976. Applications of microcapsules developed at Fuji. In *Microencapsulation,* ed. J. R. Nixon, pp. 31–38. New York: Marcel Dekker.

Welty, J. R., C. E. Wicks, and R. E. Wilson. 1984. *Fundamentals of Momentum, Heat, and Mass Transfer,* 3rd ed. pp. 471–500. New York: John Wiley & Sons.

Widder, K. J., and A. E. Senyei. 1983. Magnetic microspheres: a vehicle for selective targeting of drugs. *Pharm. Ther.* 20:377.

Wong, P., B. Barclay, J. Deters, and F. Theeuwes. 1986. Osmotic device with dual thermodynamic activity. U.S. Patent 4,612,008.

Wright, S., and L. Huang. 1989. Antibody-directed liposomes as drug-delivery vehicles. *Adv. Drug Deliv. Rev.* 3:343–389.

Zentner, G. M., G. S. Rork, and K. J. Himmelstein. 1985. The controlled-porosity osmotic pump. *J. Contr. Rel.* 1:269–282.

Index

Acetone, 150
Acetylene, 857
N-Acetyl-L-tyrosine ethyl ester (ATEE), 816–828
Acid and base production, 260, 261
Acid gas removal, 88–92, 770, 792, 794, 802–804, 851–855, 898, 899. See also Carbon dioxide separation
Activated carbon filter, 318–320
Activated sludge, 335
Activation energy, 37
Activity, 117
Advancing front model, 630–636, 644–647, 679–682
Affinity membranes, 447, 450
Air flotation, 335
Air gap membrane distillation, 900, 901, 903, 904
Air separation, 85–87, 100, 843–851
Air stripping, 12, 318, 320, 896
Albumin, 165
Alcohol reduction of beverages, 148, 210, 211
Ammonia/hydrogen separation, 853
Ammonia/nitrogen separation, 853
Ammonia removal from wastewater, 598, 672, 712, 899
Analytical applications of microfiltration, 549–565
Analytical solutions in reverse osmosis, 301–303
Angiotensins I and III, 170
Angular velocity, 467
Anion-exchange membranes, 10, 210, 219, 233–238, 242, 243
Antibiotic, 448, 449, 516, 535, 536, 581, 583, 584, 586
Anticoagulant, 209, 213
Antiscalants, 308, 318, 319, 437
Apple juice concentration, 442
Aqueous-filled pores, 170, 186, 728, 729, 733, 734, 737, 738, 747, 748
Aromatics extraction, 729, 757

Arrhenius activation energy, 37
Arterial pressure, 163
Artificial gill, 897
Artificial kidney, 163, 164, 190
Aseptic processing, 516, 532, 539, 540
Asphaltene, 450
ASTM standards for microfiltration, 513, 560
Asymmetric bead, 929
Asymmetric membranes. See Membranes, asymmetric
Azeotrope separation, 135, 142, 155, 905, 908

Back-diffusion, 464, 484–487
Back extraction, 743, 744, 747, 748, 758
Back-filtration, 193, 208
Bacteria adhesion, 305
Bacterial grow-through, 522, 524
Bacterial integrity tests for membranes, 522, 523
Batch concentration, 441, 442, 480–482, 576, 583, 586
Beer, 148, 206, 210, 537, 538
Benzene-cyclohexane separation, 905
Beverage processing, 210, 211, 536
Biochemical applications, 598, 714–716, 816, 825
Bioerodible polymers for controlled release, 916, 917
Biogas processing, 91, 802–804
Biomedical applications, 516, 549, 581, 587, 598. See also Controlled release
Biopharmaceutical processing, 215
Biot number, 635
Bipolar membranes, 10, 238, 239, 260, 261
Blood oxygenation, 10, 886, 890, 897
Blotting, 560–563
Boltzmann's constant, 400
Bolus, 930
Born repulsion, 471
Boundary layers, 172–174, 278, 401, 917

Boundary layer thickness, 401
Bovine serum albumin (BSA), 403, 404, 436, 438
Brackish water desalination, 255, 256, 266, 295–297, 313–325, 356–361, 363–366, 369–373
Breakthrough pressure, 728, 745, 746, 795, 796, 886, 895
Brownian diffusivity, 483–485, 497
Bubble point, methanol, 517, 521, 522
Bubble point pressure, 426, 509, 520–522, 819
Bubble point, water, 517, 520
Bubble-pressure breakthrough, 426, 509, 513, 520, 521, 886, 895
Burst effect, 921, 922

Cake compressibility, 463, 464
Cake drying, 466
Cake formation, 462, 466, 490
Cake layer resistance
    crossflow microfiltration, 491, 492, 495, 496
    deadend microfiltration, 463
    ultrafiltration, 396, 401
Cake permeability, 463
Cake washing, 466
Capital costs, 99, 137–139, 147, 213, 251, 356, 357, 360, 451, 452, 540, 541, 578, 586, 587, 589, 593, 718, 719, 721
Capital recovery costs, 366
Carbon adsorption, 307, 309
Carbon dioxide separation, 88–92, 98–100, 770, 802, 803, 851–855, 898
Carbon filtration, 306
Carbon monoxide separation, 804, 857, 858
Carman-Kozeny equation, 463
Carrier saturation, 834
Cascade/staged operation
    dialyzer, 203, 204
    emulsion liquid membranes, 672, 673, 675–677, 679–682, 706
    gas permeation, 72, 73, 90, 91
    membrane-based gas absorption/stripping, 897
    membrane-based solvent extraction, 755, 756
    microfiltration, 547, 581, 583
    perstraction, 905, 906
    pervaporation, 135, 136
    reverse osmosis, 320–322
    ultrafiltration, 443
Case II transport, 927
Catalytic membrane reactor-based processes, 164. See also Membrane reactors
Catalytic/reactive membranes, 809, 811, 812, 817
Cation-exchange membranes, 10, 210, 219, 226, 232–237, 242, 243
Cationic nylon membrane, 562
Cell blotting, 560–564
Cell growth on microporous membranes, 563–565
Cell harvesting, 581
Cellophane, 106, 107, 191, 233

Cellulose, 106, 163, 165, 186–190, 194, 208, 214, 408, 414, 420, 438, 746
Cellulose acetate (CA), 45, 82, 83, 88, 89, 98, 99, 107, 113, 121, 134, 155, 186, 187, 189, 282, 283, 295–298, 306, 326, 327, 329, 330, 332, 335, 336, 340, 341, 408, 411, 414, 415, 418–420, 422–424, 510, 511, 524, 736, 746, 844, 853, 854
Cellulose diacetate, 283
Cellulose ester, 128, 194, 284, 415, 817, 854
Cellulose film, 163
Cellulose nitrate, 420
Cellulose steeping liquors, 163
Cellulose triacetate, 134, 283, 284, 324, 419, 425
Cellulosic membranes, 181, 186, 187, 189, 194, 931
Central gas filters, 542
Centrifugation, 437, 580, 582
Centripetal force, 440
Cephalosporin C, 448, 449, 581
Ceramic membranes. See Membranes, ceramic
Ceramic membrane modules, 436, 437, 510, 511, 572–578, 584–587, 744
Cerini dialyzer, 188
Channel spacer, 293, 294, 301, 303, 433, 435
Channel thickness
    brine, 303
    permeate, 303
Channeling, 176, 199, 751, 753, 755, 891, 893, 896
Charge capacity of membrane, 275
Cheese whey, 257, 258, 346, 446, 447, 452
Chemical compatibility of membranes
    dialysis, 187, 192
    gas permeation, 57–59
    membrane-based solvent extraction, 748, 749
    reverse osmosis, 305, 306, 319, 320
    ultrafiltration, 424
Chemical complexation, 450. See also Facilitated transport
Chemical filters, 544, 545
Chemical oxidation, 307, 309
Chemical potential
    dialysis, 165, 168
    gas permeation, 23
    gradient, 4, 117, 168, 266
    membrane-based solvent extraction, 735
    pervaporation, 117, 133
Chitisan, 113
Chlorine, 318, 319, 330
Chlorine-alkaline electrolysis, 259, 260
Chloroform, 150
Chromium extraction, 598, 622, 623, 626, 627, 628, 669
Chymotrypsin-catalyzed hydrolysis, 816, 825
Circular dichroism spectra, 525
Clarification, 507, 515, 536, 581, 582, 587, 589–591
Clausius-Clapeyron equation, 902
Clean-in-place (CIP), 129, 440, 441, 532, 585
Clean-out-of-place (COP), 532
Clearance, 178, 187, 194, 195, 207
Coagulation, 307
Coagulation cascade, 208

Coagulation process in membrane fabrication, 188
Coal liquefaction wastewater, 327
Cobalt extraction, 598, 627, 628, 669, 868, 869, 872–876
Cocurrent flow (permeators/contactors), 61, 179, 180, 195, 196, 672, 675, 677–679, 682, 683, 774, 790, 801, 826, 827
Co-ions, 210, 226, 227, 248, 274, 275
Coliform, 553
Colloidal substance in wine, 538
Column extractor, 672, 673, 678, 681, 683
Composite membranes. *See* Membranes, composite/thin-film-composite (TFC)
Concavalin A, 450
Concentration boundary layer, 163, 173, 174, 300, 301
Concentration factor, 589
Concentration gradient, 8, 9, 12, 13, 25, 168, 277, 659, 663, 837, 870
Concentration polarization
    control strategies in ultrafiltration, 437
    dialysis, 181
    electrodialysis, 246, 249
    osmotic pressure model, 403
    pervaporation, 125, 126
    resistances in series model, 409
    reverse osmosis, 277, 300, 301
    ultrafiltration, 395, 400, 401
Concentric cylinder devices, 439
Conductive heat flux, 902
Contact angle, 427, 428, 436, 746, 819, 895, 903
Contacting/contactors
    gas-liquid, 885
    liquid-liquid, 727
Continuous membrane column, gas permeation, 73, 74
Continuum hydrodynamic theory, 167, 169
Controlled release, 915
    active ingredient, 915
    applications, 927
    asymmetric membrane, 929, 930
    burst effect, 921, 922
    core, 916, 925, 926
    dense (nonporous) membrane, 916–918
    diffusivity, 917–919
    dispersed system, 922, 923
    dissolved system, 922, 923
    driving forces, 916, 925
    kinetics, 915, 916, 919–924
    matrix systems, 916, 922–924, 930
    membrane-coated reservoir, 920
    microporous membrane, 916, 929
    nondiffusion-based technologies, 916, 917
    osmotic pump, 916, 925, 926, 931
    partition coefficient, 917, 918
    permeability, 917
    pheromones, 929, 930
    rate-controlling membrane, 916, 920
    reservoir systems, 916, 918, 920–922
    swelling-controlled systems, 916, 926, 927
    technologies, 916, 917
    time lag, 921, 922
    toxic effects, 915, 916
Convective flux, 181, 182
Convective transport through membrane, 165, 167, 181, 182
Cooling-tower blowdown, 335
Copper extraction/recovery, 598, 625, 627, 641, 670, 800, 801
Copper leaching solution, 163
Corrosion in reverse osmosis, 323
Cost estimates (economics)
    air separation, 100
    carbon dioxide separation, 98–100
    crossflow microfiltration, 578, 586, 592, 593
    deadend microfiltration, 533, 534, 540, 541, 544–546, 549, 558, 559
    dialysis, 212–215
    electrodialysis, 246, 249–253
    emulsion liquid membranes, 718–723
    gas permeation, 95–100
    hydrogen separation, 96–98
    pervaporation, 132, 135, 137–139, 146, 147
    reverse osmosis, 294, 295, 297, 298, 355–388
    ultrafiltration, 451, 452
Countercurrent flow (permeators/contactors), 61, 65, 68, 178–180, 183, 189, 195–197, 200, 672, 673, 676–678, 680–683, 730, 739, 774, 780, 781, 792, 793, 821, 825, 826, 828, 885, 893, 904
Counter-ions, 210, 226, 274
Coupling coefficient/parameter, 119–121, 267, 271
Creatinine, 170
Critical surface tension, 903
Crossflow (permeators/contactors), 61, 62, 65–68, 179, 459, 480, 571, 575, 750, 751
Crossflow microfiltration, 480, 571
    application window, 578, 579
    asymmetric filter, 575, 577
    batch concentration, 481, 482, 576, 577, 583, 586
    biotechnology process applications, 578, 581
    ceramic microfilter, 572–578, 584, 585
    clarification, 581, 582, 587, 589–591
    cleaning, 577, 585, 588
    concentration polarization, 483–485, 576
    costs/economics, 578, 586, 593
    diafiltration, 481, 482, 582–584
    feed-and-bleed, 481, 483, 586
    flux, 584
    flux decline, 483, 495, 499, 576, 577
    general design standards, 592
    harvesting of mammalian cells, 587, 590
    hollow fiber membranes, 572
    materials of construction, 592
    membranes, 576
    modules, 572, 574, 575, 577, 585, 587, 590, 591
    mycelia concentration, 581
    operational configurations, 481
    particle size, 497, 501, 502, 571, 574

Crossflow microfiltration (*continued*)
   perfusion, 587
   plasmapheresis, 484, 500, 581
   process design, 582, 584, 589
   process parameters, 584, 590
   rejection, 583
   service life, 577, 578
   solid-phase recovery applications, 581
   system design, 575, 585, 590, 592
   system maintenance, 585, 590
   theory, 480
   typical crossflow system, 572, 574, 575
   yeast suspension/concentration, 500, 581
Cuprammonium process, 186
Cuprophan®, 121, 170, 189, 734, 736, 746, 749, 777

Dairy applications
   cheese whey, 257, 258, 446, 447
   egg whites, 447
   milk concentration, 447, 448
   reverse osmosis, 345
   ultrafiltration, 446
   whey protein concentrate, 447
Damköhler number, 836
Darcy's constant, 198
Darcy's law, 462, 463, 495, 529
DC18C6 (dicyclohexano-18-crown-6), 663, 670, 671
Deadend microfiltration, 461, 506
   analytical applications, 549–565
   application categories, 507
   beverage applications, 536–541
   biological process systems, 534–537
   blotting, 560–563
   clarification, 507, 515, 536
   costs/economics, 533, 540, 541, 544–546, 549, 558, 559
   design considerations, 515, 529, 531
   membrane modules, 507, 508, 512, 520, 526, 546, 548, 550, 551
   membrane pore size, 516, 517, 520–522, 527, 534–536, 538, 542, 543, 545, 549, 550, 554, 556, 557, 565
   membrane selection, 520, 522, 524, 550, 551, 554
   membranes, 507–513, 516, 517, 520–525, 527, 532, 534–536, 538, 542–544, 546, 548, 550, 551, 554, 556, 564, 565
   microbiological assay, 552–557
   particle size, 506
   permeability, 512, 551
   pharmaceutical applications, 515, 518
   process parameters, 515
   semiconductor manufacturing, 541
   sterilization, 515–517, 527, 532, 533, 535
   theory, 461, 529–531
   throughput, 516
Deasphalting, 447
Dechlorination, 306, 319

Dehydration
   ethanol, 107, 133–140
   isopropyl alcohol, 144–146
   organics, 140–148
Deionized (DI) water filters, 542, 547–549
Demulsification, 689–692
Dense (nonporous) membrane, 25, 269, 270, 907, 916
Depth filtration, 458, 461, 467, 471
Depth filtration mechanisms, 467–477
Desalination
   brackish water, 255, 256, 266, 295–297, 313–325, 356–361, 363–366, 369–373
   design of reverse osmosis plant, 314, 315
   pretreatment methods, 318
   reverse osmosis plant locations, 314
   seawater, 266, 295–297, 308, 309, 313–325, 356–361, 363–366, 369–373
Design (process/system)
   crossflow microfiltration, 575, 582, 584, 585, 589, 590, 592
   deadend microfiltration, 515, 516, 531, 538
   dialysis, 186–204
   electrodialysis, 246–253
   emulsion liquid membranes, 656–692
   gas permeation, 54–74
   hollow-fiber contained liquid membrane, 796–799
   membrane-based gas absorption/stripping, 895–897
   membrane-based solvent extraction, 752–756
   pervaporation, 123–130
   reverse osmosis, 281–309
   ultrafiltration, 432–444
Device clearance. *See* Clearance
Dewatering applications, 346
Dewaxing, 447
Dextrans, 411, 413–415, 423, 427, 428, 436
Diafiltration, 209, 396, 443, 447, 481, 482, 582, 583
   discontinuous, 444
   sequential, 444
Dialysance, 178, 182, 183, 203
Dialysate, 164, 165, 172, 173
Dialysate-side resistance, 173, 175, 200, 201
Dialysis, 6–8, 163, 482
   applications, 206–211
   batch operation, 202
   continuous operation, 202
   cost estimates, 212–215
   definitions, 163–166
   design, 186–204
   dialyzers, 188–194, 198, 207–209
   Donnan dialysis, 7, 209, 210
   driving force, 165
   end-stage renal disease, 193, 207
   hemodiafiltration, 209
   hemodialyzers, 190, 207
   hemofiltration, 209
   high-efficiency membrane, 187
   industrial dialyzers, 188
   mass transfer, membrane, 167
   mass transfer, overall, 172

Dialysis (continued)
   materials selection, 202
   membrane fabrication, 187, 188
   membrane function and structure, 165
   membrane material, 186, 214
   membrane module, 188
   membrane selection, 186
   membrane suppliers, 214
   process, 163, 206
   solute permeability, 169–172, 181, 182, 188, 189, 198–200
   solvent permeability, 170, 171
   theory, 167–183
   See also Membranes, dialysis
Dichloroethylene, 143
Dicyclohexano-18-crown-6 (DC18C6), 663, 670, 671
Diethanolamine (DEA), 99, 100, 771, 795, 802–804, 854
Di(2-ethylhexyl)dithiophosphoric acid (DTPA), 657, 658, 670, 671, 703, 704, 710–713
Di(2-ethylhexyl)phosphoric acid (D2EHPA), 625, 657–659, 665, 669–671, 703, 708, 874, 875, 879
Di(2-ethylhexyl)thiophosphoric acid (MTPA), 703, 704, 712, 714
Diffusive capture, 468–470, 472
Diffusion
   Knudsen, 20, 890, 901
   surface, 20
Diffusion coefficient (diffusivity)
   Brownian diffusivity, 483–485, 497
   carbon dioxide in natural rubber, 38
   carbon dioxide in poly(ethylene terephthalate) (PET), 33, 39
   carbon dioxide in silicone rubber, 35
   concentration and pressure dependence, 32–35
   controlled release, 916–918, 924
   dialysis, 168, 170
   effective diffusivity, 170, 632, 634, 641, 679, 685, 734, 924
   gas diffusivity-free volume correlation for polycarbonates, 44
   gas permeation, 25–27, 32–39
   hydrogen in natural rubber, 38
   methane in natural rubber, 38
   mutual diffusivity, 36
   plasticization effect in gas permeation, 34, 35
   shear-induced hydrodynamic diffusivity, 484, 486, 489, 493, 497
   temperature dependence, 37
   thermodynamic diffusivity, 36
   ultrafiltration, 402
Diffusion dialysis, 259
Diffusion test for microfiltration, 517, 522, 523, 532
Diffusive permeability, 164, 169, 170
Diffusivity. See Diffusion coefficient
Diluents, 667–669, 704
Dilution factor, 444
Dioctylphthalate, 470
Dispersion coefficient, 395, 681, 684

Disposal of concentrate in reverse osmosis, 322
Dissolved oxygen, 320
Distribution coefficient, 632–634, 679, 684, 731, 735–737, 739, 741, 742, 748, 750, 752, 754, 786–788, 790, 791, 793, 794, 797, 917
Dithiophosphinic acid, 657, 658, 703, 711
DLVO theory, 471
Donnan dialysis, 7, 209, 210
Donnan (exclusion) equilibrium, 209, 227, 267, 274
Donnan potential, 219, 227, 274
Downstream processing, 448, 449, 581
Drinking water standards, 338, 339
Driving force, 4, 8, 9, 12, 13
   controlled release, 916, 917, 925, 926
   dialysis, 165
   electrodialysis, 221
   electrostatic pseudo-liquid-membrane, 870
   emulsion liquid membrane, 601, 637, 659, 662, 663
   facilitated transport, 792, 837
   gas permeation, 19, 22–24
   hollow-fiber contained liquid membrane, 792, 793
   membrane-based solvent extraction, 731, 735
   microfiltration, 457
   pervaporation, 113, 117, 133
   reverse osmosis, 266
   ultrafiltration, 393
Drug, 915, 919, 920
Dual-mode model, 29, 30

Economics. See Cost estimates
Eddy constant, 303, 304
Effective diffusivity, 170, 632, 634, 641, 679, 685, 734, 924
Effluent/wastewater treatment
   ammonia wastewater by emulsion liquid membranes (ELMs), 598, 672, 712
   blast-furnace scrubber water by reverse osmosis (RO), 336
   condensate streams by RO, 336
   electroplating industry by electrodialysis, 257
   electroplating industry (nickel) by ELMs, 597, 708–711
   hazardous wastewater by RO, 326
   incineration plants by ELMs, 598, 713, 714
   landfill leachate by RO, 335
   metallurgical plants by ELMs, 598, 712, 713
   municipal wastewater by RO, 328
   nickel processing plants by electrostatic pseudo-liquid-membrane, 881
   oil and grease from can washing by ultrafiltration (UF), 448
   paper and pulp industry by RO, 332, 333, 448, 449
   petrochemical wastewater by RO, 336
   petroleum industry by RO, 335
   phenol wastewater by ELMs, 597, 599–602, 711
   photographic processing waste stream by RO, 336

Index 941

Effluent/wastewater treatment (*continued*)
    power generation industry by RO, 335
    radioactive wastewaters by RO, 336
    textile industry by RO, 334, 335
    textile industry by UF, 448
    trace organics wastewater by pervaporation, 148
    viscose rayon industry (zinc) by ELMs, 597, 599, 602, 701–708, 718–723
Electrical double layer forces, 471
Electrical field, 439, 440, 690, 691, 869, 872
Electrocoat paint recovery, 446
Electroconductive membrane, 439
Electrodialysis, 6, 8, 217
    applications, 255–261
    concentration polarization, 246, 249
    cost estimates/economics, 246, 249–253
    current utilization, 246, 248, 251
    definitions, 219–222
    desalination of brackish water, 255, 256
    design, 246–253
    driving force, 221
    effective cell pair area, 251
    electromotive force, 260, 261
    energy requirements, 223–226
    Gibbs free energy change, 223, 224, 260
    hybrid processes, 256, 258
    ion-exchange membranes, 226, 230–244. *See also* Membranes, ion-exchange
    limiting current density, 246–249, 251, 252
    membrane fouling, 235, 236, 249
    permselectivity, 240–243
    production of table salt, 256, 257
    reverse polarity, 250, 251, 256
    stack design, 248, 249
    theory, 223–229
    types, 221
Electrofiltration, 439
Electromotive force, 260, 261
Electrophoretic blotting, 561
Electroplating, 331, 332
Electrostatic coalescence, 689–692
Electrostatic deposition, 468
Electrostatic force, 272, 690
Electrostatic pseudo-liquid-membrane, 13, 867
    applications, 879–881
    driving force, 870
    mass transfer process, 872–876
    scale-up, 879
    swelling, 877
    theory, 869–872
Emulsification, 656–672
Emulsion liquid membranes, 6, 7, 9, 595
    applications/separations, 597–599, 701–716
    breakage/rupture, 614, 686
    capital and operating costs, 718–723
    chemical reagent consumption, 617, 618, 622, 629, 630
    classification of extraction cases, 624–628
    definitions, 597–603
    demulsification, 689–692
    design considerations, 656–692
    diluents, 667–669, 704
    dispersion, 672
    driving force, 601, 637, 659, 662, 663
    emulsification, 656–672
    estimation of overall mass transfer coefficients, 623, 624
    estimation of physical properties, 633, 634
    extractants/stripping agents, 657–663
    extraction, 611, 675–685
    facilitated mechanisms, 601–603
    feed pretreatment, 672
    globule size, 600, 673–675
    hybrid processes, 709, 710, 713
    leakage, 628, 629, 685–687
    membrane formulation, 656–667
    membrane systems, 668–672
    phenol extraction/removal, 597, 599–602, 615, 618, 623, 626, 634, 635, 668, 711
    surfactants, 664–671, 704
    swelling, 687–689
    theory, 611–647
    zinc removal, 597, 599, 602, 625, 627, 640, 641, 671, 701–708, 718–723
Endotoxin, 209
End-stage renal disease (ESRD), 193, 207
Enzymatic reactor, 203, 759
Enzymatic resolution, 816
Enzyme immobilization, 164, 759, 817
Equilibrium stage, 727, 754
Equivalent annulus approximation, 175
Erythropoietin (EPO), 207, 208
Ethanol, 106, 107, 113, 133–140, 150, 211, 759, 816, 825–828
Ethyl acetate, 150, 151
Ethyl cellulose, 45
Ethylene-chlorotrifluoroethylene copolymer (ECTFE), 544, 545
Ethylenediaminetetraacetic acid (EDTA), 257, 441, 657
Ethylene oxide, 533, 554
Ethylene-propylene copolymer (EPR), 202
Exchange flow, 171
Extract, 601, 730, 735
Extractants, 657, 669–671
    acidic, 657–659, 703, 704
    basic, 659–661
    effect of extractant concentration, 874, 875
    neutral, 661–663
Extraction ratio, 179–181, 196, 197, 201, 204

Facilitated transport, 13, 833
    applications, 843–859
    competitive, 840, 841
    driving force, 792, 837
    emulsion liquid membranes, 601–603, 611, 636, 675, 682
    facilitation factor, 836–838

Facilitated transport (continued)
    fixed carrier membranes, 842, 847–851
    hollow-fiber contained liquid membrane, 775, 782, 788, 789–792
    hybrid processes, 856, 857
    immobilized liquid membranes, 843–847
    optimization of reaction properties, 839, 840
    temperature effects, 842
    terminology, 835, 836
    theory, 837–842
    transient effects, 842
    two-dimensional systems and modules, 782, 783, 841, 842
Facilitation factor, 836–838
Faraday's constant, 23, 224, 227, 247, 251, 252, 260, 275
Fe^{+2} EDTA, Fe^{+3} EDTA, 804
Fecal coliforms, 553
Feed-and-bleed configuration, 441, 443, 481, 483, 586
Feed pretreatment
    emulsion liquid membranes, 672
    gas permeation, 85, 88, 89, 96
    microfiltration, 540
    pervaporation, 129
    reverse osmosis, 305–309, 317–320, 324, 325
    ultrafiltration, 437
Feed water composition in desalination, 315, 316
Feed water intake systems, 316
Ferritin, 428
Ferry equation, 399
Fiber packing density, 176, 199, 201, 751, 779, 796, 892
Fick's
    first law, 25, 32, 118, 168, 169, 172, 522, 907, 916, 917, 920
    second law, 169, 172
Film theory, 278
Film thickness, 278
Filter aid, 466
Fixed carrier membranes, 842, 847–851
Flexibility of polymer chains, 919
Flow configurations
    parallel flow, 195–199, 750
    perpendicular flow, 179, 181, 196–198
    Starling's flow, 193, 208
    tangential flow, 179, 459, 753
    well-mixed dialyzate flow, 179
    See also Cocurrent flow (permeators/contactors); Countercurrent flow (permeators/contactors); Crossflow (permeators/contactors)
Flory-Huggins equation, 27
Flue gas cleanup, 804, 855
Flux, 4
    current (density) in electrodialysis, 224, 247
    dialysis, 168, 171–173, 181, 182
    emulsion liquid membranes, 681, 684, 685
    gas permeation, 23, 25, 29, 56, 57
    hollow-fiber contained liquid membrane, 785, 786, 790
    membrane-based solvent extraction, 731
    microfiltration, 462–464, 481, 483, 485, 486, 488, 491, 493–495, 497, 498, 531, 576, 584
    pervaporation, 113, 118, 119, 125, 126, 134, 142, 144–146, 149, 154
    reverse osmosis, 266, 270, 271, 274–276, 283–289, 300
    ultrafiltration, 394–396, 398, 400, 401, 403, 409–420, 423, 427
    water vapor, 901–903
Flux decline
    long-term, 459, 460, 483, 495, 576
    pore plugging, 404
    protein adsorption, 404
    short-term, 459, 460, 483, 495, 576
    ultrafiltration, 403
Flux, volumetric, 23, 56, 57, 171, 181, 182, 267, 394, 395, 459, 462
Food processing applications
    fruit juice clarification, 347, 447, 448
    fruit juice concentration, 345, 346
    fruit juice fractionation, 346, 347
    reverse osmosis, 344, 345
    See also Dairy applications
Fouling
    biological foulants, 307, 308, 318, 319
    clean-in-place procedure, 440, 532, 585
    colloids, 307, 318, 319, 433
    control strategies in ultrafiltration, 437
    electrodialysis, 235, 236, 249
    external cake fouling, 404, 460
    internal fouling, 460
    macromolecules, 433
    membrane-based solvent extraction, 752, 753
    metal oxides, 308
    organic foulants, 308, 309, 318, 320
    particulates, 306, 307, 317
    protein adsorption, 187, 209, 438, 524, 525
    reverse osmosis, 277, 305–309
    scale-forming salts, 307, 317, 318
    ultrafiltration, 396, 403, 433, 436
Fourier number, 169
Fractional penetration, 468–470, 472, 473
Free surface model, 175, 197
Free volume, 35, 42
    diffusion model/theory, 35–37
    fractional free volume, 36, 42, 44
Friction factor, 402
Frictional force
    reverse osmosis, 272
    ultrafiltration, 399
Fruit juice clarification
    apple juice, 442, 448
    cranberry, 448
    grape, 448
    microfiltration, 537
    pear, pineapple, 448
    reverse osmosis, 347
    ultrafiltration, 448

Fruit juice concentration, 345, 346
Fruit juice fractionation, 346, 347
Fulvic acid, 320

Gas adsorption, 400, 428
Gas filtration, 542–544
Gas membrane, 5, 13, 887, 888, 891, 897–899, 901
Gas permeation, 6, 8, 17
   applications, 78–93
   carbon dioxide separation, 88–92, 98–100
   commercial-scale membrane suppliers, 78, 79
   definitions, 19–24
   design of gas permeation systems, 54–74
   driving force, 19, 22–24
   economics, 95–100
   feed pretreatment, 85, 88, 89, 96
   helium recovery, 87, 88
   hybrid processes/systems, 87, 90, 91
   hydrogen recovery, 67, 68, 78–85, 96–98
   membrane design, 55–60
   membrane modules, 60–72
   membranes, 20–22, 39–48, 55–60
   oxygen/nitrogen separation, 85–87, 100
   petrochemical applications, 83–85
   refinery applications, 80–83
   theory, 25–49
Gas separation, 764, 768–773, 843–860, 885. *See also* Gas permeation
Gel concentration, 401, 402, 442
Gel layer, 395, 404
Gel-type membrane, 169, 187, 188
Gelatin concentration, 446
Gibbs free energy change, 223, 224, 260
Glass transition temperature, 27, 29, 39, 40, 60
Glassy polymer, 27, 45–47, 919
Gossyplure, 930
Graetz number, 750
Graetz problem, 174
Graetz solution, 741, 742
Graver dialyzer, 189
Gravitational settling, 468
Grow-through, 522, 524
Gutter layer, 57

Hagen-Poiseuille equation, 69, 170, 193, 398, 462, 530, 751
Hazardous waste treatment, 325
Heat flux, 902
Heat transfer coefficients
   fluid film transfer coefficients, 902, 903
   membrane transfer coefficients, 902, 903
Heavy metal removal, 598, 625, 712–714, 756, 757, 800, 801
Height of transfer unit (HTU), 740, 743
Helium recovery, 87, 88
Hematocrit, 207
Hemicellulose, 163

Hemodiafiltration, 7, 209
Hemodialysis, 164, 190, 206
Hemofiltration, 209
Henry's constant, 27, 29, 30, 522, 778, 792, 888, 889
Henry's law, 27, 522, 888, 891
Heparin, 209
Heterogeneous membranes, 231, 234, 235
Hexadiene, 856
Hexene, 856
HIGEE contactor, 887
High-efficiency dialyzers, 207, 208
High-performance liquid chromatography (HPLC), 282, 400
High pressure reverse osmosis, 266
Hindrance factor
   convective filtration, 400
   diffusion, 400
Hollow-fiber contained liquid membrane, 13, 764
   acid gas removal, 770, 792, 794, 802–804
   applications, 799
   citric acid recovery, 775, 776, 789–794, 801, 802
   concept, gas separation, 768
   concept, liquid separation, 773
   device, 796–799
   effective mass transfer coefficients, 784
   effective membrane thickness, 777–780, 795, 797
   facilitated transport, 775, 782, 788, 789, 792
   fermentation product recovery, 801, 802
   flue gas cleanup, 804
   geometrical characteristics of modules, 796, 799
   microporous hydrophilic fiber, 773, 774, 776, 777, 784, 785, 787, 788
   microporous hydrophobic fiber, 768–770, 772, 774–776, 784–788
   modes of operation, 770–772, 792
   module design, 796–798
   permeator models, 780–791
   pervaporation, 777
   pressure conditions, 770, 771, 773–776, 795, 796
   purification limits, 792, 793
   stability, 770–772, 774–776, 800–804
   stripping liquid, 773, 791, 792, 800, 801
   supported liquid membrane structure, 765–768, 775
   sweep gas/liquid, 769–771, 792, 795, 803
   theory, 777–796
   unit cell, 777–780
Homogeneous membranes, 167, 168, 230–234, 270, 507
Humic acid, 320
Hybrid processes, 7
   air separation, 87
   carbon dioxide separation, 90, 91
   dialysis, 203, 209
   electrodialysis, 256, 258
   emulsion liquid membranes, 709, 710, 713
   facilitated transport, 856, 857
   hemodialysis, 209
   pervaporation, 135, 146–148, 155, 347, 348
   reverse osmosis, 323, 324, 346–348
Hybridomas, 587

Hydraulic diameter, 301, 751
Hydraulic permeability, 164, 170, 171, 267, 303
Hydrocarbon separations, 449, 450, 628, 729, 757, 905, 906, 908
Hydrodynamics
  dialysis, 172–174, 193, 195–197
  reverse osmosis, 290
Hydrogel membranes, 165, 169, 188
Hydrogen bonding, 438, 919
Hydrogen/carbon monoxide adjustment, 79, 80
Hydrogen recovery, 67, 68, 78–85, 96–98
Hydrogen sulfide, 320, 792
Hydrogen sulfide separation, 89, 851, 852
Hydrometallurgical extraction, 598, 625, 712–714, 756, 757, 880
Hydrophilic polymers/membranes, 186, 187, 728, 729, 734, 736–738, 745–750, 752, 757, 887, 889, 898
Hydrophilicity of membranes, 422, 438
Hydrophobic interactions, 438
Hydrophobic polymers/membranes, 188, 422, 517, 521, 522, 527, 728, 729, 731–733, 735–737, 743–746, 748, 752, 755, 886–891, 898–901, 903, 904
Hydrostatic pressure, 925, 926
Hydrotrope, 905
Hydroxyoximes, 658
Hydroxyquinolines, 658
Hyperfiltration, 266

Immobilized/supported liquid membranes, 60, 765–768, 817–821, 843–847, 868, 869, 878
Immunoglobulins, 165, 437, 438, 524, 525
Impaction, 467, 468
Inertial lift theory, 484, 493, 494
Inertial migration velocity, 493, 494
Insecticide, 915, 927
  permethrin, 927–929
  pyrethroid, 927
  insecticide-releasing eartags, 928
Instantaneous chemical reaction, 631, 743, 744, 893
Insulin, 428
Integrally skinned membranes, 22
Integrated process, 450
  enzyme saccharification with ultrafiltration, 450
  extractive bioreactor, 759
Integrity test for membranes, 522, 523
Interception, 467–473
Interfacial area per unit equipment volume, 740, 748, 749, 887, 892
Interfacial reaction, 602, 639, 745, 789, 790, 793
Interfacial tension, 426, 738, 746, 747, 819
Inulin, 170, 413, 414
Ion-exchange membrane equilibria, 209, 210, 226–229
Ion-exchange membranes. *See* Membranes, ion-exchange
Ionization reactions, 894
Irreversible thermodynamics, 164, 171, 273, 274
Isopropyl alcohol dehydration, 144–146
Isotropic membrane, 424

Kaolin concentration, 447
Kedem-Katchalsky equations, 171, 172, 273, 274
Kelvin equation, 428
Kerosene, 668–671, 879
Kinetic diameter, 28
Knudsen diffusion, 20, 890, 901

Lactalbumin, 428, 446
Lactobacillus delbrueckii, 715
Lactoglobulin, 428, 446
Langmuir-Blodgett layer film, 438
Leakage, 628, 685–687, 877
Length of transfer unit (LTU), 740–743, 892, 893, 896
Lévêque solution, 174, 175, 485, 743, 750, 896
Lime softening in reverse osmosis, 318, 319
Limiting current density, 246–249, 251, 252
Limiting flux, 401
Lipophilic membrane, 187
Liposomes, 916, 917
Liquid membranes. *See* Membranes, liquid
Log reduction value (LRV), 468, 474–477, 520, 521

Magnetic systems in controlled release, 916, 917
Mass transfer coefficient, 172–178, 198–200, 401–403, 613, 731, 784, 785, 819, 871, 888–893
  length-averaged dialyzate side, 177, 198, 199
  length-averaged feed side, 174, 198
  length-averaged overall, 173, 198, 739, 740, 742, 743
  membrane, 613, 683, 731, 734, 747, 890
  overall, 172, 173, 174, 177, 198, 200, 612, 614–616, 623, 624, 682–684, 731–734, 736, 739–745, 748, 755
Mass transfer correlation
  shell-side, 175–177, 750, 751, 895
  tube-side, 173–175, 749, 750
Mass transfer in membrane, 167–170, 613, 731, 888, 890, 901, 904, 907
Mass transfer resistance
  adsorption, 276
  dialysate, 173, 177, 197, 201
  feed, 173, 177, 197, 201
  gas film, 198, 889, 890
  liquid film, 198, 889, 890
  membrane, 56, 57, 173, 177, 197, 198, 201, 396, 462, 463, 736, 737, 785, 787, 889, 890
  overall, 56, 173, 198, 736, 737
Matrix systems. *See* Controlled release
Maximum contaminant levels, 338, 339
Mean free path, 458, 890
Megasphaera, 538
Membrane, 3
Membrane-based equilibrium processes, 10–12, 209, 727, 885
  membrane gas absorption, 10, 12, 885
  membrane gas stripping, 10, 12, 885
  membrane solvent extraction, 12, 727

Membrane-based equilibrium processes (*continued*)
  vacuum membrane distillation, 12, 901, 903
Membrane-based gas absorption and stripping, 10, 12, 885
  absorbent-filled membrane pore, 886, 887, 889, 890
  aerobic fermentation, 898
  applications, 897
  blood oxygenation, 10, 886, 890, 897
  breakthrough pressure, 886, 895
  devices, 894, 895
  gas absorption, 886–891, 894, 897, 898
  gas-filled membrane pore, 886, 888, 889
  gas-liquid contractor, 886, 887
  gas-liquid interfacial areas, 887, 892
  gas membrane, 887, 888, 891, 897–899, 901
  gas stripping, 885–887, 893, 896–898
  hydrophilic membranes, 887, 889, 898
  hydrophobic membranes, 886–891, 898
  length of transfer unit (LTU), 892, 893, 896, 897
  membrane contactor, 887, 893
  membrane modules, 893–895
  modular cascade, 897
  module design, 895–897
  number of transfer units (NTU), 892, 893, 896
  pore condensation, 888
  pore wetting, 887, 889, 890
  scrubbing of flue gas, 890, 896, 898, 899
  silicone-coated microporous membrane, 890, 891
  stripping of volatile organic compounds, 896, 898
  theory, 888–894
Membrane-based pouches, 920
Membrane-based solvent extraction, 12, 727
  applications, 756
  asymmetric membrane, 747
  batch recirculation, 753, 754
  breakthrough pressure, 745–747
  chemical reaction, 743–745
  composite membrane, 729
  continuous contact, 730, 753, 754
  description, 727
  design, 752
  hollow fiber modules, 748, 749
  hydrophilic membrane, 728, 749
  hydrophobic membrane, 728, 748
  immobilized interface, 728, 729
  inorganic membrane, 747, 749, 753
  membrane module design, 752
  membrane modules and suppliers, 748, 749
  membrane selection, 752, 753
  membranes, 746–748
  microporous membranes, 728–730, 734, 746, 748
  modular cascade, 752, 755, 756
  solvent-swollen membranes, 730
  theory, 730–748
Membrane characterization
  adsorption-based methods, 273, 428
  bubble-pressure breakthrough, 426, 509, 513, 517, 520, 521, 895
  electron microscopy, 428
  high-performance liquid chromatography (HPLC), 282, 400
  ion-exchange membranes, 239–244
  mercury porosimetry, 427, 512
  morphology, 425, 426
  NMR measurement, 429
  porosity, 425, 513
  solute retention challenge, 427
  surface properties, 429
Membrane cleaning
  backflushing, 440
  clean-in-place (CIP), 129, 440, 441, 532, 585
  clean-out-of-place (COP), 532
  cleaning solutions, 441, 532, 577, 585, 590
  microfiltration, 532, 577, 585, 590
  pervaporation, 129, 130
  spongeball technique, 441
  ultrafiltration, 440, 441
Membrane compaction, 58–60, 396
Membrane contactors
  gas absorption and stripping, 886
  solvent extraction, 727
Membrane dehumidification, 898, 899
Membrane distillation, 10, 11, 13, 899
  applications, 903, 904
  devices, 903
  driving force, 901, 903
  membranes, 903
  pore size, 901, 903
  spiral-wound device, 903
  temperature polarization, 902
  theory, 901–903
  types, 900, 901
  water vapor flux, 901, 902
Membrane flux. *See* Flux
Membrane fouling. *See* Fouling
Membrane impedance, 170
Membrane lifetime
  crossflow microfiltration, 577, 578
  deadend microfiltration, 528, 534, 544, 549
  dialysis, 191–193, 214
  electrodialysis, 237, 252
  membrane-based solvent extraction, 748, 749, 758
  pervaporation, 130
  reverse osmosis, 290, 294, 363, 372
  ultrafiltration, 451
Membrane manufacturers (suppliers)
  dialysis membranes and modules, 188–190, 214
  electrodialysis membranes, 242, 243
  gas permeation, 78, 79
  membrane-based solvent extraction, gas absorption, 748, 749, 894, 895
  microfiltration, 510, 511, 550, 551, 574, 576, 585
  pervaporation, 128, 129, 156
  reverse osmosis membranes and modules, 283–289, 291, 292
  ultrafiltration membranes and systems, 408–421, 432–437

Membrane mass transfer coefficient, 613, 683, 731, 734, 747, 890
Membrane mass transfer resistance, 56, 57, 173, 177, 197, 198, 201, 396, 462, 463, 736, 737, 785, 787, 889, 890
Membrane materials
  chemical compatibility for gas permeation, 57–59
  dialysis, 186
  electrodialysis, 228–239
  gas permeation, 39–48
  membrane formulation for emulsion liquid membranes, 656–667
  microfiltration, 510, 511, 550, 573, 576, 585
  pervaporation, 128
  reverse osmosis, 282–289
  ultrafiltration, 408–420
Membrane modification, 422
  polymer blends, 422
  surface grafting, 422
Membrane module design. *See also* Design
  dialysis, 192
  electrodialysis, 248–252
  gas absorption/stripping, 892, 893, 895, 896
  gas permeation, 54, 60–72
  hollow-fiber contained liquid membrane, 796–798
  membrane-based solvent extraction, 754, 755
  microfiltration, 529, 530, 540, 576, 577, 585, 590
  modeling, gas permeation, 68–72
  pervaporation, 128, 129
  reverse osmosis, 300–305
  ultrafiltration, 432–436
Membrane modules
  arrangement of gas separators, 72–74
  cassettes, ultrafiltration, 412, 413, 415, 416
  ceramic, 424, 436, 437, 510, 511, 572–578, 584, 585
  comparison, deadend microfiltration, 507, 520, 526, 550, 551
  comparison, monolith ceramics, microfiltration, 585
  comparison, reverse osmosis, 282, 290
  comparison, ultrafiltration, 432, 434
  continuous membrane column, gas permeation, 73, 74
  disks, reverse osmosis, 299
  disks, ultrafiltration, 409, 412, 413, 415, 416
  flow patterns in gas permeation, 61–63, 66–68
  hollow-fiber, dialysis, 189, 190, 192
  hollow-fiber, gas permeation, 64–68
  hollow-fiber, gas absorption/solvent extraction, 739, 748, 749, 893–895
  hollow-fiber, reverse osmosis, 284, 290, 294, 296, 297, 304, 305
  hollow-fiber, ultrafiltration, 409, 410, 417–420, 435
  internally staged gas permeator, 74
  laminar flow thin channel, 432
  modeling, gas permeation, 68–72
  monolith, crossflow microfiltration, 572, 574, 575, 585
  monolithic tubular, ultrafiltration, 409, 436, 437
  packing density, 290, 432, 434
  plate-and-frame, crossflow microfiltration, 572, 574, 575, 591
  plate-and-frame, dialysis, 190, 191
  plate-and-frame, electrodialysis, 248–250
  plate-and-frame, pervaporation, 128–130
  plate-and-frame, reverse osmosis, 283, 286, 299
  plate-and-frame, ultrafiltration, 411–414, 418–420, 434, 435
  pleated cartridge, deadend microfiltration, 508, 512, 546, 548, 550, 551
  spiral-wound, gas permeation, 63, 64
  spiral-wound, crossflow microfiltration, 572–574, 577
  spiral-wound, reverse osmosis, 283–290, 293–296
  spiral-wound, ultrafiltration, 409–411, 413–418, 433, 434
  stacked disk, deadend microfiltration, 508, 512, 546, 549–552
  tank type, dialysis, 188
  two-membrane gas permeator, 74
  tube type, dialysis, 188
  tubular, crossflow microfiltration, 572–574, 577
  tubular, reverse osmosis, 283–290, 292, 297, 298
  tubular, ultrafiltration, 409–411, 415–417, 419, 420, 435, 436
  turbulent-flow wide bore, 432–435
Membrane morphology, 22, 55, 59, 187, 188, 423–425
Membrane permeability. *See* Permeability (coefficient)
Membrane pore size, 187, 273, 399, 517, 520–522, 527, 534–536, 538, 542, 543, 545, 549, 550, 554, 556, 557, 565, 576, 585, 734, 746, 747, 799, 894, 895, 903
Membrane porosity, 60, 425, 462, 513, 575, 576, 734, 735, 755, 797, 799, 820, 890, 902, 924, 929
Membrane process configurations, 72–74, 441
  batch concentration, 441, 442, 480, 481
  batch operation, 202, 482, 672
  batch recirculation, 754
  continuous membrane column, gas permeation, 73, 74
  continuous operation, 202, 672, 677, 678
  crossflow versus deadend microfiltration, 458
  diafiltration, 396, 443, 444, 481, 482, 582–584
  double-pass reverse osmosis, 320, 321
  feed-and-bleed, 441, 443, 481, 483, 586
  modes of operation for pervaporation, 123, 124
  multistage. *See* Cascade/staged operation
  recycle, 73, 202, 481, 679, 828
  single-pass reverse osmosis, 320
  single-stage, 72, 706
Membrane processes, 5
  characteristics of commercialized processes, 6, 8, 9
  characteristics of new processes for separation under development, 11–14
  selection of commercialized processes, 6
Membrane reactors, 11, 809
  applications, 814–816, 824

Membrane reactors (*continued*)
   catalytic/reactive membranes, 809, 811, 812, 817
   chymotrypsin-catalyzed hydrolysis, 816, 825
   definitions, 809–815
   immobilized liquid membranes, 817–821
   membrane reactor performance, 821–824
   product separation and enrichment, 815–824
   separation of multiple products, 824–827
   thermodynamic considerations, 827, 828
Membrane rejection
   batch concentration, 442
   charged membranes, 274, 275
   decline in reverse osmosis, 305
   dextran, 411, 413–415, 423
   dialysis, 165, 183
   hydrocarbon, 425
   intrinsic (true), 266, 270–273, 275, 404
   limiting salt, 267, 275, 276
   microfiltration, 509, 520, 521, 583, 584
   observed, 266, 274, 276, 283–289, 404
   pH effect in reverse osmosis, 276
   reverse osmosis, 266, 267, 270–276, 283–289, 295, 297, 298, 300, 305
   ultrafiltration, 395, 399, 404, 411, 413–415, 423, 425, 426, 428
Membrane resistance. *See* Membrane mass transfer resistance
Membrane structure. *See* Membrane morphology
Membrane tortuosity, 509, 734, 735, 755, 785, 786, 790, 797, 820, 890, 902, 924
Membrane transport
   dual-mode model, 29, 30
   finely porous model, 399
   Hagen-Poiseuille equation, 170, 398, 462
   hindered transport model, 400
   irreversible thermodynamics model, 171, 273, 274
   matrix model, 30, 31
   resistance model, 56, 57
   solution-diffusion, 21, 25, 111, 117, 168, 270, 271, 602, 631, 907, 916
   surface force-pore flow model, 271–273, 399
   See also Facilitated transport
Membranes, asymmetric
   controlled release, 929, 930
   dialysis, 188
   gas permeation, 22, 55, 56
   membrane-based solvent extraction, 734, 735, 747
   microfiltration, 460, 513, 575, 577
   pervaporation, 128
   reverse osmosis, 269, 270, 282
   ultrafiltration, 408, 423, 424
Membranes, bipolar, 10, 238, 239, 260, 261
Membranes, catalytic/reactive, 809, 811, 812, 817
Membranes, ceramic
   gas permeation, 21
   membrane-based solvent extraction, 747, 749
   microfiltration, 510, 511, 572–578, 585, 587
   ultrafiltration, 408, 409, 424, 425

Membranes, composite/thin-film-composite (TFC)
   gas permeation, 22
   microfiltration, 508
   pervaporation, 128
   resistance model, 56, 57
   reverse osmosis, 285, 287, 305, 306, 333, 341–343
   ultrafiltration, 411, 423, 424
Membranes, dialysis, 188, 189
   cellulose, 186–190, 194
   cellulose acetate, 186, 187, 189
   PAN (polyacrylonitrile), 190, 194
   PAN-co-methallylsulfonic acid, 186
   PAN-methallyl sulfate, 189
   polyethersulfone, 186
   polyethylene-co-vinylalcohol (Eval), 186
   polymethylmethacrylate (PMMA), 186, 189
   polysulfone, 189, 194
Membranes, electrodialysis. *See* Ion-exchange membranes
Membranes, gas permeation, 20–22, 39–48, 55–60
Membranes, heterogeneous, 231, 234, 235
Membranes, homogeneous, 167, 168, 230–234, 270, 507
Membranes, integrally skinned, 22
Membranes, ion-exchange, 210, 226, 230–244, 835
   anion-exchange membranes, 10, 219, 233–238, 242, 243
   cation-exchange membranes, 10, 210, 219, 226, 232–237, 242, 243
   characterization, 239–244
   electrical resistance, 241–244
   gel water content, 240, 242, 243
   heterogeneous, 231, 234, 235
   homogeneous, 230–234
   ion-exchange capacity (IEC), 240, 242, 243
   ion-exchange membrane equilibria, 209, 210, 226–229
   membrane fouling, 235, 236, 249
   perfluorocarboxylic acid, 237
   perfluorosulfonic acid (PFSA), 237, 835, 841, 851–853, 856, 857
   permselectivity, 240–243
   special property, 235–239
   sulfonated polydimethylphenyleneoxide, 856
Membranes, liquid, 5, 764, 868, 869
   electrostatic pseudo-liquid-membrane, 13, 867–881
   emulsion. *See* Emulsion liquid membranes
   hollow-fiber contained liquid membrane, 13, 764–804
   immobilized/supported, 60, 765–768, 817–821, 843–847, 868, 869, 878
   molten salt, 846, 853
Membranes, microfiltration, 507–513, 516, 517, 520–525, 527, 532, 534–536, 538, 542–544, 546, 548, 550, 551, 554, 556, 564, 565, 576
Membranes, nanofiltration, 288
   Desal-5, 289

Membranes, nanofiltration (*continued*)
  DRC-1000, 289
  HC-50, 289
  MPT-20, 289, 337
  MPT-30, 289, 337
  NF40, 288, 336, 337, 343, 344
  NF50, 288, 343
  NF70, 288, 336, 343, 344
  NF-PES-10/PP 60, 289
  NF-CA-50/PET 100, 289
  NTR-7250, 288
  NTR-7410, 288
  performance of commercial membranes, 300
  SU200HF, 289
  SU600, 289
  SU700, 289
  UTC-20HF, 289
  XP20, 288, 337
  XP45, 288, 337
Membranes, operating range
  chemical compatibility for gas permeation, 57–59
  dialysis, 170, 193, 195, 196, 207, 208
  maximum pressure/temperature for deadend microfiltration, 550, 551
  membrane-based solvent extraction, 748, 749
  pressure limit for crossflow microfiltration, 585
  pressure/temperature effects for gas permeation, 58–60
  reverse osmosis, 266, 275, 276, 296
  ultrafiltration, 409–420, 424, 425
Membranes, pervaporation, 128
Membranes, reverse osmosis
  aromatic polyamide, 282, 284, 297, 306, 327, 329, 331
  categories, 282
  cellulose acetate, 282, 283, 295–298, 306, 326, 327, 329, 330, 332, 335, 336, 340, 341
  cellulose esters, 284
  cellulose triacetate, 283, 284, 324
  characteristics, 283, 300
  desalination performance, 300
  dynamic membranes, 334
  FT30, 284, 285, 327, 328, 332, 336, 340, 341
  high temperature, 305
  nanofiltration, 288. *See also* Membranes, nanofiltration
  performance characteristics, 300
  polyamide, 286, 306, 327, 341, 342
  polyetherurea, 282
  polyfuran, 286
  polypiperazinamide, 282
  polyurea, 286, 306
  polyvinylalcohol, 285
  sulfonated polysulfone, 305, 306
  suppliers, 291, 292
  thin-film-composite (TFC), 285, 287, 305, 306, 333, 341–343

Membranes, solid, 4
  fixed carrier, 842, 847–851
Membranes, ultrafiltration, 394, 408
  acrylic vinyl polymer, 410
  alumina, 408, 409, 424
  borosilicate glass, 408, 424
  carbon-based, 419, 425
  cellulose acetate (CA), 408, 411, 414, 415, 418–420, 422–424
  cellulose nitrate, 420
  cellulose triacetate, 419, 425
  dynamically formed, 425, 449
  inorganic, 424
  polyacrylonitrile (PAN), 408, 410, 411, 415, 418, 419, 424, 426
  polyamide, 408, 410, 412, 414, 415, 426
  polyethersulfone (PES), 408, 411–416, 419, 423–424, 426, 432, 438
  polyimide, 417
  polyolefin, 417
  polysulfone, 408–412, 414, 415, 417–420, 422–425, 432, 438, 447
  polytetrafluoroethylene, 413
  polyurea/polyurethane, 423
  polyvinylchloride, 422
  polyvinylidenefluoride (PVDF), 408, 414, 416–419, 422, 424–426, 432, 438, 448
  regenerated cellulose, 408, 414, 420, 438
  sulfonated polystyrene, 438
  sulfonated polysulfone, 418, 422
  thin-film-composite (TFC), 411, 423, 424
  track-etched, 426
  zirconia, 408, 425
  ZrO_2/carbon composite, 419
Membranology, 187
Mercury intrusion, 427, 512
Metal extraction. *See* Hydrometallurgical extraction
Metal finishing, 331, 332
Methanol, 155, 156
Methyl-tert-butyl ether (MTBE), 155, 156
Methyltrioctylammonium chloride, 657
Mevinolinic acid, 747, 758
Micelles, 450, 687, 688, 715, 716, 758, 759
Microbial growth control, 306
Microcapsules, 920, 930
Microorganisms, 281, 319, 538, 552
Microfiltration, 6, 9, 457
  definitions, 457–460
  driving force, 457, 458
  See also Crossflow microfiltration; Deadend microfiltration
Microporous beads, 929, 930
Microporous membranes, 10
  controlled release, 916, 929
  gas permeation, 21
  hollow-fiber contained liquid membrane, 768–770, 773–776

Microporous membranes (*continued*)
  immobilized/supported liquid membrane, 60, 765–768, 817–821, 843–847, 868, 869, 878
  membrane-based absorption/stripping/distillation, 885–888
  membrane-based solvent extraction, 727–730
  membrane distillation, 899–901
  microfiltration, 509
  reverse osmosis, 271
  ultrafiltration, 393, 408
Microporous reservoir, 928, 929
Middle molecules, 208
Mixer-settler, 672, 673, 675–677, 679, 680, 682, 685, 727
Mobility, 36, 37
Mobility of carbon dioxide in silicone rubber, 37
Molecular diffusion, 901
Molecular sieving, 21
Molecular weight cutoff (MWCO), 393, 409–420
Molten salt membranes, 846, 853
Momentum boundary layer, 173, 174
Momentum entrance region, 174
Monoclonal antibodies, 916, 917
Monocytes, 209
Monoliths, ceramic, 434, 436, 510, 572, 575, 576, 585
Morantel tartrate, 930
Most penetrating particle size, 470
Multi-leaf, 293, 294
Multistaging. *See* Cascade/staged operation
Municipal wastewater treatment, 313, 328, 329, 340

Nafion®, 134, 210, 236, 237, 260, 812, 856
Nanofiltration, 266, 267, 269, 275, 288, 289, 300, 336–338, 343, 344
Nanofiltration membranes. *See* Membranes, nanofiltration
Natural rubber, 38, 45
Nernst equation, 241
Nernst-Planck equation, 275
Neutrofils, 209
Nickel extraction/recovery, 597, 625, 627, 670, 708–711, 881
Nifedipine, 931
Nitric oxide (NO), 855
Nitrile-butadiene rubber (NBR), 120
Nitrocellulose, 562
Nitrogen-enriched air (NEA), 85, 86, 100
Nondispersive operation
  gas-liquid, 886
  liquid-liquid, 728
Nonideal effects in dialysis, 172
Nonpolar organic-polar organic systems, 729, 732, 738, 746, 757
Normalized diffusivity, 170
Northern blotting, 561
Number of transfer units (NTU), 179–181, 201, 204, 740–742, 755, 892, 893, 896
Nylon membranes, 510, 511, 543, 562, 746, 749

Oil-shale retort operation wastewater, 335
Oil/water emulsions, 447–449, 600
Oily wastewater, 335, 447, 449
Olefins/saturates separations, 855–857
Onsager's law, 171
Operating costs, 99, 252, 253, 361, 451, 452, 533, 541, 546, 587, 588, 718–723
Operating range of membranes, *See* Membranes, operating range
Opthalmic solutions, 516
Organic-filled pore, 728, 729, 731, 732, 738, 744, 747, 748, 752
Organic/organic separations, 152–156
Organic pollutant extraction, 757
Organics removal from water, 148–152
Osmosis, 163, 266
Osmotic agent, 925, 926
Osmotic distillation, 11, 13, 900, 901, 904
Osmotic flow, 171
Osmotic pressure, 401, 403, 457
  osmotic pressure difference, 171, 267, 270, 271, 276, 395, 396, 401, 403, 925, 926
  seawater, 315, 316
Oxygen carriers, 844–850
Oxygen-enriched air (OEA), 85, 100
Oxygen/nitrogen separation, 85–87, 100, 843–851
Ozonation-membrane process, 328, 329

Packing density of fibers, 176, 197, 199, 751, 779, 795, 796
Packing density of polymer chains, 918
Parallel-plate configuration, 175, 277, 300
Parchment membrane, 163
Parenterals, 515, 516, 518, 533, 534
Partition coefficient, 168, 907, 917, 918
Peclet number, 181, 400, 469, 684
Pectinatus, 538
Penicillin G, 581
Peptide C, 170
Perfluorocarboxylic acid, 237
Perfluorosulfonic acid (PFSA), 237, 835, 841, 851–853, 856, 857
Perfusion, 587
Permeability (coefficient)
  carbon dioxide in commercial carbon dioxide separation membranes, 57
  carbon dioxide in natural rubber, 38, 45
  carbon dioxide in polycarbonates, 28, 30, 34, 42, 45
  carbon dioxide in polyestercarbonates, 46
  carbon dioxide in poly(ethylene terephthalate), 39
  carbon dioxide in polyimides, 42, 46, 47
  carbon dioxide in silicone rubber, 28, 34, 45
  gas permeabilities in polymers, 45–47
  gas permeabilities in silicone rubber and polycarbonate, 28
  gas permeability-free volume correlation for polycarbonates, 44

Permeability (coefficient) (*continued*)
   gas permeation, 25–32
   helium in polycarbonates, 42, 45
   helium in polyestercarbonates, 46
   helium in polyimides, 42, 44, 46, 47
   hydrogen in commercial hydrogen separation membranes, 57
   hydrogen in natural rubber, 38, 45
   hydrogen in polyaramides, 43
   hydrogen in polyimides, 44, 46, 47
   hydrogen in polysulfones, 43, 45
   membrane reactors, 819, 820
   methane in natural rubber, 38
   microfiltration, 512, 551
   oxygen in commercial air separation membranes, 57
   oxygen in polycarbonates, 42, 45
   oxygen in polyestercarbonates, 46
   oxygen in polyimides, 42, 46, 47
   perstraction, 907, 908
   reverse osmosis, 270, 271
   solute permeability in dialysis, 169–172, 181, 182, 188, 189, 198–200
   solvent permeability in dialysis, 170, 171
   ultrafiltration, 395, 403
Permeate, 19, 266, 267, 457
Permeate flux. *See* Flux
Perpendicular flow, 179, 181, 196–198
Perstraction, 13, 124, 125, 904
   applications, 908
   azeotrope separation, 905, 909
   benzene-cyclohexane separation, 905, 908, 909
   cascade, 905, 906
   close boiling point systems, 908
   driving force, 904, 905
   extracting liquid, 904, 905
   flux, 907
   hydrotrope, 905
   mass transfer, 907, 908
   membrane conditioning, 908
   membranes, 905, 906, 908
   permeability coefficient, 907, 908
   pertraction, 904, 906
   pervaporation versus perstraction, 905–907
   purge liquid, 904–906
   solution-diffusion mechanism, 907
   stripping/sweep liquid, 904–907
   toluene-*n*-heptane separation, 905, 909
Pervaporation, 6–8, 103, 777
   applications, 132–157
   concentration polarization, 125, 126
   definitions and background, 105–115
   dehydration of organics, 140–148
   design, 123–130
   driving force, 113, 117, 133
   economics, 132, 135, 137–139, 146, 147
   ethanol dehydration, 107, 133–140
   feed pretreatment, 129
   hybrid processes, 135, 146–148, 155, 347, 348

   isopropyl alcohol dehydration, 144–146
   membrane modules, 128, 129
   membranes, 128
   modes of operation, 123, 124
   multicomponent, 119–121
   organic/organic separations, 152–156
   removal of organics from water, 148–152
   temperature polarization/drop, 125–127
   theory, 117–121
Phase barrier, 12, 886, 887
Phenol extraction/removal, 597, 599–602, 615, 618, 623, 626, 634, 635, 668, 711, 775, 788, 789, 797, 800
Phenolic resins, 234
Phenomenological model, 168
Pheromones, 929, 930
Phosphinic acid, 658
Photoresist filters, 542, 545, 546
Pilot plant, 139–141, 153, 156, 584
Pink bollworm, 929, 930
Piperonyl butoxide, 927, 928
Plasma, 209
Plasmapheresis, 484, 500, 581
Plasma proteins, 208, 524, 525
Plasticization, 37, 919, 929
Plasticizer, 37, 919
Pleated cartridge, 508, 512, 546, 548–551
Point-of-use (POU) filters, 542–544, 546
Poiseuille's law. *See* Hagen-Poiseuille equation
Polarity, 919
Polyacrylic acid, 149, 403, 404
Polyacrylic acid-polycation, 134
Polyacrylonitrile (PAN), 128, 190, 194, 408, 410, 411, 415, 418, 419, 424, 426
   PAN-co-methallylsulfonic acid, 186
   PAN-co-4-vinylpyridine, 854
   PAN-methallyl sulfate, 189
Polyamide, 84, 282, 284, 286, 297, 306, 327, 329, 331, 341, 342, 408, 410, 412, 414, 415, 426
Polyamine, 665, 668–670
Polyaramides, 43
Polybutylmethacrylate (PBMA), 847–849
Polycarbonate (PC), 28, 34, 35, 40–45, 194, 511, 565, 843, 845
Polydimethylphenyleneoxide (PPO), 45, 856
Polydimethylsiloxane (PDMS) (silicone rubber), 28, 34, 35, 37, 45, 128, 148–150, 890
Polyestercarbonates, 46
Polyetheramides, 149
Polyetherimide (PEI), 45
Polyethersulfone (PES), 186, 238, 408, 411–416, 419, 423–424, 426, 432, 438
Polyetherurea, 282
Polyethylene, 106, 107, 118, 154, 228, 233, 234, 908
Polyethylene-co-vinylalcohol (Eval), 186
Polyethyleneglycol, 417, 427, 428, 705, 759, 814, 856, 931
Poly(ethylene oxide), 395
Poly(ethylene terephthalate) (PET), 33, 39, 40, 919

Polyethylenimine, 286
Polyethylmethacrylate (PEMA), 40
Polyfuran, 286
Polyhexylmethacrylate-co-4-vinylpyridine, 849
Polyhydroxyethylacrylate-co-hydroxyethylmethacrylate, 848
Polyimide, 41–44, 46, 47, 59, 417
Polymethylmethacrylate (PMMA), 32, 40, 186, 189
Poly(4-methyl-1-pentene), 45
Polyoctylmethacrylate, 848
Polyoctylmethacrylate-co-1-vinylimidazole, 848, 849
Polyoctylmethacrylate-co-2-vinylpyridine, 848
Polypiperazinamide, 282
Polypropylene, 107, 149, 154, 423, 426, 510, 511, 544, 545, 746, 748, 799, 819, 847, 894, 903
Poly(pyrrolone), 46
Polysaccharides, 427, 450
Polystyrene, 219, 228, 233, 850
Polystyrene, sulfonated, 438
Polystyrene-co-acrylonitrile, 192
Polysulfone, 43, 45, 128, 134, 189, 194, 228, 238, 408–412, 414, 415, 417–420, 422–425, 432, 438, 447, 510, 511
Polysulfone, sulfonated, 234, 305, 306, 418, 422
Polytetrafluoroethylene (PTFE), 154, 237, 413, 473, 475, 510, 511, 535, 543–545, 746, 819, 825, 898, 903, 904
Polytrimethylsilylpropyne (TMSP), 45, 853
Polytrimethylvinylsilane, 847, 858
Polyurea, 286, 306
Polyurea/polyurethane, 423
Polyurethane, 853
Polyvinylalcohol (PVA), 113, 128, 134, 143, 154, 233, 285
Polyvinylammonium thiocyanate, 853
Polyvinylchloride, 107, 234, 422
Polyvinylfluoride/acrylic acid, 134
Polyvinylidenefluoride (PVDF), 238, 408, 414, 416–419, 422, 424–426, 432, 438, 448, 470, 473, 476, 477, 510, 511, 543, 544, 549, 563, 847, 903
Polyvinylidenefluoride-N-vinylimidazole, 134
Polyvinylpyridine, 236
Polyvinylpyrrolidone (PVP), 154, 186, 413–415, 426
Pore size. *See* Membrane pore size
Pore size distribution
    dialytic membrane, 165, 170, 187
    microfiltration membrane, 512, 513, 516, 521
    reverse osmosis membrane, 272, 273
    ultrafiltration membrane, 399, 400, 427
Pores
    cylindrical, 272, 398–400, 462, 509
    slit shaped, 400
Porosigens, 924
Porosity. *See* Membrane porosity
Porous membranes, 21, 167, 169, 507, 509
Positively charged membranes, 219, 438. *See also* Anion-exchange membranes
Post-treatment, 321, 322
Potential function, 272, 399, 400

Power recovery, 322
Precipitation, 307, 317, 536
Precoat, 466
Preconcentration of milk, 446
Prefiltration, 437, 528, 529, 531, 532, 534, 539, 540
Pressure-driven membrane processes, 22, 266, 393, 457
Pressure drop
    shell side, 197
    tube side, 69, 70, 751
Pressure-sensitive controlled release device, 916, 917
Pressure vessel, 294
Pretreatment. *See* Feed pretreatment
Process configurations. *See* Membrane process configurations
Process design. *See* Design
Process economics. *See* Cost estimates
Product separation and enrichment, 815–824
Productivity, 295, 297, 298, 302, 304, 305
Protein adsorption, 165, 187, 209, 403, 404, 428, 438, 524, 525, 561
Protein aggregation, 404
Protein retention, 409–420, 427, 428
Pseudo-binary approach, 168
Pseudomonas diminuta, 517, 521–523, 532
Pulp and paper industry effluent treatment, 332–334
Pure water permeability constant, 270, 271

Raffinate, 601, 730, 735
Raffinose, 170
Rare earths, 627, 628, 671, 880, 881
Recombinant organisms, 535
Recombinant proteins, 519
Recycle ratio, 483, 679
Red cells, 208
Reflection coefficient, 171, 181, 182, 267, 274
Regenerated cellulose, 106, 165, 186, 189, 190, 194, 208, 214, 408, 414, 420, 438, 746
Rejection. *See* Membrane rejection
Release kinetics. *See* Controlled release kinetics
Resistance model, 56, 57
Retentate (concentrate), 19, 266, 267, 457
Reverse osmosis (RO), 6, 7, 9, 265
    applications, 312–348
    concentration polarization, 277, 300
    cost estimates, 294, 295, 297, 298, 355–388
    definitions, 265–267
    desalination, 266, 295–297, 308, 309, 313–325, 356–361, 363–366, 369–373
    design, 281–309
    double-pass RO, 320, 321
    driving force, 266
    extended Nernst-Planck model, 275
    feed pretreatment, 305–309, 317–320, 324, 325
    flux, 266, 270, 271, 274–276, 283–289, 300
    fouling, 277, 305–309
    hybrid processes, 323, 324, 346–348
    irreversible thermodynamics model, 273, 274
    loose RO, 266

Reverse osmosis (RO) (*continued*)
  low pressure RO, 266, 294, 295, 300, 356
  membranes. *See* Membranes, reverse osmosis
  membrane module suppliers, 291, 292
  membrane modules, 290
  module design, 300
  nanofiltration, 266, 267, 269, 275, 288, 289, 300, 336–338, 343, 344
  operating variables, 275
  preferential sorption-capillary flow model, 271
  process control, 323
  process design, 281, 324
  process schematic, 267, 320–322
  single-pass RO, 320, 321
  solute flux, 267, 270, 271, 274
  solute rejection, 266, 267, 270–276, 283–289, 295, 297, 298, 300, 305
  solution-diffusion model, 270, 271
  solution-diffusion-imperfection model, 271
  solvent flux, 266, 267, 283–289
  surface force-pore flow model, 271–273
  systems, 308, 309, 315, 321, 322, 324, 325, 330, 334, 348
  theory, 269–278, 300–305
  total dissolved solids, 316, 324, 327, 329, 330, 333, 336
  total organic carbon, 326–329, 335, 336, 340
Reversed micelles, 687, 688, 715, 716, 758, 759
Reversible chemical reaction, 639, 783, 788, 789, 837
Reynolds number, 176, 177, 196, 197, 199, 247, 402, 675, 742, 743, 750, 751, 789
Rotary drum filter, 466, 467, 582
Rotating cylindrical device, 439, 730, 748, 753
Rotating drum artificial kidney, 163, 190
Rubbery polymers, 27, 36, 45, 92, 919
Rubow's theory, 468–470

Saccharomyces cerevisiae, 450
Salt distribution coefficient, 270, 271, 274, 275
Salt rejection, 266, 267, 270, 271, 275, 283–289, 295, 297, 298, 300, 305
Sanitary design, 435, 531, 592
Sanitary processing, 441, 592
Sauter mean diameter, 614, 624, 674, 675
Scale formation, 317, 318
Scaling factors, 214, 359, 360
Schmidt number, 176, 177, 199, 742, 743, 750, 751
Seawater desalination, 266, 295–297, 308, 309, 313–325, 356–361, 363–366, 369–373
Selectivity (separation)
  acetic acid/hydrochloric acid, 858
  ammonia/hydrogen, 853
  ammonia/nitrogen, 853
  butanol/cyclohexane, 154
  butene/butane, 857
  1-butene/other butenes, 857
  carbon dioxide/carbon monoxide, 853
  carbon dioxide/hydrogen, 853
  carbon dioxide/methane in cellulose acetate, 89
  carbon dioxide/methane in commercial carbon dioxide separation membranes, 57
  carbon dioxide/methane in polycarbonates, 42, 43
  carbon dioxide/methane in polyimides, 42, 43
  carbon dioxide/nitrogen, 794, 795, 854
  carbon dioxide/oxygen, 854
  carbon monoxide/nitrogen, 857, 858
  chloroform/*n*-hexane, 154
  citric acid/glucose, 801, 858
  ethylene/ethane, 855–857
  fumaric acid/L-maleic acid, 858
  gas selectivities in polymers, 45–47
  helium/methane in polycarbonates, 42, 43
  helium/methane in polyimides, 42, 43
  hydrogen/methane in commercial hydrogen separation membranes, 57
  hydrogen/methane in polyamides, 43
  hydrogen/methane in polysulfones, 43
  hydrogen sulfide/carbon dioxide, 89
  hydrogen sulfide/methane, 89, 852
  isopropyl alcohol/*n*-hexane, 154
  isopropyl alcohol/toluene, 154
  organic/organic, 152–156
  organics/water, 148–152
  oxygen/nitrogen in commercial air separation membranes, 57
  oxygen/nitrogen in polycarbonates, 42, 43
  oxygen/nitrogen in polyimides, 42, 43
  oxygen/nitrogen via facilitated transport, 843–851
  1-pentene/other pentenes, 857
  propylene/nitrogen, 856
  propylene/propane, 855–857
  styrene/ethylbenzene, 856
  sulfur dioxide/carbon dioxide, 804, 855
  water/ethanol, 133–139
  water/isopropyl alcohol, 144–146
  water/organics, 140–148
Semiconductor manufacturing, 541
  process gas filtration, 542–544
  process liquids filtration, 544–549
Semipermeable membrane, 163, 266, 925, 926
Sensitivity studies
  capital costs for reverse osmosis, 369
  operating costs for reverse osmosis, 369, 370
Separation factor, 4. *See also* Selectivity (separation)
  dialysis, 179
  facilitated transport, 836
  gas permeation, 26, 42–47
  hollow-fiber contained liquid membrane, 794, 795
  perstraction, 907, 908
  pervaporation, 113, 114
Separation of multiple products, 824–827
Serum, 516, 519
Sewage sludge, 439
Shear-induced hydrodynamic diffusivity, 484, 486, 489, 493, 497
Shear rate, 403, 438, 485, 486, 489, 493, 498
Shedding, 524–526

Sherwood number, 173, 176, 177, 196, 199, 247, 402, 741, 750, 751, 836, 838, 895
    feed-side, 174, 175
    laminar flow, 278
    length-averaged overall, 174, 177
    spiral-wound module, 301
    turbulent flow, 278
    wall, 174, 200
Sieving, 8, 9
    kinetic sieving dimensions of penetrants, 28
    molecular, 21
Silicone rubber. *See* Polydimethylsiloxane (PDMS)
Single pore channel model, 170
Size-exclusion, 393
Skim milk, 439, 447
Slipcasting, 424, 439
SO_2/NO_x, 804, 855
Sodium bisulfite treatment, 306, 318–320
Sodium hexametaphosphate, 308, 318, 319, 441
Solid membranes. *See* Membranes, solid
Solubility (coefficient)
    carbon dioxide in natural rubber, 38
    carbon dioxide in poly(ethylene terephthalate), 39
    gas solubility correlation with Lennard-Jones well potential for polycarbonates, 44
    hydrogen in natural rubber, 38
    methane in natural rubber, 38
    temperature dependence, 37–39
Solubility parameter, 919
Solute chemical potential gradient, 266
Solute flux, 171, 181, 270, 271
Solute permeability, 169–172, 181, 182, 188, 189, 198–200, 270
Solute preferential sorption, 266, 271
Solute rejection. *See* Membrane rejection
Solute transport parameter, 270
Solution-diffusion theory, 25, 168, 270, 271, 631, 907
Solvent extraction, 603, 657, 867, 868. *See also* Membrane-based solvent extraction
Solvent flux, 266, 270, 271
Solvent permeability, 170, 171, 270, 271
Solvent preferential sorption, 266
Sorbitan monooleate (SPAN 80), 664, 666, 668–671
Sorption
    carbon dioxide sorption in polymethylmethacrylate, 32
    dual-mode model, 29, 30
    Flory-Huggins equation, 27
    Henry's law, 27
    Langmuir, 29, 40
    matrix model, 30, 31
    temperature dependence, 37–39
Southern blotting, 561
SPAN 80 (sorbitan monooleate), 664, 666, 668–671
Specific cake resistance, 463, 465, 466
Spent sulfite liquor, 333, 443
Stacked disk cartridge, 508, 512, 546, 549–552
Starling's flow, 193, 208
Staverman coefficient, 164, 171, 267

Steam-in-place, 202, 532, 533
Stearic acid, 438
Stefan-Maxwell formulation, 903
Steric hindrance, 272, 399
Sterile filtration, 533, 534
Sterilization, 516, 532, 533
Sterilization of filters, 532, 533
Sterilizing grade filter, 517
Stokes-Einstein relation, 483, 485, 497
Stokes radius, 187, 272
Stripper, 885–887, 893, 896–898
Sucrose, 170, 428
Sulfonated polystyrene, 438
Sulfonated polysulfone, 234, 305, 306, 418, 422
Sulfite pulp liquor processing, 443
Supported liquid membranes. *See* Immobilized/supported liquid membranes
Surface properties, membrane, 429
Surface tension, 427, 428, 895, 903
Surface water treatment by membrane, 337–343
Surfactant treatment of membrane, 438
Surfactants, 664–671, 704
Suspended solids, 306, 307, 317, 318
Sweep gas membrane distillation, 901
Swelling
    electrostatic pseudo-liquid-membrane, 877
    emulsion liquid membranes, 687–689
    polymeric membranes, 118, 239, 908

Table salt production, 256, 257
Tangential flow, 179, 459, 480, 487, 571, 730, 753
Taylor vortices, 439
Temperature polarization, 125–127, 902
Terminal sterilization, 516
Ternary phase diagram, 422
Textile industry effluent treatment, 334, 335, 447, 448
Textile sizing agents, 447
Thermoporometry, 429
Thin-film-composite (TFC) membranes. *See* Membranes, composite/thin-film-composite (TFC)
Time lag, 921, 922
Tortuosity. *See* Membrane tortuosity
Transdermal drug delivery systems, 920
Translumenal pressure drop, 193, 195
Transmembrane pressure, 590
Transmembrane pressure difference, 171, 207, 267
Transport
    co-transport, 603, 660
    counter-transport, 603, 660
    in emulsion liquid membranes, 601, 611, 636, 675, 682
    in glassy polymers, 27–31
    in rubbery polymers, 27
    mixed gas transport, 31, 32
    numbers, 240, 241, 247
    solution-diffusion theory, 25, 168, 270, 271, 631, 907

Transport (*continued*)
    See also Facilitated transport; Membrane transport
Triangular array of fibers, 197
1,1,2-Trichloroethane, 150, 152, 153
Triethanolamine, 99
Trihalomethanes, 319, 338
Trilaminate, 930
Trilaurylamine, 775, 858
Tri-*n*-octylamine (TNOA), 659, 661, 670, 671
Trioctylamine (TOA), 659, 669, 775, 776, 801
Trioctylphosphine oxide (TOPO), 625, 669, 858
Tubular pinch effect, 484, 493
Turbulence inducers, 438
    pulsatile flow, 438
    surface corrugation, 438
    Taylor vortices, 439
Turbulence promoters, ultrafiltration, 433, 438
Two-dimensional effects, 731, 735
Two-phase aqueous extraction systems, 729, 759

Ultrafiltration, 6, 9, 393
    applications, 307, 318, 346, 446, 548
    batch concentration, 441–443
    concentration polarization, 395, 400–404
    costs/economics, 451, 452
    definitions, 393–396
    diafiltration, 396, 443, 444
    feed clarification, 394, 446, 448
    feed pretreatment, 437
    flux, 394, 395, 398, 400
    flux decline, 403
    fractionation, 394
    gel layer formation, 401, 402
    membrane characterization, 425–429
    membrane fabrication, 408, 422
    membrane fouling, 396, 403, 433, 436
    membrane permeability, 395, 403
    membrane rejection, 395, 399, 404, 409, 423, 425, 426, 428
    membrane resistance, 396
    membranes. See Membranes, ultrafiltration
    module and process configuration, 432–444
    pressure differential, 395
    solute retention, 395, 399, 409–420, 428
    theory, 398–404

Ultrafiltration coefficient, 171, 208
Ultrapure water, 258, 355, 366, 548
Uranium extraction, 598, 625–627, 671, 682, 683
Urea, 170, 206

Vaccines, 516, 533, 535, 587
Vacuum distillation, 11, 12, 211, 901, 903
Vacuum stripping, 898
Validation guides, 531–533
van der Waals force, 272, 399, 400, 471
van't Hoff equation, 171, 403
Vapor permeation, 92, 156, 157
Vents, 518, 526, 527, 533
Vinegar brewing, 446
Vitamin B-12, 170, 189
Volatile organic compound stripping, 896, 898
Volumetric flux, 23, 56, 57, 171, 267, 394, 395, 462

Wastewater treatment. *See* Effluent/wastewater treatment
Water factory, 330
Water flux, 266, 270, 271, 276, 531, 584, 590
Water for injection (WFI), 516, 518, 536
Water purification, 258, 313, 507, 534, 535
Water recovery, 266, 283–287, 321, 370, 371
Water softening, 343, 344, 346, 356
Weber number, 674, 675, 869
Well control fluid, 599
Western blotting, 561
Wettability, 186, 509, 522, 527
Whey protein concentrate, 447
Whey treatment, 346, 446, 447
Wilke-Chang correlation, 624
Wilson plot, 177
Wine, 148, 345, 347, 447, 537, 538
Wine coolers, 537

Young-Laplace equation, 509, 512, 746

Zinc removal, 597, 599, 602, 625, 627, 640, 641, 671, 701–708, 718–723
Zirconia membranes, 408, 419, 425
Zirconium oxide-polyacrylate membrane, 334